Phosphorus(III) Ligands in Homogeneous Catalysis: Design and Synthesis

Phosphorus(III) Ligands in Homogeneous Catalysis: Design and Synthesis

Edited by

PAUL C. J. KAMER
EaStCHEM, School of Chemistry, University of St. Andrews, UK

and

PIET W. N. M. VAN LEEUWEN
Institute of Chemical Research of Catalonia (ICIQ), Spain

WILEY

A John Wiley & Sons, Ltd., Publication

This edition first published 2012

Registered Office
John Wiley & Sons, Ltd, The Atrium, Southern Gate, Chichester, West Sussex, PO19 8SQ, United Kingdom

For details of our global editorial offices, for customer services and for information about how to apply for permission to reuse the copyright material in this book please see our website at www.wiley.com.

Library of Congress Cataloging-in-Publication Data

Kamer, Paul C. J.
Phosphorus(III) ligands in homogeneous catalysis : design and synthesis / edited by Paul C. J. Kamer & Piet W. N. M. van Leeuwen.
p. cm.
Includes bibliographical references and index.
ISBN 978-0-470-66627-2 (hardback)
1. Phosphorus compounds. 2. Ligands. 3. Catalysis. I. Leeuwen, P. W. N. M. van (Piet W. N. M.) II. Title.
QD181.P1K36 2012
546'.712595–dc23

2012000246

A catalogue record for this book is available from the British Library.

HB ISBN: 9780470666272

Set in 10/12pt Times by SPi Publisher Services, Pondicherry, India

Contents

List of Contributors

Jérôme Bayardon
Institut de Chimie Moléculaire
Université de Bourgogne
9 Avenue A
Savary-Dijon-21078
France

Duncan Carmichael
Laboratoire "Hétéroéléments et Coordination"
Ecole Polytechnique, CNRS
Route de Saclay
F-91128 Palaiseau cedex
France

Sergio Castillón
Departament de Química Analítica I Química Orgànica
Universitat Rovira i Virigili
C/ Marcel·lí Domingo s/n
43007 Tarragona
Spain

David J. Cole Hamilton
EaStCHEM, School of Chemistry,
University of St Andrews
Purdie Building, North Haugh
St Andrews, Fife KY16 9ST
United Kingdom

Eamonn D. Conrad
Department of Chemistry
University of British Columbia
2036 Main Mall
Vancouver, BC V6T 1Z1
Canada

Gregory R. Dake
Department of Chemistry
University of British Columbia
2036 Main Mall
Vancouver, BC V6T 1Z1
Canada

Verónica de la Fuente
Departament de Química Analítica I Química Orgànica
Universitat Rovira i Virigili
C/ Marcel·lí Domingo s/n
43007 Tarragona
Spain

René den Heeten
Huntsman Holland BV
Huntsman Polyurethanes
Rozenburg Works, Merseyweg 10
3197 KG Rotterdam
The Netherlands

Johannes G. de Vries
DSM Resolve and DSM Innovative Synthesis BV
A Unit of DSM Pharma Chemicals,
PO Box 18
6160 MD Geleen
The Netherlands

Julien Dugal-Tessier
Department of Chemistry
University of British Columbia
2036 Main Mall
Vancouver, BC V6T 1Z1
Canada

Derek P. Gates
Department of Chemistry
University of British Columbia
2036 Main Mall
Vancouver, BC V6T 1Z1
Canada

Jason A. Gillespie
EaStCHEM, School of Chemistry,
University of St Andrews
Purdie Building, North Haugh
St Andrews, Fife KY16 9ST
United Kingdom

Cyril Godard
Departament de Química Analítica I Química
Orgànica
Universitat Rovira i Virigili
C/ Marcel·lí Domingo s/n
43007 Tarragona
Spain

Aitor Gual
Departament de Química Analítica I Química
Orgànica
Universitat Rovira i Virigili
C/ Marcel·lí Domingo s/n
43007 Tarragona
Spain

Sylvain Jugé
Institut de Chimie Moléculaire
Université de Bourgogne
9 Avenue A
Savary-Dijon-21078
France

Paul C. J. Kamer
EaStCHEM, School of Chemistry,
University of St Andrews
Purdie Building, North Haugh
St Andrews, Fife KY16 9ST
United Kingdom

Wouter Laan
EaStCHEM, School of Chemistry,
University of St Andrews
Purdie Building, North Haugh
St Andrews, Fife KY16 9ST
United Kingdom

Laurent Lefort
DSM Resolve and DSM Innovative
Synthesis BV
A Unit of DSM Pharma Chemicals, PO Box 18
6160 MD Geleen
The Netherlands

Wei Li
Department of Chemistry and Chemical Biology
Rutgers, The State University of New Jersey
610 Taylor Road
Piscataway, NJ 08854
United States

Angelica Marson
Clariant International Ltd.
Rothausstrasse 61
4132 Muttenz
Switzerland

Christian Müller
Freie Universität Berlin
Institute of Chemistry and Biochemistry
Fabeckstraße 34-36
14195 Berlin
Germany

Andreas Pfaltz
Department of Chemistry
University of Basel
St. Johanns-Ring 19
CH-4056 Basel
Switzerland

Paul G. Pringle
School of Chemistry
University of Bristol
Bristol BS8 1TS
United Kingdom

T. V. (Babu) RajanBabu
Department of Chemistry
The Ohio State University
100 West 18th Avenue
Columbus, OH 43210
United States

Joost N. H. Reek
Supramolecular and Homogeneous Catalysis
van 't Hoff Institute for Molecular Sciences
University of Amsterdam
Science Park 904
1098 XH Amsterdam
The Netherlands

Christopher A. Russell
School of Chemistry
University of Bristol
Bristol BS8 1TS
United Kingdom

Michiel C. Samuels
EaStCHEM, School of Chemistry,
University of St Andrews
Purdie Building, North Haugh
St Andrews, Fife KY16 9ST
United Kingdom

J. Chris Slootweg
Department of Chemistry and Pharmaceutical Sciences
VU University Amsterdam
De Boelelaan 1083
1081 HV Amsterdam
The Netherlands

Martin B. Smith
Department of Chemistry
University of Loughborough
Loughborough LE11 3TU
United Kingdom

Bert H. G. Swennenhuis
Department of Chemistry
Texas A&M University at Qatar
Doha
Qatar

René Tannert
Department of Chemistry
University of Basel
St. Johanns-Ring 19
CH-4056 Basel
Switzerland

Nell S. Townsend
School of Chemistry
University of Bristol
Bristol BS8 1TS
United Kingdom

Jarl Ivar van der Vlugt
Supramolecular and Homogeneous Catalysis
van 't Hoff Institute for Molecular Sciences
University of Amsterdam
Science Park 904
1098 XH Amsterdam
The Netherlands

Piet W. N. M. van Leeuwen
Institute of Chemical Research of Catalonia (ICIQ)
Av. Països Catalans 16
43007 Tarragona
Spain

Paul B. Webb
Sasol Technology UK Ltd
Purdie Building, North Haugh
St Andrews, Fife KY16 9ST
United Kingdom

Xumu Zhang
Department of Chemistry and Chemical Biology
Rutgers, The State University of New Jersey
610 Taylor Road
Piscataway, NJ 08854
United States

Erik Zuidema
SABIC Technology & Innovation
Urmonderbaan 22
6167 RD Geleen
The Netherlands

Preface

Phosphorus was named "The Devil's Element" by John Emsley in his book *The Shocking History of Phosphorus*. Indeed, ignorance and inexperience can lead to severe problems in the handling and application of phosphorus compounds. From a synthetic point of view, the chemistry of phosphorus is challenging and does not always follow the predicted routes. Nevertheless, the synthesis of tervalent phosphorus compounds has been well developed and the rich chemistry of phosphorus not only is crucial in biological systems but also has led to numerous useful applications and materials. Tervalent phosphorus compounds have proven to be extremely powerful in the field of homogeneous catalysis, both in academic research and in industrial applications.

This book covers the important area of the design and synthesis of P(III) donor ligands in homogeneous catalysis. The reactivity of organotransition metal complexes is dependent on the ligand environment of the metal. Consequently, optimizing the catalytic center by varying the ligand properties is a powerful tool in homogeneous catalysis. Impressive results have been obtained in both small-scale (asymmetric) catalytic preparation of fine chemicals and industrial production of bulk chemicals. In the introductory Chapter 1, common methods for using quantitative ligand parameters are discussed, illustrating how rational ligand design can be achieved.

Even so, while increasing knowledge about organotransition–metal compounds and computational chemistry has provided fundamental knowledge of the factors influencing elementary reaction steps, catalyst development is often hampered by synthesis of ligands with the appropriate structure. This is probably the reason that the majority of catalytic studies are still being performed employing traditional phosphine and phosphite ligands, with the exception of sophisticated bidentate ligands such as those employed in asymmetric catalysis. Moreover, these ligands are very often relatively simple structures; over the last decades, the most commonly used ligands for homogeneous catalysis are based on triphenylphosphine derivatives. To develop new catalysts for as yet uncatalyzed reactions, existing catalysts must undergo major improvements. Discovery of new catalytic transformations requires the exploration of new transition metal complexes with very diverse and totally new ligand systems. In this book we aim to address the design and synthesis of a comprehensive compilation of P(III) ligands for homogeneous catalysis. We will not only focus on the well-known traditional ligands that have been explored by catalysis researchers, but also include promising ligand types that have traditionally been ignored mainly because of their challenging synthesis. To promote the application of phosphine ligands in catalysis and to increase the practical value of the book, we have included detailed and reliable synthetic procedures in the book, provided by leading experts in the field. This may be of particular importance for the less traditional phosphine ligands, and their synthesis should no longer form a barrier for their use. In our opinion, this renders the book an invaluable reference book for researchers in the fields of catalysis and organic synthesis.

Naturally, the first chapters will cover the syntheses and applications of traditional tervalent phosphorus donor ligands that have successfully been applied in homogeneous catalysis. Triarylphosphine ligands have been widely investigated after the seminal discovery of Wilkinson that rhodium phosphine complexes are excellent hydrogenation catalysts. This has triggered a tremendous boost in synthetic and catalytic studies of phosphine ligands, which is reviewed in Chapter 2. Soon after the first applications of phosphines, catalytic studies based on related ligand structures have been reported, such as phosphites (Chapter 3), phosphoramidites (Chapter 4), and phosphonites and phosphinites (Chapter 5).

In organometallic chemistry, phosphorus donor ligands are known for their large trans labilizing effects. By combining phosphorus donors with other donor atoms, resulting in mixed-donor ligand systems, the different trans effects can be exploited to steer regioselectivity of catalytic transformations, which is described in Chapter 6.

Parallel to developments in the synthesis of traditional phosphorus donor ligands, there has been great progress in the field of phosphorus chemistry leading to a plethora of new compounds which have not been intensively studied as ligands in homogeneous catalysis, but do have potential.

Promising examples are P-containing heterocycles such as phospholes (Chapter 7) and phosphinines (Chapter 8). Other relatively unexploited ligand systems are highly strained phosphorus compounds (Chapter 9), low-valent phosphorus donors such as phosphaalkenes (Chapter 10), or even phosphaalkynes (Chapter 11). In contrast to its congener nitrogen, the barrier for a Walden inversion of phosphorus is high, allowing for the synthesis of P-stereogenic phosphorus ligands (Chapter 13) and bringing the chiral center in close proximity to the metal. Extremely bulky cage compounds (Chapter 12) and well-organized hemispherical calixarene-based ligands (Chapter 14) can lead to surprising steric effects and shape selectivity, respectively. Many of these phosphorus compounds have been regarded as chemical curiosities in the past and their investigations have mainly been addressed from a synthetical and theoretical point of view. These phosphorus compounds possess defined electronic, steric, and coordination properties that differ from those of the classical triphenylphosphine derivatives and can be tuned for optimal selectivity of important catalytic reaction steps. In general a catalytic transformation consists of several elementary steps that will be influenced in different ways by ligand modifications. These relatively scarcely explored compounds might provide many (pleasant or unpleasant) surprises in catalysis research.

Several approaches can be anticipated to facilitate the synthesis of libraries of phosphorus donor ligands; examples are the supramolecular assembly of bidentate ligands (Chapter 15) or the solid-phase synthesis of immobilized catalyst systems (Chapter 16).

The oxygen sensitivity of most tervalent phosphorus compounds renders them unsuitable as cofactors in biological systems. Therefore, the use of powerful phosphorus donor ligands in protein-based catalysts is restricted to the realm of artificial metalloenzymes (Chapter 17).

Finally, as most phosphorus donor ligand systems are being applied in homogeneous catalysis, ample research has been dedicated to the important problem of catalyst separation and recycling. The present state-of-the-art methods are reviewed in Chapter 18.

The recognition of the catalytic power of phosphorus donor ligands in catalysis is beyond doubt, but in spite of this, relatively few classes have been thoroughly investigated. We believe that the synthetic difficulties of the less conventional phosphorus ligands are most likely the reason for this. In this book, we want to discuss the chemistry of both well-established and unconventional ligand classes and make them accessible to catalysis researchers by including experimental procedures for representative examples of each ligand class provided by renowned experts in the field. The presentation of the established ligands will certainly lead to new ideas as regards new structures as well as new applications.

1

Phosphorus Ligand Effects in Homogeneous Catalysis and Rational Catalyst Design

Jason A. Gillespie[1], Erik Zuidema[2], Piet W. N. M. van Leeuwen[3], and Paul C. J. Kamer[1]

[1]EaStCHEM, School of Chemistry, University of St Andrews, St Andrews, United Kingdom
[2]SABIC Technology & Innovation, Geleen, The Netherlands
[3]Institute of Chemical Research of Catalonia (ICIQ), Tarragona, Spain

1.1 Introduction

Over the last 60 years, the increasing knowledge of transition metal chemistry has resulted in an enormous advance of homogeneous catalysis as an essential tool in both academic and industrial fields. The continuously growing importance of transition metal catalysis is well illustrated by the recent awards of three Nobel Prizes in 2001, 2005 and 2010 to this field of chemistry. Remarkably, phosphorus(III) donor ligands have played an important role in several of the acknowledged catalytic reactions [1–5]. The positive effects of phosphine ligands in transition metal homogeneous catalysis have contributed largely to the evolvement of the field into an indispensable tool in organic synthesis and the industrial production of chemicals.

An astounding diversity of ligand types and structures is known in literature: mono-, bi- and polydentates, ligands based on single donoratoms (such as phosphorus or nitrogen) or multiple donoratoms (such as P–N or P–O), achiral or chiral ligands, and ligands with exotic steric or electronic constraints. This extensive ligand library is in part the result of the fast developments in organometallic chemistry leading to a wide variety of ligand structures which have been exploited in transition metal complexes. Furthermore, the urge to optimise transition metal complex properties such as catalytic performance triggered an evolutionary type of growth of ligand libraries. Systematic variation and combination of successful ligand structures, intended to optimise ligand performance, inevitably led to new and unprecedented properties, in addition to the expected optimised catalytic systems.

Figure 1.1 represents an extremely tiny sampling of phosphorus donor ligands successful in diverse catalytic reactions, displaying an incredibly large variety in their structure. A striking feature is that ligands of

Phosphorus(III) Ligands in Homogeneous Catalysis: Design and Synthesis, First Edition.
Edited by Paul C. J. Kamer and Piet W. N. M. van Leeuwen.

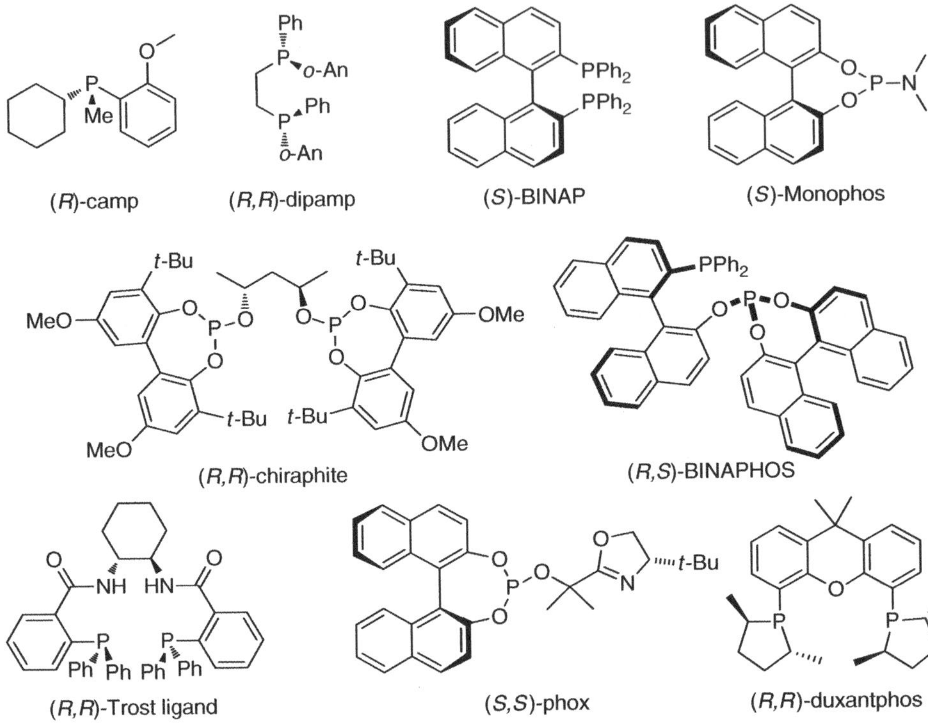

Figure 1.1 Selection of phosphorus(III) ligands.

very different structures can provide similarly high efficiency in the same catalytic reaction while ligands with very similar structures can behave very differently. The first row in Figure 1.1 shows successful ligands in asymmetric hydrogenation. The monodentate ligand *R*-camp provided good enantioselectivity in the asymmetric hydrogenation of dehydro amino acids [6], but Knowles *et al.* showed that bidentate ligands like dipamp performed superior compared to monodentates [7]. Unexpectedly, Feringa and de Vries showed two decades later that monodentates like *S*-monophos can outperform bidentate ligands [8]. The ligands on the second row are structurally quite different and showed different coordination modes in hydridorhodium carbonyl complexes; *R,R*-chiraphite coordinates in bisequatorial fashion in the trigonal bipyramidal rhodium complex [9], whereas binaphos occupies an apical and an equatorial site [10]. Nevertheless, both ligands perform well in asymmetric hydroformylation of styrene. All three ligands of the third row perform well in palladium-catalyzed asymmetric allylic substitution, although the ligands are based on different types of chirality and/or donor atom type [11–13]. Finally, ligands such as BINAP (2,2′-bis(diphenylphosphino)-1,1′-binaphthyl) show excellent performance in several catalytic reactions [1], which instigated Jacobsen to coin such ligands ‘privileged’ ligands [14].

When designing a new catalyst, the choice of the metal is naturally of utmost importance. This choice is usually dictated by the envisaged catalytic reaction and based on pre-existing knowledge or by screening via trial and error. Although most transition metals are capable of facilitating all elementary steps which constitute a catalytic cycle, several catalytic reactions are dominated by specific metals such as palladium for allylic substitutions and rhodium for the hydroformylation of alkenes. The next step is, in general, adjusting the reactivity of the metal by adding donor ligands. It is not surprising that the nature of the donor atom is pivotal in influencing the reactivity of the metal. The σ-donor and π-acceptor properties as well as the steric congestion imposed on the metal strongly influence the catalyst performance. In the case of bidentate ligands, the

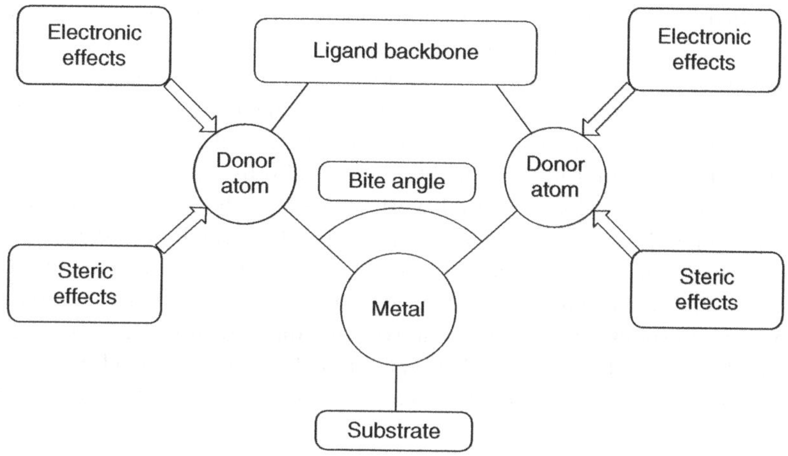

Figure 1.2 *Schematic representation of parameters used to design new and optimize existing catalysts.*

bite angle enforced on the metal also has a profound effect on the steric and electronic properties of the metal (Figure 1.2). Ligand effects are very powerful, and in fact different combinations of transition metals and donor ligands can result in very similar reactivity. Figure 1.2 shows which points of variation may be considered when designing new bidentate catalysts.

Phosphorus has often been the donor atom of choice, and it has a long history as a soft, strongly ligating atom for late-transition metals, which is easily rationalised by hard/soft acid base theory. Moreover, in-depth understanding of the effects of phosphorus ligands on the properties of transition metal complexes is greatly advanced by facile analysis by ^{31}P NMR spectroscopy; in fact, the many successes of phosphorus ligands in the field of homogeneous catalysis might very well be mainly due to this easy structural (in situ) analysis by NMR.

One of the first examples of the crucial effect of phosphine ligands on catalyst performance was reported in the 'Reppe' chemistry, to form acrylic esters from alkynes, alcohols and carbon monoxide [15]. A homogeneous triphenylphosphine nickel iodide complex was the catalyst responsible for the production of acrylic acid from acetylene, carbon monoxide and water. In the same period, another important discovery in the field of homogeneous catalysis was made by Otto Roelen, who accidentally discovered that cobalt carbonyl complexes could convert alkenes into valuable aldehydes by applying high pressures of CO and H_2 [16]. Although commercialisation of these processes was accomplished years later, these discoveries revealed the potential power of homogeneous catalysis and initiated the development of a prosperous field of research in both academia and industry.

The nickel-catalysed hydrocyanation of butadiene as developed and applied by Du Pont is an excellent example of an industrial application in which ligand parameters proved to be crucial [17]. The catalytic reaction employs a nickel catalyst modified by aryl phosphite ligands for the atom economic synthesis of adiponitrile from butadiene and hydrogen cyanide. This seminal application of transition metal catalysis clearly illustrates the level of sophistication which can be achieved in fine tuning the activity and selectivity of the metal centre. Key to the development of this process was a systematic study of 'ligand effects' by Tolman which led to the first systematic description of ligand properties in the field of homogeneous catalysis using organometallic complexes [18].

The Tolman steric and electronic parameters proved powerful tools for quantitative and qualitative understanding of the influence of first coordination sphere ligands on transition metal complex properties. The effects of these ligand parameters upon metal–ligand and ligand–substrate interactions are the basis for a rational design approach in catalyst development. The quantification of the steric and electronic contributions of phosphorus(III) ligands to M–P bonding and reactivity has contributed to a large extent to the discovery and improvement of catalytic activity of new phosphine transition metal complexes.

An important boost in the application of phosphorus donor ligands in homogeneous catalysis was triggered by the discovery of $RhCl(PPh_3)_3$ as a catalyst for alkene hydrogenation by Wilkinson and co-workers in the mid-1960s [19]. It is fair to say that the mechanism of this reaction is now among the best studied in catalysis research. Studies of electronic and steric ligand effects led to detailed understanding of the elementary reaction steps of the catalytic cycle. Increasing the donor capacity of the phosphine ligands by introducing donating substituents at the aromatic rings attached to the phosphorus atom resulted in higher reaction rates [20]. This work was soon followed by the initial work of Vaska, who showed that oxidative addition of hydrogen to Rh and Ir is one of the important elementary steps of the catalytic cycle [21].

Soon thereafter, phosphorus donor ligands proved to be very effective in rhodium-catalyzed hydroformylation as well. Wilkinson *et al.* showed that rhodium carbonyl complexes modified by alkylphosphines and arylphosphines were active catalysts for the hydroformylation of alkenes under mild conditions (70°C and 100 bar) [22]. Workers at Union Carbide Corporation found that phosphites were also very effective ligands for this reaction and that the catalyst performance was strongly dependent on the type of phosphite [23].

Bidentate ligands lead to increased stability of organometallic complexes due to the chelate effect, which has strong impact on the chemistry at the metal centre. The synthesis of the archetypical diphosphine dppe was already reported in 1959 [24], and Chatt and Hieber [25] explored the coordination chemistry of several diphosphines with an ethane bridge. Slaugh explored the use of dppe in cobalt-catalysed hydroformylation [26], but it did not result in significant changes in catalyst performance compared to the phosphine-free cobalt carbonyl catalyst. The use of alkylphosphines led to the first ligand-modified (cobalt) hydroformylation catalyst (the Shell HydroFormylation process, or SHF). A strong beneficial effect of bidentate ligands was discovered by Keim and coworkers at Shell Development in the late 1960s [27]. Certain bidentates containing an oxygen and a phosphorus donor atom formed excellent nickel catalysts for the oligomerization of ethene. Most efficient ligands were diphenylphosphinoacetic acid or 2-diphenylphosphinobenzoic acid, which was named SHOP ligand after the resulting Shell Higher Olefins Process that came on stream in 1977 (Figure 1.3).

In 1966 Iwamoto and Yuguchi described the first *advantageous* results for a range of diphosphine ligands with varying bridge lengths in the co-dimerisation of butadiene and ethene using iron catalysts [28]. In many other cases, the activity of catalysts containing dppe instead of PPh_3 was lowered due to the strong chelating power of the diphosphine. This anticipated higher stability of chelating diphosphine complexes was seen as a drawback in the development of more active catalysts.

Theoretical work of Thorn and Hoffmann [29] corroborated that migration reactions were slow in complexes containing chelating ligands such as dppe. During the migration process, the neighbouring phosphine ligand was shown to have a tendency to enlarge the P–M–P bite angle, which is prevented by the constrained C_2 bridge of the ligand.

Asymmetric hydrogenation is arguably one of the most mature fields showing beneficial use of bidentate phosphine ligands. In 1971 Kagan reported the use of DIOP for the rhodium diphosphine–catalysed hydrogenation of N-acetylphenylalanine [30,31]. This first indication of improved performance of bidentate ligands compared to monodentates was soon followed by the report of Knowles *et al.* that the P-chiral diphosphine DIPAMP led to excellent enantioselectivities compared to the monodentates CAMP and PAMP [6,7]. This discovery led to the first industrial applications of bidentate phosphines in the production of L-DOPA via the rhodium-catalysed asymmetric hydrogenation by Monsanto. Selke developed the sugar-based bisphosphonite Phenyl-β–GLUP for the same process (Figure 1.4) [32], which has been applied for many years by the German company VEB-Isis. Ever since, many other chiral diphosphines have shown wide applicability in the area of asymmetric hydrogenation, and several applications have been developed. Important new ligands that have been introduced comprise (Figures 1.1 and 1.4) Noyori's BINAP [33], DuPhos (Burk) [34], Takaya's BINAPHOS [10] and C_1-symmetric ferrocene-based ligands ('Josiphos') introduced by Togni [35].

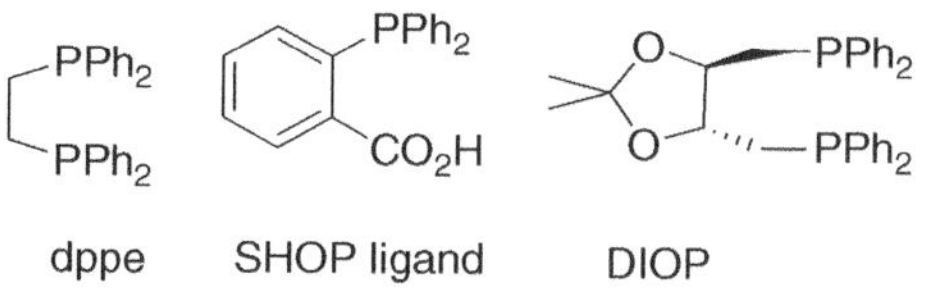

Figure 1.3 *Structures of early bidentate phosphine ligands dppe, SHOP ligand and DIOP.*

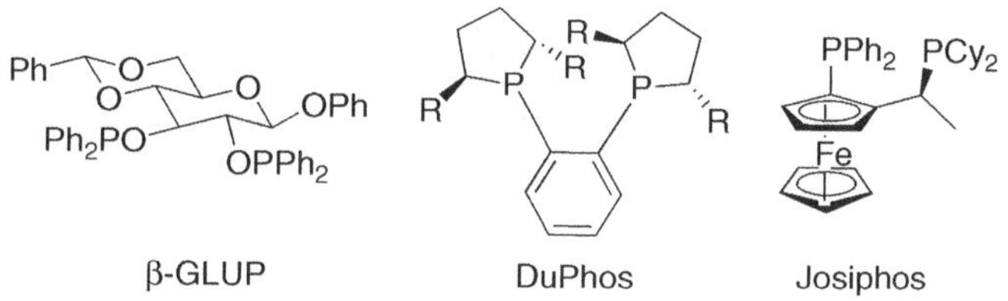

Figure 1.4 *Structures of Phenyl-β-GLUP, DuPhos and Josiphos.*

Late-transition-metal-catalysed coupling reactions, such as palladium- or nickel-catalysed C–C, C–O and C–N bond formation, have proven to be very sensitive to ligand effects [36]. The amount of data available on ligand effects for these reactions is extensive, although rationalization of trends often remains a challenging task [37].

The efficacy of the Trost ligand in asymmetric allylic alkylation was originally explained by the 'embracing' effect of the bidentate ligand [38]. By simply reasoning that larger bite angles would lead to more efficient catalysts, the wide bite angle ligand Duxantphos (Figure 1.1) was successfully designed and applied [13]. Detailed mechanistic studies by Lloyd–Jones showed that hydrogen bonding of the ligand amide N–H to the carbonyl group of the leaving group activates the allylic ester, which is crucial for the selectivity of the reaction [39]. This clearly illustrates that successful ligand design can be achieved even when it is founded on an imperfect or incomplete mechanistic basis.

The ligand size seems to be a dominant factor in the palladium-catalysed Heck reaction, as bulky phosphines [40], phosphites [41] and phosphoramidites [42] were found to lead to highly effective catalysts. For the most efficient phosphoramidite, it was shown that the steric congestion of the ligand led to mono-ligated complexes, which are more prone to substrate coordination than bis-ligated complexes. This effect had been observed before in rhodium-catalysed hydroformylation using bulky phosphite ligands (Figure 1.5) [43,44].

Hydroformylation of alkenes is a reaction that is extremely sensitive to ligand effects as well as to specific reaction conditions. Since Shell reported the beneficial use of alkylphosphines in cobalt-catalysed hydroformylation [45], many industries started applying phosphine ligands in the rhodium process as well [46]. While strongly σ-donating alkylphosphines are the ligands of choice for cobalt, they lead to low reaction rates when applied in rhodium catalysis. The application of arylphosphines reported by Wilkinson in the mid-1960s resulted in very active rhodium catalysts under very mild conditions [19].

The discovery of the rhodium triphenylphosphine hydroformylation catalysts was the basis for several industrial processes developed by Celanese (1974), Union Carbide Corporation (1976) and Mitsubishi Chemical Corporation (1978), all using this catalyst system. The UCC (now Dow Chemical) method has been licensed to many other users and it is often referred to as the low-pressure oxo (LPO) process. Major advantages of the rhodium catalysts over the cobalt ones are the higher activities, which are translated into milder reaction conditions, and higher product selectivities, resulting in better feedstock utilization. A drawback of the rhodium phosphine catalysts is the low reactivity for internal alkenes, rendering them unsuitable for higher alkene feedstocks, which are used for detergent alcohol production.

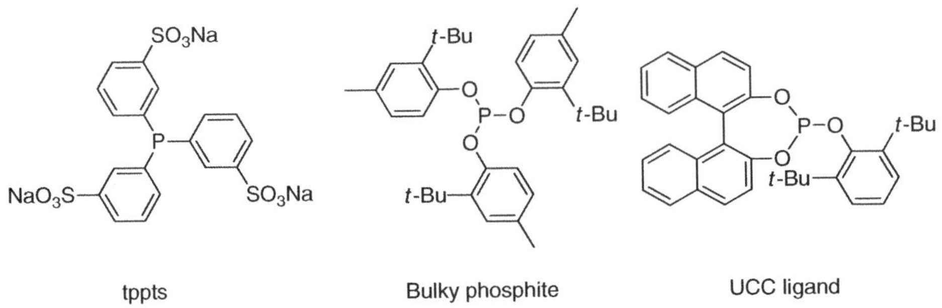

Figure 1.5 Structures of ttpts, van Leeuwen's 'bulky phosphite' and a highly stable, bulky phosphite from UCC.

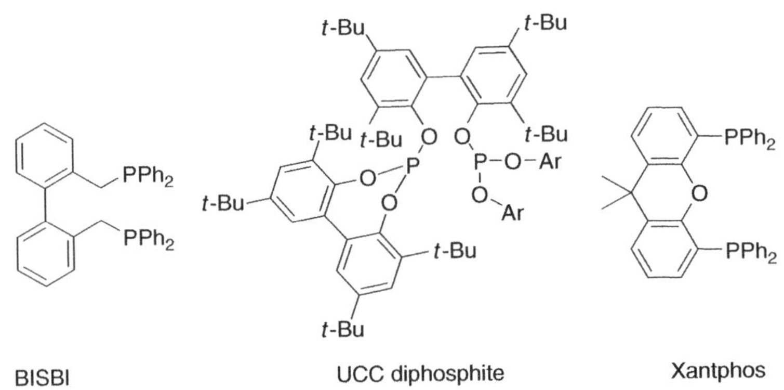

Figure 1.6 Eastman's BISBI, typical diphosphite from Union Carbide Corporation, and Xantphos.

An elegant solution for the catalyst separation problem is provided in the Ruhrchemie–RhonePoulenc process that utilises a two-phase system containing water-soluble rhodium-tppts (Figure 1.5) in the aqueous phase and the product butanal in the organic phase. The process has been in operation in Oberhausen since 1984 by Ruhrchemie (Celanese).

Almost two decades after the initial report by Pruett and Smith [23], a renewed interest in phosphites as ligands in the hydroformylation reaction was triggered after van Leeuwen and Roobeek reported that very high rates could be obtained by applying bulky monophosphites [44]. Bryant and coworkers at Union Carbide improved this system further by making more stable bulky monophosphites [47]. The high reaction rates were obtained at the expense of a reduced chemoselectivity due to the formation of unwanted isomerised alkenes and branched aldehydes. By changing to chelating diphosphites (Figure 1.6), they succeeded in combining very high linearities with still higher reaction rates compared to the triphenylphosphine system [48]. The 'bulky monophosphite' is applied commercially in a small-scale operation for the hydroformylation of 3-methylbut-3-en-1-ol by Kuraray [49].

Following up on the improved performance of diphosphites, diphosphines also experienced a renewed interest in the rhodium-catalysed hydroformylation. Casey and co-workers reported that the chelate angle of bidentate diphosphines affected the regioselectivity of the rhodium-catalysed hydroformylation of 1-alkenes dramatically [50]. They reported that a ligand developed by workers at Eastman, 2,2′-bis((diphenylphosphino) methyl)-1,1′-biphenyl (BISBI; Figure 1.6), coordinated preferentially in a bis-equatorial fashion in the trigonal bipyramidal resting state of the catalyst. This resulted in a linear-to-branched aldehyde ratio as high as 66:1 compared to a ratio of only 2:1 for equatorially-apically coordinating dppe (1,2-bis(diphenylphosphino)ethane).

Applying the modelling methodology developed by Casey and Whiteker [51], van Leeuwen and Kamer started a search for organic backbones for diphosphine ligands that would enforce coordination modes in between *cis* and *trans* coordination, leading to the new series of Xantphos ligands (Figure 1.6) [52]. Their studies showed that the ligand-induced bite angle had a distinct effect not only on the selectivity and activity of the rhodium hydroformylation catalyst but also in several other reactions such as the hydrocyanation of alkenes and cross-coupling reactions [53].

1.2 Properties of phosphorus ligands

The key role of phosphorus(III) ligands in late-transition metal catalysis has prompted many attempts to quantify the factors that influence the bonding of phosphorus donor ligands to transition metals and the reactivity of the complexes. For monodentate phosphine ligands, metal–ligand bonding is affected by both the electronic and the steric properties of the ligand. In order to rationalise the reactivity of transition metal–phosphine complexes, several quantitative ligand parameters have been developed. The quantitative and qualitative understanding of these ligand parameters is central to a rational design methodology, which considers their effect upon metal–ligand and ligand–substrate interactions retroactively. Moreover, these parameters have been used in linear free-energy relationships with the aim of predicting the physical properties and catalytic activity of new phosphine transition metal complexes. Unfortunately, this has proven to be less straightforward than one might expect. In order to rationalise catalytic results as functions of the ligand structures, structure–activity relationships usually explore small families of ligands which ideally vary only in a single ligand variable, such as bite angle size, steric bulk, electronic contribution and so on. Intrinsically it is difficult to affect one such parameter without affecting others; nonetheless the quantification and manipulation of ligand parameters are undoubtedly principal tools in structure–activity relationship analysis and the design of new ligands.

1.2.1 Electronic ligand parameters

The overall electronic properties of phosphorus(III) ligands have been studied using a variety of methods. One approach is to use mixed ligand complexes and assess the response of the other ligands upon electronic changes at the metal centre caused by the phosphine ligand. Strohmeier already showed that the IR carbonyl frequencies of metal complexes were a measure of the electronic properties of ligands [54]. Measuring the stretching frequencies of coordinated carbonyl ligands in transition metal complexes has become the most common way to assess the electronic effects of phosphorus ligands [55,56]. Tolman defined an electronic parameter (χ) for phosphorus ligands based on IR spectra of $NiL(CO)_3$ complexes, using $P(t\text{-}Bu)_3$ as a reference ligand [55]. The χ parameter of differently substituted phosphorus ligands is defined as the difference between the A_1 stretching frequencies of $LNi(CO)_3$ and $P(t\text{-}Bu)_3Ni(CO)_3$ in cm^{-1}. Tolman showed that individual contributions of the substituents are often additive (i.e. the overall χ-value of a ligand can be taken as the sum of the χ_i-values of the individual substituents).

This approach leads to a measure of the overall electronic-donating ability of the phosphine. However, it is often necessary to consider the electronic properties of phosphines as arising from two contributions, σ-donation and π-acidity. σ-Donation is the effective dative electron donation of the phosphorus lone pair towards empty metal orbitals, whilst π-acidity refers to the acceptance (back donation) of electron density from filled metal orbitals to empty ligand orbitals (Figure 1.7). Which empty ligand orbitals are involved in back donation is still a matter of debate, but the current prevailing view is that back donation occurs from the metal d-orbitals into the σ*-orbitals of the phosphorus ligand [57].

The separation of these two entwined electronic parameters is not trivial, and several studies have been devoted to this. One way to measure the σ-donation is via the Brønsted acidity of phosphonium salts, as

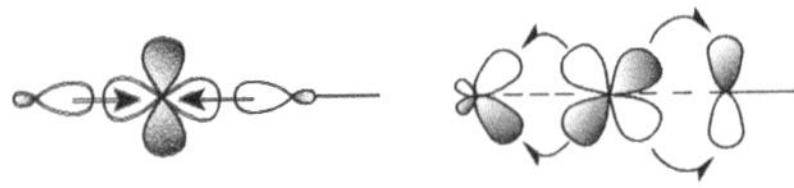

Figure 1.7 σ-donation (left) and π-back donation (right) contributions to the metal–ligand bond.

protons are not capable to participate in π–back donation and, therefore, the electron-donating potential of phosphorus can be estimated via their pK_a values. Although the interaction between the phosphorus ligand and the hard acid H^+ is very different from the interaction between the phosphorus ligand and a soft transition metal complex, a number of organometallic reactions do show a linear relationship between pK_a values of phosphorus ligands and the log k of the reaction. Considering the extreme sensitivity of the pK_a values of phosphorus bases to solvation energies, these good correlations are remarkable.

Drago postulated that σ-donation itself may be more complicated and should be defined by two parameters E_B and C_B, representing electrostatic and covalent contributions to the phosphine reactivity [58,59]. Drago's model is based on the enthalpies of adduct formation between a phosphorus σ-donor and a number of different π-acceptors, measured in poorly solvating solvents, which are used to quantify the σ-donating property of a phosphine donor. However, this method showed significant steric and π-acidity effects for bulky acceptor ligands [59].

The development of a quantitative measure for the π-acidity of phosphorus ligands has also proven to be difficult. Graham and Treichel discussed almost simultaneously the use of CO-stretching frequencies in octahedral $LM(CO)_5$ complexes for separation of the σ-donation and π–back donation contributions to the metal–ligand bond [60]. They reasoned that the π-acceptor capacity of ligand L should mainly affect the stretching frequency of the *trans*-coordinated carbon monoxide ligand by competing for the electron density of metal d-orbitals of the correct symmetry that are used in the M–CO back donation (Figure 1.7). In contrast, the σ-donating property of ligand L would affect the stretching frequencies of both the *cis*- and *trans*-coordinated carbon monoxide ligands in a similar fashion. Unfortunately, the method proved to be not very straightforward, as it predicted, for instance, that tributylphosphine is more π-acidic than triphenylphosphine.

One of the most comprehensive studies of ligand effects, known as quantitative analysis of ligand effects (QALE), was described by Giering and Poe [61]. The QALE model is based on the analysis of several physical properties of metal complexes: υ_{CO} for $(\eta^5\text{-cp})(CO)(L)Fe(COMe)$, E^0 and H^0 for the $(\eta^5\text{-cp})(CO)(L)Fe(COMe)^{+/0}$ couple, the pK_a value of HPR_3^+ and the ionisation energy of PR_3. Giering showed that the electronic properties of a ligand should be described by three instead of two electronic parameters: their σ-donor capacity (χ_d), π-acidity (π_p) and an aryl effect (E_{ar}). This last parameter is now believed to be more fundamental and not limited to aryl-containing phosphines, but its physical meaning is still unclear [61].

Computational methods offer an excellent platform to study electronic effects in metal–phosphorus bonding in individual complexes and in some catalytic reactions [62,63]. The effects of σ-donation and π–back-donation in transition metal complexes can be distinguished on the basis of the symmetry of the orbitals involved in the metal–phosphorus bonding. Pacchioni *et al.* studied the metal–PR_3 interaction in several palladium complexes and found only minor variations in the σ-donor strength of PR_3 ligands, while large differences in π-acidity were observed [64]. Computational studies by Branchadell and co-workers revealed similar behaviour in $Fe(CO)_4PR_3$ complexes [65]. Based on these results, they proposed to divide ligands into pure σ-donor ligands and σ-donor–π-acceptor ligands. Landis *et al.* studied the bond dissociation energies in rhodium–phosphine complexes and showed that the relative importance of σ-donation and π–back donation in the metal–phosphorus bond is highly dependent on the other ligands coordinated to the metal [66]. Landis argued that this σ–π synergy leads to very subtle effects at the metal centre, which makes prediction of ligand effects *a priori* difficult.

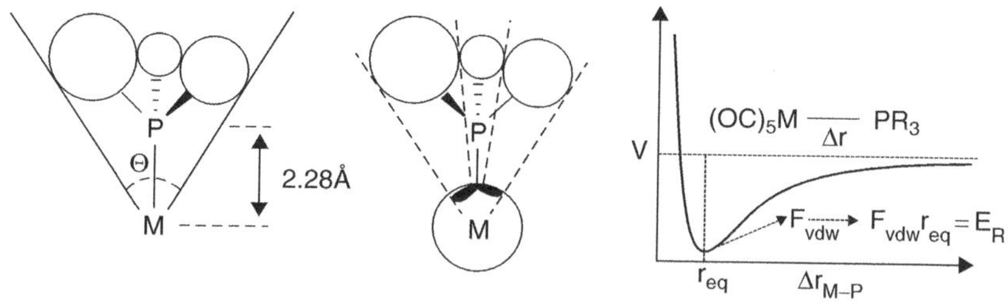

Figure 1.8 *Tolman's definition of the cone angle (left), White's definition of the solid angle (middle) and Brown's definition of E_R.*

1.2.2 Steric ligand parameters

Complementary to the development of reliable electronic parameters, several steric parameters have been introduced. One of the most widely applied concepts for quantifying the steric bulk of monophosphines is the cone angle (θ) (Figure 1.8), which was introduced by Tolman [18]. From a metal centre located 2.28 Å from the phosphorus atom of the ligand (a typical Ni–P bond distance), a cone that embraces all the atoms in the ligand based on CPK models is constructed (Figure 1.8). The resulting cone angle θ is a measure for the steric bulk of a ligand. Unfortunately, ligands rarely form a perfect cone and multiple ligands coordinated to one metal centre can mesh, creating less steric bulk around the metal than would be expected based on the sum of their individual cone angles. In many square-planar complexes, the sum of the individual cone angles is found to be much more than the available 360°. The increasing use of computational methods has resulted in several more elaborate approaches to alleviate these shortcomings [67]. White *et al.* introduced the concept of solid angles [68]. Based on either crystal structures or calculated structures, the van der Waals radii of the ligand atoms are projected onto the van der Waals surface of the metal centre (Figure 1.8). The amount of coverage, given by the solid angle, is a measure of the steric bulk around the metal.

Most measures for steric bulk are geometric in nature, but the actual physical meaning of steric congestion is a repulsive energy induced between different ligands in a transition metal complex. Brown *et al.* defined a steric repulsive energy parameter, E_R, that is based on a molecular mechanics computational model of $LCr(CO)_5$ [67,69]. The van der Waals repulsive force between the metal complex and the ligand (F_{vdw}) is calculated starting from the optimised structure of $LCr(CO)_5$ by varying the Cr–L distance. E_R is defined as the product of this van der Waals repulsive force and the metal–ligand bond distance (Figure 1.8). This method has a distinct advantage over the purely ligand-based methods as the conformation of the ligand corresponds better to the real conformation of the ligand in transition metal complexes. A drawback is that intimate knowledge of all force constants of the complex is required.

Barron *et al.* developed another semiquantitative approach to describe the steric properties of bidentate ligands. The 'pocket angle' concept is based on X-ray crystal structure data of palladium(II)–diphosphine complexes and estimates the size of the active site at palladium by calculating the interior cone angle of the bidentate ligand [70].

Leitner *et al.* proposed the 'accessible molecular surface' as an alternative approach based on molecular modelling to describe the interplay between the PR_2 fragment and the metal. The accessible molecular surface is determined by calculating the conformational space of the active fragment (i.e. [(P-P)Rh]). These conformations are then superimposed to provide a 'pseudo-dynamic' structure which can be visually interpreted for qualitative understanding or quantified by AMS analysis [71]. This method requires substantial knowledge of force-field parameters, but in this case these could be obtained from X-ray crystallographic data.

In recent years, the development of new powerful computational techniques has enabled the calculation of larger catalytic systems and the detailed study of steric effects in catalysis [72]. Although most catalyst

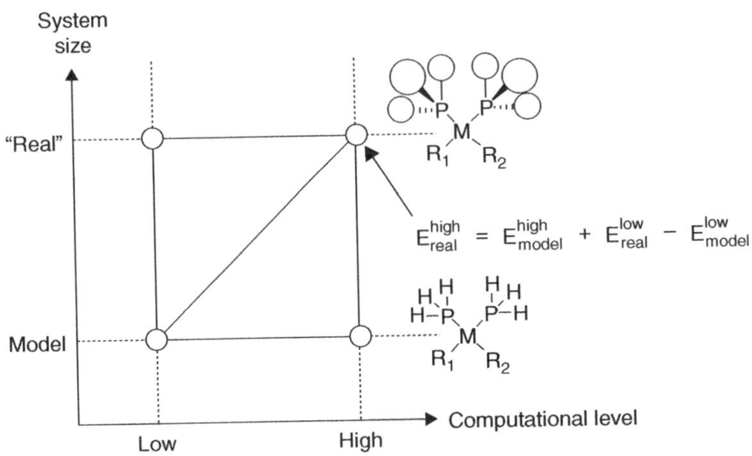

Figure 1.9 *General computational scheme for the calculation of large molecular systems.*

systems are still too large for the use of accurate, high-level *ab initio*– or density functional theory–based methods, the new techniques allow the separation of the catalytic complex into a reactive centre and a non-reactive periphery (Figure 1.9). While the reactive centre is treated at a high level of theory, the nonreactive periphery, which is mostly responsible for steric effects, is treated at a lower (and therefore less costly) level of theory. This also allows the straightforward separation of steric and electronic effects. These techniques have now been used to calculate key transition metal complexes, as well as determine pure steric effects in some catalytic processes such as the hydrogenation of enamides [73,74], the hydroboration of arenes [75], the hydroformylation of alkenes [76], the hydrosilylation of alkenes [77], the cyclodimerisation of 1,3-butadiene [62] and early- and late-transition metal-catalysed olefin polymerisation reactions [78]. One of the most important conclusions from these calculations is that not only is the amount of steric bulk important, but also the geometrical placement of the bulk around the metal centre can have a dramatic influence on the activity and selectivity of transition metal complexes in catalysis.

1.2.3 Bite angle effects

Bidentate phosphine ligands have found wide application in catalytic reactions since they can lead to increased stability and often an increased regio- and stereoselectivity of the catalyst system. The chelating effect of bidentate ligands reduces the tendency for ligand dissociation during the catalytic cycle, resulting in better defined catalytic species under catalytic conditions. Classification of bidentate ligands with the use of Tolman's electronic and steric parameters has proven to be difficult. Next to the traditional steric and electronic effects caused by the substituents at the phosphorus centre in monodentate ligands, the phosphorus–phosphorus distance in bidentate ligands, induced by the ligand backbone, proved to be an important ligand parameter [79,80].

To be able to quantify the effect of strain imposed by bidentate phosphine ligands, Casey and co-workers developed the concept of the natural bite angle [51]. This ligand parameter is based on simple molecular mechanics simulations. Similar to Tolman's cone angle concept, a dummy metal atom is introduced that coordinates to both phosphorus donor atoms at a distance of 2.315 Å. The function of this dummy metal is to ensure that bidentate coordination is preserved by allowing only ligand geometries in which the donor atoms have the proper orientation. In order to exclude contributions of the metal to the P–M–P angle, the P–M–P force constant is set to zero during geometry optimisation. The result is a purely ligand-induced P–M–P angle that can be used to compare different bidentate ligands. These calculations require prior knowledge of the remaining force constants in the metal complex, including force constants describing M–P stretching, M–P–X bending, dihedral bending and torsional deformations. These are often ignored, and changes in these parameters may cause large differences in the calculated natural bite angles. Despite this, the natural bite angle parameter has been correlated successfully with bite angle effects in a number of reactions involving transition metal complexes, including industrially important processes such as rhodium-catalysed hydroformylation, nickel-catalysed hydrocyanation and palladium-catalysed alkoxy- and hydroxycarbonylation.

The bite angle of the diphosphine ligand affects the properties of the metal complex in two fundamentally different ways; via an electronic effect on the metal complex and via the steric bulk induced by the ligand [79,81]. While in some reactions the electronic effect seems dominant, in other reactions the steric bulk of the bidentate ligand determines the overall bite angle effect. In many reactions, however, the physical origin of the observed overall bite angle effect remains unclear. Especially when multiple reaction steps in the catalytic cycle are affected by the bite angle of the bidentate ligands, rationalisation of the observed (overall) effect can be difficult. Clearly, separation of electronic and steric bite angle effects in catalytic reactions would increase our understanding of transition metal-catalysed reactions and allow for rational design of new catalyst systems.

1.2.3.1 Electronic bite angle effect

The term *electronic bite angle effect* refers to the electronic changes at the metal centre as a function of the natural bite angle of the bidentate ligand. Dierkes *et al.* introduced the concept of metal-preferred bite angle [79]. The *metal-preferred bite angle* is defined as the lowest energy P–M–P angle of the metal complex in the absence of steric effects. The better the ligand's natural bite angle matches this metal-preferred angle, the better the ligand is able to stabilise the metal complex. In a rough approximation, ligands exhibiting a natural bite angle of around 90° stabilise square planar and octahedral complexes, ligands exhibiting a natural bite angle of around 109° stabilise tetrahedral complexes and ligands exhibiting natural bite angles around 120° stabilise trigonal bipyramidal complexes. However, other ligands coordinated to the metal do affect the metal-preferred bite angle. More accurate metal-preferred bite angles can be obtained from *ab initio* and density functional theory (DFT) calculations on model complexes, which also allow the determination of metal-preferred bite angles of transition states.

Deviations of the metal-preferred bite angle of a complex induced by a bidentate ligand lead to electronic destabilisation of the metal complex and consequently to different reactivity. Especially in reactions where the geometry at the metal centre changes during the reaction, large differences in the reactivity and relative thermodynamic stability of the reaction intermediates are observed. Figure 1.10 depicts the orbital changes associated with the change of the L–Pt–L angle in the PtL_2 fragment [82]. For Pt(0), all d-orbitals at the metal centre are filled and the optimal orbital stabilisation is obtained for a L–Pt–L angle of 180°. For Pt(II), only four d-orbitals are filled, resulting in an empty δ_g*/2b_1 orbital. Consequently optimal stabilisation for the Pt(II) centre is obtained for a L–Pt–L angle of 90°.

Theoretical methods have been used to investigate electronic bite angle effects in several transition metal-mediated reactions. Thorn and Hoffman used extended Hückel calculations to show that, for hydride migration to a platinum-coordinated alkene, the bite angle of the spectator diphosphine ligand increases from 95° in the reactant to 110° in the transition state [29]. They argued that ligands with a natural bite angle of around 110° stabilise the transition state and destabilise the ground state and consequently accelerate the reaction. Indeed, a β-agostic intermediate between an alkene–hydride and an alkyl structure was observed for PtH(ethene)(dppp-Bu), which in the solid state showed a P–M–P angle of around 105°. The corresponding PtH(ethene)(dppe-Bu) complex, exhibiting a smaller bite angle of 90°, did not show this β-agostic intermediate.

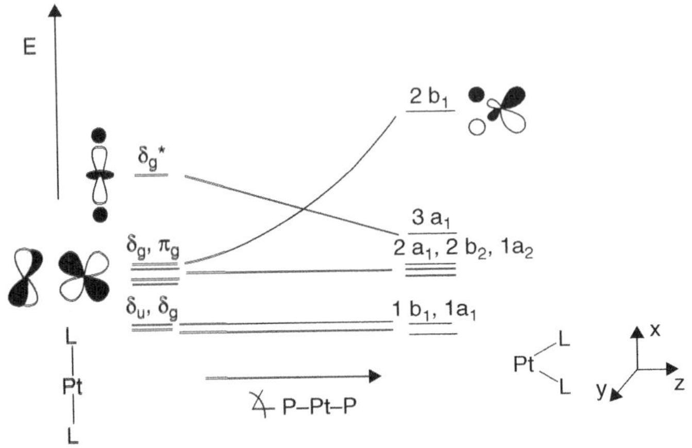

Figure 1.10 *Electronic bite angle effect in platinum complexes containing a predominantly σ-donor bidentate ligand [82]. Reprinted from S. Otsuka, 1980 with permission of Elsevier.*

Electronic bite angle effects have also been observed in other experimental studies. Dubois and co-workers found a decrease of the half-wave potentials of Ni(II/I) and Pd(II/0) couples in square planar M(diphosphine)$_2$(BF$_4$)$_2$ with an increase of the natural bite angle of the ligand [83]. Similarly, both the hydride donor strength and pKa of HM(diphosphine)$_2$X (M=Ni, Pd, Pt) decrease as the natural bite angle of the diphosphine ligand increases [84]. Both observations were attributed to a larger tetrahedral distortion of the reduced product for large bite angle ligands. Also, Angelici observed that the basicity of trigonal bipyramidal Fe(diphosphine)(CO)$_3$ decreases as the bite angle of the diphosphine ligand increases [85], large bite angle ligands favouring the trigonal bipyramidal geometry of the reactant complex over the octahedral geometry of the protonated product.

1.2.3.2 Steric bite angle effect

The steric bite angle effect is based on the change in steric interactions around the metal complex when the backbone of the bidentate ligand is modified, while keeping the substituents at the donor atoms constant. Since ligand–ligand and ligand–substrate interactions influence the relative energies of stable intermediates and transition states of the catalytic cycle, the change in bite angle has a direct influence on the activity and regioselectivity of the catalyst system (Figure 1.11).

Purely steric bite angle effects in catalysis are uncommon. Recently, we argued that the observed bite angle effect on the regioselectivity of the rhodium–diphosphine catalysed hydroformylation of alkenes, and the product selectivity in the CO–ethylene copolymerisation is largely steric in origin [81]. In the allylic alkylation reaction, both the regio- [86] and stereoselectivity [38,87] of the reaction are influenced by the steric properties of the bidentate ligands. The origin of both effects was attributed to the increased embracing of the metal by wide bite angle ligands. It should be noted that in this reaction also significant electronic effects were observed.

1.2.3.3 Steric versus electronic bite angle effects

Rational catalyst design relies on proper understanding of individual steps during a catalytic cycle. This process can be complicated, however, as each step will contain an associated transition state which cannot be observed directly. Therefore, investigating steric and electronic contributions to the transition state is often

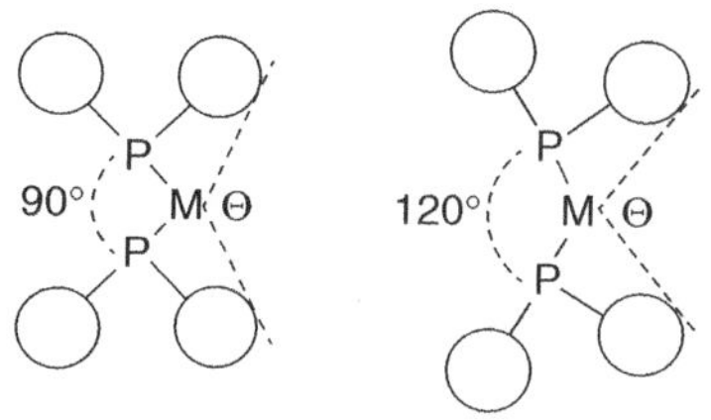

Figure 1.11 Steric bite angle effect. *θ* represents the pocket angle, as defined by Barron and co-workers [70].

approached computationally. Bickelhaupt *et al.* [88] have attempted to distinguish between steric and electronic bite angle effects on the oxidative addition of CH_3-X (where X=H, CH_3 and Cl) to palladium complexes with a series of bridged primary diphosphine ligands of the general structure $PH_2(CH_2)_nPH_2$ (where n=1–6) by performing DFT calculations. For comparison, the diphosphine complex $Pd(PH_3)_2$ was included in the study. They examined the strain energies of deformation experienced by both the catalyst and the substrate upon coordination and oxidative addition, the substrate–catalyst interaction energy, and the transition state bite and twist angles. Bickelhaupt *et al.* suggested that ligands enforcing narrow bite angles could alleviate steric interactions and lower the activation barrier, thereby essentially activating the metal complex to oxidative addition [88] (i.e. the bite angle effect is steric in nature). However, experimental studies by the group of van Leeuwen *et al.* revealed both steric and electronic effects on the activity of nickel-catalysed hydrocyanation of styrene [89,90]. These ligands were designed to activate the catalyst to undergo reductive elimination (the reverse of oxidative addition). Bickelhaupt's assertion would predict that a larger bite angle would give a higher activity for the reductive elimination. However, the observation was that the system had an optimum activity at a ligand bite angle of 105–106° with a loss in activity as the bite angle was further increased to 109° [89]. DFT calculations on the reductive elimination of alkyl cyanides from palladium–diphosphine complexes revealed that ligands favouring a bite angle of 105–106° optimally stabilise the transition state structure while destabilizing the square planar reactant [91]. This effect was observed for both methyl- and phenyl-substituted diphosphines, ligands with different steric properties. QM–MM calculations showed that the bite angle effect could be reproduced only if the electronic influences of the substituents at phosphorus were accounted for, demonstrating that for these ligands in this reaction the electronic bite angle effect dominates. Moreover, experimental studies showed that electron-withdrawing ligands further enhanced catalyst activity for a series of ligands with an effectively constant bite angle [90]. Therefore, evidence suggests that the bite angle effect is not solely steric in origin and does include electronic and orbital contributions, at least for Ni-catalysed hydrocyanation.

1.2.4 Molecular electrostatic potential (MESP) approach

The parameters for steric and electronic properties described until now are based on experimental data. Actual prediction of the properties of newly designed ligands would be very advantageous [92]. The use of molecular electrostatic potential (MESP) for the simultaneous evaluation of steric and electronic properties of phosphines was recently investigated by Suresh *et al.* [93,94].

The MESP approach was used to calculate the most negatively charged point in the lone pair region (i.e. the global minimum). The MESP is a physical property that can be determined either experimentally (by X-ray diffraction) or computationally from the electron density distribution ($\rho(r)$) (see Figure 1.12; Z_A is the charge of nucleus A located at R_A). Normally, lone pair regions on a molecule show minimum values of V due to a larger value of the electronic term in the equation as compared to the nuclear term.

MESP is defined as the energy required to bring a unit test positive charge from infinite to r. V_{min} points represent centres of negative charge on a molecule, and could be related to the 'location' of the lone pairs

$$V(r) = \sum_{A}^{N} \frac{Z_A}{|r - R_A|} - \int \frac{\rho(r')\, d^3 r'}{|r - r'|}$$

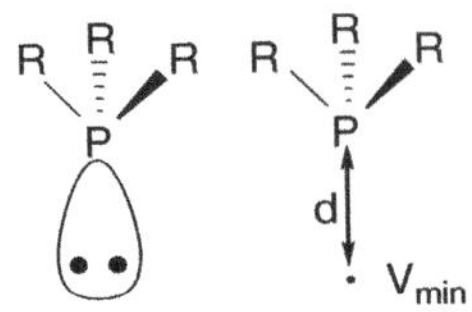

R	V_{min} (kcal/mol)	d (Å)
H	–28.22	1.850
Cy	–44.49	1.741
Ph	–34.85	1.816

Figure 1.12 *Calculation of the minimum electrostatic potential (V_{min}).*

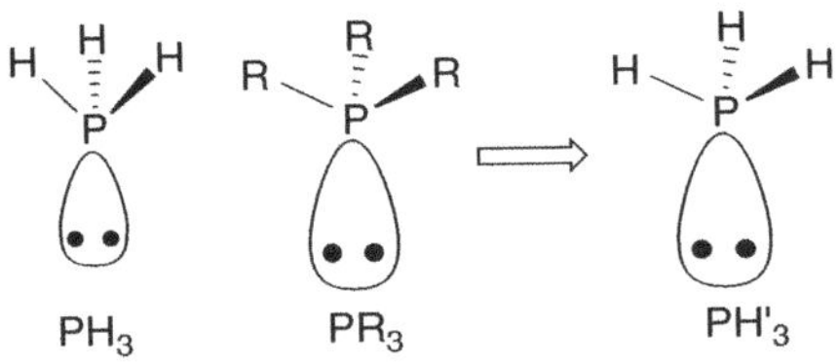

$$MESP_{steric} = V_{min}(PH_3) - V_{min}(PH'_3)$$

$$\theta_{MESP} = -8.8464\,(MESP_{steric}) + 87$$

Figure 1.13 *Schematic representation of the approach used to calculate $MESP_{steric}$ and the associated cone angle (θ_{MESP}).*

(Figure 1.13). The value of V_{min} is a quantitative measure of the 'lone pair strength', and consequently of the electron-donating properties of the phosphine [93]. The accuracy of this measure was proved by obtaining acceptable linear correlations with previously defined electronic parameters as pK_a, ν(CO), ΔE, ΔH° and E.

The V_{min} obtained by the MESP approach is inherently dependent on the steric properties of the ligand as well; the bulkier the substituents, the wider the R–P–R angle, and consequently the more p-character on the sp^3-hybridised lone pair orbital. This subtle change is reflected in an increased Lewis basicity of the phosphine and absolute value of V_{min}.

This fact can be used to evaluate steric parameters by a two-layer calculation, similar to the approach depicted in Figure 1.9. The intervalence R–P–R bond angles of the PR_3 ligand are determined by optimising the ligand structures using molecular mechanics (MM) calculations employing a universal force field. In the model, the R groups, representing the outer layer of the model, are connected to the P via imaginary link H atoms. These H atoms define an inner layer consisting of a PH_3 ligand, which is optimised sterically by the MM calculation as if H were the R group. The thus obtained (sterically influenced) optimal geometry of the inner layer is then evaluated using DFT calculations to determine V_{min}. The difference between this V_{min} and that of a fully optimised PH_3 structure is defined as the $MESP_{steric}$ parameter (i.e. the steric effect of R on the ligands electronic properties) (Figure 1.13).

A linear correlation was found between this parameter (MESPsteric) and Tolman's cone angle, and from this correlation, a new steric parameter (θ_{MESP}) can be defined. This methodology constitutes *the first quantum mechanically derived electronic quantity that interpreted the cone angle data* [94].

Fey and Harvey *et al.* [95,96] have recently approached the problem by developing ligand knowledge bases using DFT calculated ligand and ligand complex descriptors, including, but not limited to, descriptors for electronic properties (e.g. frontier molecular orbital energies and proton affinities), steric properties (e.g. bite angles and He_8-wedges) and other chemically relevant information such as metal–ligand bond lengths and changes in ligand geometry upon complexation [96]; the authors used their calculated descriptors to produce the first map of ligand space for bidentate ligands via principal component analysis (PCA). This approach to visualise correlations between ligand descriptors is useful but limited in its ability to quantitatively analyse ligand sets. Nevertheless such maps can prove very useful in qualitative analysis of ligand clustering, identifying ligand similarities and perhaps aiding in intuitive ligand design.

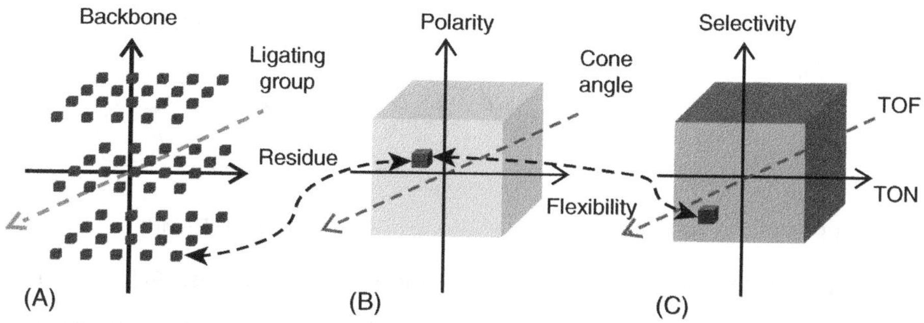

Figure 1.14 Simplified representation of the three multidimensional spaces A, B, and C. Reprinted from A. Maldonado, et al., 2009 with permission of Wiley-VCH Verlag GmbH & Co. KGaA.

Consequently, the authors have successfully demonstrated a computational approach to describing ligand properties in a computationally, chemically and statistically robust way. Currently, the limitation of such an approach, even by the authors' own admission, is in the availability of experimental data to test and challenge the calculated models and maps. However, they are optimistic that more even and extensive sampling of ligand structures, especially of more exotic systems and not just alkyl–aryl phosphines, could improve reliability and the predictive power of such maps, thus providing a valuable tool for ligand design and catalyst discovery [95,96].

Recently they reported an interesting application of their computational approach in which they detected fluorophosphines as promising ligands for rhodium-catalysed hydroformylation of alkenes [97]. This unexplored class of ligands was found to be close to phosphites in ligand space in their computational studies and indeed showed good performance in this catalytic reaction. This nicely illustrates that computational methods can have predictive power, leading to a new class of potentially useful ligands that probably would not have been explored otherwise.

Rothenberg and co-workers have developed another approach, which they called *in silico design*, aiming at identifying new active catalysts [98–102]. Instead of trying to understand ligand parameter control, they attempt to derive quantitative structure–activity relationships based upon topological descriptors, using a virtual library of known ligands in a multidimensional approach consisting of three spaces (Figure 1.14). Space A contains a library of catalysts, space B contains the catalyst and reaction descriptors (e.g. backbone flexibility and temperature) and space C contains figures of merit such as measures of catalyst performance (e.g. TON, TOF and ee), and also real-world concerns such as cost.

The relationships between B and C are quantified using quantitative structure–activity relationship (QSAR) and quantitative structure–property relationship (QSPR) models. New catalyst designs based on this computational method may be achieved using the validated model. Depending upon the size of the virtual library, millions of catalysts can be screened *in silico*, which could lead to new particularly interesting catalyst structures.

1.3 Asymmetric ligands

The use of chiral phosphine ligands for asymmetric catalysis has seen a spectacular growth in recent decades. The development of better asymmetric catalysts has been driven by the industrial need for enantiopure compounds that are not available in the chiral pool, and the costs and difficulties associated with the resolution of racemates. From the first report in 1968 by Knowles and Sabacky [103] on asymmetric hydrogenation using a known chiral phosphine (–)PMePh(*i*-Pr), a plethora of chiral mono- and diphosphines have appeared in the literature. Some early outstanding examples are shown in Figure 1.15 [7,104].

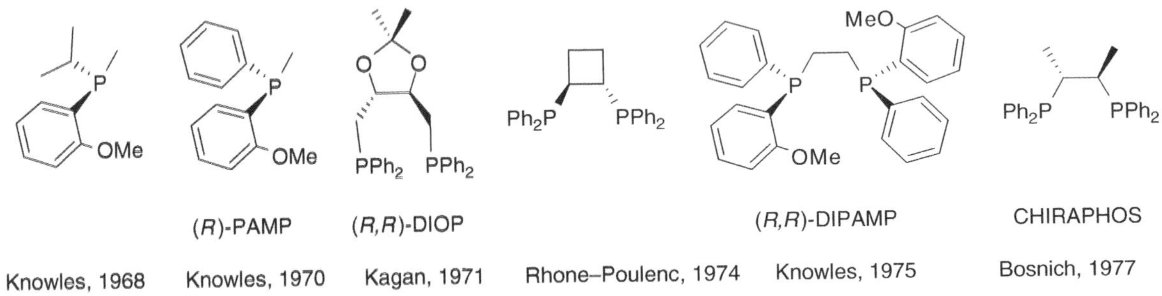

Figure 1.15 *Some chiral mono- and diphosphines relevant to asymmetric hydrogenation.*

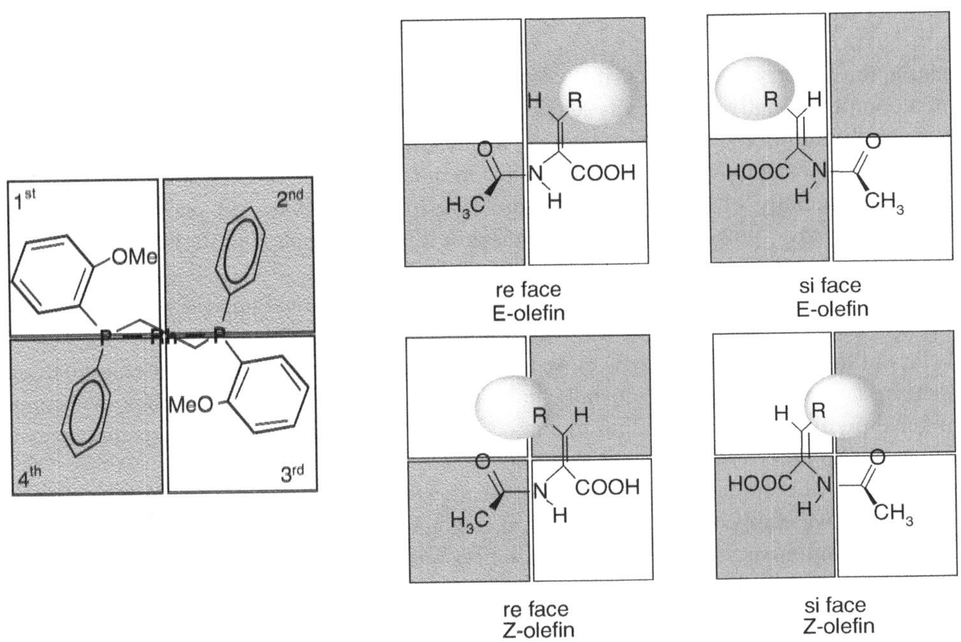

Figure 1.16 *Quadrants model for Rh-(R,R)-DIPAMP.*

In particular, chelating diphosphine ligands provided efficient catalysts for the hydrogenation of Z-enamides to form α-amino acids with high enantioselectivities (over 95% ee). Already in the early days, it was noted that all these ligands create a C_2-symmetric environment around the metal centre when coordinating to a square planar rhodium complex. In the case of DIPAMP, the X-ray structure of the catalyst precursor $[Rh(DIPAMP)(diene)]^+$ showed that enantioselectivity is achieved thanks to a restrained disposition of the phenyl groups, two of them being face exposed and the other two edge exposed (see Figure 1.16). A similar orientation is taken by other C_2-symmetric diphosphines, which have stereogenic centres on the backbone rather than on the phosphorus. This special arrangement divides the space available for substrate coordination into "four quadrants" of which two are closer to edge-exposed phenyls (second and fourth) and consequently more hindered than the ones which are closer to face-exposed phenyls (first and third). Coordination of the enamide to the metal centre takes place via the substituted alkene and the carbonyl oxygen as an additional donor atom. The preferred coordination face of the alkene (*re* or *si*, for short) will be the one that minimises the steric interaction with the phenyls of the ligand. The importance of steric interactions was soon corroborated by the observation that Z-enamides were more efficiently hydrogenated (faster and with higher ee) than E-enamides [105].

Although a direct correlation of the major adduct formed and the stereoinduction of the reaction could be attractive, detailed studies by Halpern [106] and Brown [107] proved that the major enantiomeric product does not derive from the most stable of the alkene adducts. In fact, the less stable intermediate alkene complex reacts faster in the subsequent reactions, showing the so called *anti-lock-and-key* behaviour.

This early example established the first attempted rationalization of the relationship between 'chirality of the ligand' and asymmetric induction. Although the final understanding required a better knowledge of the specific reaction mechanism, the quadrants model constitutes nowadays a useful tool to rationalise the stereoinduction observed when using many other C_2-symmetric ligands (some examples are cited in [74,108]). Historically, it has been accepted that C_2-symmetric ligands that effectively create a chiral environment around the metal centre are most efficient in asymmetric catalysis and consequently this kind of ligand has proliferated enormously in recent decades.

The quadrant model is a qualitative descriptor based on steric encumbrance. It intrinsically does not provide any quantitative predictions of enantioselectivity, but merely predicts the sign of the stereochemical outcome. Recently, the quadrants model has also been applied to describe the chiral environment created for some C_1-symmetric diphosphines [109], which effectively block three of the four quadrants by the substituents of the phosphorus atom (Figure 1.17). Also several C_1-symmetric phosphines have been very successful in many catalytic reactions. In most of the cases, not only steric but also electronic factors play a key role in the enantiodiscrimination [10,110].

Although the quadrants model is a very intuitive method based on simple symmetry principles, it is still one of the few successful tools used to design a chiral diphosphine. Nevertheless, several new methods are being developed by theoretical groups in the last years in an attempt to change this situation.

Similar to applying steric and electronic parameters for rational catalyst design, a parameter to quantify the chirality content of a molecule could facilitate similar design of chiral ligands and later on relate it with the enantiomeric outcome of the reaction. The definition in 1995 by Avnir of *chirality* as a continuous molecular structural property can be considered a milestone [111]. He defined a new parameter called *continuous chirality measure* (CCM, *S*), as shown in Figure 1.18.

Applied to molecules, P_i represent the positions of the *n* atoms on a molecule and P'_i the position that they would occupy in the 'nearest' achiral conformation. D is a normalization factor to make S independent of the

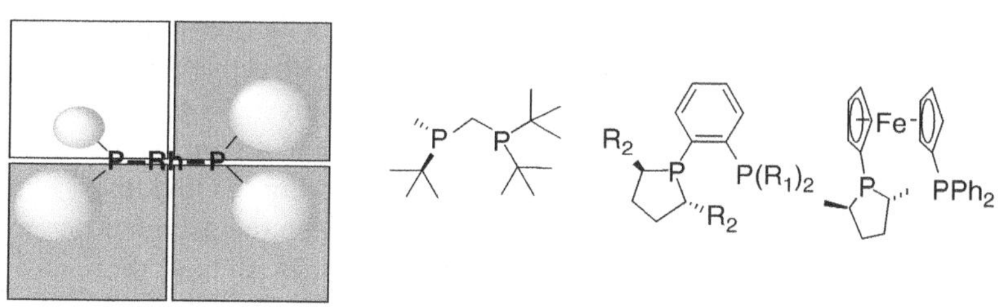

Figure 1.17 *Diphosphines fulfilling the three-full-quadrants model.*

$$S = \frac{100}{n\,D^2} \sum_{i=1}^{n} (P_i - P'_i)$$

Figure 1.18 *Continuous chirality measure.*

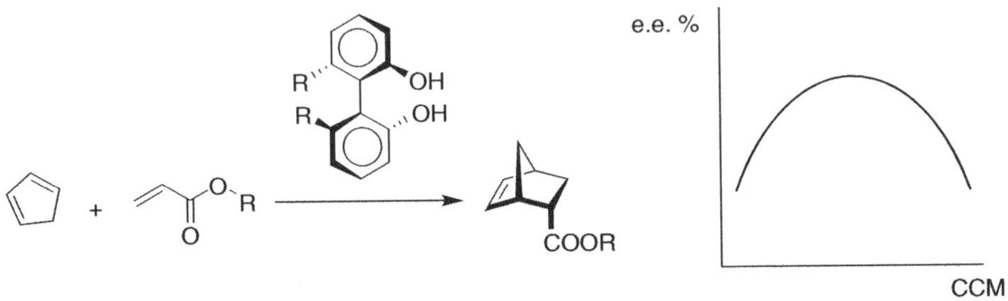

Figure 1.19 *Correlation between the ee obtained and CCM value for a 2,2′-biaryldiol catalysed Diels–Alder reaction.*

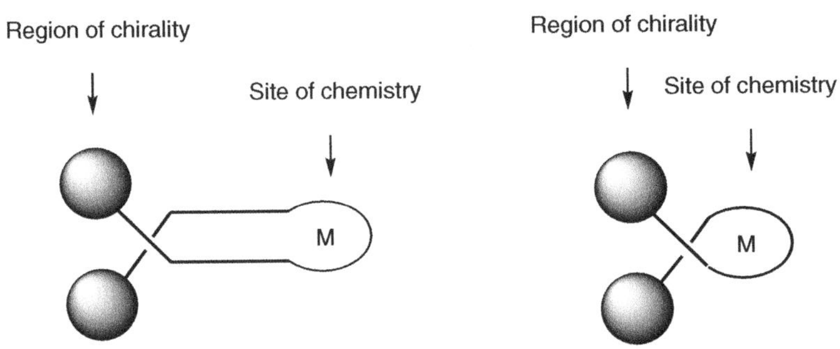

Figure 1.20 *Efficient and inefficient chiral catalyst as defined by Lipkowitz.*

size of the molecule. The S parameter varies from 0 for nonchiral molecules to 100, which is the highest possible value.

Later publications relating this new parameter with the enantiomeric excesses experimentally obtained showed encouraging success [112], but one example concerning Diels–Alder condensations catalysed by 2,2′-biaryldiols evidenced that it could not be generalised so easily [113].

They observed a nonlinear correlation between the computed CCM values and the experimental enantioselectivity. In fact, the correlation observed was basically the same as the one observed between the twist angle of the aryl rings and the ee (Figure 1.19). These results demonstrate that quantifying chirality does not automatically provide predictive power by itself, as an effective ligand should also create an asymmetric environment in the vicinity of the 'active site' of the catalyst. As formulated by Lipkowitz, 'chiral catalysts that are efficient at inducing asymmetry will have their region of maximum stereoinduction spatially congruent with the site of chemistry (bond making-breaking)'. Obviously, one of the important factors influencing asymmetric induction will be the distance between the asymmetric environment and the active site. If they are too far apart, the diastereomeric transition states will have the same or similar energies and no asymmetric induction will be observed (Figure 1.20).

In order to create a more effective methodology, Lipkowitz developed a procedure called *stereocartography* [114]. It comprises mapping of the most stereo-inducing regions around a catalyst. The calculations are done in the following manner: the centre of mass of the catalyst is placed at the centre of a Cartesian coordinates system, and a three-dimensional (3D) fine grid is defined around it. The molecule(s) reacting with it in their transition state conformation are placed sequentially at each grid point in all the possible orientations, and the intermolecular energy force is computed. The same procedure is followed for the pro-R and pro-S transition states. The results obtained for both enantiomeric transition state "substrates" are

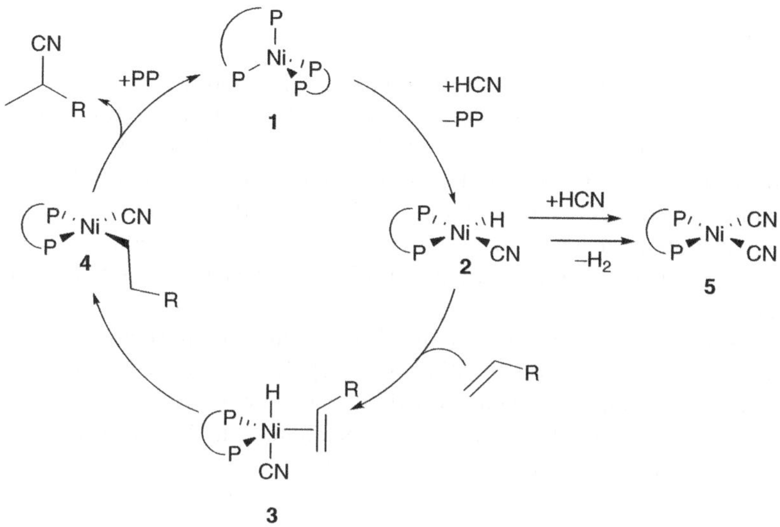

Scheme 1.1 *Simplified mechanism for nickel catalysed hydrocyanation.*

compared, locating the grid points of maximal energy difference which are considered as the points of maximal stereodifferentiation. Out of a set of 18 catalysts evaluated by using this methodology, 17 showed a correlation between the distance linking these maximal stereoinduction points to the region where the chemistry is actually taking place and the observed enantioselectivity.

1.4 Rational ligand design in nickel-catalysed hydrocyanation

1.4.1 Introduction

Despite advances in computational modelling and fundamental understanding of the influence of ligand parameters on organotransition metal chemistry, the predominant approaches to catalyst design employ a cycle of crude design, synthesis, testing and refinement of the design. Invariably, ligand design is an iterative process, building upon the performance of previous ligands and the understanding of the mechanism of the catalytic reaction. Sometimes this informed iterative approach can lead to remarkably predictable results, but on the other hand, surprising results are also often encountered. The nickel-catalysed hydrocyanation of alkenes, briefly presented here, is an example of an important catalytic reaction where detailed mechanistic knowledge and understanding have allowed for successful design of effective ligands.

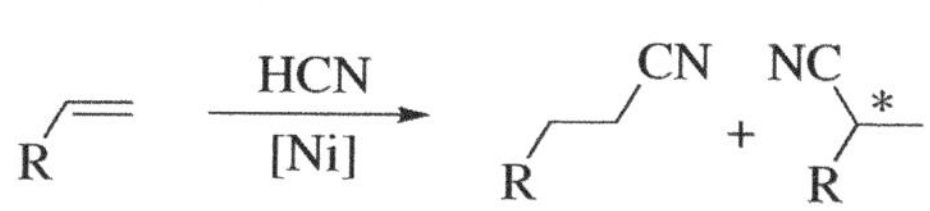

(Eq. 1.1)

Catalytic hydrocyanation (Eq. 1.1), the introduction of HCN across a double bond, was long ago established and industrialised in the well-known DuPont adiponitrile process. This process utilises a Ni/P(O-o-tolyl)$_3$/Lewis acid system. Since the commercialisation of the DuPont process, many groups have carried out detailed mechanistic studies [115–122], and Scheme 1.1 shows a simplified mechanism based on that proposed by McKinney and Roe [120,122].

1.4.2 Mechanistic insights

The catalytic cycle starts with a tetrahedral Ni(0) complex: (1) ligand dissociation followed by oxidative addition of HCN provides the intermediate square planar Ni(II) species; (2) coordination of the substrate olefin forms the intermediate trigonal bipyramidal π-olefin complex (3), which rapidly undergoes hydride migration to form the square planar σ-alkyl Ni(II) species; and (4) the final step is the reductive elimination of the product alkylnitrile and regeneration of the tetrahedral starting complex 1. A competing catalyst decomposition pathway exists when a second equivalent of HCN oxidatively adds to the square planar Ni(II) species 2, losing H_2 gas and generating a square planar biscyano Ni(II) species 5. The latter is inactive and cannot be easily regenerated.

Mechanistic studies of McKinney and Roe showed that reductive elimination is the rate-determining step in the Ni-catalysed hydrocyanation of alkenes [122]. Recent work by Vogt *et al.* showed that, also for the isomerisation of 2-methyl-3-butenenitrile to 3-pentenenitrile, reductive elimination was the rate-limiting step, and they observed a zero-order dependence on the substrate concentration [123]. The advantage of their study was that the reductive elimination step could be investigated without the complication of catalyst deactivation caused by HCN via the production of a bis cyano Ni species (5). Also, this model reaction facilitated the assessment of the effects of ligand parameters on the reductive elimination step.

1.4.3 Rational design

The first example of rational catalyst design for the hydrocyanation reaction was reported in the early 1990s. Pringle and co-workers explored chelating ligands with the intention to stabilise the nickel(0) complex and to reduce bis cyano complex formation, a disadvantage of the DuPont system. In order to retain similar electronic properties as the DuPont catalysts, they applied new chiral diphosphite ligands based upon bisphenol (L1) and bisnaphthol backbones (L2) [124,125] (Figure 1.21).

Although the 2,2′-bisphenol-based diphosphite ligand was less selective than the DuPont system for hydrocyanation of butadiene, a fourfold increase in turnover number was observed for the bidentate ligand system, and the catalyst stability was strongly enhanced [125]. Similar higher stabilities were observed for complexes of bisnaphthol diphosphite (L2) compared to the monodentate bisnaphthol phosphite (L3) and bidentate phosphines. Also much higher yields of up to 70% and ee's of up to 38% (with BPh_3 added) were

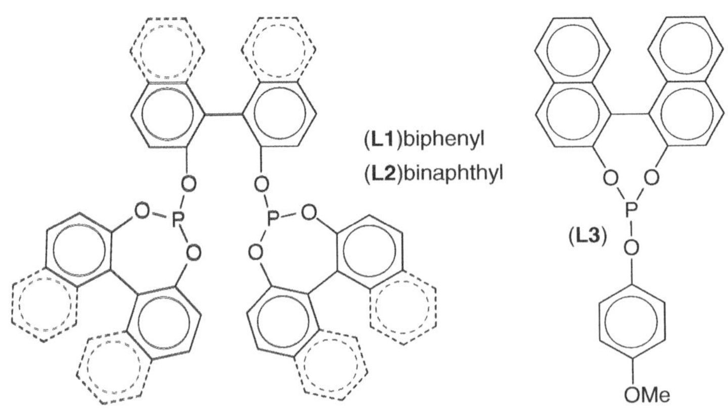

Figure 1.21 Biphenyl and binaphthyl.

observed in the hydrocyanation of norbornene using the nickel complex of ligand L2, which outperformed both the Ni–DIOP and Ni–monophosphite catalysts [124].

A further example of successful ligand design in hydrocyanation based on both mechanistic knowledge of the catalytic reaction and fundamental understanding of ligand parameters on the reactivity of metal complexes was reported by the groups of van Leeuwen and Vogt [89]. Prior to their report, application of phosphines as ligands had failed to produce results that could compete with phosphites [124]. The obvious reason is that the rate-limiting reductive elimination step is retarded by strongly σ-donating ligands such as phosphines.

The group of Kamer and van Leeuwen had prepared a series of ligands designed to enforce wide bite angles (β_n: 101–131°) (Figure 1.22), which was anticipated to destabilise the Ni(II) square planar resting state and stabilise the tetrahedral Ni(0) complex formed after reductive elimination. This was expected to compensate for the strongly donating effect of phosphines, and indeed greatly enhanced yields (up to 95%) and regioselectivities (up to 99%) compared to other diphosphines in the hydrocyanation of styrene (Table 1.1) [89]. The enhanced stabilization of the Ni(0) complexes formed after reductive elimination has a tremendous effect on the reaction rate, which is clearly illustrated in Figure 1.23, depicting a plot of the yield of hydrocyanation product versus the calculated natural bite angle.

The electronic ligand effects were further explored by testing a series of electronically tuned Thixantphos ligands (Figure 1.24, L4–L10) in the hydrocyanation of styrene (Table 1.2) [90].

The extremes of this series behaved as expected; ligand L4 gave the poorest conversion and yield (16% and 12%, respectively) whilst the most electron-withdrawing ligand, L10, gave nearly total conversion (98%) and a yield of 90%. Moreover, a high degree of regioselectivity was maintained across the series (>99% in all cases).

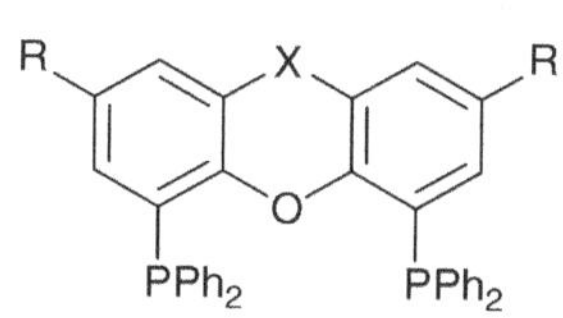

	X	R	β_n
DPEphos	H,H	H	101
Sixantphos	$SiMe_2$	H	105
Thixantphos	S	Me	106
Xantphos	$C(Me)_2$	H	109
DBFphos	Bond	H	131

Figure 1.22 *Series of wide bite angle diphosphine ligands.*

Table 1.1 *Nickel–phosphine catalysed hydrocyanation of styrene.*

Ligand	β_n	Yield (%)	Branched product (%)
DPEphos	101	35–41	88–91
Sixantphos	**105**	**94–95**	**97–98**
Thixantphos	106	69–92	96–98
Xantphos	109	27–75	96–99
DBFphos	131	0.7	83
PPh_3	—	0	—
DPPE	78	<1	ca. 40
DPPP	87	4–11	ca. 90
DPPB	98	3–8	92–95
BINAP	85	4	29

Source: Reproduced from [89].

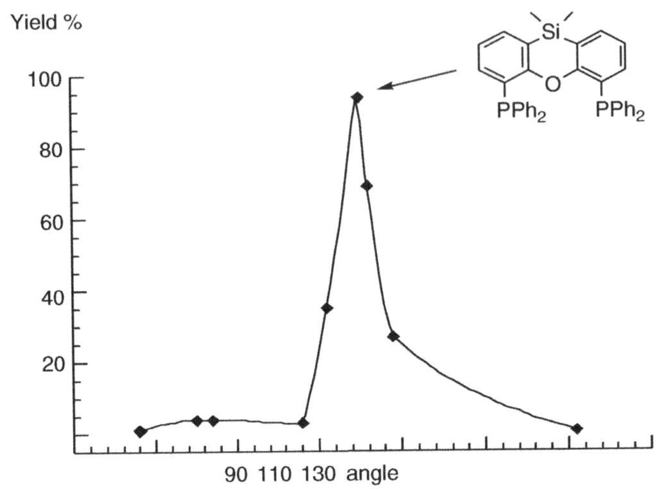

Figure 1.23 Effect of ligand bite angle on product yield in nickel-catalysed hydrocyanation of styrene.

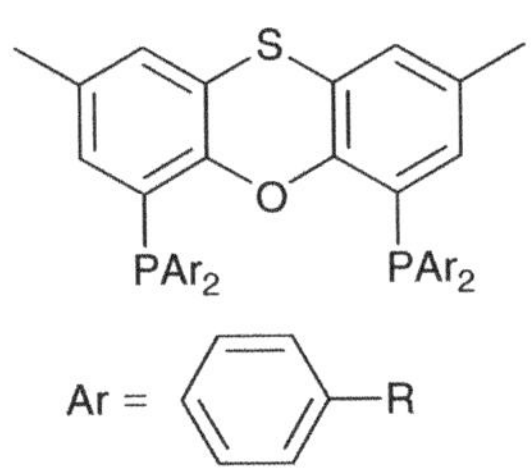

Ligand	R	β_n
L4	Me_2N	118.6
L5	MeO	113.8
L6	Me	112.9
L7	H	112.8
L8	F	112.8
L9	Cl	113.4
L10	CF_3	114.5

Figure 1.24 Electronically tuned Thixantphos ligands.

Table 1.2 Catalytic hydrocyanation of styrene.

Ligand	Conversion (%)	Yield (%)	Selectivity (%)	Regioselectivity (%)
L4	16	12	76	>99
L5	33	22	66	99.5
L6	55	48	87	99.8
L7	77	70	92	99.4
L8	46	38	83	99.4
L9	75	52	70	99.3
L10	98	90	92	99.0

Source: Reproduced from [90].

1.5 Conclusions

The development of organotransition metal chemistry has contributed significantly to the enormous growth of homogeneous catalysis. Knowledge about bonding and reactivity in organometallic chemistry has been of great support to catalysis. The reactivity of organotransition metal complexes is to an astounding extent dependent on the ligand environment around the metal. By changing the ligands, the catalysis can be directed and sometimes the catalytic behaviour can even be predicted. Still, systematic studies involving small ligand families are the best way to elucidate detailed mechanistic information and investigate subtle ligand effects. The development of new and the optimisation of existing catalytic reactions can be achieved via the rational design of ligands. The development of qualitative and quantitative ligand parameters has proven to be an increasingly useful tool in this respect.

Computational methods have become increasingly important in catalyst development. Not only have advanced higher level theoretical methods contributed enormously to mechanistic understanding of catalytic reactions, but also proper development of ligand descriptors has contributed to the discovery of completely new families of ligands, catalysts derived from them and even new catalytic transformations. The rapid growth of knowledge and understanding of catalytic reactions in combination with advances in computational design will undoubtedly lead to further improvement of rational ligand design.

Arguably, phosphorus donor compounds are among the most successfully applied ligands in homogeneous catalysis. This might be due to the vast amount of detailed structural information that is readily available by, for example, *in situ* ^{31}P NMR. This has definitely contributed to proper understanding of ligand effects and the development of reliable ligand parameters. Computational approaches such as ligand knowledge bases have already led to the discovery of unexpected ligand structures that were less likely to be explored by experimental catalysis researchers. In this book, we decided to include not only ligands which have shown ample successful applications in homogeneous catalysis, such as phosphines and phosphites, but also ligand classes which have been considered as chemical curiosities for a long time. There are many phosphorus compounds out there which have been studied only by fundamental main group chemists and are considered to be difficult to prepare and handle. These compounds, however, might possess the right steric and electronic ligand properties required for novel catalysts to be discovered. To facilitate the introduction of less studied ligand groups in new areas, we have included representative experimental procedures provided by experts in the field for each class of ligands described in the following chapters.

References

[1] R. Noyori, *Angew. Chem. Int. Ed.* **2002**, *41*, 2008.

[2] W. S. Knowles, *Angew. Chem. Int. Ed.* **2002**, *41*, 1998.

[3] R. H. Grubbs, *Angew. Chem. Int. Ed.* **2006**, *45*, 3760.

[4] E.-i. Negishi, *Angew. Chem. Int. Ed.* **2011**, *50*, 6738.

[5] A. Suzuki *Angew. Chem. Int. Ed.* **2011**, *50*, 6722.

[6] W. S. Knowles, M. J. Sabacky, D. Vineyard, *J. Chem. Soc., Chem. Commun.* **1972**, 10.

[7] W. S. Knowles, M. J. Sabacky, B. D. Vineyard, D. J. Weinkauff, *J. Am. Chem. Soc.* **1975**, *97*, 2567.

[8] M. van den Berg, A. J. Minnaard, E. P. Schudde. J. van Esch, A. H. M. de Vries, J. G. de Vries, B. L. Feringa, *J. Am. Chem. Soc.* **2000**, *122*, 11539.

[9] (a) J. E. Babin, G. T. Whiteker, W.O. 93 03839, US Pat. 911 518, **1992** (to Union Carbide Corporation); *Chem. Abstr.* **1993**, *119*, 159872; (b) G. J. H. Buisman, E. J. Vos, P. C. J. Kamer, P. W. N. M. van Leeuwen, *J. Chem. Soc. Dalton Trans.* **1995**, 409; and (c) G. J. H. Buisman, L. A. van der Veen, A. Klootwijk, W. G. J. de Lange, P. C. J. Kamer, P. W. N. M. van Leeuwen, D. Vogt, *Organometallics* **1997**, *16*, 2929.

[10] N. Sakai, S. Mano, K. Nozaki, H. Takaya, *J. Am. Chem. Soc.* **1993**, *115*, 7033.

[11] B. M. Trost, D. L. Van Vranken, *Angew. Chem. Int. Ed.* **1992**, *31*, 228.

[12] A. Pfaltz, R. Prétôt, *Angew. Chem. Int Ed.* **1998**, *37*, 323.

[13] (a) P. Dierkes, S. Ramdeehul, L. Barloy, A. De Cian, J. Fischer, P. C. J. Kamer, P. W. N. M. van Leeuwen, J. A. Osborn, *Angew. Chem., Int. Ed.* **1998**, *37*, 3116; and (b) S. Ramdeehul, P. Dierkes, R. Aguado, P. C. J. Kamer, P. W. N. M. van Leeuwen, J. A. Osborn, *Angew. Chem. Int. Ed.* **1998**, *37*, 3118.

[14] T. P. Yoon, E. N. Jacobsen, *Science* **2003**, *299*, 1691.

[15] W. Reppe, W. J. Schweckendiek, *Annalen* **1948**, *104*, 560.

[16] O. Roelen, Chemische Verwertungsgesellschaft Oberhausen m.b.H., **1938/1952**, DE 849548, Germany; O. Roelen, *Chem. Exp. Didakt.* **1977**, *3*, 119; and O. Roelen, Chemische Verwertungsgesellschaft Oberhausen m.b.H., **1943**, US 2327066.
[17] W. C. Drinkard, R. V. Lindsey, Du Pont, **1970**, US 3655723; E. S. Brown, In *Aspects of Homogeneous Catalysis*; R. Ugo, Ed., Reidel, **1974**, *Vol. 2*, 57–78.
[18] C. A. Tolman, *Chem. Rev.* **1977**, *77*, 313.
[19] J. F. Young, J. A. Osborn, F. A. Jardine, G. Wilkinson *J. Chem. Soc., Chem. Comm.* **1965**, 131.
[20] C. O'Connor, G. Wilkinson, *Tetrahedron Lett.* **1969**, *18*, 1375.
[21] L. Vaska, R. E. Rhodes, *J. Am. Chem. Soc.* **1965**, *87*, 4970.
[22] D. Evans, J. A. Osborn, G. Wilkinson, *J. Chem. Soc. A.* **1968**, 3133; and F. H. Jardine, J. A. Osborn, G. Wilkinson, J. F. Young, *Chem. and Ind. (London)* **1965**, 560.
[23] R. L. Pruett, J. A. Smith, *J. Org. Chem.* **1969**, *34*, 327.
[24] K. Issleib, D. W. Müller, *Chem. Ber.* **1959**, *92*, 3175.
[25] W. Hieber, W. Freyer, *Chem. Ber.* **1960**, *93*, 462; J. Chatt, W. J. Hart, *J. Chem. Soc.* **1960**, 1378.
[26] L. G. Cannel, L. H. Slaugh, R. D., Mullineaux, Shell Development, **1960**, DE 1186455.
[27] R. S. Bauer, P. W. Glockner, W. Keim, H. van Zwet, H. Chung, Shell Development, **1972**, US 3644563.
[28] M. Iwamoto, S. Yuguchi, *J. Org. Chem.* **1966**, *31*, 4290.
[29] D. L. Thorn, R. J. Hoffmann, *Am. Chem. Soc.* **1978**, *100*, 2079.
[30] T. P. Dang, H. B. Kagan, *J. Am. Chem. Soc.* **1972**, *92*, 6429.
[31] J. C. Poulin, T. P. Dang, H. B. Kagan, *J. Organomet. Chem.* **1975**, *84*, 87.
[32] R. Selke, H. Pracejus, *J. Mol. Catal.* **1986**, *37*, 213.
[33] A. Miyashita, A. Yasuda, H. Tanaka, K. Toriumi, T. Ito, T. Souchi, R. Noyori, *J. Am. Chem. Soc.* **1980**, *102*, 7932.
[34] M. J. Burk, *J. Am. Chem. Soc.* **1991**, *113*, 8518.
[35] A. Togni, *Angew. Chem.* **1993**, *108*, 1581.
[36] *Metal-catalyzed Cross-coupling Reactions*, F. Diederich, P. J. Stang, Eds., Wiley-VCH, **1998**.
[37] M-N. Birkholz (née Gensow), Z. Freixa, P. W. N. M. van Leeuwen, *Chem. Soc. Rev.*, **2009**, *38*, 1099.
[38] B. M. Trost, D. J. Murphy, *Organometallics* **1985**, *4*, 1143.
[39] C. P. Butts, E. Filali, G. C. Lloyd-Jones, P.-O. Norrby, D. A. Sale, Y. Schramm, *J. Am. Chem. Soc.* **2009**, *131*, 9945.
[40] M. Beller, T. H. Riermeijer, S. Haber, H. J. Kleiner, W. A. Herrmann, *Chem. Ber.* **1996**, *129*, 1259.
[41] D. A. Albisson, R. B. Bedford, S. E. Lawrence, P. N. Scully, *J. Chem. Soc., Chem. Commun.* **1998**, 2095.
[42] G. P. F. van Strijdonck, M. D. K. Boele, P. C. J. Kamer, J. G. de Vries, P. W. N. M. van Leeuwen, *Eur. J. Inorg. Chem.* **1999**, 1073.
[43] P. W. N. M. van Leeuwen, C. F., Roobeek, Shell Oil, GB 2068377 and US 4467116.
[44] A. van Rooy, E. N. Orij, P. C. J. Kamer, P. W. N. M. van Leeuwen, *Organometallics* **1995**, *14*, 34; T. Jongsma, B. Challa, P. W. N. M. van Leeuwen, *J. Organomet. Chem.* **1991**, *421*, 121; and P. W. N. M. van Leeuwen, C. F. Roobeek, *J. Organomet. Chem.* **1983**, *258*, 343.
[45] L. H. Slaugh, R. D. Mullineaux, *J. Organomet. Chem.* **1968**, *13*, 469; and J. L. Van Winkle, S. Lorenzo, R. C. Morris, R. F. Mason, US Patent, 3,420,898, **1969**, US Appl. 1965-443703, *Chem. Abstr.* **1967**, *66*, 65101.
[46] T. Onoda, *Chem. Tech.* **1993**, *Sept.*, 34.
[47] E. Billig, A. G. Abatjoglou, D. R. Bryant, R. E. Murray, J. M., Maher, Union Carbide Corp., **1986**, US 4599206.
[48] E. Billig, A. G. Abatjoglou, D. R., Bryant, Union Carbide Corp., **1987**, US 4769498, US 4668651 and US 4748261.
[49] N. Yoshinura, Y. Tokito, Kuraray, EP 223103.
[50] C. P. Casey, G. T. Whiteker, M. G. Melville, L. M. Petrovich, J. A. Gavney Jr., D. R. Powell, *J. Am. Chem. Soc.* **1992**, *114*, 5535.
[51] C. P. Casey, G. T. Whiteker, *Isr. J. Chem.* **1990**, *20*, 299.
[52] M. Kranenburg, Y. E. M. van der Burgt, P. C. J. Kamer, P. W. N. M. van Leeuwen, K. Goubitz, J. Fraanje, *Organometallics* **1995**, *14*, 3081.
[53] P. C. J. Kamer, P. W. N. M. van Leeuwen, J. N. H. Reek, *Acc. Chem. Res.* **2001**, *34*, 895.
[54] W. Strohmeier, F. J. Müller, *Chem. Ber.* **1967**, *100*, 2812.
[55] C. A. Tolman, *J. Am. Chem. Soc.* **1970**, *92*, 2953.
[56] A. Roodt, S. Otto, G. Steyl, *Coord. Chem. Rev.* **2003**, *245*, 121.

[57] S. X. Xiao, W. C. Trogler, D. E. Ellis, Z. Berkovitch-Yellin, *J. Am. Chem. Soc.* **1983**, *105*, 7033; D. S. Marynick, *J. Am. Chem. Soc.* **1984**, *106*, 4064; J. A. Tossel, J. H. Moore, J. C. Giordan, *Inorg. Chem.* **1985**, *24*, 1100; and A. G. Orpen, N. G. Connelly, *Organometallics* **1990**, *9*, 1206.

[58] R. S. Drago, S. Joerg, *J. Am. Chem. Soc.* **1996**, *118*, 2654.

[59] S. Joerg, R. S. Drago, J. Sales, *Organometallics* **1998**, *17*, 589.

[60] W. A. G. Graham, *Inorg. Chem.* **1968**, *7*, 315; and R. P. Stewart, P. M. Teichel, *Inorg. Chem.* **1968**, *7*, 1942.

[61] M. N. Golovin, M. M. Rahman, J. E. Belmonte, W. P. Giering, *Organometallics* **1985**, *4*, 1981; and A. Fernandez, M. R. Wilson, D. C. Woska, A. Prock, W. P. Giering, C. M. Haar, S. P. Nolan, B. M. Foxman, *Organometallics* **2002**, *21*, 2758.

[62] S. Tobisch, T. Ziegler, *J. Am. Chem. Soc.* **2002**, *124*, 13290.

[63] D. Gleich, J. Hutter, *Chem. Eur. J.* **2004**, *10*, 2435.

[64] G. Pacchioni, P. S. Bagus, *Inorg. Chem.* **1992**, *31*, 4391.

[65] O. Gonzalez-Blanco, V. Branchadell, *Organometallics* **1997**, *16*, 5556.

[66] C. R. Landis, S. Feldgus, J. Uddin, C. E. Wozniak, *Organometallics* **2000**, *19*, 4878.

[67] T. L. Brown, K. J. Lee, *Coord. Chem. Rev.* **1993**, *128*, 89.

[68] D. White, B. C. Tavener, N. J. Coville, P. W. Wade, *J. Organomet. Chem* **1995**, *495*, 41.

[69] T. L. Brown, *Inorg. Chem.* **1992**, *31*, 1286.

[70] Y. Koide, S. G. Bott, A. R. Barron, *Organometallics*, **1996**, *15*, 2213.

[71] K. Angermund, W. Baumann, E. Dinjus, R. Fornika, H. Goerls, M. Kessler, C. Krueger, W. Leitner, F. Lutz, *Chem.-Eur. J.*, **1997**, *3*, 755.

[72] F. Maseras, K. Morokuma, *J. Comput. Chem.* **1995**, *16*, 1170; and M. Svensson, S. Humbel, R. D. J. Froese, T. Matsubara, S. Sieber, K. Morokuma, *J. Phys. Chem.* **1996**, *100*, 19357.

[73] C. R. Landis, S. Feldgus, *Angew. Chem. Int. Ed.* **2000**, *39*, 2863; S. Feldgus, C. R. Landis, *J. Am. Chem. Soc.* **2000**, *122*, 12714.

[74] S. Feldgus, C. R. Landis, *Organometallics* **2001**, *20*, 2374.

[75] A. M. Segarra, E. Daura-Oller, C. Claver, J. M. Poblet, C. Bo, E. Fernandez, *Chem. Eur. J.* **2004**, *10*, 6456.

[76] D. Gleich, R. Schmid, W. A. Herrmann, *Organometallics* **1998**, *17*, 4828; J. J. Carbo, F. Maseras, C. Bo, P. W. N. M. van Leeuwen, *J. Am. Chem. Soc.* **2001**, *123*, 7630; S. A. Decker, T. R. Cundari, *J. Organomet. Chem.* **2001**, *635*, 132; and C. R. Landis, J. Uddin, *J. Chem. Soc., Dalton Trans.* **2002**, 729.

[77] A. Magistrato, A. Togni, U. Rothlisberger, *Organometallics* **2006**, *25*, 1151.

[78] K. Vanka, Z. T. Xu, T. Ziegler, *Organometallics* **2005**, *23*, 419; and L. Q. Deng, T. K. Woo, L. Cavallo, P. M. Margl, T. Ziegler, *J. Am. Chem. Soc.* **1997**, *119*, 6177.

[79] P. Dierkes, P. W. N. M. van Leeuwen, *J. Chem. Soc., Dalton Trans.* **1999**, 1519.

[80] P. W. N. M. van Leeuwen, P. C. J. Kamer, J. N. H. Reek, P. Dierkes, *Chem. Rev.* **2000**, *100*, 2741.

[81] Z. Freixa, P. W. N. M. van Leeuwen, *J. Chem. Soc., Dalton Trans.* **2003**, 1890.

[82] S. Otsuka, *J. Organomet. Chem.* **1980**, *200*, 191.

[83] A. Miedaner, R. C. Haltiwanger, D. L. DuBois, *Inorg. Chem.* **1991**, *30*, 417.

[84] D. E. Berning, B. C. Noll, D. L. DuBois, *J. Am. Chem. Soc.* **1999**, *121*, 11432; and J. W. Raebiger, A. Miedaner, C. J. Curtis, S. M. Miller, O. P. Anderson, D. L. DuBois, *J. Am. Chem. Soc.* **2004**, *126*, 5502.

[85] R. J. Angelici, *Acc. Chem. Res.* **1995**, *28*, 51.

[86] R. J. van Haaren, H. Oevering, B. B. Coussens, G. P. F. van Strijdonck, J. N. H. Reek, P. C. J. Kamer, P. W. N. M. van Leeuwen, *Eur. J. Inorg. Chem.* **1999**, 1237.

[87] B. M. Trost, D. L. van Vranken, C. Bingel, *J. Am. Chem. Soc.* **1992**, *114*, 9327.

[88] W. J. van Zeist, R. Visser, F. M. Bickelhaupt, *Chem.-Eur. J.*, **2009**, *15*, 6112.

[89] M. Kranenburg, P. C. J. Kamer, P. W. N. M. van Leeuwen, D. Vogt, W. Keim, *J. Chem. Soc., Chem. Commun.*, **1995**, 2177.

[90] W. Goertz, W. Keim, D. Vogt, U. Englert, M. D. K. Boele, L. A. van der Veen, P. C. J. Kamer, P. W. N. M. van Leeuwen, *J. Chem. Soc., Dalton Trans.*, **1998**, 2981.

[91] E. Zuidema, P. W. N. M. van Leeuwen, C. Bo, *Organometallics*, **2005**, *15*, 3703.

[92] O. Kühl, *Coord. Chem. Rev.* **2005**, *249*, 693.

[93] C. H. Suresh, N. Koga, *Inorg. Chem.* **2002**, *41*, 1573.

[94] C. H. Suresh, *Inorg. Chem.* **2006**, *45*, 4982.
[95] N. Fey, A. G. Orpen, J. N. Harvey, *Coord. Chem. Rev.* **2009**, *253*, 704.
[96] N. Fey, J. N. Harvey, G. C. Lloyd-Jones, P. Murray, A. G. Orpen, R. Osborne, M. Purdie, *Organometallics* **2008**, *27*, 1372.
[97] P. G. Pringle, *Angew. Chem. Int. Ed.* **submitted**.
[98] E. Burello, G. Rothenberg, *Int. J. Mol. Sci.* **2006**, *7*, 375.
[99] E. Burello, P. Marion, J. C. Galland, A. Chamard, G. Rothenberg, *Adv. Synth. Catal.* **2005**, *347*, 803.
[100] E. Burello, G. Rothenberg, *Adv. Synth. Catal. 2005*, 347, **1969**.
[101] J. A. Hageman, J. A. Westerhuis, H. W. Fruhauf, G. Rothenberg, *Adv. Synth. Catal.* **2006**, *348*, 361.
[102] A. G. Maldonado, J. A. Hageman, S. Mastroianni, G. Rothenberg, *Adv. Synth. Catal.* **2009**, *351*, 387.
[103] W. S. Knowles, M. J. Sabacky, *Chem. Commun.* **1968**, 1445.
[104] M. D. Fryzuk, B. Bosnich, *J. Am. Chem. Soc.* **1979**, *101*, 3043; and T. P. Dang, H. B. Kagan, *J. Chem. Soc. Chem. Commun.* **1971**, 481.
[105] W. S. Knowles, *Acc. Chem. Res.* **1983**, *16*, 106; and B. D. Vineyard, W. S. Knowles, M. J. Sabacky, G. L. Bachman, D. J. Weinkauff, *J. Am. Chem. Soc.* **1977**, *99*, 5946.
[106] C. R. Landis, J. Halpern, *J. Am. Chem. Soc.* **1987**, *109*, 1746; J. Halpern, *Science* **1982**, *217*, 401; and A. S. C. Chan, J. J. Pluth, J. Halpern, *J. Am. Chem. Soc.* **1980**, *102*, 5952.
[107] J. M. Brown, P. A. Chaloner, *J. Am. Chem. Soc.* **1980**, *102*, 3040.
[108] R. Noyori, H. Takaya, *Acc. Chem. Res.* **1990**, *23*, 345; T. Imamoto, J. Watanabe, Y. Wada, H. Masuda, H. Yamada, H. Tsuruta, S. Matsukawa, K. Yamaguchi, *J. Am. Chem. Soc.* **1998**, *120*, 1635; E. Fernandez, A. Gillon, K. Heslop, E. Horwood, D. J. Hyett, A. G. Orpen, P. G. Pringle, *Chem. Commun.* **2000**, 1663; M. J. Burk, J. E. Feaster, W. A. Nugent, R. L. Harlow, *J. Am. Chem. Soc.* **1993**, *115*, 10125; and M. J. Burk, *Accounts Chem. Res.* **2000**, *33*, 363.
[109] K. Matsumura, H. Shimizu, T. Saito, H. Kumobayashi, *Adv. Synth. Catal.* **2003**, *345*, 180; G. Hoge, H. P. Wu, W. S. Kissel, D. A. Pflum, D. J. Greene, J. Bao, *J. Am. Chem. Soc.* **2004**, *126*, 5966; and S. Basra, J. G. de Vries, D. J. Hyett, G. Harrison, K. M. Heslop, A. G. Orpen, P. G. Pringle, K. von der Luehe, *Dalton Trans.* **2004**, 1901.
[110] A. Togni, C. Breutel, A. Schnyder, F. Spindler, H. Landert, A. Tijani, *J. Am. Chem. Soc.* **1994**, *116*, 4062; T. V. RajanBabu, A. L. Casalnuovo, *J. Am. Chem. Soc.* **1996**, *118*, 6325; K. W. Kottsieper, U. Kuhner, O. Stelzer, *Tetrahedron: Asymmetry* **2001**, *12*, 1159; P. Kocovsky, S. Vyskocil, M. Smrcina, *Chem. Rev.* **2003**, *103*, 3213; T. Hayashi, M. Kumada, *Acc. Chem. Res.* **1982**, *15*, 395; S. Castillon, C. Claver, Y. Diaz, *Chem. Soc. Rev.* **2005**, *34*, 702; and K. Achiwa, *J. Am. Chem. Soc.* **1976**, *98*, 8265.
[111] H. Zabrodsky, D. Avnir, *J. Am. Chem. Soc.* **1995**, *117*, 462.
[112] K. B. Lipkowitz, S. Schefzick, D. Avnir, *J. Am. Chem. Soc.* **2001**, *123*, 6710; and J. W. Handgraaf, J. N. H. Reek, L. Bellarosa, F. Zerbetto, *Adv. Synth. Catal.* **2005**, *347*, 792.
[113] D. Q. Gao, S. Schefzick, K. B. Lipkowitz, *J. Am. Chem. Soc.* **1999**, *121*, 9481.
[114] K. B. Lipkowitz, T. Sakamoto, J. Stack, *Chirality* **2003**, *15*, 759; and K. B. Lipkowitz, C. A. D'Hue, T. Sakamoto, J. N. Stack, *J. Am. Chem. Soc.* **2002**, *124*, 14255.
[115] J. E. Backvall, O. S. Andell, *J. Chem. Soc., Chem. Commun.*, **1981**, 1098–1099.
[116] W. R. Jackson, C. G. Lovel, *Tetrahedron Lett.*, **1982**, *23*, 1621–1624.
[117] J. E. Backvall, O. S. Andell, *J. Chem. Soc., Chem. Commun.*, **1984**, 260–261.
[118] C. A. Tolman, W. C. Seidel, J. D. Druliner, P. J. Domaille, *Organometallics*, **1984**, *3*, 33–38.
[119] J. D. Druliner, *Organometallics*, **1984**, *3*, 205–208.
[120] R. J. McKinney, D. C. Roe, *J. Am. Chem. Soc.*, **1985**, *107*, 261–262.
[121] J. E. Backvall, O. S. Andell, *Organometallics*, **1986**, *5*, 2350–2355.
[122] R. J. McKinney, D. C. Roe, *J. Am. Chem. Soc.*, **1986**, *108*, 5167–5173.
[123] J. Wilting, C. Muller, A. C. Hewat, D. D. Ellis, D. M. Tooke, A. L. Spek, D. Vogt, *Organometallics*, **2005**, *24*, 13–15.
[124] M. J. Baker, P. G. Pringle, *J. Chem. Soc., Chem. Commun.*, **1991**, 1292–1293.
[125] M. J. Baker, K. N. Harrison, A. G. Orpen, P. G. Pringle, G. Shaw, *J. Chem. Soc., Chem. Commun.*, **1991**, 803–804.

2
Chiral Phosphines and Diphosphines

Wei Li and Xumu Zhang

Department of Chemistry and Chemical Biology, Rutgers, The State University of New Jersey, Piscataway, United States

2.1 Introduction

In all the homogeneous catalytic methodologies, trivalent phosphorus compounds play a particularly important role. Since 1968 a vast number of catalytic systems have been developed and studied in the enantioselective reactions, having a significant impact on the fast-growing area of asymmetric catalysis. Among these developed chiral catalysts, phosphines and diphosphines occupy a key category due to their ligating properties to form catalytically active transition metal complexes, since the seminal discoveries from Knowles [1], Kagan [2], and Horner [3].

Driven by both academic curiosity and industrial applications, hundreds of chiral phosphines and disphosphines have been developed based on different scaffolds and via different preparation methods, providing important accesses to chiral products with excellent enantioselectivies.

This chapter reviews the up-to-date achievement of developing chiral diphosphines and their applications in homogeneous transition metal catalyzed-asymmetric transformations. We intend to emphasize the preparation of various phosphine ligand families and attempt to trigger new insights and inventiveness in the development of new ligands and new methodologies. Readers should be aware that only the most important and representative results are summarized, due to the word limit and large diversity of ligand families.

2.1.1 Early developments

The first essential discovery in asymmetric catalysis was made in the area of asymmetric hydrogenation soon after the milestone breakthrough made by Wilkinson *et al.* [4]. The first chiral ligand introduced, methyl(isopropyl)phenylphosphine, was applied in the hydrogenation of α-substituted acrylic acids. Although

Phosphorus(III) Ligands in Homogeneous Catalysis: Design and Synthesis, First Edition.
Edited by Paul C. J. Kamer and Piet W. N. M. van Leeuwen.

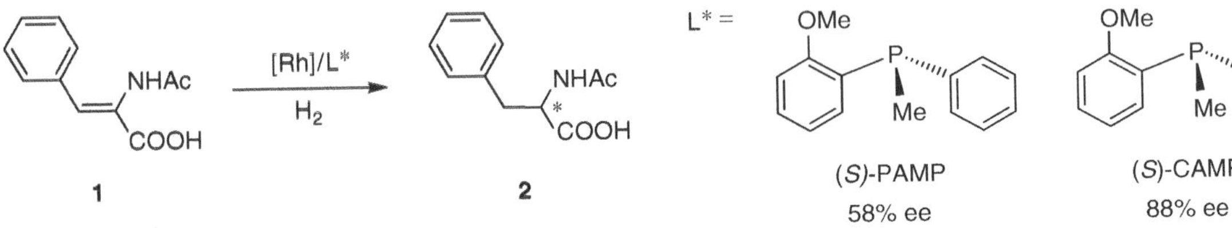

Scheme 2.1 *Asymmetric hydrogenation of α-dehydroamino acid substrate with a Rh–phosphine complex.*

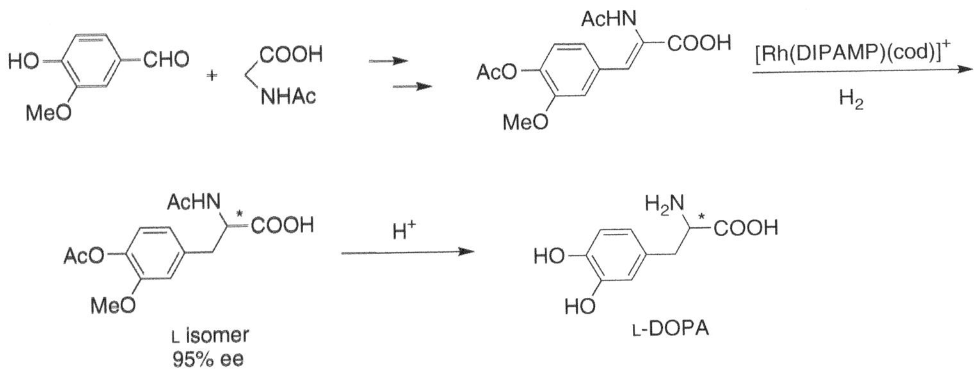

Scheme 2.2 *Monsanto L-DOPA process.*

the modest enantioselectivity (15% ee with α-phenylacrylic acid) was of no preparative value, it indicated that this hydrogenation process can have a definite asymmetric bias with the assistance of an ideal ligand. Horner also reported even more modest results in hydrogenation using the same phosphine.

In search of a phosphine of better performance in an innovative asymmetric hydrogenation methodology to synthesize L-DOPA, which was discovered to have activity in treating Parkinson's disease, modifications were made to vary the substituents directly on the phosphorus atom. Significant improvement of ee was made by the introduction of an *o*-anisyl group to provide more steric hindrance as well as a possible hydrogen-bonding site. Methylphenyl-*o*-anisylphosphine (PAMP) and methylcyclohexyl-*o*-anisylphosphine (CAMP) provided *N*-acetylphenyl-alanine **2** in 58% and 88% ee, respectively, when reducing the model substrate α-dehydroamino acid **1** (Scheme 2.1) [5a].

Furthermore, by dimerizing PAMP a chelating diphosphine DIPAMP was subsequently proved to be more effective in the commercial process of L-DOPA at Monsanto (Scheme 2.2) [5].

In the improved synthetic procedure, both the synthesis of CAMP (**6**) and of DIPAMP (**7**) were accomplished via Mislow's menthyl ester and a common intermediate **5** after introducing the last aryl group (Scheme 2.3) [5].

Another milestone was the invention of the first chelating diphosphine DIOP (**8**), in which inexpensive naturally occurring tartaric acid containing the stereogenic carbons served as the building scaffold and connected the two phosphine moieties. Despite the lack of a stereogenic P-center, the well-defined stereogenic centers in the backbone and the converged five-membered cyclic acetal provide a rigid conformation of P–Rh–P chelating pattern, thus give a good stereoselectivity. DIOP afforded up to 72% ee in the hydrogenation of dehydroamino acid derivatives (Scheme 2.4) [2a]. The synthesis of DIOP was straightforward from tartaric acid ester within four steps (Scheme 2.5) [2b].

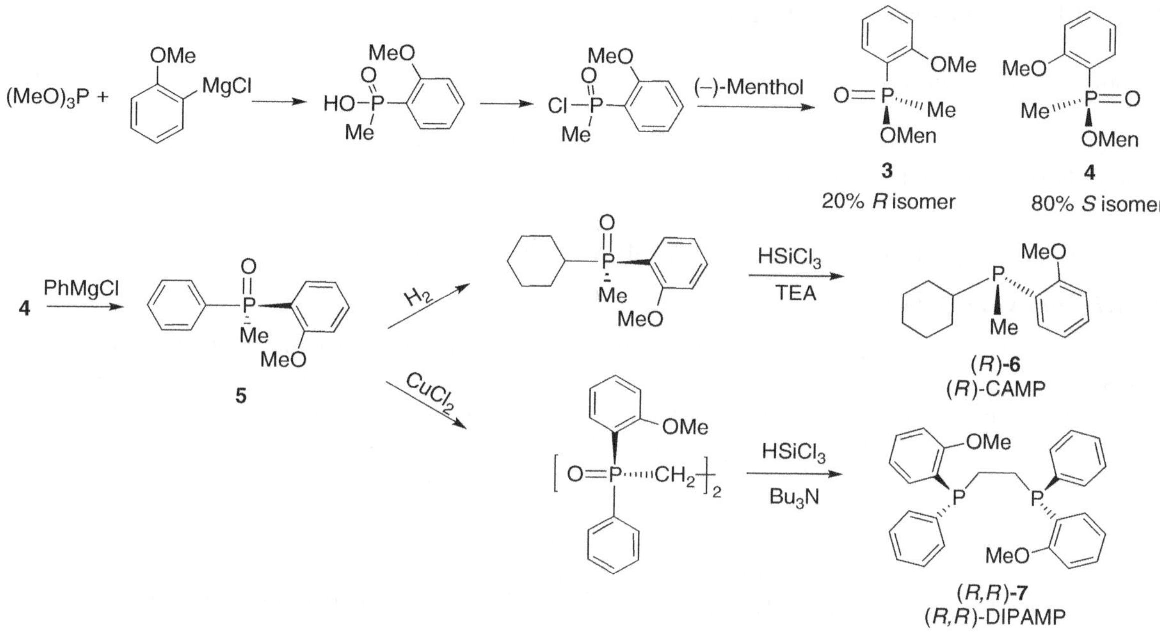

Scheme 2.3 Synthetic procedures for CAMP and DIPAMP.

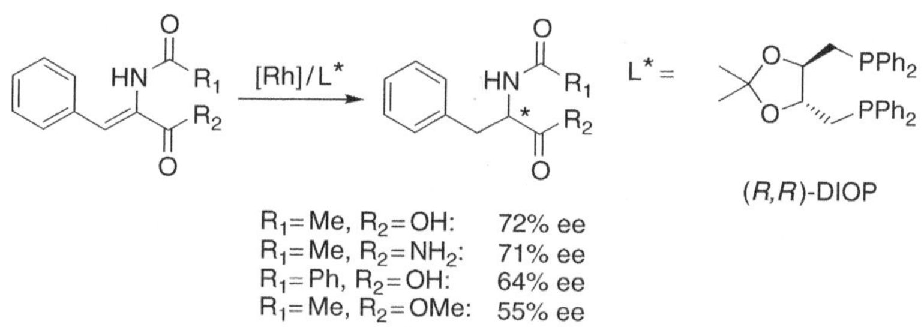

Scheme 2.4 Asymmetric hydrogenation of dehydroamino acid derivatives by Rh–DIOP.

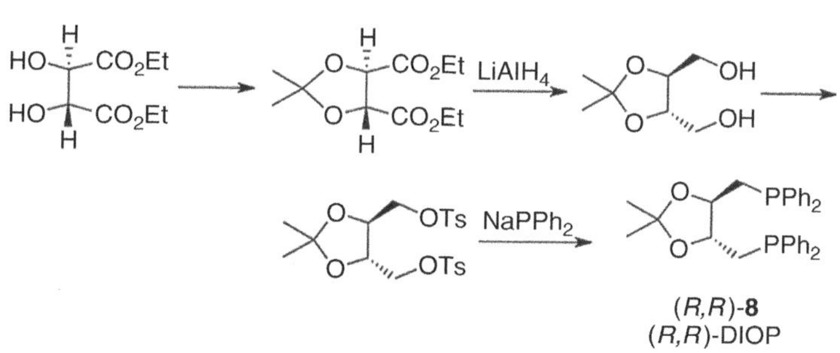

Scheme 2.5 Synthesis of (R,R)-DIOP.

2.2 Chiral chelating diphosphines with a linking scaffold

In addition to a suitable transition metal species, chiral phosphine ligands play a crucial role in asymmetric catalysis. The good enantioselectivity and high activity of the catalyst often result from well-defined structures of chiral ligands. Continuing novel ligand development and modifications of the current ligand families have been significant in this field.

2.2.1 Building chiral backbones from naturally available materials

2.2.1.1 Early development

The success of DIOP introduced a few concepts such as C_2-symmetric ligands and the installation of phosphine groups directly to or in close proximity of stereogenic centers, which are well set in a chiral backbone. Enlightened by the pioneering work by Kagan, the early study on phosphorus ligands produced a number of chiral ligands such as Bosnich's CHIRAPHOS [6] and PROPHOS [7], Rhone Poulenc's CBD [8], and Achiwa's BPPM [9]. In synthesis of these chiral ligands, one most important basic synthetic strategy involves the construction a well-defined chiral backbone, followed by the incorporation of phosphino group to form the P–C bonds in the last step(s). Substitution reactions of a chiral ditosylate, dimesylate, ditriflate, or dihalide with disubstituted phophide became the most widely adopted procedure for the introduction of phosphino groups.

Bosnich and Fryzuk successfully designed and synthesized CHIRAPHOS (**9**) and PROPHOS (**10**), which were the first members of the 1,2-diphosphine ligand family. The design of C_2-symmetric CHIRAPHOS aimed at easy accessibility and at ligands that would give consistently high ee toward a wide range of substrates [6]. The straightforward method of preparation involved tosylation of (*R*,*R*)-butane-2,3-diol, and substitution reaction with lithium diphenylphosphide (Scheme 2.6). However, even under optimized conditions, the occurrence of side reactions of elimination and loss of configurational integrity due to neighboring group participation diminished the isolation yield and made the purification process troublesome. Nickel(II) salt and cyanide ions were applied to obtain optically pure ligand production via forming an insoluble nickel complex before the final recrystallization step.

Aiming at a similar substrate scope as that of CHIRAPHOS, including mostly (*Z*)-*N*-acylaminoacrylic acids, a series of a C_1-symmetric family of ligands such as PROPHOS (**10**) [7], PHENPHOS (**11**) [10], and CYCPHOS

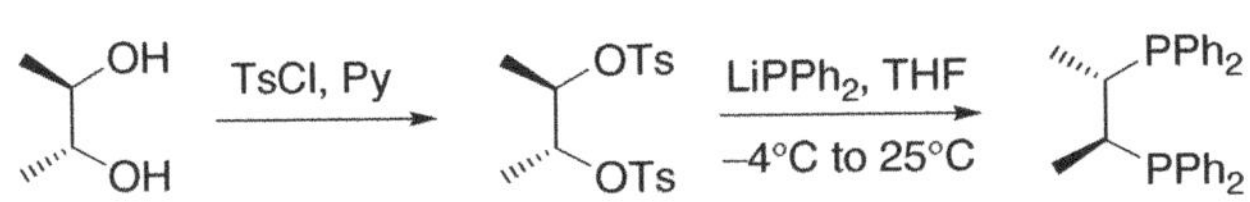

Scheme 2.6 *Synthesis of (S,S)-CHIRAPHOS.*

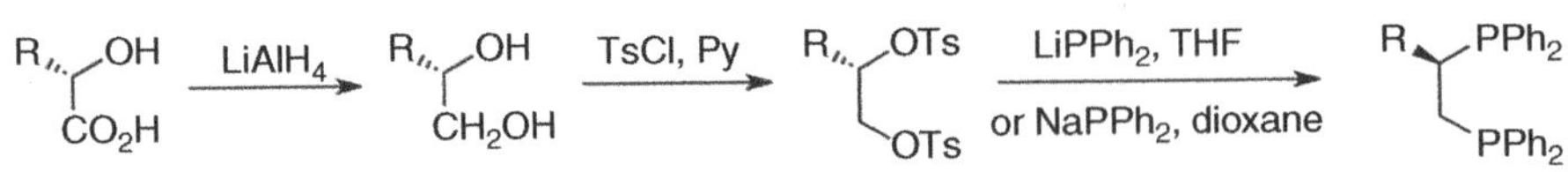

Scheme 2.7 *Synthesis of C_1-symmetric 1,2-diphosphine family.*

(**12**) [11] was synthesized following very similar procedures to that of CHIRAPHOS, albeit from different but analogous optically active natural product starting materials: (*S*)-lactic acid, (*S*)-mandelic acid, and (*S*)-hexahydromandelic acid, respectively (Scheme 2.7). The higher melting point of **11** brought advantageous ease of isolation. In the catalytic asymmetric hydrogenation studies, the enantioselectivity performance of **12** appeared to be the best among this series, while **11** tended to give slightly lower ee's than **10** [10,11].

2.2.1.2 Syntheses of DIOP variants

The great success of DIOP led to a continuous development of related structural studies long after its pioneering discovery, taking DIOP as the model ligand. By switching to different aryl groups on the P-center, such as *ortho*- or *meta*-substituted phenyl groups [12] and 4-dimethylamino group or 4-methoxy-3,5-dimethyl group [13], an array of DIOP analogues **13** were synthesized and the effects of substituent were studied. Furthermore, the conversion of ditosylate to corresponding difluoride species **14** made the reactions with lithium dialkyl-phosphide accessible with modest to good yields (Scheme 2.8) [14].

Comparing to other 1,2- or 1,3-diphosphine ligands, the higher degree of conformational flexibility of DIOP could contribute to the erosion of enantioselectivity, forming a larger seven-membered ring in the form of metal–ligand complexes. To reduce the structural flexibility and also to move the backbone chirality closer to the phosphine moiety that coordinates to the metal, Kagan attempted to incorporate two methyl groups as well as new stereogenic centers next to the phosphino groups and synthesized ligand **16**, which nevertheless gave less satisfying results even than DIOP. Zhang and coworkers rationalized this outcome and proposed that by inverting the configuration of the two stereogenic centers in **16**, the two methyl groups could locate in the favorable equatorial orientation in the metal complex and transition state [15a]. Thus, the diastereomeric ligand DIOP* (**17a**) and its analogues were synthesized from *D*-mannitol and the matching configuration of the two equatorial-positioned α-substituting groups leads to high enantioselectivities in enamide hydrogenations (Scheme 2.9) [15].

In contrast to the simple introduction of stereogenic carbon center in the acetal, modification of the acetal ring size could play a more important role, since it could have remarkable influence on the equatorial orientation of all substituents in the metal–ligand complex. Taking advantage of “BDA” and “Dispoke” methodologies, a more sterically hindered 1,4-dioxane ring could be introduced to prepare ligands **18** and **19**, which appeared to be highly effective toward substrates such as β-substituted enamides and MOM-protected β-hydroxyl enamides (up to 98% and >99% ee) [16] (Scheme 2.10).

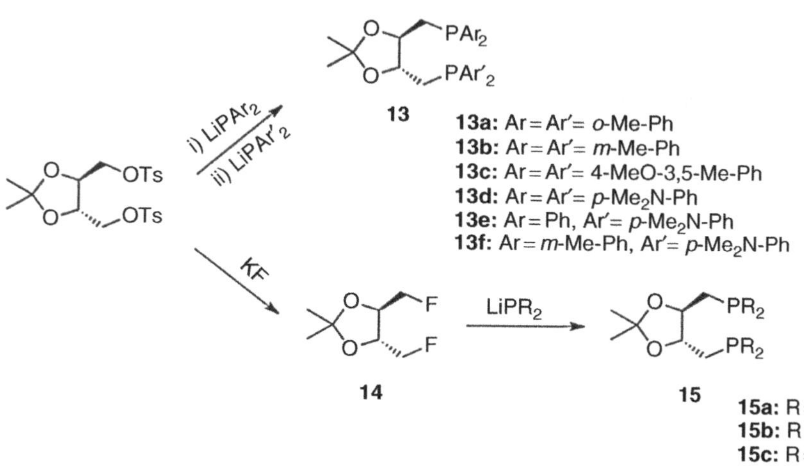

13a: Ar = Ar′ = *o*-Me-Ph
13b: Ar = Ar′ = *m*-Me-Ph
13c: Ar = Ar′ = 4-MeO-3,5-Me-Ph
13d: Ar = Ar′ = *p*-Me_2N-Ph
13e: Ar = Ph, Ar′ = *p*-Me_2N-Ph
13f: Ar = *m*-Me-Ph, Ar′ = *p*-Me_2N-Ph

15a: R = Et
15b: R = *i*-Pr
15c: R = Cy

Scheme 2.8 *Synthetic procedure for the DIOP analogue family.*

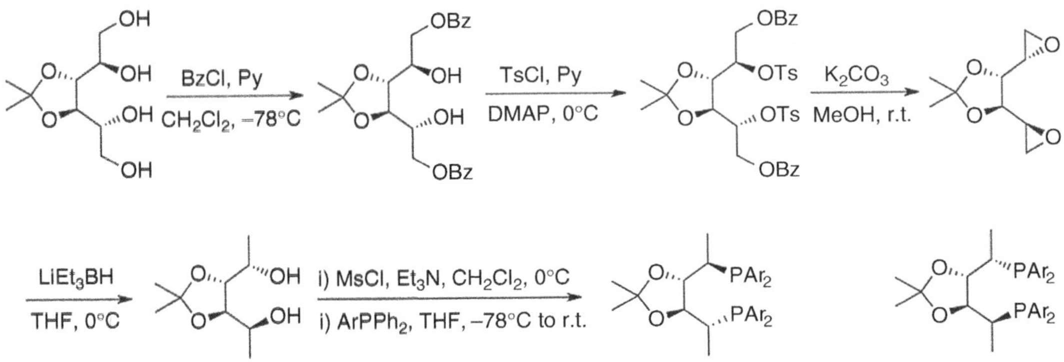

(*R*,*S*,*S*,*R*)-**17a**: Ar = Ph
(*R*,*S*,*S*,*R*)-**17b**: Ar = 3,5-Me-Ph
(*R*,*S*,*S*,*R*)-**17c**: Ar = 3,5-CF_3-Ph
(*R*,*S*,*S*,*R*)-**17d**: Ar = *p*-MeO-Ph

16

Scheme 2.9 *Synthesis of DIOP* ligands.*

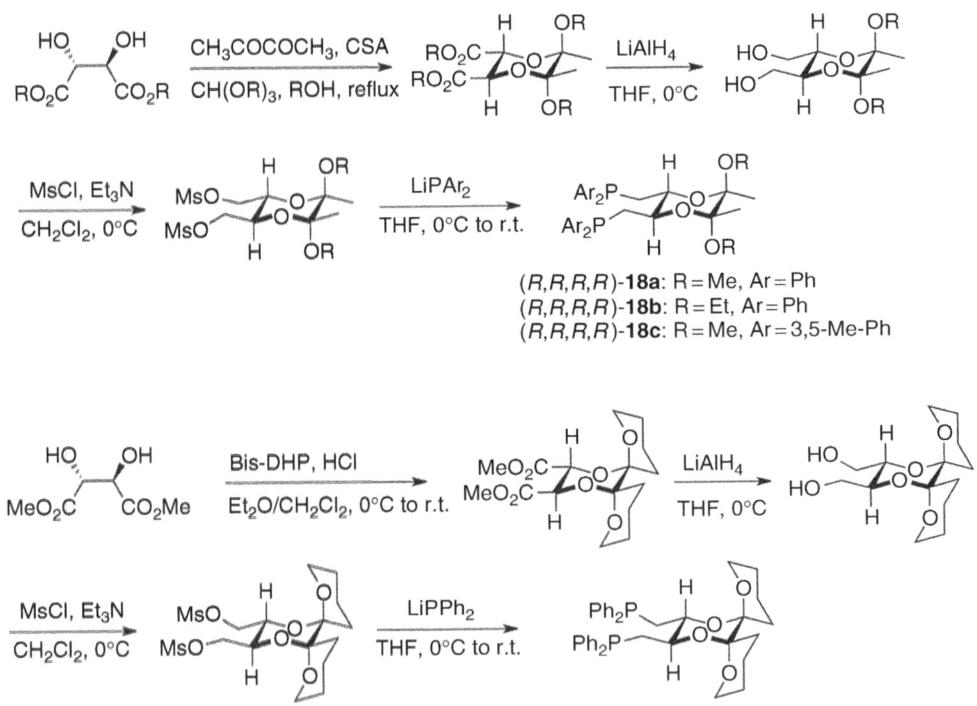

(*R*,*R*,*R*,*R*)-**18a**: R = Me, Ar = Ph
(*R*,*R*,*R*,*R*)-**18b**: R = Et, Ar = Ph
(*R*,*R*,*R*,*R*)-**18c**: R = Me, Ar = 3,5-Me-Ph

(*R*,*R*,*R*,*R*)-**19**
(*R*,*R*,*R*,*R*)-SK-Phos

Scheme 2.10 *Syntheses of T-Phos and SK-Phos.*

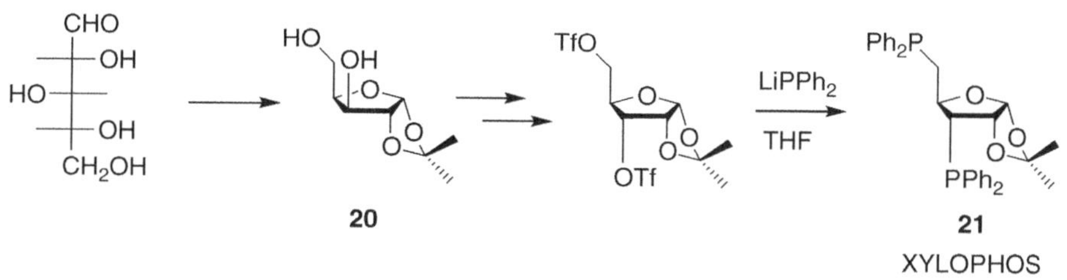

Scheme 2.11 *Synthesis of XYLOPHOS.*

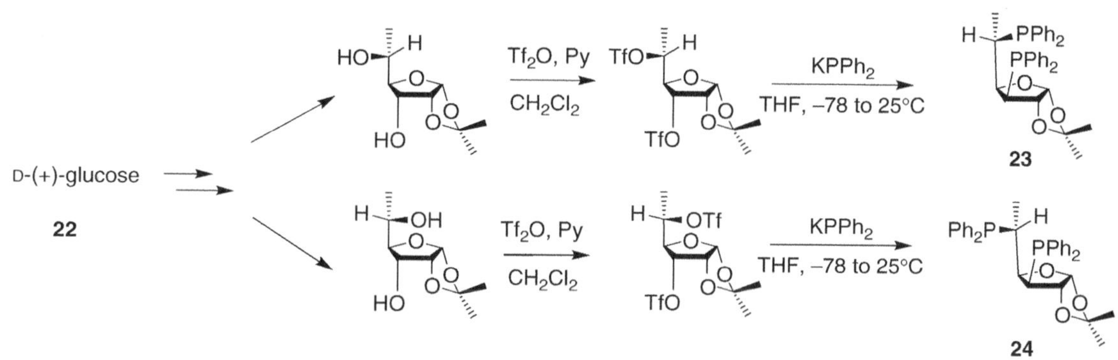

Scheme 2.12 *Synthesis of ligands from* D*-(+)-glucose.*

2.2.1.3 Synthesis from other natural chiral backbones

Carbohydrates and their derivatives bearing chiral centers could be good candidates for building chiral phosphines, because they are readily available, functionalized, and convenient starting materials to install phosphine moieties [17]. Pamies *et al.* developed a series of C_1-symmetric ligands based on their study on the furanose and glucose skeletons. Starting from commercially available D-(+)-xylose **20**, ligand **21** was prepared and isolated in high yields and named XYLOPHOS (Scheme 2.11) [18a].

Moreover, starting from D-(+)-glucose **22**, Diéguez *et al.* developed two forms of diphosphines **23** and **24** after introducing another stereocenter into the backbone with fine tuning of the two diastereomers. In the Rh-catalyzed hydrogenation of unsaturated carboxylic derivatives, ligand **24** showed better enantioselectivities due to its matching configuration (Scheme 2.12) [18b]. One of the main limitations of carbohydrate backbones is that often only the D-(+)-configuration is readily available.

Another family of 1,4-diphosphine ligands, BPPM and its derivatives, was designed and synthesized by Achiwa and coworkers based upon a chiral scaffold of L-hydroxyproline [9,19]. In the synthesis of (*S*,*S*)-**26**, esterification and Boc-protection treatments of the chiral starting material 4-hydroxy-(*S*)-proline **25** were followed by the reduction and inversion of stereogenic configuration in the last step (Scheme 2.13) [9]. This C_1-symmetric ligand featuring two phosphino groups in *cis* positions could provide excellent chelating coordination to the metal. Moreover, the other enantiomer (*R*,*R*)-**26** could be easily accessible according to the subsequent report (Scheme 2.13) [20], and modifications of the *N*-substituent and different phosphino groups could also be achieved to fine-tune the structures and thus reactivities of the BPPM ligand families [19].

Pyrrolidine-based ligands can be made from more readily accessible naturally occurring structures, such as tartaric acid. For example, by converting tartaric acid into a five-membered pyrrolidine backbone (*R*,*R*)-**27** via cyclic imide formation, a series of commercially available phosphine ligands, (*R*,*R*)-**28** and related

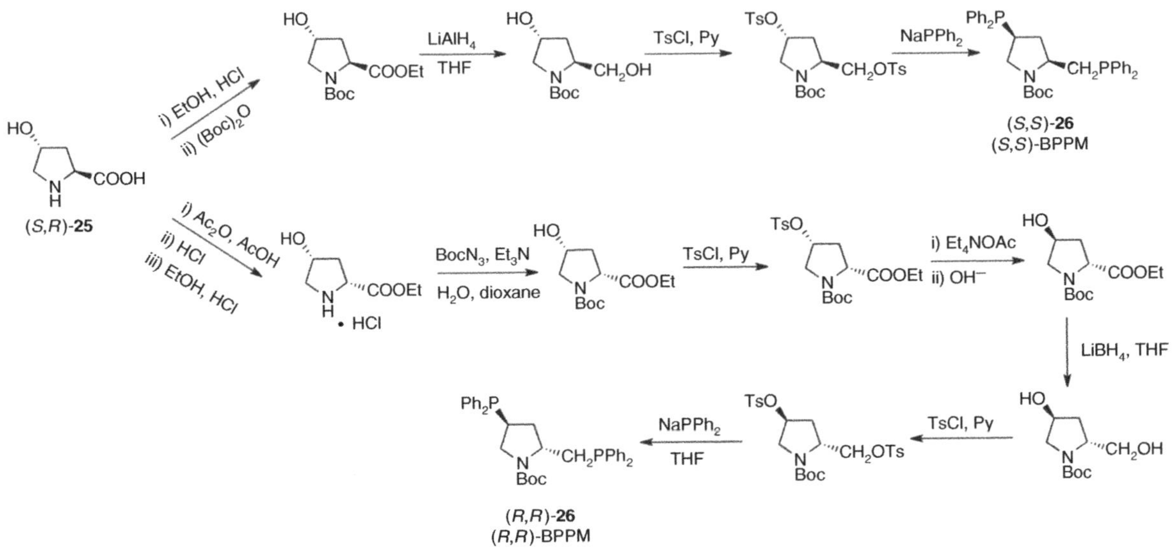

Scheme 2.13 Synthesis of BPPM ligands.

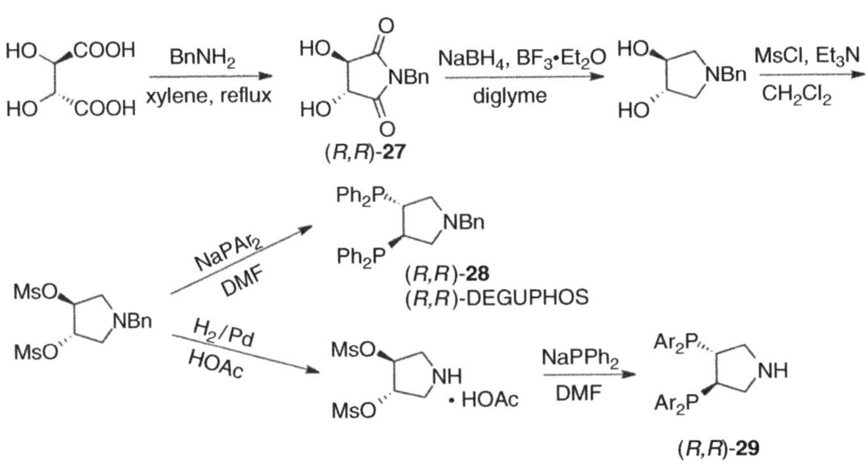

Scheme 2.14 Synthesis of DEGUPHOS and related ligands.

structures, can be formed within a few practical steps (Scheme 2.14) [21]. Another similar derivatization methodology was reported by Lee and coworkers in which a C_2-symmetric vicinal diamine (*S*,*S*)-**30** bearing two benzyl-protected hydroxyl groups was prepared straightforwardly from L-tartaric acid. A subsequent cyclization with carbonyldiimidazole led to the key five-membered ring backbone as well as the possibility of *N*-alkylation or *N*-arylation via the Hartwig–Buchwald method toward various ligand derivatives (Scheme 2.15) [22]. Structural diversity and optimization for different substrates were allowed due to the modular synthesis strategy. >99% ee was achieved in the Rh-catalyzed asymmetric hydrogenation of *N*-acetyl α-arylenamides by using ligand (*S*,*S*)-**31c**.

Furthermore, many other natural homochiral molecules such as chiral terpenes could be utilized as an inexpensive and reliable source for the synthesis of chiral ligands. Knochel *et al.* reported the synthesis

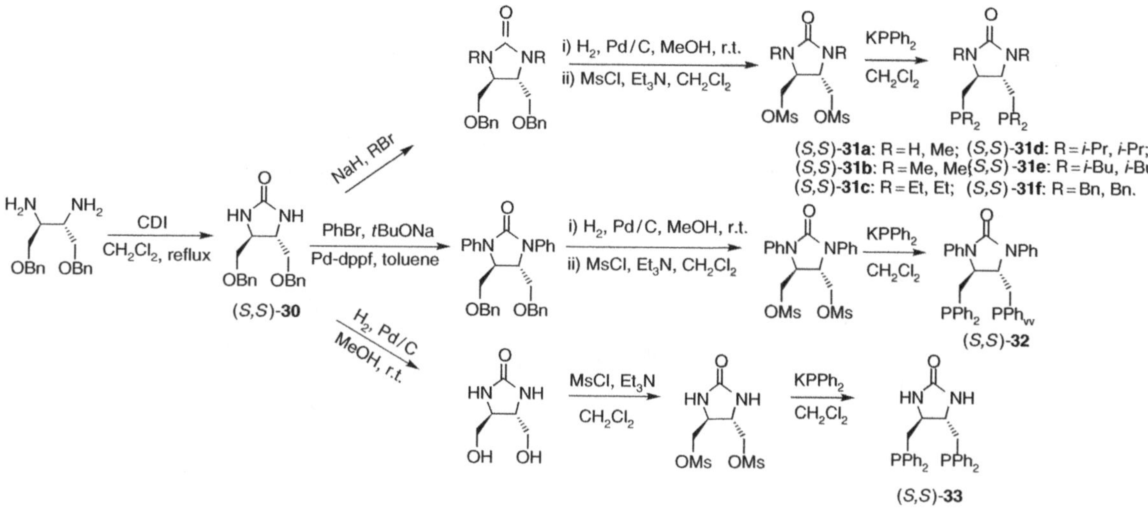

Scheme 2.15 *Synthesis of imidazolidin-2-one scaffold-based ligands.*

of a series of diphosphine ligands containing a pinene framework, and their good performance in catalytic asymmetric hydrogenations [23].

2.2.2 Design and synthesis of chiral backbones

Instead of taking advantage of the naturally occurring chiral molecule as the starting materials or precursors, structures containing stereogenic centers serving as building scaffolds could be prepared based on rational designs and asymmetric synthesis methodologies.

2.2.2.1 Chiral backbones synthesized through asymmetric catalysis

Due to the great success of DIOP and many related derivatives in asymmetric hydrogenation, and also due to the established practical synthetic steps to convert unprotected hydroxyl groups into various phosphine groups, synthesis of the key intermediate chiral diols that are not naturally available became important transformations. Although resolution of available racemic diols could often provide the desired enantiomer(s) with satisfactory optical purity, the tedious procedures and low efficiency could always limit the development of different enantiopure diol intermediates.

When studying the performance of (*R*)-PROPHOS ((*R*)-**10**) in asymmetric catalysis, Bosnich developed the concept of a "self-breeding" catalytic system, which described the production of enantiomeric intermediate that leads to the corresponding ligand enantiomer by the ligand itself. In the asymmetric hydrogenation study of (*R*)-**10**, the enol acetate derivative of the cheap readily available ethyl pyruvate could be efficiently hydrogenated to (*S*)-ethyl *O*-acetyllactate (*S*)-**34** in 84% ee. **34** could then be readily reduced to the corresponding propane diols and ditosylated to give chiral ditosylate **35** (80% ee), which, in turn, underwent a single recrystallization before it was further converted to generate the (*R*)-enantiomer of **10** (Scheme 2.16) [7]. It was also noteworthy that (*S,S*)-CHIRAPHOS ((*S,S*)-**9**) could also serve as a source for breeding more (*S*)-**10**, since the Rh catalyst of (*S,S*)-**9** could be optically enriched in the hydrogenation of the same enol acetate.

The successful synthesis of CHIRAPHOS and PROPHOS encouraged Bosnich and coworkers to prepare SKEWPHOS ((*S,S*)-**37**) by reacting chiral disulfonate derivative of (*R,R*)-2,4-pentanediol ((*R,R*)-**36**) with lithium diphenylphosphide under established conditions, but the chiral camphorsulfonate ester was obtained

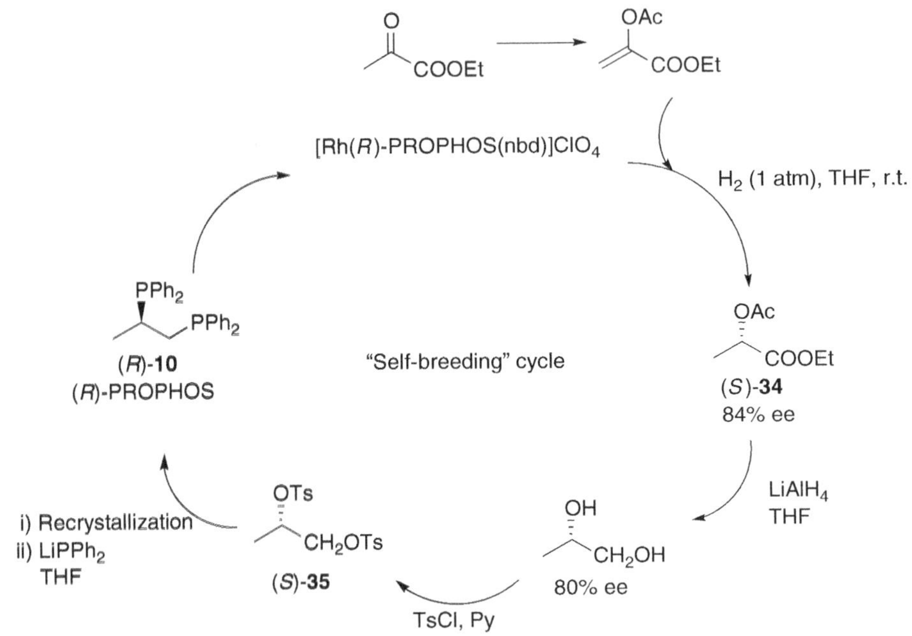

Scheme 2.16 *"Self-breeding" catalytic system of PROPHOS.*

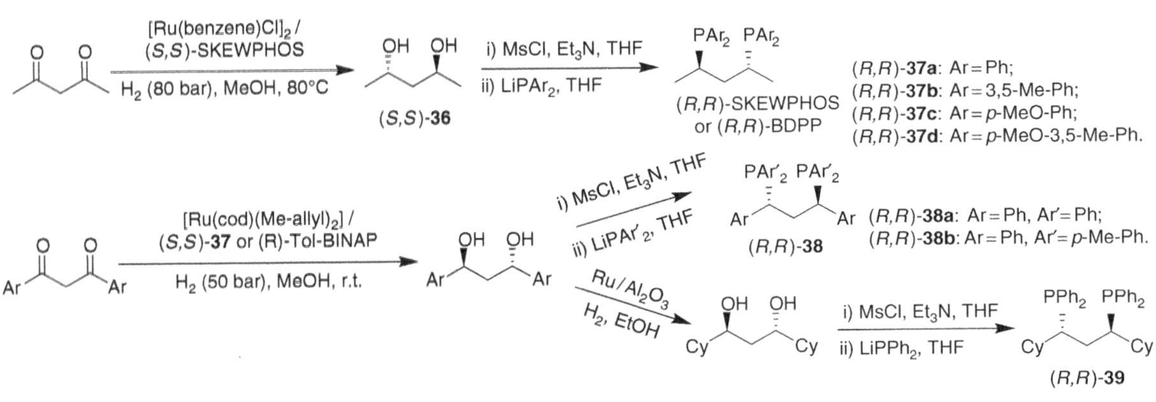

Scheme 2.17 *Synthesis of SKEWPHOS (BDPP) and related ligands via a homogeneous asymmetric hydrogenation method.*

from optical resolution using camphorsulfonic ester as the resolving agent (70% resolution yield). The purification of the ligand also required formation of a nickel complex with cyanide ions, from which the phosphine was liberated [24]. Interestingly, the development of asymmetric hydrogenation of acetylacetone in the presence of Ru precursor and SKEWPHOS could efficiently make chiral 2,4-pentanediol available, which remarkably accelerated the development of SKEWPHOS or also named BDPP ligand families sharing the same 2,4-pentane-2,4-diyl skeleton but with various P-substituents (Scheme 2.17) [25]. Brunner and Terfort named this production of chiral precursor for the opposite enantiomer "cross breeding" [26]. Similar approaches were also available for the synthesis of DuPhos and *i*-Pr-BPE [27]. The SKEWPHOS ligand family was successfully applied in the Rh-catalyzed asymmetric hydrogenation of dehydroamino

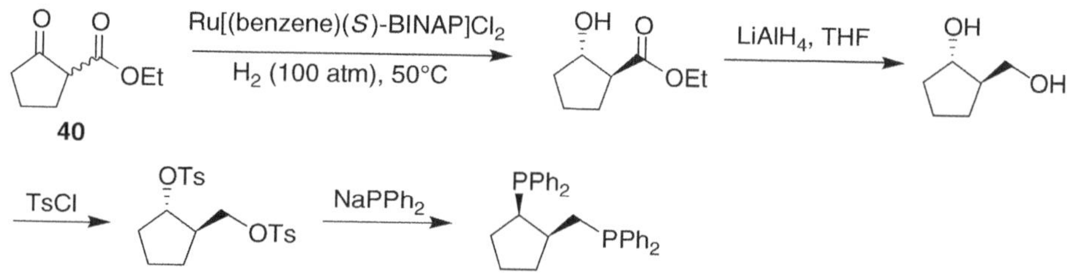

Scheme 2.18 *Synthesis of (R,S)-PPCP via asymmetric hydrogenation.*

acid derivatives, acetophenone and acetophenonbenzylimine [24,28]. A very similar cross self-breeding system was reported by Börner and coworkers. Ru-diphosphine catalyst using (*S*,*S*)-**38** or Tol-BINAP afforded chiral 1,3-diarylpropane-1,3-diols with excellent diastereoselectivity and enantioselectivity from corresponding diketones. The obtained enantiopure diols could be subsequently converted to chiral diphosphine products in an economic and advantageous way (Scheme 2.17) [29]. A desymmetrized analogue of BDPP, BDPBzP, which contains a 3-benzyl group, was documented by Bianchini *et al.* [30]. The precursor acetylacetone was modified at the 3-position before it was asymmetrically reduced to chiral diols using Ru – BINAP as the catalyst.

Along with the great breakthroughs in the development of asymmetric hydrogenation methodology, wider prochiral substrate spectra provide more and more chiral key intermediates for building chiral phosphine ligands. In another example reported by Inoguchi, a designed ligand (*R*,*S*)-PPCP ((*R*,*S*)-**41**) with favorable skew conformation of the six-membered chelate ring with rhodium metal was synthesized for the hydrogenation of (*Z*)-2-acetamidocinnamic acids [31]. The well-defined stereogenic centers in the backbone originated from the asymmetric hydrogenation of a cyclic α-keto ester **40** by Ru–BINAP catalyst (Scheme 2.18).

2.2.2.2 Design and synthesis of ligands containing spiro backbones

The successes of many C_2-symmetric phosphine ligands inspired continuing research on developing structural rigid C_2-symmetric motifs for the preparation of new chiral ligands. Biaryls, most notably 1,1′-BINOL derivatives, serve as a prominent member in the C_2-symmetric chiral ligand family in asymmetric synthesis. However, to further enhance the configurational stability and completely restrict the rotational racemization around the central axial bond, spiranes containing a quaternary center were rationally designed.

The application of *cis*,*cis*-spiro[4.4]nonae-1,6-diols (*cis*,*cis*-**43**) in asymmetric reduction of ketone became one of the early examples employing chiral spiro auxiliaries to obtain enantioselectivies [32]. To improve the previous synthesis of the spiro dione (**42**) and diol (**43**) [33], Keay *et al.* reported an efficient synthesis and resolution method of *cis*,*cis*-**43** to give an optical enantiomer scaffold (Scheme 2.19) [34], which were later used by Chan and Jiang in the synthesis of the chiral phosphinite ligand SpirOP [35].

Although the tetrahedral structure of the spiro carbon atom and the perpendicular orientation of the two rings can be perfect rigidity-enhancing features, the difficulty of the enantiopure spiro framework impeded the development of spiro ligands before the last decade. A breakthrough of synthesizing 1,1′-spirobiindane-7,7′-diol (spinol, **45**) in a six-step sequence was reported by Birman *et al.* (Scheme 2.20) [36]. Based on this key spinol scaffold, Zhou and coworkers reported the synthesis of a series of ligands including diphosphine ligand SDP and its derivatives (**46**) and other related ones such as SIPHOS and SITCP (Scheme 2.20) [37]. The excellent performance of these spiro skeleton-based rigid ligands in the enantioselective hydrogenation of a broad scope

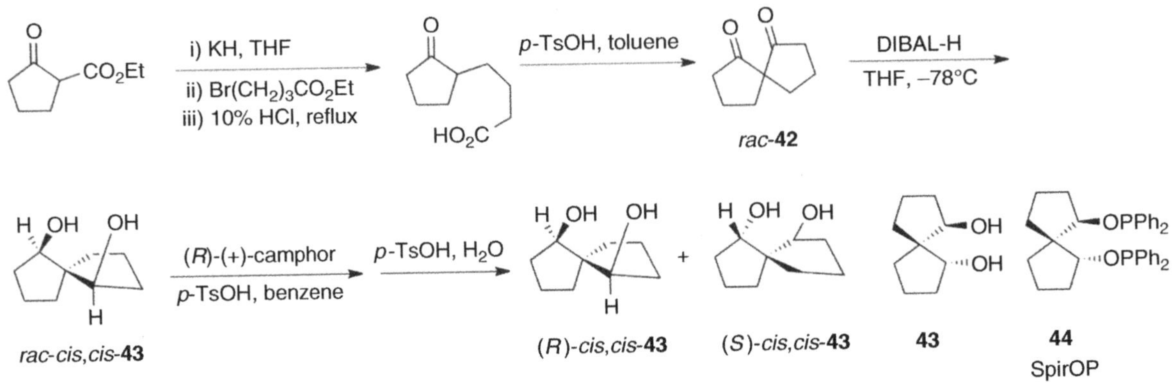

Scheme 2.19 *Synthesis of spiro chiral diol cis,cis-(+)- and cis,cis-(−)-**43** and structure of SpirOP (**44**).*

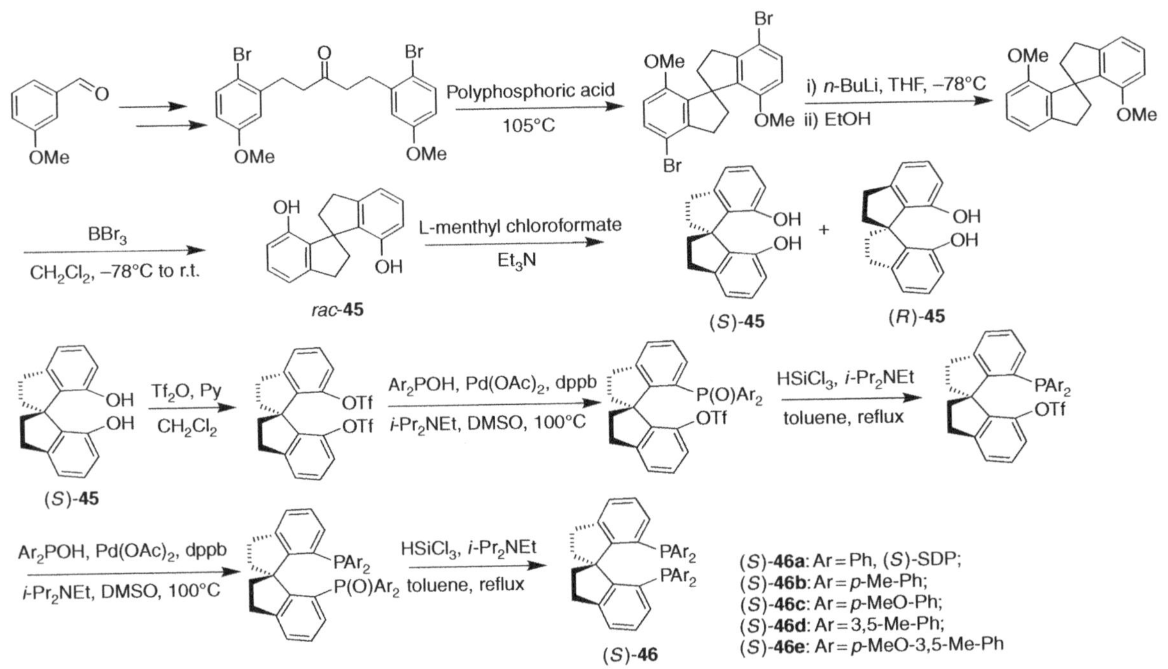

Scheme 2.20 *Synthesis of 1,1′-spirobiindane backbone **45** and chiral diphosphine SDP family ligands **46**.*

of unsaturated substrates such as dehydroamino acid derivatives, cyclic and acyclic enamides, unfunctionalized ketones, and unsaturated acid revealed the prominent privilege of the spiro backbone for chiral ligand design.

Another scaffold bearing the similar privileged rigid spiro quaternary center was also reported by Zhou *et al.* [38]. The more rigid and bulkier spirobifluorene backbone SFDP was successfully synthesized following a continuous ring-closing route to give racemic 9,9′-spirobifluorene-1,1′-diol (**47**) and hence both its enantiomers taking advantage of inclusion crystallization technique using 2,3-dimethoxy-*N,N,N′,N′*-tetracyclohexyl succinamide as the differentiating reagent. Furthermore, the available access to **47** led to the production of chiral diphosphine SFDP (**48**) via the same synthetic route from SDP synthesis (Scheme 2.21) [39]. The large dihedral

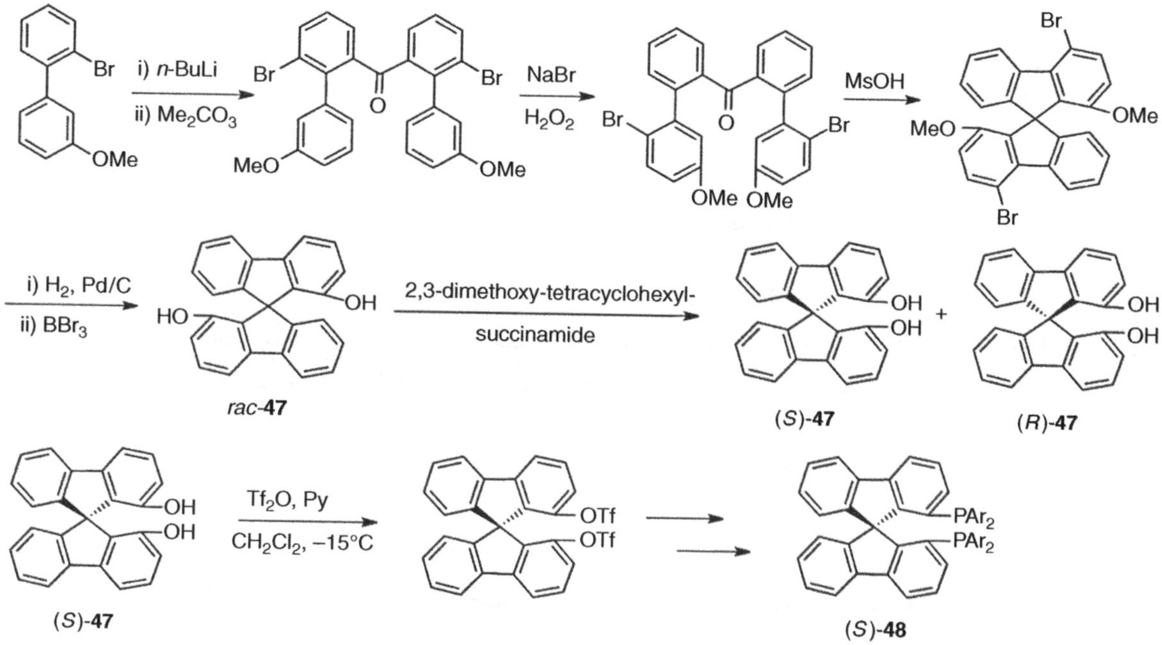

Scheme 2.21 *Synthesis of 9,9′-spirobifluorene-1,1′-diol **47** and chiral diphosphine SFDP ligand **48**.*

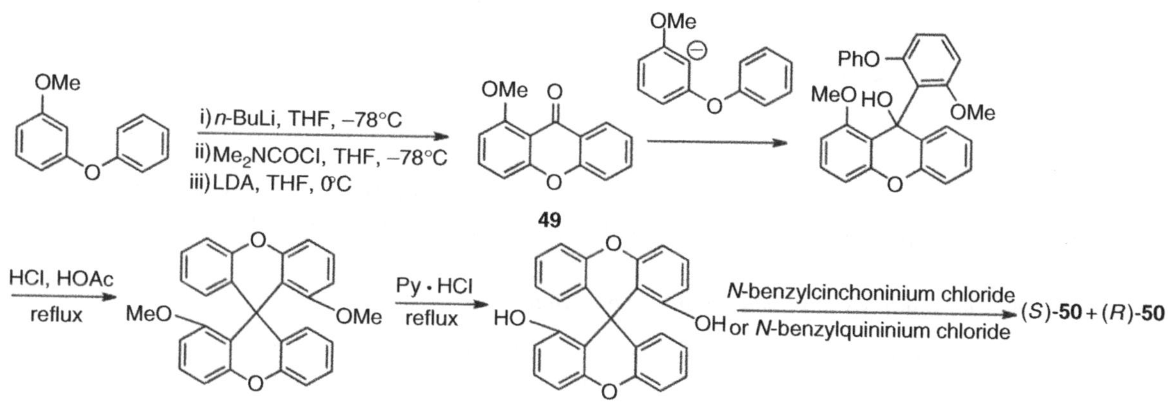

Scheme 2.22 *Synthesis of spiro diol backbone **49**.*

angle from the spiro backbone and the further enhanced rigidity rendered the SFDP ligand great performance in Ru-catalyzed asymmetric hydrogenation of a wide range of α-methylcinnamic and tiglic acid derivatives.

Attempts to synthesize a similar, highly sterically hindered spiro backbone were also reported by Zhang *et al.* [40]. Tandem directed *ortho* metalation (D*o*M) and directed remote metalation (DreM) could convert 3-phenoxyanisole into intermediate **49** with high yields. After the ring closure to form the quaternary spiro center, chiral diol product **50** was achieved after refluxing in pyridine·HCl (Scheme 2.22). Although both enantiomers of 48 were obtained through optical resolution, the diphosphine ligand based on this 9,9′-spirobixanthene-1,1′-diol framework was not successfully achieved.

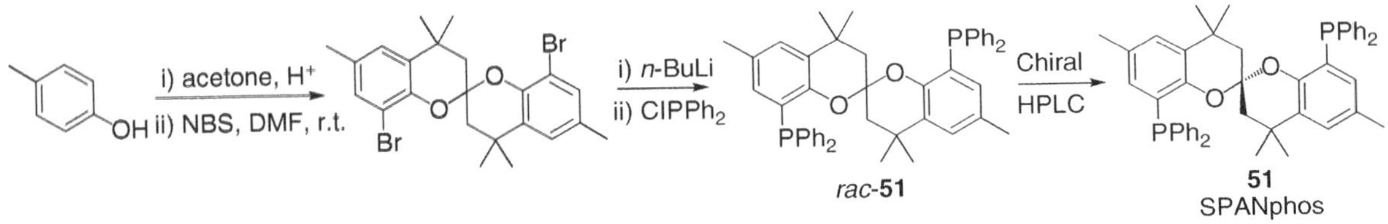

Scheme 2.23 *Synthesis of SPANphos.*

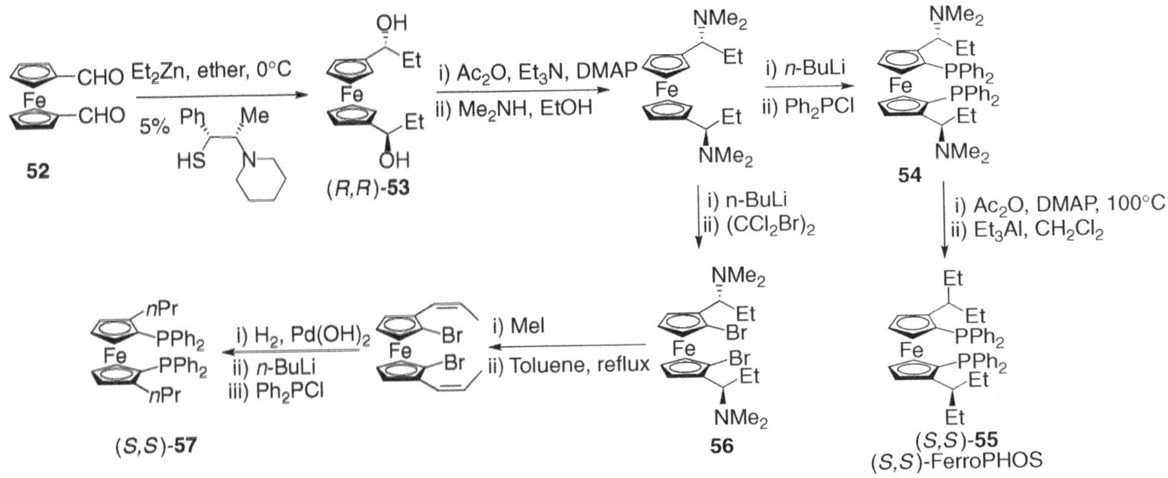

Scheme 2.24 *Syntheses of (S,S)-FerroPHOS and (S,S)-nPr-FerroPHOS.*

Van Leeuwen and coworkers documented the successful preparation of another C_2-symmetric diphosphine ligand containing the spiro backbone, SPANphos (**51**), which featured *trans*-coordinating to transition metals [41]. The synthetic route was very efficient from cheap starting materials. However, it involved chiral HPLC separation of the final diphosphine product instead of a chiral spiro diol (Scheme 2.23).

2.2.2.3 Design and synthesis of chiral ferrocene backbones

To explore more candidate frameworks to synthesize chiral diphosphines, structures carrying high rigidity and an ability to be easily converted to diverse derivatives have been extensively studied. Ferrocene-based structures are one of the successful structure categories utilized in ligand design and synthesis.

Kang *et al.* designed and reported the synthesis of FerroPHOS family ligands. The available 1,1-ferrocenedicarboxaldehyde **52** was converted to chiral diol (*R*,*R*)-**53** in the presence of thiazazincolidine catalyst. The resulting diol (*R*,*R*)-**53** was subsequently transformed into diamino diphosphine **54** and also the air-stable triaryl ferrocenyl ligand (*S*,*S*)-FerroPHOS ((*S*,*S*)-**55**) following established procedures (Scheme 2.24) [42]. This ligand of planar chirality was approved to be highly efficient in the Rh-catalyzed asymmetric hydrogenation of dehydroamino acid derivative and Rh-catalyzed hydroborations of styrene and indene. The variations of **55** bearing other substituents such as the methyl or phenyl group could be achieved by replacing Et_3Al with Me_3Al or $PhMgBr/ZnBr_2/BF_3{\cdot}OEt_2$ in the final step [43]. The prepared chiral backbone can also undergo a modified synthetic route in which dibromide intermediate **56** formed and dimethyl amino groups were eliminated before hydrogenation to finally give (*S*,*S*)-*n*Pr-FerroPHOS product ((*S*,*S*)-**57**) (Scheme 2.24) [43]. All FerroPHOS ligands are highly air stable.

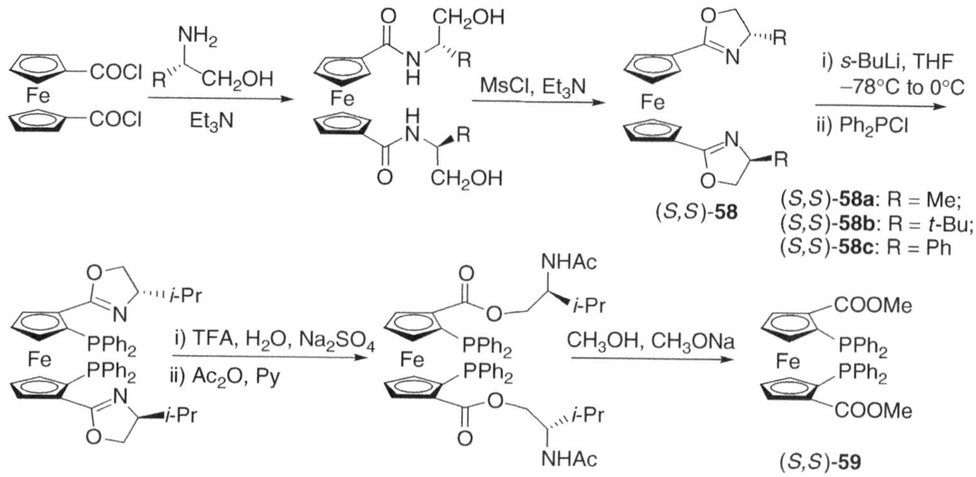

Scheme 2.25 *Synthesis of ferrocene-based ligand (S,S)-**59**.*

Another example of similar ferrocene-based backbone construction was the synthesis of chiral 1,1′-bis(oxazolinyl)ferrocenes ((*S,S*)-**58**), which could serve as a key intermediate for the synthesis of a series of diphosphine ligands. Once the chiral bisoxazolinylferrocene backbone was achieved via an established two-step method with high yield (85% and 90%, respectively), the diastereoselective control of the dilithiation process became a key issue. *sec*-Butyllithium was carefully chosen to minimize the formation of unexpected *meso* and other side products (22% *meso* product for **58a**). Diphosphine ligand (*S,S*)-**59** was then prepared by Meyer's method, in which the oxazoline rings were opened under acidic conditions, and *in situ* acetylated before the final transesterification step took place (Scheme 2.25) [44]. Ligand A58 was tested to produce good enantioselectivies in Pd-catalyzed allylic alkylation of 1,3-diphenyl-2-propenyl acetate.

Zhang and coworkers reported the synthesis of ligand (*R,R*)-**61** from chiral bis-sulfoxide ferrocene scaffold. The ligand was screened for asymmetric reactions such as hydrogenations of dehydroamino acid ester and allylic alkylation with good ee's (Scheme 2.26) [45].

2.2.2.4 Design and synthesis of other chiral backbones

Based upon the success of DIOP, BINAP, and other phosphines ligands, Zhang and coworkers designed (*R,R*)-BICP ((*R,R*)-**64**) [46]. The interconnected two cyclopentane rings and four well-defined stereogenic carbon centers provide conformational rigidity and restrict the orientation of the *P*-phenyl substituents, and thus lead to high enantioselectivity. The chiral diol scaffold **63** was synthesized from readily available 1,1′-dicyclopentene (**62**), followed by hydroboration by monoisopinocamphenylborane ((+)-IpcBH2) and oxidation by H_2O_2 (Scheme 2.27). The CAChe molecular modeling of BICP showed that the skewed conformation and protruding equatorial phenyl rings can form a defined chiral environment in the proximity of transition metal with two diagonal quadrants blocked and thus lead to enantioselectivities. The excellent enantiocontrol of the synthesized ligand was demonstrated in the asymmetric hydrogenation of α-acetamidocinnamic acid and other dehydroamino acid derivatives [40] and also unfunctionalized ketones [47].

The success story of PhanePhos in the area of asymmetric catalysis proved the advantageous structural properties of the planar chiral paracyclophane scaffold. Its excellent performance in the enantiomeric transformations such as asymmetric hydrogenation of dehydroamino acids, allylic acid, β-ketoesters,

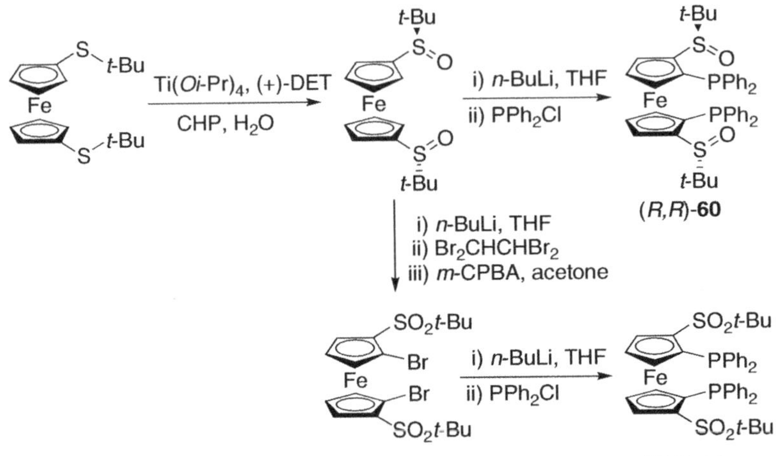

Scheme 2.26 Syntheses of ferrocene-based ligand (R,R)-**60** and (R,R)-**61**.

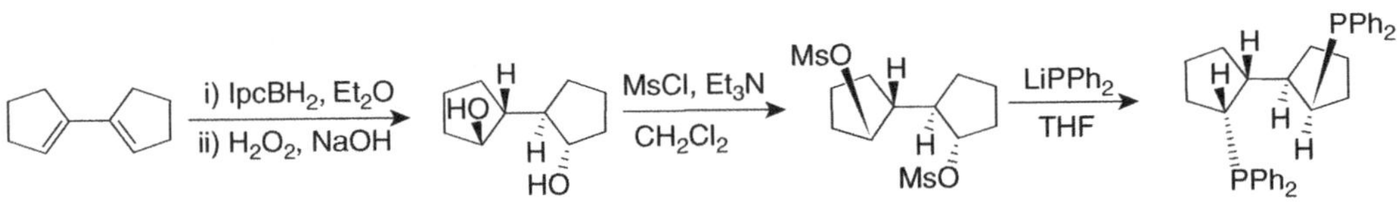

Scheme 2.27 Synthetic route of (R,R)-BICP.

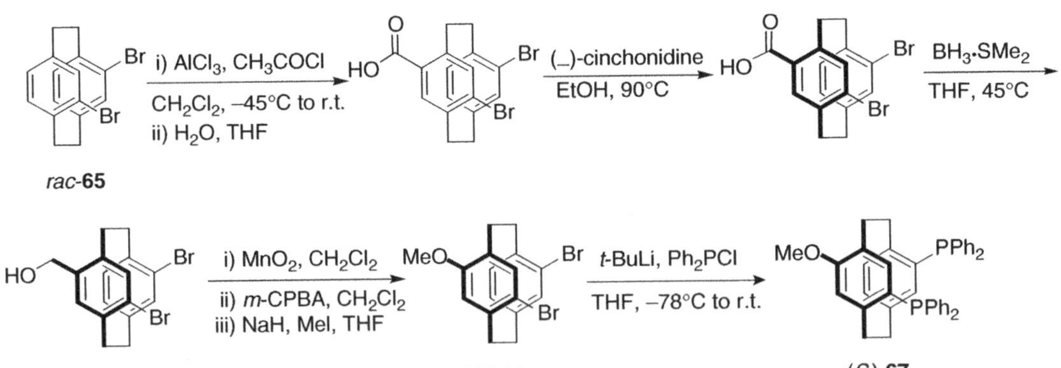

Scheme 2.28 Synthesis of paracyclophane backbone (S)-**65** and (S)-ParaPhos.

unfunctionalized ketones, and palladium-catalyzed amination [48,49] promoted the studies on the paracyclophane-based phosphine ligands. Hems *et al*. reported the synthesis of electrophilically substituted enantiomeric paracyclophane backbone after investigating the reactivity of the dibromoparacyclophane **65** in Lewis-acid catalyzed substitutions. The functionalized dibromide intermediate (*S*)-**66** could then be transformed into corresponding diphosphine **67** (Scheme 2.28) [50]. The development of this synthetic

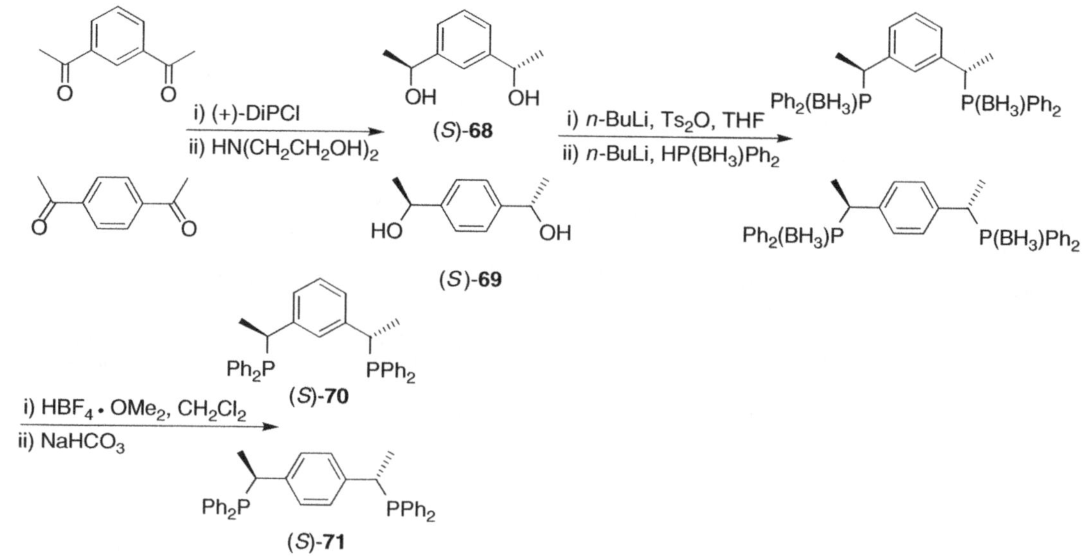

Scheme 2.29 *Synthesis of aromatic backbone-based ligands (S)-**70** and (S)-**71**.*

methodology could facilitate the fine-tuning and accessibility of the chiral backbone (*S*)-**66** and could lead to explore more ligands featuring planar chirality.

Rigid aromatic backbones can provide efficient transfer of chirality from the metal–ligand catalyst to the product, and large ligand bite angle would definitely favor the enantioselective transformation of some substrate type. Zhang and coworker reduced the readily accessible 1,3- and 1,4-diacetylbenzene to obtain enantiopure rigid chiral diols (*S*)-**68** and (*S*)-**69**. However, the established procedure of forming dimesylates or ditosylates intermediate failed at benzylic positions, and racemization could occur in the following nucleophilic substitution step. This major problem of installing the phosphine group was resolved by *in situ* tosylating the hydroxyl groups and replacing the nucleophile to borane-protected phosphines. The enantiomerically pure chiral phosphines (*S*)-**70** and (*S*)-**71** were obtained after the final deprotection step with $HBF_4{\cdot}OMe_2$ (Scheme 2.29) [51].

Many other stereoselective methodologies have been employed to construct the chiral backbone for various diphosphines and other phosphorus ligands. Achiwa and Morimoto exemplified the application of the diastereoselective Diels–Alder reaction and Wagner–Meerwin-type rearrangement reaction in the synthesis of chiral norbornene skeletons for ligand synthesis [52]. To prepare this chiral scaffold (**72**), trimethylsilylcyclopentadiene with dimentyl fumarate underwent an asymmetric Diels–Alder reaction with 97% *de*, and then selective bromination and silver ion–promoted solvolysis in methanol gave the dicarboxylate ester **73**. Then corresponding diol (*S*,*S*)-**72** was achieved by protecting the hydroxyl group and reduction using $LiAlH_4$. Then the introduction of the phosphine group could be fulfilled by established procedures to give ligands with a protected –OH (*S*,*S*)-**73** or unprotected –OH group (*S*,*S*)-**74** (Scheme 2.30) [51,53]. The stereospecific Diels–Alder and rearrangement reaction ensured the optical purity of the BHMP ligands synthesized.

2.2.3 Synthesis from optical resolution of phosphine precursors or intermediates

Although the optical resolution could be tedious and relatively unpredictable, it is still one of the important strategies applied in the synthesis of chiral phosphines, especially under the circumstances that the design and synthesis of the ligand do not involve a natural chirality source and also building a chiral

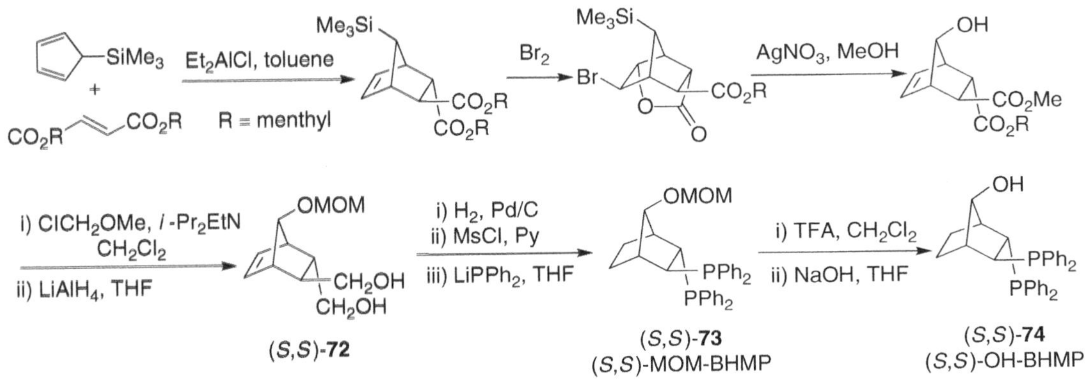

Scheme 2.30 *Synthesis of BHMP ligands.*

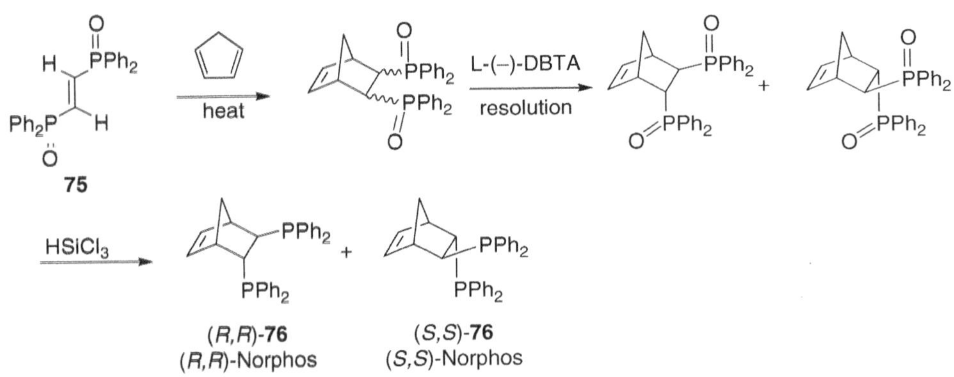

Scheme 2.31 *Synthesis of Norphos via a Diels–Alder reaction and optical resolution.*

scaffold could not be easily realized before installing a phosphino moiety. Phosphine oxide is often considered a good intermediate to be resolved in the synthetic route, while in some examples other phosphorus compound intermediates could also be resolved if the proper resolving reagent and appropriate approach are selected.

In the example reported by Brunner *et al.*, a racemic chiral diphosphine oxide linked on a norbornene backbone was synthesized highly efficiently as the adduct produced from the Diels–Alder reaction of *trans*-1,2-bis(diphenylphosphino)ethylene **75** with norbornadiene [54]. To separate both enantiomers of the diphosphine oxide, L-(–)-di-*O*-benzoyltartaric acid ((–)-DBTA) was used to form diastereomer with (–)-Norphos oxide which were separated from the mixture. Then the reduction of both enantiomers of the resolved oxides with $HSiCl_3$ allowed the preparation of both (*S,S*)- and (*R,R*)-Norphos (**76**) (Scheme 2.31).

In the development of PHANEPHOS (**81**), Merck chemists also applied the resolution technique in the separation of the enantiomers of the diphosphine oxide species. The synthesis of the C_2-symmetric planar chiral PHANEPHOS started with iron-catalyzed dibromination of [2.2]paracyclophane **77**. The racemic *pseudo ortho* diastereomer **79** was enriched through crystallization and thermal isomerization, and separated from highly insoluble *pseudo para* diastereomer **78**. The racemic phosphine oxide **80** was afforded by transmetallation and reaction with $Ph_2P(O)Cl$. The enantiomerically pure isomers of **80** were then obtained using D-(+)-di-*O*-benzoyl-tartaric acid ((+)-DBTA) as the reagent, and subsequently reduced to diphosphine ligand PHANEPHOS (**81**) (Scheme 2.32) [48].

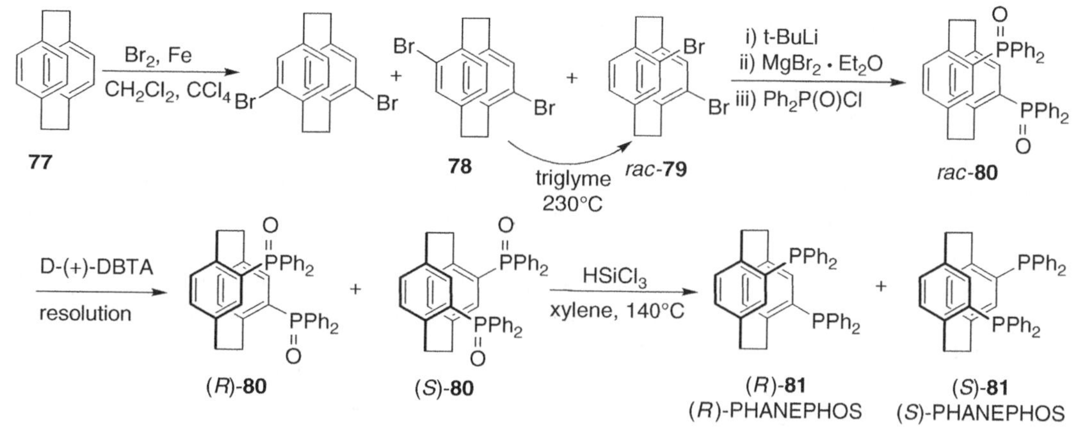

Scheme 2.32 *Synthesis of PHANEPHOS.*

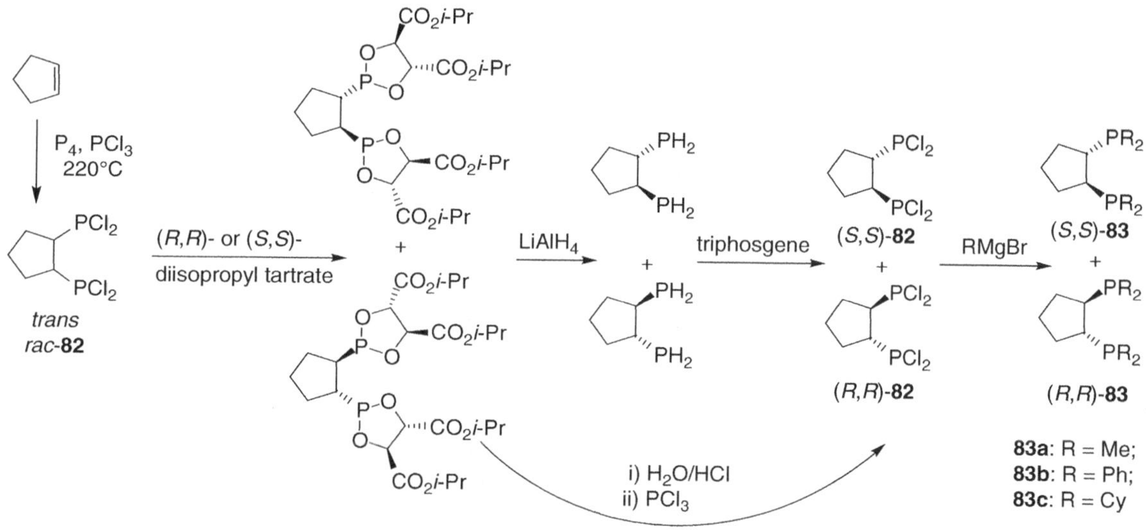

Scheme 2.33 *Resolution of rac-**82** and synthesis of chiral ligand **83**.*

Resolution of phosphine oxides has been widely applied as an effective strategy of preparing chiral phosphines. However, the resolution step is limited to not only phosphine oxides but also other phosphorus-containing intermediates such as phosphine dichlorides. Dahlenburg and coworkers reported an important approach of resolving the racemate of *trans* bis(phosphine dichloride) *rac*-**82** by forming diastereomeric phosphonites with enantiopure diisopropyl tartrate [55]. Both pure diastereomers were then reduced with $LiAlH_4$ to give enantiomeric primary bisphosphine and then reacted with triphosgene to accomplish the resolution providing the (*R*,*R*)- and (*S*,*S*)- enantiomer of **82** (Scheme 2.33). As "covalently" resolved by tartaric acid esters, chiral intermediate **82** has the great potential to play the role of key synthon for modular synthesis of diphosphine with different substituents and also more phosphorus ligands such as phosphinites, aminophosphines, and many others [56].

2.3 Chiral atropisomeric biaryl diphosphines

During the fast development of asymmetric catalysis in recent decades, different categories of well-defined chiral ligands have been designed and investigated to pursue high enantioselectivities to meet the increasing demand for optically active compounds. Chelating C_2-symmetric atropisomeric diphosphine ligands are one important class of ligands that have fundamental significance and influence on the asymmetric catalysis area. The most remarkable milestone of the development history of this ligand class was BINAP, which was first reported by Noyori and Takaya in 1980, decades after the discovery of practical asymmetric catalysis [57]. BINAP became one of the most successful chiral diphosphine ligands, and its great success in a wide range of asymmetric catalysis involving various metal–ligand complexes is attributed to its unique structural features, including the rigid 1,1′-binaphthyl backbone of high stability and steric influence, as well as its conformational flexibility from the (partially) revolving axial single bond [58]. Although BINAP was originally designed for Rh-catalyzed asymmetric hydrogenation of α-(acylamino)acrylic acids, more and more breakthroughs have been made in other asymmetric transformations such as Rh-catalyzed asymmetric isomerization and Ru-catalyzed asymmetric hydrogenation of functionalized ketones and simple ketones since then. Inspired by Noyori's work on the BINAP chemistry, other research groups developed many excellent atropisomeric biaryl diphosphine ligands, such as BICHEP, BIPHEMP, TunePhos, and MeO-BIPHEP. Moreover, many examples of modifications of BINAP backbone structure have been developed to achieve better performances.

2.3.1 Synthesis of BINAP and its derivatives

In Noyori and Takaya's original work, they presented the successful synthesis of BINAP starting from the racemic atropisomers of 1,1′-binaphthyl-2,2′-diol (BINOL) (*rac*-**84**) [57]. The racemates were first converted to dibromo-1,1′-binaphthyl under heated bromination conditions. The racemic BINAP (*rac*-**85**) was subsequently obtained by treatment of *t*-BuHgLi and Ph_2PCl. It was noteworthy that one step of optical resolution was performed by using the chiral palladium(II) complex through fractional recrystallization to obtained the final optically pure **85** (Scheme 2.34). Noyori and coworkers then improved the resolution process to find a more practical access to BINAP (Scheme 2.34) [59]. In this method, racemic bisdiphenylphosphine oxide (BINAPO) was prepared and then resolved using (+)-camphorsulfonic acid or DBTA as the resolving reagent.

Shortly after BINAP's initial synthesis report and excellent performance in asymmetric hydrogenation, other improved synthetic routes were developed to prepare enantiopure BINAP on a large scale, even on an industrial scale [60]. Merck Inc. developed a short synthetic route starting with optical resolution of BINOL and involving a Ni(II)-facilitated coupling with diphenylphosphine. It was reported that excessive diphenylphosphine acted as a reducing agent to generate a Ni(0) species *in situ* which is the active catalyst [61]. Another invention modified this short route by adding Zn metal and changing diphenylphosphine to diphenylchlorophosphine to improve the ease of handling and production (Scheme 2.35) [62]. Due to the great improvement of BINAP synthesis, BINAP became one of the rare chiral ligands produced on an industrial scale, and several large-scale asymmetric syntheses are now carried out with the commercially available BINAP.

Modifications of the BINAP ligand have been extensively studied, including the (1) replacement of the phenyl group on the phosphorus atoms to various substituted phenyl groups or aliphatic or heteroaromatic groups, (2) introduction of substituents on the binaphthyl backbone, and (3) partial reduction of the binaphthyl backbone to H_8-BINAP family ligands (Scheme 2.36) [60]. All these modifications systematically investigated the BINAP family ligands by manipulating the electronic and steric properties of both the backbone and phosphine groups. Moreover, some modifications also altered the bite angles of the binaphthyl scaffold.

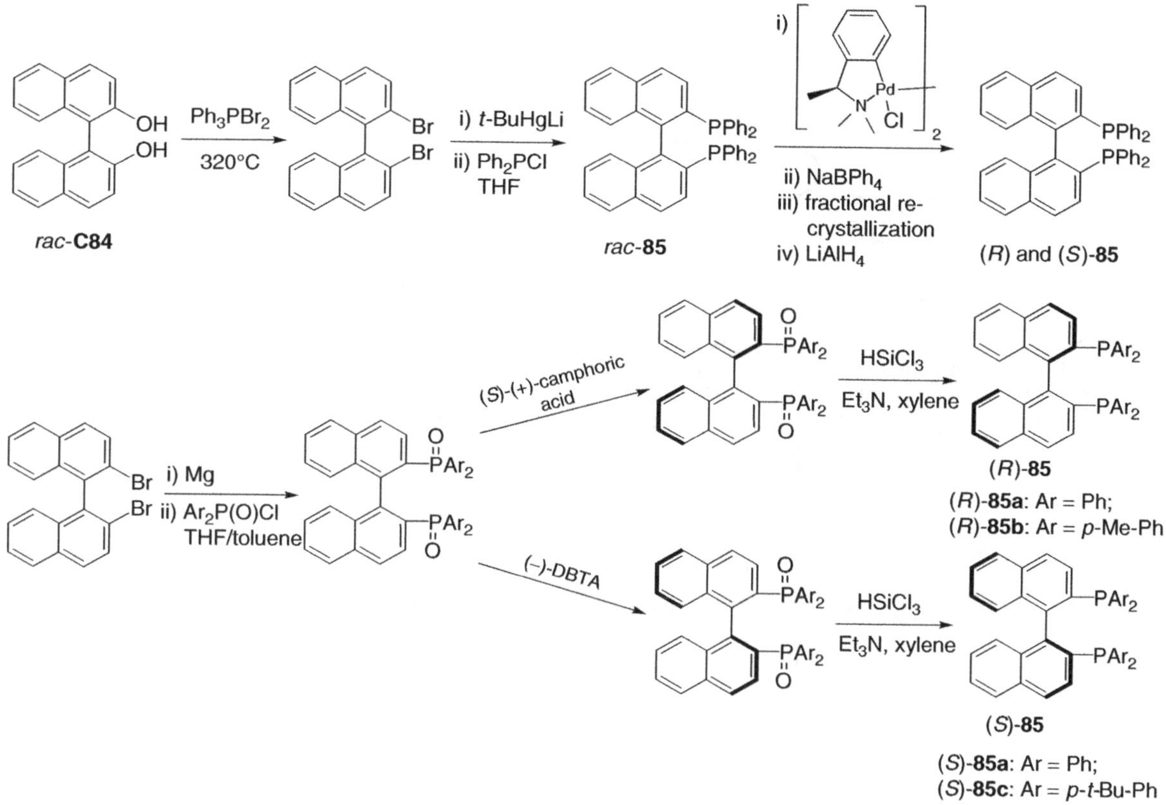

Scheme 2.34 *Synthesis of BINAP via optical resolution.*

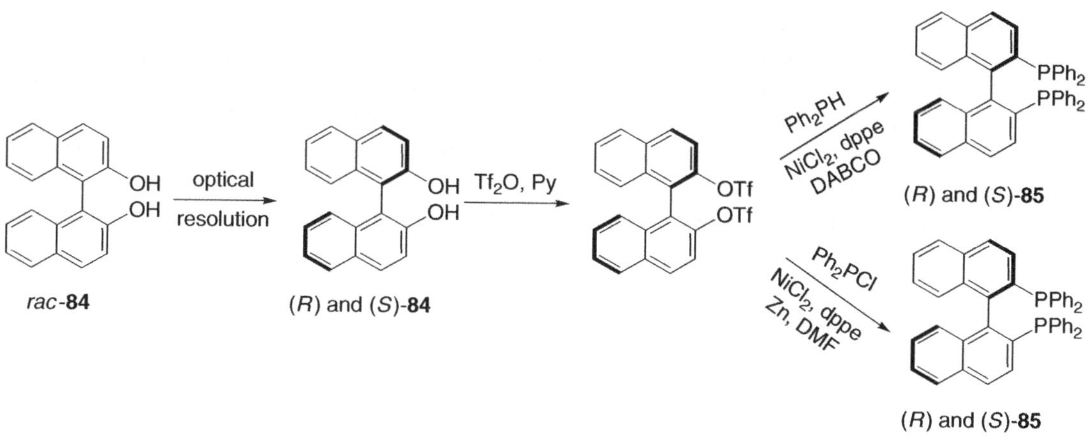

Scheme 2.35 *Modified synthetic route of BINAP.*

Many BINAP derivatives were synthesized and studied by the Takasago group by the improved method of taking advantage of Ni-catalyzed coupling of disubstituted phosphine oxide with a binaphthol bistriflate **86** [63]. Another method describing the coupling of **86** with phosphine–borane complexes to afford ligands with more complex substituents was reported [64]. The synthesis of heterotopic binaphthyldiphosphine (BINAPP′)

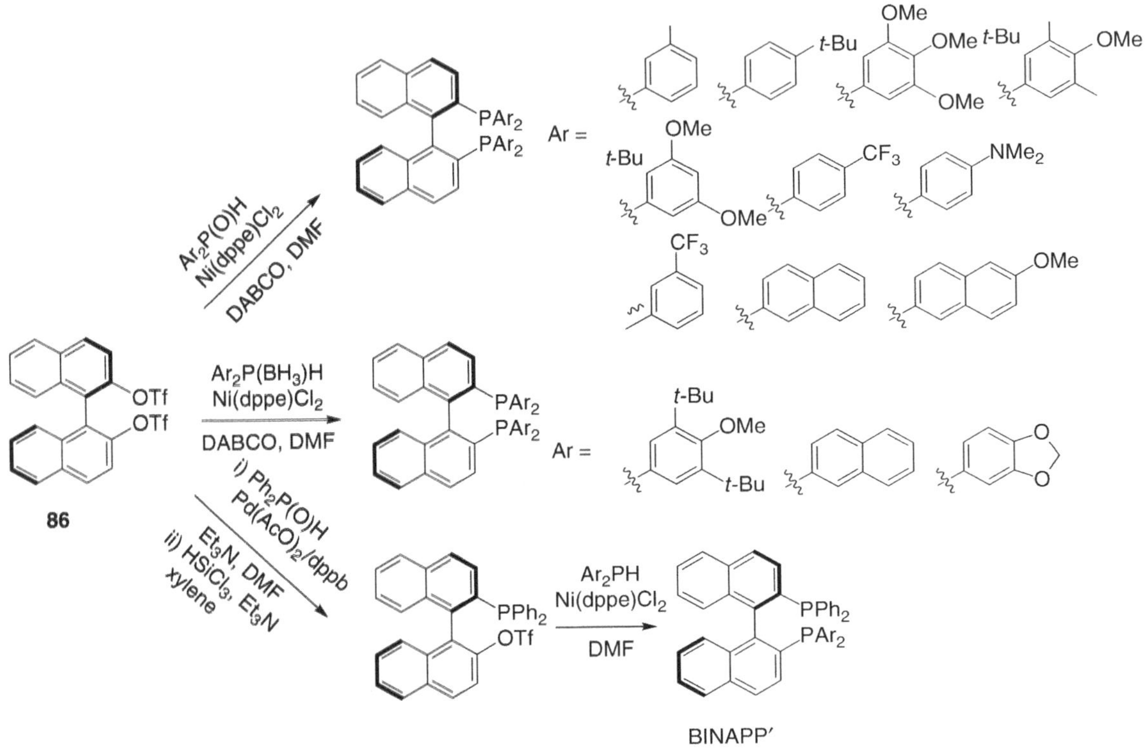

Scheme 2.36 *Syntheses of modified BINAP with aryl substituents on the phosphorus atoms.*

with different substituents in the two phosphine groups was also reported using the similar Takasago method but in a stepwise fashion to install the two different phosphine groups [65].

The syntheses of BINAP variants with aliphatic and heterocyclic substituents were not as successful as with the aryl counterpart. Takaya *et al.* exemplified the first synthesis of Cy-BINAP (**89**) by incorporating the dicyclohexyl phosphine oxide groups onto a binaphthyl backbone (Scheme 2.37) [66]. Taking a very similar synthetic strategy, Kumobayashi *et al.* synthesized cP-BINAP (**90**) with cyclopentyl groups as the substituents (Scheme 2.37) [67]. The Keay group reported the preparation of furyl-substituted BINAP, which contains fewer electron-rich phosphorus atoms [68]. Racemic TetFu-BINAP (**91**) was treated with (*S*)-camphorsulfonyl azide derivative to provide a mixture of diastereomeric phosphinimines that could be separated to give both enantiomers of the ligand (Scheme 2.37).

In another attempt to modify the structural properties and thus catalytic characteristics of BINAP, the binaphthyl skeleton was partially hydrogenated to afford more efficient ligand H_8-BINAP (**92**). The more electron-rich H_8-BINAP possesses more steric bulk and forms a larger dihedral angle than the original BINAP [69]. The H_8-BINAP was prepared via partial hydrogenation of 1,1′-binapthyl-2,2′-dibromide (**87**). Following the similar reaction procedures of BINAP synthesis, racemic H_8-BINAPO was synthesized and resolved with DBT before the final reduction step to give enantiopure **92** (Scheme 2.38) [66]. Optical resolution procedure with DBT was involved in separating the two enantiomers of the phosphine oxides H_8-BINAPO. The reduction step was accomplished with an excessive amount of $HSiCl_3$–Et_3N. H_8-BINAP exhibited excellent enantioselectivities in Ir-catalyzed reduction of ketones, Rh-catalyzed conjugated additions to α,β-unsaturated compounds, and other metal-mediated cycloaddition reactions [70].

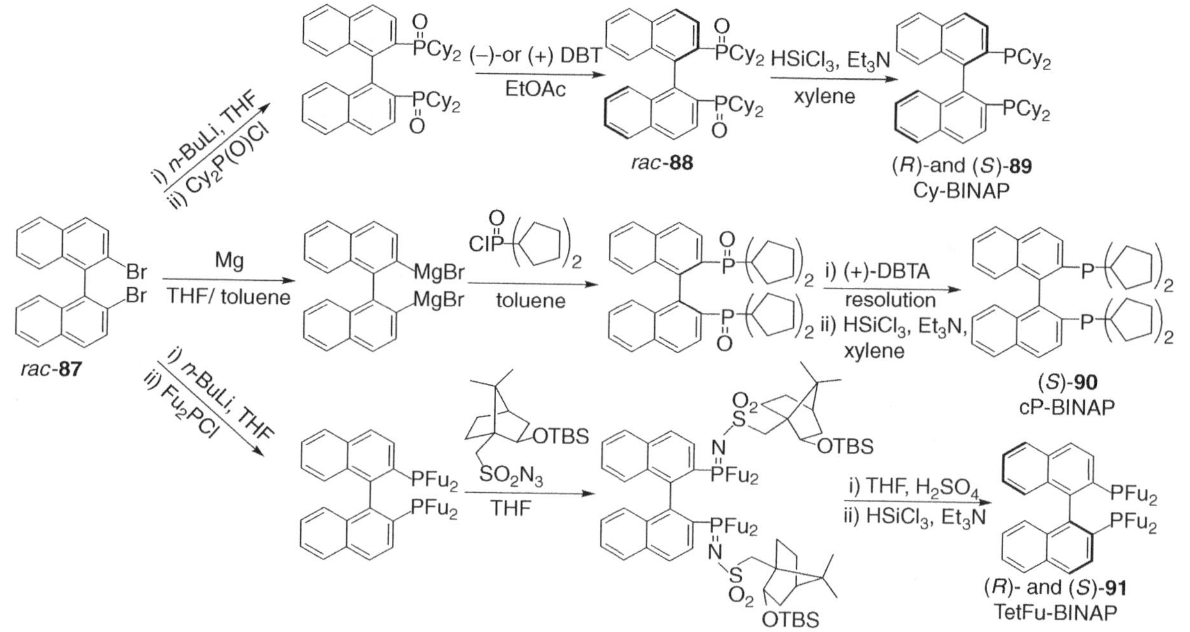

Scheme 2.37 Syntheses of modified BINAP with alkyl or heteroaromatic substituents on the phosphorus atoms.

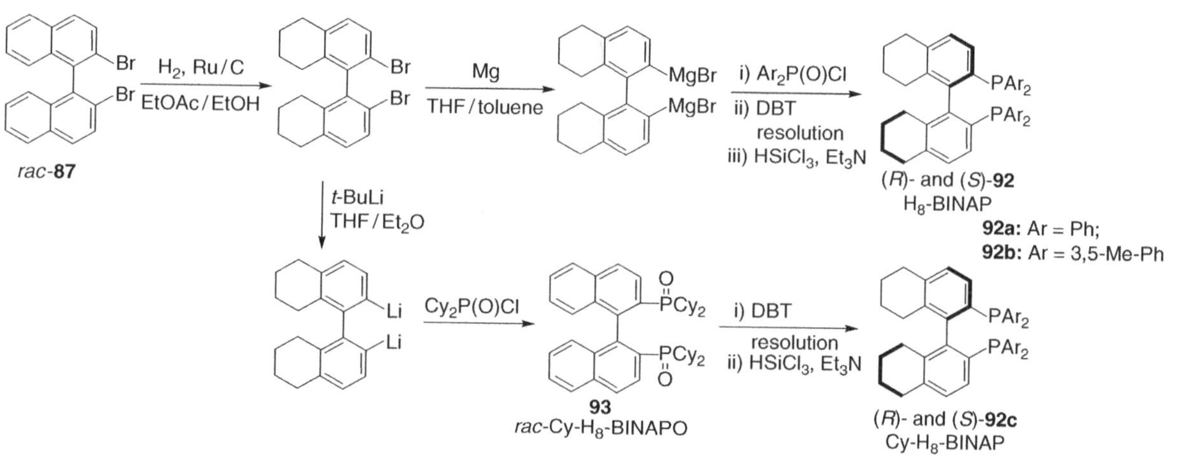

Scheme 2.38 Synthesis of H_8-BINAP.

2.3.2 Synthesis of atropisomeric biaryl ligands

Inspired by the tremendous success achieved in the use of Rh- and Ru-BINAP-catalyzed asymmetric, many atropisomeric C_2-symmetric biaryl diphosphine ligands were synthesized and extensively studied. Roche chemists reported the earliest examples of preparation of axial chiral biaryl diphosphine BIPHEMP and MeO-BIPHEP. The common steps in the two independent synthetic routes are the Ullmann coupling reaction step to form the biaryl scaffold from two monomers and optical resolution to obtain both enantiomers of the key intermediates (Scheme 2.39) [71].

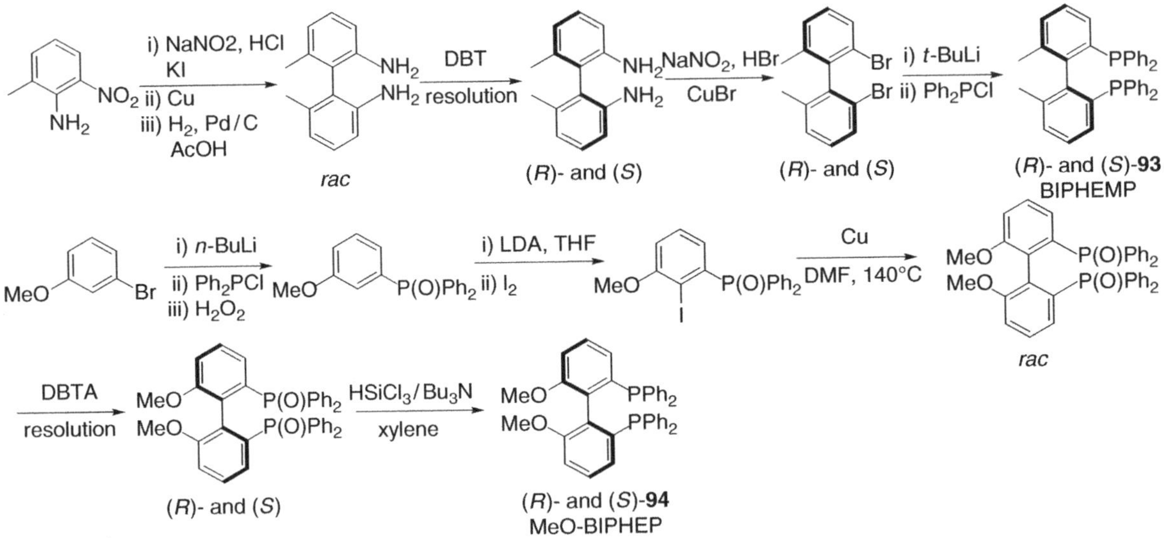

Scheme 2.39 Synthetic routes of BIPHEMP and MeO–BIPHEP.

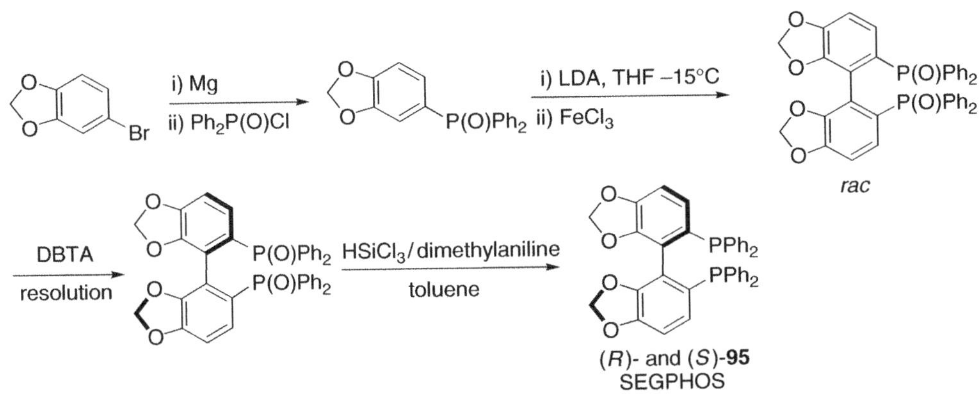

Scheme 2.40 Synthesis of SEGPHOS via an oxidative coupling method.

An alternative oxidative coupling method for aryl–aryl coupling between two lithiated phosphine oxides was developed for the synthesis of SEGPHOS (**95**) [72]. Use of $FeCl_3$ instead of Cu could lead to the desired diphosphine oxide in a racemic mixture. Further optical resolution with DBTA and reduction with $HSiCl_3$–dimethylaniline afforded the final product (*R*)- and (*S*)-**95** (Scheme 2.40).

The aryl–aryl coupling step could also be accomplished with enantiocontrol. Although introduction of chiral auxiliaries to the monomeric aryl ring prior to the coupling reaction gave diastereomeric biaryl products that could be well separated by chromatography [73], a tethered chiral linker-directed intramolecular coupling could avoid the tedious optical resolution with excellent diastereoselectivity. This attractive strategy was first reported by Chan and coworkers to synthesize PQ-Phos (**96**) [74]. The Zhang group demonstrated a more general and divergent synthetic route to prepare a modular C3-TunePhos-type of phosphine ligands using this strategy (Scheme 2.41) [75]. A highly efficient transfer of central chirality to axial chirality was the key transformation in this synthetic route, by which a substrate consisting two aryl phosphine oxides attached to a chiral linker were coupled to afford the single diastereomeric biaryl product.

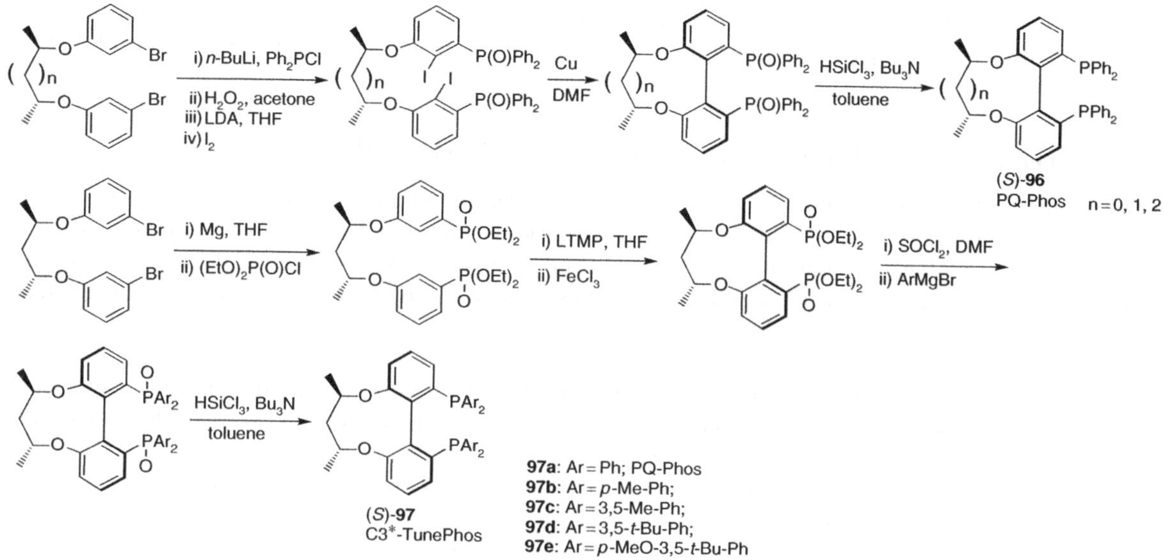

Scheme 2.41 *Syntheses of PQ-Phos and C3*-TunePhos.*

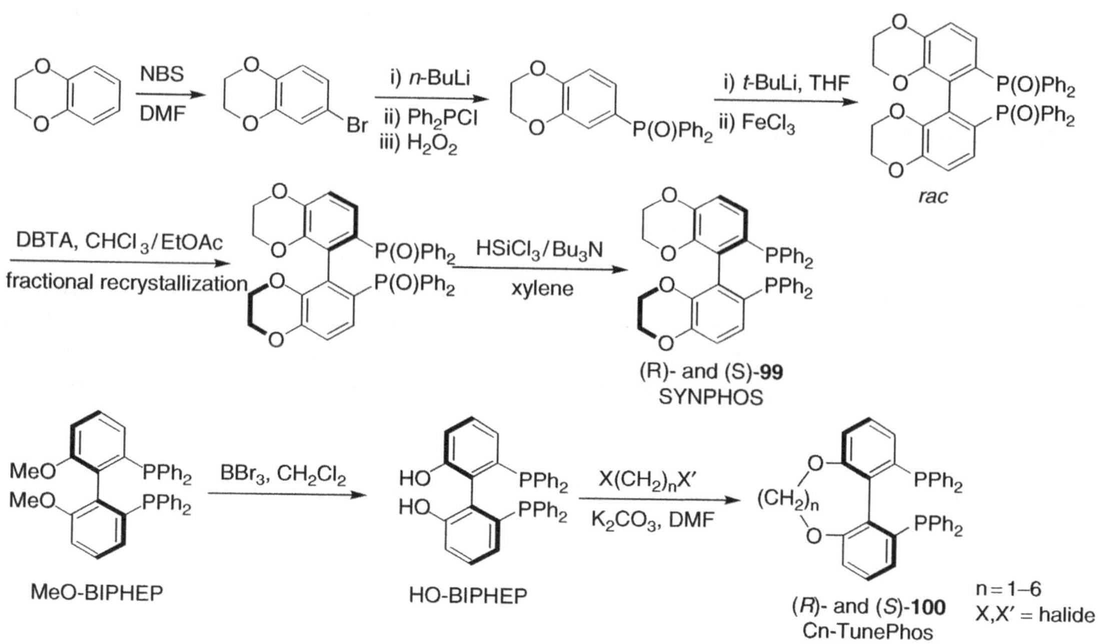

Scheme 2.42 *Syntheses of SYNPHOS and Cn-TunePhos.*

Other modifications led to the preparations of atropisomeric variants of MeO-BIPHEP. Chan and Genêt independently reported the closely related bisbenzodioxanPhos, SYNPHOS (**98**) (Scheme 2.42). The synthesis route was very similar to that of SEGPHOS, which relied on Cu-catalyzed Ullmann coupling and the resolution with DBTA [76]. The Zhang group developed a series of TunePhos ligands **99** with a tunable dihedral angle. The TunePhos ligands were synthesized by treatment of MeO-BIPHEP with BBr_3 and then tethering the two HO- groups with various linkers via etherification with corresponding alkyl dihalide [77] (Scheme 2.42).

2.3.3 General strategies of synthesizing of atropisomeric biaryl ligands

Although the synthetic routes of different atropisomeric biaryl ligands vary from case to case, they all are generally based upon the same elementary steps for constructing chiral diphosphine, yet applied in different sequences of the combination. Introduction of the phosphine moiety (**a**), introduction of axial chirality (**b**), coupling of two aryl fragments (**c**), and reduction of phosphine oxide (**d**) are the basic elements. The general synthesis strategies could be summarized:

1. Starting with a racemic biaryl backbone, followed by introduction of directly diphosphine moieties or bisphosphine oxides (reduction required in the final step), enantiomeric ligands (or oxide derivatives) are obtained from optical resolution (**b** or **c**→**b**→**d**).
2. Starting with an enantiomeric biaryl backbone, diphosphine moieties or bisphosphine oxides (reduction required in the final step) are incorporated (**b**→**a** or **b**→**a**→**d**).
3. Starting with a monomeric disubstituted aryl phosphine oxide, racemic bisphosphine oxide is formed from nonchiral coupling (Cu(0)-, Cu(II)- or Fe(III)-catalyzed), and then resolved and reduced to afford enantiomeric ligands (**a**→**c**→**b**→**d**).
4. Starting with a monomeric disubstituted aryl phosphine oxide, diastereoselective coupling (Cu(0)-, Cu(II)- or Fe(III)-catalyzed) with a tethered chiral linker affords chiral biaryl phosphine oxide that is further reduced (**a**→**bc**→**d**).

2.4 Chiral phosphacyclic diphosphines

In the continuous quest for high enantioselectivies, highly rigid and well-defined structures have been suggested for favoring the achievement of high enantioselectivies [78]. As a result, chiral ligands with phosphorus-containing heterocycles have emerged as an active area in phosphine ligand development [79]. Cyclic structures significantly reduce the flexibility, and the resulting, well-defined chiral environment of the catalyst often led to superior selectivities compared with their noncyclic counterparts. One well-known example is DuPhos developed by Burk and his DuPont colleague chemists. Its excellent selectivities, especially in asymmetric hydrogenation, have led to directing significant efforts to extend both the design and application of this ligand as well as the entire ligand class.

2.4.1 Fundamental discovery and syntheses of BPE and DuPhos

From 1990 to 1991, Burk and coworkers demonstrated the first syntheses and applications of a series of chiral *trans*-2,5-disubstituted bisphospholane ligands including BPE and DuPhos [80–82], although similar but less effective chiral phospholane ligands were previous reported by Brunner. BPE and DuPhos contain stereogenic centers close to the phosphorus atoms and also the metal catalytic center. Therefore, they tend to have the ability to deliver excellent chiral inductions in their application in metal-catalyzed catalysis, especially in Rh-catalyzed asymmetric hydrogenations. Generally good to excellent enantioselectivities were observed for a wide variety of substrates such as α- and β-dehydroamino-acid derivatives, enol acetates, *N*-acyl hydrazones, enamides, enol esters, itaconic acids, and β-keto ester derivatives [83]. This was the first series during the 1990s of chiral phosphacycle-based ligands that are highly active and stereoselective for a large variety of substrates in the asymmetric hydrogenation. The exceptional features of

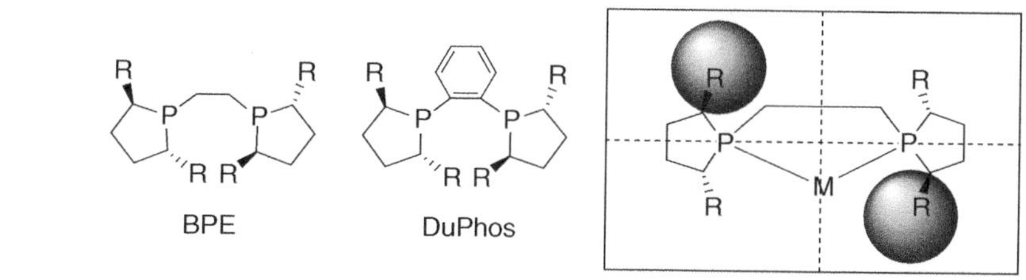

Figure 2.1 Structures of BPE and DuPhos, and quadrant diagram explaining the origin of stereoselectivity.

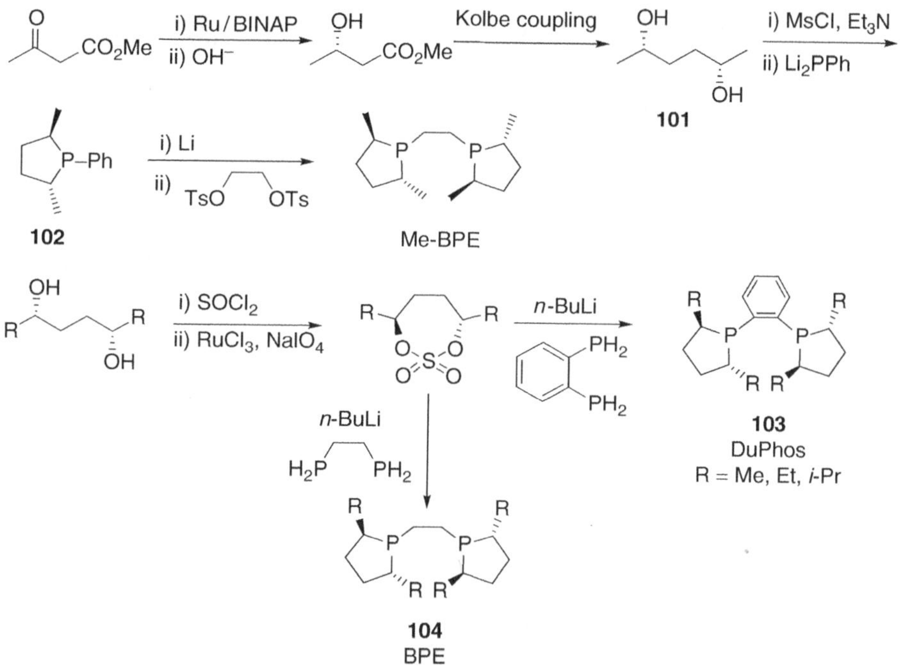

Scheme 2.43 *Syntheses of BPE and DuPhos.*

these ligands were soon employed in a series of industrial applications [84]. In many cases, the origin of the high enantioselectivities can be explained by a quadrant model (Figure 2.1) [85]. With two of the coordination quadrants blocked by the bulky substituents within the ligands, the reaction pathway is strongly influenced by steric congestions, leading to enantiomerically enriched products. The BPE and DuPhos ligand series was a fundamental discovery in phosphine ligand development and had a significant impact on the ligand design thereafter.

The first reported BPE ligand synthesis starts with a *cis*-1,4-diol **101** prepared by either a biochemical method or asymmetric hydrogenation followed by electrochemical Kolbe coupling (Scheme 2.43). Treatment with MsCl followed by Li_2PPh affords the enantiomerically pure 1-phenyl-2,5-dialkyl substituted phospholanes **102**. Cleavage of the phenyl group generates the phosphorus nucleophiles, and subsequent reaction with ditosylates or dichloride linkers yields the final ligand products [80]. This procedure, however, was problematic and was soon modified to give a more versatile route by which a series of ligands with various 2,5-dialkyl substituents can be easily accessed [81]. The new synthetic route was also applied in the

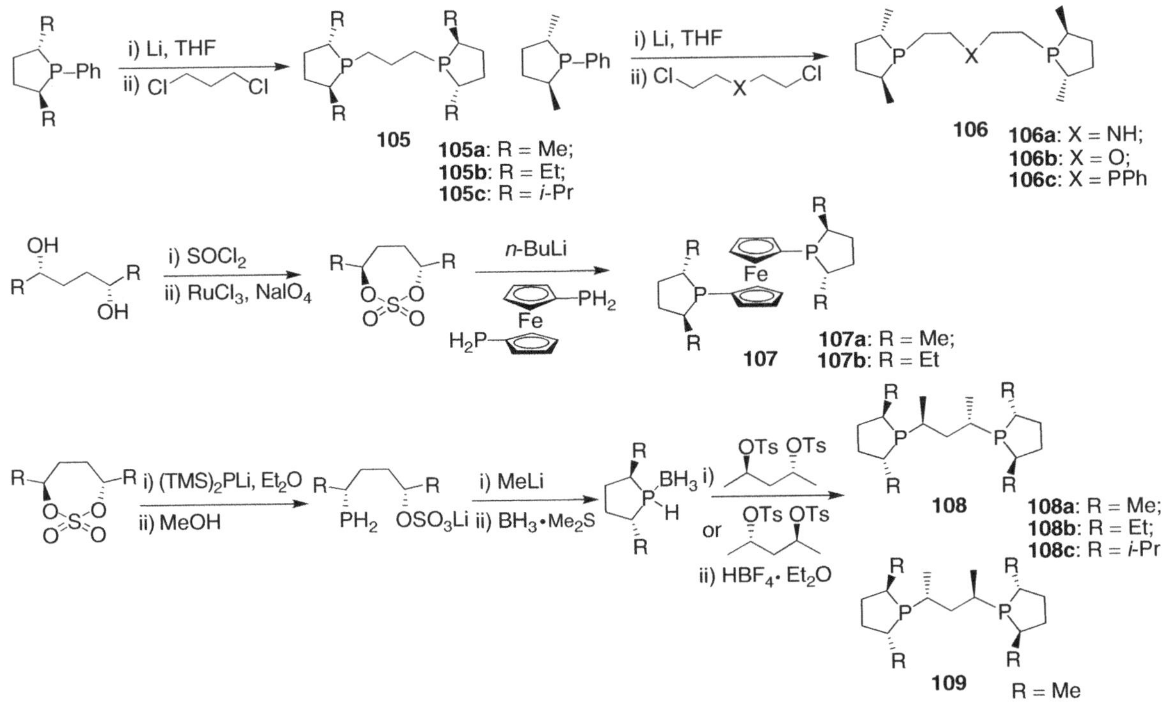

Scheme 2.44 *Syntheses of BPE analogues.*

preparation of the more rigid 1,2-benzenediyl-based DuPhos. Using a method developed by Sharpless, the *cis*-1,4-diol was converted into a cyclic sulfate as a key intermediate. Hence, the cyclic sulfate could react with deprotonated diphosphine nucleophiles to generate the bisphospholane ligand product containing phosphacycles. This methodology, involving the preparation of a cyclic sulfate derivative of chiral diols, has continued to be the most effective strategy of phosphacycle-based ligand synthesis.

Prompted by the excellent results achieved by BPE and DuPhos, and also for reasons of comparison, Burk and coworkers synthesized a series of 2,5-disubstituted analogous phospholanes of BPE via the same or slightly modified synthetic routes (Scheme 2.44) [80,82a,82b,86–88]. By switching to the linkers of longer carbon chains, **105** and **106** were successfully prepared [80,82b]. Bisphospholane ligands **107** possessing a ferrocenyl backbone could also be formed via the convenient route [87]. Ligand **108** and **109** that are tethered by chiral 2,4-pentane backbone structures were obtained by reacting the ditosylates of the chiral pentanediol and the phospholane borane adduct followed by the borane removal step. The influence of the two stereogenic centers on the linker bridge on the performance of these ligands, including the matching and mismatching effects, was studied in the Rh-catalyzed asymmetric hydrogenations of enamides and α-keto esters [86]. Indeed, matching and mismatching were observed, and, notably, the matching pair gave also higher enantioselectivity than the nonchiral pentanyl linked counterpart.

Although the above-mentioned ligand family could be all prepared via similar synthetic routes, the synthesis of analogous 2,5-diarylphospholanes had remained unsuccessful. Elimination occurred on the cyclic sulfate or dimesylates when strong basic conditions were applied in a conventional synthesis [89]. Several groups have tried to solve this problem. Fiaud and coworkers reported an epimerization process of *cis*-1,2,5-triphenylphospholane oxide in 1991 [90]. The resulting two enantiomers were originally separated by chiral phase chromatography. A new route was later developed to avoid the tedious separation procedures (Scheme 2.45) [91]. Reaction of 1,4-diphenyl-1,3-butadiene with aminophosphine dichloride, followed by

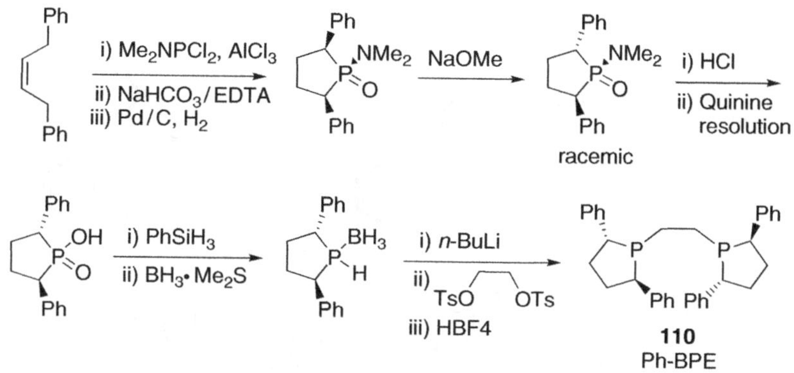

Scheme 2.45 *Synthesis of Ph–BPE.*

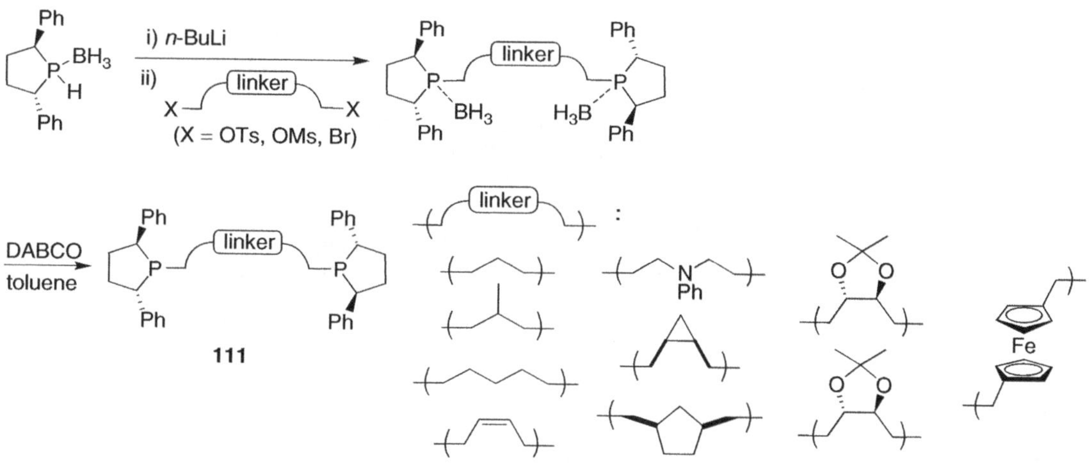

Scheme 2.46 *Synthesis of Ph–BPE analogues.*

reduction, epimerization, and hydrolysis, gave the *trans*- diphenylphospholanic acid, which was then easily resolved into the optically pure forms. Additional steps finally yielded the *trans*-1,2,5-triphenylphospholane which was used as a monodentate ligand in asymmetric hydrogenation. Utilizing this methodology developed by Fiaud, Pilkington successfully synthesized the 2,5-diphenyl BPE [89]. Replacement of the alkyl substituents with phenyl groups resulted in significantly higher enantioselectivities, in many cases examined in asymmetric hydrogenation. In 2005, Shibasaki *et al.* expanded this Ph–BPE to a ligand family **111** where various linkers connecting two 2,5-diphenyl phospholane fragments were applied (Scheme 2.46) [92]. This series of ligands was tested in copper-catalyzed enantioselective aldol reactions, and a strong dependence of the selectivities on the linker structures was concluded from the results.

The BPE and DuPhos ligand family, representing a milestone in chiral ligand development, opened a new field of research into chiral phosphine ligands. Its ability for versatile modifications has inspired many groups to develop analogous ligands to further improve the catalytic performance. Although some of these were initially designed to circumvent patent protection and, in many cases, several ligands were designed for the same catalytic process, they indeed significantly expanded the ligand family and proved an excellent opportunity to compare, evaluate, and study their structural and electronic properties. The large number of this ligand family also allows systematic screening of catalysts for one specific process.

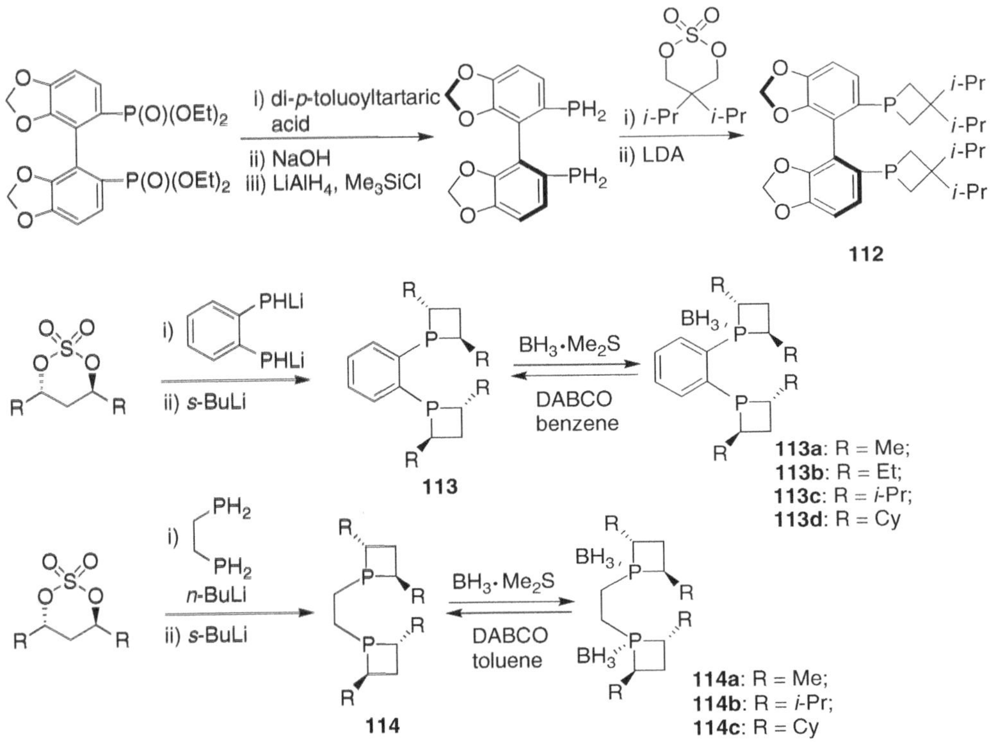

Scheme 2.47 *Synthesis of a bisphosphetanes ligand from cyclization of cyclic sulfate.*

2.4.2 Design and synthesis of bisphosphetanes

The backbones of chiral bisphosphetanes that contain a four-membered ring can provide a relatively rigid framework for asymmetric catalysis that requires a chiral environment. The synthesis of chiral bisphosphetanes could be similar to that of BPE and DuPhos, with the difference that now a phosphorus derivative reacts with a chiral three-carbon moiety to give the phosphacycles.

Taking advantage of the mesylate or cyclic sulfate methodology, many chiral bisphosphetanes were prepared based on the different moieties designed. Saito and coworkers reported the synthesis of ligand **112**, which bears a chiral biphenyl moiety (Scheme 2.47) [93]. A more sterically strained analogue of DuPhos (**113**) also synthesized using the cyclic sulfate to give **113**. Bulky substituents such as isopropyl and cyclohexyl were successfully introduced onto the four-membered rings, favoring high enantioselectivities. The borane complexation process was taken to protect the phosphorus atom in order to facilitate the purification step (Scheme 2.47) [94].

Another notable example of phosphetane synthesized is FerroTANE **115** (Scheme 2.48) [95,88]. The C_2-symmetric 1,1′-bis(phosphino)ferrocene ligand family applied was obtained via the analogous cyclization approach and became commercially available later. This ligand family exhibits high efficiency in Rh-catalyzed asymmetric hydrogenation of monoamido itaconates in terms of both reactivity and enantioselectivity.

Other phosphetanes were synthesized using a different strategy, in which the bridge connecting the two preformed phosphetane rings **116** was formed by a coupling approach. However, in this approach, enantioselective *ortho* deprotonation of the borane-protected phosphine was necessary with a chiral auxiliary reagent such as (–)-sparteine before the Cu-catalyzed oxidative couple afforded the diastereomeric dimers,

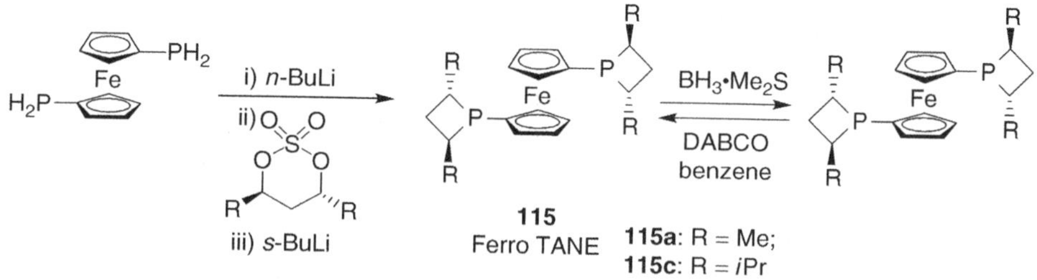

Scheme 2.48 Synthesis of FerroTANE ligands.

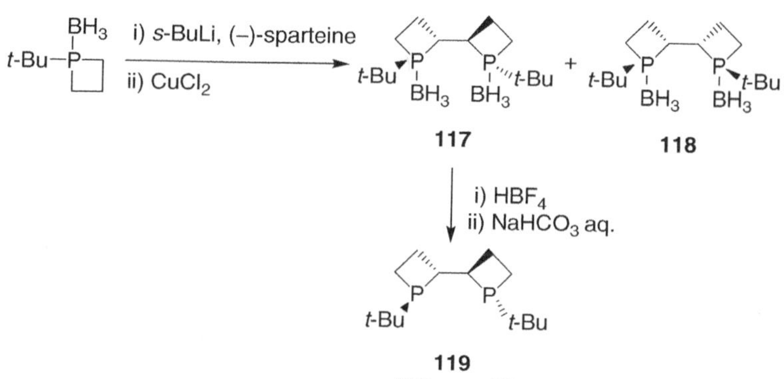

Scheme 2.49 Synthesis of ligand DiSquareP* via enantioselective oxidative coupling.

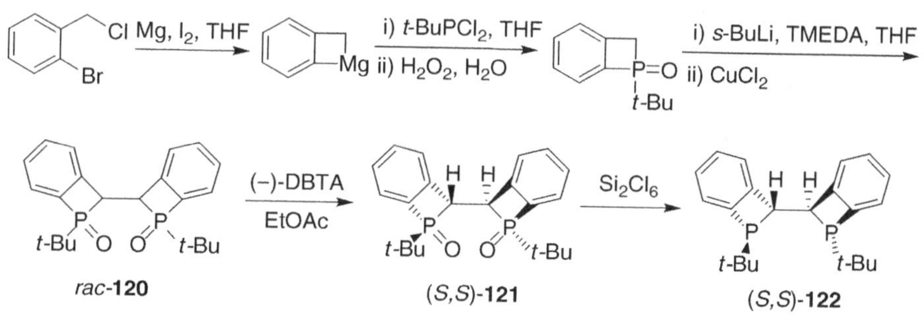

Scheme 2.50 Synthesis of bisbenzophosphetane ligand **122**.

the expected chiral borane–ligand adduct **117** and *meso* product **118** (Scheme 2.49) [96]. The removal of borane gives the ligand product DiSquareP* (**119**), which is highly air sensitive but showed excellent performance in the Rh-catalyzed asymmetric hydrogenation of α-(acetylamino)acrylates and α-substituted acetyl enamides. This oxidative coupling served as another important synthetic strategy for the construction of bisphosphacycles.

Imamoto and coworkers reported a similar oxidative synthesis of bisphosphetanes without the assistance of (−)-sparteine. However, optical resolution of the phosphine oxide by DBTA was necessary to obtain the enantiomers of ligand **122** (Scheme 2.50) [97].

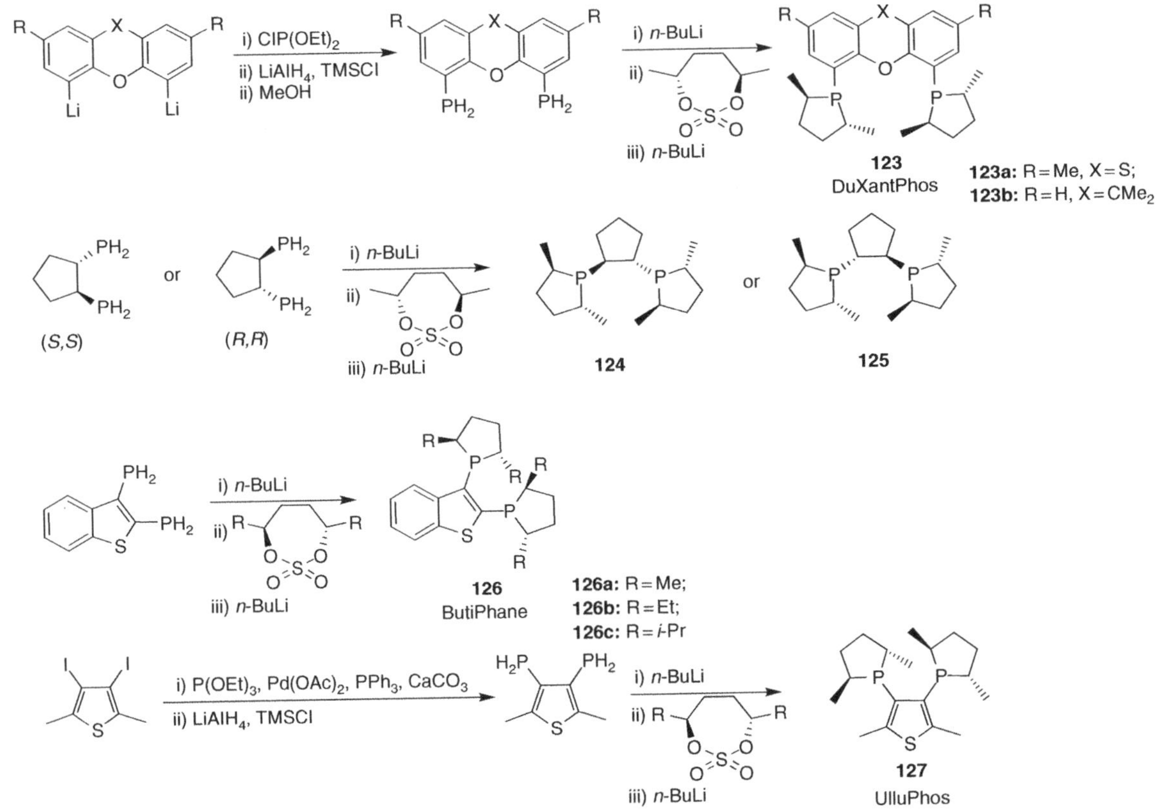

Scheme 2.51 *Synthesis of BPE and DuPhos analogues with modifications on the linker.*

2.4.3 Design and synthesis of bisphospholanes

2.4.3.1 *BPE and DuPhos analogue ligands*

The chiral phospholane ligands summarized in this chapter all exhibit some degree of similarity to BPE and DuPhos, in terms of the design motifs as well as the main synthetic strategy. Various modifications were applied to the linker, the phospholane ring, or both.

Mechanistically, analogous ligands with various linker structures connecting the two phospholane fragments have different bite angles, which may be an important factor for catalyst selectivity. They can, though, be prepared in a similar fashion using cyclic sulfate and the various primary diphosphine following the BPE and DuPhos method. In addition to the initial modifications of BPE and DuPhos by Burk and coworkers, several other analogues with different linker structures were reported (Scheme 2.51). Van Leeuwen and Osborn developed DuXantPhos (**123**), with large bite angles for an asymmetric allylic alkylation reaction [98]. Up to 97% ee was obtained with *O*-acetyl-1,3-diphenylprop-1-en-2-ol. To carry out a detailed study on the origin of the enantioselectivity of BPE in asymmetric hydrogenation, Pringle and Orpen synthesized ligands **124** and **125** with a chiral 1,2-cyclopentanyl linker [99]. The two additional chiral centers on the cyclopentanyl structure force the linker to stay exclusively in the desired conformation when the ligand coordinates with the metal. To study the effect of larger bite angles and different electronic properties of the analogous DuPhos, Berens, Salzer and Sannicolò synthesized two bisphospholane ligands with sulfur

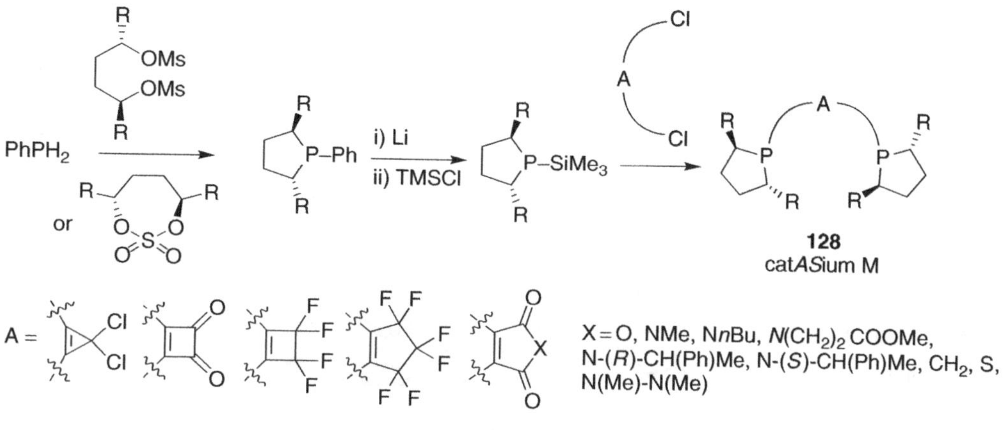

Scheme 2.52 *Synthesis of catASium® M series ligands.*

heterocycles as the linkers, ButiPhane (**126**) and UlluPhos (**127**) [100]. They were applied in asymmetric hydrogenation of dehydroamino acids, itaconic acids, keto esters, and their derivatives. In most cases, comparable selectivities with DuPhos were obtained.

Although numerous analogous ligand motifs have been developed, families of ligands with closely related structures and properties have been scarce, making it difficult to study correlations between the structure and catalytic performance of the ligands. A major contribution toward solving this problem has been made by Börner and coworkers. A library of 16 2,5-disubstituted bisphospholanes ligands cat*AS*ium® M (**128**) was prepared with various cyclic structures as the linker [101,102]. In the synthesis of **128**, the treatment of phenylphosphine dichloride with Li followed by the addition of 2,5-disubstituted dimesylates or cyclic tosylate forms the 1-phenyl-2,5-disubstituted phospholane intermediate. Replacing the phenyl group with TMS and reaction with various dichloride linkers yields the final product ligand (Scheme 2.52). When used in the asymmetric hydrogenation of dehydroamino acids and itaconic acids, a clear trend of selectivities in terms of the steric and electronic properties of the ligands was observed under the same reaction conditions, with ligands bearing maleic anhydride or maleimide linkers being the best [102]. Different synthetic methods were used to prepare these ligands in the laboratory and on an industrial scale since some of the primary diphosphines used in the traditional method were hard to obtain.

Although they usually encountered more synthetic difficulties, several ligands have been prepared with modifications on the phospholane moiety. In 1997, Knochel *et al.* reported the synthesis of two chiral borane-protected bisphospholane ligands **129** and **130** (Scheme 2.53) [103]. Unlike the traditional synthesis, the phosphines were used as electrophiles instead of nucleophiles.

In the quest for an easy route to build chiral phospholanes, Börner and coworkers prepared several bisphospholane ligands using D-mannitol as starting materials. Two 2,5-dimethyl-3,4-dialkoxy substituted RoPhos ligands **131** and **132**, combining the structural motifs of Brunner's ligands and DuPhos, were synthesized and proved to be highly selective for dehydroamino acids and unsaturated phosphonates (Scheme 2.54) [104]. A similar strategy was used to synthesize BasPhos **133** and **134** [105]. Removal of two hydroxyl groups on the 3- and 4-positions followed by formation of the cyclic sulfate and reaction with primary diphosphines yielded the 2,5-disubstituted BasPhos. The convenience and flexibility of using D-mannitol to build phospholanes were also utilized in 2000 by Zhang's and RajanBabu's groups independently to synthesize KetalPhos **135** [106,107]. Deprotection of the two hydroxyl groups on the 3- and 4-positions generates the water-soluble tetrahydroxyl ligand **136**. RajanBabu also reported two

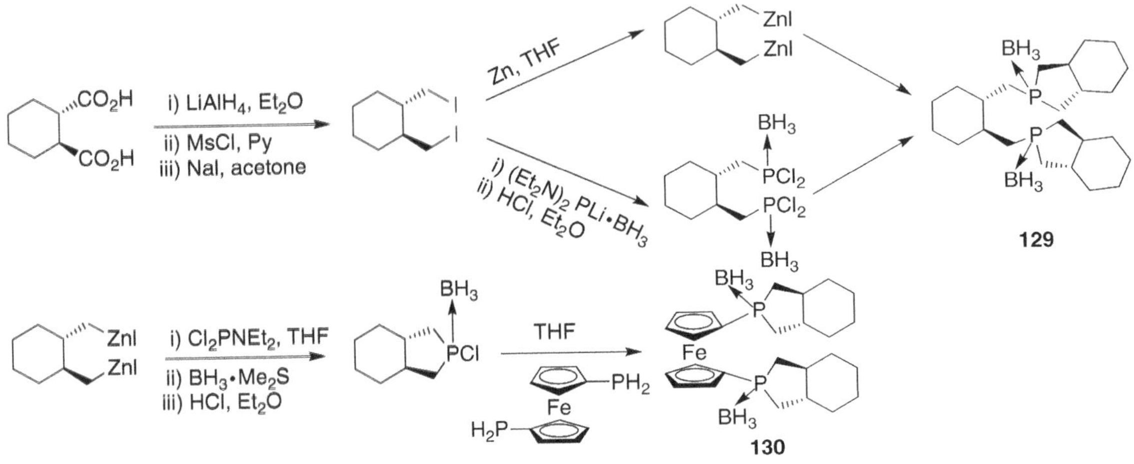

Scheme 2.53 Synthesis of ligand **129** and **130**.

diastereomeric ligands **137** and **138** (Scheme 2.54). Slightly lower selectivity than KetalPhos was observed in the Pd-catalyzed asymmetric allylation reaction. Furthermore, Zhang and coworkers reported a ferrocene-linked analogue ligand f-KetalPhos [108].

The Zhang group synthesized the highly rigid bisphospholane ligand PennPhos (**139**), a DuPhos analogue containing the phosphabicyclo[2.2.1]heptane moiety (Scheme 2.55) [109]. The fused ring system eliminates the conformational flexibility of the phospholanes in DuPhos. The bulky and electron-donating PennPhos was applied in the Rd-catalyzed asymmetric hydrogenation of several challenging substrates such as simple ketones, cyclic enol acetates, and cyclic enamides [109,110].

2.4.3.2 P-stereogenic bisphospholane ligands

Initially designed for Ru catalysis, the P-chirogenic benzophospholane ligand BeePhos (**140**) was developed by Saito in 2003 (Scheme 2.56) [111]. Interestingly, this ligand provided high selectivities in Rh-catalyzed asymmetric hydrogenation and was inactive in Ru catalysis. Hoge synthesized a series of P-chirogenic bisphospholane ligands in 2003 [112]. Inspired by the DuPhos geometry in the quadrant diagram, these ligands kept two of the coordination quadrants blocked while leaving the other two even less hindered than those of DuPhos. These P-chirogenic ligands also offered the potential advantage of attaching the chirality directly to the metal center. The synthesis of these ligands apparently required new methods. Originally, two synthetic routes were used to prepare ligand **141** (Scheme 56). Exclusively *cis*-α-alkylation of the 1-alkoxyphospholane with the assistance of (–)-sparteine/butyllithium and subsequent nucleophilic displacement with retention on the phosphorus atom gave the intermediate **143** in good yield. However, only one enantiomer of the ligand was obtained due to the lack of available (+)-sparteine in nature. The second route started from 2,3-epoxypropyl benzene, and the dimesylate was produced in three steps. Subsequent reaction with primary phosphine gave the key intermediate. Copper-catalyzed homo-oxidative coupling followed by deprotection of phosphine yielded the ligand. These two procedures were soon modified by Hoge, taking the advantage of an epimerization process to synthesize **141** and **142** [112b,c]. Starting from the chiral 1,4-diol that can be prepared on a larger scale, the cyclic sulfate was obtained according to literature-reported procedures. Reaction with primary diphosphines gave exclusively the *cis-cis* product. The desired *trans-trans* ligands were achieved by a kinetic epimerization process under heating conditions.

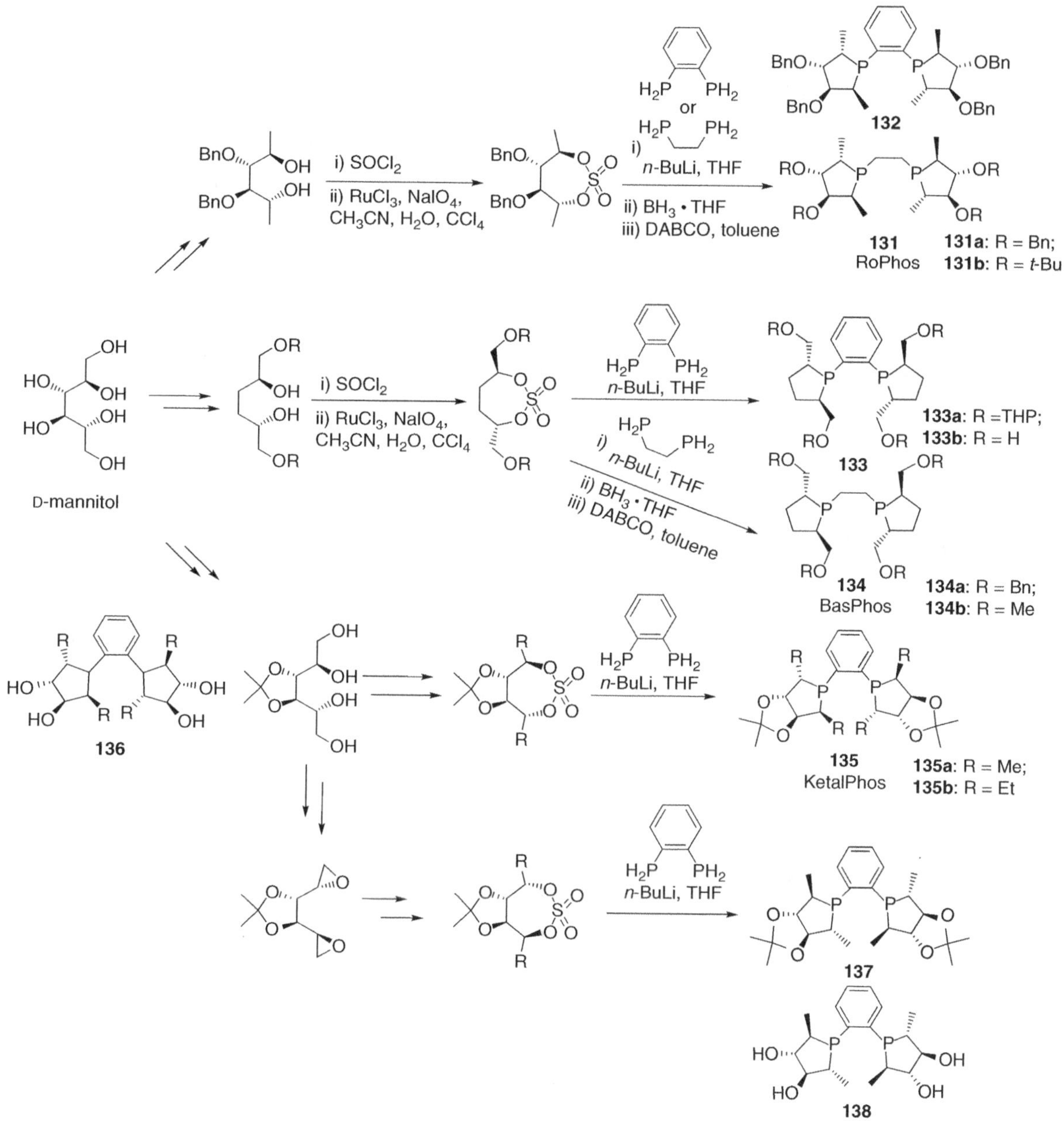

Scheme 2.54 Ligand syntheses using D-mannitol.

In addition to the vast number of DuPhos analogues, a few other structural motifs have been applied to the phospholane ligands. The research group of Zhang has developed three P-chirogenic bisphospholane ligands, TangPhos (**143**), DuanPhos (**144**), and ZhangPhos (**145**). These ligands showed outstanding performance in asymmetric hydrogenations toward a wide range of substrate types and also in many other asymmetric reactions. TangPhos, being highly rigid and strongly electron donating, proved to be an excellent ligand in

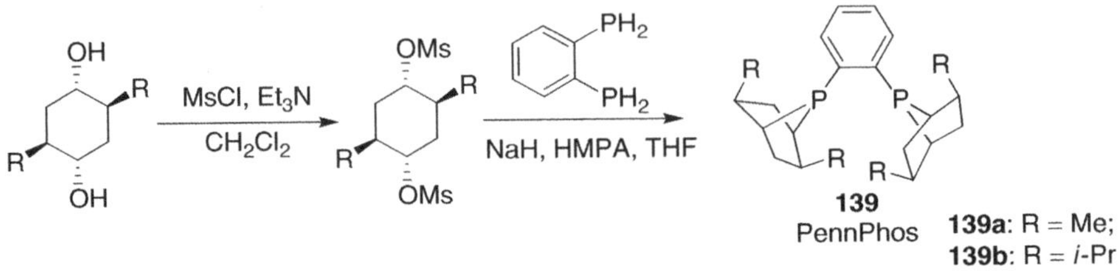

Scheme 2.55 *Synthesis of PennPhos.*

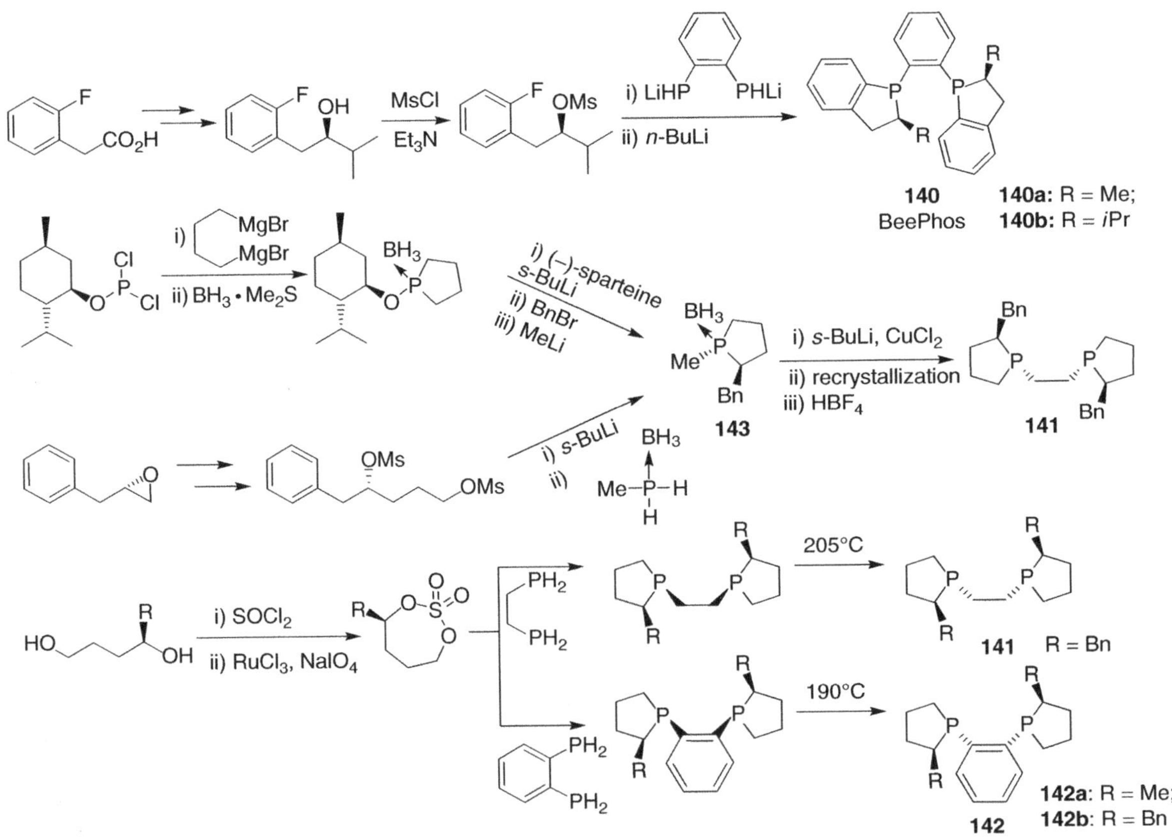

Scheme 2.56 *Syntheses of P-chirogenic ligands BeePhos and **141**, **142**.*

Rh- and Pd-catalyzed asymmetric hydrogenation of dehydroamino acids, α-arylenamides, β-(acylamino) acrylates, itaconic acids, enol acetate, and functionalized imines [113]. The synthesis of TangPhos included cyclization with a bis-Grignard reagent, enantioselective deprotonation with (–)-sparteine/butyllithium, and oxidative homocoupling. However, only one enantiomer was accessible for TangPhos using this synthetic route. To solve this problem, DuanPhos was developed; its phosphine oxide could be resolved into both optically pure forms using DBTA. While maintaining the generally excellent performance of TangPhos, the Rh–DuanPhos complex was also found to be a highly active catalyst for Rh-catalyzed hydrogenation of primary and secondary amino ketones [114]. The third generation of this series was further reported in 2010.

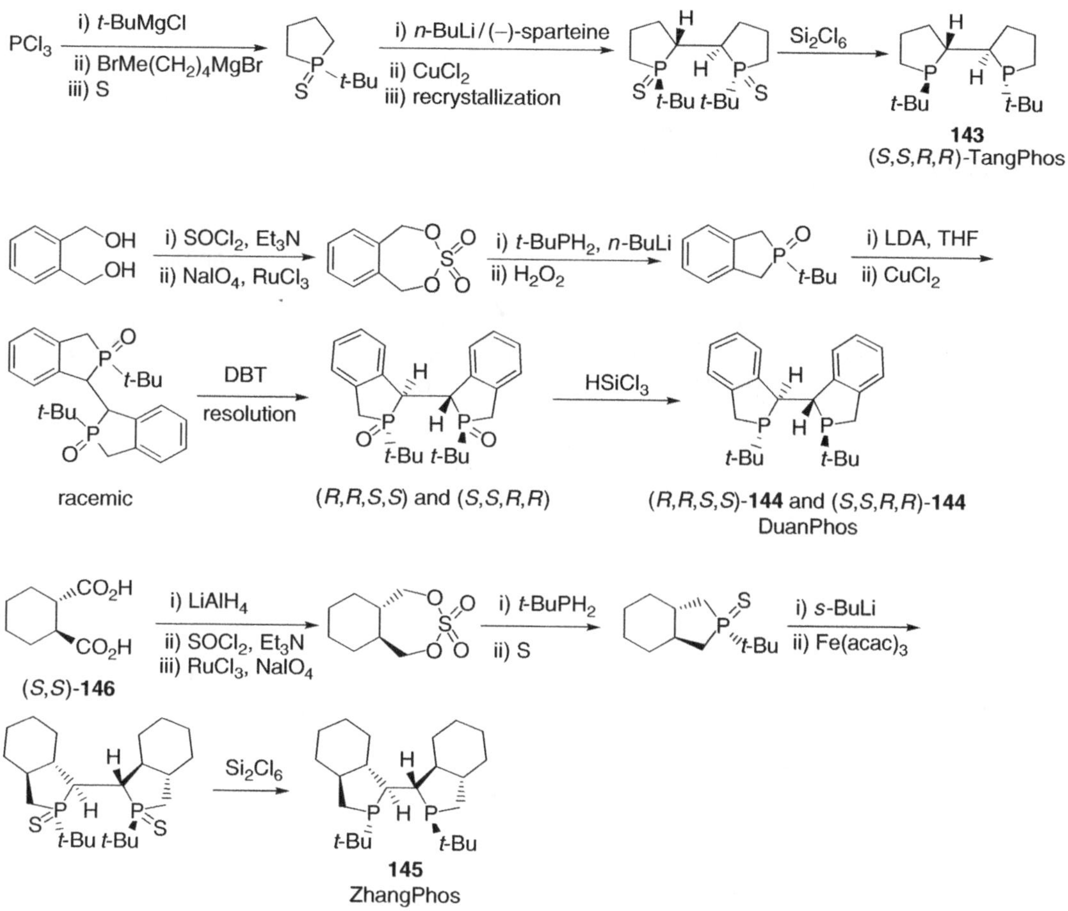

Scheme 2.57 *Syntheses of TangPhos, DuanPhos, and ZhangPhos.*

In order to further enhance the rigidity and electron-donating ability of the phosphine moieties, and to make both enantiomers accessible with resolution in the final step, ZhangPhos was designed and synthesized via a new approach, starting from enantiomerically pure cyclohexanedicarboxylic acid **146** (Scheme 2.57) [115].

2.4.4 Design and synthesis of bisphospholes

Phospholes are the phosphorus analogue of pyrrole and furan. The first example was discovered in 1953 in a dibenzophosphole form, and then the pentaphenylphosphole was synthesized in 1959. Not until 1987, though, was the parent secondary phosphole synthesized [116]. The chiral phospholes, reported as ligands in catalytic asymmetric transformations, were significantly less abundant [79a]. This is due to the fast phosphorus inversion within the phosphole ring at room temperature, making it difficult to define the P-chirality and the axial chirality in some cases. Important solutions to this problem included placing the phosphorus atom on the bridgehead of the bicyclic structures and incorporating the phosphole ring into phosphametallocene structures.

Chiral phosphole ligands with dibenzophosphole (DBP) or binaphthophosphole (BNP) structures form a large family. The double bonds of these phospholes are incorporated into the aromatic system and do not

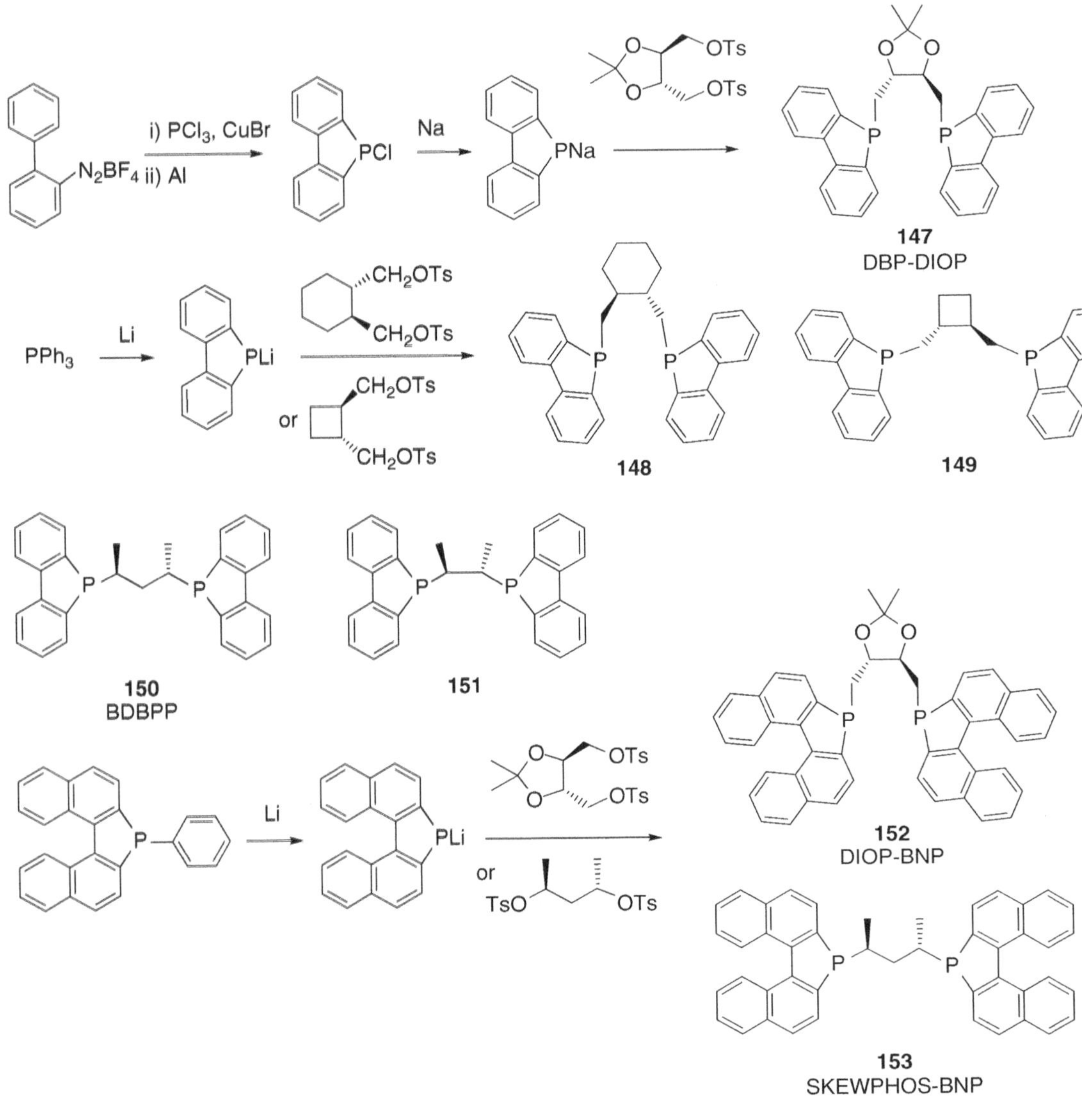

Scheme 2.58 Structures and syntheses of phosphole ligands possessing DBP or DNP structures.

exhibit the same characteristics as typical phospholes in many reactions. These ligands all have chiral substituents on the phosphorus atoms and all show some degree of similarities to existing diphosphine ligands. The DBP–DIOP (**147**) was the first of this family developed by Kagan and Tanaka (Scheme 2.58) [117]. The preliminary asymmetric hydrogenation of olefins and asymmetric hydroformylation of styrene showed moderate selectivities. This ligand was further tested in asymmetric catalysis. However, generally moderate or comparable results to DIOP were observed. Two analogous ligands with 1,2-cyclohexyl (**148**) and 1,2-cyclobutyl (**149**) backbones were synthesized by Hayashi in 1979 (Scheme 2.58) [118]. The preliminary Rh-catalyzed hydroformylation of styrene showed an improved selectivity of 1,2-cyclobutyl-DBP over the DBP–DIOP ligand. However, the enantioselectivity still remained low. The BDPP analogue BDBPP

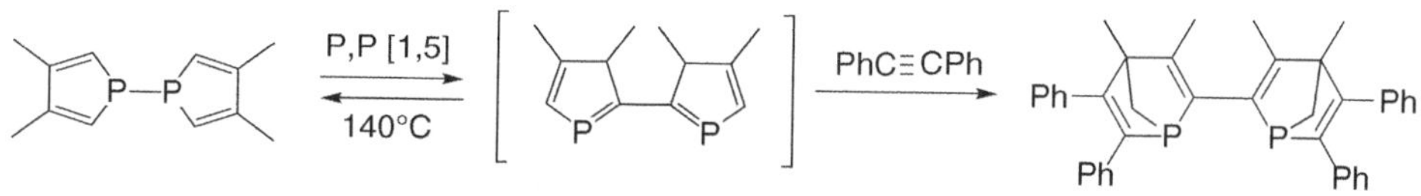

154
BIPNOR

Scheme 2.59 *Synthesis of BIPNOR.*

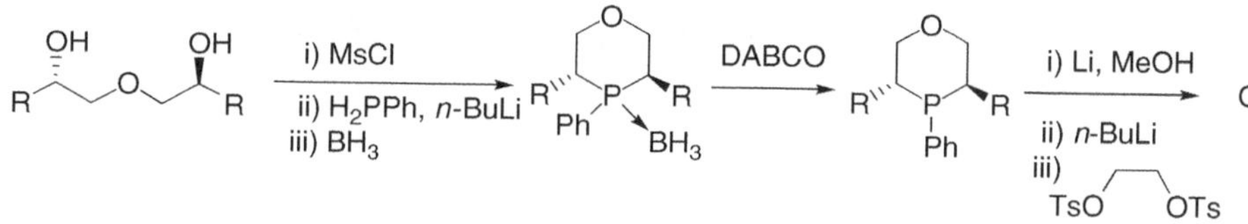

160 **160a**: R = Me; **160b**: R = *i*-Pr

Scheme 2.60 *Syntheses of bisphosphinane ligands **160**.*

(**150**) was first used by Sinou in a mechanistic study of the hydrogenation of olefins and was then tested by Tóth [119]. Detailed study on the complexations of the phosphole ligands with the transition metals showed that the favorable symmetric catalytic species and the planar structure of the phosphole ring were the reasons for the decreased enantioselectivies when compared with the diphosphine counterparts. Schenk also synthesized a similar ChiroPhos analogue (**151**), but no catalytic performance was demonstrated [120]. Gladiali reported two chiral ligands, DIOP–BNP (**152**) and SKEWPHOS–BNP (**153**), with binaphthophospholes (BNP) instead of DBP (Scheme 2.58) [121]. Unlike BINAP and many other atropisomeric diphosphine ligands, the two BNP ligands do not present axial chirality due to the rapid atropisomerization of two conformations at room temperature.

Many chiral ligands with independent phosphole structures and chiral substituents on the phosphole ring or the phosphorus atoms were developed by several research groups. The BIPNOR ligand (**154**) was prepared through a double [1,5] shift around the diphosphole ring and subsequent Diels–Alder reaction (Scheme 2.59) [122]. This is the first diphosphine ligand featuring the 1-phosphanorbornadiene structures. The meso and racemic products can be separated by silica–gel chromatography in the Pd complex form. Enantiopure ligands were obtained by coordination with a chiral palladacycle and subsequent chromatographic separation.

2.4.5 Design and synthesis of bisphosphinanes

In contrast to the vast number of phospholane ligands, chiral ligands bearing phosphinane structures are extremely rare. Helmchen and coworkers reported the first known oxaphosphinane ligand and bisphosphinane in 2002 [123]. Following a similar BPE procedure, the bisphosphinane **160** was prepared (Scheme 2.60). These ligands exhibited promising results in the asymmetric hydrogenation of dehydroamino acids and itaconic acid derivatives.

Zhang's group described the synthesis of a bis(azaphosphorinane)ligand **161** (Scheme 2.61) [124]. Starting from a readily available chiral epoxide, ring opening followed by the formation of cyclic sulfate and reaction with primary diphosphine produced the ligand in good overall yield. This ligand achieved excellent enantioselectivities in the hydrogenation of β-alkyl-β-(acylamino)-arylates and α-arylenamides.

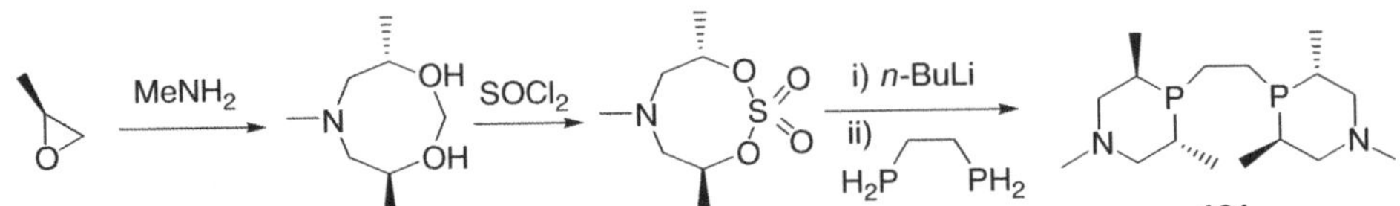

161

Scheme 2.61 *Syntheses of ligand **161**.*

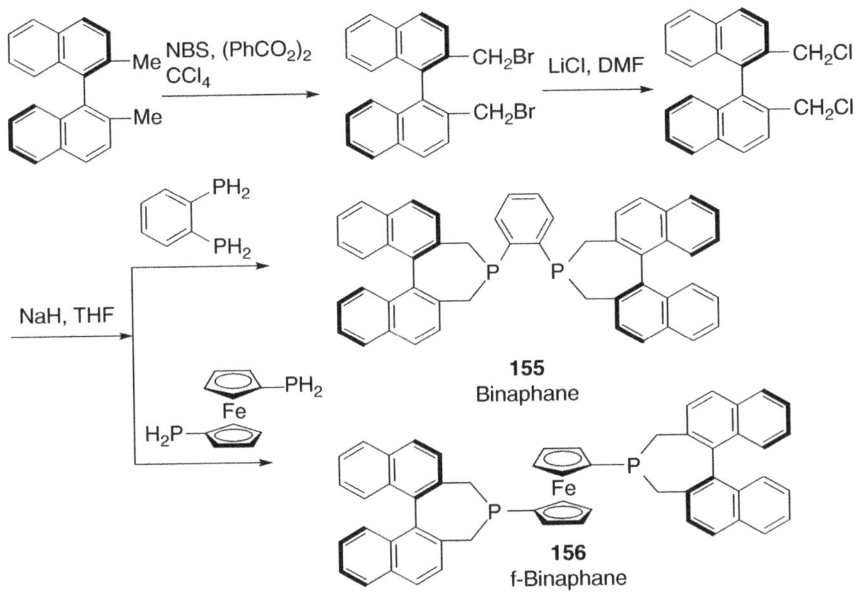

Scheme 2.62 *Syntheses of Binaphane and f-Binaphane.*

2.4.6 Design and synthesis of bisphosphepines

The chiral bisphosphepine ligands are based on the BINEPINE structures that contain an endocyclic phosphorus donor with a substituent inserted in a seven-membered ring. The structures are highly rigid with two binaphthalene scaffolds which carry axial chirality as the chiral element.

Although the first monodentate BINEPINE was synthesized in 1994, the bidentate versions of BINEPINE were not available until 2000. The Zhang group reported the synthesis of Binaphane (**155**) and f-Binaphane (**156**), in which two vicinal binaphthophosphepine fragments were linked together by a benzene or ferrocene backbone (Scheme 2.62) [125,126]. A key step of the synthetic route of the seven-membered ring relied on the transformation of the dibromo species into the corresponding less reactive dichloride intermediate before the cyclization reaction with the primary phosphines in the presence of NaH. This required reactivity decrease helped to reduce the possible intermolecular side reaction that competed with the desired intramolecular counterpart. **155** showed excellent enantioselectivities regarding the Rh-catalyzed asymmetric hydrogenation of β-substituted-α-arylenamides [125], and Ir–**156** has been proven to be highly effective in asymmetric hydrogenation toward imine substrates in particular, including *N*-aryl imines and unprotected *N*-H imines [126,127].

Bidentate ligands with two binaphthophosphepine units connected by an ethane bridge have also been prepared via two different synthetic approaches (Scheme 2.63). Moberg and coworkers reported a stepwise synthesis of ligand **158** starting from chiral bis-(bromomethyl)-1,1′-binaphthyl (**157**) with borane as the

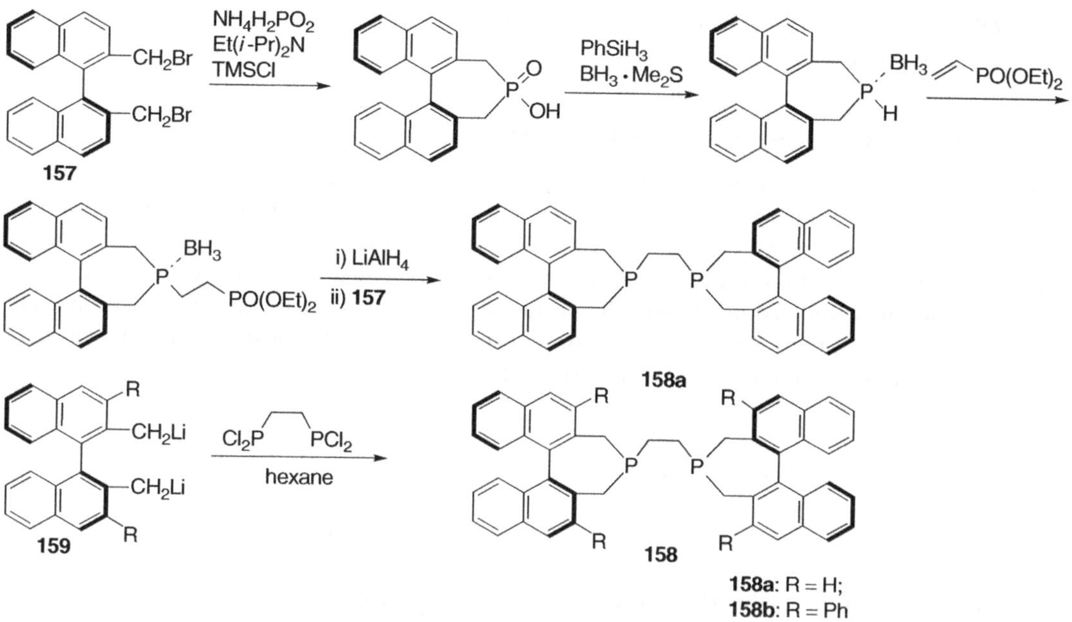

Scheme 2.63 Synthesis of ligand **158**.

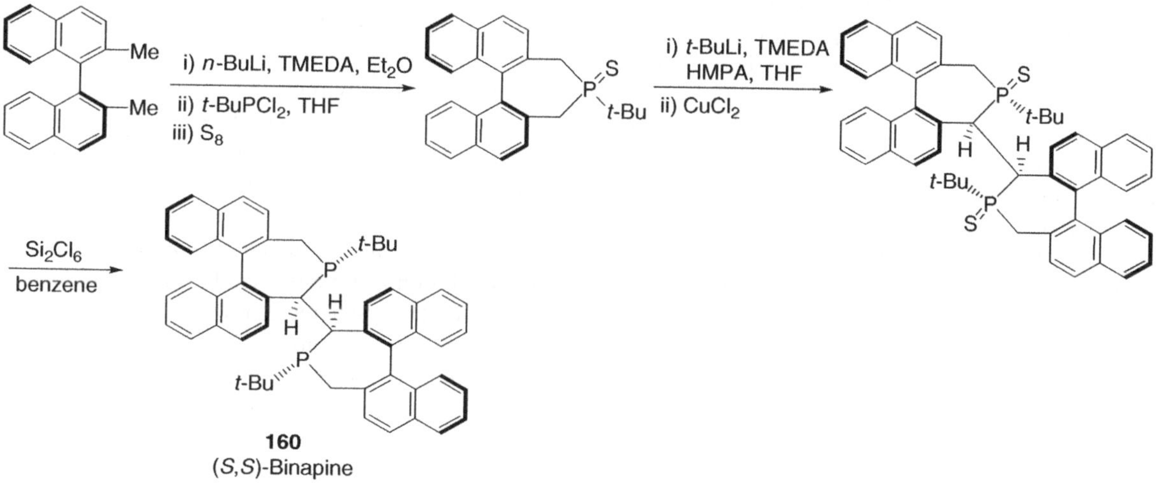

Scheme 2.64 Synthesis of (s,s)-Binapine.

protecting reagent [128]. Alternatively, a more straightforward strategy was reported by Zhang and coworkers according to which di-lithiated species **159** was reacted with 1,2-bis(dichlorophosphine)ethane [129]. The introduction of phenyl substituents on the 3- and 3′-positions on the binaphthyl rings was achievable through this route. However, the relative yield appeared to be a limitation.

An example of enantioselective coupling of the two BINEPINE units at their C(3) positions was also reported by the Zhang group, taking advantage of the similar oxidative Cu-catalyzed homocoupling strategy from the synthesis of TangPhos (**143**) and DuanPhos (**144**) (Scheme 2.64) [130]. This P-chirogenic ligand,

Binapine (**160**), combining both the axial chirality and P chirality elements, was particularly active and selective in the Rh-catalyzed hydrogenation of (*Z*)-β-(acylamino)-arylates.

2.4.7 Summary of synthetic strategies of phosphacycles

Although numerous phosphacyclic diphosphines have been synthesized since the remarkable success of BPE and DuPhos, the synthetic strategies can be summarized as follows. The dominant strategy comes from the landmark ligands, BPE and DuPhos. Cyclic sulfate, dimesylate, or ditosylate based upon chiral diols react with primary phosphines to afford the chiral diphosphine products. In another strategy, the two components having reverse roles are combined. Phosphine dichloride serves as the electrophile, and the chiral backbone reacts as the nucleophile. A third strategy involves the construction of single phosphacycles prior to stereoselective oxidative coupling affording the enantiomeric diphosphines.

2.5 P-stereogenic diphosphine ligands

The study of P-stereogenic ligands has attracted the research interest of chemists since the early stage of the history of asymmetric catalysis. The discovery of the most notable example of P-stereogenic ligands, DIPAMP, was regarded as the landmark of the homogenous asymmetric catalysis development [5]. However, relatively less attention has been paid to P-chiral diphosphine ligands, which is largely due to the remarkably synthetic difficulty of highly enantiomerically enriched P-chiral phosphines and the configurational instability such as racemization at high temperatures [131].

One of the early methodologies for the synthesis of P-stereogenic diphosphines involves the usage of a chiral palladium complex to give two diastereomers that then could be separated (Scheme 2.65) [132].

However, enantioselective deprotonation of an achiral hydrocarbyldimethylphosphine with the assistance of (–)-sparteine substantially expanded the scenario of P-stereogenic ligand synthesis. The selective deprotonation of alkyl– or aryldimethylphosphine–boranes was first reported for mono phosphine–boranes to introduce substituents on the phosphorus atoms [133]. By combining the enantioselective deprotonation with oxidative dimerization, a series of P-stereogenic diphosphines possessing a methyl group, BisP*, became accessible [133b,134,135]. The Cu(II) salt oxidized the enantiomerically enriched *ortho*-lithiated phosphine–borane complex to give the corresponding borane protected diphosphine in excellent enantioselectivity. *t*-Bu-BisP* (**162a**) was the most notable example in this ligand class (Scheme 2.66) [134]. **162a** showed excellent enantioselectivities in transition-metal catalyzed asymmetric reactions [133b,134,136]. Utilizing the similar methodology, highly rigid and electron-rich P-stereogenic bisphospholane DiSquareP* and TangPhos were synthesized subsequently [96,113a].

However, due to the availability of only one naturally occurring enantiomer of sparteine, only one enantiomer was obtained for these ligands. Interestingly, O'Brien *et al.* reported their successful attempts of

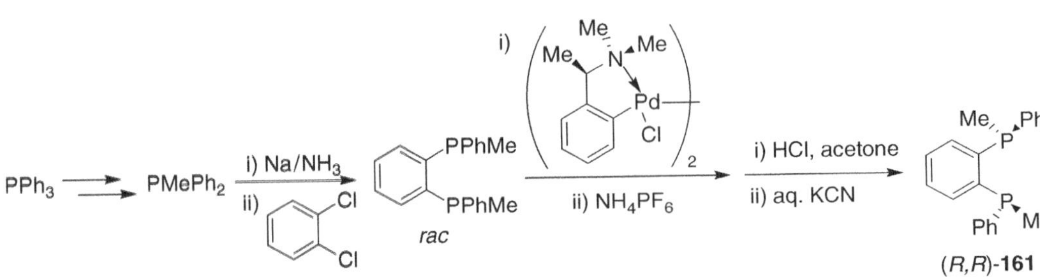

Scheme 2.65 *Synthesis of ligand **161** via optical resolution with pallacycle complex.*

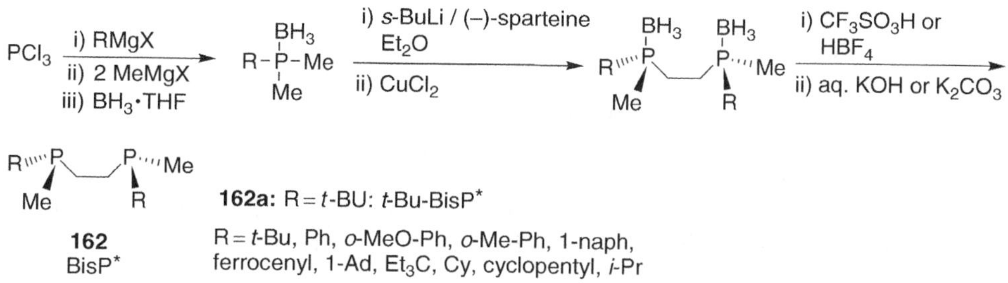

Scheme 2.66 *Synthesis of BisP*.*

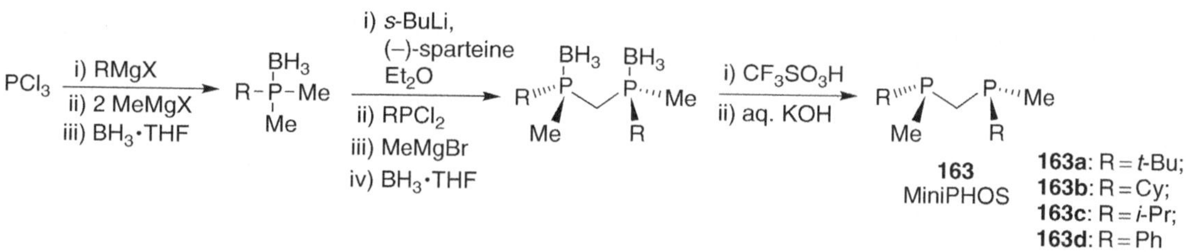

Scheme 2.67 *Synthesis of MiniPHOS.*

finding a (+)-sparteine surrogate for the synthesis of both enantiomers [137]. Excellent enantioselectivities were obtained in the different *ortho*-metalation reactions tested.

Another analogously structured but also highly electron-rich and sterically hindered diphosphine ligand was prepared by Imamoto and coworkers in 1999 (Scheme 2.67) [138].

2.6 Experimental procedures for the syntheses of selected diphosphine ligands

2.6.1 Synthesis procedure for DIOP* ligand (16)

Preparation of chiral diol bis(methanesulfonate): The enantiopure chiral diol 3,4-*O*-isopropylidene-(3*S*,4*S*)-dihydroxy-(2*S*,5*S*)-hexanediol was prepared according to literature procedures [15]. To a solution of the chiral diol (2.2 g, 11.6 mmol) and triethylamine (4.9 mL, 34.8 mmol) in CH_2Cl_2 (30 mL) was added dropwise a solution of methanesulfonyl chloride (2.0 mL, 25.8 mmol) in CH_2Cl_2 (10 mL) at 0°C. After 30 min at 0°C, the reaction mixture was stirred for an additional 30 min at room temperature, and then quenched by saturated aqueous NH_4Cl solution. The aqueous layer was extracted with CH_2Cl_2, and the combined organic solution was dried over Na_2SO_4. After evaporation of the solvent, the residue was purified by flash chromatography on silica gel eluting with CH_2Cl_2/Et_2O (9/1) to give a colorless oil 3.85 in 96% yield: $[\alpha]^{24}_D = -1.3°$ ($c = 1.04$, $CHCl_3$); ^1H NMR ($CDCl_3$, 360 MHz) δ 4.82–4.76 (m, 2 H), 3.99–3.96 (m, 2 H), 3.03 (s, 6 H), 1.45 (d, $J = 6.6$ Hz, 6 H), 1.37 (s, 6 H); ^{13}C NMR ($CDCl_3$, 90 MHz) δ 110.14, 78.19, 76.26, 38.53, 26.75, 17.63; HRMS calculated for $C_{11}H_{23}O_8S_2$ (MH^+) 347.0834 and $C_{11}H_{22}O_8S_2Na$ (MNa^+) 369.0654 found 347.0834 and 369.0654, respectively.

Preparation of DIOP*: To a solution of diphenylphosphine (1.15 mL, 6.6 mmol) in THF (50 mL) was added *n*-BuLi in hexane (4.0 mL, 6.4 mmol) at –78°C; the previous prepared chiral diol dimesylate (1.04 g, 3.0 mmol) in THF (20 mL) was added over 20 min. The resulting orange solution was warmed to room temperature and stirred overnight. The white suspension was hydrolyzed with saturated aqueous NH_4Cl solution. The aqueous layer was extracted with CH_2Cl_2, and the combined organic solution was dried over anhydrous Na_2SO_4. After removal of the solvents under reduced pressure, the residue was purified by flash chromatography on silica gel eluting with hexanes/EtOAc (95/5) to give a colorless oil 1.06 g in 67% yield: $[\alpha]^{24}_D = +41.8°$ ($c = 0.88$, toluene); ^1H NMR ($CDCl_3$, 360 MHz) δ 7.56–7.52 (m, 8 H), 7.38–7.33 (m, 12 H), 3.78–3.76 (m, 2 H), 2.50–2.46 (m, 2 H), 1.44 (s, 6 H), 0.91 (d, $J = 7.0$ Hz, 3 H), 0.87 (d, $J = 6.9$ Hz, 3 H); ^{13}C NMR (CD_2Cl_2, 90 MHz) δ 137.46 (d, $J = 15.9$ Hz), 137.03 (d, $J = 15.5$ Hz), 134.14 (d, $J = 3.7$ Hz), 133.91 (d, $J = 4.0$ Hz), 129.30 (d, $J = 8.6$ Hz), 128.75 (d, $J = 3.7$ Hz), 133.91 (d, $J = 4.0$ Hz), 129.30 (d, $J = 8.6$ Hz), 128.75 (d, $J = 7.1$ Hz), 108.27, 77.2 (dd, $J_1 = 12.0$ Hz, $J_2 = 6.8$ Hz), 31.33 (d, $J = 14.3$ Hz), 27.05, 10.74 (d, $J = 17.6$ Hz); ^{31}P NMR ($CDCl_3$) δ = −6.3 ppm; HRMS calculated for $C_{33}H_{37}O_2P_2$ (MH^+) 527.2269 found 527.2271.

2.6.2 Synthesis procedure of SDP ligands (46)

Preparation of ditriflate: The enantiopure spinol (**45**) was prepared according to a literature report [36]. To a solution of (S)-45 (5.0 g, 19.8 mmol) in 100 mL of CH_2Cl_2 was added pyridine (7.0 mL, 86.7 mmol), followed by dropwise addition of triflic anhydride (8.2 mL, 43.7 mmol) at 0°C. The mixture was stirred at room temperature overnight. After removal of the solvent, the residue was diluted with EtOAc (80 mL), and then washed with 5% aqueous HCl, saturated $NaHCO_3$, and brine (once for each). The organic layer was dried over anhydrous sodium sulfate, concentrated under reduced pressure, and passed through a silica gel plug (eluted with CH_2Cl_2) to give the ditriflate (9.9 g, 97%).

Introduction of a monophosphino group: To a mixture of (*S*)-enantiomer of the ditriflate (4.0 g, 7.75 mmol), diphenylphosphine oxide (3.13 g, 15.5 mmol), palladium acetate (87 mg, 0.39 mmol), and 1,4-bis(diphenylphosphino)butane (dppb, 166 mg, 0.39 mmol) was added 25 mL of degassed DMSO and diisopropylethylamine (4.1 g, 32 mmol), and the mixture was heated with stirring at 100°C for 6 h. After cooling to room temperature, the reaction mixture was diluted with EtOAc, and then washed with 5% aqueous HCl and saturated $NaHCO_3$. The organic phase was dried over anhydrous Na_2SO_4 and concentrated under reduced pressure. The residue was chromatographed on a silica gel column eluted with petroleum ether–EtOAc (3:1) to give (*S*)-7-diphenylphosphinyl-7′-trifluoro-methane-sulfonyloxy-1,1′-spirobiindane (4.0 g, 90%). The mixture of the product (1.4 g, 2.5 mmol) and diisopropylethylamine (13.2 g, 102 mmol) in toluene (25 mL) was added $HSiCl_3$ (4.0 mL, 39 mmol) at 0°C. The reaction mixture was stirred at 110°C for 5 days. After cooling to room temperature, the mixture was diluted with Et_2O and quenched with a small amount of saturated $NaHCO_3$. The resulting suspension was filtered through Celite, and the solid was washed with Et_2O. The combine organic layer was dried over $MgSO_4$ and concentrated under reduced pressure. The crude product was purified by silica gel column chromatography with petroleum ether–EtOAc (30:1) as eluent to give the intermediate (1.26 g, 91%) for the next step.

Preparation of SDP(46a): To a mixture of previous product (1.2 g, 2.17 mmol), diphenylphosphine oxide (0.87 g, 4.3 mmol), palladium acetate (22.4 mg, 0.1 mmol), and 1,40bis(diphenylphosphino)butane (dppb, 42.6 mg, 0.1 mmol) was added 15 mL of degassed DMSO and diisopropylethylamine (1.0 g, 8.7 mmol), and the mixture was heated with stirring at 100°C for 10 h. After cooling to room temperature, the reaction mixture was diluted with EtOAc, and then washed with 5% aqueous HCl and saturated $NaHCO_3$. The organic phase was dried over anhydrous Na_2SO_4 and concentrated under reduced pressure. The residue was chromatographed on a silica gel column eluted with petroleum ether–EtOAc (3:1) (1.2 g, 92%). The product (0.36 g, 0.6 mmol) was then dissolved in toluene (10 mL) with diisopropylethylamine (3.1 g, 24 mmol), before $HSiCl_3$ (0.9 mL, 9 mmol) was added at 0°C. The reaction mixture was stirred at 110°C for 3 days. After cooling to room temperature, the mixture was diluted with Et_2O and quenched with a small amount of saturated $NaHCO_3$. The resulting suspension was filtered through Celite, and the solid was washed with Et_2O. The combine organic layer was dried over $MgSO_4$ and concentrated under reduced pressure. The crude product was purified by silica gel column chromatography with petroleum ether–EtOAc (30:1) as eluent to give (*S*)-**45** as a white solid (0.3 g, 86%), m.p. 206–208°C. $[\alpha]^{24}_{D}=-81°$ ($c=0.5$, CH_2Cl_2); ^1H NMR ($CDCl_3$, 300 MHz) δ 7.30–7.08 (m, 14 H), 6.96 (m, 2 H), 2.96–2.80 (m, 2 H), 2.80–2.62 (m, 2 H), 2.00–1.80 (m, 4 H); ^{13}C NMR ($CDCl_3$, 75 MHz) δ 154.8, 154.7, 154.5, 144.6, 144.5, 138.4, 134.6, 134.5, 134.4, 133.5, 133.4, 133.3, 128.5, 128.2, 127.9, 127.5, 126.9, 125.3, 64.1, 40.1, 30.7; ^{31}P NMR ($CDCl_3$, 121 MHz) δ=−17.26 ppm; MS (EI) m/z 588 (M^+).

2.6.3 Synthesis procedure of (*R,R*)-BICP (64)

Preparation of bis(methanesulfonate): Compound **63** was synthesized by asymmetric hydroboration of bi-1-cyclopenten-1yl using (+)-monoisopinocampheylborane ((+)-$IpcBH_2$) according to the literature procedure [139]. The absolute configuration of the diol was assigned based on the asymmetric hydroboration of trisubstituted olefins using (+)-$IpcBH_2$. To a solution of 63 (0.8 g, 4.65 mmol) and triethylamine (1.68 mL, 12.09) in CH_2Cl_2 (30 mL) was added dropwise a solution of methanesulfonyl chloride (0.76 mL, 9.92 mmol) in CH_2Cl_2 (2 mL) at 0°C. After 30 min at 0°C, the reaction mixture was stirred for an additional 2 h at room temperature, and then quenched by saturated aqueous ammonium chloride solution (25 mL). The aqueous layer was extracted with CH_2Cl_2 (3×20 mL), and the combined organic solution was dried over Na_2SO_4. After evaporation of the solvent, a white solid was obtained which was used directly for the next step. ^1H NMR ($CDCl_3$, 200 MHz) δ 5.01 (m, 2 H), 3.94 (s, 6 H), 2.17 (m, 2 H), 2.15–1.65 (m, 10 H), 1.52–1.43 (m, 2 H); ^{13}C NMR δ 86.8, 48.2, 38.4, 32.8, 27.4, 22.5.

Preparation of (*R,R*)-64: To a solution of diphenylphosphine (1.25 mL, 7.0 mmol) in THF (80 mL) was added *n*-BuLi in hexane (4.1 mL, 6.6 mmol) at −78°C for 5 min via syringe. The resulting orange solution was warmed to room temperature and stirred for 30 min. After cooling the mixture to −78°C, the bis(methanesulfonate) (1.01 g, 3.1 mmol) from the previous step in THF (20 mL) was added over 20 min. The resulting orange solution was warmed to room temperature and stirred overnight. The white suspension solution was hydrolyzed with saturated aqueous NH_4Cl solution. The aqueous layer was extracted with CH_2Cl_2 (2×20 mL), and the combined organic solution was dried over anhydrous Na_2SO_4. After removal of the solvents under reduced pressure, the residue was dissolved in CH_2Cl_2 (50 mL), and then treated with BH_3·THF (10 mL, 10 mmol) at room temperature, and the mixture was stirred overnight. The reaction mixture was added to NH_4Cl aqueous solution, and extracted with CH_2Cl_2 (2×20 mL). The combined organic solution was dried over anhydrous Na_2SO_4. After evaporation of the solvent under reduced pressure, the residue was subjected to column chromatography on silica gel, eluting with CH_2Cl_2–hexane (1:5) and then CH_2Cl_2–hexane (2:3) affording the product as a white solid (0.36 g, 21%). To a solution of the borane complex (Section 2.6.3) of the phosphine (0.24 g, 0.45 mmol) in CH_2Cl_2 (4.5 mL) was added a tetrafluoroboric acid–dimethyl ether complex (0.55 mL, 4.5 mmol) dropwise via syringe at −5°C. After the addition, the reaction mixture was allowed to warm slowly to room temperature, and stirred for 20 h. The mixture was diluted with CH_2Cl_2, and neutralized with a saturated aqueous $NaHCO_3$ solution. The aqueous layer was extracted with CH_2Cl_2. The combined organic solution was washed with brine, followed by water, and dried over Na_2SO_4. Evaporation of the solvent gave the pure diphosphine (0.21 g, 93%). ^1H NMR ($CDCl_3$) δ 7.52–7.27 (m, 20 H), 2.53 (m, 2 H), 2.27 (m, 2 H), 1.93 (m, 2 H), 1.72 (m, 2 H), 1.70–1.43 (m, 8 H); ^{13}C NMR ($CDCl_3$) δ 139–127 (Ph), 45.9 (d, $J=12.1$ Hz), 45.8 (d, $J=12.0$ Hz), 40.34 (d, $J=14.0$ Hz), 30.9 (m), 23.8 (m); ^{31}P NMR ($CDCl_3$) δ=−14.6 ppm.

2.6.4 Synthesis procedure of SEGPHOS (95) [72]

Preparation of racemic diphosphine oxide: Under a nitrogen atmosphere, 0.7 M LDA in THF (400 mL, 280 mmol) was added dropwise to a solution of 5-diphenylphosphinyl-1,3-benzodioxole (75.22 g, 233 mmol) in THF (300 mL) at −15°C. The mixture was added to a suspension of $FeCl_3$ (45.79 g, 282 mmol) in THF (300 mL) at 0°C. After evaporation of THF, the residue was dissolved in CH_2Cl_2 (500 mL), washed with 10% HCl and water, and then dried over Na_2SO_4. The solvent was evaporated, and the residue was washed with hot ethyl acetate (200 mL) to give racemic diphosphine oxide (56.08 g, 75%).

Resolution of racemic diphosphine oxide: A solution of (+)-DBT (11.68 g, 32.6 mmol) in methanol (30 mL) was added to a solution of the previous prepared racemic diphosphine oxide (20.73 g, 32.3 mmol) in methanol (60 mL). The mixture was stirred at reflux for 5 min. The precipitates were washed with methanol to give the phosphine oxide-(+)-DBT complex as colorless crystals. The complex was stirred in a mixture of CH_2Cl_2 (90 mL) and 1.5 N NaOH (50 mL) at room temperature for 30 min. The organic layer was washed with water, evaporated, and dried under vacuum to give (*R*)-(+)-(4,4′-bi-1,5-benzodioxole)-5,5′-diylbis(diphenylphosphine oxide) (9.12 g, 44%, >99% ee). mp: 158–159°C, $[\alpha]^{24}_{D}$=+161.9° (*c*=0.063, $CHCl_3$); ^1H NMR ($CDCl_3$): δ 7.72–7.28 (m, 20 H), 6.77 (dd, J_1=14.1 Hz, J_2=8.1 Hz, 2 H), 6.65 (dd, J_1=8.1 Hz, J_2=2.1 Hz, 2 H), 5.72 (d, *J*=1.6 Hz, 2 H), 5.26 (d, *J*=1.5 Hz, 2 H). ^{31}P NMR ($CDCl_3$): δ=29.6 ppm.

Preparation of SEGPHOS (95): The mixture of (*R*)-(+)-(4,4′-bi-1,5-benzodioxole)-5,5′-diylbis-(diphenylphosphine oxide) (1.50 g, 2.34 mmol), *N,N*-dimethylaniline (3.11 g, 25.6 mmol), and $HSiCl_3$ (3.22 g, 23.3 mmol) was stirred in toluene (25 mL) at 110°C for 4 h. After the reaction mixture was cooled to 5°C with an ice-water bath, 15% aqueous NaOH (30 mL) was added. The mixture was stirred at room temperature for 30 min, and then the aqueous layer was extracted with toluene (15×2 mL). The organic layers were washed with water, 1 N HCl (30×2 mL) and water. Evaporation of the solvent, and purification by silica gel column chromatography, gave (*R*)-**95** (1.35 g, 95%). mp: 215–217°C, ç° (*c*=0.502 $CHCl_3$); ^1H NMR ($CDCl_3$): δ 7.21–7.11 (m, 20 H), 6.66 (d, *J*=8.1 Hz, 2 H), 6.51 (dd, J_1=7.9 Hz, J_2=3.1 Hz, 2 H), 5.66 (d, *J*=1.6 Hz, 2 H), 5.03 (d, *J*=1.6 Hz, 2 H). ^{31}P NMR ($CDCl_3$): δ=–12.6 ppm.

2.6.5 Synthesis procedure of Ph-BPE (110)

Preparation of 2,5-*trans*-diphenylphospholane-borane adduct: Both (*R,R*)- and (*S,S*)-1-hydroxy-1-oxo-2,5-*trans*-diphenylphospholane were prepared according to the methods described by the literature [140]. For a stirred suspension of (*R,R*)-1-hydroxy-1-oxo-2,5-*trans*-diphenylphospholane (3.50 g, 15.5 mmol) dropwise via syringe, once the addition was complete the reaction was heated to 110°C. After 16 h at 110°C, the reaction was allowed to cool to room temperature and then concentrated *in vacuo* to give an opaque gum. The gum was placed under an atmosphere of nitrogen and taken up in degassed THF (20 mL), and the resulting solution cooled to 0°C. A borane-methyl sulfide complex (4.88 mL, 51.4 mmol) was added dropwise via syringe, and once the addition was complete the reaction was allowed to warm to room temperature. After 16 h at room temperature, the reaction was concentrated *in vacuo* to yield a white solid. The solid was purified by flash column chromatography (9:1 heptane:EtOAc) to give the product as a white solid (3.12 g, 95%). ^1H NMR ($CDCl_3$, 400 MHz): δ 7.35 (m, 6 H), 7.25 (m, 4 H), 4.82 (dq, J_{HP}=361 Hz, J_{HH}=11.1 Hz, 1 H), 3.95 (m, 1 H), 3.52 (m, 1 H), 2.65–2.55 (m, 2 H), 2.25–2.15 (m, 2 H), 0.9–0.1 (br q, 3 H). ^{31}P NMR ($CDCl_3$, 162 MHz): δ=30.13 (d, J_{PH}=359 Hz); ^{13}C NMR (CDC13, 100.6 MHz) δ 137.73 (d, *J*=4.8 Hz), 136.51, 129.02, 128.67, 128.62, 128.52, 127.36, 127.31, 44.47 (d, J_{CP}=32.7 Hz), 40.66 (d, J_{CP}=29.3 Hz), 34.66 (d, J_{CP}=3.7 Hz), 33.86.

Preparation of 1,2-bis[(*R,R*)-2,5-diphenylphospholano]ethane-borane adduct: To a stirred solution of (*R,R*)-2,5-*trans*-diphenylphospholane-borane adduct (2.00 g, 7.87 mmol) in THF (20 mL) at –78°C, under an atmosphere of nitrogen, was added a 1.6 M solution of *n*-BuLi in hexanes (4.86 mL, 7.87 mmol) dropwise via syringe. The reaction was then allowed to warm to –20°C and after stirring for 1 h a solution of 1,2-ethylene ditosylate (1.45 g, 3.94 mmol) in THF (20 mL) that was added dropwise via cannula. Once the addition was complete the reaction was allowed to warm to room temperature and stirred for a further 50 h. The reaction was quenched at 0°C by the slow addition of a 1 M aqueous solution of HCl (15 mL). The reaction was concentrated *in vacuo*, and the resulting white solid residue partitioned between water

(30 mL) and CH_2Cl_2 (30 mL). The organic layer was separated, and the aqueous layer extracted with CH_2Cl_2 (3×30 mL). The organic fractions were combined, dried over $MgSO_4$, filtered, and then concentrated *in vacuo* to give a white solid. Purification by flash column chromatography (9:1 heptane:EtOAc) yielded the product as a white solid (1.44 g, 69%). ^{1}H NMR ($CDCl_3$, 400 MHz): δ 7.35–7.25 (m, 8 H), 7.20–7.10 (m, 8 H), 7.05 (m, 4 H), 3.48 (m, 2 H), 3.03 (m, 2 H), 2.40–2.30 (m, 4 H), 2.10–2.00 (m, 4 H), 1.29 (m, 4 H), 0.7–0.2 (br m, 6 H); ^{31}P NMR ($CDCl_3$, 162 MHz): δ=45.22 (br, s); ^{13}C NMR (CDCl3, 100.6 MHz) δ 136.93, 134.81, 128.86, 128.57, 128.47, 127.26, 46.91 (m), 33.62, 30.61, 17.19 (m); MS (APCI) m/z 534 (M^+).

Preparation of Ph-BPE (110): To a stirred solution of 1,2-bis[(*R*,*R*)-2,5-diphenylphospholano]ethane-borane adduct (1.10 g, 2.06 mmol) in CH_2Cl_2 (30 mL) at room temperature, under an atmosphere of nitrogen, was added $HBF_4{\cdot}OMe_2$ (2.51 mL, 20.6 mmol) dropwise via syringe. After 16 h at room temperature, a degassed saturated aqueous solution of $NaHCO_3$ (30 mL) was added via syringe. Once effervescence had ceased, the biphasic mixture was stirred for a further 15 min. The organic layer was removed via syringe and stored under nitrogen. The aqueous layer was extracted with degassed CH_2Cl_2 (4×20 mL), and all organic fractions were combined, dried over $MgSO_4$, and filtered under nitrogen. Concentration *in vacuo* gave a white solid that was recrystallized from degassed *i*-PrOH (40 mL) to give the product as white crystals (0.96 g, 92%). $[\alpha]^{24}_{D}$=−174.9° (*c*=0.3, CH_2Cl_2); ^{1}H NMR ($CDCl_3$, 400 MHz): δ 7.30 (m, 4 H), 7.21 (m, 4 H), 7.08 (m, 6 H), 3.59 (m, 2 H), 2.95 (m, 2 H), 2.48 (m, 2 H), 2.27 (m, 2 H), 2.15–2.05 (m, 2 H), 1.86–1.76 (m, 2 H), 0.96 (m, 2 H), 0,57 (m, 2 H); ^{31}P NMR ($CDCl_3$, 162 MHz): δ=16.01 ppm; ^{13}C NMR ($CDCl_3$, 100.6 MHz) δ 144.65 (m), 138.33(m), 128.53, 128.30, 127.86, 127.25, 125.83, 125.75, 50.75 (m), 46.15 (m), 37.38, 31.92, 21.45 (dd, J_{CP}=36 Hz, J_{CCP}=27.1 Hz). HRMS (EI) m/z calculated for $C_{34}H_{36}P_2$ 506.2292 found 506.2269.

2.6.6 Synthesis procedure of TangPhos (143)

Preparation of 1-*tert*-butylphospholane-1-sulfide: To a suspension of magnesium turning (7.92 g, 0.33 mol) in 300 mL of dry THF was added dropwise 1,4-dibromobutane (23.7 g, 0.11 mol) in 50 mL of THF at room temperature. The reaction was very exothermic during the addition. After the addition was complete (1 h), the resulting dark solution was kept at room temperature for 2 h. The solution of $BrMg(CH_2)_4MgBr$ was used directly for the next reaction. To a solution of phosphorus trichloride (13.7 g, 0.10 mol) in THF (300 mL) was added dropwise a solution of *t*-BuMgCl in THF (100 mL, 1.0 M) at −78°C. The addition was complete in 2 h. After the mixture was allowed to stand at −78°C for 1 h, the above solution of $BrMg(CH_2)_4MgBr$ in THF was added dropwise over 2 h. The mixture was then allowed to warm to room temperature over 2 h and stirred overnight. Sulfur power (4.8 g, 0.15 mol) was added to the reaction mixture at room temperature in one portion. The resulting solution was then stirred at room temperature for 2 h. Water (300 mL) and EtOAc (500 mL) were then added. The organic layer was washed sequentially with water (300 mL) and brine (300 mL), dried over Na_2SO_4, and concentrated. The resulting oil was purified by column chromatography (hexanes/EtOAc, 9:1), followed by recrystallization from ether to give 1-tert-butylphospholane-1-sulfide as a white crystalline solid (8 g, 45%). ^{1}H NMR ($CDCl_3$, 360 MHz): δ 2.17 (m, 4 H), 1.82 (m, 4 H), 1.25 (d, J_{HP}=16.1 Hz, 9 H); ^{31}P NMR ($CDCl_3$, 145 MHz): δ=83.5 ppm; ^{13}C NMR ($CDCl_3$, 75 MHz) δ 33.7 (d, J_{CP}=44.3 Hz), 30.4 (d, J_{CP}=47.9 Hz), 27.1 (d, J J_{CP}=5.1 Hz), 25.4 (d, J_{CP}=1.8 Hz). APCI MS 177 (M^++H); HRMS m/z calculated for $C_8H_{18}PS$ 177.0867 found 177.0858.

Preparation of (1*R*,1′*R*,2*R*,2′*R*)-1,1′-di-*tert*-butyl-[2,2′]-diphospholanyl-1,1′-disulfide: At −78°C, to a solution of (−)-sparteine (7.83 mL, 34 mmol) in ether (200 mL) was added *n*-BuLi (21.3 mL, 34 mmol, 1.6 M in hexane) dropwise. The resulting solution was kept at −78°C for 30 min. Then at this temperature, to the solution was added dropwise a solution of 1-tert0bytylphospholane-1-sulfide (5.0 g, 28.4 mmol) in ether (100 mL) over 1 h. The resulting mixture was kept at –78°C and stirred further for 8 h. Dry $CuCl_2$ (5.73 g, 42.6 mmol) was added to the solution in one portion. The resulting suspension was vigorously stirred and allowed to warm to room temperature over 4 h. To this mixture was added 150 mL of concentrated ammonium hydroxide. The water layer was washed with EtOAc (2 × 100 mL). The combined organic phase was further washed sequentially with 5% ammonia solution (100 mL), 1 N HCl solution (100 mL), water (100 mL), and brine (100 mL). The organic layer was dried over Na2SO4 and concentrated under vacuum to give an oily solid, which was subsequently purified by column chromatography to give a solid mixture (4 g) of the product (95% ee) and the *meso* compound (1*R*,1′*R*,2*S*,2′*S*)-1,1′-di-*tert*-butyl-[2,2′]-diphospholanyl-1,1′-disulfide (product:*meso* = 83:17). The ratio and ee were determined by chiral HPLC (Chiralpak AS, 1.0 mL/min, 1% *i*-PrOH/hexane, 34.5 min (1*R*,1′*R*,2*R*,2′*R*), 62.6 min (1*S*,1′*S*,2*S*,2′*S*), and 42.4 min (*meso*)). Recrystallization of the mixture from EtOAc and ethanol provided pure product (1 g, 20%, >99% ee) as colorless needles. $[\alpha]^{20}_D = 5.0°$ ($c = 0.9$, $CHCl_3$); ^1H NMR ($CDCl_3$, 360 MHz) δ 2.87 (dd, $J_1 = 16.2$ Hz, $J_2 = 7.3$ Hz, 2 H), 2.34 (m, 2 H), 2.19 (m, 2 H), 1.93 (m, 6 H), 1.69 (m, 2 H), 1.24 (d, $J_{HP} = 15.3$ Hz, 18 H); ^{13}C NMR ($CDCl_3$, 90 MHz) δ 35 (m), 34.4 (m), 30.3 (m), 30.1 (dd, $J_1 = 2.1$ Hz, $J_2 = 4.3$ Hz), 25.3 (d, $J = 1.2$ Hz), 24.6 (dd, $J_1 = 2.1$ Hz, $J_2 = 2.5$ Hz); ^{31}P NMR ($CDCl_3$, 121 MHz) δ = 89.6 ppm; APCI MS m/z 351 (M^++H); HRMS calculated for $C_{16}H_{33}P_2S_2$ 351.1499 found 351.1524.

Preparation of TangPhos (143): To a solution of (1*R*,1′*R*,2*R*,2′*R*)-1,1′-di-*tert*-butyl-[2,2′]-diphospholanyl-1,1′-disulfide (440 mg, 1.26 mmol) in 25 mL of benzene was added dropwise hexachlorodisilane (3.25 mL, 5.08 g, 18.9 mmol). The mixture was stirred at reflux for 4 h. After the solution was cooled to room temperature, 50 mL of degassed 30% (w/w) NaOH solution was slowly added to the reaction mixture in an ice-water bath. The resulting mixture was then stirred at 60°C until the aqueous layer became clear. The two phases were separated. The water phase was washed twice with degassed benzene (2 × 30 mL). The combine benzene was dried over Na_2SO_4 and concentrated. The solid residue was purified by passing through a basic Al_2O_3 plug (Et_2O:hexanes, 1:10) under nitrogen to give pure crystalline product (320 mg, 88%). ^1H NMR ($CDCl_3$, 360 MHz) δ 2.25 (m, 2 H), 2.00 (m, 2 H), 1.87 (m, 2 H), 1.64 (m, 8 H), 1.01 (d, $J_{HP} = 11.5$ Hz, 18 H); ^{13}C NMR ($CDCl_3$, 90 MHz) δ 42.9 (m), 34.6 (t, $J_{CP} = 5.1$ Hz), 28.9, 28.8, 28.5 (dd, $J_1 = 7.0$ Hz, $J_2 = 7.5$ Hz), 23.6 (t, $J_{CP} = 6.6$ Hz); ^{31}P NMR ($CDCl_3$, 121 MHz) δ = 14.5 ppm; APCI MS m/z 287 (M^++H); HRMS calculated for $C_{16}H_{33}P_2$ 287.2058 found 287.2035.

2.6.7 Synthesis procedure of Binaphane (155)

Preparation of (*R*)-2,2′-bistriflate-1,1′-binaphthyl: To a solution of (*R*)-BINOL (40.3 g, 140.7 mmol) in 900 mL of CH_2Cl_2 was added pyridine (400 mL), followed by dropwise addition of triflic anhydride (50.5 mL, 300 mmol) at 0°C. The mixture was stirred at room temperature for 6 h. After removal of the solvent, the residue was diluted with EtOAc (500 mL) and then washed with 5% aqueous HCl (100 mL), saturated $NaHCO_3$ (100 mL), and brine (100 mL). The organic layer was dried over anhydrous Na_2SO_4, concentrated, and passed through a silica gel plug (eluted with CH_2Cl_2) to give the (*R*)-bistriflate (77 g, 99%).

Preparation of (*R*)-2,2′-dimethyl-1,1′-binaphthyl: To a solution of (R)-bistriflate (77 g, 140 mmol) and $NiCl_2$·dppp (3.8 g, 7 mmol) in ether (1000 mL) was added dropwise the methyl magnesium bromide (3.0 M, 280 mL) at 0°C. The reaction mixture was heated to refluxing for 24 h. The reaction was quenched by addition of water (200 mL) slowly at 0°C and then diluted with 5% aqueous HCl (200 mL). The aqueous layer was extracted with ether (3 × 100 mL). The combined organic layer was washed with $NaHCO_3$ (100 mL), dried over anhydrous Na_2SO_4, and concentrated to afford the product as a light yellow solid (39.2 g, 99%).

Preparation of (*R*)-2,2′-dibromomethyl-1,1′-binaphthyl: A mixture of (*R*)-2,2′-dimethyl-1,1′-binaphthyl (39.2 g, 138.8 mmol), *N*-bromosuccinimide (52.4 g, 291.5 mmol), and benzoylperoxide (1.25 g) in tetrachlorocarbon (900 mL) was heated at refluxing and irradiated under a sunlight for 3 days. The mixture was cooled to room temperature and filtered. The filtrate was concentrated and passed through a silica gel plug. After removal of the solvent, the residue was recrystallized from CH_2Cl_2–hexanes to afford the product (41.1 g, 67.3%).

Preparation of (*R*)-2,2′-dichloromethyl-1,1′-binaphthyl: (*R*)-2,2′-Dibromomethyl-1,1′-binaphthyl (40 g, 90.8 mmol) and LiCl (30 g, 707 mmol) in DMF (800 mL) were mixed together and stirred at room temperature for 6 h. To this mixture was added carefully 5% aqueous HCl (300 mL) (exothermic reaction occurred). The mixture was then extracted with ether (4 × 400 mL). The organic layer was dried over anhydrous Na_2SO_4, concentrated, and recrystallized from CH_2Cl_2–hexanes to afford the product as a white solid (30 g, 93%).

Preparation of Binaphane (155): To a solution of (*R*)-2,2′-dichloromethyl-1,1′-binaphthyl (0.57 g, 1.62 mmol) and NaH (0.2 g, 8.3 mmol) in THF (20 mL) was added 1,2-bis(phosphino)benzene (0.109 mL, 0.812 mmol) at –78°C under nitrogen. The mixture was kept stirring at room temperature for 24 h and was heated at refluxing for 24 h. After the reaction was completed (monitored by ^{31}P NMR), the solvent was removed via vacuum and the residue was washed with ether (3 × 15 mL). The organic phase was filtered through a silica gel plug to give the fairly pure product. Further purification by recrystallization from ether afforded the product (0.31 g, 55%). $[\alpha]_D^{20} = 845°$ ($c = 0.22$, $CHCl_3$); ^{1}H NMR ($CDCl_3$, 360 MHz): δ 7.81–7.76 (m, 8 H), 7.57–7.54 (m, 4 H), 7.31–7.13 (m, 10 H), 7.00–6.90 (m, 2 H), 6.80–6.70 (m, 2 H), 6.64–6.62 (d, $J = 8.34$ Hz, 2 H), 2.97–2.74 (m, 8 H); ^{31}P NMR ($CDCl_3$): δ = –6.87 ppm; ^{13}C NMR ($CDCl_3$, 100.6 MHz) δ 141.70, 134.71, 134.21, 133.40, 133.27, 132.73, 132.54, 132.34, 131.20, 128.83, 128.69, 128.62, 128.08, 127.80, 127.13, 127.10, 126.32, 125.46, 125.26, 32.50, 29.83; MS m/z 698 (M^+).

2.7 Concluding remarks

The rapid development of diphosphine ligands is a major contribution to asymmetric catalysis. Their exceptional catalytic activities and outstanding enantioselectivities have extended their applications far beyond the standard substrates and laboratory application. A vast number of important intermediates and building blocks have been obtained on commercial scales through asymmetric catalysis by the metal complex of these phosphorus-based ligands. This, in turn, prompted the investigation of new efficient ligands. The large family of diphosphine ligands not only offers a great number of choices for catalysts, but also provides chemists a good opportunity to exploit the details of the catalytic processes. We are going to witness more important practical applications of these ligands and further extensive development in the future.

References

[1] Knowles, W. S., Sabacky, M. J. *Chem. Commun.* **1968**, 1445.

[2] (a) Kagan, H. B., Dang, T. P. *Chem. Commun.* **1971**, 481; and (b) Kagan, H. B., Dang, T. P. *J. Am. Chem. Soc.* **1972**, *94*, 6429.

[3] Horner, L., Siegel, H., Büthe, H. *Angew. Chem. Int. Ed. Engl.* **1968**, *7*, 941.

[4] Osborn, J. A., Jardine, F. S., Young, J. F., Wilkinson, G. *J. Chem. Soc. A* **1966**, 1711.

[5] (a) Knowles, W. S. *Angew. Chem. Int. Ed.* **2002**, *41*, 1998; and (b) Knowles, W. S., Sabacky, M. J., Vineyard, B. D., Weinkauff, D. J. *J. Am. Chem. Soc.* **1975**, *97*, 2567.

[6] (a) Fryzuk , M. B., Bosnich, B. *J. Am. Chem. Soc.* **1977**, *99*, 6262; (b) Fryzuk, M. B., Bosnich, B. *J. Am. Chem. Soc.* **1979**, *101*, 3043.

[7] Fryzuk, M. D., Bosnich, B. *J. Am. Chem. Soc.* **1978**, *100*, 5491.
[8] (a) Rhone Poulenc S. A. Fr. Patent 2230654, **1974**; and (b) Glaser, R., Geresh, S., Twaik, M. *Isr. J. Chem.* **1980**, *20*, 102.
[9] Achima, K. J. *Am. Chem. Soc.* **1976**, *98*, 8265.
[10] King, R. B., Bakos, J., Hoff, C. D., Markó, L. *J. Org. Chem.* **1979**, *44*, 1729.
[11] Riley, D. P., Shumate, R. E. *J. Org. Chem.* **1980**, *45*, 5187.
[12] Dang, T.-P., Poulin, J.-C., Kangan, H. *J. Organomet. Chem.* **1975**, *91*, 105.
[13] (a) Morimoto, T., Chiba, M., Achiwa, K. *Tetrahedron Lett.* **1988**, *29*, 4755; and (b) Morimoto, T., Chiba, M., Achiwa, K. *Tetrahedron Lett.* **1989**, *30*, 735.
[14] Tani, K., Suwa, K., Tanigawa, E., Ise, T., Yamagato, T., Otsuka, S. *J. Organomet. Chem.* **1989**, *370*, 203.
[15] (a) Li, W., Waldkirch, J. P., Zhang, X. J. *Org. Chem.* **2000**, *65*, 5871; and (b) Yan, Y.-Y., RajanBabu, T. V. *Org. Lett.* **2000**, *2*, 4137.
[16] Li, W., Waldkirch, J. P., Zhang, X. *J. Org. Chem.* **2002**, *67*, 7618.
[17] (a) Diéguez, M., Pàmies, O., Claver, C. *Chem. Rev.* **2004**, *104*, 3189; and (b) Castillón, S., Claver, C., Díaz, Y. *Chem. Soc. Rev.* **2005**, *34*, 702.
[18] (a) Pàmies, O., Net, G., Ruiz, A., Claver, C. *Eur. J. Inorg. Chem.* **2000**, 2011; and (b) Diéguez, M., Pàmies, O., Ruiz, A., Castillón, S., Claver, C. *Tetrahedron: Asymmetry* **2000**, *11*, 4701.
[19] (a) Takeda, H., Tachinami, T., Hosokawa, S., Aburatami, M., Inoguchi, K., Achiwa, K. *Chem. Pharm. Bull.* **1991**, *39*, 2706; (b) Takahashi, H., Hattori, M. Chiba, M., Morimoto, T., Achiwa, K. *Tetrahedron Lett.* **1986**, *27*, 4477; (c) Morimoto, T., Kakahashi, T., Fujii, K., Chiba, M., Achiwa, K. *Chem. Lett.* **1986**, 2061; (d) Takahashi, H., Achiwa, K. *Chem. Lett.* **1989**, 305; and (e) Inoguchi, K., Morimoto, T., Achiwa, K. *J. Organomet. Chem.* **1989**, *370*, C9–C12.
[20] Baker, G. L., Fritschel, S. J., Stille, J. R. Stille, J. K. *J. Org. Chem.* **1981**, *46*, 2954.
[21] (a) Nagel, U., Kinzel, E, Andrade, J. Prescher, G. *Chem. Ber.* **1986**, *119*, 3326; and (b) Nagel, U., Rieger, B. *Chem. Ber.* **1988**, *121*, 1123.
[22] Lee, S., Zhang, Y. J., Song, C. E., Lee, J. K., Choi, J. H. *Angew. Chem. Int. Ed.* **2002**, *41*, 847.
[23] Gavryushin, A., Polborn, K., Knochel, P. *Tetrahedron: Asymmetry* **2004**, *15*, 2279.
[24] MacNeil, P. A., Roberts, N. K., Bosnich, B. *J. Am. Chem. Soc.* **1981**, *103*, 2273.
[25] Herseczki, Z., Gergely, I. Hegedüs, C., Szöllősy, Á., Bakos, J. *Tetrahedron: Asymmetry* **2004**, 15, 1673.
[26] Brunner, H., Terfort, A. *Tetrahedron: Asymmetry* **1995**, *6*, 919.
[27] Burk, M. J., Harper, T. G. P., Kalberg, C. S. *J. Am. Chem. Soc.* **1995**, *117*, 4423.
[28] Bakos, J., Tóth, I., Heil B., Szalontai, G. Párkányi, L., Fülöp, V. *J. Organomet. Chem.* **1989**, *370*, 263.
[29] (a) Dubrovina, N. V., Tararov, V. I., Monsees, A., Kadyrov, R., Fischer, C., Börner, A. *Tetrahedron: Asymmetry* **2003**, *14*, 2739; (b) Dubrovina, N. V., Tararov, V. I., Monsees, A., Spannenberg, A., Kostas, I. D., Börner, A. *Tetrahedron: Asymmetry* **2005**, *16*, 3640.
[30] Bianchini, C., Barbaro, P., Scapacci, G., Zanobini, F. *Organometallics* **2000**, *19*, 2450.
[31] Inoguchi, K., Achiwa, K. *Synlett* **1991**, *49*.
[32] Srivastava, N., Mital, A., Kumar, A. *J. Chem. Soc., Chem. Commun.* **1992**, *493*.
[33] (a) Cram, D. J., Steinberg, H. *J. Am. Chem. Soc.* **1959**, *81*, 2753; (b) Hardegger, E., Maeder, E., Semarne, H. M., Cram, D. J. J. *Am. Chem. Soc.* **1959**, *81*, 2729; (c) Carruthers, W., Orridge, A. *J. Chem. Soc., Perkin Trans. I* **1977**, 2411; (d) Gerlach, H. *Helv. Chim. Acta.* **1968**, *51*, 1587; and (e) Harada, N., Ochiai, N., Takada, K., Uda, H. *J. Chem. Soc., Chem. Commun.* **1977**, 495.
[34] Nieman, J. A., Parvez, M., Keay, B. A. *Tetrahedron: Asymmetry* **1993**, *4*, 1973.
[35] Chan, A. S. C., Hu, W., Pai, C.-C., Lau, C.-P., Jiang, Y., Mi, A., Yan, M., Sun, J., Lou, R., Deng, J. *J. Am. Chem. Soc.* **1997**, *119*, 9570.
[36] Birman, V. B., Rheingold, A. L., Lam, K.-C. *Tetrahedron: Asymmetry* **1999**, *10*, 125.
[37] (a) Xie, J.-H., Wang, L.-X., Fu, Y., Zhu, S.-F., Fan, B.-M., Duan, H.-F., Zhou, Q.-L. J. *Am. Chem. Soc.* **2003**, *125*, 4404; (b) Fu, Y., Xie, J.-H., Hu, A.-G., Zhou, H., Wang, L.-X., Zhou, Q.-L. *Chem. Commun.* **2002**, *480*; (c) Hu, A.-G., Fu, Y., Xie, J.-H., Zhou, H., Wang, L.-X., Zhou, Q.-L. *Angew. Chem. Int. Ed.* **2002**, *41*, 2348;(d)Zhu, S.-F., Yang, Y., Wang, L.-X., Liu, B., Zhou, Q.-L. *Org. Lett.* **2005**, *7*, 2333; (e) Duan, H.-F., Xie, J.-H., Shi W.-J., Zhang, Q., Zhou, Q.-L. *Org. Lett.* **2006**, *8*, 1479; and (f) Duan, H.-F., Jia, Y.-X., Wang L.-X., Zhou, Q.-L. *Org. Lett.* **2006**, *8*, 2567.

[38] Cheng, X., Hou, G.-H., Xie, J.-H., Zhou, Q.-L. *Org. Lett.* **2004**, *6*, 2381.
[39] Cheng, X., Zhang, Q., Xie, J.-H., Wang, L.-X., Zhou, Q.-L. *Angew. Chem. Int. Ed.* **2005**, *44*, 1118.
[40] Zhang, W., Wu, S., Zhang, Z., Yennawar, H., Zhang, X. *Org. Biomol. Chem.* **2006**, *4*, 4474.
[41] Freixa, Z. Beentjes, M. S., Batema, G. D., Dieleman, C. B., van Strijdonck, G. P. F., Reek, J. N. H., Kamer, P. C. J., Fraanje, J. Goubitz, K., van Leeuwen, P. W. N. M. *Angew. Chem. Int. Ed.* **2003**, *42*, 1284.
[42] Kang, J., Lee, J. H., Ahn, S. H., Choi, J. S. *Tetrahedron Lett.* **1998**, *39*, 5523.
[43] Kang, J., Lee, J. H., Kim, J. B., Kim, G. J. *Chirality* **2000**, *12*, 378.
[44] (a) Zhang, W., Adachi, Y., Hirao, T., Ikeda, I. *Tetrahedron: Asymmetry* **1996**, *7*, 451; and (b) Zhang, W., Kida, T., Nakatsuji, Y., Ikeda, I. *Tetrahedron Lett.* **1996**, *37*, 7995.
[45] Raghunath, M., Gao, W., Zhang, X. *Tetrahedron: Asymmetry* **2005**, *16*, 3676.
[46] Zhu, G., Cao, P., Jiang, Q., Zhang, X. *J. Am. Chem. Soc.* **1997**, *119*, 1179.
[47] Genov, D. G., Ager, D. *J. Angew. Chem. Int. Ed.* **2004**, *43*, 2816.
[48] Pye, P. J., Rossen, K., Reamer, R. A., Tsou, N. N., Volante, R. P., Reider, P. J. *J. Am. Chem. Soc.* **1997**, *119*, 6207.
[49] (a) Boulton, L. T., Lennon, I. C., McCague, R. *Org. Biomol. Chem.* **2003**, *1*, 1094; (b) Rossen, K., Pye, P. J., Maliakal, A., Volante, R. P. J. *Org. Chem.* **1997**, *62*, 6462; (c) Pye, P. J., Rossen, K., Reamer, R. A., Volante, R. P., Reider, P. J. *Tetrahedron Lett.* **1998**, *39*, 4441; and (d) Burk, M. J., Hems, W. P., Herzburg, Herzburg, D., Malan, C., Zanotti-Gerosa, A. *Org. Lett.* **2000**, *2*, 4173.
[50] Dominguez, B., Zanotti-Gerosa, A., Hems, W. *Org. Lett.* **2004**, *6*, 1927.
[51] Zhang, X., Longmire, J. M. *Tetrahedron Lett.* **1997**, *38*, 1725.
[52] Achiwa, K. Morimoto, T., Yamazaki, A. *Tetrahedron: Asymmetry* **1993**, *4*, 2287.
[53] Morimoto, T., Yamazaki, A., Achiwa, K. *Chem. Pharm. Bull.* **2004**, *52*, 1367.
[54] Brunner, H., Pieronczyk, W., Schönhammer, B., Streng, K., Bernal, I., Korp, U. J. *Chem. Ber.* **1981**, *114*, 1137.
[55] (a) Eckert, C., Dahlenburg, L., Wolski, A. Z. *Naturforsch. B: Chem. Sci.* **1995**, *50*, 1004;(b) Dahlenburg, L., Kaunert, *Eur. J. Inorg. Chem.* **1998**, *885*; and (c) Brunner, H., Stefaniak, S., Zabel, M. *Synthesis* **1999**, 1776.
[56] (a) Dahlenburg, L. *Eur. J. Inorg. Chem.* **2003**; 2733.(b) Dahlenburg, L., Mertel, S. J. *Organomet. Chem.* **2001**, *630*, 221; and (c) Dahlenburg, L., Eckert, C. *J. Organomet. Chem.* **1998**, *563*, 227.
[57] Miyashita, A., Yasuda, T., Toriumi, K., Ito, T., Souchi, T., Noyori, R. *J. Am. Chem. Soc.* **1980**, *102*, 7932.
[58] Noyori, R., Ohkuma, T. *Angew. Chem. Int. Ed.* **2001**, *40*, 40.
[59] Takaya, H., Mashima, K., Koyano, K., Yagi, M., Kumobayashi, H., Taketomi, T., Akutagawa, S., Noyori, R. *J. Org. Chem.* **1986**, *51*, 629.
[60] Berthod, M., Mignani, G., Woodward, G., Lemaire, M. *Chem. Rev.* **2005**, *105*, 1801.
[61] (a) Cai, D., Payack, J. F., Bender, D. R., Hughes, D. L., Verhoeven, T. R., Reider, P. J. J. *Org. Chem.* **1994**, *59*, 7180; and (b) Cai, D. (Merck). US Patent 252,306, **1994**, US Patent 5,399,771, 1995.
[62] (a) Laneman, S. A. *Chem. Commun.* **1997**, 2359; and (b) Laneman, S. A. (Monsanto). US Patent 5,902,904, 1997.
[63] Sayo, N., Zhang, X., Oh, T., Yoshida, A., Hiratsuka-Shi, K.-K., Yokozawa, T. (Takasago International Corp.). European Patent Appl. EP 0771812, **1997**, US Patent 5,693,868, 1997.
[64] (a) Goto, M, Mano, M. (Takeda Chemical Industries). PCT Int. Appl. WO 03048174, **2003**; and (b) Goto, M, Mano, M. (Takeda Chemical Industries). European Patent Appl. EP 1452537, **2004.**
[65] Gladiali, S., Dore, A., Fabbri, D., Medici, S., Pirri, G., Pulacchini, S. *Eur. J. Org. Chem.* **2000**, 2861.
[66] Zhang, X., Mashima, K., Koyano, K., Sayo, N., Kumobayashi, H., Akutagawa, S., Takaya, H. *Tetrahedron Lett.* **1991**, *32*, 7283.
[67] Hori, Y., Kumobayashi, H. (Takasago International Corp.). European Patent Appl. EP 0466405, **1992**.
[68] Andersen, N. G., McDonald, R., Keay, B. A. *Tetrahedron: Asymmetry* **2001**, *12*, 263.
[69] Au-Yeung, T. T.-L., Chan, S. S., Chan, A. S. C. *Adv. Synth. Catal.* **2003**, *345*, 537.
[70] (a) Zhang, X., Mashima, K., Koyano, K., Sayo, N., Kumobayashi, H., Akutagawa, S., Takaya, H. *J. Chem. Soc., Perkin Trans. 1* **1994**, 2309; (b) Sayo, N., Cho, K., Imai, T., Miura, T. (Takasago International Corp.). Japanese Patent 10120670, **1998**; (c) Zhang, X., Kumobayashi, H., Takaya, H. *Tetrahedron: Asymmetry* **1994**, *5*, 1179; (d) Senda, T., Ogasawara, M., Hayashi, T. J. *Org. Chem.* **2001**, *66*, 6852; (e) Shintani, R., Duan, W.-L., Hayashi, T. J. *Am. Chem. Soc.* **2006**, *128*, 5628; (f) Oderaotoshi, Y., Cheng, W., Fujitomi, S., Kasano, Y., Minakata, S., Komatsu, M. *Org. Lett.* **2003**, *5*, 5043; and (g) Shibata, T., Takami, K., Kawachi, A. *Org. Lett.* **2006**, *8*, 1343.

[71] (a) Schmid, R., Cereghetti, M., Heiser, B., Schönholzer, P., Hansen, H.-J. *Helv. Chim. Acta* **1988**, *71*, 897; and (b) Schmid, R., Roricher, J., Cereghetti, M., Schönholzer, P. *Helv. Chim. Acta* **1991**, *74*, 370.

[72] Saito, T., Yokozawa, T., Ishizaki, T., Moroi, T., Sayo, N., Miura, T., Kumobayashi, H. *Adv. Synth. Catal.* **2001**, *343*, 264.

[73] (a) Enev, V., Ewers, C. L. J., Harre, M., Nickisch, K. Mohr, J. T. *J. Org. Chem.* **1997**, *62*, 7092; (b) Enev, V., Harre, M., Nickisch, K., Schneider, M., Mohr, J. T. *Tetrahedron: Asymmetry* **2000**, *11*, 1767; (c) Qiu, L., Qi, J., Pai, C.-C., Chan, S., Zhou, Z., Choi, M. C. K., Chan, A. S. C. *Org. Lett.* **2002**, *4*, 4599; and (d) Gorobets, E., Wheatley, B. M. M., Hopkins, J. M., McDonald, R., Keay, B. A. *Tetrahedron Lett.* **2005**, *46*, 3843.

[74] (a) Qiu, L., Wu, J., Chan, S., Au-Yeung, T. T.-L., Ji, J.-X., Guo, R., Pai, C.-C., Zhou, Z., Li, X., Fan, Q.-H., Chan, A. S. C. *Proc. Natl. Acad. Sci. USA* **2004**, *101*, 5815; and (b) Qiu, L., Kwong, F. Y., Wu, J., Lam, W. H., Chan, S., Yu, W.-Y., Li, Y.-M., Guo, R., Zhou, Z., Chan, A. S. C. *J. Am. Chem. Soc.* **2006**, *128*, 5955.

[75] Sun, X., Zhou, L., Li, W., Zhang, X. J. *Org. Chem.* **2008**, *73*, 1143.

[76] (a) Pai, C.-C., Li, Y. M., Zhou, Z.-Y., Chan, A. S. C. *Tetrahedron Lett.* **2002**, *43*, 2789; (b) de Paule, S. D., Jeulin, S., Ratovelomanana-Vidal, V., Genêt, J.-P., Champion, N., Dellis, P. *Tetrahedron Lett.* **2003**, *44*, 823; and (a) de Paule, S. D., Jeulin, S., Ratovelomanana-Vidal, V., Genêt, J.-P., Champion, N., Dellis, P. *Eur. J. Org. Chem.* **2003**, 1931.

[77] Zhang, Z., Qian, H., Longmire, J., Zhang, X. *J. Org. Chem.* **2000**, *65*, 1931.

[78] (a) Pietrusiewicz, K. M., Zablocka, M. *Chem. Rev.* **1994**, *94*, 1375; (b) Holz, J., Quirmbach, M., Börner, A. *Synthesis* **1998**, 1391; (c) Marinetti, A., Carmichael, D. *Chem. Rev.* **2002**, *102*, 201; and (d) Tang, W., Zhang, X. *Chem. Rev.* **2003**, *103*, 3029.

[79] (a) Holz, J., Gensow, M.-N., Zaya, O., Börner, A. *Curr. Org. Chem.* **2007**, *11*, 61; (b) Clark, T. P., Landis, C. R. *Tetrahedron: Asymmetry* **2004**, *15*, 2123.

[80] Burk, M. J., Feaster, J. E., Harlow, R. L. *Organometallics* **1990**, *9*, 2653.

[81] Burk, M. J. *J. Am. Chem. Soc.* **1991**, *113*, 8518.

[82] (a) Burk, M. J., Harlow, R. L. *Angew. Chem. Int. Ed. Engl.* **1990**, *29*, 1462; (b) Burk, M. J., Feaster, J. E., Harlow, R. L. *Tetrahedron: Asymmetry* **1991**, *2*, 259; and (c) Burk, M. J., Feaster, J. E., Nugent, W. A., Harlow, R. L. *J. Am. Chem. Soc.* **1993**, *115*, 10125.

[83] (a) Burk, M. J., Gross, M. F., Martinez, J. P. J. *Am. Chem. Soc.* **1995**, *117*, 9375; (b) Burk, M. J., Allen, J. G., Kiesman, W. F. *J. Am. Chem. Soc.* **1998**, *120*, 657; (c) Burk, M. J., Feaster, J. E. J. *Am. Chem. Soc.* **1992**, *114*, 6266; (d) Burk, M. J., Wang, Y. M., Lee, J. R. *J. Am. Chem. Soc.* **1996**, *118*, 5142; (e) Burk, M. J., Kalberg, C. S., Pizzano, A. *J. Am. Chem. Soc.* **1998**, *120*, 4345; (f) Burk, M. J., Bienewald, F., Harris, M., Zanotti-Gerosa, A. *Angew. Chem. Int. Ed.* **1998**, *37*, 1931; and (g) Burk, M. J., Harper, T. G. P., Kalberg, C. S. *J. Am. Chem. Soc.* **1995**, *117*, 4423.

[84] (a) Blaser, H.-U., Spindler, F., Studer, M. *Appl. Catal. A: General* **2001**, *221*, 119; (b) Blaser, H.-U., Malan, C., Pugin, B., Spindler, F., Steiner, H., Studer, M. *Adv. Synth. Catal.* **2003**, *345*, 103; and (c) Cobley, C. J., Johnson, N. B., Lennon, I. C., McCague, R., Ramsden, J. A., Zanotti-Gerosa, A. in *Asymmetric Catalysis, on an Industrial Scale: Challenges, Approaches and Solutions*, ed. Blaser, H.-U., Schmidt, E. Wiley-VCH, Weinheim, **2004**.

[85] Burk, M. J. *Acc. Chem. Res.* **2000**, *33*, 363.

[86] Burk, M. J., Pizzano, A., Martín, J. A. *Organometallics* **2000**, *19*, 250.

[87] Burk, M. J., Gross, M. F. *Tetrahedron Lett.* **1994**, *35*, 9363.

[88] Berens, U., Burk , M. J., Gerlach, A., Hems, W. *Angew. Chem. Int. Ed.* **2000**, *39*, 1981.

[89] Pilkington, C. J., Zanotti-Gerosa, A. *Org. Lett.* **2003**, *5*, 1273.

[90] Fiaud, J.-C., Legros, J.-Y. *Tetrahedron Lett.* **1991**, *32*, 5089.

[91] (a) Guillen, F., Fiaud, J.-C. *Tetrahedron Lett.* **1999**, *40*, 2939; (b) Guillen, F., Rivard, M., Toffano, M., Legros, J.-Y., Daran, J.-C., Fiaud, J.-C. *Tetrahedron* **2002**, *58*, 5895.

[92] Oisaki, K., Zhao, D., Suto, Y., Kanai, M., Shibasaki, M. *Tetrahedron Lett.* **2005**, *46*, 4325.

[93] Shimizu, H., Ishizaki, T., Fujiwara, T., Saito, T. *Tetrahedron: Asymmetry* **2004**, *15*, 2169.

[94] (a) Marinetti, A., Jus, S., Genêt, J. P., Ricard, L. *Tetrahedron* **2000**, *56*, 95; and (b) Marinetti, A., Jus, S., Genêt, J. P., Ricard, L. *J. Organomet. Chem.* **2001**, *624*, 162.

[95] Marinetti, A., Labrue, F., Genêt, J. P. *Synlett* **1999**, 1975.

[96] Imamoto, T., Oohara, N., Takahashi, H. *Synthesis* **2004**, 1353.

[97] Imamoto, T., Crepy, K. V. L., Katagiri, K. *Tetrahedron: Asymmetry* **2004**, *15*, 2213.

[98] Dierkes, P., Ramdeehul, S., Barloy, L., De Cian, A., Fischer, J., Kamer, P. CJ., van Leeuwen, P. W. N. M., Osborn, J. A. *Angew. Chem. Int. Ed.* **1998**, *37*, 3116.

[99] Fernandez, E., Gillon, A., Heslop, K., Horwood, E., Hyett, D. J., Orpen, A. G., Pringle, P. G. *Chem. Commun.* **2000**, 1663.

[100] (a) Berens, U., Englert, U., Geyser, S., Runsink, J., Salzer, A. *Eur. J. Org. Chem.* **2006**, 2100; and (b) Beniocori, T., Pilati, T., Rizzo, S., Sannicolò, F., Burk, M. J., de Ferra, L, Ullucci, E., Piccolo, O. *J. Org. Chem.* **2005**, *70*, 5436.

[101] (a) Holz, J., Monsees, A., Jiao, H., You, J., Komarov, I. V., Fischer, C., Drauz, K., Börner, A. *J. Org. Chem.* **2003**, *68*, 1701; (b) Almena, J., Monsees, A., Kadyrov, R., Riermeier, T. H., Gotov, B., Holz, J., Börner, A. *Adv. Synth. Catal.* **2004**, *346*, 1263; (c) Börner, A., Holz, J., Monsees, A., Riermeier, T., Kadyrov, R., Schneider, C. A., Dingerdissen, U., Drauz, K. (Degussa AG), WO 03/084971 A1, 2003.

[102] Holz, J., Zayas, O., Jiao, H., Baumann, W., Spannenberg, A., Monsees, A, Riermeier, T. H., Almena, J., Kadyrov, R., Börner, A. *Chem. Eur. J.* **2006**, *12*, 5001.

[103] Lougneau, A., Durand, S., Knochel, P. *Tetrahedron: Asymmetry* **1997**, *8*, 987.

[104] Holz, J., Quirmbach, M., Schmidt, U., Heler, D., Stürmer, R., Börner, A. *J. Org. Chem.* **1998**, *63*, 8031.

[105] (a) Holz, J., Zayas, O, Jiao, H., Baumann, W., Spannenberg, A., Monsees, A., Riermeier, T. H., Almena, J., Kadyrov, R., Börner, A. *Chem. Eur. J.* **2006**, *12*, 5001; (b) Holz, J., Stürmer, R., Schmit, U., Drexler, H.-J., Heller, D., Krimmer, H.-P., Börner, A. *Eur. J. Org. Chem.* **2001**, 4615.

[106] (a) Li, W., Zhang, Z., Xiao, D., Zhang, X. *Tetrahedron Lett.* **1999**, *40*, 6701; and (b) Li, W., Zhang, Z., Xiao, D., Zhang, X. *J. Org. Chem.* **2000**, *65*, 3489.

[107] (a) Yan, Y.-Y., RajanBabu, T. V. J. *Org. Chem.* **2000**, *65*, 900; and (b) RajanBabu, T. V., Yan, Y.-Y., Shin, S. *J. Am. Chem. Soc.* **2001**, *123*, 10207.

[108] Liu, D., Li, W., Zhang, X. *Org. Lett.* **2002**, *4*, 4471.

[109] Jiang, Q., Jiang. Y., Xiao, D., Cao, P., Zhang, X. *Angew. Chem. Int. Ed.* **1998**, *37*, 1100.

[110] (a) Jiang, Q., Xiao, D., Zhang, Z., Cao, P, Zhang, *X. Angew. Chem. Int. Ed.* **1999**, *38*, 516; and (b) Zhang, Z., Zhu, G., Jiang, Q., Xiao, D., Zhang, X. *J. Org. Chem.* **1999**, *64*, 1774.

[111] Shimizu, H., Saito, T., Kumobayashi, H. *Adv. Synth. Catal.* **2003**, *345*, 185.

[112] (a) Hoge, G. *J. Am. Chem. Soc.* **2003**, *125*, 10219; (b) Hoge, G. *J. Am. Chem. Soc.* **2004**, *126*, 9920; (c) Hoge, G., Samas, B. *Tetrahedron: Asymmetry* **2004**, *15*, 2155.

[113] (a) Tang, W., Zhang, X. *Angew. Chem. Int. Ed.* **2002**, *41*, 1612; (b) Tang, W., Zhang, X. *Org. Lett.* **2002**, *4*, 4159; (c) Tang, W., Liu, D., Zhang, X. *Org. Lett.* **2003**, *5*, 205; (d) Yang, Q., Shang, G., Gao, W., Deng, J., Zhang, X. *Angew. Chem. Int. Ed.* **2006**, *45*, 3832; (e) Shang, G., Yang, Q., Zhang, X. *Angew. Chem. Int. Ed.* **2006**, *45*, 6360; and (f) Chen, J., Zhang, W., Geng, H., Li, W., Hou, G., Lei, A., Zhang, X. *Angew. Chem. Int. Ed.* **2009**, *48*, 800.

[114] Liu, D., Zhang, X. *Eur. J. Org. Chem.* **2005**, 646.

[115] Zhang. X, Huang, K., Hou, G., Cao, B., Zhang, X. *Angew. Chem. Int. Ed.* **2010**, *49*, 6421.

[116] (a) Quin, L. D. *A Guide to Organophosphorus Chemistry*, John Wiley & Sons, New York, **2000**; and (b) Mathey, F. *Chem. Rev.* **1988**, *88*, 429.

[117] (a) Daug, T. P., Poulin, J.-C., Kagan, H. B. J. *Organomet. Chem.* **1975**, *91*, 105; and (b) Tanaka, M., Ikeda, Y., Ogata, I. *Chem. Lett.* **1975**, 1115.

[118] Hayashi, T., Tanaka, M., Ikeda, Y., Ogata, I. *Bull. Chem. Soc. Jpn.* **1979**, *52*, 2605.

[119] (a) Bakos, J., Karaivanov, R., Laghmari, M., Sinou, D. *Organometallics* **1994**, *13*, 2951; and (b) Tóth, I., Elsevier, C. J., de Vries, J. G., Bakos, J., Smeets, W. J. J., Spek, A. L. J. *Organomet. Chem.* **1997**, *540*, 15.

[120] Schenk, W. A., Stubbe, M., Hagel, M. *J. Organomet. Chem.* **1998**, *560*, 257.

[121] Gladiali, S., Fabbri, D., Kollàr, L. *J. Organomet. Chem.* **1995**, *491*, 91.

[122] Mathey, F., Mercier, F., Robin, F., Ricard, L. *J. Organomet. Chem.* **1998**, *557*, 117.

[123] Ostermeier, M. Prieβ, J., Helmchen, G. *Angew. Chem. Int. Ed.* **2002**, *41*, 612.

[124] Yan, Y., Zhang, X. *Tetrahedron Lett.* **2006**, *47*, 1567.

[125] Xiao, D., Zhang, Z., Zhang, X. *Org. Lett.* **1999**, *1*, 1679.

[126] Xiao, D., Zhang, X. *Angew. Chem. Int. Ed.* **2001**, *40*, 3425.

[127] (a) Hou, G., Gosselin, F., Li, W., McWilliams, J. C., Sun. Y., Weisel, M., O'Shea, P. D., Chen, C.-Y., Davies, I. W., Zhang, X. *J. Am. Chem. Soc.* **2009**, *131*, 9882; and (b) Hou, G., Li, W., Ma, M., Zhang, X., Zhang, X. *J. Am. Chem. Soc.* **2010**, *132*, 12844.

[128] Vasse, J.-L., Stranne, R., Zalubovskis, R., Gayet, C., Moberg, C. J. *Org. Chem.* **2003**, *68*, 3258.
[129] Chi, Y., Zhang, X. *Tetrahedron Lett.* **2002**, *43*, 4849.
[130] Tang, W., Wang, W., Chi, Y., Zhang, X. *Angew. Chem. Int. Ed.* **2003**, *42*, 3509.
[131] (a) Pietrusiewicz, K. M, Zablocka, M. *Chem. Rev.* **1994**, *94*, 1375; (b) Imamoto, T. In *Handbook of Organophosphorus Chemistry*, Engel, R., Ed., Marcel Dekker: New York, **1992**, Chapter 1; and (c) Kagan, H. B., Sasaki, M. In *Chemistry of Organophosphorus Compounds*, Hartley, F. R., Ed., Wiley & Sons: New York, **1990**, Vol. 1, Chapter 3.
[132] Roberts, N. K., Wild, S. B. *J. Am. Chem. Soc.* **1979**, *101*, 6254.
[133] (a) Muci, A. R., Campos, K. R., Evans, D. A. J. *Am. Chem. Soc.* **1995**, *117*, 9075; (b) Gridnev, I. D., Yamanoi, Y., Higashi, N., Tsuruta, H., Yasutake, M., *Imamoto. T. Adv. Synth. Catal.* **2001**, *343*, 118; and (c) Nagata, K., Matsukawa, S., Imamoto, T. *J. Org. Chem.* **2000**, *65*, 4185.
[134] Imanoto, T., Watanabe, J., Wada, Y., Masuda, H., Yamada, H., Tsuruta, H., Matsukawa, S., Yamaguchi, K. *J. Am. Chem. Soc.* **1998**, *120*, 1635.
[135] (a) Crépy, K. V. L., Imamoto, T., Seidel, G., Fürstner, A. *Org. Synth.* **2005**, *82*, 22; and (b) Oohara, N., Katagiri, K., Imamoto, T. *Tetrahedron: Asymmetry* **2003**, *14*, 2171.
[136] (a) Gridnev, I. D., Yasutake, M., Higashi, N., Imamoto, T. *J. Am. Chem. Soc.* **2001**, *123*, 5268; (b) Yasutake, M., Gridnev, I. D., Higashi, N., Imamoto, T. *Org. Lett.* **2001**, *3*, 1701; (c) Crépy, K. V. L., Imamoto, T. *Adv. Synth. Catal.* **2003**, *345*, 79; (d) Gridnev, I. D., Yasutake, M., Imamoto, T., Beletskaya, I. P. *Proc. Natl. Acad. Sci. USA* **2004**, *101*, 5385; and (e) Imamoto, T., Iwadate, N., Yoshida, K. *Org. Lett.* **2006**, *8*, 2289.
[137] Genet, C., Canipa, S. J., O'Brien, P., Taylor, S. *J. Am. Chem. Soc.* **2006**, *128*, 9336.
[138] Yamanoi, Y., Imamoto, T. *J. Org. Chem.* **1999**, *64*, 2988.
[139] Brown, H. C., Jadhav, P. K., Mandal, A. K. *J. Org. Chem.* **1982**, *47*, 5074.
[140] Guillen, F., Rivard, M., Toffano, M., Legros, J.-Y., Daran, J.-C., Fiaud, J.-C. *Tetrahedron* **2002**, *58*, 5895.

3

Design and Synthesis of Phosphite Ligands for Homogeneous Catalysis

Aitor Gual, Cyril Godard, Verónica de la Fuente, and Sergio Castillón

Departament de Química Analítica I Química Orgànica, Universitat Rovira i Virigili, Tarragona, Spain

3.1 Introduction

Over recent decades, phosphite ligands [1] of general formula $P(OR)_3$, have attracted much interest in the field of catalysis, and have been particularly successful in asymmetric reactions. One of the main advantages of these compounds is that they can be readily prepared from alcohols [2], which allows the rapid synthesis of ligand libraries, and their screening in the search for catalysts provides high activities and selectivities [3]. They are less prone to oxidation than phosphines but can decompose by hydrolysis or alcoholysis and the Arbuzov reaction, although the rate of hydrolysis is often low for bulky phosphites.

Phosphites have a very rich coordination chemistry, and examples of phosphite-containing complexes have been reported with most transition metals. They are strong π acceptors and form stable complexes with electron-rich transition metals. The strong binding of phosphites suggests that π-back bonding may be dominating over σ donation in these complexes. The question arises as to which orbitals of phosphorus are responsible for π-back donation. The current view is that the anti-bonding σ^* orbitals of phosphorus to oxygen play the role of the π acceptor orbital on phosphorus [4]. The steric properties are defined by the cone angle in monophosphites and by the bite angle in diphosphites.

As phosphite function is present in many ligands, alone or combined with other coordinating heteroatoms such as nitrogen or sulfur, or other phosphorus functions such as phosphines, phosphinites or phosphoramidites, this chapter will deal only with ligands that exclusively contain the phosphite function.

Phosphorus(III) Ligands in Homogeneous Catalysis: Design and Synthesis, First Edition.
Edited by Paul C. J. Kamer and Piet W. N. M. van Leeuwen.

3.2 Synthesis of phosphites

According to their coordination properties, phosphites can be classified as mono-, di-, tri and tetraphosphites, although very few examples of triphosphites have been reported to date (Figure 3.1). Symmetrically substituted phosphite ligands (**A**) are the simplest components of this family of ligands and have been used in catalysis since the 1960s, just in the dawn of homogenous catalysis, and the number of new examples of ligands belonging to this family keeps growing continuously. Asymmetrically substituted ligands are usually prepared by combining an alcohol and a diol (**C**), and very successful ligands with this type of structure have been reported. Less common are phosphites of type **B**, containing two distinct alkoxy substituents. No examples of monophosphites with three different substituents (**D**) have been reported.

Diphosphite ligands contain two $P(OR)_2$ units bonded to a diol backbone. There are only a few examples of ligands with the structure **E**. Most of the reported diphosphite ligands have the structure **F**, in which the diol backbone is linked to dioxaphospho-cyclic units. There are also a few examples of the tridentate ligands of type **G** (or related).

3.2.1 Monophosphites

3.2.1.1 Symmetrically substituted monophosphites

Symmetrically substituted monophosphites can be readily prepared by reaction of an alcohol (**A**, R=alkyl, aryl; Figure 3.1) with phosphorus trichloride (PCl_3) in the presence of a base and a solvent (i, Scheme 3.1). The base is used to trap the generated HCl in the presence of which the phosphite can react to afford the corresponding dialkyl hydrogen phosphite (ii, Scheme 3.1) [5].

Symmetrically substituted trialkyl [6–30] and triaryl [6,20,21,25,27–30] monophosphites have been extensively used in homogeneous catalysis. One of the reasons is that they can be prepared in only one step, and their electronic and steric properties can be easily tuned by varying the R substituents (Figure 3.2). Alkyl

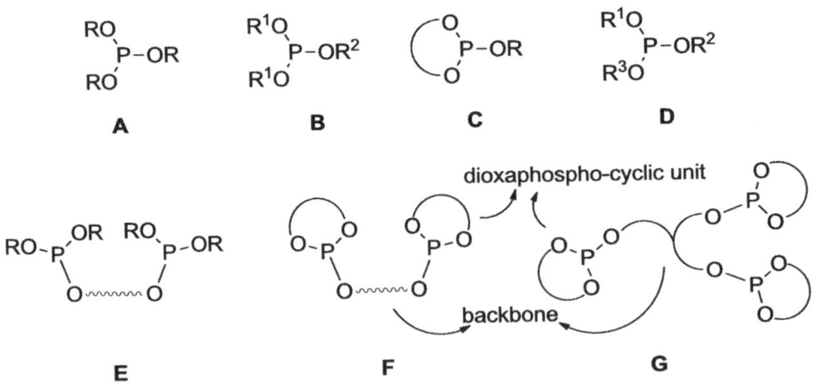

Figure 3.1 General structures for phosphites.

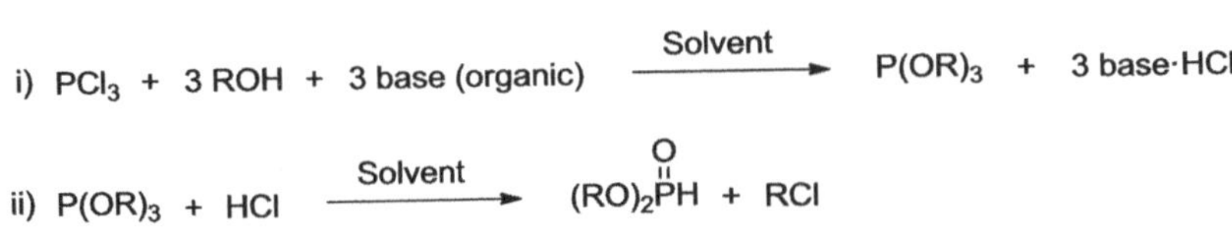

Scheme 3.1 General procedure for the synthesis of symmetrically substituted monophosphites.

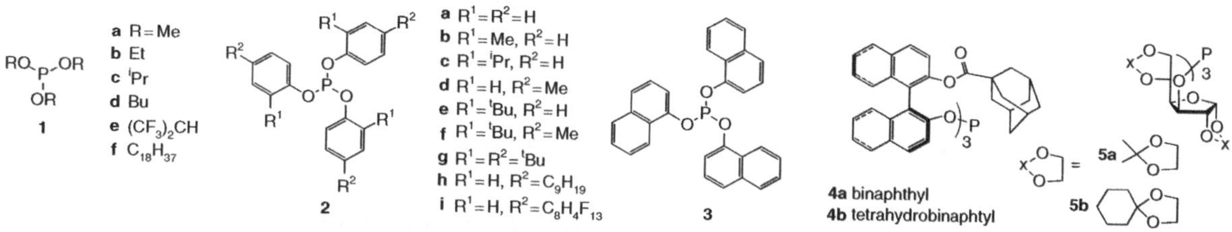

Figure 3.2 Representative symmetrically substituted monophosphites used in homogeneous catalysis.

monophosphites with various alkyl chains (**1**), aryl monophosphites with substituents at positions 2, 4 or both (**2**) or naphthyl groups (**3**) have been prepared [6]. Binaphthol-containing phosphites such as **4** were prepared by monoacylation of the biphenol unit and further reaction with PCl_3 in the presence of triethylamine (Et_3N) [7]. The phosphites **5**, which contain three units of glucofuranoside, were prepared by treating the corresponding alcohols with butyllithium to form the lithium alkoxydes and reaction with PCl_3 [8].

This methodology has been applied for a wide range of alkyl and aryl alcohols even bearing 18 carbon atoms. A representative example is the synthesis of $P(OPh)_3$ presented in this chapter.

Common problems in this synthesis are (1) the reaction is exothermic and, thus, it is necessary to cool down the reaction mixture, (2) an equimolar amount of amine with respect to the alcohol is necessary, (3) the separation of the salts formed during the reaction is sometimes ineffective and (4) the product needs to be purified due to the contamination with amine and salts. This issue can be circumvented by the use of potassium phosphate as a base and under phase transfer conditions [9]. Other procedures were reported using acetonitrile as solvent and heating the solution at 80°C for 2 or 4 h [10,11].

Using similar procedures to that of $P(OPh)_3$, various symmetrically substituted trialkyl, triaryl and mixed monophosphites were synthesised. The reaction conditions vary depending on the electronic and steric properties of the alcohol. However, caution must always be taken during the addition of PCl_3 since the reaction is strongly exothermic and efficient cooling is required. An interesting variation of these ligands (**1e**, **2i**) contains fluorinated alkyl chains which make it soluble in supercritical CO_2. This approach led to the formation of efficient biphasic catalytic systems which provide a solution to one of the main drawbacks of homogeneous catalysis, namely the recycling of the catalyst [12].

The synthesis of sterically hindered aryl-monophosphites is more difficult than the preparation of conventional phosphites. Phosphites having bulky substituents at the 2,4 positions have been widely used in catalysis and their synthesis required either harsher reaction conditions or the use of catalysts [38]. Highly substituted 2,4,5-trisubstituted phosphites with *t*-butyl, *t*-amyl, *t*-octyl, *t*-nonyl, *t*-dodecyl or dimethyl benzyl groups at the aryl ring were prepared using 2-mercaptobenzothiazole as catalysts [13].

3.2.1.2 Non-symmetrically substituted monophosphites

Nonsymmetrically substituted monophosphite ligands of type **B** (Figure 3.1), containing two distinct substituents such as compounds **6, 7a–f**, are synthesised by reacting an alcohol (OR^2) and $(R^1O)_2PCl$ in the presence of a base (i, Scheme 3.2). The key of the process is the synthesis of the phosphochloridite $(R^1O)_2PCl$ since mixtures containing trisubstituted and monosubstituted derivatives can also be obtained [14,15]. This issue can be partially circumvented using a slight excess of PCl_3 and by an appropriate selection of the alcohol. The second step is carried out under standard reaction conditions.

Generally, nonsymmetrically substituted phosphites $(R^1O)_2POR^2$ were synthesised by reacting the monochlorophosphites $(R^1O)_2PCl$, where R^1 can be alkyl or aryl substituent, with the corresponding alcohol in presence of a base. An interesting alternative procedure for the synthesis of mixed phosphites consists in a transesterification by reaction of a symmetrically substituted phosphite with a lithium alkoxide (ii, Scheme 3.2)

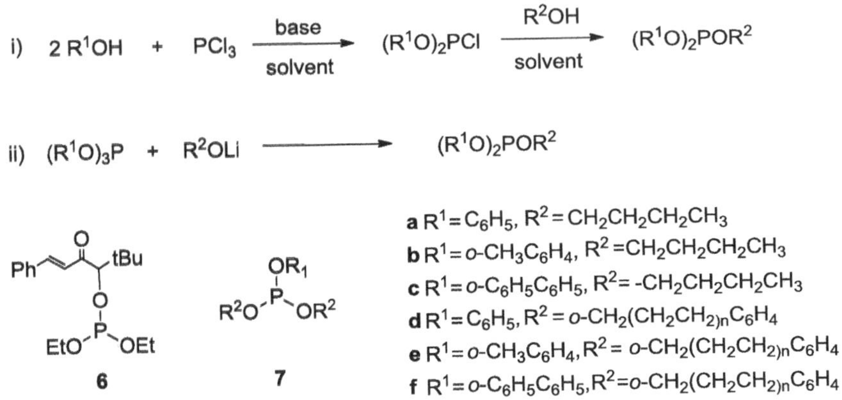

Scheme 3.2 *Synthesis of nonsymmetrically substituted phosphite ligands.*

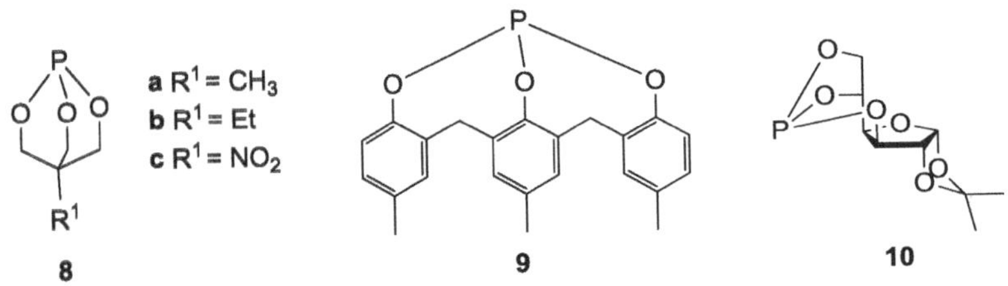

Figure 3.3 *Caged monophosphite ligands.*

[49]. However, with this procedure, mixtures of products can be obtained and the stoichiometry of the reagents needs to be carefully controlled.

The mixed phosphite ligands can also be synthesised by phase transfer catalysis using potassium phosphate as a base, as previously described for trialkyl or triaryl phosphites [43]. To the best of our knowledge, mixed phosphite ligands containing three distinct substituents have not been reported to date.

3.2.1.3 Caged monophosphites

Cage phosphite ligands were discovered 40 years ago, and although they present interesting features such as their strong π acceptor nature (exceeding that of $P(OPh)_3$) and their very small cone angle (Figure 3.3), they have been used in only a limited number of applications [16]. Their synthesis involves two intramolecular processes in the presence of triols and PCl_3. Compounds derived from alkyl, aryl or sugar-triols have been prepared (Figure 3.3).

Compound **10** was prepared from the corresponding triol by reaction with hexa-alkylphosphorus triamide [17a] or with PCl_3 in the presence of triethylamine [51b].

3.2.1.4 Monophosphites bearing dioxaphospho-cyclic units

A large variety of mixed monophosphites bearing dioxaphospho-cyclic units (type **C**, Figure 3.1) have been prepared. The most general synthetic procedure for these compounds first involves the formation of a cyclic phosphochloridite, which reacts in a second step with an alcohol (Scheme 3.3). The most common diols used for the synthesis of cyclic phosphochloridites are shown in Scheme 3.3. These species are 1,2- and 1,3-diol derivatives, catechol, naphthalene-1,8-diol, biphenol, binaphthol and taddol derivatives.

As an alternative procedure to that shown in Scheme 3.3, glycols undergo ester interchange with $P(OPh)_3$ to yield cyclic phosphites (type **C**) [18]. The initial experiments were carried out with ethylene glycol and

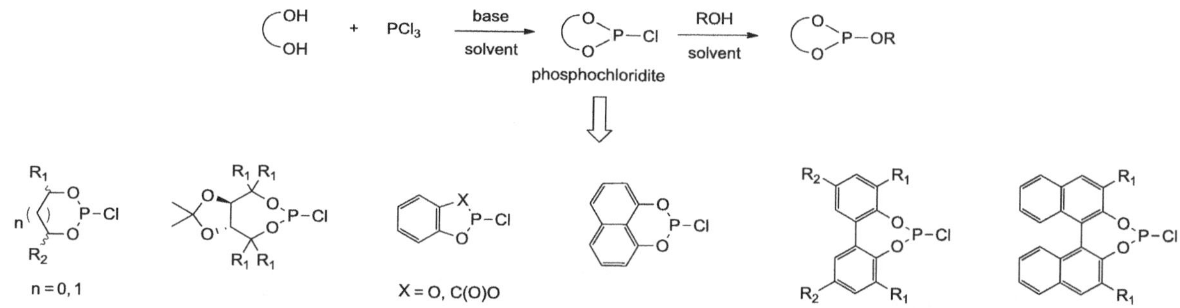

Scheme 3.3 *Most common dioxaphospho-cyclic units used in the synthesis of monophosphites.*

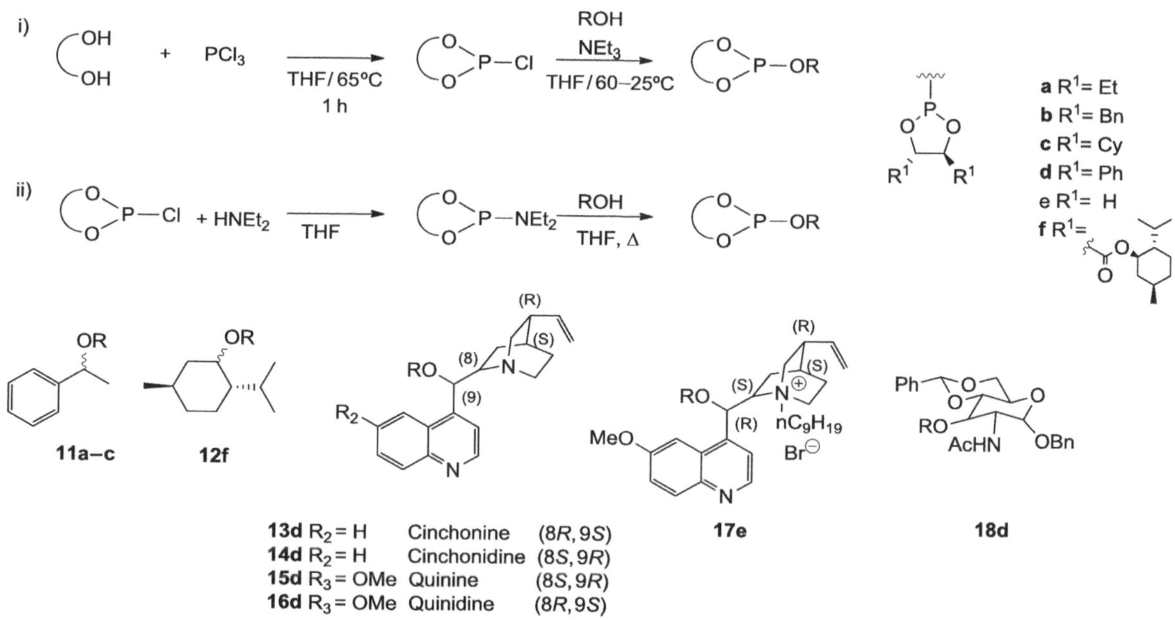

Scheme 3.4 *Monophosphites bearing a 1,3,2-dioxaphospholane unit.*

trimethylene glycol, both of which react with triphenylphosphite **2a** without an added acid catalyst with liberation of two moles of phenol. It is noteworthy that the former reacts at room temperature, while the latter requires heating at 100°C. When aliphatic diols are used, the yields of phosphites containing a phospho-cyclic unit depend on the number of atoms separating the reacting hydroxyl groups. The yield of steadily decreases when the resulting phosphite contains from a five- to an eight-membered ring, and the 9-membered cyclic phosphite is not even formed at all; a similar effect was observed in the analogous reaction of glycols with phosphorus trichloride.

2-chloro-1,3,2-dioxaphospholanes (**a**-Cl – **f**-Cl, Scheme 3.4) are prepared from PCl_3 and dialkyl tartrates, or 1,2-diols (i, Scheme 3.4). [19] The synthesis of 2-chloro-1,3,2-dioxaphospholanes was preferentially performed in refluxing THF using a 7–8-fold excess of PCl_3 in order to ensure selective mono-adduct formation [20]. Alternative procedures employed an equimolar amount of phosphorus reagent and diethyl tartrate in the presence of $PhNEt_2$ [21]. These phosphochloridites were air stable for months if stored at 20°C

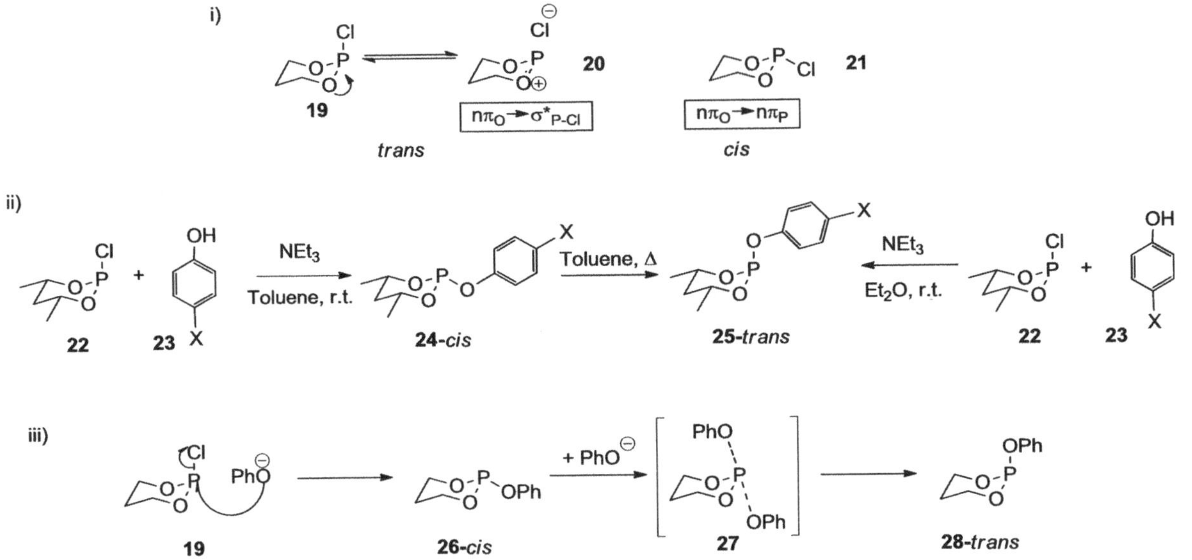

Scheme 3.5 *Synthesis of monophosphites bearing a 1,3,2-dioxaphosphinane unit.*

in a stop-cocked round-bottomed flask [54]. Further reactions of these intermediates with the corresponding alcohols readily yield the phosphite ligands **11–12** in good yields (Scheme 3.4).

Ligands **13d–17d** can be prepared in 'one pot' (Scheme 3.4) from (*S,S*)-1,2-di-phenyl-1,2-ethane diol and cinchona alkaloids. Initially, (*S,S*)-1,2-diphenyl-1,2-ethanediol (hydrobenzoin) was added to a solution of PCl_3 in dry THF to give the phosphochloridite intermediate, which was directly used with cinchona alkaloids in the presence of triethylamine as an acid scavenger. After stirring overnight at 60°C, ligands **13–16d** were obtained [22]. These ligands were shown to successfully induce chirality in Pd-catalysed allylic alkylation processes (enantiomeric excess (ee) up to 94%) [56a]. (*N,N*-diethyl)amido-1,3,2-dioxaphospholanes $P(OR)_2NEt_3$ was used in the synthesis of ligand **17e** (ii, Scheme 3.4) [23].

Phosphochloridite derived from the diols (*S,S*)- and (*R,R*)-hydrobenzoin were also synthesised in pure PCl_3 at 75°C (**d**-Cl, Scheme 3.4) [24]. Ligand **18d** was synthesised by dropwise addition of a dichloromethane solution containing an equimolar amount of alcohol and 4-dimethylaminopyridine (DMAP) to the phosphochloridite previously dissolved in the same solvent, and maintaining the reaction overnight at room temperature [58,25]. This ligand was used in the Ni-catalysed hydrovinylation of styrene and provided moderate ee (up to 30%) [59].

Lucas *et al.* described the synthesis of alkyl phosphites bearing a 1,3,2-dioxaphosphinane unit using glycols and phosphorus trichloride, followed by the addition of an alkoxy group [26]. The *trans*-phosphochloridite **19** was produced from the reaction between 1,3-butanediol and phosphorus trichloride [27]. The *trans* isomer is thermodynamically more stable than the *cis* analogue because of the stabilizing (O–P–Cl) anomeric interactions (i, Scheme 3.5) [28] [29]. The synthesis of both series of aryl phosphites *cis* and *trans* was accomplished through the stereoselective formation of the phosphite intermediates from the phosphochloridite and aryl alcohols as shown in Scheme 3.5 [30].

The addition of phosphochloridite **22**-*trans* to a solution of phenol and triethylamine at room temperature led exclusively to the phenyl phosphite **25**-*trans*. In spite of the fact that the configuration of the phenyl phosphite **25**-*trans* suggests a reaction with retention of configuration, the results were explained via a S_N2P mechanism [31].

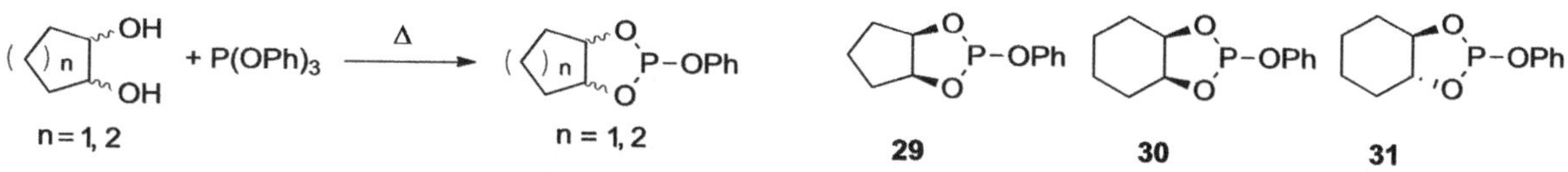

Scheme 3.6 Synthesis of compounds **29**, **30**, and **31** via transesterification.

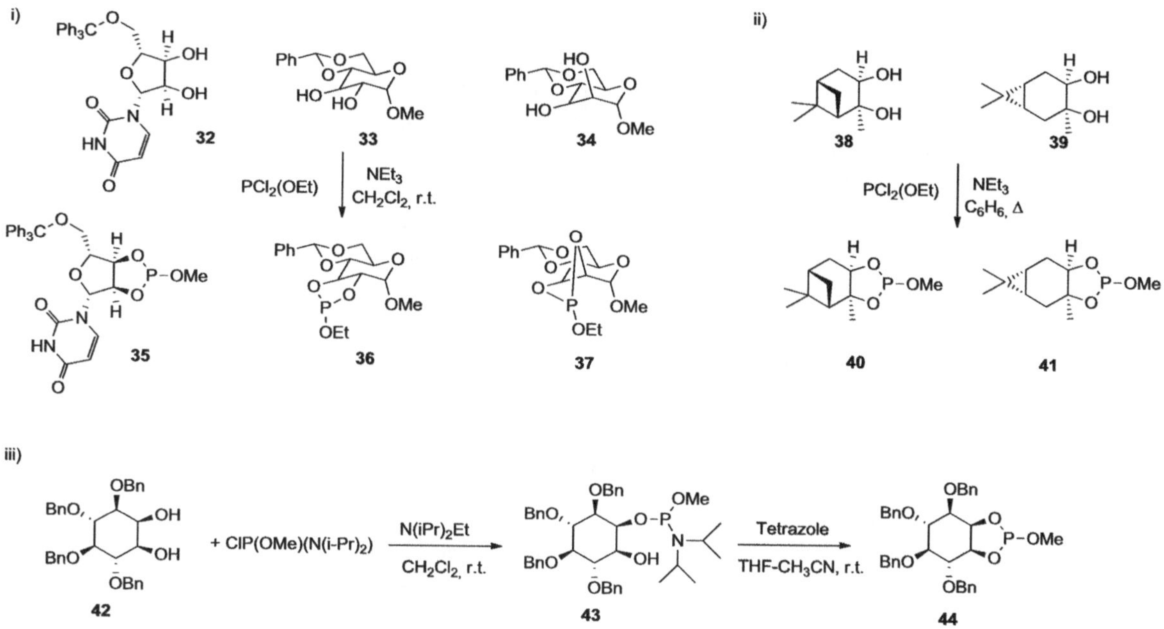

Scheme 3.7 Synthesis of monophosphites derived from natural products.

To avoid acidic medium, the synthesis of **26**-*cis* was performed with sodium phenoxide (generated from phenol and sodium) in anhydrous ethyl ether at room temperature; however, the phenyl phosphite **28**-*trans* was again obtained as a single product (iii, Scheme 3.5) [61]. Further experiments showed that an excess of phenoxide also excludes the formation of the *cis* phosphite via a second attack of the phenoxide ion on phenyl phosphite **26**-*cis* (iii, Scheme 3.5) [61,32].

The synthesis of compounds **29**, **30** and **31** was carried out by a transesterification reaction from $P(OPh)_3$ (Scheme 3.6) [52]. As expected, the reaction is controlled by the stereochemical configurations of the reacting hydroxyl groups. With cyclopentane-1,2-diols, only the *cis* isomer gives a cyclic phosphite (1,3,2-dioxaphospholane). However, both *cis*- and *trans*-cyclohexane-1,2-diol give the corresponding cyclic phosphite, as neither of the two stereoisomeric cyclic phosphites is greatly strained and their formation requires little or no distortion of the cyclohexane ring from its preferred chair conformation.

The phosphites **35**, **36**, and **37** derived from uridine and methyl 4,6-*O*-benzylidene-α-D-gluco- and mannopyranosides, respectively, were synthesised by reacting the corresponding diols **32–34** with $PCl_2(OMe)$ or $PCl_2(OEt)$ in CH_2Cl_2 (i, Scheme 3.7) [33a,b]. The phosphites **40** and **41** were synthesised from diols **38** and **39** using $PCl_2(OEt)$ in refluxing benzene (ii, Scheme 3.7) and used in Pd-catalysed allylic substitutions with various nucleophiles, but provided poor results in terms of conversion and enantioselectivity [34]. D/L-1,4,5,6-tetra-*O*-benzyl-myo-inositol **42** was initially monophosphorylated by

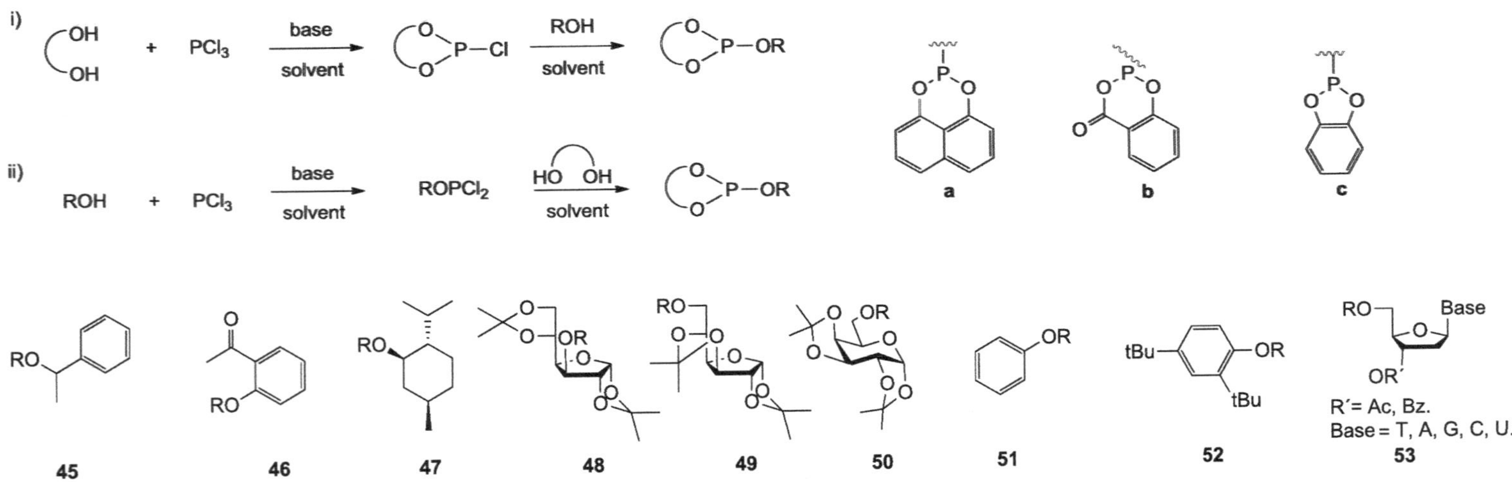

Scheme 3.8 Monophosphites bearing a 1,3,2-dioxaphospholane and phosphinane units derived from phenols and naphthols.

reaction with $ClP(OCH_3)N(iPr)_2$ to give **43**, which was then treated with tetrazole in $THF–CH_3CN$ to produce **44** via intramolecular cyclization (iii, Scheme 3.7) [35].

o-Phenylene phosphochloridite (**c**-Cl, Scheme 3.8) was first prepared by adding PCl_3 to a solution of catechol in diethyl ether, but later, and more conveniently for large-scale synthesis, the reaction was carried out without solvent (i, Scheme 3.8) [36,37]. Compounds **45c–46c** were synthesised by adding *o*-phenylene phosphochloridite dropwise to a cooled (−5°C to 0°C) solution of an equimolar mixture of pyridine (or triethylamine) and the corresponding alcohol in hexane–ether (4:1) [38,39] . Ligand **45c** was tested in the Rh-catalysed hydroformylation of styrene and provided promising results in terms of activity and regioselectivity [72]. Similar procedures were described for the synthesis of phosphites **47c–49c** using triethylamine as a base and benzene (or toluene) as solvent [40].

Phosphochloridite derived from naphthalene-1,8-diol was prepared in a similar manner [41] and was then reacted with phenol to afford the phosphite **51a** (Scheme 3.10a) [75b]. This compound was also prepared by reacting $P(OPh)Cl_2$ with naphthalene-1,8-diol and triethylamine as a base (Scheme 3.8b) [75a]. Ligand **52b** was synthesised by reaction of the commercially available 2-chloro-4H-1,3,2-benzo[*d*][1,3,2]dioxaphosphinin-4-one (**b**-Cl, Scheme 3.8) with 2,4-di-*tert*-butylphenol and triethylamine as base in refluxing toluene (i, Scheme 3.8) [42]. In the presence of a Pd precursor, the orthometallated complex is formed, which was successfully used as precatalysts in Suzuki–Miyaura reactions. The presence of the carbonyl function in backbone **b** was shown to enhance the π-acceptor properties of these phosphite ligands. Compound **53b** was synthesised by a similar procedure from the corresponding partially protected nucleosides using pyridine as a base and DMF as a solvent [43]. This species were used as an intermediate in the synthesis of nucleosides-(α-*p*-borano)triphosphites.

The synthesis of ligands **54a** and **54j** was carried out in two steps. First, a solution of PCl_3 in THF was added to a mixture of biphenol or binaphthol in the presence of triethylamine to form the phosphochloridite [44]. Then, triethylamine and phenol were added to a solution of the phosphochloridite in THF (i, Scheme 3.9). The synthesis of ligands **54a** and **54j** was also described by addition of the corresponding phosphochloridite in THF to a solution of the deprotonated phenol at −78°C using *n*-BuLi as a base [45]. These ligands were successfully used in asymmetric *O*-nitroso aldol reactions of silyl enol ethers with ee up to 94%. Similar procedures were employed for the synthesis of the phosphite **54q**, which was synthesised at room temperature using toluene as solvent and trietylamine as a base [46]. This ligand provided the highest ee reported using monodentate phosphite ligands in the Rh-catalyzed hydroformylation of several mono-substituted alkenes (ee up to 44%).

Phosphites **55c** and **55j** were synthesised using an alternative procedure [47]. A toluene solution of 2,4-di-*tert*-butylphenol was first treated with PCl_3 and later with Et_3N, and the resulting mixture was heated at 90°C for 18 h to give the dichlorophosphite (ii, Scheme 3.9). To a solution of the dichlorophosphite in toluene, the corresponding biphenol or binaphtol and Et_3N were subsequently added. The resulting mixture was stirred at 90°C for 16 h, affording **55c** and **55j**. The complexes **Pd/55** were used as catalysts in the Rh-catalysed conjugated 1,4-addition of boronic acid to α,β-unsaturated ketones and in the addition of tributiltin hydride to benzaldehyde, but poor ee were obtained.

The phosphites **56a**, **56j** and **57i** were prepared by first synthesising the phosphochloridite derived from the diol and further reaction with the corresponding alcohol (i, Scheme 3.9a) [48,49]. Ligand **57i** provided interesting results in Rh-catalysed hydroboration of styrene with ee up to 80%.

Two different routes were considered for the synthesis of cationic phosphites **58–62** (Scheme 3.9). The first method consists in the synthesis of an aminoaryl phosphite and subsequent quaternization of the amino group with an appropriate reagent, and the second includes the preparation of a phenol containing an ammonium group followed by reaction with $(OR)_2P$-Cl [50]. The phosphites **58a**, **60a** and **58q** were synthesised using the first method [84], while the phosphites **59j**, **61j** and **62j** were produced via the second one. In both cases, NEt_3 was used as the base and toluene as the solvent [51]. These ligands were designed for their solubility in ionic liquids and provided good regioselectivity to the linear product in the Ni-catalysed hydrocyanation of 3-pentonitrile, and high enantioselectivity in the hydrogenation of dehydroaminoacids.

The chiral monophosphites derived from (1*R*,2*S*,5*R*)-(−)-menthol **63,a,c,j**, (1*R*,2*R*,3*R*,5*S*)-(−)-isopinocamphenol **64a,c,j**, (−)-borneol **65a,b,j** and (1*R*)-*endo*-(+)-fenchol **66a,b,j** were synthesised following route ii, Scheme 3.9 [52]. First, the dichlorophosphite was formed by reacting PCl_3 with the alcohols (**63–66**, R=H) dissolved in CH_2Cl_2 at room temperature for 2 h. After removal of solvent and the excess of PCl_3, the resulting residue was dissolved in THF and a solution of the biphenol and Et_3N in THF was added. A particular example of C2-symmetric monophosphite bearing a bihelicenol unit is **63s**, which was also prepared by route i, Scheme 3.9 [53,54]. An easy way to produce chiral phosphites is the use of a chiral alcohol such as menthol, which yields diastereomeric mixtures by reaction with the phosphochloridite derived from racemic binols or helicenols, which can be separated by chromatography. In the Rh-catalysed 1,4-addition of boronic acids to enones and the hydrogenation of prochiral olefins, the use of ligands **63–66** combined with an equimolar amount of phosphoramidite provided a high level of enantioselectivity (ee up to 99%). However, when either the phosphite or phosphoramidite ligands alone were used, moderate to low ee were obtained (*ca.* 30%) [86a–d].

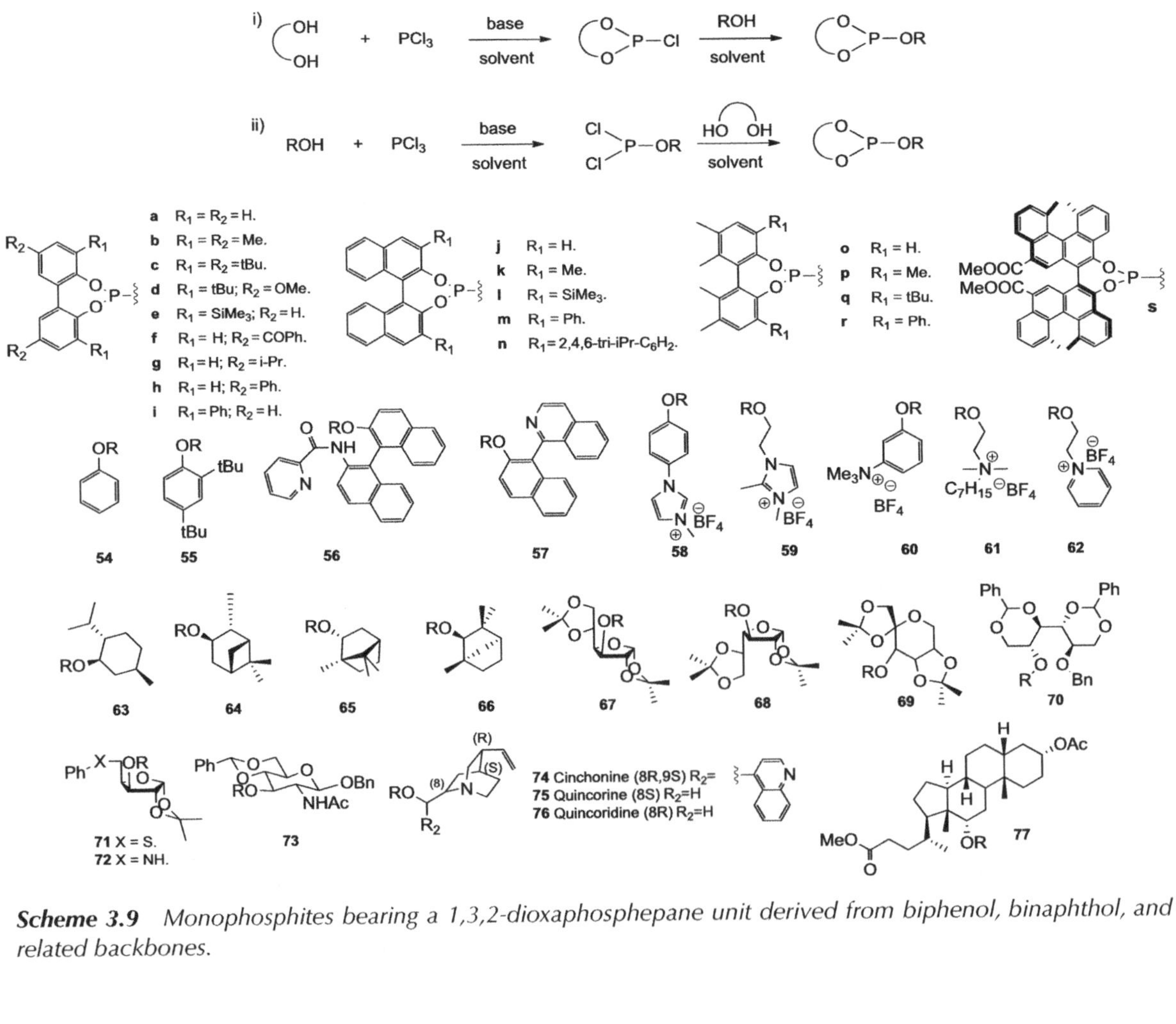

Scheme 3.9 Monophosphites bearing a 1,3,2-dioxaphosphepane unit derived from biphenol, binaphthol, and related backbones.

Carbohydrate-derived monophosphites such as those derived from D-glucose (**67a,c,j**), D-galactose (**68a–e,j**), D-fructose (**69a–e,j**) and D-mannitol (**70a,c,j**) were synthesised by forming first the dichlorophosphite derived from alcohols **67–70** (R=H), and adding afterwards the corresponding biphenol or binaphthol (ii, Scheme 3.9) [42,55a–c,56]. Monophosphites **67a,c** were synthesised by first forming the biphenol or binapthol containing phosphochloridite and further reaction with the 1,2;3,5-di-*O*-isopropyliden-D-glucofuranose using Et_3N as a base and toluene as solvent (i, Scheme 3.9) 89d,e. These ligands were tested in many asymmetric processes such as Rh-catalysed hydrogenation of dehydroamino acids, esters and 1,2- and 1,4-addition of organometallic reagents to aldehydes and enones, respectively, and provided high levels of enantioselectivity [89].

For the synthesis of ligands **71a,c** and **72a,c,d**, the corresponding alcohols were reacted with the *in situ* formed phosphochloridite dissolved in toluene in the presence of pyridine [57]. Ligand **73d** was synthesised by dropwise addition of an equimolar solution of alcohol (**73**, R=H) and DMAP in dichloromethane to a solution of phosphochloridite [58,59]. The resulting mixture was stirred overnight at room temperature.

The phosphite **74a** was synthesised by the addition of dichlorophosphite derived from cichonine, which was synthesised by reacting an excess of PCl_3 with the alcohol at room temperature and using $CHCl_3$ as solvent, to a solution of biphenol and Et_3N in $CHCl_3$ [56b]. Phosphites **75–76j** were obtained by the addition of a solution of quincorine and quincoridine in Et_3N to an ether or toluene solution of phosphochloridite [58a,b].

Further progress in the design of inexpensive and readily accessible ligands is represented by the asymmetric activation of tropos species (Scheme 3.10) [59]. According to this strategy, a configurationally fluxional (tropos) phosphorus ligand can be obtained from an achiral biphenol and a configurationally stable chiral amine or alcohol. The chirality of this substituent favours a prevalent sense of twist on the biphenyl moiety in the phosphite or phosphoramidite species. Coordination of the ligand to a metal centre was also shown to slow down the equilibrium (Scheme 3.10) and thus favour one of the atropoisomeric forms of the biphenol unit. For instance, the cholestanic backbone **77** is able to induce a prevalent sense of twist on the biphenyl moiety of tropos biphenylphosphites linked at the 1,2-position of the deoxycholic acid.

Phosphites **77a,f–j** containing the cholestanic backbone were synthesised following method i, Scheme 3.9 [60] These ligands provided excellent results in the Rh-catalysed asymmetric addition of phenylboronic acid to cyclohexenone and the asymmetric Rh hydrogenation of dimethylitaconate.

An interesting strategy for the synthesis of monophosphites bearing axial chirality such as **84** was described by Rampf *et al.* [61]. Their synthetic approach employs the achiral 2,2-6,6-tetrahydroxybiphenyl (**80**) as starting material. Following a procedure published by Harada *et al.* [62], axial chirality is induced by reacting two of the four hydroxy groups of **80** with the chiral auxiliary **81** to yield the chiral biphenol **82** (i, Scheme 3.11). This strategy avoids the locking of the biphenyl backbone conformation by bulky substituents in *ortho* positions and the separation of the enantiomers. The synthesis of the phosphites was carried out by reacting PCl_3 with the enantiomerically pure diol **82** dissolved in toluene at –78°C and using Et_3N as a base to form the phosphochloridite **83**. Further reaction with isopropanol yielded the phosphite **84**, which was successfully used in the asymmetric Rh-catalysed hydrogenation of dehydroaminoacid esters and enamines.

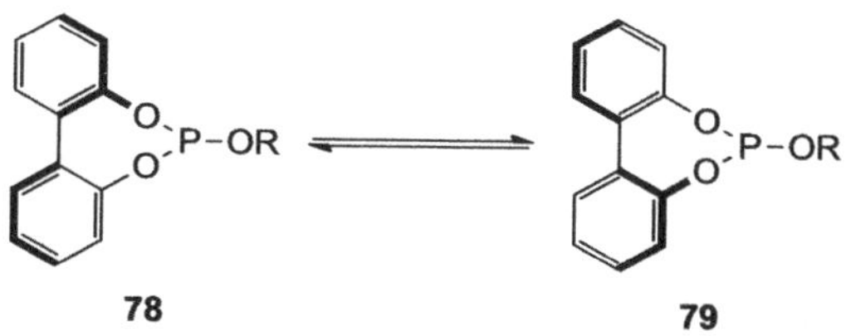

Scheme 3.10 *Atropoisomerism equilibrium in chiral phosphite ligands based on a biphenol unit.*

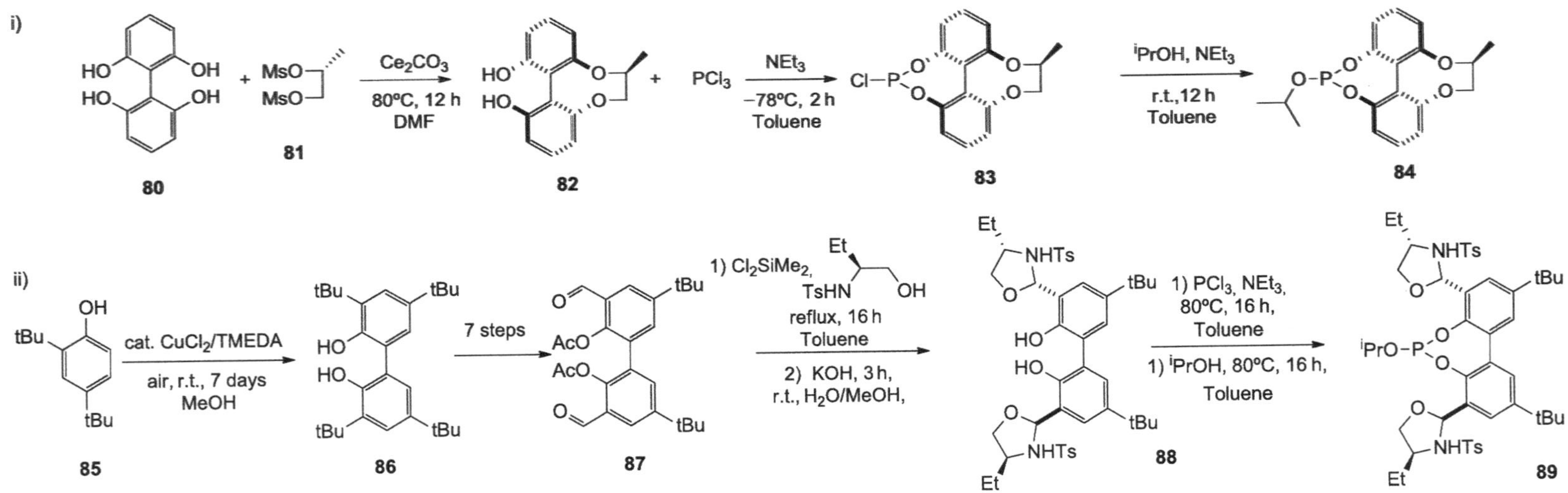

Scheme 3.11 *Synthesis of chiral monophosphites bearing a 1,3,2-dioxaphosphepane unit by using chiral auxiliaries.*

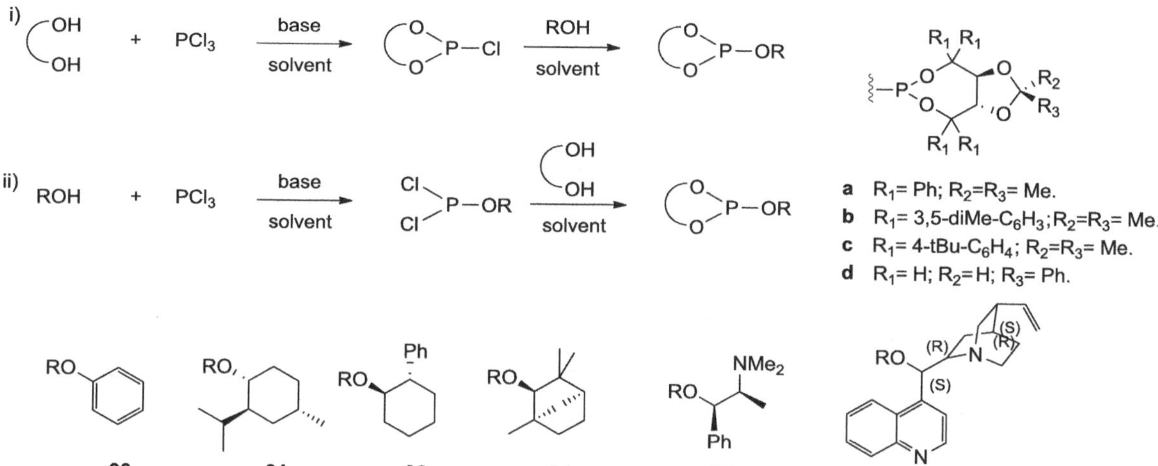

Scheme 3.12 *Monophosphites bearing a 1,3,2-dioxaphosphepane unit derived from TADDOL.*

Hoppe *et al.* [63] reported the synthesis of the phosphite **89** bearing axial chirality using aminoalcohols. The synthetic route started with the coupling of two 2,4-di-tertbutyl-phenol **85** to form the biphenol **86** (ii, Scheme 3.11) [64]. After seven steps, the biphenol **87** containing two aldehyde functions in *ortho* positions was obtained. This compound reacts with N-arenesulfonylamino alcohols in the presence of dichlorodimethylsilane in refluxing toluene to give bis-1,3-oxazolidines **88** in excellent yields after hydrolysis. Finally, the phosphite **89** was obtained by reaction of **88** with PCl_3 to give the phosphochloridite and further reaction with an alcohol ROH. These ligands were successfully used in the asymmetric Cu-catalysed 1,4-addition of diethylzinc to cyclohexenone with ee up to 83%.

TADDOL and analogues were prepared by the addition of an excess of aryl Grignard reagent to the corresponding acetals or ketals of tartrate esters [65]. The synthesis of the monophosphites **90a–c** was carried out in two steps (i, Scheme 3.12). First, phosphochloridite was formed by adding PCl_3 to a cooled solution (−78°C) of TADDOL-derived diol and triethylamine in THF and maintaining the reaction at room temperature for 12 h [66]. After filtration and removal of volatiles, the residue was dissolved in THF and the resulting solution added to a mixture of phenol and Et_3N in THF and stirred for 12 h. These ligands were successfully applied in the amine-directed Rh-catalysed asymmetric hydroboration of trisubstituted alkenes providing high levels of enantioselectivity (ee up to 97%).

The monophosphites **90–94a** were prepared by reaction of the di-lithium salt of TADDOL in THF at −70°C with the neat dichlorophosphites derived from the corresponding alcohols (**90–94**, R=H) (ii, Scheme 3.12) [67]. These ligands were tested in the Pd-allylic alkylation of 1,3-diphenylallyl acetate with dimethyl malonate and in the Cu-catalysed 1,4-addition of Et_2Zn to enones providing ee up to 80% and 96%, respectively. The monophosphite **95d** was synthesised by similar procedures using Et_3N as a base and $CHCl_3$ as solvent [56].

The synthesis of the monophosphite **96a** was carried out by addition of Et_3N to a solution of 2,2′-methylenebis(4-methyl-6-tert-butylphenyl)chlorophosphite and 2,4-di-tertbutylphenol in THF (i, Scheme 3.13) [68]. This bulky ligand forms orthopalladated complexes which provided high activity in the Suzuki and Stille coupling reactions.

Ligands **100a,b** were synthesised in 'one pot' by treating PCl_3 in toluene at 0°C with the corresponding diol in the presence of Et_3N overnight, followed by slow addition of the corresponding alcohol (i, Scheme 3.13) [89e]. The synthesis of the monophosphites **97–99c** was carried out in two steps by method i (Scheme 3.9): first, reaction of *R*-1,10-spirobiindane-7,70-diol with PCl_3 to form the phosphochloridite, and further reaction

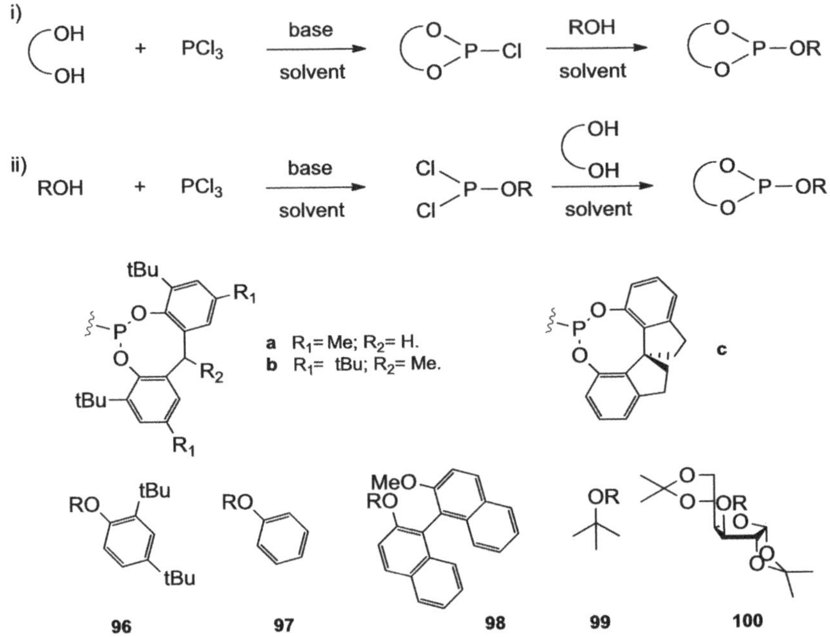

Scheme 3.13 Monophosphites bearing a 1,3,2-dioxaphosphocyclooctane unit.

with lithium phenolate [69]. Ligands **97–99c** provided excellent asymmetric induction in the Pd-catalysed hydrovinylation of vinylarenes with ethene with ee up to 92%.

Ligand **97c** was alternatively synthesised by addition of $(PhO)PCl_2$ to a solution of (*S*)-1,1′-spirobiindane-7,7′-diol in CH_2Cl_2 in the presence of diisopropylethylamine at 0°C (ii, Scheme 3.13) [70]. This ligand was successfully applied in the Rh-catalysed 1,2-addition of arylboronic acids to aldehydes with ee up to 87%.

Kamer and co-workers reported the solid-phase synthesis of monophosphite ligands based on the synthesis of supported phosphoramidites (Scheme 3.14) [71]. Thus, phosphoramidites linked to the solid support via amido functionality (**102**) were prepared by reaction of prolinamide-modified polymer **101** with a phosphochloridite. The acid-catalysed substitution of the prolinamide moiety by alcohols affords monophosphites liberated from the resin. This route allows the combinatorial solid-phase synthesis of phosphites. In the absence of an acid activator, no reaction with methanol was observed for any of the phosphoramidites. Activators like 1H-tetrazole, N-methyl-imidazolinium trifluoroacetate and pyridiniumtetrafluoroborate were tested but proved unsatisfactory, as conversion to product was low or incomplete and the prolonged reaction times promoted side product formation via oxidation and hydrolysis. *N*-methylanilinium trifluoroacetate (TAMA) was found to be the most suitable activator, resulting in the fast formation of products **103–107** in good yields.

3.2.2 Diphosphite ligands

3.2.2.1 *Diphosphites not containing a dioxaphospho-cyclic unit*

Similarly to monophosphites of type **B**, in which two identical substituents are present, scarce attention has been devoted to diphosphites of type **E** (Figure 3.1). These diphosphites can be prepared by reaction of a phosphochloridite, $(RO)_2PCl$, with a diol. The diol backbone bridges the two phosphite moieties and

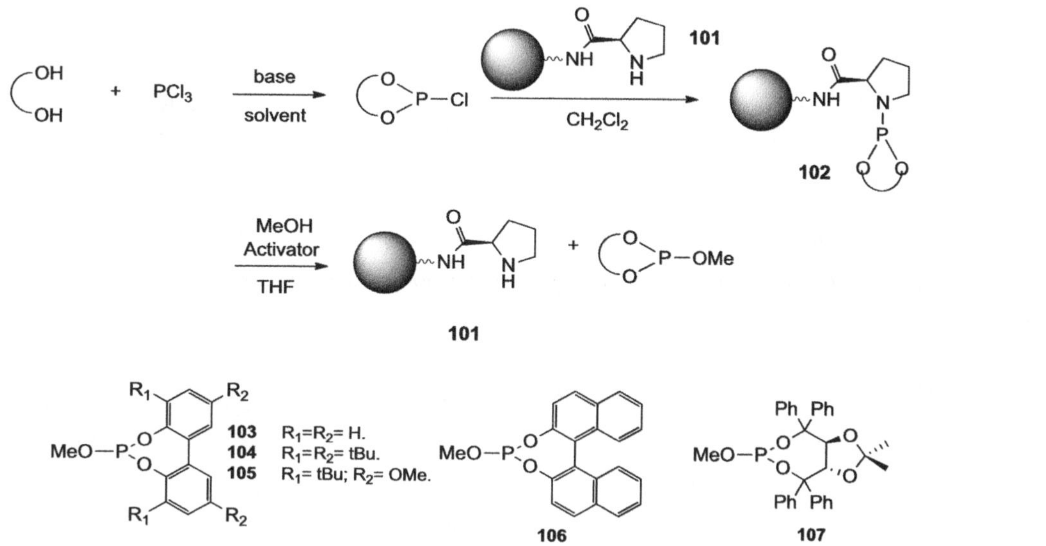

Scheme 3.14 Solid-phase synthesis of phosphoramidites as intermediates in the synthesis of phosphites (the balls represent the polystyrene support (DVB-PS), and the waved lines represent poly(ethylene glycol) linkers).

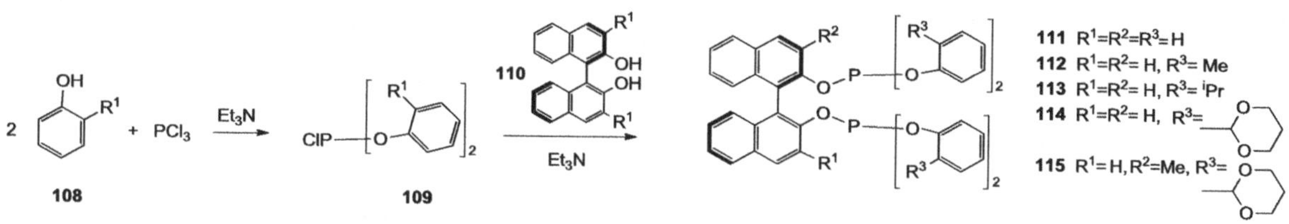

*Scheme 3.15 Synthesis of the diphosphite ligands **111–115** not containing a dioxaphospho-cyclic unit.*

determines the bite angle of the resulting ligand. Ligands **111–115** were prepared by reaction of two moles of phenol or *ortho*-substituted phenols with 1 mol of PCl_3 in the presence of Et_3N, and subsequent treatment with 0.5 mol of (*R*)-(1,1′)-binaphthalenyl-2,2′-diol. Ligands **111–115** were used in the Ni-catalysed hydrocyanation of alkenes with HCN and provided 86% ee for 1,3-cyclohexadiene with ligand **115** [72].

Ligands **116a,b** and **117a,b** (Figure 3.4) are related to ligands **111–115** shown in Scheme 3.15, and were prepared following a similar procedure. These ligands were used in the Ni-catalysed hydrocyanation of butadiene to produce adiponitrile [73].

3.2.2.2 Diphosphites bearing dioxaphospho-cyclic units

As previously mentioned, diphosphites bearing two dioxaphospho-cyclic units (ligands of type **F**, Figure 3.1) are the widest family of diphosphite ligands reported in the literature. In this case, the synthesis consists of the reaction of two equivalents of 2-chloro1,3,2-dioxaphospho-cyclic (phosphochorhidite) unit with one equivalent of the corresponding diol. These 2-chloro1,3,2-dioxaphospho-cyclic fragments, combined with the large variety of diols available from natural or artificial sources, have provided a huge number of diphosphites of type **F**.

Most diphosphites contain a chiral moiety either in their diol backbone or in the dioxaphospho-cyclic unit, but there are also many examples where both moieties are chiral, at times having cooperative effects in catalytic processes.

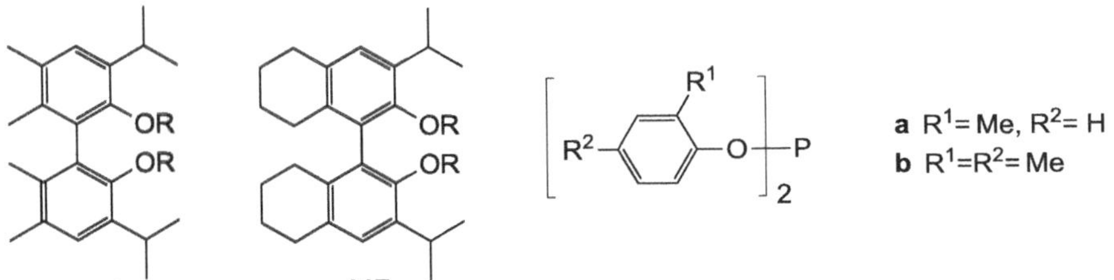

Figure 3.4 Relevant examples of diphosphites not containing a dioxaphospho-cyclic unit.

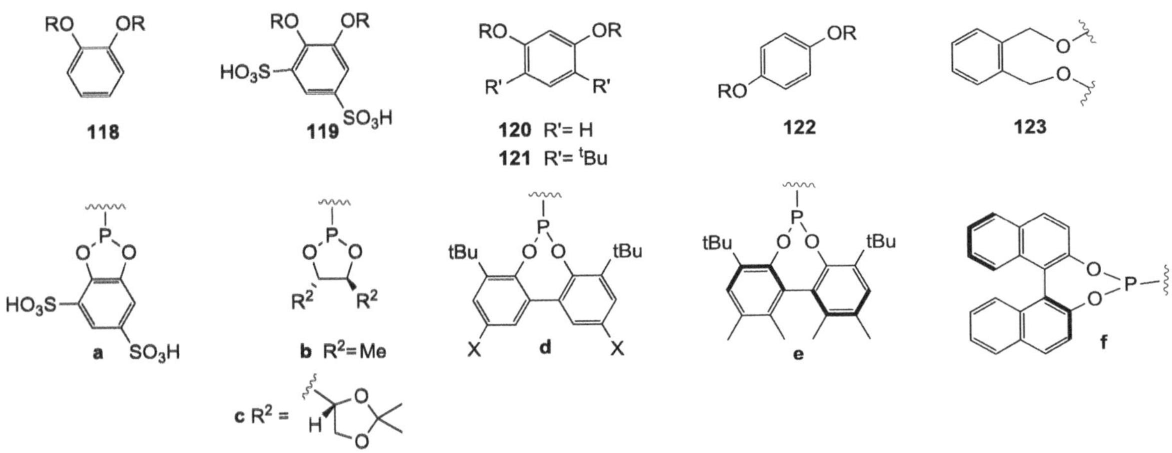

Figure 3.5 Diphosphites with a benzene backbone.

The diphosphite **118f** was prepared from phosphochloridite **f** (R=H) and tested in the rhodium-catalysed styrene hydroformylation with moderate results [74]. Polysulphonated phosphites such as **119b** and related were prepared from triphenylphosphite **2a** by reaction with disulfonated catechol and similar sulfonated phenols [75]. The diphosphites **120** and **121** (Figure 3.5) contain a resorcinol backbone and were prepared to investigate their coordination modes to transition metals. Thus, upon coordination, the phosphite **120 g** forms rhodium(I) dimers having a 'cis-A-frame' structure [76], while ligands **120e,f** [77] and **121c,d,g** [78] give pincer complexes due to the presence of an acidic proton at the 2 position. These ligands **120e,f,g** and **121g** were prepared from the corresponding phosphochloridite (Scheme 3.16i), while **121c,d** were prepared from the phosphodichloridite (ii, Scheme 3.16).

Diphosphites containing a 'pseudo-crown ether' in the diol backbone such as **126d** [79] or in the phosphite moiety such as **124b** [80] were prepared to study potential interactions with cations and to investigate the effects of secondary interactions on the selectivity of catalytic processes (Figure 3.6). Diphosphites containing chiral backbones such as **127–131** (Figure 3.6) are commonly combined with achiral dioxaphospho-cyclic units such as **d–f** or with the chiral **c** and **h** (Figure 3.6). Ligands **127c–129c** were prepared by reaction of the diols (**127**, **129**, R=H) with the corresponding 2-chloro-dioxaphospho-cyclic derivatives (phosphochloridites) [81].

The diphosphites **130e–g** [82] **131f** and **131j** [83] (Figure 3.6) were prepared by reaction of the spirodiols with the corresponding phosphochloridites (Scheme 3.16i). In the case of ligands **131**, the starting diol was

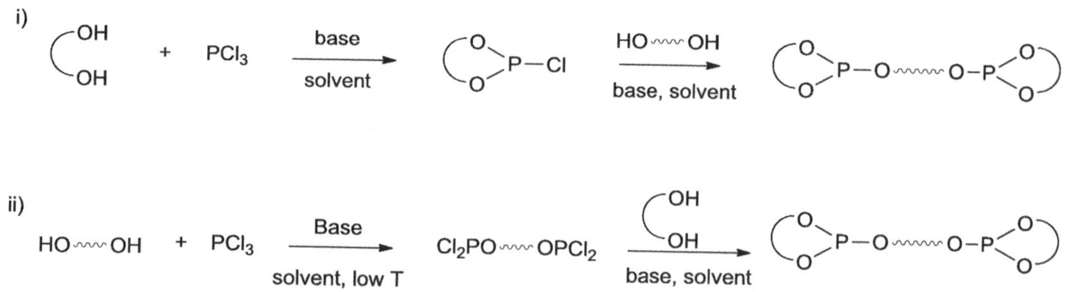

Scheme 3.16 General methods for the synthesis of diphosphite ligands bearing a dioxaphospho-cyclic unit.

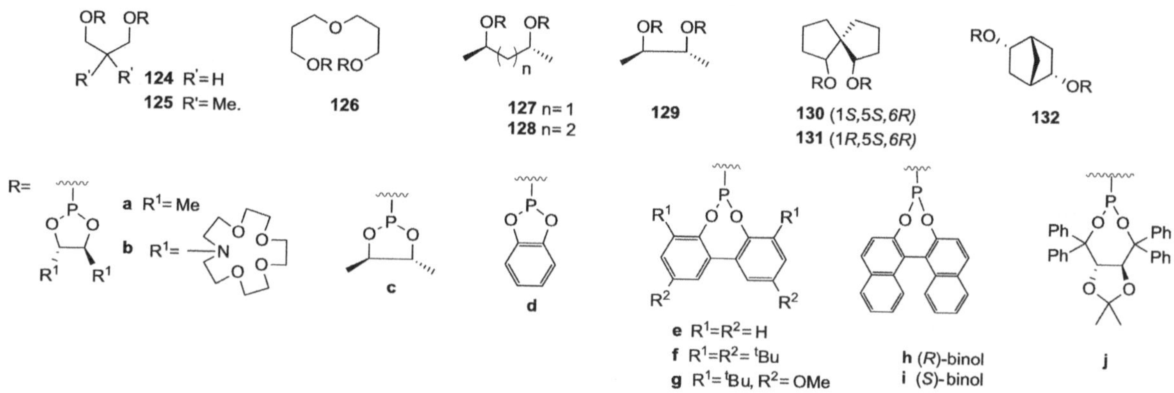

Figure 3.6 Diphosphites with an alkylic backbone.

enantioselectively obtained by asymmetric reduction of the spiro[4,4]nonane-1,6-dione using a chiral oxazaborolidine as catalyst. These ligands were tested in the Rh-catalysed hydroformylation of styrene and the use of ligand **130f** provided up to 69% ee, while **131f** provided lower enantioselectivity. However, in both cases the major isomer exhibited the *S* configuration.

The diphosphites **132e,h,i** (Figure 3.6) were obtained from the diol **132** (R=H) (i, Scheme 3.16), which in turn was obtained from norbornene by hydroxylation, ketone formation and enantioselective reduction of diketone. Ligand **132e** (Figure 3.6) provided low ee in the hydrogenation of enamidoesters and dimethyl itaconate. However, when ligands **132h,i** (Figure 3.6) were employed, enantioselectivities higher than 99% were achieved. The enantioselectivity was shown to be governed by the chirality of the binaphthol unit.

Carbohydrates are chiral polyhydroxylated compounds that are ubiquitous in nature. The combination of different configurations together with the large number of protecting groups available provide a huge number of potential structures that are appropriate for the synthesis of chiral ligands. In addition, the presence in these compounds of hydroxyl groups facilitates the synthesis of oxygenated phosphorus functions such as phosphinites, phosphonites and particularly phosphites [84]. Pentoses and hexoses are the most frequently used carbohydrates that can be converted into furanosides (five-membered rings) or pyranosides (six-membered rings) by the appropriate choice of synthetic pathways. As such, both pentoses and hexoses react with acetone in acidic medium to form 1,2-*O*-isopropylidene-furanosides. Furthermore, hexoses also react with methanol and provide methyl pyranosides. In these reactions, one issue is the control of the configuration at the anomeric position.

In Figure 3.7, the most relevant diphosphite ligands derived from carbohydrates are described. Ligand **133a**, derived from 1,2:5,6-di-*O*-cyclohexylidene-D-mannitol, was the first sugar-based diphosphite ligand reported and tested in asymmetric hydrogenation [85]. The related cyclohexylidene derivative was also

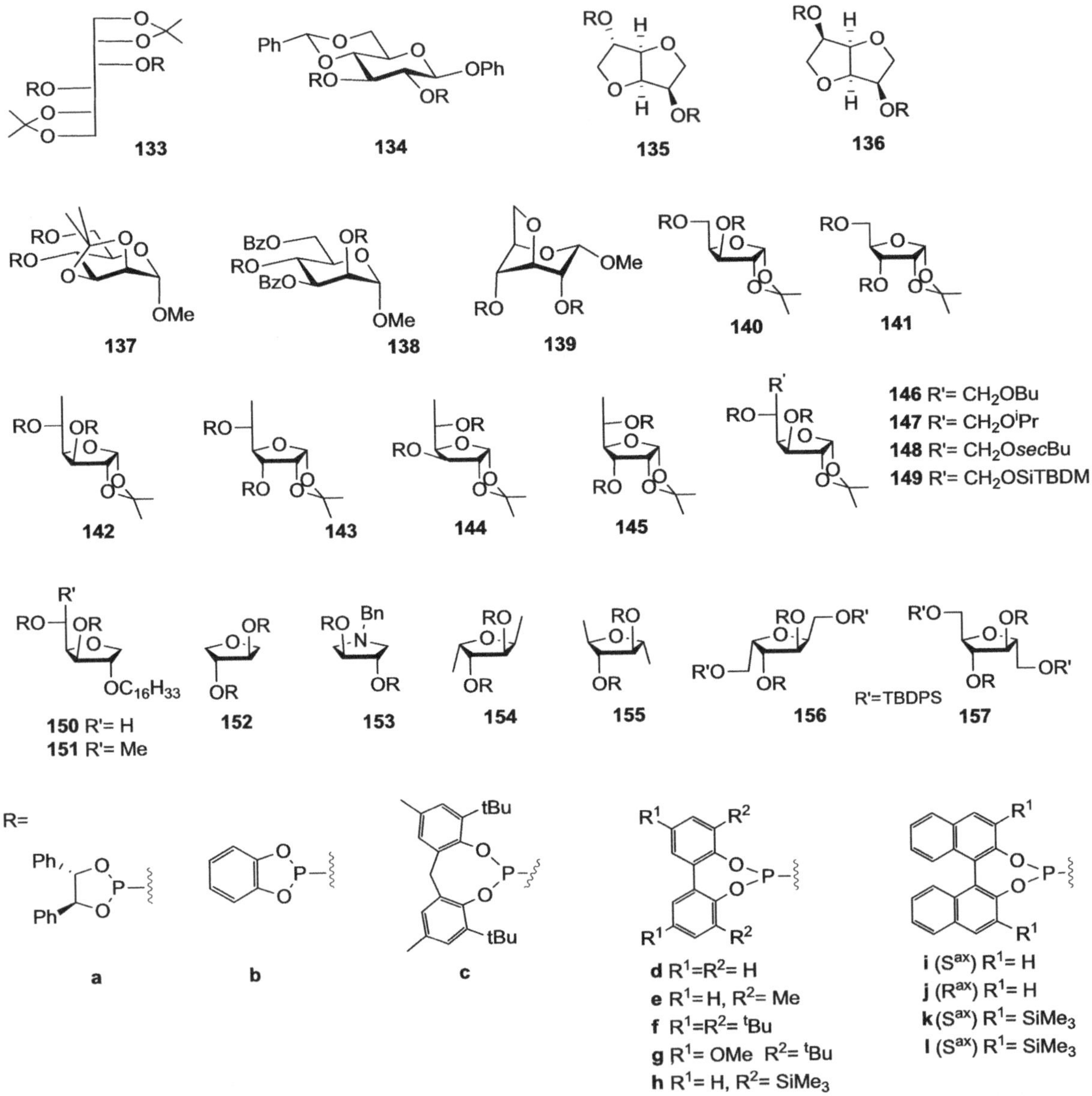

Figure 3.7 *Relevant diphosphites containing a sugar backbone.*

synthesised and tested in the diethyl zinc addition to enones [86]. Phosphite ligands with a pyranoside backbone such as **134** were prepared [55b], although the corresponding phosphinites have been more widely studied in catalytic processes such as Ni-catalysed hydrocyanation. The chiral diols for the backbone of ligands **135** and **136** are the commercially available 1,4:3,6-dianhydro-D-glucitol and -mannitol, respectively. The diphosphites **135f** and **136f** [87] were tested in the Pd-catalysed allylic alkylation, and the diphosphites **136c,d,e,f,i,j** [88] in asymmetric Rh-catalysed hydrogenation of dimethylitaconate and *N*-acetamidoacrylate. Significantly, ligand **136e** with the conformationally flexible biphenol unit provided the highest ee (up to 98%).

The pyranosides **137f,g** were prepared from methyl 2,3-*O*-isopropylidene-α-D-mannopyranoside by reaction with the corresponding phosphochloridites (**f**-Cl, **g**-Cl, Figure 3.7) providing ee up to 64% in the Rh-catalysed hydroformylation of styrene. In this reaction, the disphosphites **138g** and **139g**, prepared in a similar manner, afforded very low enantioselectivities [89].

Diphosphites containing a furanoside backbone have been much more investigated. The first diphosphites of this type that were reported were ligands **140d–g**, which were prepared from the commercially available 1,2-*O*-isopropylidene-α-D-xylofuranoside by reaction with phosphochloridites (**d**-Cl, **g**-Cl) [124]. This strategy was followed for synthesising the different diphosphites with a sugar backbone (commented in this chapter). Ligands **140d–g** were tested in the Rh-catalysed asymmetric hydroformylation of styrene providing ee up to 62%, and in the Cu-catalysed 1,4-addition of diethylzinc to cyclohexenone affording ee up to 41%. Ligands **141d–g**, with a configuration of ribose, can also be obtained from the *xylo* derivative by inversion of the configuration at position C3. A significant improvement of this type of ligands was the introduction of an additional stereogenic centre at C5, as in ligands **142d–g** which were obtained from 1,2-*O*-isopropylidene-glucofuranose and deoxygenation of position C6. These ligands provided an increase in enantioselectivity in the Rh-catalysed hydroformylation of vinylarenes (ee up to 93% with **142g**) [90].

Ligands **143–145** were obtained by systematically inverting the configuration at positions C3 and C5 of the carbohydrate backbone. These modifications, combined with various binol structures, led to the production of a library of new diphosphite ligands. Although the binol units **d–h** are not chiral, they usually adopt a preferred atropoisomeric form, which is controlled by the neighbouring stereogenic centres within the ligand (Scheme 3.10). Ligands containing these units are called 'adaptive ligands'. Remarkably, ligand **145g** behaves as a pseudoenantiomer affording 89% ee in Rh-catalysed hydroformylation of styrene. Furthermore, ligand **142g**, which differs from **145g** only by the configuration of the C3 and C5 atoms of the carbohydrate, yielded the opposite enantiomer with comparable ee value. Ligands **143** and **144** provided the lowest ee of this ligand series. These results were explained by the presence of several Rh hydride species [$RhH(CO)_2$(P-P)] under catalytic conditions, where these ligands are coordinated to the Rh centre of the resting state of the catalytic cycle in eq-eq and eq-ax manners [125c].

Ligands **140** and **142** were used in various asymmetric processes. Ligands **140d,f** were used in the Ir-catalysed asymmetric hydrogenation of benchmark imines (up to 57% ee) [91]. Ligands **140–145** were tested in various processes such as Cu-catalysed conjugate 1,4-addition to cyclohexanone (up to 84% ee, with **142k**) [92], Rh-catalysed hydroformylation of dihydrofurans (up to 75% ee, ligand **142g**) [93], Rh-catalysed hydrosilylation of ketones (up to 45% ee, ligand **140f**), Rh-catalysed hydrogenation of substrates such as dimethyl itaconate, (*N*)-acetylamidoacrylate and (*Z*)-(*N*)-acetylaminocynnamate (up to 99% ee, ligand **145h**) [94] and Pd-catalysed allylic substitutions of allyl acetates providing ee up to 99% [95].

Recently, some modifications were carried out to improve the enantioselectivity of several processes. The diol precursors were modified to incorporate a bulky group at position C6 to afford ligands **146–149**, which can be prepared in one step from 1,2:5,6-di-*O*-isopropylidene-D-glucofuranose or from 1,2-*O*-isopropylidene-D-glucofuranose. Ligands **150** and **151**, from which the 1,2-*O*-isopropylidene group has been removed, were prepared to test the greater flexibility of the structure. Ligands **146–148** and **150–151** provided slightly lower ee than that obtained with ligand **142** in the Rh-catalysed hydroformylation of styrene. From this series, the best results were obtained with ligand **151g** (83% ee) [96]. However, when these ligands were applied in the Rh-catalysed hydroformylation of dihydrofurans, ee up to 88% was achieved for 2,5-dihydrofuran using ligand **151d** and 84% ee was obtained with 2,3-dihydrofuran using ligand **147d** [97]. Some of these diphosphite ligands were used as stabilisers for Pd, Ir, Ru and Rh nanoparticles, which were subsequently applied in catalysis. The resulting Pd nanoparticles stabilised by ligand **140d** provided excellent ee and kinetic resolution in allylic alkylation of 3-acetoxy-1,3-diphenyl-propene [98]. Furthermore, the Ru, Rh and Ir nanoparticles stabilised with ligands **140**, **150** and **151** were used in the hydrogenation of arenes. Interestingly, ligands **150** and **151**, which contain a long alkylic chain, provide smaller and thus more active nanoparticles than those stabilised by ligand **140** [99].

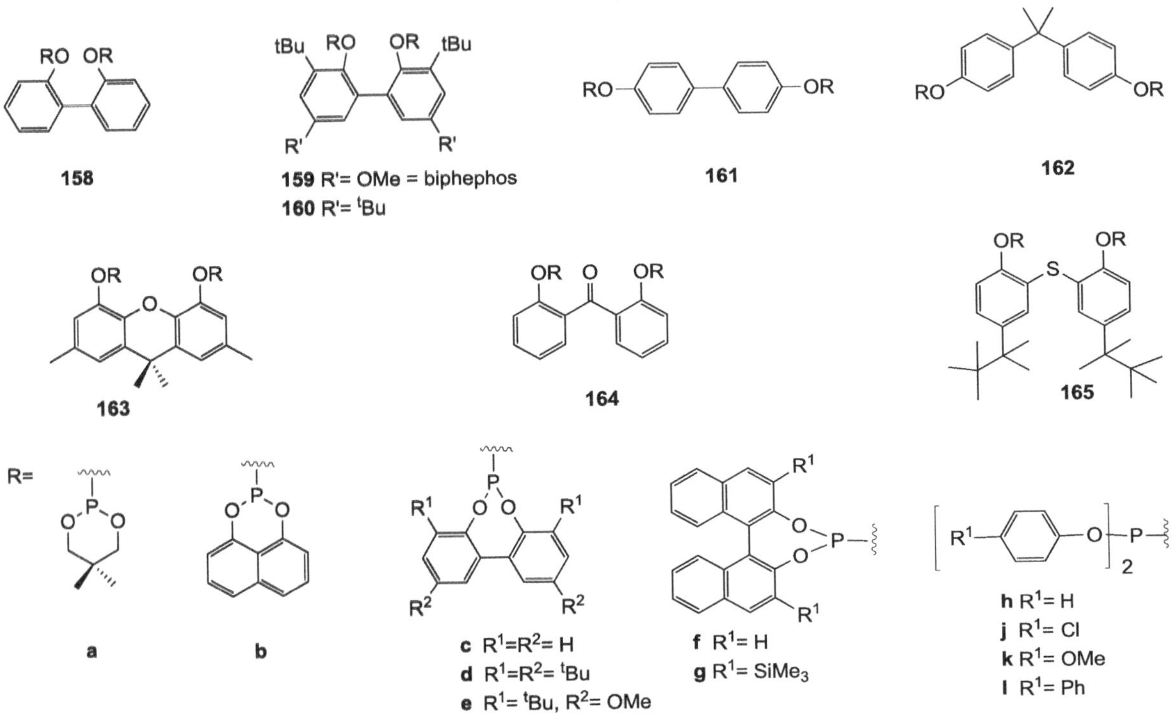

Figure 3.8 Diphosphites with a biphenol backbone.

Most of the sugar-based diphosphite ligands described above have C1 symmetry. C2-symmetrical diphosphite libraries can be prepared from D-mannitol. As an alternative, the C2 symmetry ligand **157**, which contains a tetrahydrofuran backbone, can be readily prepared from glucosamine by activation of the amino group as a diazonium salt and subsequent ring contraction reaction to generate a 1-formyl derivative. After reduction of the formyl group, the tetrol **157** (R=R′=H) is obtained and can be selectively protected at the primary alcohols, while the secondary alcohols can be derived to form the corresponding phosphite (or other phosphorus functions) [100]. The tetrol precursor of the ligand **156** can be obtained in modest yield from D-glucitol. Ligands **156** and **157** have identical configuration for the secondary alcohols and opposite configuration for the alkoxymethylene groups. For comparative purposes, ligands **152** and **154–155** were also prepared. Closely related ligands such as **153** were prepared from tartaric acid [122]. The diphosphites **152, 154–157** were used in the Rh-catalysed hydroformylation of styrene derivatives affording moderate ee (up to 60% with ligand **157f**) [101], and reduction of oxosteroids with very high activity and diastereoselectivity [102]. Recently, the catalytic system Pd/**157f** showed unprecedented activities in the Pd-catalysed allylic alkylation and amination, providing TOFs up to 22000 h^{-1} and 400 h^{-1}, respectively, and ee up to 99%. This catalytic system was active even at substrate–Pd ratios of 10000:1, providing an efficient kinetic resolution under these conditions [103].

Since the discovery of ligand **159c** (biphephos, Figure 3.8) in 1987, diphosphite ligands with biphenol units have attracted much interest [104]. This ligand provides a very high linear-to-branched ratio in the Rh-catalysed hydroformylation of alkenes [105]. This behaviour was explained by the steric hindrance introduced at the *ortho* position of the backbone. The related ligands with structures **158d**, **159c** and **160c** (Figure 3.8), together with the mixed phosphite ligands **160** (R=**c**, R=**h–l**), were prepared and tested in the Rh-catalysed hydroformylation of 1-octene. The catalysts bearing the rigid diphosphites **159, 160** (Figure 3.8) were shown to provide high activity and selectivity to the linear products. The high regioselectivity was

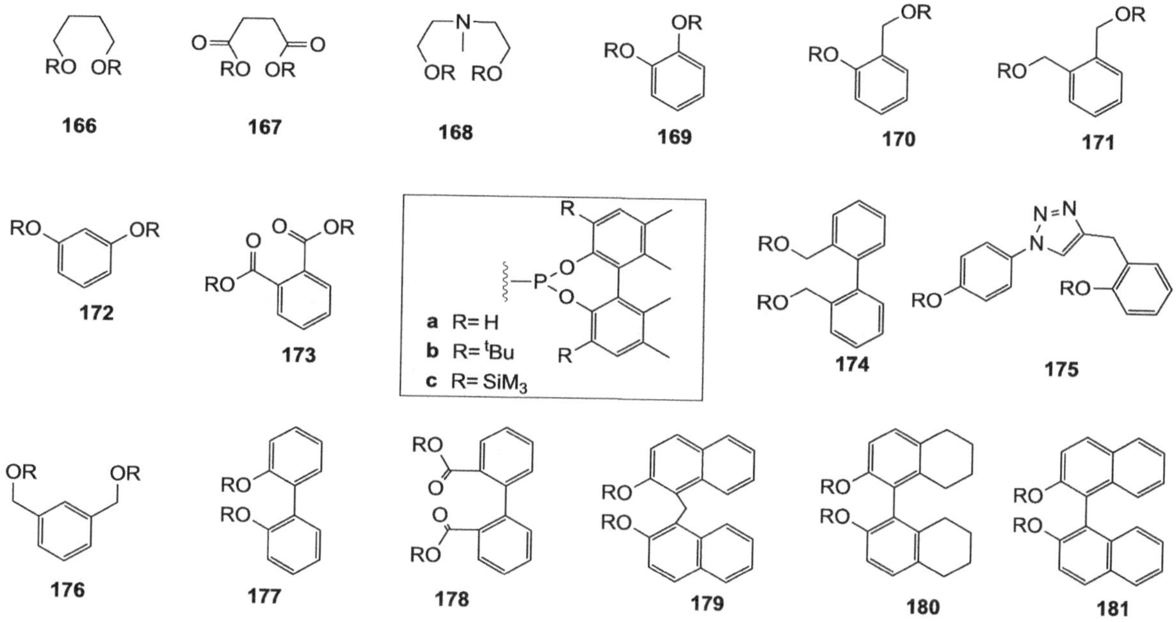

Figure 3.9 Diphosphite analogues of kelliphite.

explained by means of steric repulsions between the diphosphites coordinated to the rhodium centre in a bisequatorial fashion and the approaching substrate [106].

A series of diphosphite ligands **163c,d** (Figure 3.8) related to Xantphos, a diphosphine successfully used in homogeneous catalysis and containing a 9*H*-xhantene skeleton, was prepared from the diol by reaction with the corresponding phosphochloridite (**c**-Cl, **d**-Cl) [107]. In this case, ligands **163c,d** (Figure 3.8) were shown to exclusively coordinate to the Rh centre in axial-equatorial mode in the $[RhH(CO)_2(P–P)]$ complexes, the resting state in the hydroformylation process. The synthesis of the related ligands **161a** and **162f** (Figure 3.8) was also described [108,109].

Most of the diphosphite ligands used in asymmetric catalysis contain an optically active backbone. A different approach in the synthesis of chiral ligands consists in the introduction of a chiral $(RO)_2P$ phosphite moiety and achiral diol backbone. The diphosphite ligands **166–181** (Figure 3.9) were prepared from the commercially available chiral (*S*)-3,3′-di.*tert*-butyl-5,5′,6,6′-tetramethyl-biphenyl-2,2′-diol (BIPHEN-H_2) and an achiral diol. In some cases (**180**, **181**, R=H), a chiral diol was used [80,110]. These diphosphite ligands can be prepared by reaction of (*S*)-(BIPHEN)PBr with the corresponding diols in the presence of Et_3N (1.1 equiv.) as a base in toluene at room temperature. The use of phosphorobromidite, or phosphoroiodite, allows the synthesis to be performed under milder conditions. These ligands were tested in the Rh-catalysed hydroformylation of allyl cyanide and vinylacetate. The best results were achieved using ligand **177b** with up to 79% ee for allyl cyanide and 88% ee for vinyl acetate. Similar ligands containing a binaphthol as chiral moiety, or with a chiral diol bridge, did not improve the enantioselectivity [111]. A new family of diphosphites was obtained by linking two monophosphite units by a *v*-triazol unit **175b** [112].

Despite the facts that monophophosphites and diphosphites with binol units have been largely explored, and that diphosphites with a binol backbone were reported quite early [113], ligands **182–188** (Scheme 3.17) have been applied relatively infrequently as ligands in asymmetric catalysis.

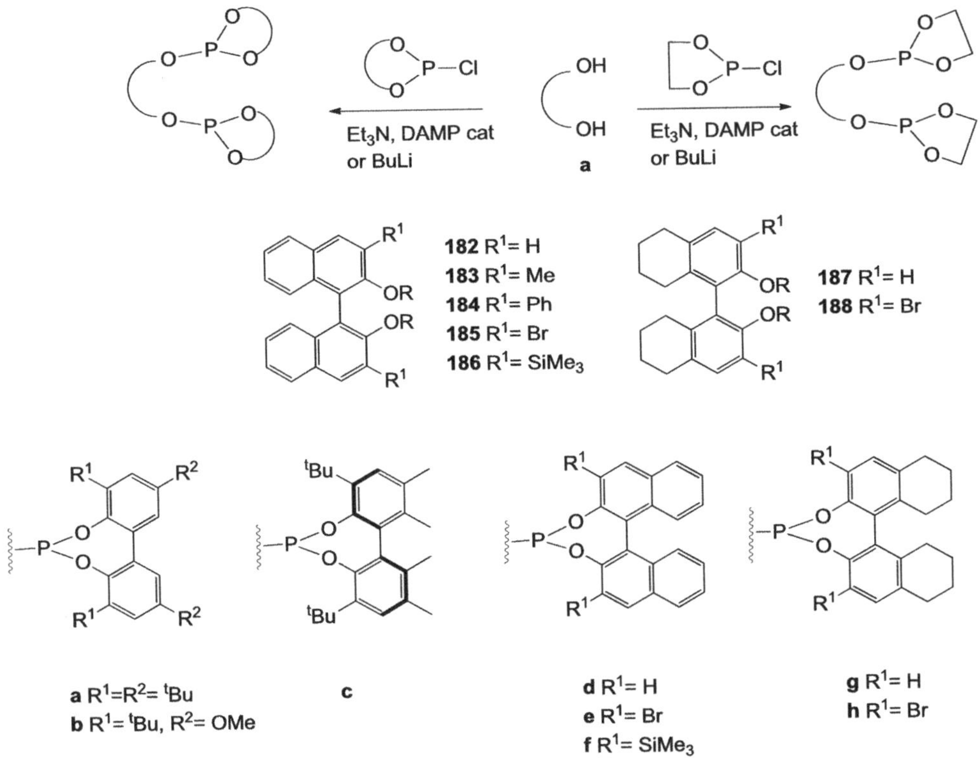

Scheme 3.17 *Diphosphites containing a binaphthol backbone.*

The synthesis of these ligands starts from a binaphthol that reacts with PCl_3, affording a phosphochloridite which reacts another binaphthol unit in the presence of Et_3N (or previous treatment with BuLi to form the lithium salt). The resulting ligands thus contain three identical binol units in their structure. To obtain mixed ligands, a binaphthol must be reacted in the presence of a phosphochloridite bearing a biphenol or another binol unit.

A Ni(0) complex bearing the diphosphite (*S,S,S,*)-**182a** (Scheme 3.17) was used in the hydrocyanation of norbornene with acetone cyanohydrine to give the corresponding product with up to 38% ee [148]. $Cu(OTf)_2$/(*S,S,S,*)-**182a** was tested in the conjugate 1,4-addition of diethylzinc to enones, and enantioselectivities up to 92% were achieved [114a,b]. The ligands (*S,R,S,*)-**182g**, (*S,S,S,*)-**182g** and (*S,R,S,*)-**187g** were also tested in the same reaction using $AlMe_3$ as nucleophile, and ee up to 96% was achieved [149c]. All these ligands were mainly tested in the rhodium-catalysed asymmetric hydroformylation of various substrates such as styrene, vinyl acetate and allyl cyanide [108,146,115,116]. Furthermore, the regioselective hydroformylation of the terminal carbon in alkenes such as 1-pentene, 2-pentene and 1-octene was also reported [117,118].

The catalytic system Co/(*S,S,S*)-**182c** was tested in the Pauson–Khand reaction to obtain cyclopentanones with ee up to 75% [119]. The conjugate addition of arylboronic acids to dehydroalanine derivatives catalysed by Rh/(*R,R,R*)-**182c** afforded unnatural amino acid esters with up to 72% ee [120]. The ligand **182d** and related phosphepines were tested in the Rh-catalysed hydrogenation of α-acetamidocinamate, achieving up to 98% ee [121]. The ligand **182i** was also tested in the Pd-catalysed allylic amination of 3-acetoxy-1,3-diphenylpropene (81% ee) and in the Rh-catalysed hydrogenation of dimethylitaconate with poor ee [122].

Pyrophosphites (Scheme 3.18) represent an interesting class of organophosphorus compounds that have been scarcely used as ligands in transition metal-catalysed processes [123a]. The binol derivatives **189** and

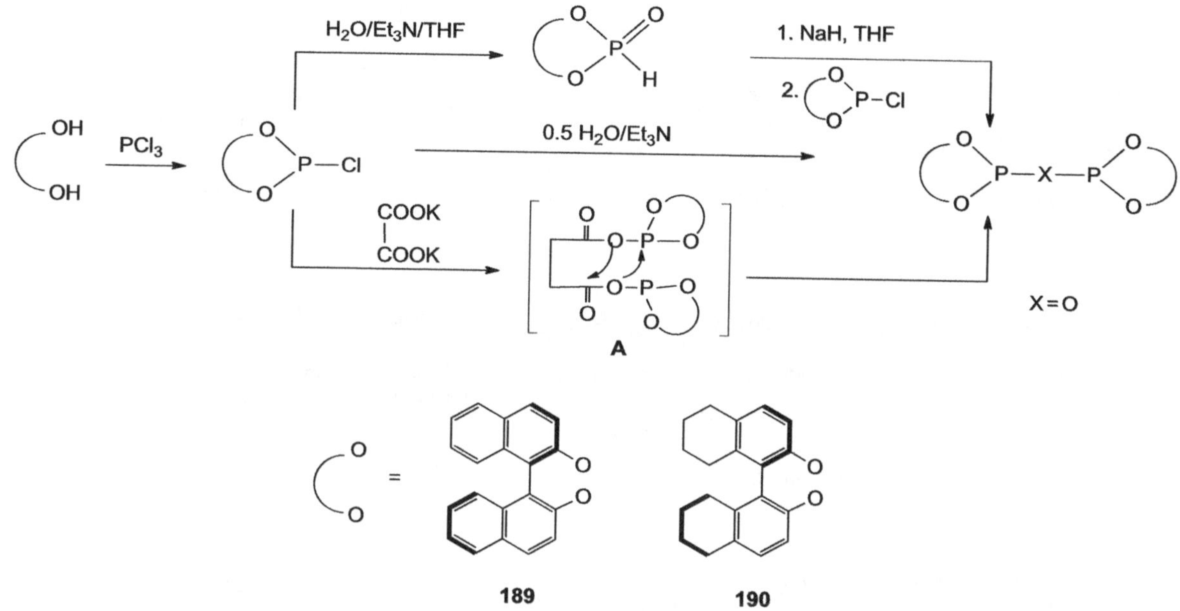

Scheme 3.18 *Synthetic methods for the synthesis of pyrophosphites.*

190 were prepared from phosphochloridite by various procedures (Scheme 3.18). The most direct method consists in treating the phosphochloridite with 0.5 equivalent of water in the presence of tributyl amine. These compounds can be purified by flash chromatography and are air stable. The binaphthol derivatives **189** and **190** were prepared in *ca.* 75% yield, but the reaction was unsuccessful when bulky substituents were present at the *ortho* positions [158b].

Otherwise, pyrophosphites can also be prepared from phosphochloridites by reaction with potassium oxalate (Scheme 3.18), and the reaction is believed to evolve through intermediate **A**. Alternatively, the reaction of the phosphochloridite with a stoichiometric amount of water affords the phosphonate, which can be treated with sodium hydride to afford the pyrophosphate [158c]. These ligands were used in Rh-catalysed hydrogenation of dimethylitaconate, and methyl *N*-acetylaminoacrilates with moderate ee, and in Pd-catalysed allylic substitutions (ee up to 57%). Analogues containing a carbon or nitrogen bridge have also been reported.

The diphosphites **193** and **196** are structurally distinct from those previously presented, since in this case both phosphorus atoms are bonded to a bicyclic or tricyclic moiety (Scheme 3.19). The diphosphite **193** was prepared from the pentaerythritol **191** by slow addition of PCl_3 to form the bis-phosphorochoridite **192** [124]. By slowly adding PCl_3, the reaction temperature can be maintained below 50°C. 2,4-di-*tert*-butylphenol was then slowly added, maintaining the temperature at 60–80°C to obtain **193** in good yield after removal of the volatiles under vacuum. The diphosphite **196** was obtained from terephthalic acid by initial silylation to yield **195**. Further reaction with $EtOPCl_2$ yields the product **196** [125].

The carboranyl phosphites (Scheme 3.20) **201** and **203** were prepared in high yields (92–96%) by one-step phosphorylation of the corresponding *ortho*-**198** and *meta*-carboranediols **200** with binaphthylphosphochloridite in the presence of triethylamine. The phosphochloridite **199** was synthesised from the diol **198**, and further reaction with binol afforded the carboranyl diphosphite **202** [126]. These ligands provided moderate enantioselectivity in Rh-catalysed hydrogenation of dimethylitaconate.

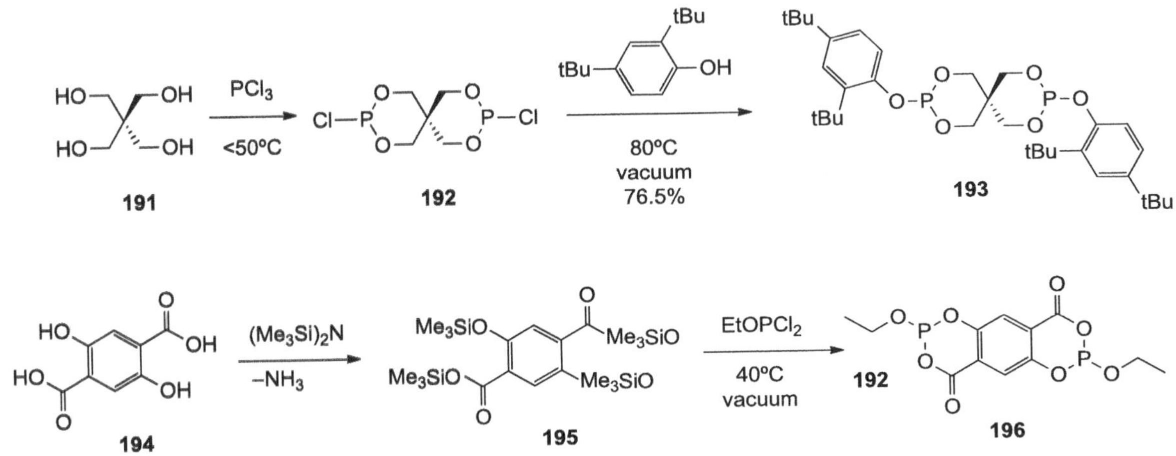

Scheme 3.19 *Bicyclic and tricyclic diphosphites.*

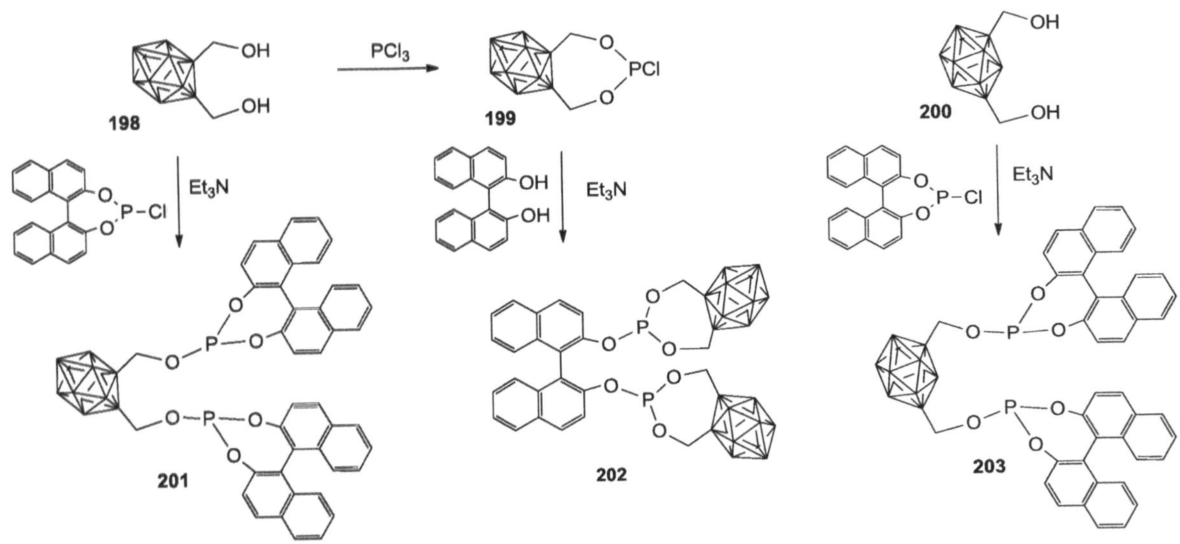

Scheme 3.20 *Syntheses of carboranylphosphites.*

The structure of most diphosphites previously presented consists in a diol backbone bonded to two identical 1,3,2-dioxaphospho-cyclic units. There is also a category of diphosphites where the diol backbone is bonded to two distinct phophocyclic or phosphoacyclic units (Scheme 3.21). Many combinations of diol backbones and different phosphocyclic moieties can be considered, although diphosphites containing two biphenol units are the most commonly studied. In their synthesis, a biphenol such as **204** or **205** reacts with an equimolar amount of their corresponding phosphochloridites **206**, **207** to provide the monophosphites **208**, **209**. The reaction of these phosphites containing a free hydroxyl group with a distinct phosphochloridite unit affords the diphosphites **210b,c–211a–g** [127].

The related diphosphites **219–221** were prepared in a slightly distinct way since the monophosphites **212–214** were reacted with PCl_3 to give the dichloro derivatives **215–217**. Further reaction with the diol **218**

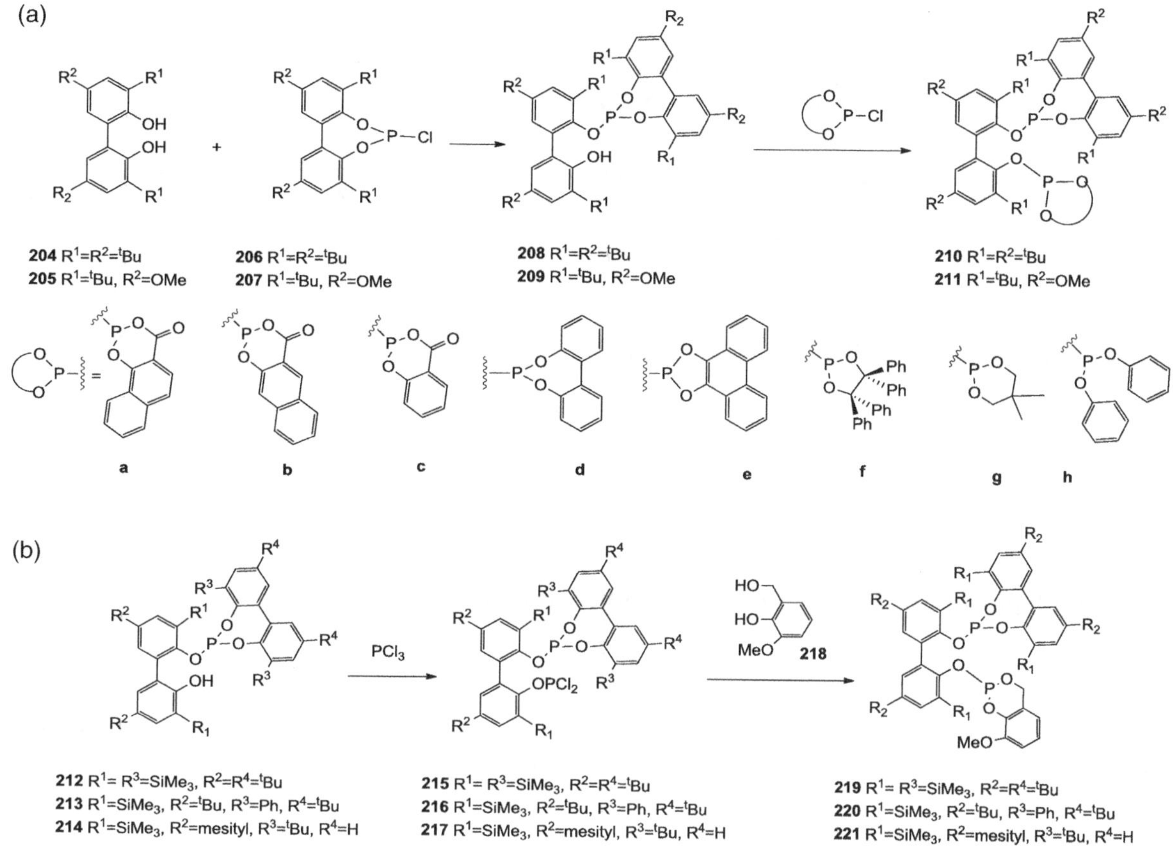

Scheme 3.21 *Diphosphites bearing distinct substitutions at the phosphorus.*

afforded the diphosphites **219–221** [128]. All these unsymmetrical diphosphites were tested in the Rh-catalysed hydroformylation of terminal and internal alkenes, providing moderate to good selectivity to the lineal aldehydes.

3.2.3 Triphosphites

Triphosphite ligands have been much less studied than mono- and diphosphites, although the first report was published in 1962. In 1984, Volcko *et al.* reported the triphosphites **222a** and **222b** [50b]. Recently, the triphosphites **222a**, **223a** and **224a** were synthesised starting from triols and the corresponding phosphochloridite (**a**-Cl) (Figure 3.10) [129]. The triphosphite ligands **225c**, **226c** (*S,S,S*)-TRISPHOS were prepared by reaction of phosphochloridite (c-Cl) with triethanolamine and (*S,S,S*)-triisopropanolamine, respectively, in the presence of triethylamine. Ligand **224c** was used in the Rh-catalysed hydrosilylation of acetophenone, achieving 81% ee [130].

Tetraphosphorus derivatives with the general structure **227**, where each phosphorus is bonded to two pyrrol units **227d**, were recently described [131]. Similar tetraphosphite derivatives could be prepared by reacting the tetrol with distinct phosphochloridites [132].

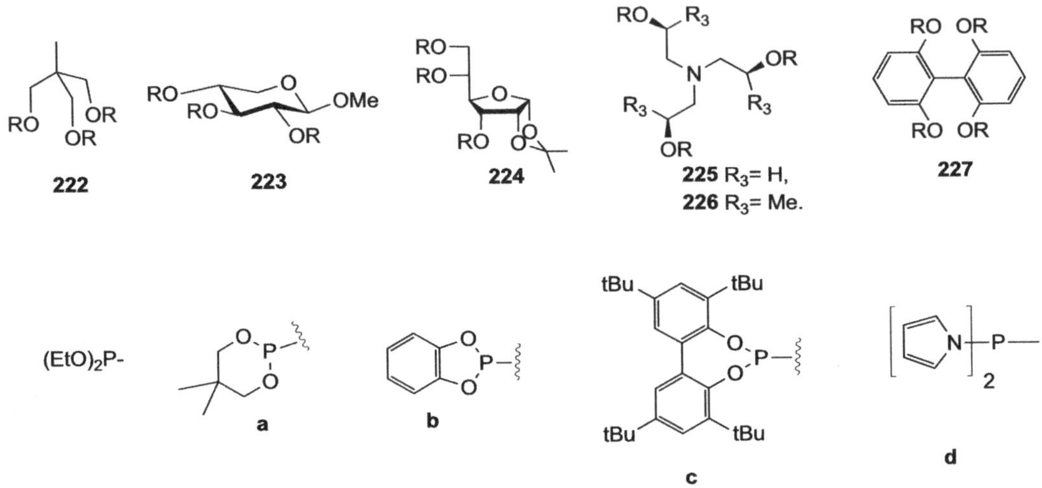

Figure 3.10 *Tri- and tetraphosphites bearing cyclic moiety.*

3.3 Highlights of catalytic applications of phosphite ligands

To achieve the highest levels of reactivity and selectivity in catalysis, several reaction parameters must be optimised, the most crucial of which is perhaps the design of the chiral ligand [1]. Phosphite ligands are extremely attractive for catalysis as they are easy to prepare from alcohols and highly tunable. In this section, the most relevant applications in catalysis of phosphite ligands are described.

3.3.1 Hydrogenation reactions

The development of effective catalysts for hydrogenation began with the concept of modifying the Wilkinson catalyst by replacing triphenylphosphine by more sophisticated ligands. Nowadays, the asymmetric hydrogenation of prochiral compounds catalysed by chiral transition-metal complexes is intensively studied to design more efficient catalytic systems for stereoselective organic synthesis. Several of them were successfully applied in industrial processes [1]. Although significant progress has been made in this area in recent years, asymmetric hydrogenations remain a challenge for many of these substrates. To date, diphosphines are the ligands of choice for the metal-catalysed hydrogenation of alkenes, but phosphite ligands recently received considerable attention due to their potential in asymmetric induction.

For the hydrogenation of functionalised olefins, Rh complexes containing phosphorus and nitrogen chiral ligands have proved to be the best catalysts. Excellent activities and enantioselectivities were achieved in recent decades for the asymmetric hydrogenation of dehydroamino acids and other functionalised alkenes (Scheme 3.22). The first successful results in the Rh-asymmetric hydrogenation of pro-chiral alkenes were reported by Reetz and co-workers [123]. They described the synthesis of the diphosphite **136i–j**, bearing binaphthol moieties, and their application in the hydrogenation of **S1** and **S3** with ee up to 89% and 98%, respectively. The same authors also reported the first successful application of the monophosphite ligands **223i–j**, related to **136i–j**, in these reactions [79b]. The initial success of monophosphite ligands bearing binaphthol moieties in the Rh-catalysed hydrogenation of prochiral alkenes encouraged several authors to perform studies on the synthesis of this type of ligands and apply them in asymmetric catalysis. To date, monophosphite ligands derived from commercially available nonchiral (**224, 226, 228**) or chiral alcohols (**67, 70, 225, 227, 229–230**) were applied with success in these reactions [86e,89c,90a,133]. Recently, several examples of polymer (**231–233**) [134] and ionic liquids (**58–62**) [85] which incorporated phosphite functions

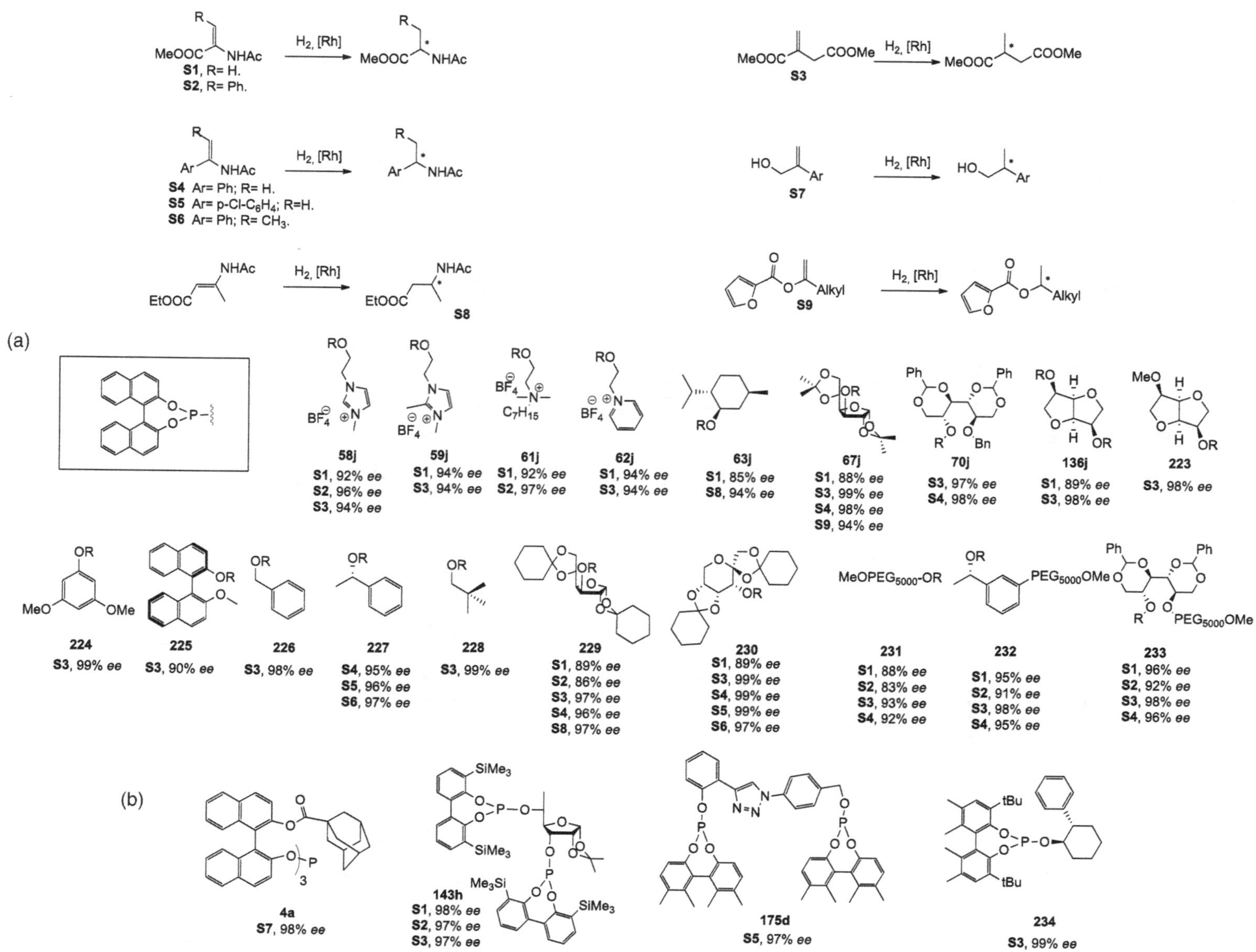

Scheme 3.22 *Rh-catalysed asymmetric hydrogenation of pro-chiral alkenes: (a) with ligands containing dinaphtho[1,3,2]-dioxaphosphepine units; and (b) other ligands.*

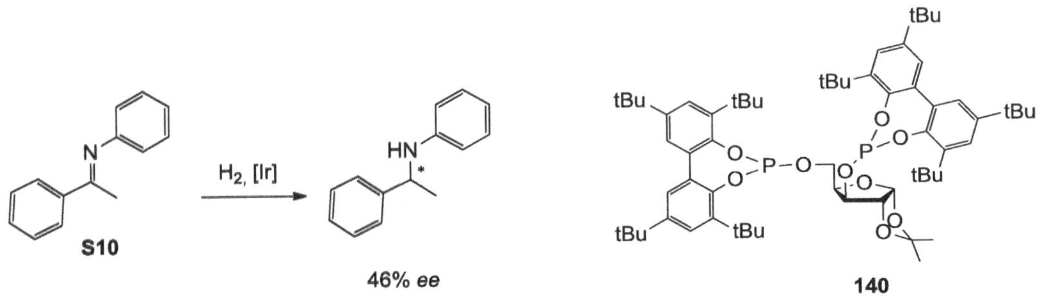

Scheme 3.23 Ir-catalysed asymmetric hydrogenation of imines.

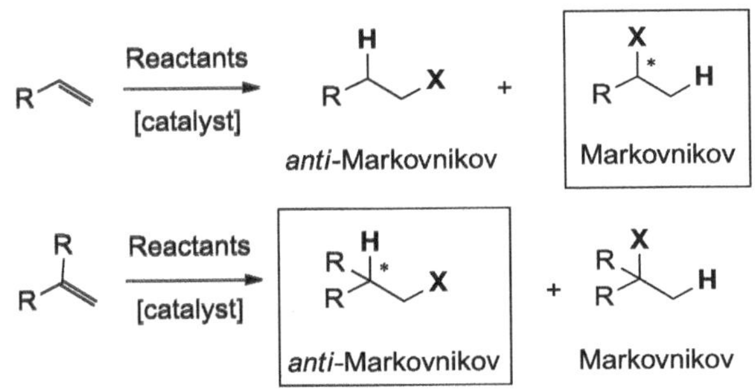

Scheme 3.24 Functionalisation of monosubstituted and 1,1-disubstituted alkenes: hydroformylation (X = CHO) and hydrocyanation (X = CN).

in their structure were reported (Scheme 3.22). These strategies allow the recovery and reuse of the catalyst. Concerning other types of phosphites, several reports described the application of ligands bearing nonchiral (**143**) [129,135] or chiral (**175** and **234**) bisphenol moieties [147,136] and the C3-symmetrical monophosphite ligand **4** (Scheme 3.22) in hydrogenation reactions [41].

In the Ir-catalysed asymmetric hydrogenation of imines, phosphites such as ligand **140** produced moderate to low enantioselectivities in this reaction (Scheme 3.23) [126].

The ability of diphosphite ligands to stabilise metal (Rh-, Ru- and Ir-) nanoparticles for their use as nanocatalysts in the hydrogenation of disubstituted arenes was probed using 1,3-diphosphite ligands such as **140** [134]. These nanocatalysts were active in the hydrogenation of o- and m-methylanisole, producing excellent *cis*-selectivity (*cis*-selectivity up to 100% and 82%, respectively), although no asymmetric induction was achieved.

3.3.2 Functionalization of alkenes: hydroformylation and hydrocyanation

The functionalization of alkenes to produce aldehydes (hydroformylation) or nitriles (hydrocyanation) has attracted much attention due to the high versatility of these compounds in organic synthesis. For substrates other than ethene, the regioselectivity becomes an important issue due to the formation of nonchiral compounds (Scheme 3.24).

3.3.2.1 *Hydroformylation*

The hydroformylation of alkenes is nowadays one of the most important industrial applications of homogeneous catalysis [3,137,138]. Today, over 9 million tons of so-called oxo-products are produced per

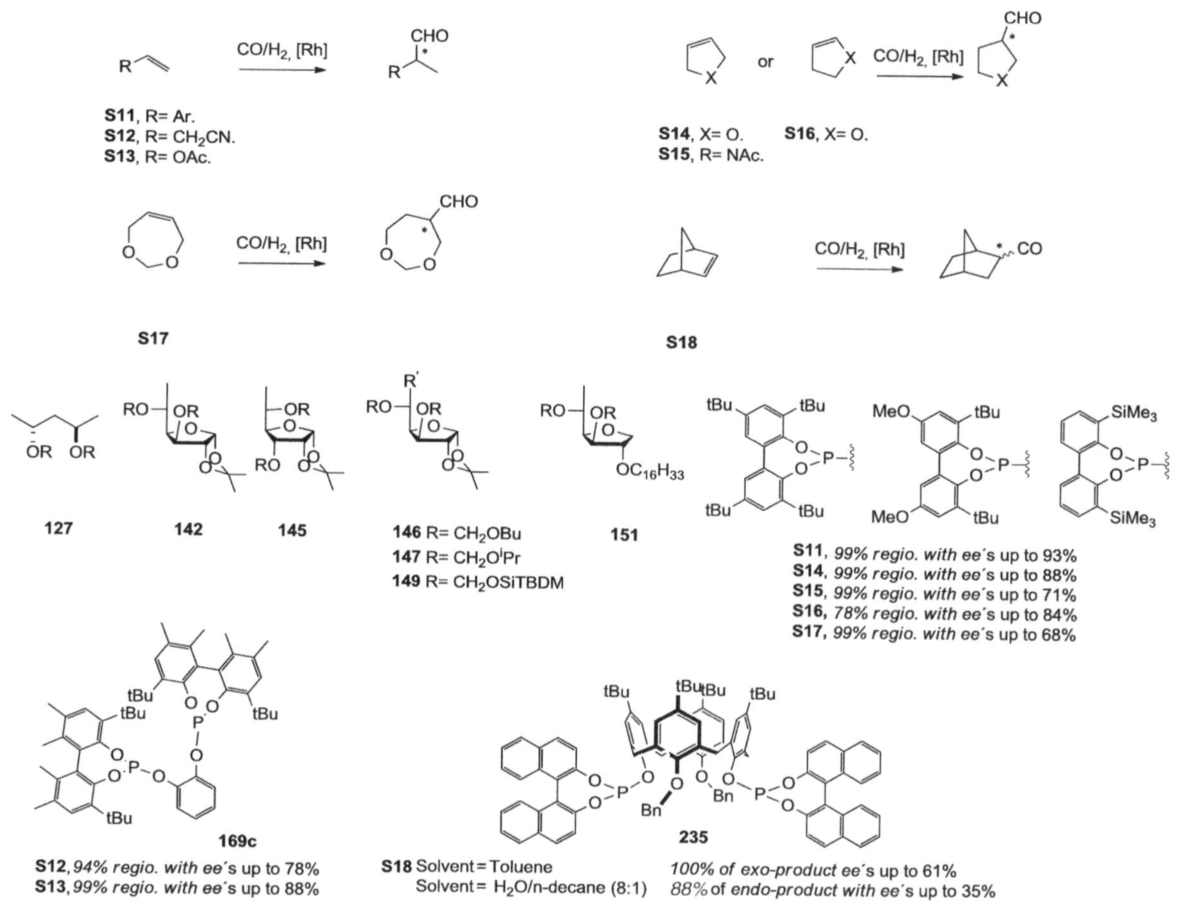

Scheme 3.25 *Rh-catalysed asymmetric hydroformylation of pro-chiral alkenes.*

year, a number which is still rising. The majority of these oxo-products are obtained from the hydroformylation of propene, which is a fraction of the steam-cracking process. The resulting products, n-butanal and iso-butyraldehyde, are important intermediates for the production of esters, acrylates and 2-ethylhexanol [172a]. Several diphosphites such as **159c** (biphephos) [140] and binapthol derivatives **182–187** [152,153] provide a very high linear-to-branched ratio in the Rh-catalysed hydroformylation of alkenes.

The Rh-catalysed asymmetric hydroformylation of alkenes has attracted much attention as a potential tool for preparing enantiomerically pure aldehydes [172,173]. The hydroformylation of monosubstituted vinylarenes (**S11**, Scheme 3.25) is used as a model for the synthesis of 2-aryl propionaldehydes, which can be used as intermediates in the synthesis of 2-aryl propionic acids, the Profen class of nonsteroidal anti-inflammatory drugs. The Rh-catalysed asymmetric hydroformylation of other monosubstituted substrates such as allyl cyanide, vinyl acetate or cyclic 1,2-disubstituted olefins was successfully carried out in recent years [174–179]. 1,3-diphosphites (**127**, **142**, **145–147**, **149**, **151**) bearing a nonchiral biphenol unit were shown to be efficient ligands in the Rh-catalysed asymmetric hydroformylation of vinylarenes **S11**, dihydrofurans **S14** and **S16**, 2,5-dihydropyrrolidone **S15** and dioxapines **S7** [116b,139,140,141]. The use of carbohydrate-derived ligands was revealed to be beneficial due to the possibility of easy modulation of the ligand structure [124,125c,131,132,142]. Concerning other types of substrates, the diphosphite **169** bearing a chiral biphenol unit was revealed to be the

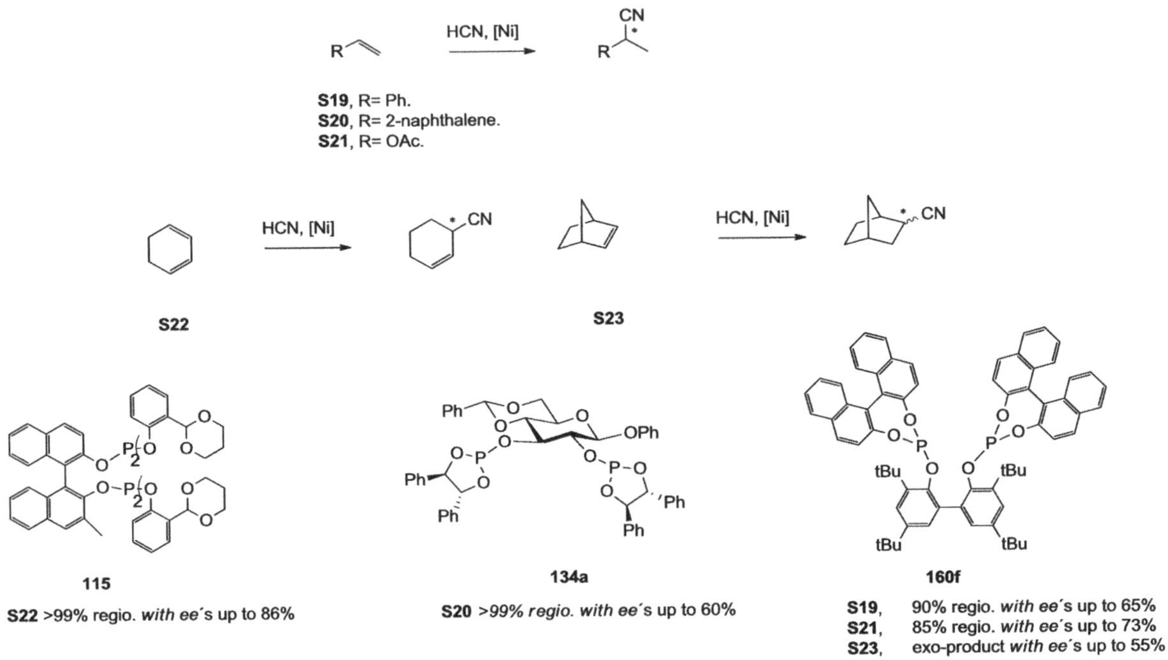

Scheme 3.26 Ni-catalysed asymmetric hydrocyanation of pro-chiral alkenes.

most efficient ligand so far for the Rh-catalysed asymmetric hydroformylation of allyl cyanide **S12** and vinyl acetates **S13** [80,145,143]. Recently, the use of diphosphite ligand **235** in the Rh-catalysed asymmetric hydroformylation of norbornene **S18** was reported, with moderate to low enantioselectivity [144].

3.3.2.2 Hydrocyanation

Although the toxicity of HCN requires a special experimental setup, the 100% atom economy and the inexpensive feedstocks of the reaction make this process of interest. The reaction has tremendous industrial impact, mainly because of the adiponitrile production by DuPont via hydrocyanation of butadiene using aryl phosphite-modified nickel catalysts [145]. The industrial process is among the early success stories of homogeneous catalysis. The actual catalyst is a Ni(0) complex which contains tris-*o*-tolylphosphite as ligands and Lewis acids as promoters [146]. The superior catalyst performance of the ligand tris-*o*-tolylphosphite can be explained by the result of an advantageous combination of a relatively large cone angle (141°) and strong π acidity [147].

More recently, diphosphites have also been reported to show good performance in the hydrocyanation of various substrates (Scheme 3.26) [148]. In this reaction, the most promising results in terms of regio- and enantioselectivity were obtained using the diphosphites **115** (substrate **S22**) [106a], **160** (substrate **S19, S21** and **S23**) [149] and **134** (substrate **S20**) [150].

3.3.3 Addition of nucleophiles to carbonyl compounds and derivatives

The catalytic asymmetric addition of nucleophiles to carbonyl compounds is one of the most powerful C–C and C–X bond-forming reactions in organic synthesis (Scheme 3.27).

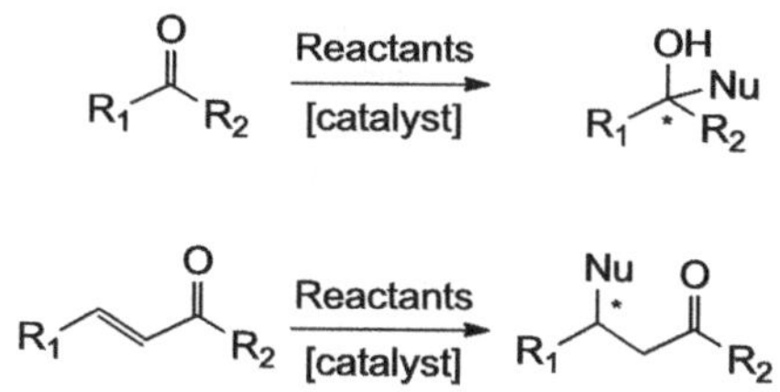

Scheme 3.27 *Metal-catalysed asymmetric 1,2- and 1,4-addition of nucleophiles to carbonyl compounds.*

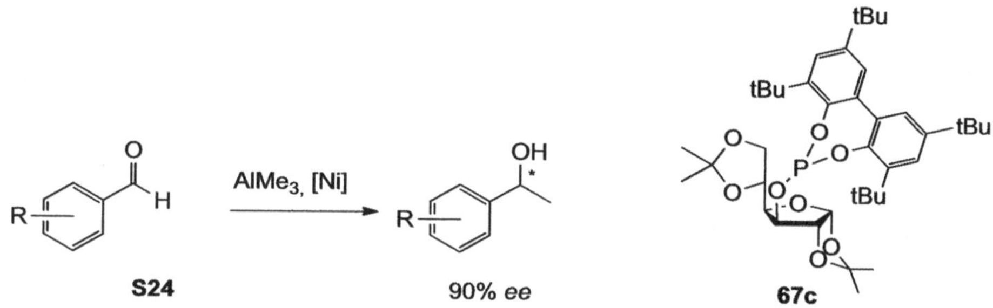

Scheme 3.28 *Ni-catalysed 1,2-addition to aldehydes.*

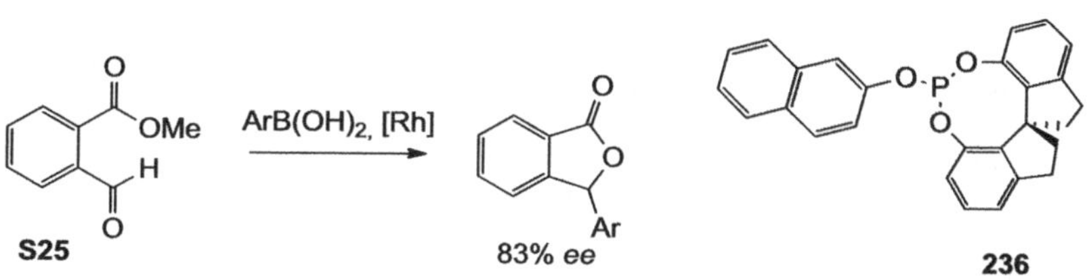

Scheme 3.29 *Rh-catalysed 1,2-addition of aryl boronic acids to aldehydes.*

3.3.3.1 1,2-addition

The nickel-catalysed 1,2-addition of trialkylaluminium to aldehydes constitutes a powerful method for the production of chiral alcohols [151]. The monophosphite ligand **67** gave high enantioselectivities (up to 94%) in the Ni-catalysed trialkylaluminium addition to benzaldehydes **S24** (Scheme 3.28) [89b].

The Rh(I)-catalysed asymmetric addition of arylboronic acids to aldehyde **S25** was carried out using SPINOL-based phosphites as ligands such as **236**, and up to 83% ee was achieved (Scheme 3.29) [152]. This reaction provides an efficient method to synthesise chiral 3-substituted phthalides.

3.3.3.2 1,4-addition

Over the first decade of the twenty-first century, several breakthroughs were reported in the area of Cu-catalysed asymmetric 1,4-addition of alkyl organometallic nucleophiles to enones [188–196]. Most of the successful asymmetric versions of this process used organozinc reagents, especially $ZnEt_2$ [153]. The inherently low reactivity of organozinc reagents towards unsaturated carbonyl compounds has facilitated

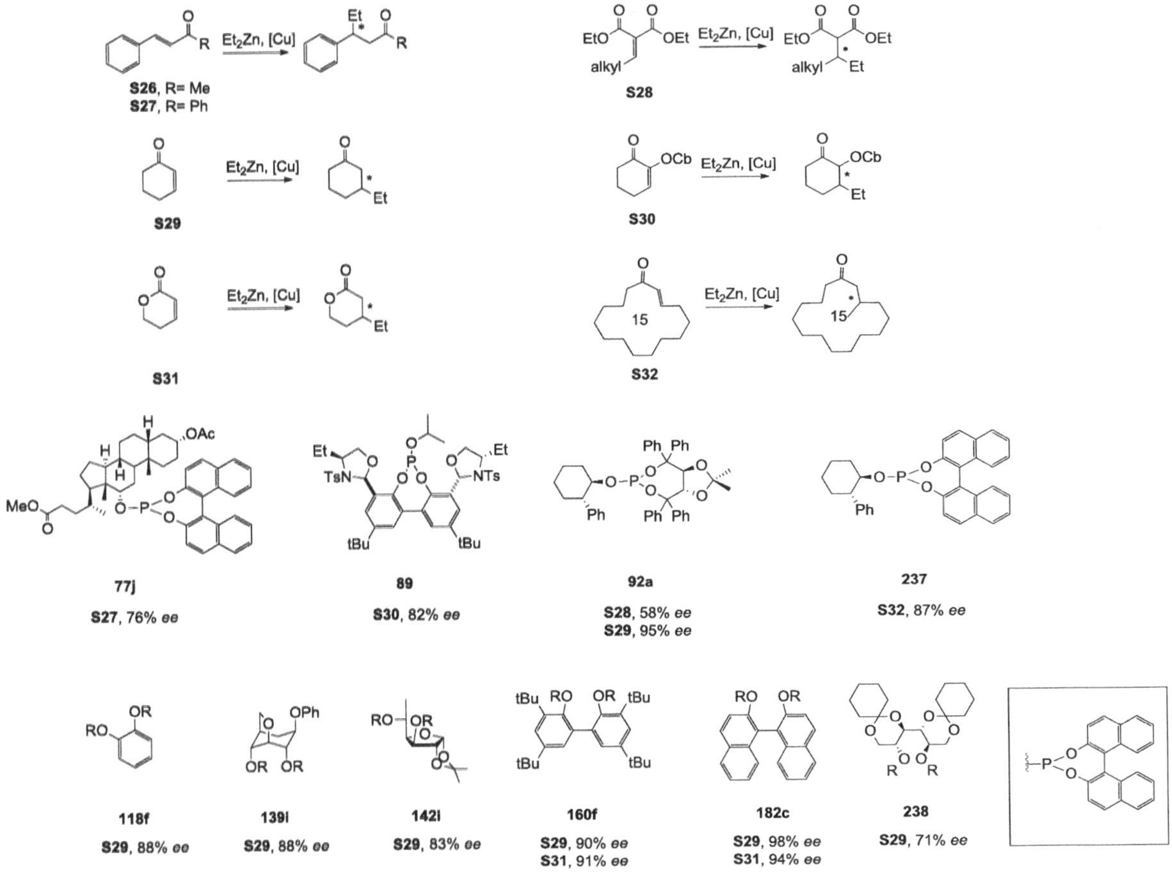

Scheme 3.30 *Cu-catalysed asymmetric 1,4-addition of $ZnEt_2$ to enones (**S26–S32**).*

the development of a plethora of chiral phosphorus-based ligands providing excellent regio- and enantioselectivities (>95% ee) to acyclic and cyclic enones, lactones and malonates [154].

The monophosphites bearing the taddol **92**, binaphthol **77** and **237** and biphenol **89** units were successfully applied in this reaction and provided excellent results in terms of enantioselectivity for disubstituted acyclic enones **S27** [94,155], diethyl malonates **S28** [156], disubstituted cyclic enones **S29** [157] and **S32** [158], and trisubstituted cyclic enones **S30** [159] (Scheme 3.30). Concerning diphosphite ligands, the most promising results in this reaction were achieved using the cyclic enones **S29** and lactones **S31** as substrates and the diphosphites **118, 139, 142, 160, 182** and **238**, which are bearing a binaphthol phosphite unit (Scheme 3.30) [121,127,149,160].

Trialkylaluminum reagents have recently emerged as an interesting alternative to organozinc reagents since they are also commercially available and can be easily prepared (Scheme 3.31) [161]. Their higher reactivity allows the Cu-catalysed 1,4-addition to very challenging substrates such as trisubstituted enones, which are inert to organozinc methodologies. Nowadays, only moderate to low enantioselectivities are achieved using monophosphites with cyclic and acyclic enones and nitroolefins [89a].

The asymmetric Rh-catalysed conjugate addition of organoborane reagents to enones has become the method of choice for introducing an aryl or vinyl group in β position (Scheme 3.32) [162]. This approach therefore complements the Cu-catalysed conjugate addition that usually provides the highest levels of enantioselectivity when alkyl groups are introduced. Gennari reported that the heterocombination of a biphenyl-based phosphite

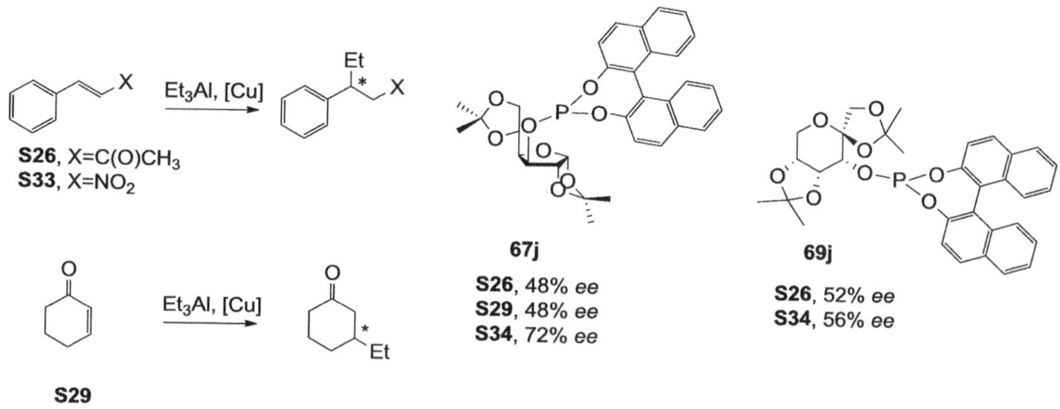

Scheme 3.31 Cu-catalysed 1,4-addition of $AlEt_3$ to enones.

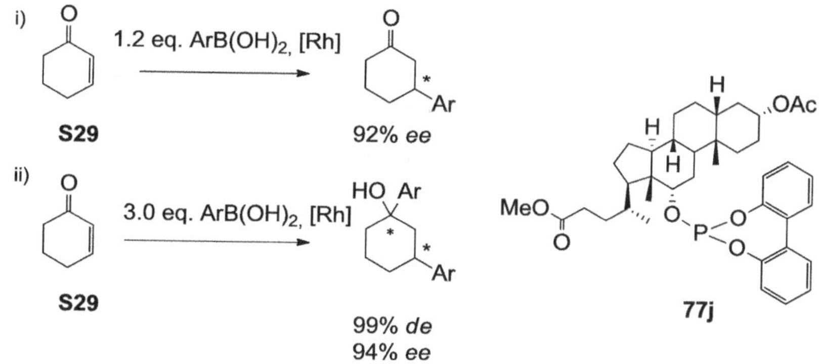

Scheme 3.32 Rh-catalysed 1,4-addition of organoborane reagents to enones.

derived from (*1R,2S*)-2-(2-phenylpropan-2-yl)cyclohexanol and a biphenyl-based phosphoramidite derived from (*2R,5R*)-2,5-diphenylpyrrolidine gave the most enantioselective catalyst (ee up to 99%) [86,163]. More recently, the above-mentioned biphenyl-based tropos monophosphite ligands derived from deoxycholic acid **77** were also used in the Rh-catalysed addition of arylboronic acids to cyclic enone **S29**, giving ee up to 92% (i, Scheme 3.32) [164]. Furthermore, this ligand also provided excellent results in the double addition (1,2- and 1,4-) of arylboronic acids to cyclic enones (diastereomeric excess (de) up to 99% and ee up to 94%) (ii, Scheme 3.32).

3.3.4 Allylic substitution reactions

The Pd-catalysed asymmetric allylic substitutions constitute a powerful synthetic tool for preparing enantiomerically enriched compounds through the enantioselective formation of carbon–carbon and carbon–heteroatom bonds [165].

The 1,3 diphosphites **127**, **140**, **142** and 1,2-diphosphites **157** bearing biphenol phosphite units were revealed to be very effective ligands in the Pd-allylic alkylation and amination of 1,3-substituted substrates, both acyclic and cyclic substrates (Scheme 3.33) [122,130,138,161b,166]. Additionally, with these ligands, a kinetic resolution of the substrate often takes place when the reaction is performed at high substrate-to-metal ratio. The monophosphites **239** [168b,167], **91** [101] and **59** [85b] were also revealed to be efficient ligands in the Pd-catalysed allylic alkylation, amination and sulphonation of 3-acetoxy-1,3-diphenyl-propene (Scheme 3.33).

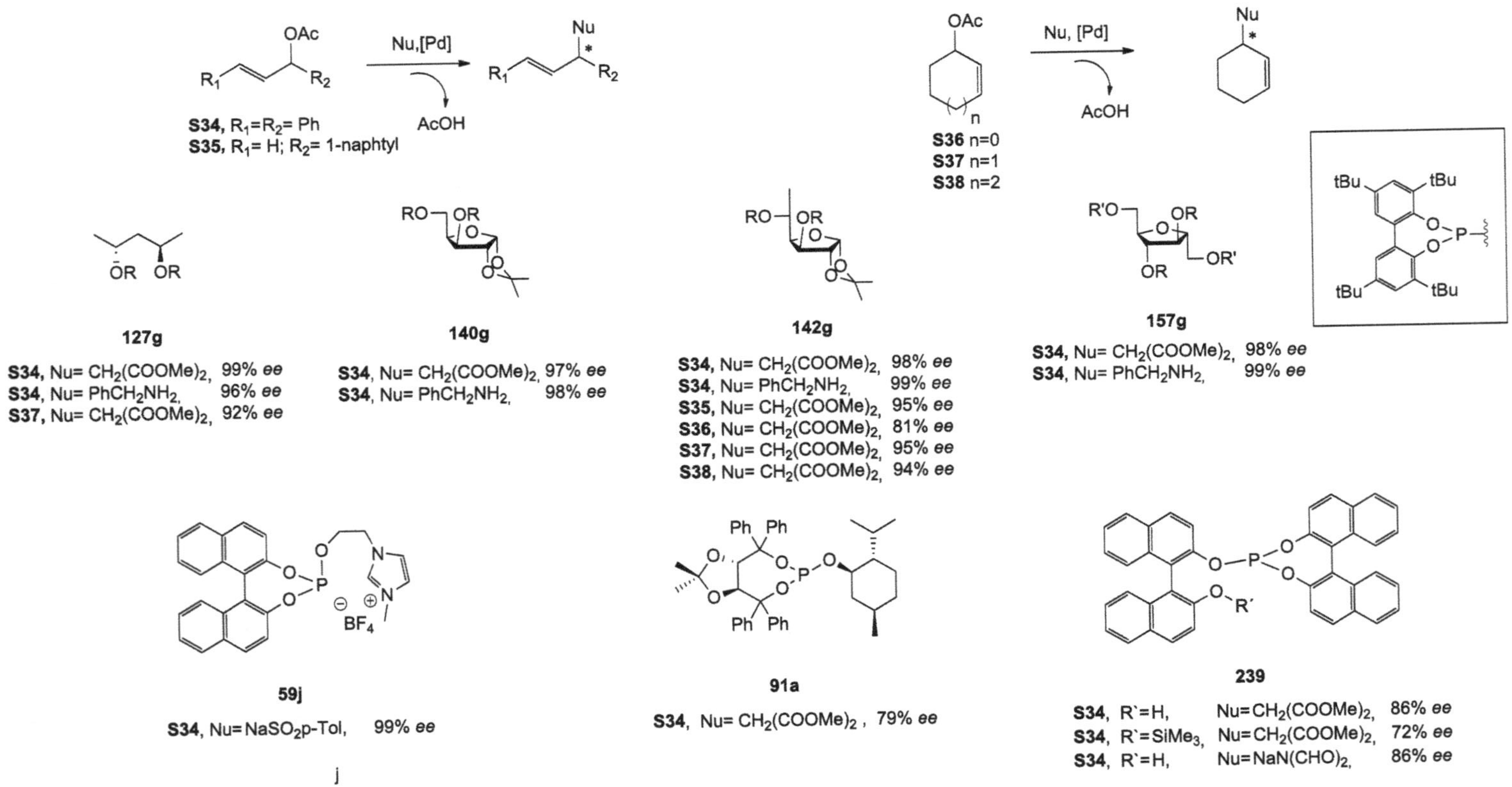

Scheme 3.33 *Pd-catalysed asymmetric allylic substitution.*

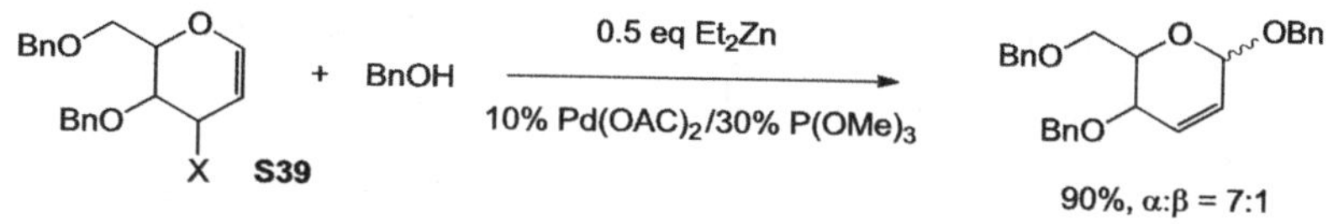

Scheme 3.34 *Pd-catalysed allylic substitution of glycals.*

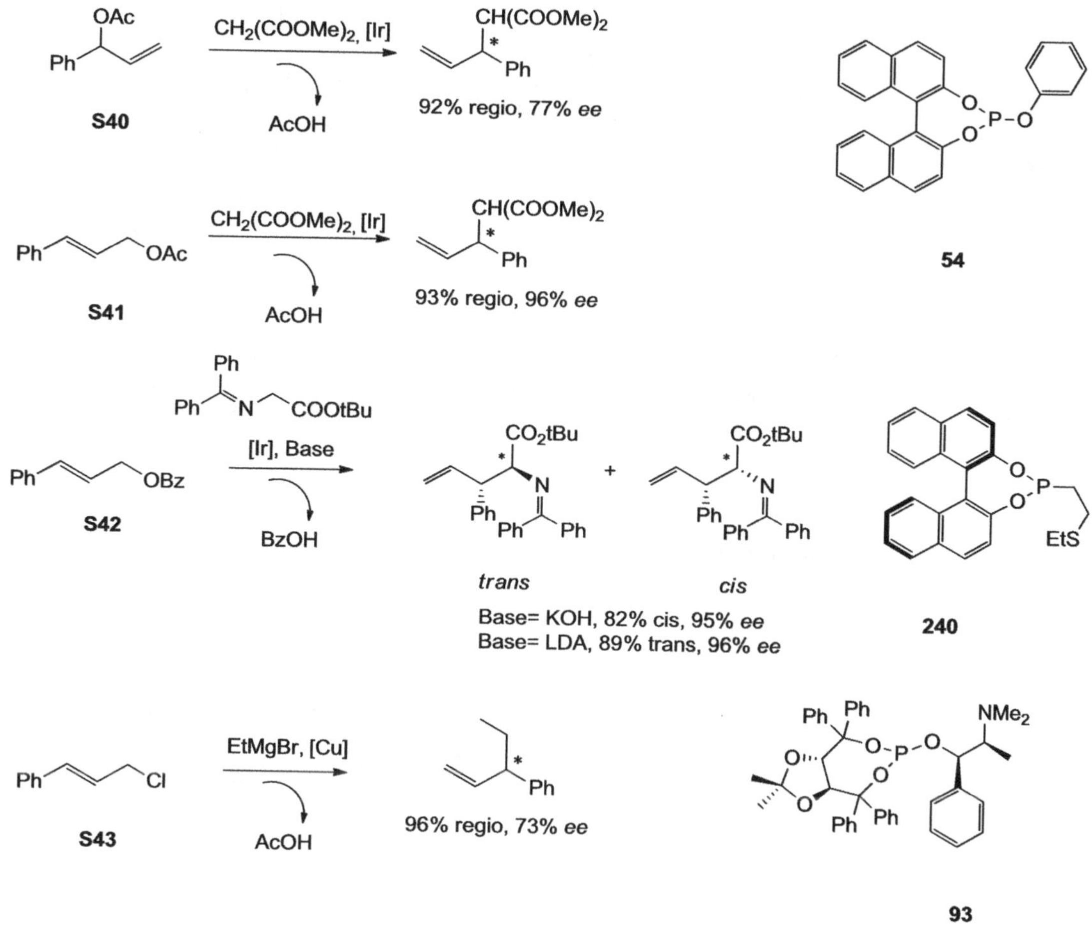

Scheme 3.35 *Ir- and Cu-catalysed asymmetric allylic substitution.*

The Pd-catalysed allylic substitution of glycals has been explored mainly in relation to the synthesis of C-glycosides (Scheme 3.34). A Pd-catalysed O-glycosylation method affording 2,3-anhydroglycosides was reported [168]. This procedure allows the direct use of glycals **S39** as glycosyl donors, and the anomeric stereochemistry is controlled by the ligand. Thus, $P(OMe)_3$ affords mainly α-glycoside (α:β=7:1), while β-glycosides are obtained by using DTBBP (α:β=1:25).

Monophosphites were also shown to be efficient ligands in the Ir- and Cu-catalysed allylic substitution of 3-acetoxy-3-phenyl-propene **S40** [169], 3-acetoxy-1-phenyl-propene **S41** [170], 3-benzyloxy-1-phenyl-propene **S42** [171] and 3-chloro-1-phenyl-propene **S43** [172], producing high selectivity to give the chiral product (Scheme 3.35).

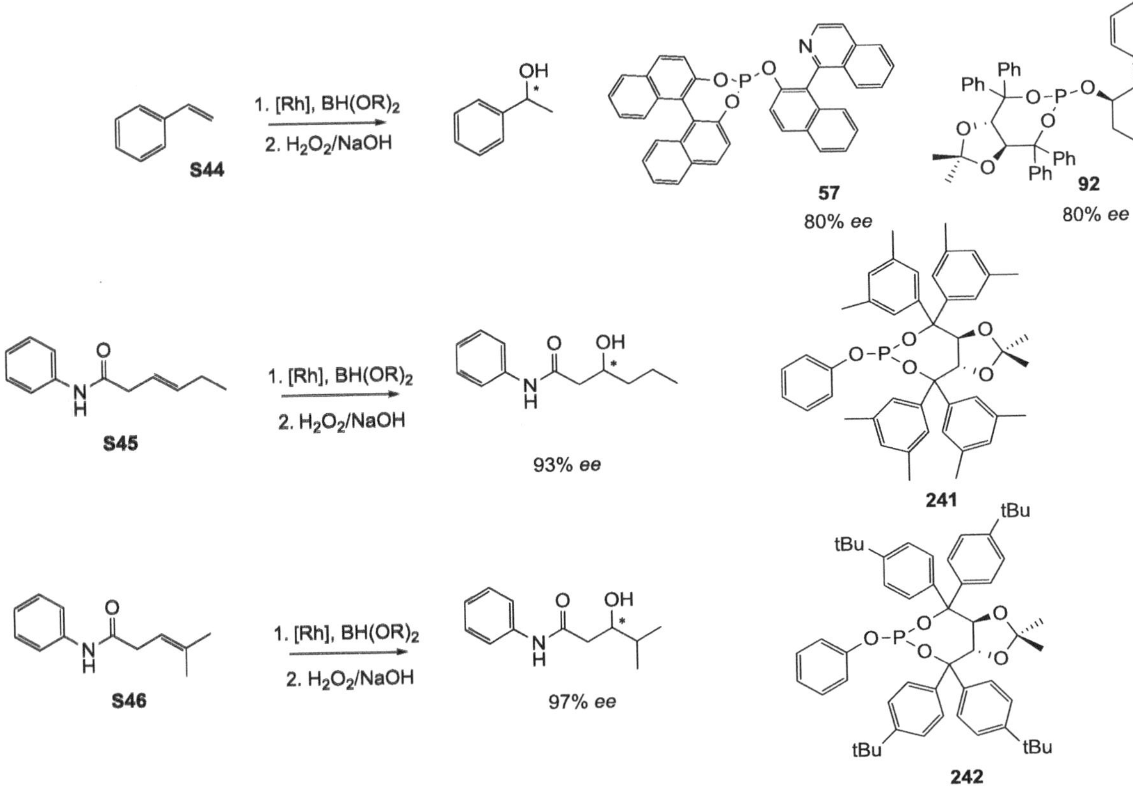

Scheme 3.36 Rh-catalysed asymmetric hydroboration.

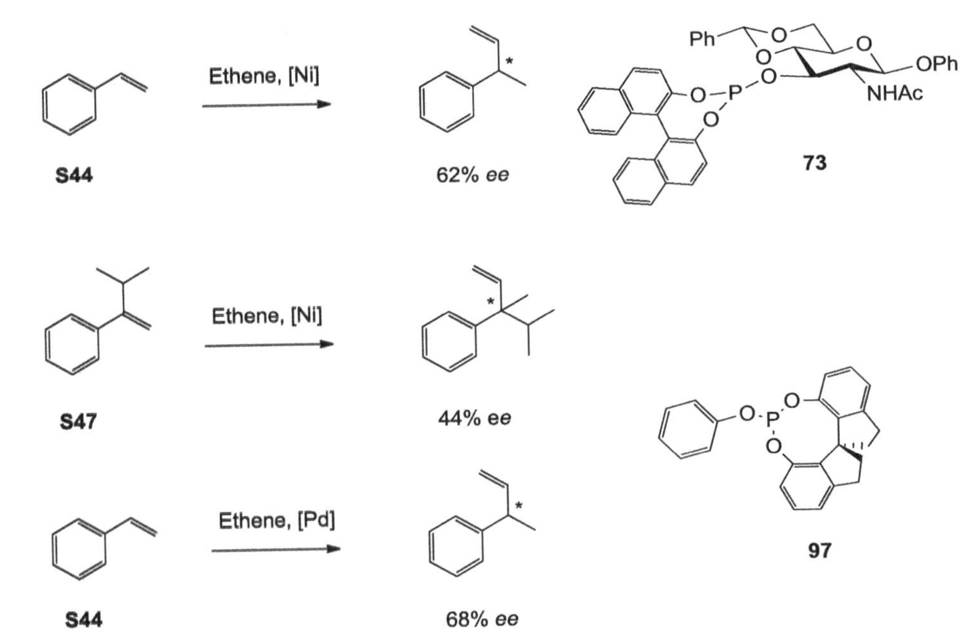

Scheme 3.37 Ni- and Pd-catalysed asymmetric hydrovinylation.

3.3.5 Miscellaneous reactions

Due to the large variety of structures, phosphite ligands have been used in a huge number of processes catalysed by transition metal complexes. In this section, a few examples of successful applications of phosphite ligands are described.

In the Rh-catalysed asymmetric hydroboration of alkenes **S44–S46** (Scheme 3.36), the use of monophosphites **57**, **92**, **241** and **242**, the three last bearing TADDOL-modified units, provided highly active catalysts with excellent enantioselectivities (up to 97%) [83,100,173].

In the Ni- and Pd-catalysed hydrovinylation of **S44** and **S47** (Scheme 3.37), moderate enantioselectivities were obtained using the monophosphites **73** and **97**, respectively [59,103a,104a].

In the Rh-catalysed asymmetric hydrosilylation of acetophenone (**S48**) (Scheme 3.38), moderate enantioselectivities were obtained using the monophosphite **5** (up to 58%) [42]. However, the use of the triphosphite ligand **226** provided enantioselectivities up to 81% [165].

A nice example of the influence of the ancillary ligands in catalysis is the gold-catalysed cyclization of the dienyl allenes **S49** and **S50** (Scheme 3.39). Thus, compound **S49**, in the presence of a cationic gold complex containing the phosphine **243**, affords the product resulting of a [4+3] cyclization. However, when a gold

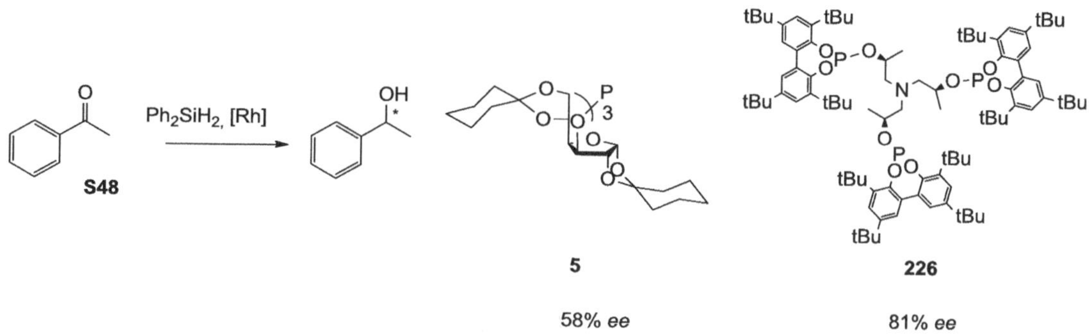

Scheme 3.38 Rh-catalysed asymmetric hydrosylilation.

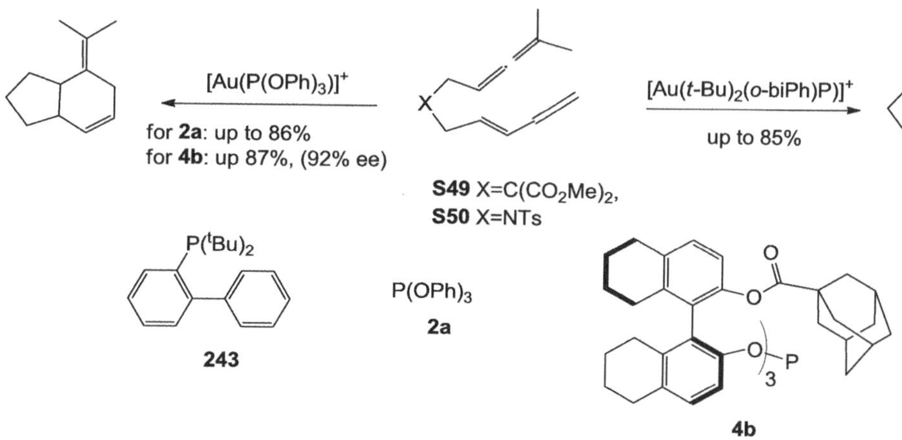

Scheme 3.39 Au-catalysed cycloaddition reactions.

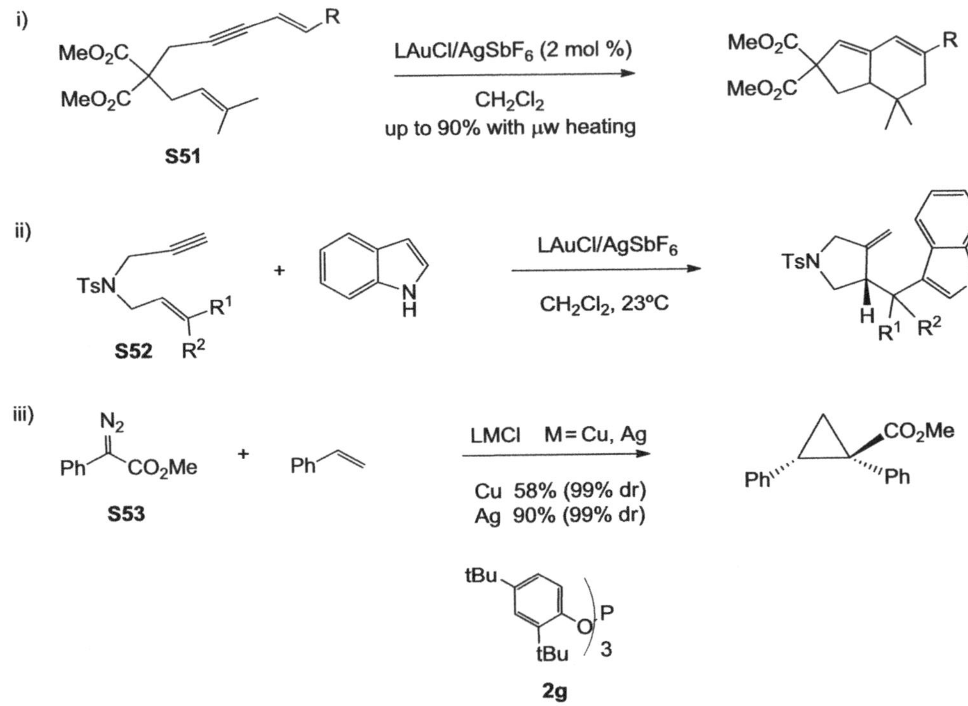

Scheme 3.40 Au-, Cu, and Ag-catalysed cycloaddition reactions.

complex bearing triphenylphosphite **2a** is used, a [4+2] cyclization takes place [174]. An enantioselective version of this process has been reported, achieving the product of cyclization [4+2] ($X{=}C(CO_2Me)_2$) with 92% ee when the C3-symmetric phosphite **4b** was used [175].

Gold complexes bearing tris-(2,4-di-*terc*-butylphenyl)phosphite **2g** as ligand are highly active in several inter- and intramolecular processes. Thus, when the enyne **S51** was treated with catalytic amounts of $[LAu]^+$, the compound formed through a [4+2] cyclization process was obtained (i, Scheme 3.40) [176]. The same catalytic system was used in the reaction of 1,6-enynes **S52** with electron-rich alkenes or heteroarenes such as indoles, affording the product resulting from the attack of indol to alkene with concomitant cyclization (ii, Scheme 3.40) [177]. Copper and silver catalysts bearing ligand **2g** also catalyse the cycloaddition of a stabilised carbene to styrene to give cyclopropane derivatives (iii, Scheme 3.40) [178].

Phosphites have also been used as ligands in palladium-catalysed inter- and intramolecular cycloaddition reactions involving acetylenes and alkenes. For instance, alk-5-ynyidenecyclopropanes **S56**, which can be easily formed by addition of the appropriate nucleophiles to allylcyclopropyltosylate, undergo Pd-catalysed intramolecular [3+2] cycloaddition to yield bicyclo[3.3.0]octenes (iii, Scheme 3.41). The most appropriate catalytic systems were $[Pd_2(dba)_3]$/tris-isopropylphosphite ($P(O\textit{i}Pr)_3$) [179] and $[Pd_2(dba)_3]$/tris-(3,5,-di-*tert*-butylphenyl)phosphite (**244**, Scheme 3.41) [180]. Bicyclo[4.3.0]nonanes can also be obtained using Pd/**244** as a catalytic system. Alkylidenecyclopropanes linked to 2-pyridyl unit **S54** undergoes a Pd/**2g** catalysed regioselective hydroalkynylation. The reaction also evolves when the 2-pyridyl unit is replaced by an ethylcarboxylate moiety (63% yield) (i, Scheme 3.41) [181]. Tri- and penta-substituted benzenes can be obtained by a cycloaddition involving three components: an allyl tosylate **S57** and two alkynes (iv,

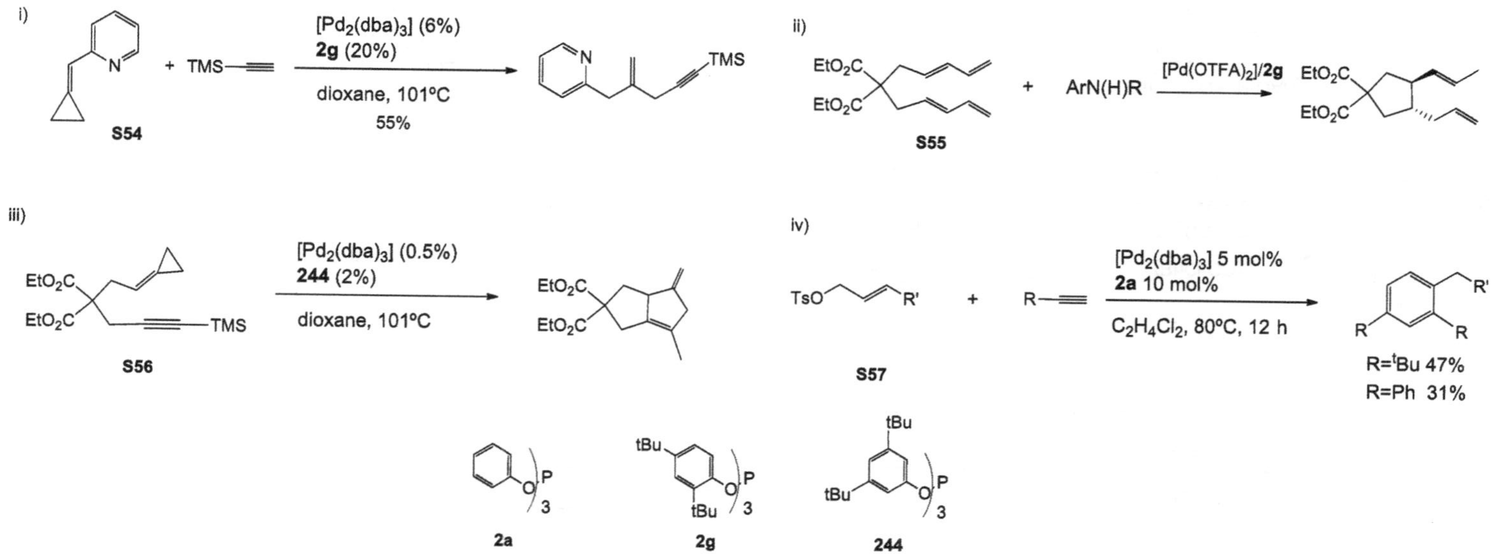

Scheme 3.41 *Pd-catalysed cycloaddition reactions.*

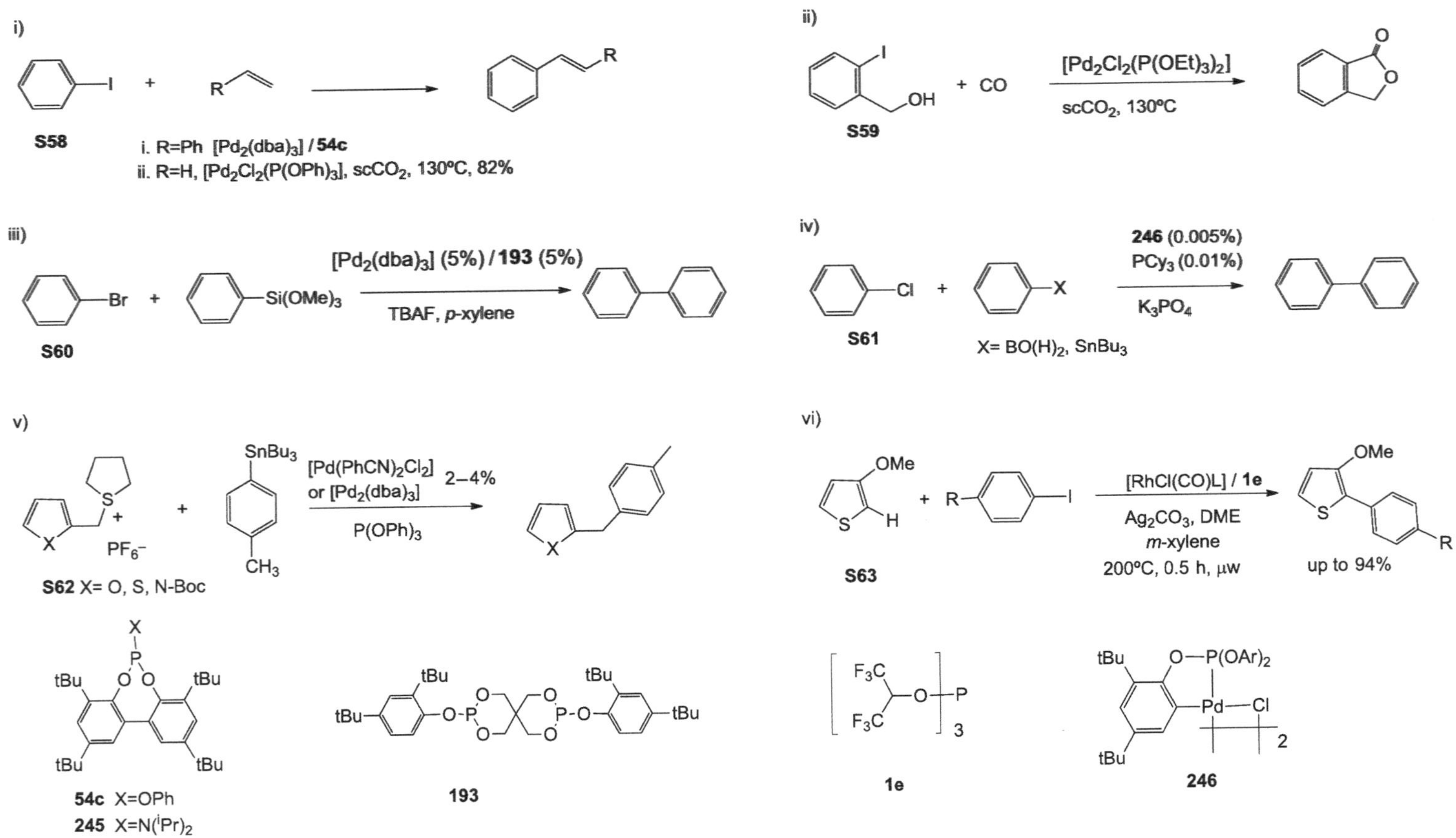

Scheme 3.42 *C–C bond formation reactions using phosphite ligands.*

Scheme 3.41). Doubly substituted alkynes (internal alkynes) react in the presence of Pd–PPh_3 with allyl tosylates to afford pentasubstituted benzenes. However, this catalytic system affords a mixture of isomers when terminal alkynes are used. The use of triphenylphosphite **2a** improved the regioselectivity and increased the yield up to 47% yield for R=^{t}Bu (Scheme 3.41) [182].

The Pd-catalysed bisdiene cyclization in the presence of amines was studied following a combinatorial approach, in which several palladium precursors and ligands were systematically tested (ii, Scheme 3.41). The catalytic system $Pd(OTFA)_2$/**2g** was among the most active catalysts that were tested [168e].

Although phosphines are usually the ligand of choice for cross-coupling reactions, phosphites have also been successfully used [183]. To study the effect of ligand properties in the Heck reaction using iodobenzene **S58** and styrene as model substrates, it was found that $Pd(acac)_2$–phosphoramidite **245** (i, Scheme 3.42) was the most reactive, followed by the $Pd(acac)_2$–phosphite **54c** system [184]. A similar reaction was explored using iodobenzene and ethylene as substrates to give styrene using $[PdCl_2(P(OPh)_3)_2]$ as catalyst and working under supercritical carbon dioxide ($scCO_2$) conditions. At 100 bar of CO_2 and 10 bar of ethylene, 97% of conversion and 88% selectivity were achieved [185]. The alkoxycarbonylation of aryl halides **S59** was also found to be efficiently catalysed by $[PdCl_2(P(OEt)_3)_2]$ in supercritical $scCO_2$ (ii, Scheme 3.42) [186]. The Hiyama coupling between bromobenzene **S60** and trimethoxysilylbenzene was studied using $Pd(acac)_2$ as a catalyst precursor in the presence of a series of phosphine and phosphite ligands. The catalytic system containing ligand **193** was the most active, even more than the more widely studied tris-(2,4-di-*tert*-butylphenyl)phosphite (iii, Scheme 3.42) [187].

Triarylphosphite ligands $P(OAr)_3$ undergo facile orthometallation reactions with transition metals, and in some cases orthometallated triaryl phosphites can provide active catalytic systems. For instance, complex **246** has shown to be active in the Heck arylation of alkenes [188] and in both Stille and Suzuki couplings, although PCy_3 is also usually used as additive (iv, Scheme 3.41) [102a]. A similar platinum complex was also shown to be very active in Suzuki couplings [189].

Heterobenzylic sulfonium salts (derived from furan, thiophene and pyrrole), which can be easily prepared from the corresponding alcohols, undergo palladium-catalysed cross-coupling with a variety of organostannanes, organoboronic acids and organozinc halides. Sulphonium salts derived from all heterocycles reacted with organostannanes, but organozinc and organoboron reagents reacted only with furan and thiophene derivatives. The high reactivity of sulphonium salts precludes the use of nucleophilic phosphines as ligands; however, triphenylphosphite **2a** was effective in heterobenzylic cross-coupling (v, Scheme 3.42) [190].

Iodoarenes can be directly coupled with electron-rich arenes and heteroarene involving a C–H activation using $[RhCl(CO)(P(OCH(CF_3)_2)_3)_2]$ (**1e**) and silver carbonate, which provide good to excellent yields (iv, Scheme 3.42) [191].

A Pd(0) catalyst bearing the sterically hindered phosphite ligand **2g** is capable of promoting a tandem ring-opening coupling process involving terminal alkynes and alkylidenecyclopropanes [216]. Importantly, this study demonstrated that the presence of an appropriate directing group, such as a 2-pyridyl unit or a carboxylate moiety, leads to a regioselective hydroalkynylation (i, Scheme 3.43). The Pt-catalysed intramolecular silaboration of borylsilanyl homoallyl ethers was reported and the diastereoselectivity in the products was highly dependent on the type and structure of the phosphorus ligand used (ii, Scheme 3.43) [192]. They observed a direct relationship between the steric hindrance of the phosphite and the *cis-trans* selectivity: a platinum catalyst bearing a sterically less hindered phosphite ligand, such as $P(OMe)_3$, showed moderate *trans* preference, while the *cis* selectivity was dramatically improved using a catalyst containing a bulkier ligand such as $P(OPh)_3$. So far, the sterically hindered phosphite **2g** is the ligand of choice for *cis*-selective silaboration. Catalysts formed *in situ* from mixtures of palladium acetate, indium(III) triflate and the chiral nonchelating bis(phosphite) **162** yielded good to excellent conversions and reasonable enantioselectivity in the asymmetric dimerisation of styrenes (iii, Scheme 3.43) [193].

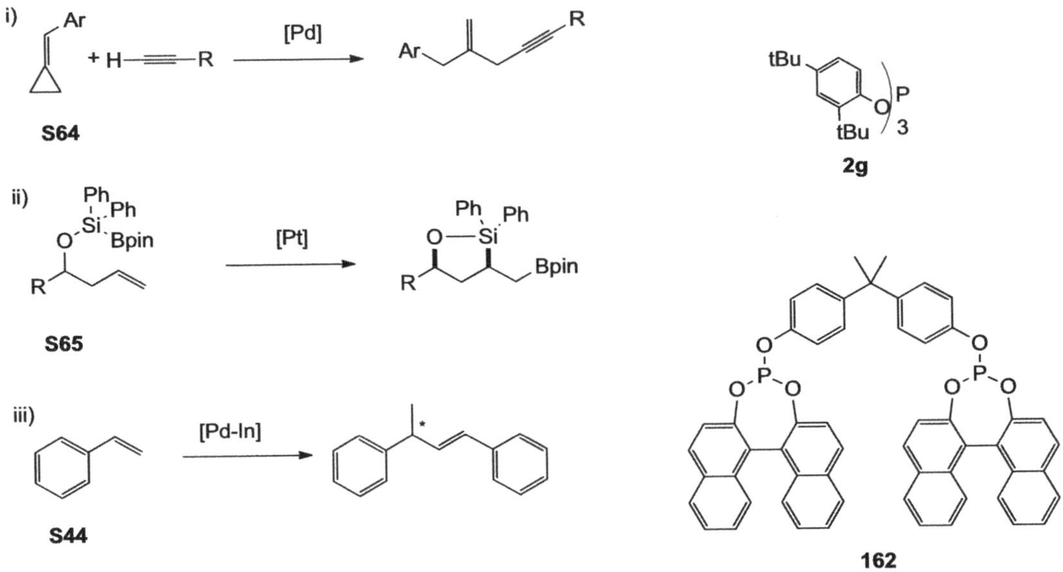

Scheme 3.43 Miscellaneous reactions.

3.4 General synthetic procedures

3.4.1 Symmetrically substituted phosphites

As representative examples of the synthesis of symmetrically substituted phosphites, we collected above the synthesis of triphenylphosphite (**2a**) and the most sterically hindered tris(2,4-di-tert-butylphenyl) phosphite (**2g**).

Synthesis of triphenylphosphite (2a) [39]: In a 500 mL flask, fitted a with a dropping funnel, stirring bar and rubber septum, were added a solution of phenol (34.2 mmol) and Et_3N (37.0 mmol) in dry and deoxygenated tetrahydrofuran (THF) (200 mL). Then PCl_3 (11.4 mmol) were added by syringe. The solution was stirred at room temperature for 24 h. The solution was then filtered to remove the salts formed during the reaction. The solid was washed with dry and deoxygentated THF (2 × 15 mL), and the solvent was removed under vacuum to afford the compound **2a** in 80% yield.

Cautions

- It is extremely important that all the reactants are free of water to avoid the formation of phosphoric acid and chlorhidric acid in the presence of PCl_3 which decreased the reaction yield. Therefore, phenol has to be distilled and Et_3N and THF dried before reaction. It is also important to avoid the oxidation of the phosphorus (III) species in the presence of molecular oxygen, and thus it is recommendable to carry out the reaction and the purification process under inert atmosphere using deoxygenated solvents.
- Generally, commercially available PCl_3 is not stored under inert atmosphere and therefore the purification by distillation under vacuum is recommendable before use.
- If a highly exothermic acid base reaction occurs, then in order to avoid damages, PCl_3 has to be carefully added dropwise to the solution of phenol and Et_3N in THF at 0°C.
- TLC and ^{31}P NMR are useful to check the product formation.
- Traces of acid are present in some deuterated solvent ($CDCl_3$) and thus it has to be purified before use.

Synthesis of tris(2,4-di-tert-butylphenyl)phosphite ***(2g)*** [38]: (a) A solution of 2,4-di-tert-butylphenol (0.03 mmol) in dried and deoxygenated THF (35 mL) was carefully added dropwise to a suspension of NaH (0.04 mmol) in THF (25 mL) at room temperature over 30 min, and the stirring was continued for 2 h. *Cautions*: (a) H_2 is generated; and (b) see observations in the synthesis of triphenylphosphite **2a** (Section 3.3.5). Then, the mixture was filtered off, PCl_3 (0.009 mmol) was added to the resulting solution of sodium phenolate, and the mixture was stirred for 24 h at 65°C. Then, the solvent was removed, hexane was added, and the suspension of NaCl obtained was filtered off. The solution was concentrated under reduced pressure and the residue was crystallised at −15°C, affording the compound **2g** with a 90% yield.

3.4.2 Nonsymmetrically substituted phosphites

As a representative example of the synthesis of nonsymmetrically substituted phosphites, we collected here the synthesis of diethyl 5,5-dimethyl-3-oxo-1-phenyl-1-hexen-4-yl phosphite (**6**).

Synthesis of diethyl 5,5-dimethyl-3-oxo-1-phenyl-1-hexen-4-yl phosphite ***(6)*** [48]: (a) Synthesis of diethylchlorophosphite: Ethanol (0.1 mol) was charged under nitrogen in a Schlenk tube with a reflux condenser and PCl_3 (55 mmol) was added. The top of the condenser was fitted with a drying tube of $CaCl_2$ and Drierite, and the mixture was stirred at 120°C for 6 h. The solution was then cooled to room temperature. The unreacted PCl_3 was removed under vacuum, and the residue was distilled to afford $(EtO)_2PCl$. (b) Synthesis of phosphite **6**: $(EtO)_2PCl$ (1.3 mmol) was added to a solution of 5,5-dimethyl-3-oxo-1-phenyl-1-hexen-4-ol (1.2 mmol) and Et_3N (11 mmol) in dichloromethane (20 mL) at −78°C. The reaction mixture was allowed to warm to 25°C and stirred for 4 h. The solution was concentrated and the residue was washed with pentane, the solution was filtered through celite, and the filtrate was concentrated to give the desired product in 91% yield. The compound can be purified by column chromatography on silica gel.

Cautions

- See observations in the synthesis of triphenylphosphite **2a**.
- It is recommendable to collect PCl_3 in an intermediate trap and destroy it using a Na_2CO_3-saturated water solution.
- TLC and ^{31}P NMR are useful to check the product formation.
- Filtration though celite can be excluded if the product purification by column chromatography is required.
- Silica gel has to be dried under vacuum and deoxygenated before its use. For that, silicagel is heated at 100–150°C for 2–4 h, and then cooled to room temperature under inert atmosphere.
- Solvents have to be dried and deoxygenated before their use.
- Column chromatrography has to be carried out under nitrogen atmosphere to avoid product decomposition.
- Flash chromatography procedures are carried out using a flow of inert gas, and the fractions have to be collected under inert atmosphere.

3.4.3 Phosphites bearing dioxaphospho-cyclic units

As a representative example of the synthesis of phosphites bearing dioxaphospho-cyclic units, we collected here the synthesis of **127f**.

Synthesis of diphosphite ***127f*** *[194]*: [a] Synthesis of phosphochloridite (**f**-Cl, Figure 3.6): 4,4′,6,6′-Tetra-ter t-butyl-2,2′-bis(phenol) (7.0 mmol, 3.32 g), azeotropically dried with toluene (3×5 mL), was dissolved in toluene (25 mL) and pyridine (25 mmol, 2 mL). This solution was added dropwise to a cooled solution (0°C) of PCl_3 (8 mmol, 0.70 mL) and pyridine (25 mmol, 2 mL). The reaction mixture was refluxed for 2 h. The solvent and excess of PCl_3 were removed under vacuum to obtain the phosphochloridite that was directly used in the next reaction.

- See observations in the synthesis of triphenylphosphite **2a** (Section 3.3.5).
- It is recommendable to collect PCl_3 in an intermediate trap and destroy it using a Na_2CO_3-saturated water solution.
- ^{31}P NMR is useful to check the phosphochloridite formation before its subsequent use.

[b] Synthesis of diphosphite **127f**: The crude phosphochloridite (2.2 mmol) generated previously was dissolved in toluene (10 mL) and pyridine (20 mL, 1.62 mL). 2R,4R-pentane-2,4-diol (2.0 mmol, 0.208 g) was azeotropically dried with toluene (3×1 mL) and dissolved in toluene (15 mL). The diol solution was transferred slowly to the solution of phosphochloridite over 30 min at 0°C. The reaction mixture was stirred overnight at room temperature, and the pyridine salts were filtered off. Evaporation of the solvent gave a white foam, which was purified by flash chromatography (eluent: 5% ethylacetate-hexane, Rf 0.46). Yield 1.76 g, 90% of a white powder.

Cautions

- TLC and ^{31}P NMR are useful to check the phosphite formation before its purification.
- Silica gel has to be dried under vacuum and deoxygenated before its use. For that silicagel is heated at 100–150°C for 2–4 h, and the cooled to room temperature under inert atmosphere.
- Solvents have to be dried and deoxygenated before their use.
- Column chromatrography has to be carried out under nitrogen atmosphere to avoid product decomposition.
- Flash chromatography procedures are carried out using a flow of inert gas, and the fractions have to be collected under inert atmosphere.

References

[1] For recent reviews see, for example, (a) I. Ojima, *Catalytic Asymmetric Synthesis*, John Wiley & Sons, Inc., **2010**; (b) A. Börner, *Phosphorus Ligands in Asymmetric Catalysis*, Wiley-VCH, **2008**; (c) M. Christmann, S. Bräse, *Asymmetric Synthesis: The Essentials*, Wiley-VCH, **2007**; (d) H. U. Blaser, E. Schmidt, in *Asymmetric Catalysis on Industrial Scale*, John Wiley & Sons, Inc., **2004**; and (e) E. N. Jacobsen, A. Pfaltz, H. Yamamoto, *Comprehensive Asymmetric Catalysis*, Springer, **1999**.

[2] D. Peña, A. J. Minnaard, J. G. de Vries, B. L. Feringa, *J. Am. Chem. Soc.* **2002**, *124*, 14553.

[3] (a) P. W. M. N. van Leeuwen, *Homogeneous Catalysis: Understanding the Art*, Kluwer, **2004**; (b) P. W. M. N. van Leeuwen *Appl. Catal. A: Gen.* **2001**, *212*, 61.

[4] (a) D. S. Marynick, *J. Am. Chem. Soc.* **1984**, *106*, 4064; and (b) A. G. Orpen, N. G. Connelly, *J. Chem. Soc. Chem. Commun* **1985**, 1310.

[5] N.-S. Fusao Tada, O.-S. Seiei Inamine, S.-S. Tsutomu Hatanada, US Patent 3674897.

[6] A. A. Dabbawala, H. C. Bajaj, R. V. Jasra, *Journal of Molecular Catalysis A: Chemical* **2009**, *302*, 97.

[7] M. T. Reetz, H. Guo, J.-A. Ma, R. Godard, R. J. Mynott, *J. Am. Chem. Soc.*, **2009**, *131*, 4136.

[8] A. Suárez, A. Pizzano, I. Fernández, N. Khiar, *Tetrahedron: Asymmetry* **2001**, *12*, 633.

[9] G. Ilia, S. Iliescu, L. Macarie, A. Popa, *Heteroatom Chemistry* **2008**, *19*, 4, 360.

[10] E. Pietrusewicz, M. Sienczyk, J. Oleksyszyn, *J. Enzyme Inhibition and Medicinal Chemistry* **2009**, *24*, 6, 1229.

[11] R. Grzywa, A. M. Sokol, M. Sienczyk, M. Radziszewicz, B. Kosciolek, M. P. Carty, J. Oleksyszyn, *Bioorg. Med. Chem.* **2010**, *18*, 2930.

[12] T. Mathivet, E. Monflier, Y. Castanet, A. Mortreux, J.-L. Couturier, *Tetrahedron* **2002**, *58*, 3877.

[13] B. A. Hunter, US Patent 5254709.

[14] E. Bergbreiter, R. Chandran, *J. Org. Chem.* **1986**, *51*, 4754.

[15] W.-Y. Zhang, D. J. Jakiela, A. Maul, C. Knors, J. W. Lauher, P. Helquist, D. Enders, *J. Am. Chem. Soc.* **1988**, *110*, 4652.

[16] (a) J. R. Cole, M. E. Dellinger, T. J. Johnson, B. A. Reinecke, R. D. Pike, Robert, W. T. Pennington, M. Krawiec, A. L. Rheingold, *J. Chem. Cryst.* **2003**, *33*, 341; (b) E. J. Volcko, J. G. Verkade, *Phosphorus Sulfur* **1984**, *21*, 111; (c) W. S. Wadsworth Jr., W. D. Emmons, *J. Am. Chem. Soc.* **1962**, *84*, 610; (d) O. Neunhoeffer, W. Maiwald, *Chem. Berichte* **1962**, *95*, 108; (e) A. B. Kudryavtsev, A. T. Teleshev, E. E. Nifant'ev, W. Linert, *Russ. Chem. Bull.* **2003**, *52*, 2740; (f) A. B. Koudriavtsev, A. T. Teleshev, A. A. Zhdanov, E. E. Nifant'ev, W. A. Linert, *Monatsh. Chem.* **2001**, *132*, 1001; and (g) D. Weber, W. D. Habicher, E. E. Nifantev, A. T. Teleshev, A. A. Zhdanov, V. K. Belsky, *Phosphorus Sulfur*, **1999**, *149*, 143.

[17] (a) N. K. Kochetkov, E. E. Nifant'ev, M. P. Koroteev, Z. K. Zhane, A. A. Borisenko, *Carbohydr. Res.* **1976**, *47*, 221; and (b) E. E. Nifantyev, M. P. Koroteev, S. A. Lysenko, S. A. Rumyantseva, N. K. Kochetkov, *Carbohydr. Res.* **1982** , *111*, 67.

[18] D. C. Ayres, H. N. Rydon, *J. Chem. Soc.* **1957**, *108*, 1109.

[19] (a) T. Maki, K. Ishihara, H. Yamamoto, *Org. Lett.* **2005**, *7*, 5047; (b) M. Lautens, E. Tayama, C. Herse, *J. Am. Chem. Soc.* **2005**, *127*, 72; (c) K. Furuta, Q. Gao, H. Yamamoto, *Organic Syntheses*, John Wiley & Sons, *Inc.*, **1998**, *9*:722; and (d) M. Egli, M. Dobler, *Helv. Chim. Acta* **1989**, *72*, 1136.

[20] M. Amberg, U. Bergsträsser, G. Stapf, J. Hartung, *J. Org. Chem.* **2008**, *73*, 3907.

[21] (a) J.-M. Brunel, O. Pardigon, M. Maffei, G. Buono, *Tetrahedron: Asymmetry* **1992**, *3*, 1243; and (b) R. Kadyrov, D. Heller, R. Selke, *Tetrahedron: Asymmetry* **1998**, *9*, 329.

[22] (a) Q.-F. Wang, W. He, X.-Y. Liu, H. Chen, X.-Y. Qin, S.-Y. Zhang, *Tetrahedron: Asymmetry* **2008**, *19*, 2447; and (b) Q.-F. Wang, X.-Y. Liu, X.-Y. Qin, X.-L. Li, S.-Y. Zhang, *Chin. Chem. Lett.* **2009**, *20*, 797.

[23] (a) K. N. Gavrilov, I. S. Mikhel, D. V. Lechkin, G. I. Timofeeva, *Phosphorus Sulfur* **1996**, *108*, 285; and (b) A. I. Polosukhin, K. N. Gavrilov, O. G. Bondarev, A. V. Korostylev, P. V. Petrovskii, V. A. Davankov, *J. Organomet. Chem.* **2000**, *608*, 89.

[24] H. Park, T. V. RajanBabu, *J. Am. Chem. Soc.* **2002**, *124*, 734.

[25] H. Park, R. Kumareswaran, T. V. RajanBabu, *Tetrahedron* **2005**, *61*, 6352.

[26] W. J. Lucas, F. W. Mitchell, C. I. Scully, *J. Am. Chem. Soc.* **1960**, *72*, 5491.

[27] (a) B. Gordillo, C. Garduiio, G. Guadarrama, J. Hernández, *J. Org. Chem.* **1995**, *60*, 5180; and (b) E. L. Eliel, S. Chandrasekaran, L. E. Carpenter, J. G. Verkade, *J. Am. Chem. Soc.* **1986**, *108*, 6651.

[28] (a) J. A. Mosbo, J. G. Verkade, *J. Am. Chem. Soc.* **1973**, *95*, 4659; (b) R. V. Hodges, F. A. Houle, J. L. Beauchamp, R. A. Montag, J. G. Verkade, *J. Am. Chem. Soc.* **1980**, *102*, 932; (c) J. G. Verkade, *Phosphorus Sulfur* **1976**, *2*, 251; and (d) R. F. Hudson, J. G. Verkade, *Tetrahedron Lett.* **1975**, 3231.

[29] K. Taira, D. G. Gorenstein, *J. Am. Chem. Soc.* **1984**, *106*, 7825.

[30] J. Hernández, R. Ramos, N. Sastre, R. Meza, H. Hommer, M. Salas, B. Gordillo, *Tetrahedron* **2004**, *60*, 10927.

[31] For similar types of mechanisms, see (a) R. F. Hudson, J. L. Keay, *J. Chem. Soc.* **1956**, 2463; and (b) D. W. White, R. D. Bertrand, G. K. MeEwen, J. G. Verkade, *J. Am. Chem. Soc.* **1970**, *92*, 7125.

[32] (See the stereoelectronic effects in transition states, for instance, in the glycoside hydrolysis: P. Deslongschamps, *Pure Appl. Chem.* **1993**, *65*, 1161.

[33] (a) N. J. Zhang, X. Chen, H. Y. Lu, Y. P. Feng, Y. F. Zhao, *Synlett* **1997**, *4*, 373; and (b) J.-J. Hu, Y. Ju, Y.-F. Zhao, Y.-X. Cui, *Synth. Commun.* **2000**, *30*, 4087.

[34] E. B. Benetsky, S. V. Zheglov, T. B. Grishina, F. Z. Macaev, L. P. Bet, V. A. Davankov, K. N. Gavrilov, *Tetrahedron Lett.* **2007**, *48*, 8326.

[35] G. Lin, M.-D. Tsai, *J. Am. Chem. Soc.* **1989**, *111*, 3099.

[36] L. Anschütz, W. Broeker, R. Neher, A. Ohnheiser, *Ber. Dtsch. Chem. Ges.* **1943**, *76*, 218.

[37] P. C. Crofts, J. H. H. Markes, N. D. Rygon, *J. Chem. Soc.* **1958**, 4250.

[38] I. Odinets, T. Kégl, Elena Sharova, O. Artyushin, E. Goryunov, G. Molchanova, K. Lyssenko, T. Mastryukova, G.-V. Röschenthaler, G. Keglevich, L. Kollár, *J. Organomet. Chem.* **2005**, *690*, 3456.

[39] L. M. Abdrakhmanova, V. F. Mironov, T. A. Baronova, D. B. Krivolapov, I. A. Litvinov, M. N. Dimukhametov, R. Z. Musin, A. I. Konovalov, *Russ. Chem. Bull., Int. Ed.* **2008**, *57*, 1559.

[40] L. H. Cao, H. I. Gao, C. J. Zhou, Y. T. Liu, *Chin. Chem. Lett.* **2004**, *15*, 171.

[41] (a) E. E. Nifantév, E. E. Milliaresi, N. G. Ruchkina, L. M. Druyan-Poleshchuk, *Zh. Obshch. Khim.*, **1979**, *49*, 2390; and (b) E. E. Nifantév, E. E. Milliaresi, N. G. Ruchkina, L. M. Druyan-Poleshchuk, L. K. Vasyanina, E. A. Tyan, *Zh. Obshch. Khim*, **1981**, *51*, 1528.

[42] R. B. Bedford, S. L. Hazelwood, M. E. Limmert, *Chem. Commun.* **2002**, 2610.

[43] (a) K. He, A. Hasan, B. Krzyzanowska, B. R. Shaw, *J. Org. Chem.* **1998**, *63*, 5769; (b) B. K. Krzyzanowska, K. He, A. Hasan, B. R. Shaw, *Tetrahedron* **1998**, *54*, 5119; and (c) J. Lin, B. R. Shaw, *Chem. Commun.* **2000**, 2115.

[44] (a) V. E. Albrow, A. J. Blake, R. Fryatt, C. Wilson, S. Woodward, *Eur. J. Org. Chem.* **2006**, 2549; and (b) S. Woodward, PCT Int. Appl. WO 2007036701.

[45] (a) P. H. Dussault, K. R. Woller, *J. Org. Chem.* **1997**, *62*, 1556; (b) M. T. Reetz, G. Mehler, *Angew. Chem. Int. Ed.* **2000**, *39*, 3889; and (c) M. Kawasaki, P. Li, H. Yamamoto, *Angew. Chem. Int. Ed.* **2008**, *47*, 3795.

[46] C. J. Cobley, K. Gardner, J. Klosin, C. Praquin, C. Hill, G. T. Whiteker, A. Zanotti-Gerosa, J. L. Petersen, K. A. Abboud, *J. Org. Chem.* **2004**, *69*, 4031.

[47] (a) R. B. Bedford, H. Dumycz, M. F. Haddow, L. T. Pilarski, A. G. Orpen, P. G. Pringle, R. L. Wingad, *Dalton Trans.* **2009**, 7796; (b) V. V. Dunina, P. A. Zykov, M. V. Livantsov, I. V. Glukhov, K. A. Kochetkov, I. P. Gloriozov, Y. K. Grishin, *Organometallics* **2009**, *28*, 425; and (c) J. Li, M. A. Siegler, M. Lutz, A. L. Spek, R. J. M. Klein Gebbink, G. van Koten, *Organometallics* **2009**, *28*, 5323.

[48] Y. Hu, X. Liang, J. Wang, Z. Zheng, X. Hu, *J. Org. Chem.* **2003**, *68*, 4542.

[49] A. Korostylev, I. Gridnev, J. M. Brown, *J. Organomet. Chem.* **2003**, *680*, 329.

[50] (a) C. Vallée, Y. Chauvin, J. M. Basset, C. C. Santini, J.-C. Galland, *Adv. Synth. Catal.* **2005**, *347*, 183; and (b) J.-C. Galland, Y. Chauvin, C. Vallée, FR 2873696 A1 20060203.

[51] (a) S. E. Lyubimov, V. A. Davankov, P. M. Valetskii, P. V. Petrovskii, M. G. Maksimov, K. N. Gavrilov, *Russ. Chem. Bull. Inter. Ed.* **2006**, *55*, 1448; (b) K. N. Gavrilov, S. E. Lyubimov, O. G. Bondarev, M. G. Maksimova, S. V. Zheglov, P. V. Petrovskii, V. A. Davankov, M. T. Reetz, *Adv. Synth. Catal.* **2007**, *349*, 609; and (c) S. E. Lyubimov, P. V. Petrovskii, V. A. Davankov, *Russ. Chem. Bull. Inter. Ed.*, **2009**, *58*, 528.

[52] (a) C. Monti, C. Gennari, U. Piarulli, *Chem. Eur. J.* **2007**, *13*, 1547; (b) C. Monti, C. Gennari, U. Piarulli, *Chem. Commun.* **2005**, 5281; (c) C. Monti, C. Gennari, U. Piarulli, J. G. de Vries, A. H. M. de Vries, L. Lefort, *Chem. Eur. J.* **2005**, *11*, 6701; (d) C. Monti, C. Gennaria, U. Piarulli, *Tetrahedron Lett.* **2004**, *45*, 6859; and (e) W. Chen, J. Xiao, *Tetrahedron Lett.* **2001**, *42*, 8737.

[53] D. Nakano, M. Yamaguchi, *Tetrahedron Lett.* **2003**, *44*, 4969.

[54] D. Nakano, R. Hirano, M. Yamaguchi, C. Kabuto, *Tetrahedron Lett.* **2003**, *44*, 3683.

[55] (a) E. Raluy, O. Pàmies, M. Diéguez, S. Rosset, A. Alexakis, *Tetrahedron: Asymmetry* **2009**, *20*, 2167; (b) Y. Mata, M. Diéguez, O. Pàmies, S. Woodward, *J. Org. Chem.* **2006**, *71*, 8159; (c) H. Huang, X. Liu, S. Chen, H. Chen, Z. Zheng, *Tetrahedron: Asymmetry 2004*, 15, **2011**; and (d) J. E. Babiarz, S. D. Pastor, *Phosphorus Sulfur* **1994**, *91*, 263.

[56] (a) H. Huang, X. Liu, H. Chen, Z. Zheng, *Tetrahedron: Asymmetry* **2005**, *16*, 693; and (b) H. Huang, Z. Zheng, H. Luo, C. Bai, X. Hu, H. Chen, *J. Org. Chem.* **2004**, *69*, 2355.

[57] (a) M. Diéguez, O. Pàmies, G. Net, A. Ruiz, C. Claver, *J. Mol. Catal. A: Chem.* **2002**, *185*, 11; and (b) M. Diéguez, A. Ruiz, C. Claver, *Tetrahedron: Asymmetry* **2001**, *12*, 2861.

[58] (a) C. Saluzzo, J. Breuzard, S. Pellet-Rostaing, M. Vallet, F. Le Guyader, M. Lemaire, *J. Organomet. Chem.* **2002**, *643–644*, 98; and (b) K. N. Gavrilov, O. G. Bondarev, R. V. Lebedev, A. I. Polosukhin, A. A. Shyryaev, S. E. Lyubimov, P. V. Petrovskii, S. K. Moiseev, V. N. Kalinin, N. S. Ikonnikov, *J. Organomet. Chem.* **2002**, *655*, 204.

[59] For a review on tropos catalysts, see K. Mikami, M. Yamanaka, *Chem. Rev.* **2003**, *103*, 3369.

[60] (a) A. Iuliano, S. Facchetti, T. Funaioli, *Chem. Commun.* **2009**, 457; (b) A. Iuliano, D. Losi, S. Facchetti, *J. Org. Chem.* **2007**, *72*, 8472; (c) S. Facchetti, D. Losi, A. Iuliano, *Tetrahedron: Asymmetry* **2006**, *17*, 299; (d) A. Iuliano, S. Facchetti, G. Uccello-Barretta, *J. Org. Chem.* **2006**, *71*, 4943; and (e) A. Iuliano, P. Scafato, *Tetrahedron: Asymmetry* **2003**, *14*, 611.

[61] P. Hannen, H.-C. Militzer, E. M. Vogl, F. A. Rampf, *Chem. Commun.* **2003**, 2210.

[62] T. Harada, T. M. T. Tuyet, A. Oku, *Org. Lett.* **2000**, *2*, 1319.

[63] S. Wulnnemann, R. Fröhlich, D. Hoppe, *Org. Lett.* **2006**, *8*, 2455.

[64] (a) G. J. H. Buisman, P. C. J. Kamer, P. W. N. M. van Leeuwen, *Tetrahedron: Asymmetry* **1993**, *4*, 1625; and (b) T. Yamato, K. Hasegawa, Y. Saruwatari, L. Doamekpor, *Chem. Ber.* **1993**, *126*, 1435.

[65] D. Seebach, R. Dahinden, R. E. Marti, A. K. Beck, D. A. Plattner, F. N. M. Kuhnle, *J. Org. Chem.* **1995**, *60*, 1788.

[66] (a) S. M. Smith, J. M. Takacs, *J. Am. Chem. Soc.* **2010**, *132*, 1740; and (b) S. M. Smith, N. C. Thacker, J. M. Takacs, *J. Am. Chem. Soc.* **2008**, *130*, 3734.

[67] (a) I. S. Mikhel, G. Bernardinelli, A. Alexakis, *Inorg. Chim. Act.* **2006**, *359*, **1826**; (b) A. Alexakis, J. Burton, J. Vastra, C. Benhaim, X. Fournioux, A. van den Heuvel, J.-M. Levêque, F. Mazé, S. Rosset, *Eur. J. Org. Chem.* **2000**, 401; and (c) J.-M. Brunel, G. Buono, *J. Org. Chem.* **1993**, *58*, 7313.

[68] (a) R. B. Bedford, S. L. Hazelwood (*neé* Welch), M. E. Limmert, D. A. Albisson, S. M. Draper, P. N. Scully, S. J. Coles, M. B. Hursthouse, *Chem. Eur. J.* **2003**, *9*, 3216; and (b) R. B. Bedford, S. L. Hazelwood, M. E. Limmert, J. M. Brown, S. Ramdeehul, A. R. Cowley, S. J. Coles, M. B. Hursthouse, *Organometallics* **2003**, *22*, 1364.

[69] (a) W.-J. Shi, J.-H. Xie, Q.-L. Zhou, *Tetrahedron: Asymmetry* **2005**, *16*, 705; and (b) W.-J. Shi, L.-X. Wang, Y. Fu, S.-F. Zhu, Q.-L. Zhou, *Tetrahedron: Asymmetry* **2003**, *14*, 3867.

[70] (a) W.-J. Shi, Q. Zhang, J.-H. Xie, S.-F. Zhu, G.-H. Hou, Q.-L. Zhou, *J. Am. Chem. Soc.* **2006**, *128*, 2780; and (b) H.-F. Duan, J.-H. Xie, W.-J. Shi, Q. Zhang, Q.-L. Zhou, *Org. Lett.* **2006**, *8*, 1479.

[71] B. H. G. Swennenhuis, R. Chen, P. W. N. M. van Leeuwen, J. G. de Vries, P. C. J. Kamer, *Org. Lett.* **2008**, *10*, 989.

[72] (a) J. Wilting, M. Janssen, C. Müller, M. Lutz, A. L. Spek, D. Vogt *Adv. Synth. Catal.* **2007**, *349*, 350; and (b) L. Bini, E. A. Pidko, C. Müller, R. A vanSanten, D. Vogt, *Chem. Eur. J.* **2009**, *15*, 8768.

[73] T. Foo, J. M. Garner, R. Ozer, for Invista Technologies, WO 2008008930.

[74] M. Yan, X. Li, A. S. C. Chan, *Chin. Sc. Bull.* **2003**, *48*, 2188.

[75] H. Bahrmann, B. Fell, G. Papadogianakis, for Hoechst AG, DE 3942787, **1991**.

[76] S. Ini, A. G. Oliver, T. D. Tilley, R. G. Bergman, *Organometallics* **2001**, *20*, 3839.

[77] (a) M. Rubio, A. Suárez, D. del Río, A. Galindo, E. Alvarez, A. Pizzano, *Dalton Trans.* **2007**, 407; and (b) M. Rubio, A. Suárez, D. del Río, A. Galindo, E. Alvarez, A. Pizzano, *Organometallics*, **2009**, *28*, 547.

[78] R. Angharad-Baber, R. B. Bedford, M. Betham, M. E. Blake, S. J. Coles, M. F. Haddow, M. B. Hursthouse, A. G. Orpen, L. T. Pilarsky, P. G. Pringle, R. L. Wingad, *Chem. Commun.* **2006**, 3880.

[79] J. Powell, A. Kuksis, C. J. May, P. E. Meindi, S. J. Smith, *Organometallics* **1989**, *8*, 2933.

[80] D. K. MacFarland, C. R. Landis, *Organometallics* **1996**, *15*, 483.

[81] P. Krüger, B. Weberndörfer, H. Werner, *Z. Anorg. Allg. Chem.* **2000**, *626*, 2228.

[82] Y. Jiang, S. Xue, Z. Li, J. Deng, A. Mi, A. S. C. Chan, *Tetrahedron Asymmetry* **1998**, *9*, 3185.

[83] (a) J. Jiang, S. Xue, K. Yu, Z. Li, J. Deng, A. Mi, A. S. C. Chan, *J. Organmet. Chem.* **1999**, *586*, 159; and (b) S. Xue, *Y-Z. Jiang Chin. J. Chem.* **2004**, *22*, 1456.

[84] For the more common procedures of preparation of protected carbohydrates affording precursors of chiral ligands, see S. Castillón, C. Claver, Y. Díaz, *Chem. Soc. Rev.* **2005**, *34*, 702.

[85] H. Brunner, W. Pieronczyk, *J. Chem. Res. Synop.* **1980**, 74.

[86] Q.-L. Zhao, L.-L. Wang, F. Y. Kwong, A. S. C. Chan, *Tetrahedron: Asymmetry* **2007**, *18*, 1899.

[87] M. Diéguez, O. Pàmies, C. Claver, *J. Org. Chem.* **2005**, *70*, 3363.

[88] M. T. Reetz, T. Neugebauer, *Angew. Chem. Int. Ed.* **1999**, *38*, 179.

[89] G. J. H. Buisman, M. E. Martin, E. J. Vos, A. Klootwijk, P. C. Kamer, P. W. N. M. van Leeuwen, *Tetrahedron: Asymmetry* **1995**, *6*, 719.

[90] (a) M. Diéguez, O. Pàmies, A. Ruiz, S. Castillón, C. Claver, *Chem. Commun.* **2000**, 1607; (b) M. Diéguez, O. Pàmies, A. Ruiz, C. Claver, *New J. Chem.* **2002**, *26*, 826; and (c) M. Diéguez, O. Pàmies, A. Ruiz, S. Castillón, C. Claver, *Chem. Eur. J.*, **2001**, *7*, 3086.

[91] E. Guiu, B. Muñoz, C. Claver, S. Castillón, *Adv. Synth. Catal.* **2003**, *345*, 169.

[92] M. Diéguez, A. Ruiz, C. Claver, *Tetrahedron: Asymmetry* **2001**, *12*, 2895

[93] (a) M. Diéguez, O. Pàmies, C. Claver, *Chem. Commun.* **2005**, 1121; and (b) J. Mazuela, M. Coll, M. Diéguez, O. Pàmies, *J. Org. Chem.* **2009**, 5440.

[94] M. Diéguez, A. Ruiz, C. Claver, *J. Org. Chem.* **2002**, *67*, 3796.

[95] M. Diéguez, O. Pàmies, C. Claver, *Adv. Synth. Catal.* **2005**, *347*, 1257.

[96] A. Gual, C. Godard, C. Claver, S. Castillón, *Eur. J. Org. Chem.* **2009**, 1191.

[97] A. Gual, C. Godard, S. Castillón, C. Claver, *Adv. Synth. Catal.* **2010**, *352*, 463.

[98] S. Jansat, M. Gómez, K. Philippot, G. Muller, E. Guiu, C. Claver, S. Castillón, B. Chaudret, *J. Am. Chem. Soc.* **2004**, *126*, 1592.
[99] (a) A. Gual, R. Axet, K. Philippot, B. Chaudret, A. Denicourt-Nowicky, A. Rocoux, S. Castillón, C. Claver, *Chem. Commun.* **2008**, 2759; and (b) A. Gual, C. Godard, K. Philippot, B. Chaudret, A. Denicourt-Nowicky, A. Rocoux, S. Castillón, C. Claver, *ChemSusChem* **2009**, *2*, 769.
[100] M. Aghmiz, A. Aghmiz, Y. Díaz, A. Masdeu-Bultó, C. Claver, S. Castillón *J. Org. Chem.* **2004**, *69*, 7502.
[101] R. Axet, J. Benet-Buchholz, C. Claver, S. Castillón, *Adv. Synth. Catal.* **2007**, *349*, 1983.
[102] R. M. D. Nunes, A. F. Peixoto, R. M. Axet, M. M. Pereira, M. J. Moreno, L. Kollar, C. Claver, S. Castillón, *J. Mol. Catal. A* **2006**, *247*, 275.
[103] A. Balanta-Castillo, I. Favier, E. Teuma, S. Castillón, C. Godard, A. Aghmiz, C. Claver, M. Gómez, *Chem. Commun.* **2008**, 6197.
[104] (a) E. Billig, A. G. Abatjoglou, D. R. Bryant, to Union Carbide Corp., EP 213639; and (b) E. Billig, A. G. Abatjoglou, D. R. Bryant, R. E. Murray, J. E. Maher to Union Carbide Corp., US Patent 4717775.
[105] G. D. Cuny, S. L. Buschwald *J. Am. Chem. Soc.* **1993**, *115*, 2066.
[106] (a) E. Billig, A. G. Abatjoglou, D. R. Bryant, to Union Carbide Corp., US Patent US1988/4769498; and (b) A. van Rooy, P. C. J. Kamer, P. W. N. M. van Leeuwen, K. Goubitz, J. Fraanje, N. Veldman, A. L. Spek, *Organometallics* **1996**, *15*, 835.
[107] C. B. Dieleman, P. C. J. Kamer, J. N. H. Reek, P. W. N. M. van Leeuwen, *Helv. Chim. Acta* **2001**, *84*, 3269.
[108] S. D. Pastor, J. L. Hyun, P. A. Odorisio, R. K. Rodebaugh, *J. Am. Chem. Soc.* **1988**, *110*, 6547.
[109] C. G. Arena, D. Drommi, F. Faraone, C. Graiff, A. Tiripicchio, *Eur. J. Inor. Chem.* **2001**, 247.
[110] G. T. Whiteker, J. Klosin, K. J. Gardner to *Dow Chemical Comp.*, US Pat. Appl. US2004/01990023.
[111] C. J. Cobley, R. D. Froese, J. Klosin, C. Qin, G. T. Whiteker, *Organometallics.* **2007**, *26*, 2986.
[112] Q. Zhang, J.M. Takacs, *Org. Lett.* **2008**, *10*, 545.
[113] M. J. Baker, P. G. Pringle, *Chem. Commun.* **1991**, 1292.
[114] (a) M. Yan, L.-W. Yang, K.-Y. Wong, A. S. C. Chan, *Chem. Commun.* **1999**, 11; (b) M. Yan, Z.-Y. Zhou, A. S. C. Chan, *Chem. Commun* **2000**, 115; and (c) L. Su, X. Li, W. L. Chan, X. Jiz, A. S. C. Chan, *Tetrahedron: Asymmetry 2003*, 14, **1865**.
[115] S. Cserepi-Szucs, G. Huttner, L. Zsolnai, A. Szolosy, C. Hegedus, J. Bakos, *Inorg. Chim. Acta* **1999**, *296*, 222.
[116] Y. Zou, Y. Yan, X. Zhang, *Tetrahedron Lett.* **2007**, *48*, 4781.
[117] D. Selent, W. Baumann, K.-D. Wise, A. Börner, *Chem. Commun.* **2008**, 6023.
[118] I. Piras, R. Jennerjahn, R. Jackstell, W. Baumann, A. Spannenberg, R. Franke, K.-D. Wiese, M. Beller, *J. Organomet. Chem.* **2010**, *695*, 479.
[119] S. J. Sturla, S. L. Buchwald, *J. Org. Chem.* **2002**, *67*, 3398.
[120] C. J. Chapman, K. J. Wadsworth, C. G. Frost, *J. Organomet. Chem.* **2003**, *680*, 206.
[121] K. Junge, B. Hagemann, S. Enthaler, G. Erre, M. Beller, *Arkivok*, **2007**, *V*, 50.
[122] S. E. Lyubinov. R. P. Zhuravskii, V. I. Rozenber, A. S. Safroov, P. V. Petrovskii, V. A. Davankov, *Russ. Chem. Bull. Int. Ed.* **2008**, *57*, 137.
[123] (a) R. A. Smith, S. D. Pastor, for Ciba Specialty Chemicals Corp, US Pat. 5858905, **1999**; (b) A. Korostylev, D. Selent, A. Monsees, C. Borgmann, A. Börner, *Tetrahedron:Asymmetry 2003*, 14, **1905**; and (c) G/ P. Calabrò, D. Drommi, G. Bruno, F. Faraone, *Dalton Trans.* **2004**, 81.
[124] G. Marlin, for General Electric Comp., US Patent US1999/5919966
[125] L. M. Abdrakhamova, V. F. Mironov, T. A. Branova, R. Z. Musin, *Russ. J. Org. Chem.* **2007**, *43*, 1090.
[126] (a) S. E. Lyubimov, A. S. Safranov, A. A. Tyutyunov, V. N. Kalinin, E. E. Saod-aliev, A. R. Khokhlov, P.V. Petrovskii, P.M. Valetskii, V. A. Davankov, *Russ. Chem. Bull. Int. Ed.* **2008**, *57*, 345; and (b) S. E. Lyubimov, A. A. Tyutyunov, P. A. Vologzhanin, A. S. Safranov, P. V. Petrovskii, V. N. Kalinin, K. N. Gavrilov, V. A. Davankov, *J. Organomet. Chem.* **2008**, *693*, 3321.
[127] (a) D. Hess, K. D. Wiese, C. Borgmann, A. Börner, D. Selent, R. Schmutler, C. Kunze, for Oxeno Olefinchemie, EP 2000/1201675; (b) D. Selent, A. Börner, B. Kreidler, D. Hess, K.-D. Wiese, for Evonik Oxeno, WO 2008/071508; (c) R. Paciello, L. Siggel, H. J. Kneuper, N. Walker, M. Röper, *J. Mol. Catal. A* **1999**, *143*, 85; (d) E. Billing, A. G. Abatjoglu, D. R. Bryant, for Union Carbide, US Patent 4769498; and (e) P. Lorz, W. Bertleff, M. Röper, D. Koeffer, for BASF, EP 472071.

[128] W. J. Peng, J. E. Holladay, for Union Carbide Chemical, WO 2004/035595.

[129] N. M. Pugashova, A. M. Koroteev, I. S. Bushmarinov, K. A. L. Lysenko, E. En Andriutse, M. P. Koroteev, E. E. Nifantév, *Russ. J. Gen. Chem.* **2008**, *78*, 883.

[130] S. D. Pastor, S. P. Shun, *Tetrahedron: Asymmetry* **1998**, *9*, 543.

[131] (a) Y. Yan, X. Zhang, *J. Am. Chem. Soc.* **2006**, *128*, 16056; and (b) S. Yu, Y. Chie, Z. Guan, X. Zhang, *Org. Lett.* **2008**, *10*, 3469.

[132] X. Zhang, Y. Yan, for the Penn State Research Foundation),WO2007/078859

[133] (a) T. Jerphagnon, J.-L. Renaud, P. Demonchaux, A. Ferreira, C. Bruneau, *Adv. Synth. Catal.* **2004**, *346*, 33; (b) K. N. Gavrilov, S. E. Lyubimov, S. V. Zheglov, E. B. Benetsky, P. V. Petrovskii, E. A. Rastorguev, T. B. Grishina, V. A. Davankov, *Adv. Synth. Catal.* **2007**, *349*, 1085; (c) M. T. Reetz, J.-A. Ma, R. Godard, *Angew. Chem. Int. Ed.* **2005**, *44*, 412; (d) M. T. Reetz, A. Meiswinkel, G. Mehler, K. Angermund, M. Graf, W. Thiel, R. Mynott, D. G. Blackmond, *J. Am. Chem. Soc.* **2005**, *127*, 10305; (e) M. T. Reetz, L. J. Goossen, A. Meiswinkel, J. Paetzold, J. F. Jensen, *Org. Lett.* **2003**, *5*, 3099; (f) H. Huang, Z. Zheng, H. Luo, C. Bai, X. Hu, H. Chen, *Org. Lett.* **2003**, *5*, 4137; (g) W. Chen, J. Xiao, *Tetrahedron Lett.* **2001**, *42*, 2897; (h) M. T. Reetz, G. Mehler, A. Meiswinkel, T. Sell, *Tetrahedron Lett.* **2002**, *43*, 7941; and (i) B. Lynikaite, J. Cvengros, U. Piarulli, C. Gennari, *Tetrahedron Lett.* **2008**, *49*, 755.

[134] W. Chen, S. M. Roberts, J. Whittall, *Tetrahedron Lett.* **2006**, *47*, 4263.

[135] (a) M. Diéguez, A. Ruiz, C. Claver, *Dalton Trans.* **2003**, 2957; (b) O. Pàmies, G. Net, A. Ruiz, C. Claver, *Eur. J. Inorg. Chem.* **2000**, 1287.

[136] B. D. Chapsal, Z. Hua, I. Ojima, *Tetrahedron: Asymmetry* **2006**, *17*, 642.

[137] K. Weissermel, J. P. Arpe, in *Industrial Organic Chemistry*, Wiley-VCH, **2003**.

[138] See, for example: (a) A. Gual, C. Godard, S. Castillon, C. Claver, *Tetrahedron: Asymmetry* **2010**, *21*, 1135; (b) J. Klosin, C. R. Landis, *Acc. Chem. Res.* **2007**, *40*, 1251; (c) B. Breit, *Top. Curr. Chem.* **2007**, *279*, 139; (d) F. Ungvári, *Coord. Chem. Rev.* **2007**, *251*, 2087; (e) F. Ungvári, *Coord. Chem. Rev.* **2007**, *251*, 2072; and (f) K. D. Wiese, D. Obst, *Top. Organomet. Chem.* **2006**, *18*, 1.

[139] (a) J. E. Babin, G. T. Whiteker, Asymmetric Synthesis, World Patent, WO 1993/9303839; and (b) G. T. Whiteker, J. R. Briggs, J. E. Babin, G. A. Barne, in *Asymmetric Catalysis Using Biphosphite Ligands in Chemical Industries*, Marcel Dekker, vol. *89*, **2003**.

[140] (a) G. J. H. Buisman, L. A. van der Veen, A. Klootwijk, W. G. J. de Lange, P. C. J. Kamer, P. W. N. M. van Leeuwen, D. Vogt, *Organometallics* **1997**, *16*, 2929; and (b) G. J. H. Buisman, L. A. van der Veen, P. C. J. Kamer, P. W. N. M. van Leeuwen, *Organometallics* **1997**, *16*, 5681.

[141] (a) S. Cserépi-Szûcs, I. Tóth, L. Párkanyi, J. Bakos, *Tetrahedron: Asymmetry* **1998**, *9*, 3135; and (b) R. Abdallah, J. A. J. Breuzard, M. C. Bonet, M. Lemaire, *J. Mol. Cat.: A Chem.* **2006**, *249*, 218.

[142] (a) O. Pàmies, G. Net, A. Ruiz, C. Claver, *Tetrahedron: Asymmetry* **2000**, *11*, 1097; and (b) M. Diéguez, O. Pàmies, A. Ruiz, C. Claver, *New. J. Chem.* **2002**, *26*, 827.

[143] C. Cobley, J. Klosin, C. Qin, G. T. Whiteker, *Org. Lett.* **2004**, *6*, 3277.

[144] (a) D. Sémeril, D. Matt, L. Toupet, *Chem. Eur. J.* **2008**, *14*, 7144; and (b) L. Monnereau, D. Sémeril, D. Matt, L. Toupet, *Adv. Synth. Catal.* **2009**, *351*, 1629.

[145] S. Krill, in *Applied Homogeneous Catalysis with Organometallic Compounds*, 2nd ed., B. Cornils, W. A. Herrmann, Eds., Wiley-VCH Verlag GmbH, **2002**, 468–490.

[146] (a) W. C. Seidel, C. A. Tolman, to DuPont, US Patent US1973/3850973; *Chem. Abstr.* **1975**, *82*, 97704; (b) C. M. King, W. C. Seidel, C. A. Tolman, to DuPont, US Patent US1975/3925445; *Chem. Abstr.* **1976**, *84*, 88921; and (c) M. Rapoport, to DuPont, US Patent US1983/4371474, **1983**; *Chem. Abstr.* **1983**, *98*, 125452.

[147] (a) C. A. Tolman, R. J. McKinney, W. C. Seidel, J. D. Druliner, W. R. Stevens, *Homogeneous Nickel-Catalysed Olefin Hydrocyanation. Advances in Catalysis*, Academic Press, vol. *33*, 1, **1985**; and (b) R. J. McKinney, in *Homogeneous Catalysis*, G. W. Parshall, Ed., John Wiley & Sons, Inc., **1992**, 42.

[148] (a) K. A. Kreutzer, W. Tam, to DuPont, US Patent US1996/5512696; (b) A. I. Breikss, to DuPont, US Patent US996/5523453; (c) K. A. Kreutzer, W. Tam, to DuPont, US Patent US1997/5663369; and (d) W. Tam, K. A. Kreutzer, R. J. McKinney, to DuPont, US Patent US1997/5,688,986; and (e) M. J. Baker, K. N. Harrison, A. G. Orpen, P. G. Pringle, G. Shaw, *J. Chem. Soc., Chem. Commun.* **1991**, 803.

[149] M. Yan, Q.-Y. Xu, A. S. C. Chan, *Tetrahedron: Asymmetry* **2000**, *11*, 845.

[150] (a) T. V. RajanBabu, A. L. Casalnuovo, to DuPont, US Patent US1992/5,175,335; and (b) T. V. RajanBabu, A. L. Casalnuovo, T. A. Ayers, in *Ligand Tuning in Asymmetric Catalysis: Hydrocyanation and Hydrogenation Reactions*, M. P. Doyle, Ed., *Advances in Catalytic Processes*, vol. 2, 1–41, JAI Press, **1997**.
[151] K. Biswas, O. Prieto, P. J. Goldsmith, S. Woodward, *Angew. Chem. Int. Ed.* **2005**, *44*, 2232.
[152] C.-H. Xing, Y.-X. Liao, P. He, Q.-S. Hu, *Chem. Commun.* **2010**, *46*, 3010.
[153] A. Alexakis, J. C. Frutos, P. Mangeney, *Tetrahedron: Asymmetry* **1993**, *4*, 2427.
[154] For a recent review, see A. Alexakis, J. E. Bäckvall, N. Krause, O. Pàmies, M. Diéguez, *Chem. Rev.* **2008**, *108*, 2796.
[155] S. Facchetti, D. Losi, A. Iuliano, *Tetrahedron: Asymmetry* **2006**, *17*, 2993.
[156] A. Alexakis, C. Benhaim, *Tetrahedron: Asymmetry* **2000**, *12*, 1151.
[157] (a) A. Alexakis, J. Burton, J. Vastra, C. Benhaim, X. Fournioux, A. van den Heuvel, J.-M. Levêque, F. Mazé, S. Rosset, *Eur. J. Org. Chem.* **2000**, 4011; and (b) A. Alexakis, J. Vastra, J. Burton, C. Benhaim, P. Mangeney, *Tetrahedron Lett.* **1998**, *39*, 7869.
[158] A. Alexakis, C. Benhaïm, X. Fournioux, A. van der Heuvel, J.-M. Levêque, S. March, S. Rosset, *Synlett* **1999**, 1811.
[159] Y.-L. Chen, R. Fröhlich, D. Hoppe, *Tetrahedron: Asymmetry* **2009**, *20*, 1144.
[160] (a) L. Wang, Y.-M. Li, C.-W. Yip, L. Qiu, Z. Zhou, A. S. C. Chan, *Adv. Synth. Catal.* **2004**, *346*, 947; (b) O. Pàmies, G. Net, A. Ruiz, C. Claver, *Tetrahedron: Asymmetry 1999*, 10, **2007**; (c) O. Pàmies, M. Diéguez, G. Net, A. Ruiz, C. Claver, *Tetrahedron: Asymmetry* **2000**, *11*, 4377; (d) L. Liang, M. Yan, Y.-M. Li, A. S. C. Chan, *Tetrahedron: Asymmetry* **2004**, *15*, 2575; and (e) M. Yan, A. S. C. Chan, *Tetrahedron Lett.* **1999**, *40*, 6645.
[161] F. A. Cotton, G. Wilkinson, *Advanced Inorganic Chemistry*, 5th ed., John Wiley & Sons, Inc., **1988**.
[162] T. Hayashi, K. Yamasaki, *Chem. Rev.* **2003**, *103*, 2829.
[163] C. Marelli, C. Monti, C. Gennari, U. Piarulli, *Synlett*, **2007**, 2213.
[164] S. Facchetti, I. Cavallini, T. Funaioli, F. Marchetti, A. Iuliano, *Organometallics* **2009**, *28*, 4150.
[165] For reviews, see J. Tsuji, in *Palladium Reagents and Catalysis, Innovations in Organic Synthesis*, John Wiley & Sons, Inc., **1995**; (b) B. M. Trost, D. L. van Vranken, *Chem. Rev.* **1996**, *96*, 395; (c) M. Johannsen, K. A. Jorgensen, *Chem. Rev.* **1998**, *98*, 1689; (d) A. Pfaltz, M. Lautens, in *Comprehensive Asymmetric Catalysis*, E. N. Jacobsen, A. Pfaltz, H. Yamamoto, Eds., vol. *2*, Springer-Verlag, **1999**; (e) G. Helmchen, A. Pfaltz, *Acc. Chem. Res.* **2000**, *33*, 336; (f) A. M. Masdeu-Bultó, M. Diéguez, E. Martín, M. Gómez, *Coord. Chem. Rev.* **2003**, *242*, 159; (g) B. M. Trost, M. L. Crawley, *Chem. Rev.* **2003**, *103*, 2921; (h) E. Martín, M. Diéguez, *C. R. Chem.* **2007**, *10*, 188; and (i) Z. Lu, S. Ma, *Angew. Chem. Int. Ed.* **2008**, *47*, 258.
[166] (a) M. Diéguez, S. Jansat, M. Gomez, A. Ruiz, G. Muller, C. Claver, *Chem. Commun.* **2001**, 1132; (b) O. Pàmies, G. P. F. van Strijdonck, M. Diéguez, S. Deerenberg, G. Net, A. Ruiz, C. Claver, P. C. J. Kamer, P. W. N. M. van Leeuwen, *J. Org. Chem.* **2001**, *66*, 8867; (c) I. Favier, M. Gómez, G. Muller, M. R. Axet, S. Castillón, C. Claver, S. Jansat, B. Chaudret, K. Philippot, *Adv. Synth. Catal.* **2007**, *349*, 2459; and (e) A. Gual, S. Castillón, O. Pàmies, M. Diéguez, C. Claver, *Dalton Trans.* **2011**, doi:10.1039/c0dt01067g.
[167] K. N. Gavrilov, S. E. Lyubimov, P. V. Petrovskii, S. V. Zheglov, A. S. Safronov, R. S. Skazov, V. A. Davankov, *Tetrahedron* **2005**, *61*, 10514.
[168] H. Kim, H. Men, C. Lee, *J. Am. Chem. Soc.* **2004**, *126*, 1336.
[169] N. Kinoshita, K. H. Marx, K. Tanaka, K. Tsubaki, T. Kawabata, N. Yoshikai, E. Nakamura, K. Fuji, *J. Org. Chem.* **2004**, *69*, 7960.
[170] K. Fuji, N. Kinoshita, K. Tanaka, T. Kawabata, *Chem. Commun.* **1999**, 2289.
[171] T. Kanayama, K. Yoshida, H. Miyabe, Y. Takemoto, *Angew. Chem. Int. Ed.* **2003**, *42*, 2054.
[172] A. Alexakis, K. Croset, *Org. Lett.* **2002**, *4*, 4147.
[173] S. A. Moteki, D. Wu, K. L. Chandra, D. S. Reddy, J. M. Takacs, *Org. Lett.* **2006**, *8*, 3097.
[174] P. Mauleón, R. M. Zeldin, A. Z. González, F. Dean Toste, *J. Am. Chem. Soc.* **2009**, *131*, 6348.
[175] A. Z. González, F. Dean Toste, *Org. Lett.* **2010**, *121*, 200.
[176] C. Nieto-Oberhuber, P- Pérez-Galán, E. Herrero-Gómez, T. Lauterbach, C. Rodríguez, S. López, C. Bour, A. Rosellón, D. J. Cárdenas, A. M. Echavarren. *J. Am. Chem. Soc.* **2008**, *130*, 269.
[177] C. H. M. Amijs, C. Ferrer, A. M. Echavarren, *Chem. Commun.* **2007**, 698.
[178] A. Prieto, M. R. Fructos, M. M. Díaz-Requejo, P. J. Pérez, P. Pérez-Galán, N. Delpont, A. M. Echavarren, *Tetrahedron* **2009**, *65*, 1790.
[179] A. Delgado, J. R. Rodríguez, J. L. Mascareñas, L. Castedo, *J. Am. Chem. Soc.* **2003**, *125*, 9282.

[180] J. Durán, M. Gulías, L. Castedo, J. L. Mascareñas, *Org. Lett.* **2005**, *7*, 5693.
[181] L. Vilariño, F. López, L. Castedo, J, L. Mascareñas, *Chem. Eur. J.* **2009**, *15*, 13308.
[182] N. Tsukada, S. Sugawara, Y. Inoue, *Org. Lett.* **2000**, *2*, 655.
[183] (a) M. Beller, A. Zapf, *Synlett*, **1998**, 792; (b) D. A. Albinson, R. B. Bedford, S. E. Lawrence, P. N. Scully, *J. Chem. Soc. Chem. Commum.* **1998**, 2095.
[184] G. P. F. van Strijdonck, M. D. K. Boele, P. C. J. Kamer, J. G. de Vries, P. W. N. M. van Leeuwen, *Eur. J. Inorg. Chem.* **1999**, 1073.
[185] Y. Kasaki, Y. Noguchi, T. Ikariya, *Chem. Commun.* **2000**, 2245.
[186] Y. Kayaki, Y. Noguchi, S. Iwasa, T. Ikariya, R. Noyori, *Chem. Commun.* **1999**, 1235.
[187] J. Ju, H. Nam, H.-M. Jung, S. Lee, *Tetrahedron Lett.* **2006**, *47*, 8673.
[188] R. B. Bedford, S. L. Welch, *Tetrahedron Lett.* **1998**, *39*, 9793.
[189] R. B. Bedford, S. L. Hazelwood, *Organometallics* **2003**, *21*, 2599.
[190] S. Zhang, D. Marshall, L. S. Liebeskind, *J. Org. Chem.* **1999**, *64*, 2796.
[191] (a) S. Yanagisawa, T. Sudo, R. Noyori, K. Itami, *Tetrahedron* **2008**, *64*, 6073; and (b) K. Itami, D. Yamazaki, J. Yoshida, *J. Am. Chem. Soc.* **2004**, *126*, 15396.
[192] T. Ohmura, H. Furukawa, M. Suginome, *J. Am. Chem. Soc.* **2006**, *128*, 13366.
[193] R. B. Bedford, M. Betham, M. E. Blake, A. Garcés, S. L. Millara, S. Prashar, *Tetrahedron* **2005**, *61*, 9799.
[194] (a) J. E. Babin, G. T. Whiteker, WO Patent 93/03830, **1992**; and (b) G. J. H. Buisman, E. J. Vos, P. C. J. Kamer, P. W. N. M. van Leeuwen, *J. Chem. Soc. Dalton Trans.* **1995**, 409.

4

Phosphoramidite Ligands

Laurent Lefort and Johannes G. de Vries

DSM Innovative Synthesis BV, P.O. Box 18, 6160 MD Geleen, The Netherlands

4.1 Introduction

Phosphoramidites are trivalent phosphorus compounds with 2 P–O bonds and 1 P–N bond. The substituents around the phosphorus can either be independent or form a heterocycle with the phosphorus atom as shown in Scheme 4.1. Compounds of type II are obtained from diols, whereas compounds of type III are synthesised from amino-alcohols.

In most of the cases, the N atom is derived from a primary or secondary (eventually cyclic) amine. In a few cases, phosphoramidites have been prepared from amides [1], sulfonamides [2], ureas [3] or hydrazines [4] leading to the structures shown in Scheme 4.2.

Although both the P and the N atoms have a free pair of electrons, phosphoramidites bind to a metal centre via a dative bond involving the phosphorus atom. These ligands are less σ donating than phosphines but better π-back acceptors. Several strategies have been followed to obtain bis-phosphoramidites as shown in Scheme 4.3.

The use of a bridging diamine is the most common way to prepare bis-phosphoramidites. Chiral diols can also be used as bridge in conjunction with amino-alcohols. There are only a few examples in the literature based on tethered diols [5–7] or polyols [8]. Pincer-type bisphosphoramidites have also been made [9].

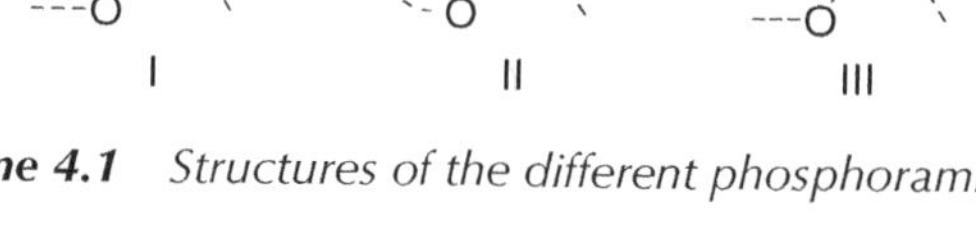

Scheme 4.1 *Structures of the different phosphoramidites.*

Phosphorus(III) Ligands in Homogeneous Catalysis: Design and Synthesis, First Edition.
Edited by Paul C. J. Kamer and Piet W. N. M. van Leeuwen.

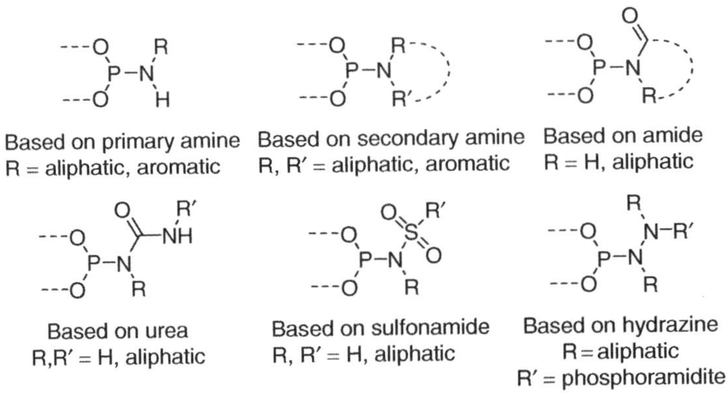

Scheme 4.2 Different N moieties found in phosphoramidites.

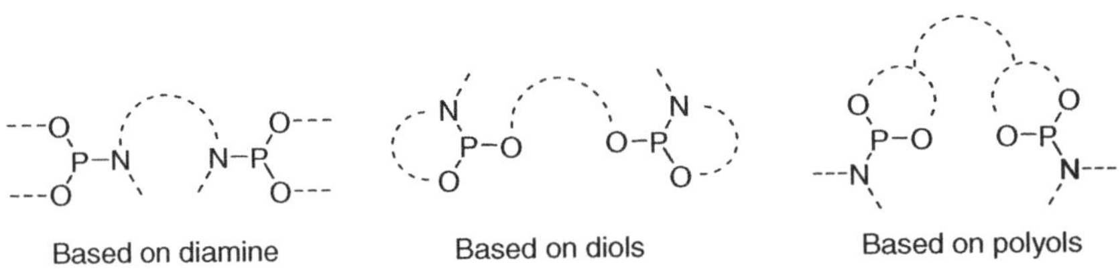

Scheme 4.3 General structure of bis-phosphoramidites.

4.1.1 History

Up until the 1990s, phosphoramidites were primarily used as phosphitylating agents in the syntheses of various biomolecules, such as phospholipids, sugar phosphates and oligonucleotides [10]. In particular, the automated solid-support DNA synthesis involved the repeated reaction of a nucleoside phosphoramidite with the 5-hydroxy function of the growing support-bound oligonucleotide chain. The displacement of the amino moiety by the incoming hydroxy group is usually catalysed by weak nitrogen acids, such as amine hydrohalides and azoles, tetrazole being the most extensively used activator.

Chiral phosphoramidites have been investigated as chiral derivatising agents for determining the enantiomeric excess (ee) of alcohols and amines by nuclear magnetic resonance (NMR), but they were found to be unreactive with both alcohols and amines [11].

Probably the first successful nonchiral application of phosphoramidites as ligands for catalytically active transition metal complexes is in rhodium-catalysed hydroformylation [12,16b]. Chiral phosphoramidites have been used quite early in asymmetric catalysis with little to moderate success [13]. The first real breakthrough in asymmetric catalysis was the invention of the BINOL-based phosphoramidites as ligands for the copper-catalysed 1,4-addition of Et2Zn [14]. Thereafter, many applications have been found for chiral and nonchiral phosphoramidite ligands [15].

4.2 Synthesis of phosphoramidites

Phosphoramidites are generally prepared in two steps from PCl_3 and alcohols and amines (Scheme 4.4, routes I and II). Because both consecutive reactions lead to the release of HCl, a base is normally used in each step. The reaction of BINOLs with PCl_3 is an exception. This reaction can advantageously be carried out

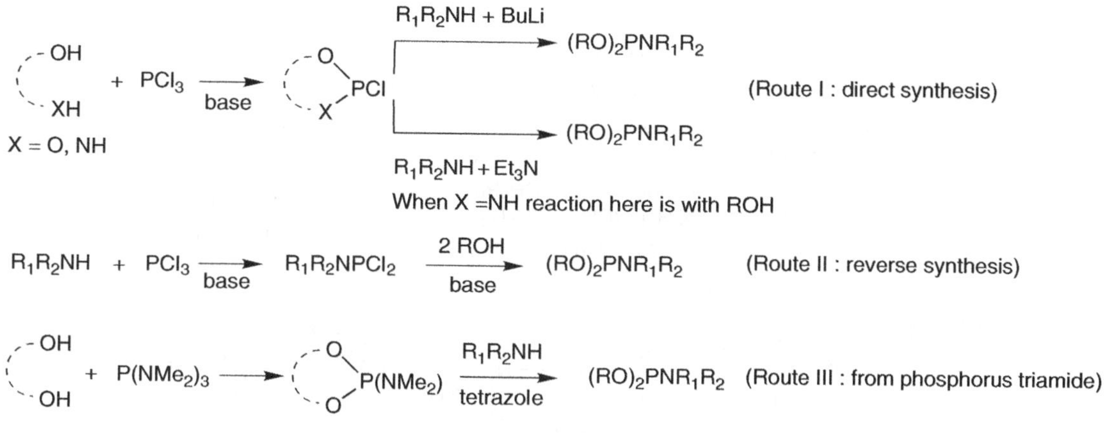

Scheme 4.4 *Synthesis of phosphoramidites.*

without base in excess refluxing PCl_3. As illustrated in Scheme 4.4 (compare routes I and II), the order of the two steps can be reversed. This versatility of the synthetic strategy appears to be particularly beneficial for the preparation of phosphoramidites containing a bulky secondary amine [16]. In this case, the amine is reacted first, leading almost exclusively to the mono-aminated phosphorus derivative to which the alcohols can be added.

After the first step, both routes I and II yield chlorophosphite intermediates that are rather sensitive to hydrolysis and have to be handled under inert atmosphere. No data could be found on their air sensitivity. The phosphoramidites are usually purified by column chromatography or by crystallisation. In reactions with butyllithium as a base, the reader is well advised to titrate the butyllithium solution before use. If the strength is less then assumed, formation of the amine. HCl salt is the inevitable result. In several cases, this salt was reported to co-elute with the ligand in the chromatographic purification step [17]. This problem can be avoided altogether by using an aqueous workup after the second step. It is advised to keep the pH at or above 7 during this procedure. The aqueous workup is mandatory when preparing large amounts (>5 g) of ligand. We have noticed that bringing the ligand solution containing the $Et_3N.HCl$ salt on the column leads to evolution of heat, which may destroy the ligand.

It is also possible to prepare phosphoramidites in one step by reaction of a diol with a phosphorus triamide (route III) [11]. We have commonly carried out this reaction for the preparation of 100–1000 g amounts of MonoPhos. This synthesis is particularly efficient since the isolation of the desired product is a simple filtration of the solid that crystallised out of the toluene solution. Via a transamination reaction catalysed by weak acid, new phosphoramidites can be formed.

4.3 Reactivity of the phosphoramidites

Only a few papers treat the reactivity of phosphoramidites. Lönnberg [18] and co-workers published a number of them in the framework of procedures used in automated oligonucleotide synthesis and explored the reactivity and the mechanism of the reaction between phosphoramidites and alcohols in the presence of acids. Interestingly, the authors claim that weak acids are more efficient catalysts for this reaction since they protonate the N atom rather than the P atom.

Our experience is that BINOL and TADDOL phosphoramidites based on secondary amines are stable in air as solids while those based on primary amines are much more labile and prone to hydrolysis. Hydrolysis

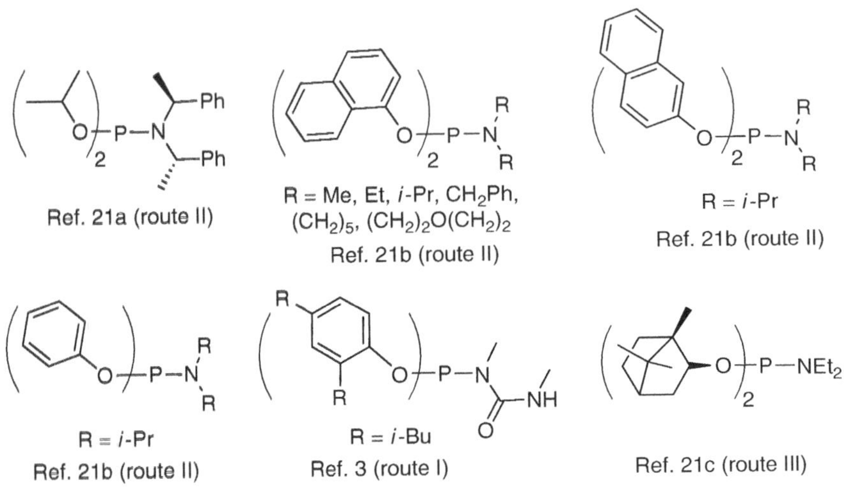

Scheme 4.5 *Acyclic phosphoramidites reported in the literature (with the corresponding reference and the synthetic route used according to the nomenclature of Scheme 4.4).*

occurs more readily under acidic conditions [19]. Hence, performing the synthesis of phosphoramidites with a slight excess of base avoids the formation of undesired hydrolysed by-products [3e]. When bound to a metal centre, phosphoramidites can resist very harsh conditions such as 100°C in the presence of water [20].

4.4 Types of phosphoramidite ligands

4.4.1 Acyclic monodentate phosphoramidites (type I, Scheme 4.1)

Only a few papers report the preparation of acyclic monodentate phosphoramidites (Scheme 4.5) [21]. Acyclic phosphoramidites comporting three different substituents around the P centre are extremely rare [12]. These P-chiral compounds could potentially be interesting for asymmetric catalysis, although their preparation in enantiopure form may be difficult and racemisation of the P-chiral centre may occur. Consequently, almost all the prepared acyclic phosphoramidites bear two identical alkoxy groups. As shown in Scheme 4.5, they have been prepared using all three different routes.

Worth mentioning is the work of Kühl *et al.* [3] using an unconventional amino counterpart: by reacting N,N′-dimethyl urea with one equivalent of (2,4-di-*t*Bu-$C_6H_3O)_2$ PCl in the presence of Et_3N with tetrahydrofuran (THF) as a solvent, they could prepare the mono-phosphoramidite in high yield (97%). In the attempted reaction of N,N′-dimethylurea with two equivalents of the chlorophosphite in toluene, the desired bis-phosphoramidite was not obtained. Instead, a mixture of the mono-phosphoramidite and unreacted chlorophosphite was recovered from the reaction mixture. The authors argue that the bulky chlorophosphite ligand is sterically too demanding for the formation of the bis-phosphoramidite urea derivative.

Mikel *et al.* [21c] report the preparation of a chiral acyclic phosphoramidite based on [(1 S)-endo]-(−)-borneol using the phosphorus triamide route [22].

4.4.2 Cyclic monodentate phosphoramidites based on diols (type II, Scheme 4.1)

The general structures of the different diols used to prepare phosphoramidites are presented in Scheme 4.6. Binaphthol and biphenol (and their analogs) are the most commonly used diols and will be discussed in

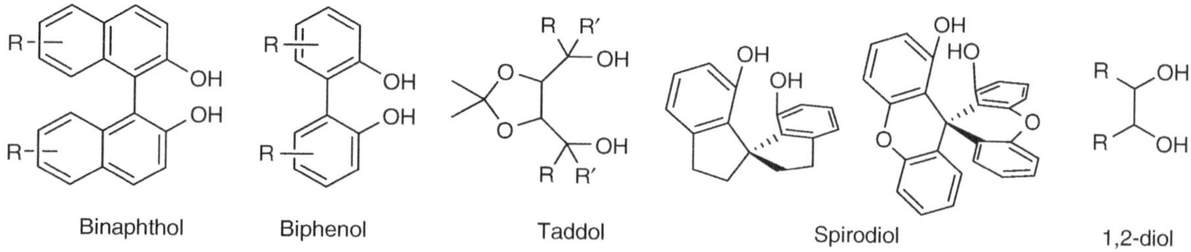

Scheme 4.6 *Main diol moieties used to prepare phosphoramidites.*

Section 4.4.2.1. In Section 4.4.2.2, TADDOL-, spirodiol- and 1,2-diol-based phosphoramidites will be discussed. Finally, phosphoramidites based on uncommon diol moieties will be discussed in Section 4.4.2.3.

4.4.2.1 Synthesis of binaphthol- and biphenol-based phosphoramidites

Several reviews have been written about enantiopure BINOL- and biphenol-based phosphoramidites and their use as ligands in asymmetric catalysis [15,23]. In general, the synthesis of these ligands is considered easy and high-yielding. The preferred method follows route I (Scheme 4.4) with PCl_3 as the phosphorus source. For bulky phosphoramidites, the reverse synthesis (route II, Scheme 4.4) is preferred. At DSM, we perform the synthesis of MonoPhos™ (Scheme 4.7) using hexamethyl phosphorus triamide (HMPT) as the starting material (see the procedure in this chapter).

Many different types of substituted biphenols and BINOLs, and primary and secondary alkyl and aryl amines, have been used to prepare phosphoramidites. Ojima *et al.* prepared phosphoramidites based on enantiopure 6,6′-dimethylbiphenols [24]. Up to five different substituents at the 3,3′-positions (i.e. H, Me, Br, *t*-Bu and Ph) were introduced onto the biphenol skeleton. Chan *et al.* were the first to prepare phosphoramidites based on partially hydrogenated binaphthol (H_8-MonoPhos, Scheme 4.7) [25]. For asymmetric catalysis, many studies have demonstrated the importance of the 3,3′-substituents on the binaphthol or biphenol unit in the chiral ligand. Bulky substituents in these positions are keys for high ee [26]. Phosphoramidites based on binaphthol with bulky substituents in the 3,3′-positions were prepared by Fan *et al.* using the direct route via the chlorophosphites, prepared in neat PCl_3 [27]. NMP was used as a base. The reaction with dibenzyl amine was performed in refluxing THF in the presence of Et_3N. The yields over the two steps varied between 46% and 86% but did not correlate with the size of the 3,3′-substituents. Beller *et al.* described an efficient method to prepare bulky phosphoramidite ligands with 3,3′-substituents relying on the use of partially hydrogenated H_8-binaphthol [28]. Indeed, this diol can be selectively brominated at the 3,3′-positions with Br_2. Direct bromination of binaphthol would occur in the 6,6′-position, and thus it would require the use of *ortho*-directing groups to prepare the 3,3′-disubstituted diol, which adds two additional steps. Palladium-catalysed coupling reactions with arylboronic acids allows the introduction of the bulky group in the 3,3′-position. The resulting diols can be treated with $P(NMe_2)_3$ in refluxing toluene to obtain ligands in moderate to good yields. Instead of $P(NMe_2)_3$, $PCl_2(NEt_2)$ in the presence of triethylamine can also be used to vary the amino moiety.

The use of monosubstituted enantiopure BINOLs allows the preparation of C_1-symmetric phosphoramidite ligands with a P* stereocentre. Reetz *et al.* were the first to use this concept with eight different BINOLs monosubstituted in the 3 position [29]. In three cases (for R=CH_2OBn, $SiMe_3$ and $SiPh_3$), phosphorylation with $P(NMe_2)_3$ led to only one diastereomer. Gavrilov *et al.* synthesised (*S*)-6-bromo-BINOL which upon reaction with $P(NEt_2)_3$ led to only the (S_a, S_p) epimer [30].

In addition to alkyl or aryl groups, uncommon substituents have been introduced in the 3,3′-position of the binaphthol unit of phosphoramidites. Lloyd-Jones *et al.* prepared a 3,3′-trifluorosulfonyl substituted phosphoramidite in just three steps from enantiopure BINOL (Scheme 4.8A) [31]. The substituted BINOL was

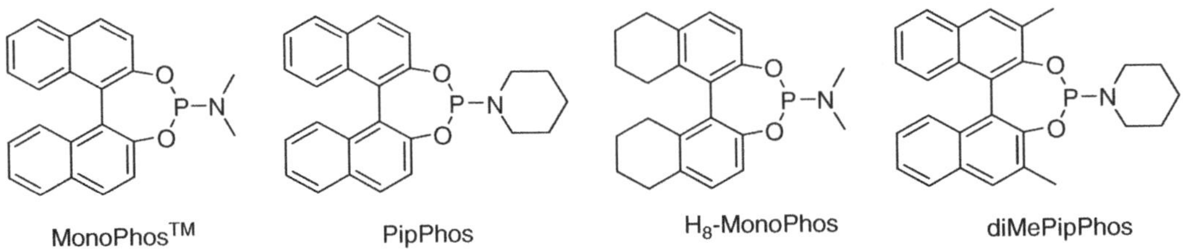

Scheme 4.7 *Names of some phosphoramidites commonly used at DSM.*

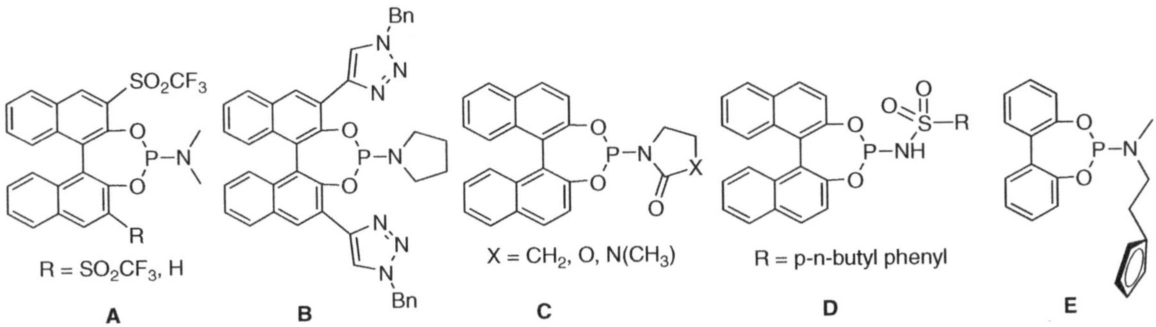

Scheme 4.8 *Phosphoramidites with less common diol or amino moieties.*

synthesised by way of an anionic thia-Fries rearrangement of the bis O–SO_2RF esters of BINOL. Using one or two equivalents of LDA in this reaction allowed the preparation of the mono- or di-substituted BINOL, respectively. McErlean *et al.* prepared 3,3′-bispyridyl-BINOL but could not obtain the corresponding phosphoramidite [32]. Use of either PCl_3 or phosphorus triamide only led to the oxidised phosphoramidite. The author reasoned that another nitrogen-containing heterocycle may reduce the propensity towards oxidation and turned towards the 3,3′-bistriazolyl substituted compound synthesised via azide–alkyne cycloaddition. For this purpose, a 3,3′-di-acetylenyl-BINOL-based phosphoramidite was needed. This intermediate was prepared either by phosphorylation of the corresponding BINOL via the PCl_3 route or, interestingly, via a Sonogashira reaction performed on the 3,3′-diiodo-phosphoramidite. The click reaction to form the triazolyl was difficult and after optimisation of the reaction conditions, the authors could obtain the desired product, although with a low yield (17%), together with the oxidised compound (Scheme 4.8B).

Primary and secondary alkyl and aryl amines are the most common building blocks used to prepared phosphoramidites. However, a few groups used other N-containing molecules. Takeuchi *et al.* prepared BINOL-based phosphoramidites with an amide moiety expecting that such a ligand could act as a bidentate ligand through coordination of both the phosphorus and the carbonyl oxygen of the amide group (Scheme 4.8C) [33]. The synthesis involved reaction of the amide and the phosphochloridite in anhydrous THF at room temperature for 40 h in the presence of Et_3N. The products were obtained in 40–50% yield after recrystallisation. Phosphoramidites with a sulfonamide moiety were prepared in the same way by Reek and co-workers (Scheme 4.8D) [34]. Upon treatment with Et_3N, these ligands can be deprotonated at the nitrogen atoms of the amido group, thus forming an anion paired with Et_3NH^+. Esteruelas *et al.* used an amine with a tethered cyclopentadienyl (Cp) group to form a chelating ligand with the metal (Scheme 4.8E). Interestingly, the phosphoramidite was synthesised from the Os and Ru complex where the Cp group was already bound to the metal centre whilst the amino group was free. Piperidinomethyl polystyrene was used as a base [35].

An additional chiral centre can be introduced in the ligand by using a chiral amine in addition to the axially chiral biphenol and binaphthol. In this case, match–mismatch effects have to be considered and both

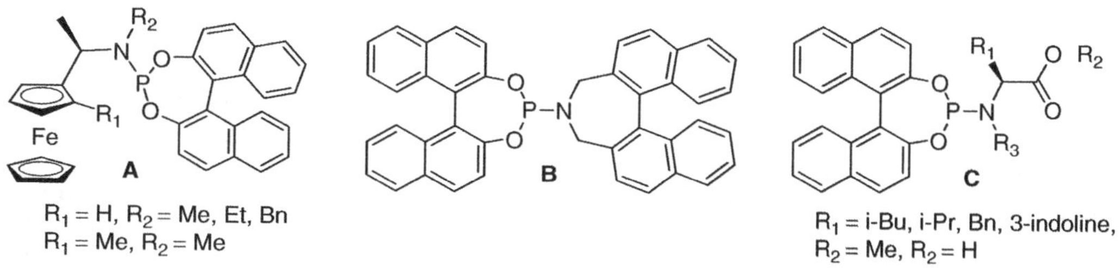

Scheme 4.9 *Phosphoramidites based on chiral amines.*

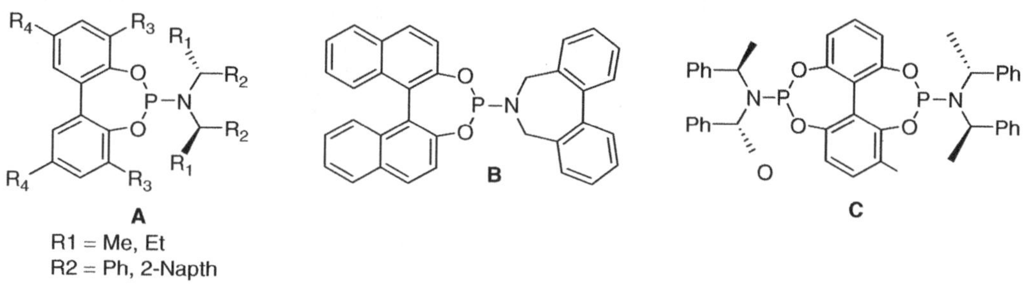

Scheme 4.10 *Atropos phosphoramidites.*

diastereomers have to be prepared. Zheng *et al.* prepared the so-called FAPhos phosphoramidites based on a chiral ferrocene amine (Scheme 4.9A) [36]. Matt *et al.* and Rosini *et al.* used the enantiopure binaphtho-azepine (Scheme 4.9B) [37]. The former group noticed a significant difference in the ^{31}P chemical shifts ($\delta = 146.0$ and 143.0 ppm, respectively) for the two pairs of enantiomers ($S_{azepine}$,S_{binol})/($R_{azepine}$,R_{binol}) and ($R_{azepine}$,S_{binol})/($S_{azepine}$,R_{binol}). Reek *et al.* prepared 12 BINOL-based phosphoramidites using enantiopure amino acid esters (Scheme 4.9C). The synthesis of these ligands could be accomplished either by the chlorophosphite route or by the transamination route from the -NEt_2 precursors. Better yields were obtained using the former route (i.e. a *ca.* 90% yield after purification) [38]. Similar ligands were also prepared by Matt and co-workers [39].

In the case of biphenol, the absence of substituents in the 6,6′ position allows axial rotation of the two aryl moieties (atropisomerisation) and renders the ligand achiral [40]. Several groups used bulky chiral amines to block this rotation and induce the axial chirality. Alexakis made the most significant contribution to this field preparing a large number of tropos ligands based on biphenol with substituents (or not) in the 3,3′- and 5,5′-positions (Scheme 4.10A) [41]. In an inverted approach, Rosini *et al.* used enantiopure BINOL to induce asymmetry in the tropos biphenylazepine moiety of the ligand (Scheme 4.10B) [42]. Zhang *et al.* used achiral tetrahydroxybiphenyls (Scheme 4.10C), obtaining four isomers with the configuration of axial R–axial S and *trans-cis* relating to the relative stereochemistry of the two phosphorus atoms [43]. Also worth citing is the contribution of Gennari and Piarulli in this field. These authors used mixtures of tropos ligands L1 with atropos ligands L2 in catalysts of the type ML^1L^2 [44].

Because of their ease of synthesis, two groups have embarked on designing protocols for the preparation of libraries of chiral phosphoramidite ligands in order to speed up the discovery of the right ligand for a given asymmetric transformation. Waldman *et al.* prepared libraries using solid-phase synthesis on functionalised polystyrene beads [45]. The ligands and catalysts were tested on the solid support. At DSM, we designed a protocol for the synthesis of a library of ligands in solution in order to avoid the common problems related with heterogenising the catalyst (mass transfer limitations, and non-innocent interaction with the polymeric

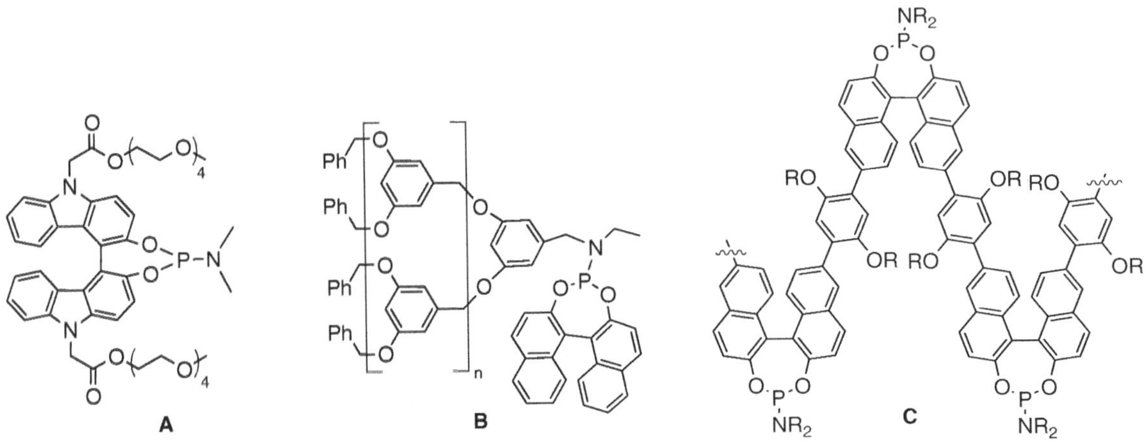

Scheme 4.11 Water-soluble phosphoramidite, dendritic, and polymeric phosphoramidites.

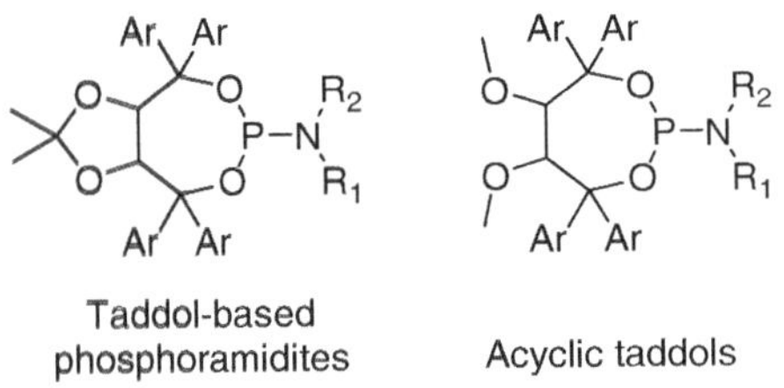

Scheme 4.12 TADDOL-based phosphoramidites.

surface) [46]. Additionally, phosphoramidites have been deposited on inorganic solids [47], modified to be water soluble [48] (Scheme 4.11A) and incorporated into dendritic [49] (Scheme 4.11B) and polymeric structures (Scheme 4.11 C) [50].

4.4.2.2 Synthesis of TADDOL-based phosphoramidites

Since the discovery of the DIOP ligand by Kagan and co-worker, much work has been performed on the development of similar chiral backbones. Seebach and co-workers developed the synthesis of TADDOL, which became a privileged structure in the field of asymmetric catalysis [51]. Thus, it was not surprising that TADDOLs were also used as chiral backbones for the synthesis of chiral phosphoramidites [52]. The preparation of these compounds is entirely similar to the one based on BINOLs. Recently, Oestreich *et al.* reported an improved synthesis of TADDOL-based phosphoramidites [53]. The synthesis begins with the reaction of TADDOL with PCl_3 in toluene using N-methylmorpholine as base which appeared to give higher yields than Et_3N. The HCl salt was removed by filtration at –78°C. After a solvent switch from toluene to THF, the chlorophosphite was added to a freshly prepared solution of a lithium amide. With this procedure, the authors were able to introduce a sterically demanding 2,2,6,6-tetramethylpiperidinyl (TMP) group at phosphorus with 50% yield. Fürstner prepared phosphoramidites based on acyclic TADDOLs in order to increase the secondary interactions between the metal centre and the pending aromatic groups (Scheme 4.12) [54]. The synthesis was performed according to the direct route from PCl_3 with Et_3N in toluene (in the presence of 4 Å mol sieves). The second step involved the reaction of the phosphochloridite with lithiated amine in THF. The yields were 60–70%.

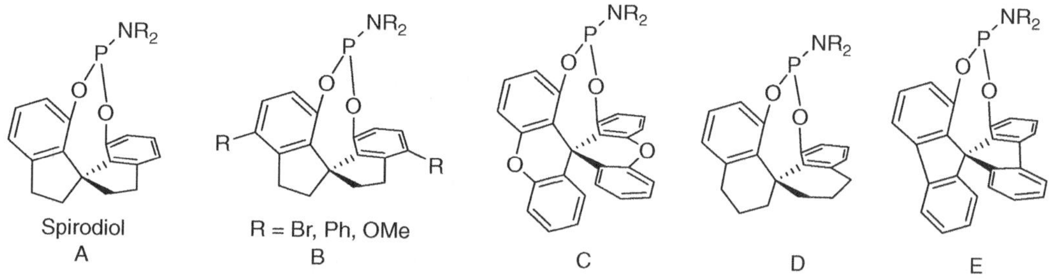

Scheme 4.13 *Spiro phosphoramidites.*

4.4.2.3 *Synthesis of spiro-based phosphoramidites*

Monodentate spiro phosphoramidites rely on 1,1′-spirobiindane-7,7′-diol. These ligands were first introduced by Zhou *et al.* (Scheme 4.13A) [55]. The dihedral angle of the two aromatic planes is larger in this class of ligands than in the one based on unsubstituted BINOLs. Equally, the synthesis of these phosphoramidites is identical to the one reported in this chapter. For example, in a procedure extracted from reference [56c], a mixture of (R)-1,1′-spirobiindane-7,7′-diol (200 mg, 0.8 mmol), HMPT (0.2 mL, 1 mmol) and dry toluene (2 mL) was heated at reflux under argon for 2 h. After cooling to room temperature, the mixture was concentrated and purified by chromatography on a silica gel column with 16:1 petroleum ether–EtOAc to give the desired phosphoramidite as a white solid (237 mg, 92% yield).

The introduction of substituents on the spiro-diols backbone via bromination leads to substitution in the 4,4′-position (Scheme 4.13B) [36b]. Zhang and co-workers reported a related spiro phosphoramidite with a 9,9′-spirobixanthene backbone [56] (Scheme 4.13 C), and Zhou reported ligands containing a 1,1′-spirobitetraline backbone (Scheme 4.13D) [36d] and a 9,9′-spirobifluorene backbone (Scheme 4.13E) [36e].

4.4.2.4 *Synthesis of 1,2-diol-based phosphoramidites*

Phosphoramidites based on 1,2-diols have not been extensively prepared (Scheme 4.14). Feringa and co-workers used the achiral cathechol as the diol backbone. However, since their interest was in asymmetric catalysis, they rendered the ligand chiral via the use of a chiral enantiopure amine [57]. Benetsky *et al.* prepared a P*-chiral phosphoramidite ligand based on a diol obtained from 3-carene [58]. One equivalent of diol was reacted with $P(NEt_2)_3$ at 120°C for 40 min in the absence of solvent. The desired product was obtained in 72% yield after distillation as a single diastereomer. Protected sugars have been used as diols by Nifantyef and co-workers [13,59].

4.4.2.5 *Phosphoramidites based on unusual diols*

Scheme 4.15 shows a set of unusual diols used for the preparation of phosphoramidites [60]. Ding *et al.* prepared the chiral diphenol intermediate from a chiral diamine, either (*R,R*)-1,2-di(2-dimethoxyphenyl)-1,2-ethylenediamine [60a] or (1*R*,2*R*)-1,2-diaminocyclohexane [60b]. Mikami *et al.* used *tropos* benzophenone-like diphenol backbones [60c,d]. Goldfuss constructed phosphoramidites based on BIFOL (biphenylbisfenchol) containing a rigid, terpene-based bicyclo[2.2.1]heptane unit [60e]. Putala prepared bis-*ortho*-phenylene and a bis-*meta*-phenylene homologue of BINOL via Negishi cross-coupling of the chiral binaphthyl diiodide [60f]. Rieger *et al.* started from mannitol to prepare a diol which was further elaborated into a phosphoramidite [60g,h]. Reek *et al.* used a bicarbazole skeleton (BICOL) and took advantage of the easy functionalisation of the nitrogen atoms of BICOL to synthesise dendrimer-functionalised ligands [60i].

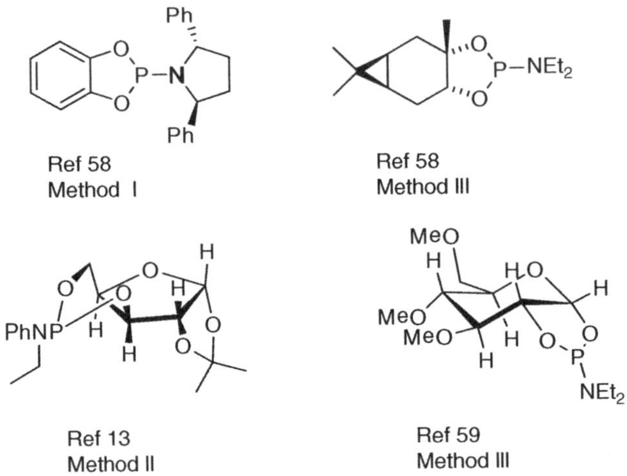

Scheme 4.14 1,2- and 1,3-diol-based phosphoramidites.

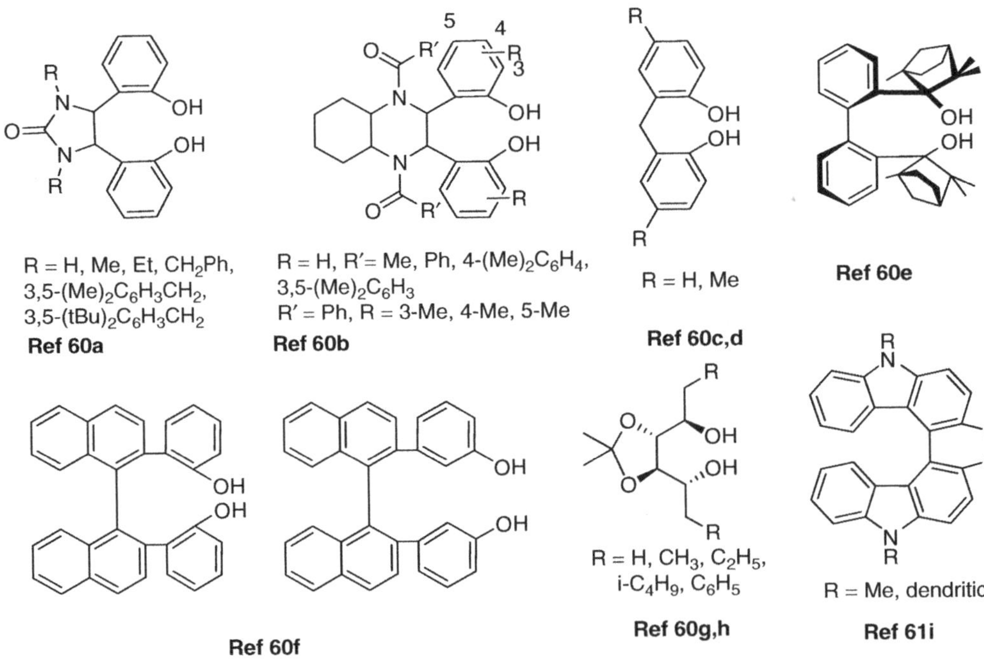

Scheme 4.15 Unusual diol skeletons used for the preparation of phosphoramidites.

4.4.3 Cyclic phosphoramidites based on amino alcohols (type III, Scheme 4.1)

Amino alcohols are attractive building blocks for phosphoramidites as they are available and structurally diverse (Scheme 4.16). Gavrilov and co-workers made monodentate phosphoramidites from prolinol, using a range of alcohols and phenols as the third substituent on phosphorus [61]. Bondarev and co-workers used (S)-α,α-diphenylprolinol which upon reaction with PCl_3 led exclusively to the (R)-P-chiral chlorophosphite. Further reaction with an alcohol in the presence of Et_3N allowed the formation of the phosphoramidite [62]. Reaction with a diol such as glycol, diethylene glycol or butane-diol led to the preparation of bidentate phosphoramidites [63].

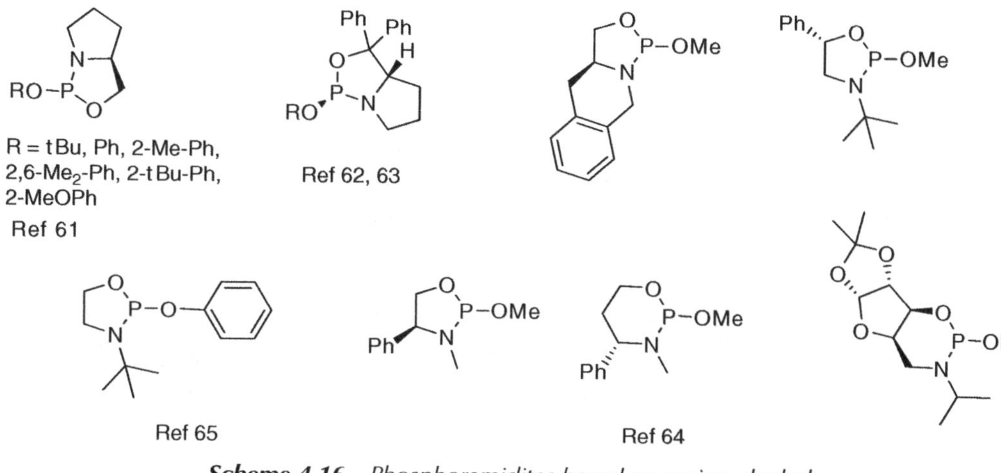

Scheme 4.16 Phosphoramidites based on amino alcohols.

Benetsky *et al.* synthesised a library of monodentate *P**-chiral phosphoramidite ligands having five- and six-membered phosphacycles and a methoxy as exocyclic substituent. These compounds were prepared by direct phosphorylation of optically active amino alcohols derived from (S)-2-amino-2-phenylacetic acid, (S)-1,2,3,4-tetrahydroisoquinoline-3-carboxylic acid, malonic acid and D-xylofuranose. A solution of the amino alcohol in benzene was added dropwise to a solution of PCl_2OMe and Et_3N in benzene at 0°C for 20 min. The mixture was then briefly heated to the boiling point. The solid $Et_3N.HCl$ was filtered off. The product was obtained by distillation *in vacuo* (1 Torr) with yield from 61% to 85% as a mixture of the two epimeric forms at the P*-stereocentre. The 2 six-membered phosphacycles contained small amounts (around 2%) of the minor epimer [64].

Pastor, Rodebaugh and co-workers prepared mono-phosphoramidites based on an amino alcohol and phenols. The phoshorochloramidite was prepared by the reaction of 2-(*N*-tert-butylamino)ethanol with PCl_3 and triethylamine (yield = 67%, distilled) or without (yield = 57%, distilled). The reaction of the phoshorochloramidite with phenolates led to the formation of the phosphoramidites in good yields. For less sterically hindered phenols, the reaction could be performed directly from the phenol in presence of Et_3N. Interestingly, the reaction with phenol gave a mixture of the desired phosphoramidite and the ring-opened phosphite [65].

4.4.4 Bis-phosphoramidites

Although in some particular cases it was shown that phosphoramidites can interact with the metal centre via the phosphorus atom and via another part of the molecule (e.g. η_2 interaction of a pending arene, and cyclometallation via CH activation), these ligands are primarily monodentate in the sense that they bind to the metal centre only via the phosphorus atoms. Consequently, due to the large success of bidentate ligands in asymmetric catalysis, many groups have prepared bis-phosphoramidites by linking two phosphoramidites together, the linker originating from either a diamine or a diol. A range of bisphosphoramidites, both diol bridged as well as diamine bridged, was described by Wissing *et al.* [12].

4.4.4.1 Bis-phosphoramidites based on diamines (Scheme 4.17)

Waldmann used bicyclic diamines such as bispidine to prepare C2-symmetric bisphosphoramidites [66]. Feringa prepared nine new bidentate phosphoramidites from 1,2- or 1,3-diamines [67]. Faraone prepared two new bidentate phosphoramidites based on DACH [68]. Matt *et al.* prepared bis-phosphoramidites based on hydrazine. The ligands were prepared in *ca.* 30–40% yield by reacting a dialkylhydrazine hydrochloride,

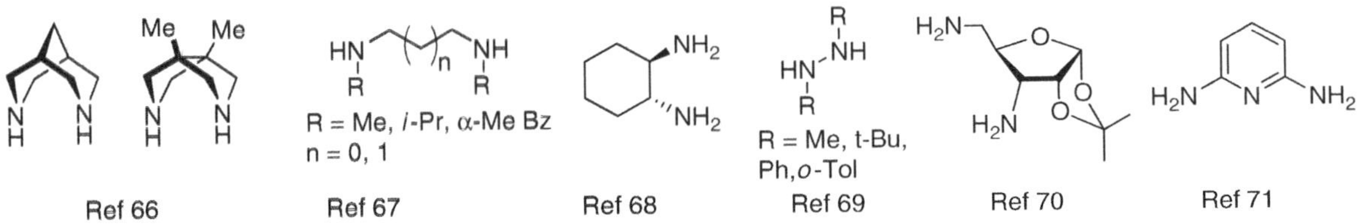

Scheme 4.17 *Diamines used for the preparation of bis-phosphoramidites.*

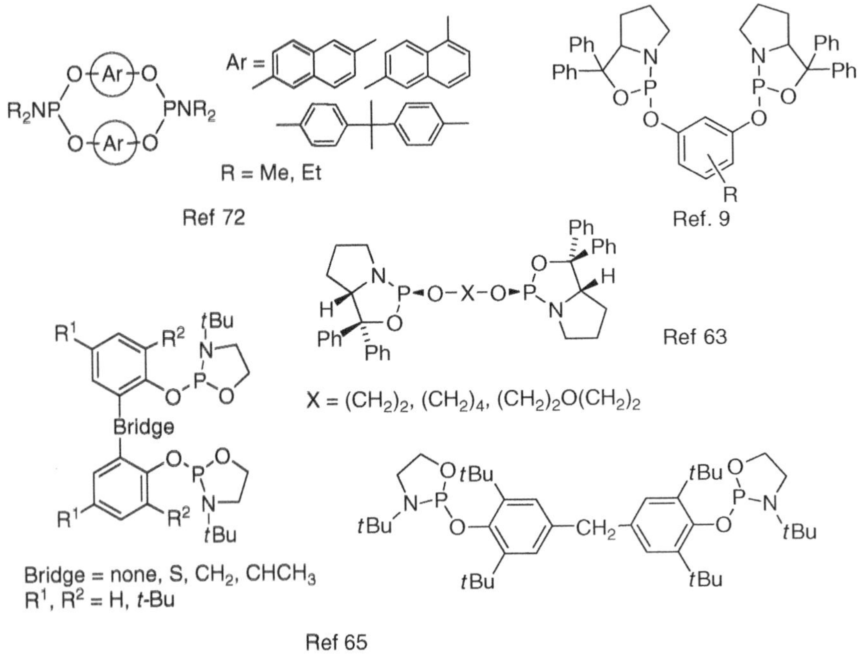

Scheme 4.18 *Bis-phosphoramidites based on diol linkers.*

[R[1]NH–NHR[1]].HCl, with three equivalents of diisopropylethylamine and two equivalents of the appropriate enantiopure phosphochloridite. However, arylhydrazine derivatives did not react following this protocol. These ligands had to be prepared by reduction of an azobenzene with sodium in THF followed by reaction with the corresponding optically pure phosphochloridite (yields *ca.* 50%) [69]. Pamies prepared bisphosphoramidites from a diamine synthesised from inexpensive D-(+)-xylose [70]. Kirchner and co-workers prepared a series of bis-phosphoramidite ligands based on 2,6-diaminopyridine. Their diol backbones included ethylene glycol, tartaric acid esters and 1,2-diphenyl-ethyleneglycol [71].

4.4.4.2 *Bis-phosphoramidites based on diols (Scheme 4.18)*

Nifantyev *et al.* prepared bis-phosphoramidites by reacting one equivalent of phosphorus triamide with one equivalent of noncyclisable diols such as bis-phenol or naphthodiols. Such diols cannot form a ring with the phosphorus and, consequently, two of them act as linkers between two different PNR_2 units. A route to prepare this compound involved the disproportionation of the bis-phosphorylated diol with one equivalent of phosphorus triamide. Several days were needed to obtain yields >50% [72].

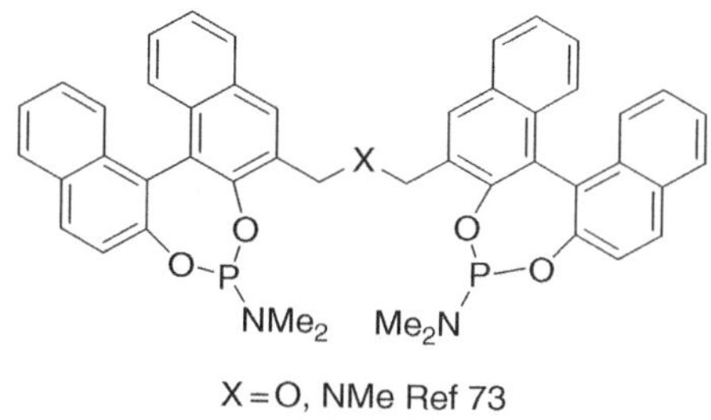

Figure 4.1 *Miyaura's bis-phosphoramidite.*

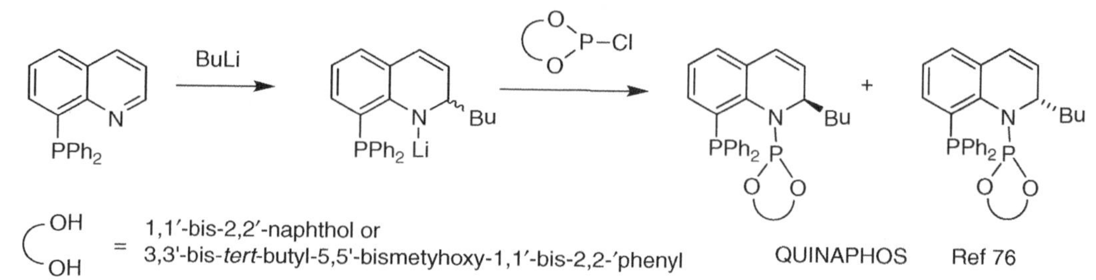

Scheme 4.19 *Synthesis of QuinaPhos.*

Rodebaugh *et al.* prepared bidentate phosphoramidites by linking two phoshochloramidites based on a centre with a bisphenol. When using 3,5-*t*-butyl-biphenol, they obtained a single diastereomer with two non-equivalent phosphorus atoms showing a through-space $^{7}J_{(P,P)}$-coupling constant of 30.3 Hz [65]. In the same vein, Reetz and Bondarev used aliphatic diols to attach two centre-based chlorophosphites [63]. Klein-Gebbink and Song followed a similar approach to prepare pincer ligands using meta-catechol as a linker [9].

4.4.4.3 Other bidentate phosphoramidites

Miyaura developed a bidentate phosphoramidite based on the dimeric ether-bridged BINOL that was earlier developed by Shibasaki. Later, he also developed the analogous nitrogen-bridged bidentate phoshoramidite (Figure 4.1) [73].

4.4.5 Mixed bidentate ligands

Next to bidentate phosphoramidites, quite a large number of other mixed bidentate phosphoramidites have been made. The area has been reviewed twice [74,75].

4.4.5.1 Phosphoramidite–phosphines

In all cases, the synthetic strategy is formation of the aminophosphine moiety first, followed by reaction of the amine moiety with a suitable phosphochloridite. In many cases researchers have chosen for a chiral bridge, to allow for more variation. The first example of this class was QuinaPhos, developed by Leitner, Faraone and co-workers (Scheme 4.19) [76].

Crévisy and co-workers synthesised a chiral amine from enantiopure BOC-phenylalaninol, which was substituted by $LiPPh_2$ via its O-tosylate. This aminophosphine was reacted with the phosphochloridite

[77]. Zheng, Hu and co-workers used an *ortho*-phosphinylated α-methylbenzylamine, which was reacted with the phosphochloridite [78]. Kostas and co-workers prepared Me-Anilaphos by the reaction of *o*-methylamino-phenyl-diphenylphosphine with the BINOL phosphochloridite [79]. Franciò, Leitner and co-workers prepared similar ligands with chiral substituents on the nitrogen atom [80]. Another linker developed by Zheng and co-workers is 1,8-naphthyl-bridged aminophosphine [81]. Boaz and co-workers [82], Zheng [83] and also Li, Yeung, Chan and co-workers [84] developed ferrocene-bridged

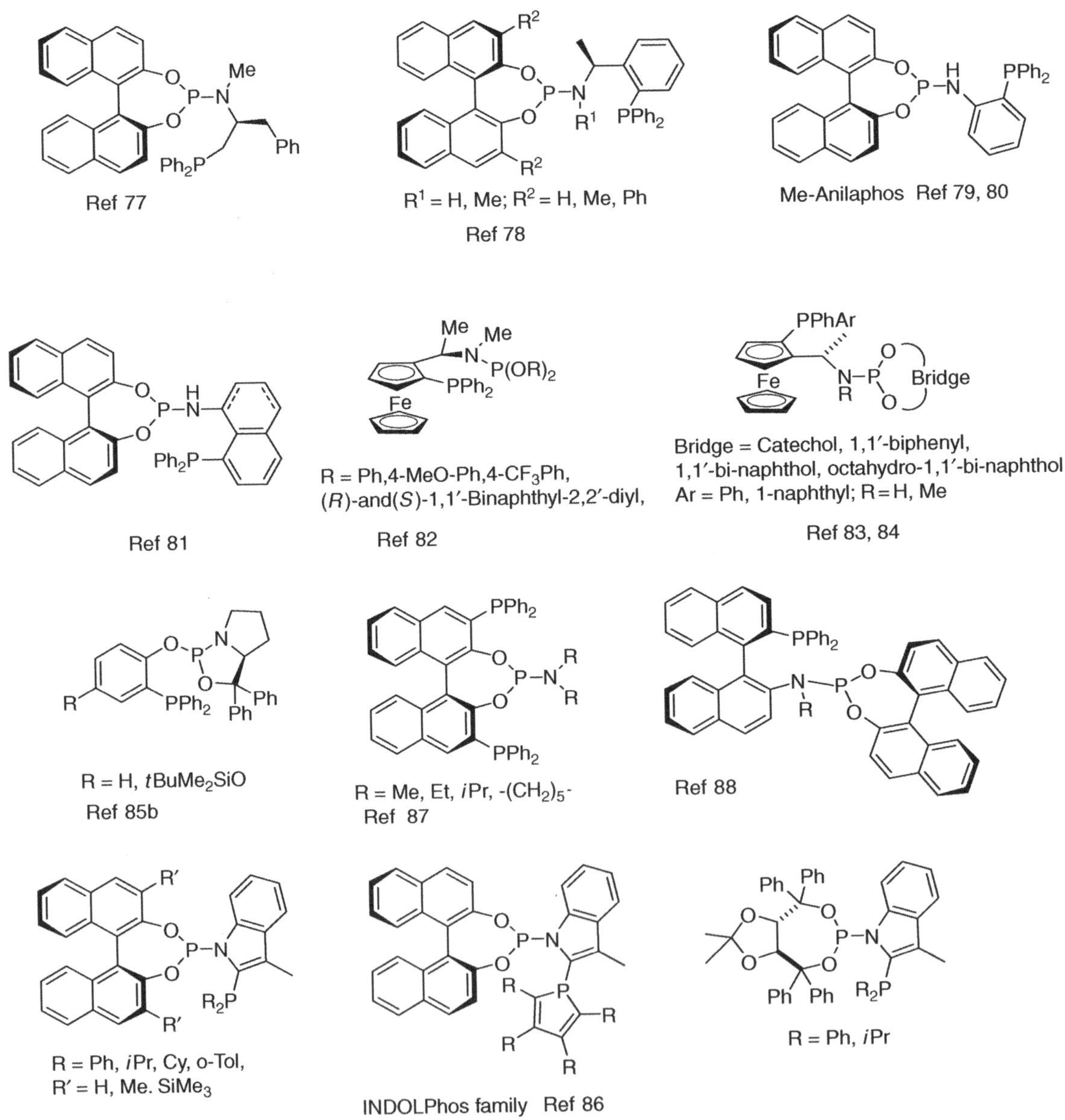

Scheme 4.20 *Phosphoramidite–phosphines.*

phosphine-phosphoramidites based on amino-phosphines made from the Ugi amine (enantiopure N, N-dimethyl-1-ferrocenylethylamine). Schmaltz and co-workers developed a useful synthetic strategy to synthesise a library of bidentate ligands starting from differentially protected hydroquinone. *Ortho*-lithiation allowed introduction of the phosphine moiety after which the adjacent hydroxyl group was deprotected and functionalised as phosphite or phosphoramidite [85]. A similar strategy was employed by Reek and co-workers, who used *ortho*-lithiation of 3-methyl-indole to introduce a PAr_2-group or a phosphole in the 2 position. Treatment with a BINOL or TADDOL phosphochloridite gave a class of bidentate ligands which they named INDOLPhos [86]. Zhang *et al.* introduced two identical -PPh_2 substituents in the 3,3′-positions of chiral BINOL, which upon treatment with HMPT generated a bidentate ligand, phosphine-phosphoramidite, since all three phosphorus atoms cannot bind simultaneously to the same metal centre [87]. The same group also prepared the phosphoramidite analogue of BINAPHOS and a number of analogous compounds [88]. Finally, although outside the scope of this chapter, worth mentioning is the supramolecular approach of Reek and co-workers which allows the formation of a bidentate phosphine-phosphoramidite by self-assembly [89].

4.4.5.2 *Phosphoramidite–phosphite*

Probably the first phosphoramidite–phosphite ligand was made by DSM researchers together with van Leeuwen (Scheme 4.21) [21]. They used bisphenol or bisnaphthyls as bridge. Laschat and co-workers used tropinone as a scaffold for bidentate ligands using the hydroxyl and amino groups for reaction with the phosphochloridite. The ketone was either transformed in a cyclic acetal or reduced to create another attachment point [90]. Kostas, Börner and co-workers were able to prepare a phosphoramidite–phosphite ligand in a single step by phosphitylating N-phenyl-aminoethanol [91]. Probably most work in this area was performed by Pàmies in collaboration with the groups of Diéguez, Claver and Alexakis (Scheme 4.22). They made a library of phosphoramidite–phosphites using either commercially available amino alcohols [92] (mostly ephedra alkaloids and reduced amino acids) or furanosides as scaffolds [70,93].

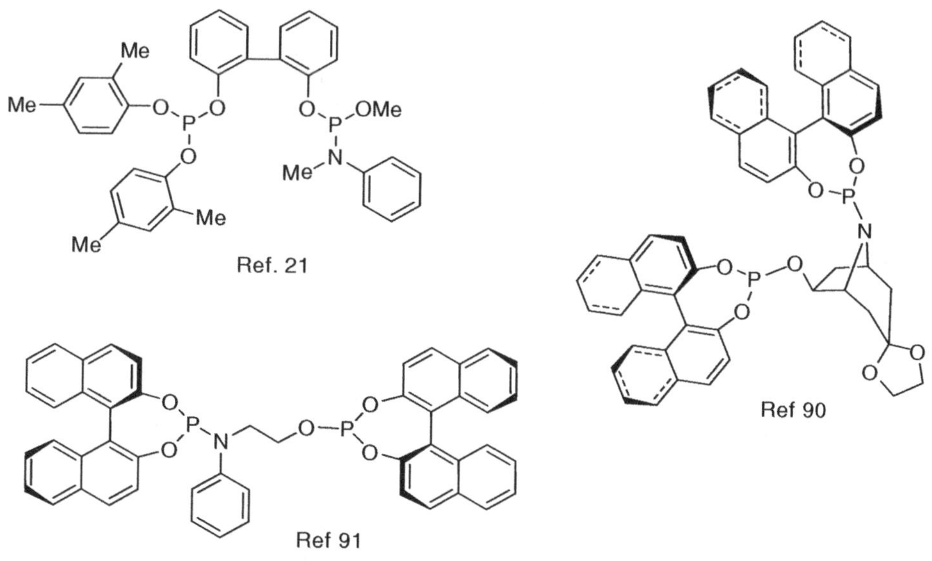

Scheme 4.21 *Phosphoramidite–phosphites.*

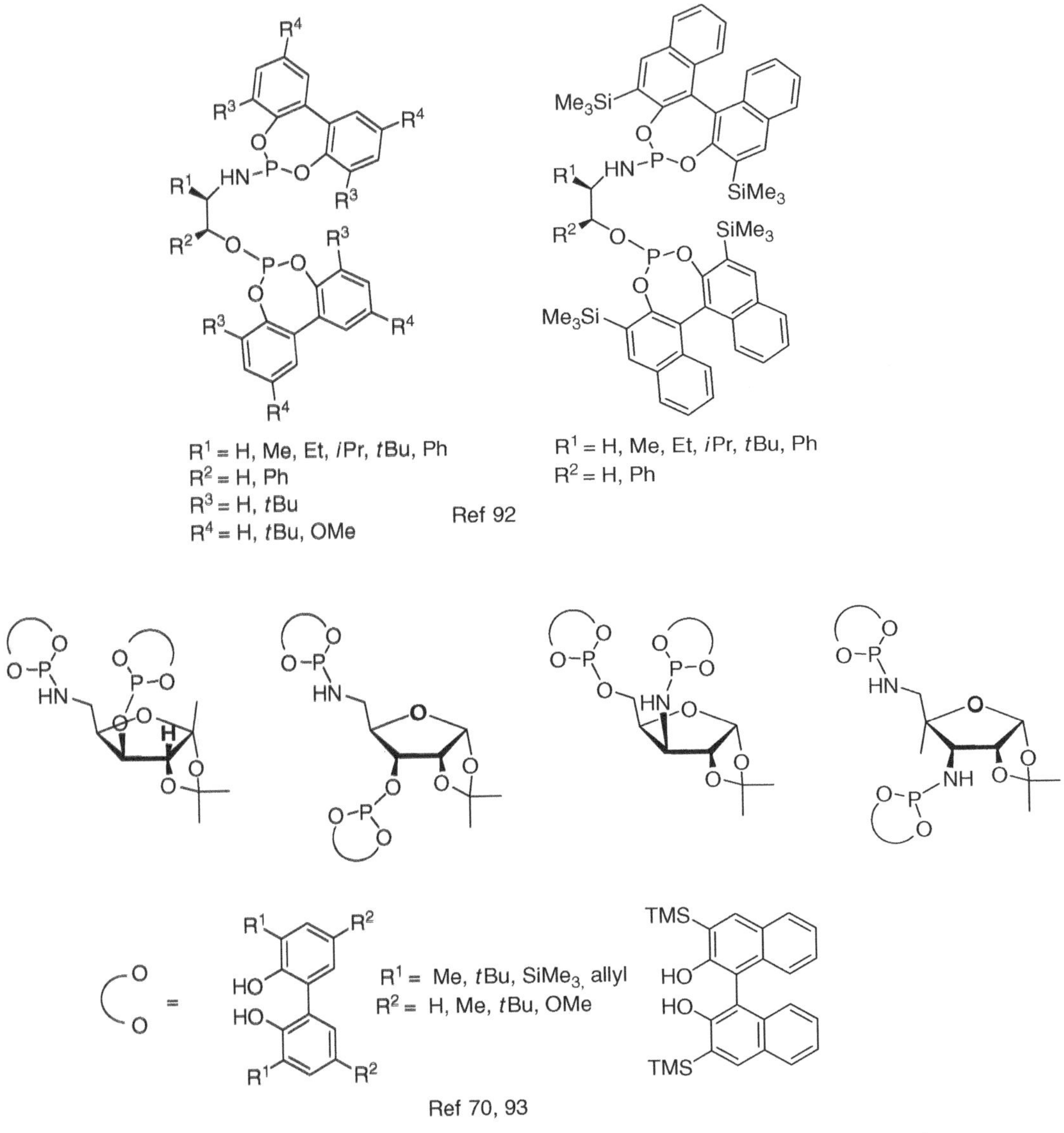

Scheme 4.22 *Phosphoramidite–phosphites developed by Pàmies and coworkers.*

4.4.5.3 Phosphoramidite–amines

Several groups have developed synthetic protocols for the synthesis of bidentate phosphoramidite–amine ligands (Scheme 4.23). Two different synthetic strategies are possible depending on if the starting material is an amino alcohol or a diamine. Both strategies have been used. Gavrilov and co-workers made these ligands by reacting a dialkylaminoethanol with a substituted 2-chloro-1,3,2-oxazaphospholidine [94]. A bicyclic ligand was obtained by the same group by reacting the 2-chloro-1,3,2-oxazaphospholidine made from (*S*)-prolinol with a dialkylaminoethanol [95]. Guiry and co-worker prepared phosphoramidite-oxazolines by phosphitylating amino-oxazolines with the phosphochloridite made from TADDOL [96]. Buono reacted prolinol and ephedrine-type amino alcohols with HMPT, and the resulting 2-dimethylamino-1,3,2-oxazaphospholidine was reacted

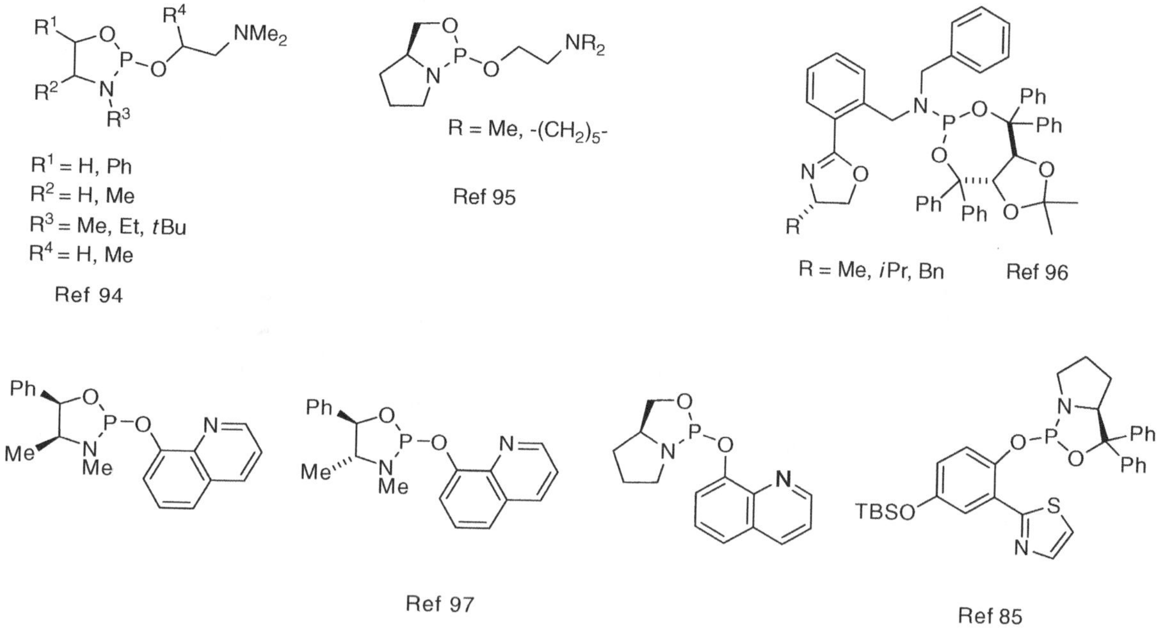

Scheme 4.23 *Phosphoramidite–amines.*

with 8-hydroxy-quinoline to obtain the bidentate ligands in moderate yields [97]. Schmalz and co-workers prepared a bidentate phosphoramidite–thiazole using his dihydroquinone-based synthesis scheme discussed in this chapter [85].

4.4.5.4 Other bidentate phosphoramidite ligands (Scheme 4.24)

Crévisy prepared a phosphoramidite–thioether ligand [77]. The precursor amino thioethers were prepared by reaction of 4-tolyl-thiol with N-protected mesylated amino alcohols that were prepared from amino acids or from 1-amino-2-indanol. Zeng and co-workers subjected alpha-methyl-benzylamine to *ortho*-lithiation after protecting the amine with TMS and quenched the aryllithium with PhSSPh. The ensuing amino thioether was methylated and reacted with the BINOL phosphochloridite to produce the desired ligands [98]. Schmalz and co-workers used their strategy based on protected hydroquinone to prepare a prolinol-based phosphoramidite–thioether ligand [85].

Douthwaite and co-workers prepared phosphoramidite–imidazole ligands that were converted into phosphoramidite–NHC ligands (Scheme 4.25) [99].

4.4.6 Polydendate phosphoramidites (Scheme 4.26)

Nifantyef *et al.* were able to prepare linear polymers of phosphoramidite by condensing bifunctional hydroxyl-containing compounds HOR′OH (R′=alkylene, arylene such as 1,6-hexanediol, 2,2′-(4-hydroxyphenyl)-propane, hydroquinone and 4,4′-dihydroxybiphenyl) with phosphorous triamides $P(NR_2)_3$ (R=Me, Et) under mild conditions (T=20–45°C, reduced pressure). The reactions were carried out with equimolar amounts of reactants without a solvent or in diethylene glycol dimethyl ether with removal of the liberated amine by distillation. The resulting polymeric materials were obtained in yields of 76–99% and due to their good swelling in organic solvent could be used as ligands in Rh-catalysed hydrogenation [100].

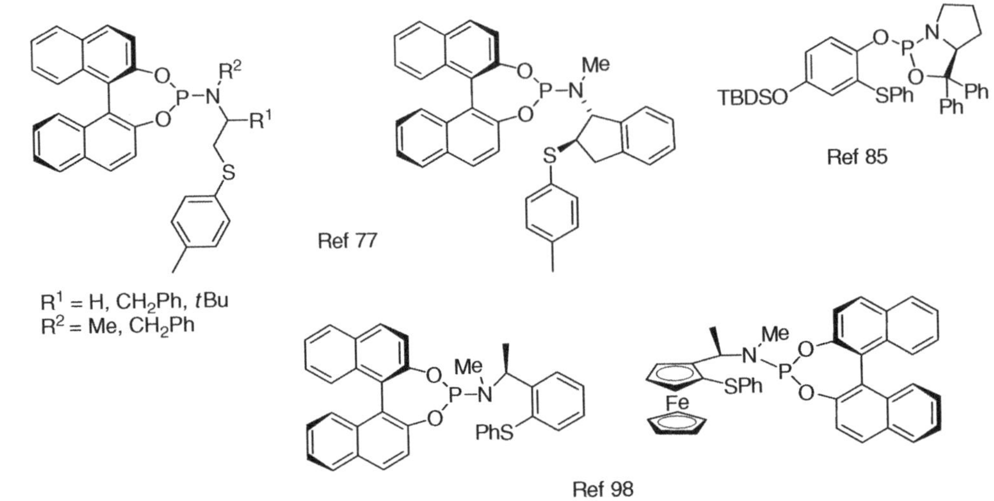

Scheme 4.24 *Other mixed bidentate phosphoramidite ligands.*

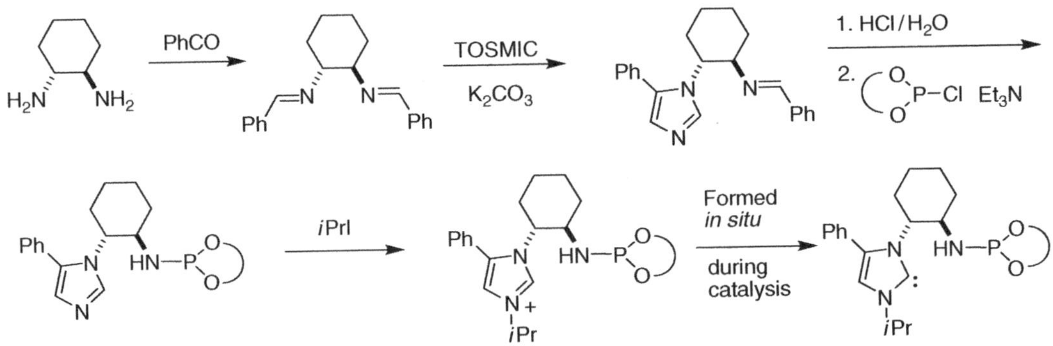

Scheme 4.25 *Synthesis of bidentate phosphoramidite–NHC ligands.*

Nifantyev *et al.* used phosphorus triamides ($P(NR_2)_3$ with R=Me, Et and piperidyl) to achieve the cyclophosphorylation of resorcinarenes, thus obtaining a tetrakis–phosphoramidite [101]. In most cases, the reactions proceeded under mild conditions and the desired phosphocavitands crystallised either directly from the reaction mixtures or upon partial evaporation of the solvent. The stereoselectivity depended on the nature and the bulk of substituents at the phosphorus atom in the phosphorylating reagent. Symmetrical phosphoramidite cavitands could be obtained with yields of 72–96%. Suprisingly, they also obtained phosphoramidites when using phosphorus ester diamide, $P(OAr)(NR_2)_2$ instead of the expected phosphite. They attribute this behaviour to stacking interactions between the aromatic group on the phosphorus and those of the calixarene, leading to an intermediate prone to trans-esterification.

Doherty *et al.* prepared two different vinyl-substituted phosphoramidites. Upon co-polymerisation of these compounds with styrene, they obtained two functionalised resins having good solubility in organic solvent but being isolable by precipitation with MeOH. In Rh-catalysed hydrogenation, one of the resins performed the same as its homogeneous analogue [102].

Several groups prepared immobilised phosphoramidites by reacting phosphochloridites on polymers functionalised with amines [103,104]. Jiang *et al.* used commercially available polystyrene resins substituted

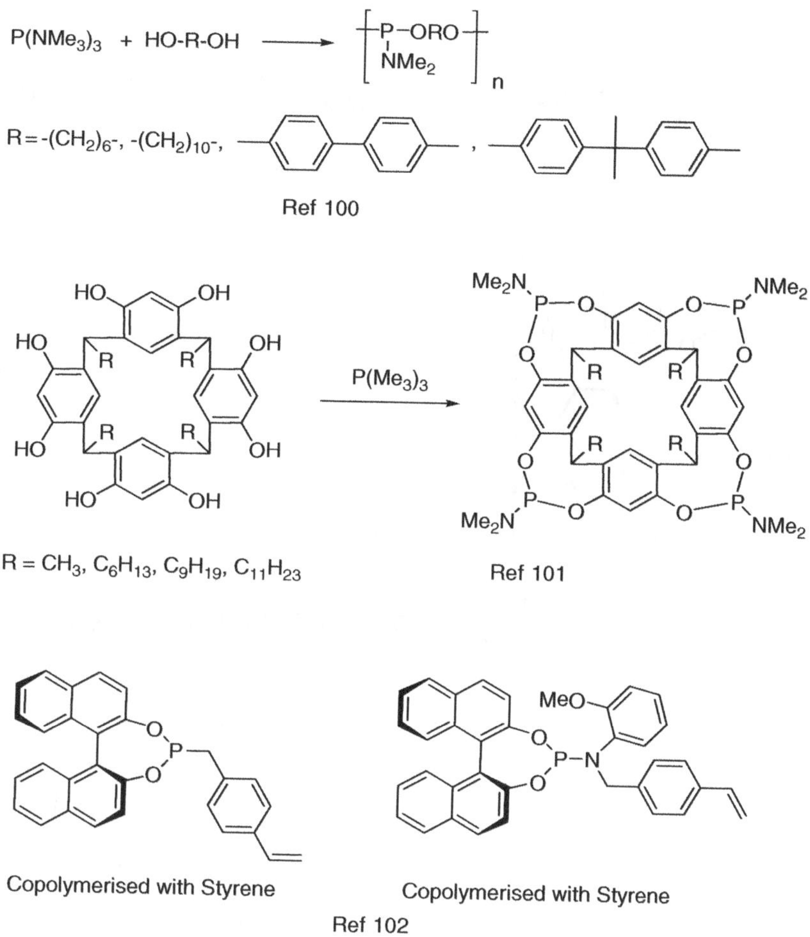

Scheme 4.26 *Polydentate phosphoramidites.*

with *N*-methylaminomethyl and pyrrolidine (Scheme 4.27) [105]. Reactions of the resins with the TADDOL- and BINOL-based phosphochloridites in the presence of DIPEA (*N,N*-diisopropylethylamine) at room temperature in THF for 1 h followed by washing afforded the immobilised ligands.

Kamer and co-workers prepared a library of immobilised phosphoramidites, which upon treatment with an alcohol in the presence of catalytic amounts of N-methylanilinium trifluoroacetate (TAMA) delivered highly pure phosphites. Thus, this allowed the preparation of not only a library of immobilised phosphoramidites but also a library of soluble phosphites [106].

For the preparation of a metal-organic framework, Ding *et al.* prepared four types of multitopic MonoPhos ligands with different bridging linkers between the BINOL subunits [107]. Reaction of these multitopic phosphoramidites (di-, tri- and tetra-meric ligands were prepared) with $[Rh(COD)_2]BF_4$ led to the formation of metal-organic frameworks that were used as catalysts. Not only could the catalysts be reused, but also it was possible to use them as stationary catalysts in a flow system for a continuous period of 144 h. (See Scheme 4.28 [108].)

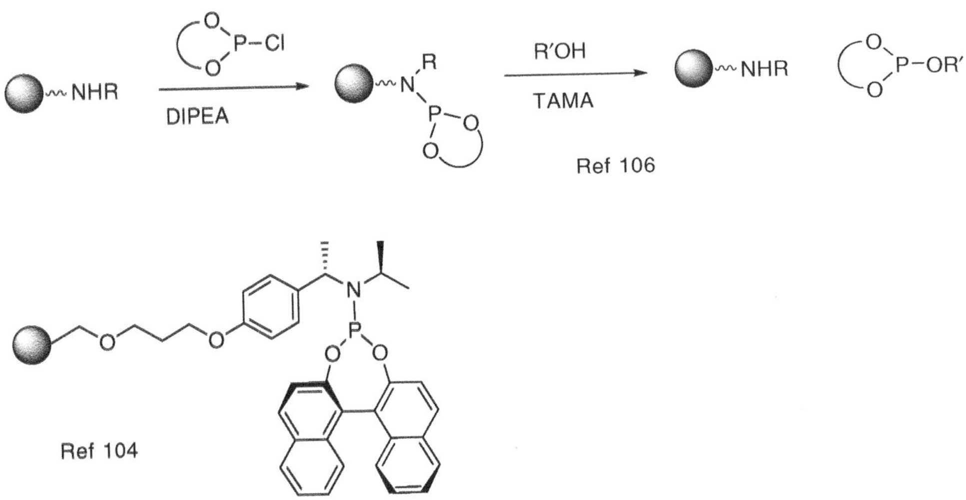

Scheme 4.27 *Immobilized phosphoramidite ligands.*

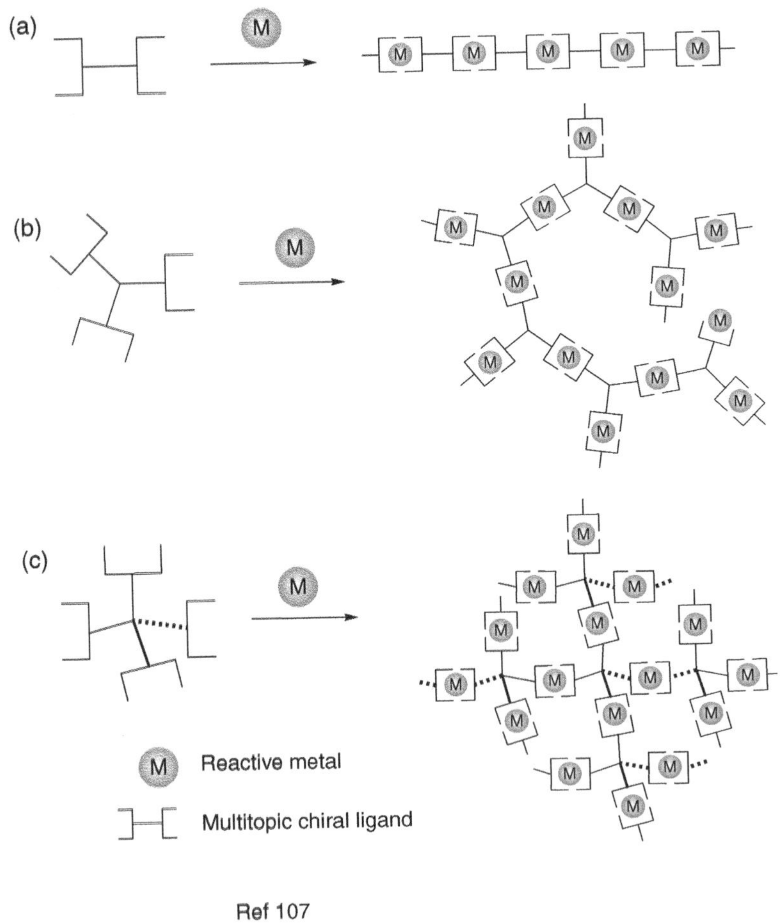

Scheme 4.28 *Metal-organic frameworks based on multitopic chiral ligands.*

4.5 Conclusion

The synthesis of new phosphoramidite ligands has blossomed tremendously in the past 15 years. This is due to a number of developments. Firstly, their synthesis is relatively easy, often in only two steps from commercially available starting materials. Secondly, many interesting applications in asymmetric synthesis have culminated in an industrial application by DSM for the asymmetric hydrogenation of a pharma intermediate [26b]. For an overview of all catalytic applications, the reader is referred to the review by Teichert and Feringa [15]. An interesting feature of monodentate ligands is the possibility to use complexes that have two different ligands [108]. Although at first view this approach would not seem to be very attractive, due to the formation of a mixture of two homo-ligated complexes and only one mixed ligated complex, in practice the approach can work if the mixed complex is formed preferentially or if the mixed complex is kinetically superior. In practice, both cases have been found. In addition, it is possible by changing the relative amounts of the two ligands to completely suppress formation of one of the two homo complexes. Finally, due to the relative ease of their synthesis it is possible to prepare a library of ligands by parallel synthesis in a robot [46]. This allows high-throughput screening, which has tremendously aided their application in both academia and industry.

We expect many more new applications of this interesting class of ligands in the years to come.

4.6 Synthetic procedures

Route I [109]. (*S*)-Bis-ß-naphthol (10 g, 35 mmol) in 40 ml of PCl_3 was heated under reflux for 8 h. Excess of PCl_3 was removed by distillation *in vacuo* (20 mbar). The residual solid was subjected to azeotropic distillation with toluene (2 × 10 ml) and dried *in vacuo* until a white foam resulted (12.2 g, 100%). The residue was dissolved in toluene to afford 50 ml of a phosphochloridite stock solution. A 10 ml aliquot of the above prepared phosphochloridite stock solution was added at 0°C to a solution of 2.13 ml (15.4 mmol, 2.2 eq) of triethylamine and 7.7 mmol (1.1 eq) of the corresponding amine in 15 ml of dry THF. The reaction mixture was allowed to warm to RT and stirred overnight. The mixture was diluted with diethyl ether (50 ml) and filtered over a plug of silica, then washed with 50 ml of diethyl ether, and the solvent was removed *in vacuo*. Column chromatography over silica with pentane–ethyl acetate as eluent or recrystallisation from diethyl ether gave the pure phosphoramidites as foams or solids.

Route II [16b]. *3,3′-Di-tert-butyl-5,5′-dimethoxy-2,2′-diyl, N,N-diisopropyl-phosphoramidite* A solution of 3,3′-di-*tert*-butyl-5,5′-dimethoxy-biphenyl-2,2′-diol (8.96 g, 25.4 mmol) and Et_3N (15 ml, 108.2 mmol) in toluene was added dropwise to solution of $Cl_2PN(i\text{-}Pr)_2$ (4.6 g, 22.8 mmol) in toluene at 0°C. When addition was complete, the cooling bath was removed and the mixture was refluxed for 6 h. After cooling to RT and filtration, the solvent and excess Et_3N were evaporated, leaving a yellow precipitate. The product was washed twice with cold hexane and dried *in vacuo*. Yield 9.16 g (18.8 mmol, 83%) of white crystals.

Route III [11]. *(R)-(-)-MonoPhos*™. (*R*)-(+)-1,1′-bi(2-naphthol) (2.00 g, 7.5 mmol), hexamethylphosphorustriamide (1.4 g, 9.5 mmol), NH_4Cl (0.01 g) and toluene (10 ml; the original recipe called for the use of benzene) were heated to reflux for 12 h. The mixture was concentrated under reduced pressure, affording an oil. The oil was stirred with 25 g of ether, resulting in the formation of white crystals. The crystals were isolated and, if necessary, recrystallised from dry diethyl ether.

References

[1] G. Onodera, K. Watabe, M. Matsubara, K. Oda, S. Kezuka, R. Takeuchi, *Adv. Synth., Catal.*, **2008**, *350*, 2725.

[2] F. W. Patureau, S. de Boer, M. Kuil, J. Meeuwissen, P. A. R. Breuil, M. A. Siegler, A. L. Spek, A. J. Sandee, B. de Bruin, J. N. H. Reek, *J. Am. Chem. Soc.* **2009**, *131*, 6683.

[3] O. Kühl, *Dalton Trans.* **2003**, 949.
[4] L. Eberhardt, D. Armspach, D. Matt, B. Oswald, L. Toupet, *Org. Biomol. Chem.* **2007**, *5*, 3340.
[5] K. Kurihara, Y. Yamamoto, N. Miyaura, *Tetrahedron Lett.* **2009**, *50*, 3158.
[6] K. Kurihara, Y. Yamamoto, N. Miyaura, *Adv. Synth. Catal.* **2009**, *351*, 260.
[7] L. Shi, X. W. Wang, C. A. Sandoval, Z. Wang, H. J. Li, J. Wu, L. T. Yu, K. L. Ding, *Chem. Eur. J.* **2009**, *15*, 9855.
[8] M. Yang, J. Zhang, N. Wang, X. Q. Yu, *Arkivoc* **2008**, 279.
[9] J. Li, M. Lutz, A. L. Spek, G. P. M. van Klink, G. van Koten, R. Klein Gebbink, *Organometallics* **2010**, *29*, 1379.
[10] (a) É. E. Nifant'ev, M. K. Grachev, *Russ. Chem. Rev.* (Engl. Transl.), **1994**, *63*, 575; *Usp. Khim.*, 1994, **63**, 602; and (b) S. L. Beaucage and R. P. Iyer, *Tetrahedron*, **1992**, *48*, 2223.
[11] R. Hulst, N. K. de Vries, B. L. Feringa, *Tetrahedron: Asymmetry*, **1994**, *5*, 699.
[12] E. Wissing, A. J. J. M. Theunissen, C. B. Hansen, P. W. N. M. van Leeuwen, A. van Rooy, D. Burgers, to DSM N.V. and E.I. Dupont de Nemours and Company, World Patent WO9616923, **1996**.
[13] E. E. Nifantyev, S. A. Rumyantseva, M. P. Koroteyev, E. M. Abbasov, A. T. Teleshev, V. A. Pavlov, E. I. Klabunovsky, *Phosphorus, Sulfur Rel. Elem*, **1981**, *12*, 27.
[14] (a) A. H. M. de Vries, A. Meetsma, B. L. Feringa, *Angew. Chem. Int. Ed. Engl.* **1996**, *35*, 2374; and (b) B. L. Feringa, M. Pineschi, L. A. Arnold, R. Imbos, A. H. M. de Vries, *Angew. Chem. Int. Ed. Engl.* **1997**, *36*, 2620.
[15] J. F. Teichert, B. L. Feringa, *Angew. Chem. Int. Ed.* **2010**, *49*, 2486.
[16] (a) Y. H. Choi, J. Y. Choi, H. Y. Yang, Y. H. Kim, *Tetrahedron: Asymmetry* **2002**, *13*, 801; (b) A. van Rooy, D. Burgers, P. C. J. Kamer, P. W. N. M. van Leeuwen, *Recl. Trav. Chim. Pays-Bas* **1996**, *115*, 492; and (c) K. Tissot-Croset, D. Polet, S. Gille, C. Hawner, A. Alexakis, *Synthesis* **2004**, 2586.
[17] (a) A. Duursma, Asymmetric catalysis with chiral monodentate phosphoramidite ligands, thesis, University of Groningen, **2004**; (b) N. Mrsic, Asymmetric hydrogenation of imines, enamines and N-heterocycles using phosphoramidite ligands, thesis, University of Groningen, **2010**.
[18] (a) E. J. Nurminen, J. K. Mattinen, H. Lonnberg, *J. Chem. Soc. Perkin Trans. II* **1998**, 1621; (b) E. J. Nurminen, J. K. Mattinen, H. Lonnberg, *J. Chem. Soc. Perkin Trans. II* **2000**, 2238; (c) E. J. Nurminen, J. K. Mattinen, H. Lonnberg, *Phosphorus Sulfur Silicon Rel. Elem.* **2002**, *177*, 2209; (d) E. J. Nurminen, J. K. Mattinen, H. Lonnberg, *Helv. Chim. Acta* **2003**, *86*, 2005; and (e) L. K. Sal'keeva, M. T. Nurmaganbetova, and O. Sh. Kurmanaliev, *Russ. J. Gen. Chem.* **2003**, *73*, 533.
[19] N. S. Kumar, S. Kumaraswamy, M. A. Said, K. C. K. Swamy, *Org. Proc. Res. Dev.* **2003**, *7*, 925.
[20] J.-G. Boiteau, R. Imbos, A. J. Minnaard, B. L. Feringa, *Org. Lett.* **2003**, *5*, 681.
[21] (a) J. Duran, M. Gulias, L. Castedo, J. L. Mascarenas, *Org. Lett.* **2005**, *7*, 5693; (b) Z. J. Zhang, J. C. Mao, R. L. Wang, F. Wu, H. L. Chen, B. S. Wan, *J. Mol. Catal. A Chem.* **2006**, *243*, 239; and (c) I. S. Mikhel, K. N. Gavrilov, A. I. Polosukhin, A. I. Rebrov *Russ. Chem. Bull.* **1998**, *47*, 1585.
[22] K. N. Gavrilov, I. S. Mikhel, K. A. Lysenko, M. Y. Antipin, G. I. Timofeeva, A. I. Polosukhin, A. V. Korostylev, *Russ. J. Inorg. Chem.*, **1997**, *42*, 520.
[23] (a) A. J. Minnaard, B. L. Feringa, L. Lefort, J. G. de Vries, *Acc. Chem. Res.* **2007**, *40*, 1267; (b) T. Jerphagnon, M. G. Pizzuti, A. J. Minnaard, B. L. Feringa, *Chem. Soc. Rev.* **2009**, *38*, 1039; and (c) L. Eberhardt, D. Armspach, J. Harrowfield, D. Matt, *Chem. Soc. Rev.* **2008**, *37*, 839.
[24] B. D. Chapsal, Z. H. Hua, I. Ojima, *Tetrahedron: Asymmetry* **2006**, *17*, 642.
[25] (a) Q. Zeng, H. Liu, X. Cui, A. Q. Mi, Y. Z. Jiang, X. S. Li, M. C. K. Choi, A. S. C. Chan, *Tetrahedron: Asymmetry* **2002**, *13*, 115; (b) Q. Zeng, H. Liu, A. Mi, Y. H. Jiang, X. S. Li, M. C. K. Choi, A. S. C. Chan, *Tetrahedron* **2002**, *58*, 8799; and (c) X. S. Li, X. Jia, G. Lu, T. T. L. Au-Yeung, K. H. Lam, T. W. H. Lo, A. S. C. Chan, *Tetrahedron: Asymmetry* **2003**, *14*, 2687.
[26] See, for example, (a) F. Giacomina, A. Meetsma, L. Panella, L. Lefort, A. H. M. de Vries, J. G. de Vries, *Angew. Chem. Int. Ed.* **2007**, *46*, 1497; and (b) J. A. F. Boogers, U. Felfer, M. Kotthaus, L. Lefort, G. Steinbauer, A. H. M. de Vries, J. G. de Vries. *Org. Proc. Res. Dev.* **2007**, *11*, 585–591. Also in G. J. H. Buisman, P. C. J. Kamer, P. W. N. M. van Leeuwen, *Tetrahedron: Asymmetry* **1993**, *4*, 1625.
[27] F. Zhang, Q. H. Fan, *Org. Biomol. Chem.* **2009**, *7*, 4470.
[28] S. Enthaler, G. Erre, K. Junge, K. Schröder, D. Addis, D. Michalik, M. Hapke, D. Redkin, M. Beller, *Eur. J. Org. Chem.* **2008**, 3352, and references therein. The bromination procedure was reported earlier: D. J. Cram, R. C. Helgeson, S. C. Peacock, L. J. Kaplan, L. A. Domeier, P. Moreau, K. Koga, J. M. Mayer, Y. Chao, *J. Org. Chem.* **1978**, *43*, 1930.

[29] M. T. Reetz, J. A. Ma, R. Goddard, *Angew. Chem. Int. Ed.* **2005**, *44*, 412.

[30] K. N. Gavrilov, E. B. Benetsky, V. E. Boyko, E. A. Rastorguev, V. A. Davankov, B. Schaffner, A. Börner, *Chirality* **2010**, *22*, 844.

[31] M. Kruck, M. P. Munoz, H. L. Bishop, C. G. Frost, C. J. Chapman, G. Kociok-Köhn, C. P. Butts, G. C. Lloyd-Jones, *Chem. Eur. J.* **2008**, *14*, 7808.

[32] C. Recsei, C. S. P. McErlean, *Tetrahedron: Asymmetry* **2010**, *21*, 149.

[33] G. Onodera, K. Watabe, M. Matsubara, K. Oda, S. Kezuka, R. Takeuchi, *Adv. Synth. Catal.* **2008**, *350*, 2725.

[34] F. W. Patureau, S. de Boer, M. Kuil, J. Meeuwissen, P. A. R. Breuil, M. A. Siegler, A. L. Spek, A. J. Sandee, B. de Bruin, J. N. H. Reek, *J. Am. Chem. Soc.* **2009**, *131*, 6683.

[35] M. Batuecas, M. A. Esteruelas, C. Garcia-Yebra, E. Onate, *Organometallics* **2010**, *29*, 2166.

[36] (a) Q. H. Zeng, X. P. Hu, Z. C. Duan, X. M. Liang, Z. Zheng, *J. Org. Chem.* **2006**, *71*, 393; (b) Z. C. Duan, X. P. Hu, D. Y. Wang, J. D. Huang, S. B. Yu, J. Deng, Z. Zheng, *Adv. Synth. Catal.* **2008**, *350*, 1979; and (c) Z. C. Duan, X. P. Hu, C. Zhang, D. Y. Wang, S. B. Yu, Z. Zheng, *J. Org. Chem.* **2009**, *74*, 9191.

[37] (a) L. Eberhardt, D. Armspach, D. Matt, L. Toupet, B. Oswald, *Eur. J. Org. Chem.* **2007**, 5395; and (b) P. Scafato, S. Superchi, C. Rosini, *Arkivoc* **2008**, *116*.

[38] P. A. R. Breuil, J. N. H. Reek, *Eur. J. Org. Chem.* **2009**, 6225.

[39] L. Eberhardt, D. Armspach, D. Matt, L. Toupet, B. Oswald, *Eur. J. Inorg. Chem.* **2007**, 4153.

[40] A. Iuliano, *Tetrahedron: Asymmetry* **2010**, *21*, 1943.

[41] (a) A. Alexakis, S. Rosset, J. Allamand, S. March, F. Guillen, C. Benhaim, *Synlett* **2001**, 1375; (b) A. Alexakis, D. Polet, C. Benhaim, S. Rosset, *Tetrahedron: Asymmetry* **2004**, *15*, 2199; (c) A. Alexakis, D. Polet, S. Rosset, S. March, *J. Org. Chem.* **2004**, *69*, 5660; and (d) E. Raluy, O. Pamies, M. Dieguez, S. Rosset, A. Alexakis, *Tetrahedron: Asymmetry* **2009**, *20*, 2167.

[42] P. Scafato, G. Cunsolo, S. Labano, C. Rosini, *Tetrahedron* **2004**, *60*, 8801.

[43] (a) F. Fang, H. Zhang, F. Xie, G. Q. Yang, W. B. Zhang, *Tetrahedron* **2010**, *66*, 3593; and (b) H. Zhang, F. Fang, F. Xie, H. Yu, G. Q. Yang, W. B. Zhang, *Tetrahedron Lett.* **2010**, *51*, 3119.

[44] (a) C. Monti, C. Gennari, U. Piarulli, *Tetrahedron Lett* **2004**, *45*, 6859; (b) C. Monti, C. Gennari, U. Piarulli, J. G. de Vries, A. H. M. de Vries, L. Lefort, *Chem. Eur. J.* **2005**, *11*, 6701.

[45] (a) O. Huttenloch, E. Laxman, H. Waldmann, *Chem. Eur. J.* **2002**, *8*, 4767; and (b) O. Huttenloch, E. Laxman, H. Waldmann, *Chem. Commun.* **2002**, 673.

[46] L. Lefort, J. A. F. Boogers, A. H. M. de Vries, J. G. de Vries, *Org. Lett.* **2004**, *6*, 1733.

[47] C. Simons, U. Hanefeld, I. Arends, A. J. Minnaard, T. Maschmeyer, R. A. Sheldon, *Chem. Commun.* **2004**, 2830.

[48] R. Hoen, S. Leleu, P. N. M. Botman, V. A. M. Appelman, B. L. Feringa, H. Hiemstra, A. J. Minnaard, J. H. van Maarseveen, *Org. Biomol. Chem.* **2006**, *4*, 613.

[49] (a) W-J. Tang, Y-Y. Huang, Y.-M. He, Q.-H. Fan, *Tetrahedron: Asymmetry* **2006**, *17*, 536; and (b) F. Zhang, Y. Li, Z. W. Li, Y. M. He, S. F. Zhu, Q. H. Fan, Q. L. Zhou, *Chem. Commun.* **2008**, 6048.

[50] M. Yang, X. B. Yang, X. Q. Yu, *Arkivoc* **2008**, 84.

[51] D. Seebach, A. K. Beck, A. Heckel, *Angew. Chem. Int. Ed.* **2001**, *40*, 92.

[52] E. Keller, J. Maurer, R. Naasz, T. Schader, A. Meetsma, B. L. Feringa, *Tetrahedron: Asymmetry* **1998**, *9*, 2409.

[53] M. Mewald, A. Weickgenannt, R. Fröhlich, M. Oestreich, *Tetrahedron: Asymmetry* **2010**, *21*, 1232.

[54] H. Teller, S. Flugge, R. Goddard, A. Fürstner, *Angew. Chem. Int. Ed.* **2010**, *49*, 1949.

[55] (a) A.-G. Hu, Y. Fu, J.-H. Xie, H. Zhou, L.-X. Wang, Q.-L. Zhou, *Angew. Chem., Int. Ed.* **2002**, *41*, 2348; (b) S. F. Zhu, Y. Fu, J. H. Xie, B. Liu, L. Xing, Q. L. Zhou, *Tetrahedron: Asymmetry* **2003**, *14*, 3219; (c) H. Zhou, W.-H. Wang, Y. Fu, J.-H. Xie, W.-J. Shi, L.-X. Wang, Q.-L. Zhou, *J. Org. Chem.* **2003**, *68*, 1582; (d) X. H. Huo, J. H. Xie, Q. S. Wang, Q.-L. Zhou, *Adv. Synth. Catal.* **2007**, *349*, 2477; and (e) J. H. Xie, Q. L. Zhou, *Acc. Chem. Res.* **2008**, *41*, 581.

[56] (a) S. L. Wu, W. C. Zhang, Z. G. Zhang, X. M. Zhang, *Org. Lett.* **2004**, *6*, 3565; and (b) W. C. Zhang, C. J. Wang, W. Z. Gao, X. M. Zhang, *Tetrahedron Lett.* **2005**, *46*, 6087.

[57] R. Hoen, M. van den Berg, H. Bernsmann, A. J. Minnaard, J. G. de Vries, B. L. Feringa, *Org. Lett.* **2004**, *6*, 1433.

[58] E. B. Benetsky, S. V. Zheglov, T. B. Grishina, F. Z. Macaev, L. P. Bet, V. A. Davankov, K. N. Gavrilov, *Tetrahedron Lett.* **2007**, *48*, 8326.

[59] E. E. Nifantyev, S. A. Rumyantseva, K. Sisneros, M. P. Koroteev, A. I. Lutsenko, N. K. Kochetkov, *Zh. Obshch. Khim.*, **1983**, *53*, 2619.

[60] (a) Y. Liu, K. L. Ding, *J. Am. Chem. Soc.* **2005**, *127*, 10488; (b) B. G. Zhao, Z. Wang, K. L. Ding, *Adv. Synth. Catal.* **2006**, *348*, 1049; (c) K. Wakabayashi, K. Aikawa, S. Kawauchi, K. Mikami, *J. Am. Chem. Soc.* **2008**, *130*, 5012; (d) K. Walkabayashi, K. Aikawa, K. Mikami, *Heterocycles* **2009**, *77*, 927; (e) T. Kop-Weiershausen, J. Lex, J. M. Neudörfl, B. Goldfuss, *Beilstein J. Org. Chem.* **2005**, *1*; (f) N. Miklasova, O. Julinek, M. Meskova, V. Setnicka, M. Urbanova, M. Putala, *Tetrahedron Lett.* **2010**, *51*, 1966; (g) A. Bayer, P. Murszat, U. Thewalt, B. Rieger, *Eur. J. Inorg. Chem.* **2002**, 2614; (h) A. Bayer, U. Thewalt, B. Rieger, *Eur. J. Inorg. Chem.* **2002**, 199; and (i) P. N. M. Botman, A. Amore, R. van Heerbeek, J. W. Back, H. Hiemstra, J. N. H. Reek, J. H. van Maarseveen, *Tetrahedron Lett.* **2004**, *45*, 5999.

[61] A. I. Polosukhin, O. G. Bondarev, S. E. Lyubimov, A. A. Shiryaev, P. V. Petrovskii, K. A. Lysenko, K. N. Gavrilov, *Russ. J. Coord. Chem.* **2001**, *27*, 591.

[62] O. G. Bondarev, R. Goddard, *Tetrahedron Lett.* **2006**, *47*, 9013.

[63] M. T. Reetz, G. Mehler, O. Bondarev, *Chem. Commun.* **2006**, 2292.

[64] E. B. Benetsky, S. V. Zheglov, T. B. Grishina, F. Z. Macaev, L. P. Bet, V. A. Davankov, K. N. Gavrilov, *Tetrahedron Lett.* **2007**, *48*, 8326.

[65] (a) S. D. Pastor, J. L. Hyun, P. A. Odorisio and R. K. Rodebaugh, *J. Am. Chem. Soc.*, **1988**, *110*, 6547; and (b) S. D. Pastor, A. Togni, *Inorg. Chim. Acta*, **1989**, *159*, 3.

[66] O. Huttenloch, J. Spieler, H. Waldmann, *Chem. Eur. J.* **2001**, *7*, 671.

[67] A. Mandoli, L. A. Arnold, A. H. M. de Vries, P. Salvadori, B. L. Feringa, *Tetrahedron: Asymmetry* **2001**, *12*, 1929.

[68] C. G. Arena, V. Casilli, F. Faraone, *Tetrahedron: Asymmetry* **2003**, *14*, 2127.

[69] L. Eberhardt, D. Armspach, D. Matt, B. Oswald, L. Toupet, *Org. Biomol. Chem.* **2007**, *5*, 3340.

[70] E. Raluy, M. Diéguez, O. Pàmies, *J. Org. Chem.* **2007**, *72*, 2842.

[71] D. Benito-Garagorri, K. Mereiter, K. Kirchner, *Eur. J. Inorg. Chem.* **2006**, 4374.

[72] (a) E. N. Rasadkina, P. V. Slitikov, M. S. Mel'nik, A. I. Stash, V. K. Bel'skii, E. E. Nifant'ev, *Russ. J. Gen. Chem.* **2004**, *74*, 1080; and (b) E. E. Nifant'ev, E. N. Rasadkina, S. V. Suvorkin, D. V. Gusev, *Russ. J. Gen. Chem.* **2001**, *71*, 1325.

[73] (a) Y. Yamamoto, K. Kurihara, N. Sugishita, K. Oshita, D. Piao, N. Miyaura, *Chem. Lett.* **2005**, *34*, 1224; (b) K. Kurihara, Y. Yamamoto, N. Sugishita, K. Oshita, D. Piao, N. Miyaura, *J. Organomet. Chem.* **2007**, *692*, 428; (c) K. Kurihara, Y. Yamamoto, N. Miyaura, *Tetrahedron Lett.* **2009**, *50*, 3158; and (d) K. Kurihara, Y. Yamamoto, N. Miyauraa, *Adv. Synth. Catal.* **2009**, *351*, 260.

[74] F. Boeda, T. Beneyton, C. Crévisy, *Mini-Rev. Org. Chem.* **2008**, *5*, 96.

[75] L. Eberhardt, D. Armspach, J. Harrowfield, D. Matt, *Chem. Soc. Rev.* **2008**, *37*, 839.

[76] (a) G. Franciò, F. Faraone, W. Leitner, *Angew. Chem. Int. Ed.* **2000**, *39*, 1428; and (b) T. Pullmann, B. Engendahl, Z. Y. Zhang, M. Hölscher, A. Zanotti-Gerosa, A. Dyke, G. Franciò, W. Leitner, *Chem. Eur. J.* **2010**, *16*, 7517.

[77] F. Boeda, D. Rix, H. Clavier, C. Crévisy, M. Mauduit, *Tetrahedron: Asymmetry* **2006**, *17*, 2726.

[78] (a) J. D. Huang, X. P. Hu, Z. C. Duan, Q. H. Zeng, S. B. Yu, J. Deng, D. Y. Wang, Z. Zheng, *Org. Lett.* **2006**, *8*, 4367; and (b) X. M. Zhou, J. D. Huang, L. B. Luo, C. L. Zhang, X. P. Hu, Z. Zheng, *Org, Biomol. Chem.* **2010**, *8*, 2320.

[79] K. A. Vallianatou, I. D. Kostas, J. Holz, A. Börner, *Tetrahedron Lett.* **2006**, *47*, 7947.

[80] M. Eggenstein, A. Thomas, J. Theuerkauf, G. Franció, W. Leitner, *Adv. Synth. Catal.* **2009**, *351*, 725.

[81] (a) D. Y. Wang, X. P. Hu, J. D. Huang, J. Deng, S. B. Yu, Z. C. Duan, X. F. Xu, Z. Zheng, *Angew. Chem. Int. Ed.* **2007**, *46*, 7810; (b) M. Qiu, X. P. Hu, D. Y. Wang, J. Deng, J. D. Huang, S. B. Yu, Z. C. Duan, Z. Zheng, *Adv. Synth. Catal.* **2008**, *350*, 1413; (c) S. B. Yu, J. D. Huang, D. Y. Wang, X. P. Hu, J. Deng, Z. C. Duan, Z. Zheng, *Tetrahedron: Asymmetry* **2008**, *19*, 1862; and (d) M. Qiu, D. Y. Wang, X. P. Hu, J. D. Huang, S. B. Yu, J. Deng, Z. C. Duan, Z. Zheng, *Tetrahedron: Asymmetry* **2009**, *20*, 210.

[82] N. W. Boaz, J. A. Ponasik, S. E. Large, *Tetrahedron: Asymmetry* **2005**, *16*, 2063.

[83] (a) X. P. Hu, Z. Zheng, *Org. Lett.* **2004**, *6*, 3585; (b) X. P. Hu, Z. Zheng, *Org. Lett.* **2005**, *7*, 419; (c) Q. H. Zeng, X. P. Hu, Z. C. Duan, X. M. Liang, Z. Zheng, *Tetrahedron: Asymmetry* **2005**, *16*, 1233; (d) S. B. Yu, X. P. Hu, J. Deng, D. Y. Wang, Z. C. Duan, Z. Zheng, *Tetrahedron: Asymmetry* **2009**, *20*, 621.

[84] M. Jia, X. S. Li, W. S. Lam, S. H. L. Kok, L. J. Xu, G. Lu, C. H. Yeung, A. S. C. Chan, *Tetrahedron: Asymmetry* **2004**, *15*, 2273.

[85] (a) R. Kranich, K. Eis, O. Geis, S. Mühle, J. W. Bats, H.-G. Schmalz, *Chem. Eur. J.*, **2000**, *6*, 2847; and (b) F. Blume, S. Zemolka, T. Fey, R. Kranich, H.-G. Schmalz, *Adv. Synth. Catal.* **2002**, *344*, 868.
[86] (a) J. Wassenaar, J. N. H. Reek, *Dalton Trans.* **2007**, 3750; (b) J. Wassenaar, J. N. H. Reek, *J. Org. Chem.* **2009**, *74*, 8403; and (b) J. Wassenaar, S. van Zutphen, G. Mora, P. Le Floch, M. A. Siegler, A. L. Spek, J. N. H. Reek, *Organometallics* **2009**, *28*, 2724.
[87] (a) W. C. Zhang, X. M. Zhang, *Angew. Chem. Int. Ed.* **2006**, *45*, 5515; (b) W. C. Zhang, X. M. Zhang, *J. Org. Chem.* **2007**, *72*, 1020.
[88] X. W. Zhang, B. N. Cao, Y. J. Yan, S. C. Yu, B. M. Ji, X. M. Zhang, *Chem. Eur. J.* **2010**, *16*, 871.
[89] P. E. Goudriaan, M. Kuil, X. B. Jiang, P. W. N. M. van Leeuwen, J. N. H. Reek, *Dalton Trans.* **2009**, 1801.
[90] N. Cramer, S. Laschat, A. Baro, *Organometallics* **2006**, *25*, 2284.
[91] L. D. Kostas, K. A. Vallianatou, J. Holz, A. Börner, *Tetrahedron Lett.* **2008**, *49*, 331.
[92] (a) O. Pàmies, M. Diéguez, C. Claver, *Adv. Synth. Catal.* **2007**, *349*, 836; and (b) O. Pàmies, M. Diéguez, *Chem. Eur. J.* **2008**, *14*, 944.
[93] (a) E. Raluy, O. Pàmies, M. Diéguez, S. Rosset, A. Alexakis, *Tetrahedron: Asymmetry*, **2009**, *20*, 1930; and (b) E. Raluy, O. Parmies, M. Dieguez, *Adv. Synth. Catal.* **2009**, *351*, 1648.
[94] A. V. Korostylev, O. G. Bondarev, S. Y. Lyubimov, A. Y. Kovalevsky, P. V. Petrovskii, V. A. Davankov, K. N. Gavrilov, *Inorg. Chim. Acta* **2000**, *303*, 1.
[95] A. I. Polosukhin, O. G. Bondarev, S. E. Lyubimov, A. A. Shiryaev, P. V. Petrovskii, K. A. Lysenko, and K. N. Gavrilov, *Russ. J. Coord. Chem.* **2000**, *27*, 591.
[96] R. P. J. Bronger, P. J. Guiry, *Tetrahedron: Asymmetry* **2007**, *18*, 1094.
[97] G. Delapierre, J. M. Brunel, T. Constantieux, G. Buono, *Tetrahedron: Asymmetry* **2001**, *12*, 1345.
[98] L. Zeng, S. B. Yu, Z. Cao, D. W. Yang, *Catal. Lett.* **2010**, *136*, 243.
[99] (a) R. Hodgson, R. E. Douthwaite, *J. Organomet. Chem.* **2005**, *690*, 5822; and (b) L. G. Bonnet, R. E. Douthwaite, B. M. Kariuki, *Organometallics* **2003**, *22*, 4187.
[100] (a) W. Tie, A. T. Teleshev, V. Yu. Mishina, E. E. Nifantyev, *Russ. J. Gen. Chem.* **2008**, *78*, 152; and (b) V. Y. Mishina, I. V. Abrashina, A. T. Teleshev, L. K. Vasyanina, E. E. Nifant'ev, *Russ. J. Org. Chem.*, **2002**, *38*, 977.
[101] E. E. Nifantyev, V. I. Maslennikova, R. V. Merkulov, *Acc. Chem. Res.* **2005**, *38*, 108.
[102] S. Doherty, E. G. Robins, I. Pal, C. R. Newman, C. Hardacre, D. Rooney, D. A. Mooney, *Tetrahedron: Asymmetry* **2003**, *14*, 1517.
[103] B. H. G. Swennenhuis, R. F. Chen, P. van Leeuwen, J. G. de Vries, P. C. J. Kamer, *Eur. J. Org. Chem.* **2009**, 5796.
[104] A. Mandoli, M. Calamante, B. L. Feringa, P. Salvadori, *Tetrahedron: Asymmetry* **2003**, *14*, 3647.
[105] D. Jiang, Z. H. Meng, *Chin. J. Chem.* **2007**, *25*, 542.
[106] B. H. G. Swennenhuis, R. Chen, P. van Leeuwen, J. G. de Vries, P. C. J. Kamer, *Org. Lett.* **2008**, *10*, 989.
[107] L. Shi, X. W. Wang, C. A. Sandoval, Z. Wang, H. J. Li, J. Wu, L. T. Yu, K. L. Ding, *Chem. Eur. J.* **2009**, *15*, 9855.
[108] (a) D. Peña , A. J. Minnaard , J. A. F. Boogers , A. H. M. de Vries, J. G. de Vries, B. L. Feringa, *Org. Biomol. Chem.*, **2003**, *1*, 1087; and (b) R. Hoen, J. A. F. Boogers, H. Bernsmann, A. J. Minnaard, A. Meetsma, T. D. Tiemersma-Wegman, A. H. M. de Vries, J. G. de Vries, B. L. Feringa, *Angew, Chem., Int. Ed.* **2005**, *44*, 4209.
[109] H. Bernsmann, M. van den Berg, R. Hoen, A. J. Minnaard, G. Mehler, M. T. Reetz, J. G. de Vries, B. L. Feringa, *J. Org. Chem.* **2005**, *70*, 943.

5

Phosphinite and Phosphonite Ligands

T. V. (Babu) RajanBabu

Department of Chemistry, The Ohio State University, Columbus, United States

5.1 Introduction

Transition metal complexes containing tertiary phosphines (R_3P) and phosphites [$P(OR)_3$] had a long history in inorganic and organometallic chemistry [1–3] much before phosphinites [R_2POR] and phosphonites [$RP(OR)_2$] began to receive increased attention. In the series R_3P, $R_2P(OR)$, $RP(OR)_2$, and $P(OR)_3$, the cone angle decreases from 132° (phosphine) to 109° (phosphite) for R=Et, and from 118° to 107° for R=Me, with intermediate values expected for the phosphinite and phosphonite [4–6]. The π acceptor ability of these ligands would be expected to decrease in the order $P(OR)_3 > RP(OR)_2 > R_2P(OR) > R_3P$, especially for small R groups where steric effects are not important. Electron-withdrawing OR groups should lower the energies of the empty 3d orbitals of phosphorus, which enhances π acceptor properties. Since the OR groups should stabilize the σ donor orbitals, σ-bonding tendencies should have the series order in reverse, with the R_3P ligands being the best. Add to this the steric effects, and a complex picture emerges for all but the simplest ligands, and generalizations are often unreliable. For each metal–ligand combinations, both the σ- and π-bonding and steric effects should be considered explicitly, and structures and properties of closely related and structurally analogous complexes examined, before planning applications such as in homogeneous catalysis.

This review deals with the important developments in the syntheses and applications of phosphinite and phosphonite ligands [7]. Wherever possible, significant applications of these ligands in homogeneous catalysis are included. The coverage reflects the fact that the majority of these ligands were developed for selective catalysis, and it is intended for an audience of chemists interested in this line of research. This review, which covers the literature until July 2011, is meant to be illustrative more than exhaustive.

Phosphorus(III) Ligands in Homogeneous Catalysis: Design and Synthesis, First Edition.
Edited by Paul C. J. Kamer and Piet W. N. M. van Leeuwen.

5.2 General methods for synthesis of complexes

Early systematic investigations of phosphinite and phosphonite ligands were conducted primarily by Robinson, who examined the neutral [8] and cationic [9,10] complexes of copper, silver, gold, iridium, ruthenium, osmium, and platinum metals. The initially reported procedures for the preparation of the metal complexes are still being used widely. Typical examples of the synthesis of neutral complexes [8] are listed in Equations 5.1, 5.2, 5.3, 5.4, 5.5, 5.6, 5.7, 5.8, 5.9, 5.10, and 5.11. These involve (1) direct reaction of the ligands with halides (Eqs. 5.1, 5.2, 5.3, 5.4, and 5.5) or carbonyl halides (Eq. 5.7), or (2) substitution of a weekly coordinated alkene by the ligand (Eqs. 5.8, 5.9, 5.10, and 5.11). Well-explored cobalt (Eqs. 5.2 and 5.3) [11] and nickel (Eqs. 5.4, 5.5, and 5.6) [12] complexes are prepared by the procedure (i).

$$AgCl \xrightarrow[\text{2. hexane}]{\text{1. } Ph_2POEt\text{, MeOH}} \{(Cl)Ag[Ph_2POEt]_2\}_2 \qquad \text{(Eq. 5.1)}$$

Also prepared by a similar route:
$(Cl)Ag[Ph_2POMe]$, $[(Cl)Ag(PhP(OEt)_2]$, $[ClAg(PhP(OMe)_2]$

$$CoCl_2 + \underset{\text{(2 equiv.)}}{EtP(OEt)_2} \xrightarrow{C_6H_6\text{, 15 d}} (Cl)_2Co[EtP(OEt)_2]_2 \qquad \text{(Eq. 5.2)}$$

$$CoCl_2 + \underset{\text{(3.6 equiv.)}}{Et_2P(OEt)} \xrightarrow{\text{neat, 12 d}} (Cl)_2Co[Et_2P(OEt)]_2 \qquad \text{(Eq. 5.3)}$$

$$Ni(CN)_2 + \underset{\text{(3 equiv.)}}{Me_2POMe} \xrightarrow{\text{acetone, rt}} \left\{ \begin{array}{c} (CN)_2Ni[Me_2P(OMe)]_3 \\ + \\ (CN)_2Ni[Me_2P(OMe)]_2 \end{array} \right. \qquad \text{(Eq. 5.4)}$$

$$Ni(CN)_2 + \underset{\text{(2.5 equiv.)}}{MeP(OMe)_2} \xrightarrow[\text{(high vacuum)}]{\text{acetone}} \begin{array}{l} (CN)_2Ni[(MeP(OMe)_2]_3 \\ (CN)_2Ni[(MeP(OMe)_2]_2 \end{array} \qquad \text{(Eq. 5.5)}$$

$$Ni(ClO_4)_2.6H_2O + \underset{\text{(5 equiv.)}}{MeP(OMe)_2} \longrightarrow \{Ni[MeP(OMe)_2]_5\}\ (ClO_4)_2 \qquad \text{(Eq. 5.6)}$$

Also prepared by a similar route: $(CN)_2Ni[(Me_2P(OMe)]_3$ and $(CN)_2Ni[(Me_2P(OMe)]_2$

(i) MeO 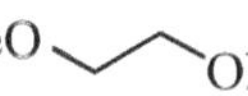 OH, CO, reflux

$$RuCl_3{\cdot}3H_2O \xrightarrow[\text{(ii) } \mathbf{L}\text{, CO, heat}]{\text{(i), CO, reflux}} (Cl)_2Ru(CO)_2(\mathbf{L})_2 \qquad \text{(Eq. 5.7)}$$

L = [PhP(OMe)$_2$], [PhP(OEt)$_2$], Ph$_2$POMe, Ph$_2$POEt

$$[(COD)IrX]_2 \xrightarrow{\text{1. } \mathbf{L}\text{ (4 equiv.)}} [IrX(COD)]\mathbf{L}_2 \xrightarrow{(CO)} [IrX(CO)]\mathbf{L}_2 \qquad \text{(Eq. 5.8)}$$

X = Cl, I; **L** = PhP(OEt)$_2$, Ph$_2$POEt

(Cl)(H)(COD)Ir 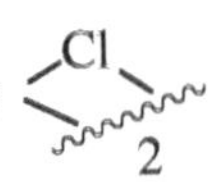

$$\xrightarrow[\text{2. hexane}]{\text{1. } \mathbf{L}\text{, } C_6H_6\text{, reflux}} \{(Cl)_2(H)Ir[\mathbf{L}]_3\} \qquad \text{(Eq. 5.9)}$$

L = PhP(OMe)$_2$, PhP(OEt)$_2$, Ph$_2$POMe, Ph$_2$POEt

$(COD)MCl_2$, M = Pd, Pt $\xrightarrow[\text{2. hexane}]{\text{1. } Ph_2POEt,\ C_6H_6}$ $Cl_2M(Ph_2POEt)_2$ (Eq. 5.10)

L = $PhP(OMe)_2$, $PhP(OEt)_2$, Ph_2POMe, Ph_2POEt

$(Cl)Au(cyclooctene)$ $\xrightarrow{\text{1. } Ph_2P(OEt),\ C_6H_6}$ $(Cl)Au(Ph_2POEt)$ (Eq. 5.11)

The cationic complexes are typically prepared by addition of the free ligand to a solution or suspension of a labile transition metal precursor in an alcoholic solvent [9,10]. The mixture is warmed as needed, and subsequently sodium tetraphenylborate is added to precipitate the complex. Examples are shown in Equations 5.12, 5.13, 5.14, 5.15, 5.16, 5.17, 5.18, 5.19, and 5.20. Other specific examples are listed throughout this chapter.

$[Cl_2Ru(NBD)]_n$ + **L** $\xrightarrow[\text{2. } Na^+\ Ph_4B^-]{\text{1. alcohol solvent reflux}}$ cationic complexes

$[Ru]^{2+}$ $[\mathbf{L}]_6$ 2 $[Ph_4B]^-$

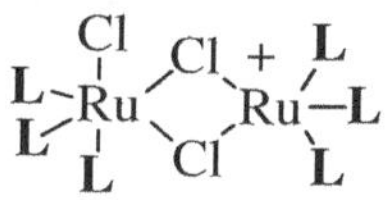

$[Ph_4B]^-$ (Eq. 5.12)

L = $[PhP(OMe)_2]$, $[PhP(OEt)_2]$

$(Cl)(CO)(H)Ru(PPh_3)_3$ + **L** $\xrightarrow[\text{2. } Na^+\ Ph_4B^-]{\text{1. } MeOH/C_6H_6\text{, reflux}}$ $[(CO)(H)Ru(L)_4]^+$ Ph_4B^- (Eq. 5.13)

L = Ph_2POMe, Ph_2POEt

$[(COD)IrCl]_2$ $\xrightarrow[\text{2. } Na^+\ Ph_4B^-]{\text{1. } Ph_2POMe\text{, MeOH}}$ $[Ir(Ph_2POMe)_4]^+$. CH_2Cl_2 Ph_4B^- (Eq. 5.14)

Also prerpared by a similar route: $\{[Ir[PhP(OEt)_2]_4\}^+$ Ph_4B^-

$[(Cl)(CO)Ir(PPh_3)_2]_2$ $\xrightarrow[\text{2. } Na^+\ Ph_4B^-]{\text{1. } PhP(OMe)_2\text{, MeOH}}$ $[(H)Ir(PhP(OMe)_2]_5{}^{2+}$ • CH_2Cl_2 2 Ph_4B^- (Eq. 5.15)

(Cl)(H)(COD)Ir

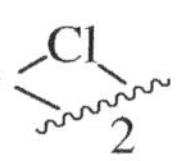

$\xrightarrow[\text{2. } \mathbf{L}\text{, } Na^+\ Ph_4B\text{, MeOH or EtOH}]{\text{1. } \mathbf{L}\text{, } C_6H_6\text{, reflux}}$ *trans*-$\{(Cl)(H)Ir[\mathbf{L}]_4\}^+$ Ph_4B^- (Eq. 5.16)

L = $PhP(OMe)_2$, $PhP(OEt)_2$, Ph_2POMe, Ph_2POEt

$(COD)MCl_2$, M = Pd, Pt $\xrightarrow[\text{2. } Na^+\ Ph_4B^-]{\text{1. } Ph_2POMe\text{, MeOH}}$ $[M(Ph_2POMe)_4]^{2+}$ 2 Ph_4B^- (Eq. 5.17)

Also prepared by a similar route: $[Pt(PhP(OEt)_2]_4{}^{2+}$ 2 Ph_4B^-

$[Pt(Ph_2P(OMe)]_4{}^{2+}$ 2 Ph_4B^-

$$\mathrm{CuCl} \xrightarrow[\text{2. Na}^+\ \mathrm{Ph_4B^-}]{\text{1. Ph}_2\text{P(OMe), MeOH}} [\mathrm{Cu(Ph_2POMe)_4}]^+\ \mathrm{Ph_4B^-} \qquad \text{(Eq. 5.18)}$$

$$\mathrm{AgNO_3} \xrightarrow[\text{2. Na}^+\ \mathrm{Ph_4B^-}]{\text{1. PhP(OMe)}_2\text{, MeOH}} \{\mathrm{Ag[PhP(OMe)_2]_4}\}^+\ \mathrm{Ph_4B^-} \qquad \text{(Eq. 5.19)}$$

Also prepared by a similar route:

$[Ag(Ph_2POMe)_4]^+\ Ph_4B^-$ $[Ag(Ph_2POEt)_4]^+\ Ph_4B^-$

$$\text{(Cl)Au(cyclooctene)} \xrightarrow[\text{2. Na}^+\ \mathrm{Ph_4B^-}]{\text{1. }\mathbf{L}\text{, MeOH}} \{\mathrm{Au}[\mathbf{L}]_4\}^+\ \mathrm{Ph_4B^-} \qquad \text{(Eq. 5.20)}$$

$\mathbf{L} = PhP(OMe)_2, PhP(OEt)_2, Ph_2P(OMe), Ph_2P(OEt)$

Some of the simple phosphinites (alkyl dialkylphosphinites, alkyldiarylphosphinites, aryl diarylphosphinites, and aryl dialkylphosphinites) and phosphonites (dialkyl alkylphosphonites, dialkyl arylphosphonites, diaryl alkylphosphonites, and diaryl arylphosphonites) and the corresponding P(III)-chlorides are commercially available. Others can be prepared by modifications of procedures available in the literature [9,11,13,14]. More recent methods used for the syntheses of complex phosphinite and phosphonite ligands under different classes are described in this chapter. Several of these methods can also be employed for the syntheses of simpler analogs.

5.3 Syntheses and applications of phosphinite ligands

5.3.1 Early studies

One of the earliest applications of phosphinite ligands in catalysis has been the use of diphenylphosphinite derived from (*S,S*)-*trans*-1,2-cyclohexanediol in the Rh(I)-catalyzed asymmetric hydrogenation of alkenes (Scheme 5.1) [15]. The synthesis of the ligand and its use in hydrogenations of alkenes shown in Scheme 5.1 are prototypes of many such ligands synthesized from various diols, most notably those derived from readily available carbohydrates [7c,7e]. Complexes of these modular ligands, which can be easily tuned for steric and electronic variations, have been used in a variety of C–H, C–C, and C–heteroatom bond-forming reactions.

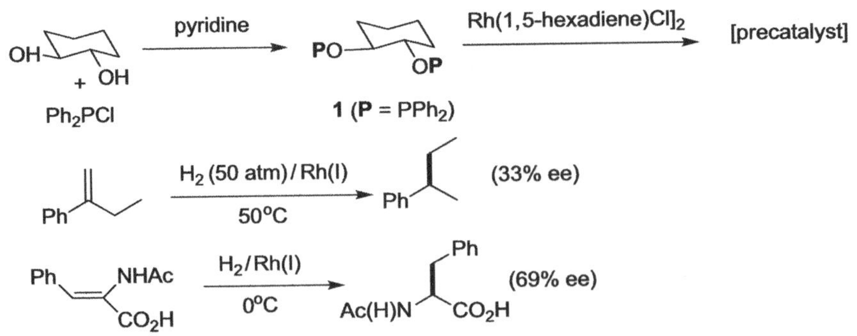

Scheme 5.1 *Synthesis and use of a chelating 1,2-bis-diarylphosphinite ligand in Rh-catalyzed hydrogenations.*

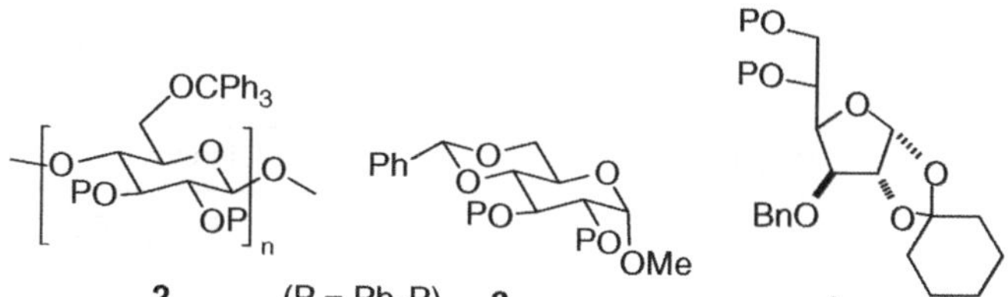

2 (P = Ph_2P) 3 4

Figure 5.1 *Some early examples of phosphinite ligands in Rh-catalyzed hydrogenations.*

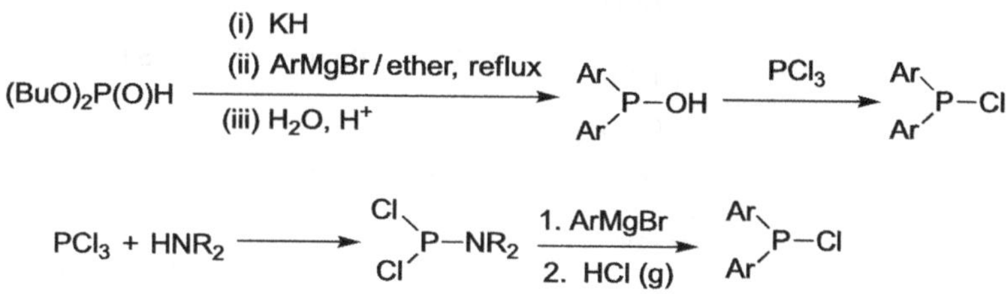

Scheme 5.2 *Synthesis of diarylchlorophosphines.*

5.3.2 Phosphinite ligands from carbohydrates

Following the pioneering contributions (see Figure 5.1 for some typical examples) of Kawabata [16], Cullen [17], Selke [18], Thompson [19], and Johnson [20], a systematic investigation of various sugar-derived 1,2-diarylphosphinites for various reactions was performed by the RajanBabu group. These included Ni(0)-catalyzed asymmetric hydrocyanation [21–24] Rh(I)-catalyzed hydrogenation of 2-*N*-acylacrylic acid and cinnamic acid derivatives [25–27], Pd(0)- [28,29] and Ni(0)- [30] catalyzed allylations, and Ni(II)-catalyzed asymmetric hydrovinylation [31,32].

A typical synthesis from a glucose-derived diol and a chlorophosphine is shown in Equation 5.21 [22,26]. Syntheses of the various chlorophosphines can be accomplished by one of two ways (Scheme 5.2), starting from either dialkylphosphite [33–36] or phosphorus trichloride [37].

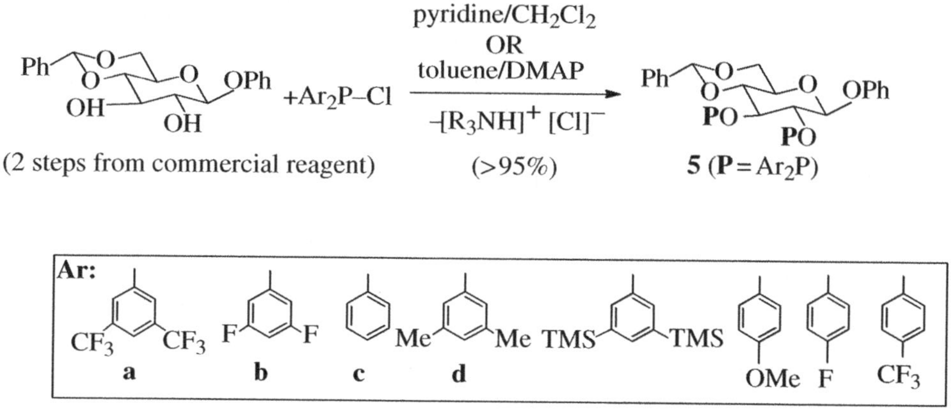

(Eq. 5.21)

Table 5.1 *Electronic effects on Rh-catalyzed hydrogenation: making L-amino acids in high optical purity*[a].

Entry	Substrate	% ee (Ligand)			
		5d	5c	5b	5a
1.	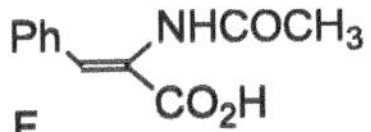	99.0	94.0	60.0	-
2.	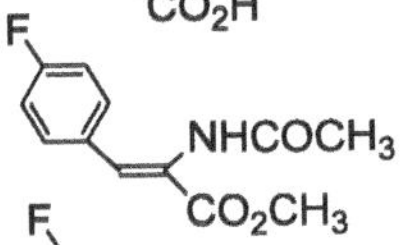	97.2	85.0	13.0	9.0
3.	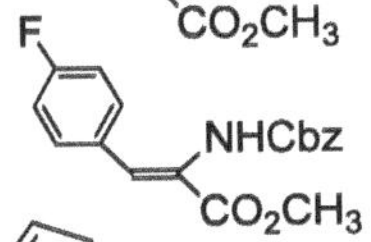	95.7	62.0	<3.0	<5.0
4.	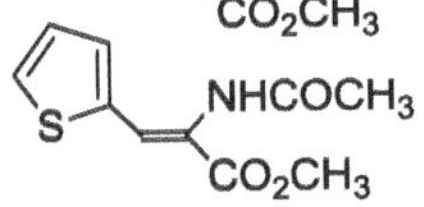	96.7	86.6	--	--

[a] reaction conditions: 0.05-0.1 mol% $LRh^+(COD)SbF_6^-$, 30-40 psi H_2, in THF, rt.

5.3.2.1 *Rh-catalyzed asymmetric hydrogenation of dehydroaminoacids*

Electronically tuned phosphinites for highly enantioselective hydrogenations of dehydroaminoacids. In his pioneering work with sugar phosphinites, Selke [38] reported that simple diphenylphosphinites such as **5** (Eq. 5.21, **P** = Ph_2P) were good ligands for the Rh-catalyzed hydrogenation of dehydrophenylalanine derivatives (Eq. 5.22, R = Ph, R' = Me, H, and R" = Me). However, two limitations became apparent in this approach [25]: (1) the scope of the ligand with the diphenylphosphino moiety (**5**, **P** = Ph_2P) was limited, and selectivities for substituted phenylalanines and the corresponding heteroaromatic derivatives, and for amino acids carrying *N*-protecting groups other than the acyl group, were low; and (2) the synthesis of D-amino acids would require the more expensive L-sugars if one were to use the 2,3-diarylphophinites. During an examination of the electronic effect of ligands [25,39] on this venerable reaction, it was found that electron-rich phosphinites produced highly active catalysts that gave high enantioselectivities (Table 5.1). For example, in the synthesis of methyl 4-fluorophenylalaninate (entry 2), the enantiomeric excess (ee) drops from 85% for the diphenylphosphinite **5c** (see Eq. 5.21 for structures of ligands) to 9% for the di[(bis-3,5-trifluoromethyl)phenyl] phosphinite **5a**. Remarkably, the use of phosphinite **5d** with comparatively more electron-donating bis-(3,5-dimethylphenyl) groups on phosphorus increased the ee to over 97%. Excellent yields (~100%) and selectivities are obtained over a wide range of substrates when **5d** is used as the ligand [26]. Successful asymmetric hydrogenations of dehydroamino acid derivatives with easily removable protecting groups on nitrogen have been rare. It was found that with the use of the bis-(3,5-dimethylphenyl)phosphinite **5d**, even the carboxybenzyl (CBZ)–protected derivative of 4-fluorophenyl alanine can be prepared in 96% ee (entry 3). An extensive list of amino acids prepared by this route can be found in reference [26].

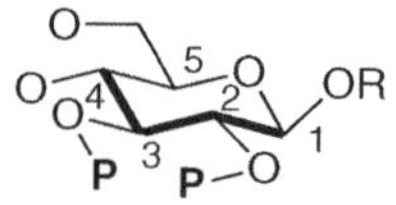

6
2,3-Di-O-**P**-D-*gluco*

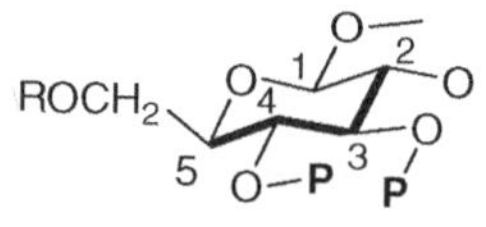

7
3,4-Di-O-**P**-D-*gluco*

P = diarylphosphino-

Figure 5.2 *D-glucopyranosides with 2,3- and 3,4-disubstitution pattern are approximately mirror images.*

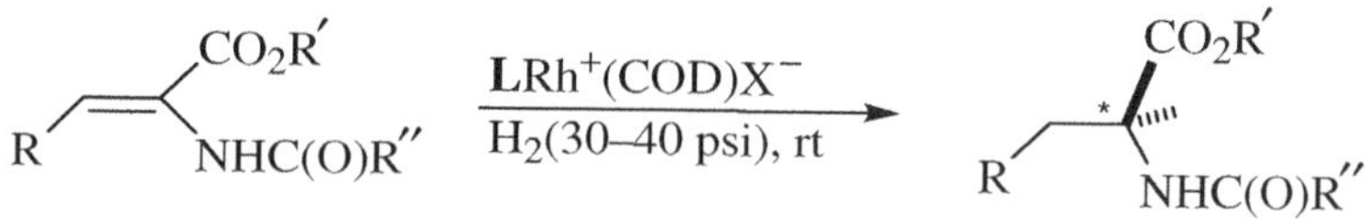

(Eq. 5.22)

R = Aryl, alkyl, R′ = H, Me, R″ = Me, OBn

Making L- or D-amino acids using D-sugars as ligand precursors: pseudo-enantiomeric 2,3- versus 3,4-bis-O-diarylphosphinoxy-D-gluco ligands. As pointed out earlier, the preparation of D-amino acids would normally require the enantiomers of the ligands that must be derived from prohibitively expensive L-glucose bis-2,3-diarylphosphinite. A cheaper and more practical option in the context of the D-glucopyranoside backbone is to use a ligand that is pseudo-enantiomeric, for example with a 3,4-diarylphosphinite motif (see Figure 5.2, **6** and **7**). Since the enantioselectivity of the reaction is largely dictated by the chirality of the two carbons to which the diarylphosphinoxy moieties are attached, one should expect D-amino acids from ligand **7**. In the event, it was found that 3,4-diarylphosphinite ligands (**8** and **9**) from methyl α-D-glucopyranoside and methyl β-D-2-deoxy-2-acetamidoglucose provided the D-amino acid derivatives in excellent ee. As before, the electron-rich 3,5-dimethylphenylphosphinites gave the highest ee (>96%) for a variety of substituted phenylalanine derivatives (Table 5.2) [26]. From a commercial perspective, these results mean that both D- and L-amino acids can now be prepared in excellent enantioselectivities by using ligands derived from two of the most abundantly available sugars, D-glucose and D-2-acetamido-2-deoxyglucose. This concept of generating pseudo-enantiomeric ligands from D-sugars has since been used by other workers [40].

While the sugar-derived ligands are among the most inexpensive ones for aromatic amino acid synthesis, for *aliphatic* amino acids these ligands are still not competitive with some of the best ligands available (e.g., the DUPHOS series) except for alanine, and further optimizations are needed before they can be applied in this context.

Incidentally, the phosphinite **10**, derived from methyl *D*-rhamnopyranoside, is one of the early examples of a monophosphinite that gave exceptionally high ee in the reduction of dimethyl itaconate (the use of other monodentate ligands is also discussed later in this chapter) [41,42]. A remarkable effect of a spectator ligand (1,5-COD vs. cyclooctene) on the enantioselectivity of this reaction is also noteworthy in this study.

Table 5.2 Synthesis of D- or L-arylalanines using 3,4- or 2,3-diarylphosphinites from D-sugars (**5d**, **8**, or **9**).

Entry	R in Eq. 22	Ligand[a]	%ee
1.	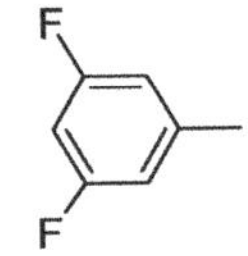	8 5d	96.2 (*R*) 98.4 (*S*)
2.	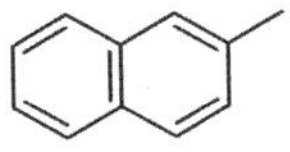	8 5d	96.0 (*R*) 97.1 (*S*)
3.	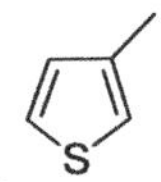	8 5d	97.0 (*R*) 96.7 (*S*)
4.	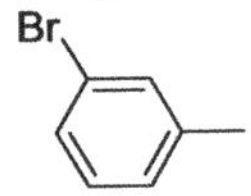	8 5d	96.4 (*R*) 96.8 (*S*)
5.	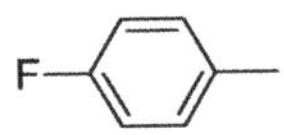	9 8 5d	97.8 (*R*) 96.2 (*R*) 97.2 (*S*)
6.	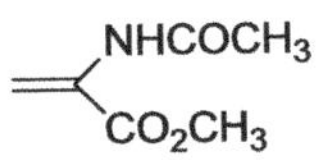	9 5d	95.0 (*R*) 96.9 (*S*)

[a] See Eq. 21 for structure of 5d.

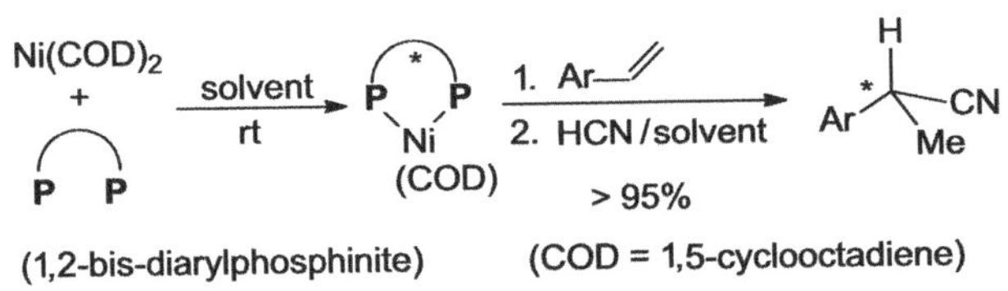

Scheme 5.3 Hydrocyanation of vinyl arenes.

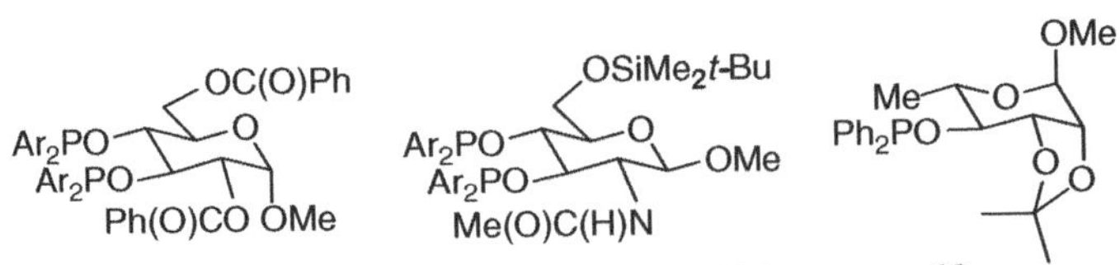

8 (Ar = 3,5-Me_2-C_6H_4) **9** (Ar = 3,5-Me_2-C_6H_4) **10**

Figure 5.3 3,4-Diarylphosphinite ligands for the synthesis of D-amino acids.

5.3.2.2 Ni(0)-catalyzed asymmetric hydrocyanation

Another outstanding application of the highly tunable carbohydrate-derived 1,2-bis-diarylphosphinites is in the Ni(0)-catalyzed asymmetric hydrocyanation of alkenes, a challenging reaction (Scheme 5.3) with no successful asymmetric variant known when the work in the area started [21,43]. Having established bis-diarylphosphinites as viable ligands for this reaction, it is relatively easy to make structural changes in this system since the phosphinites are readily assembled from the diols and the chlorophosphines. While the

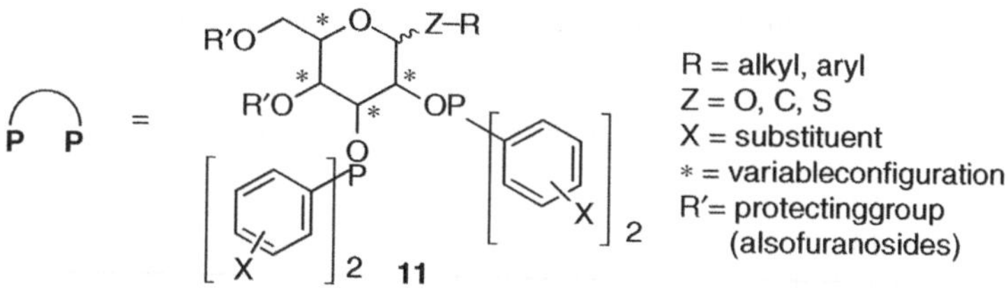

Figure 5.4 *Tunable sites on a typical sugar-derived ligand.*

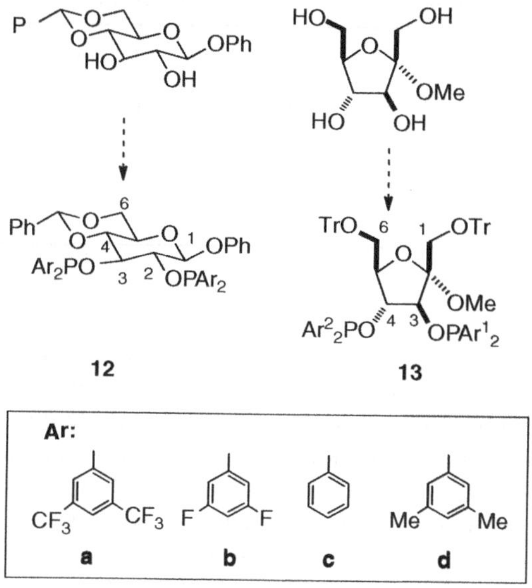

Figure 5.5 *Prototypical phosphinite ligands from cheap sugars.*

carbohydrate backbone provided the necessary stereochemical diversity, substitution patterns around phosphorus are used to vary the steric and electronic properties of the ligating atom (Figure 5.4).

The hydrocyanation reaction (Scheme 5.3) is typically carried out with 0.1–1.0 mol% of the Ni-catalyst derived from the phosphinites and $Ni(COD)_2$ in a hydrocarbon solvent. For vinylarenes, a near quantitative reaction ensued at room temperature (RT) to give the Markovnikov addition product.

Effect of the sugar backbone. Over 100 1,2- and 1,3-diarylphosphinite ligands were synthesized from various sugars, including glucose, galactose, fructose, 2-acetamido-2-deoxy-glucose, lactose, and trehalose [43]. Two of these scaffoldings along with the substituents on the *P*-aryl groups that were examined are shown in Figure 5.5. The results of hydrocyanation of 6-methoxy-2-vinylnaphthalene (MVN), a potential precursor for the popular anti-inflammatory agent naproxen, using these ligands are shown in Table 5.3. Inspection of these results suggests that the local chirality defined by the phosphinoxy-bearing carbons controls the absolute configuration of the major product. Thus, ligands of the type 2,3-bis-O-disubstituted D-gluco-diarylphosphinite (**12**) gave the (*S*)-nitrile as the major product, while the 3,4-disubstituted α-D-fructofuranoside (**13**) was found to give the (*R*)-nitrile.

Table 5.3 *Hydrocyanation of 2-methoxy-6-vinylnaphthalene: effect of the sugar backbone.*

Entry	Ligand	Phosphinite	Attachment site	% ee
1	**12**	Ph_2P	2,3	40 (*S*)
2	**13**	Ph_2P	3,4	30 (*R*)

Table 5.4 *Ligand electronic effects on asymmetric hydrocyanation of vinylarenes.*

Substrate	Ligand (% ee)			
	12a	**12b**	**12c**	**12d**
2-Methoxy-6-vinylnaphthalene	85–91	78	35	16
Acenaphthene	59	—	0	—
4-isobutylstyrene	56	38	6	—

Steric and electronic effects of the phosphorus substituents. While the steric and electronic tuning of the aglycone (Z-R in Figure 5.4) gave a modest, yet discernible, improvement in the selectivity of the reaction [43], substituents on the ligating phosphorus had the most pronounced effect. Hydrocyanation of a number of vinylarenes using *o*-, *m*-, and *p*-substituted gluco-diarylphosphinite ligands (**12a–12d** and others) led to the unexpected discovery that the electronic characteristics of the phosphorus–aryl substituents had a much greater impact on the enantioselectivity than the inherent size of the group. This electronic effect can be seen clearly from the results obtained using *m*-disubstituted ligands (**12a–d**; Figure 5.5), as shown in Table 5.4. The ee increased dramatically (from 16% to>85% for (6-methoxy-2-vinylnaphthalene, MVN), and from 6% to 56% for 4-isobutylstyrene) as the electron-withdrawing power of the *P*-aryl substituent increased (for **12a–d**, σ_m = 0.43, 0.34, 0.0, and –0.07, respectively). Other examples are included in reference [22]. A mechanistic rationale based on kinetic and deuterium-labeling studies is also included in this chapter.

Effect of electronic asymmetry: fine tuning of a ligand for high enantioselectivity [23]. During the initial scouting experiments, it was discovered that 3,4-bis-O-diphenylphosphino-α-D-fructofuranoside (**13**, Ar = Ph) gave a modest selectivity (30–40%, *R*), but *opposite to that of the D-gluco-1,2-phosphinite*, in the asymmetric hydrocyanation of MVN (Table 5.3, entry 2). The methyl α–fructofuranoside is an especially attractive scaffolding for further fine tuning of the ligand by virtue of the fact that the two hydroxyl groups (C_3 and C_4) are placed in completely different steric environments. Accordingly, a series of electronically "symmetrical" and "unsymmetrical" bis-(diaryl)phosphinite ligands from this sugar (Figure 5.6) was prepared, and their effects on the hydrocyanation were examined. Several of these ligands have the unique feature that the steric effects around the two chelating phosphorus atoms are largely kept constant, while the relative electron densities on these atoms are systematically varied. Results of the hydrocyanation of MVN are shown in Figure 5.6 [23].

As anticipated from previous work, electron-donating substituents on the phosphorus aryl groups (**14a**) give the lowest ee. Electron-deficient phosphinites (**14b**) increase the selectivity to some extent. But the

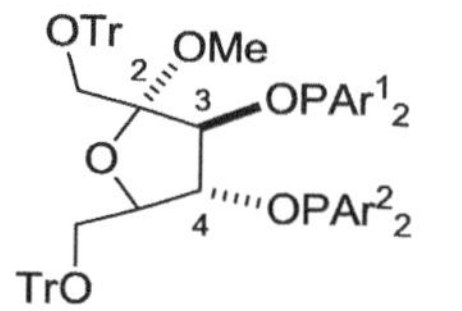

	(% ee)
14a $Ar^1 = Ar^2 = Ph$	(43)
14b $Ar^1 = Ar^2 =$ 3,5-*bis*-CF_3-C_6H_3	(56)
14c $Ar^1 =$ 3,5-*bis*-CF_3-C_6H_3; $Ar^2 = Ph$	(58)
14d $Ar^1 = Ph$; $Ar^2 =$ 3,5-*bis*-CF_3-C_6H_3	(95 at 0°C)

Figure 5.6 *Electronically unsymmetrical ligands from fructose.*

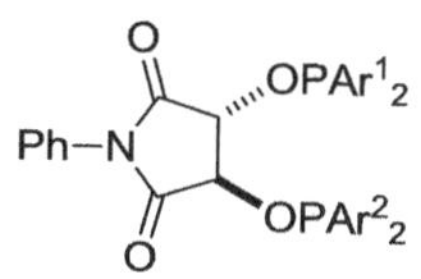

	(% ee)
15a $Ar^1 = Ar^2 = Ph$	(54)
15b $Ar^1 = Ar^2 =$ 3,5-*bis*-CF_3-C_6H_3	(70)
15c $Ar^1 = Ph$; $Ar^2 =$ 3,5-*bis*-CF_3-C_6H_3	(77)

Figure 5.7 *Electronically unsymmetrical ligands from tartranil.*

highest ee (89% at RT, 95% at 0°C) was obtained with C_4 carrying the electron-deficient P, and C_3 the electron-rich one (**14d**). On the other hand, electron-rich P on C_4 and electron-deficient P on C_3 gave only marginal ee. Several other examples of electronically unsymmetrical phosphinites in the fructose series corroborate this general trend [23].

A similar observation was also made using (*3S, 4S*)-tartranil phosphinites (**15a–c**, Figure 5.7); the highest ee, 77% (*S*), was obtained with a mixed phenyl/3,5-bis(trifluoromethyl)phenyl derivative (**15c**), whereas the C_2-symmetric phenyl and 3,5-bis(trifluoromethyl)phenyl derivatives **15a** and **15b** gave 54% and 70% ee, respectively.

Thus the selective manipulation of the electronic properties of ligands from two readily available monosaccharides (D-glucose and D-fructose) allows the preparation of either the (*R*)- or the (*S*)-enantiomer of an important 2-arylpropionitrile in over 90% ee.

From a practical standpoint, turnover numbers of 700–800 in $Ni(COD)_2$ and up to 4500 in the ligand have been achieved for the production of naproxen nitrile (Scheme 5.3, Ar=2-(6-methoxy)naphthyl), and this material can be recrystallized to optical purity (>99% ee). The reaction has not been optimized for large-scale production or for other 2-arylpropionitriles. The ee is only modest in most other cases studied so far (e.g., 56% for ibuprofen nitrile, Scheme 5.3, Ar=4-*i*-butylphenyl). Mild reaction conditions, high turnover frequency of the catalyst, and prospects for ligand tuning to improve the enantioselectivity should provide ample incentives for further research in this area.

Room-temperature hydrocyanation of 1,3-dienes. 1,2-Bis-diarylphosphinites derived from cyclohexanediol (**1**) and phenyl β-glucopyranoside (**12**) are good ligands for Ni-catalyzed hydrocyanation of various 1,3-dienes (Eq. 5.23) at RT [24]. The enantioselectivities are the highest known for any 1,3-dienes reported thus far.

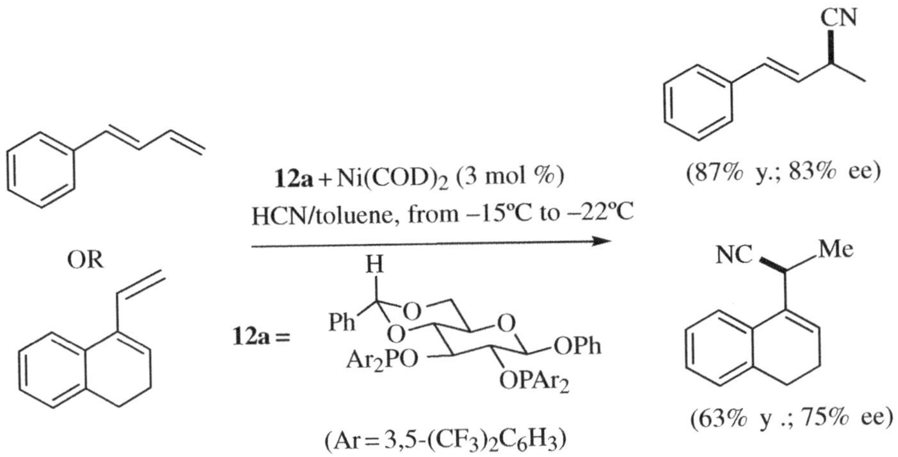

(Eq. 5.23)

5.3.2.3 *Ni(0)- and Pd(0)-catalyzed allylic substitution by carbon nucleophiles*

(i) Ni(0)-catalyzed allylic substitution using a Grignard reagent. The sugar-derived 1,2-bis-diarylphosphinites are excellent ligands for Ni(0)-catalyzed substitution of an allylic phenoxy group by a Grignard reagent, even though the enantioselectivity is only modest [30].

The sugar-derived phosphinites and those from 2,2′-binaphthol are excellent ligands for the Pd(0)-catalyzed addition of stabilized carbanions to allylic electrophiles (Tsuji–Trost reaction), even though only moderate enantioselectivities are realized (see later, Eq. 5.41) [29]. Hybrid thioether–phosphinites give the best selectivities for this reaction [106] (see later, Eqs. 5.42 and 5.43).

A nucleophilic allylation of cinnamaldehyde using allyltributyltin catalyzed by a (bis-phosphinite)$PtCl_2$ complex derived from chloramphenicol gives very high yield and selectivity using 1 mol% catalyst (Eq. 5.24B) [85].

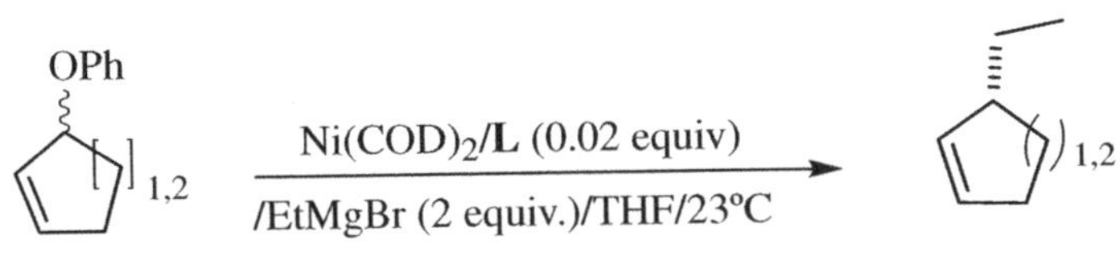

n	L	yield / % ee
1	**12a**	88/12 (*S*)
1	**12d**	78/0 (–)
2	**12a**	62/79 (*S*)
2	**12d**	99/36 (*S*)
2	**8**	55/36 (*R*)

(Eq. 5.24A)

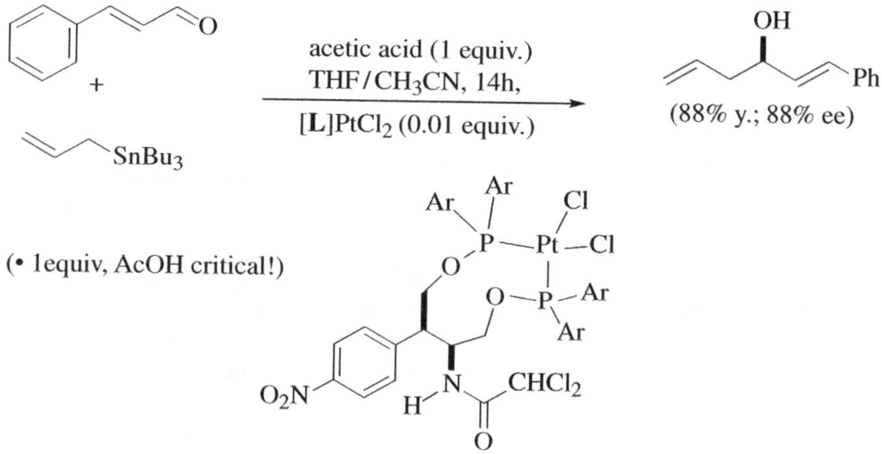

(Eq. 5.24B)

5.3.2.4 *Rh(I)-catalyzed hydroformylation of vinylarenes*

Ligands **12a–12d** have been used for Rh-catalyzed hydroformylation of 2-vinylnaphthalene (Eq. 5.25). Using the ligand **12a** (0.0015 equiv.), the reaction proceeds at RT (18 h) in hexane at 500 psi (1:1 CO–H_2) to give a quantitative yield of the product aldehydes with an unusually high branched-to-linear ratio of 19:1) for this reaction [44]. The enantioselectivity was a modest 50% ee. Most notably, the electron-rich bis-phosphinites **12c** and **12d** under comparable reaction conditions gave lower conversion and with less than 10% enantioselectivity.

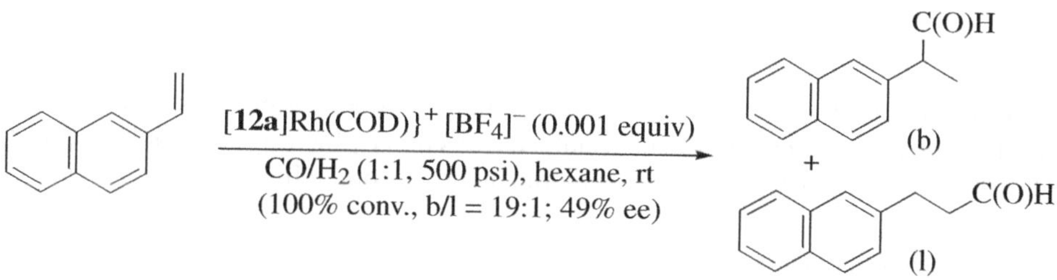

(Eq. 5.25)

5.3.2.5 *Ni(II)-catalyzed asymmetric hydrovinylation of alkenes*

Hydrovinylation of alkenes (Eq. 5.26, Scheme 5.4) is an important C–C bond-forming reaction where very few ligands have given practical levels of efficiency and selectivity [45,46]. Vogt *et al.* [47] identified a (–)-menthol-derived *P*-chiral phosphinite (**16**),which gave good selectivity and yield in a Pd(II)-catalyzed hydrovinylation of styrene (Eq. 5.26). In continued efforts to improve the enantioselectivity of the corresponding Ni-catalyzed reaction, our group [31,32] screened a large number of carbohydrate (a generic structure: **18**) and other monophosphinites, and discovered that easily accessible β-acetamido-derivative **17** (prepared in five steps from readily available 2-deoxy-2-acetamidoglucose) served as an excellent ligand for this exacting reaction (Scheme 5.4). Bromostyrene gives >98% isolated yield with >99% selectivity for the desired 3-arylbutene and 89% ee. The resulting product has been transformed into (*R*)-ibuprofen.

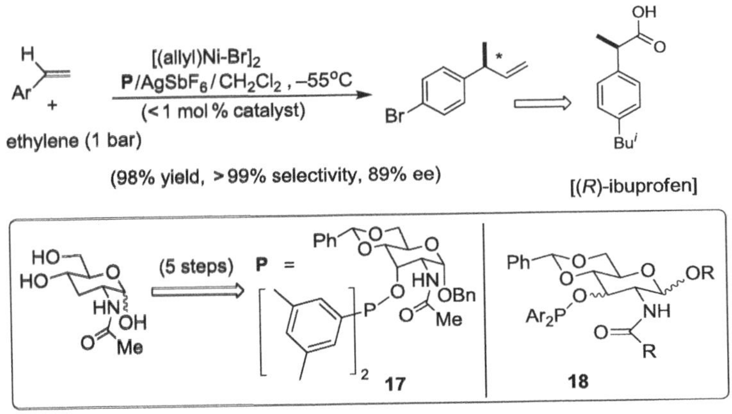

Scheme 5.4 Phosphinite ligands for Ni-catalyzed hydrovinylation.

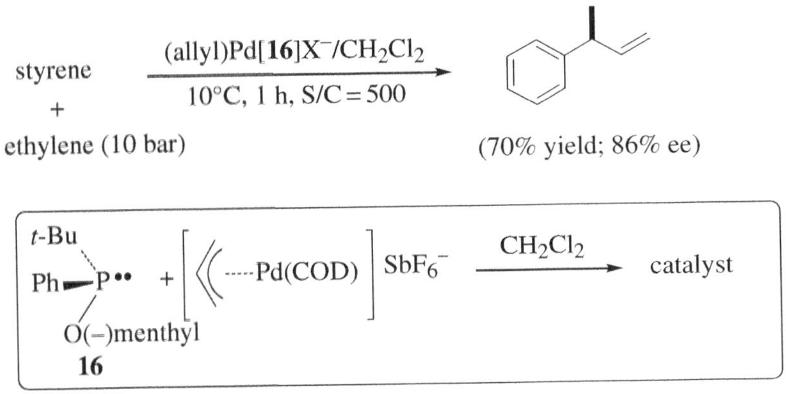

(Eq. 5.26)

Several other ligands based on various pyranose and furanose carbohydrate diols have been prepared and applied to a wide variety of transition-metal catalyzed reactions [48–52]. A notable application is the Rh-catalyzed hydrogenation of 100% ee (*S*)(!) using the monophospinite ligand **10** (Figure 5.3) [41].

5.3.2.6 *Ligands for homogeneous catalysis in water*

Water-soluble versions of sugar-derived phosphinites (Figure 5.8: **19** [53], **20** [54,55], **21** [55,56], and **22** [57]) as ligands for catalyzed reactions in aqueous media have also been reported. Use of surfactants (e.g., dodecyl sulfate) to promote better reactions of hydrophobic substrates is a highlight in this area [53–59]. Carbohydrate-derived phosphinites (e.g., **23**) have also been used for Pd(0)-catalyzed allylation reactions in water [60]. While the biphasic conditions do allow some catalyst recycling, the enantioselectivities in aqueous reactions, except in rare cases [55], are generally lower compared to those obtained in the reactions conducted in organic solvents.

5.3.3 Phosphinite ligands from other alcohols

In addition to the carbohydrate-derived bis-diarylphosphinites, several others from readily available diols, among them dimethyl tartrate, (*R*)-pinenediol, and *t*-butyl 3-ethoxycarbonylcholate, were prepared and evaluated in Ni(0)-catalyzed asymmetric hydrocyanation [43]. While these ligands provided good turnover numbers for the reactions, the enantioselectivity seldom exceeded 50%. Several mono-diarylphosphinites have also been found useful in the hydrovinylation (Scheme 5.4) reactions [32].

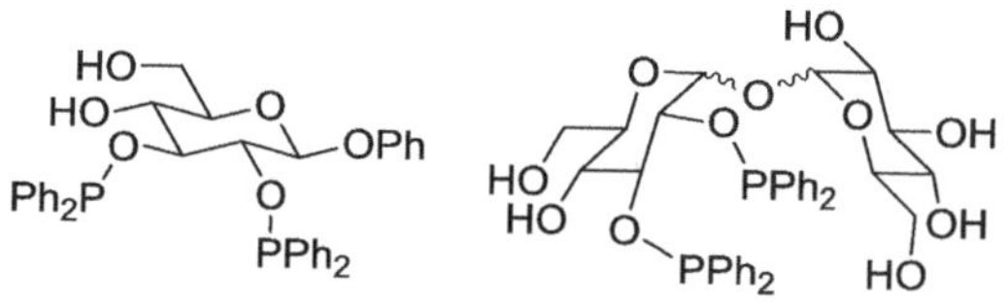

19 **20**(α,α) **21**(β,β)

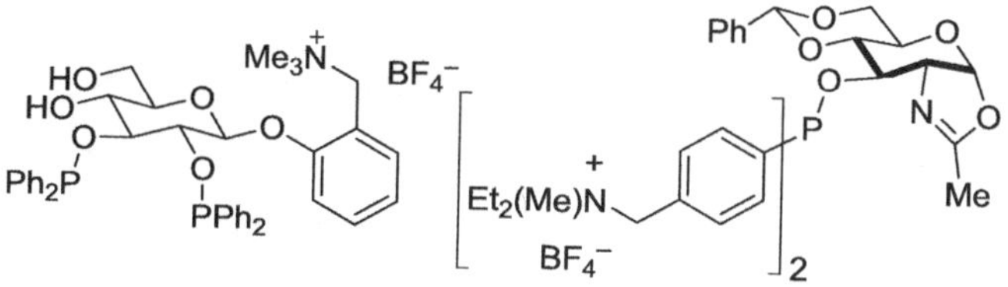

22 **23**

*Figure 5.8 Water-soluble ligands for Rh-catalyzed hydrogenations (**19–22**) and Pd-catalyzed allylations (**23**).*

5.3.4 Phosphine–phosphinite and amine–phosphinite ligands

Mixed ligands like phosphine–phosphinites derived from xylose [48] and L-rhamnose [61] were among the first chelating systems tried for Rh-catalyzed hydrogenation (Eq. 5.22), even though the selectivities were poor. Most useful ligands in this series are based on the BINAP scaffolding. Nozaki [62] has reported the use of the so-called BIPNITE ligands (e.g., **24**) for the hydroformylation of a β-lactam (**25**, Eq. 5.27) with excellent diastereoselectivity. A Rh complex of a 3-phenyl derivative of **29** (prepared from **28** [63,64]) gives >99% ee for the reduction of a wide range of 2-N-acetylcinnamates [65]. Synthesis of **29** (Scheme 5.5) is typical for these types of ligands with 3-substituents.

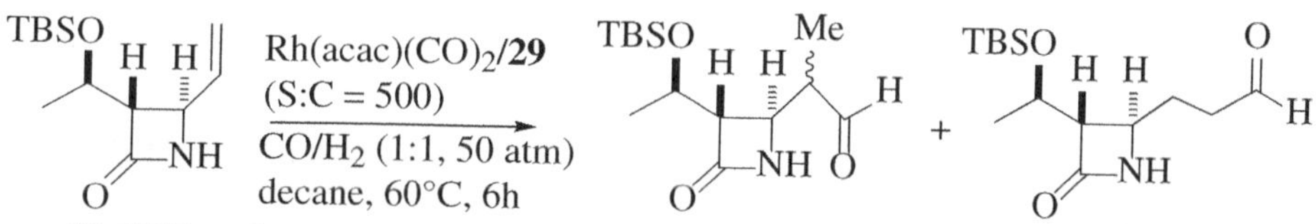

25 (95% y.; branched:linear = 74:26) **26** (β:α 96:4) **27**

(Eq. 5.27)

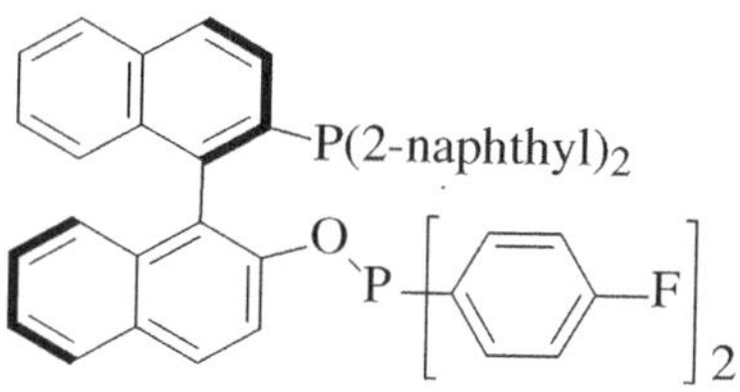

24

Two other noteworthy ligands in this general class are phosphine–phosphinite (**31**) [66] and phosphine–aminophosphine (**33**) [67], both prepared from ferrocenylamine **30** (Scheme 5.6) [68].The ligand **33** is easily prepared (Scheme 5.6) and remarkably stable in air, and its Rh complex is very active (turnover frequency (TOF) up to 77,000 h^{-1}) in reduction of N-acylcinnamates, the corresponding β-alkyl substituted acrylates, and even those containing tert-butyloxycarbonyl (BOC)– and CBZ-protecting groups on nitrogen (Eq. 5.28).

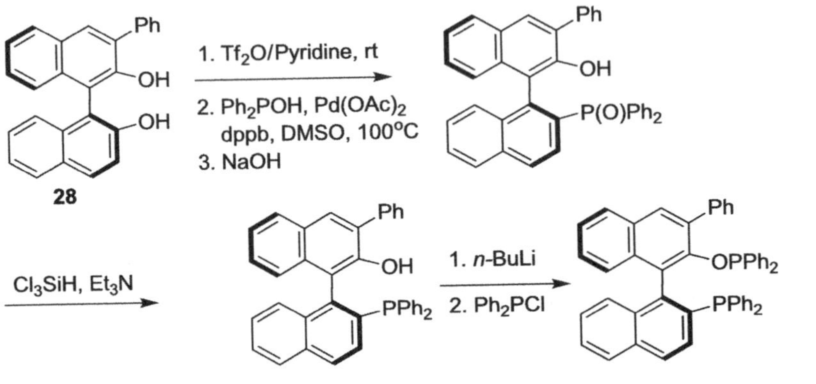

(25% from **28**) (*S*)-**29** (*S*)-3-Ph-BIPNITE (R = Ph)

Scheme 5.5 *Synthesis of a 3-aryl-2,2′-binaphthol phosphinite.*

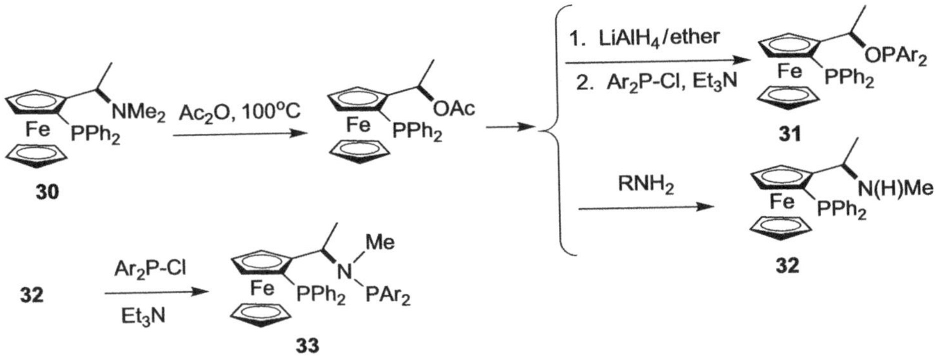

Scheme 5.6 *Highly active ligands for Rh-catalyzed hydrogenation.*

The full paper describing this work reports a direct side-by-side comparison of several popular ligands for the reduction of the prototypical substrate, (Z)-methyl N-acetylcinnamate [67].

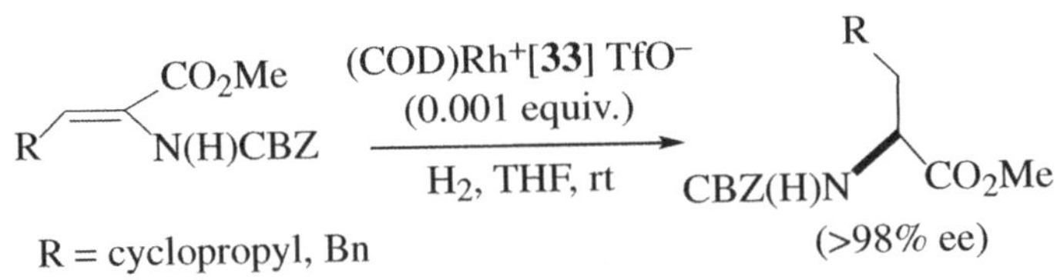

(Eq. 5.28)

5.3.5 Phosphinites from amines, amino alcohols, and amino acids

5.3.5.1 *Aminophosphines*

The first use of diarylphosphino-1,2-bis-amines as ligands dates back to 1976, when Fiorini reported the hydrogenation of *N*-acetylcinnamic acid using the complex [Rh(**34**)Cl]$_2$ (Scheme 5.7) in 84% ee [69,70]. This was quickly followed by equally impressive results for Rh-catalyzed hydrogenation of a more challenging substrate using ligand **35** (Scheme 5.8) [71]. Miyano introduced axially dissymmetric biaryl-based

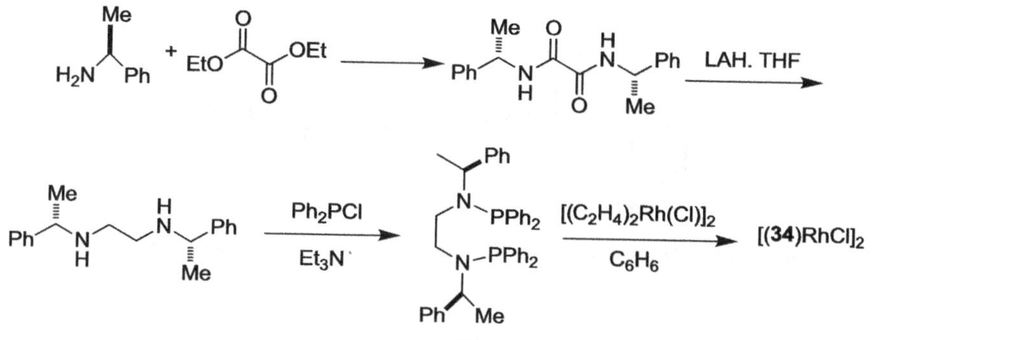

Scheme 5.7 Synthesis of a 1,2-bis-diphenylaminophosphine ligand and a Rh complex.

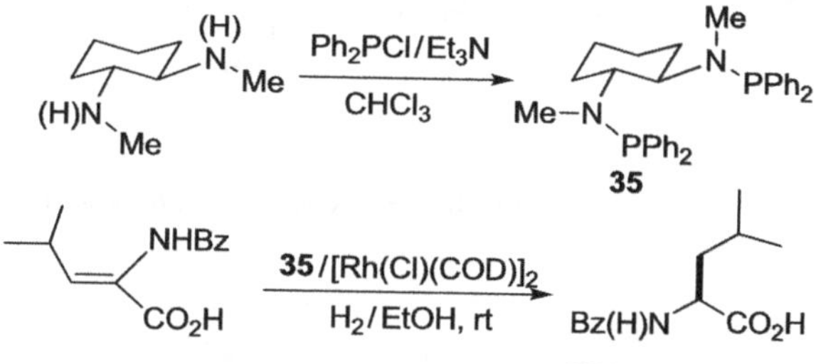

Scheme 5.8 An early example of the synthesis of an aliphatic amino acid using a 1,2-bis-diphenylaminophosphine ligand.

diaminophosphinites like **36a** and **36b** to the Rh-catalyzed hydrogenation area and obtained moderate selectivity in the reductions of typical dehydroamino acids (up to 90%) (Eq. 5.30; Figure 5.9) [72a]. A hydrogenated version of this binaphthyl ligand (**37**) was found to be an excellent ligand for the hydrogenation of 1-arylethylen-1-caboxamide (Eq. 5.30) [72b]. Zhang recently reported further applications of a related ligand **38** in Rh-catalyzed hydrogenation of *N*-acetylcinnamates [73].

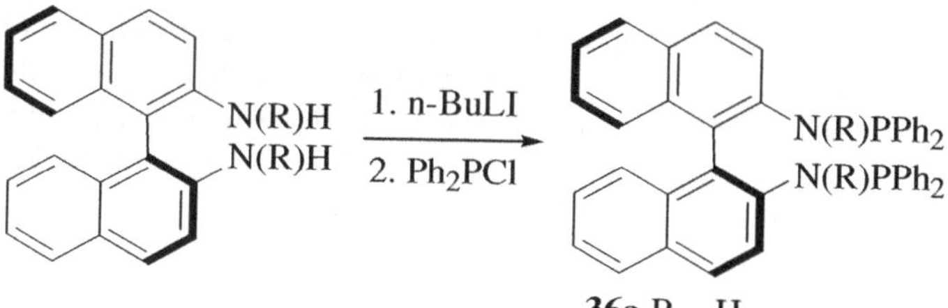

36a R = H,
36b R = Me

(Eq. 5.29)

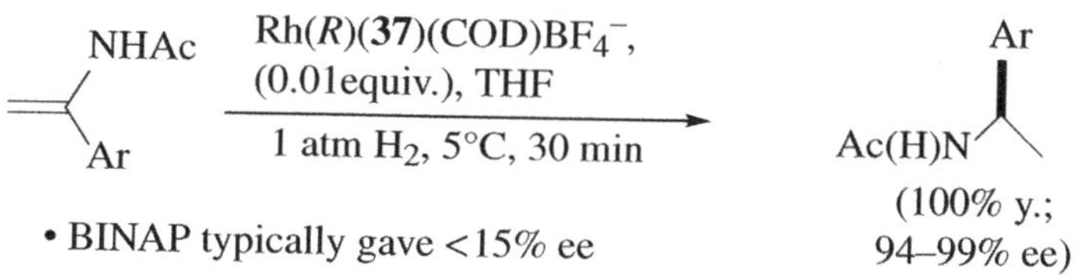

• BINAP typically gave <15% ee

(Eq. 5.30)

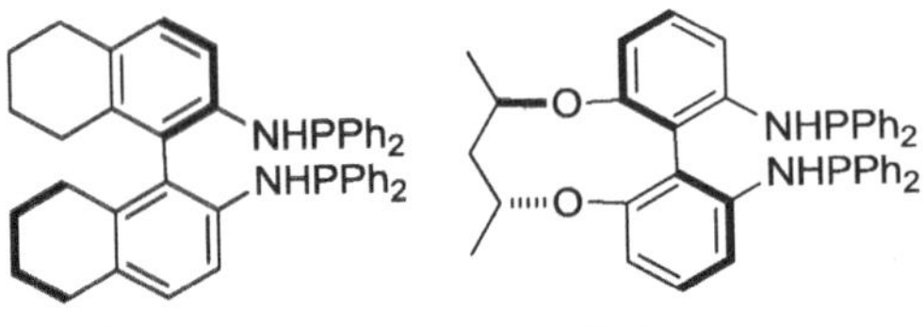

Figure 5.9 Aminophosphine ligands derived from axially chiral biaryls.

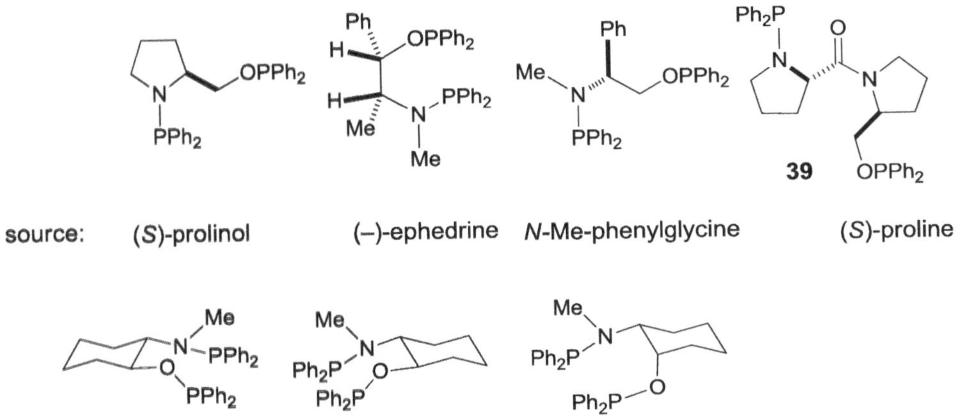

Figure 5.10 Aminophosphine–phosphinite ligands.

5.3.5.2 Aminophosphine–phosphinite (AMPP) ligands

Syntheses of aminophosphine–phosphinites Chelating aminophosphine–phosphinites have received more attention than the corresponding aminophosphines, especially in connection with Rh-catalyzed hydrogenations, because of the ease of their syntheses from widely available amino acids, and amino alcohols, many of which are either components of pharmaceuticals or intermediates involved in their syntheses. Extensive reviews on aminophosphine–phosphinites [74–76] have been published. The discussion in this subsection is limited to typical syntheses of these types of ligands and some of their applications that result in high selectivities.

Rhodium complexes of aminophosphine–phosphinites derived from prolinol [77], ephedrines, *N*-methylphenylglycinol [78–80], and the hydroxyamide **39** derived from (*S*)-proline [81] (Figure 5.10) were found to give excellent yields in hydrogenation of *N*-acylaminoacrylates even though enantioselectivities observed were modest by today's standards. Ligands derived from *cis*- and *trans*-2-*N*-methylaminocyclohexanols **40a–40c** [82] are noteworthy in that high enantioselectivity was observed for the reduction of *N*-acetylvaline precursor using a Rh complex of **40c**.

Equations 5.31, 5.32, 5.33, 5.34, and 5.35 illustrate various strategies for the synthesis of aminophosphine–phosphinites of various structural types. Except when the acidity of the N–*H* is low, treatment with a chlorophosphine and a base usually provides good to excellent yields of the product, which can be purified by column chromatography on basic alumina or, in some cases, silica gel under nitrogen. These compounds are generally quite stable in an anaerobic and moisture-free environment, especially in the solid state. Metathesis reaction with a metal alkene complex is by far the most common method for the generation of the metal complexes.

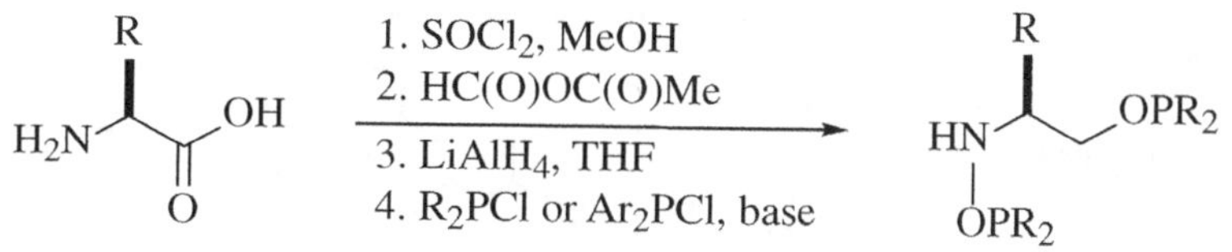

(Eq. 5.31)

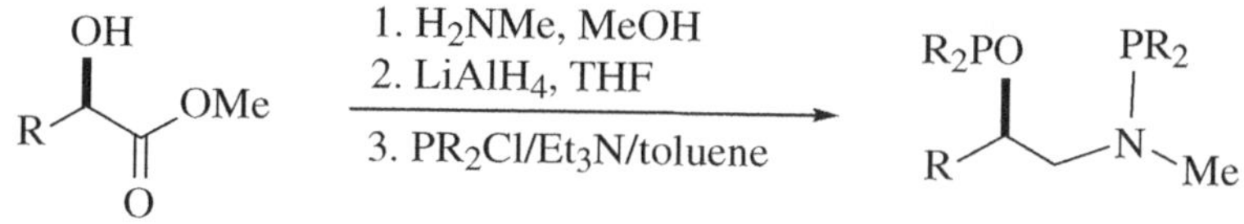

(Eq. 5.32)

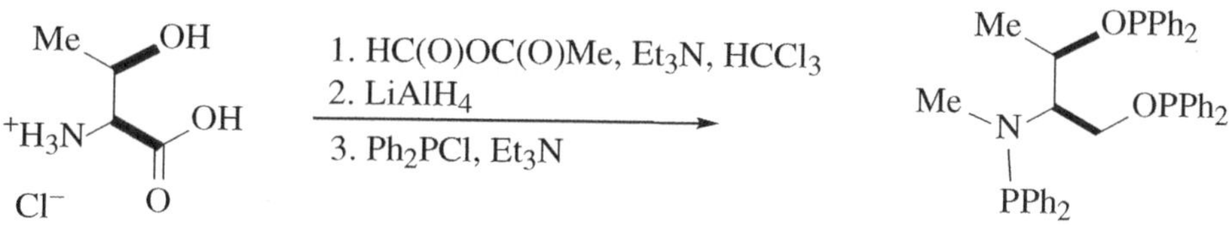

(Eq. 5.33)

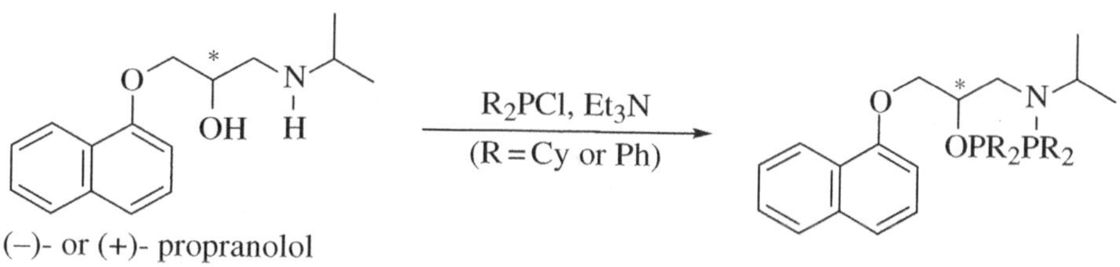

(Eq. 5.34)

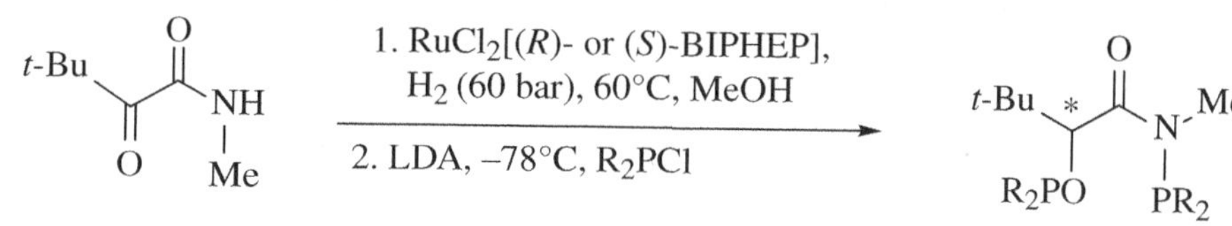

(Eq. 5.35)

An especially attractive way of generating the complexes of air-sensitive phosphinites uses the so-called template synthesis [83–85] in which a chlorophosphine complex is directly reacted with the diol or amino alcohol (Scheme 5.9).

Applications of aminophosphine–phosphinite ligands A notable early application of an aminophosphine–phosphinite ligand, derived from threonine (Eq. 5.33), is for an asymmetric hydrovinylation of a 1,3-diene (Eq. 5.36) [86]. Other applications of AMPP ligands in olefin dimerizations have also been recorded [7f,76].

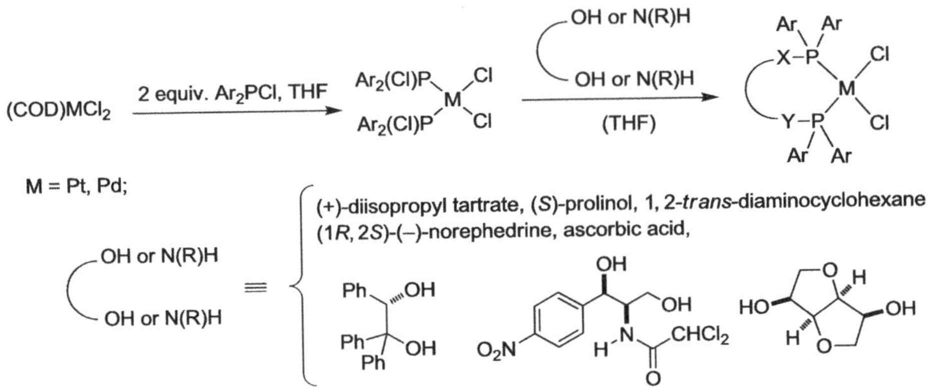

Scheme 5.9 *"Metal–template" synthesis of aminophosphine and phosphinite complexes.*

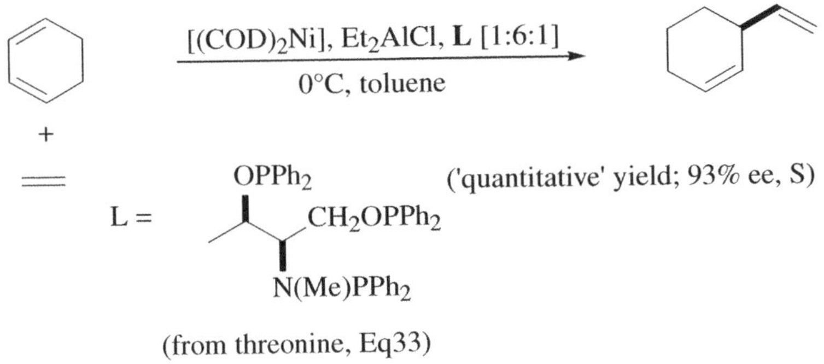

(Eq. 5.36)

Even though complexes of AMPP ligands have been used to effect Pd(0)-catalyzed allylation of 1,3-diphenylallyl acetate (Pd) [87], hydroformylation of vinylarenes (Pt [88] and Rh [89]), Co-catalyzed [2+2+2]-cycloaddition between norbornadiene and acetylenic derivatives (Eq. 5.37) [75], and hydrogenation of enolacetates (Rh) [90], the most notable results are in the area of Rh-catalyzed hydrogenation of functionalized ketones (Eqs. 5.38, 5.39, and 5.40) [76,91–93]. Significant results have also been achieved in the synthesis of amino acids via hydrogenation of the corresponding enamide derivatives [94]. Neutral dicycloalkylphosphinite complexes in nonpolar solvents are best for hydrogenation of functionalized ketones, whereas cationic diarylphosphinite complexes seem best for dehydroaminoacids, especially when the reaction is conducted in polar solvents.

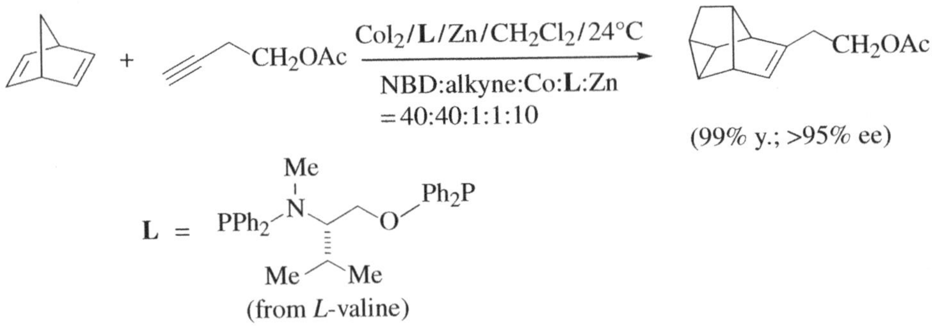

(Eq. 5.37)

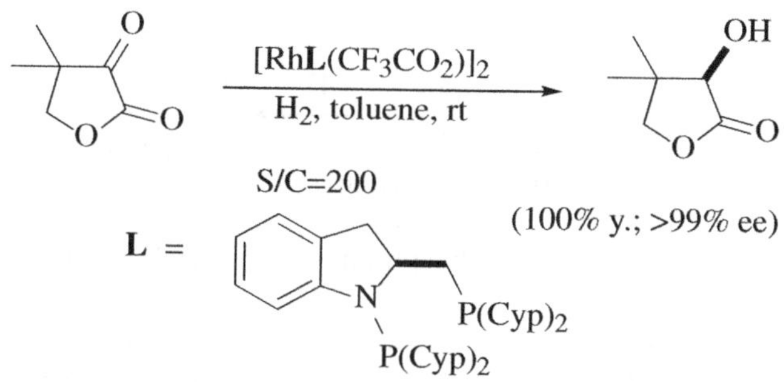

(Eq. 5.38)

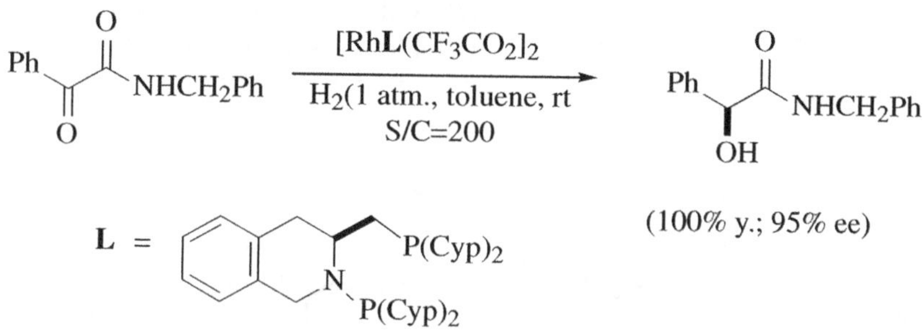

(Eq. 5.39)

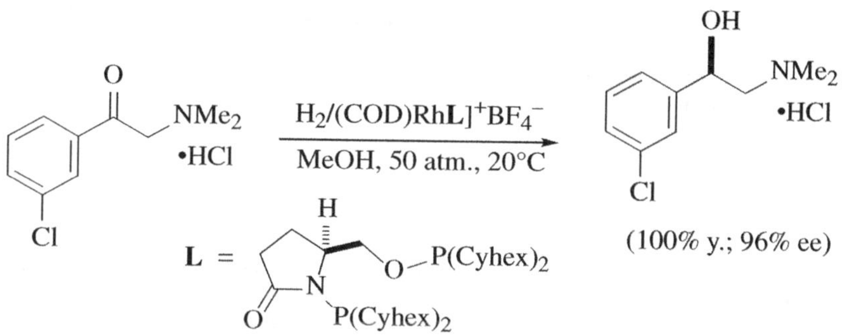

(Eq. 5.40)

5.3.6 Bisphosphinite ligands with other scaffoldings

Apart from the carbohydrate and amino alcohol backbones, two other scaffoldings have been found to be useful in diaryl- and dialkylphosphinite ligand synthesis. These are the spirononane and biaryl systems shown in Figure 5.11. Thus, the spirocyclic bisphosphinite **41** [95] in a side-by-side comparison with other well-known ligands matches, or sometimes exceeds, the best selectivities seen in the literature for the Rh-catalyzed hydrogenation of α-acetamidoacrylates and the corresponding cinnamates. A 2,2′-binaphthol-derived phosphinite was among the first of these ligand classes to be synthesized [96], and similar ligands have been used for Rh-catalyzed hydrogenation [97,98] and Pd(0)-catalyzed allylation of malonates [99]. Seebach has reported the use of a tartrate-derived 1,2-bis-diarylphosphinite **44** for this reaction, resulting in a modest enantioselectivity [100]. A detailed study of electronic effects on the latter reaction (Eq. 5.41) has revealed that the enantioselectivity of the reaction can be reversed by electronic tuning of the phosphinite substituents [29].

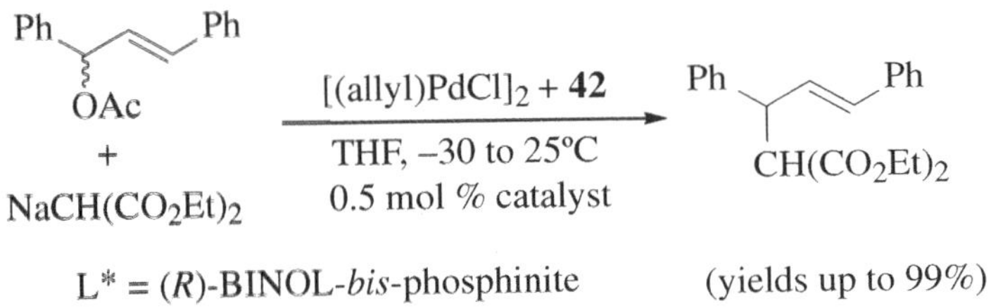

(Eq. 5.41)

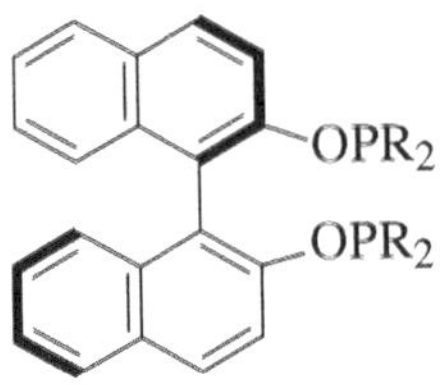

	R	% ee
(*R*)-**42d**	C_6H_5	80 (*R*)
(*R*)-**42e**	C_6H_{11}	87 (*S*)

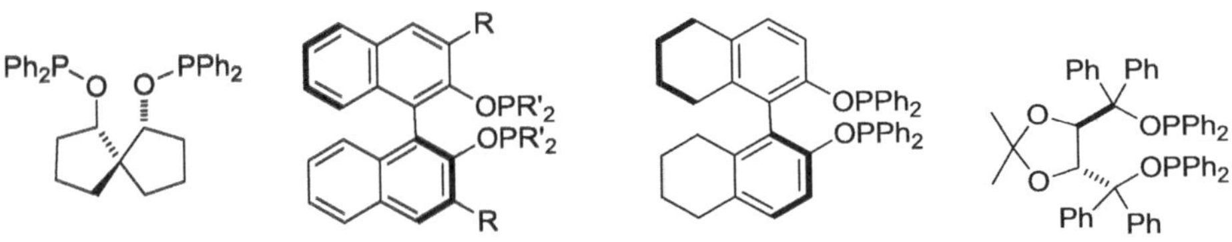

41

42a R = 3,5-$Me_2C_6H_3$; R' = Ph
42b R = Ph; R' = 3,5-$Me_2C_6H_3$
42c R = Ph, R' = Ph
42d R = H, R' = Ph
42e R = H, R' = CyHex

43 (*S*)-H8-BINAPO

44

Figure 5.11 *Bisphosphinites derived from spirononanediol, 2,2′-binaphthols, and TADDOL.*

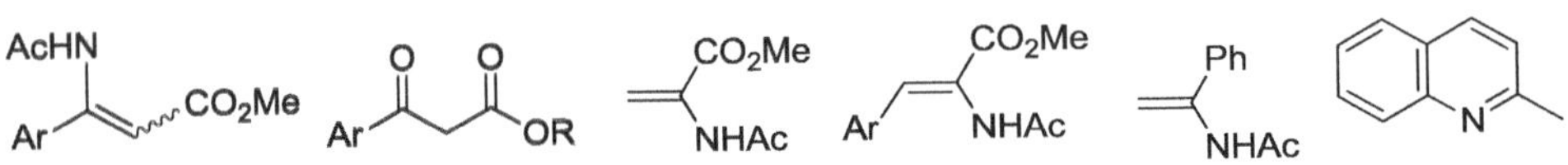

45 **46** **47** **48** **49** **50**

Figure 5.12 *Prototypical prochiral substrates for enantioselective hydrogenation using BINAP–phosphinites.*

3,3′-Disubstituted BINAPO ligands **42a** and **42b** are exceptionally good ligands for the Ru-catalyzed hydrogenation of difficult-to-reduce substrates such as β-aryl-substituted β-acylaminoacrylates (**45**) and β-keto esters (**46**) (Figure 5.12) [101]. Ligand **42c** gives acceptable selectivities for the Rh-catalyzed reduction of popular substrates, methyl α-*N*-acetylacrylate (**47**), a variety of *N*-acetylcinnamates (**48**), and 1-phenyletheneamide (**49**) [102]. The partially hydrogenated analog **43** is a good ligand for the Ir-catalyzed hydrogenation of 2-methylquinoline (**50**) [103].

5.3.7 1,1′-Diaryl-2,2′-phosphinites and dynamic conformational control in asymmetric catalysis

The exquisite selectivity realized by many enzymes relies on the dynamic conformational features produced by molecular folding to communicate structural information over large distances to the active site. Synthetic

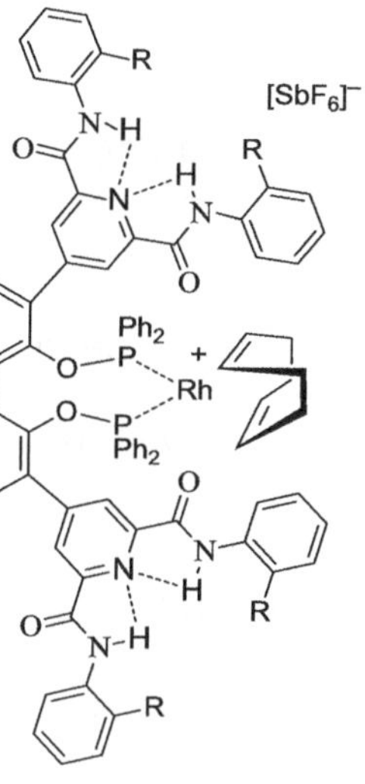

51a-c

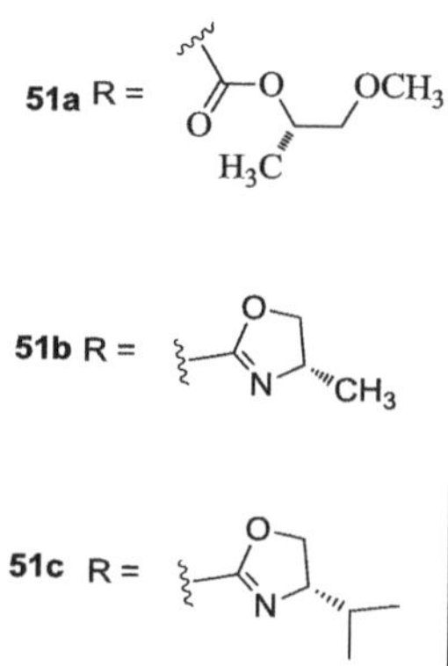

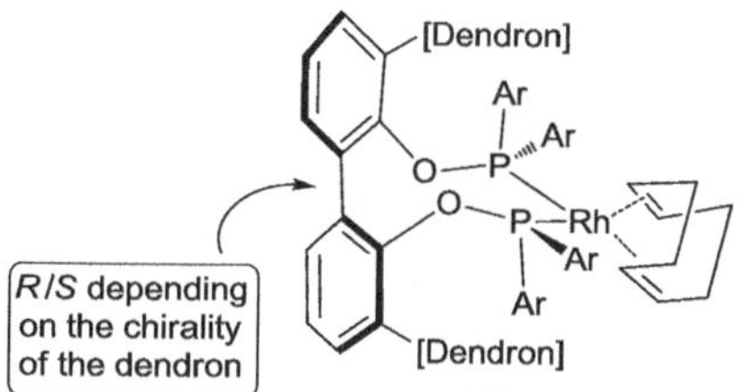

53

Ligand	ΔG‡ (P/M isomer.[1])[2]	% ee[3]
51a	<5	18 (*R*)
51b	13	57 (*S*)
51c	>20	87 (*S*)
52a	>20	97 (*R*)
52b	>20	95 (*S*)

1. For the dendron attached at 3/3′ positions
2. In kcal/mol
3. See Eq 22

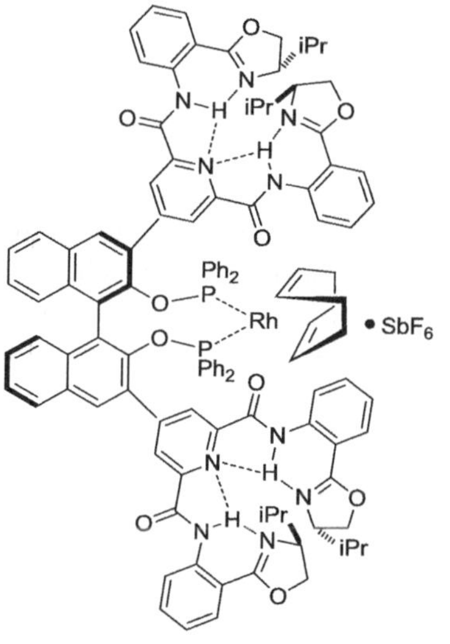

52a backbone: (*R*)-BINAPO (shown)
52b backbone: (*S*)-BINAPO

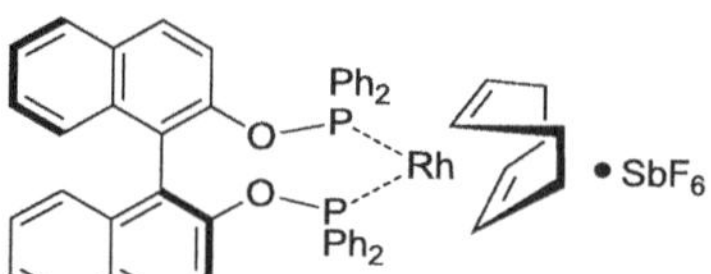

[57% ee (*S*)]

52c [(*R*)-BINAPO]Rh(I)(COD) with no dendron

Figure 5.13 *Design of a dynamic catalyst for Rh(I)-catalyzed asymmetric hydrogenation.*

catalysts generally depend on static, proximal structural information for selectivity, including enantioselectivity. The quest for truly biomimetic molecules that exploit a chiral, folded secondary structure in asymmetric catalysis remained an unrealized objective in abiological systems until recently. In 2008, dendritic catalysts (**51a–c**, Figure 5.13) based on highly fluxional biphenyl bisphosphinite were used [104] to demonstrate the idea that the dynamic conformational properties of a remote dendron (capable of low-energy *P*/*M* interconversion) can affect molecular folding, and this in turn can be used to communicate structural information, and hence exert stereochemical control, over large distances to the active site. Thus the remote chirality originating in the bis-oxazoline dendron within the catalyst **51c** (Figure 5.13) controls the

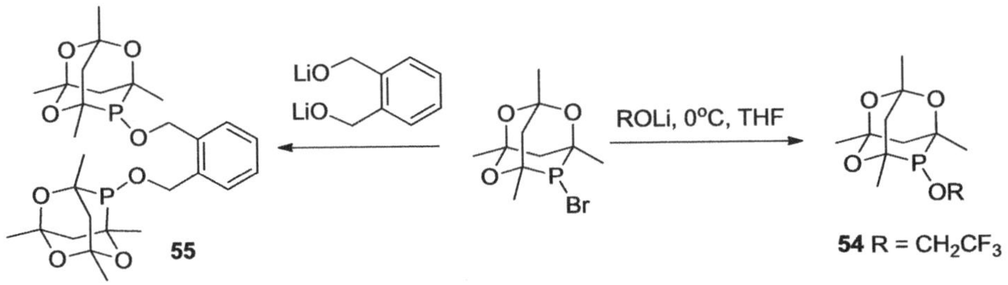

Scheme 5.10 Hydrolytically stable cage phosphinites similar to $P(O\text{-}2\text{-}tol)_3$ in Ni-catalyzed hydrocyanation of butadiene.

enantioselectivity of hydrogenation of methyl *N*-acetylcinnamate (Eq. 5.22) efficiently by relaying on chirality transfer over 14 bonds (**53**) to the catalytic site. Control experiments with atropisomeric binaphthyl systems **52a–52c** and dendritic fragments with lower barriers for the *P*/*M* interconversion of the tether (**51a** and **51b**) demonstrate that the dendrons themselves *do not* affect the selectivity of the reaction, but chirality transfer to the biphenyl core is essential for the selectivity. Circular dichroism studies support the correlation between the helical bias for *P* chirality in the dendrons (**51a**<**51b**<**51c**) and the extent of enantioselectivity observed for the hydrogenation reaction.

5.3.8 Monophosphinite ligands

Chiral monophosphinites have been found to be excellent ligands for a number of reactions like Rh-catalyzed hydrogenation of enamides [41,42] and Ni(II)-catalyzed hydrovinylations [32], but enantioselectivities are generally low except in rare cases [31,41]. Most recent additions to monophosphinites, which resemble the well-known triarylphosphite ligands (Scheme 5.10) in their Ni(0)-catalyzed hydrocyanation activity but are more hydrolytically stable, are the so-called cage phosphinites (e.g., **54**) [105]. A corresponding bisphosphinite (**55**) is also known.

5.3.9 Hybrid ligands containing phosphinites

5.3.9.1 Thioether–phosphinite ligands

Evans has developed a family of diarylphosphinite–thioether ligands that give good to excellent selectivities for a number of prototypical C–C, C–H, and C–N bond-forming reactions. These include Pd-catalyzed allylation reactions (Eqs. 5.42 and 5.43) [106], Rh-catalyzed asymmetric hydrogenations (Eq. 5.44) [107], and ketone hydrosilylations using 1-naphthylphenylsilane. (Eq. 5.45) [107]. Particularly noteworthy in this study is a model proposed for the asymmetric induction in the Rh-catalyzed hydrogenation reactions that does not follow the well-established Curtin–Hammett-driven mechanistic hypothesis [108] for hydrogenation of enamides using chelating diphosphines.

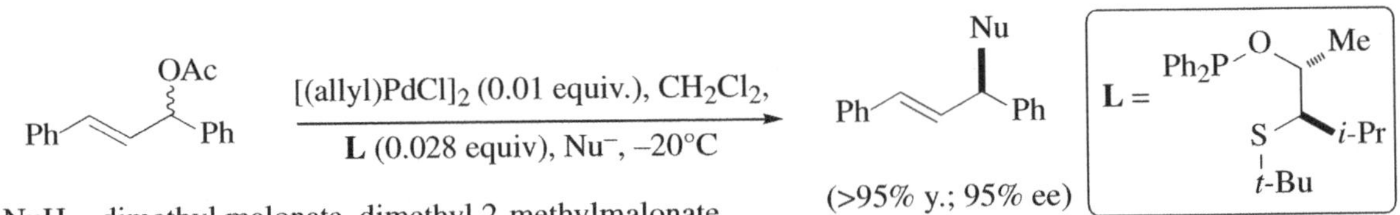

NuH = dimethyl malonate, dimethyl 2-methylmalonate
benzyl amine

(Eq. 5.42)

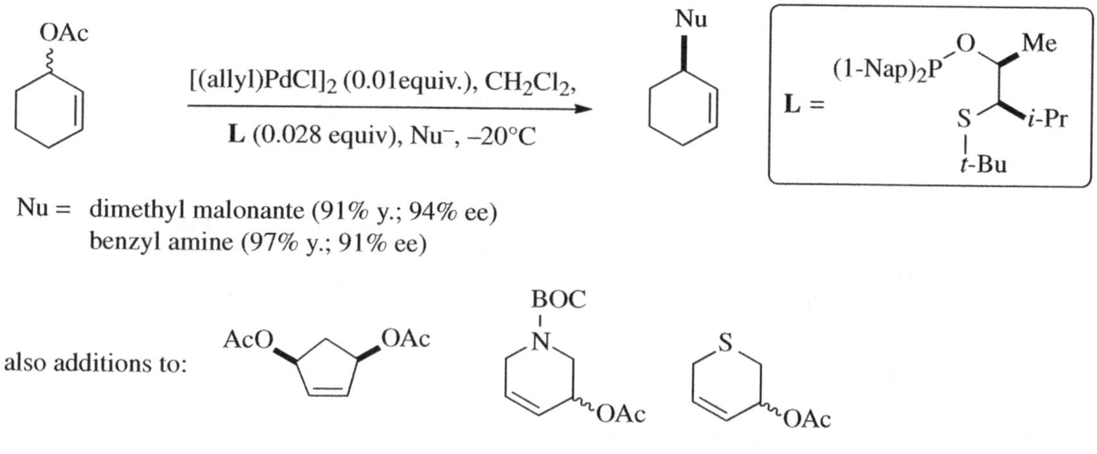

(Eq. 5.43)

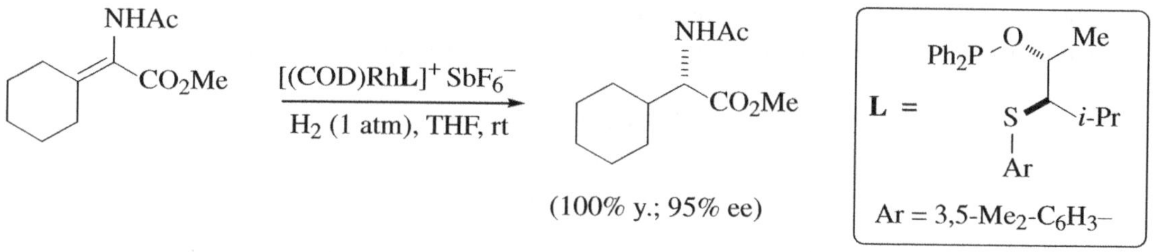

(Eq. 5.44)

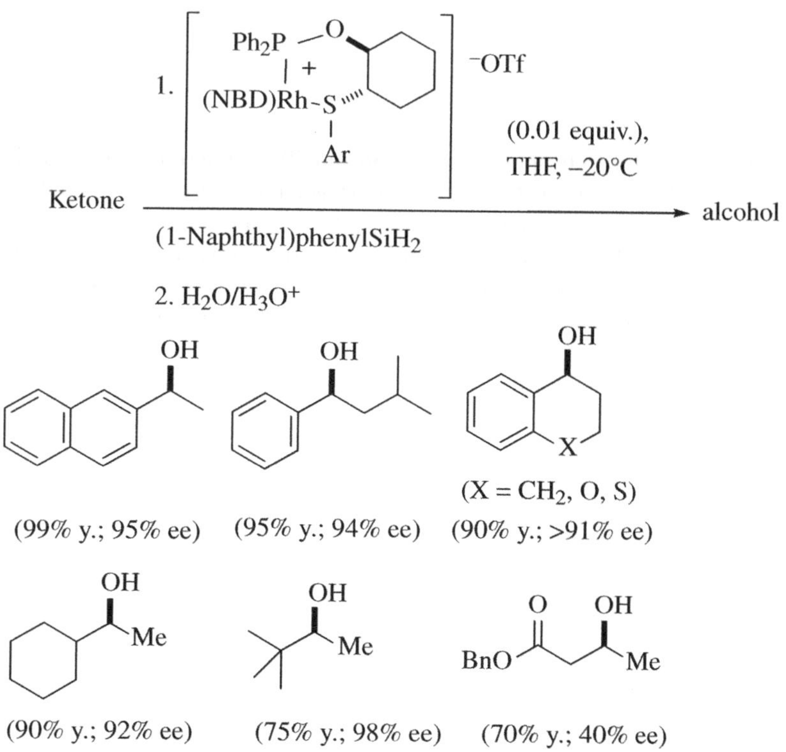

(Eq. 5.45)

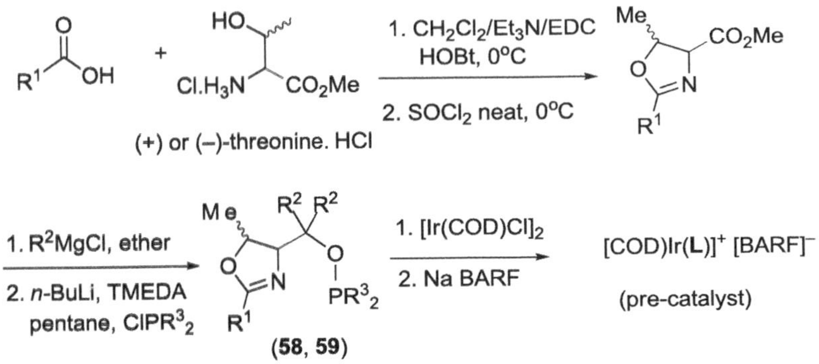

Scheme 5.11 *Typical synthesis of an oxazoline–phosphinite ligand and an Ir complex.*

5.3.9.2 *Oxazoline–phosphinite and pyridine–phosphinite ligands*

A carbohydrate-derived oxazoline–phosphinite was mentioned earlier (**23**; Figure 5.8) in connection with water-soluble phosphinite ligands [60]. An especially noteworthy use of a similar ligand is in a highly substrate-specific intermolecular Heck reaction of phenyltriflate with dihydrofuran (Eq. 5.46) [109].

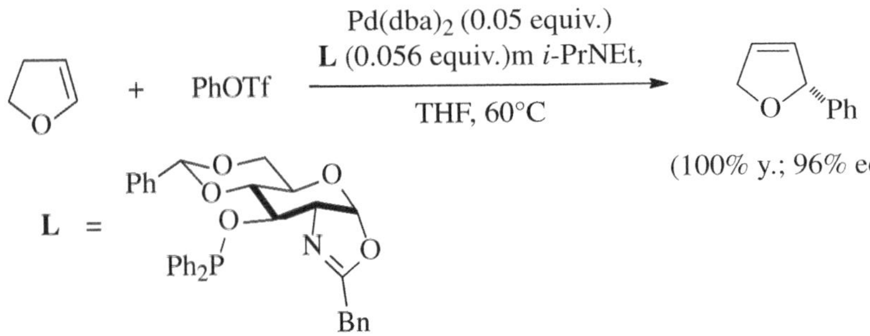

(Eq. 5.46)

Pfaltz prepared various oxazoline–phosphinites (e.g., **58** and **59**) and pyridine–phosphinites (**62**) and found them to have broad applicability in various Ir-catalyzed hydrogenations of alkenes including unfunctionalized alkenes, which are the most demanding substrates [110]. These ligands are easily synthesized from readily available amino acids and amino alcohols and have considerable flexibility with respect to steric and electronic tuning. The enantioselectivities reported are often the highest achieved for the respective substrates shown here. A typical example of the synthesis of an oxazoline–phosphinite is shown in Scheme 5.11 [111,112].

Equations 5.47 and 5.48 and Figure 5.14 show some typical applications. Ligand **60** shown in Figure 5.14 is among the most accessible (two steps from *t*-leucinol), yet one of the most powerful, for several enantioselective hydrogenations [113].

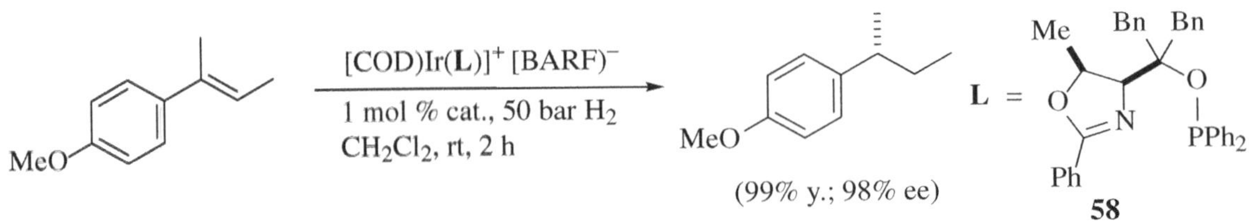

(Eq. 5.47)

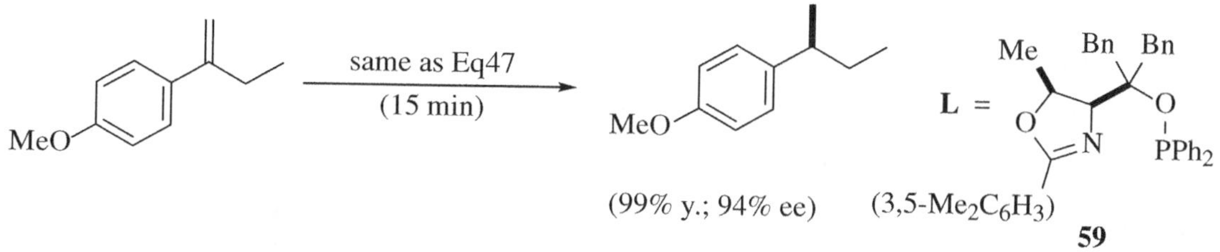

(Eq. 5.48)

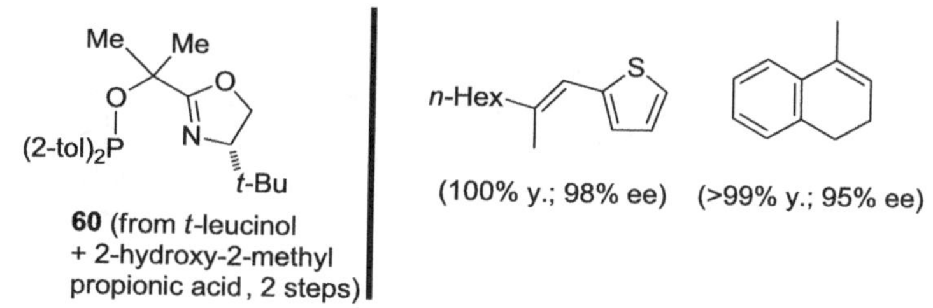

Figure 5.14 *A simple oxazoline–phosphinite for Ir-catalyzed hydrogenation.*

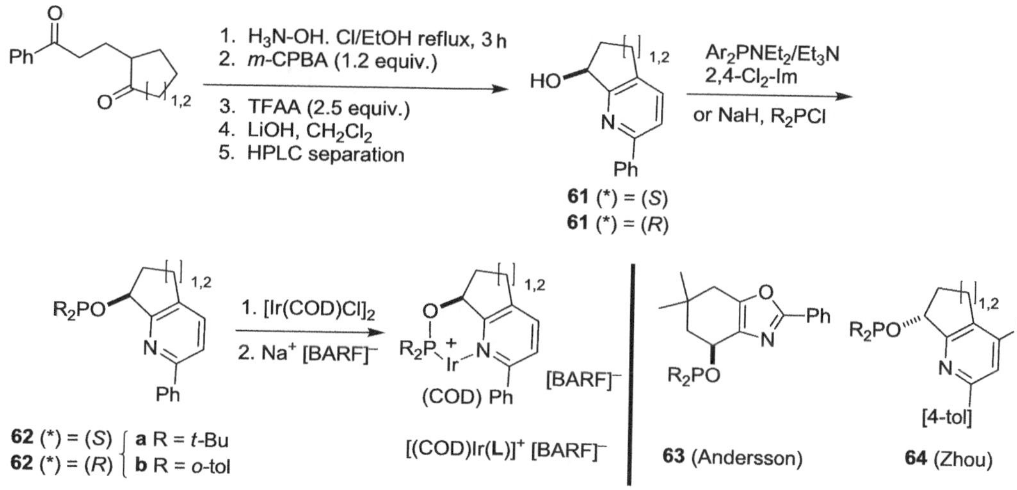

Scheme 5.12 *Synthesis of pyridine–phosphinites and an Ir hydrogenation catalyst.*

The pyridine-bearing phosphinites (**62**) shown in Scheme 5.12 [114,115] are the best ligands for Ir-catalyzed hydrogenation of unfunctionalized alkenes shown in Figure 5.15. These ligands also reduce furan derivatives including benzofurans in good to excellent enantioselectivities. Andersson [116] and Zhou [117] have also developed phosphinite ligands (**63** and **64**, respectively) that fall into this structural type.

Pyridine–phosphinite ligand **65**, readily synthesized from (–)-menthone and 2-bromopyridine (Scheme 5.13), has been used for the Rh-catalyzed hydroformylation of vinyl derivatives. Even though the enantioselectivity in this reaction is modest, an unusually high branched-to-linear ratio of the aldehyde products (near 100% for 2-vinylnaphthalene, methyl acrylate, and vinyl acetate) has been recorded for this reaction [118].

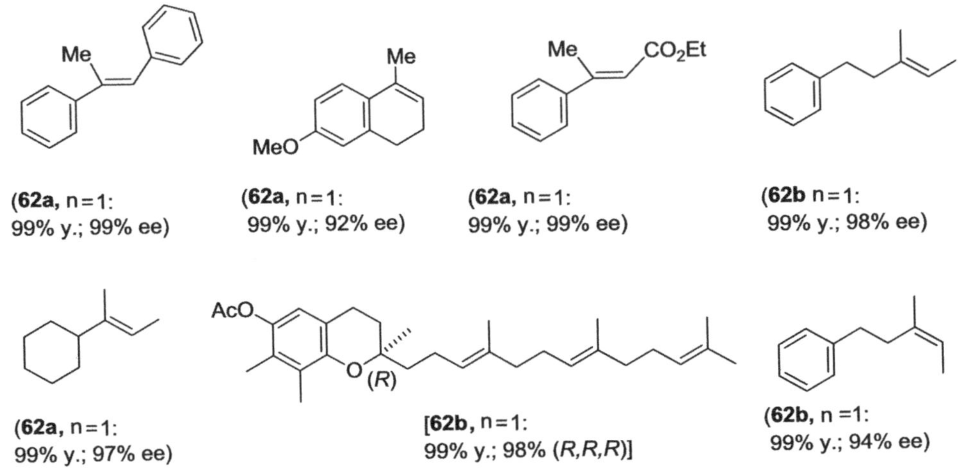

Figure 5.15 Substrates for hydrogenations with pyridine–phosphinite complexes $[(COD)Ir(L)]^+$ $[BARF]^-$. Structures of ligands shown in brackets.

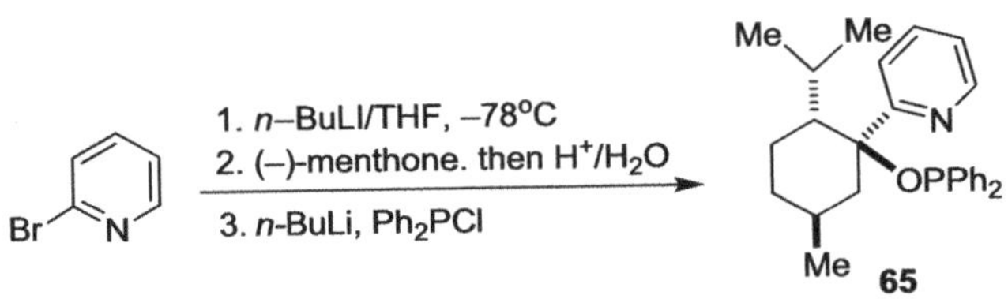

Scheme 5.13 Synthesis of a pyridine–phosphinite useful for Rh-catalyzed hydroformylation.

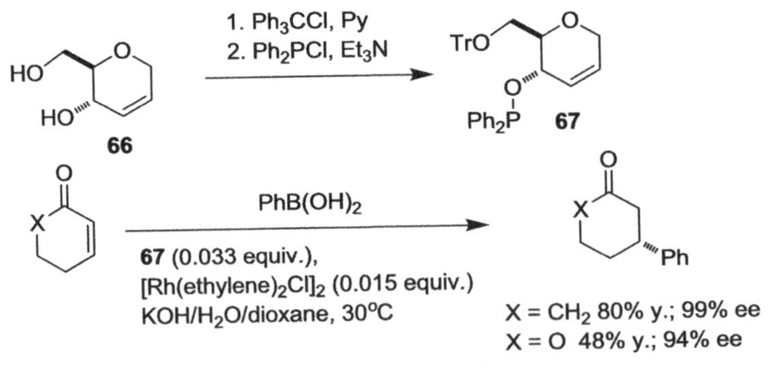

Scheme 5.14 Synthesis of an alkene–phosphinite and an application.

5.3.9.3 *An alkene–phosphinite ligand*

Latest among the hybrid ligands containing a phosphinite moiety is the alkene–phosphinite **67** derived from the glycal **66** in two steps. This ligand promotes reaction of phenyl boronic acid with cyclic enones and an α,β-unsaturated lactone with surprisingly good selectivity (Scheme 5.14) [119].

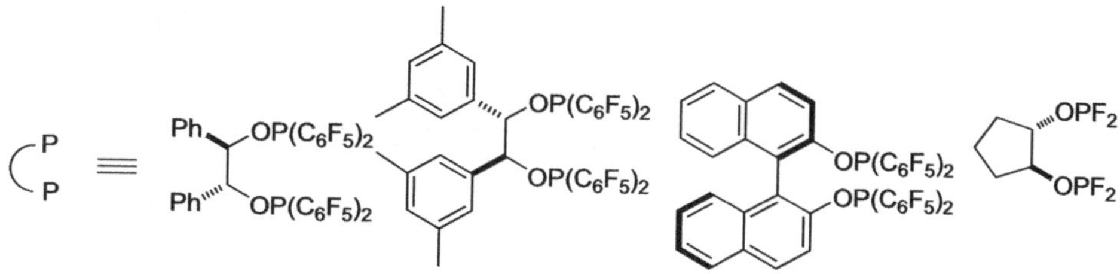

A B C D

Figure 5.16 Highly electron-deficient phosphinites for synthesis of transition-metal Lewis acids.

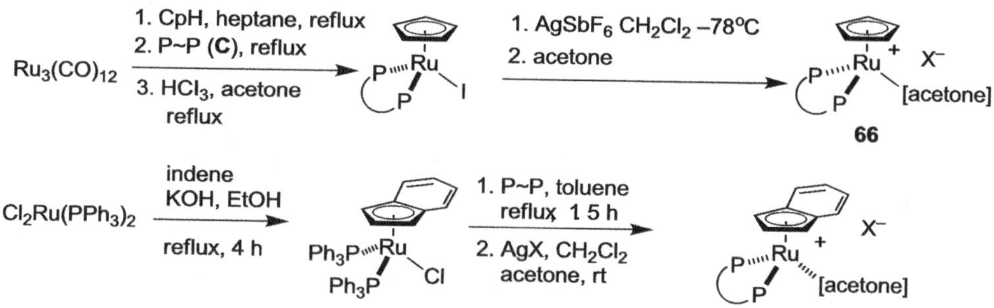

66

67

Scheme 5.15 Synthesis of cationic Ru–phosphinite complexes.

5.3.9.4 *Chiral transition metal Lewis acids bearing electron-withdrawing phosphinites*

The electron-withdrawing ability of fluorinated bis-1,2-diarylphosphinite ligands was firmly established by examining the IR-stretching frequencies of $(OC)_2Ni(\mathbf{L})$ complexes, where **L** represents β-phenyl glucoside-derived ligand **5** (Eq. 5.21) [22,120]. Kündig *et al.* prepared a series of chiral Fe [121] and Ru [122] complexes of 1,2- and 1,4-bis-di-pentaflurophenylphosphinites [123] (Figure 5.16) and successfully tested them as Lewis acids in various reactions.

The Ru complexes are more useful in synthesis, and preparations of two of these complexes from readily available metal precursors are shown in Scheme 5.15 [124,125]. The indenyl moiety makes the substitution of the Ph_3P ligands by a chelating ligand more facile, and the complex **67** can be synthesized in high yield starting from $Cl_2Ru(PPh_3)_2$. The reactivity of the cationic complex is affected by the counter anion, with the complexes carrying the most dissociated ions (SbF_6^- and $BARF^-$) showing the highest reactivity. Various salts are easily prepared from the corresponding neutral metal–halide complexes by metathesis reactions [122].

The use of the cationic Ru and Fe complexes has been investigated mostly in the [4+2]- and [3+2]-cycloaddition reactions. Diels–Alder reactions between α,β-unsaturated aldehydes or ketones and cyclopentadiene are challenging reactions (Eqs. 5.49 and 5.50). Catalysis by the Ru-complexes gives very high yield, selectivity, and ee in several of these reactions [125,126]. An intramolecular Diels–Alder reaction shown in Equation 5.51 proceeds under mild conditions to afford endo products in 82–100% yield and very high enantioselectivity [127]. Likewise, *N*-methylisoxazolidines are produced via asymmetric [3+2]-cycloadditions between nitrones and α,β-unsaturated aldehydes in high regio-, diastereo-, and enantioselectivity in the presence of a chiral cationic Ru–Lewis acid **66A** (Eq. 5.52) [128].

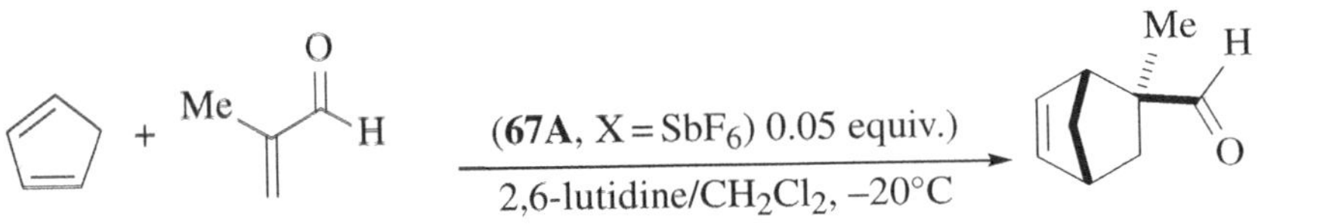

(85% y.; exo/endo=332; 88% ee)

(Eq. 5.49)

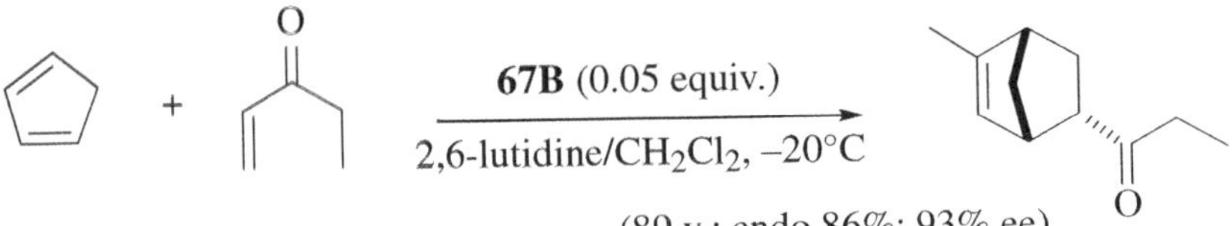

(89 y.; endo 86%; 93% ee)

(Eq. 5.50)

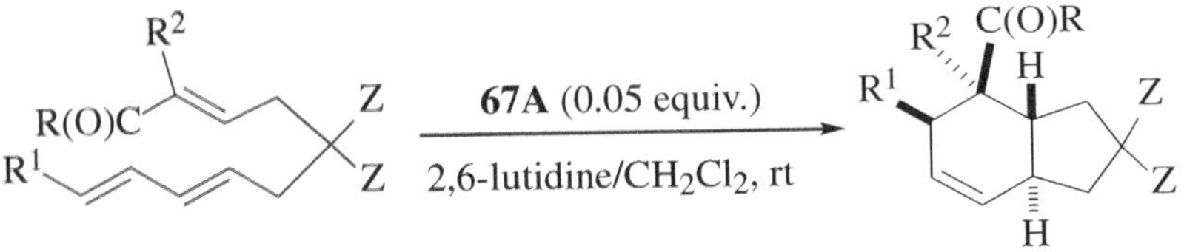

(82–100% y.; endo major; up to 92% ee)

(Eq. 5.51)

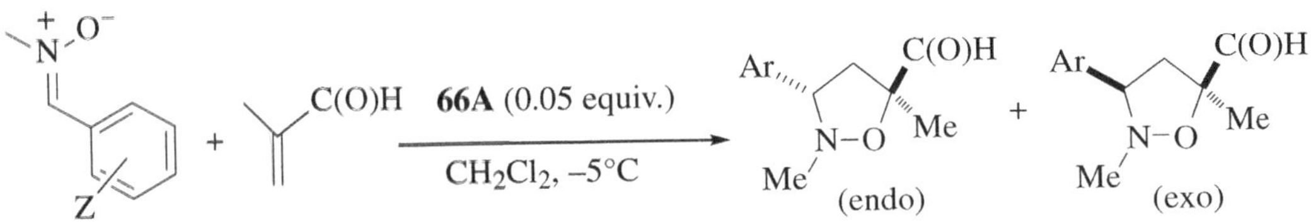

(60–85% y.; endo major; up to 97% ee)

(Eq. 5.52)

5.4 Synthesis and applications of phosphonite ligands

5.4.1 Early studies

Early studies on phosphonites were mostly concerned about the syntheses and structures of their complexes [3]. Robinson and coworkers prepared a variety of Ag and Au phosphonite and phosphinite complexes using the general procedures outlined in Section 5.1 [8,9]. The Ag complex is prepared from AgCl or $AgNO_3$ and the appropriate ligand in an alcoholic solvent (Eq. 5.1). The Au analog is prepared from [Au(cyclooctene)Cl] (Eq. 5.11) [8]. The preparation of Co and Ni phosphonite complexes is shown in Equation 5.2 [11] and Equations 5.5 and 5.6 [12], respectively. Co(II) complexes of $(EtO)_2PPh$, $[CoL_4X]^+$ $[BPh_4]^-$ (X=Cl, Br, I, N_3) have been prepared by reaction of CoX_2, $NaBPh_4$ in ethanol [129]. [Palladium, $[L_4]$ where **L**=$Ph_2P(OMe)$, is formed from $Pd(COD)Cl_2$ and the ligand, but the phosphonite analogue is unstable [9]. Many mixed ligands of the type **LL'**$PdCl_2$ (**L**, **L'**=combinations of phosphine, phosphinite, and phosphonite) have also been described in connection with a highly illuminating investigation of the thermodynamics of various equilibria of complexes containing these ligands [130]. The paper describing this work is a treasure trove of information on ligand parameters such as Tolman's $\Sigma\chi_i$ and nuclear magnetic resonance (NMR) chemical shifts. Rhodium(I) diene complexes readily give the $[RhL_4]^+$ or $[RhL_5]^+$ cations upon reaction with the ligands in cold alcohol, whereas Ir complexes are formed only under more forcing conditions [131–133]. Larger phosphonite and phosphinite ligands $(EtO)_2PPh$, $(EO)PPh_2$ in ethanol yield $[L_3RuCl_3RuL_3]^+$ cations upon reaction with $[Ru(NBD)Cl_2]_n$ (Eq. 5.12) [9,133]. Triphenylphosphine complexes (e.g., $(Cl)(CO)(H)Ru(PPh_3)_3$) can also be used for the synthesis of ethyl and methyl diphenylphosphonites (Eq. 5.13) [10].

5.4.2 Phosphonites from TADDOL and related compounds

Apart from a stoichiometric reaction probing the stereoelectronic effect of ligands on H_2 addition to a Vaska's-type Ir complex $(X)[(EtO)_2PPh]Ir(CO)$ (X=Cl, I) [134], the first notable application of a phosphonite complex involved the use of the so-called TADDOL–phosphonites (**69**, TADDOL=α,α,α′,α′-tetraaryl-2,2-dimethyl-1,3-dioxolan-4,5-dimethanol) prepared by Seebach from dialkyl tartrates (Scheme 5.16) [135,136]. An improved procedure for the synthesis of phosphonites and phosphoramidites uses the reaction of TADDOL with PCl_3 in toluene containing *N*-methylmorpholine followed by reaction of the resulting chlorophosphite with a lithium reagent (Scheme 5.17) [137].

Initial applications [138] of the TADDOL ligands for Rh-catalyzed hydroformylation of styrene (CO (40 bar), H_2 (80 bar), 0.005 equiv. **69**, 0.0005 equiv. Rh(acac)(COD), benzene, 40-70°C, 5 h) gave 93–95% of the branched product (R >75% yield, ~20% ee). The same ligand in combination with $[Rh(COD)Cl]_2$ [(Rh:**L**=1)] catalyzes hydroboration of styrene with 1.1 equiv. catecholborane to give 1-phenylethanol (70–80% yield, 16% ee, *S*) after oxidative workup. Highly regioselective hydrosilylation with Cl_3SiH catalyzed by (allyl)Pd[**69b**] at 0°C gave a 70% yield of the secondary trichlorosilane in 34% ee (*S*-adduct). This product is converted into 1-phenylethanol by H_2O_2–KF–K_2CO_3–THF-MeOH.

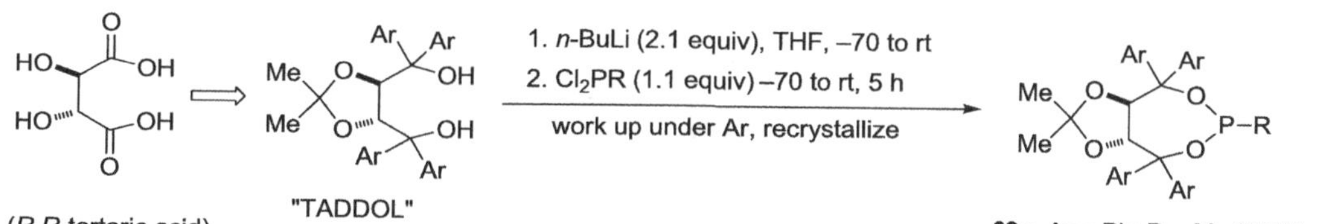

68a Ar = Ph
68b Ar = 2-Nap

69a Ar = Ph; R = Me (76%)
69b Ar = Ph; R = Ph (47%)
69c Ar = 2-Nap; R = Ph (75%)
69d A = R = 2-nap

Scheme 5.16 *of TADDOL-derived phosphonites.*

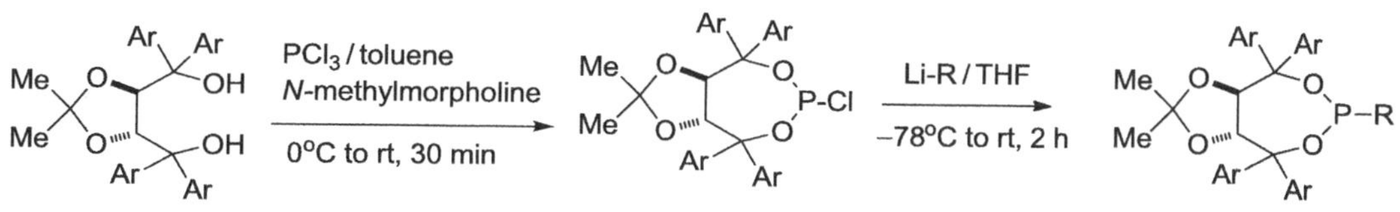

69d Ar = R = 2-Nap 62%
69e Ar = 2-Nap; R = *t*-Bu 78%
69f Ar = Ph; R = (2-pyridylmethyl) 26%

Scheme 5.17 *An improved synthesis of TADDOL-derived phosphonites.*

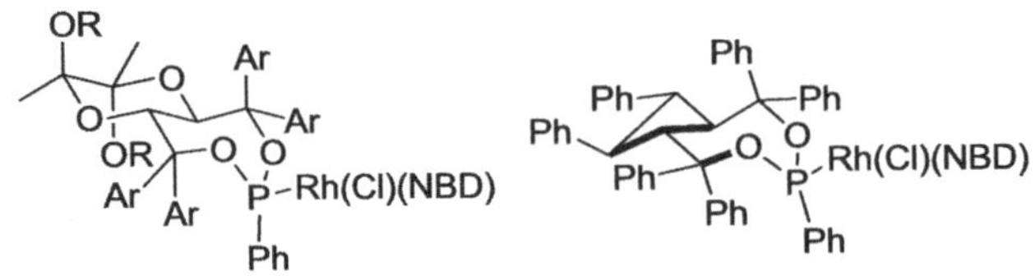

70 (R = Et, Ar = Ph) **71**

Figure 5.17 *Assorted complexes of TADDOL-like ligands.*

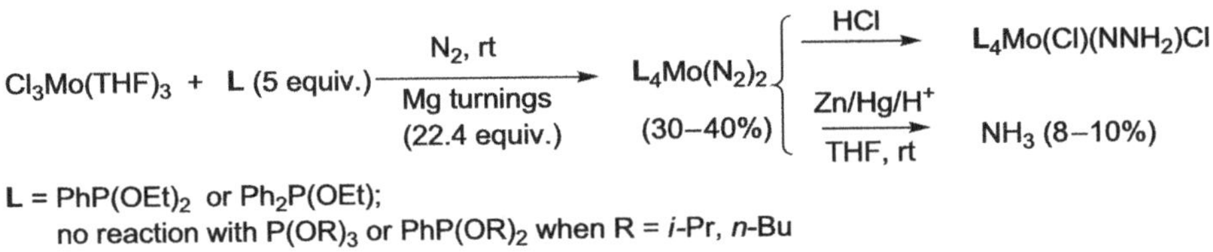

Scheme 5.18 *Formation and reactions of a molybdenum–nitrogen complex.*

Other ligands patterned after TADDOL but with 1,4-dioxane and 1,3-dioxalane backbones (**70** and **71**) gave lower selectivities compared to complexes of **69** in the hydrosilylation reactions of acetophenone [139]. Alexakis has investigated a large number of TADDOL-type phosphonite, phosphite, and phosphoramidite ligands for Cu(I)-catalyzed asymmetric conjugate addition of diethyl zinc to α,β-unsaturated ketones [140], esters [141], and nitroolefins [142a]. While the phosphonite(s) ligand **69** enables the reaction to be run at low temperature and yields are generally excellent, the selectivity is only marginal in most cases except for nitroolefins (e.g., 98% yield, 33% ee (*S*) for addition to cyclohexenone) (Figure 5.17; Eqs. 5.53 and 5.54).

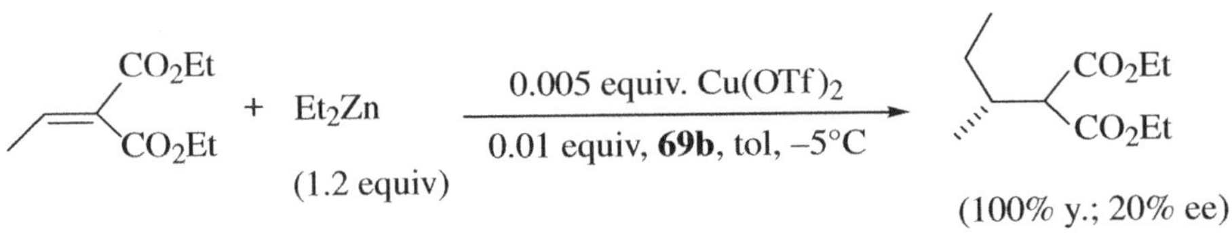

(Eq. 5.53)

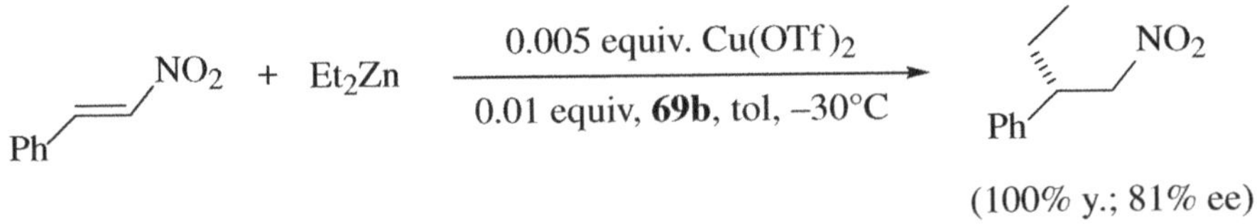

(Eq. 5.54)

Phosphonite ligands impart unique reactivity to a Mo complex, $\mathbf{L}_4Mo(N_2)_2$ (**L**=$PhP(OEt)_2$, which upon reaction with Zn–Hg in the presence of lutidine, HCl, and excess acid gives ammonia in 8–10% yield. Acid alone gives a hydrazido(2-)[$MoCl(NNH_2)[PhP(OEt)_2]Cl$ derivative (Scheme 5.18).

An innovative enantio- and diastereoselective synthesis of β-hydroxyketones from a vinyl ketone and an aldehyde via hydrogenative aldol coupling (Eq. 5.55) uses modified TADDOL-like phosphonites (**72**) [142b]. Systematic variation of the different structural elements (Ar, R^1, and R^2) leads to **73**, a ligand which gave the highest selectivity in this Rh-catalyzed reaction. As illustrated by the examples, the reaction conditions are quite tolerant to a broad set of substrates, and selectivities are synthetically useful.

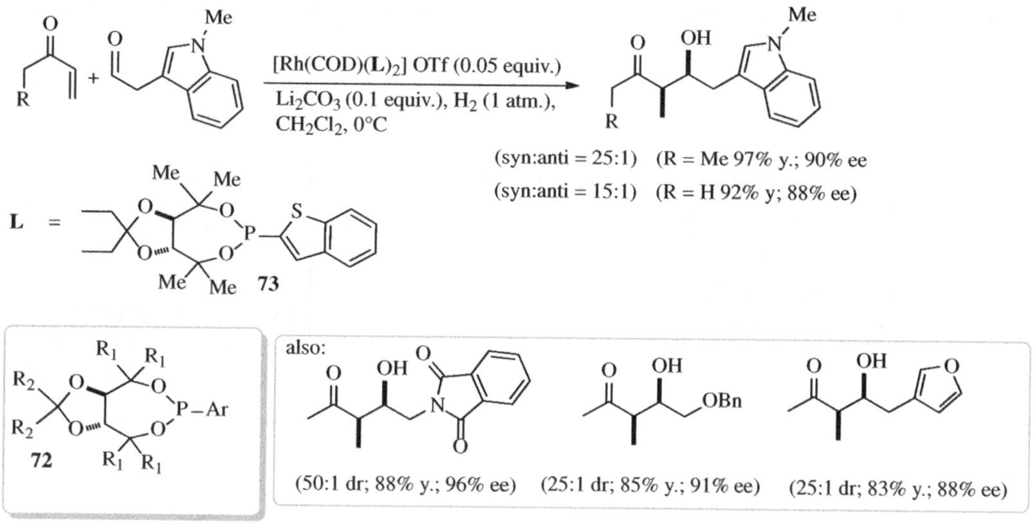

(Eq. 5.55)

Asymmetric 1,4-hydroxylation of dienes is accomplished by Pt-catalyzed diboration using bis(pinacolato)diboraon [B_2(pin)$_2$] followed by oxidative workup of the intermediate diboron species (Eqs. 5.56 and 5.57) [143a].

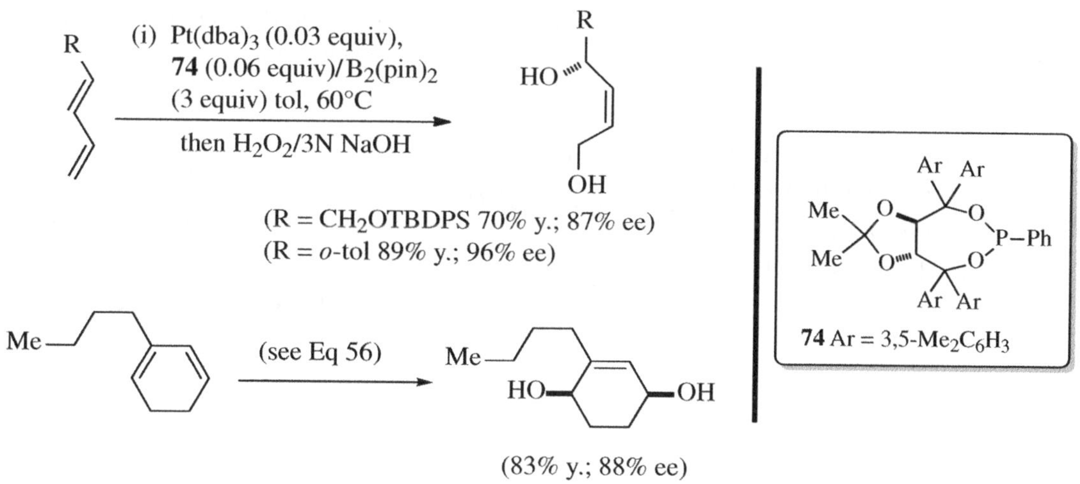

(Eq. 5.56)
(Eq. 5.57)

Morken *et al.* have also reported [143b,c]enantioselective allylation of unsymmetrical dialkylidene ketones (Eq. 5.58) and dienals (Eq. 5.59) using allylboronic acid pinacol ester [allylB(pin)] in a Ni(0)-catalyzed reaction that uses phenylphosphonite derivatives **75** and **69b** as ligands. Presumably these reactions proceed through conjugate addition of the B–Ni species followed by an unusual 3,3′-reductive elimination, the discussion of which is beyond the scope of this chapter. The original paper should be consulted for a highly readable discussion of this aspect of the reaction.

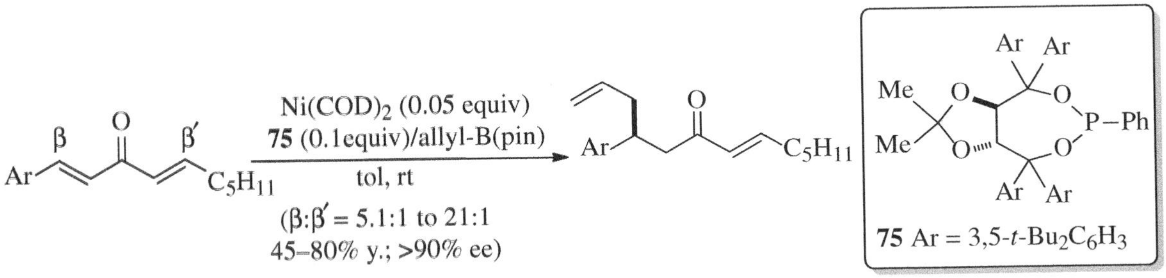

(Eq. 5.58)

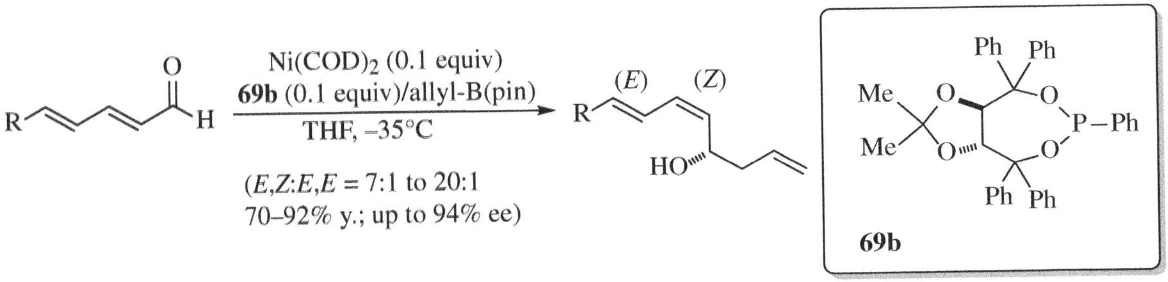

(Eq. 5.59)

Diborylaton of alkenes is catalyzed by a Pt complex of the phosphonite **76** (Eq. 5.60) [144]. Alkyl- and aryl-substituted alkenes take part in this reaction, and the products are readily converted into 1,2-diols under oxidative conditions.

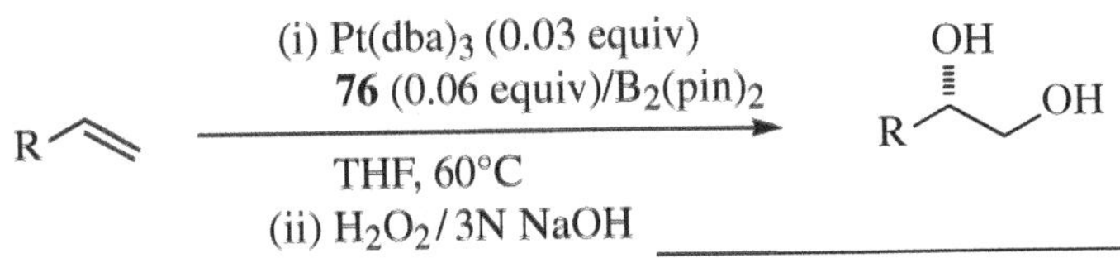

R	% yield	% ee
n-hex	83	92
t-Bu	46	90
Ph	93	90
TBDPSO-CH_2	84	86

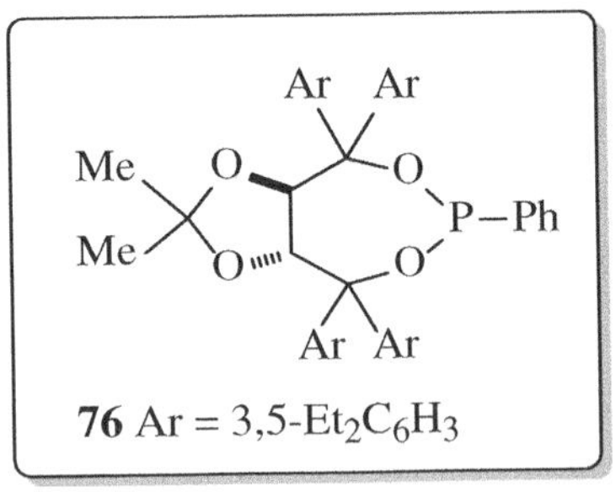

(Eq. 5.60)

Yet another important development that demonstrates advantages of the tunablility of the TADDOL-derived ligand for a "challenging" reaction is shown in Eq. 5.61. The dehydrogenative silicon–oxygen coupling proceeds with moderate to excellent kinetic selectivity (s = up to 35) when the appropriate silane and phosphonite ligands [for Cu(I)] are employed. In this instance, the generally superior phosphoramidites give poor results. Even though the reaction is of a narrow scope, is limited to pyridine-bearing substrates that facilitate coordination to Cu, and suffers from low turnovers, it signifies an important conceptual advance.

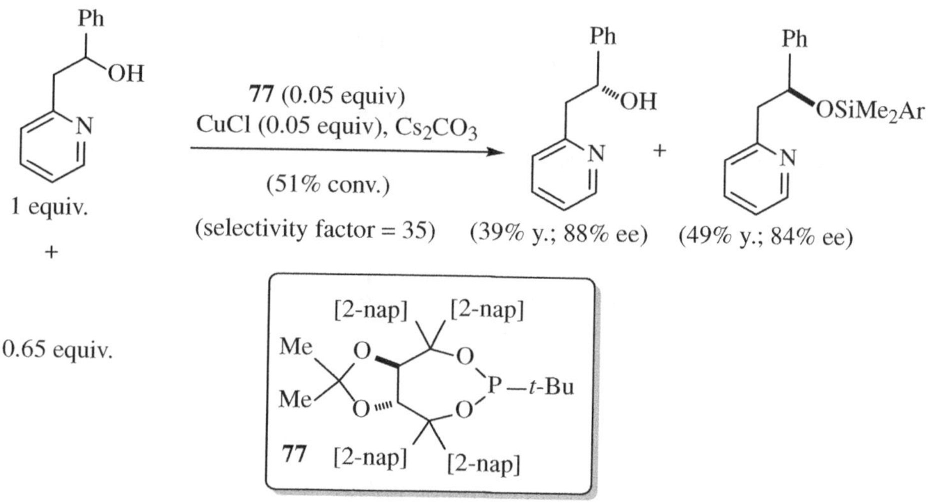

(Eq. 5.61)

5.4.3 Phosphonites derived from 2,2′-hydroxybiaryls and related compounds

Phosphonites derived from 2,2′-hydroxybiaryls form yet another popular class of ligands that have found wide applications. Various strategies for the syntheses of four types of ligands in this class that use well-known reactions are shown in Scheme 5.19 [145,146], Scheme 5.20 [146,147], Scheme 5.21 [145,148], and Scheme 5.22 [149,150].

Ferrocene can be dilithiated with *n*-BuLi/TMEDA in hexane [146], and the dilithium salt formed reacts with $ClP(NEt_2)_2$ to give a bis-diethylaminophosphine derivative, which can be converted into a bis-dichlorophosphine. This compound is useful for the synthesis of a wide variety of phosphonites (Scheme 5.19) [145,146]. Hydrogenation of dimethyl itaconate and methyl 2-acetamidoacrylate

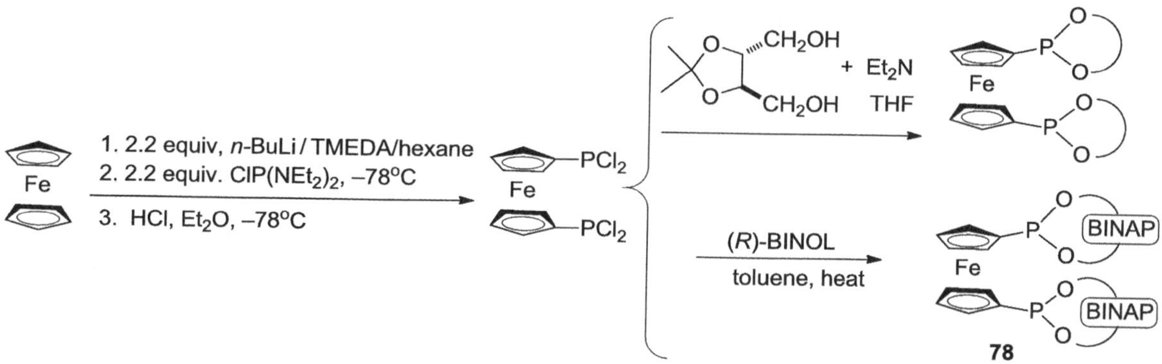

Scheme 5.19 *Phosphonites from ferrocene via a bis-dichlorophosphine route.*

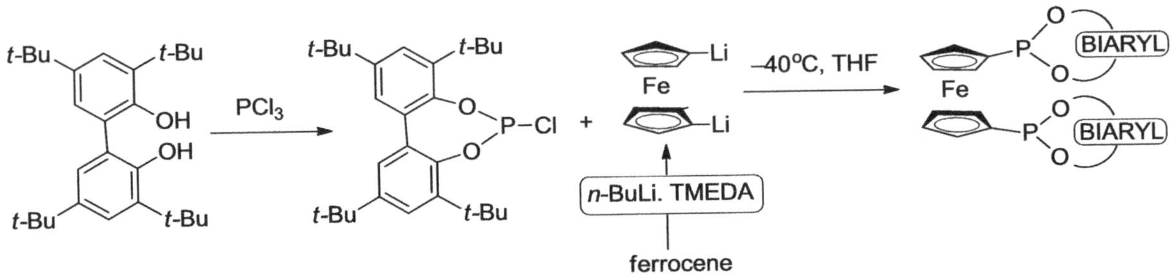

Scheme 5.20 Sterically demanding phosphonites from ferrocene via a dilithio derivative.

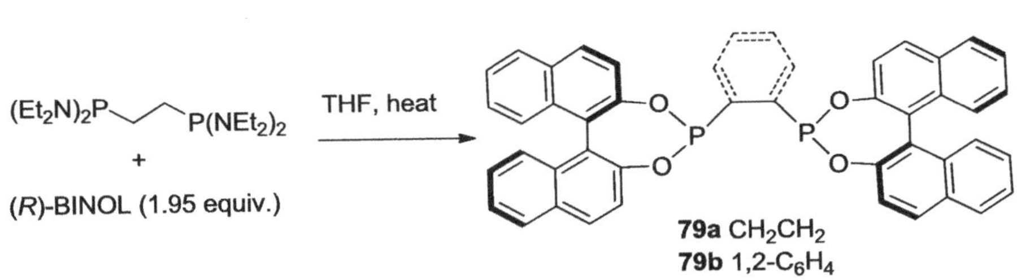

Scheme 5.21 Synthesis of bis-phosphonites from BINOL.

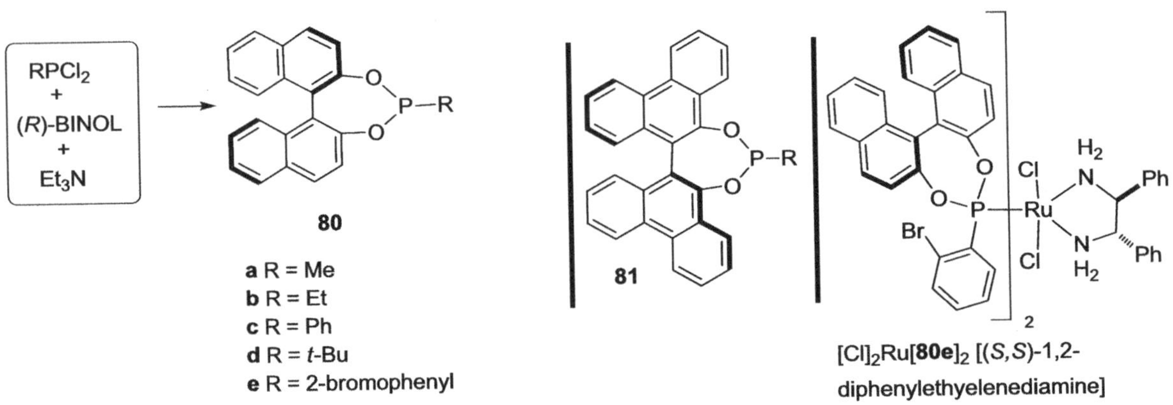

Scheme 5.22 Synthesis of monophosphonites from BINOL, BIANTHRACENOL, and a Ru complex.

catalyzed by $[(COD)Rh(\mathbf{78})]^+ BF_4^-$ (0.001 equiv., 42 psi H_2, CH_2Cl_2) proceeds in quantitative yield and in 99.5% ee [145].

The bulky biarylphosphonite depicted in Scheme 5.20 can be synthesized directly from the dilithiated ferrocene using a chlorophosphite. This ligand has been successfully tested for Rh-catalyzed hydrogenation of dehydroaminoacids, Rh-catalyzed hydrosilylation of ketones, and Pd-catalyzed Suzuki coupling reaction [147].

As in the synthesis of phosphinites [96], a diethylaminophosphine can also be used directly in a metathesis reaction (usually by heating in toluene or THF with a diol) to generate a phosphonite (Scheme 5.21).

The monophosphonites **80a–80d**, synthesized directly from $RPCl_2$ and a chiral diol (Scheme 5.22), are also surprisingly efficient (e.g., up to 93% ee for the reduction of methyl 2-acetamidoacrylate with **80d**) [148,150].

Monophosphonites **80a–d** (and other similar ligands) have been used to demonstrate a new principle in combinatorial asymmetric catalysis where optimization of selectivity has been accomplished by combining two different monodentate ligands in a reaction that is known to proceed through intermediates carrying at least two ligands. The original test case was Rh-catalyzed asymmetric hydrogenation of methyl *N*-acetylcinnamate using combinations of phosphonites **80** [151] and corresponding phosphites. In several instances, heterocombinations gave higher ee as compared to the homocombinations. Similar combinations of phosphonites with other ligands with fluxional behavior [152] or even with ones possessing no chirality can work in these systems as long as a diastereomeric or other **MLL'** combination is catalytically competent and is more efficient vis-à-vis the $\mathbf{ML_2}$ or $\mathbf{ML'_2}$ homocombinations [153,154].

Only marginal differences are observed between binaphthol and biphenanthrol-derived phosphonites (e.g., **81**, Scheme 5.22) in Rh-catalyzed hydrogenations [150] and Cu-catalyzed conjugate additions of organozinc reagents to enones [155]. Neutral Rh complexes of **79a** and **80c** are excellent ligands (100% conversion at 70 psi, RT, 0.01 equiv. cat., MeOH–CH_2Cl_2) for hydrogenation of benzylimine of acetophenone even though the ee obtained is low [156]. Complexes of the type η^6-arene-Ru(**80c**)Cl_2 (arene = benzene, *p*-cymene, and mesitylene) are good catalysts for isomerization of allyl alcohols to ketones, and a modest kinetic resolution is seen in this process [157].

Other applications of ligands such as **78–81** include the Cu(I)-catalyzed conjugate addition of Et_2Zn to enones (up to 99% ee for cyclohex-2-enone, 0.01 equiv. Cu(OTf), ligand 3,3′-Me_2-**78**, THF, –30°C) [155,158], conjugate addition of arylboronic acids to cyclohex-2-enone (>99% ee, 100% conv., 0.03 equiv. Rh(acac)(ethylene)$_2$), **79b** (C_6H_4 backbone, 100°C, dioxane–H_2O 10:1, 5 h) [159], Rh-catalyzed hydrogenation of enamines [(COD)Rh[**80c**] BF_4] [160], Pd-catalyzed intramolecular cyano-amidation (**L** = **80c**, cf. phosphoramidites are the best ligands for this reaction) [161], and Ru(II)-catalyzed transfer hydrogenation of aryl methyl ketones (Scheme 5.22; $[Cl]_2Ru[\mathbf{80e}]_2$ [(*S*,*S*)-1,2-diphenylethyelenediamine], 0.0005 equiv., 2-propanol, 0.005 equiv. *t*-BuOK: up to 99% ee for *o*-bromoacetophenone) [162].

Incidentally, Ding *et al.* has prepared a series of highly modular phosphonites **83** starting from *bis*-dichlorophosphinylferrocene and bis-phenols **82** (Eq. 5.62), and tested them for hydroformylation of vinyl acetate and styrene [163]. Moderate to good selectivities are observed. Structurally related bidentate phosphonites have also been prepared from combinations of **82**,**84** and 1,2-bis-diclorophosphinobenzene (**85**) and 1,2-bis-dichlorophosphinoethane (**86**, Figure 5.18) [164]. Complexes of phosphine–phosphonite ligands (**87a** and **87b**) based on ferrocene have been used for Heck and hydroformylation reactions [165]. In the hydroformylation of 1-octene using **87b** (0.005 equiv. catalyst, 80°C, 10 atm CO/H_2, 16 h), a surprisingly high proportion of products from olefin isomerization followed by hydroformylation (nonanal:methyloctanal:2-ethylheptanal:2-propylhexanal = 52:37:7:4) was observed.

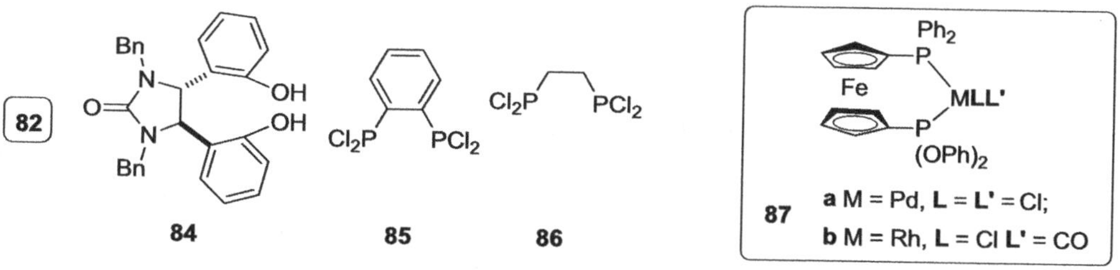

Figure 5.18 *Components of modular phosphonite ligands useful for Rh-catalyzed hydroformylation.*

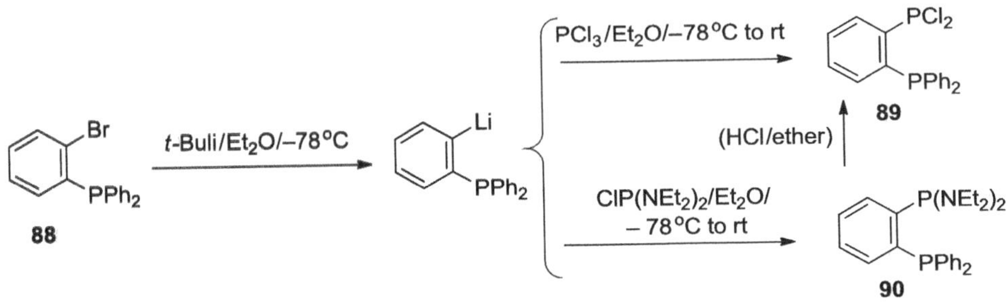

Scheme 5.23 Synthesis of dichlorophosphino- or bis(dimethylaminophosphino)phenyl-diphenylphosphine.

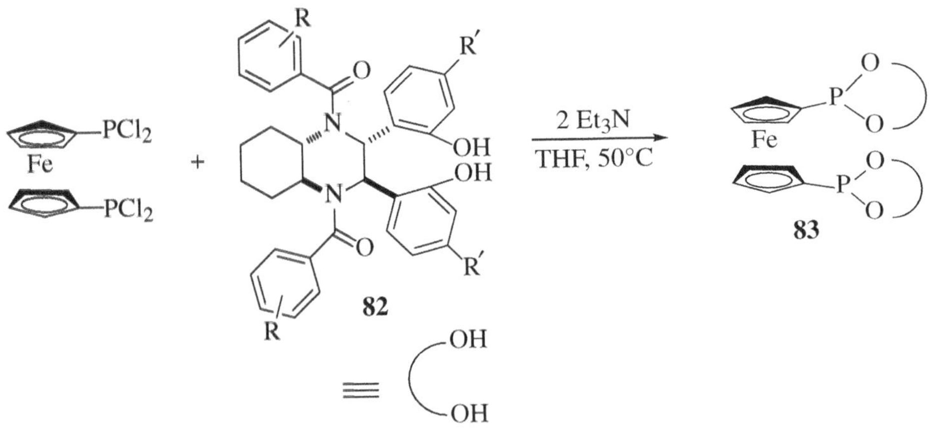

(Eq. 5.62)

5.4.4 Phosphine–phosphonite ligands

Phosphine–phosphonites including those derived from binaphthols have been prepared starting with either Ar-PCl_2 (**89**) [166] or ArP$(NMe_2)_2$ [90] [167], as shown in Scheme 5.23. These valuable precursors are readily prepared from bromophenyldiphenylphosphine **88** [166] by lithium–bromine exchange with *t*-BuLi [168] followed by reaction with the appropriate phosphorus chloride. Several phosphonite–phosphine ligands were prepared from either **89** or **90** and a series of mono- or diols (Scheme 5.24). Rhodium complexes of ligand **91a** are moderately effective for hydrogenation of dimethyl itaconate [169].

5.4.5 Phosphonites with paracyclophane backbone

Enantiomerically pure *pseudo-ortho*-dibromoparacyclophane has been converted into a series of phosphonites following the chemistry described in Scheme 5.25 [170].

Bis(dichlorophosphino)paracyclophane (**93**) is readily converted into a variety of phosphonites (**94**), and the resulting ligands were successfully tested for Rh-catalyzed asymmetric hydrogenation of standard substrates in MeOH (up to 99% ee with substrate:catalyst ratio ~1000 for methyl 2-acetamidoacrylate and methyl 2-acetamidocinnamate using **94e**). The achiral biphenol derivative **94a** gave 96% ee. Both **94d** (with *S*-BINOL) and **94e** (with *R*-BINOL) gave the (*S*) aminoacid (the former with lower ee), suggesting that the paracyclophane core controls the selectivity. Low selectivity was observed with the appendages 2,6-dimethylphenol, TADDOL, and 1,2-diphenylethanediol.

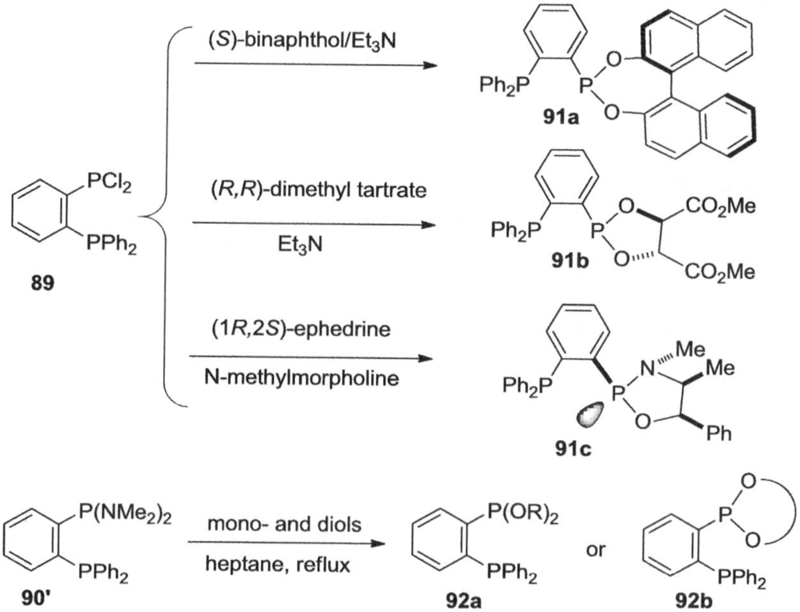

alcohols: 2-propanol, (–)-menthol, isopinocampheol, (1*R*,2*R*)-cyclohexane diol, (–)-pinanediol, BINOL, 1,2:5,6-diisopropylidene-D-mannitol, or (1*R*,2*R*)-1,2-diphenyl-1,2-ethane diol

Scheme 5.24 *Synthesis of tunable phosphine–phosphonites.*

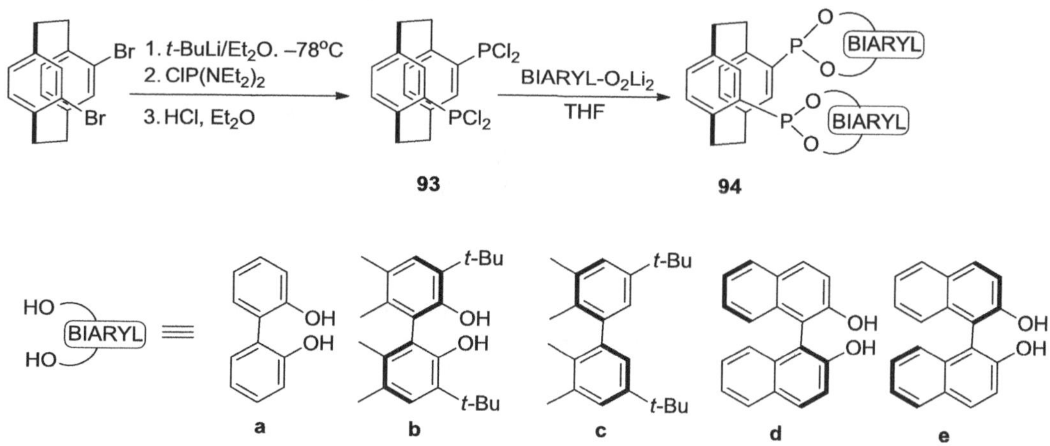

• low selectivity in hydrogenations using ligands with: 2,6-dimethylphenol, TADDOL, 1,2-diphenylethanediol

Scheme 5.25 *Paracyclophane-based bis-phosphonites.*

5.4.6 Phosphonites with a spirobisindane backbone

Zhou has prepared (Eq. 5.63) phosphonite ligands based on spirobisindane scaffolding (from **95** [171]) and found them to be efficient for Rh(I)-catalyzed hydrogenation of not only methyl *N*-acetamidocinnamates [172] but also enamines of the type 1-dialkylamino-1,2-diarylethylenes (Eq. 5.64) [160]. As in the case of the

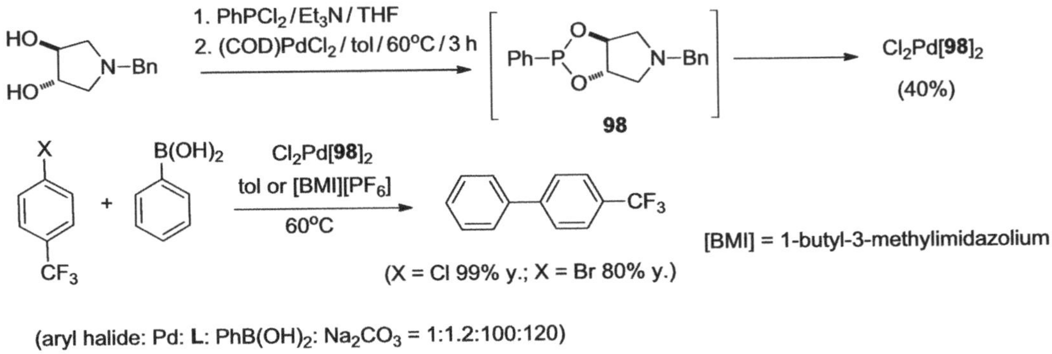

Scheme 5.26 *Phosphonite Pd complexes as catalyst for Suzuki reactions in ionic media.*

phosphinites [25–27], a strong electronic effect was observed in the hydrogenation of the dehydroamino-acids. Palladium(II) complex derived from **96** (X=O; Z=H) also serves as a modest catalyst for Et_3B-mediated allylation of benzaldehyde [173]. Not unexpectedly, a corresponding phosphine (**97**, X=CH_2; Z=H) is a much superior ligand in this reaction.

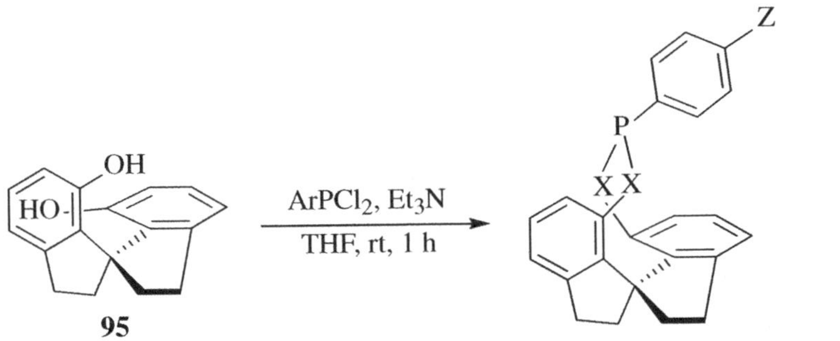

96 X = O; Z = H, Me, OMe, Cl, CF_3
97 X = CH_2; Z = H

(Eq. 5.63)

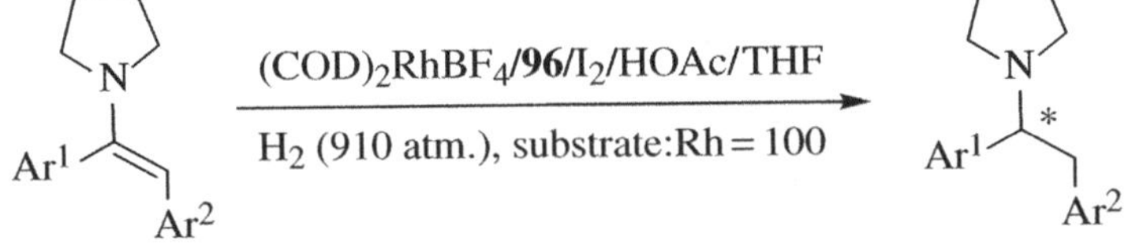

(Eq. 5.64)

5.4.7 Miscellaneous phosphonite ligands

Palladium complexes of ligand **98** are good catalysts for Suzuki coupling between bromo- and chloroarenes and phenylboronic acid in both toluene and an ionic liquid, (BMI)(PF_6) [BMI=1-butyl-3-methylimidazolium] [174]. A typical example is shown in Scheme 5.26. Coupling between sterically demanding substrates such as 2-bromotoluene and 2,6-dimethylphenylboronic acid gives low conversions, the reaction mostly leading to deboronation.

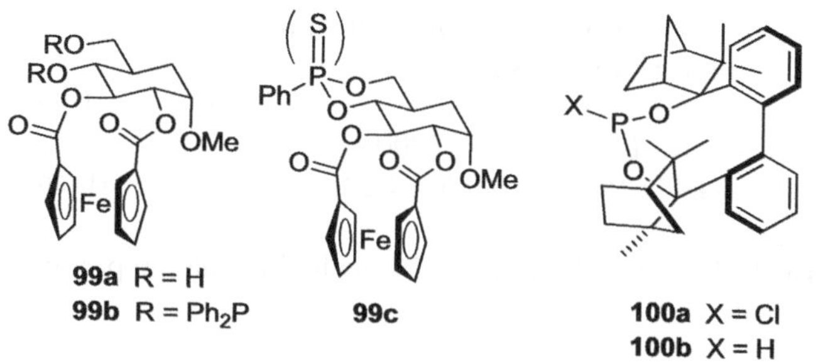

Figure 5.19 Miscellaneous phosphinites from ferrocenylsugars and terpenes.

Syntheses of an unusual ferrocenylglucose-derived phosphonite **99c** (and a bis-diphenylphosphinite **99b**) have been accomplished, starting from the corresponding 4,6-sugar diol, **99a** (Figure 5.19) [175]. Only a Pd(0)-catalyzed allylation of dimethylmalonate using the phosphinite ligand **99b** (up to 39% ee) has been reported. The phosphonite was characterized as a sulfide.

A series of highly hindered phosphonites have been prepared from readily available terpenic ketones and organolithium derivatives [176]. Examples are shown in Figure 5.19. The chlorophosphine **100a** is highly resistant to nucleophilic displacement including with aqueous KOH, but a dialkyl-H-phosphonite **100b** can be prepared using $LiAlH_4$ at high temperature. A Cu complex of this unusual P–H phosphonite ligand yields up to 65% ee in catalyzed addition of Et_2Zn to cyclohexenone [176]. Allylation of dimethyl malonate with 1-phenyl-2-propenyl acetate in the presence of $[(allyl)PdCl]_2$ and **100b** mostly leads to linear products (l:br = 80:20, 68% yield), with the ee for the branched product reaching up to 65% [177].

5.4.8 Development of phosphonite ligands for industrially relevant processes

The quest for improved selectivity in large commercial processes that employ homogeneous catalysis continues to generate ever more complex ligands including phosphonite and phosphinite ligands. Three examples relate to reactions of alkenes, Rh-catalyzed hydroformylation, Ni(0)-catalyzed hydrocyanation, and Ni(II)-, Co(II)-, and Fe(II)-catalyzed oligomerization (mostly dimerization) of ethylene and propylene. Unique aspects of phosphonite ligands developed for these and other related reactions are discussed in this section.

5.4.8.1 Phosphonite ligands in hydroformylation

Formations of linear aldehydes via hydroformylation of terminal alkenes (Eq. 5.65) as well as from internal alkenes through tandem isomerization followed by hydroformylation (Eq. 5.66) continue to be a significant challenge in this industrially relevant area. Börner *et al.* [178] found a partial solution to this problem as illustrated by the hydroformylation of mixtures of *n*-octene isomers (with only 3.3% 1-octene) to produce up to 47% of the desired linear 1-nonanal using the ligand **101c** (Scheme 5.27A), which carries an uncommon phenolic moiety on the backbone. Unusually low temperatures and pressures at which these reactions proceed are also noteworthy. Ligand **103** (Figure 5.20) with bulky aryl groups is also useful for hydroformylation of a mixture of C_8 alkenes, even though the selectivity for the desired product appears to be poor [179].

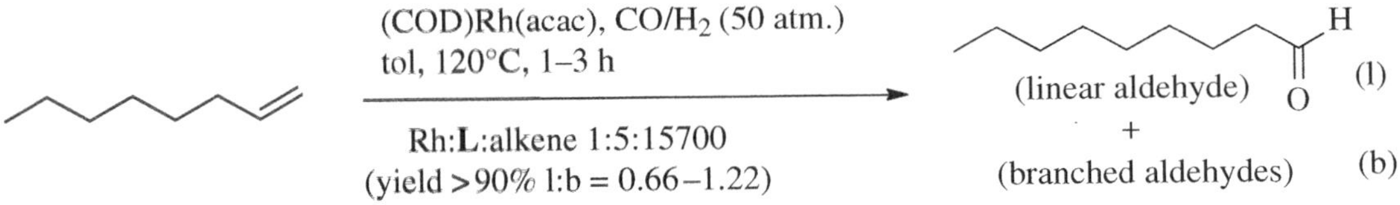

• at 80°C with **L = 101a**: TOF >83000 h^{-1}; l:b = 1.57; no hydrogenation product; only 1.6% 2-ethylheptanal (5.65)

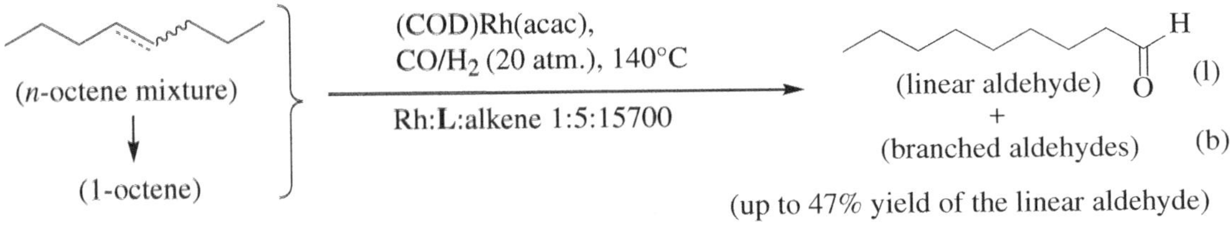

(5.66)

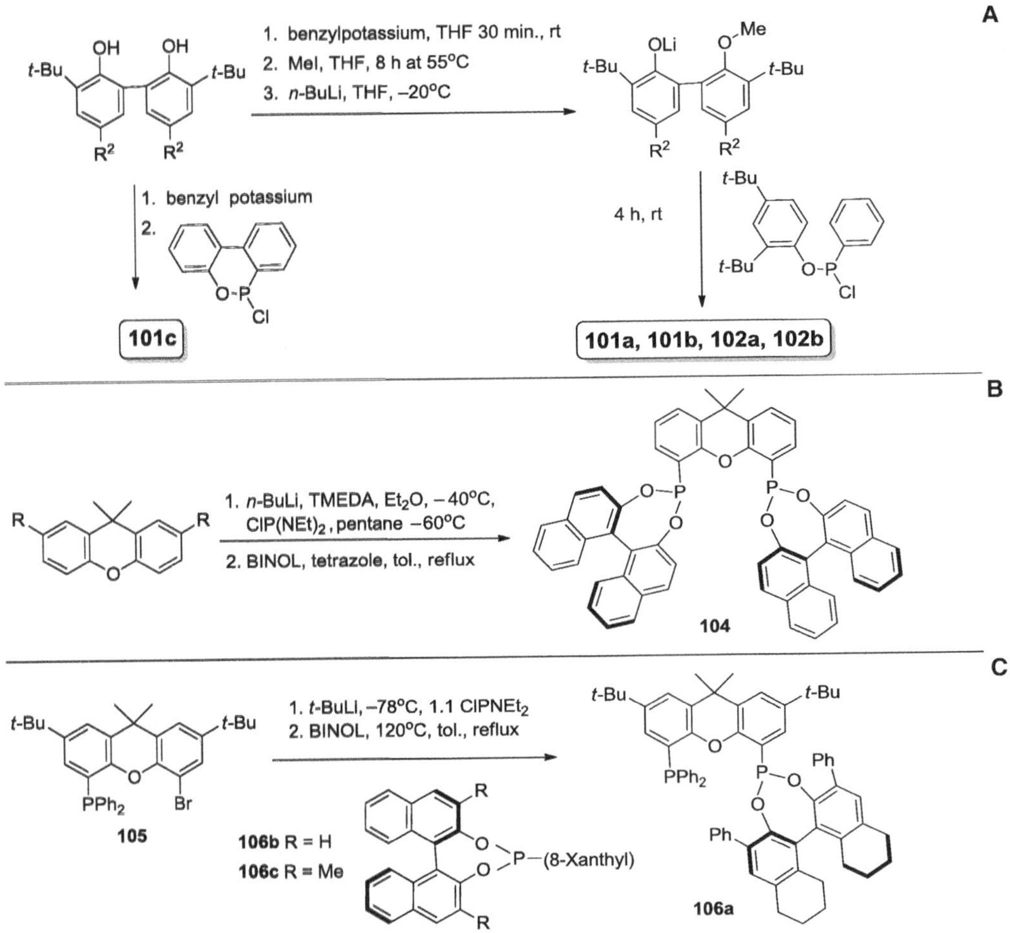

Scheme 5.27 *Syntheses of ligands for hydroformylation.*

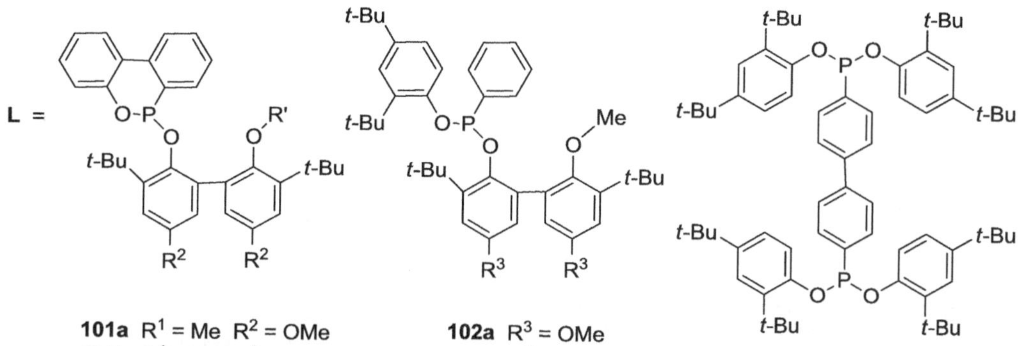

101a R^1 = Me R^2 = OMe
101b R^1 = Me R^2 = *t*-Bu
101c R^1 = H R^2 = OMe

102a R^3 = OMe
102b R^3 = *t*-Bu

103

Figure 5.20 *Phosphonite ligands for Rh(I)-catalyzed hydroformylation.*

In the hydroformylation of 1-octene, phosphonite ligands prepared from calix[4]-arenes also give a high proportion of linear aldehyde (up to 63%, with the rest being mostly 2-methyl-octanal) [180].

Ligand **104** (Scheme 5.27B), readily assembled from 9,9-dimethylxanthene [181], is moderately effective for the hydroformylation of styrene (b:l=9:1, 33% ee), vinyl acetate (b:l=90:1, 51% ee), and Rh-catalyzed hydrogenation of methyl *N*-acetamidocinnamate (54% ee). Coordination chemistry and structural studies on Pd, Pt, and Rh chloride complexes of related highly hindered biphenol-derived xanthene ligands have also been published [182].

A related phosphine–phosphonite ligand **106a** with a xanthene backbone (prepared from **105** [183]; Scheme 5.27C) is an excellent ligand for hydroformylation of dihydrofurans (Eq. 5.67) [184]. Nearly perfect regioselectivities and ee up to 91% for **108** seen in this reaction are unprecedented. Under slightly modified conditions, 1,2-dihydrofuran (**107b**) gave the opposite enantiomer of the product **108** along with a regio-isomer **109**. The regioselectivity for this substrate (**107b**) is relatively low with up to 25% the 2-carboxaldehyde being formed. Ligands **106b** and **106c** performed poorly in these reactions.

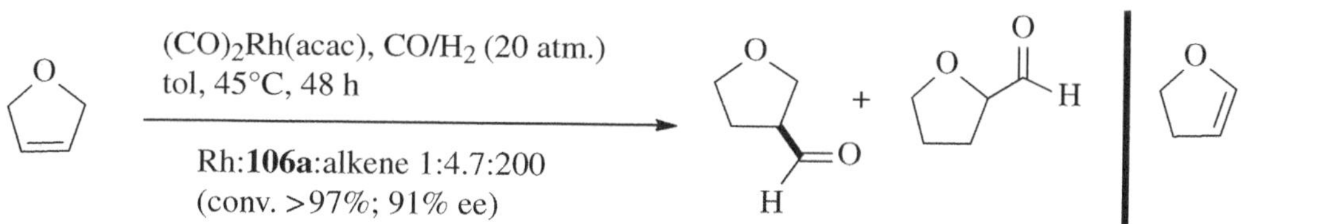

107a **108** **109** (not formed) **107b** (5.67)

5.4.8.2 Phosphonite ligands in Ni(0)-catalyzed hydrocyanation

Production of adiponitrile from 1,3-butadiene (BD) depends on the initial hydrocyanation of BD to form a mixture of 2-methyl-3-butenenitrile (2M3BN, ~30%) and 3-pentenenitrile (3PN, ~70%), followed by isomerization of the 2M3BN to 3PN, subsequent double bond migration, and a second hydrocyanation (Scheme 5.28). A great deal of work has been done in optimizing the ligands for the isomerization of 2M3BN to 3PN and 4-pentenenitrile (4PN).

Phosphonite ligand **114** prepared by Hofmann *et al.* is among the best to accomplish this transformation and compares favorably with the best phosphite ligands developed by the DuPont workers [185]. This ligand and others used in this study (e.g., **112**) were readily synthesized by well-established routes as shown in

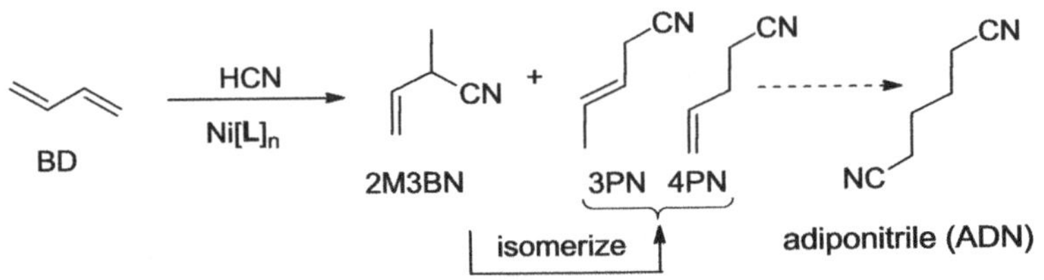

Scheme 5.28 *Ni(0)-catalyzed hydrocyanation of butadiene and synthesis of adiponitrile.*

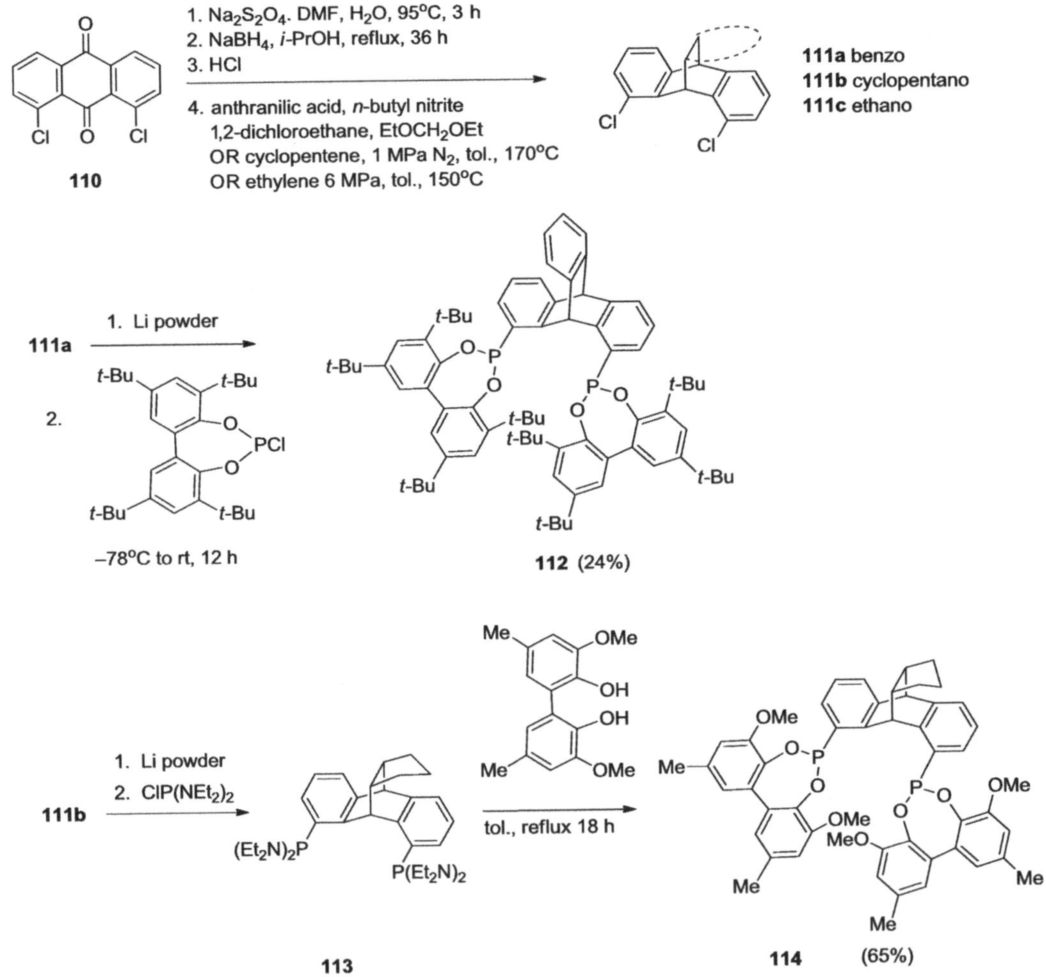

Scheme 5.29 *Synthesis of 9,10-bridged anthracene phosphonite ligands.*

Scheme 5.29 [186]. Other ligands (**115** [187], **116** [187], **117a,b** [188], and **118** [188]) developed for this reaction are shown in Figure 5.21. Ligand **117a** promotes Ni(0)-catalyzed hydrocyanation of styrene with excellent regioselectivity (99.9% branched) and yield (>99%) using 1 mol% of the corresponding Ni(0) complex. Ligand **118** is exceptionally active for hydrocyanation of butadiene (0.001 equiv. Ni, 120°C, 3h, 2M3BN 41%, *t*-3PN 49%, TON 644, TOF 215 h^{-1}) [188].

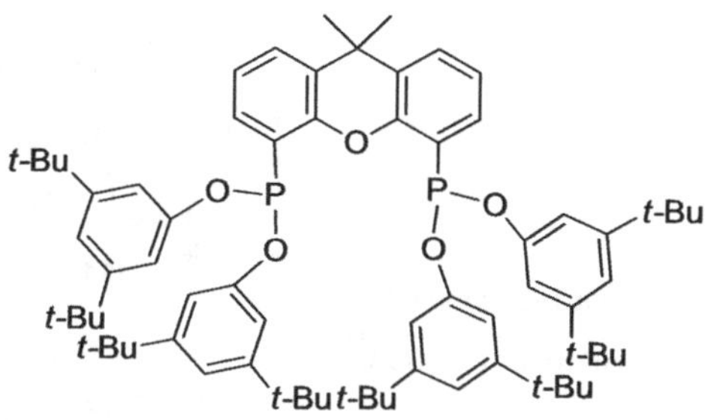

115

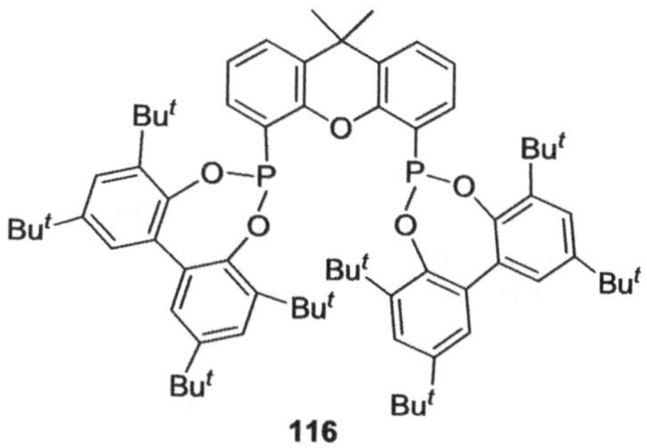

116

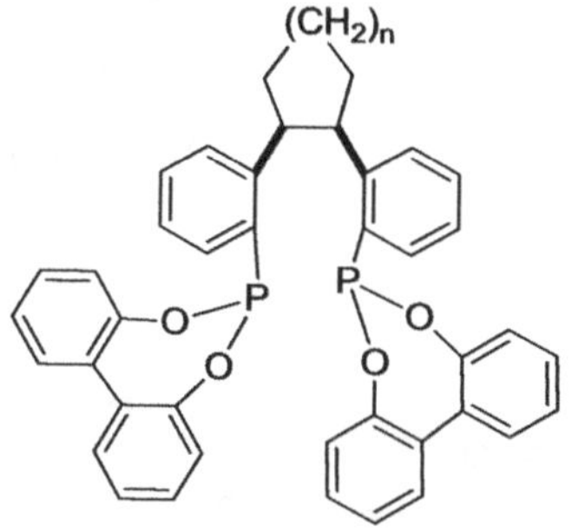

117a n = 1

117b n = 2

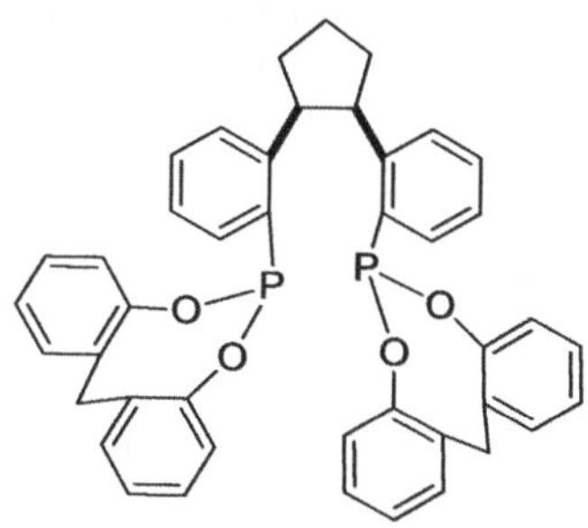

118

Figure 5.21 *Assorted phosphonite ligands developed for adiponitrile synthesis from 1,3-butadiene.*

5.4.8.3 Oxazoline–phosphonite ligands and olefin dimerization

Bis-oxazoline–phosphonite ligands have been prepared as shown in Scheme 5.30. Complexation with [Cl_2Ru (*p*-cymene)]$_2$ gives either a coordinately unsaturated monomeric complex (**121**) with the more hindered dimethyloxazoline or an 18-e dimer (**122**) with the isopropyl derivative [189]. Complex **121** forms a CO adduct, but does not react with ethylene, Ph_3P, or $(PhO)_3P$. Complex **121** catalyzes transfer hydrogenation of acetophenone in the presence of 2-propanol and excess sodium *i*-propoxide, albeit in low ee. Attempted cyclopropanation of styrene with ethyl diazoacetate in the presence of **121** mostly leads to diethyl fumarate and maleate [189].

A phosphinite **123** derived from **119a** reacts with [Pd(η^3-allyl)(μ-Cl)]$_2$ to form a rare η^1-allyl-Pd complex **124**, challenging the conventional notion that formation of such mono-hapto species requires strong tridentate chelation around the metal. Note that the η^1-allyl unit is *trans* to nitrogen, the weaker of the two σ donors, nitrogen and phosphorus. A similar η^1 coordination (**125a**) ensues with the bisoxazoline–phosphonite ligand **120a** [190]. In this case, the second oxazoline acts as a noncoordinating dangling ligand, although NMR evidence suggests a dynamic behavior for the complex. Yet another example of a similar complex (**125b**) has also been reported [191]. The η^1 complexes undergo facile insertion (upon bubbling CO through a toluene solution) of CO to form an acyl–Pd complex, providing some support for the mechanistic hypothesis that η^3–η^1 slippage and CO coordination to the metal precede CO migratory insertion (Scheme 5.31).

Studies of Ir complexes (Scheme 5.32) with the bis-oxazoline–phosphonite ligand **120a** have led to an unusual η^2 coordination mode for COD (**126**) and an equally uncommon Ir–polyhydride adduct (**128**) upon reaction of a cationic derivative **127** with hydrogen [192].

The phosphonite **120a** forms a distorted tetrahedral complex **129** upon reaction with $CoCl_2$ with a dangling uncoordinated oxazoline [193], like the corresponding Pd(II) complex [190], which is square planar. In sharp

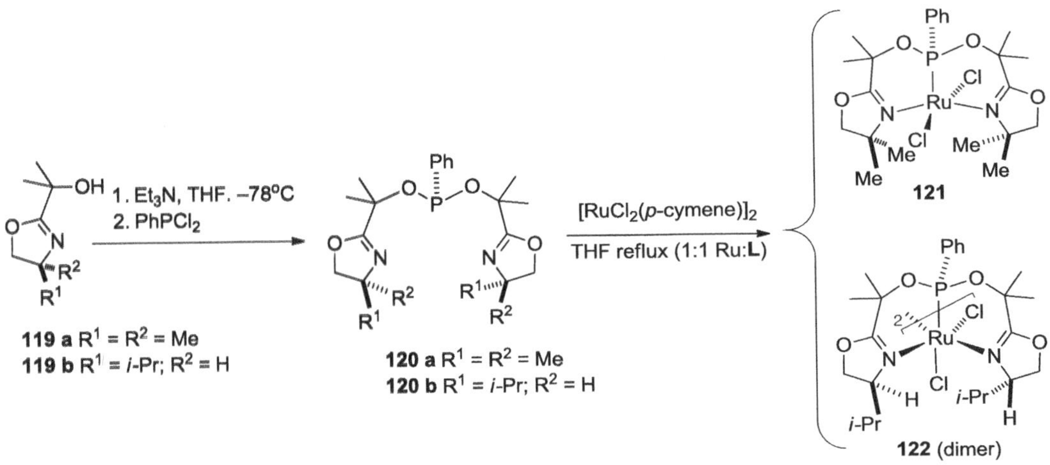

Scheme 5.30 Synthesis of a bisoxazoline–phosphonite and Ru complexes.

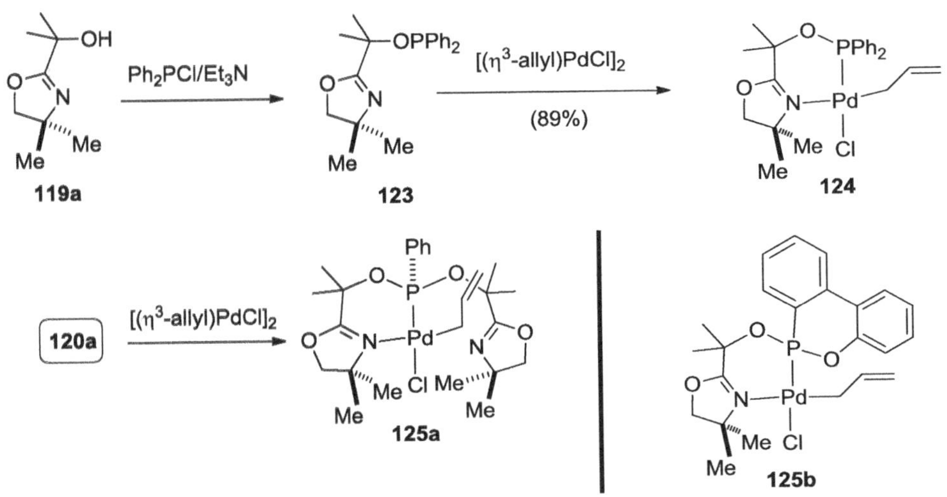

Scheme 5.31 Uncommon η^1 coordination in an allyl–palladium oxazoline–phosphonite complex.

contrast, the corresponding oxazoline–phosphine ligand (**130**) with Co and Fe forms unprecedented 8-membered *N,N*-chelates without involving the phosphorus atom (**131**, Scheme 5.33). Complex **129** in the presence of $EtAlCl_2$ (6 equiv.) catalyzes ethylene polymerization (30 atm ethylene; 80°C; 17700 h^{-1}).

For the dimerization of ethylene in the presence of MAO, the pyridine–phosphinite Ni complex **132** is more active than **133** (Figure 5.22); but **133** is more selective for 1-butene (up to 72% with 200 equiv. MAO) [194,195]. In the presence of MAO, complex **132** serves as one of the most active catalysts for dimerization of propylene (3.29×10^4 [propylene][Ni]$^{-1}$ h^{-1}).

The phosphonite **134**, synthesized from the oxazoline **119a**, reacts with (COD)PdCl(Me) to give a neutral complex, which upon reaction with AgOTf produces a labile triflate intermediate (Scheme 5.34). This complex undergoes sequential insertion of CO and ethylene to give **138**, validating two of the elementary steps in the Pd-catalyzed polyketone synthesis [196].

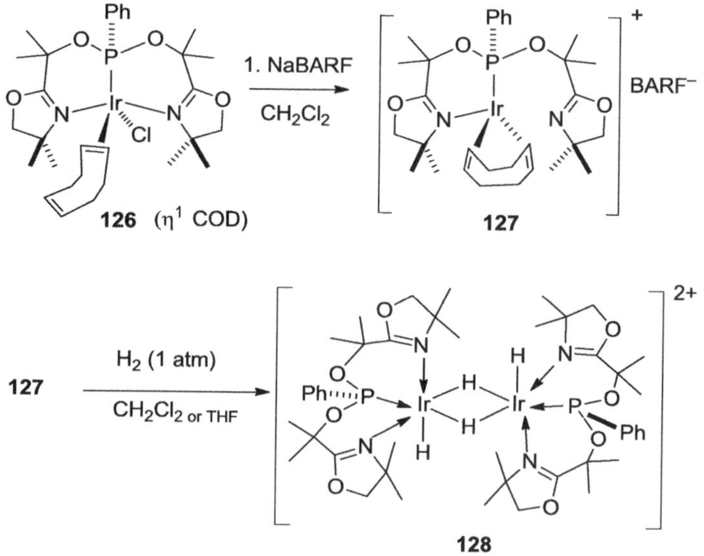

• Only hydrides as bridging ligands
• Two 3c-2e bonds in the $Ir_2(\mu\text{-}H_2)$

Scheme 5.32 *Unusual Ir complexes with oxazoline–phosphonite ligands.*

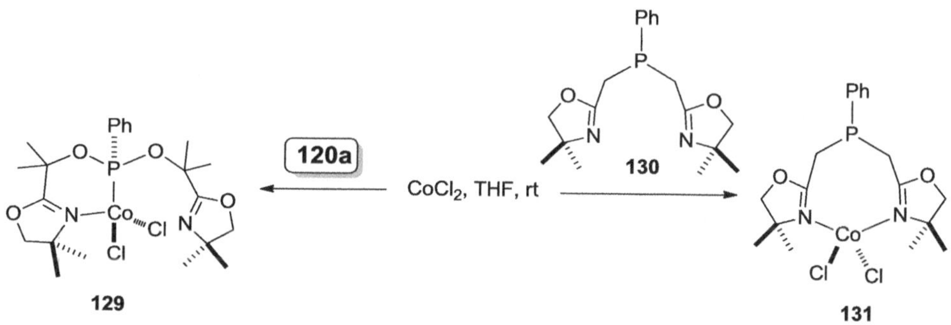

• $FeCl_2.4H_2O/CH_2Cl_2$ + **130** also gives a similar 8-membered*N,N*-chelate
• $PdCl_2.(NCCH_3)_2$ + **130** gives a square planar 5-membered *P,N*-chelate

Scheme 5.33 *Contrasting bonding modes of oxazoline–phosphine versus oxazoline–phosphinite ligands.*

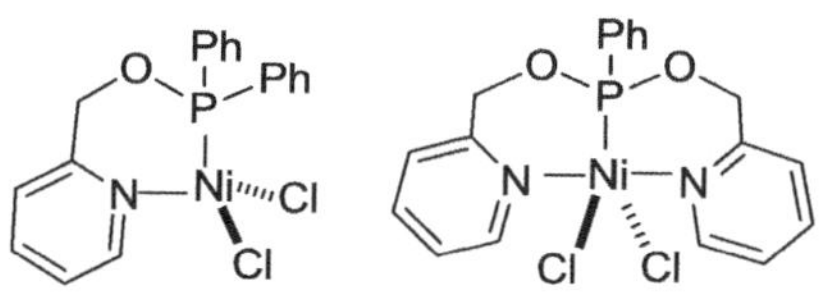

132 **133**

Figure 5.22 *Pyridine–phosphinite and phosphonite complexes for selective ethylene and propylene dimerization.*

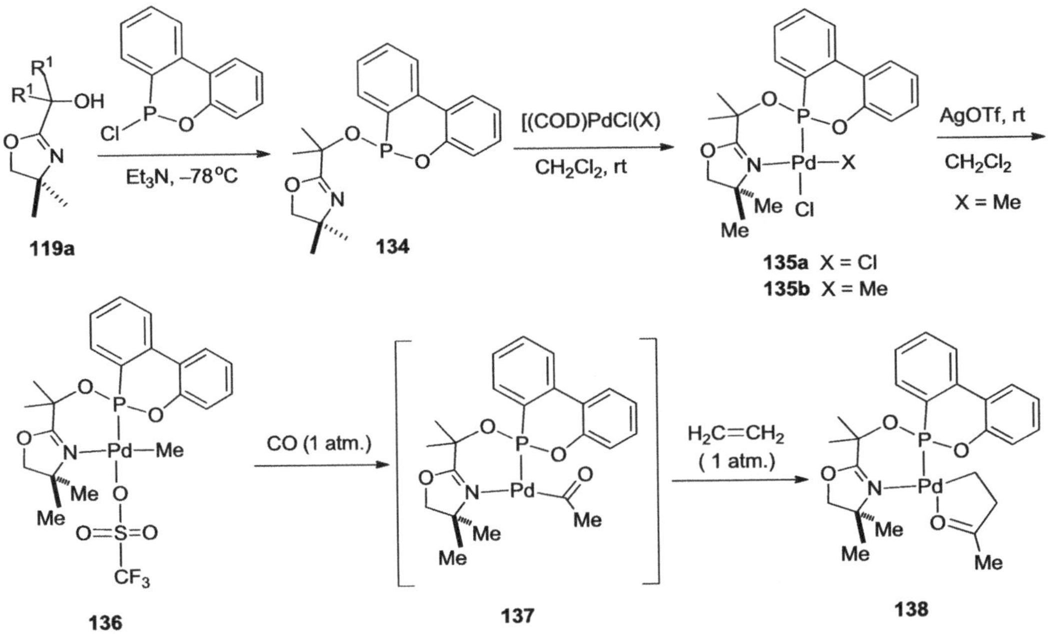

Scheme 5.34 Insertion reactions of oxazoline Pd(II) complexes. Elementary steps in polyketone synthesis.

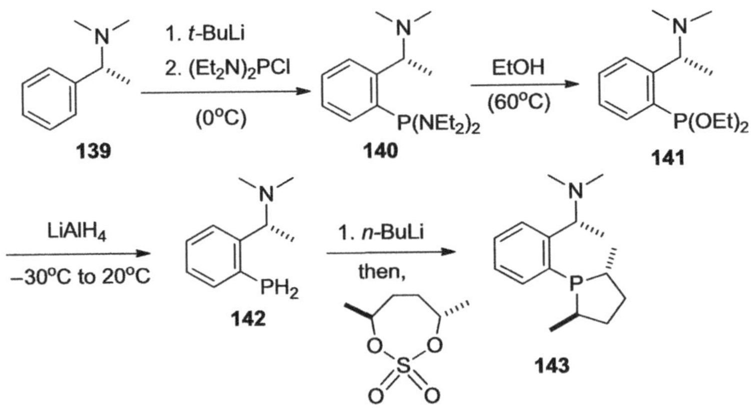

Scheme 5.35 Use of a phosphonite intermediate for the synthesis of an arylphosphine and a phospholane.

5.4.9 Use of the phosphonite functionality to synthesize other ligands

Aminophosphonites. The electrophilic phosphonite functionality has been used for the synthesis of other unusual ligands. Several examples are shown in Schemes 5.35, 5.36, 5.37, and 5.38. The phosphonite **141**, easily synthesized from the amine **139** in two steps, can be reduced to the phosphine **142**, which in turn can be converted into the phospholane **143**, analogous to one of the ligands in the DUPHOS series (Scheme 5.35) [197].

Synthesis of π-acidic ligands. Synthesis of π-acidic ligands such as **147** and **148** involves intermediacy of **146**, a diphenylalkyl bis-phosphonite, which undergoes facile Grignard addition to give the

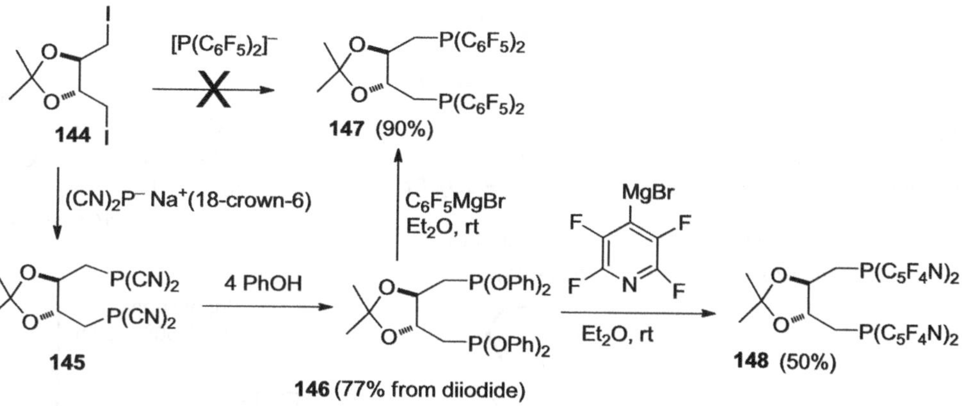

Scheme 5.36 Use of a phosphonite intermediate for the syntheses of π-acidic ligands.

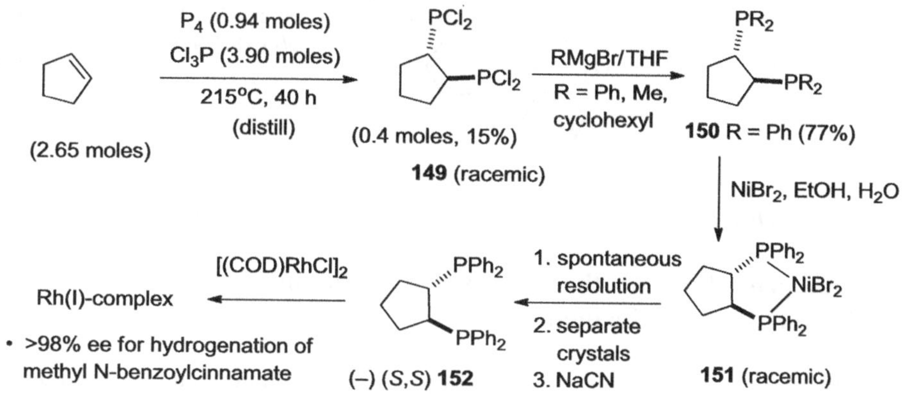

Scheme 5.37 Synthesis of trans-1,2-dichlorophosphinylcyclopentane and its reactions.

perfluroarylphosphines **147** and **148** (Scheme 5.36) [198]. Direct displacement of iodide from **144** with pentaflouorophenylphosphide in the presence of Lewis acids failed to produce the expected product **147**.

Synthesis and applications of enantiopure 1,2-bis-dichlorophosphines. Trans-bis-1,2-dichlorophosphinocyclopentane (**149**) is easily synthesized [199] from white phosphorus, PCl_3, and cyclopentene at high temperature (Scheme 5.37). This compound can be converted into diarylphosphines (**150**, Scheme 5.37) or to various phosphonites via the reactions shown in Scheme 5.38 [200,201].

The 1,2-bis-dichlorophosphine **149** can be resolved through a series of reactions starting with a diastereoselective reaction with (*R*,*R*)-isopropyl tartrate to produce **153** (Scheme 5.38). Reduction of **153** with $LiAlH_4$ (to give **154**) [200] followed by reaction with triphosgene gives (*S*,*S*)-**149**, which serves as a precursor for a wide variety of phosphonites including the BINAPO derivative **156** [202]. Synthetic applications of these ligands have not been reported. It should be noted that the original paper [199] that reports the synthesis of the bis-dichlorophosphine **149** also includes the preparation and resolution of a *1*,2-bis-diphenylphosphinocyclopentane, which appears to be an exceptionally good ligand for the Rh-catalyzed hydrogenation of dehydroamino acids. This result portends well for possible applications of ligands with similar scaffolding like **155** and **156**.

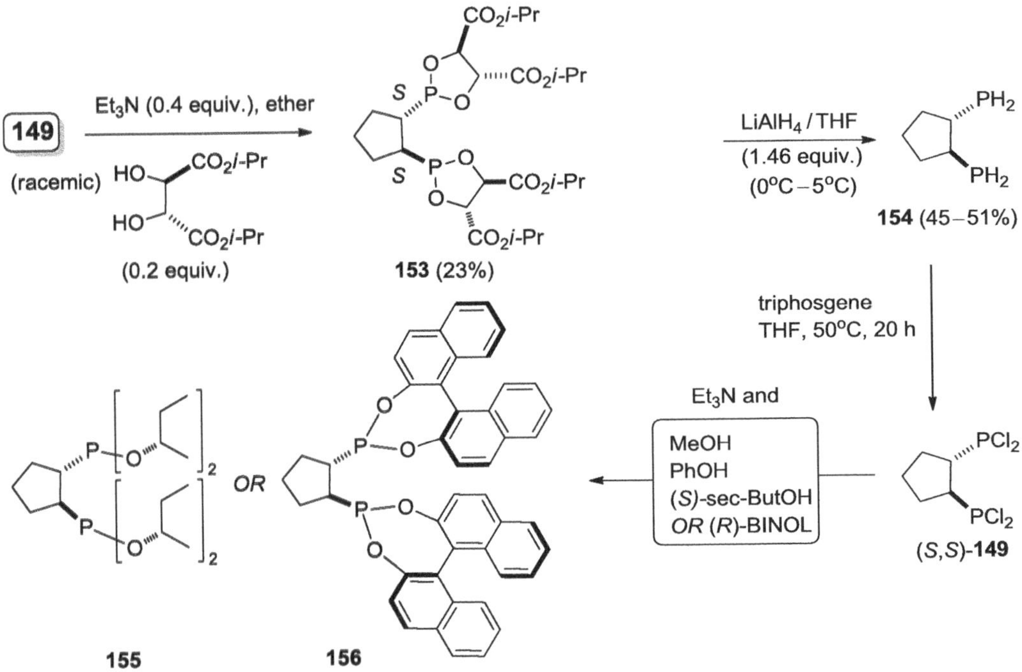

Scheme 5.38 Synthesis of cycloalky-2,2-diphospinato ligands.

5.5 Experimental procedures for the syntheses of prototypical phosphinite and phosphonite ligands

Many of the phosphorus precursors for the synthesis of simple phosphinites and phosphonites are commercially available. These include $(C_6H_5)_2PCl$, (cyclohexyl)$_2$PCl, Me_2PCl, Et_2PCl, $MePCl_2$, $EtPCl_2$, and $C_6H_5PCl_2$. Et_2PCl, $PhP(OEt)_2$, $Me_2P(O)OH$, $Me_2P(O)Cl$, and others are easily synthesized by metathesis reactions between the halides and alcohols or amines. Two widely used precursors $[(i\text{-Pr})_2N]_2PCl$ [203] and $[(i\text{-Pr})_2N]PCl_2$ [204] are readily prepared by stoichiometry-controlled reactions of the diisopropyl amine and freshly distilled PCl_3. Hydrolytic and oxidative stability of the phosphinites and phosphonites varies widely depending on the *P*-substituents. Many are quite stable in crystalline form, and may be exposed to air briefly; in general they undergo facile oxidation and hydrolysis in solution when exposed to oxidants (including oxygen) and moisture. Appropriate care should also be exercised in handling electrophilic phosphorus compounds, many of which are very toxic.

5.5.1 Phosphinite ligands

5.5.1.1 $Me_2P(OMe)$ [12]

The ligand Me_2POMe was prepared by heating a 1:1 molar mixture of Me_2PNMe_2 and MeOH at 65° for 1 h and then distilling the product at bp 58–59° (760 mm).

The reactant Me_2PNMe_2 is prepared by heating Me_2PCl with 2 mol of Me_2NH in Na-dried diethyl ether at −78° under a nitrogen atmosphere. The Me_2NH was slowly added through an addition funnel with a pressure-equalizing side arm. The reaction flask was also equipped with a very efficient stirrer and a dry ice condenser. After complete addition of Me_2NH, the reaction mixture was allowed to come slowly to RT with stirring, the amine hydrochloride was filtered off under suction, and the product was distilled (bp 95–100°C, 760 mm). The direct reaction of Me_2PCl or $MePCl_2$ with MeOH in the presence of base gave very poor yields of the phosphinite and phosphonite. The spectrum of Me_2POMe consists of two doublets at 1.19 ppm (CH_3, J=6.0 Hz) and at 3.32 ppm (OCH_3, J=1 3.5 Hz) with intensities of 2:1, respectively.

5.5.1.2 *Et_2POEt and $EtP(OEt)_2$* [11]

Diethyl ethylphosphonite and ethyl diethylphosphinite are prepared by a modification of the triethyl phosphite synthesis [205]. A mixture of 70.0 g (0.534 mol) of ethyldichlorophosphine and 50 mL of hexane is added dropwise with stirring under nitrogen to a mixture of 62.4 mL (1.068 mol) of ethanol, 170.2 mL (1.068 mol) of *N,N*-diethylaniline, and 200 mL of hexane at 0°C. The reaction mixture is then refluxed for 3 h and filtered under nitrogen. The *N,N*-diethylaniline hydrochloride precipitate is washed with hexane, and the hexane is removed from the filtrate by distillation under 1 atm of nitrogen. The residue is distilled through a Vigreux column under nitrogen to give diethyl ethylphosphonite (74%, bp 43°C, 15 Torr).

In the same way, ethyl diethylphosphinite is prepared (44%, bp 65–66°C, 73 Torr; 83–87°C, 151 Torr) using equimolar quantities of diethylchlorophosphine, ethanol, and *N,N*-diethylaniline. Throughout the work, infrared spectra were taken of the phosphorus ligands and of their complexes to ensure the absence of hydrolysis or of Arbuzov isomerization to the phosphoryl compounds, for which characteristic infrared frequencies have been reported. Their ^{31}P (chemical shift δ from H_3PO_3) and ^1H NMR (δ from TMS) data are reported below: Et_3P: ^{31}P: +19.5 (m), ^1H:1.13 (m); Et_2POEt: ^{31}P: −182 (m), ^1H:1.32 (m), 3.94 (m); $EtP(OEt)_2$: ^{31}P: −135 (m), ^1H:1.28 (m), 3.84 (m); $P(OEt)_3$: ^{31}P: −142 (sept), ^1H:1.19, 1.19, 3.85.

5.5.1.3 *Synthesis of methyl 3,4-bis-O-[bis(3,5-dimethylphenyl)phosphino]-2,6-di-O-benzoyl–α–D-glucopyranoside (Ligand 8) (Scheme 5.39)* [26]

Diisopropylphosphoramidous dichloride [$(i\text{-}Pr)_2NPCl_2$] [204,206]. A dry 1 L Morton flask equipped with an overhead stirrer, alcohol thermometer, nitrogen bubbler, and 250 mL addition funnel is purged with nitrogen, charged with 500 mL of toluene and 68.5 g (43.5 mL, 0.50 mol) of phosphorus trichloride (CAUTION: EXTREMELY MOISTURE SENSITIVE, CORROSIVE, AND IRRITATING), and cooled to 5°C with an ice–acetone bath. With the vent closed, the addition funnel is charged with 101 g of diisopropylamine, which is then added over 1 h, maintaining the internal temperature at 15–25°C (single peak in the ^{31}P NMR at this time). After the addition is complete, the cooling bath is removed, and the mixture is stirred at 35°C for 2 h. The product is filtered in a dry box (CAUTION: THE PRODUCT IS EXTREMELY MOISTURE

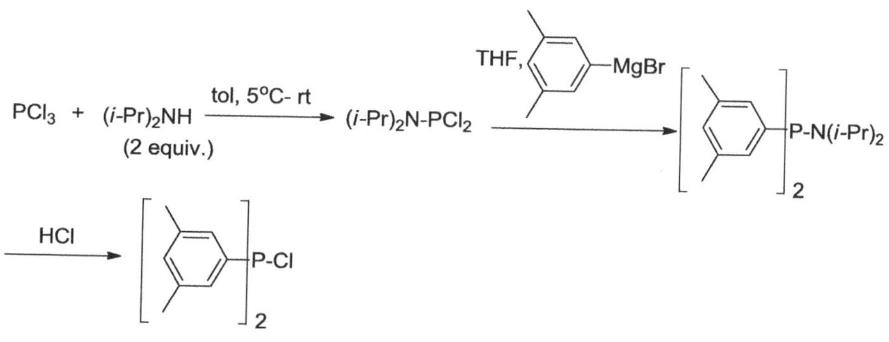

Scheme 5.39

SENSITIVE) into a 3L three-necked flask to remove the solid amine hydrochloride, which is pressed dry with a rubber dam and washed with three 50mL portions of toluene. The filtrate is used in the next step without purification.

N,N-diisopropyl bis(3,5-dimethylphenyl)phosphinamide [37,206]. A 2L three-necked flask equipped with an overhead stirrer, condenser, thermometer, vented addition funnel, and nitrogen bubbler is dried under nitrogen and charged with 30g (1.23mol) of Mg turnings and 70mL of anhydrous tetrahydrofuran (THF). A small portion (about 15mL) of a solution of 194g (143mL, 1.05mol) of 5-bromo-*m*-xylene in 250mL of anhydrous THF and 350mL of toluene is added, and the reaction is initiated with 0.5mL of 1,2-dibromoethane. When the exotherm subsides, the flask is cooled with a 20–25°C bath while the remaining solution is added over 1.5–2.0h at such a rate to maintain the temperature at 30–35°C. When the addition is complete, the cooling bath is removed, and the mixture is allowed to stir at 30–35°C for 1h. Meanwhile, the solution from step A is cooled to 5–10°C and the warm Grignard solution is cannulated in over 0.5–1.0h, maintaining the reaction temperature at 40–50°C. When the transfer is complete, the mixture is heated for an additional 1h at 45–50°C. Complete conversion of the intermediate Ar(Cl)PN(i-Pr)$_2$ to the product may be confirmed by gc or ^{31}P-NMR (C_6D_6). The intermediate appears at δ 135ppm (^{31}P), while the product resonates at δ 38ppm. The product may also be analyzed by gas chromatography on a 5m HP-1 column: 100°C, then 10°C/min to 280°C (2min): product @ 9.5min, Ar_2P(O)N(i-Pr)$_2$ @ 11.7min, Ar_2P(O)Par$_2$ @ 18min. The mixture is distilled under reduced pressure (~20mm) with a nitrogen bleed to remove most of the THF. The magnesium salts are filtered and washed with two 100mL portions of toluene; the filtrate is concentrated on the rotary evaporator and cooled to ambient temperature. Traces of toluene are removed by adding 100mL of cold methanol and concentration on the rotary evaporator at 25°C. The solid residue is reslurried with 300mL of cold (0–5°C) methanol, filtered, washed with 50mL of cold MeOH, and suction-dried to afford 111g (65%) of product as a white solid, mp 111–112°C. ^{31}P-NMR (C_6D_6) δ 38ppm; ^{1}H-NMR (C_6D_6) δ 1.2 (d, 6H), 2.2 (s, 6H), 3.5 (m, 1H), 6.85 (br s, 1H), 7.45 (br d, 2H).

Bis(3,5-dimethylphenyl)phosphinous chloride (Scheme 5.40) [206]. The product from the previous step (34.1g, 0.100mol) is dissolved in 300mL of cyclohexane in a dry flask equipped with a thermometer, gas inlet tube, and nitrogen bubbler connected to a trap and caustic scrubber. The flask is purged with nitrogen and cooled to 5°C, and dry HCl gas (ca. 8g) is fed subsurface with efficient stirring at 5–15°C until the exotherm subsides and the system begins to off-gas HCl (two molar equivalents are required for complete conversion). Excess HCl is purged with a subsurface nitrogen stream, and the mixture is filtered in a drybox to remove diisopropylamine hydrochloride. The filtrate is concentrated to dryness *in vacuo* to afford the crude phosphinous chloride in nearly quantitative yield as an extremely moisture-sensitive white crystalline solid. ^{31}P-NMR δ 84ppm. The product may be purified by vacuum distillation (bp ~150°C/5Torr), but is generally suitable for coupling with diols to form bis-phosphinite ligands.

Bis-phosphinite ligand (8) [26,206]. A dry 200mL flask equipped with a thermometer and magnetic stirring bar is charged in a drybox with 9.33g (33.8mmol) of the chlorodiarylphosphine from the previous step, 75mL of anhydrous toluene, and 4.52g (37.0mmol) of 4-dimethylaminopyridine (DMAP). To this solution is then added 6.47g (16.1mmol) of the diol in portions over 40min, maintaining the internal temperature

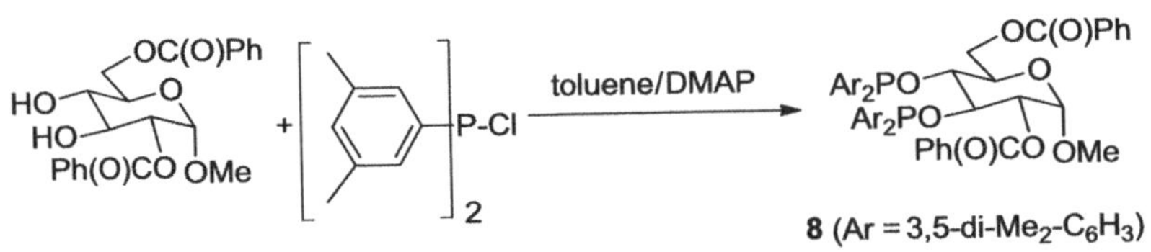

8 (Ar = 3,5-di-Me$_2$-C_6H_3)

Scheme 5.40

at 25–30°C. The mixture is stirred for an additional 16 h, the precipitated DMAP hydrochloride is filtered under nitrogen and washed with two 15 mL portions of toluene, and the filtrate is concentrated to dryness *in vacuo* to provide 14 g of crude ligand. The yield of recovered DMAP hydrochloride is 5.66 g (97%). The crude product may be purified to remove traces of polar impurities by dissolving in 20 mL of acetonitrile, cooling to −30°C, filtering rapidly, and drying *in vacuo* to provide 10.0 g (70%) of pure ligand as a white crystalline solid, mp 162–164°C. The solid ligand is reasonably air stable, but is readily degraded in solution by atmospheric oxygen. ^{1}H-NMR (C_6D_6) δ 1.73 (s, 6), 1.88 (s, 6), 1.91 (s, 6), 2.02 (s, 6), 2.75 (s, 3), 3.90 (dd,1 J=4, 12 Hz), 4.04 (dd, J=4, 10 Hz), 4.42 (d, 1 ,J=12 Hz), 4.89 (m, 1), 5.00 (d, 1, J=3 Hz), 5.23 (m, 1), 5.51 (dd, 1 J=4, 10 Hz), 6.03 (s, 1), 6.32 (s, 1), 6.46 (s, 1), 6.63 (s, 1), 6.7–7.3 (m, 14), 7.80 (m, 2), 8.13 (m, 2); ^{31}P-NMR (C_6D_6) δ 118.8 and 124.7 ppm, $J_{P\text{-}P}$=2.4 Hz. CH analysis C 72.30; H 6.39 (calculated C 72.10; H 6.39).

5.5.1.4 *Preparation of phenyl 2,3-bis-*O*-[bis[3,5-bis(trifluoromethyl)phenyl]-phosphino)-4,6-*O*-benzylidene-glucopyranoside (5a) (Scheme 5.41)* [22]

Bis[3,5-bis(trifluoromethyl)phenyl]chlorophosphine from (Et_2N)PCl_2. A 1.0 M solution of [3,5-bis(trifluoromethyl)phenyl]magnesium bromide was prepared by slow addition of 18.5 g (60 mmol) of 3,5-bis-(trifluoromethy1)bromobenzene in 40 mL of THF to a slurry of Mg turnings in 20 mL of THF. After 1 h, this solution was added slowly to a solution of 5.0 g (29 mmol) of (Et_2N)PCl_2 in 30 mL of THF at 0°C. After 2 h, the mixture was concentrated *in vacuo*. Cyclohexane (100 mL) was added, and the mixture was filtered through celite to provide a solution of bis[3,5-bis(trifluoromethyl)phenyl](diethylamino)phosphine. Dry HC1 was passed through this solution for 1 h. After filtration under a nitrogen atmosphere (in some instances, it was necessary to degas the solution to precipitate the amine hydrochloride) and concentration, 12.4 g (88%) of the chlorophosphine was collected as a white solid.
Ligand **5a**. A solution of $(CF_3)_2C_6H_3)_2PCl$ (1.246 g, 2.53 mmol) from the previous step in 5 mL of CH_2Cl_2 was cooled to 0°C and added dropwise to a solution of phenyl 4,6-*O*-benzylidene-D-glucopyranoside(0 .397 g, 1.15 mmol) and DMAP (0.030 g, 0.25 mmol) in 15 mL of CH_2Cl_2/pyridine (1:1) at 0°C. The reaction mixture was warmed to RT, stirred overnight, and concentrated to dryness *in vacuo*. The resulting solids were slurried in hot benzene and filtered, and the filtrate concentrated *in vacuo* to a dry white solid. These white solids were washed with about 2 mL of cold hexane and dried *in vacuo* to give 1.14 g (79%) of the ligand **5a**.

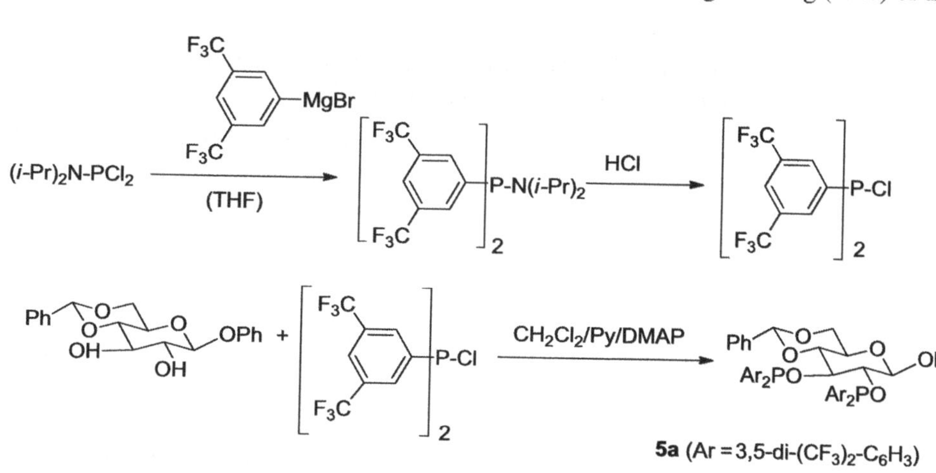

Scheme 5.41

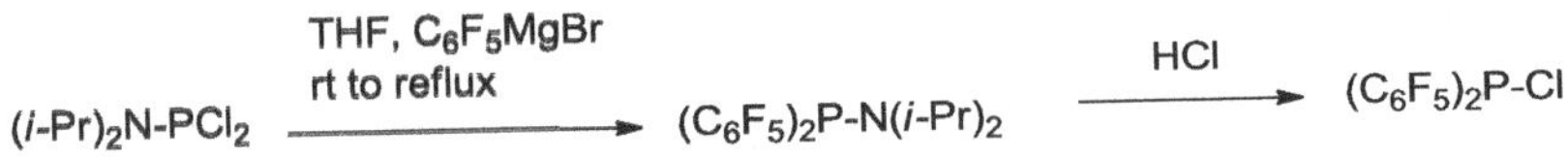

Scheme 5.42

^{31}P(^{1}H) NMR: δ 108.4 (s, lP), 107.5 (s, 1P).^{1}H NMR: δ 2.98 (m, lH), 3.15 (t, J=9.2 Hz, lH), 3.27 (t,10.2 Hz, lH), 3.88 (dd, 4.9, 10.4 Hz, lH), 4.19 (m, 2 H), 4.69 (m, lH),4.94 (s, lH), 6.33 (d, 7.9 Hz, 2 H), 6.69 (t, 7.3 Hz, lH), 6.88 (t, 8.0 Hz,2 H), 6.97 (m, 2 H), 7.05 (m, 3 H), 7.35 (s, 2 H), 7.39 (s, 2 H), 7.82 (m, 8 H). Anal. C, H, P.

5.5.1.5 *Preparation of bis-(pentafluorophenyl)chlorophosphine (Scheme 5.42)* [27]

Pentafluorobromobenzene (8.20 mL, 66.0 mmol) in dry THF (15 mL) is added dropwise to freshly crushed magnesium turnings (24.31 g, 69.0 mmol) in THF (60 mL). After 2 h, this solution is added slowly to a solution of (i-Pr$_2$N)PCl$_2$ (5.76 g, 30.0 mmol) in THF (20 mL) at RT. The reaction mixture was heated to reflux. After 13 h, the mixture was cooled to RT, and saturated aqueous NH_4Cl (80 mL) is added. The aqueous layer is extracted with hexane (80 mL × 3), and the combined organic extract is dried (Na_2SO_4) and concentrated *in vacuo* to give $(C_6F_5)_2$PN(i-Pr) as a colorless oil. Crystallization from methanol gives white crystals (11.1 g, 79.8%): mp 79–82°C; ^{1}H NMR ($CDCl_3$) δ 3.58 (m, 1 H), 1.19 (d, J) 6.1 Hz, 6 H; ^{31}P NMR ($CDCl_3$) δ 1.61 (quintet, J) 29.8 Hz. Dry HCl is passed through a solution of the aminophosphine (2.45 g, 5.29 mmol) in dry cyclohexane (40 mL) at RT for 20 min. After filtration under argon atmosphere and concentration, 1.97 g (93.1%) of $(C_6F_5)_2$PCl was collected as a colorless oil: ^{31}P ($CDCl_3$) δ 35.37 (quintet, J) 38.5 Hz.

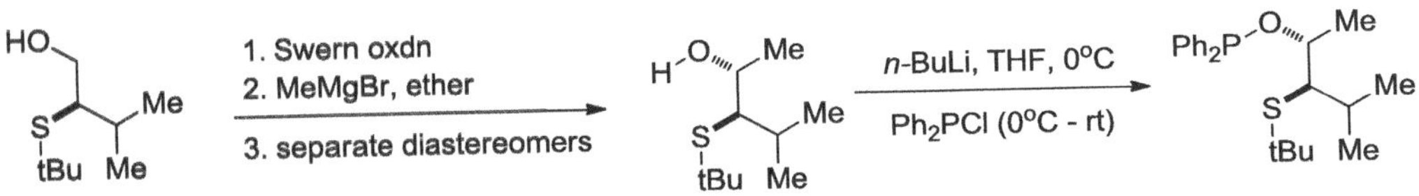

Scheme 5.43

5.5.1.6 *An alternate general procedure for phosphinite incorporation. [(2S,3R)-3-phenylthio-4-methylpent-2-oxy]diphenylphosphine (Scheme 5.43)* [106]

To a Schlenck flask containing β-hydroxysulfide (57.0 mg, 0.275 mmol) in THF (0.28 M) at 0°C is added n-BuLi (2.50 M in hexanes, 110 µL, 0.275 mmol). The reaction is warmed to RT and stirred for1 h. The reaction is then cooled to 0°C, and chlorodiphenylphosphine(49 µL, 0.275 mmol) is added via syringe. The reaction is allowed to warm to RT and stirred for 3 h. The solvent is removed *in vacuo*, and the reaction is diluted with 95:5 hexane–ethylacetate (0.5 mL, dried over sieves and degassed with argon). The resulting slurry is loaded onto a plug of silica and purified by flash chromatography (1 × 3 cm, 95% hexane, 5% ethyl acetate) to yield the product as an oil (101 mg, 93% yield). $[\alpha]_d^{23}$ +27.1° (c 0.28 CH_2Cl_2); IR(CH_2Cl_2) 2965, 1480, 1436, 1046, 978, 882 cm^{-1}; ^{1}H NMR (500 MHz,C_6D_6) δ7.62–7.57 (m, 4 H), 7.40 (d, 2 H, J) 7.2 Hz, 7.13–7.08 (m, 4 H), 7.07–7.01 (m, 2 H), 6.98–6.95 (m, 2 H), 6.90 (t, 1 H, J) 7.1 Hz, 4.33 (qdd, 1 H, J) 8.2, 7.5, 6.3 Hz, 3.17 (dd, 1 H, J) 7.5, 4.3 Hz, 2.28 (dsept, 1 H, J) 6.7, 4.3 Hz, 1.40 (d, 3 H, J) 6.2 Hz, 0.99(d, 3 H, J) 6.8 Hz, 0.97 (d, 3 H, J) 6.8 Hz; ^{13}C NMR (125 MHz,C_6D_6) δ143.8 (d, J) 18 Hz, 143.0 (d, J) 17 Hz, 138.5, 131.4,131.3, 131.2, 130.7, 130.4, 129.5, 129.3, 129.2, 129.1, 128.5, 128.4,126.3, 79.2 (d, J) 21 Hz, 64.7 (d, J) 6 Hz, 29.1,

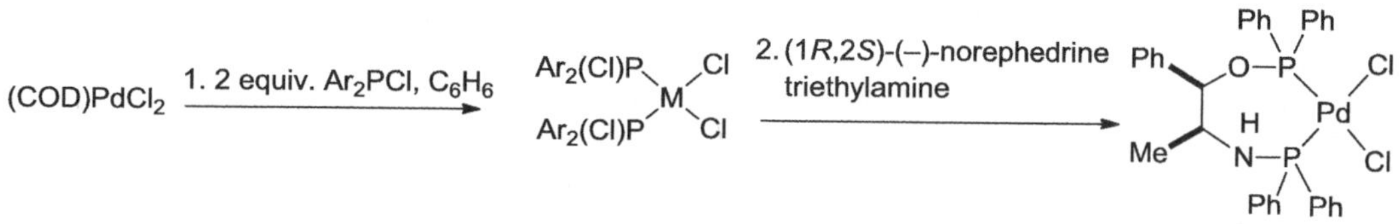

Scheme 5.44

21.7, 20.7 (d, *J*) 5 Hz, 18.3; ^{31}P NMR (202 MHz, C_6D_6) δ 109.6 (s); TLC *Rf* 0.6 (95% hexane, 5% ethyl acetate); exact mass calculated for $C_{24}H_{28}O_1P_1S_1$-H requires *m/z* 395.1599, found *m/z* 395.1590 (CI–NH_3).

5.5.1.7 *Metal – template synthesis of an amino1,2-diarylphosphine-diarylphophinite complex (Scheme 5.44)* [84]

[$PdCl_2$(1,5-COD)] [207] (0.150 g, 0.525 mmol) is suspended in distilled benzene (25 mL), and PPh_2Cl (189 µL, 1.050 mmol) is added to give the immediate formation of the yellow soluble intermediate. Triethylamine (147 µL, 1.050 mmol) and (1*R*,2*S*)-(–)-norephedrine (0.080 g, 0.525 mmol) are added in sequence, followed by the precipitation of NEt_3.HCl. After 18 h of stirring, the solution containing the product complex is separated by filtration and taken to dryness. The yellow-orange solid product is purified by recrystallization from CH_2Cl_2 and diethyl ether. Yield: 0.215 g (59%). $C_{33}H_{31}Cl_2NOP_2Pd$ (696.9): calculated C 56.88, H 4.48, N 2.01; found C 56.54, H 4.48, N 1.91; ^{1}H NMR δ (200 MHz, $CDCl_3$, 25°C): δ 0.95 (d, $J_{H,H}$ 7.3 Hz, 3 H, CH_3), 2.0 (br, s, 1 H, NH), 3.9 (br. s, 1 H, CH–N), 5.2 (m, 1 H, CH–O), 6.8–8.2 (m, 25 H, Ph). ^{31}P NMR (81.01 MHz, $CDCl_3$, 25°C): δ 70.5 (d, P_A, J_{PP} 35 Hz), 123.6 (d, P_B, J_{PP} 35 Hz) ppm. $[\alpha]^D_{25}$ = –13.2 (*c* 0.53, CH_2Cl_2). Recrystallization of crude complex from dichloromethane–diethyl ether gives crystals suitable for a diffractometric study.

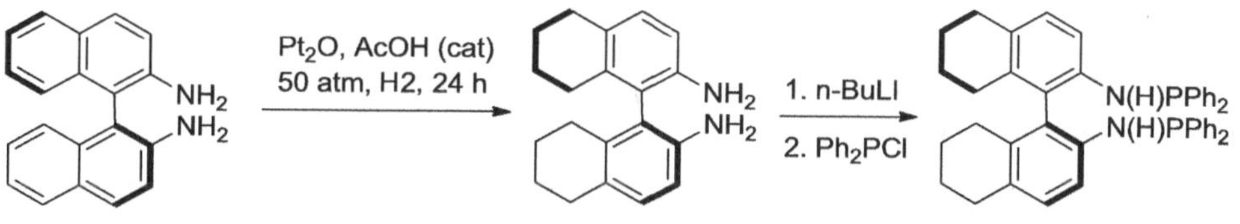

37

Scheme 5.45

5.5.1.8 *Procedure for the preparation of a bis-aminodiaryphosphine (R)-37 (Scheme 5.45)* [72]

The procedure for the preparation of (R)-octahydro-2.2′-binaphthyldiamine [72b]. A 50 mL autoclave equipped with a magnetic stirring bar is charged with 200 mg (*R*)-2.2’-binaphthyldiamine (purchased from Aldrich Chemical Co.), 20 mg of PtO_2, 20 mL of glacial acetic acid, and 2 mL of water. The autoclave is closed, and 3 atm of H_2 is charged. The solution is stirred for 24 h at 50°C. After releasing the hydrogen gas and removing the solid catalyst by filtration, the mixture is neutralized with aqueous $NaHCO_3$ solution followed by extraction with 20 mL of ethyl acetate three times. The combined extracts are dried with sodium sulfate, and the solvent is removed with a rotary evaporator to give 200 mg of crude product, (*R*)-5,5, 6 6′,7,7′,8,8′-octahydro-1,1′-binaphthyl-2,2′-diamine. The crude product is purified by crystallization with 5 mL of ethyl acetate and 15 mL of hexane to give 180 mg of crystals of the reduced product (88% of theoretical yield). The analytical data are as follows: mp: 210°C dec; $[\alpha]_D$ 133 (*c* 1.0, pyridine); ^{1}HNMR

(400 MHz, $CDCl_3$) δ: 6.90 (d, *J* 8.0 Hz, 2 H); 6.60 (d, *J*8.2 Hz, 2 H); 3.07 (s, 4 H); 2.70 (m, 4 H); 2.22 (m, 4 H); 1.67 (m, 8 H). ^{13}CNMR (101 MHz, $CDCl_3$) δ 141.9, 136.6, 129.6, 128.0, 122.4, 113.5, 29.7, 27.4, 23.6 ppm. Anal. calculated for $C_{20}H_{24}N_2$: C 82.15, H 8.27, N 9.58. Found C 82.34, H 8.06, N 9.61.

Ligand 37 [72b]. (*R*)-5,5, 6,6′,7,7′,8,8′-Octahydro-1,1′-binaphthyl-2,2′-diamine (200 mg, 0.7 mmol) in THF (20 mL) is charged to a 50 mL flask under a nitrogen atmosphere. This flask is cooled to –30°C, and into the solution is added a solution of *n*-butyllithiumin hexane (0.88 mL of a 1.6 M solution, 1.4 mmol) in a dropwise manner. The mixture is stirred for 2 h at –30°C with a magnetic stirrer. Then a solution of chlorodiphenylphosphine (0.32 mL, 1.8 mmol) in THF (5 mL) is added dropwise. The system is allowed to stir for 5 h, and the temperature is raised to about 25°C. The mixture is filtered to remove the solid. The THF solvent is removed *in vacuo* to give 420 mg of the bisphosphine. The crude product is purified by recrystallization in diethyl ether solvent at –30°C for 24 h to afford 390 mg of white, needlelike crystals of (*R*)-**37** (84.0% of theoretical yield). The analytical data of (*R*)-**37** were as follows: mp: 137–139°C. $[\alpha]_D$ –47 (*c* 1.0, CH_2Cl_2). ^{1}H NMR (400 MHz, $CDCl_3$) δ 7.24 (m, 22 H); 6.98 (d, *J* 8.34 Hz, 2 H); 4.27 (d, J_{PH} 7.0 Hz, 2 H); 2.67 (m, 4 H); 2.10 (m, 4 H); 1.58 (m, 8 H). ^{13}C NMR (101 MHz, $CDCl_3$) δ: 141.9, 141.7, 141.2, 141.0, 140.6, 140.5, 136.1, 131.0, 130.8, 130.4, 130.2, 129.6, 128.9, 128.7, 128.4, 128.3, 128.2, 123.2, 112.7, 112.5, 29.3, 27.3, 23.1, 23.0; ^{31}P NMR (162 MHz, $CDCl_3$) δ 27.25 ppm. Anal. calculated for $C_{44}H_{42}N_2P_2$: C 79.97, H 6.41, N 4.24. Found: C 79.78, H 6.40, N 4.24.

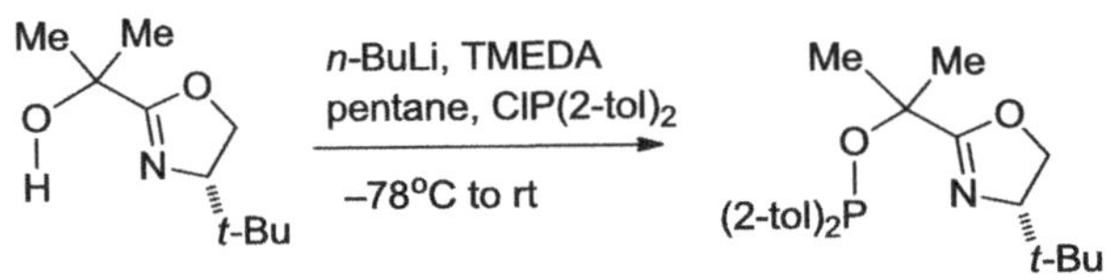

Scheme 5.46

5.5.1.9 *(–)-(S)-4-tert-butyl-2-{1-di(2′-methylphenyl)phosphinite-1-methyl-ethyl}-4,5-dihydro-oxazole (60) (Scheme 5.46)* [113]

2-(4-*tert*-butyl-4,5-dihydro-oxazol-2-yl)-propan-2-ol (500 mg, 2.7 mmol) [208,209] is dissolved in 50 mL dry pentane, and at –78°C 2.2 mL (3.51 mmol, 1.3 equiv) *n*-BuLi solution (~1.6 M in hexanes) is added dropwise. After 10 min, 0.82 mL (5.4 mmol, 2 equiv.) TMEDA is added and a pink colored suspension is obtained. After 1.5 h stirring at –78°C, the mixture is allowed to warm to RT over 1 h. To this white suspension, a solution of 873 mg (3.51 mmol, 1.3 equiv) di(2-methylphenyl)-chloro-phosphine in 10 mL pentane is added dropwise at 0°C. The mixture is allowed to warm to RT overnight, and the solvent volume is reduced to 2 mL. This suspension is transferred directly on a column prepared under argon (silica, pentane:Et_2O:NEt_3=95:5:1). After chromatography, a colorless oil is obtained. Yield: 536 mg (1.35 mmol, 50%). $C_{24}H_{32}NO_2P$ (397.49) ^{1}H NMR (400.1 MHz, C_6D_6, 300 K): δ=0.76 (s, 9 H, *t*-Bu CH_3), 1.69 (s, 3 H, P–O–C–CH_3), 1.72 (s, 3 H, P–O–C–CH_3), 2.37 (s, 3 H, *o*-Tol CH_3), 2.42 (s, 3 H, *o*-Tol CH_3), 3.60 (dd, ^3J≈8 Hz, 8 Hz, 1 H, oxaz CH), 3.65 ("t", *J*≈8 Hz, 1 H, oxaz CH_2), 3.68 ("t", *J*(H,H)≈8 Hz, 1 H, oxaz CH_2), 6.90–6.94 (m, 2 H, arom. CH), 7.01–7.15 (m, 4 H, arom. CH), 7.95 (mc, 2 H, ortho arom. CH); ^{31}P{^{1}H} NMR(162.0 MHz, C_6D_6, 300 K): δ=81.2 (s); ^{13}C{^{1}H} NMR(100.6 MHz, C_6D_6, 300 K): δ=20.7 (d, $3J_{(P,C)}$=3.8 Hz, oxaz CCH3), 20.9 (d, $3J_{(P,C)}$=4.2 Hz, oxaz CCH_3), 26.0 (s, *t*Bu CH_3), 27.6 (d, $3J_{(P,C)}$=8.8 Hz, *o*Tol-CH_3), 27.7 (d, $3J_{(P,C)}$=8.4 Hz, *o*Tol-CH_3), 33.8 (s, *t*Bu quat. C), 69.0 (oxaz CH_2), 76.0 (oxaz CCH_3), 76.2 (oxaz CH), 126.2 ("t", $3J_{(P,C)}$=3.1 Hz, 2 C, meta arom. C), 129.3, 129.4, 130.2 (d, $J_{(P,C)}$=3.4 Hz), 130.3 (d, $J_{(P,C)}$=3.4 Hz), 131.8 (d, $J_{(P,C)}$=8.4 Hz), 132.2 (d, $J_{(P,C)}$=8.8 Hz), 141.0 (superimposed d, 4 C, quat. arom.

C), 168.8 (oxaz quat. C); IR (neat, on NaCl): ν ~ = 3055 m, 2957vs, 2904 m, 2868 m, 1920vw, 1730w, 1666vs, 1590w, 1566vw, 1508vw, 1467 s, 1454 s, 1394w, 1380 m, 1364 m, 1271 m, 1208 m, 1157vs, 1131vs, 1068w, 1031w, 979 s, 958 s, 921 m, 877 s, 800w, 749vs, 720 m, 673w cm^{-1};+EI MS, 70 eV, m/e: 168 ([M – OP*o*-$Tol_2]^+$, 100%), 215 ([M – *o*$Tol_2]^+$, 29%), 229 ([OP*o*-$Tol_2]^+$,88%), 397 (M+, 3%); $[\alpha]_D^{20}$ –38.2 (*c* 0.7, $CHCl_3$); TLC: R_f = 0.32 (pentane:Et_2O:NEt_3 = 95:5:1 on silica).

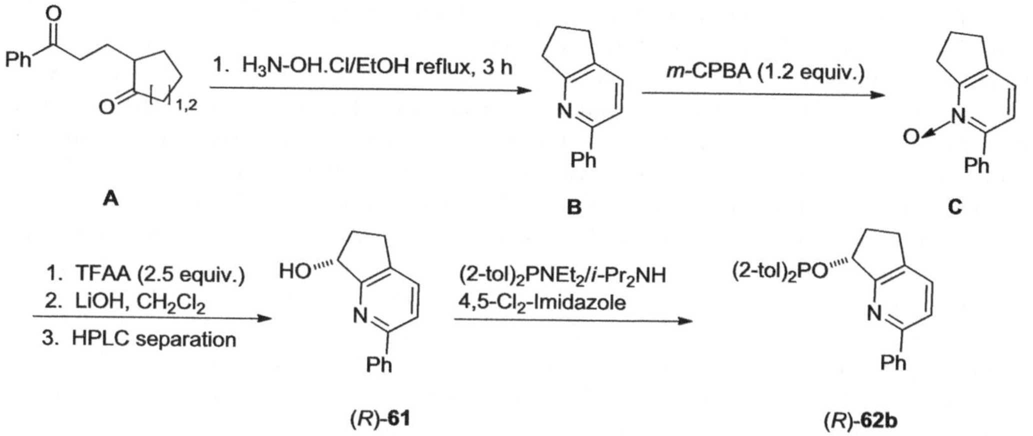

Scheme 5.47

5.5.1.10 (R)-7-(2-phenyl-6,7-dihydro-5 H-[1]pyrindin)-di-(2'-methylphenyl)-phosphinite (62b) (Scheme 5.47) [114]

2-Phenyl-6,7-dihydro-5 H-[1]*pyrindine* (**B**). 2-(3-Oxo-3-phenyl-propyl)-cyclopentanone (**A**) (2.00 g, 9.25 mmol) and hydroxylamine hydrochloride (0.643 g, 9.25 mmol) dissolved in 10 mL of ethanol are placed in a two-necked flask with a reflux condenser. After 3 h at reflux, saturated sodium carbonate solution and 40 mL of water are added to neutralize the reaction mixture. The water phase is extracted with 150 mL of methylene chloride. The combined organic layers are dried ($MgSO_4$) and concentrated to yield a crude brown oil which is purified by flash column chromatography (silica gel; hexanes:CH_2Cl_2:triethylamine, 5:1:1). Colorless crystals are obtained (1.53 g, 85%, $C_{14}H_{13}N$, 195.26 g mol^{-1}). TLC: *Rf* = 0.61 (hexanes:CH_2Cl_2:triethylamine (5:1:1); mp: 79°C; elemental analysis [found (calc.)]: C 86.03 (86.12), H 6.73 (6.71), N 7.03% (7.17); MS(EI): 195 (100.0, M+), 115 (6.5), 91 (7.3), 83 (6.8%); IR(KBr): 3051w, 2916 s, 2838 s, 1568 s,1431br, s, 1315 m, 1220 s, 842 cm^{-1} s; ^{1}H-NMR (400 MHz, $CDCl_3$): δ 7.95 (2 H, d, J_{HH} = 6.8 Hz, H_{Ar}), 7.55 (1 H, d, J_{HH} = 7.8 Hz, H_{Ar}), 7.45 (3 H, mc, H_{Ar}), 7.37 (1 H, tt, J_{HH} = 7.3 Hz, J_{HH} = 1.4 Hz, H_{Ar}), 3.09 (2 H, t, J_{HH} = 7.7 Hz, CH_2), 2.97 (2 H, t, J_{HH} = 7.5 Hz, CH_2), 2.17 ppm (2 H, quintet, J_{HH} = 7.6 Hz, CH_2); ^{13}C-NMR(100 MHz, $CDCl_3$): δ 165.9 (s, C_{Ar}), 155.9 (s, C_{Ar}), 140.0 (s, C_{Ar}), 135.4 (s, C_{Ar}), 132.5 (s, HC_{Ar}), 128.6 (s, HC_{Ar}), 128.3 (s, HC_{Ar}), 126.9 (s, HC_{Ar}), 118.0 (s, HC_{Ar}), 34.4 (s, CH_2), 30.5 (s, CH_2), 23.3 ppm (s, CH_2).

2-Phenyl-6,7-dihydro-5 H-[1]*pyrindine-N-oxide* (**C**). A two-necked flask is charged with *m*CPBA (0.530 g, 3.07 mmol) suspended in 10 ml of methylenechloride. The suspension is cooled to 0°C, and 2-phenyl-6,7-dihydro-5 *H*-[1]pyrindine (**B**) (0.500 g, 2.56 mmol) is added dropwise. After removing the ice bath, the reaction mixture is stirred at RT overnight. An additional 0.6 equiv. *m*CPBA are added and stirred for 2 h. Then the organic phase is washed several times with diluted NaOH solution to remove all remaining peroxides. The solvent is evaporated, and the yellow solid was purified by flash column chromatography (silica gel; he xanes:CH_2Cl_2:triethylamine, 5:5:1) to yield colorless crystals (0.473 g, 87%, $C_{14}H_{13}NO$,211.26 g mol^{-1}). TLC: Rf=0.41 (hexanes:CH_2Cl_2:triethylamine (5:5:1); mp: 125°C; Elemental analysis [found (calc.)]: C 79.11 (79.59), H 6.11 (6.20), N 6.49 (6.63), O 7.82% (7.57); MS(EI): 211 (57.9, M$^+$), 210 (>100.0, M-H$^+$), 195 (60.7), 194 (64.4), 183 (28.2), 157 (26.1), 115 (11.6), 91 (9.5), 77 (12.0%); IR (KBr): 3061 s, 2965 m, 2922 m, 1473 s, 1439 m, 1375 s, 1260br, s, 768 cm^{-1} m; ^{1}HNMR(400 MHz, $CDCl_3$): δ 7.81 (2 H, mc, *H*Ar), 7.45 (2 H, mc, *H*Ar), 7.41 (1 H, mc, *H*Ar), 7.24 (1 H, d, *J*=7.8 Hz, *H*Ar), 7.14 (1 H, d, *J*=7.6 Hz, *H*Ar), 3.22 (2 H, t, *J*=7.7 Hz, CH_2), 3.06 (2 H, t, *J*=8.0 Hz, CH_2), 2.22 ppm (2 H, quintet, J_{HH}=7.6 Hz, CH_2);^{13} CNMR(100 MHz, $CDCl_3$): δ 153.8 (s, *C*Ar), 147.4 (s, *C*Ar), 140.5 (s, *C*Ar), 133.0 (s, *C*Ar), 129.4 (s, H*C*Ar), 129.0 (s, H*C*Ar), 128.0 (s, H*C*Ar), 125.3 (s, H*C*Ar), 122.0 (s, H*C*Ar), 31.6 (s, CH_2), 30.1 (s, CH_2), 22.3 ppm (s, CH^2).

*7-Hydroxy-2-phenyl-6,7-dihydro-5H-[1]pyridine (R)-***61** *and (S)-***61**. A two-necked flask is charged with 2-phenyl-6,7-dihydro-*5H*-[1]pyridine-*N*-oxide (**C**) (0.400 g, 1.89 mmol) and dissolved in 10 ml of methylene chloride. After cooling the solution to 0°C with an ice bath, trifluoroacetic anhydride (0.660 ml, 0.994 g, 4.73 mmol) is slowly added. The reaction solution is allowed to warm up to RT and is stirred 4 h while the color changes from yellow to green to brown. The volatile compounds are removed under reduced pressure, and the remaining brown oil is dissolved in 10 ml of methylene chloride and 10 mL of a 2 M LiOH solution is added. After stirring for 3 h, the water phase is extracted five times with 40 mL of methylene chloride. The combined organic extracts were washed with water, then brine, then dried ($MgSO_4$) and evaporated. The crude product is purified by flash column chromatography (silica gel; EA:hexanes, 1:1), and colorless crystals were obtained (0.36 g, 89%, $C_{14}H_{13}NO$, 211.3 g mol^{-1}). The racemic alcohol is separated into its component enantiomers by chiral semipreparative HPLC; Daicel Chiralcel OD, 10% isopropanol in hexanes, 6 mL/min, 34.6 min (–), (*R*), 46.3 min (+), (*S*). TLC: R_f=0.65 (EA:hexanes (1:1); mp: 119°C; $[\alpha]^{20}_D$: +31 (*c* 1.01, $CHCl_3$),–31 (*c* 0.95,$CHCl_3$). Elemental analysis [found (calc.)]: C 79.29 (79.59), H 6.25 (6.20), N 6.63 (6.63), O 7.86% (7.57); MS(EI): 211 (58.4, M$^+$), 182 (100.0), 155 (40.7%); IR(KBr): 3170br, s, 3050w, 2938 s, 2726 m, 1571 s, 1452 s, 1309 s, 1051 s, 765 cm^{-1} m; ^{1}H-NMR(500 MHz, CDC13): δ 7.96 (2 H, mc, *H*Ar), 7.62 (1 H, d, J_{HH}=7.9 Hz, *H*Ar), 7.58 (1 H, d, J_{HH}=7.9 Hz, *H*Ar), 7.46 (2 H, mc, *H*Ar), 7.40 (1 H, tt, J_{HH}=7.3 Hz, J_{HH}=2.2 Hz, *H*Ar), 5.25 (1 H, t, J_{HH}=6.7 Hz, *C*H(OH)), 3.22 (1 H, s, O*H*), 3.05 (1 H, mc, CH_2), 2.87 (1 H, mc, CH_2), 2.58 (1 H, mc, CH_2CH(OH)), 2.07 ppm (1 H, mc, CH_2CH(OH)); ^{13}C-NMR(125 $[\alpha]^{20}_D$ 16 MHz, $CDCl_3$): δ 164.7 (s, *C*Ar), 156.6 (s, *C*Ar), 139.5 (s, *C*Ar), 134.7 (s, *C*Ar), 133.7 (s, H*C*Ar), 128.7 (s, H*C*Ar), 127.0 (s, H*C*Ar), 120.1 (s, H*C*Ar), 74.9 (s, *C*H(OH)), 33.1 (s, CH_2CH(OH)), 27.2 ppm (s, CH_2).

General procedure for the formation of diaryl phosphinites. The appropriate alcohol (1.0 equiv.) and 4,5-dichloroimidazole (2.0 equiv.) are dissolved in methylene chloride (10 mL) to give a white suspension. To this is added diisopropylamine and triethylamine, respectively (2.0 equiv.), causing the solution to clarify. The mixture is cooled to 0°C, and the diethylaminodiaryl phosphines (2.0 equiv.) added. The cooling bath is removed and the solution stirred for several hours at RT. The mixture is then concentrated and directly chromatographed under argon on degassed silica using degassed solvents for elution (Et_2O:pentane, 1:1). Phosphinites are rather sensitive moieties and are readily decomposed by activated silica gel. As such, the authors recommend degassing silica gel by loading it dry into the column and passing argon backward through it for 10–20 min immediately prior to use. This provides the phosphinites as colorless oils that were used directly in complexation reactions.

(R)-7-(2-Phenyl-6,7-dihydro-5H-[1]pyrindin)-di-(2'-methylphenyl)-phosphinite (**62b**). (*R*)-(–)-2-Phenyl-6,7-dihydro-*5H*-[1]pyrindin-7-ol (**61**) (100 mg, 0.473 mmol) is converted with diethylamino-di-2-methylphenylphosphine (270.1 mg, 0.947 mmol) in the presence of 4,5-dichloroimidazole (130 mg, 0.947 mmol) diisopropylamine (66.3 µl, 47.9 mg, 0.473 mmol) and triethylamine (65.8 µl, 47.9 mg, 0.473 mmol) to (*R*)-7-(2-phenyl-6,7-dihydro-*5H* [1]pyrindin)-di-(2'-methylphenyl)-phosphinite (**62b**) after 138 h (90.6 mg, 45%, C28H26NOP, 423.49 g mol^{-1}). TLC: R_f=0.74 (diethyl ether:pentane (1:1); ^{31}P-NMR(161 MHz, C6D6): δ 102.1 ppm (s).

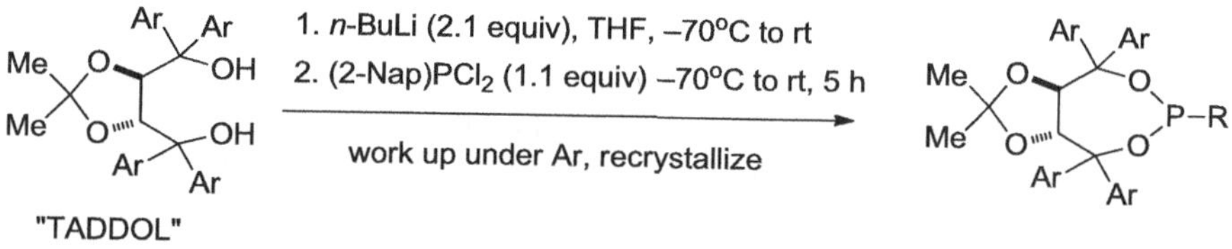

68b Ar = 2-Nap

69d Ar = R = 2-nap

Scheme 5.48

5.5.2 Phosphonite ligands

5.5.2.1 *(IR,7R)-9,9-dimethyl-2,2,4,6.6-penta(2-naphthyl)-3,5,8,l0-tetraoxa-4-phosphabicyclo[5.3.0]-decane (69d) (Scheme 5.48)* [138]

Chlorobis(diethylamino)phosphine [203]. A solultion of Et_2NH (70 g, 0.96 mol) in Et_2O (100 mL) is dropped into a solution of PCl_3 (32.9 g, 0.24 mol) in Et_2O (500 mL) at –70°C over 2 h. The mixture is allowed to warm up to RT. The suspension is filtered and washed several times with Et_2O. The filtrate and washings are concentrated *in vacuo*. The residue is distilled under reduced pressure to give the title compound (quant.). b.p. 78–79°/(0.5 Torr).

(2-Naphthyl)phosphonous dichloride. (2-Naphthy1)magnesium bromide (prepared from 2-bromonaphthalene (11.4 g, 55 mmol) and Mg turnings (1.34 g, 55 mmol)) solution of Et_2O (80 mL) is dropped into a solution of chlorobis (diethy1amino)phosphine (7.63 g, 50 mmol) in Et_2O (100 mL) at –70°C over 1.5 h. The mixture is slowly warmed up to RT and stirred overnight, then refluxed for 0.5 h. HCl (7.3 g, 0.2 mol) in Et_2O (100 mL) is slowly added under ice cooling. The mixture is stirred for 2 h at RT and filtered. The filtrate is concentrated *in vacuo* and distilled under reduced pressure to give a title compound (2.88 g, 25%). b.p. 100° (0.2 Torr).

Preparation of phosphorus TADDOL phosphonites (general procedure). *n*-BuLi (42 mmol, 1.55 M solution in hexane) is added to a stirred solution of the diol (20 mmol) in 80 mL of THF under Ar at –70°C. During this addition, the temperature rises up to –50°C. The mixture is cooled again to –70°C and stirred at this temperature for 5 min, and then warmed up to RT within 1 h. The mixture is cooled again to –70°C, and dichlorophosphorus reagent (22 mmol) is added slowly without allowing the temperature to rise. The mixture is warmed up again to RT within 1 h and stirred for 5 h at this temperature. After evaporation of the solvent under reduced pressure by high-vacuum pump, 50 mL of pentane are added to the residue. The suspension is stirred for 1 h and filtered through a pressure filter funnel under Ar. To this solid residue were added 100 mL of toluene. The suspension was stirred for 10 h and filtered through a pressure filter funnel again. The filtrate is concentrated *in vacuo* to give a solid, which is purified as described for the individual compound. The ligands obtained are stored under Ar.

The general procedure is applied with 2.0 g (3 mmol) of **68b** and 824 mg (3.6 mmol) of (2-naphth)PCl_2. The dichloride is added as a solution of THF (2 mL). After the addition of (2-naphth)PCl_2, the mixture is stirred at RT for 15 h. Then, THF is removed under reduced pressure and 20 mL of toluene are added to the residue directly. The suspension is stirred for 3 h and filtered through a pressure filter funnel. The filtrate is concentrated *in vacuo*, and the residue is chromatographed (silica gel; pentane:Et_2O, 15:1) to give a pale yellow solid. To a solution of this solid in 20 mL of Et_2O is added 1.0 g of activated charcoal. The mixture is stirred for 3 min at RT and filtered off. The filtrate is evaporated *in vacuo*, and the residue is purified by silica-gel column chromatography (pentane:Et_2O, 15:l) again to 1.1 g (45%) of **69d** as a colorless solid.

M.p. 171–173.5°C (dec.). $[\alpha]_D = -133.3$; ^{1}H-NMR ($CDCl_3$400 MHz): 0.18 (s, Me); 1.68 (s, Me); 5.18 (*d*, *J*=8.54, CH); 6.01 (*dd*, *J*=8.54,4.84, CH); 7.40–8.81 (*m*, 35 arom. H). [13]C-NMR ($CDCl_3$, 100 MHz): 25.27; 28.03; 82.63; 82.87; 82.91; 83.66; 83.73; 83.96; 84.00; 111.94; 125.54; 125.70; 125.81; 125.93; 125.96; 126.00; 126.04; 126.08; 126.10; 126.13; 126.21; 126.36; 126.51; 127.15; 127.18; 127.28; 127.32; 127.37; 127.40; 127.49; 127.60; 127.88; 127.94; 128.19; 128.40; 128.45; 128.58; 128.67; 128.69; 128.78; 128.81; 131.34; 131.67; 132.50; 132.54; 132.63; 132.73; 132.78; 132.81; 132.84; 134.71; 138.39; 138.50; 138.73; 138.96; 142.94; 142.98; 143.73. [31]P-NMR ($CDCl_3$) 158.3. EI-MS: 822 (<1, M^+), 746 (<1), 631 (30), 630 (59), 572 (26), 545 (40), 544 (92), 366 (29), 365(91), 308 (36), 307 (89), 281 (27), 280 (43), 279 (84), 278 (48), 277 (23), 276 (30), 268 (28), 267 (93), 266 (71), 265 (100), 264 (27), 263 (33), 50 (23). HR-MS: $C_{57}H_{43}O_4P$ (M^+: calc.: 822.2899; found: 822.2707).

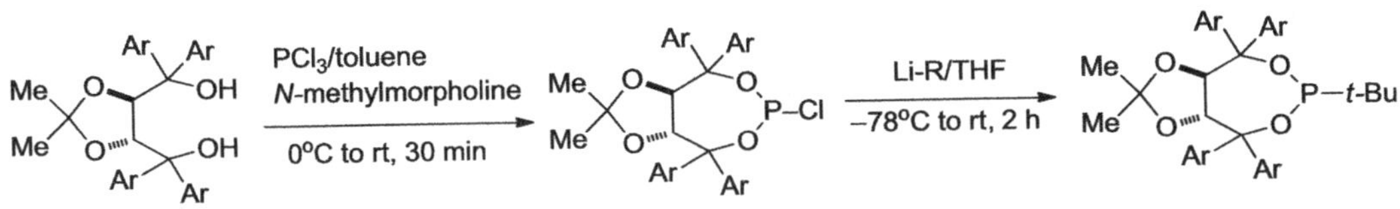

68b Ar = 2-Nap

69e Ar = 2-Nap

Scheme 5.49

5.5.2.2 *(1R,7R)-4-tert-Butyl-9,9-dimethyl-2,2,6,6-tetra(naphthalen-2-yl)-3,5,8,10-tetraoxa-4-phosphabicyclo-[5.3.0]decane* (**69e**) *(Scheme 5.49)* [137]

General procedure for the preparation of TADDOL-derived phosphoramidites and phosphonites. To an ice-cold solution of PCl_3 (1.0 equiv.) in toluene (0.4 M) is added a solution of TADDOL **68b** (1.0 equiv.) and freshly distilled *N*-methylmorpholine (2.0 equiv.) in toluene (0.2 M based on I). The resulting suspension is stirred at RT for 30 min, cooled to −78°C, and filtered under argon. The solvent of the filtrate is removed under reduced pressure, and the pale-yellow residue is dissolved in THF (0.3 M). The solution is cooled to −78°C and added dropwise to the lithium amide (1.0 equiv.) in THF (for phosphoramidites), or the appropriate organolithium solution (1.0 equiv.) is added to the dissolved crude intermediate (for phosphonites). After stirring for an additional 2 h at RT, the mixture was concentrated to a slurry and subjected directly to flash column chromatography on silica gel using cyclohexane–Et_3N mixtures as eluent. Ligands are obtained as white or pale-yellow solids.

(1R,7R)-4-tert-Butyl-9,9-dimethyl-2,2,6,6-tetra(naphthalen-2-yl)-3,5,8,10-tetraoxa-4-phosphabicyclo-[5.3.0]decane (**69e**). According to the general procedure, a solution of [(4*R*,5*R*)-2,2-dimethyl-1,3-dioxolan-4,5-diyl]bis(dinaphthalen-2-ylmethanol)(2.57 g, 3.85 mmol) and *N*-methylmorpholine (779 mg, 7.70 mmol) in toluene (20 mL) is added to an ice-cold solution of PCl_3 (529 mg, 3.85 mmol) in toluene (10 mL). The resulting suspension is stirred for 30 min at RT, cooled to −78°C, and filtered under argon. The solvents of the filtrate are removed under reduced pressure. The pale-yellow residue is dissolved in THF (25 mL) and *tert*-butyl lithium (1.9 M in n-pentane, 2.07 mL, 3.85 mmol) is added dropwise at −78°C. The solution is maintained at RT for 2 h and concentrated under reduced pressure. The residue is subjected to flash column chromatography on silica gel (cyclohexane/Et_3N 4:1), yielding the ligand **69e** as a pale-yellow solid (2.22 g, 78%). R_f=0.15(cyclohexane–*tert*-butylmethylether 97:3).

Mp 130°C (decomp.). ^{1}H NMR (300 MHz, C_6D_6): δ 0.18 (s, 3 H), 1.38 (d, J=12.8 Hz, 9 H), 1.63 (s, 3 H), 5.51 (d, J=8.6 Hz, 1 H), 6.50 (dd, J=8.6, 4.9 Hz, 1 H),7.05–7.74 (m, 20 H), 7.82–7.94 (m, 2 H), 8.08 (m, 2 H), 8.35 (m, 1 H),8.48 (s, 1 H), 8.58 (m, 1 H), 9.10 (m, 1 H) ppm. ^{13}C NMR (75 MHz,C_6D_6): δ 24.1 (d, J_{C-P}=16.4 Hz), 25.4, 28.2, 35.4 (d, J_{C-P}=13.2 Hz),82.8 (d, J_{C-P}=2.2 Hz), 83.0 (d, J_{C-P}=23.5 Hz), 83.9

(d, J_{C-P} =5.4 Hz), 84.9 (d, J_{C-P}=3.6 Hz), 112.0, 126.2, 126.3, 126.3, 126.4, 126.5, 126.5, 126.5, 126.6, 126.7, 127.1, 127.1, 127.7–128.5 (m), 128.9–129.3 (m), 133.1, 133.2, 133.3, 133.4, 133.6, 140.3 (m),144.4, 144.4 ppm. ^{31}P NMR (121 MHz, C_6D_6) δ 172.0 ppm. IR(ATR): 2926 (m), 1032 (m), 802 (s), 741 (s) cm^{-1}; HRMS (ESI) calculated for $C_{51}H_{45}O_4PH$ ([M+H]$^+$): 753.3131. Found: 753.3128. Anal. calculated for $C_{51}H_{45}O_4P$ (752.87): C, 81.36; H, 6.02. Found: C, 81.68; H, 5.99. $[\alpha]_D^{20}$ −95:2 (c 1.05, $CHCl_3$).

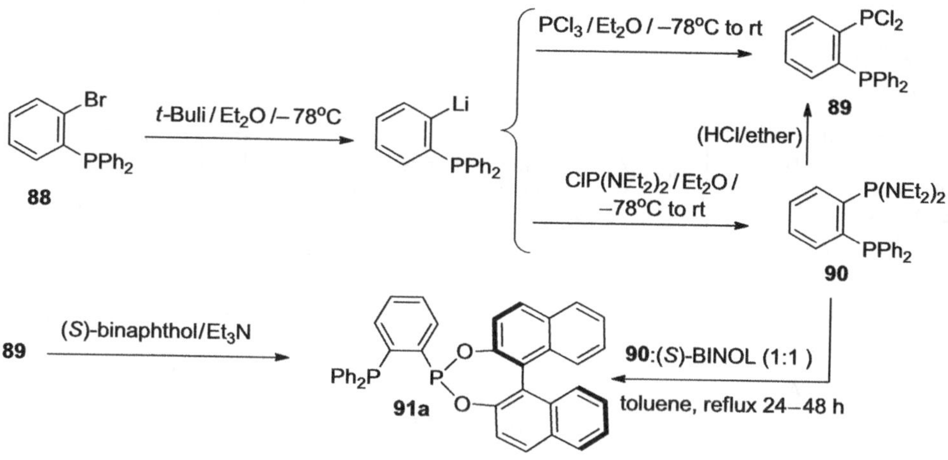

Scheme 5.50

5.5.2.3 *Synthesis of (S)-2-[2-(diphenylphosphino)phenyl]-1,3,2-dinaphtho[d^{1,2},f^{1,2}]dioxaphosphe-pine (91a) (Scheme 5.50)* [166]

A Schlenk flask is charged with 2-bromophenyldiphenylphosphine (**88**) (0.413 g, 1.14 mmol) [166b,169], ether (15 mL), and a stirrer bar and cooled to −78°C (2-propanol/CO_2). Then *t*-BuLi (1.5 mL, 2.3 mmol, 1.5 Min pentane) is added dropwise with stirring. The resulting suspension is stirred for 10 min and transferred slowly via catheter to a stirred solution of PCl_3 (1.0 mL, 11 mmol) in ether (20 mL) cooled to −78°C (2-propanol–CO_2). The cold bath is removed and the reaction mixture allowed to warm to RT. The resulting pale-yellow suspension is filtered, and the solvent is removed under oil-pump vacuum to give a yellow oil (**89**) which is taken up in CH_2Cl_2 (10 mL). Then NEt_3 (0.42 mL, 3.0 mmol) and (*S*)-binaphthol (0.143 g, 0.499 mmol) are added, and the solution is stirred overnight. The solvent is removed under oil-pump vacuum, and the residue was taken up in benzene (15 mL). The solution is chromatographed on a 12 mm diameter silica gel column (10 g, benzene). The solvent is removed from the eluate under oil-pump vacuum to give (*S*)-**91a** as a white foam (0.257 g, 90%). ^{1}H NMR ($CDCl_3$): δ 7.94 (d, *J*=8.8 Hz, 1 H), 7.87 (d, *J*=8.1 Hz, 1 H), 7.74 (d, *J*=8.1 Hz, 1 H), 7.56–6.95 (m, 22 H), 6.14 (d, *J*=8.8 Hz, 1 H). ^{31}P{^{1}H}NMR ($CDCl_3$): δ −19.1 (d, *J*=207 Hz, *P*Ar_2), 175.3 (d, *J*=207 Hz, *P*O_2). MS: *m/e* 577 (M+1).

This compound has also been prepared from the bis-diethylaminophosphine **90** [166].

5.5.2.4 *4,5-Bis{di[(2-tert-butyl)phenyl]phosphonito}-9,9-dimethylxanthene (115) (Scheme 5.51)* [187]

4,5-(Bis[bis-diethylamido]phosphonito)-9,9-dimethylxanthene (**B**). 9,9-Dimethylxanthene (3.00 g, 14.3 mmol) and TMEDA (4.14 g, 35.7 mmol) are dissolved in 50 mL of diethyl ether and cooled to −40°C. *n*-BuLi (14.3 mL of a 2.5 M solution in hexanes, 35.7 mmol) is added via a syringe over a period of 10 min.

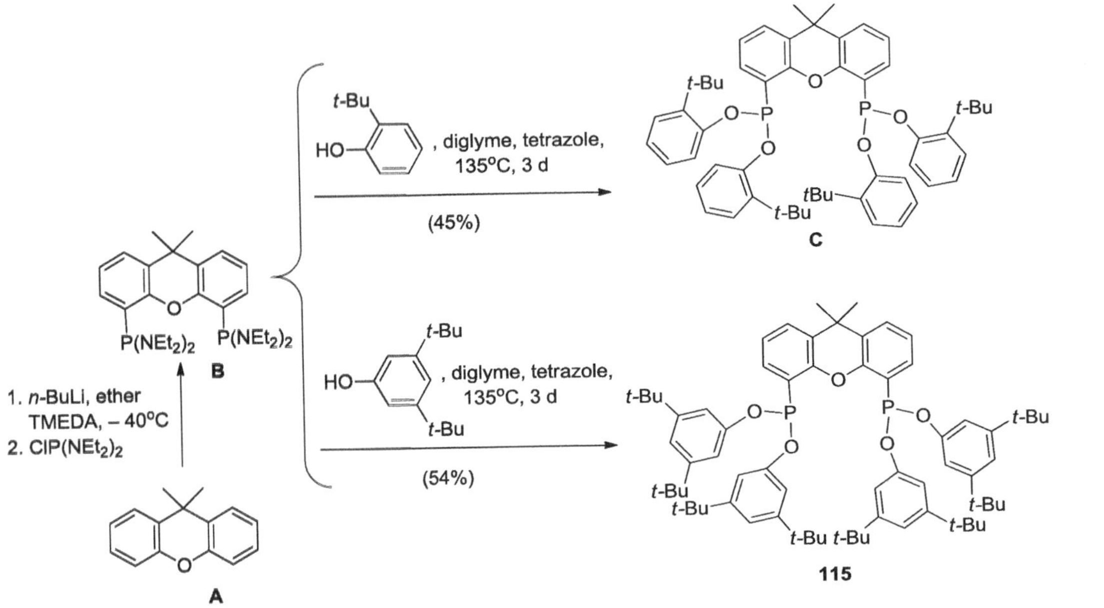

Scheme 5.51

The mixture is allowed to warm to up to RT and is stirred overnight. Subsequently, $ClP(NEt_2)_2$ (6.32 g, 30.0 mmol) is dissolved in 30 mL of pentane and cooled to –60°C. The dilithioxanthene solution is slowly added via a cannula. Subsequently, the reaction mixture is allowed to warm to RT and stirred overnight. Solvents are removed under vacuum to leave a yellow solid. 15 mL pentane is added, and the suspension filtered. Remaining TMEDA is removed by stripping three times with 10 mL of toluene. Upon cooling to –25°C, colorless crystals are obtained. Yield: 5.03 g (63%); ^{1}H NMR (300 MHz, C_6D_6): δ 7.60 (d, 2H, J = 7.5 Hz), 7.22 (d, 2H, J = 7.5 Hz), 7.05 (t, 2H, J = 7.5 Hz), 3.19 (m, 16H, CH_2), 1.55 (s, 6H, CH_3), 1.06 (t, 12H, CH_3, J = 7.1 Hz); ^{13}C NMR (75 MHz, C_6D_6): δ 151.0, 130.6, 130.1, 129.5, 126.0, 122.4, 43.3 (CH_2), 33.9 (CMe_2), 32.7 (CMe_2), 14.5 (CH_3); ^{31}P{^{1}H} NMR (162 MHz, C_6D_6): δ 92.3 (s); anal. calculated for $C_{31}H_{51}N_4OP_2$: C 66.6, H 9.4, N10.0; found: C 66.3, H 9.3, N 9.9.

4,5-Bis{di[(2-tert-butyl)phenyl]phosphonito}-9,9-dimethylxanthene (**C**). **B** (2.11 g, 3.58 mmol), 2-*tert*-butylphenol (2.15 g, 14.3 mmol), and 2 mol% of tetrazole are dissolved in 20 mL of di(2-methoxyethyl) ether. The reaction mixture is heated to 135°C for 3 days, while the diethylamine formed during reaction is removed under vacuum twice a day. Solvent is removed under vacuum to give a light yellow solid. It is dissolved in 10 mL of dichloromethane and after layering of the solution with 10 mL of ethanol a pure white powder precipitated; yield of **C**: 1.40 g (45%); ^{1}H NMR (400 MHz, C_6D_6): δ 8.15 (dd, 2H, J 7.2 Hz, J 1.2 Hz), 7.26 (dd, 4H, J 8 Hz, J 1.6 Hz), 7.17 (dd, 2H, J 8 Hz, J 1.6 Hz), 7.14 (d, 4H, J 8 Hz, 2 Hz), 7.0 (t, 2H, 1 7.6 Hz), 6.86 (t, 4H, J 8 Hz, J 2.4 Hz), 6.81 (t, 4H, J 7.6 Hz, J 1.6 Hz), 1.35 (s, 6H, CH_3), 1.32 (s, 36H, *t*-Bu); ^{13}C NMR (100 MHz, C_6D_6): δ 155.3, 139.4, 130.5, 129.0, 128.7, 127.4, 127.1, 124.1, 122.7, 119.2, 119.1, 118.9, 34.7, 31.4, 30.1; ^{31}P{^{1}H} NMR (162 MHz, C_6D_6): δ 147.8 (s); anal. calculated for $C_{55}H_{64}O_5P_2$: C 76.2, H 7.4; found: C 76.3, H 7.4. MS (Maldi-TOF): m/z (%) 899.45 (100) $[M+2O]^+$.

4,5-(Bis{di[(3,5-di-tert-butyl)phenyl]phosphonito}-9,9- dimethylxanthene (**115**). Starting from **B** (4.32 g, 7.73 mmol) and 3,5-bis(*tert*-butyl)phenol(6.35 g, 30.9 mmol), compound **115** is obtained as a white powder in analogy to the synthetic procedure described for ligand **C**. Yield: 4.57 g (54%); ^{1}H NMR(400 MHz, C_6D_6): δ 8.15 (dd, 2H, J 7.6 Hz, J 1.2 Hz), 7.32 (d, 8H, J 1.6 Hz), 7.15 (t, 4H, J 1.6 Hz), 7.07 (dd, 6H, J 7.6 Hz, J 1.6 Hz), 6.92 (t, 2H, J 7.6 Hz), 1.25 (s, 6H, CH_3), 1.14 (s, 72H, *t*-Bu); ^{13}C NMR (100 MHz,

C_6D_6): d 155.8, 152.2, 130.0, 129.5, 128.5, 128.3, 117.3, 115.5, 115.4, 34.8, 31.8, 31.4; $^{31}P\{^1H\}$ NMR(162 MHz, C_6D_6): δ 148.4 (s); anal. calculated for $C_{71}H_{96}O_5P_2$: C78.1, H 8.9; found: C 78.0, H 8.6; MS (Maldi-TOF): m/z (%) 1123.72 (100) $[M+2O]^+$.

5.6 Acknowledgments

I am grateful for a highly fruitful collaboration with Drs. Albert Casalnuovo and Timothy Ayers during the early stages of our exploration of chelating phosphinites as highly tunable ligands for asymmetric hydrocyanation and hydrogenation reactions. Since then, many talented students and postdoctoral fellows in my group have made significant contributions to our research in this and related areas. These include Drs. Branko Radetich, Haengsoon Park, Nobuyoshi Nomura, and Kumareswaran Ramaiah. Support for our recent research at the Ohio State University comes primarily from the US National Science Foundation (CHE-1057818) and the US National Institutes of Health (R01 GM075107).

Abbreviations

acac	acetylacetonate
ADN	adiponitrile
AMPP	aminophosphine–phosphinite (ligand)
BARF	*tetrakis*[3,5-bis-(trifluoromethyl)phenyl]borate
BD	1,3-butadiene
BINAP	2,2-bis-(diphenylphosphino)-1,1-binaphthyl
BINAPO	2,2′-bis-oxy-1,1-binaphthyl-
BINOL	2,2′-dihydroxy-1,1′-binaphthyl
BMI	1-butyl-3-methylimidazolium (ionic liquid component)
cat.	In catalytic amounts
C_2H_4	ethylene
C_6H_6	benzene
$C_{10}H_7$	naphthyl
COD	1,5-cyclooctadiene
CpH	cyclopentadiene
Cyhex	cyclohexyl
DUPHOS	1,2-(1-(2,5-dialkylphospholano)-benzene
dba	dibenylideneacetone
dppb	1,4-bis-diphenylphosphinobutane
dr	diastereomeric ratio
ee	enantiomeric excess
EI	electron impact (in mass spectrometry)
equiv.	equivalent
Eq.	equation
HR-MS	high-resolution mass spectrometry
Im	imidazole
LA	Lewis acid
2M3BN	2-methyl-3-butenenitrile
MCPBA	*m*-chloroperbenzoic acid
Mes	mesityl, 2,4,6-trimethylphenyl

MVN	6-methoxy-2-vinylnaphthalene
NBD	norbornadiene
oxaz-	oxazoline
PN	pentenenitrile
Py	pyridine
rac.	racemic
RT	room temperature
S/C	substrate:catalyst molar ratio
TADDOL	(*R*,*R*)- or (*S*,S)-α,α,α',α'-tetraaryl-1,3-dioxalane-4,5-methanol
TBDPS	*tert*-butyldiphenylsilyl
Tf or tf	trifluromethylsulfonyl
TFAA	trifluoroacetic anhydride
THF	tetrahydrofuran
TMEDA	tetramethylethylenediamine
TMS	tetramethylsilane
TMSX	trimethylsilyl halide
TOF	turnover frequency (in catalysis)
TOF	time of flight (in mass spectrometry)
tol	tolyl or toluene
VN	vinylnaphthalene
y.	yield

References

[1] O. Stelzer, Transition metal complexes of phosphorus ligands, in *Topics in Phosphorus Chemistry*, Vol. 9, E. J. Griffith, M. Grayson (Eds.), John Wiley, **1977**, 1–229.

[2] R. Mason, D. W. Meek, Versatility of tertiary phosphine ligands in coordination and organometallic chemistry, *Angew. Chem. Int. Ed. Engl.* **1978**, *17*, 183.

[3] C. A. McAuliffe, W. Levason, *Studies in Inorganic Chemistry. Phosphine, Arsine and Stibene Complexes of the Tranistion Elements*, Vol. 1, Elsevier, **1979**, 1–546 (a review with over 2700 references containing mainly literature between 1971 and 1978). For earlier work, see K. K. Chow, W. Levason, C. A. McAuliffe, Transition metal complexes containing phosphine ligands, in *Transition-Metal Complexes of Phosphorus, Arsenic and Antimony Donor Ligands*, C. A. McAuliffe (Ed.), McMillan, **1973**.

[4] C. A. Tolman, Phosphorus ligand exchange equilibria on zero valent nickel: a dominant role for steric effects, *J. Am. Chem. Soc.* **1970**, *92*, 2956.

[5] C. A. Tolman, Steric effects of phosphorus ligands in organometallic chemistry and homogeneous catalysis, *Chem. Rev.* **1977**, *77*, 313.

[6] D. White, N. J. Coville, Quantification of steric effects in organometallic chemistry, *Adv. Organomet. Chem.* **1994**, *36*, 95.

[7] For recent reviews on the synthesis and applications of different types of ligands, see (a) A. Börner (Ed.), *Phosphorus Ligands in Asymmetric Catalysis*, Vols. 1–3, Wiley-VCH, **2008**; (b)H. Fernandez-Perez, P. Etayo, A. Panossian, A. Vidal-Ferran, Phosphine-phosphinite and phosphine-phosphite ligands: preparation and applications in asymmetric catalysis, *Chem. Rev.* **2011**, *111*, 2119; (c) S. Woodward, M. Diéguez, O. Pamies, Use of sugar-based ligands in selective catalysis: recent developments, *Coord. Chem. Rev.* **2010**, *254*, 2007; (d) M. T. Reetz, Combinatorial transition-metal catalysis: mixing monodentate ligands to control enantio-, diastereo-, and regioselectivity, *Angew. Chem. Int. Ed.* **2008**, *47*, 2556; (e) M. Diéguez, O. Pamies, C. Claver, Ligands derived from carbohydrates for asymmetric catalysis, *Chem. Rev.* **2004**, *104*, 3189; (f) F. Agbossou-Niedercorn, I. Suisse, Chiral aminophosphine-phosphinite ligands and related auxiliaries, *Coord. Chem. Rev.* **2003**, *242*, 145; (g) F. Agbossou, J.-F. Carpentier, F. Hapiot, I. Suisse, A. Mortreux, The aminophosphine-phosphinites and related ligands: synthesis, coordination chemistry and enantioselective catalysis, *Coord. Chem. Rev.* **1998**, *178–180*, 1615.

[8] D. A. Couch, S. D. Robinson, J. N. Wingfield, Transition-metal complexes containing phosphorus ligands. XIII. Dialkyl phenylphosphonite and alkyl diphenylphosphinite derivatives of silver, gold, and platinum metals, *J. Chem. Soc., Dalton Trans.* **1974**, 1309.

[9] D. A. Couch, S. D. Robinson, Transition metal complexes containing phosphorus ligands. XII. Cationic dialkyl phenylphosphonite and alkyl diphenylphosphinite derivatives of copper, silver, gold, and the platinum metals, *Inorg. Chem.* **1974**, *13*, 456, and references cited therein. This paper also lists the contributions of earlier workers, primarily Gray, Muetterties, and Haines.

[10] C. J. Creswell, S. D. Robinson, A. Sahajpal, Complexes containing phosphorus ligands. XVI. New phosphinite, phosphonite and phosphite derivatives of ruthenium, osmium and iridium, *Polyhedron* **1983**, *2*, 517.

[11] I. Joedicke, H. V. Studer, J. T. Yoke, Four-coordinate high-spin and five-coordinate low-spin complexes of cobalt(II) chloride with ethyl phosphite, phosphonite, and phosphinite ligands: refinement of the Evans nuclear magnetic resonance method for the study of spin equilibria, *Inorg. Chem.* **1976**, *15*, 1352.

[12] E. J. Lukosius, K. J. Coskran, Four- vs. five-coordination of nickel(II) with certain tertiary phosphorus ligands, *Inorg. Chem.* **1975**, *14*, 1922.

[13] R. Rabinowitz, J. Pellon, Phosphorus-containing monomers 1. Synthesis of vinyl phosphines, oxides, sulfides, and phosphonium compounds, *J. Org. Chem.* **1961**, *26*, 4623.

[14] L. Maier, Organische Phosphorverbindungen. IX. Über Einige Reaktionen des Methyldibromphosphins [CH_3PBr_2], *Helv. Chim. Acta* **1963**, *46*, **2667**.

[15] M. Tanaka, I. Ogata, Asymmetric hydrogenation by chiral diphosphinite rhodium complexes, *J. Chem. Soc., Chem. Commun.* **1975**, 735.

[16] Y. Kawabata, M. Tanaka, I. Ogata, Asymmetric hydrogenation by a rhodium catalyst complexed with a phosphinite derived from cellulose, *Chem. Lett.* **1976**, 1213.

[17] W. R. Cullen, Y. Sugi, Asymmetric hydrogenation catalyzed by diphosphinite Rh complexes derived from a sugar, *Tetrahedron Lett.* **1978**, 1635.

[18] R. Selke, Phosphinites of carbohydrates as chiral ligands for asymmetric synthesis catalyzed by complexes, I. Hydrogenation of aminoacid precursors by rhodium(I) chelates of methyl 4,6-*O*-benzylidene-2,3-bis-*O*-(diphenylphosphino)-*a*-*D*-glucopyranoside, *React. Kinet. Catal. Lett.* **1979**, *10*, 135.

[19] R. Jackson, D. J. Thompson, Asymmetric hydrogenation using readily prepared chiral sugardiphosphinite-rhodium complexes, *J. Organomet. Chem.* **1978**, *159*, C29.

[20] T. H. Johnson, G. Rangarajan, Chiral phosphinites from *a*-*D*-glucofuranose and α-*L*-idofuranose: a comparison to chiral phosphines in asymmetric hydrogenation, *J. Org. Chem.* **1980**, *45*, 62. For the synthesis of bis-diarylphosphinites derived from tartaric acid derivatives, see J. Bourson, L. Oliveros, Asymmetric hydrogenations catalyzed by diphosphinite rhodium complexes derived from natural tartaric acid, *J. Organomet. Chem.* **1982**, *229*, 77.

[21] T. V. RajanBabu, A. L. Casalnuovo, Tailored ligands for asymmetric catalysis: the hydrocyanation of vinylarenes, *J. Am. Chem. Soc.* **1992**, *114*, 6265.

[22] A. L. Casalnuovo, T. V. RajanBabu, T. A. Ayers, T. H. Warren, Ligand electronic effects in asymmetric catalysis: enhanced enantioselectivity in asymmetric hydrocyanation of vinylarenes, *J. Am. Chem. Soc.* **1994**, *116*, 9869.

[23] T. V. RajanBabu, A. L. Casalnuovo, Role of electronic asymmetry in the design of new ligands: the asymmetric hydrocyanation reaction, *J. Am. Chem. Soc.* **1996**, *118*, 6325.

[24] B. Saha, T. V. RajanBabu, Nickel(0)-catalyzed asymmetric hydrocyanation of 1,3-dienes, *Org. Lett.* **2006**, *8*, 4657.

[25] T. V. RajanBabu, T. A. Ayers, A. L. Casalnuovo, Electronic amplification of selectivity in Rh-catalyzed hydrogenations: *D*-glucose-derived ligands for the synthesis of *D*- or *L*-amino acids, *J. Am. Chem. Soc.* **1994**, *116*, 4101.

[26] T. V. RajanBabu, T. A. Ayers, G. A. Halliday, K. K. You, J. C. Calabrese, Carbohydrate phosphinites as practical ligands in asymmetric catalysis: electronic effects and dependence of backbone chirality in Rh-catalyzed asymmetric hydrogenations: synthesis of *R*- or *S*-amino acids using natural sugars as ligand precursors, *J. Org. Chem.* **1997**, *62*, 6012.

[27] T. V. RajanBabu, B. Radetich, K. K. You, T. A. Ayers, A. L. Casalnuovo, J. C. Calabrese, Electronic effects in asymmetric catalysis: structural studies of precatalysts and intermediates in Rh-catalyzed hydrogenation of dimethyl itaconate and acetamidocinnamic acid derivatives using C_2-symmetric diarylphosphinite ligands, *J. Org. Chem.* **1999**, *64*, 3429.

[28] N. Nomura, Y. C. Mermet-Bouvier, T. V. RajanBabu, Unprecedented electronic and steric effects in palladium-catalyzed asymmetric allylation: switching of enantioselectivity with a single chiral backbone, *Synlett.* **1996**, 745.

[29] D. S. Clyne, Y. C. Mermet-Bouvier, N. Nomura, T. V. RajanBabu, Substituent effects on ligands on asymmetric induction in a prototypical palladium-catalyzed allylation reaction: making both enantiomers of a product in high optical purity using the same source of chirality, *J. Org. Chem.* **1999**, *64*, 7601.

[30] T. A. Ayers, T. V. RajanBabu, A. L. Casalnuovo, Sugar diphosphinites as ligand for allylation reactions: quasi-enantiomeric relationship of ligands derived from *D*-glucose, *Proceedings of Organometallic Chemistry Directed toward Organic Synthesis (OMCOS 8)* **1995**, Santa Barbara, 89.

[31] H. Park, T. V. RajanBabu, Tunable ligands for asymmetric catalysis: readily available carbohydrate-derived diarylphosphinites induce high selectivity in the asymmetric hydrovinylation of styrenes, *J. Am. Chem. Soc.* **2002**, *124*, 734.

[32] H. Park, R. Kumareswaran, T. V. RajanBabu, Tunable phosphinite, phosphite and phosphoramidite ligands for the asymmetric hydrovinylation reactions, *Tetrahedron* **2005**, *61*, 6352.

[33] G. M. Kosolapoff, R. M. Watson, The anhydrides of dimethyl- and diethylphosphinic acids, *J. Am. Chem. Soc.* **1951**, *73*, 5466.

[34] R. H. Williams, L. A. Hamilton, Di-normal alkylphosphine oxides. 1. Synthesis, *J. Am. Chem. Soc.* **1952**, *74*, 5418.

[35] B. B. Hunt, B. C. Saunders, Esters containing phosphorus. Part XV. Preparation and reactions of diphenylphosphine oxide, *J. Chem. Soc.* **1957**, 2413.

[36] L. D. Quin, H. G. Anderson, Some new reactions of functions containing secondary phosphorus, *J. Org. Chem.* **1966**, *31*, 1206.

[37] J. D. Unruh, J. R. Christenson, A study of the mechanism of rhodium-phosphine-catalyzed hydroformylation: use of 1,1′-bis(diarylphosphino)ferrocene ligands, *J. Mol. Catal.* **1982**, *14*, 19.

[38] R. Selke, H. Pracejus, Phosphinites of carbohydrates as chiral ligands for asymmetric synthesis catalyzed by complexes part II. Superiority of phenyl 4,6-O-(*R*)-benzylidene-2,3-O-bis(diphenylphoshino)-beta-D-glucopyranoside for rhodium(I)-hydrogenation of amino acid precursors, *J. Mol. Catal.* **1986**, *37*, 213.

[39] Few examples of electronic tuning of asymmetric catalysts had been reported in the literature at the time this work was originally published: (a) E. N. Jacobsen, W. Zhang, M. L. Guler, Electronic tuning of asymmetric catalysis, *J. Am. Chem. Soc.* **1991**, *113*, 6703;(b) T. V. RajanBabu, A. L. Casalnuovo, Tailored ligands for asymmetric catalysis: the hydrocyanation of vinylarenes, *J. Am. Chem. Soc.* **1992**, *114*, 6265;(c) K. Inoguchi, S. Sakuraba, K. Achiwa, Design concepts for highly efficient chiral biphosphine ligands, *Synlett.* **1992**, 169; and (d) H. Nishiyama, S. Yamaguchi, M. Kondo, K. Itoh, Electronic substituent effect of nitrogen ligands in catalytic asymmetric hydrosilylation of ketones: chiral 4-substituted bis(oxazolinyl)pyridines, *J. Org. Chem.* **1992**, *57*, 4306.

[40] N. Khiar, R. Navas, B. Suarez, E. Alvarez, I. Fernadez, Asymmetric enamide hydrogenation using phosphinite thioglycosides: synthesis of D- and L-aminoesters using D-sugars as catalyst precursors, *Org. Lett.* **2008**, *10*, 3697.

[41] M. Yamashita, M. Kobayashi, M. Sugiura, K. Tsunekawa, T. Oshikawa, S. Inokawa, H. Yamamoto, Homogeneous asymmetric hydrogenation using a chiral phosphinite derivative of carbohydrates as ligand, *Bull. Chem. Soc., Jpn.* **1986**, *59*, 175.

[42] M. Yamashita, M. Naoi, H. Imoto, T. Oshikawa, Asymmetric hydrogenation using diphenylphosphinite derivatives of carbohydrates as chiral ligands of rhodium catalysts, *Bull. Chem. Soc., Jpn.* **1989**, *62*, 942.

[43] T. V. RajanBabu, A. L. Casalnuovo, T. A. Ayers, Ligand tuning in asymmetric catalysis: hydrocyanation and hydrogenation reactions, in *Advances in Catalytic Processes*, M. Doyle (Ed.), Vol. 2, JAI Press, **1998**, 1.

[44] T. V. RajanBabu, T. A. Ayers, Electronic effects in asymmetric catalysis: hydroformylation of olefins, *Tetrahedron Lett.* **1994**, *35*, 4295.

[45] T. V. RajanBabu, Asymmetric hydrovinylation, *Chem. Rev.* **2003**, *103*, 2845.

[46] T. V. RajanBabu, In pursuit of an ideal carbon-carbon bond-forming reaction: development and applications of the hydrovinylation of olefins, *Synlett.* **2009**, 853.

[47] R. Bayersdörfer, B. Ganter, U. Englert, W. Keim, D. Vogt, Asymmetric hydrovinylation of styrene applying cationic palladium complexes of a P-chiral ligand, *J. Organomet. Chem.* **1998**, *552*, 187.

[48] H. Brunner, W. Pieronczyk, Asymmetric catalyses. Part III. Asymmetric hydrogenation of (*Z*)-α-acetamidocinnamic acid with phosphorus derivatives of sugars, tartaric acid, menthol, and 1-phenylethylamine as ligands in rhodium catalysts, *J. Chem. Res. (S)* **1980**, 76.

[49] I. Habus, Z. Raza, V. Sunjic, Chiral bidentate ligands from the most widespread monosaccharides for enantioselective hydrogenation: diphenylphosphinites from *D*-glucose and *D*-xylose, *J. Mol. Catal.* **1987**, *42*, 173.

[50] M. Diéguez, O. Pamies, C. Claver, Ligands derived from carbohydrates for asymmetric catalysis, *Chem. Rev.* **2004**, *104*, 3189.

[51] C. Claver, O. Pamies, M. Diéguez, Chiral diphosphinites, diphosphonites, diphosphites and mixed P,O-ligands, in *Phosphorus Ligands in Asymmetric Catalysis*, Vol. 2, A. Börner (Ed.), Wiley-VCH, **2008**, 506.

[52] S. Woodward, M. Diéguez, O. Pamies, Use of sugar-based ligands in selective catalysis: recent developments, *Coord. Chem. Rev.* **2010**, *254*, 2007.

[53] A. Kumar, G. Oehme, J. P. Roque, M. Schwarze, R. Selke, Increase in enantioselectivity of asymmetric hydrogenation in water influenced by surfactants or polymerized micelles, *Angew. Chem. Int. Ed. Engl.* **1994**, *33*, 2197.

[54] S. Shin, T. V. RajanBabu, Water-soluble organometallic catalysts from carbohydrates 1. diphosphinite-Rh complexes, *Org. Lett.* **1999**, *1*, 1229.

[55] K. Yonehara, K. Ohe, S. Uemura, Highly enantioselective hydrogenation of enamides and itaconic acid in water in the presence of water-soluble rhodium (I) catalyst and dodecyl sulfate, *J. Org. Chem.* **1999**, *64*, 9381.

[56] K. Yonehara, T. Hashizume, K. Ohe, S. Uemura, Novel water-soluble bisphosphinite chiral ligands derived from α,α- and β,β-trehalose: application to asymmetric hydrogenation of dehydroamino acids and their esters in water or an aqueous/organic biphasic medium, *J. Org. Chem.* **1999**, *64*, 5593.

[57] Y. Y. Yan, T. V. RajanBabu, Water-soluble organometallic catalysts from carbohydrates 2: a strategy for the preparation of catalysts with pendant quaternary ammonium groups using *D*-salicin, *J. Org. Chem.* **2001**, *66*, 3277.

[58] I. Grassert, E. Paetzold, G. Oehme, Influence of different types of amphiphiles on rhodium (1) complex-catalyzed asymmetric hydrogenation of (Z)-methyl-alpha-acetamidocinnamate in aqueous medium, *Tetrahedron* **1993**, *49*, 6605.

[59] R. Selke, M. Ohff, A. Riepe, Asymmetric hydrogenation – influence of the structure of carbohydrate derived catalysts on the relative enantioselectivity $Q_{H/Me}$ regarding acid and ester substrates and its inversion – selectivity increase in water by amphiphiles, *Tetrahedron* **1996**, *52*, 15079.

[60] T. Hashizume, K. Yonehara, K. Ohe, S. Uemura, A novel amphiphilic chiral ligand derived from *D*-glucosamine: application to palladium-catalyzed asymmetric allylic substitution reaction in an aqueous or an organic medium, allowing for catalyst recycling, *J. Org. Chem.* **2000**, *65*, 5197.

[61] M. Yamashita, K. Hiramatsu, M. Yamada, N. Suzuki, S. Inokawa, Synthesis of sugar derivatives of tervalent phosphorus compounds and their application to homogeneous asymmetric hydrogenation, *Bull. Chem. Soc., Jpn.* **1982**, *55*, 2917.

[62] K. Nozaki, W.-G. Li, T. Horiuchi, H. Takaya, T. Saito, A. Yoshida, K. Matsumura, Y. Kato, T. Imai, T. Miura, H. Kumobayashi, A new stereocontrolled approach to 1-β-methylcarbapenem: asymmetric hydroformylation of 4-vinyl β-lactams catalyzed by Rh(I) complexes of chiral phosphine-phosphites and phosphine-phosphinites, *J. Org. Chem.* **1996**, *61*, 7658.

[63] K. Ishihara, H. Kurihara, M. Matsumoto, H. Yamamoto, Design of Brönsted acid-assisted chiral Lewis acid (BLA) catalysts for highly enantioselective Diels-Alder reactions, *J. Am. Chem. Soc.* **1998**, *120*, 6920.

[64] B. Saha, T. V. RajanBabu, Syntheses and applications of 2-phosphino-2'-alkoxy-1,1′-binaphthyl ligands. Development of a working model for asymmetric induction in hydrovinylation reactions, *J. Org. Chem.* **2007**, *72*, 2357.

[65] Y. Yan, Y. Chi, X. Zhang, Novel phosphine-phosphite and phosphine-phosphinite ligands for highly enantioselective asymmetric hydrogenation, *Tetrahedron: Asymmetry* **2004**, *15*, 2173.

[66] X. Jia, X. Li, W. S. Lam, S. H. L. Kok, L. Xu, G. Lu, C.-H. Yeung, A. S. C. Chan, The synthesis of new chiral phosphine-phosphinites, phosphine-phosphoramidite, and phosphine-phosphite ligands and their applications in asymmetric hydrogenation, *Tetrahedron: Asymmetry* **2004**, *15*, 2273.

[67] N. W. Boaz, E. B. Mackenzie, S. D. Debenham, S. E. Large, J. A. Ponasik, Synthesis and application of phosphinoferrocenylaminophosphine ligands for asymmetric catalysis, *J. Org. Chem.* **2005**, *70*, 1872.

[68] D. Marquard, H. Klusacek, G. Gokel, P. Hoffmann, I. Ugi, Stereoselective syntheses 5. Correlation of central and planar chirality in ferrocene derivatives, *J. Am. Chem. Soc.* **1970**, *92*, 5389.

[69] M. Fiorini, G. M. Giongo, F. Marcati, W. Marconi, Asymmetric hydrogenation by chiral aminophosphine-rhodium complexes, *J. Mol. Catal.* **1976**, *1*, 451.

[70] G. Pracejus, H. Pracejus, Chiral amino and diamino phosphines as ligands for asymmetric hydrogenation catalysts, *Tetrahedron Lett.* **1977**, *18*, 3497.

[71] K. Hanaki, K. Kashiwabara, J. Fujita, Asymmetric hydrogenation of α-acylaminoacrylic acids by rhodium(I) complex of (*1R, 2R*)-*bis*(*N*-diphenylphosphinomethylamino)cyclohexane, *Chem. Lett.* **1978**, 489. For other related

results, see (b) K. Onuma, T. Ito, A. Nakamura, Asymmetric hydrogenation of olefins with rhodium complexes of (1*S*,2*S*)-1,2-bis(diphenylphosphino)cyclohexane and (2*S*,3*S*)-2,3-bis(diphenylphosphinamino)butane, *Bull. Chem. Soc., Jpn.* **1980**, *53*, 2012; and (c) K. Onuma, T. Ito, A. Nakamura, Asymmetric hydrogenation with chiral aminophosphine-rhodium complexes and chiral recognition by bisphosphine-rhodium complexes in the asymmetric hydrogenation of olefins through the chiral helical conformation of phenyl groups on the phosphorus atom, *Bull. Chem. Soc., Jpn.* **1980**, *53*, 2016.

[72] (a) S. Miyano, M. Nawa A., Mori, H. Hashimoto, Axially disymmetric bis-aminophosphines derived from 2,2′-diamino-1,1-biphenyl: synthesis and application to rhodium(i)-catalyzed asymmetric hydrogenations, *Bull. Chem. Soc. Jpn.* **1984**, *57*, 2171; and (b) F.-Y. Zhang, C.-C. Pai, A. S. C. Chan, Asymmetric synthesis of chiral amine derivatives through enantioselective hydrogenation with a highly effective rhodium catalyst containing a chiral bisaminophosphine ligand, *J. Am. Chem. Soc.* **1998**, *120*, 5808.

[73] X. Sun, W. Li, L. Zhou, X. Zhang, Enantioselective hydrogenation of α-dehydroamino acid esters catalyzed by rhodium complexes with chiral bisaminophosphine ligands, *Adv. Synth. Catal.* **2010**, *352*, 1150.

[74] F. Agbossou, J.-F. Carpentier, F. Hapiot, I. Suisse, A. Mortreux, The aminophosphine-phosphinites and related ligands: synthesis, coordination chemistry and enantioselective catalysis, *Coord. Chem. Rev.* **1998**, *178–180*, 1615.

[75] O. Pardigon, A. Tenaglia, G. Buono, Enantioselective syntheses of monofunctionalized deltacyclenes using a [CoI_2/Zn] catalytic system, *J. Org. Chem.* **1995**, *60*, 1868.

[76] F. Agbossou-Niedercorn, Chiral aminophosphine-phosphinites (AMPP), in *Phosphorus Ligands in Asymmetric Catalysis*, Vol. 2, A. Börner (Ed.), Wiley-VCH, **2008**, 477.

[77] E. Cesarotti, A. Chiesa, G. D'Alfonso, Asymmetric hydrogenation by chiral aminophosphine-phosphinite rhodium complexes, *Tetrahedron Lett.* **1982**, *23*, 2995.

[78] M. Petit, A. Mortreux, F. Petit, G. Buono, G. Peiffer, Preparation of chiral aminophosphine-phosphinites. applications in homegeneous asymmetric catalysis, *Nou. J. Chim.* **1983**, *7*, 593.

[79] G. Pracejus, H. Pracejus, Chiral *O,N-bis*(diphenylphosphino)aminoalkanols as ligands for enantioselective metal complex hydrogenation catalysts, *J. Mol. Catal.* **1984**, *24*, 227.

[80] A. Karim, A. Mortreux, F. Petit, G. Buono, G. Peiffer, C. Siv, Synthese d'aminophosphinephosphinites chiraux. Utilisation en reduction asymetrique catalytique, *J. Organomet. Chem.* **1986**, *317*, 93.

[81] E. Cesarotti, A. Chiesa, L. Prati, L. Colombo, Asymmetric hydrogenation catalyzed by aminophosphine-phosphinite rhodium complexes derived from natural amino alcohols, *Gazz. Chim. Ital.* **1987**, *117*, 129.

[82] H. Pracejus, G. Pracejus, B. Costisella, *O,N*-Bis(diphenylphosphino) derivatives of chiral *trans*- and *cis*-2-aminocyclohexanols: synthesis and enantioselective behavior as ligands in Rh-based homogeneous hydrogenation catalysts, *J. Prakt. Chem.* **1987**, *329*, 235.

[83] P. Bergamini, V. Bertolasi, M. Cattabriga, V. Ferretti, U. Loprieno, N. Mantovani, L. Marvelli, Template synthesis of chiral vicinal diphosphinites as their Pt(II) and Pd(II) complexes, *Eur. J. Inorg. Chem.* **2003**, *2003*, 918.

[84] P. Bergamini, V. Bertolasi, F. Milani, The construction of P-O/P-N ligands on platinum and palladium, *Eur. J. Inorg. Chem.* **2004,** 1277.

[85] R. K. Sharma, A. G. Samuelson, Asymmetric allylation of aldehydes with chiral platinum phosphinite complexes, *Tetrahedron: Asymmetry* **2007**, *18*, 2387.

[86] G. Buono, C. Siv, G. Peiffer, C. Triantaphylides, P. Denis, A. Mortreux, F. Petit, A. Threophos, A new chiral aminophosphine phosphinite (AMPP) ligand highly efficient in asymmetric hydrovinylation of cyclohexa-1,3-diene catalyzed by nickel complexes, *J. Org. Chem.* **1985**, *50*, 1781.

[87] L. Z. Gong, G. S. Chen, A. Q. Mi, Y. Z. Jiang, F. M. Fu, X. Cui, A. S. C. Chan, Chiral aminophosphine-phosphinite-palladium catalysts in asymmetric allylic alkylations, *Tetrahedron-Asymmetry* **2000**, *11*, 4297.

[88] S. Naieli, J.-F. Carpentier, F. Agbossou, A. Mortreux, G. Nowogrocki, J.-P. Wignacourt, Platinum(II) aminophosphine- and amidophosphine-phosphinite complexes: synthesis, structure, and use in catalytic asymmetric hydroformylation of styrene. crystal structure of [(*s*)-1-(diphenylphosphino)-2-(((diphenylphosphino)oxy)methyl) pyrrolidine]dichloroplatinum(II), *Organometallics* **1995**, *14*, 401.

[89] Y. Pottier, A. Mortreux, F. Petit, Utilization of aminophosphine-phosphinite ligands in hydroformylation catalysis using $RhH(CO)_2$**L***, *J. Organomet. Chem.* **1989**, *370*, 333.

[90] E. A. Broger, W. Burkart, M. Hennig, M. Scalone, R. Schmid, New amidophosphine-phosphinites (tLANOPs) as chiral ligands for asymmetric hydrogenation reactions, *Tetrahedron: Asymmetry* **1998**, *9*, 4043.

[91] A. Roucoux, L. Thieffry, J.-F. Carpentier, M. Devocelle, C. Méliet, F. Agbossou, A. Mortreux, A. J. Welch, Amidophosphine-phosphinites: synthesis and use in rhodium-based asymmetric hydrogenation of activated keto compounds: crystal structure of bis[μ-chloro)(*S*)-2-((diphenylphosphino)oxy)-2-pheny 1-*N*-(diphenylphosphino)-*N*-methylacetamide)rhodium(I)], *Organometallics* **1996**, *15*, 2440.

[92] C. Pasquier, Enantioselective hydrogenation of functionalized ketones: synthesis and application of new chiral aminophosphine-phosphinite ligands, *Synlett.* **1998**, *1998*, 1162.

[93] M. Devocelle, A. Mortreux, F. Agbossou, J.-R. Dormoy, Alternative synthesis of the chiral atypical beta-adrenergic phenylethanolaminotetraline agonist SR58611A using enantioselective hydrogenation, *Tetrahedron Lett.* **1999**, *40*, 4551.

[94] H. J. Kreuzfeld, C. Döbler, U. Schmidt, H. W. Krause, Synthesis of non-proteinogenic (D)- or (L)-aminoacids by asymmetric hydrogenation, *Amino Acids* **1996**, *11*, 269, and references cited therein.

[95] A. S. C. Chan, W. H. Hu, C. C. Pai, C. P. Lau, Y. Z. Jiang, A. Q. Mi, M. Yan, J. Sun, R. L. Lou, J. G. Deng, Novel spiro-phosphinite ligands and their application in homogeneous catalytic hydrogenation reactions, *J. Am. Chem. Soc.* **1997**, *119*, 9570. For reduction of 1-aryletheneamide (86–97% ee with 1 mol% catalyst), see W. H. Hu, M. Yan, C. P. Lau, S. M. Yang, A. S. C. Chan, Y. Z. Jiang, A. Q. Mi, A highly effective rhodium spirocyclic phosphinite catalyst for the asymmetric hydrogenation of enamides, *Tetrahedron Lett.* **1999**, *40*, 973.

[96] R. H. Grubbs, R. A. DeVries, Asymmetric hydrogenation by an atropisomeric diphosphinite rhodium complex, *Tetrahedron Lett.* **1977**, *18*, 1879.

[97] R. Guo, T. T.-L. Au-Yeung, J. Wu, M. C. K. Choi, A. S. C. Chan, Modified BINAPO ligands for Rh-catalysed enantioselective hydrogenation of acetamidoacrylic acids and esters, *Tetrahedron: Asymmetry* **2002**, *13*, 2519.

[98] I. Gergely, C. Hegedus, A. Szollosy, A. Monsees, T. Riermeier, J. Bakos, Electronic and steric effects of ligands as control elements for rhodium-catalyzed asymmetric hydrogenation, *Tetrahedron Lett.* **2003**, *44*, 9025.

[99] B. M. Trost, D. J. Murphy, A model for metal-templated catalytic asymmetric induction via π-allyl fragments, *Organometallics* **1985**, *4*, 1143.

[100] D. Seebach, E. Devaquet, A. Ernst, M. Hayakawa, F. N. M. Kühnle, W. B. Schweizer, B. Weber, Preparation of the $PdCl_2$ complexes of TADDOP, the bis(diphenylphosphinite) of TADDOL: use in enantioselective 1,3-diphenylallylations of nucleophiles and discussion of the mechanism, *Helv. Chim. Acta* **1995**, *78*, 1636.

[101] Y. G. Zhou, W. J. Tang, W. B. Wang, W. G. Li, X. M. Zhang, Highly effective chiral ortho-substituted BINAPO ligands (*o*-BINAPO): applications in Ru-catalyzed asymmetric hydrogenations of β-aryl-substituted β-(acylamino)acrylates and β-keto esters, *J. Am. Chem. Soc.* **2002**, *124*, 4952.

[102] Y. G. Zhou, X. M. Zhang, Synthesis of novel BINOL-derived chiral bisphosphorus ligands and their application in catalytic asymmetric hydrogenation, *Chem. Commun.* **2002**, 1124.

[103] K. H. Lam, L. J. Xu, L. C. Feng, Q. H. Fan, F. L. Lam, W. H. Lo, A. S. C. Chan, Highly enantioselective iridium-catalyzed hydrogenation of quinoline derivatives using chiral phosphinite H8-BINAPO, *Adv. Synth. Catal.* **2005**, *347*, 1755.

[104] J. Yu, T. V. RajanBabu, J. R. Parquette, Conformationally driven asymmetric induction in a catalytic dendrimer, *J. Am. Chem. Soc.* **2008**, *130*, 7845.

[105] I. S. Mikhel, M. Garland, J. Hopewell, S. Mastroianni, C. L. McMullin, A. G. Orpen, P. G. Pringle, Cage phosphinites: ligands for efficient nickel-catalyzed hydrocyanation of 3-pentenenitrile, *Organometallics* **2011**, *30*, 974.

[106] D. A. Evans, K. R. Campos, J. S. Tedrow, F. E. Michael, M. R. Gagné, Chiral mixed phosphorus/sulfur ligands for palladium-catalyzed allylic alkylations and aminations, *J. Am. Chem. Soc.* **2000**, *122*, 7905. For the first report of a similar ligand, see E. Hauptman, R. Shapiro, W. Marshall, Synthesis of chiral bis(phosphinite) ligands with a tetrahydrothiophene backbone: use in asymmetric hydrogenation, *Organometallics* **1998**, *17*, 4976.

[107] D. A. Evans, F. E. Michael, J. S. Tedrow, K. R. Campos, Application of chiral mixed phosphorus/sulfur ligands to enantioselective rhodium-catalyzed dehydroaminoacid hydrogenation and ketone hydrosilylation processes, *J. Am. Chem. Soc.* **2003**, *125*, 3534. Carbohydrate-based thioether ligands have since appeared in the literature; see (b) E. Guimet, M. Diéguez, A. Ruiz, C. Claver, Asymmetric hydrogenation of prochiral olefins catalysed by furanoside thioether-phosphinite Rh(I) and Ir(I) complexes, *Dalton Trans.* **2005**, 2557; and (c) N. Khiar, R. Navas, B. Suarez, E. Alvarez, I. Fernadez, Asymmetric enamide hydrogenation using phosphinite thioglycosides: synthesis of *D*- and *L*-aminoesters using *D*-sugars as catalyst precursors, *Org. Lett.* **2008**, *10*, 3697.

[108] C. R. Landis, J. Halpern, Origin of enantioselection in asymmetric hydrogenation of MAAC, *J. Am. Chem. Soc.* **1987**, *109*, 1746.

[109] K. Yonehara, K. Mori, T. Hashizume, K.-G. Chung, K. Ohe, S. Uemura, Palladium-catalyzed asymmetric intermolecular arylation of cyclic or acyclic alkenes using phosphinite-oxazoline ligands derived from D-glucosamine, *J. Organomet. Chem.* **2000**, *603*, 40.

[110] S. J. Roseblade, A. Pfaltz, Iridium-catalyzed asymmetric hydrogenation of olefins, *Acc. Chem. Res.* **2007**, *40*, 1402.

[111] J. Blankenstein, A. Pfaltz, A new class of modular phosphinite-oxazoline ligands: Ir-catalyzed enantioselective hydrogenation of alkenes, *Angew. Chem. Int. Ed.* **2001**, *40*, 4445.

[112] F. Menges, A. Pfaltz, Threonine-derived phosphinite-oxazoline ligands for the Ir-catalyzed enantioselective hydrogenation, *Ad. Synth. Catal.* **2002**, *344*, 40.

[113] S. P. Smidt, F. Menges, A. Pfaltz, SimplePHOX, a readily available chiral ligand system for iridium-catalyzed asymmetric hydrogenation, *Org. Lett.* **2004**, *6*, 2023.

[114] S. Kaiser, S. R. Smidt, A. Pfaltz, Iridium catalysts with bicyclic pyridine-phosphinite ligands: asymmetric hydrogenation of olefins and furan derivatives, *Angew. Chem. Int. Ed.* **2006**, *45*, 5194.

[115] S. Bell, B. Wustenberg, S. Kaiser, F. Menges, T. Netscher, A. Pfaltz, Asymmetric hydrogenation of unfunctionalized, purely alkyl-substituted olefins, *Science* **2006**, *311*, 642.

[116] K. Kallstrom, C. Hedberg, P. Brandt, A. Bayer, P. G. Andersson, Rationally designed ligands for asymmetric iridium-catalyzed hydrogenation of olefins, *J. Am. Chem. Soc.* **2004**, *126*, 14308.

[117] Q.-B. Liu, C.-B. Yu, Y.-G. Zhou, Synthesis of tunable phosphinite-pyridine ligands and their applications in asymmetric hydrogenation, *Tetrahedron Lett.* **2006**, *47*, 4733.

[118] C. G. Arena, F. Nicolo, D. Drommi, G. Bruno, F. Faraone, Enantioselective hydroformylation with the chiral bidentate *P,N*-ligand 2-[1-(1*S*,2*S*,5*R*)-(−)-menthoxydiphenylp hosphino]-pyridine cationic rhodium(I) complexes, *J. Chem. Soc., Chem. Commun.* **1994,** *1994*, 2251.

[119] T. Minuth, M. M. K. Boysen, Novel, efficient alkene-phosphinite hybrid ligand based on D-glucose, *Org. Lett.* **2009**, *18*, 4212.

[120] T. V. RajanBabu, A. L. Casalnuovo, Electronic effects in asymmetric catalysis: enantioselective carbon-carbon bond-forming processes, *Pure Appl. Chem.* **1994**, *66*, 1535.

[121] E. P. Kündig, B. Bourdin, G. Bernardinelli, Asymmetric Diels-Alder reactions catalyzed by a chiral iron Lewis acid, *Angew. Chem. Int. Ed.* **1994**, *33*, 1856.

[122] E. P. Kundig, C. M. Saudan, F. Viton, Chiral cyclopentadienyl-iron and -ruthenium Lewis acids containing the electron-poor BIPHOP-F ligand: a comparison as catalysts in an asymmetric Diels-Alder reaction, *Adv. Synth. Catal.* **2001**, *343*, 51.

[123] R. Ali, K. B. Dillon, Some novel phosphoranides containing pentafluorophenyl group, *J. Chem. Soc. Dalton Trans.* **1990**, 2593.

[124] V. Alezra, G. Bernardinelli, C. Corminboeuf, U. Frey, E. P. Kundig, A. E. Merbach, C. M. Saudan, F. Viton, J. Weber, CpRu((*R*)-Binop-F)(H_2O) SbF_6, a new fluxional chiral Lewis acid catalyst: synthesis, dynamic NMR, asymmetric catalysis, and theoretical studies, *J. Am. Chem. Soc.* **2004**, *126*, 4843.

[125] E. P. Kundig, M. C. Saudan, V. Alezra, F. Viton, G. Bernardinelli, (Indenyl)Ru(biphop-F)(+): a Lewis acid catalyst that controls both the diene and the dienophile facial selectivity in Diels-Alder reactions, *Angew. Chem. Int. Ed.* **2001**, *40*, 4481.

[126] J. Rickerby, M. Vallet, G. Bernardinelli, F. Viton, E. P. Kündig, Ruthenium-Lewis acid catalyzed asymmetric Diels-Alder reactions between dienes and α.α-unsaturated ketones, *Chem. Eur. J.* **2007**, *13*, 3354.

[127] S. Thamapipol, G. Bernardinelli, C. Besnard, E. P. Kündig, Chiral ruthenium Lewis acid catalyzed intramolecular Diels-Alder reactions, *Org. Lett.* **2010**, *12*, 5604, and references cited therein.

[128] A. Badoiu, G. Bernardinelli, E. P. Kundig, Ruthenium Lewis acid catalyzed asymmetric 1,3-dipolar cycloadditions between *N*-methylnitrones and enals, *Synthesis* **2010**, 2207.

[129] A. Bertacco, U. Mazzi, A. A. Orio, New five-coordinate complexes of cobalt(II) with phenyldiethoxyphosphine, *Inorg. Chem.* **1972**, *11*, 2547.

[130] A. W. Verstuyft, D. A. Redfield, L. W. Cary, J. H. Nelson, Mixed-ligand bis-monodentate phosphorus donor ligand complexes of palladium(II), **LL'**$PdCl_2$: a comprehensive investigation, *Inorg. Chem.* **1976**, *15*, 1128, and references cited therein.

[131] L. M. Haines, Cationic complexes of rhodium(I) and their reactivity toward air, *Inorg. Chem.* **1971**, *10*, 1685.

[132] L. M. Haines, E. Singleton, Cationic complexes of iridium(I) containing phosphites, phosphines, or arsines as ligands, *J. Chem. Soc., Dalton Trans.* **1972**, 1891.

[133] D. A. Couch, S. D. Robinson, Transition metal complexes containing phosphorus ligands. XI. Cationic dialkyl phenylphosphonite and alkyl diphenylphosphinite derivatives of gold, silver and the platinum metals, *Inorg. Chem. Acta* **1974**, *9*, 39.

[134] X. L. Luo, D. Michos, R. H. Crabtree, M. B. Hall, Synthesis and characterization of some new complexes of the Vaska-type *trans*-[IrX(CO)$\mathbf{L}_2$] (X=Cl or IL=phosphite, phosphinite or phosphonite) and the direction of dihydrogen addition to these species, *Inorg. Chimica Acta* **1992**, *198*, 429.

[135] D. Seebach, M. Hayakawa, J. Sakaki, W. B. Schweizer, Derivatives of α,α,α',α'-tetraaryl-2,2-dimethyl-1,3-dioxolan-4,5-dimethanol (TADDOL) containing nitrogen, sulfur, and phosphorus atoms: new ligands and auxiliaries for enantioselective reactions, *Tetrahedron* **1993**, *49*, 1711.

[136] H. Lam, TADDOL-derived phosphonites, phosphites, and phosphoramidites in asymmetric catalysis, *Synthesis* **2011**, 2011.

[137] M. Mewald, A. Weickgenannt, R. Frohlich, M. Oestreich, Expeditious synthesis of TADDOL-derived phosphoramidite and phosphonite ligands, *Tetrahedron: Asymmetry* **2010**, *21*, 1232.

[138] J. Sakaki, W. B. Schweizer, D. Seebach, Catalytic enantioselective hydrosilylation of aromatic ketones using rhodium complexes of TADDOL-derived cyclic phosphonites and phosphites, *Helv. Chim. Acta* **1993**, *76*, 2654.

[139] D. Haag, J. Runsink, H. D. Scharf, [**L***Rh(NBD)Cl] (**L***=chiral cyclic monophosphonite): a novel class of rhodium(I) complexes and their evaluation in the asymmetric hydrosilylation of ketones: investigations of the effects of temperature and ligand backbone, *Organometallics* **1998**, *17*, 398.

[140] A. Alexakis, J. Burton, J. Vastra, C. Benhaim, X. Fournioux, A. van den Heuvel, J. M. Leveque, F. Maze, S. Rosset, Synthesis and application of chiral phosphorus ligands derived from TADDOL for the asymmetric conjugate addition of diethyl zinc to enones, *Eur. J. Org. Chem.* **2000**, 4011.

[141] A. Alexakis, C. Benhaim, Asymmetric conjugate addition to alkylidene malonates, *Tetrahedron: Asymmetry* **2001**, *12*, 1151.

[142] (a) A. Alexakis, C. Benhaim, Enantioselective copper-catalyzed conjugate addition of dialkyl zinc to nitro-olefins, *Org. Lett.* **2000**, *2*, 2579. (b) Bee, C.; Han, S. B.; Hassan, A.; Iida, H.; Krische, M. J. Diastereo- and enantioselective hydrogenative aldol coupling of vinyl ketones: Design of effective monodentate TADDOL-Like phosphonite ligands, *J. Am. Chem. Soc.* **2008**, *130*, 2746.

[143] (a) H. E. Burks, L. T. Kliman, J. P. Morken, Asymmetric 1,4-dihydroxylation of 1,3-dienes by catalytic enantioselective diboration, *J. Am. Chem. Soc.* **2009**, *131*, 9134. (b) Sieber, J. D.; Morken, J. P. Asymmetric Ni-Catalyzed Conjugate Allylation of Activated Enones", *J. Am. Chem. Soc.* **2008**, *130*, 4978. (c) Zhang, P.; Morken, J. P. "Catalytic Enantioselective Allylation of Dienals through the Intermediacy of Unsaturated pi-Allyl Complexes, *J. Am. Chem. Soc.* **2009**, *131*, 12550.

[144] L. T. Kliman, S. N. Mlynarski, J. P. Morken, Pt-catalyzed enantioselective diboration of terminal alkenes with B-2(pin)(2), *J. Am. Chem. Soc.* **2009**, *131*, 13210.

[145] M. T. Reetz, A. Gosberg, R. Goddard, S.-H. Kyung, Diphosphonites as highly efficient ligands for enantioselective rhodium-catalyzed hydrogenation, *Chem. Commun.* **1998**, 2077.

[146] I. E. Nifantev, L. F. Manzhukova, M. Y. Antipin, Y. T. Struchkov, E. E. Nifantev, Ferrocenylenediphosphonites as ligands in the synthesis of oligonuclear heterometal complexes, *Russian J. Gen. Chem.* **1995**, *65*, 682. Original article in Russian: *Zh. Obshch. Khim.* **1995**, *65*, 756.

[147] S. P. Shum, S. D. Pastor, G. Rihs, Ferrocene bis(phosphonite)s: synthesis and characterization of a novel class of sterically congested ligands, *Inorg. Chem.* **2001**, *41*, 127.

[148] M. T. Reetz, T. Sell, Rhodium-catalyzed enantioselective hydrogenation using chiral monophosphonite ligands, *Tetrahedron Lett.* **2000**, *41*, 6333.

[149] K. Tani, T. Yamagata, K. Nagata, (+/–)-8-Phenyldinaphtho[2,1-d-1',2'-f[1,3,2]-dioxaphosphepine, *Acta Crystallogra. Sect. C.: Cryst. Struct. Commun.* **1994**, *C50*, 1274.

[150] C. Claver, E. Fernandez, A. Gillon, K. Heslop, D. J. Hyett, A. Martorell, A. G. Orpen, P. G. Pringle, Biarylphosphonites: a class of monodentate phosphorus(III) ligands that outperform their chelating analogues in asymmetric hydrogenation catalysis, *Chem. Commun.* **2000**, 961. For a detailed mechanistic study of various

mono- versus bidentate asymmetric hydrogenation phosphonite catalysts, see D. W. Norman, C. A. Carraz, D. J. Hyett, P. G. Pringle, J. B. Sweeney, A. G. Orpen, H. Phetmung, R. L. Wingad, Bidentates versus monodentates in asymmetric hydrogenation catalysis: synergic effects on rate and allosteric effects on enantioselectivity, *J. Am. Chem. Soc.* **2008**, *130*, 6840.

[151] M. T. Reetz, T. Sell, A. Meiswinkel, G. Mehler, A new principle in combinatorial asymmetric transition-metal catalysis: mixtures of chiral monodentate P ligands, *Angew. Chem. Int. Ed.* **2003**, *42*, 790.

[152] M. T. Reetz, X. Li, Mixtures of configurationally stable and fluxional atropisomeric monodentate P ligands in asymmetric Rh-catalyzed olefin hydrogenation, *Angew. Chem. Int. Ed.* **2005**, *44*, 2959.

[153] M. T. Reetz, X. G. Li, Combinatorial approach to the asymmetric hydrogenation of β-acylamino acrylates: use of mixtures of chiral monodentate P-ligands, *Tetrahedron* **2004**, *60*, 9709.

[154] M. T. Reetz, Combinatorial transition-metal catalysis: mixing monodentate ligands to control enantio-, diastereo-, and regioselectivity, *Angew. Chem. Int. Ed.* **2008**, *47*, 2556.

[155] A. Martorell, R. Naasz, B. L. Feringa, P. G. Pringle, Phosphonite ligands for enantioselective copper(I)-catalysed conjugate addition of diethylzinc to enones, *Tetrahedron: Asymmetry* **2001**, *12*, 2497.

[156] A. Martorell, C. Claver, E. Fernandez, The influence of chiral phosphite and phosphonite ligands on the hydrogenation of imines with rhodium and iridium complexes, *Inorg. Chem. Commun.* **2000**, *3*, 132.

[157] M. A. Fernandez-Zumel, B. Lastra-Barreira, M. Scheele, J. Diez, P. Crochet, J. Gimeno, Chiral phosphonite, phosphite and phosphoramidite η^6-arene-ruthenium(II) complexes: application to the kinetic resolution of allylic alcohols, *Dalton Trans.* **2010**, *39*, 7780.

[158] M. T. Reetz, A. Gosberg, D. Moulin, Copper-catalyzed enantioselective conjugate addition of diethylzinc to α,β-unsaturated carbonyl compounds using diphosphonites as chiral ligands, *Tetrahedron Lett.* **2002**, *43*, 1189.

[159] M. T. Reetz, D. Moulin, A. Gosberg, BINOL-based diphosphonites as ligands in the asymmetric Rh-catalyzed conjugate addition of arylboronic acids, *Org. Lett.* **2001**, *3*, 4083.

[160] G.-H. Hou, J.-H. Xie, L.-X. Wang, Q.-L. Zhou, Highly efficient Rh(I)-catalyzed asymmetric hydrogenation of enamines using monodente spiro phosphonite ligands, *J. Am. Chem. Soc.* **2006**, *128*, 11774.

[161] Y. Yasui, H. Kamisaki, Y. Takemoto, Enantioselective synthesis of 3,3-disubstituted oxindoles through Pd-catalyzed cyanoamidation, *Org. Lett.* **2008**, *10*, 3303.

[162] Y. Xu, N. W. Alcock, G. J. Clarkson, G. Docherty, G. Woodward, M. Wills, Asymmetric hydrogenation of ketones using a ruthenium(II) catalyst containing BINOL-derived monodonor phosphorus-donor ligands, *Org. Lett.* **2004**, *6*, 4105.

[163] X. G. Peng, Z. Wang, C. G. Xia, K. L. Ding, Ferrocene-based bidentate phosphonite ligands for rhodium(I)-catalyzed enantioselective hydroformylation, *Tetrahedron Lett.* **2008**, *49*, 4862.

[164] B. G. Zhao, X. G. Peng, Z. Wang, C. G. Xia, K. L. Ding, Modular chiral bidentate phosphonites: design, synthesis, and application in catalytic asymmetric hydroformylation reactions, *Chem. Eur. J.* **2008**, *14*, 7847.

[165] M. Laly, R. Broussier, B. Gautheron, Ferrocene-based phosphonite-phosphine ligands, Pd and Rh complexes, *Tetrahedron Lett.* **2000**, *41*, 1183.

[166] (a) T. L. Schull, A. Knight, New chiral phosphine-phosphonites derived from (2R,3R)-dimethyl tartrate, (*S*)-binaphthol and (1*R*,2*S*)-ephedrine, *Tetrahedron: Asymmetry* **1999**, *10*, 207; and for a synthesis of 2-bromophenyldiphenylphosphine, see (b) T. L. Schull, J. C. Fettinger, D. A. Knight, Synthesis and characterization of palladium(II) and platinum(II) complexes containing water-soluble hybrid phosphine-phosphonate ligands, *Inorg. Chem.* **1996**, *35*, 6717.

[167] K. W. Kottsieper, U. Kuhner, O. Stelzer, Synthesis of enantiopure C-1 symmetric diphosphines and phosphino-phosphonites with ortho-phenylene backbones, *Tetrahedron: Asymmetry* **2001**, *12*, 1159.

[168] T. L. Schull, J. C. Fettinger, D. A. Knight, Synthesis and characterization of palladium(II) and platinum(II) complexes containing water-soluble hybrid phosphine-phosphonite ligands, *Inorg. Chem.* **1996**, *35*, 6717.

[169] M. T. Reetz, A. Gosberg, New non-C-2-symmetric phosphine-phosphonites as ligands in asymmetric metal catalysis, *Tetrahedron: Asymmetry* **1999**, *10*, 2129.

[170] A. Zanotti-Gerosa, C. Malan, D. Herzberg, Phosphonites based on the paracyclophane backbone: new ligands for highly selective rhodium-catalyzed asymmetric hydrogenation, *Org. Lett.* **2001**, *3*, 3687.

[171] V. B. Birman, A. L. Rheingold, K.-C. Lam, 1,1′-Spirobiindane-7,7′-diol: a novel, C_2-symmetric chiral ligand, *Tetrahedron: Asymmetry* **1999**, *10*, 125.

[172] Y. Fu, G. H. Hou, J. H. Xie, L. Xing, L. X. Wang, Q. L. Zhou, Synthesis of monodentate chiral spiro phosphonites and the electronic effect of ligand in asymmetric hydrogenation, *J. Org. Chem.* **2004**, *69*, 8157.

[173] S.-F. Zhu, Y. Yang, L.-X. Wang, B. Liu, Q.-L. Zhou, Synthesis and application of chiral spiro phospholane ligand in Pd-catalyzed asymmetric allylation of aldehydes with allylic alcohols, *Org. Lett.* **2005**, *7*, 2333.

[174] M. V. Escarcega-Bobadilla, E. Teuma, A. M. Masdeu-Bulto, M. Gomez, New bicyclic phosphorous ligands: synthesis, structure and catalytic applications in ionic liquids, *Tetrahedron* **2011**, *67*, 421.

[175] A. A. Nazarov, C. G. Hartinger, V. B. Arion, G. Giester, B. K. Keppler, Synthesis of ferrocenylglucose phosphonite and pisphosphinite: Pd(II) and Pt(II) complexes,Pd-catalyzed allylic alkylation, *Tetrahedron* **2002**, *58*, 8489.

[176] T. Kop-Weiershausen, J. Lex, J. M. Neudorfl, B. Goldfuss, An exceptional P-H phosphonite: biphenyl-2,2′-bisfenchylchlorophosphite and derived ligands (BIFOPs) in enantioselective copper-catalyzed 1,4-additions, *Beilstein J. Org. Chem.* **2005**, *1*:6, 1.

[177] B. Goldfuss, T. Loschmann, T. Kop-Weiershausen, J. Neudorfl, F. Rominger, A superior P-H phosphonite: asymmetric allylic substitutions with fenchol-based palladium catalysts, *Beilstein J. Org. Chem.* **2006**, 2:7, 1.

[178] D. Selent, K.-D. Wiese, D. Röttger, A. Börner, Novel oxyfunctionalized phosphonite ligands for the hydroformylation of isomeric *n*-olefins, *Angew. Chem. Int. Ed.* **2000**, *39*, 1639.

[179] J. K. Jeon, Y. K. Park, J. M. Kim, Hydroformylation of mixed octenes using rhodium-bulky phosphonite complexes with excellent catalytic activity and stability, *Chem. Lett.* **2004**, *33*, 174.

[180] C. Kunze, D. Selent, I. Neda, M. Freytag, P. G. Jones, R. Schmutzler, W. Baumann, A. Börner, Calix[4]arene-based bis-phosphonites, bis-phosphites, and bis-*o*-acyl-phosphites as ligands in the rhodium(I)-catalyzed hydroformylation of 1-octene, *Z. Anorg. Allg. Chem.* **2002**, *628*, 779.

[181] J. I. van der Vlugt, J. M. J. Paulusse, E. J. Zijp, J. A. Tijmensen, A. M. Mills, A. L. Spek, C. Claver, D. Vogt, Coordination chemistry and asymmetric catalysis with a chiral diphosphonite, *Eur. J. Inorg. Chem.* **2004**, 4193.

[182] J. I. van der Vlugt, R. Sablong, A. M. Mills, H. Kooijman, A. L. Spek, A. Meetsma, D. Vogt, Coordination chemistry and X-ray studies with novel sterically constrained diphosphonite ligands, *Dalton Trans.* **2003**, 4690.

[183] L. A. van der Veen, P. K. Keeven, P. C. J. Kamer, P. W. N. M. van Leeuwen, Novel arsine ligands for selective hydroformylation of alk-1-enes employing platinum/tin catalysts, *Chem. Commun.* **2000**, 333.

[184] S. H. Chikkali, R. Bellini, G. Berthon-Gelloz, J. I. van der Vlugt, B. Bruin, J. N. H. Reek, Highly enantioselective hydroformylation of dihydrofurans catalyzed by hybrid phosphine-phosphonite rhodium complexes, *Chem. Commun.* **2010**, *46*, 1244.

[185] M. E. Tauchert, T. R. Kaiser, A. P. V. Gothlich, F. Rominger, D. C. M. Warth, P. Hofmann, Phosphonite ligand design for nickel-catalyzed 2-methyl-3-butenenitrile isomerization and styrene hydrocyanation, *Chem Catchem* **2010**, *2*, 674.

[186] For a direct comparison of effects of different ligands on this isomerization, see reference [185].

[187] J. I. van der Vlugt, A. C. Hewat, S. Neto, R. Sablong, A. M. Mills, M. Lutz, A. L. Spek, C. Müller, D. Vogt, Sterically demanding diphosphonite ligands: synthesis and application in nickel-catalyzed isomerization of 2-methyl-3-butenenitrile, *Adv. Synth. Catal.* **2004**, *346*, 993.

[188] A. P. V. Gothlich, M. Tensfeldt, H. Rothfuss, M. E. Tauchert, D. Haap, F. Rominger, P. Hofmann, Novel chelating phosphonite ligands: syntheses, structures, and nickel-catalyzed hydrocyanation of olefins, *Organometallics* **2008**, *27*, 2189.

[189] P. Braunstein, F. Naud, A. Pfaltz, S. J. Rettig, Ruthenium complexes with novel tridentate N,P,N ligands containing a phosphonite bridge between two chiral oxazolines. catalytic activity in cyclopropanation of olefins and transfer hydrogenation of acetophenone, *Organometallics* **2000**, *19*, 2676.

[190] P. Braunstein, F. Naud, A. Dedieu, M.-M. Rohmer, A. DeCian, S. J. Rettig, Palladium complexes with a new hemilabile bis(oxazoline)phenylphosphonite ligand: characterization of an unprecedented chloro palladium(II)-η^1-(1-allyl) complex, *Organometallics* **2001**, *20*, 2966.

[191] J. Zhang, P. Braunstein, R. Welter, Stable, chloride-induced monohapto-bonding mode for an allyl ligand in a Pd(II) complex bearing a new bidentate phosphonite-oxazoline ligand, *Inorg. Chem.* **2004**, *43*, 4172.

[192] R. Peloso, R. Pattacini, C. S. J. Cazin, P. Braunstein, Structure and reactivity of new iridium complexes with bis(oxazoline)-phosphonito ligands, *Inorg. Chem.* **2009**, *48*, 11415.

[193] A. Kermagoret, P. Braunstein, Contrasting bonding modes of a tridentate bis(oxazoline)phosphine ligand in cobalt and iron vs. palladium complexes: unprecedented *N-N*-coordination for a N,P,N ligand, *Dalton Trans.*

2008, *2008*, 585. For another similar example of varying coordination modes between a phenylbis(pyridylmethyl) *phosphine* and bis(pyrid-2-ylmethyl)*phenylphosphonite*, see F. Hung-Low, A. Renz, K. K. Klausmeyer, The variable binding modes of phenylbis(pyrid-2-ylmethyl)phosphine and bis(pyrid-2-ylmethyl) phenylphosphonite with Ag-I and Cu-I, *Eur. J. Inorg. Chem.* **2009**, 2994.

[194] A. Kermagoret, F. Tomicki, P. Braunstein, Nickel and iron complexes with *N,P,N*-type ligands: synthesis, structure and catalytic oligomerization of ethylene, *Dalton Trans.* **2008**, *2008*, 2945.

[195] P. Chavez, I. G. Rios, A. Kermagoret, R. Pattacini, A. Meli, C. Bianchini, G. Giambastiani, P. Braunstein, Nickel complexes with phosphinito-oxazoline ligands: temperature-controlled formation of mono- or dinuclear complexes and catalytic oligomerization of ethylene and propylene, *Organometallics* **2009**, *28*, 1776.

[196] M. Agostinho, P. Braunstein, R. Welter, Phosphinito- and phosphonito-oxazoline Pd(II) complexes as CO/ethylene insertion intermediates: synthesis and structural characterization, *Dalton Trans.* **2007**, 759.

[197] D. J. Brauer, K. W. Kottsieper, S. Rossenbach, O. Stelzer, Novel P,N ligands derived from (*R*)- and (*S*)-1-phenylethylamine with (2*R*,5*R*)-2,5-dimethylphospholanyl groups (DuPHAMIN) for asymmetric catalysis, *Eur. J. Inorg. Chem.* **2003**, *2003*, 1748.

[198] B. Hoge, P. Panne, Synthesis of a functional C-2-symmetrical bidentate diphenylphosphonite DIOP derivative and its conversion into the corresponding π-acidic bis(pentafluorophenyl) and bis(p-tetrafluoropyridyl) compounds, *Chem. Eur. J.* **2006**, *12*, 9025.

[199] D. L. Allen, V. C. Gibson, M. L. H. Green, J. F. Skinner, J. Bashkin, P. D. Grebenik, A new route to chiral *bis*-tertiary phosphine ligands: synthesis, resolution, and crystal structure of *trans*-bis-1,2-(diphenylphosphino)-cyclopentane and the nickel adduct $NiBr_2$[*trans*-1,2-$(PPh_2]_2$ C_8H_8, *J. Chem. Soc., Chem. Commun.* **1983**, *1983*, 895.

[200] C. Eckert, L. Dahlenburg, A. Wolski, Oligophosphine ligands 39. (1*S*,2S)-*Trans*-cyclopentane-1,2-diyl-bis(phosphine), $C_5H_8(PH_2)_2$, by reduction of a diastereomerically pure cyclopentanediyl-*bis*(dioxaphospholane), *Z. Naturforsch. B: Chem. Sci.* **1995**, *50*, 1004.

[201] H. Brunner, S. Stefaniak, M. Zabel, Enantioselective catalysis 130: optically active expanded ligands based on the *trans*-1,2-substituted cyclopentane skeleton, *Synthesis* **1999**, 1776.

[202] L. Dahlenburg, A. Kaunert, (1*S*,2*S*)-Cyclopentane-1,2-diyl-bis(phosphonous dichloride, (1*S*,2*S*)-$C_5H_8(PCl_2)_2$: a versatile optically active reagent, *Eur. J. Inorg. Chem.* **1998**, *1998*, 885.

[203] P. G. Chantrell, C. A. Pearce, C. R. Toyer, R. Twaits, Preparation of some novel diphosphorus compounds. I. 1,4-phenylene-diphosphorus compounds, *J. Appl. Chem.* **1964**, *14*, 563.

[204] J. Perich, Di-*tert*-butyl *N,N*-diethylphosphoramidite: a new phosphitylating agent for the efficient phosphorylation of alcohols, *Synthesis* **1988**, *1988*, 142.

[205] A. H. Ford-Moore, B. J. Perry, Triethyl phosphite, *Org. Synth., Coll. Col.* **1963**, *4*, 955.

[206] This modified protocol, based on procedures reported in the literature, was developed by T. V. RajanBabu, A. L. Casalnuovo, R. Shapiro, and T. A. Ayers in 1994 at the Ohio State University and DuPont Company.

[207] D. Drew, J. R. Doyle, A. G. Shaver, Cyclic-diolefin complexes of platinum and palladium, *Inorg. Synth.* **1972**, *13*, 47.

[208] J. V. Allen, J. M. J. Williams, Enantiomerically pure oxazolines tethered to alcohols: preparation and use in asymmetric catalysis, *Tetrahedron-Asymmetry* **1994**, *5*, 277.

[209] L. N. Pridgen, G. Miller, Synthesis of 2-(α-hydroxyalkyl)-1,3-heterocyclic alcohols and aryl carbamates, *J. Heterocyl. Chem.* **1983**, *20*, 1223.

6
Mixed Donor Ligands

René Tannert and Andreas Pfaltz

Department of Chemistry, University of Basel, Basel, Switzerland

6.1 Introduction: general design principles

The history of phosphorus(III) ligands in homogeneous catalysis is closely linked to the development of asymmetric hydrogenation. The starting point was Wilkinson's catalyst [$Rh(PPh_3)_3Cl$] [1], which was transformed to a chiral catalyst by replacing triphenylphosphine by a chiral analog. Although the enantioselectivities in the asymmetric hydrogenation with such complexes were low, these initial studies constituted an important proof of principle that opened up new directions in asymmetric catalysis (Scheme 6.1) [2].

Kagan's seminal work then shifted the focus to bidentate chelating diphosphines, which in most cases proved to be superior to monophosphines. With the DIOP ligand (**1**), Kagan introduced several fundamental design principles that strongly influenced the development of asymmetric catalysis (Figure 6.1) [3].

He recognized that bidentate ligands offer various advantages over monodentate ligands: their conformationally more rigid ligand backbones allow a better control of the metal–substrate binding modes resulting in a more efficient discrimination of the enantiotopic faces of the prochiral substrate. Furthermore, the use of bidentate ligands significantly reduces the tendency for the formation of complexes of unwanted stoichiometry and preferentially produces the desired 1:1 complexes of metal and ligand.

The high enantioselectivities of Rh(diop) catalysts demonstrated that ligands with chirality in the backbone can be as efficient as P-chiral ligands. Although the distance between stereogenic centers and coordination site of the substrate increases, the overall effect is a single conformation of the ligand around the metal allowing a spatially defined coordination of the substrate.

DIOP was also the first example of a C_2-symmetric diphosphine. An apparent advantage of C_2-symmetric ligands is the resulting relative simplicity regarding the mechanistic analysis of the catalytic reaction since the number of putative transition states is halved compared to ligands devoid of symmetry. Also from the

Phosphorus(III) Ligands in Homogeneous Catalysis: Design and Synthesis, First Edition.
Edited by Paul C. J. Kamer and Piet W. N. M. van Leeuwen.

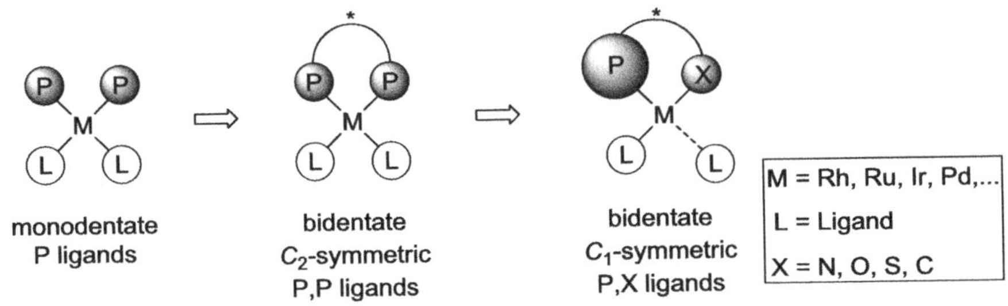

Scheme 6.1 *Development from monodentate over C_2-symmetric bidentate P,P ligands to unsymmetrical P,X ligands.*

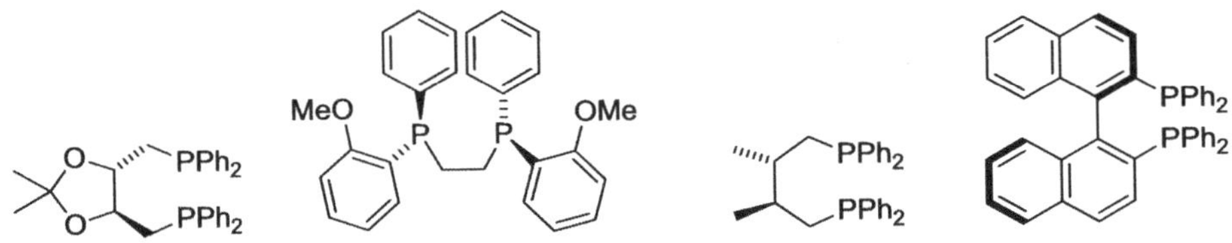

1 (DIOP) **2** (DIPAMP) **3** (CHIRAPHOS) **4** (BINAP)

Figure 6.1 *Structures of important bidentate P,P- and N,N ligands.*

synthetic point of view, C_2-symmetry can offer advantages as it often simplifies the analysis of ligands and intermediates as well as the synthetic planning, for instance considering dimerization reactions.

Subsequently, many other highly effective C_2-symmetric diphosphines such as DIPAMP (**2**) [4], CHIRAPHOS (**3**) [5] or BINAP (**4**) [6] were developed. As a consequence, C_2-symmetric ligands have been dominating asymmetric catalysis for a long time. However, there is no fundamental reason why C_2-symmetric ligands should be necessarily superior. Indeed, many nonsymmetric ligands have been found that induce higher enantioselectivities in asymmetric hydrogenations and other reactions than C_2-symmetric diphosphines. Moreover, for certain reactions mechanistic arguments can be made that nonsymmetrical ligands with two sterically and electronically different donor atoms should, in principle, allow better enantiocontrol than C_2-symmetric analogues. This has been realized by Achiwa for the hydrogenation of olefins with Rh-diphosphine catalysts [7]. Because the intermediates in the catalytic cycle are nonsymmetric, the Rh-bound substrate interacts with the two P-donor atoms in a different manner. As a consequence, Achiwa postulated that the steric and electronic properties of the two donor atoms should be optimized individually, to achieve optimal reactivity and enantioselectivity. He was able to demonstrate the utility of what he called the 'respective control concept' by desymmetrizing the DIOP ligand. By replacing one of the diphenylphosphino groups by a dicyclohexylphosphino group, he observed a significant increase of both catalytic activity and enantioselectivity.

An even more effective way to desymmetrize a P,P-ligand was to switch to a mixed donor ligand system. In this regard, nitrogen donors seemed especially promising as they display electronic properties as rather 'hard' σ donors that differ significantly from the behavior as 'soft' π acceptors attributed to phosphorus donor atoms. Following this line of thought, the phosphinooxazolines were developed by Pfaltz [8], Helmchen [9], and Williams [10] as a new, highly versatile class of P,N-ligands.

The reaction, for which these ligands were developed, was the Pd-catalyzed asymmetric allylic substitution (Scheme 6.2). Starting from a racemic mixture of allylic acetates **5** and *ent*-**5**, an allyl complex is formed. Since both allylic acetates are converted to the same intermediate (**6**), the stereochemical information is lost at this point. It is the subsequent nucleophilic addition to the allyl system that determines which enantiomer

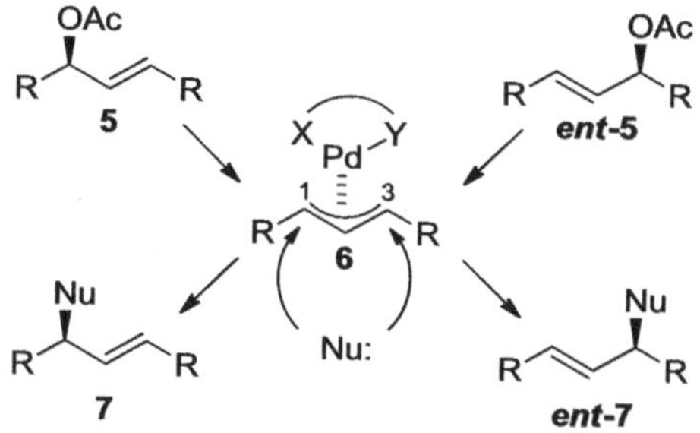

Scheme 6.2 Palladium-catalyzed allylic substitution of racemic substrates.

of the product is formed: nucleophilic attack at C1 leads to product **7**, whereas attack at C3 results in formation of its enantiomer ***ent*-7**. Consequently, the problem of inducing enantioselectivity is equivalent to controlling the regioselectivity of the nucleophilic attack.

The complexation by P,N-ligands should result in effective electronic differentiation of the allylic termini because of the different *trans*-influence of the two electronically dissimilar heteroatoms (Scheme 6.1). Indeed, Pd–PHOX complexes were found to be highly reactive and selective catalysts providing high enantiomeric excess with a range of C-terminal and N-terminal nucleophiles. The selectivities could be rationalized by a combination of steric and electronic effects, and the preference of nucleophilic attack at the allylic carbon atom located *trans* to the phosphino was identified as a key factor [11].

When compared to phosphines, their oxygen-substituted analogs (i.e., phosphinites, phosphonites, and phosphites) exhibit decreasing σ donor and increasing π acceptor properties. As ligands, they effect a more cationic character at the corresponding metal center. Thus, an effective electronic fine-tuning of the P-donor atom in bidentate ligands leading to different catalytic reactivity of the metal complex can be realized through a switch from phosphines to analogous oxygen-substituted phosphorus compounds. A successful example for the application of this concept is provided again by an allylic substitution reaction: by changing from phosphine ligands to phosphites, a switch of the preference for the linear allylation product toward the branched product could be effected with cinnamyl-type substrates [12].

The examples of asymmetric hydrogenation and allylic substitution were chosen to illustrate the potential advantages of mixed donor ligands. Since opportunities for structural modifications of the catalysts depend primarily on the synthetic accessibility of the corresponding ligands, the following sections of this chapter intend to provide an overview of existing mixed donor ligand systems and include selected synthetic procedures to efficient ligands starting from commercially available precursors.

6.2 Synthesis of bidentate P,X-ligands

6.2.1 P,N-ligands

The most prominent class of the mixed bidentate ligands is composed by P,N-ligands. An attractive feature of these ligands is their easy accessibility from members of the chiral pool, in particular from the corresponding chiral amino acids.

6.2.1.1 Oxazoline-based P,N-ligands

Ligand systems based on oxazolines as N-donor groups represent the most prominent subclass of P,N-ligands. The classical members of this bidentate ligand family are the phosphinooxazolines **12** (PHOX) independently discovered in 1993 by Pfaltz [8], Helmchen [9], and Williams [10] (Scheme 6.3). Until

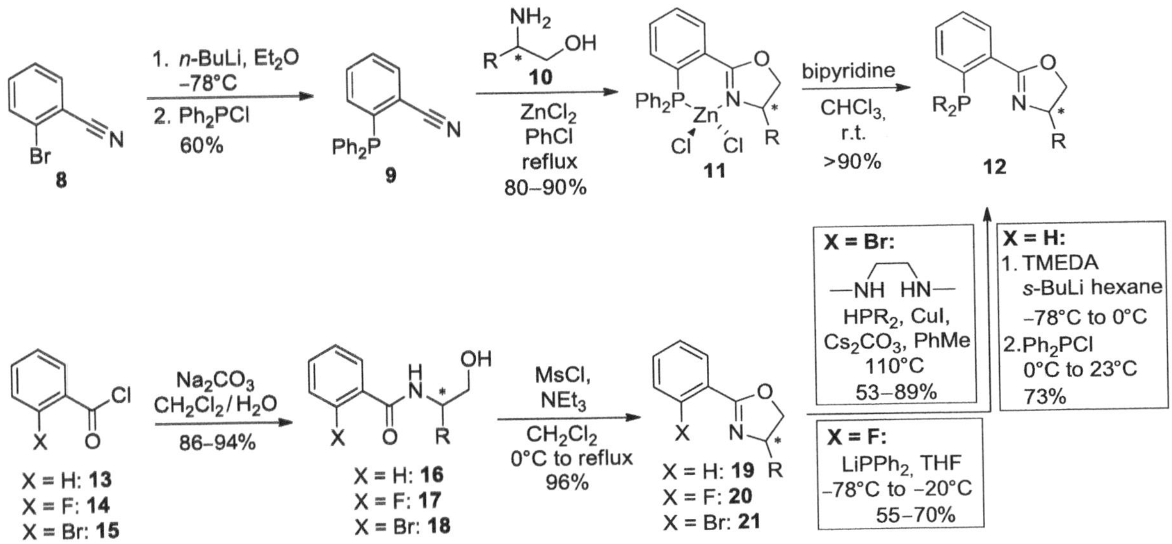

Scheme 6.3 *Synthetic approaches to PHOX ligands bearing an aryl backbone.*

today, they have proven great utility in a wide range of asymmetric transformations including allylic substitutions, Heck reactions, hydrogenations, Diels–Alder reactions, conjugate additions, and Pauson–Khand reactions [11,13–15].

Their highly modular character renders phosphinooxazolines synthetically easily accessible: depending on whether one wants to vary the substitution pattern on oxazoline or phosphine, either of the two groups can be formed in the last step of the ligand synthesis.

One synthetic approach for the synthesis of PHOX ligands **12** involves the conversion of 2-bromobenzonitrile (**8**) into phosphine **9** by halogen–metal exchange and subsequent nucleophilic substitution reaction upon treatment with chlorodiphenylphosphine (Scheme 6.3). Oxazoline formation with amino alcohols **10** in the presence of a metal salt such as zinc chloride followed by decomplexation of the resulting metal–PHOX complex **11** affords the desired ligands **12** differing in the substitution pattern of the oxazoline [16]. Another viable approach that facilitates variation of the phosphine moiety involves the amide coupling of an acid chloride **13–15** under Schotten–Baumann conditions [17,18] followed by intramolecular condensation [19,20] of the resulting β-hydroxy amide **16–18** upon activation with mesyl chloride.

Unfunctionalized arenes **19** (X=H) can be converted into the corresponding PHOX ligand **12** by a sequence of *ortho*-lithiation and subsequent nucleophilic attack to a chlorodiarylphosphine. Other arene substitution patterns allow milder methods of functionalization to be performed: fluoroarene **20** can be transformed into the desired phosphines by means of nucleophilic aromatic substitution [21]. Alternatively, the C–P bond can be formed by means of an Ullmann-type coupling of a secondary phosphine with bromoarene **21** [22]. The latter method outperformed the former ones in particular for the coupling of electron-deficient bromoarenes.

Since the synthetic opportunities for oxazoline formation are manifold [23–25], several other methods have found widespread application in the synthesis of PHOX ligands. Worth mentioning are certainly the conditions described by Vorbrüggen and Krolikiewicz [26,27] to perform both amide coupling and cyclization, under Appel conditions [28] (i.e., PPh3, CCl4, and amine base) and the use of Burgess's reagent [29,30] for the cyclization of β-hydroxy amides.

Using the abovementioned synthetic strategies, an array of PHOX derivatives containing an aromatic backbone has been prepared and tested for their performance in transition metal catalysis (Figure 6.2). Their synthesis included variation of the amino alcohol part derived from commercially available bulky α-amino

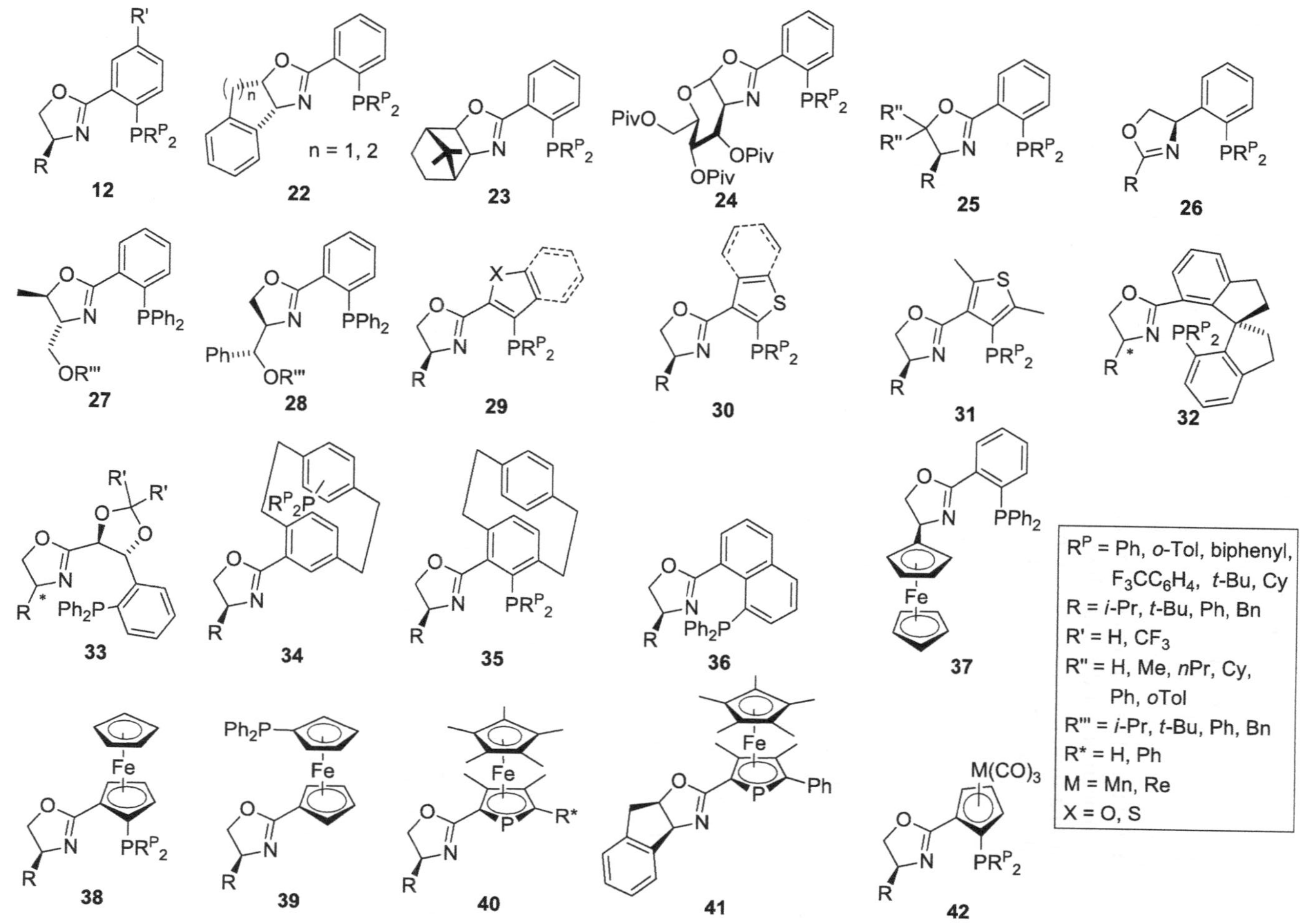

Figure 6.2 *Representative collection of PHOX ligands with (hetero)aryl backbone.*

acids (valine, *tert.*-leucine, phenylalanine, and phenylglycine). Most of the phosphinooxazolines contain aromatic phosphines, but occasionally alkyl phosphines were also employed when very electron-rich phosphines were demanded. As mentioned, the original PHOX ligands **12** were described by Pfaltz, Williams, and Helmchen [8–10]. The introduction of ligands **12** being electronically modified in the aromatic ligand backbone was realized more recently by Stoltz and coworkers (see Section 6.4) [31]. Saigo [32] and Helmchen [33] reported on ligands **22** bearing tricyclic oxazolines that were derived from indan-1,2-diol and tetralin-1,2-diol, respectively. Additionally, the synthesis of **23** from (*2S,3R*)-3-hydroxy-bornylamine was described by Helmchen [33]. The use of glucosamine-derived ligands **24** was described by Kunz and coworkers [34], and Pfaltz demonstrated the utility of ligands **25** carrying two additional geminal alkyl or aryl groups [35]. Changing the usual connectivity of the oxazoline moiety, Zhang and colleagues reported on ligands **26** that were synthesized starting from phenylglycinol [36].

Moberg showed that ligands **27** [37] and **28** [38] carrying an additional free hydroxyl or methyl ether group in the oxazoline part could be accessed by reduction and subsequent derivatization of serine. The groups of Tietze [39] and Guiry [40,41] described the synthesis of heteroaryl ligands **29–31** which were named HETPHOX. Zhou reported the four-step synthesis of SIPHOX ligands **32** [42] starting from 1,1′-spirobiindan-7,7′-diol (SPINOL) which, in turn, can be prepared from *meta*-anisaldehyde in six chemical operations [43] followed by separation of the enantiomers [44] using *N*-benzylcinchonidinium chloride. The Morken group contributed a group of ketal-containing ligands (**33**) they termed StePHOX accessed by ring opening of coumarin followed by Sharpless asymmetric dihydroxylation, ketalization, and oxazoline formation [45]. Paracyclophane-based ligands **34** displaying a "*pseudo*-geminal" substitution pattern have been reported by the groups of Pelter [46], Hou [47], and Bolm, who additionally described the synthesis of "*pseudo*-ortho" and *ortho*-substituted (**35**) derivatives [48]. Ma and coworkers published on phosphinonaphtyloxazolines **36** which were synthesized from 8-bromo-1-naphthoyl chloride [49].

Oxazoline-based P,N ligands containing elements of planar chirality have also been described in the literature. Among these ligands are a number of phosphinooxazolines that carry a metallocene unit. Patti and coworkers reported the synthesis of PHOX ligands **37** having a ferrocenyl side chain at the oxazoline [50]. The groups of Uemura [51] and Richards [52] independently described the synthesis of *ortho*-substituted ferrocene–PHOX ligands **38** accessed by (directed) *ortho*-lithiation of ferrocenyloxazoline [53,54] and subsequent treatment with chlorophosphines. Ikeda's group later reported the synthesis of ligands **39** carrying substituents at different cyclopentadienyl units [55]. The Fu group published on *phospha*-ferrocene–oxazoline ligands **40** and **41** [56,57]. Helmchen and coworkers published the synthesis of ligands **42** bearing a tricarbonylcyclopentadienylmanganese (cymantrene) unit and an additional *P*-stereogenic center. Their ligand synthesis commenced with amide coupling of cymantrenecarbonyl chloride and a chiral amino alcohol. After cyclization and oxazoline-directed *ortho*-lithiation, final treatment with chlorodiarylphosphines afforded the desired ligands **42** [58]. Later, Bolm reported on the analogous synthesis of tricarbonylcyclopentadienylrhenium (cyrhetrene) ligands [59].

The synthesis of axially chiral BINAP-containing ligands **46** was independently described by the groups of Ikeda [60] and Hayashi [61]. The synthetic approaches of both work groups started with the known preparation of binaphthyl methyl ester **44** via quantitative conversion of binaphthol **43** into the corresponding *bis*-triflate [62] and subsequent Pd-catalyzed carboxylation [63] (Scheme 6.4). After Pd-mediated coupling of diphenylphosphine oxide to the remaining aryl triflate moiety, Ikeda described the reduction of the phosphine oxide giving **45** followed by amide bond formation with an enantiopure amino alcohol (via transesterification and an O→N acyl shift). A chromatographic separation of the resulting diastereomers and subsequent oxazoline formation delivered the ligand diastereomers **46a** and **46b**, respectively. Hayashi reported on a slightly different procedure where phosphine oxide reduction represented the last step of the synthesis and where an additional saponification step preceded a carbodiimide-mediated coupling of the amino alcohol to acid **47**. Following a very similar synthetic strategy, also configurationally flexible

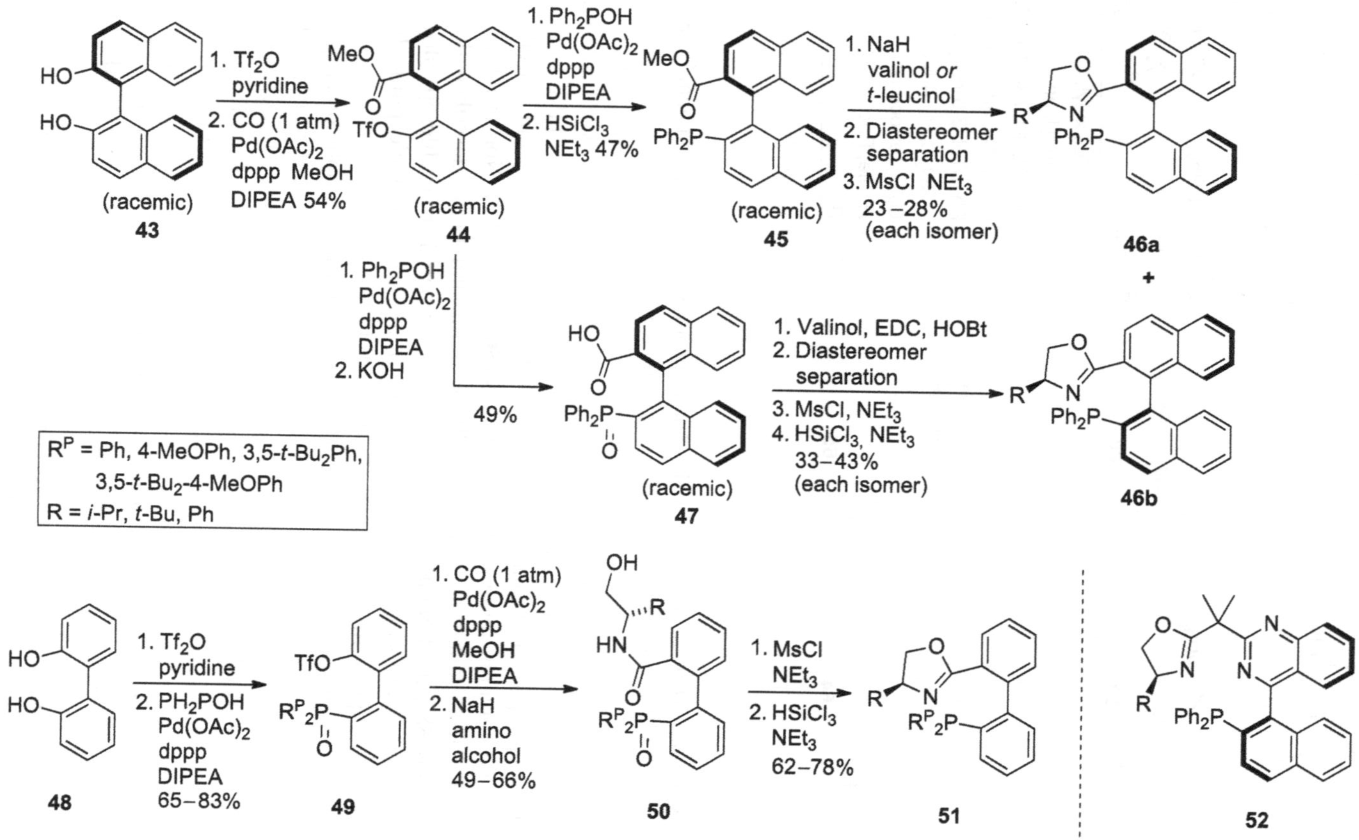

Scheme 6.4 *Synthetic approaches to PHOX ligands bearing a biaryl backbone as described by Ikeda (top), Hayashi (middle), and Zhang (bottom); structure of the heterobiaryl ligand* ***52*** *introduced by Guiry.*

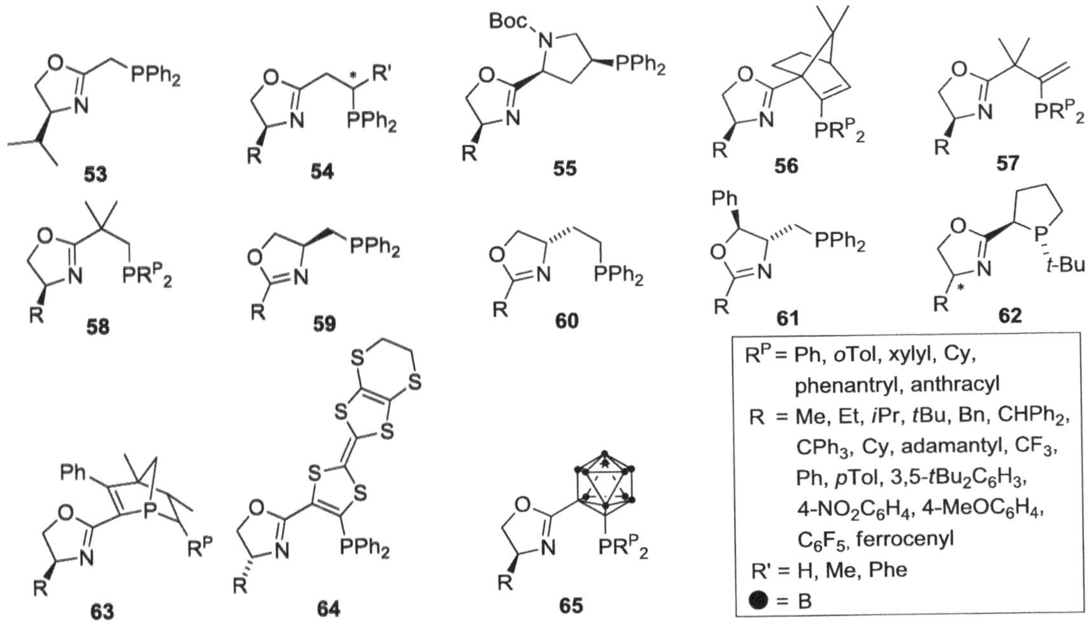

Figure 6.3 *Representative collection of oxazoline-based P,N-ligands with alkyl backbone.*

biphenyl-based PHOX ligands **51** were prepared by the Zhang group starting from biphenol (**48**) [64]. Guiry *et al.* published the synthesis of heterobiaryl phosphinooxazolines **52**.

Besides the abovementioned aromatic backbones, many oxazoline-based ligands that carry alkyl backbones have been described in the literature. In 1993, Sprinz and Helmchen described the synthesis of **53**, the first phosphinooxazoline with an alkyl group tethering the two donor atoms (Figure 6.3) [9]. In analogy to this ligand, the Gilbertson group reported on simplified ligands **54** [65,66]. Later, they introduced proline-based compounds **55** [67,68] and vinyl phosphinooxazoline ligands (**56** and **57**) [69]. Norbornyl ligands **56** could be prepared in four steps from ketopinic acid. In 2009, Schrems and Pfaltz described the synthesis of structurally related ligands **58** called NeoPHOX. These can be prepared from chloropivaloyl chloride by means of consecutive amide bond formation, oxazoline ring closure, and final nucleophilic displacement of a neopentyl chloride using a diphosphine anion (see Section 6.4) [70].

Burgess published the combinatorial synthesis of a number of simplified phosphinomethyl- (**59**) [71] and phosphinoethyloxazoline (**60**) [72,73] ligands derived from serine and aspartic acid, respectively. As reported by James and coworkers, commercially available (*1S,2S*)-(+)-2-amino-1-phenyl-1,3-propanediol can be transformed into similar ligands **61** that bear an additional phenyl group [74]. The Zhang group described *P*-chiral phospholane–oxazoline ligands **62** traced back to 1-tert-butyl-phospholane 1-sulfide [75]. Gilbertson demonstrated the use of phosphanorbornene oxazoline ligands **63** [76]. Agbossou-Niedercorn, Avarvari, and coworkers prepared ligands composed of redox-active tetrathiafulvalene (TTF) derivatives **64** [77,78]. The Kang group described the synthesis of *ortho*-carborane-based phosphinooxazolines (CABPHOX) **65** [79].

Bidentate oxazoline-based ligands consisting of trivalent P-donor moieties other than phosphines Several ligand families consisting of phosphinites have been successfully employed in asymmetric catalysis. In 1999, Uemura described the synthesis of glucosamine-derived phosphinite–oxazoline ligands **66** (Figure 6.4) [80,81]. Somewhat later, the same group published an amphiphilic modification of these ligands (**67**) where

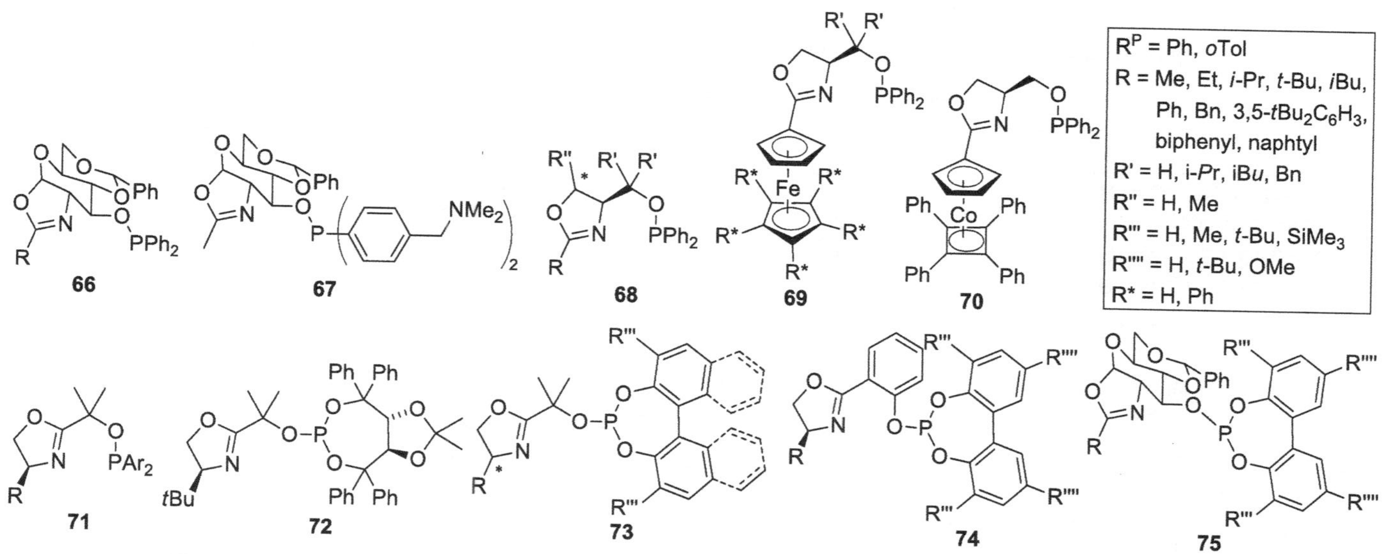

Figure 6.4 *Structures of bidentate oxazoline ligands with phosphinites or phosphites as P-donor unit.*

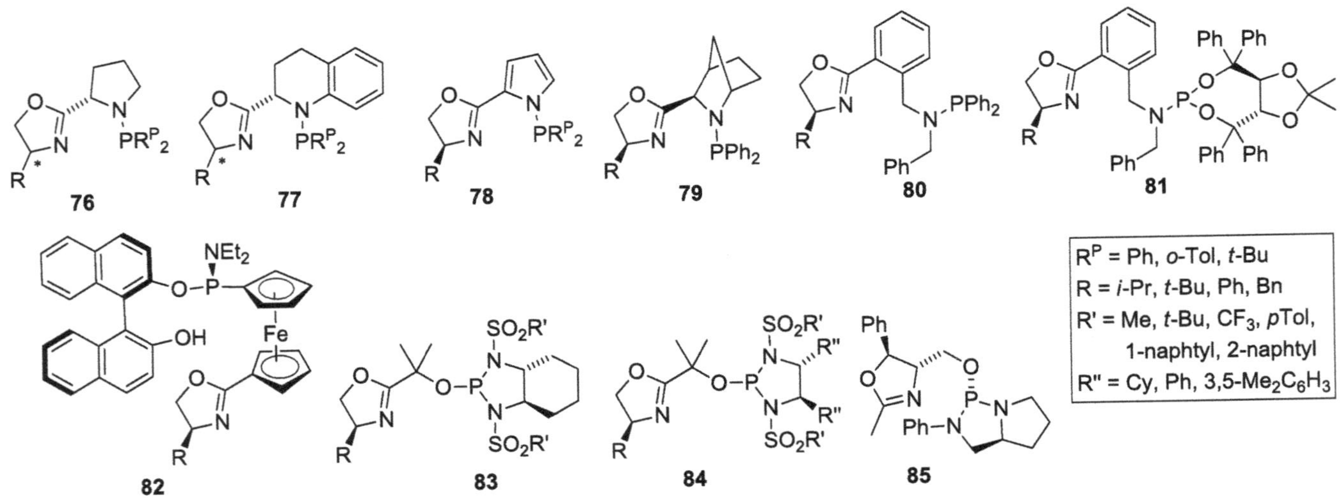

Figure 6.5 *Structures of bidentate oxazoline ligands carrying nitrogen-substituted P-donor units.*

both aryl groups at phosphorus carry an additional tertiary amine function. Such ligands allow application in biphasic or aqueous media when the corresponding metal complexes are methylated (e.g., by treatment with Meerwein salt) affording a charged quaternary ammonium species [82].

Phosphite–oxazoline ligands with a similar alkyl backbone have been reported independently by the groups of Pfaltz (**68** and **69**) [83,84] and Richards (**69** and **70**) [85] which were derived from either serine (SerPHOX) or threonine (ThrePHOX). Ligands **71** which display a reversed connectivity of the oxazoline moiety could be synthesized from commercial amino alcohols and 2-hydroxyisobutyric acid in only two steps (that is the reason why ligands **71** were termed SimplePHOX) [86]. The research group of Pfaltz was the first to disclose binaphthol-containing phosphite–oxazolines. They introduced TADDOL-derived ligand **72** and biaryl phosphite **73** [12]. In 2005, Pàmies and Diéguez reported on biphenol-derived phosphites **74** as close analogs of the classical PHOX ligands [87]. They also described the synthesis of similar glucosamine-derived phosphite ligands **75** [88,89].

Besides oxygen-substituted P-donor ligands, one can also use the pendant nitrogen compounds. In general, the higher polarity of the N–H compared to C–H bond renders amines more acidic and, therefore, easier to deprotonate for the formation of the desired P–N bond. On the other hand, that bond is also more susceptible toward hydrolysis, typically resulting in a more difficult reaction workup. Nevertheless, nitrogen-substituted phosphorus ligands are frequently employed in catalysis. Gilbertson was the first to report the synthesis of proline-derived ligands **76** (Figure 6.5) [67,90,91]. Later, Agbossou-Niedercorn also reported the preparation of **76**, its diastereomers and 6-membered tetrahydroquinoline analogs **77** [92]. Cozzi, Pfaltz, and coworkers reported on PyrPHOX (**78**). As mentioned, the synthesis of these phosphinopyrrolyl-oxazolines involves the use of milder base for the coupling of the chlorophosphine as compared to PHOX (**12**), but a more careful workup had to be performed in order to suppress hydrolysis [93]. Andersson described bicyclic aminophosphine ligands **79** [94]. In 2007, the Guiry group reported on aminophosphine **80** and TADDOL-derived phosphoramidite oxazolines **81** forming eight-membered metallacycles [95] Hou and Dai described ferrocene-containing aminophosphinite oxazoline ligands **82** [96–98]. Pfaltz and colleagues reported on tosylaminophosphine-oxazolines **83** and **84** [99,100]. The Bondarev group contributed the bicyclic ligand **85** which was prepared in three steps from enantiopure (S)-anilinomethylpyrrolidine (accessible from *L*-glutamic acid) [101].

6.2.1.2 Imidazoline-based P,N-ligands

Phosphinoimidazolines were first studied as ligands for Heck reactions and hydrogenations at Boehringer Ingelheim by Busacca and colleagues, with patents filed in 2001 [102]. Throughout recent years, the Busacca group described the preparation of a diverse library of their so-called BIPI ligands **84** and **85** allowing the in-depth study of electronic effects exerted by the imidazoline (and phosphine) substituents (Figure 6.6) [103–105].

Also the research group of Claver and Castillón contributed certain members of ligand **84** that bear unsubstituted imidazoline moieties (R′=H) [106]. Pfaltz and coworkers described the synthesis of phosphinophenylimidazolines 85 [107] and 86 [108,109] as nitrogen analogs of the PHOX ligands and SimplePHOX ligands, respectively.

6.2.1.3 Oxazole-, thiazole-, and imidazole-based P,N-ligands

In 2004, Andersson *et al.* reported on a first generation of bicyclic oxazole-based ligands describing the phosphinites **90** (Figure 6.7) [110]. Later, they reported on phosphinothiazoles **91** as highly efficient ligands for the asymmetric Ir-catalyzed hydrogenation [111]. More recently, the Andersson group, in collaboration with Diéguez and colleagues, described the construction of a biarylphosphite ligand library (**92–94**) consisting of

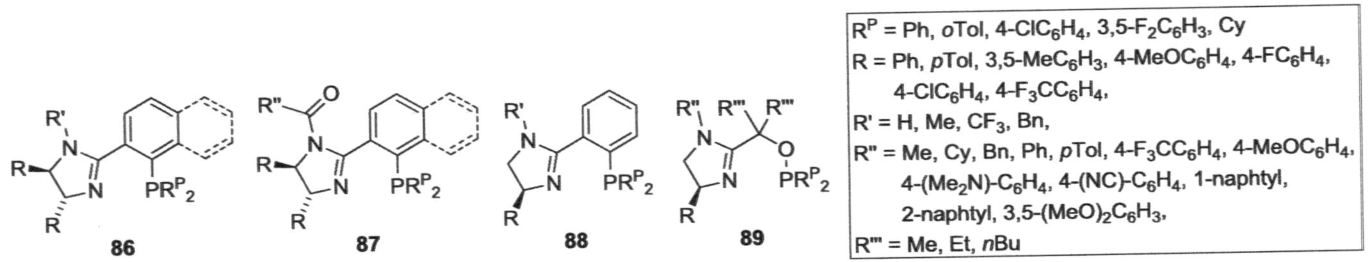

Figure 6.6 *Structures of bidentate phosphine–imidazoline ligands.*

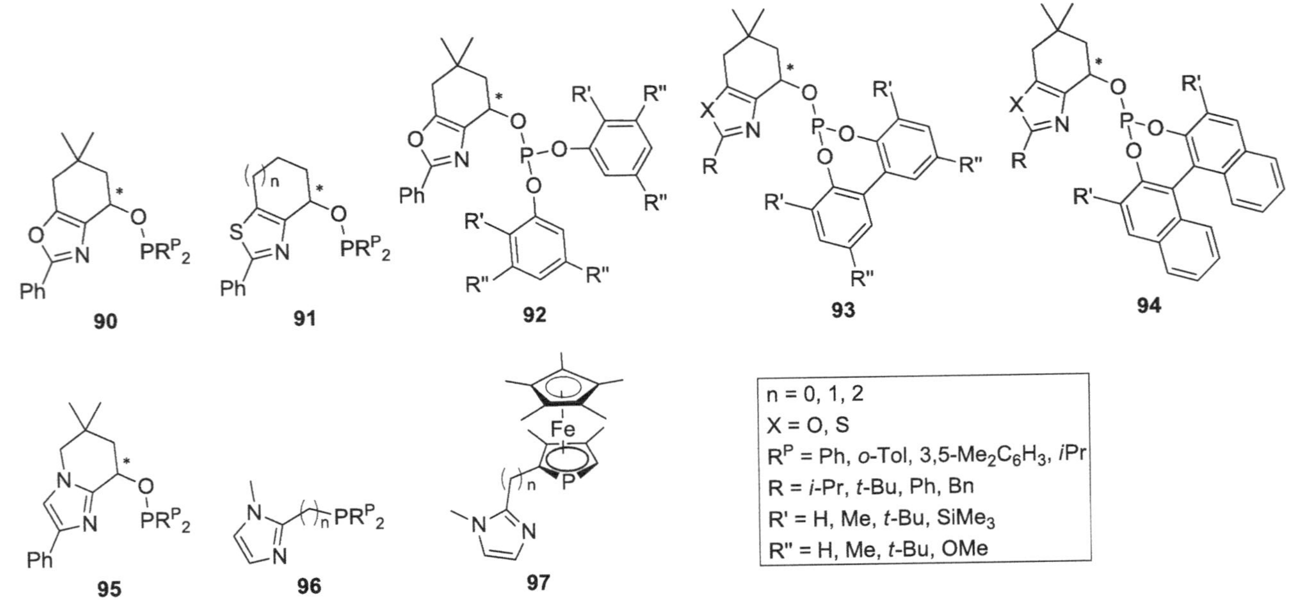

Figure 6.7 *Structures of bidentate oxazole-, thiazole- and imidazole-based P,N-ligands.*

more than 50 compounds [112]. The Andersson group also reported on imidazole-based ligands **95** [113]. Whereas Messerle *et al.* prepared similar simplified achiral phosphine–imidazolyl ligands **95** [114], the Ganter group contributed imidazole ligands **97** that contain a phosphole as a P-donor moiety [115].

6.2.1.4 Pyridine-based P,N-ligands

A large number of bidentate P,N-ligands contains a pyridine as N-donor [116–118]. Since complexes of the formula [Ir(py)(cod)PR_3)]PF_6 which were first studied by Crabtree *et al.* had shown high catalytic activity in the hydrogenation of tri- and tetrasubstituted olefins [119,120], the interest in chiral analogs of Crabtree's (pre-)catalyst has prompted many research groups to investigate metal complexes that bear pyridine-based chiral bidentate P,N-ligands [118].

Among the first examples of chiral pyridine-based P,N-ligands were compounds **98a** prepared by the group of Mathey using a Diels–Alder reaction to construct the bicyclic core (Figure 6.8) [121]. Relying on a similar strategy, the *exo*-selective synthesis of P-chiral bidentate pyridine phosphine ligands **98b** could be achieved by Leung *et al.* [122]. Ito, Katsuki and coworkers have introduced bicyclic 2-(phosphinoaryl) pyridines **99** for asymmetric allylic alkylations. The ligands were prepared by Suzuki–Miyaura cross-coupling of the bicyclic 2-chloropyridine derivative and 2-hydroxyphenylboronic acid followed by transformation of the phenol group into the desired phosphine. The Wilson group reported on similar compounds **100** [123]. Chelucci *et al.* [124] and Malkov, Kočovský, and colleagues [125] have independently described ligands **101–104** as synthetically more easily accessible yet similarly efficient alternatives to **99**. The chiral precursors (−)-pinocarvon, (−)-isopinocampheol, (−)-β-pinene, and (+)-camphor were converted into the corresponding exo-methylene ketones, and Kröhnke's method [126] using pyridinium salts for the synthesis of pyridines was applied [124]. The Chelucci group also reported on the synthesis of pyridyl ligands **105** and their bipyridyl analogs, applying a strategy that started from tartaric acid and relied on Co-catalyzed cyclotrimerization [127,128] as key step for pyridine formation [129].

Pyridine phosphinite ligands **106** were formed from 2-hydroxypyridine (or 2-hydroxymethylpyridine) *tris*(dimethylamino)phosphine and (*S*)-2-anilinomethylpyrrolidine. For the synthesis of QUIPHOS ligands **107-111**, *tris*(dimethylamino)phosphine was first treated with a chiral diamine, amino alcohol (ephedrine, pseudoephedrine or prolinol) or diol (2,2′-binaphthol), and then reacted with 8-hydroxyquinoline to afford the desired ligands [101,130–132].

By lithiation of 8-bromoquinoline and subsequent coupling to a chlorodiaminophosphine, ligands **112**, having a more inert P–C bond in place of a hydrolizable P–O bond when compared to **107–111**, were synthesized by Buono [133]. Mathieu and colleagues described the synthesis of *P*-chiral ligands **113** that could be separated by fractionated crystallization as diastereomeric palladium complexes followed by liberation of the ligand using KCN. Bulky members of this ligand family carrying *P*-anisyl groups showed hemilabile character and apparently bind to metals not exclusively in a P,N-bidentate fashion, but also to some extent as P,O-chelate ligands [134]. Lithiation of pyridine followed by treatment first with menthone or camphor, then with chlorodiphenylphosphine, delivered ligands **114** and **115**, respectively, as reported by Faraone and coworkers [135].

Simplified chiral phosphinopyridine ligands **116** have been described more recently by the groups of Elsevier and Kamer, who also incorporated very bulky substituents in the 2-position of the pyridine [136,137]. P-chiral variants of such ligands were prepared by Leung *et al.* [138]. Similar phosphinite ligands **117** with *gem*-dimethyl substitution patterns were reported by the Braunstein group [139]. Pyridine–phosphines **118** and –phosphinites **119** and **120** were developed by Pfaltz and coworkers [140,141]. The synthesis of **120** involved the Furukawa–Boekelheide rearrangement [142,143] of bicyclic pyridine N-oxide to give a mixture of the corresponding pyridine alcohol enantiomers which can be separated by chromatography or by enzymatic kinetic resolution for further elaboration to the desired ligands [144].

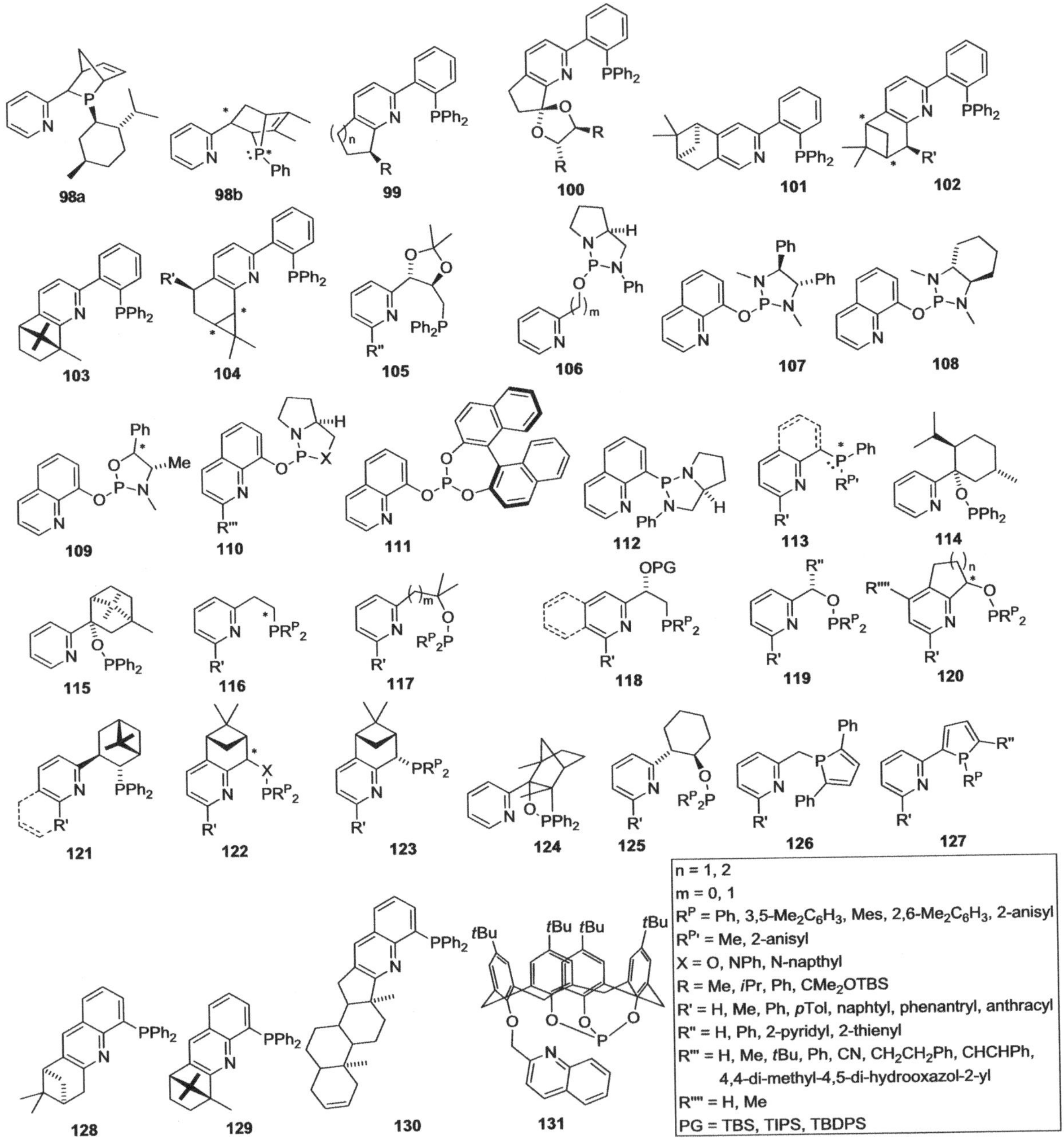

Figure 6.8 *Representative set of structures of pyridine-based P,N-ligands.*

Ligands **118**, **119**, and especially **120** proved to be highly effective ligands for Ir-catalyzed hydrogenation. Ir-complexes with ligands **120** are the only catalysts known that allow the highly enantioselective hydrogenation of purely alkyl-substituted alkenes. Compounds **120** were also reported independently by Zhou and colleagues [145].

Knochel reported on terpene-derived ligands **121** [146]. Verendel and Andersson described pinene-derived pyridyl phosphine(it)e ligands **122** that form six-membered metallacycles upon chelation [147], and, later, the Li group prepared binaphthyl phosphite variants of **122** [148] and analogous compounds **123** [149] that form five-membered metallacycles. Starting from (–)-fenchone, Rominger *et al.* reported on the synthesis of pyridine-substituted fenchyl-phosphinite (FENOP) ligands **124** [150]. Liu and Zhou described the preparation of cyclohexane-based ligands **125** [145]. Pyridyl-phosphole ligands **126** [151] and **127** [152] were reported by the groups of Mathey and Réau, respectively. Chelucci and colleagues published the synthesis of further terpene-derived ligands **128** and **129**, as well as steroid-derived ligand **130**. The group of van Leeuwen and Kamer demonstrated the synthesis of the calixarene-derived quinoline ligand **131** [153].

Phosphinoarylquinoline **132**, published in 1991 by Brown *et al.*, constitutes an early example of an axially chiral pyridine-based P,N ligand (Figure 6.9). However, it was found that this compound easily racemized, rendering it of limited use in asymmetric catalysis [154]. This drawback was overcome by the same authors with the synthesis of quinoline–naphthyl (QUINAP) ligands **133** which could be obtained by Suzuki cross-coupling of 2-chloroquinoline to methoxynaphthylboronic acid. After methyl ether cleavage and conversion of the free phenol into the pendant triflate, a Pd-mediated coupling of diphenylphosphine oxide followed by reduction to the free phosphine delivered racemic QUINAP ligands **133**. Upon formation of diastereomeric Pd complexes and fractionized crystallization, enantiopure ligands could be liberated using 1,2-bisdiphenylphosphinoethane (DPPE) [155–157]. It was later shown by Carreira *et al.* that analogous quinazine ligands (PINAP ligands) are synthetically more accessible yet equally efficient in asymmetric catalysis (see Section 3.1.6) [158].

Phenanthrene-naphthyl (PHENAP) ligands **134** were prepared in a manner similar to the synthesis of QUINAP, and it was found upon complexation with Pd that only one of the two diastereomeric forms behaved as chelating ligand, whereas the other binds in a monodentate fashion via *P*-coordination [159]. Brown demonstrated that metal complexes of the easily racemizable indole-derived ligand **135** can be used to gain mechanistic insights in the Pd-catalyzed allylic substitution [160]. Kwong and Chan described the synthesis of an array of atropoisomerically pure phosphinoarylpyridines **136** [161]. Zhang and coworkers reported on binaphthyl ligands **137** that have a large bite angle; they were prepared in a straightforward manner from 2-pyridinecarboxylic acid and 2-amino,2′-phosphinobinaphtol through amide coupling [162]. The Ganter group published pyridyl-substituted *phospha*-ferrocenes **138** [163]. The Fu group contributed ferrocenylpyridine ligands **139** [164], and Chung and colleagues described chromocene-based ligand **140** [165]. Ruzziconi *et al.* reported on quinolinophane–phosphines **141** [166].

6.2.1.5 Amine- and imine-based P,N-ligands

For catalytic hydrogenations, allylic substitutions, Michael additions, and other applications, aminophosphines are attractive ligands due to their structural simplicity and the synthetic accessibility resulting therefrom [167]. A variety of very simple aminophosphines such as **142** can be accessed by reduction and subsequent nucleophilic substitution of the corresponding naturally occurring amino acids as described by Achiwa more than a decade ago (Figure 6.10) [168,169]. Conformationally more rigid ligands **143** are readily accessible from naturally occurring chiral amino alcohols like ephedrine and its derivatives [170].

The nucleophilic ring opening of aziridines as reported by Yudin represents a viable method to access *β*-aminophosphines like **144**. The synthesis can be performed with achiral reagents as the racemic products can be resolved via the tartaric acid salts [171]. The treatment of amino alcohols with thionyl chloride followed by oxidation delivers cyclic sulfamidates which, in turn, can be subjected to nucleophilic substitution at the carbon center bound to oxygen. Using this sulfamidate route, access to structures of the general formula **145** and **146** suitable for the formation of both five- and six-membered metallacycles can be obtained

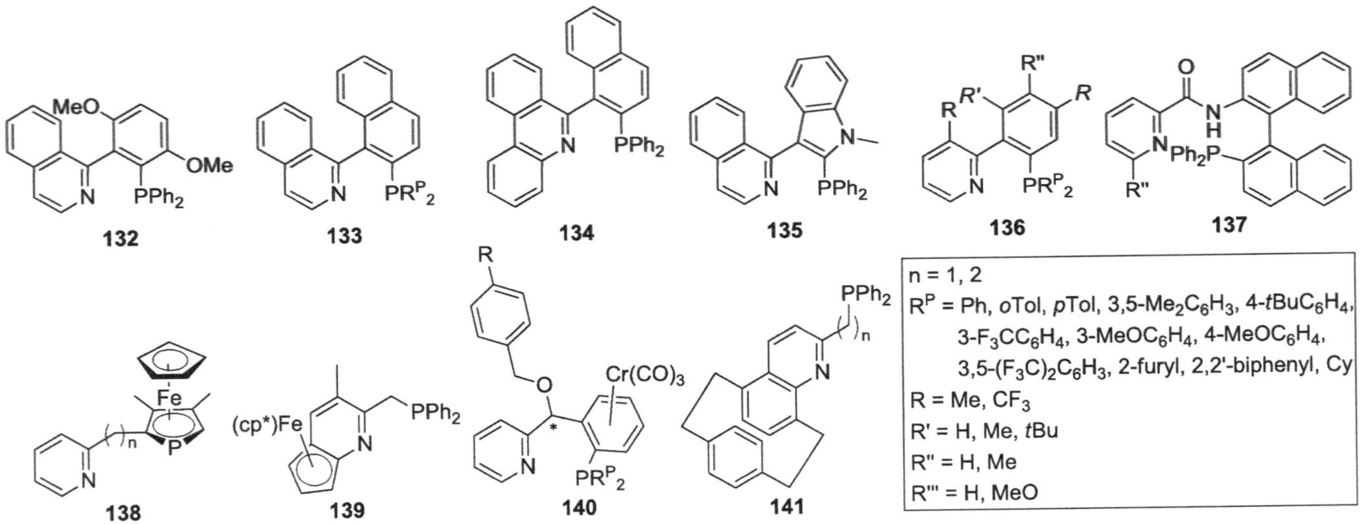

Figure 6.9 *Pyridine-based P,N-ligands bearing elements of axial or planar chirality in the backbone.*

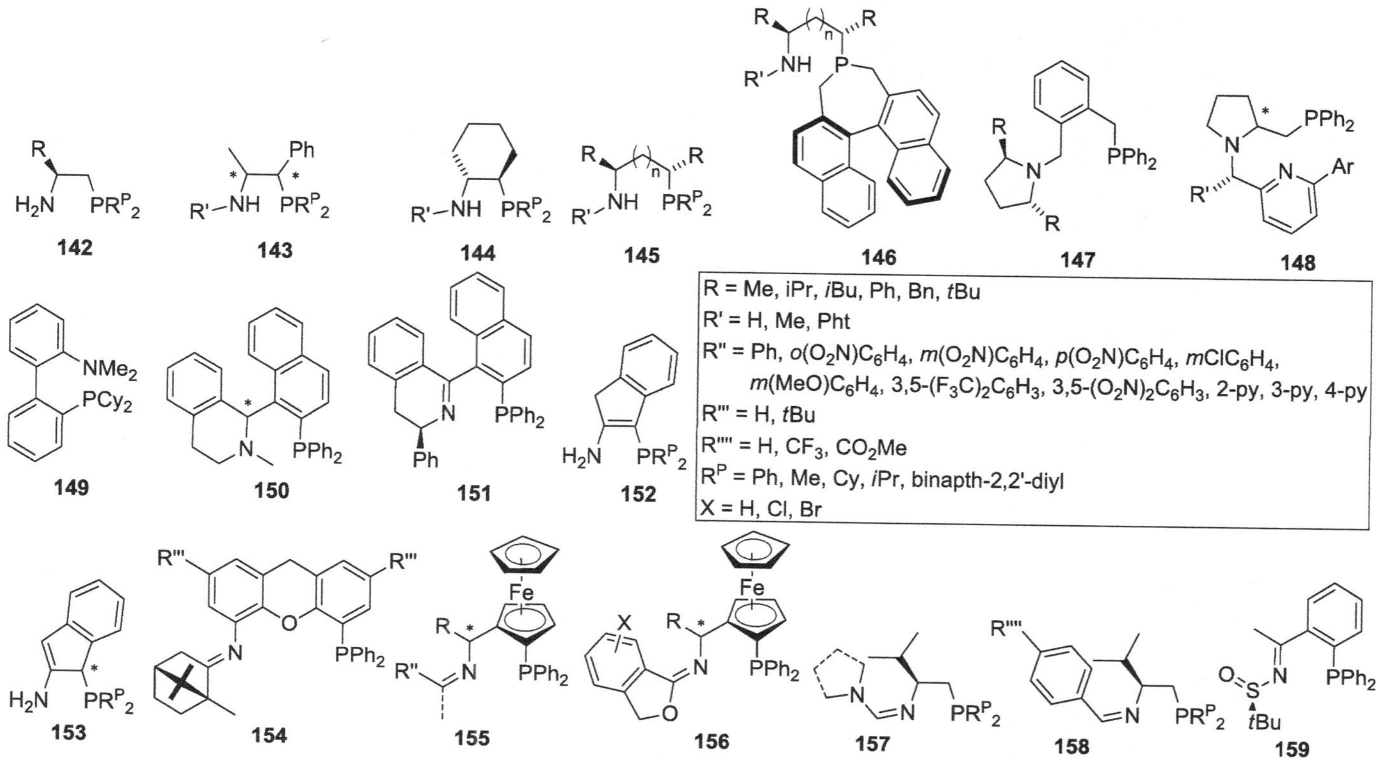

Figure 6.10 *Structures of published P,N-ligands incorporating either an amine or an imine as the N-donor group.*

[172,173]. Alternative synthetic approaches to aminophosphines include the hydroamination of vinylphosphine oxides followed by phosphorus reduction [174,175], the amide coupling of borane-protected β-phosphinoacids followed by amide reduction and phosphine deprotection [176,177], and the asymmetric guanidine-catalyzed *phospha*-Michael addition of secondary phosphine oxides to nitroalkenes followed by reduction of the nitro group and phosphine oxide [178].

Guiry published the synthesis of pyrrolidine–phosphine **147** [179]. Uenishi and Hamada reported on pyrrolidine-centered ligands **148** with an additional pyridine moiety, which, however, was shown not to be involved in metal binding [180]. The Buchwald group developed ligand **149** (DavePhos), an air-stable crystalline solid that is very suitable for Suzuki cross-couplings and aromatic aminations [181]. Dihydroisoquinoline and 1-naphthol were the starting materials for the synthesis of ligand **150** (named mTHIQ-NAP) by Li and coworkers. The same group also reported on configurationally unstable FliNAP ligands **151** [182]. Stradiotto *et al.* introduced achiral (**152**) [183] and chiral (**153**) [184] indenyl-based amino phosphines which upon metal complexation are converted to anionic ligands. The Pfeffer group reported on terpene-derived phosphine–imine ligands **154** with a large bite angle due to the xanthene backbone [185]. The groups of Zheng and van der Eycken described the synthesis of ferrocene ligands **155** [186,187] and **156** [188], respectively. Achiwa *et al.* studied amidine ligands **157** and **158**, readily available from the corresponding amino alcohols [168,169]. Schenkel and Ellman reported on S-chiral phosphine–sulfonylimine ligands **159** prepared in one step by Ti-mediated condensation of *t*-butanesulfinamide with (commercially available) 2-(diphenylphosphino)-benzaldehyde [189].

6.2.1.6 *Other P,N-ligands*

A few additional P,N-ligands that do not belong to any of the aforementioned categories are worth mentioning in the context of this chapter: among those are the PINAP ligands developed by Carreira *et al.* (Figure 6.11) [158]. The Lewis acid–mediated addition of 1,4-dichlorophthalazine to 2-naphthol followed by coupling of either (*R*)-phenylethanol or (*R*)-phenylethylamine, conversion of the phenol to the triflate, and subsequent Ni-catalyzed phosphine coupling delivers a mixture of diastereomers of **160** that can be readily separated by silica gel chromatography, rendering these ligands a convenient alternative to QUINAP which is more difficult to prepare. Similar compounds **161** and **162** based on a 1,3-quinazine scaffold were reported by Guiry *et al.* with **162** being susceptible to racemization [190,191,192]. Quinazolinone ligands **163** were reported by Virgil and coworkers [193,194]. Thiel and coworkers reported *ortho-* and *meta*-substituted pyrazole–phosphine ligands **164** [195], as well as phosphites **165** tethered to either a pyrazole [196] or a triazole [197].

6.2.2 P,O-ligands

Many important transition metal-catalyzed transformations make use of hemilabile bidentate P,O-ligands, some even on an industrial scale. The Shell Higher Olefin Process (SHOP), for instance, is run using simple phosphinocarboxylic acid ligands of type **166**, and phosphinosulfonic acid such as **167** serves as ligands for the palladium-catalyzed alternating co-polymerization of alkenes and carbon monoxide (Figure 6.12) [198–200]. Keim and Schulz systematically screened bidentate ligand families **166–169** for studies of their ability to chelate nickel in the oligomerization of ethylene [201]. The groups of Pfaltz [202] and Rieger [203] reported independently on the synthesis of phospinoarylsulfonic acid **170** and analogs by *ortho*-lithiation of phenylsulfonic acid and subsequent coupling of chlorophosphines [198].

The combinatorial synthesis of achiral or racemic acetal ligands **171–173** for applications in the Pd-catalyzed Suzuki–Miyaura cross-couplings [204] and Buchwald–Hartwig aminations [205] was reported by the industrial research team of Guram. Again, the ligand synthesis involved halogen–lithium exchange of halogenated arenes followed by coupling of chlorophosphines [206]. Clayden and coworkers found that the

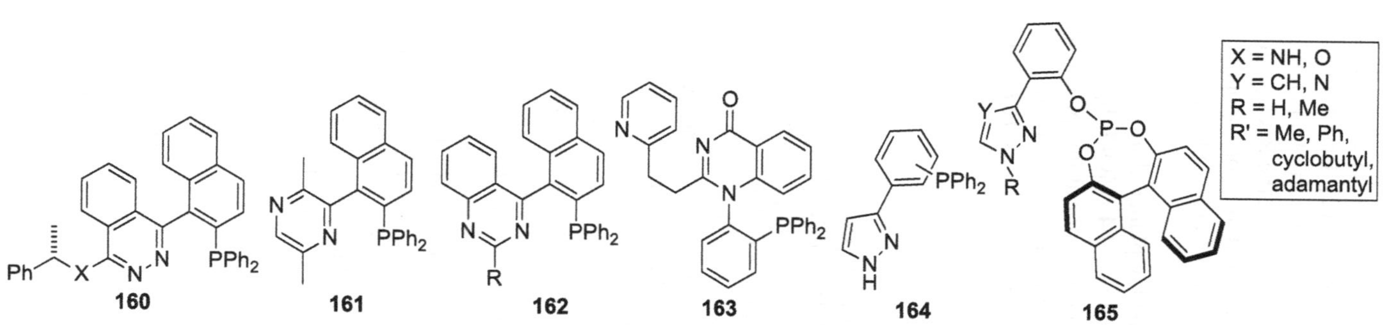

Figure 6.11 *Structures of P,N-ligand classes not discussed so far.*

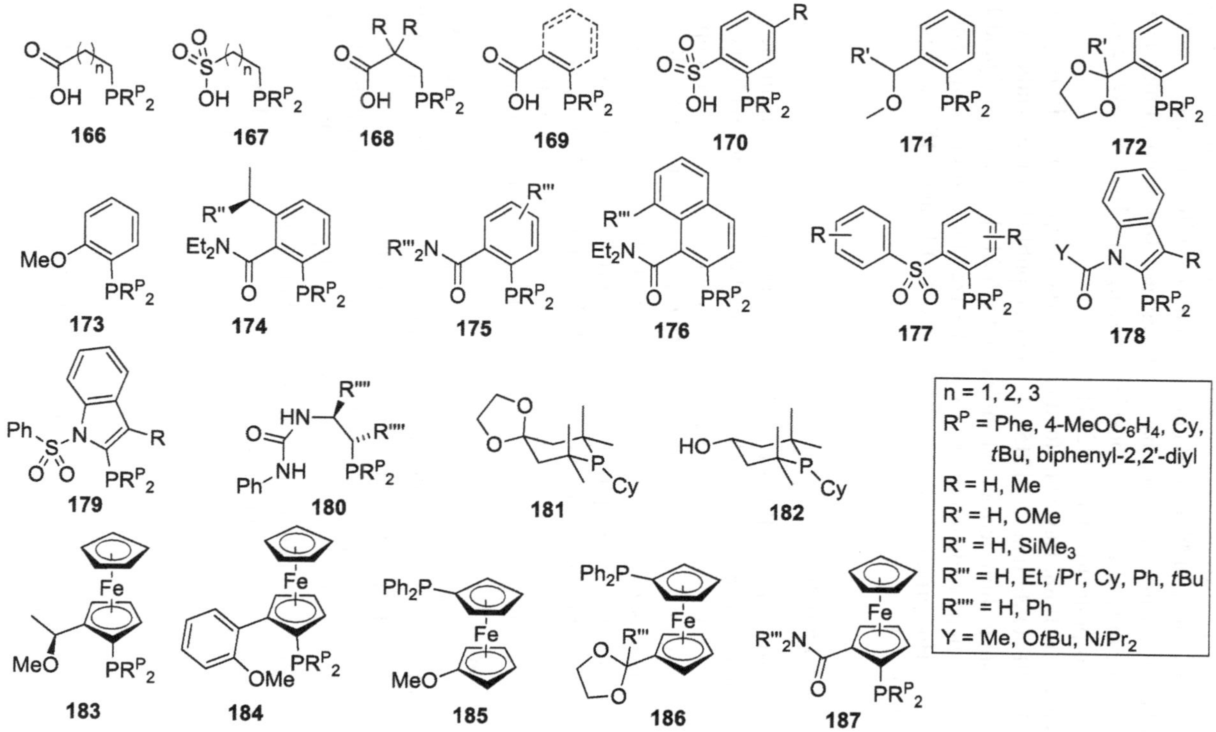

Figure 6.12 *Structures of important bidentate P,O-ligands.*

installation of sterically demanding substituents in phosphinoaryl amides **174** leads to suitable chiral ligands due to suppressed interconversion of the two atropoisomers [207]. The groups of Kwong and Chan [208] and Dai [209–211] independently prepared a variety of air-stable Bphos ligands **175** and **176**. The industrial research team of Tom reported on readily available achiral phosphinoarylsulfones **177** [212]. Based on the indole scaffold, Kwong and coworkers reported on P,O-ligands where the O donor either is part of an amide, carbamate, or urea (**178**) or was embedded into a sulfonamide moiety (**179**) [213]. Recently, the Reek group described the synthesis of urea-based P,O-ligands **180** by coupling phenyl isocyanate with the corresponding amines. McNulty introduced phosphorinane ligands **181** and **182** accessed by double Michael addition of a primary phosphine to phorone [214].

As early as 1988, Hayashi, Ito, and coworkers reported on P,O-ligands that display planar chirality, such as **183** (PPF–OMe ligands) [215]. Johannsen published the synthesis of aryl analogs **184** (MOPF ligands) starting from a chiral ferrocenyl sulfoxide [216,217]. Ferrocene ligand **185** with both donor atoms placed at different cyclopentadienyl units was synthesized by Gibson, Long, and colleagues from 1,1′-dibromoferrocene. The Hor group prepared acetal-based P,O-ligands **186**, and phosphinoferrocenyl amides **187** were published by Kwong and Chan [213].

6.2.3 P,S-ligands

Starting from his well-known chiral nonracemic oxazolidinones, Evans developed a synthesis of α-bromoamide **188** that can undergo nucleophilic substitution with thiols (giving **189**), followed by conversion to aldehyde **190** (Scheme 6.5). Subsequent Grignard addition and coupling of a chlorodiarylphosphine afford phosphinites **191** which, together with **192**, proved to be viable ligands for a number of metal-catalyzed transformations (e.g., allylic substitution, asymmetric hydrogenation, and hydrosilylation) [218,219]. At about the same time, Hauptman [220] and Dervisi [221] described the synthesis of phosphine homologs **193** forming five- instead of six-membered chelate rings, and Imamoto demonstrated P-chiral variants of these ligands [222]. Molander introduced cyclopropane-based P,S-ligands **194** and **195** [223].

Several P,S-ligands based on sugar scaffolds have been reported. Khiar and coworkers obtained both anomers of **196** in five steps from galactose pentaacetate and prepared its *quasi*-enantiomer **197** from D-arabinose tetraacetate [224,225]. RajanBabu introduced phosphine ligands **198** derived from D-mannitol [226]. The groups of Net and Ruiz [227], as well as Diéguez *et al.* [228,229], reported on xylofuranose-derived ligands **199** bearing either a phosphite or a phosphinite unit.

Other P,S-ligands rely on terpenes as the primary source of chirality. Starting from pulegone, Pericàs and coworkers described the synthesis of bicyclic P,S-ligands **200** named PuPhos. Having a very short tether between the two donor atoms (only one carbon), they turned out to be very efficient chelators of both metal centers in bimetallic complexes such as those formed by $CO_2(CO)_6$ and alkynes. Thus, these ligands have been successfully applied in Pauson–Khand processes [230]. Later, Verdaguer, Riera, and colleagues introduced CamPhos (**201**) the camphor-derived analog of PuPhos [231,232]. Similar to ligands **199–201**, Nakano, Hongo, and coworkers reported on ligands **202** (from pulegone) [233,234], **203** and **204** ((iso)borneol) [233,234], and **205** (xylofuranose-based) [235], where the tether between the mixed donor units contains an additional aryl moiety. Later, Nakano also reported on polymer-supported analogs **206** [236].

Hiroi *et al.* were the first to explore amino acids as chiral building blocks for the synthesis of P,S-ligands such as the proline-derived compounds **207–209** [237–241]. They also reported on S-chiral ligands including phosphinoarysulfoxides **209** [237–241] and **210** [237–241], the pendant aminophosphine sulfoxides **211** [242], and sulfoxide aminophosphines derived from pyrrole (**212**) [243] or indole (**213** and **214**) [243]. Riera and Verdaguer introduced minimally tethered phosphino sulfoxides **215** and **216** [244,245]. In 2007, phosphino sulfonamides (PNSO ligands) [246,247] **217** and **218** were published as nitrogen analogs which showed a considerably higher tendency to undergo hydrolytic cleavage of the P–N bond. Toru and colleagues

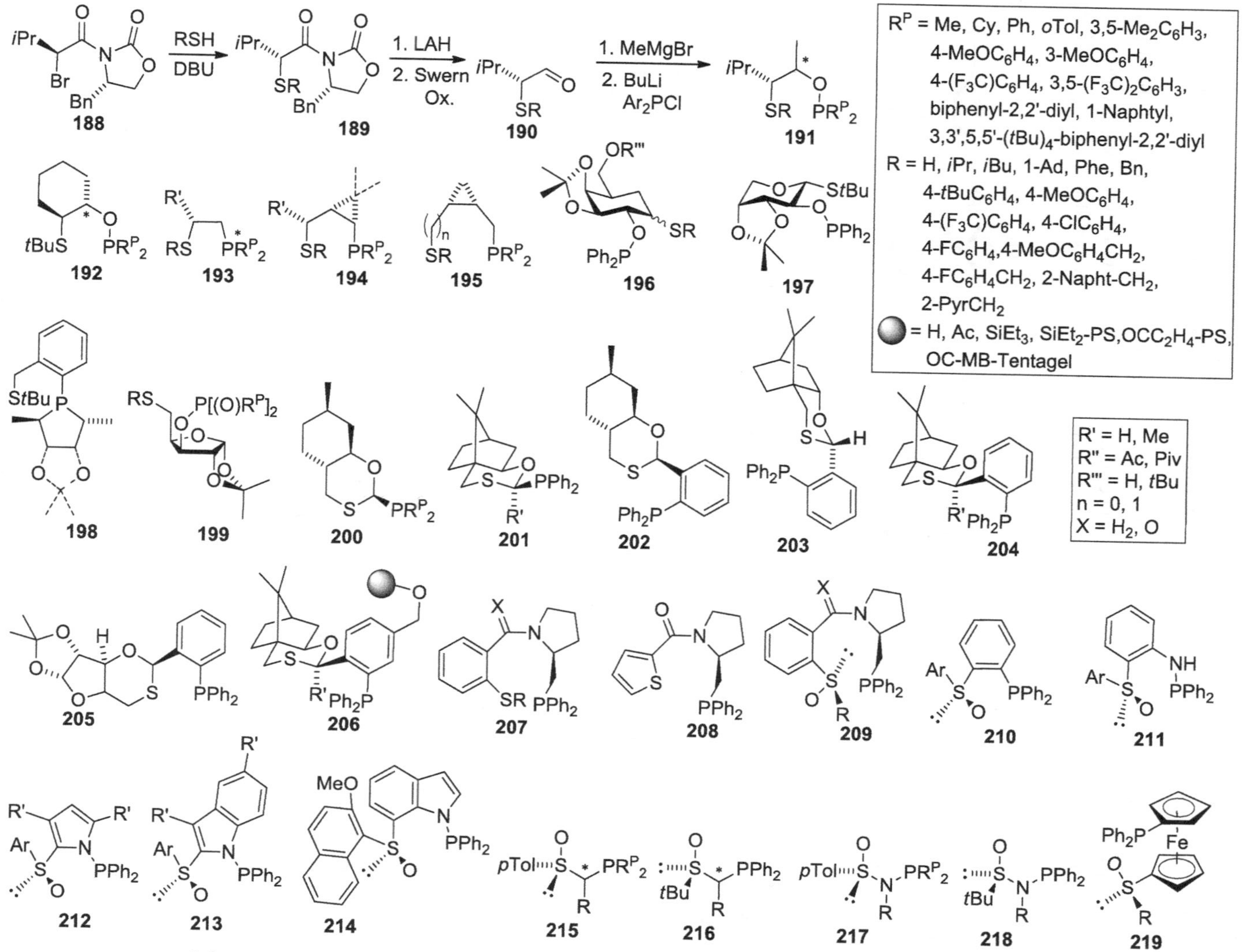

Scheme 6.5 Structures of the most prominent P,S-ligands with elements of central chirality.

combined central chirality at sulfur and planar chirality of ferrocenes focusing on nonracemic phosphinoferrocenyl sulfoxides **219** [248].

P,S-ligands solely based on elements of planar chirality have found applications in metal catalysis that span from allylic substitutions over ring-opening reactions of strained alkenes to [4+2]- and [3+2]- cycloadditions [249].

In 2002, for example, Carretero introduced ferrocene-based P,S-ligands **222** termed FesulPHOS (Scheme 6.6). They are readily accessible by a sequence of diastereoselective lithiation of enantiopure sulfenylferrocene **220**, coupling with a chlorophosphine, and final reduction of the resulting sulfoxide **221** with

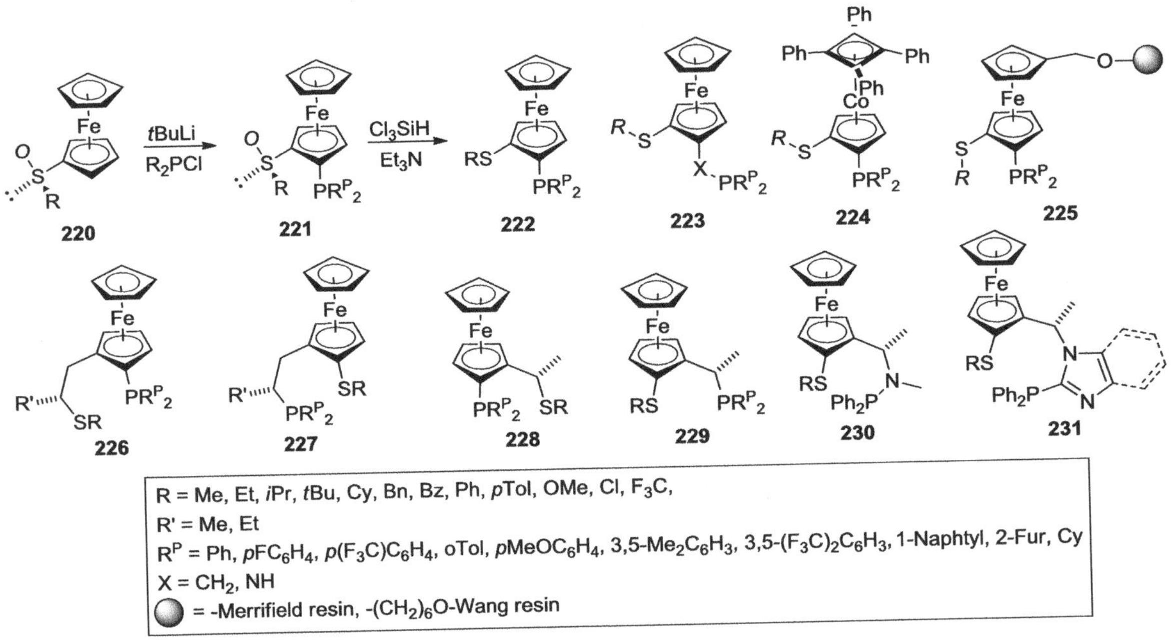

Scheme 6.6 *Structures of the most prominent P,S-ligands with planar chirality.*

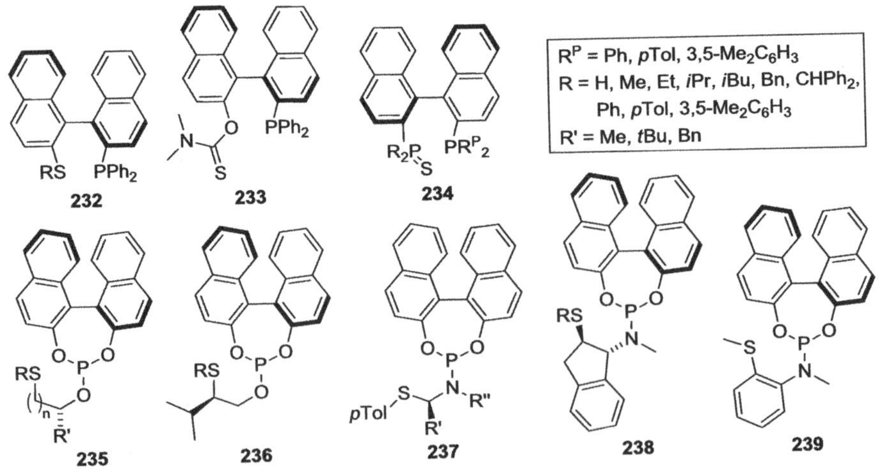

Figure 6.13 *Structures of common P,S-ligands with axial chirality.*

chlorosilane [250]. Later they published homologous (amino)phosphine compounds **223** [251], cobaltocene complex **224** [252], and polymer-supported ligands **225** that allow facile separation and recycling of the corresponding metal complex [253]. In 1999, the Enders group had already come up with ligands **226** and **227** [254], and Hou, Dai, and coworkers published later similar ligands that allow the formation of six-membered metallacycles [255]. Zeng and Zhong described a systematic study of the electronic effects displayed by sulfide and phosphine within ligands **228** and **229** [256]. Chan *et al.* explored aminophosphine ligands **230** [257] and (benz)imidazole ligands **231** [258].

Axially chiral P,S-ligands **232** were first reported in 1994 by the group of Gladiali [259], and shortly thereafter by Kang and coworkers (Figure 6.13) [260]. Shi and colleagues described the synthesis of **233** and thiocarbamate analogs **234**, and Faller *et al.* introduced mixed phosphite sulfide ligands **235**. The groups of Pregosin [261] (**236**) and Takemoto [262,263] (**236–237**) published binaphthol-derived phosphite ligands, whereas Crévisy, Mauduit, and coworkers focused on phosphoramidites **238–239** [264].

6.2.4 P,C-ligands

To some extent, N-heterocyclic carbenes (NHC) share with phosphorus donors the ability to act as π acceptors. The pronounced σ-donating character of NHC units, on the other hand, is believed to render the metal–carbene bond more stable than a metal–phosphorus bond. Thus, in line with the concept of hemilability, the P donor of a bidentate P,C-ligand can dissociate freeing a coordination site at the metal center. This electronically unique situation makes ligands consisting of phosphorus and carbene donors attractive research targets in catalysis, and successful applications include Heck reactions, hydrogenations, and Suzuki couplings [265,266].

The first carbene-based bidentate P,C-ligand was reported by Herrmann *et al.* in 1996 [267]. The synthesis starts with the reaction of lithiated imidazole (**240**) with (diphenylphosphoryl)ethyl chloride, followed by N-quaternization using methyl iodide (Scheme 6.7). The phosphine oxide group of the resulting imidazolium salt **241** is then reduced with *in situ*-generated methylethoxysilanes affording free phosphine **242**. This, in turn, is first complexed with a metal salt before the carbene functionality of **243** is generated by treatment with strong base.

Subsequently, a number or research groups described further analogs of ligands **244** and applied them in Heck reactions and hydrogenations [268–271]. The Zhou group reported the synthesis of phosphinobenzylcarbenes **245** [272], and Bappert and Helmchen disclosed rigid naphthyl-bridged ligands **246** [273]. Nanchen

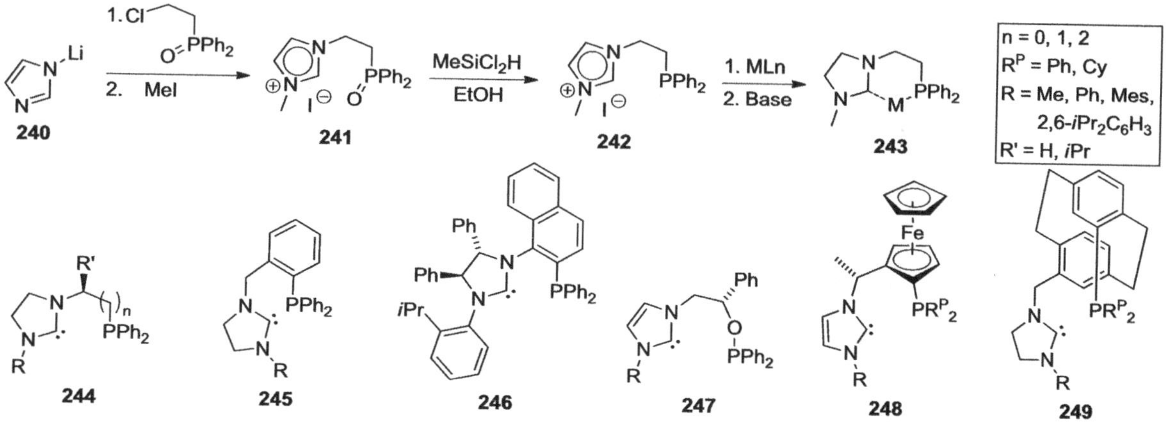

Scheme 6.7 *Synthesis of NHC-based P,C-complex **243** and structures of the most common P,C-ligands.*

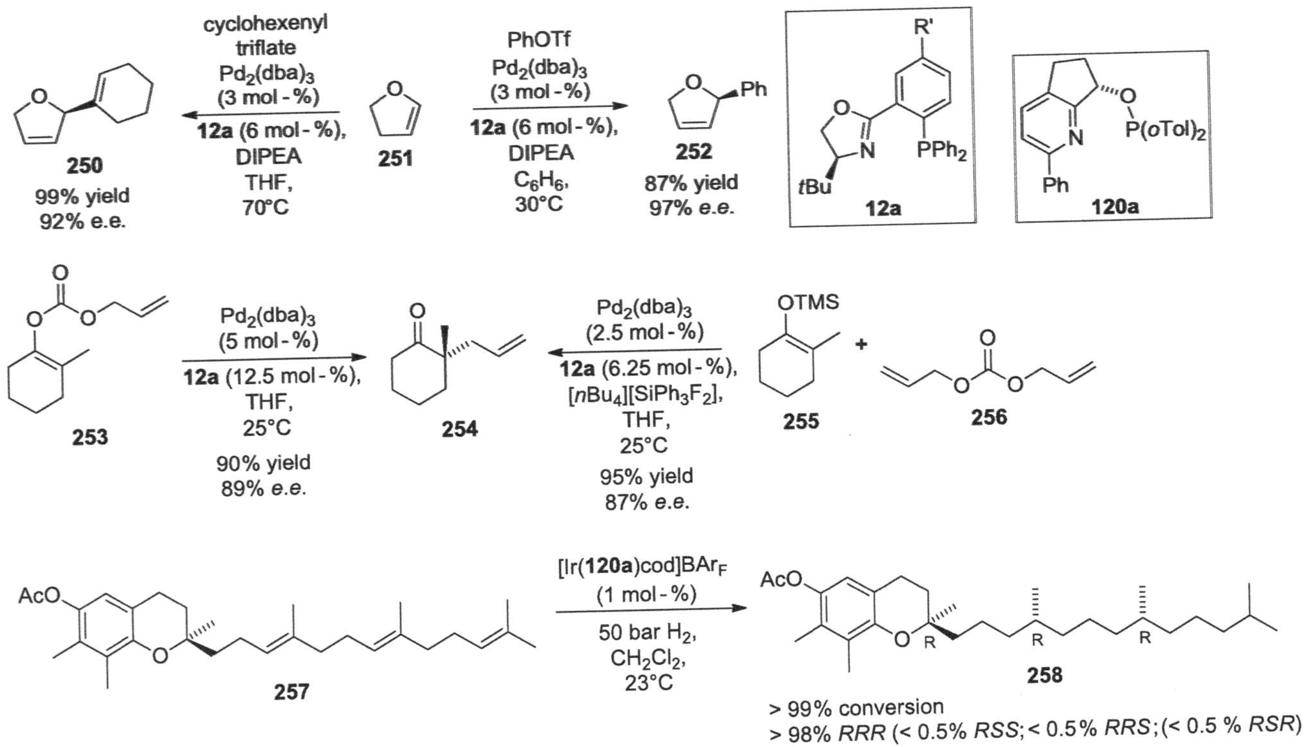

Scheme 6.8 *Useful applications of the most versatile ligands* ***12*** *and* ***120***.

and Pfaltz introduced phosphinite–carbene analogs **247** readily accessible by nucleophilic ring opening of chiral nonracemic epoxides [270]. Chiral ferrocene- and cyclophane-based ligands **248** and **249** were developed by the Chung [274] and the Bolm [275] groups, respectively.

6.3 Conclusion

The numerous examples of P,N-, P,O-, P,S-, and P,C-ligand families described in this chapter demonstrate that the electronic and steric differentiation of donor atoms of bidentate ligands for applications in catalysis has become common practice. Given the sheer number of ligand families, it seems advisable at this point to highlight the most versatile mixed donor ligand systems and the applications achieved with them.

Among the most versatile ligands is the PHOX system **12a**. It allows effective enantiocontrol of Pd-catalyzed Heck reactions, for example the conversion of 2,3-dihydrofuran (**251**) with aliphatic (such as **250**) or aromatic substrates (like **252**). This reaction occurs without migration of the double bond, which is often observed with the BINAP complexes (Scheme 6.8) [276].

PHOX ligands **12a** also showed excellent performance in the palladium-catalyzed allylic substitutions. A recent example is represented by the (intra- and intermolecular) decarboxylative allylation developed by Stoltz *et al.* who demonstrated the construction of quaternary carbon centers using Pd(phox) complexes [277].

Iridium-catalyzed hydrogenation has emerged as a particularly fruitful area of application for P,X-ligands [278–281]. Iridium complexes with P,N- and P,C-ligands have strongly enhanced the range of substrates that can be hydrogenated with high enantioselectivity. In contrast to rhodium and ruthenium diphosphine complexes, they do not require a coordinating group near the C=C bond and, therefore, allow highly enantioselective hydrogenations of a wide range of unfunctionalized tri- and even tetrasubstituted alkenes. In addition, they have also been successfully used for the hydrogenation of various functionalized olefins, heterocycles such as furans and indoles, and imines. The potential of these catalysts is demonstrated by the highly enantio- and diastereoselective introduction of the two stereocenters of the side chain of vitamin E by simultaneous hydrogenation of three C=C bonds using pyridine-phosphinte ligand **120a** (Scheme 6.8) [282].

The wealth of mixed donor ligands described in this chapter illustrates the almost unlimited structural variations that are possible and leaves no doubt that many further applications will be found for metal catalysts with P,X-ligands.

6.4 Experimental procedures

6.4.1 Synthesis of PHOX ligand 12a [22] (Scheme 6.9)

Under argon atmosphere, an oven-dried 500 mL three-neck flask equipped with a magnetic stirring bar, a reflux condenser, a pressure-equalizing addition funnel, and an internal thermometer was charged with (L)-*t*-leucine (**259**, 5.00 g, 38.1 mmol) and absolute tetrahydrofuran (100 mL, 0.38 M). The resulting suspension was cooled to 4°C in an ice-water bath, and $NaBH_4$ (2.40 eq, 91.5 mmol, 3.46 g) was added in one portion. The addition funnel was charged with a solution of I_2 (1.00 eq, 38.1 mmol, 9.67 g) in absolute tetrahydrofuran (25 mL) via syringe and added dropwise to the suspension over a period of 30 min. The reaction mixture was warmed to reflux (oil bath temperature 80°C) for 18 h. After cooling to ambient temperature, MeOH (50 mL) was added slowly resulting in a slightly turbid solution. After stirring for 30 min the solution was transferred to a 500 mL flask with MeOH (50 mL) and concentrated to give a white semisolid which was dissolved in aqueous KOH (20% *w/w*; 75 mL) and stirred for 5 h at ambient temperature. The aqueous layer was extracted

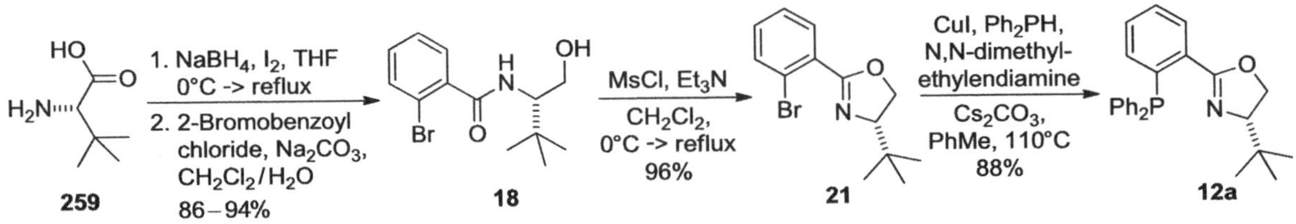

Scheme 6.9 *Synthesis of PHOX ligand* ***12a***.

with CH_2Cl_2 (6×60 mL), and the combined organic extracts were dried over Na_2SO_4, filtered, concentrated, and dried under vacuum to yield crude (*S*)-*tert*-leucinol (4.42–4.45 g, 37.7–37.9 mmol, 99% yield) as a colorless oil. This material was used in the following step without purification.

A 500 mL flask equipped with a magnetic stirring bar was charged with crude (*S*)-*tert*-leucinol (1.00 eq, 37.7 mmol, 4.42 g), CH_2Cl_2 (125 mL, 0.30 M), and then a solution of Na_2CO_3 (3.00 eq, 113.1 mmol, 11.98 g) in H_2O (95 mL) was added at ambient temperature. The biphasic mixture was stirred vigorously, and 2-bromobenzoyl chloride (1.15 eq, 43.3 mmol, 5.67 mL) was added dropwise to the emulsion via syringe over a period of 15 min. The reaction mixture was stirred for 10 h; the layers were separated, the aqueous layer was extracted with CH_2Cl_2 (4×50 mL), and the combined organic extracts were stirred with 1 N KOH solution in MeOH (19 mL) for 30 min at ambient temperature and then acidified to neutral pH with 1 N HCl (16 mL). After the addition of H_2O (25 mL), the layers were separated and the aqueous layer extracted with CH_2Cl_2 (4×35 mL). The combined organic extracts were washed with brine (75 mL), dried over Na_2SO_4, filtered, and concentrated to give an off-white solid which was dissolved in a minimal amount of hot acetone (ca. 10 mL) and treated with hexanes until a turbid solution was obtained (ca. 45 mL). The crystals formed upon cooling and aging for 3 h at 0°C were collected, washed with hexanes, and dried under vacuum to afford bromobenzamide **18** (9.23–9.61 g, 30.8–32.0 mmol) as white blocks. The mother liquor was concentrated and the remaining solid recrystallized from acetone–hexanes (2 mL/10 mL) to afford additional product (0.62–1.13 g, 2.1–3.76 mmol) as white blocks, for a combined yield of 9.85–10.74 g (32.8–35.8 mmol, 86–94% yield over two steps).

Under argon atmosphere, an oven-dried 500 mL three-neck flask was equipped with a magnetic stirring bar, a reflux condenser, and an internal thermometer and charged with bromobenzamide **18** (9.85 g, 32.8 mmol), CH_2Cl_2 (170 mL, 0.19 M), and Et_3N (2.40 eq, 78.6 mmol, 11.0 mL). The resulting colorless solution was cooled to 4°C in an ice-water bath, and $MeSO_2Cl$ (1.15 eq, 37.7 mmol, 2.92 mL) is added dropwise via syringe over a period of 3 min. The slightly yellow solution was warmed to reflux (oil bath temperature 50°C) while monitoring conversion by TLC. Upon completed cyclization, the reaction was allowed to cool to ambient temperature and saturated aqueous $NaHCO_3$ solution (60 mL) was added with vigorous stirring for 5 min. The layers were separated, and the aqueous phase extracted with CH_2Cl_2 (2×35 mL). The combined organic layers were washed with saturated brine (75 mL), dried over $MgSO_4$, filtered, and concentrated to afford a red-brown semisolid. The residue was dissolved in a minimal amount of CH_2Cl_2 (ca. 35 mL), dry-loaded onto silica gel (8 g), and purified by silica gel chromatography to afford bromophenyloxazoline **21** (8.85–8.86 g, 31.4 mmol, 96% yield) as a pale yellow oil solidifying when placed in a −20°C freezer.

Under an argon atmosphere, a dried 150 mL Schlenk flask equipped with a magnetic stirring bar was charged with CuI (0.005 eq, 0.10 mmol, 19.0 mg), Ph_2PH (1.25 eq, 25.0 mmol, 4.35 mL), *N,N'*-dimethylethylenediamine (0.025 eq, 0.50 mmol, 53 μL), and toluene (20 mL). The flask was sealed with a glass stopper, and the colorless contents were stirred at ambient temperature for 20 min. Under a positive pressure of argon, the flask was charged with bromophenyloxazoline **21** (1.00 eq, 20.0 mmol, 5.64 g), Cs_2CO_3 (1.50 eq, 30.0 mmol, 9.78 g), and toluene (20 mL, 0.50 M total) to wash the neck and walls of the flask. The flask was equipped with a reflux condenser with a two-tap Schlenk adapter connected to a bubbler and an

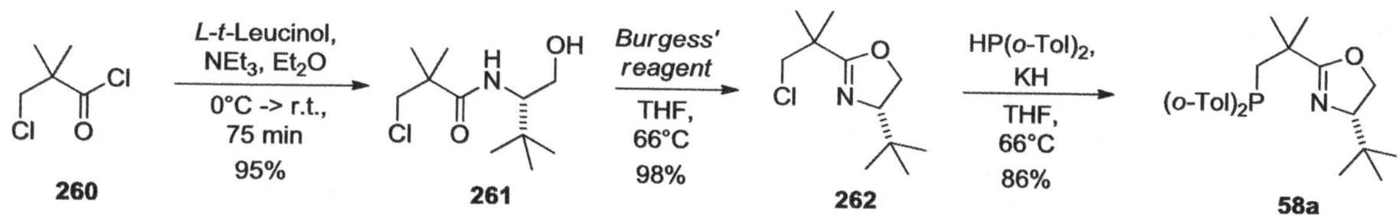

Scheme 6.10 *Synthesis of NeoPHOX ligand **58a**.*

argon–vacuum manifold. The yellow suspension was placed in a 110°C oil bath and vigorously stirred under argon atmosphere. Following consumption of starting material, the reaction was allowed to cool to ambient temperature, it was filtered through a pad of celite, and the filter cake was washed with CH_2Cl_2 (2×40 mL). The filtrate was concentrated to a pale yellow semisolid which was dissolved in a minimal amount of CH_2Cl_2 (ca. 40 mL), dry loaded onto silica gel (10 g), and purified by silica gel chromatography eluting with 4% Et_2O in hexanes until excess Ph_2PH elutes, then with 10% Et_2O in CH_2Cl_2. The combined fractions were concentrated to give a viscous, pale yellow oil which was layered with MeCN (ca. 5 mL) to facilitate crystallization. The flask was swirled while crystals formed within seconds. After 15 min, the flask was placed under high vacuum to remove volatiles affording (*S*)-*t*-butyl-PHOX **12a** (6.81 g, 17.6 mmol, 88% yield) as white blocks.

6.4.2 Synthesis of NeoPHOX ligand 58a [70] (Scheme 6.10)

Under an argon atmosphere, L-*tert*.-leucinol (1.17 g, 10.0 mmol) and NEt_3 (3.30 eq, 33.0 mmol, 3.39 g) were dissolved in absolute Et_2O (60 mL). A solution of chloropivaloyl chloride (**260**, 1.00 eq, 10.0 mmol, 1.55 g) in absolute Et_2O (10 mL) was added dropwise at 0°C to this mixture, leading to the precipitation of HCl·NEt_3. After the addition of chloropivaloyl chloride, the solution was stirred for 2 h at room temperature. Saturated aqueous $NaHCO_3$ solution (50 mL) was added, and the layers were separated. The aqueous layer was extracted with Et_2O (1×50 mL, 2×25 mL). The combined organic layers were washed with brine (50 mL), dried with $MgSO_4$, filtered, and concentrated, furnishing hydroxyamide **261** as a colorless solid (2.24 g, 9.50 mmol, 95% yield) which needed no further purification.

Hydroxyamide **261** (4.62 g, 19.6 mmol) and Burgess's reagent (1.30 eq, 25.5 mmol, 6.07 g) were dissolved in absolute tetrahydrofuran (200 mL). The mixture was heated to reflux for 4 h and concentrated. To the residue was added Et_2O (100 mL) to dissolve the soluble components of the reaction mixture. The organic layer was decanted and concentrated. Bulb-to-bulb distillation (100°C, 0.15 mbar) of the crude product gave a colorless oil which was filtered through a plug of silica (h=4.5 cm, d=3.5 cm) eluting with 10% Et_2O in pentane. The desired chloroalkyl oxazoline **262** was obtained as a colorless oil (4.18 g, 19.2 mmol, 98% yield).

In a nitrogen-filled glove box, an oven-dried Schlenk flask equipped with a magnetic stirring bar was charged with (*o*-Tol)$_2$PH (221 mg, 1.03 mmol) and KH (2,14 eq, 2.02 mmol, 81.0 mg). The flask was removed from the glove box, and chloroalkyl oxazoline **262** (1.07 eq, 1.10 mmol, 240 mg) and tetrahydrofuran (2 mL) were added under positive pressure of argon. The orange-red solution was heated to reflux for 4 h. A ^{31}P-NMR spectrum recorded in C_6D_6 indicated full conversion. The NMR sample was added to the reaction mixture followed by *t*-BuOMe (4 mL) and saturated aqueous NH_4Cl solution (4 mL). The layers were separated, and the aqueous layer diluted with H_2O (2 mL). The aqueous layer was extracted with *t*-BuOMe (3×4 mL), and the combined organic layers washed with brine (10 mL), dried over Na_2SO_4, filtered, and dry loaded onto silica gel (3 g). After silica gel chromatography (h=11 cm, d=2.5 cm; 2–4% AcOEt in pentane) the desired phosphinooxazoline **58a** was obtained as a colorless oil (352 mg, 0.89 mmol, 86% yield).

References

[1] J. A. Osborn, F. H. Jardine, J. F. Young, G. Wilkinson, *J. Chem. Soc. A* **1966**, 1711.
[2] W. S. Knowles, M. J. Sabacky, *Chem. Commun. (London)* **1968**, 1445.
[3] H. B. Kagan, T. P. Dang, *J. Am. Chem. Soc.* **1972**, *94*, 6429.
[4] W. S. Knowles, M. J. Sabacky, B. D. Vineyard, D. J. Weinkauff, *J. Am. Chem. Soc.* **1975**, *97*, 2567.
[5] M. D. Fryzuk, B. Bosnich, *J. Am. Chem. Soc.* **1977**, *99*, 6262.
[6] A. Miyashita, A. Yasuda, H. Takaya, K. Toriumi, T. Ito, T. Souchi, R. Noyori, *J. Am. Chem. Soc.* **1980**, *102*, 7932.
[7] K. Inoguchi, S. Sakuraba, K. Achiwa, *Synlett.* **1992**, 169.
[8] P. von Matt, A. Pfaltz, *Angew. Chem. Int. Ed. Engl.* **1993**, *32*, 566.
[9] J. Sprinz, G. Helmchen, *Tetrahedron Lett.* **1993**, *34*, 1769.
[10] G. J. Dawson, C. G. Frost, J. M. J. Williams, S. J. Coote, *Tetrahedron Lett.* **1993**, *34*, 3149.
[11] G. Helmchen, A. Pfaltz, *Acc. Chem. Res.* **2000**, *33*, 336.
[12] R. Prétôt, A. Pfaltz, *Angew. Chem. Int. Ed.* **1998**, *37*, 323.
[13] W. Tang, X. Zhang, *Chem. Rev.* **2003**, *103*, 3029.
[14] H. A. McManus, P. J. Guiry, *Chem. Rev.* **2004**, *104*, 4151.
[15] G. C. Hargaden, P. J. Guiry, *Chem. Rev.* **2009**, *109*, 2505.
[16] H. Witte, W. Seeliger, *Liebigs Ann. Chem.* **1974**, 996.
[17] C. Schotten, *Ber. Deutsch. Chem. Ges.* **1884**, *17*, 2544.
[18] E. Baumann, *Ber. Deutsch. Chem. Ges.* **1886**, *19*, 3218.
[19] J. Attenburrow, D. F. Elliott, G. F. Penny, *J. Chem. Soc.* **1948**, 310.
[20] G. W. Moersch, D. P. Hylander, *J. Am. Chem. Soc.* **1954**, *76*, 1703.
[21] M. Peer, J. C. de Jong, M. Kiefer, T. Langer, H. Rieck, H. Schell, P. Sennhenn, J. Sprinz, H. Steinhagen, B. Wiese, G. Helmchen, *Tetrahedron* **1996**, *52*, 7547.
[22] M. R. Krout, J. T. Mohr, B. M. Stoltz, *Org. Synth.* **2009**, *86*, 181.
[23] J. A. Frump, *Chem. Rev.* **1971**, *71*, 483.
[24] T. G. Gant, A. I. Meyers, *Tetrahedron* **1994**, *50*, 2297.
[25] D. J. Ager, I. Prakash, D. R. Schaad, *Chem. Rev.* **1996**, *96*, 835.
[26] H. Vorbrüggen, K. Krolikiewicz, *Tetrahedron Lett.* **1981**, *22*, 4471.
[27] H. Vorbrüggen, K. Krolikiewicz, *Tetrahedron* **1993**, *49*, 9353.
[28] R. Appel, *Angew. Chem. Int. Ed. Engl.* **1975**, *14*, 801.
[29] P. Wipf, C. P. Miller, *Tetrahedron Lett.* **1992**, *33*, 907.
[30] S. Burckhardt, *Synlett.* **2000**, *2000*, 559.
[31] K. Tani, D. C. Behenna, R. M. McFadden, B. M. Stoltz, *Org. Lett.* **2007**, *9*, 2529.
[32] A. Sudo, K. Saigo, *J. Org. Chem.* **1997**, *62*, 5508.
[33] B. Wiese, G. Helmchen, *Tetrahedron Lett.* **1998**, *39*, 5727.
[34] B. Gläser, H. Kunz, *Synlett.* **1998**, 53.
[35] R. Stohler, F. Wahl, A. Pfaltz, *Synthesis* **2005**, 1431.
[36] D. Liu, W. Tang, X. Zhang, *Org. Lett.* **2004**, *6*, 513.
[37] A. Frölander, S. Lutsenko, T. Privalov, C. Moberg, *J. Org. Chem.* **2005**, *70*, 9882.
[38] A. Frölander, C. Moberg, *Org. Lett.* **2007**, *9*, 1371.
[39] L. F. Tietze, J. K. Lohmann, *Synlett.* **2002**, *2002*, 2083.
[40] T. G. Kilroy, P. G. Cozzi, N. End, P. J. Guiry, *Synlett.* **2004**, 106.
[41] T. G. Kilroy, P. G. Cozzi, N. End, P. J. Guiry, *Synthesis* **2004**, 1879.
[42] S.-F. Zhu, J.-B. Xie, Y.-Z. Zhang, S. Li, Q.-L. Zhou, *J. Am. Chem. Soc.* **2006**, *128*, 12886.
[43] V. B. Birman, A. L. Rheingold, K.-C. Lam, *Tetrahedron: Asymmetry* **1999**, *10*, 125.
[44] J.-H. Zhang, J. Liao, X. Cui, K.-B. Yu, J. Zhu, J.-G. Deng, S.-F. Zhu, L.-X. Wang, Q.-L. Zhou, L. W. Chung, T. Ye, *Tetrahedron: Asymmetry* **2002**, *13*, 1363.
[45] S. Trudeau, J. P. Morken, *Tetrahedron* **2006**, *62*, 11470.
[46] A. Marchand, A. Maxwell, B. Mootoo, A. Pelter, A. Reid, *Tetrahedron* **2000**, *56*, 7331.
[47] X.-W. Wu, K. Yuan, W. Sun, M.-J. Zhang, X.-L. Hou, *Tetrahedron: Asymmetry* **2003**, *14*, 107.

[48] D. K. Whelligan, C. Bolm, *J. Org. Chem.* **2006**, *71*, 4609.
[49] S.-M. Ma, D.-H. Duan, *Chinese J. Chem.* **2002**, *20*, 1363.
[50] A. Patti, M. Lotz, P. Knochel, *Tetrahedron: Asymmetry* **2002**, *12*, 3375.
[51] Y. Nishibayashi, S. Uemura, *Synlett.* **1995**, *1995*, 79.
[52] C. J. Richards, T. Damalidis, D. E. Hibbs, M. B. Hursthouse, *Synlett.* **1995**, *1995*, 74.
[53] T. Sammakia, H. A. Latham, D. R. Schaad, *J. Org. Chem.* **1995**, *60*, 10.
[54] T. Sammakia, H. A. Latham, *J. Org. Chem.* **1995**, *60*, 6002.
[55] W. Zhang, Y.-I. Yoneda, T. Kida, Y. Nakatsuji, I. Ikeda, *Tetrahedron: Asymmetry* **1998**, *9*, 3371.
[56] R. Shintani, M. M. C. Lo, G. C. Fu, *Org. Lett.* **2000**, *2*, 3695.
[57] R. Shintani, G. C. Fu, *Org. Lett.* **2002**, *4*, 3699.
[58] S. Kudis, G. Helmchen, *Angew. Chem. Int. Ed.* **1998**, *37*, 3047.
[59] C. Bolm, L. Xiao, M. Kesselgruber, *Org. Biomol. Chem.* **2003**, *1*, 145.
[60] Y. Imai, W. Zhang, T. Kida, Y. Nakatsuji, I. Ikeda, *Tetrahedron Lett.* **1998**, *39*, 4343.
[61] M. Ogasawara, K. Yoshida, H. Kamei, K. Kato, Y. Uozumi, T. Hayashi, *Tetrahedron: Asymmetry* **1998**, *9*, 1779.
[62] Y. Uozumi, A. Tanahashi, S. Y. Lee, T. Hayashi, *J. Org. Chem.* **1993**, *58*, 1945.
[63] Y. Uozumi, N. Suzuki, A. Ogiwara, T. Hayashi, *Tetrahedron* **1994**, *50*, 4293.
[64] F. Tian, D. Yao, Y. J. Zhang, W. Zhang, *Tetrahedron* **2009**, *65*, 9609.
[65] S. R. Gilbertson, C.-W. T. Chang, *Chem. Commun. (Cambridge, UK)* **1997**, 975.
[66] S. R. Gilbertson, C.-W. T. Chang, *J. Org. Chem.* **1998**, *63*, 8424.
[67] S. R. Gilbertson, D. Xie, *Angew. Chem. Int. Ed.* **1999**, *38*, 2750.
[68] S. R. Gilbertson, D. Xie, Z. Fu, *J. Org. Chem.* **2001**, *66*, 7240.
[69] S. R. Gilbertson, Z. Fu, *Org. Lett.* **2001**, *3*, 161.
[70] M. G. Schrems, A. Pfaltz, *Chem. Commun. (Cambridge, UK)* **2009**, 6210.
[71] A. M. Porte, J. Reibenspies, K. Burgess, *J. Am. Chem. Soc.* **1998**, *120*, 9180.
[72] D.-R. Hou, K. Burgess, *Org. Lett.* **1999**, *1*, 1745.
[73] D.-R. Hou, J. H. Reibenspies, K. Burgess, *J. Org. Chem.* **2001**, *66*, 206.
[74] M. B. Ezhova, B. O. Patrick, B. R. James, F. J. Waller, M. E. Ford, *J. Mol. Catal. A: Chem.* **2004**, *224*, 71.
[75] W. Tang, W. Wang, X. Zhang, *Angew. Chem. Int. Ed.* **2003**, *42*, 943.
[76] S. R. Gilbertson, D. G. Genov, A. L. Rheingold, *Org. Lett.* **2000**, *2*, 2885.
[77] C. Rethore, M. Fourmigue, N. Avarvari, *Chem. Commun. (Cambridge, UK)* **2004**, 1384.
[78] C. Réthoré, I. Suisse, F. Agbossou-Niedercorn, E. Guillamón, R. Llusar, M. Fourmigué, N. Avarvari, *Tetrahedron* **2006**, *62*, 11942.
[79] J.-D. Lee, T. T. Co, T.-J. Kim, S. O. Kang, *Synlett.* **2009**, *2009*, 771.
[80] K. Yonehara, T. Hashizume, K. Mori, K. Ohe, S. Uemura, *Chem. Commun. (Cambridge, UK)* **1999**, 415.
[81] K. Yonehara, T. Hashizume, K. Mori, K. Ohe, S. Uemura, *J. Org. Chem.* **1999**, *64*, 9374.
[82] T. Hashizume, K. Yonehara, K. Ohe, S. Uemura, *J. Org. Chem.* **2000**, *65*, 5197.
[83] J. Blankenstein, A. Pfaltz, *Angew. Chem. Int. Ed.* **2001**, *40*, 4445.
[84] F. Menges, A. Pfaltz, *Adv. Synth. Cat.* **2002**, *344*, 40.
[85] G. Jones, C. J. Richards, *Tetrahedron Lett.* **2001**, *42*, 5553.
[86] S. P. Smidt, F. Menges, A. Pfaltz, *Org. Lett.* **2004**, *6*, **2023**.
[87] O. Pàmies, M. Diéguez, C. Claver, *J. Am. Chem. Soc.* **2005**, *127*, 3646.
[88] Y. Mata, M. Diéguez, O. Pàmies, C. Claver, *Adv. Synth. Cat. 2005*, 347, **1943**.
[89] M. Diéguez, J. Mazuela, O. Pàmies, J. J. Verendel, P. G. Andersson, *J. Am. Chem. Soc.* **2008**, *130*, 7208.
[90] G. Xu, S. R. Gilbertson, *Tetrahedron Lett.* **2002**, *43*, 2811.
[91] G. Xu, S. R. Gilbertson, *Tetrahedron Lett.* **2003**, *44*, 953.
[92] C. Blanc, J. Hannedouche, F. Agbossou-Niedercorn, *Tetrahedron Lett.* **2003**, *44*, 6469.
[93] P. G. Cozzi, N. Zimmermann, R. Hilgraf, S. Schaffner, A. Pfaltz, *Adv. Synth. Cat.* **2001**, *343*, 450.
[94] A. Trifonova, J. S. Diesen, P. G. Andersson, *Chem. – Eur. J.* **2006**, *12*, 2318.
[95] R. P. J. Bronger, P. J. Guiry, *Tetrahedron: Asymmetry* **2007**, *18*, 1094.
[96] S.-L. You, X.-Z. Zhu, Y.-M. Luo, X.-L. Hou, L.-X. Dai, *J. Am. Chem. Soc.* **2001**, *123*, 7471.
[97] L.-X. Dai, T. Tu, S.-L. You, W.-P. Deng, X.-L. Hou, *Acc. Chem. Res.* **2003**, *36*, 659.

[98] X.-L. Hou, N. Sun, *Org. Lett.* **2004**, *6*, 4399.
[99] R. Hilgraf, A. Pfaltz, *Synlett.* **1999**, 1814.
[100] R. Hilgraf, A. Pfaltz, *Adv. Synth. Cat.* **2005**, *347*, 61.
[101] J. M. Brunel, T. Constantieux, G. Buono, *J. Org. Chem.* **1999**, *64*, 8940.
[102] C. A. Busacca (Boeringer Ingelheim Pharmaceuticals, Inc.), US Patent 6316620, **2001**.
[103] C. A. Busacca, D. Grossbach, R. C. So, E. M. O'Brie, E. M. Spinelli, *Org. Lett.* **2003**, *5*, 595.
[104] C. A. Busacca, D. Grossbach, S. J. Campbell, Y. Dong, M. C. Eriksson, R. E. Harris, P.-J. Jones, J.-Y. Kim, J. C. Lorenz, K. B. McKellop, E. M. O'Brien, F. Qiu, R. D. Simpson, L. Smith, R. C. So, E. M. Spinelli, J. Vitous, C. Zavattaro, *J. Org. Chem.* **2004**, *69*, 5187.
[105] C. A. Busacca, J. C. Lorenz, N. Grinberg, N. Haddad, H. Lee, Z. Li, M. Liang, D. Reeves, A. Saha, R. Varsolona, C. H. Senanayake, *Org. Lett.* **2008**, *10*, 341.
[106] E. Guiu, C. Claver, J. Benet-Buchholz, S. Castillón, *Tetrahedron: Asymmetry* **2004**, *15*, 3365.
[107] F. Menges, M. Neuburger, A. Pfaltz, *Org. Lett.* **2002**, *4*, 4713.
[108] F. Menges, PhD thesis, University of Basel, Switzerland, **2004**.
[109] F. Menges, A. Pfaltz (Solvias AG), European Patent EP1191030, **2005**.
[110] K. Källström, C. Hedberg, P. Brandt, A. Bayer, P. G. Andersson, *J. Am. Chem. Soc.* **2004**, *126*, 14308.
[111] C. Hedberg, K. Källström, P. Brandt, L. K. Hansen, P. G. Andersson, *J. Am. Chem. Soc.* **2006**, *128*, 2995.
[112] J. Mazuela, A. Paptchikhine, P. Tolstoy, O. Pàmies, M. Diéguez, P. G. Andersson, *Chem. – Eur. J.* **2010**, *16*, 620.
[113] P. Kaukoranta, M. Engman, C. Hedberg, J. Bergquist, P. G. Andersson, *Adv. Synth. Cat.* **2008**, *350*, 1168.
[114] L. D. Field, B. A. Messerle, K. Q. Vuong, P. Turner, *Dalton Trans.* **2009**, 3599.
[115] H. Willms, W. Frank, C. Ganter, *Organometallics* **2009**, *28*, 3049.
[116] G. R. Newkome, *Chem. Rev. 1993*, 93, **2067**.
[117] P. Espinet, K. Soulantica, *Coord. Chem. Rev.* **1999**, *193–195*, 499.
[118] G. Chelucci, G. Orrù, G. A. Pinna, *Tetrahedron* **2003**, *59*, 9471.
[119] R. H. Crabtree, G. E. Morris, *J. Organomet. Chem.* **1977**, *135*, 395.
[120] R. Crabtree, *Acc. Chem. Res.* **1979**, *12*, 331.
[121] R. De Vaumas, A. Marinetti, L. Ricard, F. Mathey, *J. Am. Chem. Soc.* **1992**, *114*, 261.
[122] G. He, S.-K. Loh, J. J. Vittal, K. F. Mok, P.-H. Leung, *Organometallics* **1998**, *17*, 3931.
[123] M. P. A. Lyle, A. A. Narine, P. D. Wilson, *J. Org. Chem.* **2004**, *69*, 5060.
[124] G. Chelucci, A. Saba, F. Soccolini, *Tetrahedron* **2001**, *57*, 9989.
[125] A. V. Malkov, M. Bella, I. G. Stará, P. Kocovský, *Tetrahedron Lett.* **2001**, *42*, 3045.
[126] F. Kröhnke, *Synthesis* **1976**, 1.
[127] H. Bönnemann, *Angew. Chem. Int. Ed. Engl.* **1985**, *24*, 248.
[128] G. Chelucci, *Tetrahedron: Asymmetry* **1995**, *6*, 811.
[129] G. Chelucci, M. A. Cabras, C. Botteghi, C. Basoli, M. Marchetti, *Tetrahedron: Asymmetry* **1996**, *7*, 885.
[130] J.-M. Brunel, T. Constantieux, A. Labande, F. Lubatti, G. Buono, *Tetrahedron Lett.* **1997**, *38*, 5971.
[131] T. Constantieux, J.-M. Brunel, A. Labande, G. Buono, *Synlett.* **1998**, 49.
[132] G. Delapierre, J. M. Brunel, T. Constantieux, G. Buono, *Tetrahedron: Asymmetry* **2001**, *12*, 1345.
[133] G. Delapierre, M. Achard, G. Buono, *Tetrahedron Lett.* **2002**, *43*, 4025.
[134] H. Yang, N. Lugan, R. Mathieu, *Organometallics* **1997**, *16*, 2089.
[135] C. G. Arena, F. Nicolo, D. Drommi, G. Bruno, F. Faraone, *J. Chem. Soc., Chem. Commun.* **1994**, 2251.
[136] J. Flapper, P. Wormald, M. Lutz, A. L. Spek, P. W. N. M. van Leeuwen, C. J. Elsevier, P. C. J. Kamer, *Eur. J. Inorg. Chem.* **2008**, *2008*, 4968.
[137] J. Flapper, H. Kooijman, M. Lutz, A. L. Spek, P. W. N. M. van Leeuwen, C. J. Elsevier, P. C. J. Kamer, *Organometallics* **2009**, *28*, 3272.
[138] S. Chen, J. K.-P. Ng, S. A. Pullarkat, F. Liu, Y. Li, P.-H. Leung, *Organometallics* **2010**, *29*, 3374.
[139] F. Speiser, P. Braunstein, L. Saussine, R. Welter, *Inorganic Chemistry* **2004**, *43*, 1649.
[140] S. Kaiser, S. P. Smidt, A. Pfaltz, *Angew. Chem. Int. Ed.* **2006**, *45*, 5194.
[141] D. H. Woodmansee, M.-A. Müller, M. Neuburger, A. Pfaltz, *Chemical Science* **2010**, *1*, 72.
[142] G. Kobayashi, S. Furukawa, *Pharm. Bull.* **1953**, *1*, 347.
[143] V. Boekelheide, W. J. Linn, *J. Am. Chem. Soc.* **1954**, *76*, 1286.

[144] M. Maywald, A. Pfaltz, *Synthesis* **2009**, *2009*, 3654.
[145] Q.-B. Liu, Y.-G. Zhou, *Tetrahedron Lett.* **2007**, *48*, 2101.
[146] T. Bunlaksananusorn, K. Polborn, P. Knochel, *Angew. Chem. Int. Ed.* **2003**, *42*, 3941.
[147] J. J. Verendel, P. G. Andersson, *Dalton Trans.* **2007**, 5603.
[148] X. Meng, Y. Gao, X. Li, D. Xu, *Catalysis Communications* **2009**, *10*, 950.
[149] X. Meng, X. Li, D. Xu, *Tetrahedron: Asymmetry* **2009**, *20*, 1402.
[150] B. Goldfuss, T. Löschmann, F. Rominger, *Chem. – Eur. J.* **2004**, *10*, 5422.
[151] C. Thoumazet, M. Melaimi, L. Ricard, F. Mathey, P. Le Floch, *Organometallics* **2003**, *22*, 1580.
[152] F. Leca, C. Lescop, L. Toupet, R. Réau, *Organometallics* **2004**, *23*, 6191.
[153] A. Marson, J. E. Ernsting, M. Lutz, A. L. Spek, P. W. N. M. van Leeuwen, P. C. J. Kamer, *Dalton Trans.* **2009**, 621.
[154] N. W. Alcock, J. M. Brown, M. Pearson, S. Woodward, *Tetrahedron: Asymmetry* **1992**, *3*, 17.
[155] N. W. Alcock, J. M. Brown, D. I. Hulmes, *Tetrahedron: Asymmetry* **1993**, *4*, 743.
[156] N. W. Alcock, D. I. Hulmes, J. M. Brown, *J. Chem. Soc., Chem. Commun.* **1995**, 395.
[157] H. Doucet, J. M. Brown, *Tetrahedron: Asymmetry* **1997**, *8*, 3775.
[158] T. F. Knöpfel, P. Aschwanden, T. Ichikawa, T. Watanabe, E. M. Carreira, *Angew. Chem. Int. Ed.* **2004**, *43*, 5971.
[159] J.-M. Valk, T. D. W. Claridge, J. M. Brown, D. Hibbs, M. B. Hursthouse, *Tetrahedron: Asymmetry* **1995**, *6*, 2597.
[160] T. D. W. Claridge, J. M. Long, J. M. Brown, D. Hibbs, M. B. Hursthouse, *Tetrahedron* **1997**, *53*, 4035.
[161] F. Y. Kwong, K. S. Chan, *Organometallics* **2001**, *20*, 2570.
[162] X. Hu, H. Chen, X. Zhang, *Angew. Chem. Int. Ed.* **1999**, *38*, 3518.
[163] C. Ganter, C. Glinsböckel, B. Ganter, *Eur. J. Inorg. Chem.* **1998**, *1998*, 1163.
[164] B. Tao, G. C. Fu, *Angew. Chem. Int. Ed.* **2002**, *41*, 3892.
[165] J. W. Han, H.-Y. Jang, Y. K. Chung, *Tetrahedron: Asymmetry* **1999**, *10*, 2853.
[166] R. Ruzziconi, C. Santi, S. Spizzichino, *Tetrahedron: Asymmetry* **2007**, *18*, 1742.
[167] D. Amoroso, T. W. Graham, R. Guo, C.-W. Tsang, K. Abdur-Rashid, *Aldrichim. Act.* **2008**, *41*, 15.
[168] A. Saitoh, T. Morimoto, K. Achiwa, *Tetrahedron: Asymmetry* **1997**, *8*, 3567.
[169] A. Saitoh, K. Achiwa, K. Tanaka, T. Morimoto, *J. Org. Chem.* **2000**, *65*, 4227.
[170] L. Dahlenburg, R. Götz, *J. Organomet. Chem.* **2001**, *619*, 88.
[171] A. Caiazzo, S. Dalili, A. K. Yudin, *Org. Lett.* **2002**, *4*, 2597.
[172] R. Guo, X. Chen, W. Jia, K. Abdur-Rashid (Kanata Chemical Technologies, Inc.), US Patent Application 60/942,699, **2007**.
[173] P. Rönnholm, M. Södergren, G. Hilmersson, *Org. Lett.* **2007**, *9*, 3781.
[174] A. M. Maj, K. M. Pietrusiewicz, I. Suisse, F. Agbossou, A. Mortreux, *Tetrahedron: Asymmetry* **1999**, *10*, 831.
[175] A. M. Maj, K. M. Pietrusiewicz, I. Suisse, F. Agbossou, A. Mortreux, *J. Organomet. Chem.* **2001**, *626*, 157.
[176] M. Léautey, G. Castelot-Deliencourt, P. Jubault, X. Pannecoucke, J.-C. Quirion, *Tetrahedron Lett.* **2002**, *43*, 9237.
[177] M. Léautey, P. Jubault, X. Pannecoucke, J.-C. Quirion, *Eur. J. Org. Chem.* **2003**, *2003*, 3761.
[178] X. Fu, Z. Jiang, C.-H. Tan, *Chem. Commun.* (Cambridge, UK) **2007**, 5058.
[179] J. P. Cahill, A. P. Lightfoot, R. Goddard, J. Rust, P. J. Guiry, *Tetrahedron: Asymmetry* **1998**, *9*, 4307.
[180] J. I. Uenishi, M. Hamada, *Tetrahedron: Asymmetry* **2001**, *12*, 2999.
[181] D. W. Old, J. P. Wolfe, S. L. Buchwald, *J. Am. Chem. Soc.* **1998**, *120*, 9722.
[182] J. Feng, S. Dastgir, C.-J. Li, *Tetrahedron Lett.* **2008**, *49*, 668.
[183] R. J. Lundgren, M. A. Rankin, R. McDonald, M. Stradiotto, *Organometallics* **2007**, *27*, 254.
[184] D. Wechsler, G. Schatte, M. Stradiotto, *J. Organomet. Chem.* **2009**, *694*, 1943.
[185] G. Malaisé, J.-B. Sortais, L. Barloy, M. Pfeffer, N. Kyritsakas, *Polyhedron* **2006**, *25*, 3349.
[186] X. Hu, C. Bai, H. Dai, H. Chen, Z. Zheng, *J. Mol. Catal. A: Chem.* **2004**, *218*, 107.
[187] X. Hu, H. Dai, C. Bai, H. Chen, Z. Zheng, *Tetrahedron: Asymmetry* **2004**, *15*, 1065.
[188] T. Noël, K. Bert, E. Van der Eycken, J. Van der Eycken, *Eur. J. Org. Chem.* **2010**, *2010*, 4056.
[189] L. B. Schenkel, J. A. Ellman, *Org. Lett.* **2003**, *5*, 545.
[190] M. McCarthy, R. Goddard, P. J. Guiry, *Tetrahedron: Asymmetry* **1999**, *10*, 2797.
[191] M. McCarthy, P. J. Guiry, *Tetrahedron* **1999**, *55*, 3061.
[192] P. M. Lacey, C. M. McDonnell, P. J. Guiry, *Tetrahedron Lett.* **2000**, *41*, 2475.
[193] X. Dai, A. Wong, S. C. Virgil, *J. Org. Chem.* **1998**, *63*, 2597.

[194] X. Dai, S. Virgil, *Tetrahedron Lett.* **1999**, *40*, 1245.
[195] Y. Sun, W. R. Thiel, *Inorg. Chim. Acta* **2006**, *359*, 4807.
[196] C. K. Seubert, Y. Sun, W. R. Thiel, *Dalton Trans.* **2009**, 4971.
[197] C. May, Y. Sun, G. Wolmershäuser, W. R. Thiel, *Z. Naturforsch., B: J. Chem. Sci.* **2009**, *64*, 1438.
[198] J. A. Van Doorn, E. Drent, P. W. N. M. Van Leeuwen, N. Meijboom, A. B. Van Oort, R. L. Wife (Shell Internationale Research Maatschappij B.V.), European Patent EP0280380, **1988**.
[199] W. Keim, *Angew. Chem. Int. Ed. Engl.* **1990**, *29*, 235.
[200] E. Drent, P. H. M. Budzelaar, *Chem. Rev.* **1996**, *96*, 663.
[201] W. Keim, R. P. Schulz, *J. Mol. Cat.* **1994**, *92*, 21.
[202] T. Schultz, A. Pfaltz, *Synthesis* **2005**, *2005*, 1005.
[203] A. K. Hearley, R. J. Nowack, B. Rieger, *Organometallics* **2005**, *24*, 2755.
[204] X. Bei, T. Crevier, A. S. Guram, B. Jandeleit, T. S. Powers, H. W. Turner, T. Uno, W. H. Weinberg, *Tetrahedron Lett.* **1999**, *40*, 3855.
[205] X. Bei, A. S. Guram, H. W. Turner, W. H. Weinberg, *Tetrahedron Lett.* **1999**, *40*, 1237.
[206] X. Bei, T. Uno, J. Norris, H. W. Turner, W. H. Weinberg, A. S. Guram, J. L. Petersen, *Organometallics* **1999**, *18*, **1840**.
[207] J. Clayden, P. Johnson, J. H. Pink, M. Helliwell, *J. Org. Chem.* **2000**, *65*, 7033.
[208] F. Y. Kwong, W. Har Lam, C. Hung Yeung, K. Shing Chan, A. S. C. Chan, *Chem. Commun. (Cambridge, UK)* **2004**, 1922.
[209] W.-M. Dai, Y. Li, Y. Zhang, K. Wah Lai, J. Wu, *Tetrahedron Lett.* **2004**, *45*, 1999.
[210] W.-M. Dai, K. K. Y. Yeung, Y. Wang, *Tetrahedron* **2004**, *60*, 4425.
[211] W.-M. Dai, Y. Zhang, *Tetrahedron Lett.* **2005**, *46*, 1377.
[212] R. A. Singer, N. J. Tom, H. N. Frost, W. M. Simon, *Tetrahedron Lett.* **2004**, *45*, 4715.
[213] F. Y. Kwong, A. S. C. Chan, *Synlett.* **2008**, *2008*, 1440.
[214] E. Ullah, J. McNulty, A. Robertson, *Tetrahedron Lett.* **2009**, *50*, 5599.
[215] T. Hayashi, K. Hayashizaki, T. Kiyoi, Y. Ito, *J. Am. Chem. Soc.* **1988**, *110*, 8153.
[216] H. L. Pedersen, M. Johannsen, *Chem. Commun. (Cambridge, UK)* **1999**, 2517.
[217] J. F. Jensen, M. Johannsen, *Org. Lett.* **2003**, *5*, 3025.
[218] D. A. Evans, K. R. Campos, J. S. Tedrow, F. E. Michael, M. R. Gagne, *J. Org. Chem.* **1999**, *64*, 2994.
[219] D. A. Evans, K. R. Campos, J. S. Tedrow, F. E. Michael, M. R. Gagné, *J. Am. Chem. Soc.* **2000**, *122*, 7905.
[220] E. Hauptman, P. J. Fagan, W. Marshall, *Organometallics* **1999**, *18*, **2061**.
[221] A. Dervisi, R. L. Jenkins, K. M. A. Malik, M. B. Hursthouse, S. Coles, *Dalton Trans.* **2003**, 1133.
[222] H. Sugama, H. Saito, H. Danjo, T. Imamoto, *Synthesis* **2001**, 2348.
[223] G. A. Molander, J. P. Burke, P. J. Carroll, *J. Org. Chem.* **2004**, *69*, 8062.
[224] N. Khiar, B. Suárez, M. Stiller, V. Valdivia, I. Fernández, *Phosphorus, Sulfur Silicon Relat. Elem.* **2005**, *180*, 1253.
[225] N. Khiar, B. Suárez, V. Valdivia, I. Fernández, *Synlett.* **2005**, 2963.
[226] Y.-Y. Yan, T. V. RajanBabu, *Org. Lett.* **1999**, *2*, 199.
[227] O. Pàmies, M. Diéguez, G. Net, A. Ruiz, C. Claver, *Organometallics* **2000**, *19*, 1488.
[228] E. Guimet, M. Diéguez, A. Ruiz, C. Claver, *Tetrahedron: Asymmetry* **2005**, *16*, 959.
[229] M. Diéguez, O. Pàmies, C. Claver, *Tetrahedron: Asymmetry* **2005**, *16*, 3877.
[230] X. Verdaguer, A. Moyano, M. A. Pericàs, A. Riera, M. A. Maestro, J. Mahía, *J. Am. Chem. Soc.* **2000**, *122*, 10242.
[231] X. Verdaguer, M. A. Pericàs, A. Riera, M. A. Maestro, J. Mahía, *Organometallics* **2003**, *22*, 1868.
[232] J. Solà, A. Riera, X. Verdaguer, *J. Organomet. Chem.* **2010**, *695*, 2377.
[233] H. Nakano, Y. Okuyama, H. Hongo, *Tetrahedron Lett.* **2000**, *41*, 4615.
[234] H. Nakano, Y. Okuyama, M. Yanagida, H. Hongo, *J. Org. Chem.* **2001**, *66*, 620.
[235] H. Nakano, J.-I. Yokoyama, R. Fujita, H. Hongo, *Tetrahedron Lett.* **2002**, *43*, 7761.
[236] H. Nakano, K. Takahashi, Y. Suzuki, R. Fujita, *Tetrahedron: Asymmetry* **2005**, *16*, 609.
[237] K. Hiroi, Y. Suzuki, *Tetrahedron Lett.* **1998**, *39*, 6499.
[238] K. Hiroi, Y. Suzuki, I. Abe, *Tetrahedron: Asymmetry* **1999**, *10*, 1173.
[239] K. Hiroi, Y. Suzuki, I. Abe, *Chem. Lett.* **1999**, *28*, 149.
[240] K. Hiroi, Y. Suzuki, R. Kawagishi, *Tetrahedron Lett.* **1999**, *40*, 715.
[241] Y. Suzuki, I. Abe, K. Hiroi, *Heterocycles* **1999**, *50*, 89.

[242] K. Hiroi, Y. Suzuki, I. Abe, R. Kawagishi, *Tetrahedron* **2000**, *56*, 4701.
[243] K. Hiroi, I. Izawa, T. Takizawa, K.-I. Kawai, *Tetrahedron* **2004**, *60*, 2155.
[244] C. Ferrer, A. Riera, X. Verdaguer, *Organometallics* **2009**, *28*, 4571.
[245] Y. Ji, A. Riera, X. Verdaguer, *Org. Lett.* **2009**, *11*, 4346.
[246] J. Solà, M. Revés, A. Riera, X. Verdaguer, *Angew. Chem. Int. Ed.* **2007**, *46*, 5020.
[247] M. Revés, T. Achard, J. Solà, A. Riera, X. Verdaguer, *J. Org. Chem.* **2008**, *73*, 7080.
[248] S. Nakamura, T. Fukuzumi, T. Toru, *Chirality* **2004**, *16*, 10.
[249] F. L. Lam, F. Y. Kwong, A. S. C. Chan, *Chem. Commun. (Cambridge, UK)* **2010**, *46*, 4649.
[250] J. Priego, O. G. Mancheno, S. Cabrera, R. G. Arrayas, T. Llamas, J. C. Carretero, *Chem. Commun. (Cambridge, UK)* **2002**, 2512.
[251] O. G. Mancheño, J. Priego, S. Cabrera, R. G. Arrayás, T. Llamas, J. C. Carretero, *J. Org. Chem.* **2003**, *68*, 3679.
[252] R. Gomez Arrayas, O. Garcia Mancheno, J. C. Carretero, *Chem. Commun.* (Cambridge, *UK)* **2004**, 1654.
[253] B. Martín-Matute, S. I. Pereira, E. Peña-Cabrera, J. Adrio, A. M. S. Silva, J. C. Carretero, *Adv. Synth. Cat.* **2007**, *349*, 1714.
[254] D. Enders, R. Peters, R. Lochtman, G. Raabe, *Angew. Chem. Int. Ed.* **1999**, *38*, 2421.
[255] T. Tu, Y.-G. Zhou, X.-L. Hou, L.-X. Dai, X.-C. Dong, Y.-H. Yu, J. Sun, *Organometallics* **2003**, *22*, 1255.
[256] W. Zeng, Y.-G. Zhou, *Tetrahedron Lett.* **2007**, *48*, 4619.
[257] F. L. Lam, T. T. L. Au-Yeung, H. Y. Cheung, S. H. L. Kok, W. S. Lam, K. Y. Wong, A. S. C. Chan, *Tetrahedron: Asymmetry* **2006**, *17*, 497.
[258] H. Y. Cheung, W.-Y. Yu, F. L. Lam, T. T. L. Au-Yeung, Z. Zhou, T. H. Chan, A. S. C. Chan, *Org. Lett.* **2007**, *9*, 4295.
[259] S. Gladiali, D. Antonio, F. Davide, *Tetrahedron: Asymmetry* **1994**, *5*, 1143.
[260] J. Kang, S. H. Yu, J. I. Kim, H. G. Cho, *Bull. Kor. Chem. Soc.* **1995**, *16*, 439.
[261] K. Selvakumar, M. Valentini, P. S. Pregosin, A. Albinati, *Organometallics* **1999**, *18*, 4591.
[262] T. Kanayama, K. Yoshida, H. Miyabe, Y. Takemoto, *Angew. Chem. Int. Ed.* **2003**, *42*, 2054.
[263] T. Kanayama, K. Yoshida, H. Miyabe, T. Kimachi, Y. Takemoto, *J. Org. Chem.* **2003**, *68*, 6197.
[264] F. Boeda, D. Rix, H. Clavier, C. Crévisy, M. Mauduit, *Tetrahedron: Asymmetry* **2006**, *17*, 2726.
[265] K. Albert, P. Gisdakis, N. Rösch, *Organometallics* **1998**, *17*, 1608.
[266] F. E. Hahn, M. C. Jahnke, *Angew. Chem. Int. Ed.* **2008**, *47*, 3122.
[267] W. A. Herrmann, C. Köcher, L. J. Gooßen, G. R. J. Artus, *Chem. – Eur. J.* **1996**, *2*, 1627.
[268] C. Yang, H. M. Lee, S. P. Nolan, *Org. Lett.* **2001**, *3*, 1511.
[269] A. A. Danopoulos, S. Winston, T. Gelbrich, M. B. Hursthouse, R. P. Tooze, *Chem. Commun. (Cambridge, UK)* **2002**, 482.
[270] S. Nanchen, A. Pfaltz, *Hel. Chim. Acta* **2006**, *89*, 1559.
[271] C.-C. Ho, S. Chatterjee, T.-L. Wu, K.-T. Chan, Y.-W. Chang, T.-H. Hsiao, H. M. Lee, *Organometallics* **2009**, *28*, 2837.
[272] A.-E. Wang, J.-H. Xie, L.-X. Wang, Q.-L. Zhou, *Tetrahedron* **2005**, *61*, 259.
[273] E. Bappert, G. Helmchen, *Synlett.* **2004**, 1789.
[274] H. Seo, H.-j. Park, B. Y. Kim, J. H. Lee, S. U. Son, Y. K. Chung, *Organometallics* **2003**, *22*, 618.
[275] T. Focken, G. Raabe, C. Bolm, *Tetrahedron: Asymmetry* **2004**, *15*, 1693.
[276] O. Loiseleur, M. Hayashi, M. Keenan, N. Schmees, A. Pfaltz, *J. Organomet. Chem.* **1999**, *576*, 16.
[277] D. C. Behenna, B. M. Stoltz, *J. Am. Chem. Soc.* **2004**, *126*, 15044.
[278] X. Cui, K. Burgess, *Chem. Rev.* **2005**, *105*, 3272.
[279] S. J. Roseblade, A. Pfaltz, *Acc. Chem. Res.* **2007**, *40*, 1402.
[280] T. L. Church, P. G. Andersson, *Coord. Chem. Rev.* **2008**, *252*, 513.
[281] D. H. Woodmansee, A. Pfaltz, in *Topics in Organometallic Chemistry, Vol. 34, Iridium Catalysis*, 1st ed., P. G. Andersson, Ed., Springer Verlag, **2011**, 31.
[282] S. Bell, B. Wüstenberg, S. Kaiser, F. Menges, T. Netscher, A. Pfaltz, *Science* **2006**, *311*, 642.

7

Phospholes

Duncan Carmichael

Laboratoire "Hétéroéléments et Coordination", Ecole Polytechnique, CNRS, Route de Saclay, France

7.1 Introduction

Phospholes are obtained upon substitution of a cyclopentadiene CR functionality by a phosphorus atom, and this formal definition is useful in that it correctly places phospholes closer to their all-carbon (dienic) counterparts than their more aromatic congeners such as furan, thiophen and pyrrole. Well-recognised applications in catalysis, molecular materials [1–3] and recent developments in medicine [4–6] mean that phospholes are well served in reviews, with Quin's series being outstanding [7]. Simple phospholes show partial dienic and partial aromatic character, and controlling these contrasting facets of reactivity is the key to exploiting their potential as a source of new ligands. A short résumé outlining the interplay between the dienic and aromatic characteristics of phospholes is therefore necessary before treating their synthesis, coordination chemistry and catalytic applications.

The chemistry of phospholes is governed by the electropositivity of phosphorus relative to carbon (P: 2.1, C: 2.5 according to Pauling), the marked difference between λ^3- P–C and P=C bond strengths (estimated as σCP: 66.0, πCP: 49.4; compare σCC: 88.3, πCC: 69.6 kcal mol^{-1}) [8] and the very different energies and space occupancy of the phosphorus 3s and 3p orbitals [9]. Of the three possible isomers, the 1*H*-phospholes **1** are the most chemically stable, and the high-phosphorus *s*- electron density in their lone pairs combines with ring strain constraints to generate a highly pyramidalised ground state **2** (Figure 7.1). Calculations [10–13] show that for 1*H*-phospholes, (1) hyperconjugation between the P–R bond and the butadiene π system generates aromatic stabilisation energy of about 3 to 6 kcal mol^{-1} [11,14–16]; and (2) the degree of hyperconjugation is more influenced by the electronegativity of the phosphorus substituent than its conjugating effects, which are small because of the near-orthogonality of the P-substituent and butadiene π–system planes [10]. The electron-releasing P-P bond in 1,1′-biphospholes therefore confers stronger aromaticity upon **4** than simple carbon

Phosphorus(III) Ligands in Homogeneous Catalysis: Design and Synthesis, First Edition.
Edited by Paul C. J. Kamer and Piet W. N. M. van Leeuwen.

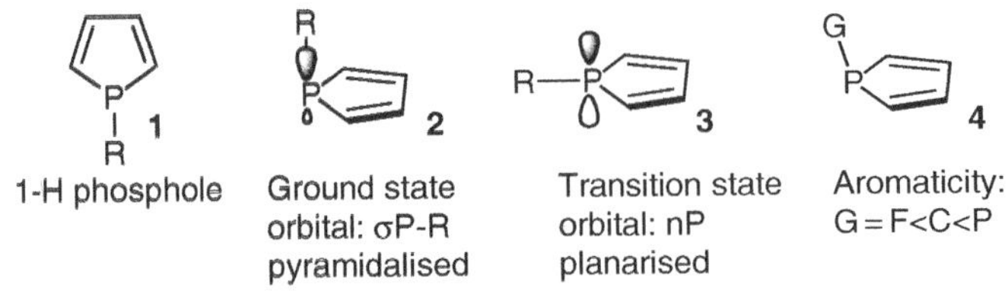

Figure 7.1 *Factors influencing the aromaticity of phospholes.*

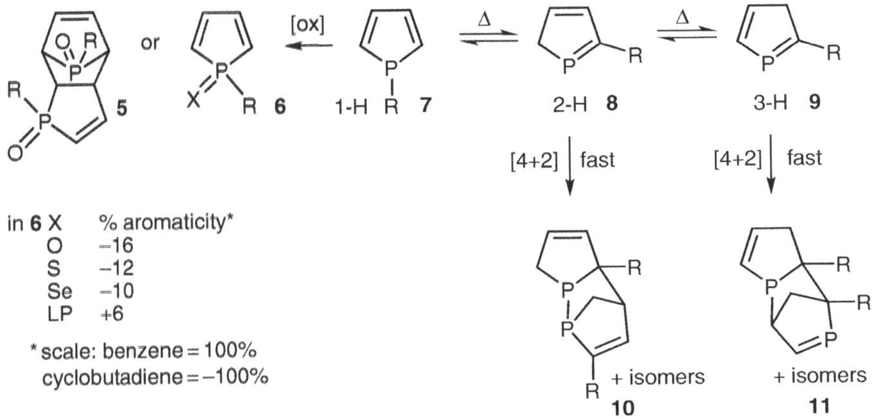

Figure 7.2 *Relative stability of phosphole isomers.*

substituents do [17,18], whilst electron-withdrawing groups generate a 4πe antiaromatic component which results from an increased positive charge at phosphorus [10,19]. Similarly, antiaromaticity rises in the oxidised $\sigma^4\lambda^5$-chalcogen derivatives **6** with increasing electronegativity of the chalcogen (Se: 10%, S: 12% and O: 16%; Figure 7.2) [19,20], or when the phosphorus lone pair in λ^3-phospholes is protonated [21]. Increased antiaromaticity raises the reactivity of diene component significantly with respect to cycloadditions and 1,4 additions, and this effect should be taken into account when phosphole ligands are used to support strongly electrophilic metal centres.

When phosphorus has carbon-based substituents, the 1*H*-phosphole isomer is often only slightly thermodynamically favoured over its shift isomer, the 2*H*-phosphole **8**; the 3*H*-compound **9** lies higher, but not inaccessibly so (Figure 7.2) [11,22]. The hyperconjugated 6π electron system generated between the dienic system and the substituent σ−bond allows these [1,5] shift isomers to interconvert through low-lying, essentially planar [23,24], transition states, provided that the migrating substituent can easily increase its coordination number by one unit. This is the case for H, Si, P, C*sp* and Csp^2 substituents but not for O, S and Csp^3 [25]. This phenomenon is important in terms of stability: a simple 1*H*-phosphole resembles cyclopentadiene in showing a moderate tendency to undergo [4+2] cycloaddition reactions though the carbon atoms of its butadiene system, with slightly higher activation energies for the phosphole because of its aromaticity [26], but the isomeric 2*H*- and 3*H*-phospholes have weak and reactive P=C double bonds which undergo rapid, asynchronous [26–28] cycloadditions with both dienes and dienophiles (including themselves) through very low-lying transition states (Figure 7.2) [23,27]. If the P-substituents in a simple phosphole ligand are conducive to undergoing shift reactions, its reactivity patterns under forcing conditions are therefore liable to become complex upon dissociation from the metal.

The electronic descriptions above are valid for pyramidalised phospholes **2**. As the extracyclic P-substituent of a simple $\sigma^3\lambda^3$-phosphole moves towards the ring plane in **3**, ring strain increases rapidly but the associated

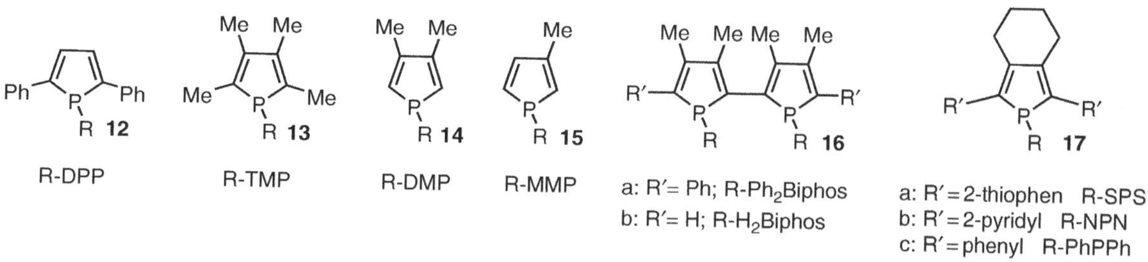

Figure 7.3 *Commonly used ligand structures based upon phospholes.*

phosphorus rehybridisation improves *p*-orbital overlap with the π system, thus increasing aromaticity (Figure 7.1) [13,14,29]. Pyramidalised phospholes have been assigned approximately 42% of the aromaticity of pyrrole, with the corresponding planarised phospholes being estimated at 108% [30]. Artificially planarised phospholes are therefore unlike pyramidalised analogues and undergo electrophilic substitutions [31], and the aromatic stabilisation associated with the planarised phosphole also means that inversion barriers for typical phospholes are lower than those in classical phosphines by about 15–20 kcal mol^{-1} [13,30,32,33]. A significant corollary pertaining to ligand design is that P-chiral phospholes will normally invert rapidly at room temperature [32] (Figure 7.3).

Finally, substituents at the carbon atoms of the phosphole nucleus affect the electronic properties of the system profoundly. Extensive studies of **17**a–c, having 2-phenyl [11], 2-pyridyl and 2-thiophenyl [34,35] or cases having 2-phospholyl [36] substituents, show that the aromatic character of the phosphole ring falls in the order: H > Py > Th, with the substituent showing more pronounced effects upon the highest occupied molecular orbital (HOMO) than the lowest unoccupied molecular orbital (LUMO) [34,37]. Such conjugation of the diene into the 2,5-aryl goups normally lowers the tendency of the phosphole to undergo cycloaddition reactions significantly and confers stability. Benzoannelation has more profound effects; the drive for conjugation within the arene is significantly larger than within the phosphole, and this disrupts the electronic properties of the phosphole nucleus to the point that dibenzo[*b*,*d*]phospholes can be classified as either phosphorus- substituted biphenyls or constrained triarylphosphines [38,39]. Consequently, the major part of this summary will be given over to the chemistry of nonbenzannelated phospholes.

7.2 Creation of phospholes for use as ligands

There are three currently preferred routes to phospholes (Figure 7.4).

7.2.1 Reactions of phosphorus dihalides with metallated dienes (Eq. 7.1)

This (first) route to the phosphole nucleus [40–42] remains the most important and versatile. Suitable dianionic dienes may be prepared through Li [40,42], Zr [43,44], Ti [45] or other transition metal-mediated reductive dimerisations of alkynes [46,47]; Ti-based systems show good functional group compatibility (e.g. primary alkynes [48]) but zirconacycles [3,49–54] are currently more widely employed and a fully optimised Zr-based procedure has appeared [55]. Diyne starting materials can be used to impose the desired product substitution pattern, as in **17**, but simple dissymmetrically substituted alkynes can often be coupled with reliable and predictable regiochemistry; for example, bulky [48,56] or silyl [57] groups feature preferentially on the carbon atoms bound to the metal. Techniques for preparing dissymmetrically substituted phospholes through zirconacyles have also been developed [58–60].

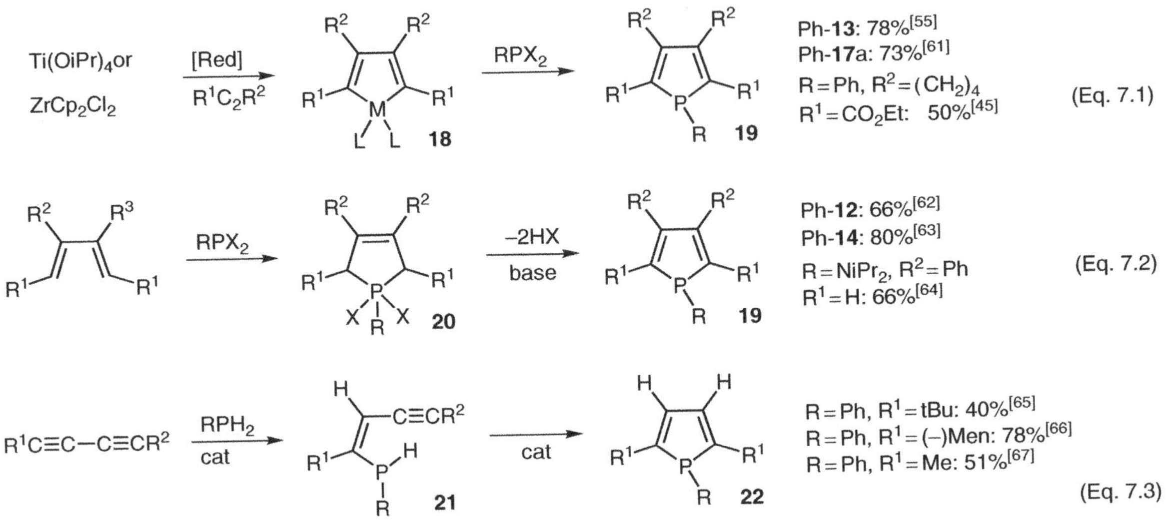

Figure 7.4 The principal synthetic routes to phospholes.

Except when extremely hindered [68,69], these zirconacycles react directly with dihalophosphines to give the corresponding phospholes which can normally be isolated by simple extraction into hexane; bromophosphines often give cleaner phosphinations than the corresponding chlorides [68]. Difunctional reagents such as [$Cl_2PCH_2CH_2PCl_2$] react well [70] and have been used to generate 1,2-dissymmetrical ligands [71]. Using PX_3 rather than PRX_2 (X=halogen) leads directly to phosphole halides [50,68], which are difficult to obtain by other synthetic pathways. Excellent yields make the zirconium approach ideal for the preparation of complex high-value products such as oligophospholes and helicene-derived phospholes [52,54,72]. For its limitations, see [52].

7.2.2 Reactions of phosphorus dihalides with dienes (Eq. 7.2)

These operationally simple routes are normally preferred when large quantities of phosphole are required. In outline, the method involves [4+1] cycloaddition of an appropriate conjugated diene with a dihalophosphine to give a λ^5-dihalophospholene that undergoes double dehydrohalogenation to provide a λ^3-phosphole (Eq. 7.2 and Figure 7.4). The original synthesis, involving refluxing 1,4-diphenylbuta-1,3-diene in dichlorophenyl phosphine, gives the widely used ligand 1,2,5-triphenylphosphole Ph-DPP **12** directly in good (66%) yield [62]. Lower temperature methods are required for less robust phospholes, and the most useful variant is Mathey's [73] dehydrohalogenation of λ^5-dihalophosphol-3-enes **20** which are prepared according to Quin [74]. Reaction of dienes with $PhPX_2$ (X=Cl, Br) and dehydrohalogenation with 2-methylpyridine or N-methylimidazole [63] gives one-pot, two-step yields of typically 75% for 1-phenyl-3-methylphosphole Ph-MMP **15** and 86% for 1-phenyl-3,4-dimethylphosphole Ph-DMP **14**; batches up to 30 kg have been described [63]. 3-ester groups are tolerated, but few functional phospholes have been obtained by this route [75]. The hindrance-sensitive λ^5-dihalophosphol-3-ene synthesis restricts the range of usable dienes [74,76], but this constraint can be partially overcome through using aminochlorophosphenium ions instead of RPX_2 as the phosphorus source [64]; a wider range of 3-phospholenes and otherwise inaccessible phospholes then becomes available. For very bulky phospholes (R=mes, $C_6H_2tBu_3$ *etc.*), more complex methodologies are necessary [77,78].

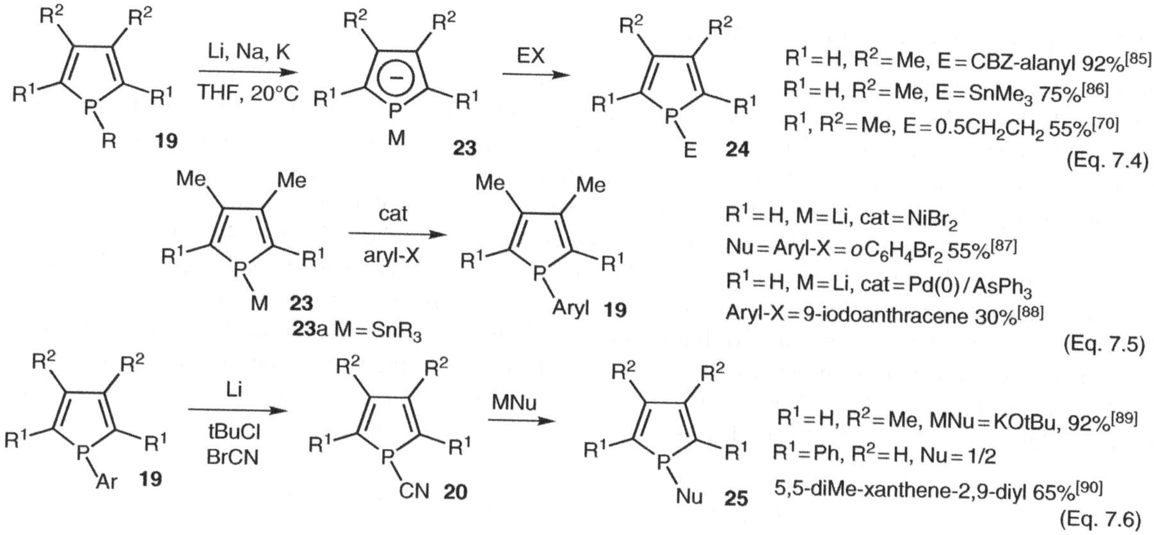

Figure 7.5 *Elaboration of phospholes at phosphorus.*

7.2.3 Michael addition of primary phosphines to dienes (Eq. 7.3) [67]

This approach is rarely used because the phosphorus source- RPH_2 (or its synthetic equivalents, $RP(TMS)_2$ or $RP(CH_2OH)_2$) presents unpleasant characteristics (extreme malodorousness, toxicity and air sensitivity) and product yields can be moderate. Catalysed by Cu salts [67,79], radicals [67,79] or more usually bases [32,67,80], this method is best adapted to phospholes having bulky substituents in the 2,5 positions [65,66] whose stability facilitates chromatographic separation from the often significant quantities of byproducts. An enantiopure atropomeric derivative of Ph-**12** has been prepared through this approach [81]. A related addition of Group 2 trimethylsilylphosphides to 1,4-diphenylbuta-1,3-diyne gives high (up to 82%) yields of 3,4-di(silyl)functionalised phospholide anions, which are potentially excellent precursors for phospholes; see section 7.3.1 [82].

Other methods, notably based on $AlCl_3$-mediated alkyne cyclisations [83] and on phosphonium ylids [84], are also available [85,86].

7.3 Postsynthetic functionalisation

Postsynthetic functionalisation of the phosphole nucleus is often necessary for sophisticated applications (Figure 7.5). The 'value' of the three different ring atom positions (P, C_α and C_β) is reflected in the synthetic methodology available for their elaboration: the phosphorus atom is by far the easiest to functionalise, and much work has also been devoted to elaborating the 2-position because of its importance in preparing chelating ligands. 3-functionalisation methodologies are less well established.

7.3.1 Functionalisation at phosphorus (Figure 7.5)

Very convenient functionalisation methodologies based upon phospholide anions **23** have been used extensively since their groundbreaking discovery by Braye [79]; approaches using the less stable formal phosphole cations are less accessible and less well developed. In exceptional cases, such as with *t*BuLi, direct S_N2 attack at the PPh bond of phenylphospholes is possible with liberation of PhLi [91].

7.3.2 At phosphorus: use of electrophiles (Eq. 7.4)

Phosphole aromaticity rises as P-substituent electronegativity falls and the polarisability of the P–R bond rises, so alkali metal 'substituents' on the phosphole ring confer a strong aromaticity which has been estimated at about 85% of the value in the cyclopentadienide anion [14,92]. Alkali metal treatment of phospholes having a range of P–R groups (R = halogen [93], Ph [79], phosphole [94] *etc.*) therefore generates the corresponding phospholide anion easily, and the stability and availability of P-phenyl substituted phospholes [49,62,63] usually make them starting materials of choice. Quenching the LiPh byproduct with tBuCl [95] and alkylation of the phospholide anion at P by simple alkyl, vinyl, benzyl, halides, tosylates, α,ω-dielectrophiles *etc.* completes what is usually a simple one-pot process [75]; products are often isolated directly, often as moderately air-sensitive liquids or, for 2,5-diaryl derivatives, air-stable solids. The borane functionality is rarely used for protecting the phosphorus lone pair during workup because of the low stability of phosphole boranes [96,97]. If not sufficiently pure for immediate use, the phosphole is usually isolated as the sulphide [98,99] or a metal complex [97] and then liberated in a subsequent step.

7.3.3 At phosphorus: use of nucleophiles and aromatics (Eqs. 7.5, 7.6 and 7.7)

S_N 2Ar P-arylation of phospholes can be effected with favourable reagents, such as aryl fluorides having directing substituents [100]. Aryl iodides can also undergo direct substitution, although reduction is usually competitive [101]. Arylations of phospholides or P-trialkylstannanes have also been palladium, nickel or copper catalysed (Eq. 7.5) [88,102–104]. However, Pd-catalysed arylations with aryl iodides give variable yields [88,102], methodologies for triflates are not available and bromides can be used only in very favourable cases such as 2-pyridyl [103]. These difficulties in cross-coupling simple phospholes contrast with their benzannelated analogues which can be classically cross-coupled through their stable secondary oxides [105], so methodologies formally involving phosphole cations are generally preferred. Such phosphole cation equivalents can be accessed from aryl phospholes through an initial conversion into the phospholide anion and subsequent umpolung using cyanogen bromide (Eq. 7.6). The quite stable P-cyanide products **20** can be purified by chromatography [89,90,106,107], and show good reactivity towards alkoxides [89,108,109], amides [106] and similar reagents, as well as to redox resistance toward aryl and alkyl anions [89,99,110,111], and they are employed extensively. Conversion of phospholides into the more attractive but highly sensitive P-halogen derivatives **25** through oxidative coupling to the 1,1′-biphosphole and further treatment with bromine [112] or iodine [101] carries a significant risk of overoxidation; organozirconium reagents allow a better and more direct route to them (Eq. 7.7) [50,113]. The P-chlorides again appear nicely resistant to redox chemistry [50].

7.3.4 At carbon: elaboration about the phosphole nucleus (Figure 7.6)

Deprotonation of phosphole oxides or sulfides in the 2-position is rarely employed because of the low thermal stability of many simple phosphole oxides [75] and the competing reactivity at other positions in the case of sulphides [114]; cleaner reactivity is usually observed with 2,-bromo [98] or 2,5-dibromophospholes **26**a–b [115] which undergo metal–halogen exchange reactions cleanly at –105°C (Eq. 7.8). Close temperature control is usually necessary for P-aryl substituents, but the P-N*i*Pr_2 derivatives may have greater stability [64]. 2- Halogen substituted phosphole oxides that are sufficiently hindered to be stable with respect to autocycloaddition reactions can be cross- coupled using classical palladium- based chemistry.[115b, c].

A second approach involves combining [1,5] sigmatropic shifts with deprotonation. The normally unstable 2*H*-phospholes resulting from sigmatropy (or their synthetically equivalent [4+2] cyclodimers) can be deprotonated *in situ* by a suitable base to form 2-functionalised phospholides (Eq. 7.9) [87,95,104,116]. The method is probably the most powerful available for the elaboration of the phospholes in their 2 positions.

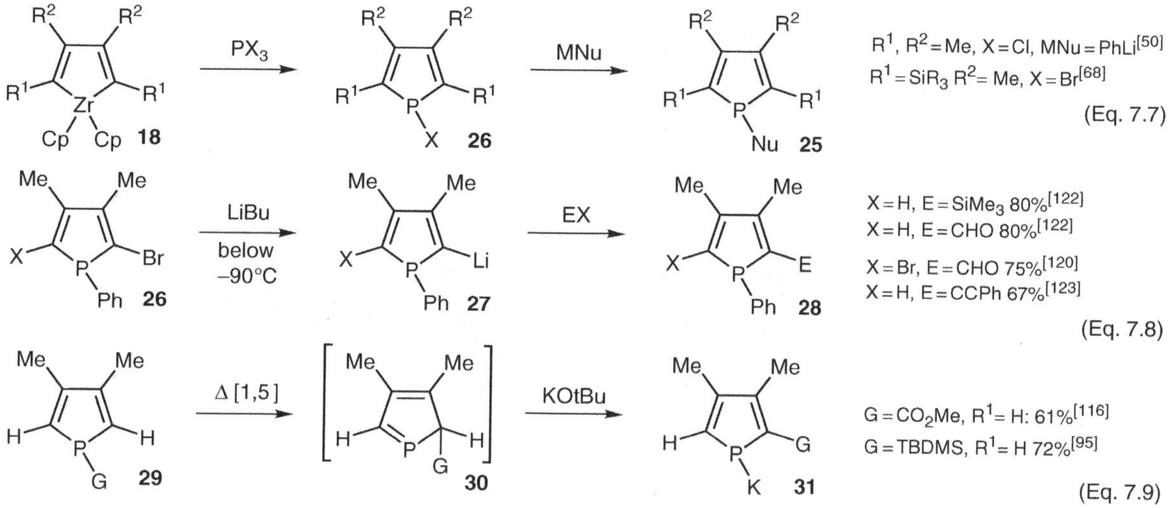

Figure 7.6 *Elaboration of phospholes at carbon.*

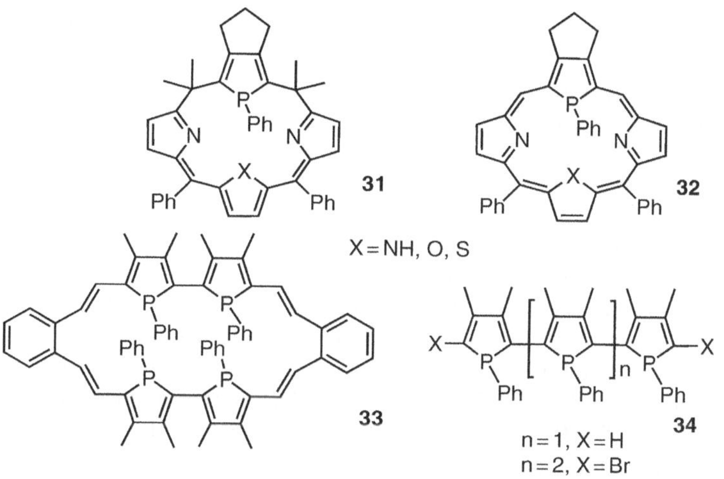

Figure 7.7 *Examples of C-elaborated phospholes* [118–121].

A similar but more restricted methodology is available for the 3 position, the constraint being the presence of strongly bound substituents in both the 2 and 5 positions [117]. A very active recent area has involved quite classical organic elaboration of phospholes in the 2 and 5 positions for preparing new ligands or molecular materials. The sophistication possible is illustrated in selected examples in Figure 7.7 [118–123].

7.4 Phosphole coordination chemistry

The fundamental properties of phospholes as ligands have been reviewed comprehensively and a wide range of coordination modes are possible [124]. As with more classical phosphines, computational and PES studies [125–127] show that the P lone pair is the preferred donor and simple η^1 complexes are usually formed; the

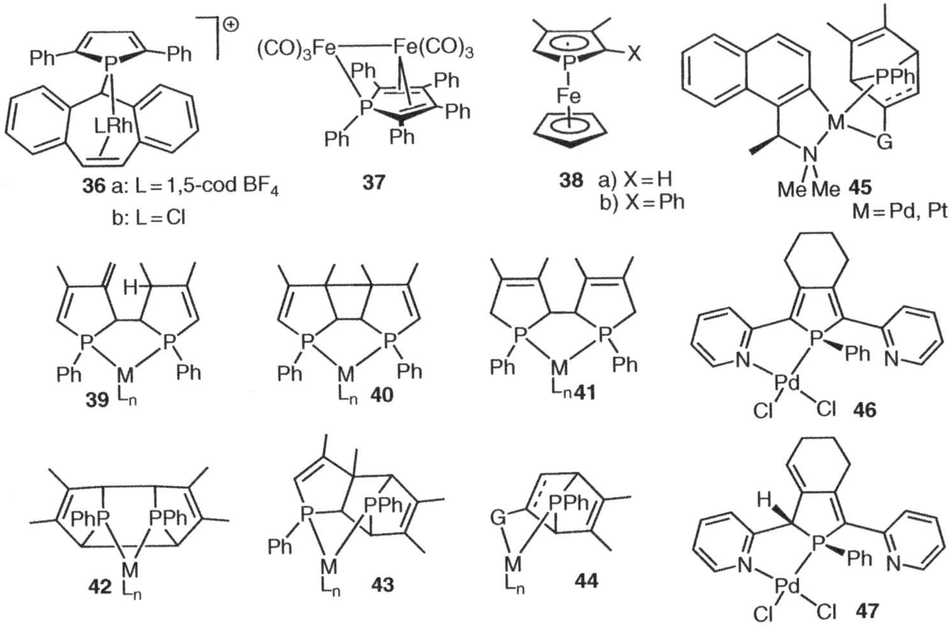

Figure 7.8 Phosphole ligands and some of their evolution products in metal coordination spheres. $ML_n = PdCl_2$ [144] $PtCl_2$ [145], $RuCp^+$ [146] and $Ru(arene)^{2+}$ [147].

low-lying σ*P–R orbital is the acceptor [128]. CDA calculations at the 6–31 G** level indicate that back donation plays a more significant role in the binding of $Ni(CO)_3$ to Ph-**12** (which shows a donor:backacceptor quotient of 4.7) than in triarylphosphines or benzophospholes (both 6.5) [128]. P-coordinated phospholes can therefore be considered to confer less electron density upon a coordinated metal centre than related classical phosphines. When expressed in more conventional Tolman [129] terms, their donor power rises in the order Ph-**12** (2071.9 cm^{1}) < pentaphenylphosphole < Ph-**14** (2067.9 cm^{1}) << tBu-**14** (2063.5 cm^{1}) [130]; these data are comparable with PPh_3: 2068.9 and PMe_3: 2064.1 cm^{1}. The steric parameter θ is estimated at 101° for 1-benzylphosphole and 145° for Ph- **12**; these values lie close to $PhPH_2$ and PPh_3 respectively [130]. Phospholes generally undergo little structural reorganisation upon coordination [131], and the straightforward access to 2- and 2,5-disubstituted phospholes makes a wide range of steric profiles available.

Reliable procedures for attaching phospholes to metal centres are available [124,131–135] and, unless the phosphole is very bulky [136,137], coordination is generally straightforward. Ligands such as **36** [138] can be designed wherein an olefin component intervenes to stabilise open-shell metal structures that can serve as catalytic resting states, and the diene component of well- designed phospholes can also be used for this purpose [90]. Note that coordination chemistry under forcing conditions should be avoided; phospholes can give rise to multinuclear species resulting from simultaneous engagement of the lone pair and diene functionality [42,124,139] (such as **37** upon thermolysis of pentaphenylphosphole with $Fe_3(CO)_{12}$ or from loss [140,141] or migration of the P-substituent [142,143] (as in **38**a,b from thermolysis of Ph-**14** with $[FeCp(CO)_2]_2$).

Stability: Phosphines are rarely perfect "spectator" ligands [148] and metal sensitivity can be acute for some phospholes (Figure 7.8). Coordinated Ph-**14** shows reactivity towards many catalytically important substrates, the most obvious being activated alkenes and alkynes [149]. This metal-induced reactivity provides excellent potential for the synthesis of new ligands: for example, reductive dimerisation at Ni(II) [150] and Pt(II) [145] centres, and [4+4], [4+2] and [2+2] cycloadditions between two phosphole ligands

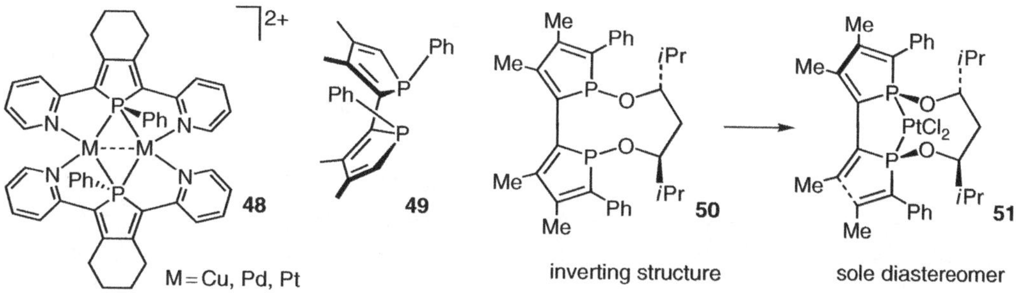

Figure 7.9 *Unusual properties associated with phospholes.*

[144–147], are also well documented, giving **39-43**. Lewis acidic Pd(II) [151] and Pt(II) [152,153] cations have been used to promote templated [4+2] cycloaddition reactions between (mainly activated G=CO_2R, C=O(R), SO_2R, SR, R, PR_2 *etc.*) -alkenes or -alkynes and the diene system of Ph-DMP (or its sulphide [154]) to give **44**; complexes bearing enantiopure auxilliaries provide clean enantiomers of analogous bicyclic phosphines **45** in reactions whose diastereoselectivity can rationalised [155,156]. Phospholes can also undergo formal [20] 1,3 proton shifts in metal coordination spheres (Pd(II), Pt(II)) in the absence of a second phosphole ligand with the diastereoselective formation of phospholene complexes such as **47** [157,158]. Complex **47**, having an *exo*- methylene ligand, is more stable in the metal coordination sphere than **46** (by 10.8 $kJmol^{-1}$) but the phosphole isomer is preferred by 18.3 $kJmol^{-1}$ when the metal is removed; the drive for the shift partially reflects the contrast between the antiaromatic nature of the coordinated phosphole ring and the weak aromaticity in the free phosphole [158]. These *exo*-methylene phospholenes are potentially very strong donors [159] and have been obtained enantiopure [160].

Two unusual characteristics distinguish phospholes from more classical phosphines, and these provide the potential for designing novel ligands (Figure 7.9). They are (1) a rare [161,162] capacity amongst $\sigma^3\lambda^3$-phosphines to adopt bridging coordination modes, and (2) their very low inversion barriers.

Bridging ligands: 2-pyridylphospholes often give $\sigma^3\lambda^3$-phosphole bridges at palladium(I), platinum(I), copper(I) and silver(I) centres [163–167] and this bridging mode is quite robust; for example, the **48** (M=Pd) core is stable to air, CO, $TMSN_3$ $TMSCHN_2$ and PR_3 [166], this stability being attributed to high lone pair *s*-character. Both symmetrical and dissymmetrical phosphole bridges can be observed, the potential surface being quite shallow. Mixed metal dimers can be prepared rationally [163]. Unusually, such bridging phospholes stabilise dimers unsupported by metal–metal bonds (as in d^{10} copper complexes) and clusters [168]. They provide attractive potential for supramolecular assembly and have been elegantly exploited, *inter alia*, in the construction of chiral π-stacked organometallic paracyclophane analogues [72].

Tropos ligands: These have been developed impressively with 2,2′-biphospholes [169]. Low-phosphorus inversion barriers (ΔH: 63.5 $kJmole^{-1}$ ΔS: 25$Jmol^{-1}$) and rapid 2,2′-bond rotation at −70°C allow different 2,2′-biphosphole diastereomers to interconvert easily [170]. One case, **49**, resolves spontaneously into large single-enantiomer conglomerate crystals which can be redissolved and coordinated to [$PdCl_2$] [171] and $[Ir(cod)]^+$ [172] centres at low temperature without loss of optical activity; clean coordination to $[Rh(1,5\text{-}cod)]^+$ is more difficult [172]. The *tropos* properties of these systems can be harnessed more universally through functionalisation with enantiopure diolates [109] or diyls [99] which dictate a preferred configuration for the coordinated biphosphole. For example, the pentane-2,4-diyl linked diphosphinite **50** shows inverting conformers in solution at temperatures above 60°C [173], but the chain linking the two phosphole groups confers a strong conformational preference upon coordination to Pt, Pd *etc.* centres and allows clean enantiopure coordination compounds (e.g. **51**) to be prepared [174]. Note that free ligand structures are not necessarily reliable predictors of chirality in the 2,2′-biphosphole component of the corresponding coordination complex [175].

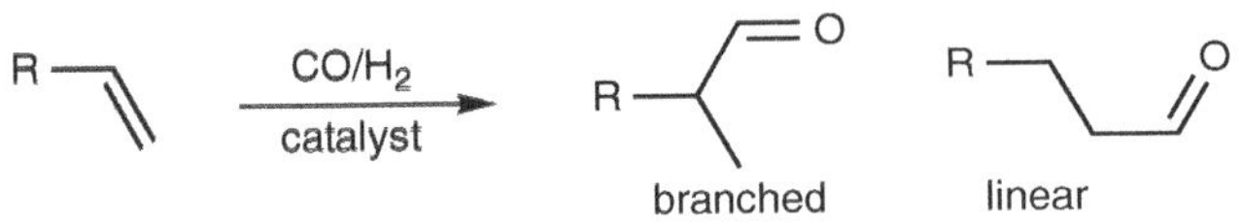

(Eq. 7.10)

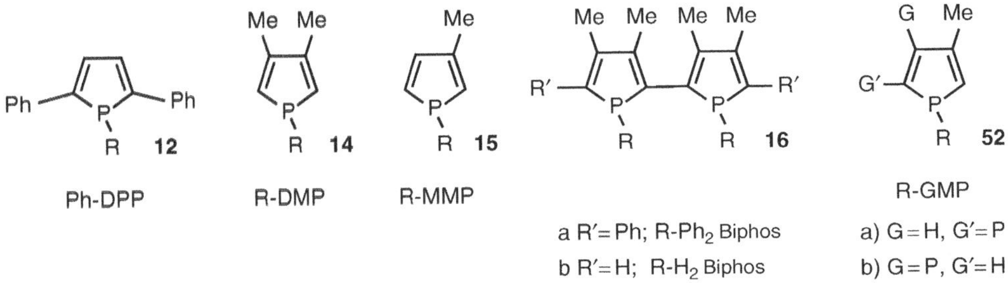

Figure 7.10 *Hydroformylation with phospholes.*

7.5 Phospholes in catalysis

In general, phospholes have been used sparingly as ligands for catalysis and it will be noted that 2,5-diarylsubstituted derivatives have received by far the most attention, presumably because of their relative stability and low propensity towards cycloaddition. The integrity of phosphole ligands under catalytic conditions is rarely verified, so their influence should be interpreted carefully, particularly if metals having significant Lewis acidity ($Pd(II)^+$ *etc.* that might provoke ligand reactivity) are involved. Phospholes have been involved in catalytic processes which include the following (Figure 7.10):

Hydroformylation: Phospholes have long been associated with hydroformylation (Eq. 7.10), and very early studies identified phosphole-modified cobalt catalysts as being highly active with terminal olefins [176]. Ph-**12** is also frequently described in Rh(I)-based processes and a seminal study showed that styrene hydroformylation is insensitive to ligand at Rh:Ph-**12** ratios greater than two [177]; the Ph-**12**-based catalyst gives 100% aldehyde (T≤80°C, P=20–30 bar), is twice as active as PPh_3-supported systems [178] and gives a thermodynamically controlled linear:branched ratio of 8.5:1 that is relatively insensitive to both t and T. Similar results were found with ethyl acrylate [179], 2,3- and 2,5- dihydrofurans [180] and 1-hexene, where the optimised Ph-**12** system was 10% more selective and 18 times more active than PPh_3 [130,135]. Experiments under deuterated syngas imply that a reversible and thermodynamically controlled insertion of styrene occurs into a RhH bond, with the probable intervention of an association complex between the styrene and [RhH(CO)(Ph-**12**)$_2$] [181]; the [RhH(CO)$_2$(Ph-**12**)$_2$] resting state exists as equilibrating stereoisomers, but the formulation of the minor isomer (10%) as having an equatorial hydride [182] has been contested [183]. Current best data imply a single hydroformylation cycle wherein the rhodium is invariantly coordinated to two Ph-**12** ligands, and give an analytical equation for the reaction rate that is consistent with the complex dependence of the product distribution upon syngas composition and pressure [184].

The influence of very bulky phosphole P-substituents (R=mes, tBu$_3$C$_6$H$_2$, iPr$_3$C$_6$H$_2$, tBu$_2$MeC$_6$H$_2$) and of ring substitution upon styrene hydroformylation has been evaluated under 40 or 50 bar syngas using 0.1 mol% of catalysts that were generated *in situ* from two equivalents [137,185,186] of phosphole (R-**14**, R-**15**, R-**52**) and [Rh(nbd)Cl)]$_2$ or four equivalents with [RhCp*Cl$_2$]$_2$ [136]. Exocyclic (P=O) groups G in the phosphole 2 or 3 positions of R-**52** were linked to higher activity [186] and branched aldehydes were generally observed. The coordination chemistry of the bulky phospholes appears to be both atypical and difficult [136,137] and differences from the classic Réau systems above may reflect this. A P-(dibenzoazepine)phosphole **36a** that can function as a

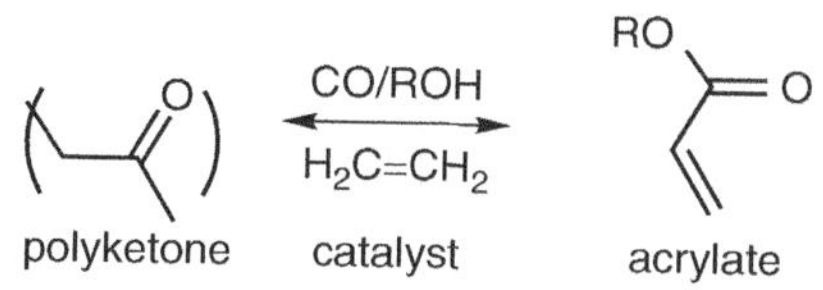

(Eq. 7.11)

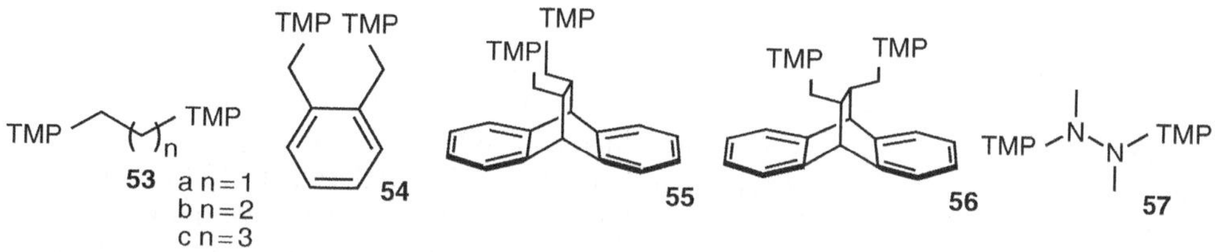

Figure 7.11 *Pd-catalysed copolymerisationation with phospholes (TMP=2,3,4,5-tetramethylphosphole,* ***13****).*

chelating 2,5-diphenylphosphole-olefin ligand gives selectivity up to 96% for branched products with styrene under 20 bar of syngas, and also showed some activity with cyclooctene and the quite difficult substrate tetramethylethylene, although TOF was low at 6 h^{-1} [138]. Hydroformylation of styrene with the biphosphole-based rhodium complexes [Rh(cod)(+/−)(Ph-**16**)]BF_4 for 20 h at room temperature using 150 atm of 1:1 syngas gave moderate yields but a good regiochemistry of 98:2 in favour of the branched isomer in MeCN; use of MeOH improved conversion to 88% but inverted the regiochemistry, giving a branched to linear ratio of 29:71 [172]. Note that small and reactive 3,4-dimethylphosphole-based ligands R-**14** (R=Ph [186], *t*Bu, *n*Bu, Ph [178]) give poor results in Rh-based systems.

A series of phospholes having bulky aryl groups to increase planarity (as discussed in this chapter) have also been compared in platinum-catalysed hydroformylation. However, the simplest ligand Ph-**14** gave the best results in L_2PtCl_2–$SnCl_2$ catalysed systems under 1:1 H_2–CO at 80 bar and showed 90% chemo- and 85.5% regioselectivity with styrene; conversions were low in all cases [137]. R-**16**a-based complexes (where R=enantiopure diyl- and diolate functions) also show excellent chemoselectivity for L_2PtCl_2–$SnCl_2$-catalysed styrene hydroformylation, and regioselectivities of 76–91% in favour of branched aldehydes at 20°C. These fell at 60°C. Rates, and enantioselectivities <20%, were low. Performance in these biphosphole-based systems was only slightly affected by the nature of the heteroatom at P [174].

Copolymerisation: The outcome of palladium-catalysed olefin carbonylations, which can give either polyketones or acrylates, is strongly influenced by the supporting ligand; for example, α,α-o-xylyl*bis*(PR_2) ligands are highly active for methylpropionate when R=*t*Bu, but give copolymer when R=Ph [187]. The performance of chelating tetramethylphosphole (R-**13**)–based ligands has been shown to be very sensitive to the form of the tether: under 20 bar CO–C_2H_4 at 90°C in MeOH, **52**a and **52**c are inactive, but **53**b and **54**–**56** having *cis*- or *trans*-fumaryl and α,α-o-xylydiyl linkers give copolymerisation with tightly defined molecular weight distributions (M_wM_n=1.7), good to excellent efficiency (a mole of catalyst giving up to 46Kg polymer h^{-1} for **54**) and higher degrees of polymerisation (to n=80 for **55**) than the corresponding *bis*-(diphenylphosphanyl)-substituted analogues. Note that *trans*-anthracene **56** gives copolymerisation, but its PPh_2 analogue gives mainly (73%w/w) methylpropanoate; this chemoselectivity may reflect the relatively two-dimensional phosphole's ability to protect the apical sites at Pd. The inactivity of **53**c appears to stem from its inability to form chelates at Pd(II) [70,188]. Predictably [189], monodentate Ph-TMP gives only propoanote, but inefficiently [188]. Independently, Fagan finds that for **53**a, acid-free copolymerisation with a preformed [$PdMeL_2$]SbF_6 catalyst gives Mw=236000 copolymer with M_wM_n=5 at 45 bar and 100°C over 18 h with a production of 28 kg polymer per mole of catalyst per hour, which fell to Mw=149000 copolymer with M_wM_n=2.4 at 45 bar; the **53**b homologue gave 76 kg at 45 bar. Hydrazine derivative **57** is less useful [71] (Figure 7.11).

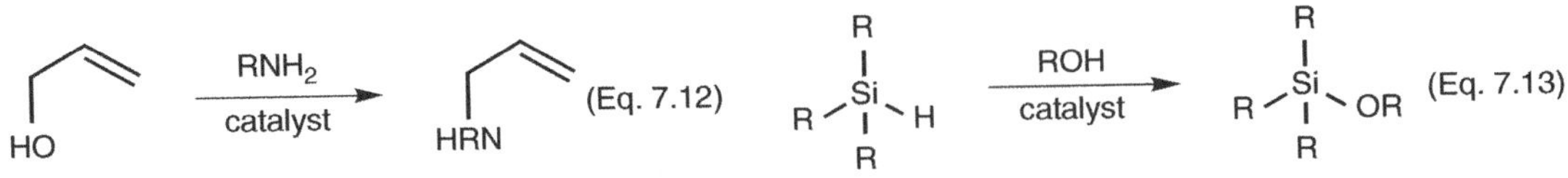

Figure 7.12 Allylic amination (left) and dehydrogenative silylation (right).

Ethylene–CO copolymerisation can also be achieved at 20 bar of CO–C_2H_4 using cationic palladium 2-pyridylphosphole complexes **58**a–c (R = Ph, Cy). The system is unstable in methanol, which rapidly causes reduction to **48** (M = Pd) but works well in dichloromethane. Whilst significantly less active than the best diphosphine-based classes, it shows good performance (TOF to 75 h^{-1} and TON to 1586 at 20 bar) relative to other PN-based systems. Norbornadiene is also copolymerised with a small *MwMn* ratio [165].

The complementary methoxycarbonylation of styrene has been effected using $Pd(OAc)_2$–L–acid catalytic systems with biphosphole-based ligands. Racemic *tropos* Ph-**16b** performed disappointingly at 50°C at 20 bar CO, giving 10% consumption of styrene and 3% conversion into 88% of the branched product methyl (2-phenyl)propanoate over 21 h; the corresponding 2,2′-biphosphole **16** linked at the P-atoms by a (*S*)-1-methylpropylidenediyl chain performed similarly and gave a low ee (13%) [190].

Palladium-catalysed reactions with aryl halides: In early studies, Ph-**14** was also found to be far superior to PPh_3 for Pd-catalysed carbonylation of organic halides [191], and recently complex phosphole-containing hybrid calixphyrins **59**ab have been found to catalyse the Heck reaction of bromoarenes with *n*-butylacrylate in DMA at 135°C with turnover numbers of 9700–12300 [192,193]. An earlier macrocyclic phosphole-based structure also gave excellent TON in Heck arylations [194]. Miyaura and Heck couplings have also been effected using pyridyl and oxazolinyl substituted ligands **60** and **61**. Iodobenzene and dihydrofuran gave 2-phenyl-2,5-dihydrofuran with 89% ee and 67% yield [195]; Pd complexes of pyridyl-based phosphole pincer ligands have also been shown to catalyse the Miyaura couplings of arylbromides and iodides [196].

Allylic amination of unactivated allylic alcohols (Eq. 7.12): Early studies showing that the electron-poor $[Pd(allyl)(Ph\text{-}\mathbf{12})_2]^+$ cation outperforms dibenzophosphole (DBP)– or PPh_3-based analogues in the amination of unactivated allylic alcohols by morpholine and aniline [128] have led to the development of second-generation palladium- [90] and platinum- [197] based allylic alcohol amination catalysts that contain the Xantphos- type phosphole **62**b. DFT-calculated catalytic cycles are substantially the same for the two metals, with the role of the relatively acceptor 2,5-diphenylphosphole ligands and the wide-angle (Xanthene-based) bridge being to create low-basicity systems that shut down a catalyst-poisoning pathway involving metal protonation and insertion of allylic amine into the M–H bond which has been shown to operate experimentally for Pt(dppe)- based cycles [197]. Whilst the palladium-based system also aminates allylic ethers [90], platinum appears to accept a wider range of amines and gives 98% amination of allyl alcohol by morpholine (4 h at 50 °C, 0.1 mol% catalyst) [197] (Figure 7.12).

Dehydrogenative silylation of alcohols: Use of a Xantphos-based ligand has allowed the formation and isolation of a cationic digold hydride cation $[Au_2(\mathbf{62}b)_2H]^+$, whose solutions induce dehydrogenative silylation of alcohols. Rates vary as a function of the counterion, with the best (triflate) giving complete triethylsilylation of 2-phenylethanol using 1 mol% of catalyst in 2 hours at 50°C. The stability of the gold hydride species appears to be marginal, and the design of the supporting ligand is crucial to its stability [198]. DFT studies rationalising the $[Au_2(\mathbf{62}b)_2H]^+$ cation and treating the catalytic pathway are available [199].

Ethylene dimerisation: $[Ni(\mathbf{58})Cl_2]/AlClEt_2$ (1:70) mixtures give good activity (TOF to 56000 h^{-1}) in the nickel-catalyzed oligomerisation of ethylene, with the phosphole 5-substituent (Ph, Th) being more influential than the group at P (Ph, Cy). Selectivities are pressure dependent: at 41 bar, highly C_4-rich fractions (up to 97%) having good selectivity for 1-butene (80% of C_4) were obtained [200]. Modification of $[NiBr_2dme]$

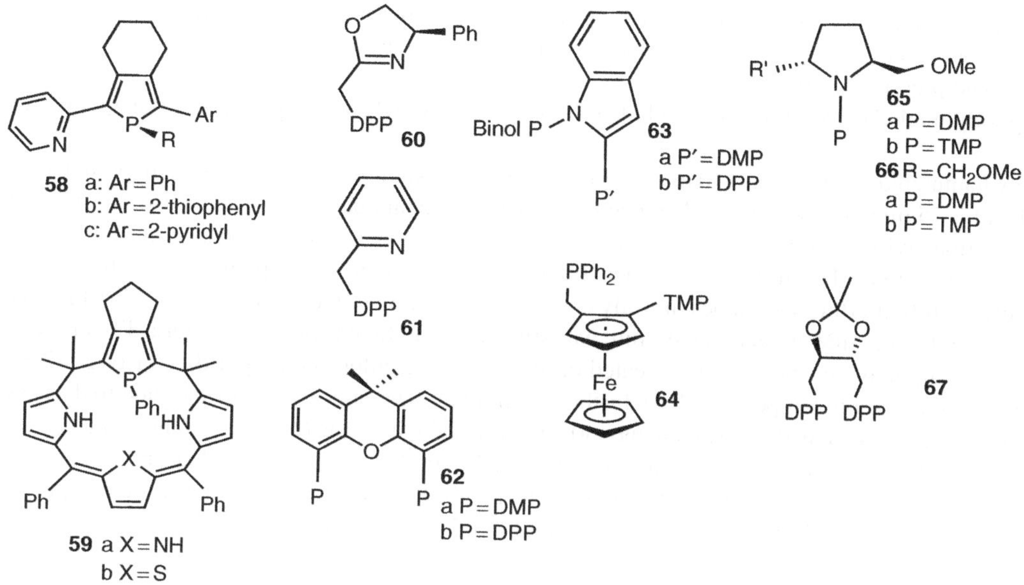

Figure 7.13 *Less common phosphole ligands used in catalysis.*

by the simple ligands Ph-**12** or Ph-**14** gives poor catalysts but, predictably [201], the wide-angle DMP–Xantphos-based complex **62**a gave far superior performance (97% of C_4 comprising 90% 1-butene with TOF=43000 h^{-1}) [202].

Hydrogenation [109,134,172,203,204] *and allylic malonation* [99,106,111,205]: These classical tests for new ligands have been extensively explored for phospholes and a wide variety of studies have been conducted, including with ligands **63–67**. The reader is directed to the primary literature for details. The data, particularly with respect to hydrogenation, are consistent with models wherein phospholes play the role of ligands having slightly lower donor capacity than the corresponding phosphines. Useful comparisons with dibenzophospholes and triphenylphosphine are available [203]. Finally, Pd-catalysed asymmetric *hydrosilylation* of 5,5-dimethyl-1-hexen-3-yne using $HSiCl_3$ by a phenyl-substituted phosphole having 2,5-menthyl substituents is less efficient and less enantioselective than with the corresponding phosphametallocene [81], and Rh-catalysed hydrosilylations has been reported using hybrid calixphyrins [192]. $[RhCl(CO)_2]_2$/**59**a catalysed acetophenone hydrosilylation with $SiPh_2H_2$ gave 91% 1-phenylethanol after hydrolysis; **59**b was inferior.

The current state of play appears to be that many quite highly elaborate and elegant phosphole derivatives are available for coordination chemistry and catalysis, but that they have not been exploited to the full. Further developments in these areas can be confidently expected (Figure 7.13).

7.6 Experimental procedures

1,2,5-triphenylphosphole (Ph-**12**): the preparation is done with protection from air.

Phenyldichlorophosphine (STENCH!, 15 mL, 115.7 mMol) and 1,4-diphenylbuta-1,3-diene (5.0 g, 16.0 mMol) were mixed under an inert atmosphere in a flask fitted with a reflux condenser and a nitrogen inlet and heated to reflux (222°C) with good stirring in an electric mantle for 8 hours. The highly moisture-sensitive mixture was stoppered carefully, and the deep yellow-orange solution was cooled overnight in a freezer.

After filtration under protection from air at a frit, the solid product was washed with ether (5×20 mL), and then petroleum ether (5×20 mL) or until odourless. The pure yellow, air- and water-stable product was dried under reduced pressure (3.0 g, 64%).

1-(2-pyridylmethyl)-2,5-diphenylphosphole ($NC_5H_4CH_2$-DPP) **61**: the preparation is done with protection from air.

Freshly hammered lithium (1 g, 140 mMol; containing 1% of sodium) was stirred with naphthalene (0.25 g, 2.3 mMol) for 15 minutes in THF that had been freshly distilled and degassed. The blue solution was treated with 1,2,5-triphenylphosphole (3.8 g, 12.2 mMol) in three portions at 20°C for 6 h or until all the solution became homogeneous. The mixture was then filtered under an inert atmosphere with a frit or canula to remove excess lithium (this should be carefully destroyed according to local safety rules). Tert-butyl chloride was added (4 mL, 36.2 mMol), and the mixture was stirred overnight at room temperature. The crude solution of lithium 2,5-diphenylphospholide was treated dropwise with 2-chloromethylpyridine (1.64 g, 12.8 mMol) at 0°C and stirred at room temperature for a further 15 minutes. The product was evaporated to dryness under reduced pressure, taken up in dichloromethane (30 mL), washed with brine (10 mL) and then water (2 × 10 mL) and dried briefly (5 minutes) over magnesium sulfate. Ethanol (50 mL) was added and the mixture was reduced in volume to 5 mL under reduced pressure. The precipitate of 1-(2-pyridylmethyl)-2,5-diphenylphosphole (3.1 g, 78%) was collected, washed with ethanol (5 mL) and dried.

References

[1] O. Fadhel, M. Gras, N. Lemaitre, V. Deborde, M. Hissler, B. Geffroy, R. Reau, *Advanced Materials* **2009**, *21*, 1261.
[2] N. Le Maitre, B. Geoffroy, O. Fadhel, M. Hissler, R. Reau, World patent WO2008/012298 to the CEA and University of Rennes 1, France, 31 Jan 2008.
[3] D. Joly, D. Tondelier, V. Deborde, B. Geffroy, M. Hissler, R. Reau, *New Journal of Chemistry* **2010**, *34*, 1603.
[4] S. Urig, K. Fritz-Wolf, R. Reau, C. Herold-Mende, K. Toth, E. Davioud-Charvet, K. Becker, *Angewandte Chemie-International Edition* **2006**, *45*, 1881.
[5] E. Viry, E. Battaglia, V. Deborde, T. Muller, R. Reau, E. Davioud-Charvet, D. Bagrel, *Chemmedchem* **2008**, *3*, 1667.
[6] K. Becker- Brandenburg, E. Daviuod- Charvet, V. Deborde, R. Réau, R. H. Schirmer, C. Herold-Mende, World patent WO 2006/024770 to CNRS, France, 9 March 2006.
[7] L. D. Quin, in *Current Organic Chemistry*, **2006**, *10*, 43–78; L. D. Quin, in *Comprehensive Heterocyclic Chemistry*, 2nd ed., Pergamon, **1996**, 2, 757; L. D. Quin, in *Phosphorus Carbon Heterocyclic Chemistry* (ed. F. Mathey), Pergamon, **2001**, 219–307; L. D. Quin, G S. Quin, in *Phosphorus Carbon Heterocyclic Chemistry* (ed. F. Mathey), Pergamon, **2001**, 307–364; L. D. Quin, in *Comprehensive Heterocyclic Chemistry*, Pergamon, **1984**, *1*, 518–522; and see also Y. Matano, T. Imahori, *Organic and Biomolecular Chemistry*, **2009**, *7*, 1258–1271; A. N Hughes, C. Srivanavit, *Journal of Heterocyclic Chemistry* **2009**; and F. Mathey, *Chemical Reviews* **1988**, *88*, 429.
[8] P. V. Schleyer, D. Kost, *Journal of the American Chemical Society* **1988**, *110*, 2105.
[9] P. Le Floch, *Coordination Chemistry Reviews* **2006**, *250*, 627.
[10] E. Mattmann, F. Mathey, A. Sevin, G. Frison, *Journal of Organic Chemistry* **2002**, *67*, 1208.
[11] D. Delaere, A. Dransfeld, M. T. Nguyen, L. G. Vanquickenborne, *Journal of Organic Chemistry* **2000**, *65*, 2631.
[12] L. Nyulaszi, *Chemical Reviews* **2001**, *101*, 1229.
[13] L. Nyulaszi, *Journal of Physical Chemistry* **1995**, *99*, 586.
[14] A. Dransfeld, L. Nyulaszi, P. V. Schleyer, *Inorganic Chemistry* **1998**, *37*, 4413.
[15] P. V. Schleyer, P. K. Freeman, H. J. Jiao, B. Goldfuss, *Angewandte Chemie-International Edition in English* **1995**, *34*, 337.
[16] M. K. Cyranski, T. M. Krygowski, A. R. Katritzky, P. V. Schleyer, *Journal of Organic Chemistry* **2002**, *67*, 1333.
[17] O. Fadhel, D. Szieberth, V. Deborde, C. Lescop, L. Nyulaszi, M. Hissler, R. Reau, *Chemistry-a European Journal* **2009**, *15*, 4914.
[18] C. Fave, M. Hissler, T. Karpati, J. Rault-Berthelot, V. Deborde, L. Toupet, L. Nyulaszi, R. Reau, *Journal of the American Chemical Society* **2004**, *126*, 6058.

[19] Z. Mucsi, G. Keglevich, *European Journal of Organic Chemistry* **2007**, 4765.
[20] L. Nyulaszi, O. Holloczki, C. Lescop, M. Hissler, R. Reau, *Organic & Biomolecular Chemistry* **2006**, *4*, 996.
[21] D. B. Chesnut, L. D. Quin, *Heteroatom Chemistry* **2007**, *18*, 754.
[22] S. M. Bachrach, L. Perriott, *Journal of Organic Chemistry* **1994**, *59*, 3394.
[23] T. C. Dinadayalane, K. Geetha, G. N. Sastry, *Journal of Physical Chemistry A* **2003**, *107*, 5479.
[24] S. M. Bachrach, *Journal of Organic Chemistry* **1993**, *58*, 5414.
[25] F. Mathey, *Accounts of Chemical Research* **2004**, *37*, 954.
[26] T. C. Dinadayalane, G. N. Sastry, *Organometallics* **2003**, *22*, 5526.
[27] T. C. Dinadayalane, G. Gayatri, G. N. Sastry, J. Leszczynski, *Journal of Physical Chemistry A* **2005**, *109*, 9310.
[28] S. M. Bachrach, L. M. Perriott, *Canadian Journal of Chemistry-Revue Canadienne De Chimie* **1996**, *74*, 839.
[29] L. Nyulaszi, *Tetrahedron* **2000**, *56*, 79.
[30] I. Alkorta, K. Zborowski, J. Elguero, *Structural Chemistry* **2006**, *17*, 13.
[31] G. Keglevich, T. Chuluunbaatar, B. Dajka, A. Dobo, A. Szollosy, L. Toke, *Journal of the Chemical Society-Perkin Transactions 1* **2000**, 2895.
[32] W. Egan, R. Tang, G. Zon, K. Mislow, *Journal of the American Chemical Society* **1971**, *93*, 6205.
[33] S. Pelzer, K. Wichmann, R. Wesendrup, P. Schwerdtfeger, *Journal of Physical Chemistry A* **2002**, *106*, 6387.
[34] D. Delaere, M. T. Nguyen, L. G. Vanquickenborne, *Journal of Physical Chemistry A* **2003**, *107*, 838.
[35] C. Hay, M. Hissler, C. Fischmeister, J. Rault-Berthelot, L. Toupet, L. Nyulaszi, R. Reau, *Chemistry-a European Journal* **2001**, *7*, 4222.
[36] D. Delaere, M. T. Nguyen, L. G. Vanquickenborne, *Physical Chemistry Chemical Physics* **2002**, *4*, 1522.
[37] J. Crassous, R. Reau, *Dalton Transactions* **2008**, 6865.
[38] E. S. Milosavljevic, S. Affandi, J.H. Nelson, *Organometallics* **1988**, *7*, 1735.
[39] T. Baumgartner, W. Bergmans, T. Karpati, T. Neumann, M. Nieger, L. Nyulaszi, *Chemistry-a European Journal* **2005**, *11*, 4687.
[40] F. C. Leavitt, T. A. Manuel, F. Johnson, *Journal of the American Chemical Society* **1959**, *81*, 3163.
[41] E. H. Braye, W. Hubel, *Chemistry & Industry* **1959**, 1250.
[42] E. H. Braye, I. Caplier, W. Hubel, *Journal of the American Chemical Society* **1961**, *83*, 4406.
[43] W. A. Nugent, D. L. Thorn, R. L. Harlow, *Journal of the American Chemical Society* **1987**, *109*, 2788.
[44] E. Negishi, F. E. Cederbaum, T. Takahashi, *Tetrahedron Letters* **1986**, *27*, 2829.
[45] Y. Matano, T. Miyajima, T. Nakabuchi, Y. Matsutani, H. Imahori, *Journal of Organic Chemistry* **2006**, *71*, 5792.
[46] Z. F. Xi, *Accounts of Chemical Research* **2011**, *43*, 1342.
[47] A. Wolan, Y. Six, *Tetrahedron* **2011**, *66*, 3097.
[48] S. Yamaguchi, R. Z. Jin, K. Tamao, F. Sato, *Journal of Organic Chemistry* **1998**, *63*, 10060.
[49] P. J. Fagan, W. A. Nugent, J. C. Calabrese, *Journal of the American Chemical Society* **1994**, *116*, 1880.
[50] P. J. Fagan, W. A. Nugent, *Journal of the American Chemical Society* **1988**, *110*, 2310.
[51] O. Fadhel, Z. Benko, M. Gras, V. Deborde, D. Joly, C. Lescop, L. Nyulaszi, M. Hissler, R. Reau, *Chemistry-a European Journal* **2010**, *16*, 11340.
[52] C. Hay, C. Fave, M. Hissler, J. Rault-Berthelot, R. Reau, *Organic Letters* **2003**, *5*, 3467.
[53] M. Ogasawara, T. Sakamoto, K. Nakajima, T. Takahashi, *Journal of Organometallic Chemistry* **2007**, *692*, 55.
[54] S. Graule, M. Rudolph, W. T. Shen, J. A. G. Williams, C. Lescop, J. Autschbach, J. Crassous, R. Reau, *Chemistry-a European Journal* **2010**, *16*, 5976.
[55] P. J. Fagan, W. A. Nugent, *Organic Syntheses* **1992**, *70*, 272.
[56] F. Nief, D. Turcitu, L. Ricard, *Chemical Communications* **2002**, 1646.
[57] M. Westerhausen, M. H. Digeser, C. Guckel, H. Noth, J. Knizek, W. Ponikwar, *Organometallics* **1999**, *18*, 2491.
[58] R. E. V. Spence, S. J. Brown, D. G. Harrison, Q. Wang, World Patent WO 98 50392 to Nova Chemical, Canada, 12 Nov 1998.
[59] F. X. Buzin, F. Nief, F. Mathey, J. Malinge, E. Deschamps, B. Deschamps, World Patent WO 02 059133 to Atofina, 1 August, 2002.
[60] J. Hydrio, M. Gouygou, F. Dallemer, J. C. Daran, G. G. A. Balavoine, *Journal of Organometallic Chemistry* **2000**, *595*, 261.
[61] C. Hay, C. Fischmeister, M. Hissler, L. Toupet, R. Reau, *Angewandte Chemie-International Edition* **2000**, *39*, 1812.

[62] I. G. Campbell, R. C. Cookson, M. B. Hocking, A. N. Hughes, *Journal of the Chemical Society* **1965**, 2184.
[63] A. Breque, F. Mathey, P. Savignac, *Synthesis-Stuttgart* **1981**, 983; note that the synthesis will fail (giving phospholene oxide) if excess acid is added at the neutralization stage.
[64] J. Hydrio, M. Gouygou, F. Dallemer, G. G. A. Balavoine, J. C. Daran, *European Journal of Organic Chemistry* **2002**, 675.
[65] D. Carmichael, L. Ricard, F. Mathey, *Journal of the Chemical Society-Chemical Communications* **1994**, 1167.
[66] M. Ogasawara, K. Yoshida, T. Hayashi, *Organometallics* **2001**, *20*, 1014.
[67] G. Markl, R. Potthast, *Angewandte Chemie-International Edition* **1967**, *6*, 86.
[68] X. Sava, N. Mezailles, N. Maigrot, F. Nief, L. Ricard, F. Mathey, P. Le Floch, *Organometallics* **1999**, *18*, 4205.
[69] D. Turcitu, F. Nief, L. Ricard, *Chemistry-a European Journal* **2003**, *9*, 4916.
[70] S. Doherty, G. R. Eastham, R. P. Tooze, T. H. Scanlan, D. Williams, M. R. J. Elsegood, W. Clegg, *Organometallics* **1999**, *18*, 3558.
[71] P. J. Fagan, G. Li, Z. Guan, L. Wang, World Patent WO 2000 21970 to E I DuPont Nemours **2000**, 20 April 2000.
[72] A. I. A. Perez, T. Biet, S. Graule, T. Agou, C. Lescop, N. R. Branda, J. Crassous, R. Reau, *Chemistry-a European Journal* **2011**, *17*, 1337.
[73] F. Mathey, *Comptes Rendus Hebdomadaires Des Seances De L Academie Des Sciences Serie C* **1969**, *269*, 1066.
[74] L. D. Quin, J. P. Gratz, T. P. Barket, *Journal of Organic Chemistry* **1968**, *33*, 1034.
[75] F. Mathey, *Chemical Reviews* **1988**, *88*, 429.
[76] W. B. McCormack, US Patent 2663736, to E I DuPont Nemours, 22 Dec 1953.
[77] T. A. Niemi, P. L. Coe, S. J. Till, *Journal of the Chemical Society-Perkin Transactions 1* **2000**, *10*, 1519.
[78] L. D. Quin, G. Keglevich, A. S. Ionkin, R. Kalgutkar, G. Szalontai, *Journal of Organic Chemistry* **1996**, *61*, 7801.
[79] E. H. Braye, I. Caplier, R. Saussez, *Tetrahedron* **1971**, *27*, 5523.
[80] E. J. M. de Boer, I. J. Gilmore, F. M. Korndorffer, A. D. Horton, A. van der Linden, B. W. Royan, B. J. Ruisch, L. Schoon, R. W. Shaw, *Journal of Molecular Catalysis a-Chemical* **1998**, *128*, 155.
[81] M. Ogasawara, A. Ito, K. Yoshida, T. Hayashi, *Organometallics* **2006**, *25*, 2715.
[82] M. Westerhausen, M. H. Digeser, H. Noth, W. Ponikwar, T. Seifert, K. Polborn, *Inorganic Chemistry* **1999**, *38*, 3207.
[83] F. Nief, F. Mathey, L. Ricard, F. Robert, *Organometallics* **1988**, *7*, 921.
[84] A. Schmidpeter, M. Thiele, *Angewandte Chemie-International Edition in English* **1991**, *30*, 308.
[85] S. van Zutphen, V. J. Margarit, G. Mora, P. Le Floch, *Tetrahedron Letters* **2007**, *48*, 2857.
[86] F. Nief, F. Mathey, *Journal of the Chemical Society-Chemical Communications* **1988**, 770.
[87] S. Holand, N. Maigrot, C. Charrier, F. Mathey, *Organometallics* **1998**, *17*, 2996.
[88] S. R. Gilbertson, D. G. Genov, A. L. Rheingold, *Organic Letters* **2000**, *2*, 2885.
[89] S. Holand, F. Mathey, *Organometallics* **1988**, *7*, 1796.
[90] G. Mora, B. Deschamps, S. van Zutphen, X. F. Le Goff, L. Ricard, P. Le Floch, *Organometallics* **2007**, *26*, 1846.
[91] F. Mathey, *Tetrahedron* **1972**, *28*, 4171.
[92] E. J. P. Malar, *Journal of Organic Chemistry* **1992**, *57*, 3694.
[93] T. Douglas, K. H. Theopold, *Angewandte Chemie-International Edition in English* **1989**, *28*, 1367.
[94] D. Carmichael, L. Ricard, F. Mathey, *Journal of the Chemical Society-Chemical Communications* **1994**, 2459.
[95] Y. Cabon, D. Carmichael, X. F. Le Goff, *Organometallics* **2009**, *28*, 370.
[96] N. H. T. Huy, F. Mathey, *Organometallics* **1994**, *13*, 925.
[97] M. Sebastian, M. Hissler, C. Fave, J. Rault-Berthelot, C. Odin, R. Reau, *Angewandte Chemie-International Edition* **2006**, *45*, 6152.
[98] E. Deschamps, F. Mathey, *Bulletin De La Societe Chimique De France* **1992**, *129*, 486.
[99] E. Robe, C. Ortega, M. Mikina, M. Mikolajczyk, J. C. Daran, M. Gouygou, *Organometallics* **2005**, *24*, 5549.
[100] X. Sava, A. Marinetti, L. Ricard, F. Mathey, *European Journal of Inorganic Chemistry* **2002**, 1657.
[101] D. Carmichael, L. Ricard, N. Seeboth, J. M. Brown, T. D. W. Claridge, B. Odell, *Dalton Transactions* **2005**, 2173.
[102] F. Bisaro, P. Le Floch, *Synlett* **2011**, 3081.
[103] K. Waschbusch, P. Lefloch, F. Mathey, *Bulletin De La Societe Chimique De France* **1995**, *132*, 910.
[104] S. Holand, M. Jeanjean, F. Mathey, *Angewandte Chemie-International Edition in English* **1997**, *36*, 98.
[105] H. Doucet, J. M. Brown, *Tetrahedron-Asymmetry* **1997**, *8*, 3775.

[106] J. Hydrio, M. Gouygou, F. Dallemer, J. C. Daran, G. G. A. Balavoine, *Tetrahedron-Asymmetry* **2002**, *13*, 1097.
[107] F. Laporte, F. Mercier, L. Ricard, F. Mathey, *Journal of the American Chemical Society* **1994**, *116*, 3306.
[108] E. Mattmann, D. Simonutti, L. Ricard, F. Mercier, F. Mathey, *Journal of Organic Chemistry* **2001**, *66*, 755.
[109] E. Robe, C. Hegedus, J. Bakos, J. C. Daran, M. Gouygou, *Dalton Transactions* **2009**, 6528.
[110] M. O. Bevierre, F. Mercier, L. Ricard, F. Mathey, *Angewandte Chemie-International Edition in English* **1990**, *29*, 655.
[111] J. Wassenaar, S. van Zutphen, G. Mora, P. Le Floch, M. A. Siegler, A. L. Spek, J. N. H. Reek, *Organometallics* **2009**, *28*, 2724.
[112] C. Charrier, H. Bonnard, F. Mathey, D. Neibecker, *Journal of Organometallic Chemistry* **1982**, *231*, 361.
[113] J. M. Slattery, C. Fish, M. Green, T. N. Hooper, J. C. Jeffery, R. J. Kilby, J. M. Lynam, J. E. McGrady, D. A. Pantazis, C. A. Russell, C. E. Willans, *Chemistry-a European Journal* **2007**, *13*, 6967.
[114] B. Deschamps, F. Mathey, *Organometallics* **1992**, *11*, 1411.
[115] E. Deschamps, F. Mathey, *Comptes Rendus De L Academie Des Sciences Serie Ii Fascicule C-Chimie* **1998**, *1*, 715; b) Y. Matano, Y. Kon, A. Saito, Y. Kimura, T. Murafuji, H. Imahori, Chemistry Letters **2011**, 40, 919; c) A. Saito, Y. Matano, H. Imahori, *Organic Letters*, 2010, *12*, 2675.
[116] D. Carmichael, X. F. Le Goff, E. Muller, *New Journal of Chemistry* **2010**, *34*, 1341.
[117] M. Clochard, J. Grundy, B. Donnadieu, F. Mathey, *Organic Letters* **2005**, *7*, 4511.
[118] Y. Matano, H. Imahori, *Accounts of Chemical Research* **2009**, *42*, 1193.
[119] Y. Matano, M. Nakashima, H. Imahori, *Angewandte Chemie-International Edition* **2009**, *48*, 4002.
[120] E. Deschamps, L. Ricard, F. Mathey, *Angewandte Chemie-International Edition in English* **1994**, *33*, 1158.
[121] E. Deschamps, L. Ricard, F. Mathey, *Journal of the Chemical Society, Chemical Communications* **1995**, 1561.
[122] B. Deschamps, F. Mathey, *Bulletin De La Societe Chimique De France* **1996**, *133*, 541.
[123] S. Holand, F. Gandolfo, L. Ricard, F. Mathey, *Bulletin De La Societe Chimique De France* **1996**, *133*, 33.
[124] F. Mathey, J. Fischer, J. H. Nelson, *Structure and Bonding* **1983**, *55*, 153.
[125] W. Schafer, A. Schweig, F. Mathey, *Journal of the American Chemical Society* **1976**, *98*, 407.
[126] L. Nyulaszi, L. Soos, G. Keglevich, *Journal of Organometallic Chemistry* **1998**, *566*, 29.
[127] L. Nyulaszi, G. Keglevich, L. D. Quin, *Journal of Organic Chemistry* **1996**, *61*, 7808.
[128] C. Thoumazet, H. Gruetzmacher, B. Deschamps, L. Ricard, P. Le Floch, *European Journal of Inorganic Chemistry* **2006**, 3911.
[129] C. A. Tolman, *Chemical Reviews* **1977**, *77*, 313.
[130] D. Neibecker, R. Reau, *Journal of Molecular Catalysis* **1989**, *53*, 219.
[131] J. J. Macdougall, J. H. Nelson, F. Mathey, J. J. Mayerle, *Inorganic Chemistry* **1980**, *19*, 709.
[132] W. A. Herrmann, W. R. Thiel, T. Priermeier, E. Herdtweck, *Journal of Organometallic Chemistry* **1994**, *481*, 253.
[133] J. J. Macdougall, J. H. Nelson, F. Mathey, *Inorganic Chemistry* **1982**, *21*, 2145.
[134] L. M. Wilkes, J. H. Nelson, L. B. McCusker, K. Seff, F. Mathey, *Inorganic Chemistry* **1983**, *22*, 2476.
[135] D. Neibecker, R. Reau, *Journal of Molecular Catalysis* **1989**, *57*, 153.
[136] I. Odinets, T. Kortvelyesi, T. Kegl, L. Kollar, G. Keglevich, *Journal of Organometallic Chemistry* **2007**, *691*, 5547.
[137] Z. Csok, G. Keglevich, G. Petocz, L. Kollar, *Journal of Organometallic Chemistry* **1999**, *586*, 79.
[138] G. Mora, S. van Zutphen, C. Thoumazet, X. F. Le Goff, L. Ricard, H. Grutzmacher, P. Le Floch, *Organometallics* **2006**, *25*, 5528.
[139] S. Holand, F. Mathey, J. Fischer, A. Mitschler, *Organometallics* **1983**, *2*, 1234.
[140] J. Bitta, S. Fassbender, G. Reiss, C. Ganter, *Organometallics* **2006**, *25*, 2394.
[141] F. Mathey, *Tetrahedron Letters* **1976**, 4155.
[142] F. Mercier, S. Holand, F. Mathey, *Journal of Organometallic Chemistry* **1986**, *316*, 271.
[143] F. Mercier, F. Mathey, *Journal of Organometallic Chemistry* **1984**, *263*, 55.
[144] W. L. Wilson, J. Fischer, R. E. Wasylishen, K. Eichele, V. J. Catalano, J. H. Frederick, J. H. Nelson, *Inorganic Chemistry* **1996**, *35*, 1486.
[145] W. L. Wilson, J. A. Rahn, N. W. Alcock, J. Fischer, J. H. Frederick, J. H. Nelson, *Inorganic Chemistry* **1994**, *33*, 109.

[146] H. L. Ji, J. H. Nelson, A. DeCian, J. Fischer, B. Li, C. Wang, B. McCarty, Y. Aoki, J. W. Kenney, L. Solujic, E. B. Milosavljevic, *Journal of Organometallic Chemistry* **1997**, *529*, 395.
[147] K. D. Redwine, J. H. Nelson, *Organometallics* **2000**, *19*, 3054.
[148] P. E. Garrou, *Chemical Reviews* **1985**, *85*, 171.
[149] F. Mathey, N. H. T. Huy, A. Marinetti, *Helvetica Chimica Acta* **2001**, *84*, 2938.
[150] F. Mercier, F. Mathey, J. Fischer, J. H. Nelson, *Inorganic Chemistry* **1985**, *24*, 4141.
[151] Y. X. Li, S. Selvaratnam, J. J. Vittal, P. H. Leung, *Inorganic Chemistry* **2003**, *42*, 3229.
[152] S. A. Pullarkat, Y. L. Cheow, Y. X. Li, P. H. Leung, *European Journal of Inorganic Chemistry* **2009**, 2375.
[153] G. S. He, Y. Qin, K. F. Mok, P. H. Leung, *Chemical Communications* **2000**, 167.
[154] Y. Qin, S. Selvaratnam, J. J. Vittal, P. K. Leung, *Organometallics* **2002**, *21*, 5301.
[155] N. Gul, J. H. Nelson, *Tetrahedron* **2000**, *56*, 71.
[156] P. H. Leung, *Accounts of Chemical Research* **2004**, *37*, 169.
[157] F. Leca, M. Sauthier, B. le Guennic, C. Lescop, L. Toupet, J. F. Halet, R. Reau, *Chemical Communications* **2003**, 1774.
[158] F. Leca, C. Lescop, L. Toupet, R. Reau, *Organometallics* **2004**, *23*, 6191.
[159] A. Ciric, F. Mathey, *Organometallics* **2010**, *29*, 4785.
[160] F. Leca, F. Fernandez, G. Muller, C. Lescop, R. Reau, M. Gomez, *European Journal of Inorganic Chemistry* **2009**, 5583.
[161] P. Braunstein, N. M. Boag, *Angewandte Chemie-International Edition* **2001**, *40*, 2427.
[162] P. H. M. Budzelaar, P. Vanleeuwen, C. F. Roobeek, A. G. Orpen, *Organometallics* **1992**, *11*, 23.
[163] F. Leca, C. Lescop, E. Rodriguez-Sanz, K. Costuas, J. F. Halet, R. Reau, *Angewandte Chemie-International Edition* **2005**, *44*, 4362.
[164] M. Sauthier, B. Le Guennic, V. Deborde, L. Toupet, J. F. Halet, R. Reau, *Angewandte Chemie-International Edition* **2001**, *40*, 228.
[165] M. Sauthier, F. Leca, L. Toupet, R. Reau, *Organometallics* **2002**, *21*, 1591.
[166] F. Leca, M. Sauthier, V. Deborde, L. Toupet, R. Reau, *Chemistry-a European Journal* **2003**, *9*, 3785.
[167] S. Welsch, B. Nohra, E. V. Peresypkina, C. Lescop, M. Scheer, R. Reau, *Chemistry-a European Journal* **2009**, *15*, 4685.
[168] E. Rodriguez-Sanz, C. Lescop, R. Reau, *Comptes Rendus Chimie* **2010**, *13*, 980.
[169] M. Gouygou, J. C. Daran, E. Robe, C. Ortega, *Comptes Rendus Chimie* **2010**, *13*, 1054.
[170] O. Tissot, M. Gouygou, J. C. Daran, G. G. A. Balavoine, *Chemical Communications* **1996**, 2287.
[171] O. Tissot, M. Gouygou, F. Dallemer, J. C. Daran, G. G. A. Balavoine, *Angewandte Chemie-International Edition* **2001**, *40*, 1076.
[172] O. Tissot, M. Gouygou, F. Dallemer, J. L. Daran, G. G. A. Balavoine, in *European Journal of Inorganic Chemistry*, **2001**, 2385.
[173] C. Ortega, M. Gouygou, J. C. Daran, *Chemical Communications* **2003**, 1154.
[174] E. Robe, C. Hegedus, J. Bakos, Y. Coppel, J. C. Daran, M. Gouygou, *Inorganica Chimica Acta* **2008**, *361*, 1861.
[175] J. Fischer, A. Mitschler, F. Mathey, F. Mercier, *Journal of the Chemical Society-Dalton Transactions* **1983**, 841.
[176] Dutch patent NL 6516164 to Shell International Research Maatschappij BV, 15 May 1966.
[177] C. Bergounhou, D. Neibecker, R. Reau, *Journal of the Chemical Society-Chemical Communications* **1988**, 1370.
[178] D. Neibecker, R. Reau, S. Lecolier, *Journal of Organic Chemistry* **1989**, *54*, 5208.
[179] D. Neibecker, R. Reau, *New Journal of Chemistry* **1991**, *15*, 279.
[180] A. Polo, C. Claver, S. Castillon, A. Ruiz, J. C. Bayon, J. Real, C. Mealli, D. Masi, *Organometallics* **1992**, *11*, 3525.
[181] C. Bergounhou, D. Neibecker, R. Reau, *Bulletin De La Societe Chimique De France* **1995**, *132*, 815.
[182] C. Bergounhou, D. Neibecker, R. Mathieu, *Organometallics* **2003**, *22*, 782.
[183] P. Cheliatsidou, D. F. S. White, B. de Bruin, J. N. H. Reek, D. J. Cole-Hamilton, *Organometallics* **2007**, *26*, 3265.
[184] C. Bergounhou, D. Neibecker, R. Mathieu, *Journal of Molecular Catalysis a-Chemical* **2004**, *220*, 167.
[185] G. Keglevich, T. Chuluunbaatar, B. Dajka, K. Ludanyi, G. Parlagh, T. Kegl, L. Kollar, L. Toke, *Journal of Organometallic Chemistry* **2002**, *643*, 32.
[186] G. Keglevich, T. Kegl, T. Chuluunbaatar, B. Dajka, P. Matyus, B. Balogh, L. Kollar, *Journal of Molecular Catalysis a-Chemical* **2003**, *200*, 131.
[187] T. Fanjul, G. Eastham, N. Fey, A. Hamilton, A. G. Orpen, P. G. Pringle, M. Waugh, *Organometallics*, **2010**, *29*, 2292.

[188] S. Doherty, E. G. Robins, J. G. Knight, C. R. Newman, B. Rhodes, P. A. Champkin, W. Clegg, *Journal of Organometallic Chemistry* **2001**, *640*, 182.
[189] Z. Freixa, P. van Leeuwen, *Dalton Transactions* **2003**, 1890.
[190] L. Diab, M. Gouygou, E. Manoury, P. Kalck, M. Urrutigoity, *Tetrahedron Letters* **2008**, *49*, 5186.
[191] B. M. Choudary, N. P. Reddy, M. Z. Jamil, *Polyhedron* **1986**, *5*, 911.
[192] Y. Matano, T. Miyajima, N. Ochi, T. Nakabuchi, M. Shiro, Y. Nakao, S. Sakaki, H. Imahori, *Journal of the American Chemical Society* **2008**, *130*, 990.
[193] Y. Matano, T. Miyajima, T. Nakabuchi, H. Imahori, N. Ochi, S. Sakaki, *Journal of the American Chemical Society* **2006**, *128*, 11760.
[194] F. Mercier, F. Laporte, L. Ricard, F. Mathey, M. Schroder, M. Regitz, *Angewandte Chemie-International Edition in English* **1997**, *36*, 2364.
[195] C. Thoumazet, M. Melaimi, L. Ricard, P. Le Floch, *Comptes Rendus Chimie* **2004**, *7*, 823.
[196] M. Melaimi, C. Thoumazet, L. Ricard, P. Le Floch, *Journal of Organometallic Chemistry* **2004**, *689*, 2988.
[197] G. Mora, O. Piechaczyk, R. Houdard, N. Mezailles, X. F. Le Goff, P. le Floch, *Chemistry-a European Journal* **2008**, *14*, 10047.
[198] A. Escalle, G. Mora, F. Gagosz, N. Mezailles, X. F. Le Goff, Y. Jean, P. Le Floch, *Inorganic Chemistry* **2009**, *48*, 8415.
[199] S. Labouille, A. Escalle-Lewis, Y. Jean, N. Mezailles, P. Le Floch, *Chemistry-a European Journal* **2011**, *17*, 2256.
[200] R. F. de Souza, K. Bernardo-Gusmao, G. A. Cunha, C. Loup, F. Leca, R. Reau, *Journal of Catalysis* **2004**, *226*, 235.
[201] M. Lejeune, D. Semeril, C. Jeunesse, D. Matt, F. Peruch, P. L. Lutz, L. Ricard, *Chemistry-a European Journal* **2004**, *10*, 5354.
[202] G. Mora, S. van Zutphen, C. Klemps, L. Ricard, Y. Jean, P. Le Floch, D. Bernache-Assollant, *Inorganic Chemistry* **2007**, *46*, 10365.
[203] D. G. Holah, A. N. Hughes, B. C. Hui, *Canadian Journal of Chemistry* **1972**, *50*, 3714.
[204] C. Thoumazet, M. Melaimi, L. Ricard, F. Mathey, P. Le Floch, *Organometallics* **2003**, *22*, 1580.
[205] E. Robe, W. Perlikowska, C. Lemoine, L. Diab, S. Vincendeau, M. Mikolajczyk, J. C. Daran, M. Gouygou, *Dalton Transactions* **2008**, 2894.

8

Phosphinine Ligands

Christian Müller

Freie Universität Berlin, Institute of Chemistry and Biochemistry, Berlin, Germany

8.1 Introduction

A milestone in the chemistry of low-coordinate phosphorus compounds was set with the first successful preparation of the fully unsaturated six-membered phosphorus heterocycle **1** by Märkl in 1966 and the parent compound C_5H_5P (**2**) by Ashe III in 1971 [1,2]. The stabilization of reactive P=C double bonds by their incorporation into aromatic systems opened up the access to formally sp^2-hybridized phosphorus(III) heterocycles with significantly different electronic and steric properties compared to classical ligands based on trivalent phosphorus (Figure 8.1 and Section 8.2).

The name "phosphinine" for such compounds was proposed in 1983 by the International Union of Pure and Applied Chemistry (IUPAC), while the term "phosphabenzene" is still frequently found in literature. More precisely, aromatic heterocycles containing a trivalent phosphorus atom that is linked to two carbon atoms (–P=) are designated $1\lambda^3\sigma^2$-phosphinines.

Phosphinines have long been regarded as "chemical curiosities," but state-of-the-art synthetic methodologies, developed mostly by the groups of Mathey and Le Floch, now allow specific derivatization and functionalization, including the introduction of additional donor-functionalities (Section 8.3). This is an important aspect for potential applications, as polydentate ligands with interesting coordination and functional properties can be obtained that are not accessible by using monodentate phosphinines. The design and preparation of (donor)functionalized, polydentate, and chiral 2,4,6-triaryl-substituted phosphinines via the original pyrylium salt route have been investigated by Müller *et al.* in recent years (Section 8.3). These derivatives often show a considerable kinetic stability and are inert toward water, oxygen, and many acids and bases, in contrast to less substituted phosphinines. This feature facilitates the preparation and functionalization of phosphinine-based ligands and their corresponding transition metal

Phosphorus(III) Ligands in Homogeneous Catalysis: Design and Synthesis, First Edition.
Edited by Paul C. J. Kamer and Piet W. N. M. van Leeuwen.

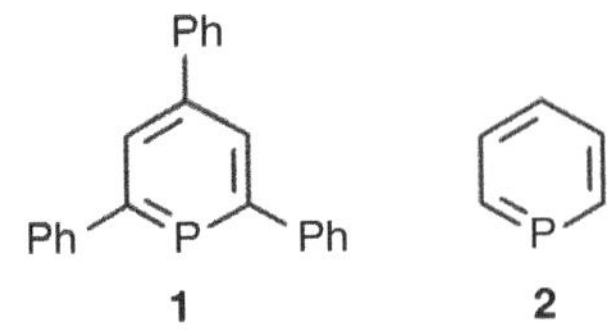

Figure 8.1 *2,4,6-triphenylphosphinine **1** and parent phosphinine **2**.*

complexes (Sections 8.4 and 8.5) and systems with tailored properties become synthetically accessible. By making use of the chelate effect of such polydentate phosphinine-based ligands, even coordination compounds with transition metal centers in medium to high oxidation states now become synthetically accessible. Despite the fact that phosphinines have been known for more than 40 years and possess rich and versatile coordination chemistry, investigations on their application as ligands in homogeneous catalysis are still comparatively rare. So far studies are limited to only a few examples, but several promising results have recently emerged (Section 8.6).

8.2 Ligand properties

The incorporation of the phosphorus atom into the aromatic moiety has a significant influence on the electronic and steric properties of phosphinines as well as on their chemical reactivity.

8.2.1 Electronic properties

Phosphinines are planar, aromatic systems. Theoretical calculations suggest their aromaticity to be as high as 88% of benzene [3,4]. The electronic properties of phosphinines differ from those of pyridines, as shown by photoelectron and electron transmission spectroscopy as well as by theoretical calculations [5–8]. The $HOMO^{-2}$ orbital has a large coefficient at the phosphorus atom and represents essentially the lone pair at the heteroatom. The phosphorus lone pair occupies a more diffuse, partly delocalized, and less directional orbital than that of pyridine. While the $HOMO^{-1}$ and HOMO orbitals contribute to π-donation, the LUMO orbital, with a large coefficient at the phosphorus atom, enables the heterocycle to act as a π-acceptor ligand, once coordinated to the metal center via the phosphorus atom (Figure 8.2). In pyridines, the LUMO is located higher in energy compared to phosphinines, while the HOMO represents the lone pair located at the nitrogen atom. As a result, phosphinines are much better π-acceptor ligands, but less good σ-donors compared to their nitrogen counterparts with significant consequences for their coordination behavior (Sections 8.4 and 8.5).

Phosphinines show a typical downfield shift of about $\delta = +200$ ppm in the $^{31}P\{^{1}H\}$ NMR spectrum, and all peripheral protons show chemical shift values downfield from benzene in the ^{1}H NMR spectrum. These phenomena can be attributed to the presence of a diamagnetic ring current typical for aromatic systems. Nucleus-independent chemical shift values (NICS) of phosphinines lead to the same conclusion [9]. In order to quantify the electronic properties of ligands, Tolman's electronic parameter χ (chi) can be used to compare different monodentate systems [10].

An ideal probe for determining the electronic properties of monodentate phosphinine ligands is the infrared (IR) stretching frequency of the CO ligand in *trans*-$L_2Rh(CO)Cl$ complexes. In fact, the synthesis of the common $LNi(CO)_3$ complexes now attracts much less attention due to the extreme toxicity and inconvenience of Ni-carbonyls and correlations between values derived from $LNi(CO)_3$ complexes, and various metal carbonyl species exist [13]. Table 8.1 shows typical values for phosphinines in comparison with phosphines and phosphites, which suggest that phosphinines are electron-withdrawing ligands (π-acceptors) with properties similar to phosphites as already anticipated from the corresponding MO scheme (Figure 8.2).

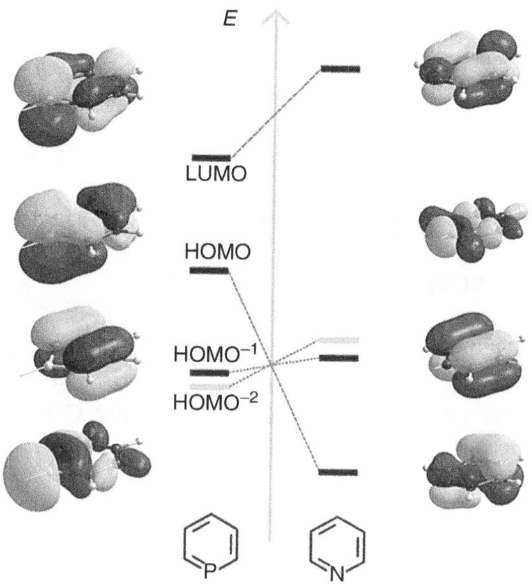

Figure 8.2 Qualitative MO diagram of the frontier orbitals of phosphinine and pyridine. Light gray: MO representing the lone pair.

Table 8.1 Selected χ-values of P-ligands.

Ligand	χ	$\tilde{\nu}_{CO(Ni)}$ [cm^{-1}]	$\tilde{\nu}_{CO(Rh)}$[b] [cm^{-1}]
Phosphinine[a]	24[15]	—	1999
PPh_3	12.9	2069	1979
PEt_3	5.6	2062	1958
$P(OPh)_3$	29.1	2085	2016 [11]
$P(OMe)_3$	23.3	2076	2006 [12]
$P[O(2\text{-t-}BuC_6H_4)]_3$	30	2086	2013

[a]2,4,6-Triphenylphosphinine. [b]*trans*-$L_2Rh(CO)Cl$ complex.

8.2.2 Structural characteristics and steric properties

Several λ^3-phosphinines have been characterized crystallographically [14]. Figure 8.3 shows part of the molecular structure of phosphinine **3** [15]. The thienyl-substituents in *ortho*-position and the phenyl group in *para*-position have been excluded for clarity, and only the corresponding *ipso*-carbon atoms (C6, C12, C16) of those substituents are shown.

The phosphorus heterocycle is essentially planar and can best be described as a distorted hexagon, in contrast to benzenes or pyridines. The P–C_α bond lengths lie in between a P–C single bond (triphenylphosphine: 1.83 Å [16]) and a P–C double bond (diphenylmethylenephosphaalkene: 1.66 Å [17]). There is no bond alternation between the C–C (~1.39 Å) and the P–C (~1.75 Å) bond lengths, which indicates delocalization of the π-system and the presence of an aromatic system. The internal C–P–C angle of approximately 100° is somewhat smaller compared to the C–N–C angle in pyridine (117°) [18]. This phenomenon is mainly attributed to the poor ability of phosphorus orbitals to hybridize, next to the lengthening of the heteroatom–carbon

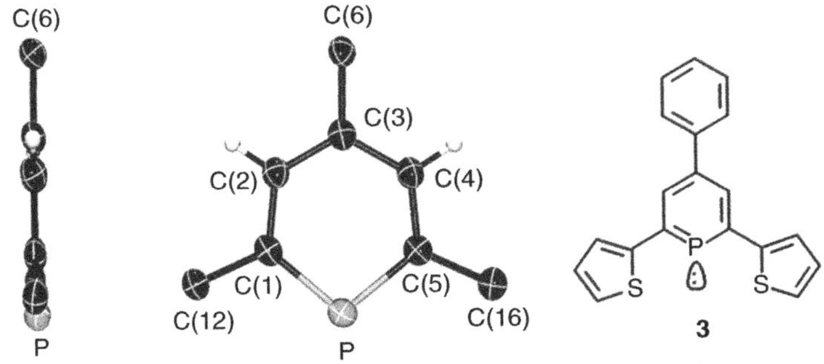

3

Figure 8.3 Molecular structure of the bis-thienyl-substituted phosphinine **3**. Only part of the structure is shown. Side view (left); front view (middle). Selected bond lengths (Å) and angles (°): P-C(1): 1.749, P-C(5): 1.750, C(1)-(C2): 1.398, C(2)-C(3): 1.394, C(3)-C(4): 1.394, C(4)-C(5): 1.392. C(1)-P-C(5): 101.37.

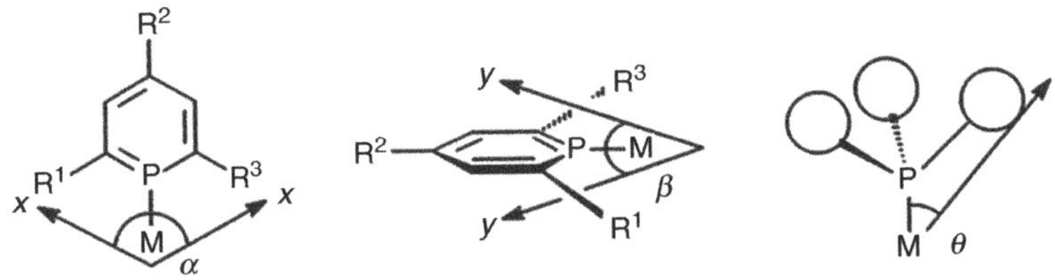

Figure 8.4 Occupancy angles *α* and *β* of a phosphinine and cone angle *θ* of a phosphine.

bond. From the planarity of the P-heterocycle, it becomes obvious that the steric demand of such ligand systems is significantly different compared to trivalent phosphines. Tolman's steric parameter θ (cone angle), which provides a measure of the volume around a metal center occupied by a ligand, is not appropriate to describe the steric properties of phosphinines. In fact, the occupancy angles α and β along the two orthogonal planes x and y show that the steric demand in the y plane is relatively small, while it is very large in the x plane. This leads to a flattened rather than a symmetrical cone, and the steric demand of phosphinines can thus be better described by the occupancy angles α and β rather than by the cone angle θ (Figure 8.4) [19].

In this respect, phosphinines are much more similar to their pyridine analogs and to N-heterocyclic carbenes. On the basis of representative conformations and calculated structures of a 2,4,6-triphenylphosphinine–nickel complex, values of $\alpha = 126.5°$ and $\beta = 54°$ have been determined (cone-angle for PPh_3: $\theta = 145°$) [20].

8.2.3 Reactivity of phosphinines

The lone pair of the phosphorus atom in λ^3-phosphinines has a strong 3*s*-orbital character (63.8% versus 29.1% in pyridine) [9]. This feature reflects the poor hybridization ability of phosphorus, leading to a very low basicity of the phosphorus center (pK_a ($C_5H_6P^+$)=−16.1 ± 1.0 in aqueous solution) [21]. As a consequence, typical reactivities of trivalent phosphorus compounds are generally not observed for phosphinines. Phosphinines can only be protonated by strong, nonoxidizing acids containing weakly nucleophilic (noncoordinating) anions. The only example reported in literature so far is the phosphininium salt **4**, which is generated by reaction of 2,4,6-tri-*tert*-butyl-1-phosphinine with $Et_3Si(CHB_{11}Me_5Cl_6)$ followed by addition of trifilic acid as reported by Reed and Nixon (Figure 8.5) [22].

In contrast, reaction of **1** with triflic acid leads after initial protonation to rearrangement reactions under formation of a P–O bond [23]. Reaction of **1** with perchloric acid or concentrated sulfuric acid causes

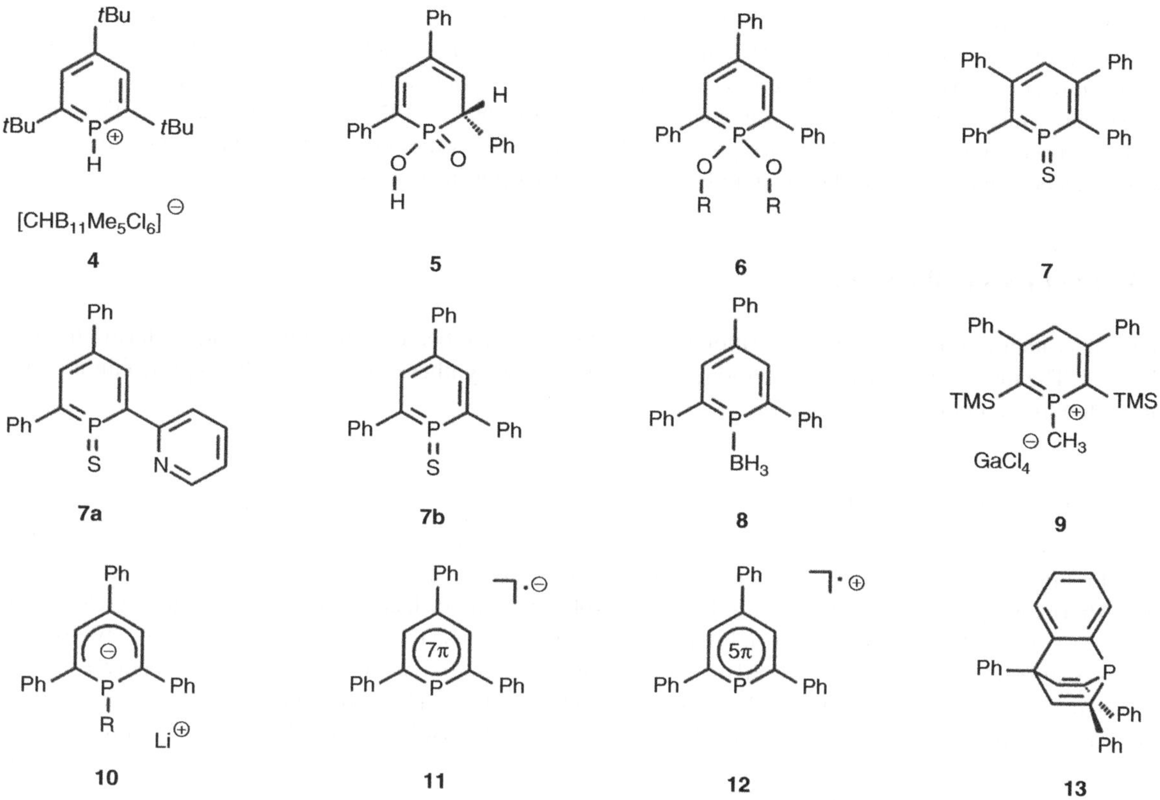

Figure 8.5 *Chemical modification of a phosphinine heterocycle.*

oxidation under formation of deep blue cationic species [24]. **1** and higher substituted phosphinines are rather stable toward air. The phosphinine-oxide has been detected but turned out to be too reactive and could not be isolated [25]. Oxidation of **1** with H_2O_2 leads to the formation of 2-hydro-phosphinic acid **5** after initial formation of a λ^5-phosphinineoxide-hydrate [24]. Oxidation of **1** with $HgCl_2$ in the presence of alcohols leads to the fairly stable and highly fluorescent λ^5-phosphinines of the type **6** [26]. Phosphinine-sulfides (**7**) can be obtained by heating toluene solutions with a slight excess of sulfur for several days at T=90°C. DFT calculations have shown that the sulfur lone pairs in these sulfides are strongly stabilized by the two σ*(P–C) orbitals and the phosphinine π* system [27]. Interestingly, the oxidation with sulfur to phosphinine-sulfides can be accelerated considerably in the presence of pyridine. The groups of Müller and Mézailles recently observed that 2-(2′-pyridyl)-4,6-diphenylphosphinine (section 8.3.1) is converted rapidly within 1 day with a slight excess of sulfur at T=90°C to the corresponding phosphinine-sulfide **7a**. Accordingly, 2,4,6-triphenylphosphinine (**1**) is converted rapidly into the corresponding phosphinine-sulfide **7b** in the presence of pyridine. This reactivity can be attributed to a donor-assisted facilitated ring opening of the S_8 molecule along with an enhancement of reactivity. Phosphinine-selenides are, on the other hand, unknown. Reaction of **1** with B_2H_6 gives only at low temperature the corresponding borane-protected phosphinine **8** in an equilibrium reaction [24]. The preparation of the 1-methylphosphininium cation **9** has been achieved only recently by reaction of λ^5-1-methyl-1-chlorophosphinines with $GaCl_3$ [28]. Strong nucleophiles, such as RMgX or RLi reagents, attack the phosphorus atom in phosphinines to form (1*R*)-phosphahexadienyl anions of the type **10** [29]. Phosphinines can easily be reduced to the radical anion **11** upon reaction with Na–K alloy, potassium metal in DME, or sodium naphthalenide and can thus be regarded as "non-innocent" ligands [24,30–32]. Dimerization under formation of a P–P bond in analogy with the dimerization of the phosphinyl radical $R_2P\cdot$ has been observed as well [33]. Reaction of **1** with the 2,4,6-triphenyoxyl radical leads to the formation of the stable radical cation **12**, which was characterized by means of ESR spectroscopy [24,34]. The [4+2] cycloaddition reaction of **1** with *in situ* generated benzyne affords the corresponding phosphabarrelene **13**, which has been identified as a bulky, mainly σ-donating phosphorus ligand [35–37].

8.3 Synthesis of Phosphinines

Several synthetic routes for the preparation of phosphinines have been reported in literature during the last decades. The most commonly used methods, especially for the synthesis of functionalized phosphinines, are presented throughout this section.

8.3.1 O^+/P exchange reaction

The first reported phosphinine was prepared from 2,4,6-triphenylpyrylium tetra-fluoroborate and $P(CH_2OH)_3$ in refluxing pyridine, and was obtained as a yellow air- and moisture stable solid in 24–30% yield (Scheme 8.1) [1]. Alternatively, $P(SiMe_3)_3$ can be used as a phosphorus source, leading to slightly higher yields [38]. Exchange of BF_4^- by I^- can lead to even higher yields (45%) because the formation of Me_3SiI as a by-product of the synthesis is preferred over Me_3SiBF_4 [39]. Despite its high toxicity, the use of PH_3 gives pure **1** in 77% yield after recrystallization and only H_2O as a by-product [38,40,41]. Performing the O^+/P exchange with PH_4I as PH_3-source in butanol at T=110–120°C gives **1** in 61% isolated yield [42]. PH_4I is, however, not commercially available anymore. Recrystallization of **1** from hot CH_3CN as the solvent of choice affords yellow needles of the highly pure phosphinine.

In contrast to arenes and pyridines, the direct functionalization of phosphinines is challenging because reagents for electrophilic substitution and metalation reactions usually attack at the phosphorus atom. The incorporation of additional donor functionalities within the heterocyclic framework is, however, an important aspect for the design of phosphinines. The chelate effect is an essential feature especially for the stabilization of metal complexes in higher oxidation states, preventing otherwise fast ligand dissociation from the metal center. The combination of "soft" and "hard" donor sites within the same molecule (hybrid ligands) leads to electronically different binding sites in the corresponding metal complexes, which can influence significantly their reactivities and selectivities in catalytic reactions. Hemilability, often associated with hybrid ligands, is an important concept for the design of homogeneous catalysts, functional molecular materials, or activation of small molecules. Additional reactive groups within the phosphorus heterocycle can also be used for further functionalization and the synthesis of polydentate ligands. In this respect, the pyrylium salt route is a highly flexible and modular procedure for the preparation of functionalized 2,4,6-triarylphosphinines. The most

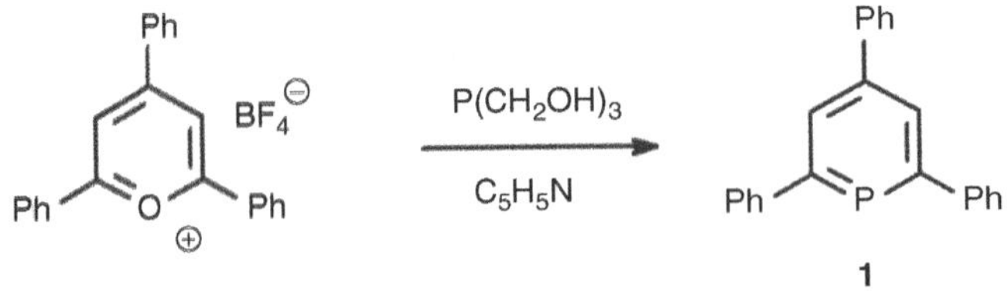

Scheme 8.1 2,4,6-triphenylphosphinine synthesis via pyrylium salt.

common route for the synthesis of triaryl-substituted pyrylium salts is the reaction of a functionalized chalcone with a functionalized acetophenone in the ratio of 2:1 and in the presence of HBF_4 (Et_2O solution) in $ClCH_2CH_2Cl$ [43]. Chalcones are accessible in high yields via classical Claisen–Schmidt condensation reactions between acetophenones and benzaldehydes. In the pyrylium salt formation reaction, one equivalent of the chalcone is consumed as it acts as hydride abstractor for the intermediate 2,4,6-triaryl-4*H*-pyrane derivative [44–47]. Alternatively, pyrylium salts can be obtained in good yields from the corresponding 1,5-diketones by reaction with benzylidene acetophenone in the presence of $BF_3{\cdot}Et_2O$ (Scheme 8.2) [43]. Symmetrical pyrylium salts are also accessible in a one-pot reaction between acetophenones and benzaldehydes in the ratio of 2:1 and in the presence of $BF_3{\cdot}Et_2O$ [43]. 2,4,6-triaryl-pyrylium salts are usually obtained as crystalline materials in 40–70% yield and show intensive fluorescence in solution.

This stepwise assembly of the pyrylium salt starting from functionalized benzaldehydes and acetophenones allows the incorporation of various (donor) substituents into specific positions of the final 2,4,6-triarylphosphinine core [15,48]. In this way, the steric and electronic properties of phosphinine-based ligands can be varied to a certain extent, and their scope is significantly extended. Selected examples of functionalized phosphinines prepared via the pyrylium salt route are illustrated in Figure 8.6, and a detailed experimental procedure for the synthesis of **1**, **15**, and **17** is given in Section 8.7. Interestingly, this synthetic

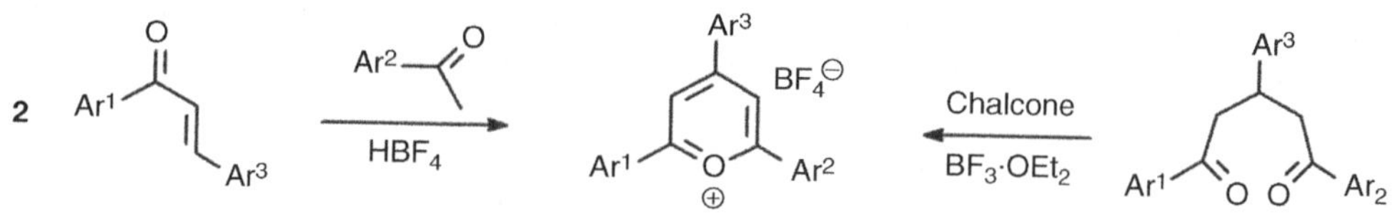

Scheme 8.2 *Synthesis of 2,4,6-triaryl-pyrylium salts.*

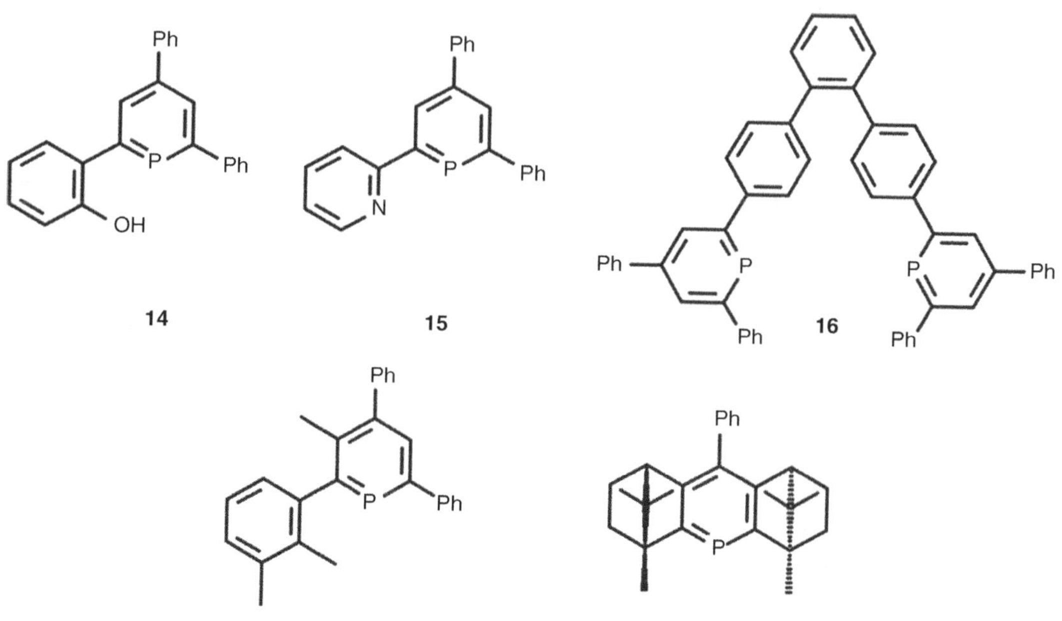

Figure 8.6 *Selected examples of phosphinines prepared via pyrylium salts.*

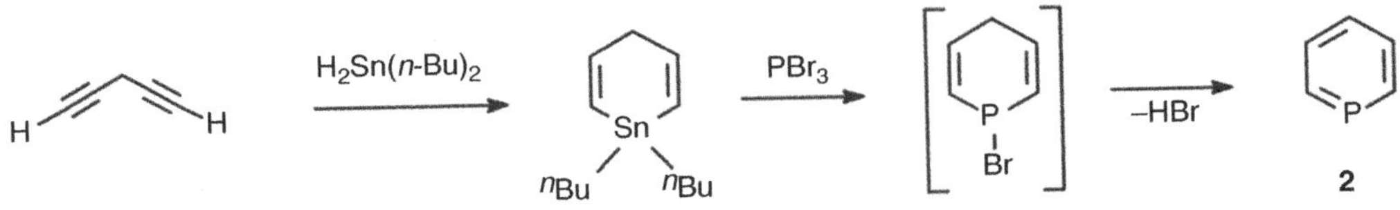

Scheme 8.3 *Synthesis of the parent phosphinine **2** from 1,4-pentadiyne.*

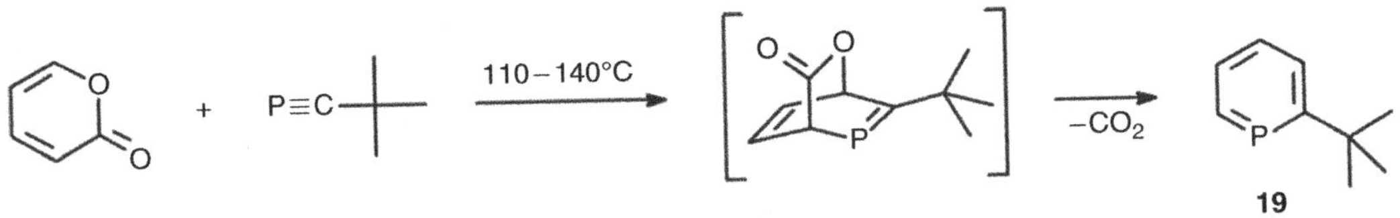

Scheme 8.4 *Synthesis of phosphinines from pyrones.*

procedure is not restricted to the preparation of mono-phosphines, as di-phosphinines (**16**) can also be prepared from the corresponding bis-pyrylium salts [49,50]. Moreover, this concept has been used to prepare the first axially chiral phosphinine **17** as well as the chiral C_2-symmetric phosphinine **18** [51–53].

The introduction of the methyl group in 3-position of the P-heterocycle in **17** is achieved by making use of propiophenone derivatives rather than acetophenones for the synthesis of the corresponding pyrylium salt [51]. The chemical stability of 2,4,6-triaryl-phosphinines becomes obvious from the synthesis of the hydroxy-functionalized phosphinine **14** [54]. While unprotected –OH functionalities might disturb the formation of pyrylium salts and phosphinines in some cases, the use of methoxy-functionalized acetophenones or benzaldehydes is initially required. The deprotection has to be performed in the last step and can be achieved by reaction of the phosphinine with BBr_3 in CH_2Cl_2 and subsequent aqueous workup or, alternatively, in refluxing HCl solution overnight [54,55].

8.3.2 Tin route

A synthetic access to the unsubstituted parent compound C_5H_5P was developed by Ashe III in 1971 [2]. Hydrostannation of 1,4-pentadiyne with $H_2Sn(n\text{-}Bu)_2$ affords 1,4-dihydro-1,1-dibutylstannabenzene in 50% yield. Reaction with PBr_3 gives C_5H_5P under loss of HBr (Scheme 8.3). **2** is a colorless, air- and moisture-sensitive liquid. 1,4-pentadiyne is not commercially available on a larger scale. It is synthesized by a Cu(I)-catalyzed coupling of propargyl magnesium chloride with acetylene and obtained as a solution in THF [56]. 1,4-dihydro-1,1-dibutylstannabenzene is a versatile precursor not only for the preparation of **2** but also for the synthesis of boratabenzenes, arsabenzene, stibabenzene, and bismabenzene [57].

8.3.3 [4+2] cycloaddition reactions

Regitz *et al.* described [4+2] cycloaddition reactions of phosphaalkynes with cyclopentadienones or pyrones at T=110–140°C under subsequent exclusion of CO and CO_2, respectively [58,59]. In this way, 2-alkyl-substituted phosphinines, such as the colorless liquid **19**, are synthetically accessible in high yields (91% for **19**) (Scheme 8.4). The substitution pattern of the heterocycle is restricted to the nature and synthetic availability of the starting materials, especially substituted phosphaalkynes. Moreover, bulky substituents in 2-position of the phosphinine core can prevent σ-coordination via the phosphorus lone pair to a metal center [60].

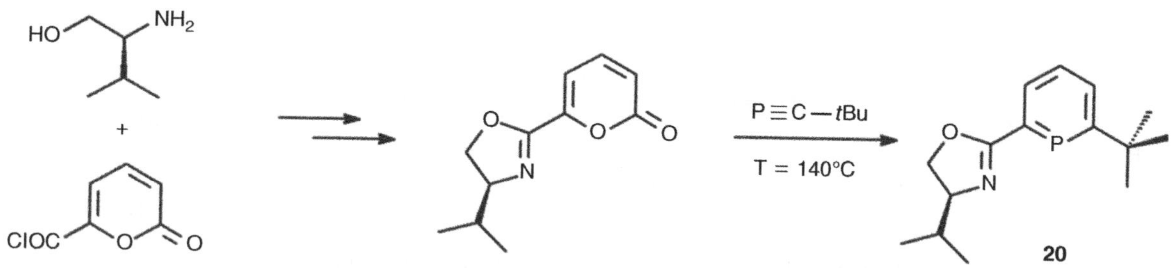

20

Scheme 8.5 *Synthesis of a chiral oxazoline-based phosphinine.*

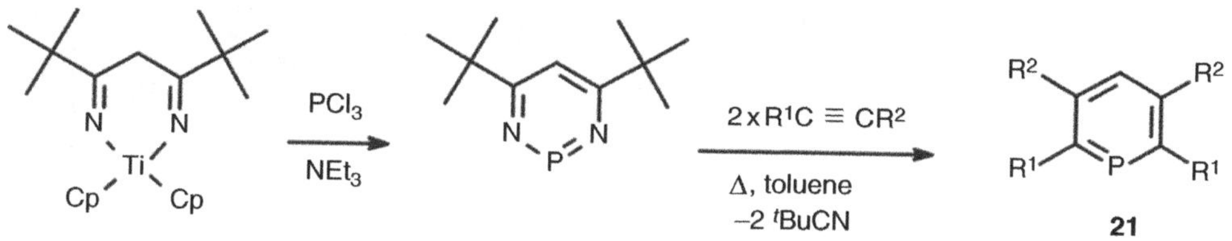

21

Scheme 8.6 *Synthesis of phosphinines from diazaphosphinines.*

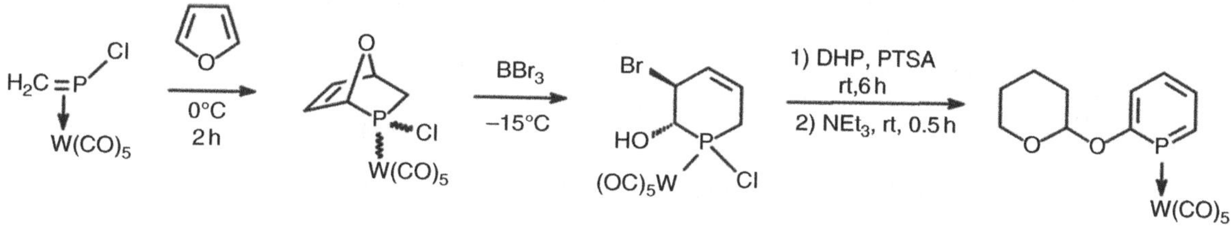

22

Scheme 8.7 *Synthesis of phosphinines from furans.*

This method was used by Breit and coworkers to prepare chiral oxazoline-substituted phosphinine-based ligands with two electronically rather different coordination sites [61] (Scheme 8.5).

The synthesis was achieved starting from L-valinol and α-pyrone-carboxylic acid chloride. The phosphinine moiety in **20** was formed by [4+2] cycloaddition reaction of the oxazoline–pyrone derivative with a phosphaalkyne under subsequent release of CO_2. The use of (di)azaphosphinines as precursors for [4+2] cycloaddition reactions with alkynes has been investigated by Le Floch and Mathey and coworkers [62] (Scheme 8.6).

This method is especially suitable for the introduction of substituents into the 2- and 6-position of the heterocyclic framework (e.g., R^1=H, $SiMe_3$, PPh_2, and Ph; Scheme 8.6). The corresponding phosphinines of the type **21** are obtained in 60–85% isolated yield. More recently, Mathey *et al.* reported on a [4+2] cycloaddition reaction starting from furans and phosphaalkenes [63] (Scheme 8.7).

Cleavage of the cycloaddition products with BBr_3 and subsequently protection of the –OH and dehydrohalogenation by triethylamine give the corresponding phosphinine-$W(CO)_5$ complex **22**. The overall yield starting from furan is 60%.

8.3.4 Ring expansion methods

Ring expansion reactions of phospholes by formal insertion of methylene fragments lead to 2-substituted phosphinines as described by Mathey *et al.* In this way, the first phosphinine derivative of 2,2′-bipyridine

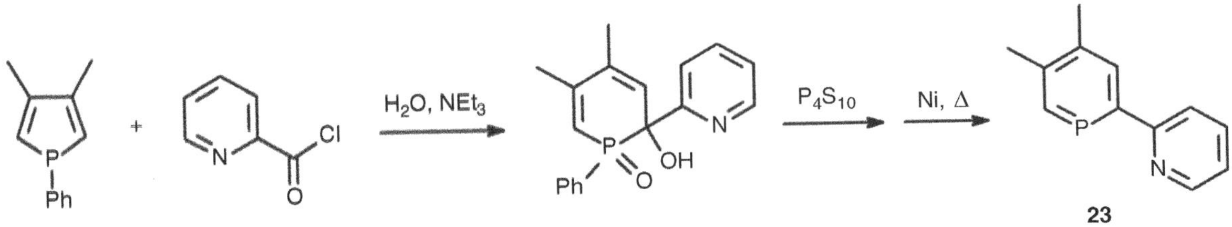

Scheme 8.8 Synthesis of phosphinines via ring expansion.

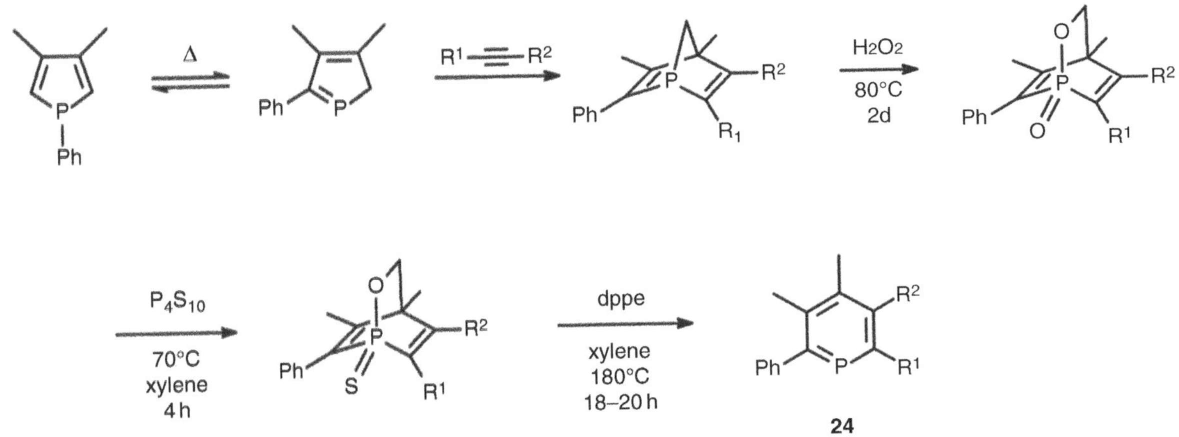

Scheme 8.9 Synthesis of phosphinines starting from phospholes.

(NIPHOS, **23**) was synthesized (Scheme 8.8) and obtained as a liquid (250°C/0.1 mmHg) in 35% overall yield [64]. This method is somewhat flexible concerning the use of the carboxylic acid chlorides [65,66]. It should be kept in mind that the precursor 1-phenyl-3,4-dimethylphosphole has to be synthesized first and the substitution pattern depends on the synthetic accessibility of this compound. Transition metal complexes containing **23** have been reported to be highly sensitive toward nucleophilic attack (*vide infra*).

A new and versatile route for the conversion of phospholes into phosphinines was recently reported by Mathey *et al.* [67]. The equilibrium between 1*H*- and 2*H*-phospholes makes 1-phosphanorbornadienes easily available through the reaction of 2*H*-phospholes with alkynes (Scheme 8.9). Oxidation with hydrogen peroxide and conversion into the corresponding P=S derivative were followed by thermal reduction–extrusion to afford phosphinine **24** in 64–82% yield, depending on the substitution pattern. This reaction provides the potential to synthesize a whole variety of differently substituted phosphinines by making use of substituted phospholes and alkynes.

8.3.5 Metal-mediated functionalizations

Donor-functionalized phosphinines are accessible via Pd-catalyzed cross-coupling reactions starting from halo-phosphinines. In this way, thienyl-, furyl-, and *N*-methyl-pyrrol-functionalized phosphinines can be prepared starting from 2,6-dibromophosphinines and the corresponding organotin derivatives (Stille coupling; Scheme 8.10) [68]. The thienyl- and furyl-functionalized phosphinines **25** and **26** were obtained as yellow solids in 60–70% yield, while the N-methyl-pyrrol-substituted compound **27** was obtained as a yellow

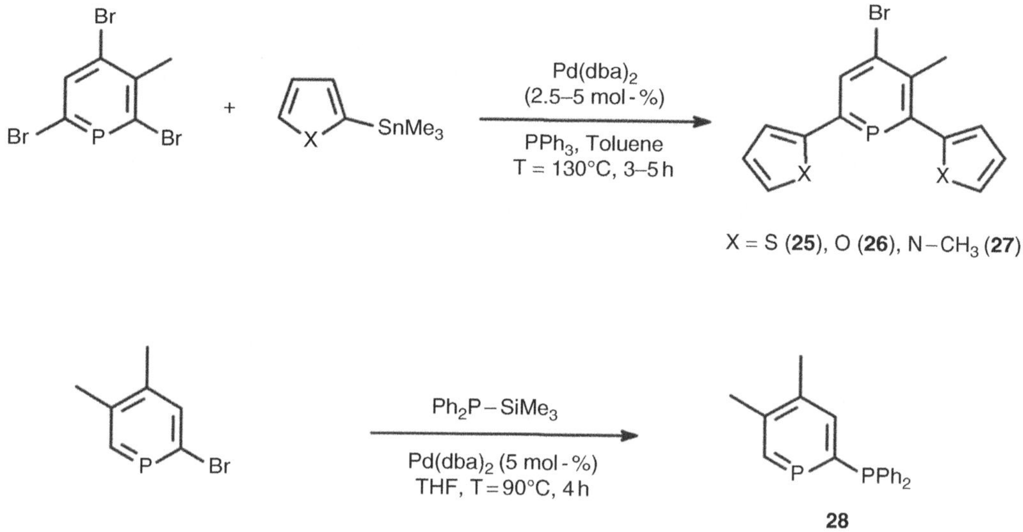

Scheme 8.10 *Metal-mediated functionalization of bromo-phosphinines.*

oil in 65–75% isolated yield. 2-Bromo-phosphinines can also be functionalized by $-PPh_2$ groups via Pd-catalyzed cross-coupling reactions using silyl-phosphines. The corresponding phosphinine–phosphine **28** was obtained as a yellow oil in 75% yield [68].

8.4 Coordination chemistry

The coordination chemistry of phosphinines has been subject to several excellent review articles published in literature within the last decade [69–73]. This section will focus on recent developments in this field only. Especially the coordination of phosphinines to transition metal centers in medium to high oxidation states has been mainly neglected and will be discussed here. Phosphinines as ambidentate ligands possess two different potential coordination sites, the lone pair of the phosphorus atom and the aromatic π-system. The $HOMO^{-2}$ of a phosphinine ligand is suitable for σ-coordination toward a metal center. Its energy is close to the energy of the $HOMO^{-1}$ and HOMO orbitals that can participate in η^6-coordination. This leads to a range of coordination modes depicted in Figure 8.7.

The coordination mode through the phosphorus atom (σ-coordination, 2e-donation) is the most common one and is generally observed with late transition metals in low oxidation states, due to the strong π-acceptor properties of the phosphinine ligand. In fact, ~95% of the published X-ray crystallographic data of phosphinine-σ-complexes contain metal centers in the formal oxidation state 0 and+I. Reduced species of the type $[(\text{Phosphinine})ML_n]^-$ and $[(\text{Phosphinine})ML_n]^{2-}$ have been reported as well. The preparation of coordination compounds containing metal centers in higher oxidation states is very challenging, as the aromaticity of the phosphinine heterocycle is significantly disrupted upon coordination. As a result the phosphinine core behaves as a cyclophosphahexatriene moiety containing a highly reactive P=C double bond, which is prone to nucleophilic attack (Section 8.5) [74]. Coordination can, however, be achieved by making use of the chelate effect. In order to form a strong bond with a Lewis acidic metal center, the lone pair at the phosphorus atom, having a high *s*-character in the free ligand, must gain considerable *p*-character upon

Figure 8.7 Common coordination modes of phosphinines.

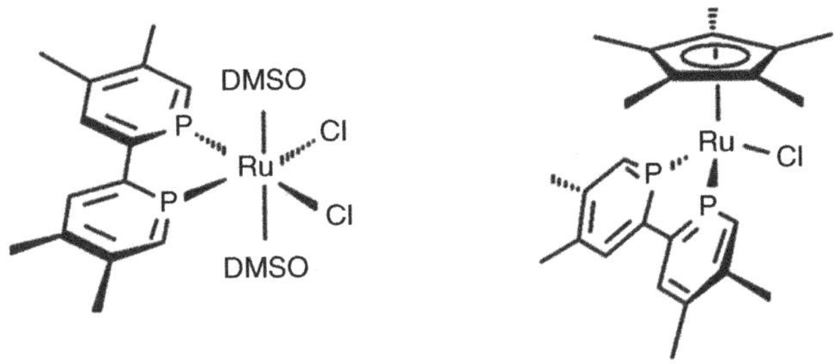

29 30

Figure 8.8 Ru(II) complexes containing the tmbp ligand.

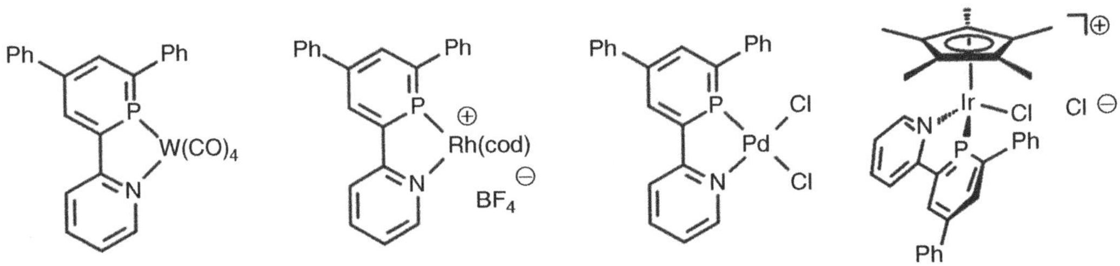

31 32 33 34

Figure 8.9 Transition metal complexes of 2-(2′-pyridyl)-4,6-diphenylphosphinine **15**.

coordination. This leads to an opening of the internal ∠C–P–C angle, which is approximately 100° in a free phosphinine. This phenomenon can thus be correlated with the electron-accepting character of a metal-fragment. Only a few isolated and crystallographically characterized M(+II) species have been reported in literature, such as the Ru(II) complexes **29** and **30**, containing the chelating ligand 4,4′,5,5′-tetramethyl-2,2′-biphosphinine (tmbp) (Figure 8.8) [75,76].

The group of Müller made use of this concept and prepared various transition metal complexes based on the bidentate P,N hybrid ligand 2-(2′-pyridyl)-4,6-diphenylphosphinine **15** (Figure 8.6) [77,78]. This ligand, which is accessible via the pyrylium salt route, allows the preparation of coordination compounds of the type [(P,N)M(0)], [(P,N)M(I)], [(P,N)M(II)], and [(P,N)M(III)], and representative examples **31–34** are depicted in Figure 8.9.

Compounds **33** represents a rare example of a phosphinine–Pd(II) species and the first isolated and crystallographically characterized phosphinine–M(III) complex **34**. In this way, a complete new set of phosphinine–metal complexes becomes synthetically accessible, leading to a much broader scope for applications of phosphinines and their corresponding metal complexes.

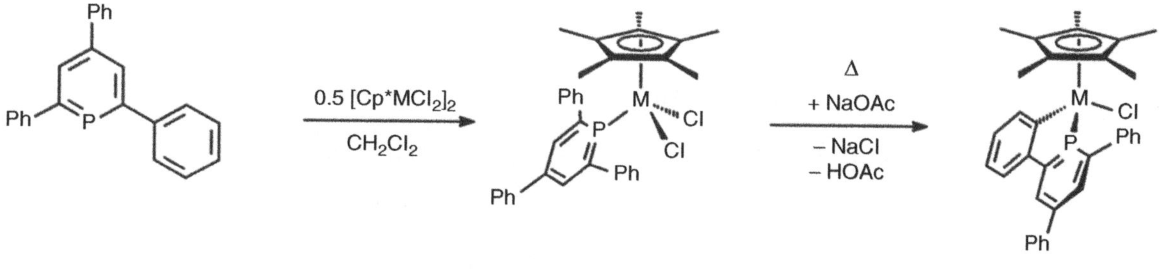

Scheme 8.11 *C–H activation of 2,4,6-triphenylphosphinine.*

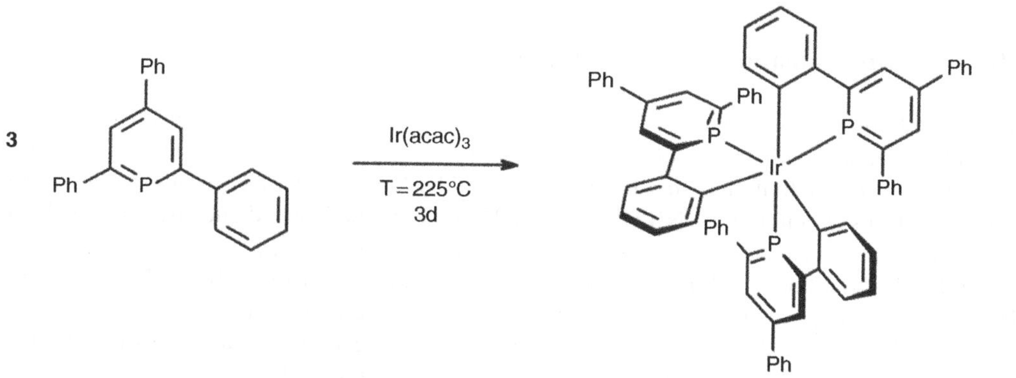

Scheme 8.12 *C-H activation of **1** under formation of a [Ir(III)(C^P)$_3$] complex **37**.*

A novel aspect in the coordination chemistry of phosphinines is the C–H activation of 2,4,6-triphenylphosphinine by late transition metal complexes. An unprecedented *ortho*-metalation of **1** by Rh(III) and Ir(III) has recently been observed by Müller *et al.* [79]. The complexes **35** and **36** could be isolated and structurally characterized (Scheme 8.11).

The reaction of **1** with Ir(acac)$_3$ in the ratio 3:1 gives exclusively and quantitatively the *fac*-isomer **37** (Scheme 8.12), which represents the first phosphorus analogue of the well-known triplet emitter [Ir(ppy)$_3$] (ppy=2-phenylpyridine). Complex **37** shows interesting photophysical properties. Apparently, exchanging nitrogen by phosphorus shifts the emission wavelength to a higher frequency as the metal-centered t_{2g} orbitals are stabilized by the π-accepting properties of the phosphinine ligand. This is particularly interesting as phosphorus-containing triplet emitters are scarce [80].

As a matter of fact, cyclometalated coordination compounds containing 2-phenylpyridines have attracted considerable interest for a wide variety of applications, such as catalysis, photochemistry, organic synthesis, and molecular material sciences (e.g., organic light-emitting diode [OLEDs]). The replacement of nitrogen by phosphorus in similar structures could lead to significantly diverse properties due to the electronic difference that exists between these heteroatoms. The C–H activation as a novel reactivity pattern of phosphinines might consequently open up new perspectives for future applications of phosphinines, especially in homogeneous catalysis and molecular material sciences.

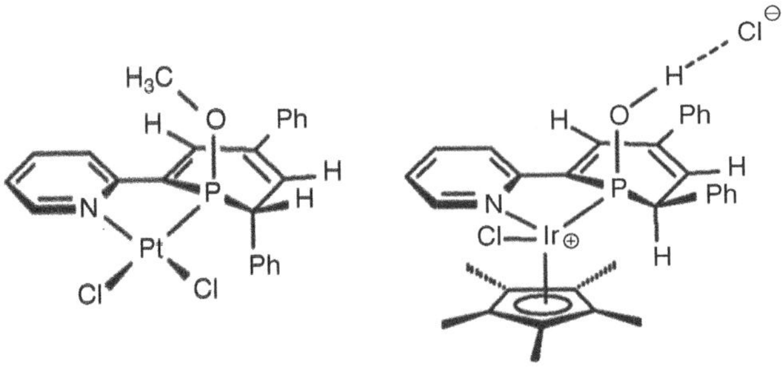

38 39

Figure 8.10 Formation of dihydrophosphinines by nucleophilic attack.

8.5 Reactivity of transition metal complexes

Transition metal complexes containing less substituted phosphinines are highly sensitivity toward nucleophilic attack and react with traces of water or alcohols under formation of dihydrophosphinine–metal complexes [74]. This reactivity has first been investigated by Venanzi *et al.* The authors showed that Pd(II) and Pt(II) complexes containing the NIPHOS ligand **23** could not be isolated and structurally characterized. Apparently, it seemed that the higher the oxidation state of the metal center, the more reactive the P=C double bond is. It should be noted here also that transition metal complexes of bipyridines undergo reaction with H_2O albeit with opposite regioselectivity due to the different charge distribution in pyridines compared to phosphinines [81]. Using 2,4,6-triarylphosphinines, however, the additional aryl-substituents contribute to a significant kinetic stabilization of the corresponding metal complex by making the P=C double bond sterically more hindered for nucleophilic attack. Nevertheless, especially at higher temperatures, the reaction with MeOH or H_2O also leads to the formation of dihydrophosphinines in this case (Figure 8.10) [78].

This is an important aspect for the application of phosphinines as ligands in homogeneous catalysis. Using the bidentate P,N ligands **15** or **23**, the reaction proceeds regioselectively. The addition of the proton exclusively to the carbon atom in 6-position might be attributed to its somewhat higher nucleophilicity compared to the one in 2-position due to the electron-withdrawing character of the adjacent pyridine ring. Interestingly, the addition of CH_3OH or H_2O to the double bond proceeds also diastereoselectively. In case of the Pt(II) complex (**15**)$PtCl_2$, the addition of CH_3OH proceeds exclusively in *syn*-fashion under the formation of **38**, while the *anti*-product **39** was observed for the addition of H_2O to the Ir(III) complex **34** [78,80]. The cause of this difference is unclear, but might be attributed to steric factors during the transfer of the proton to the carbon atom.

8.6 Application of phosphinines in homogeneous catalysis

The application of phosphinines as ligands in homogeneous catalysis has been subject to several excellent review articles, and the interested reader is referred to the corresponding literature [71–73,82,83]. Despite the fact that phosphinines have been known for more than 40 years and possess a rich and versatile coordination chemistry, investigations on their application as ligands in homogeneous catalysis were started only a decade ago. So far, studies are limited to only a few examples. Thus, Fe-catalyzed cyclotrimerizations and cyclodimerizations [84], Rh-catalyzed hydroformylations [36,38,61,85], Rh-catalyzed hydroformylation–cyclization tandem reactions [86], Ni-catalyzed cycloisomerizations [19], Rh- and Ir-catalyzed hydrogenations [54,87,88], Ir- and Pd-catalyzed hydrosilylations [53,89], Au-catalyzed cycloisomerizations [80], and Ni- and Pd-catalyzed Suzuki–Miyaura reactions [90] have been investigated.

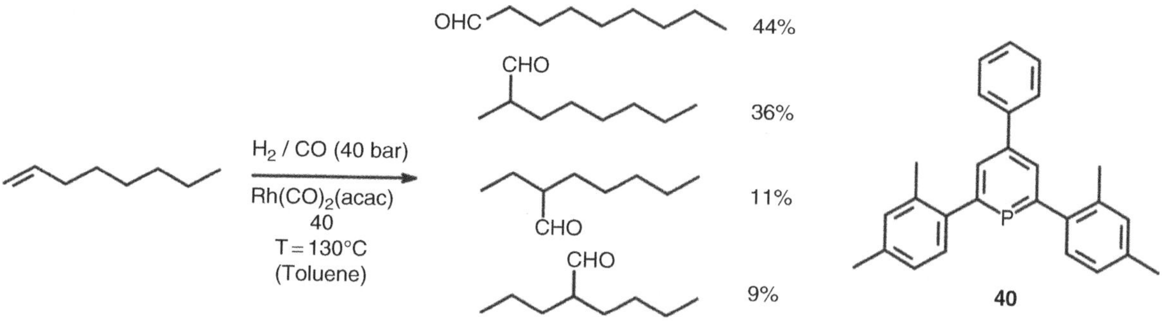

Scheme 8.13 *Hydroformylation of 1-octene with the catalyst system Rh/**40**.*

So far, the most promising results have been reported by Breit *et al.* who explored the application of phosphinines in the Rh-catalyzed hydroformylation of various alkenes [38,40,41,61,85]. While α-unsubstituted phosphinines inhibit the catalytic cycle, 2,4,6-trisubstituted ligands showed both high activity and high regioselectivity in the formation of the branched product in the Rh-catalyzed hydroformylation of styrene. In the screening of various monodentate phosphinines for the Rh-catalyzed hydroformylation of 1-octene, the substituted bulky 2,4,6-triphenylphosphinine derivative **40** (Scheme 8.13) was found to show superior performance over classical PPh_3-based catalysts. Turnover frequencies of up to $45370 \cdot h^{-1}$ have been determined (T = 130°C; p = 40 bar CO/H_2; Rh:L = 1:20), although the selectivity toward the linear product nonanal is low, as anticipated for catalysts based on monodentate ligands (l/b ~2) [38]. Moreover, complete and rapid consumption of internal octenes, which have been formed by isomerization reaction prior to hydroformylation, occurred, showing also that less reactive alkenes can be hydroformylated efficiently (Scheme 8.13).

These examples demonstrate that phosphinines have great potential as ligands in the hydroformylation reaction. As described in this chapter, the high activity in the hydroformylation of 1-octene as well as the hydroformylation activity for less reactive alkenes can be attributed to the presence of monoligated Rh-species in combination with the strong π-acceptor capabilities of the ligand. Future developments in this area will most likely focus on the design and application of bidentate phosphinines, which should lead especially to more regioselective catalysts. It should be mentioned here that $^{31}P\{^1H\}$ NMR analysis of the hydroformylation mixtures showed that no phosphinine degradation had occurred during the catalytic runs, indicating the stability of phosphinines under the applied hydroformylation conditions.

The development of chiral phosphinines with the chiral information close to the phosphorus center is very challenging and could lead to interesting developments in asymmetric homogeneous catalysis. Especially the hydroformylation of internal alkenes would lead to chiral aldehydes, which are interesting building blocks for the fine-chemicals and pharmaceutical industries. The first axially chiral atropisomeric phosphinine **17** (Figures 8.6 and 8.11) has recently been reported by the group of Müller [51,52]. Both enantiomers could be separated by means of chiral high-performance liquid chromatography (HPLC), and the absolute configuration was determined by combining circular dichroism (CD) spectroscopy and density functional theory (DFT) calculations. However, it turned out that the barrier for internal rotation ($\Delta G^{\#} = 109.5$ kJ/mol) is still too low for applications as efficient ligands in asymmetric homogeneous catalysis, especially at higher temperature. Modification of the substitution pattern for the development of suitable systems is therefore important and necessary. In this respect, the group of Müller recently designed and prepared via the pyrylium salt route the new chiral naphthyl-substituted phosphinine **41**, which exhibits a rather high calculated barrier for internal rotation of $\Delta G^{\#} = 171$ kJ/mol (Figure 8.12)!

So far, investigations on the enantioseparation of **41**-S_a and **41**-R_a as well as their use in asymmetric homogeneous catalysis have not been reported yet at this stage.

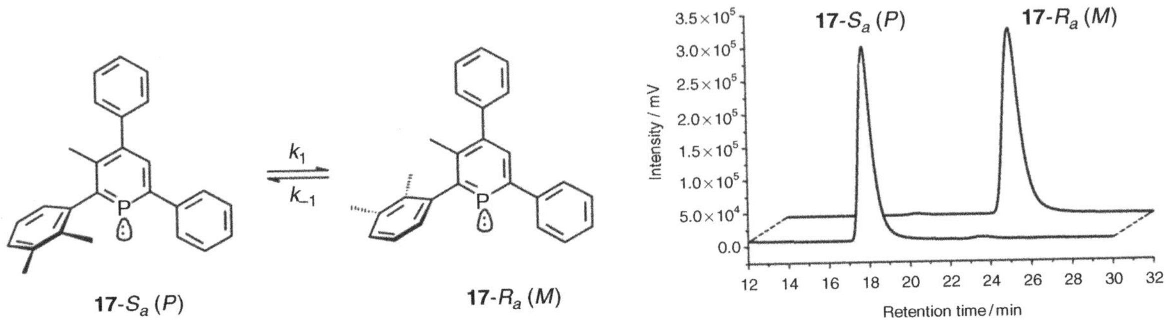

Figure 8.11 *Axially chiral phosphinine **17** (left). Separation of enantiomers by chiral HPLC and assignment of absolute configuration (right).*

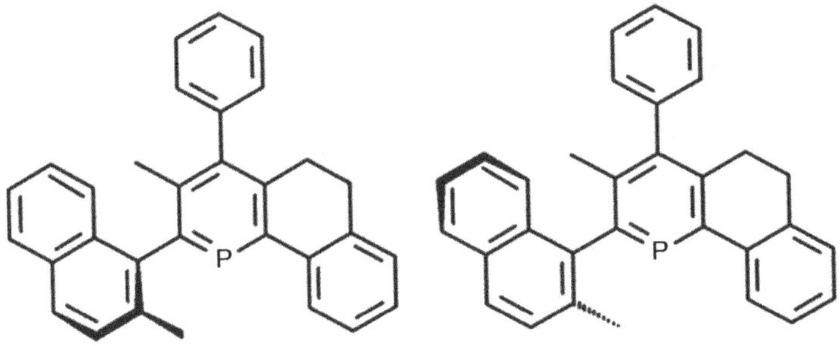

41-S_a(P) 41-R_a(M)

Figure 8.12 *Chiral naphthyl-substituted phosphinine **41**.*

In the Rh-catalyzed hydroformylation of 2-octene with a racemic mixture of the xylyl-substituted phosphinine-ligand **17**, however, it could already be demonstrated that isomerization reactions were sufficiently suppressed under the right reaction conditions and high selectivities toward branched aldehydes were indeed obtained (Figure 8.13) [80].

These results show that extensive screening of functionalized phosphinines in homogeneous catalytic reactions is an attractive research area and could lead to unexpected results in terms of activity, reactivity, and selectivity in the future. It should be kept in mind, however, that the application of phosphinines in catalysis might be restricted to specific transformations, since it has been shown in some cases that phosphinine complexes are prone to nucleophilic attack on the phosphorus atom (Section 8.5). Nevertheless, state-of-the-art synthetic methodologies now allow a straightforward derivatization and functionalization, which can also lead to a kinetic stabilization of these heterocycles as has already been shown for selected examples. Ligands with tailored properties can thus be expected, and especially the development of chiral phosphinines and their application in asymmetric homogeneous catalysis are promising. The very peculiar electronic and steric properties of phosphinines as well as the application and development of various highly interesting and promising derivatives based on phosphinines might lead to interesting results in homogeneous catalysis in the near future. A detailed exploration of these unique phosphorus-containing ligands is consequently necessary and important. Although once regarded as "laboratory curiosities," the history of phosphinines nicely demonstrates that these and related ligands can still lead to unexpected discoveries, and their further investigation remains an exciting, challenging, and rewarding research field.

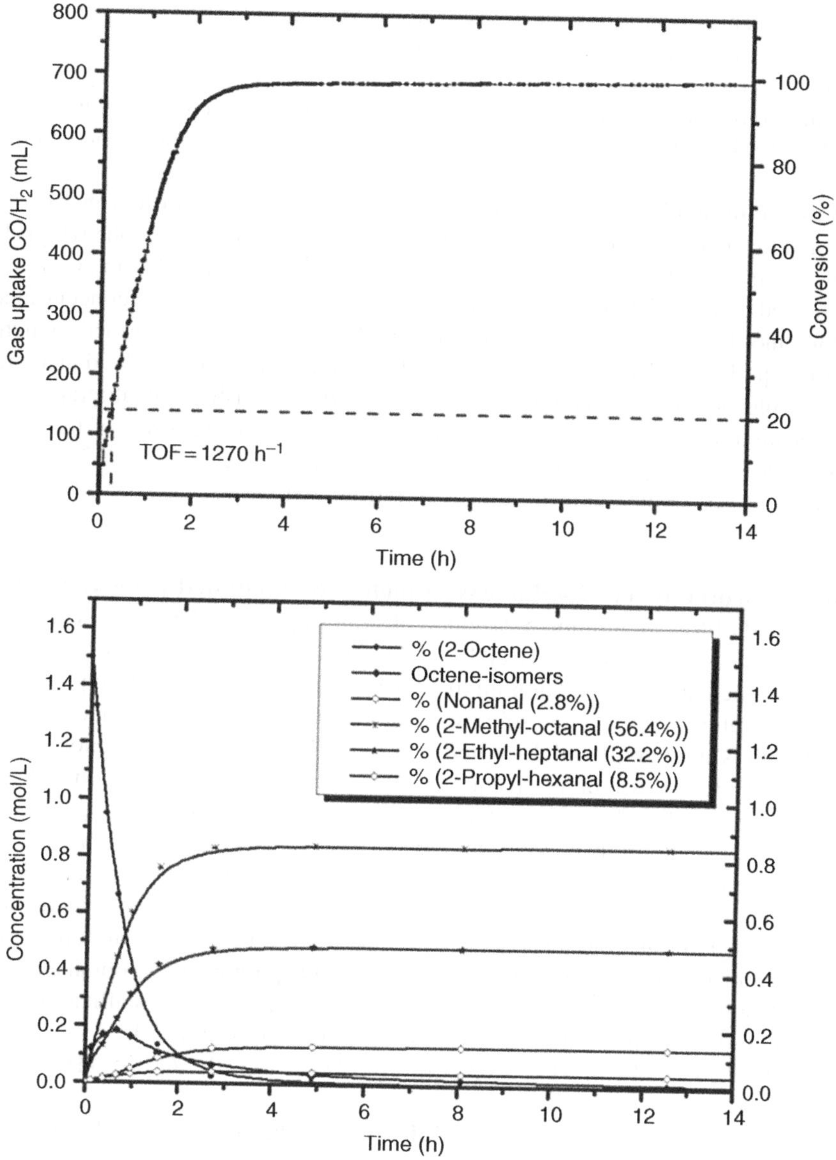

Figure 8.13 *Rh-catalyzed hydroformylation of 2-octene with a racemic mixture of **17**. Top: gas uptake curve and corresponding activity (at 20% conversion). Bottom: product distribution. Reaction conditions: T=80°C;* $p_{CO/H2}$ *=40 bar;* c_{Rh} *=1 mM; Rh:S=1:1500; Rh:L=1:20; V=8 mL (toluene). Preformation: 2 h; T=80°C; p=20 bar.*

8.7 Experimental procedure for the synthesis of selected phosphinines

Triaryl-substituted pyrylium salts can generally be synthesized either via the chalcone route (α,β-unsaturated aryl ketones) or, alternatively, by cyclization of aromatic 1,5-diketones in the presence of $HBF_4 \cdot Et_2O$ or $BF_3 \cdot Et_2O$. Both chalcones and 1,5-diketones can be prepared from the corresponding benzaldehydes and acetophenones by classical condensation reactions. Since pyrylium salts formally contain two fewer hydrogen atoms than their starting materials, an oxidant has to be added for the cyclization step, which is typically an additional equivalent of the used chalcone or benzylidene-2-acetophenone. Both synthetic protocols usually have some limitations, and the right reaction conditions have to be selected individually. The phosphorus precursors $P(SiMe_3)_3$ [91] and $P(CH_2OH)_3$ [92] are prepared according to literature procedures.

2,4,6-Triphenylphosphinine (1): The precursor 2,4,6-triphenylpyrylium tetrafluoroborate is commercially available. Dry and degassed pyridine (30 mL) is added to a mixture of 2,4,6-triphenylpyrylium tetrafluoroborate (6.6 g, 16.7 mmol) and $P(CH_2OH)_3$ (3.7 g, 29.8 mol, 1.8 equiv.) under an argon atmosphere. The suspension is heated to T = 125°C for 4 h. Subsequently, the orange solution is concentrated to 50% of its original volume, and degassed water (20 mL) is added. An orange solid is formed, which is isolated by decanting off the liquid layer. The remaining solid is dried under vacuum, dissolved in a minimum amount of a mixture of dichloromethane and petroleum ether (50:50), and flash-chromatographed over silica with the same solvent mixture to remove traces of pyridine × HBF_4. After removal of the solvents *in vacuo*, phosphinine **1** is obtained as a yellow solid in high purity. Yield: 1.20 g (3.7 mmol, 22.2%). Yellow needles of **1** are obtained by recrystallization from hot acetonitrile. (^{1}H NMR ($CDCl_3$): δ (ppm) = 7.50 (m, 9 H), 7.75 (d, $^3J_{H-H}$ = 7.5 Hz, 2 H), 7.78 (d, $^3J_{H-H}$ = 8.15 Hz, 4 H), 8.24 (d, $^3J_{H-P}$ = 5.9 Hz, 2 H, H_β). ^{31}P NMR ($CDCl_3$): δ (ppm) = 180.7.

Chalcone route – synthesis of 2,4-diphenyl-5-methyl-6-(2,3-dimethylphenyl)-phosphinine (17): Tetrafluoroboric acid (4.6 g, 27.1 mmol, 52% ethereal solution) is added at T = 70°C to a solution of benzylideneacetophenone (6.0 g, 28.6 mmol) and 2,3-dimethylpropiophenone [93] (2.32 g, 14.3 mmol) in 10 mL of 1,2-dichloroethane. The reaction mixture is heated for 6 h and is subsequently allowed to cool down to room temperature (RT). The red solution is diluted with Et_2O (150 mL) and stirred overnight. The yellow pyrylium salt is filtered off and dried under vacuum. Yield: 2.40 g (5.5 mmol, 38%). To a solution of the pyrylium salt (3.55 g, 8.1 mmol, 1 equiv.) in dry pyridine (20 mL) is added $P(CH_2OH)_3$ (1.8 g, 14.6 mmol, 1.8 equiv.) at RT. The solution is heated to T = 125°C, and stirring is continued for 3 h. Subsequently, the orange solution is concentrated to 50% of its original volume, and degassed water (5 mL) is added. An orange solid was formed, which is isolated by decanting the liquid layer. The product is dried under vacuum, dissolved in a minimum amount of a mixture of dichloromethane and petroleum ether (50:50), and flash-chromatographed over silica with the same solvent mixture to remove traces of pyridine × HBF_4. After removal of the solvents *in vacuo*, phosphinine **17** is obtained as a yellow solid. Yield: 700 mg (1.9 mmol, 24%). Yellow needles of **17** are obtained by recrystallization from hot acetonitrile. ^{1}H NMR (C_6D_6): δ (ppm) = 1.98, 2.04 (2 × s, 2 × 3 H, $-CH_3$), 2.06 (d, $^4J_{H-P}$ = 2.0 Hz, 3 H, $-CH_3$), 6.96–7.09 (m, 3 H), 7.11–7.24 (m, 8 H), 7.63–7.68 (m, 2 H), 7.95 (d, $^3J_{H-P}$ = 5.4 Hz, 1 H, H_β). ^{31}P NMR (C_6D_6): δ (ppm) = 190.5.

Diketone route – synthesis of 2-(2′-pyridyl)-4,6-diphenylphosphinine (15): $BF_3 \cdot Et_2O$ (22.3 g, 157.3 mmol, 8 equiv.) is added dropwise to a mixture of 1,3-diphenyl-5-(2-pyridyl)-1,5-pentanedione [94] (6.5 g, 19.7 mmol, 1 equiv.) and benzylidene-2-acetophenone (4.1 g, 19.8 mmol, 1 equiv.) at RT. The reaction mixture is heated to T = 70°C for 3 h. After allowing the reaction mixture to cool down to RT, a yellow solid precipitates after adding diethyl ether. The yellow solid is collected on a glass filter, washed with diethyl ether, and recrystallized from methanol to obtain the 2-(2-pyridyl)-4,6-diphenylpyrylium tetrafluoroborate as yellow needles (4.7 g, 60%). Under an argon atmosphere at room temperature, $P(SiMe_3)_3$ (2.8 g, 11.2 mmol, 2.1 equiv.) is added dropwise to a solution of the pyrylium salt (2.0 g, 5.0 mmol, 1 equiv.) in acetonitrile (12 mL). A dark reaction mixture is formed which is refluxed at T = 85°C for 6 h. Subsequently, all volatiles are removed *in vacuo* to obtain a dark solid. The crude product is purified by means of column chromatography over neutral alumina with ethyl acetate-petroleum ether (1:5) to afford the product as a yellow-orange solid (0.54 g, 30.7%). Yellow needles of **15** are obtained by recrystallization from hot acetonitrile. ^{1}H NMR (200 MHz, C_6D_6): δ (ppm): 7.21 – 7.08 (m, 8 H), 7.64 (m, 2 H), 7.47 (m, 2 H), 7.85 (m, 1 H), 8.10 (dd, 1 H, *J*=5.7 Hz, *J*=1.4 Hz), 8.55 (m, 1 H), 9.10 (dd, 1 H, *J*=5.7 Hz, *J*=1.4 Hz). ^{31}P NMR (C_6D_6): δ (ppm): 187.4.

References

[1] G. Märkl, *Angew. Chem.* **1966**, *78*, 907.
[2] A. J. Ashe III, *J. Am. Chem. Soc.* **1971**, *93*, 3293.
[3] K. K. Baldridge, M. S. Gordon, *J. Am. Chem. Soc.* **1988**, *110*, 4204.
[4] L. Nyulászi, T. Veszprémi, J. Réffi, B. Burkhardt, M. Regitz, *J. Am. Chem. Soc.* **1992**, *114*, 9080.
[5] P. D. Burrow, A. J. Ashe III, D. J. Bellville, K. D. Jordan, *J. Am. Chem. Soc.* **1982**, *104*, 425.
[6] L. Nyulászi, *Chem. Rev.* **2001**, *101*, 1229.
[7] L. Nyulászi, T. Veszprémi, *J. Phys. Chem.* **1996**, *100*, 6456.
[8] A. Modelli, B. Hajgató, J.F. Nixon, L. Nyulászi, *J. Phys. Chem. A* **2004**, *108*, 7440.
[9] G. Frison, A. Sevin, N. Avarvari, F. Mathey, P. Le Floch, *J. Org. Chem.* **1999**, *64*, 5524.
[10] C. A. Tolman, *Chem. Rev.* **1977**, *77*, 313.
[11] A. Muller, R. Meijboom, A. Roodt, *J. Organomet. Chem.* **2006**, *691*, 5782.
[12] S. Serron, S. P. Nolan, K. G. Moloy, *Organometallics* **1996**, *15*, 4301.
[13] O. Kühl, Coord, *Chem. Rev.* **2005**, *249*, 693.
[14] C. Müller, M. Lutz, A. L. Spek, D. Vogt, *Chem. Crystallogr.* **2006**, *36*, 869 and references therein.
[15] C. Müller, D. Wasserberg, J. J. M. Weemers, E. A. Pidko, S. Hoffmann, M. Lutz, A. L. Spek, S. C. J. Meskers, R. A. Janssen, R. A. van Santen, D. Vogt, *Chem. Eur. J.* **2007**, *13*, 4548.
[16] J. J. Daly, *J. Chem. Soc.*, Abstracts **1964**, 3799.
[17] J. C. J. Bart, *Angew. Chem., Int. Ed.* **1968**, *7*, 730.
[18] A. J. Ashe III, *Comprehensive Heterocyclic Chemistry II*, A.R. Katritzky, C.W. Rees E.F.V. Scriven (Eds.), Pergamon, **1996**.
[19] E. F. DiMauro, M. C. Kozlowski, *J. Chem. Soc., Perkin Trans.* **2001**, 439.
[20] M–P distance is usually set to 2.28 Å, which corresponds to a typical Ni–P distance.
[21] N.-N. Pham-Tran, G. Bouchoux, D. Delaere, M. T. Nguyen, *J. Phys. Chem. A* **2005**, *109*, 2957.
[22] Y. Zhang, F. S. Tham, J. F. Nixon, C. Taylor, J. C. Green, C. Reed, *Angew. Chem. Int. Ed. Engl.* **2008**, *47*, 3801.
[23] C. A. Reed, *Acc. Chem. Res.* **2010**, *43*, 121.
[24] K. Dimroth, *Top. Curr. Chem.* **1973**, *38*, 1.
[25] A. Hettche, K. Dimroth, *Chem. Ber.* **1973**, *106*, 1001.
[26] K. Dimroth, W. Städe, *Angew. Chem.* **1968**, *80*, 966.
[27] A. Moores, T. Cantat, L. Ricard, N. Mézailles, P. Le Floch, *New J. Chem.* **2007**, *31*, 1493.
[28] A. Moores, L. Ricard, P. Le Floch, *Angew. Chem., Int. Ed. Engl.* **2003**, *42*, 4940.
[29] G. Märkl, F. Lieb, A. Merz, *Angew. Chem.* **1967**, *79*, 59.
[30] K. Dimroth, F. W. Steuber, *Angew. Chem.* **1967**, *79*, 410.
[31] F. Gerson, P. Merstetter, S. Pfenninger, G. Märkl, *Magn. Reson. Chem.* **1997**, *35*, 384.
[32] S. Choua, C. Dutan, L. Cataldo, T. Berclaz, M. Geoffroy, N. Mézailles, A. Moores, L. Ricard, P. Le Floch, *Chem. Eur. J.* **2004**, *10*, 4080.
[33] A. Cataldo, S. Choua, T. Berclaz, M. Geoffroy, N. Mézailles, L. Ricard, F. Mathey, P. Le Floch, *J. Am. Chem. Soc.* **2001**, *123*, 6654.
[34] K. Dimroth, N. Greif, H. Perst, F. W. Steuber, *Angew. Chem.* **1967**, *79*, 58.
[35] G. Märkl, *Tetrahedron Lett.*, **1971**, *12*, 1249.
[36] B. Breit, E. Fuchs, *Chem. Commun.* **2004**, 694.
[37] C. Wallis, P. G. Edwards, M. Hanton, P. D. Newman, A. Stasch, C. Jones, R. P. Tooze, *Dalton Trans.* **2009**, 2170.
[38] B. Breit, R. Winde, T. Mackewitz, R. Paciello, K. Harms, *Chem. Eur. J.* **2001**, *7*, 3106.

[39] G. Märkl, F. Lieb, A. Merz, *Angew. Chem. Int. Ed. Engl.* **1967**, *6*, 458.
[40] R. Paciello, E. Zeller, B. Breit, M. Röper (to BASF), DE 19743197 A1, **1999**.
[41] T. Mackewitz, M. Röper (to BASF), EP 1036796 A1, **2000**.
[42] G. Märkl, F. Lieb, A. Merz, *Angew. Chem. Int. Ed. Engl.* **1967**, *6*, 944.
[43] T. S. Balaban, A. T. Balaban, *Science of Synthesis* **2003**, *14*, 11.
[44] A. R. Katritzky, M. Rezende, *J. Chem. Res.* **1980**, 312.
[45] A. R. Katritzky, S. S. Thind, *J. Chem. Soc., Perkin Trans.* **1980**, 1895.
[46] A. R. Katritzky, J. M. Lloyd, R.C. Patel, *J. Chem. Soc., Perkin Trans.* **1982**, 117.
[47] A. R. Katritzky, E. M. Elisseou, R. C. Patel, B. Plau, *J. Chem. Soc., Perkin Trans.* **1982**, 125.
[48] C. Müller, D. Vogt, *C. R. Chimie* **2010**, *13*, 1127.
[49] C. Müller, Z. Freixa, M. Lutz, A.L. Spek, D. Vogt, P. W. N. M. van Leeuwen, *Organometallics* **2008**, *27*, 834.
[50] C. Müller, E. A. Pidko, M. Lutz, A. L. Spek, D. Vogt, *Chem. Eur. J.* **2008**, *14*, 8803.
[51] C. Müller, E. A. Pidko, M. Lutz, A. L. Spek, R. A. van Santen, D. Vogt, *Dalton Trans.* **2007**, *46*, 5372.
[52] C. Müller, E. A. Pidko, A. J. P. M. Staring, M. Lutz, A. L. Spek, R. A. van Santen, D. Vogt, *Chem. Eur. J.* **2008**, *14*, 4899.
[53] J. R. Bell, A. Franken, C. M. Garner, *Tetrahedron* **2009**, *65*,9368.
[54] C. Müller, L. Guarrotxena-Lopéz, H. Kooijman, A. L. Spek, D. Vogt, *Tetrahedron Lett.* **2006**, *47*, 2017.
[55] M. D. Segarra Maset, PhD thesis, ICIQ–Universitat Rovira i Virgili, Tarragona, Spain, **2010**.
[56] A. J. Ashe III, P. Shu, *J. Am. Chem. Soc.* **1971**, *93*, 1804.
[57] A. J. Ashe III, *Acc. Chem. Res.* **1978**, *11*, 153.
[58] W. Rösch, M. Regitz, *Z. Naturforsch., B.* **1986**, *41*, 931.
[59] M. Regitz, P. Binger, *Angew. Chem.* **1988**, *100*, 1541.
[60] N. Mézailles, L. Ricard, F. Mathey, P. Le Floch, *Organometallics* **2001**, *20*, 3304.
[61] B. Breit, *J. Mol. Catal. A: Chem.* **1999**, *143*, 143.
[62] N. Avarvari, P. Le Floch, F. Mathey, *J. Am. Chem. Soc.* **1996**, *118*, 11978.
[63] Y. Mao, F. Mathey, *Org. Lett.* **2010**, *12*, 3384.
[64] J.-M. Alcaraz, A. Breque, F. Mathey, *Tetrahedron Lett.* **1982**, *23*, 1565.
[65] F. Mathey, *Tetrahedron Lett.* **1979**, 1753.
[66] N. H. T Huy, B. Donnadieu, F. Mathey, *Organometallics* **2007**, *26*, 6497.
[67] H. Wang, C. Li, D. Geng, H. Chen, Z. Duan, F. Mathey, *Chem. Eur. J.* **2010**, *16*, 10659.
[68] P. Le Floch, D. Carmichael, L. Ricard, F. Mathey *J. Am. Chem. Soc.* **1993**, *115*, 10665.
[69] P. Le Floch, F. Mathey, *Coord. Chem. Rev.* **1998**, *179–180*, 771.
[70] N. Mézailles, F. Mathey, P. Le Floch, *Prog. Inorg. Chem.* **2001**, 455.
[71] P. Le Floch, *Coord. Chem. Rev.* **2006**, *250*, 627.
[72] C. Müller, D. Vogt, *Dalton Trans.* **2007**, 5505.
[73] F. Mathey, P. Le Floch, *Science of Synthesis* **2005**, *15*, 1097.
[74] B. Schmid, L. M. Venanzi, A. Albinati, F. Mathey, *Inorg. Chem.* **1991**, *30*, 4693.
[75] D. Carmichael, P. Le Floch, L. Ricard, F. Mathey, *Inorg. Chim. Act.* **1992**, *198–200*, 437.
[76] P. Le Floch, S. Mansuy, L. Ricard, F. Mathey, *Organometallics*, **1996**, *15*, 3267.
[77] A. Campos Carrasco, E. A. Pidko, A. M. Masdeu-Bultó, M. Lutz, A. L. Spek, D. Vogt, C. Müller, *New. J. Chem.* **2010**, *34*, 1547.
[78] A. Campos-Carrasco, L. E. E. Broeckx, J. J. M. Weemers, E. A. Pidko, M. Lutz, A. M. Masdeu-Bultó, D. Vogt, C. Müller, *Chem. Eur. J.* **2011**, *17*, 2510.
[79] L. E. E. Broeckx, M. Lutz, D. Vogt, C. Müller, *Chem. Commun.* **2011**, 47, 2003.
[80] L. E. E. Broeckx, W. Delaunay, C. Latouche, M. Lutz, D. Vogt, A. Boucekkine, M. Hissler, C. Müller, *manuscript in preparation*.
[81] K. H. Al-Obaidi, R. D. Gillard, L. A. P. Kane-Maguire, P. A. Williams, *Transition Met. Chem.* **1977**, *2*, 64.
[82] L. Weber, *Angew. Chem. Int. Ed. Engl.* **2002**, *41*, 563.
[83] L. Kollár, G. Keglevich, *Chem. Rev.* **2010**, *110*, 4257.
[84] F. Knoch, F. Kremer, U. Schmidt, U. Zenneck, P. Le Floch. F. *Mathey, Organometallics* **1985**, *15*, 2713.
[85] B. Breit, *Chem. Commun. 1996*, **2071**.
[86] P. S. Bäuerlein, I. Arenas Gonzalez, J. J. M. Weemers, M. Lutz, A. L. Spek, D. Vogt, C. Müller, *Chem. Commun.* **2009**, 4944.

[87] M. T. Reetz, G. Mehler, *Tetrahedron Lett* **2003**, *44*, 4593.
[88] E. Neumann, PhD thesis, Universität Basel, Switzerland, **2006**.
[89] Y. Miyake, E. Isomura, M. Iyoda, *Chem Lett.* **2006**, *35*, 836.
[90] C. Godard, C. Müller, unpublished results.
[91] E. Niecke, H. Westermann, *Synthesis* **1988**, *4*, 330.
[92] J. W. Ellis, K. N. Harrison, P. A. T. Hoye, A. G. Orpen, P. G. Pringle, M. B. Smith, *Inorg. Chem.* **1992**, *31*, 3026.
[93] A. Shiozawa, K. Narita, G. Izumi, S. Kurashige, K. Sakitama, M. Ishikawa, *Eur. J. Med. Chem.* **1995**, *30*, 85.
[94] G. W. V. Cave, C. L. Raston, J. Chem. *Soc.*, *Perkin Trans. 1* **2001**, 3258.

9

Highly Strained Organophosphorus Compounds

J. Chris Slootweg and Koop Lammertsma

Department of Chemistry and Pharmaceutical Sciences, VU University Amsterdam, Amsterdam, The Netherlands

9.1 Introduction

Phosphorus is one of the most important main-group elements and plays a crucial role in all the fields of chemistry, of which biochemistry [1], organic synthesis [2], coordination chemistry, homogeneous catalysis [3], and material sciences [4] are just some examples. Phosphorus heterocyclic chemistry, however, has been underdeveloped for a long time and is only recently undergoing a very rapid growth [5]. Among the phosphorus rings, the inherently strained three-membered cyclopropane and cyclopropene homologues are unique because of their profoundly different electronic and chemical properties. In this chapter, a literature overview of the various reported methods to synthesize highly strained organophosphorus compounds is given. Some of these species are highly stable, while others function as reactive intermediates with exciting chemistry generating an easy access to novel organophosphorus compounds that can serve as phosphorus(III) ligands in catalysis.

The all-carbon cyclopropane **1** and the heterocycles oxirane **2** and azirane **3** (Figure 9.1) have been studied for more than a century and are even applied as valuable starting materials for the preparation of various polymeric materials in the chemical industry. In contrast, the chemistry of the phosphorus-containing three-membered rings started only in 1963 with the discovery of the parent phosphirane **4** by Wagner [6]. The next breakthrough came 20 years later with the discovery of 1*H*-phosphirene **5** [7] and 2*H*-phosphirene **6** (Figure 9.1) [8] by, respectively, Mathey and Regitz. Since then, the chemistry of the three-membered phosphacycles has evolved from laboratory curiosities to indispensable tools in organophosphorus chemistry [9]. Consequently, several reviews have covered the synthesis, structure and reactivity of these intriguing species, albeit primarily focused on the ring systems containing mainly P and C [10]. In this chapter, an overview is given of all the reported three-membered rings incorporating one or more phosphorus atoms. Special examples

Phosphorus(III) Ligands in Homogeneous Catalysis: Design and Synthesis, First Edition.
Edited by Paul C. J. Kamer and Piet W. N. M. van Leeuwen.

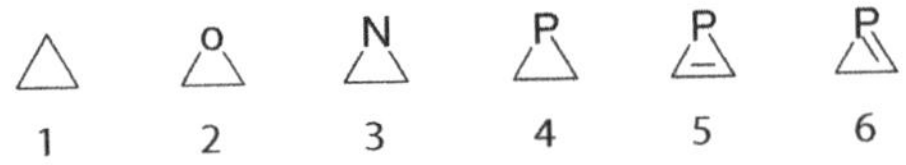

Figure 9.1 Three-membered ring systems.

Figure 9.2 Possible ways of creating a strained P-ring.

of these highly interesting species will be given, like the incorporation of these ring systems into bicyclic frameworks. The applicability of strained P-rings will become evident as they give easy access to novel organophosphorus compounds that were used as phosphine ligands in catalysis.

9.2 Three-membered rings

The inherent strain in three-membered rings gives them their unique electronic and chemical properties due to the small angles and bent bonds [11]. Incorporation of heteroatoms and/or multiple bonds into the framework influences its strain energy and thus destabilizing factors, like angle strain and π-electron–lone pair repulsion, and stabilizing factors, like σ-aromaticity, delocalization, and rehybridization, can be envisioned for a variety of hetero-elements [12]. Generally, the hetero-analogues of cyclopropane are less strained. However, the heteroatom–carbon bonds are much weaker than normal C–C bonds, which makes these ring systems more prone to rearrangements and therefore even more special from a synthetic point of view.

Various synthetic routes are available for the preparation of the phosphiranes **4** and phosphirenes **5** and **6**, the most convenient approach being the [1 + 2]-cycloaddition of reactive intermediates to olefins or acetylenes [10]. Consequently, the use of free or metal-complexed phosphinidenes (R–P) [13] is the most obvious and most developed route to obtain these desired phosphacycles, but the addition of carbenes, nitrenes, and silylenes [14] to phospha-alkenes and -alkynes is also established. Additionally, the three-membered phosphiranes **4** have also been prepared by salt elimination reactions from P and X–Y units and cyclization of P–X–Y and X–P–Y units [10] (Figure 9.2).

To give a detailed overview, all the currently known P-containing three-membered rings are represented by their parent structures in Figure 9.3. These species are structurally characterized as stable, free or metal-complexed compounds, except for 3*H*-diphosphirene **10**, which has been postulated as a transient intermediate [18].

Some specific examples of phosphirane **4** are the polycyclic dibenzophosphasemibullvalene **24** reported by Grützmacher [32] and the highly strained, thermally stable, branched phospha[7]triangulane **25** by Lammertsma [33] (Figure 9.4), both synthesized by adding low-valent phosphinidenes to the corresponding olefins.

The reverse reaction is also well established, and dissociation of phosphiranes and phosphirenes into alkenes or alkynes and low-coordinate phosphorus units is well documented for tervalent phosphiranes [34], phosphirane oxides [35], phosphiranium salts [36], and phosphirane [37] and phosphirene [23] complexes.

Although the three-membered phosphorus-containing heterocycles have been extensively studied, their cationic counterparts have not, in contrast to their corresponding all-carbon aromatic cyclopropenium salts **26** [38] (Figure 9.5). Depicted in Figure 9.5 are the two π-electron Hückel-type aromatic phosphirenylium cation **27** [39] and the σ-aromatic phosphirenium salt **28** [40], which both have been structurally characterized. The only isolable diheteroatom analogues are the recently reported diphosphirenylium **29** and diphosphirenium salt **30** [17c,41].

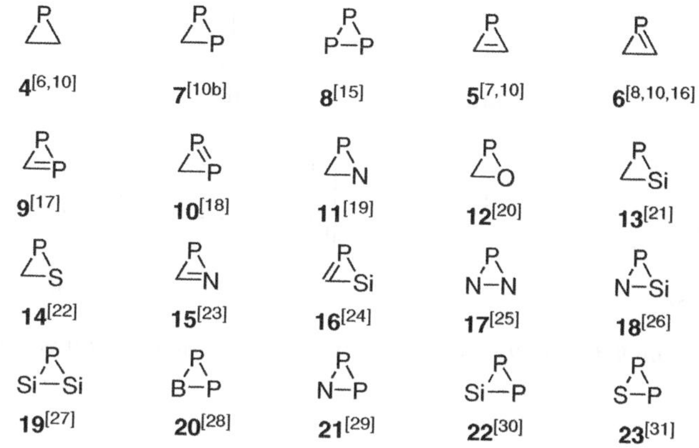

Figure 9.3 *Parent ring structures.*

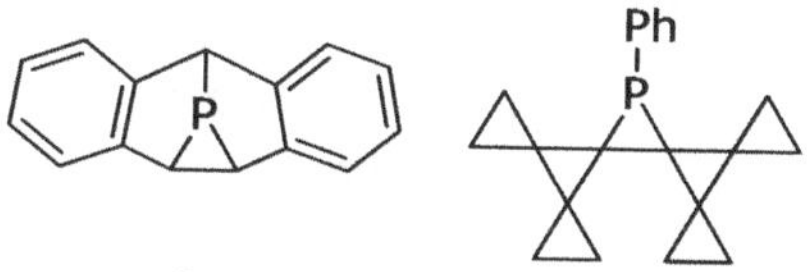

Figure 9.4 *Polycyclic dibenzophosphasemibullvalene (**24**) and branched phospha[7]triangulane (**25**).*

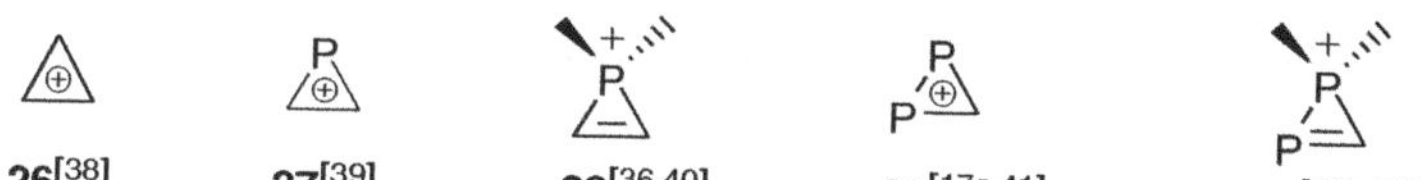

Figure 9.5 *Examples **26–30**.*

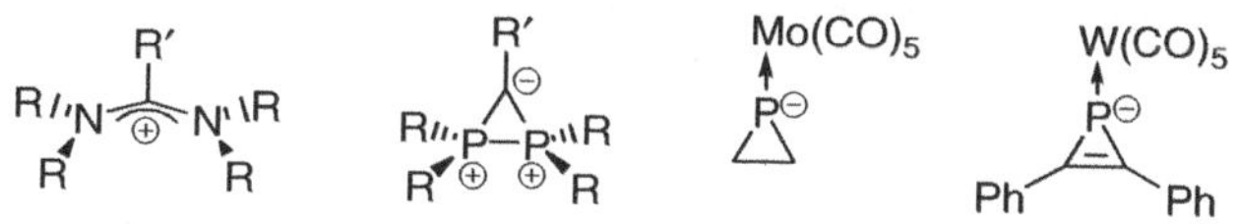

Figure 9.6 *Examples **31–34**.*

In addition, an exotic example is cation **32** that represents the diphosphorus analogue of amidinium salt **31**. **32** has been synthesized and structurally characterized as a cyclic carbanion in sharp contrast to the open planar form that is the most stable for **31** (Figure 9.6)[42].

The chemistry of the anionic three-membered rings is even more limited, and only complexed phosphiranide **33** [43] has been reported together with the corresponding unsaturated, anti-aromatic phosphirenide anion **34** (Figure 9.6) that can be generated *in situ* under phase transfer conditions [44]. Without the stabilizing effect of the transition metal, phosphiranide and phosphirenide anions appear to be very unstable and remain so far elusive.

Stable P-ring radicals are also not known. Whereas the all-carbon cyclopropenium salts **35** can be easily converted into their corresponding cyclopropenyl radicals **36** by reduction [45], diphosphorus analogue **30**

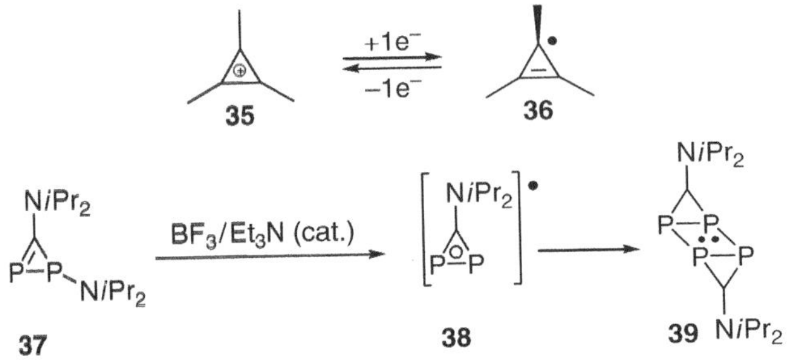

Scheme 9.1 Synthesis of phospha-benzene valence isomer **39**.

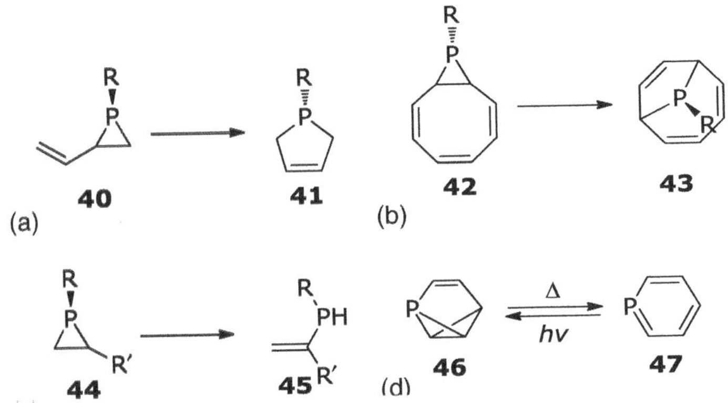

Scheme 9.2 Rearrangements involving phosphiranes.

undergoes a ring-opening reaction to yield a stable 1,3-diphosphaallyl radical [46]. However, upon cleavage of the P–N bond in 1,2-diphosphirene **37**, transient diphosphorenyl radical **38** is produced that dimerizes spontaneously to give singlet biradical **39** (Scheme 9.1). **39** was described as a tetraphosphorus valence isomer of benzene with, formally, one-electron phosphorus–phosphorus bonds, which result from the π^*–π^* interaction between the two diphosphirenyl radicals [47].

9.3 Rearrangements

The phosphorus-containing three-membered rings are more prone to rearrangements than their cyclopropane counterparts due to the weaker heteroatom–carbon bond. Consequently, a plethora of sigmatropic and electrocyclic rearrangements, involving especially phosphiranes **4**, have been observed [48].

Among the pericyclic rearrangements of the small phosphacycles, the conversion of 2-vinylphosphirane **40** into 3-phospholene **41**, which follows a suprafacial [1,3]-sigmatropic shift, has been studied in depth (Scheme 9.2) [49]. This hetero-analogue of the well-known vinylcyclopropane-cyclopentene rearrangement [50] proceeds with diradical character for the metal-complexed phosphiranes [49 g] in full analogy to the all-carbon case. In addition, the concerted antarafacial [1,5]-sigmatropic shift, which is also thermally allowed, has been established for phosphiranes like **42** (Scheme 9.2) [51].

Phosphiranes can also undergo a ring chain rearrangement to vinylphosphine **45** where simultaneously a ring opening and a [1,3]-hydrogen shift take place. [52] Phosphiranes incorporated into a bicyclo[1.1.0]butane framework, like **46**, give via a thermal ring-opening reaction the more stable phosphabenzene **47** [53], in analogy to the valence isomerization of phosphabicyclo[1.1.0]butanes. [54]

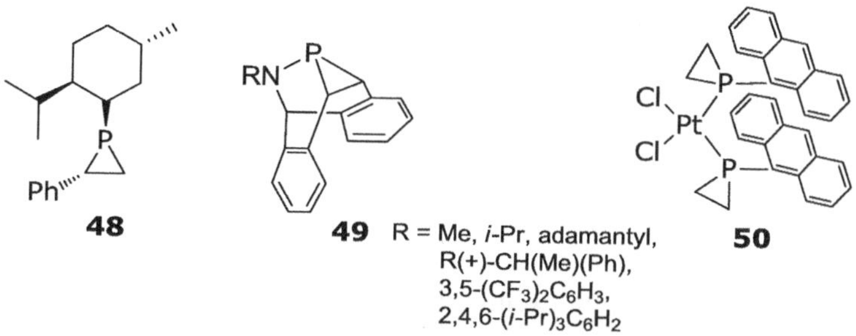

Figure 9.7 Examples **48–50**.

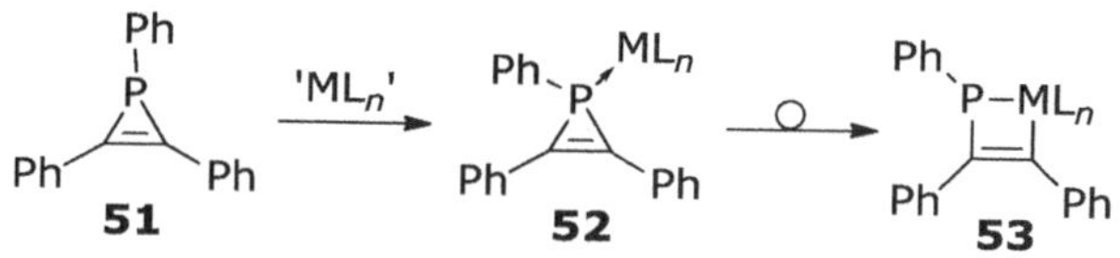

Scheme 9.3 Phosphirenes as ligands.

9.4 Homogeneous catalysis

Strained phosphorus heterocycles are not generally used as phosphine ligands in catalysis, but the few that are show unprecedented behavior. Optically active phosphiranes have been obtained from their $Mo(CO)_5$ complexes by a stereoselective decomplexation reaction with $(Ph_2PCH_2)_2$ (dppe) and used as ligands for cationic rhodium(I) complexes in the catalytic hydrogenation of olefins. Significant optical yields are obtained with P-menthylphosphirane **48** (Figure 9.7), but its use was found to be limited because of ligand decomposition [55].

Decomposition routes for the three-membered phosphacycles are generally thermal degradation, and the monocyclic phosphiranes tend to form metallaphosphetanes as the more stable products via insertion of an electron-rich metal center into one of the P–C bonds. [56] BABAR-Phos **49** (Figure 9.7) [57], on the other hand, in which the PCC ring is incorporated into a polycyclic framework, is much more robust and allows the synthesis of fairly stable Rh(I) and Pt(0) complexes which are active hydrosilylation [58] and recyclable hydroboration catalysts [59]. The metallaphosphetane formation with BABAR-Phos was found to be reversible and controllable by the co-ligands.

In a different vein, phosphiranes are also used as ligands for *cis*-$PtCl_2$ complexes, as in **50** (Figure 9.7), which possibly has cytotoxic properties and consequently can be used as an anticancer drug [60]. Interestingly, the anthracene moieties in **50** reveal π-stacking interactions in the solid state that is an important binding mode for DNA intercalation.

The parent phosphiranes **4** are weak σ-donors due to the high s-character of the lone pair on phosphorus, and as a consequence they give, for example, no stable BH_3-adducts [61]. On the other hand, phosphiranes may act as better π-acceptors as compared to other phosphines, because of the strong pyramidalization of the P-coordination sphere [62]. In contrary, the more electron-rich BABAR-Phos **49** is a relatively good electron donor, while its electron acceptor properties are not significantly different from those of other phosphines [63].

The unsaturated phosphirenes are less suited as ligands in catalysis, since complexation of 1*H*-phosphirenes, like **51**, gives phospha-metallocyclobutene **53** after rearrangement of the initially formed complex **52** (Scheme 9.3). Both metal compounds have been structurally characterized [64].

Bidentate ligands solely based on phosphiranes or phosphirenes have not been applied in catalysis yet, although these species have been reported as $W(CO)_4$-complexes, generated from transient phosphinidenes [65]. In addition, mixed phosphirane–phosphine complexes are also accessible as their $Mo(CO)_4$-adducts [66].

9.5 Conclusions

The highly strained organophosphorus compounds are fascinating species; some are highly stable, while others function as reactive intermediates with exciting chemistry, generating an easy access to novel organophosphorus species that can serve as phosphorus(III) ligands in catalysis.

9.6 Experimental procedures

9.6.1 Synthesis of BABAR-Phos 49a (R = *i*-Pr) [57a]

For Figure 9.8:

55 [57c]: Titanium chloride (1.6 mL) was slowly added to a solution of 5*H*-dibenzo[a,d]cyclohepten-5-one (**54**; 2.0 g, 9.7 mmol) in toluene (100 mL), which resulted in a dark red suspension. Subsequently, the reaction mixture was cooled to –10°C and *iso*-propylamine (5.0 g, 48.5 mmol) was added in one portion. The resulting orange mixture was allowed to warm to room temperature and was stirred overnight. Then, the reaction mixture was poured over an ice–water mixture and the light yellow organic layer was separated. The milky white aqueous layer was extracted with toluene (3 × 100 mL). The combined organic layers were dried over $MgSO_4$ and concentrated, yielding the imine analytically pure as a pale yellow, viscous oil in 97% yield (2.3 g). Recrystallization from hexane provided colorless crystals. ^{1}H-NMR ($CDCl_3$): δ 0.95 (d, $^3J_{H,H}$ = 6.2 Hz, 3 H; CH(CH_3)$_2$), 1.40 (d, $^3J_{H,H}$ = 6.2 Hz, 3 H; CH(CH_3)$_2$), 3.82 (sept, $^3J_{H,H}$ = 6.2 Hz, 1 H; *C*H(CH_3)$_2$), 6.97 (m, 2 H; *H*C=C*H*), and 7.20–7.60 (m, 8 H; Ar*H*).

56a [57c]: *N-iso*-propyl-5*H*-dibenzo[a,d]cyclohepten-5-imine (**55**; 7.0 g, 28.3 mmol) was dissolved in MeOH (250 mL) and treated five times with portions of $NaBH_4$ (0.53 g, 14.2 mmol each) at an interval of 1.5 hours. The reaction was stirred at room temperature for 20 hours and concentrated under reduced pressure, and the residue was partitioned between 130 mL of 20% Na_2CO_3 (aq) and 130 mL diethyl ether. After separation, the aqueous layer was extracted with 4 × 100 mL diethyl ether. The combined organic layers were dried over $MgSO_4$ and concentrated under reduced pressure, yielding **56a** (6.74 g, 96%; mixture of *exo*- and *endo*-isomers) as a pale yellow viscous oil. Recrystallization from hexane provided colorless crystals. ^{1}H-NMR ($CDCl_3$): δ 0.92 and 1.13 (d, $^3J_{H,H}$ = 6.2 Hz, 6 H; CH(CH_3)$_2$), 1.93 (br. s, 2× 1 H; N*H*), 2.24 and 2.91 (sept, $^3J_{H,H}$ = 6.2 Hz, 1 H; C*H*(CH_3)$_2$), 4.27 and 4.94 (s, 2× 1 H; NC*H*), 7.04 and 7.16 (s, 2 H; *H*C=C*H*), and 7.25–7.45 (m, 2× 8 H; Ar*H*).

57a [57a,c]: *n*-BuLi (9.8 mL of a 1.6 M sol. in hexanes, 15.7 mmol) was added to a solution of amine **56a** (4.0 g, 15.7 mmol; mixture of *exo*- and *endo*-isomers) in diethyl ether (100 mL) at –78°C. The reaction mixture was allowed to warm to room temperature, then added dropwise to a solution of PCl_3 (3.25 g, 23.7 mmol) in diethyl ether (20 mL) at –78°C. The obtained suspension was allowed to warm to room temperature, then all volatiles were removed in vacuo; the residue was suspended in toluene (100 mL), then filtered over Celite to remove insoluble LiCl. Subsequently the solvent was removed in vacuo to give aminodichlorophosphine **57a** (5.3 g, 96%) as a pale yellow viscous oil. Recrystallization from toluene and hexane provided colorless crystals. ^{31}P-NMR ($CDCl_3$): δ 168.7. ^{1}H-NMR ($CDCl_3$): δ 0.89 (d, $^3J_{H,H}$ = 6.9 Hz, 6 H; CH(CH_3)$_2$), 4.30 (dsept, $^3J_{H,H}$ = 6.9 Hz, $^3J_{H,P}$ = 4 Hz, 1 H; C*H*(CH_3)$_2$), 5.73 (d, $^3J_{H,P}$ = 4 Hz, 1 H; NC*H*), 7.12 (s, 2 H; *H*C=C*H*), and 7.30–7.52 (m, 8 H; Ar*H*).

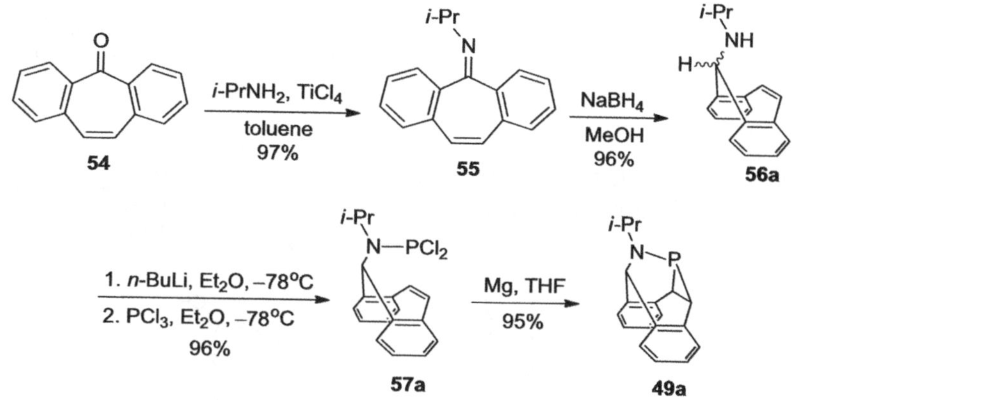

Figure 9.8 *Synthesis of BABAR-Phos* **49a** *(R=i-Pr).*

49a [57a,c]: A solution of dichlorophosphine **57a** (7.0 g, 20.0 mmol) in THF (150 mL) was stirred in the presence of Mg turnings (0.53 g, 22.0 mmol) at room temperature. After 12 h, dry dioxane (3 mL) was added to precipitate $MgCl_2$. The solvents were evaporated; the residue was dissolved in toluene (100 mL) and filtered over celite. The toluene phase was concentrated to 10% of its volume, and *n*-hexane was added to precipitate spectroscopically pure **49a** as a slightly yellow powder (95%), which was recrystallized from Et_2O–hexane. ^{31}P-NMR (CD_3CN): δ –153.5. ^{1}H-NMR (CD_3CN): δ 1.12 (d, $^3J_{H,H}$=6.6 Hz, 6 H; CH(CH_3)$_2$), 2.66 (d, $^2J_{H,P}$=20 Hz, 2 H; *HCCH*), 3.46 (dsept, $^3J_{H,H}$=6.6 Hz, $^3J_{H,P}$=27.3 Hz, 1 H; *CH*(CH_3)$_2$), 4.90 (d, $^3J_{H,P}$=4.9 Hz, 1 H; NC*H*), and 7.14–7.26 (m, 8 H; Ar*H*).

9.6.2 Synthesis of BABAR-Phos 49b (R=3,5-(CF_3)$_2$$C_6H_3$) [57a]

For Figure 9.9:

58: Step 1: A solution of NaOH (0.26 g, 6.5 mmol) and $NaBH_4$ (2.52 g, 67 mmol) in water (50 mL) was added to a solution of 5*H*-dibenzo[a,d]cyclohepten-5-one (**54**; 25.0 g, 120 mmol) in MeOH (500 mL). After stirring for 15 hours, the reaction mixture was concentrated, and the obtained white solid was dissolved in diethyl ether and washed with water. The combined organic layers were dried over $MgSO_4$, concentrated, and dried in vacuo, giving 23.05 g of the corresponding alcohol as a white solid. This material was used without further purification in the following step.

Step 2: Thionylchloride was slowly added to a cooled solution of the crude alcohol in DCM (600 mL) at 0°C, which resulted in a pink solution that was stirred overnight. After removal of all volatiles, a red solid was obtained which was stirred (20 min) in a mixture of toluene and hexane (300 mL; 1:5) with a few drops of triethylamine, yielding a white suspension. This mixture was heated, and the warm solution was filtered and slowly cooled to allow crystallization, which afforded **58** as colorless crystals in 64% yield (17.22 g). ^{1}H-NMR ($CDCl_3$): δ 6.27 (br.s, 1 H; ClC*H*), 7.16 (s, 2 H; *H*C=C*H*), and 7.38–7.47 (m, 8 H; Ar*H*).

56b [57c]: 3,5-Bistrifluoromethylaniline (5.1 g, 22.1 mmol) was added to a solution of 5-chloro-5-*H*-dibenzo[a,d]cycloheptene (**58**; 5.0 g, 22.1 mmol) in toluene (100 mL) and stirred for 3 hours at room temperature. 100 mL of a saturated Na_2CO_3 solution (aq) was added, and the mixture was stirred for another hour. All organic volatiles were removed in vacuo, and the remaining water phase was extracted with toluene (3×100 mL). The combined organic layers were dried over Na_2SO_4 and concentrated, yielding the product as an oil in >97% purity, which was crystallized from *n*-hexane affording amine **56b** as colorless

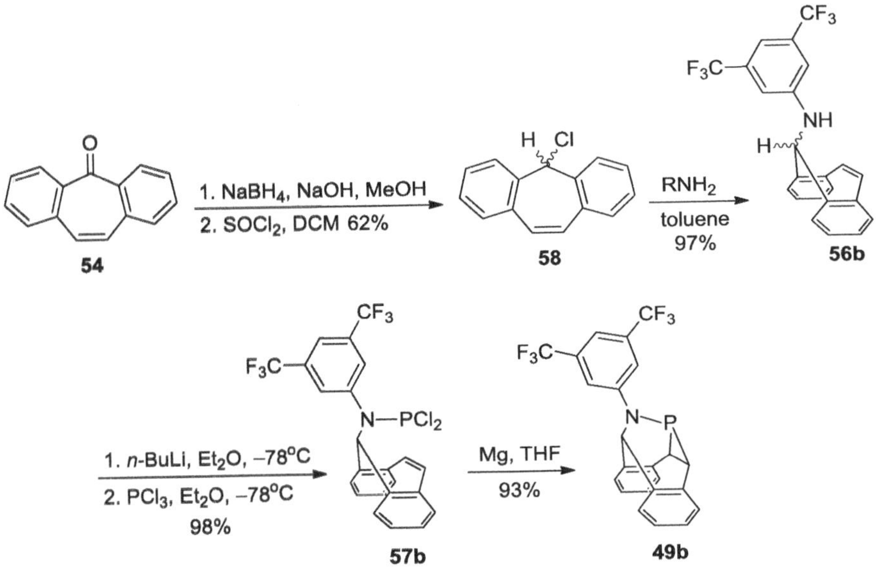

Figure 9.9 Synthesis of BABAR-Phos **49b** (R = i-Pr).

crystals in 97% yield (8.9 g). ^{1}H-NMR ($CDCl_3$): δ 4.88 (br.s, 1 H; N*H*), 5.75 (br.s, 1 H; NC*H*), 7.04 (s, 2 H; *H*C=C*H*), and 7.2–7.7 (m, 11 H; Ar*H*).

57b [57a,c]: *n*-BuLi (7.5 mL of a 1.6 M sol. in hexanes, 12.0 mmol) was added to a solution of amine **56b** (5.0 g, 12.0 mmol) in diethyl ether (100 mL) at −78°C. The reaction mixture was allowed to warm to room temperature, then added dropwise to a solution of PCl_3 (4.9 g, 36.0 mmol) in diethyl ether (20 mL) at −78°C. The obtained suspension was allowed to warm to room temperature, then all volatiles were removed in vacuo; the residue was suspended in toluene (100 mL), then filtered over Celite to remove insoluble LiCl. Subsequently the solvent was concentrated to ~10 mL to allow crystallization at 4°C, which afforded after washing (10 mL *n*-hexane) and drying aminodichlorophosphine **57b** as colorless crystals (6.1 g, 98%). ^{31}P-NMR ($CDCl_3$): δ 154.7. ^{1}H-NMR ($CDCl_3$): δ 5.97 (d, $^3J_{H,P}$ = 4.5 Hz, 1 H; NC*H*), 7.21 (s, 2 H; *H*C=C*H*), 7.37–7.56 (m, 10 H; Ar*H*), and 7.75 (br.s, 1 H; *p*-3,5-$(CF_3)_2$Ar*H*).

49b [57a,c]: A solution of dichlorophosphine **57b** (5.35 g, 10.28 mmol) in THF (100 mL) was stirred in the presence of Mg turnings (0.27 g, 33.93 mmol) at room temperature. After 5 hours, dry dioxane (3 mL) was added to precipitate $MgCl_2$. The solvents were evaporated; the residue was dissolved in toluene (100 mL) and filtered over celite. The toluene phase was concentrated to 10% of its volume, and *n*-hexane was added to precipitate spectroscopically pure **49b** as a slightly yellow powder (4.3 g, 93%). ^{31}P-NMR ($CDCl_3$): δ −146.0. ^{1}H-NMR ($CDCl_3$): δ 3.12 (d, $^2J_{H,P}$ = 22 Hz, 2 H; *H*CC*H*), 5.63 (d, $^3J_{H,P}$ = 4.9 Hz, 1 H; NC*H*), 7.31–7.48 (m, 9 H; Ar*H*), and 7.72 (s, 2 H; Ar*H*).

References

[1] (a) P. D. Boyer (Nobel lecture) *Angew. Chem.* **1998**, *110*, 2424–2436; *Angew. Chem. Int. Ed.* **1998**, 37, 2296–2307; and (b) J. E. Walker (Nobel lecture) *Angew. Chem.* **1998**, *110*, 2437–2450; *Angew. Chem. Int. Ed.* **1998**, 37, 2308–2319.

[2] G. Wittig (Nobel lecture) in *Nobel Lectures, Chemistry 1971–1980* (Ed. T. Frängsmyr), World Scientific, Singapore, **1993**, 368–376.

[3] (a) W. S. Knowles (Nobel lecture) *Angew. Chem.* **2002**, *114*, 2096–2107; *Angew. Chem. Int. Ed.* **2002**, 41, 1998–2007; and (b) R. Noyori (Nobel lecture) *Angew. Chem.* **2002**, *114*, 2108–2123; *Angew. Chem. Int. Ed.* **2002**, 41, 2008–2022.

[4] (a) A. G. MacDiarmid (Nobel lecture) *Angew. Chem.* **2001**, *113*, 2649–2659; *Angew. Chem. Int. Ed.* **2001**, 40, 2581–2590; and (b) A. J. Heeger (Nobel lecture) *Angew. Chem.* **2001**, *113*, 2660–2682; *Angew. Chem. Int. Ed.* **2001**, 40, 2591–2611.

[5] F. Mathey (Ed.), *Phosphorus-Carbon Heterocyclic Chemistry: The Rise of a New Domain*, Pergamon, **2001**.

[6] (a) R. I. Wagner, US Patents 3086053 and 3086056, **1963**; *Chem. Abstr.* **1963**, *59*, 10124; Chem. Abstr. **1964**, 60, 559; and (b)R. I. Wagner, L. V. D. Freeman, H. Goldwhite, D. G. Rowsell, *J. Am. Chem. Soc.* **1967**, *89*, 1102–1104.

[7] (a)A. Marinetti, F. Mathey, J. Fischer, A. Mitschler, J. Am. *Chem. Soc.* **1982**, *104*, 4484–4485; and (b) A. Marinetti, F. Mathey, J. Fischer, A. Mitschler, *J. Chem. Soc., Chem. Commun.* **1984**, 45–46.

[8] O. Wagner, G. Maas, M. Regitz, *Angew. Chem.* **1987**, *99*, 1328–1330; and *Angew. Chem. Int. Ed. Engl.* **1987**, 26, 1257–1259.

[9] F. Mathey, *Angew. Chem.* **2003**, *115*, 1616–1643; and *Angew. Chem. Int. Ed.* **2003**, 42, 1578–1604.

[10] (a) F. Mathey, *Pure Appl. Chem.* **1987**, *59*, 993–998; (b) F. Mathey, *Chem. Rev.* **1990**, *90*, 997–1025; (c) F. Mathey, M. Regitz in *Comprehensive Heterocyclic Chemistry II*, Vol. *1A* (Ed. A. R. Katritzky, C. W. Rees, E. F. V. Scriven), Pergamon, **1997**, 277–304; and (d) M. Regitz in *Phosphorus-Carbon Heterocyclic Chemistry: The Rise of a New Domain* (Ed. F. Mathey), Pergamon, **2001**, 17–55.

[11] (a) K. B. Wiberg in *Reactive Intermediate Chemistry* (Ed. R. A. Moss, M. S. Platz, M. Jones, Jr..), John Wiley & Sons, Inc., **2004**; 717–740; and (b) K. B. Wiberg, *Angew. Chem.* **1986**, *98*, 312–322; *Angew. Chem. Int. Ed.* **1986**, 25, 312–322.

[12] T. P. M. Goumans, Z. Benkõ, L. Nyulászi, A. W. Ehlers, K. Lammertsma, manuscript in preparation.

[13] (a) J. C. Slootweg, K. Lammertsma, in *Science of Synthesis* (Eds. B. M. Trost, F. Mathey), Vol. 42, Georg Thieme Verlag, **2009**, 15–36; (b) K. Lammertsma, *Top. Curr. Chem.* **2003**, *229*, 95–119; (c) K. Lammertsma, M. J. M. Vlaar, *Eur. J. Org. Chem.* **2002**, 1127–1138; (d) F. Mathey, N. H. Tran Huy, A. Marinetti, *Helv. Chim. Acta* **2001**, *84*, 2938–2957.

[14] Carbenes: M. Jones, Jr., R. A. Moss (Ch. 7), and G. Bertrand (Ch. 8); nitrenes: M. S. Platz (Ch. 11); and silylenes: N. Tokitoh, W. Ando (Ch. 14) in *Reactive Intermediate Chemistry* (Eds. R. A. Moss, M. S. Platz, M. Jones, Jr.), John Wiley & Sons, Inc., **2004**.

[15] (a) M. Baudler, K. Glinka, *Chem. Rev.* **1993**, *93*, 1623–1667; (b) A. –M. Caminade, J. –P. Majoral, R. Mathieu, *Chem. Rev.* **1991**, *91*, 575–612.

[16] M. Sanchez, R. Réau, C. J. Marsden, M. Regitz, G. Bertrand, *Chem. Eur. J.* **1999**, *5*, 274–279.

[17] (a) E. Niecke, R. Streubel, M. Nieger, D. Stalke, *Angew. Chem.* **1989**, *101*, 1708–1710; *Angew. Chem., Int. Ed. Engl.* **1989**, *28*, 1673–1674; (b) R. Streubel, L. Ernst, J. Jeske, P. G. Jones, *J. Chem. Soc., Chem. Commun.* **1995**, 2113–2114.; (c) D. Bourissou, G. Bertrand, *Top. Curr. Chem.* **2002**, *220*, 1–25.

[18] The transient formation of a 3*H*-diphosphirene has been postulated: E. Niecke, R. Steubel, M. Nieger, *Angew. Chem.* **1991**, *103*, 103–105; *Angew. Chem., Int. Ed. Engl.* **1991**, *30*, 90–91.

[19] R. Streubel, A. Ostrowski, H. Wilkens, F. Ruthe, J. Jeske, P. G. Jones, *Angew. Chem.* **1997**, *109*, 409–413; *Angew. Chem. Int. Ed. Engl.* **1997**, *36*, 378–382.

[20] (a) S. Bauer, A. Marinetti, L. Ricard, F. Mathey, *Angew. Chem.* **1990**, *102*, 1188–1189; *Angew. Chem. Int. Ed. Engl.* **1990**, *29*, 1166–1167; (b) F. Mathey, A. Marinetti, S. Bauer, P. le Floch, *Pure Appl. Chem.* **1991**, *63*, 855–858; (c) R. Streubel, A. Kusenberg, J. Jeske, P. G. Jones, *Angew. Chem.* **1994**, *106*, 2564–2566; *Angew. Chem. Int. Ed. Engl.* **1994**, *33*, 2427–2429.

[21] M. J. M. Vlaar, A. W. Ehlers, F. J. J. de Kanter, M. Schakel, A. L. Spek, M. Lutz, N. Sigal, Y. Apeloig, K. Lammertsma, *Angew. Chem.* **2000**, *112*, 4296–4299; *Angew. Chem. Int. Ed.* **2000**, *39*, 4127–4130.

[22] (a) M. Caira, R. H. Neilson, W. H. Watson, P. Wisian-Neilson, Z. –M. Xie, *Chem. Commun.* **1984**, 698–699; (b) G. Märkl, W. Hölzl, *Tetrahedron Lett.* **1989**, *30*, 4501–4504; (c) K. Toyota, K. Shimura, H. Takahashi, M. Yoshifuji, *Chem. Lett.* **1994**, 1927–1930; (d) R. Streubel, C. Neumann, *Chem. Commun.* **1999**, 499–500; (e) a thiaphosphasilirane and a dithiaphosphirane are also reported: I. van den Winkel, H. M. M. Bastiaans, F. Bickelhaupt, *Phosphorus, Sulfur Silicon Relat. Elem.* **1990**, *49/50*, 333–336.

[23] (a) R. Streubel, *Coord. Chem. Rev.* **2002**, *227*, 175–192; (b) R. Streubel, *Top. Curr. Chem.* **2003**, *223*, 91–109.

[24] A. Schäfer, M. Weidenbruch, W. Saak, S. Pohl, *Angew. Chem.* **1987**, *99*, 806–807; *Angew. Chem., Int. Ed. Engl.* **1987**, *26*, 776–777.

[25] (a) E. Niecke, K. Schwichtenhovel, H. –G. Schäfer, B. Krebs, *Angew. Chem.* **1981**, *93*, 1033–1034; *Angew. Chem., Int. Ed. Engl.* **1981**, *20*, 963–964. (b) E. Niecke, M. Link, M. Nieger, *Chem. Ber.* **1992**, *125*, 93–96.

[26] (a) M. Driess, H. Pritzkow, *Angew. Chem.* **1992**, *104*, 775–777; *Angew. Chem., Int. Ed. Engl.* **1992**, *31*, 751–753; (b) M. Driess, H. Pritzkow, *Phosphorus, Sulfur Silicon Relat. Elem.* **1993**, *76*, 57–60.

[27] M. Driess, S. Rell, H. Pritzkow, R. Janoschek, *Angew. Chem.* **1997**, *109*, 1384–1387; *Angew. Chem. Int. Ed. Engl.* **1997**, *36*, 1326–1329.

[28] (a) M. Baudler, C. Gruner, G. Fürstenberg, B. Kloth, F. Saykowski, V. Özer, *Z. Anorg. Allg. Chem.* **1978**, *446*, 169–176; (b) W. Köstler, G. Linti, *Eur. J. Inorg. Chem.* **2001**, 1841–1846.

[29] (a) E. Niecke, A. Nickloweit-Lueke, R. Rüger, *Angew. Chem.* **1981**, *93*, 406–407; *Angew. Chem. Int. Ed. Engl.* **1981**, *20*, 385–386; (b) R. Streubel, E. Niecke, P. Paetzold, *Chem. Ber.* **1991**, *124*, 765–767; (c) E. Niecke, B. Kramer, M. Nieger, H. Severin, *Tetrahedron Lett.* **1993**, *34*, 4627–4630; (d) D. Barion, C. Gaertner-Winkhaus, M. Link, M. Nieger, E. Niecke, *Chem. Ber.* **1993**, *126*, 2187–2196.

[30] (a) M. Weidenbruch, M. Herrndorf, A. Schäfer, K. Peters, H. G. von Schnering, *J. Organomet. Chem.* **1985**, *295*, 7–14; (b) M. Driess, H. Pritzkow, S. Rell, R. Janoschek, *Inorg. Chem.* **1997**, *36*, 5212–5217.

[31] (a) M. Baudler, H. Suchomel, G. Fürstenberg, U. Schings, *Angew. Chem.* **1981**, *93*, 1087–1088; *Angew. Chem., Int. Ed. Engl.* **1981**, *20*, 1044–1045; (b) E. Niecke, R. Rüger, *Angew. Chem.* **1983**, *95*, 154–155; *Angew. Chem., Int. Ed. Engl.* **1983**, *22*, 155–156; (c) T. Sasamori, N. Takeda, N. Tokitoh, *J. Phys. Org. Chem.* **2003**, *16*, 450–462; and references therein.

[32] J. Geier, G. Frison, H. Grützmacher, *Angew. Chem.* **2003**, *115*, 4085–4087; and *Angew. Chem., Int. Ed.* **2003**, 42, 3955–3957.

[33] J. C. Slootweg, M. Schakel, F. J. J. de Kanter, A. W. Ehlers, S. I. Kozhushkov, A. de Meijere, M. Lutz, A. L. Spek, K. Lammertsma, *J. Am. Chem. Soc.* **2004**, *126*, 3050–3051.

[34] (a) X. Li, D. Lei, M. Y. Chiang, P. P. Gaspar, *J. Am. Chem. Soc.* **1992**, *114*, 8526–8531; (b) X. Li, S. I. Weissman, T-S. Lin, P. P. Gaspar, A. H. Cowley, A. I. Smirnov, *J. Am. Chem. Soc.* **1994**, *116*, 7899–7900; and (c) G. Bucher, M. L. G. Borst, A. W. Ehlers, K. Lammertsma, S. Ceola, M. Huber, D. Grote, W. Sander, *Angew. Chem.* **2005**, *117*, 3353–3357; *Angew. Chem. Int. Ed.* **2005**, 44, 3289–3293.

[35] (a) H. Quast, M. Heuschmann, *Chem. Ber.* **1982**, *115*, 901–909; and (b) P. P. Gaspar, H. Qian, A. M. Beatty, D. A. d'Avignon, J. L-F. Kao, J. C. Watt, N. P. Rath, *Tetrahedron* **2000**, *56*, 105–119.

[36] (a) D. C. R. Hockless, M. A. McDonald, M. Pabel, S. B. Wild, *J. Chem. Soc. Chem. Commun.* **1995**, 257–258; (b) D. C. R. Hockless, M. A. McDonald, M. Pabel, S. B. Wild, *J. Organomet. Chem.* **1997**, *529*, 189–196; (c) N. E. Brasch, I. G. Hamilton, E. H. Krenske, S. B. Wild, *Organometallics* **2004**, *23*, 299–302;and (d) J. Liedke, S. Loss, C. Widauer, H. Grützmacher, *Tetrahedron*, **2000**, *56*, 143–156.

[37] (a) A. Marinetti, C. Charrier, F. Mathey, J. Fischer, *Organometallics* **1985**, *4*, 2134–2138;(b)F. Mercier, B. Deschamps, F. Mathey, *J. Am. Chem. Soc.* **1989**, *111*, 9098–9100; (c) A. Marinetti, L. Ricard, F. Mathey, *Synthesis* **1992**, 157–162; (d) F. Mercier, L. Ricard, F. Mathey, M. Regitz, *J. Chem. Soc. Chem. Commun.* **1991**, 1305–1307.

[38] (a) R. Breslow, C. Yuan, *J. Am. Chem. Soc.* **1958**, *80*, 5991–5994; and (b) R. Breslow, J. T. Groves, *J. Am. Chem. Soc.* **1970**, *92*, 984–987.

[39] (a) K. K. Laali, B. Geissler, O. Wagner, J. Hoffmann, R. Armbrust, W. Eisfeld, M. Regitz, *J. Am. Chem. Soc.* **1994**, *116*, 9407–9408;(b)calculations:A. Göller, H. Heydt, T. Clark, J. Org. Chem. **1996**, *61*, 5840–5846; (c) A. G. Avent, F. G. N. Cloke, K. R. Flower, P. B. Hitchcock, J. F. Nixon, D. M. Vickers, *Angew. Chem.* **1994**, *106*, 2406–2408; *Angew. Chem., Int. Ed. Engl.* **1994**, 33, 2330–2332; (d) J. Simon, U. Bergsträsser, M. Regitz, K. K. Laali, *Organometallics* **1999**, *18*, 817–819; and (e) X. Liu, D. M. Ivanova, D. Giblin, M. L. Gross, P. P. Gaspar, *Organometallics* **2005**, *24*, 3125–3135; for donor-stabilized phosphiranylium ions, see H. Jansen, F. B. Läng, J. C. Slootweg, A. W. Ehlers, M. Lutz, K. Lammertsma, H. Grützmacher, *Angew. Chem. Int. Ed.* **2010**, 49, 5485–5488.

[40] (a) K. S. Fongers, H. Hogeveen, R. F. Kingma, *Tetrahedron Lett.* **1983**, *24*, 643–646;and (b) J. M. Vural, S. A. Weissman, S. G. Baxter, A. H. Cowley, C. M. Nunn, *J. Chem. Soc., Chem. Commun.* **1988**, 462–463.

[41] D. Bourissou, G. Bertrand, *Acc. Chem. Res.* **1999**, *32*, 561–570.

[42] (a) T. Kato, H. Gornitzka, A. Baceiredo, W. W. Schoeller, G. Bertrand, *Science* **2000**, *289*, 754–756; and (b) T. Kato, H. Gornitzka, A. Baceiredo, W. W. Schoeller, G. Bertrand, *J. Am. Chem. Soc.* **2002**, *124*, 2506–2512.

[43] R. Bausch, E. A. V. Ebsworth, D. W. H. Rankin, *Angew. Chem.* **1971**, *83*, 111; *Angew. Chem., Int. Ed. Engl.* **1971**, 10, 125.

[44] B. Deschamps, F. Mathey, *New. J. Chem.* **1988**, *12*, 755–759.

[45] R. Breslow, M. R. Wasielewski, *J. Am. Chem. Soc.* **1976**, *98*, 4222–4229.

[46] Y. Canac, A. Baceiredo, W. W. Schoeller, D. Gigmes, G. Bertrand, *J. Am. Chem. Soc.* **1997**, *119*, 7579–7580.

[47] Y. Canac, D. Bourissou, A. Baceiredo, H. Gornitzka, W. W. Schoeller, G. Bertrand, *Science* **1998**, *279*, 2080–2082.

[48] R. E. Bulo, Ph.D. thesis, Vrije Universiteit, Amsterdam, The Netherlands, **2004**.

[49] (a)W. J. Richter, *Chem. Ber.* **1983**, *116*, 3293–3300; (b) W. J. Richter, *Chem. Ber.* **1985**, *118*, 1575–1579; (c) K. Lammertsma, J. T. Hung, P. Chand, G. M. Gray, *J. Org. Chem.* **1992**, *57*, 6557–6560; (d) B. Wang, C. H. Lake, K. Lammertsma, *J. Am. Chem. Soc.* **1996**, *118*, 1690–1695; (e) J. B. M. Wit, G. T. van Eijkel, F. J. J. de Kanter, M. Schakel, A. W. Ehlers, M. Lutz, A. L. Spek, K. Lammertsma, *Angew. Chem.* **1999**, *111*, 2716–2719; *Angew. Chem. Int. Ed.* **1999**, 38, 2596–2599;(f)M. J. van Eis, T. Nijbacker, F. J. J. de Kanter, W. H. de Wolf, K. Lammertsma, F. Bickelhaupt, *J. Am. Chem. Soc.* **2000**, *122*, 3033–3036; and (g) R. E. Bulo, A. W. Ehlers, S. Grimme, K. Lammertsma, *J. Am. Chem. Soc.* **2002**, *124*, 13903–13910.

[50] (a) J. E. Baldwin, in *The Chemistry of the Cyclopropyl Group*, Vol. 2 (Ed. Z. Rappoport), Wiley, **1995**; 469–494; (b) J. E. Baldwin, *J. Comput. Chem.* **1998**, *19*, 222–231; and (c) J. E. Baldwin, *Chem. Rev.* **2003**, *103*, 1197–1212.

[51] (a) T. J. Katz, C. R. Nicholson, C. A. Reilly, *J. Am. Chem. Soc.* **1966**, *88*, 3832–3842; (b)E. W. Turnblom, T. J. Katz, *J. Am. Chem. Soc.* **1973**, *95*, 4292–4311; (c) W. J. Richter, *Chem. Ber.* **1985**, *118*, 97–106; and (d) R. E. Bulo, H. Jansen, A. W. Ehlers, F. J. J. de Kanter, M. Schakel, M. Lutz, A. L. Spek, K. Lammertsma, *Angew. Chem.* **2004**, *116*, 732–735; *Angew. Chem. Int. Ed.* **2004**, *43*, 714–717.

[52] (a) S. Haber, P. le Floch, F. Mathey, *J. Chem. Soc., Chem. Commun.* **1992**, 1799–1800; calculations: (b) M. T. Nguyen, L. Landuyt, L. G. Vanquickenborne, *J. Chem. Soc., Faraday Trans.* **1994**, *90*, 1771–1781; (c) N-N. Pham-Tran, H. M. T. Nguyen, T. Veszprémi. M. T. Nguyen, *J. Chem. Soc., Perkin Trans 2* **2001**, 766–773; and (d) J. Mátrai, A. Dransfeld, T. Veszprémi, M. T. Nguyen, *J. Org. Chem.* **2001**, *66*, 5671–5678.

[53] (a) K. Blatter, W. Rösch, U-J. Vogelbacher, J. Fink, M. Regitz, *Angew. Chem.* **1987**, *99*, 67–68; *Angew. Chem., Int. Ed. Engl.* **1987**, 26, 85–86; derivatives: (b) Y. Kobayashi, S. Fujino, H. Hamana, Y. Hanzawa, S. Morita, I. Kumadaki, *J. Org. Chem.* **1980**, *45*, 4683–4685; and (c) M. Regitz, U. Bergsträsser, *Polish. J. Chem.* **1999**, *73*, 135–150.

[54] (a) E. Niecke, A. Fuchs, M. Nieger, *Angew. Chem.* **1999**, *111*, 3213–3216; *Angew. Chem. Int. Ed.* **1999**, 38, 3028–3031; (b) E. Niecke, A. Fuchs, F. Baumeister, M. Nieger, W. W. Schoeller, *Angew. Chem.* **1995**, *107*, 640–642; *Angew. Chem., Int. Ed. Engl.* **1995**, 34, 555–557; (c) O. Schmidt, A. Fuchs, D. Gudat, M. Nieger, W. Hoffbauer, E. Niecke, W. W. Schoeller, *Angew. Chem.* **1998**, *110*, 995–998; *Angew. Chem., Int. Ed. Engl.* **1998**, 37, 949–952; calculations: (d) W. W. Schoeller, C. Begemann, E. Niecke, D. Gudat, *J. Phys. Chem. A* **2001**, *105*, 10731–10738; 1,3-diphosphabicyclo[1.1.0]butanes: (e) P. Binger, B. Biedenbach, C. Krüger, M. Regitz, *Angew. Chem.* **1987**, *99*, 798–799; *Angew. Chem., Int. Ed. Engl.* **1987**, 26, 764–765; (f) E. Niecke, O. Altmeyer, M. Nieger, *J. Chem. Soc. Chem. Commun.* **1988**, 945–946; (g) E. Niecke, H. J. Metternich, R. Streubel, *Chem. Ber.* **1990**, *123*, 67–69; and 2-phospha-bicyclo[1.1.0]butanes: (h) J. C. Slootweg, S. Krill, F. J. J. de Kanter, M. Schakel, A. W. Ehlers, M. Lutz, A. L. Spek, K. Lammertsma, *Angew. Chem.* **2005**, *117*, 6737–6740; *Angew. Chem. Int. Ed.* **2005**, 44, 6579–6582.

[55] A. Marinetti, F. Mathey, L. Ricard, *Organometallics* **1993**, *12*, 1207–1212.

[56] (a) D. Carmichael, P. B. Hitchcock, J. F. Nixon, F. Mathey, L. Ricard, *J. Chem. Soc., Chem. Commun.* **1989**, 1389–1390; (b) D. Carmichael, P. B. Hitchcock, J. F. Nixon, F. Mathey, L. Ricard, *J. Chem. Soc., Dalton Trans.* **1993**, 1811–1822.

[57] (a) J. Liedke, S. Loss, G. Alcaraz, V. Gramlich, H. Grützmacher, *Angew. Chem.* **1999**, *111*, 1724–1727; *Angew. Chem. Int. Ed.* **1999**, 38, 1623–1626; (b) F. B. Läng, H. Grützmacher, *Chimia* **2003**, *57*, 187–190; (c) J. Liedtke, dissertation, ETH Zürich, Nr. 13688, **2000**; and (d) H. Jansen, F. B. Läng, J. C. Slootweg, A. W. Ehlers, M. Lutz, K. Lammertsma, H. Grützmacher, *Angew. Chem. Int. Ed.* **2010**, *49*, 5485–5488.

[58] J. Liedke, S. Loss, C. Widauer, H. Grützmacher, *Tetrahedron*, **2000**, *56*, 143–156.

[59] J. Liedke, H. Rüegger, S. Loss, H. Grützmacher, *Angew. Chem.* **2000**, *112*, 2596–2599; and *Angew. Chem. Int. Ed.* **2000**, 39, 2479–2481.

[60] (a) F. Yang, P. E. Fanwick, C. P. Kubiak, *Organometallics* **1999**, *18*, 4222–4225; and (b) F. Yang, P. E. Fanwick, C. P. Kubiak, *Inorg. Chem.* **2002**, *41*, 4805–4809.

[61] (a) N. Mézailles, P. E. Fanwick, C. P. Kubiak, *Organometallics* **1997**, *16*, 1526–1530; and (b) T. I. Sølling, S. B. Wild, L. Radom, *J. Organomet. Chem.* **1999**, *580*, 320–327.

[62] M. T. Nguyen, E. Van Praet, L. G. Vanquickenborne, *Inorg. Chem.* **1994**, *33*, 1153–1157.

[63] C. Laporte, G. Frison, H. Grützmacher, A. C. Hillier, W. Sommer, S. P. Nolan, *Organometallics* **2003**, *22*, 2202–2208.

[64] (a) D. Carmichael, P. B. Hitchcock, J. F. Nixon, F. Mathey, A. Pidcock, *J. Chem. Soc., Chem. Commun.* **1986**, 762–763; (b) S. S. Al-Juaid, D. Carmichael, P. B. Hitchcock, S. Lochschmidt, A. Marinetti, F. Mathey, J. F. Nixon, *J. Chem. Soc., Chem. Commun.* **1988**, 1156–1158; (c) S. S. Al-Juaid, D. Carmichael, P. B. Hitchcock, A. Marinetti, F. Mathey, J. F. Nixon, *J. Chem. Soc., Dalton Trans.* **1991**, 905–915; (d) F. A. Ajulu, D. Carmichael, P. B. Hitchcock, F. Mathey, M. F. Meidine, J. F. Nixon, L. Ricard, M. L. Riley, *J. Chem. Soc., Chem. Commun.* **1992**, 750–751.

[65] (a) C. Carrier, N. Maigrot, F. Mathey, *Organometallics* **1987**, *6*, 586–591; monodentate bisphosphiranes: (b) M. J. M. Vlaar, A. W. Ehlers, F. J. J. de Kanter, M. Schakel, A. L. Spek, K. Lammertsma, *Angew. Chem.* **2000**, *112*, 3071–3074; *Angew. Chem. Int. Ed.* **2000**, 39, 2943–2945; monodentate bisphosphirenes: (c) N. H. Tran Huy, L. Ricard, F. Mathey, *J. Chem. Soc., Dalton Trans.* **1999**, 2409–2410; and (d) N. H. Tran Huy, L. Ricard, F. Mathey, J. Organomet. *Chem.* **1999**, *582*, 53–57.

[66] (a) M. J. M. Vlaar, S. G. A. van Assema, F. J. J. de Kanter, M. Schakel, A. L. Spek, M. Lutz, K. Lammertsma, *Chem. Eur. J.* **2002**, *8*, 58–65; (b) S. G. A. van Assema, A. W. Ehlers, F. J. J. de Kanter, M. Schakel, A. L. Spek, M. Lutz, K. Lammertsma, *Chem. Eur. J.* **2006**, *12*, 4333–4340.

10

Phosphaalkenes

Julien Dugal-Tessier, Eamonn D. Conrad, Gregory R. Dake, and Derek P. Gates

Department of Chemistry, University of British Columbia, Vancouver, Canada

10.1 Introduction

The synthesis of isolable compounds with multiple bonds between phosphorus and carbon represents a landmark achievement in the field of phosphorus chemistry. Prior to this breakthrough, it was generally accepted that heavier elements of the p-block elements would not form stable (3p–2p)π bonds. This notion was reinforced by the fact that the fascinating molecules HC≡P and $HP{=}CH_2$, characterized in the gas phase using spectroscopic methods, could not be isolated in condensed form [1,2]. The first isolable compounds containing phosphorus–carbon multiple bonds were the phosphamethine cyanine cations and the phosphinines, each benefiting from the delocalization of the P=C bond over multiple centers [3–5]. The first isolable compounds with localized P=C bonds, phosphaalkenes, were prepared in 1976 (Figure 10.1) [6]. The so-called Becker phosphaalkenes (e.g., **A**) are readily accessed bearing a range of substituents by treating a silylphosphine [$RP(SiMe_3)_2$] with an acid chloride [R′COCl]. The first phosphaalkene bearing solely carbon substituents, **B** (Mes=2,4,6-trimethylphenyl), was reported by Bickelhaupt and coworkers in 1978 [7]. These early breakthroughs laid a foundation for the widespread growth that has occurred in the field of phosphaalkene chemistry as evidenced by the number of recent reviews dedicated to the subject [8–17].

From the earliest days, researchers in the field of phosphaalkene chemistry have been interested in the potential to exploit these unusual species as a novel class of ligand for transition metals [18,19]. The three basic modes of binding of phosphaalkenes to transition metals are shown in Figure 10.2. The first coordination complexes of a phosphaalkene were characterized by nuclear magnetic resonance (NMR) spectroscopy as *cis*-$M(CO)_4(\eta^1\text{-MesP=CPh}_2)_2$ (M=Cr, Mo, W), *trans*-$RhCl(PPh_3)_2(\eta^1\text{-MesP=CPh}_2)$, *trans*-$RhCl(\eta^1\text{-MesP=CPh}_2)_2(CO)$, $Rh(\eta^5\text{-}C_9H_7)(\eta^1\text{-MesP=CPh}_2)_2$, *cis*-$PtX_2(\eta^1\text{-MesP=CPh}_2)_2$ (X=Cl, I, Me), and *cis*- and *trans*-$PtCl_2(PEt_3)(\eta^1\text{-MesP=CPh}_2)$ [20]. Interestingly, each of these complexes exhibited

Phosphorus(III) Ligands in Homogeneous Catalysis: Design and Synthesis, First Edition.
Edited by Paul C. J. Kamer and Piet W. N. M. van Leeuwen.

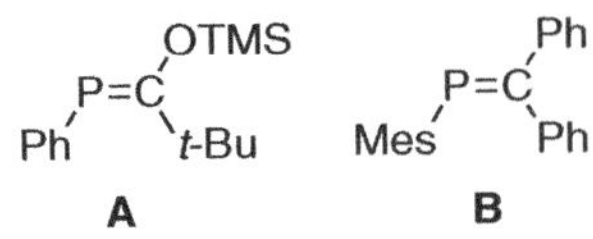

Figure 10.1 The first isolable phosphaalkenes.

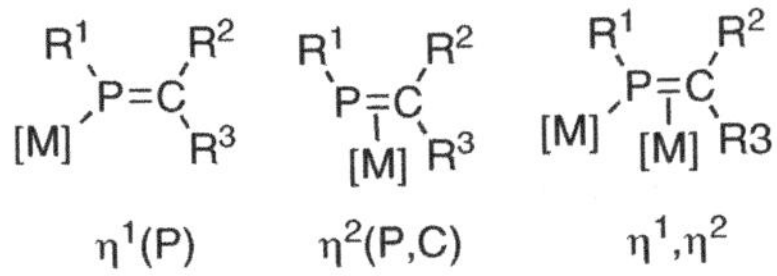

Figure 10.2 Various modes of phosphaalkene coordination to transition metals.

binding of type η^1(P) through the phosphorus center, and this was confirmed crystallographically with the structural characterization of $Cr(CO)_5(MesP{=}CPh_2)$ and *cis*-$PtCl_2(PEt_3)(MesP{=}CPh_2)$ [21,22]. Since these early developments, this field has grown considerably and now includes examples illustrating a number of different binding modes for phosphaalkenes [10,13,19,23,24]. In addition, a fascinating field of metallophosphaalkenes has been developed in which the metals are substituents at either the P or C atom of the P=C bond [12].

In this chapter, the application of phosphaalkene complexes of transition metals in homogeneous catalysis will be reviewed. This new frontier in catalysis is still in its infancy compared to the mature area of phosphine ligands. Nevertheless, exciting opportunities exist for the development of phosphaalkene complexes for catalytic applications given the poor σ-donating and good π-accepting properties of the P=C bond. The reader is also referred to three excellent reviews of low-coordinate phosphorus complexes highlighting their role in catalysis [9,10,14].

10.1.1 Frontier molecular orbitals of phosphaalkenes

Phosphaalkenes possess unique electronic properties which make them particularly attractive as ligands for application in metal catalysis. Electron donation to the metal occurs from the nonbonding phosphorus lone pair (n_P), whereas back bonding from the metal center will occur in the empty $\pi^*_{P=C}$ antibonding orbital. Calculations on $HP{=}CH_2$ have shown that the phosphorus lone pair is high in s-character (66% 3s) relative to idealized sp^2 hybridization (34% s, 66% p) (Figure 10.3) [10]. Experimentally, the high s-character of the lone pair, and therefore the low s-character of the σ-bonds, is supported by the acute R–P=C bond angle observed in phosphaalkenes [10,25]. In contrast to $HP{=}CH_2$, the lone pair of $HN{=}CH_2$ (n_N) is in an orbital that is close to idealized sp^2 hybridization (39% 2s, 61% 2p). The degree of the s-character affects the directionality of the lone pair and, consequently, its σ-donating ability toward transition metals. For phosphaalkenes, the high s-character and more spherical distribution of the lone pair electrons suggest that it would be a relatively poor σ-donor when compared to the nitrogen lone pair within the C=N function of an imine. The σ-donor properties of phosphaalkenes are also influenced by the polarity of the P–C σ-bond. A consideration of the Pauling electronegativities of phosphorus (2.1) and carbon (2.5) suggests that the P–C σ-bond is polarized slightly toward carbon. In fact, NBO calculations on $HP{=}CH_2$ agree with these predictions and suggest a positive charge at phosphorus (+0.42 e) which may further decrease the σ-donating ability of phosphaalkenes [10]. In contrast, $HN{=}CH_2$ has the reverse polarization with a build-up of negative charge on nitrogen (−0.59 e). This polarization, combined with the high p-character of n_N, imparts better σ-donating properties to $HN{=}CH_2$ when compared to $HP{=}CH_2$. Indeed, the donor properties of phosphaalkenes should be quite different from those of imines.

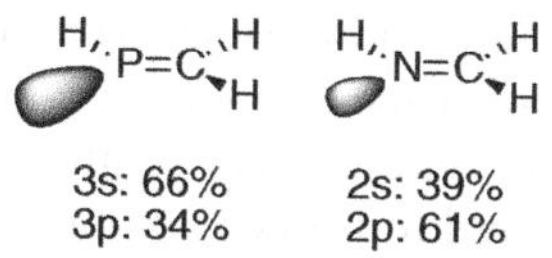

Figure 10.3 *Calculated lone pair molecular orbital compositions in HP=CH$_2$ and HN=CH$_2$.*

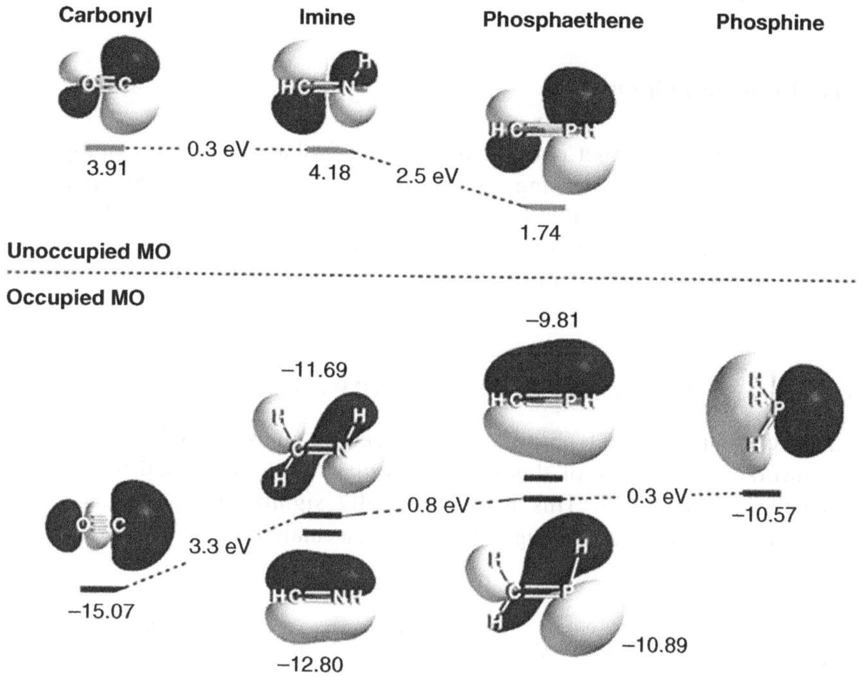

Figure 10.4 *Frontier molecular orbitals of carbon monoxide (C≡O), imine (HN=CH$_2$), phosphaalkene (HP=CH$_2$) and phosphine (PH$_3$). The orbital levels were estimated by* ab initio *MO calculations (HF/6-311G(d)). Reprinted from F. Ozawa, et al., 2004, with permission of the American Chemical Society.*

Due to the high s-character of the lone pair in phosphaalkenes, their properties as σ-donors might be better compared to carbon monoxide rather than imine. The lone pair on carbon (n_C) in C≡O is formally in an sp-hybrid orbital (50% 2s) similar in s-character to the lone pair of HP=CH$_2$ (66% 3s). In addition to the similar σ-donor properties, the analogy between phosphaalkenes and carbonyls extends to their excellent π-accepting ability [26]. The calculated energies of the frontier molecular orbitals of C≡O, HN=CH$_2$, HP=CH$_2$, and PH$_3$ are shown in Figure 10.4 [27]. These results are consistent with earlier experimental and computational investigations on HP=CH$_2$ and HN=CH$_2$ [28,29,30]. Noteworthy is the fact that the $\pi^*_{P=C}$ orbital of HP=CH$_2$ is considerably lower in energy than that of HN=CH$_2$ and C≡O. In addition, inspection of the depictions of the π* orbitals in Figure 10.4 reveals that phosphaethene, like CO, is well suited for back bonding. Therefore, the π-accepting properties of phosphaalkene ligands might potentially be comparable to those of C≡O, a well-established π-accepting ligand. Moreover, the calculated energies of the highest occupied molecular orbital (HOMO) in HP=CH_2 are the $\pi_{P=C}$ bond, with the n_P being the HOMO-1 which is consistent with the orbital energies determined experimentally [28]. The orbital energies are reversed in HN=CH_2. These facts reinforce the aforementioned poor σ-donating properties of phosphaalkenes when compared to imines, although it must also be noted that the lone pair in HP=CH_2 is close in energy to that of PH_3.

The unique π-acceptor and σ-donor properties of phosphaalkenes combined with the potential to fine-tune the steric and electronic properties of the substituents in phosphaalkenes have led to considerable excitement about the catalytic potential of P=C-based ligands. Similar ligand tuning is not possible with C≡O.

10.2 Synthesis of phosphaalkenes

The synthetic methods used to construct phosphaalkene ligands that have been employed in catalysis are described in Sections 10.2.1–10.2.5 focusing on different ligand classes. These ligand classes include diphosphinidenecyclobutenes (DPCB), other P,P-bidentate phosphaalkene ligands, P,N chelate phosphaalkenes, P,X achiral phosphaalkene ligands (X=C, O, S), and P,X chiral ligands (X=P, N). Each section describes the synthetic routes to a particular class of P=C ligand, and experimental procedures are presented in Section 10.4.1.

10.2.1 Diphosphinidenecyclobutene (DPCB) synthesis (P,P chelates)

The diphosphinidenecyclobutene (DPCB) ligand class (e.g., compounds **6–8** in Scheme 10.1) was developed by Yoshifuji and coworkers. Members of this compound class have been used to support homogeneous transition metal catalysis since 1995. This is a historically significant ligand class, as it is the first phosphaalkene-derived compound used as a ligand support in a catalytic reaction [9,31]. DPCB ligands have

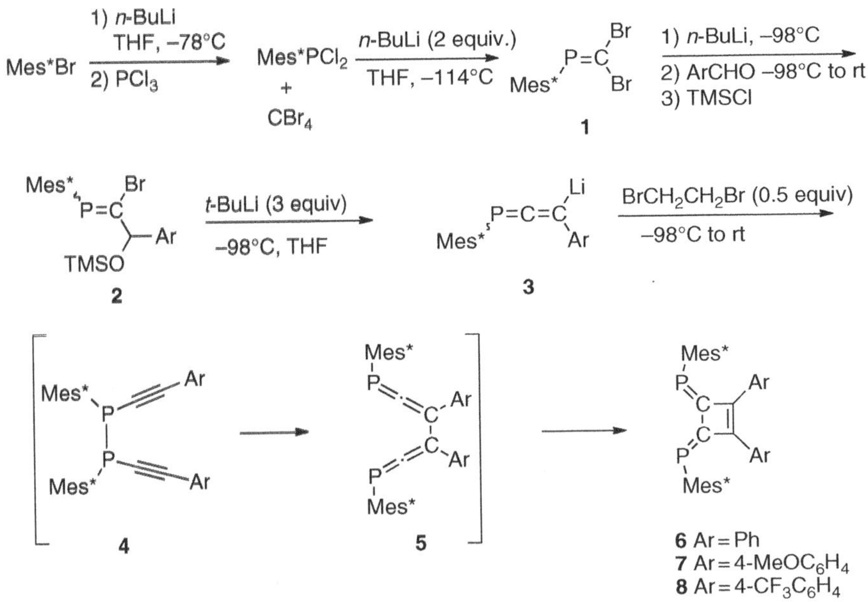

Scheme 10.1 Synthesis of DPCBs.

received considerable attention and are effective in numerous catalytic applications. A variety of different synthetic routes to DPCBs have been reported [9,32]. A detailed experimental procedure for DPCB **6** can be found in Section 4.1.

The synthesis of DPCBs involves the electrocyclization of phosphaallenes [9]. To date, DPCBs **6–8** have shown the greatest potential as ligands for catalysis and their synthesis is outlined in Scheme 10.1 [33]. Most currently known DPCBs feature the 2,4,6-tri-*tert*-butylphenyl (Mes*) group on phosphorus (rarely 2,4,6-tri-isopropylphenyl) [34] to impart kinetic and thermodynamic stability to the P=C bond. Reaction of Mes*-Br with *n*-butyllithium followed by condensation with phosphorus trichloride provides Mes*PCl_2 [35]. Treating Mes*PCl_2 and CBr_4 with *n*-BuLi at −114°C affords dibromophosphaalkene **1** [36]. Further functionalization of **1** is achieved through lithium–halogen exchange using *n*-BuLi with subsequent condensation with the appropriate aryl aldehyde to give phosphaallene precursor **2** after silylation. The use of different aldehyde electrophiles allows for the introduction of a variety of different functionalities. Phosphaalkene **2** is converted to a DPCB by treatment with three equivalents of *t*-BuLi followed by 1,2-dibromoethane. The stoichiometry of *t*-butyllithium is critical as two equivalents are required for lithium–halogen exchange, whilst the third deprotonates the putative phosphaallene. Lithiophosphaallene **3** is believed to be a crucial reaction intermediate that undergoes P–P coupling through reaction with 1,2-dibromoethane to form **4**. Subsequently, compound **4** undergoes a spontaneous Cope-type rearrangement to **5** which then electrocyclizes to the DPCB.

Interestingly, the aforementioned protocol to generate DPCBs using P–P coupling followed by sigmatropic rearrangement is amenable to the synthesis of DPCB polymers (Scheme 10.2). Compound **9**, which contains a spacer group of eight methylene carbon atoms, intermolecularly couples to afford the novel phosphaalkene polymer **10** in 53% conversion and 9% isolated yield under similar conditions to those described in Scheme 10.1 [37]. Analysis of polymer **10** by gel permeation chromatography suggested a reasonable molecular weight (M_n=58,000 g·mol^{-1}) and broad polydispersity (PDI=7.8).

10.2.2 Bidentate-chelating P,P phosphaalkene ligands

An interesting P,P-chelating 1,4-diphospha-1,3-butadiene (**13**) can be accessed in a two-step process from ethylenebis(dichlorophosphine) (**11**) (Scheme 10.3) [38]. Treating **11** with Mes*Li (2 equiv) at −80°C affords chlorophosphine **12** which upon addition of 1,5-diazabicyclo[4.3.0]non-5-ene (DBN) gives a mixture of *E,E* and *E,Z* isomers of **13**. The two isomers can be separated by fractional crystallization in toluene.

1,3-Diphosphapropenes, P,P-chelating phosphaalkenes, are accessible from dibromophosphaalkene **1** in a sequential double lithiation (Scheme 10.4) [36]. The first step involves lithiation with *n*-BuLi (1 equivalent) at −78°C followed by quenching with MeI to yield *Z*-**14** [39]. The subsequent addition of *n*-BuLi to *Z*-**14** followed by a diphenylchlorophosphine quench affords phosphinophosphaalkene *E*-**15** (66%) [40]. The

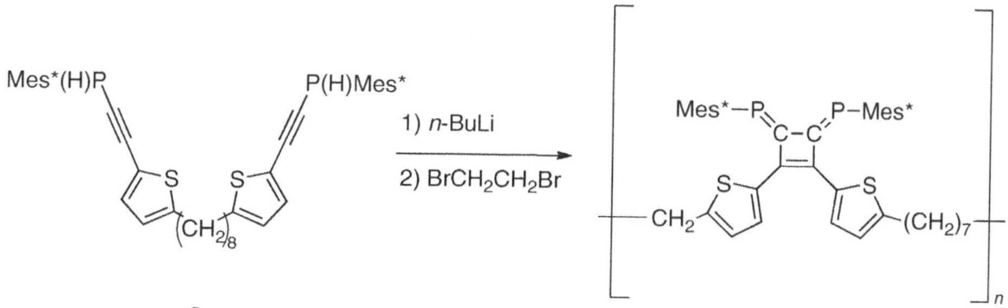

9   10

Scheme 10.2 *Synthesis of DPCB polymers.*

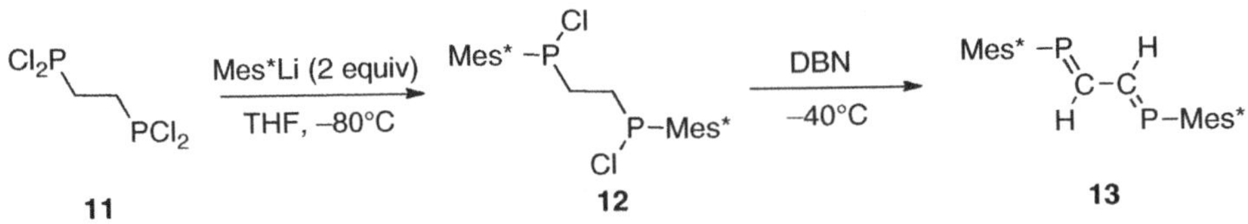

Scheme 10.3 Synthesis of diphosphabutadienes.

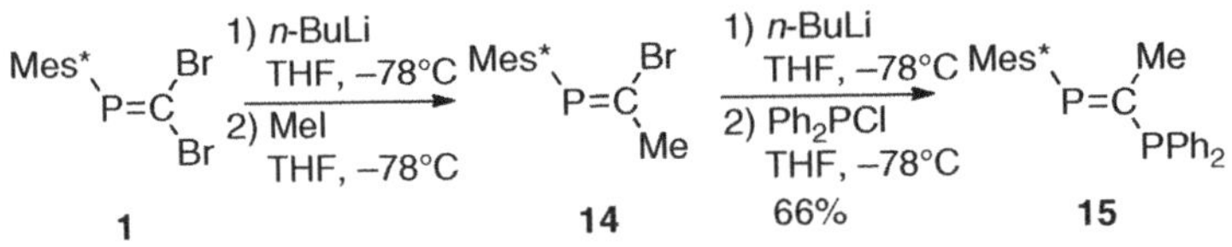

Scheme 10.4 Synthesis of diphosphapropenes.

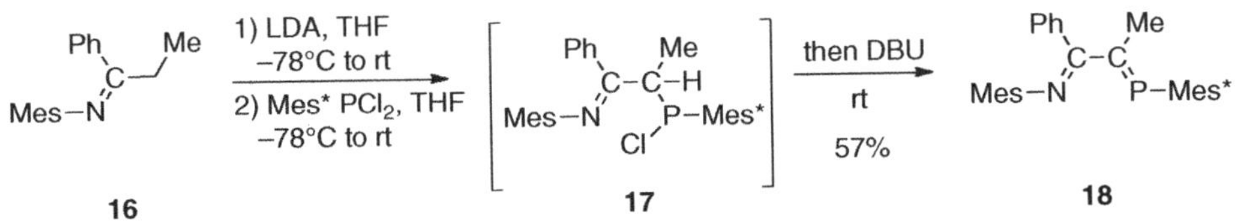

Scheme 10.5 Synthesis of phosphaalkene imines.

observed *E*-configuration is unexpected based on the *Z*-configuration of **14**. However, the observed stereochemistry may be rationalized by the steric repulsion between the Mes* substituent on the phosphorus atom and the diphenylphosphine group on the carbon atom of phosphaalkene **15**.

10.2.3 Phosphaalkenes capable of P,N-chelation to metals

P,N-chelate phosphaalkene ligand **18** can be synthesized in a one-pot, two-step process (Scheme 10.5) [41]. Treating imine **16** with lithium diisopropylamide (LDA) followed by addition of Mes*PCl_2 generates putative intermediate **17** which upon the addition of 8-diazabicyclo[5.4.0]undec-7-ene (DBU) gives phosphaalkene **18** in 57% isolated yield.

Other potential P,N-chelating phosphaalkenes containing a pyridyl group as the nitrogen donor have been synthesized following a phospha-Peterson approach (Scheme 10.6). For example, the novel tridentate PNP ligand **19** is accessed from the condensation of a pyridyldialdehyde with Mes*P(Li)$SiMe_2$(*t*-Bu) (2 equivalents) [42]. Treating MesP$(SiMe_3)_2$ with 2-benzoyl-substituted pyridine in the presence of a catalytic amount of anhydrous KOH gives bidentate phosphaalkene **20** as a 1:1 mixture of E/Z isomers [43]. Even though compound **20** bears the smaller Mes substituent on the phosphorus atom, it is stable and isolable under inert atmosphere.

10.2.4 P,X achiral phosphaalkene ligands (X=P, O, S)

Mono- and tridentate ligands have been synthesized using phospha-Wittig methodologies (Scheme 10.7) [16]. The phospha-Wittig (Mes*P=PMe_3) reagent can be accessed easily in high yields from the reaction between Mes*PCl_2, an excess of trimethylphosphine (PMe_3), and zinc dust. Mixing the phospha-Wittig reagent with the appropriate aldehyde in tetrahydrofuran (THF) at 0°C produces phosphaalkenes **21** and **22** in good isolated yields [44].

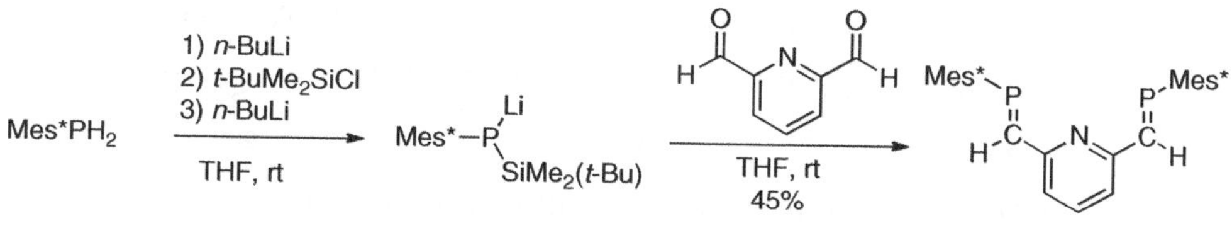

Scheme 10.6 *Synthesis of pyridyl phosphaalkenes.*

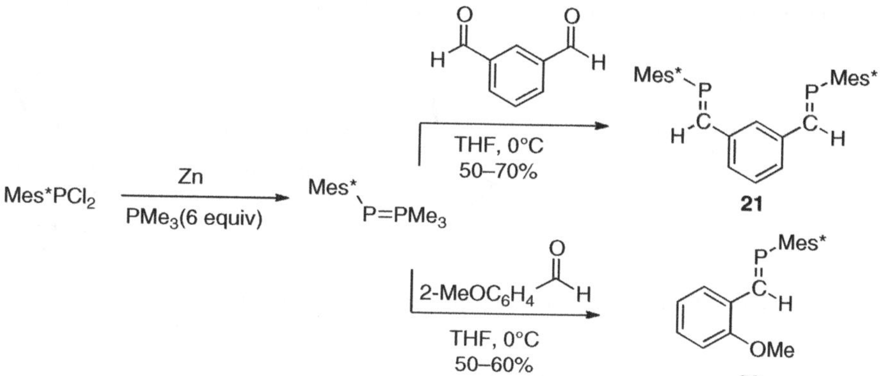

Scheme 10.7 *Synthesis of phosphaalkenes using the phospha-Wittig reaction.*

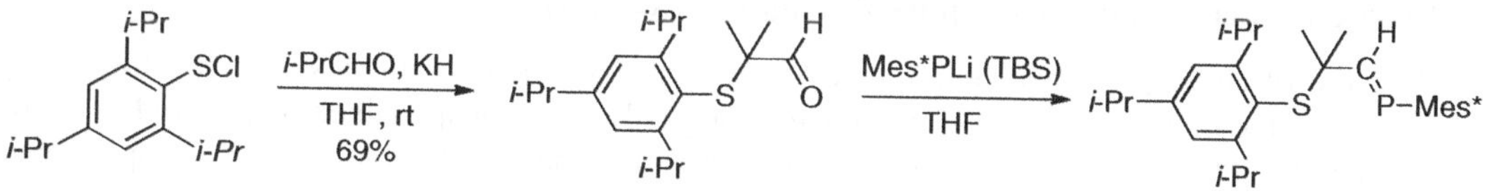

Scheme 10.8 *Synthesis of a P,S-chelate phosphaalkene.*

The interesting P,S-chelate phosphaalkene **24** can be accessed in two steps from 2,4,6-tri(isopropyl)phenylsulfenyl chloride (Scheme 10.8) [41]. The first step involves the deprotonation of isobutyraldehyde using potassium hydride followed by nucleophilic addition to 2,4,6-tri(isopropyl)phenylsulfenyl chloride affording α-arylsulfenylaldehyde intermediate **23**. This aldehyde is transformed to the P,S-chelate phosphaalkene ligand **24** using a phospha-Peterson protocol. This fascinating compound has not been isolated but is used *in situ* to form palladium complexes in high yield (87%).

Modification of the phosphine moiety of the novel 1,3-diphosphapropene **15** affords P,O- and P,S-chelate phosphaalkene ligands (Scheme 10.9). Oxidation of **15** with *meta*-chloroperbenzoic acid (*m*-CPBA) gives

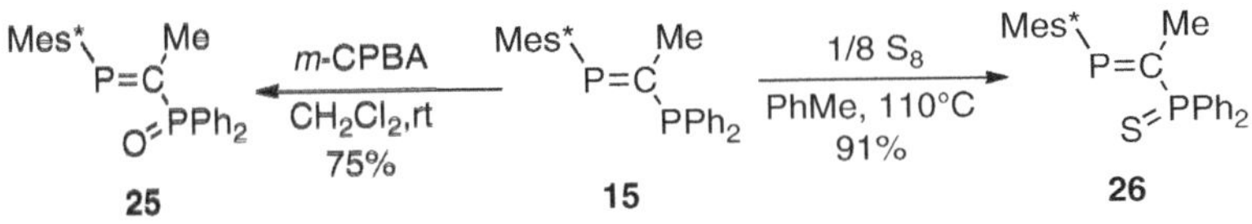

Scheme 10.9 Synthesis of P,O- and P,S-chelate phosphaalkenes.

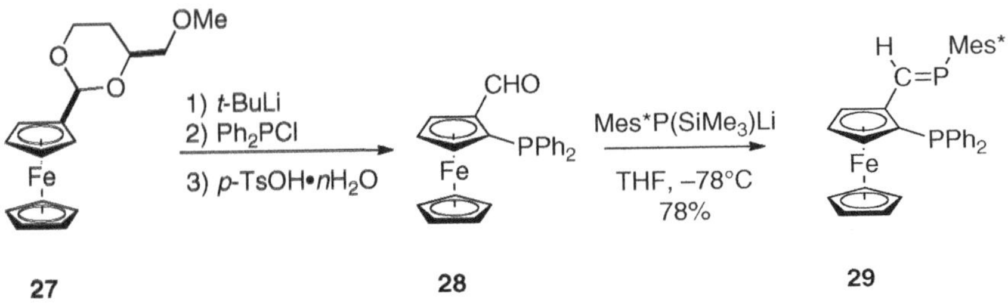

Scheme 10.10 Synthesis of chiral 1-phosphaethenyl-2-phosphanylferrocenes.

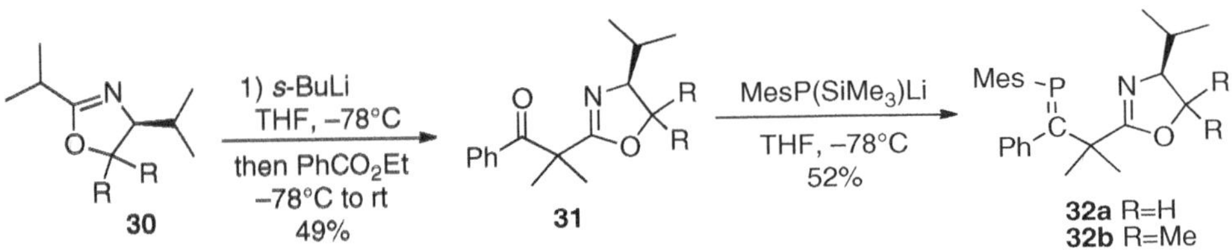

Scheme 10.11 Synthesis of enantiomerically pure PhAk–Ox ligands.

air-stable ligand **25** in 75% isolated yield [45]. Alternatively, **15** can be oxidized with elemental sulfur in refluxing toluene to produce air-stable ligand **26** in 91% yield [46,47].

10.2.5 Synthesis of enantiomerically pure P,X ligands (X=P, N)

The development of chiral phosphaalkene ligands for use in asymmetric catalysis is an exciting prospect. Only recently have the first examples of chiral, enantiomerically pure phosphaalkenes appeared in the literature. The planar chirality within substituted ferrocene derivatives has been exploited to develop an exciting class of chiral P,P-bidentate phosphaalkene compounds (Scheme 10.10) [48]. The synthetic methodology involves the site-selective lithiation of acetal **27** followed by trapping using diphenylchlorophosphine and acidic hydrolysis to afford aldehyde **28** [49]. A phospha-Peterson reaction was utilized with aldehyde **28** to generate phosphaalkene **29** in 78% yield.

A P,N-chelate chiral phosphaalkene incorporating the elements of a phosphorus–carbon double bond and a chiral oxazoline (PhAk–Ox) fragment has been prepared (Scheme 10.11) [50]. 2-Isopropyl-substituted oxazolines (**30**) are obtained from commercially available chiral amino acids in two preparative steps [51]. Deprotonation of **30** with *sec*-BuLi followed by Claisen-type condensation using ethyl benzoate generates ketones of general structure **31**. The addition of ketone **31** (R=H) to a solution of MesP(SiMe$_3$)Li at −78°C cleanly produces phosphaalkene **32a**. The synthetic route to **32a** is modular and efficient, and it can be carried out on the multigram scale without recourse to column chromatography purification. The route allows for a rapid modification of compound structure in order to modify the steric and electronic parameters of this

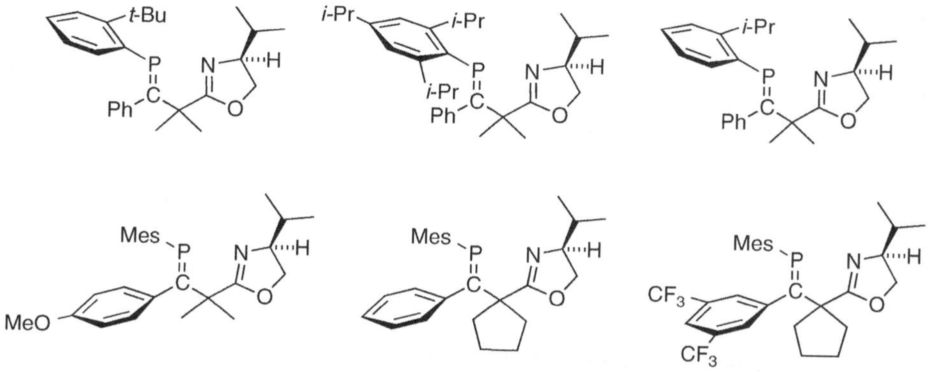

Chart 10.1 *Examples of enantiomerically pure PhAk–Ox ligands.*

ligating species. As an example, compound **32b**, which features an additional set of geminally substituted dimethyl groups on the oxazoline framework, behaves as a bulkier ligand than **32a**. Notably, compounds of general structure **32a** or **32b** are stable to atmospheric air and moisture. Following this same methodology, a number of PhAk–Ox ligands have been prepared with a range of functional substituents (Chart 10.1) [52].

10.3 Catalysis with phosphaalkene ligands

The focus of Section 10.2 centered on the synthesis of phosphaalkenes that have been utilized in catalytic applications. This section describes the use of phosphaalkene ligands in catalysis within the chemical literature (up to early 2011). The phosphaalkenes are typically utilized following one of two general methods. In the majority of instances, discrete metal complexes containing a phosphaalkene ligand were isolated and then subsequently used in a catalysis application. These metal complexes were typically formed using standard ligand substitution reactions, and consequently the synthesis of these metal complexes will not be discussed in detail. Alternatively, a phosphaalkene compound and a metal complex are mixed together to generate a putative metal–phosphaalkene complex *in situ* and a catalytic process is carried out directly. The presentation of this section will be divided into subsections based on reaction class.

10.3.1 Ethylene polymerization

As a consequence of the successful application of complexes containing α-diimine ligands in promoting ethylene polymerization [53], metal complexes of structurally related phosphaalkene ligands (**33–38**) have also been investigated for this application (Chart 10.2). Interestingly, these phosphaalkene-based complexes appear to be more stable than the comparable α-diimine complexes under typical polymerization conditions.

DPCB palladium(II) complexes **33–35** are active toward ethylene polymerization upon acid activation [$H(OEt_2)_2BAr_F$] to afford catalytically active cationic complexes [54]. Complex **33** (Ar=Ph), upon activation, is more active than **34** and **35** and polymerizes ethylene (turnover frequency (TOF)=4,600 h^{-1}, pressure=10 $kgf{\cdot}cm^{-2}$) to give polyethylene with a molecular weight (M_w) of 18,700 $g{\cdot}mol^{-1}$ and polydispersity indexes (PDIs) of 14.8. Interestingly, these DPCB complexes will polymerize ethylene at 100°C, a temperature at which α-diimine catalysts are not stable [54]. Further studies on the effect of DPCB ligands structure on catalyst performance suggested that incorporating electron-withdrawing aryl substituents into the DPCB backbone gave more active catalysts [Ar=4-$MeOC_6H_4$<Ph<3,5-$(CF_3)_2C_6H_3$<4-$CF_3C_6H_4$] [27]. The most

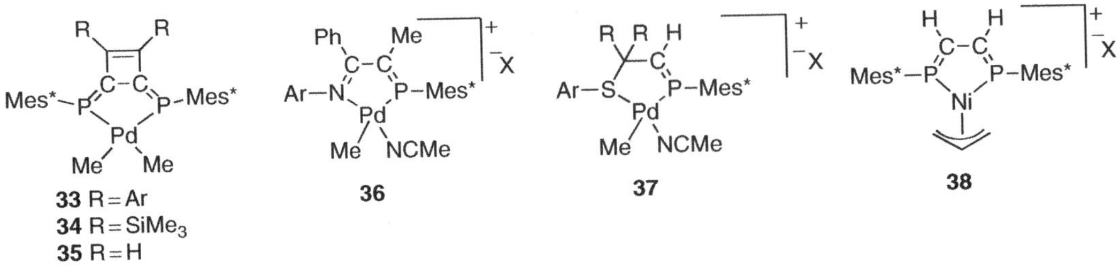

Chart 10.2 *Phosphaalkene complexes used in ethylene polymerization.*

active catalyst for ethylene polymerization is **33** (Ar=4-$CF_3C_6H_4$) with an activity of 7,500 h^{-1}, molecular weight (M_w) of 43,000 g·mol^{-1}, and PDIs of 23.7.

The mixed P,N-catalyst **36** (Ar=Mes) is also an effective catalyst for the polymerization of ethylene with low turnover frequency (TOF=94 h^{-1}) compared with α-diimine Pd(II) complexes (TOF of 4,500 h^{-1}) [41]. The low TOF was attributed to slow rates of ethylene migratory insertion, and calculations of the energy barrier supported this hypothesis. Therefore, a P,S-phosphaalkene complex **37** [Ar=2,4,6-(*i*-Pr)C_6H_2, R=H] was synthesized in an attempt to lower the migratory insertion energy barrier. Remarkably, **37** polymerizes ethylene with activities comparable to α-diimine ligands (TOF=4,300 h^{-1} for a 3 h run). Complex **37** [Ar=2,4,6-(*i*-Pr)C_6H_2, R=H] also has greater stability (TOF of 3,000 h^{-1} for a 15 h run) for the polymerization conditions than systems that featured α-diimine ligands but generated oligomers (M_n=215 g·mol^{-1}). The methyl-substituted catalyst **37** [Ar=2,4,6-(*i*-Pr)C_6H_2, R=Me] has a high activity (TOF 3,100 h^{-1}) and gives higher molecular weight polyethylene (M_n=2,300–2,500 g·mol^{-1}) with extensive branching.

Nickel(II) complexes such as **38** have also successfully been utilized as catalysts for the polymerization of ethylene to afford polymer (M_w=58,000 g·mol^{-1}; PDI=3.19) with less branching [55]. The phosphaalkene-based metal complexes seem to have a greater thermal stability than α-diimine complexes in palladium- and nickel-catalyzed polymerizations of ethylene which might be an advantage for industrial applications.

10.3.2 Cross-coupling reactions

Metal complexes containing phosphaalkene ligands have also been investigated in cross-coupling reactions. These reactions are well established and offer a useful benchmark to compare catalytic efficiencies [56]. Complexes containing phosphaalkene-based ligands are active in a variety of different cross-coupling reactions such as the Sonogashira [57], Suzuki–Miyaura [58], Stille [59], Buchwald–Hartwig [60,61], and Ullmann [62] couplings. The complexes shown in Chart 10.3 are active in the Sonogashira and Suzuki–Miyaura couplings.

Yoshifuji and coworkers were the first to use a phosphaalkene ligand in catalysis. Complex **39** effectively couples 1-bromo-4-nitrobenzene with trimethylsilylacetylene in 77% yield [31]. Interestingly, the 1,3-diphosphapropene complex **40** is a more active catalyst for the coupling of phenylacetylene and iodobenzene in high yields [40]. However, the limitations of complex **40** are its inability to couple 1-bromo-4-nitrobenzene and its sensitivity toward air and moisture [47]. To remedy the sensitivity of **40**, the oxidized complex **41** can be employed, but lower yields of cross-coupled products are obtained [45]. The more electron-rich phosphine sulfide complex **42** tolerates air and moisture and gives synthetically useful yields of diphenylacetylene [47]. Complex **42** will not cross-couple bromobenzene with phenylacetylene but rather dimerizes phenylacetylene quantitatively to afford diphenylbutadiyne (Ph–C≡C–C≡C–Ph) [46]. The authors of this work suggest that the dimerization product results from the formation of a dialkynylpalladium(II) complex. Monodentate phosphaalkene complex **43** and tridentate phosphaalkene complex **44** can effectively

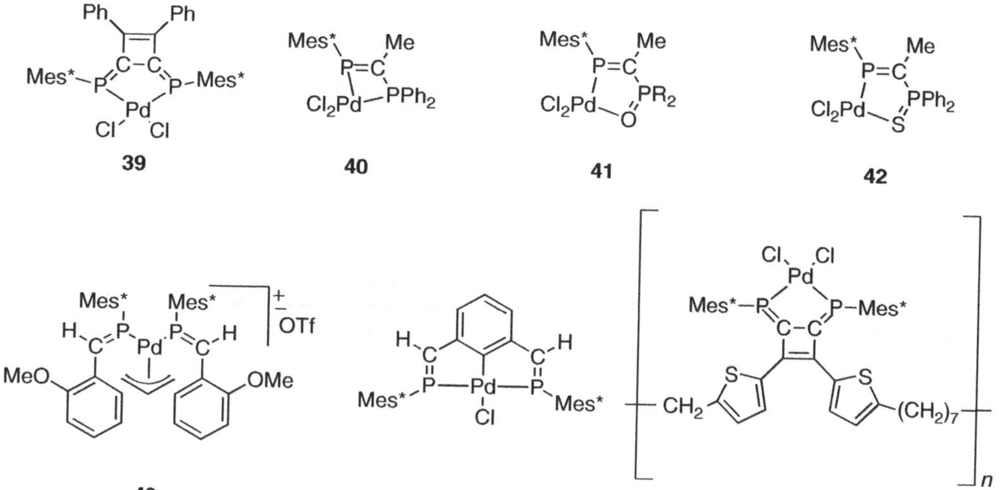

Chart 10.3 *Complexes used in Sonogashira and Suzuki–Miyaura coupling.*

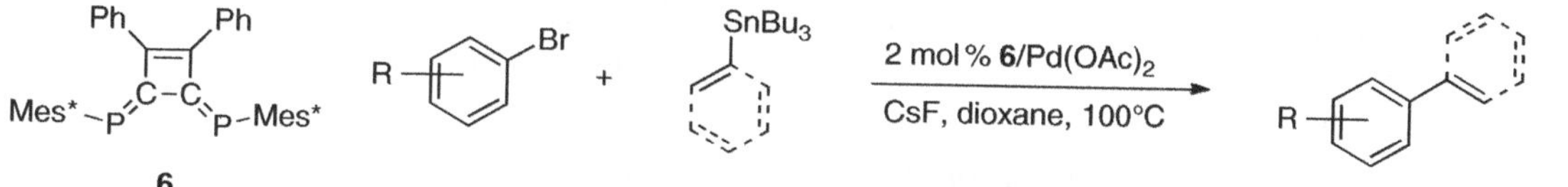

Scheme 10.12 *Stille coupling using ligand **6**.*

couple bromoarenes with phenylacetylene with low catalyst loadings (0.1 mol %) [44]. Polymer complex **45** shows good activity in cross-coupling 1-bromo-4-nitrobenzene with trimethylsilylacetylene [37]. Importantly, the polymer-supported catalyst is easily separated from the product and recycled.

The Suzuki–Miyaura coupling of aryl halides with aryl boronic acids can be catalyzed using phosphaalkene complexes **41**, **42**, **43**, and **44**. Phosphine oxide complex **41** efficiently catalyzes the cross-coupling of iodobenzene and phenylboronic acid [45]. Although the electron-rich phosphine sulfide complex **42** efficiently cross-couples iodobenzene and phenylboronic acid, its use could not be extended to less reactive electrophiles such as bromobenzene or 4-iodoanisole [47]. Remarkably, cross-coupling of bromoarenes with phenylboronic acid is achieved rapidly (2 h) and quantitatively with very low (10^{-5} mol %) loadings of complexes **43** and **44** [44].

In each of the aforementioned studies, the complexes used for catalysis were isolated prior to use. This next example features an *in situ* generated catalyst. Premixing ligand **6** and $Pd(OAc)_2$ generates a catalyst mixture that can efficiently promote the Stille cross-coupling reaction of aryl bromides with vinyl-, allyl-, and aryl- tins (Scheme 10.12) [63]. The cross-coupling conditions were general for a variety of different bromoarenes containing a variety of different electronic and steric properties. These conditions were tolerant toward unprotected alcohols and heteroaromatics.

Generating the catalyst *in situ* was also successfully used to promote the challenging cyanation of aryl bromides. A mixture of $Pd_2(dba)_3$, **6**, Zn, and $Zn(CN)_2$ effectively catalyzes the cyanation of aryl bromides at 100°C (Scheme 10.13) [64]. The catalyst loading of (2 mol % Pd) is surprisingly low, especially considering that cyanation reactions are often plagued by catalyst deactivation by cyanides [65]. However, it was

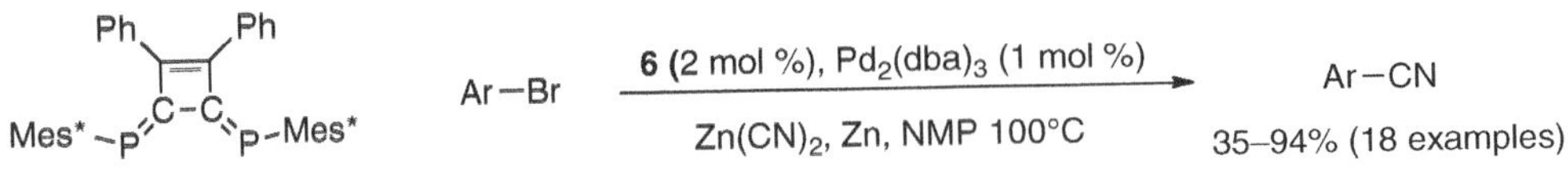

6

Scheme 10.13 *Cyanation of aryl bromides with* **6**.

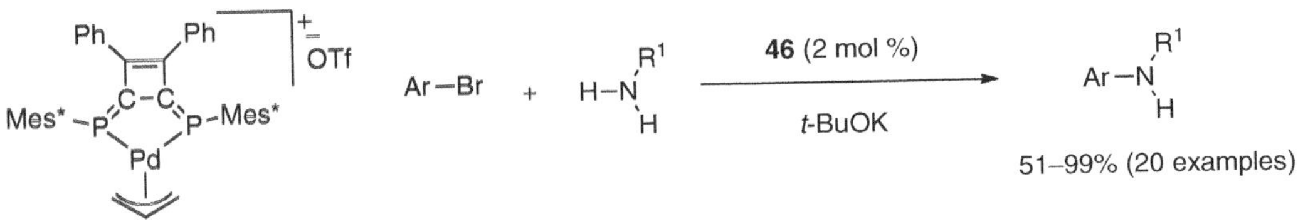

46

Scheme 10.14 *Buchwald–Hartwig coupling using* **46**.

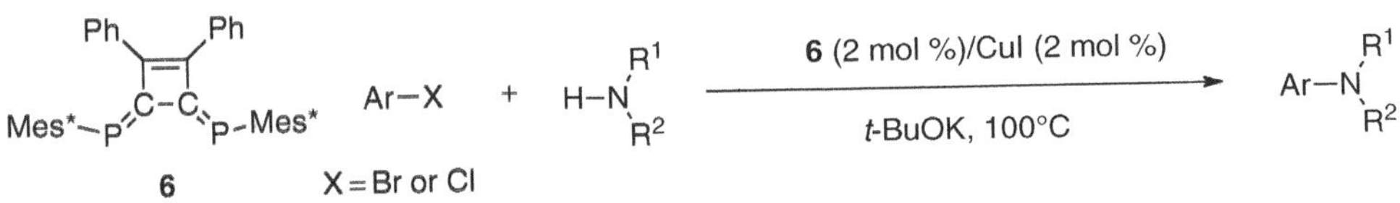

6

Scheme 10.15 *Ullman coupling using* **6**.

previously known that the presence of Zn dust as an additive permits lower catalyst loadings in cyanation reactions [66]. Cyanation is a synthetically useful reaction, and the yields obtained with **6** were generally high. The process could be employed to construct sterically hindered aryl nitriles such as 2,4,6-trimethylbenzonitrile.

Anilines can be coupled to bromoarenes (Buchwald–Hartwig coupling) in synthetically useful yields using palladium allyl complex **46** and potassium *tert*-butoxide (Scheme 10.14) [67]. It is noteworthy that the amination proceeds at room temperature without solvent. Interestingly, monoarylated anilines are the observed products at room temperature even in the presence of excess bromobenzene. Disubstitution reactions of aniline using two equivalents of bromobenzene or the cross-coupling of secondary aromatic amines such as *N*-methyl aniline could be performed but required high temperature (100°C). In contrast, aliphatic secondary amines such as piperidine and morpholine couple readily at room temperature.

Aryl bromides and chlorides can also be coupled with anilines in an Ullman coupling process using DPCB **6**, copper iodide, and potassium *tert*-butoxide (Scheme 10.15) [68]. These reactions usually require higher temperatures (100°C) relative to the palladium catalysis shown in Scheme 10.14. The yields of products for both the palladium- and copper-catalyzed processes were similar.

10.3.3 Hydro- and dehydrosilylation

The transition metal-catalyzed hydrosilylation of carbonyl compounds forming silyl ethers is a well-established process [69]. Therefore, it is quite remarkable that complex **47** (Ar=Ph) catalyzes the dehydrosilylation of ketones to silyl enol ethers (Scheme 10.16) [70]. Electronic perturbations of ligand **47** (Ar=4-$MeOC_6H_4$ or Ar=4-$CF_3C_6H_4$) lead to different reactivity profiles. For example, the reduced C=O product (up to 15%)

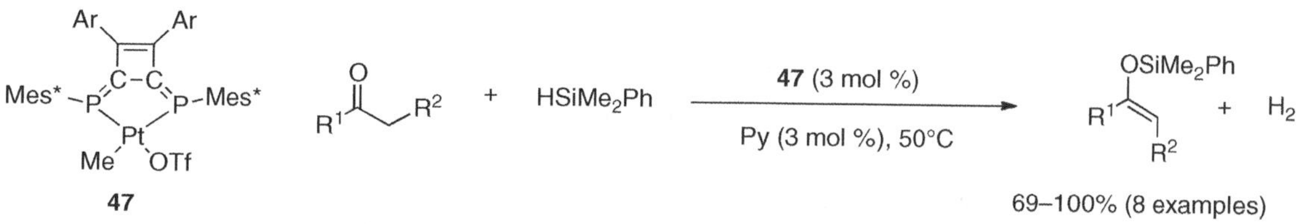

Scheme 10.16 *Dehydrogenative silylation of ketones using **47**.*

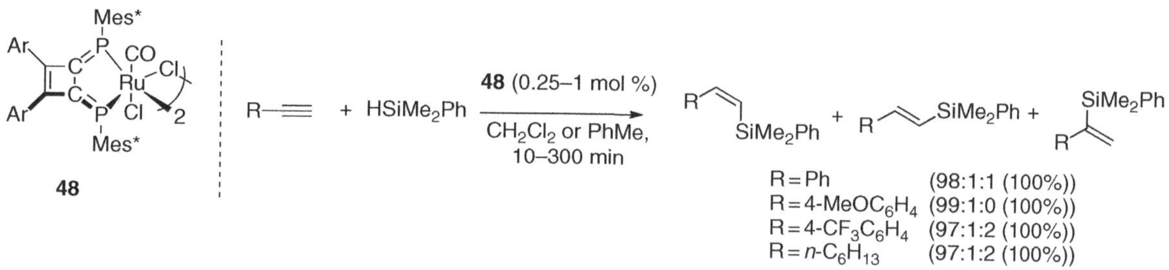

Scheme 10.17 *Z-selective hydrosilylation of alkynes.*

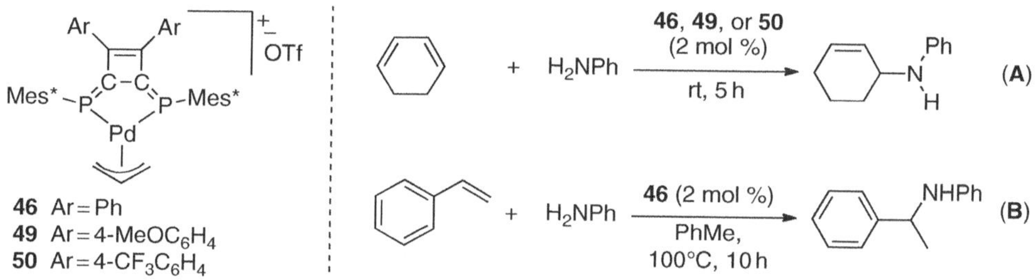

Scheme 10.18 *Hydroamination of 1,3-dienes.*

could be isolated when these electronically modified metal–ligand complexes were used. These solvent-free reactions generally form the *E*-isomer of the most substituted silyl enol ether.

In general, the hydrosilylation of alkynes mediated by a metal catalyst will generate preferentially the *E*-vinyl silane [69]. Surprisingly, hydrosilylation reactions promoted by ruthenium dimer complex **48** (Ar=4-$MeOC_6H_4$) selectively produce *Z*-vinyl silanes (Scheme 10.17) [71]. A wide range of substrates are tolerated. For example, 1-arylalkynes having either electron-donating or -withdrawing substituents, or 1-alkylalkynes, are hydrosilylated with high yields. Reactions using **48** feature low catalyst loadings and short reaction times. Mechanistic studies of the reaction suggest that the strong π-accepting properties of **48** (Ar=4-$MeOC_6H_4$) are responsible for the *Z*-selectivity [71]. Complex **48** (Ar=4-$MeOC_6H_4$) outperformed previously reported catalysts for this transformation to generate bifunctional *Z*-vinyl silanes en route to all-*cis* poly(phenylene vinylenes) (PPVs) [72].

10.3.4 Hydroamination and hydroamidation

Palladium allyl DPCB complexes **46**, **49**, and **50** are effective for the 1,4-hydroamination of 1,3-dienes with aniline (Scheme 10.18) [73]. An interesting solvent effect is observed: reactions in toluene (nonpolar) give

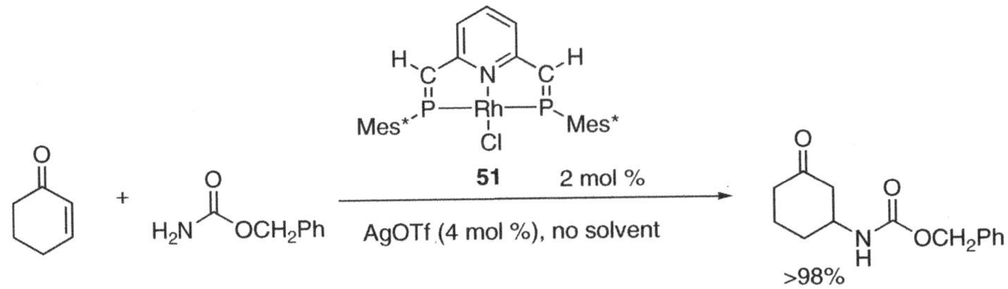

Scheme 10.19 *Conjugate addition of benzyl carbamate to enones using* ***51****.*

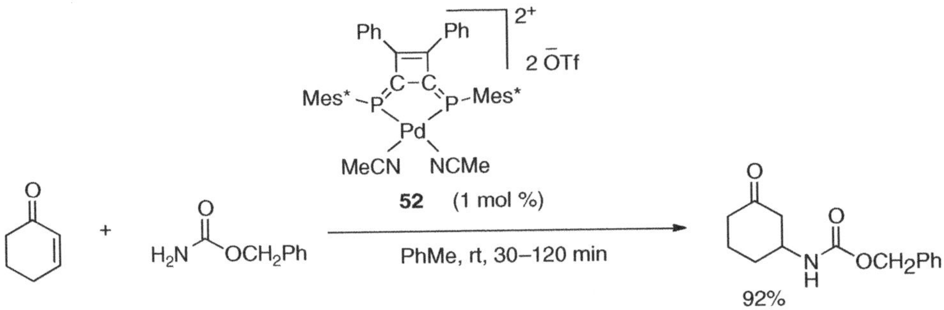

Scheme 10.20 *Conjugate addition of benzyl carbamate to enones using* ***52****.*

high yields (91%), but there is no reaction observed using otherwise identical conditions in DMF (polar) (equation **A**). The incorporation of electron-withdrawing substituents on the DPCB ligand results in lower isolated yields [**50** (58%)<**46** (89%)<**49** (91%)]. Catalyst **46** also effectively mediates the hydroamination of styrene (equation B).

P,N,P-complex **51** is active in the solvent-free conjugate addition of carbamates to enones (Scheme 10.19) in excellent yields [74]. The addition of silver triflate (AgOTf) is essential for catalytic activity. Prior exchange of the chloride counterion in **51** with a triflate counterion did not generate an active catalyst without the further addition of a catalytic amount of AgOTf. Importantly, control experiments demonstrated that AgOTf alone is not a competent catalyst for this conjugate addition, but that both **51** and AgOTf are required for activity.

Dicationic complex **52** actively catalyzes the conjugate addition of carbamates to enones (Scheme 10.20) [75]. Although complex **52** shows higher activity than $PdCl_2(MeCN)_2$, its activity was comparable to the dicationic complex $[Pd(MeCN)_4](BF_4)_2$. A possible advantage of using **52** as a catalyst is that palladium black is not observed. Palladium black is known to catalyze the undesirable disproportionation of cyclohexenone to cyclohexanone and phenol. Palladium black is generated when $[Pd(MeCN)_4](BF_4)_2$ is used as the catalyst [75].

10.3.5 Isomerization reactions

The successful gold(I)-catalyzed cycloisomerization of enynes expands the use of phosphaalkene ligands to third-row transition metals [76]. Chlorogold complexes **53**, *E*-**54**, *Z*-**54**, and **55** effectively catalyze the isomerization of enynes to cyclopentenes in high yields (Scheme 10.21) [77–79].

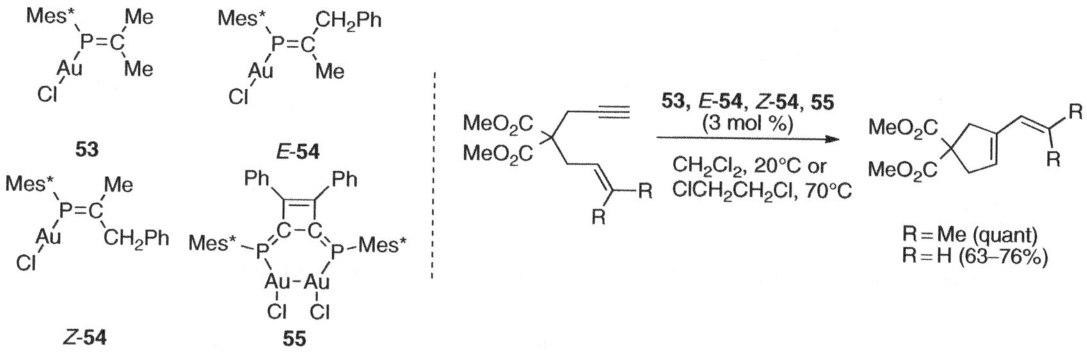

Scheme 10.21 *Gold-catalyzed cycloisomerization of enynes.*

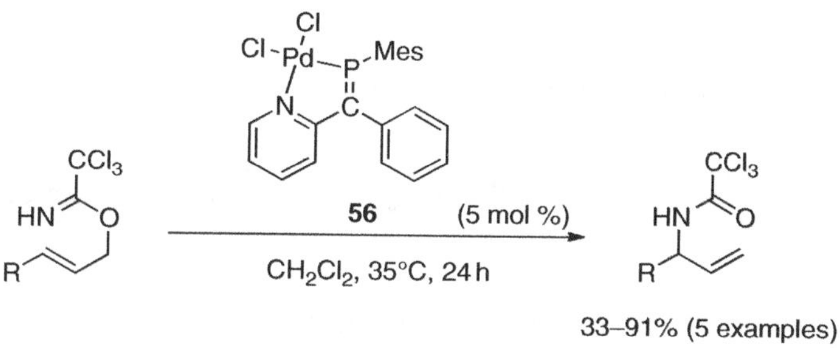

Scheme 10.22 *Overman–Claisen rearrangement using **56**.*

Interestingly, the phosphaalkene–gold complexes do not require the addition of a silver co-catalyst as is often the case in these types of reactions.

Mesityl-substituted phosphaalkene **56** is an effective catalyst for the Overman–Claisen rearrangement (Scheme 10.22) [80]. The substrate scope was investigated, and complex **56** catalyzes the conversion of linear allylic trichloroacetimidates containing linear aliphatic chains to branched allylic trichloroamidates. The purity of the trichloroacetimidate starting materials has an influence on the efficiency of the catalyzed processes.

10.3.6 Allylic substitution

Allyl alcohols usually must be converted to allyl acetates before metal-catalyzed allylic substitution can take place [81]. Therefore, the direct substitution of allylic alcohols improves synthetic efficiency as the hydroxyl activation step is rendered unnecessary. The reaction of aniline with substituted allyl alcohols is catalyzed by complexes **46** (91%), **49** (96%), **50** (82%), and **57** (63%) (Scheme 10.23) [46,47,82]. Superior results are obtained with the electron-rich complex **49**. The reaction is regioselective with nucleophilic substitution occurring at the least sterically hindered carbon atom of the Pd–allyl intermediate. The substitution of enantio-enriched allyl alcohols generates products with no erosion of enantiomeric purity. The π-acidity of the phosphaalkene ligand is critical to promote the direct substitution of allyl alcohols [83,84].

Expanding upon the direct substitution of allyl alcohols, the cyclodehydration of *cis*-2-butene-1,4-diols can be employed to synthesize dihydrofurans (Scheme 10.24) [85]. The methodology involves mixing the diol and β-dicarbonyl compounds with a catalytic amount of base and **49** in toluene at 100°C. Acetylacetone and ethyl acetoacetate generated dihydrofurans in synthetically useful yields (91% and 74%, respectively).

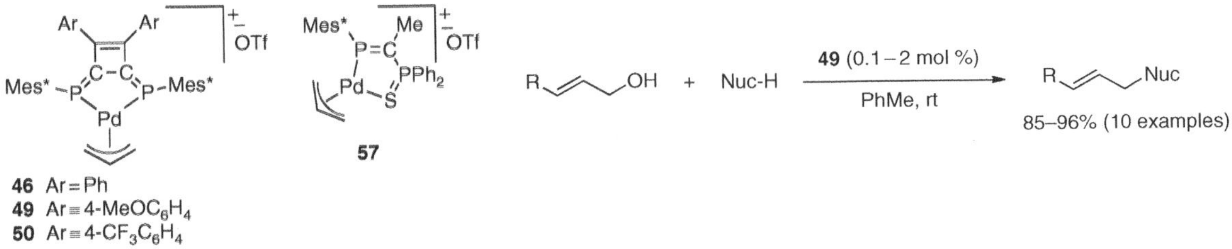

Scheme 10.23 *Allylic alkylation of allyl alcohols using phosphaalkene complexes.*

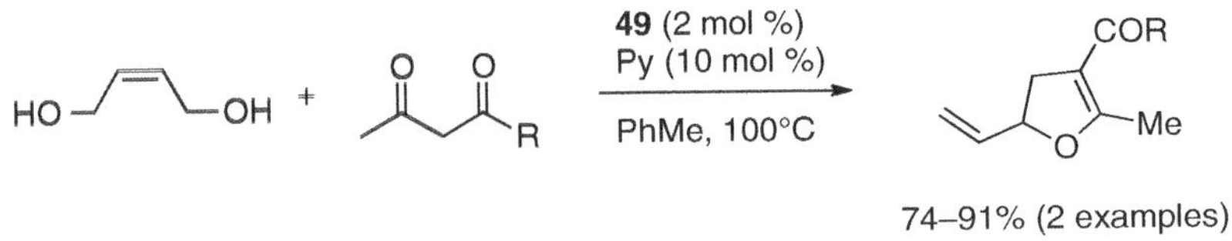

Scheme 10.24 *Cyclodehydration using* ***49****.*

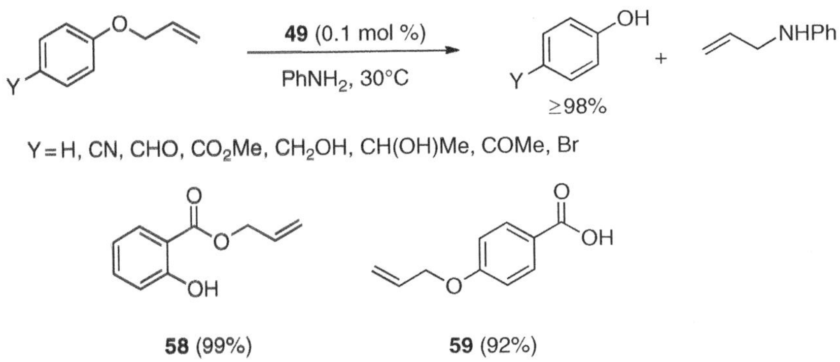

Scheme 10.25 *Deallylation reactions using* ***49****.*

Complex **49** is also an effective catalyst for the deallylation of a variety of allylic phenol ethers in the presence of aniline (Scheme 10.25). Deallylation reactions using **49** may be performed in air and in the presence of a variety of different functionalities such as vinyl, alkynyl, hydroxy, acetoxy, silyloxy, and acetonide groups [86]. An unusual selectivity is observed in the deallylation of bis-allylated hydroxybenzoic acid derivatives. Complex **49** removed only one allyl group, although two were present in the substrates. Allyl O-allyl-2-hydroxybenzoate reacts under identical conditions at only its allyl ether function to generate ester **58**. Allyl benzoate esters having allyl ether substitution at the *meta* or *para* positions react only at the ester function to afford products such as **59**. The use of $Pd(PPh_3)_4$ as the catalyst resulted in complete, global deallylation. The authors speculated that coordination effects are responsible for the observed selectivities.

10.3.7 Asymmetric catalysis

As the sophistication of the synthetic methods used to prepare phosphaalkenes continually improves, the prospect of developing chiral phosphaalkene-based ligands and utilizing them within enantioselective processes is particularly intriguing. Ozawa and coworkers employed complex **60** in an asymmetric 1,4-addition

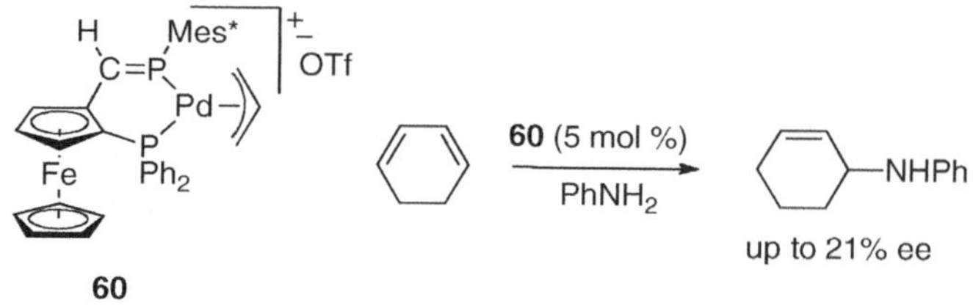

Scheme 10.26 *Enantioselective hydroamination using **60**.*

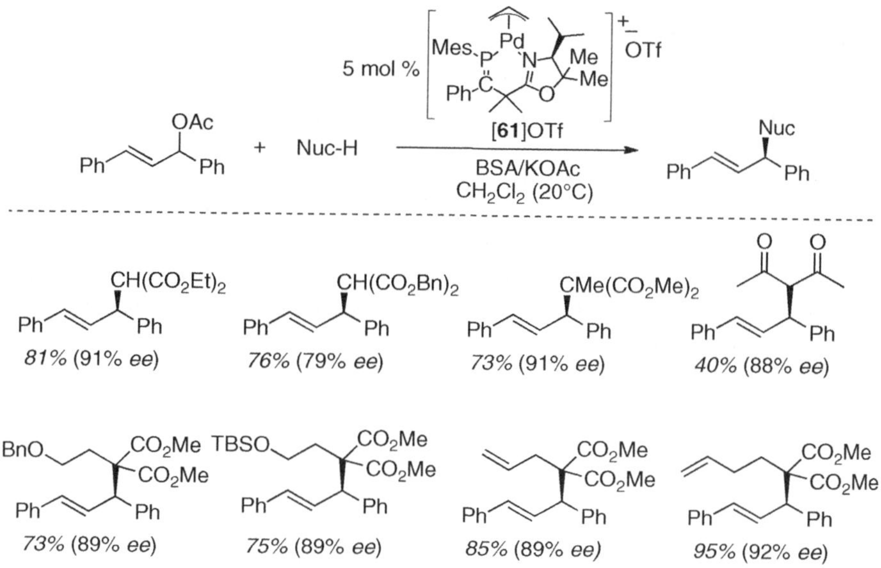

Scheme 10.27 *Allylic alkylation using complex **61**.*

of cyclobutadiene with aniline. The process suffered from low enantioselectivity (Scheme 10.26) [48]. Efforts to improve the yield generated processes having no to low enantioselectivities. Likewise, optimization to improve enantioselectivity led to low yields. Nonetheless, this report represents the first account of an enantioselective reaction using a phosphaalkene ligand.

The development of a novel class of phosphaalkene–oxazoline (PhAk–Ox) ligands for use in asymmetric catalysis has given promising results. As a proof of concept, allylic alkylation reactions can be performed in high yields and enantioselectivities using palladium(II) complexes supported by PhAk–Ox ligands (Scheme 10.27) [87]. The use of complex **[61]** OTf bearing a ligand having *gem*-dimethyl substitution on the oxazoline ring was important in obtaining enantioselectivities above 90%. A variety of substituted malonate nucleophiles can be used in reactions catalyzed by a palladium PhAk–Ox complex **[61]**OTf.

10.4 Concluding remarks

Phosphaalkene ligands display a fascinating and diverse chemistry distinct from phosphorus-containing ligands lacking a formal multiple bond at phosphorus. Improved synthetic processes to build these compounds have allowed chemists to generate phosphaalkene compounds with capabilities for mono- or multidentate coordination. Achiral and chiral compounds containing these fragments can now be constructed. Phosphaalkenes, with their novel donor–acceptor properties, have moved from fundamental curiosities to compounds with prospective applications. Their application as ligands for transition metal catalysis has been particularly fruitful with phosphaalkene complexes mediating many important organic transformations. As compounds containing a P=C bond become increasingly sophisticated, the development of fascinating catalytic applications for these species in organic chemistry will continue.

10.5 Experimental procedures for representative ligands

10.5.1 Synthesis of DPCB (6) [33]

Step 1: To a solution of 2,2-dibromo-1-(2,4,6-tri-*t*-butylphenyl)-1-phosphaethene **1** (451 mg, 1 mmol, 1 equiv.) in THF (9 mL) at –98°C was added *n*-butyllithium (1.59 M in hexanes, 0.64 mL, 1 equiv.), and the solution was stirred for 10 min. To the cooled (–98°C) solution, benzaldehyde (106 mg, 1 mmol, 1 equiv., in 1 mL THF) was added dropwise, and the solution was allowed to warm to ambient temperature. Chlorotrimethylsilane (0.25 mL, 2 equiv) was added dropwise, and the solution was stirred an additional 3 h. The reaction mixture was diluted with ether, and washed twice with water and once with brine. The organic layer was dried over $MgSO_4$, and the solvent was removed under reduced pressure to give a yellow oil. The crude product was purified by flash column chromatography on silica (hexanes and ethyl acetate, 98:2) to give **2** (333 mg, 0.61 mmol, 61% yield) as a pale yellow solid.

^{31}P NMR (121 MHz, $CDCl_3$): δ 249.9; ^{1}H NMR (300 MHz, $CDCl_3$): δ 0.23 (s, 9 H, $SiMe_3$), 1.37 (s, 9 H, *p-t*-Bu), 1.47 (s, 9 H, *o-t*-Bu), 1.55 (s, 9 H, *o-t*-Bu), 5.76 (d, 1 H, J_{PH}=13 Hz, CH), 7.2–7.7 (m, 7 H, arom.).

Step 2: To a solution of **2** (333 mg, 0.61 mmol) in THF (10 mL) at –98°C was slowly added a solution of *t*-butyllithium (1.59 M in pentane, 1.2 mL, 3 equiv.). The reaction mixture was stirred cold for 10 min, after which 1,2-dibromoethane (27 μL, 0.5 equiv.) was added dropwise. The reaction mixture was warmed to ambient temperature and stirred for an additional 6 h, during which time a color change from dark red to amber was observed. The solvent was removed under reduced pressure to give a viscous yellow oil. The crude product was purified by flash column chromatography on silica (hexanes and ethyl acetate, 98:2) to yield **6** (210 mg, 91% yield) as a yellow oil that solidified upon standing.

mp 240–243°C; ^{31}P NMR (121 MHz, $CDCl_3$): δ 169.5 (s); ^{1}H NMR (300 MHz, $CDCl_3$): δ 1.37 (s, 18 H, *p-t*-Bu), 1.54 (s, 36 H, *o-t*-Bu), 6.50 (d, 4 H, J_{HH}=7.4 Hz, *o*-Ph), 6.78 (d, J_{HH}=7.4 Hz, 4 H, *m*-Ph,), 6.97 (t, *J*=7.4 Hz, 2 H, *p*-Ph), 7.33 (s, 4 H, *m*-Mes*); ^{13}C{^{1}H} NMR (75 MHz, $CDCl_3$): δ 31.6 (s, *p*-C*Me*$_3$), 33.2 (virtual triplet, *J*=3 Hz, *o*-C*Me*$_3$), 35.0 (s, *p*-CMe$_3$), 38.8 (s, *o*-CMe$_3$), 121.8 (s, *m*-Mes*), 127.6 (s, *m*-Ph), 128.1 (s, *o*-Ph), 131.7 (s, *ipso*-Ph), 135.3 (virtual triplet, *J*=28 Hz, *ipso*-Mes*), 150.1 (s, *p*-Mes*), 154.7 (s, *o*-Mes*), 155.2 (t, J_{PC}=6 Hz, P=C-*C*), 176.2 (dd, J_{PC}=17, 10 Hz, P=*C*).

10.5.2 Synthesis of PhAk–Ox (32a) [50]

To a solution of $MesP(SiMe_3)_2$ (11.34 g, 38 mmol) in THF (25 mL) was added methyllithium (1.6 M, 24 mL, 38 mmol). The reaction mixture was heated to 55°C for 1–2 h. ^{31}P{^{1}H} NMR analysis of an aliquot removed from the reaction mixture suggested quantitative formation of MesP(Li)($SiMe_3$) (δ = –187 ppm). The reaction mixture was cooled to –78°C and treated with a solution of oxazoline **31** (9.93 g, 38 mmol) in THF (15 mL). After warming to room temperature, analysis of an aliquot removed from the reaction mixture by ^{31}P{^{1}H} NMR spectroscopy revealed a singlet resonance, which is consistent with phosphaalkene (**32a**) (δ=244). The reaction mixture was quenched with Me_3SiCl (4.14 g, 38 mmol), the solvent evaporated *in vacuo*, and the product extracted into hexanes (1×20 mL, 2×10 mL). After filtration, the hexanes were removed *in vacuo*. The product was recrystallized from *n*-pentane to afford a colorless crystalline solid, which was washed with cold *n*-pentane (2×5 mL) and dried *in vacuo* (7.87 g, 52%).

$[\alpha]_D^{22}$ −65.8 (*c* 0.50, $CDCl_3$); $^{31}P\{^1H\}$ NMR (121 MHz, $CDCl_3$): δ 244; ^1H NMR (400 MHz, $CDCl_3$): δ 7.00–6.50 (m, 7 H), 4.23–4.18 (m, 1 H), 3.99–3.95 (m, 1 H), 3.88–3.82 (m, 1 H), 2.26 (s, 3 H), 2.21 (s, 3 H), 2.09 (s, 3 H), 1.75–1.66 (m, 1 H), 1.67 (d, $^4J_{PH}$=1 Hz, 3 H), 1.64 (d, $^4J_{PH}$=1 Hz, 3 H), 0.90 (d, $^3J_{HH}$=7 Hz, 3 H), 0.81 (d, $^3J_{HH}$=7 Hz, 3 H); $^{13}C\{^1H\}$ NMR (100 MHz, $CDCl_3$): δ 203.3 (d, $^1J_{PC}$=48 Hz), 170.6 (d, $^3J_{PC}$=6 Hz), 142.8 (d, J_{PC}=15 Hz), 140.0 (d, J_{PC}=7 Hz), 139.8 (d, J_{PC}=6 Hz), 137.9, 135.9 (d, $^1J_{PC}$=41 Hz), 128.0 (d, J_{PC}=4 Hz), 126.9, 126.7, 126.6, 126.3, 72.1, 70.0, 47.5 (d, $^2J_{PC}$=25 Hz), 32.6, 28.2 (d, J_{PC}=9 Hz), 28.0 (d, J_{PC}=7 Hz), 22.6 (d, J_{PC}=7 Hz), 22.5 (d, J_{PC}=7 Hz), 21.1, 19.2, 18.0; LRMS (EI): *m/z* [%] 394, 393 [4, 15, M^+], 380, 379, 378 [12, 74, 100, $M^+ - CH_3$], 275, 274 [5, 24, M^+ − Mes]; analysis calculated for $C_{25}H_{32}NOP$: C, 76.31; H, 8.20; N, 3.56; found: C, 76.56; H, 8.17; N, 3.56.

10.6 Acknowledgments

The authors are grateful to the Natural Sciences and Engineering Research Council (NSERC) of Canada for their support of the research programs of DG and GD and scholarships to JDT. We are grateful to Professor Fumiyuki Ozawa for providing detailed procedures for the preparation of DPCB ligands, the graphic for Figure 10.4, and valuable comments on the manuscript. We thank Professor Masaaki Yoshifuji for providing valuable comments on the manuscript.

References

[1] T. E. Gier, *J. Am. Chem. Soc.* **1961**, *83*, 1769.
[2] M. J. Hopkinson, H. W. Kroto, J. F. Nixon, N. P. C. Simmons, *J. Chem. Soc., Chem. Commun.* **1976**, 513.
[3] K. Dimroth, P. Hoffman, *Angew. Chem. Int. Ed. Engl.* **1964**, *3*, 384.
[4] G. Märkl, *Angew. Chem. Int. Ed.* **1966**, *5*, 846.
[5] A. J. Ashe, *J. Am. Chem. Soc.* **1971**, *93*, 3293.
[6] G. Becker, *Z. Anorg. Allg. Chem.* **1976**, *423*, 242.
[7] T. C. Klebach, R. Lourens, F. Bickelhaupt, *J. Am. Chem. Soc.* **1978**, *100*, 4886.
[8] J. I. Bates, J. Dugal-Tessier, D. P. Gates, *Dalton Trans.* **2010**, *39*, 3151.
[9] F. Ozawa, M. Yoshifuji, *Dalton Trans.* **2006**, 4987.
[10] P. Le Floch, *Coord. Chem. Rev.* **2006**, *250*, 627.
[11] D. P. Gates, *Top. Curr. Chem.* **2005**, *250*, 107.
[12] L. Weber, *Coord. Chem. Rev.* **2005**, *249*, 741.
[13] F. Mathey, *Angew. Chem. Int. Ed.* **2003**, *42*, 1578.
[14] L. Weber, *Angew. Chem. Int. Ed.* **2002**, *41*, 563.
[15] L. Weber, *Eur. J. Inorg. Chem.* **2000**, 2425.
[16] S. Shah, J. D. Protasiewicz, *Coord. Chem. Rev.* **2000**, *210*, 181.
[17] M. Yoshifuji, *J. Chem. Soc. Dalton Trans.* **1998**, 3343.
[18] O. J. Scherer, *Angew. Chem. Int. Ed. Engl.* **1985**, *24*, 924.
[19] J. F. Nixon, *Chem. Rev.* **1988**, *88*, 1327.
[20] H. Eshtiagh-Hosseini, H. W. Kroto, J. F. Nixon, M. J. Maah, M. J. Taylor, *J. Chem. Soc., Chem. Commun.* **1981**, 199.
[21] T. C. Klebach, R. Lourens, F. Bickelhaupt, C. H. Stam, A. van Herk, *J. Organomet. Chem.* **1981**, *210*, 211.
[22] H. W. Kroto, J. F. Nixon, M. J. Taylor, A. A. Frew, K. W. Muir, *Polyhedron* **1982**, *1*, 89.
[23] F. Mathey, *Acc. Chem. Res.* **1992**, *25*, 90.

[24] K. B. Dillon, F. Mathey, J. F. Nixon, *Phosphorus: The Carbon Copy*, John Wiley & Sons, Ltd, **1998**.
[25] R. Appel, in *Multiple Bonds and Low Coordination in Phosphorus Chemistry*, M. Regitz, O. J. Scherer (Eds.), THIEME, **1990**, 157.
[26] M. Doux, A. Moores, N. Mézailles, L. Ricard, Y. Jean, P. Le Floch, *J. Organomet. Chem.* **2005**, *690*, 2407.
[27] F. Ozawa, S. Kawagishi, T. Ishiyama, M. Yoshifuji, *Organometallics* **2004**, *23*, 1325.
[28] S. Lacombe, D. Gonbeau, J. L. Cabioch, B. Pellerin, J. M. Denis, G. Pfister-Guillouzo, *J. Am. Chem. Soc.* **1988**, *110*, 6964.
[29] G. David, V. von der Gönna, E. Niecke, T. Busch, W. W. Schoeller, P. Rademacher, *J. Chem. Soc., Faraday Trans.* **1994**, *90*, 2611.
[30] W. W. Schoeller, C. Begemann, U. Tubbesing, J. Strutwolf, *J. Chem. Soc., Faraday Trans.* **1997**, *93*, 2957.
[31] K. Toyota, K. Masaki, T. Abe, M. Yoshifuji, *Chem. Lett.* **1995**, 221.
[32] F. Ozawa, M. Yoshifuji, *C. R. Chim.* **2004**, *7*, 747.
[33] K. Toyota, K. Horikawa, R. S. Jensen, K. Omori, S. Kawasaki, S. Ito, M. Yoshifuji, N. Morita, *Bull. Chem. Soc. Jpn.* **2007**, *80*, 1580.
[34] M. Yoshifuji, Y. Ichikawa, N. Yamada, K. Toyota, *Chem. Commun.* **1998**, 27.
[35] A. H. Cowley, N. C. Norman, M. Pakulski, G. Becker, M. Layh, E. Kirchner, M. Schmidt, *Inorg. Synth.* **1990**, *27*, 235.
[36] M. van der Sluis, J. B. M. Wit, F. Bickelhaupt, *Organometallics* **1996**, *15*, 174.
[37] K. Toyota, J. Ujita, S. Kawasaki, K. Abe, N. Yamada, M. Yoshifuji, *Tetrahedron Lett.* **2004**, *45*, 7609.
[38] R. Appel, J. Hünerbein, N. Siabalis, *Angew. Chem. Int. Ed. Engl.* **1987**, *26*, 779.
[39] S. Ito, K. Toyota, M. Yoshifuji, *Chem. Commun.* **1997**, 1637.
[40] H. Z. Liang, K. Nishide, S. Ito, M. Yoshifuji, *Tetrahedron Lett.* **2003**, *44*, 8297.
[41] O. Daugulis, M. Brookhart, P. S. White, *Organometallics* **2002**, *21*, 5935.
[42] A. Jouaiti, M. Geoffroy, G. Bernardinelli, *Tetrahedron Lett.* **1992**, *33*, 5071.
[43] M. Yam, J. H. Chong, C. W. Tsang, B. O. Patrick, A. E. Lam, D. P. Gates, *Inorg. Chem.* **2006**, *45*, 5225.
[44] B. Deschamps, X. Le Goff, L. Ricard, P. Le Floch, *Heteroatom Chem.* **2007**, *18*, 363.
[45] K. Nishide, H. Z. Liang, S. Ito, M. Yoshifuji, *J. Organomet. Chem.* **2005**, *690*, 4809.
[46] S. Ito, H. Z. Liang, M. Yoshifuji, *J. Organomet. Chem.* **2005**, *690*, 2531.
[47] H. Z. Liang, S. Ito, M. Yoshifuji, *Org. Lett.* **2004**, *6*, 425.
[48] R. Takita, Y. Takada, R. S. Jensen, M. Okazaki, F. Ozawa, *Organometallics* **2008**, *27*, 6279.
[49] O. Riant, O. Samuel, H. B. Kagan, *J. Am. Chem. Soc.* **1993**, *115*, 5835.
[50] J. Dugal-Tessier, G. R. Dake, D. P. Gates, *Angew. Chem. Int. Ed.* **2008**, *47*, 8064.
[51] M. J. Kurth, O. H. W. Decker, H. Hope, M. D. Yanuck, *J. Am. Chem. Soc.* **1985**, *107*, 443.
[52] J. Dugal-Tessier, S. C. Serin, E. B. Castillo-Contreras, E. D. Conrad, G. R. Dake, D. P. Gates, *Chem. Eur. J.* **2012** (In Press).
[53] S. D. Ittel, L. K. Johnson, M. Brookhart, *Chem. Rev.* **2000**, *100*, 1169.
[54] S. Ikeda, F. Ohhata, M. Miyoshi, R. Tanaka, T. Minami, F. Ozawa, M. Yoshifuji, *Angew. Chem. Int. Ed.* **2000**, *39*, 4512.
[55] A. Ionkin, W. Marshall, *Chem. Commun.* **2003**, 710.
[56] A. de Meijere, F. Diederich, *Metal-Catalyzed Cross-Coupling Reactions*, 2nd ed., Wiley-VCH, **2004**.
[57] E. Negishi, L. Anastasia, *Chem. Rev.* **2003**, *103*, 1979.
[58] N. Miyaura, A. Suzuki, *Chem. Rev.* **1995**, *95*, 2457.
[59] P. Espinet, A. M. Echavarren, *Angew. Chem. Int. Ed.* **2004**, *43*, 4704.
[60] J. P. Wolfe, S. Wagaw, J.-F. Marcoux, S. L. Buchwald, *Acc. Chem. Res.* **1998**, *31*, 805.
[61] J. F. Hartwig, *Acc. Chem. Res.* **1998**, *31*, 852.
[62] S. V. Ley, A. W. Thomas, *Angew. Chem. Int. Ed.* **2003**, *42*, 5400.
[63] A. S. Gajare, R. S. Jensen, K. Toyota, M. Yoshifuji, F. Ozawa, *Synlett* **2005**, 144.
[64] R. S. Jensen, A. S. Gajare, K. Toyota, M. Yoshifuji, F. Ozawa, *Tetrahedron Lett.* **2005**, *46*, 8645.
[65] M. Sundermeier, A. Zapf, M. Beller, *Angew. Chem. Int. Ed.* **2003**, *42*, 1661.
[66] F. Q. Jin, P. N. Confalone, *Tetrahedron Lett.* **2000**, *41*, 3271.
[67] A. S. Gajare, K. Toyota, M. Yoshifuji, F. Ozawa, *J. Org. Chem.* **2004**, *69*, 6504.

[68] A. S. Gajare, K. Toyota, M. Yoshifuji, F. Ozawa, *Chem. Commun.* **2004**, 1994.
[69] B. Marciniec, in *Advances in Silicon Science*, Vol. 1, J. Matisons (Ed.), Springer, **2009**.
[70] F. Ozawa, S. Yamamoto, S. Kawagishi, M. Hiraoka, S. Ikeda, T. Minami, S. Ito, M. Yoshifuji, *Chem. Lett.* **2001**, 972.
[71] A. Hayashi, T. Yoshitomi, K. Umeda, M. Okazaki, F. Ozawa, *Organometallics* **2008**, *27*, 2321.
[72] H. Katayama, M. Nagao, T. Nishimura, Y. Matsui, K. Umeda, K. Akamatsu, T. Tsuruoka, H. Nawafune, F. Ozawa, *J. Am. Chem. Soc.* **2005**, *127*, 4350.
[73] T. Minami, H. Okamoto, S. Ikeda, R. Tanaka, F. Ozawa, M. Yoshifuji, *Angew. Chem. Int. Ed.* **2001**, *40*, 4501.
[74] A. Hayashi, M. Okazaki, F. Ozawa, R. Tanaka, *Organometallics* **2007**, *26*, 5246.
[75] R. S. Jensen, K. Umeda, M. Okazaki, F. Ozawa, M. Yoshifuji, *J. Organomet. Chem.* **2007**, *692*, 286.
[76] Z. G. Li, C. Brouwer, C. He, *Chem. Rev.* **2008**, *108*, 3239.
[77] S. Ito, S. Kusano, N. Morita, K. Mikami, M. Yoshifuji, *J. Organomet. Chem.* **2010**, *695*, 291.
[78] S. Ito, M. Freytag, H. Liang, K. Nishide, M. Yoshifuji, *Phosphorus, Sulfur Silicon Relat. Elem.* **2008**, *183*, 555.
[79] M. Freytag, S. Ito, M. Yoshifuji, *Chem. Asian J.* **2006**, *1*, 693.
[80] J. Dugal-Tessier, G. R. Dake, D. P. Gates, *Organometallics* **2007**, *26*, 6481.
[81] B. M. Trost, M. L. Crawley, *Chem. Rev.* **2003**, *103*, 2921.
[82] F. Ozawa, H. Okamoto, S. Kawagishi, S. Yamamoto, T. Minami, M. Yoshifuji, *J. Am. Chem. Soc.* **2002**, *124*, 10968.
[83] O. Piechaczyk, C. Thoumazet, Y. Jean, P. Le Floch, *J. Am. Chem. Soc.* **2006**, *128*, 14306.
[84] F. Ozawa, T. Ishiyama, S. Yamamoto, S. Kawagishi, H. Murakami, M. Yoshifuji, *Organometallics* **2004**, *23*, 1698.
[85] H. Murakami, Y. Matsui, F. Ozawa, M. Yoshifuji, *J. Organomet. Chem.* **2006**, *691*, 3151.
[86] H. Murakami, T. Minami, F. Ozawa, *J. Org. Chem.* **2004**, *69*, 4482.
[87] J. Dugal-Tessier, G. R. Dake, D. P. Gates, *Org. Lett.* **2010**, *12*, 4667.

11

Phosphaalkynes

Christopher A. Russell and Nell S. Townsend

School of Chemistry, University of Bristol, Bristol, United Kingdom

11.1 Introduction

With hindsight, it now seems amazing to think that only 50 years ago compounds involving $p\pi$–$p\pi$ multiple bonds of the third-period elements and their heavier homologues were thought to be inaccessible, a statement that was encapsulated in the so-called double bond rule. The path from being a wholly speculative bond type to being prepared on the multigram scale is well documented but worthy of retelling. The parent HC≡P was first proposed as a reactive intermediate by Albers in 1950 [1], but it took more than a decade before this species was observed experimentally. In his ground-breaking study, Gier pyrolised PH_3 gas using rotating arc graphite electrodes, and in the somewhat unholy brew that was generated he was able to firstly chromatograph and secondly identify HC≡P by IR spectroscopy [2]. However, the compound was reported to be pyrophoric and decomposed at very low temperatures. As an addendum to this story, Mathey and Le Floch were later able to synthesise this compound via a route which did not produce highly reactive by-products; under these conditions HC≡P proved to be far more stable than was initially believed [3]. Despite the reported lack of stability, this observation served to inspire further research in the field which culminated in the 1981 report of Becker and co-workers of the preparation of the first kinetically stabilised phosphaalkyne, ^{t}BuC≡P [4]. We have revised the Becker preparation of this material to make it safer and more accessible, and this synthesis is presented in this chapter. This breakthrough represented a defining moment in phosphaalkyne chemistry, and since then literally hundreds of examples of phosphaalkynes have been prepared and their chemistry thoroughly examined. Notably it has been found that their chemistry is more closely aligned to alkynes rather than nitriles, which is also reflected in the common name being phosphaalkyne as opposed to alternatives such as phosphacyanide. Of particular relevance to this chapter is the vast number of organometallic compounds that have been synthesised covering the bulk of the transition series [5]. Photoelectron spectroscopy studies have established that the highest occupied molecular orbital (HOMO) is the π-type

Phosphorus(III) Ligands in Homogeneous Catalysis: Design and Synthesis, First Edition.
Edited by Paul C. J. Kamer and Piet W. N. M. van Leeuwen.

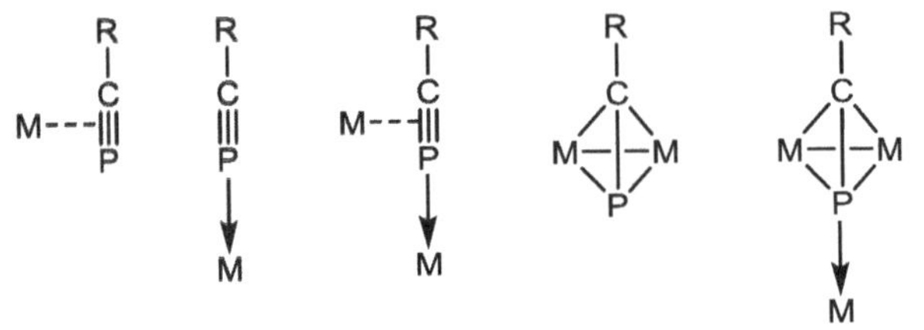

Figure 11.1 *Binding modes of phosphaalkynes at mono and dimetallic centres.*

P≡C triple bond [6], and thus the η[2]-ligation mode is preferred in binding to a metal. In common with alkynes, there are numerous modes through which the P≡C triple bond may interact with either mono- or multinuclear metal centres (Figure 11.1), but in contrast to alkynes the nominal lone pair on the P atom may in addition be used to bind to additional metal centres. If the phosphaalkyne is forced to bind into a 'pocket', then the η[1]-ligation mode may be observed [5].

Despite the wealth of publications on this topic that have appeared over the past 30 years, almost all reports have tended to focus on the intricacies of what is still regarded as a rather unusual multiple bond rather than attempt to explore the applications of such an unusual phosphorus ligand in, for example, homogeneous catalysis. It is not the aim of this chapter to set out how wonderful such species are as ligands for catalysis, but merely to redress the perception that they are inaccessible chemical exotica of no use in applications. Not only can these chemicals be accessed on a reasonable scale using conventional bench-top techniques, but also their derivitisation to provide access to a range of novel P(III) species is readily achieved. Within this class of derivatives, we should highlight the phospholide anions which are known to have the important quality of stabilising low-oxidation state transition metals by promoting δ-back bonding but whose potential as an auxiliary ligand in catalysis is untapped. We hope that this will serve to demystify this area and reveal hitherto its unrealised potential as ligands in catalysis.

11.2 General experimental

All reactions and manipulations should be performed under an inert atmosphere (dry nitrogen or argon), using standard Schlenk line and glove box techniques. Common solvents (tetrahydrofuran, toluene, n-hexane and dichloromethane) should be dried and degassed prior to use (e.g. using a commercially available solvent purification system) [7]. More specialist solvents should also be dried and distilled prior to use in the following preparation: dimethoxyethane (DME) was heated under reflux with sodium–benzophenone; C_6D_6 was stirred over molten potassium, then vacuum transferred; $CDCl_3$ and cyclohexane were dried over molecular sieves and degassed by freeze, pump and thaw methods; D_8-THF was stirred over a potassium mirror and then distilled; and pivaloyl chloride was dried by stirring over CaH_2 and distilled prior to use. All other reagents were obtained from commercial sources and used as received.

11.3 Preparation of PC^{t}Bu

The synthetic pathway described in this section is based ostensibly on the original literature procedure [4], but incorporates both slightly modified reagents and apparatus leading to numerous safety benefits. Our preparation (Scheme 11.1) replaces the hazardous mixture of white phosphorus and Na–K alloy used in the original preparation with the less hazardous red phosphorus and Na–naphthalene. During the preparation of tris(trimethylsilyl)phosphine, the previous literature procedure invoked using high-volume sinter sticks

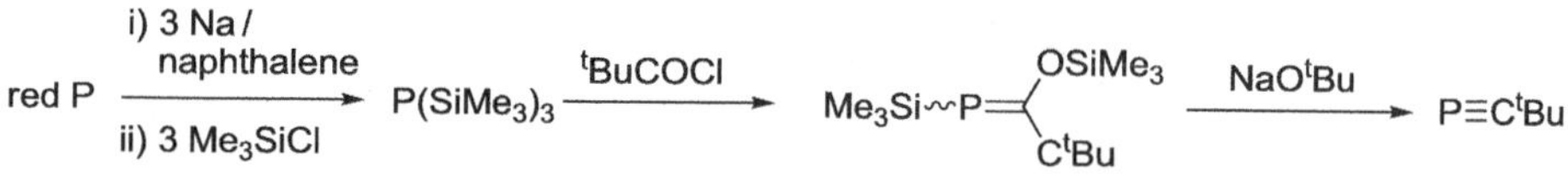

Scheme 11.1 *Synthesis of* tert-*butylphosphaalkyne.*

leading to a precarious array of large glassware being used to filter a pyrophoric material; this is replaced in the current procedure by filtration of the reaction mixture using a cannula to transfer the turgid solution through a Celite® layer over a sinter – a procedure that is both faster and safer. In the final step to generate the phosphaalkynes, the base NaO^{t}Bu proved to be more reliable than that used in the original literature report (NaOH). A revised isolation procedure was also implemented by using a column filled with Fenske helices, providing a more reliable high-volume separation procedure than that provided by trap-to-trap distillation used in the original literature method.

11.3.1 Tris(trimethylsilyl)phosphine, $P(SiMe_3)_3$ (Scheme 11.2)

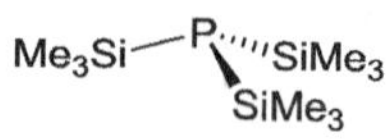

Scheme 11.2 *Tris(trimethylsilyl)phosphine.*

Although there are many reported routes towards phosphaalkynes, the vast majority of large-scale preparations of these interesting substrates start from this highly reactive bulky phosphine. Although high inherent reactivity is rarely a desirable feature in ligands, this phosphine has found application as a supporting ligand in the nickel-catalysed reaction of carbonyls and dienes using a stoichiometric amount of bis(pinacolato)diboron furnishing allyl boronic esters [8]; previously claims in the patent literature have also identified the use of this phosphine in Pd-catalysed aryl amination, although these reports have yet to appear in peer-reviewed journals [9].

WARNING: THE PRODUCT OF THIS LARGE LABORATORY-SCALE PREPARATION, TRIS (TRIMETHYLSILYL)PHOSPHINE, IS PYROPHORIC AND MUST BE HANDLED WITH EXTREME CARE. THIS SYNTHESIS MUST BE UNDERTAKEN ONLY BY SKILLED RESEARCHERS WITH EXPERIENCE OF HANDLING PYROPHORIC CHEMICALS.

To a 2 L multinecked round-bottomed flask fitted with a reflux condenser and dropping funnel and placed in a sand tray, red phosphorus (25 g, 800 mmol), naphthalene (5 g, 39 mmol) and dimethoxyethane (1 L) were added and the contents stirred by an overhead stirrer fitted with a PTFE or glass stirrer paddle. The mixture was heated to *ca.* 40°C, then sodium (55 g, 2400 mmol, chopped into *ca.* $5\times5\times5\,mm^3$ pieces) was added over 2–3 h. When all the sodium was added, the solution was heated at reflux for *ca.* 16 h. The solution was allowed to cool and the dropping funnel charged with Me_3SiCl (305 mL), which was added dropwise to the stirred solution over several hours. Extra DME (*ca.* 300 mL) was then added to reduce the viscosity of the slurry to a 'custard' consistency (i.e. mobile slurry as opposed to paste) which was then heated at reflux for 16 h. The viscous soup was filtered using large porosity 3 sinter and a 20–30 mm plug of Celite® by cannula transfer (see Figure 11.2, using PTFE tubing, internal diameter *ca.* 10 mm, as the cannula). The filtrate was washed with a total of *ca.* 0.5 L DME in several portions. The solvent was removed *in vacuo* into an external trap leaving a yellow oil. $P(SiMe_3)_3$ was isolated (*ca.* 200 mL, 86% yield) as a colourless viscous liquid

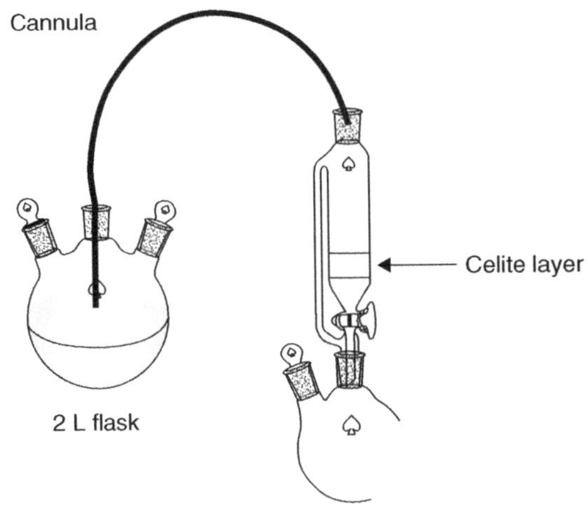

Figure 11.2 *Filtration apparatus.*

following purification by distillation at *ca.* 200°C under vacuum (10^{-2} mmHg) with the distillate being condensed using a liquid nitrogen bath.

$^{31}P\{^1H\}$NMR ($CDCl_3$, 122 MHz): δ −249 (s).
^1H NMR ($CDCl_3$, 300 MHz): δ 0.3 (d, $^3J_{PH}$ = 6 Hz).

A small amount of an impurity is often observed attributable to $HP(SiMe_3)_2$$\delta_P$ −234 (d, $^3J_{PH}$ = 192 Hz); this impurity has no detrimental effect on reactivity in the reactions described in the rest of this section.

11.3.2 *tert*-butylphosphaalkene, $Me_3SiP{=}C(OSiMe_3)^tBu$ (systematic name [2,2-dimethyl-1-(trimethylsiloxy)propylidene]–(trimethylsilyl) phosphine) (Scheme 11.3)

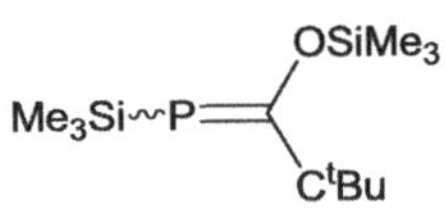

Scheme 11.3 tert-*butylphosphaalkene.*

To a Schlenk flask containing a solution of $P(SiMe_3)_3$(100 mL, 400 mmol) in cyclohexane (100 mL), pivaloyl chloride (50 mL, 400 mmol) was added and an immediate colour change from colourless to yellow was observed. The mixture was heated at reflux for 16 h, then the solvent was removed *in vacuo* to give a yellow oil, $Me_3SiP{=}C(OSiMe_3)^tBu$, in quantitative yield.

Density: 0.935 g/mL at 25°C.
$^{31}P\{^1H\}$ NMR ($CDCl_3$, 122 MHz): δ 117 (s), 132 (s) E/Z isomers.
^1H NMR ($CDCl_3$, 300 MHz): δ 1.22 (9H, d, $^4J_{PH}$ = 1.5 Hz, ^tBu) 0.35 (9H, s, $OSiMe_3$), 0.25 (9H, d, $^3J_{PH}$ = 4 Hz, $PSiMe_3$).
IR (cm^{-1}): 2959, 1253, 1190, 839, 756.

11.3.3 *tert*-butylphosphaalkyne, systematic name (2,2-dimethylpropylidyne)phosphine; $^tBuC{\equiv}P$ (Scheme 11.4)

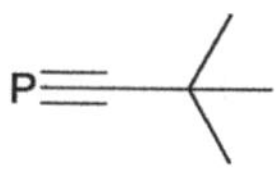

Scheme 11.4 *(2,2-dimethylpropylidyne)phosphine.*

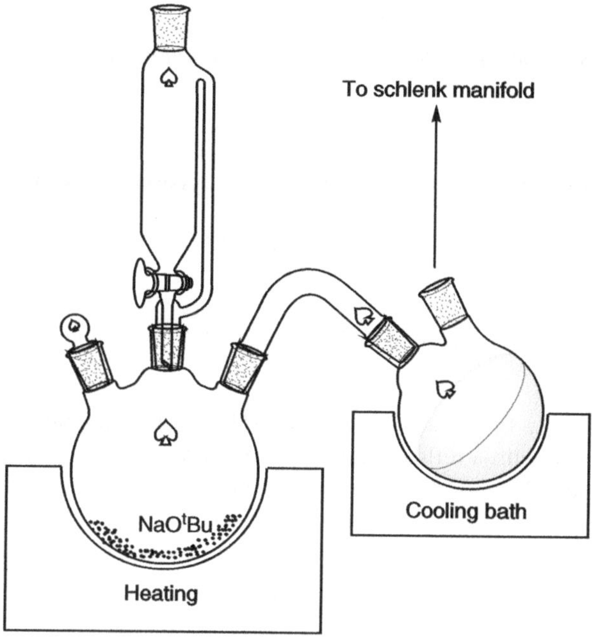

Figure 11.3 *Apparatus for the preparation of* tert-*butylphosphaalkyne.*

To a flask under a static vacuum of 1×10^{-2} mm Hg heated to 160°C (silicone oil *or* graphite bath *or* solid-state heating block) containing NaO^{t}Bu (4 g, 42 mmol), neat $Me_3SiP{=}C(OSiMe_3)^tBu$ (from the reaction above, *ca.* 100 mL) was added slowly over 2 h via a dropping funnel (see Figure 11.3). The reaction was heated for a further 30 min. The product was distilled into a second flask cooled in liquid nitrogen at *ca.* 60°C giving a yellow liquid. This was further distilled using a column filled with Fenske helices. The fraction collected between 60°C and 70°C yields $^tBuC{\equiv}P$ (*ca.* 15 mL, 40% yield) as a colourless liquid. The colourless fraction collected between 70°C and 90°C contains *ca.* 50% siloxane impurity, $Me_3SiOSiMe_3$ (^{1}H NMR $CDCl_3$ δ 0.8 (s)).

Density: 0.788 g/mL at 25°C.
^{31}P{^{1}H} NMR ($CDCl_3$, 122 MHz): δ −69 (s).
^{1}H NMR ($CDCl_3$, 300 MHz): δ 1.3 (d, $^4J_{PH}$ = 0.98 Hz, ^{t}Bu).
^{13}C{^{1}H} NMR ($CDCl_3$, 100 MHz): δ 185 (d, $^1J_{PC}$ = 38 Hz, $PCCMe_3$) 36 (d, $^2J_{PC}$ = 18 Hz, CMe_3) 31 (d, $^3J_{PC}$ = 6 Hz, CMe_3) 2 (s, $SiMe_3$).
IR (cm^{-1}): 1533 (C≡P Str.).

11.4 Adamanylphosphaalkyne, AdC≡P [10] (Scheme 11.5)

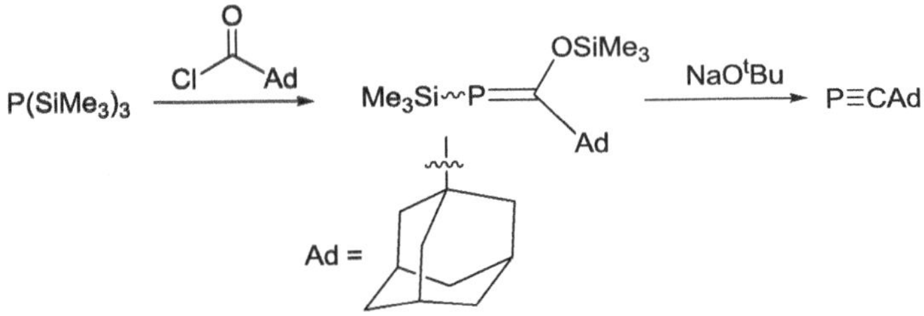

Scheme 11.5 Synthesis of adamantylphosphaalkyne.

11.4.1 Adamant-1-yl(trimethylsiloxy)methylidene (trimethylsilyl) phosphine (Scheme 11.6)

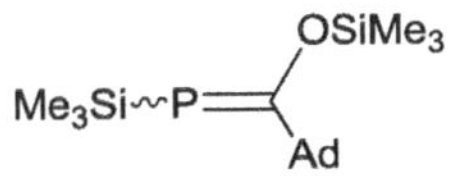

Scheme 11.6 Adamant-1-yl(trimethylsiloxy)methylidene (trimethylsilyl) phosphine.

A solution of $P(SiMe_3)_3$ (42 mL, 144 mmol) in toluene (150 mL) was prepared in a 500 mL three-necked round-bottomed flask equipped with a reflux condenser and a pressure-equalising dropping funnel. A solution of 1-adamantoyl chloride (32.3 g, 62 mmol) in toluene (130 mL) was added dropwise to the stirred solution. The mixture was heated at reflux for 24 h, and a colour change from colourless to yellow was observed. The solvent was removed *in vacuo* to yield adamant-1-yl(trimethylsiloxy)methylidene (trimethylsilyl) phosphine as a viscous yellow-orange oil (40 mL, 90%).

^{31}P {^{1}H} NMR (C_6D_6, 122 MHz): δ 122.5 (s, *Z*-isomer), 104.2 (s, *E*-isomer).
^{1}H NMR (C_6D_6, 300 MHz): δ 0.29 (9 H, s, $OSiMe_3$), 0.38 (9 H, d, $^3J_{PH}$ = 3.9 Hz, $PSiMe_3$), 1.59–2.00 (15 H, m, Ad-H).

11.4.2 (Adamant-1-ylmethylidyne)phosphine (Scheme 11.7)

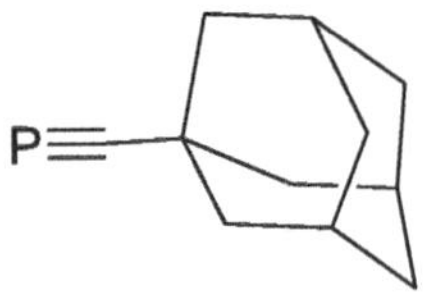

Scheme 11.7 (Adamant-1-ylmethylidyne)phosphine.

A 100 mL round-bottomed flask charged with NaO^{t}Bu (2.0 g, 208 mmol) and fitted with a dropping funnel containing adamantylphosphaalkene (15 mL, 480 mmol) was heated to *ca.* 160°C and placed under active vacuum (*ca.* 10^{-2} mmHg). The phosphaalkene was then added dropwise to the heated flask, generating adamantylphosphaalkyne, which is trapped in the receiver flask which is held at −78°C. Sublimation of the distillate (110°C, 10^{-2} mm Hg) yields adamantylphosphaalkyne as a pale yellow solid. Good yields (*ca.* 80%) have been reported for this reaction.

IR: υ_{max}(neat)/cm^{-1} 1524 (C≡P str.).
^{31}P {^{1}H} NMR (C_6D_6, 122 MHz): δ –67.1 (s).
^{1}H NMR (C_6D_6, 400 MHz): δ 1.40–1.90 (15 H, m, Ad–H).
^{13}C {^{1}H} NMR (C_6D_6, 100 MHz): δ 28.3 (s), 36.5 (s), 38.9 (d, $^2J_{CP}$ = 18.5 Hz), 43.7 (d, $^3J_{CP}$ = 5.4 Hz), 184.9 (d, $^1J_{CP}$ = 37.7 Hz).

11.5 Mesitylphosphaalkyne, MesC≡P

The method described in this section (Scheme 11.8) differs slightly from the literature-reported procedure [11], using potassium bis(trimethylsilyl)phosphide instead of the lithium salt.

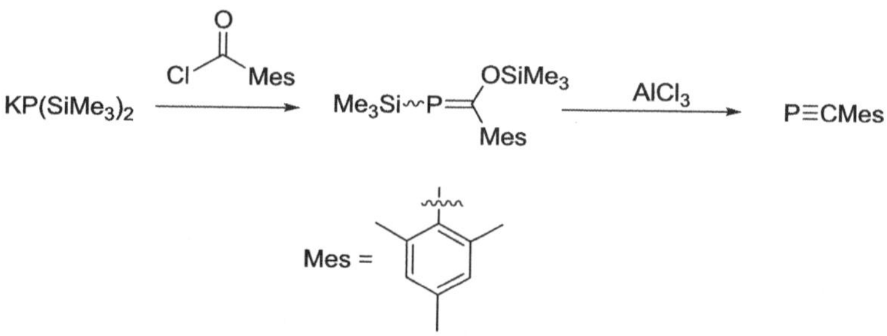

Scheme 11.8 *Synthesis of mesitylphosphaalkyne.*

11.5.1 Preparation of potassium bis(trimethylsilyl)phosphide {$KP(SiMe_3)_2$}

To a stirred solution of tris(trimethylsilyl)phosphine (11.6 mL, 40 mmol) in THF (100 mL) was added dropwise a solution of KO^{t}Bu (4.5 g, 40 mmol) in THF (50 mL). Heat was evolved, and the solution became pink-red. After stirring at room temperature for 4 h, the reaction had gone to completion (^{31}P NMR). The solvent was removed *in vacuo* to afford a pink powder, which was suspended in diethyl ether (20 mL) and again liberated of solvent *in vacuo*. The process was repeated three more times before the solid was dried under vacuum for 2 h. The pale pink solid ($KP(SiMe_3)_2$·0.25 THF·0.06 Et_2O) was isolated in 96% yield.

^{31}P{^{1}H} NMR (C_5D_5N, 121.4 MHz,): δ –295.08 (s).
^{1}H NMR (C_5D_5N, 300.0 MHz): δ 0.75 (d, $^3J_{PH}$ = 3.12 Hz, $PSi(CH_3)_3$).
Impurities: δ 1.18 (t, $^3J_{HH}$ = 6.97 Hz, Et_2O), 1.66 (m, THF), 3.39 (q, $^3J_{HH}$ = 6.97 Hz, Et_2O), 3.68 (m, THF).

11.5.2 Mesityl(trimethylsiloxy)methylene trimethylsilylphosphine (Scheme 11.9)

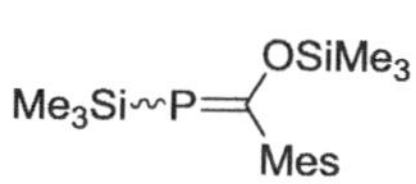

Scheme 11.9 *Mesityl(trimethylsiloxy)methylene trimethylsilylphosphine.*

To a flask containing mesitylene-2-carboxylic acid chloride (1.26 g, 6.9 mmol) in n-hexane (10 mL), potassium bis(trimethylsilyl)phosphide (2.0 g, 6.9 mmol) in n-hexane (50 mL) was added via a dropping funnel. On addition the reaction mixture turned from colourless to orange, and after 2.5 h stirring at 25°C the reaction mixture was yellow with a cream precipitate. The mixture was transferred to a sinter with a plug of Celite® via filter cannula and the solvent removed *in vacuo* to yield a yellow oil, which was purified by distillation at 150–160°C under an active vacuum using a liquid nitrogen cooling bath to obtain 1.60 g (73%).

$^{31}P\{^{1}H\}$ NMR (C_6D_6, 122 MHz): δ 126.3 (s, *E*-isomer), 122.4 (s, *Z*-isomer).
^{1}H NMR (C_6D_6, 300 MHz): E-isomer δ0.05 (d, $^{3}J_{HP}$=4.4 Hz, 9 H, $PSiMe_3$), 0.47(d, $^{5}J_{HP}$=1.1 Hz, 9 H, $OSiMe_3$), 2.10(s, 3 H, *p*-CH_3), 2.43(s, 6 H, *o*-CH_3), 6.70(s, 2 H,aryl-H) *Z*-isomer δ –0.04(s, 9 H, $OSiMe_3$), 0.5(d, $^{3}J_{HP}$=3.6 Hz, 9 H, $PSiMe_3$), 2.10(s, 3 H, *p*-CH_3), 2.36(s, 6 H, *o*-CH_3), 6.70(s, 2 H,aryl-H).

11.5.3 Mesitylphosphaalkyne (Scheme 11.10)

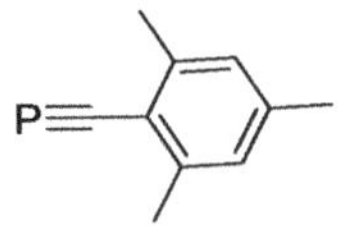

Scheme 11.10 Mesitylphosphaalkyne.

To a suspension of $AlCl_3$ (0.822 g, 6.16 mmol) in CH_2Cl_2 (15 mL), a solution of mesitylphoshaalkene (2.0 g, 6.16 mmol) in CH_2Cl_2(10 mL) was added dropwise at –40°C with vigorous magnetic stirring. The reaction was left for 2 h at this temperature, and then the solvent removed *in vacuo* (–20°C/10^{-3} mbar) to yield a dark brown residue. Immediate distillation (50–70°C/10^{-3} mbar) followed by filtration through silica using n-hexane yields the product as a colourless oil (0.299 g, 30%). The product is best stored at –78°C.

$^{31}P\{^{1}H\}$ NMR (C_6D_6, 122 MHz): δ 2.5 (s).
^{1}H NMR (C_6D_6, 300 MHz): δ 1.98 (s, 3 H, *p*-CH_3), 2.39 (s, 6 H, *o*-CH_3), 6.56 (s, 2 H, aryl-H).

11.6 Phospholide anions

The phosphaalkynes described in this chapter are an interesting class of ligand in their own right. However, the P≡C triple bond may be manipulated to give other species which have been shown to have useful properties. One key class of compound is the phospholide anions, which are phosphorus-containing analogues of the cyclopentadienyl anion. These have been shown to stabilise low-oxidation state transition metal complexes by promoting δ-back donation into the heteroatom-containing ring system. Spectacular examples of stabilisation of low-oxidation states are seen in the complexes $[Ti\{\eta^5\text{-}P_3(C^tBu)_2\}_2]$ [12], $[Ti(\eta^5\text{-}P_5)_2]^{2-}$ [13] and $[\{\eta^5\text{-}P_3(C^tBu)_2\}Sc\{\mu\text{-}\eta^6\text{-}P_3(C^tBu)_3\}Sc\{\eta^5\text{-}P_3(C^tBu)_2\}]$ [14], none of which have direct carbocyclic analogues. In the first two complexes we have Ti(II) and Ti(0), respectively; incredibly the stabilisation provided by the all-phosphorus analogue of the cyclopentadienyl anion in the latter is so pronounced that the complex is air stable. In the last of these complexes, scandium is present in the formal +1 oxidation state which is again highly unusual and testament to the stabilising characteristics of the triphospholide anion.

In addition to these complexes, there are numerous examples of phospholide complexes as ligands towards a wide variety of the s-, p- and d-block elements. The majority are synthesised from transmetallation using an appropriate alkali metal salt of the phospholide anion. An alternative route via transmetallation of the triphenylstannyl derivative of the triphospholide anion has also been reported[15]; this is often preferable as the precursor may be synthesised and stored indefinitely under an inert atmosphere prior to use – a synthesis of this tin adduct is presented in this section. In many examples the phospholide anions form complexes which merely mimic their all-carbon counterparts, albeit with the added potential to use the lone pairs on the phosphorus centres to form adducts to additional metals – such a strategy has been used to striking effect, in particular by the group of Scheer to produce supramolecular aggregates based on a fullerene-type topology [16].

Herein we present two routes towards the triphospholide anion. The second route (Sections 11.6.4–11.6.5) is preferable in terms of both yield and simplicity, but the first route (Sections 11.6.1–11.6.3) offers ready adaptability in incorporating other heteroatoms into the ring system, notably by varying reagents in Section 11.6.2.

11.6.1 Preparation of $Cp_2Zr(PC^tBu)_2$ [17] (Scheme 11.11)

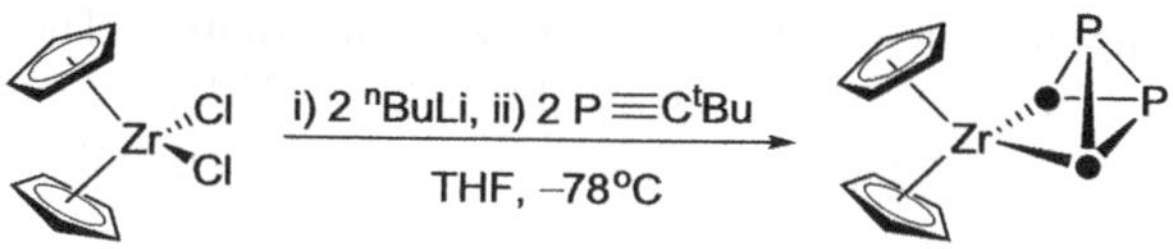

Scheme 11.11 *Preparation of $Cp_2Zr(PC^tBu)_2$.*

To a solution of Cp_2ZrCl_2 (0.84 g, 2 mmol) in THF (25 mL) at −20°C, ^nBuLi (2.5 mL of a 1.6 M solution in hexanes, 4 mmol) was added dropwise and stirred for 15 min, forming a yellow solution. $P{\equiv}C^tBu$ (0.5 mL, 4 mmol) was then added to the reaction mixture, which was warmed to room temperature and stirred for 2 h; a colour change to deep red was observed. The reaction mixture was diluted with CH_2Cl_2 (20 mL), filtered through Celite® to remove the insoluble LiCl precipitate and washed with CH_2Cl_2 (3 × 10 mL) before removing the solvent *in vacuo* to yield the product, an orange-red powder (0.95 g, 75%).

^{31}P {^1H} NMR ($CDCl_3$, 122 MHz): δ −246 (s).
^1H NMR ($CDCl_3$, 300 MHz): δ 5.9 (10 H, s, C_5H_5) 1.2 (18 H, d, $^4J_{PH}$ = 0.92, ^{t}Bu).

11.6.2 Preparation of $ClP(PC^tBu)_2$ (Scheme 11.12)

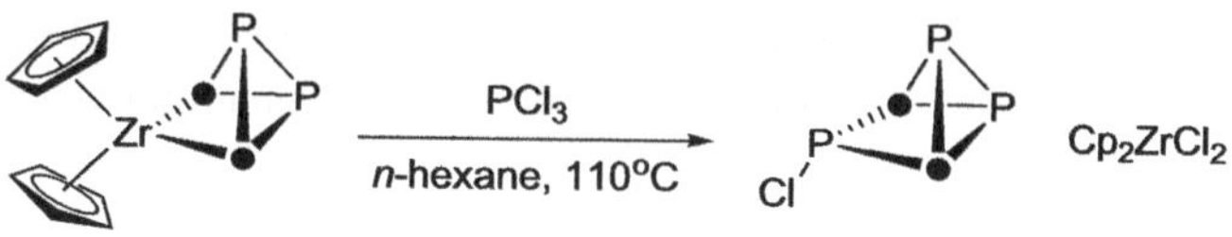

Scheme 11.12 *Preparation of $ClP(PC^tBu)_2$.*

In a large Young's flask with excess head space, phosphorus trichloride (3.4 mL, 25 mmol) was added dropwise to a solution of $Cp_2Zr(PC^tBu)_2$ (0.5 g, 1.20 mmol) in n-hexane (20 mL). The system was sealed and heated to 110°C for 2.5 h, and a mustard yellow mixture with yellow precipitate (Cp_2ZrCl_2) was formed. The reaction mixture cooled to room temperature before filtering through Celite®, washing with hexane (3 × 10 mL) and drying *in vacuo* to yield a yellow oil.

^{31}P{^1H} NMR (CD_2Cl_2, 121.42 MHz): δ 239.6 (dd, $^2J_{PP}$ 61, 54 Hz), −117.6 (dd, $^1J_{PP}$ = 145 Hz, $^2J_{PP}$ = 54 Hz), −229.8 (dd, $^1J_{PP}$ = 145 Hz, $^2J_{PP}$ = 61 Hz).

11.6.3 Preparation of the triphospholide anion and derivation to give the triphenylstannylphosphole (Scheme 11.13)

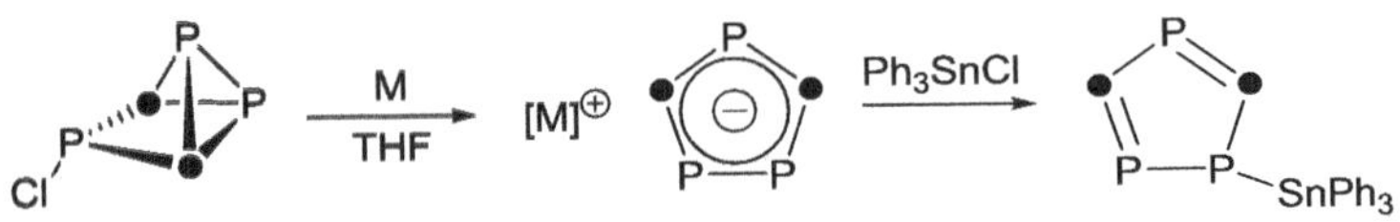

Scheme 11.13 Preparation of the triphospholide anion.

In a typical experiment, $ClP(PC^tBu)_2$ (0.172 g, 0.65 mmol) was dissolved in THF (5 mL) and added to a suspension of Na pieces (0.100 g, 4.34 mmol) in THF (5 mL). The resulting mixture was stirred for 2 h, after which time it had turned deep red indicating the formation of the triphospholide anion: a NMR spectrum recorded at this time indicated that conversion was quantitative. The solution was then filtered and cooled to −30°C, and a THF solution (10 mL) of Ph_3SnCl (0.272 g, 0.71 mmol) was added portion-wise, causing an instant colour change to yellow. Conversion was again quantitative, as shown by ^{31}P NMR spectroscopy.

The reaction with Li, K, Rb, Cs and Mg could be accomplished in the same manner using an excess of s-block metal; if required, this should be activated by sonication.

11.6.4 Preparation of $Cl_3P_3(C^tBu)_2$

Salts of the triphospholide anion can also be prepared through a more direct procedure which uses a modification of the reported procedure [18]. To a nuclear magnetic resonance (NMR) tube containing a solution of $^tBuC{\equiv}P$ (250 μL, 1.55 mmol) in d^8-toluene (0.5 mL), PBr_3(80 μL, 0.85 mmol) was added. The tube was sealed and heated to 60°C, and the resulting reaction monitored by $^{31}P\{^1H\}$spectroscopy. After 72 h, the resonance for $^tBuC{\equiv}P$ had been replaced by that for $Cl_3P_3(C^tBu)_2$ at δ 49.2 ppm. The volatile materials were removed under vacuum-affording crude $Cl_3P_3(C^tBu)_2$ that was used in the reduction reaction (see Section 11.6.5).

11.6.5 Preparation of the triphospholide anion

To a Schlenk flask containing $Cl_3P_3(C^tBu)_2$ prepared as described in this section, THF (10 mL) and magnesium turnings (*ca.* 0.42 g, 17 mmol) were added. The resulting mixture was stirred for 2 h, during which time it turned a dark red-brown colour, and ^{31}P NMR spectroscopy showed quantitative conversion to the triphospholide anion as its $[Mg_2(\mu{-}Cl)_3(THF)_6]^+$ salt.

11.7 1,3,5-Triphosphabenzene

The parent (mono)phosphabenzene (more correctly named phosphinine) and its derivatives have proven to be highly useful ligands in organometallic catalysis [19]. Related chemistry utilising triphosphabenzene, which is readily accessible by trimerisation of a phosphaalkyne, is underexplored, probably on account of the perception that it is difficult to access. Several groups have conducted studies into the chemistry of triphosphabenzene and have shown that it forms coordination complexes with transition metals where it binds in an η^6-fashion [14,20]; the corresponding η^1-ligation through a P-lone pair has also been identified [21], although no crystallographically characterised examples of this coordination mode have yet been reported [22].

11.7.1 Preparation of Cl_3VN^tBu [23] (Scheme 11.14)

Scheme 11.14 *Preparation of Cl_3VN^tBu.*

To a round-bottomed flask charged with n-hexane (100 mL) and cooled to −78°C, $VOCl_3$ (10 mL, 106 mmol) was added forming an orange solution. H_2N^tBu (6 mL, 58 mmol) was added dropwise via syringe. The solution was stirred for 30 min at −78°C, then for 2 h at room temperature. The black solution was filtered through Celite® and washed with n-hexane. The solvent was removed *in vacuo* to yield a green crystalline solid (3.84 g, 29%).

11.7.2 Preparation of 1,3,5-triphospabenzene; $P_3(C^tBu)_3$ [24] (Scheme 11.15)

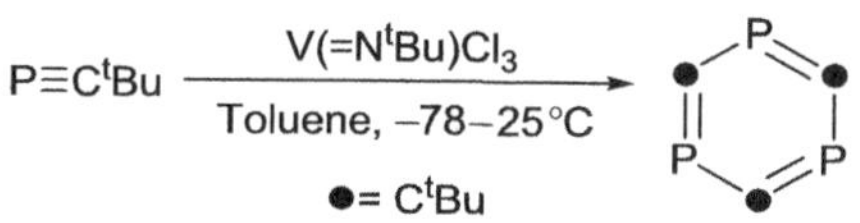

Scheme 11.15 *Preparation of 1,3,5-triphospabenzene.*

To an orange solution of $Cl_3V{=}N^tBu$ (0.10 g, 0.44 mmol) in toluene (10 mL) at −78°C, PC^tBu (0.13 mL, 1.76 mmol) was added and the now dark brown solution allowed to warm to room temperature. The solvent was removed *in vacuo* and n-hexane (10 mL) added before filtering through silica (first deactivated with 4% water). The solvent was removed *in vacuo* to yield a yellow solid (0.12 g, 68%).

^{31}P {^{1}H} NMR ($CDCl_3$, 122 MHz): δ 233 (s).
^{1}H NMR ($CDCl_3$, 400 MHz): δ 1.34 (s, 27 H, C^tBu).
^{13}C {^{1}H} NMR ($CDCl_3$, 100.6 MHz): δ 212 (s).

References

[1] H. Albers, *Angew. Chem.* **1950**, *62*, 451.
[2] T. E. Gier, *J. Am. Chem. Soc.* **1961**, *83*, 1769.
[3] P. Le Floch, F. Mathey, *J. Chem. Soc., Chem. Commun.* **1993**, 1295.
[4] G. Becker, G. Gresser, W. Uhl, *Z. Naturforsch. B* **1981**, *36*, 16.
[5] J. F. Nixon, *Coord. Chem. Rev.* **1995**, *145*, 201.
[6] J. C. T. R. Burckett-St Laurent, M. A. King, H. W. Kroto, J. F. Nixon, R. J. Suffolk, *J. Chem. Soc., Dalton Trans.* **1983**, 755.
[7] A. B. Pangborn, M. A. Giardello, R. H. Grubbs, R. K. Rosen, F. J. Timmers, *Organometallics* **1996**, *15*, 1518.
[8] H. Y. Cho, J. P. Morken, *J. Am. Chem. Soc.* **2010**, *132*, 7576.
[9] A. M. Richter, V. Lischewski, *Chem. Abstr.* **2001**, *135*, 152619; DE Patent 19,963,009, 1991.
[10] (a) T. Allspach, M. Regitz, G. Becker, W. Becker, *Synthesis* **1986**, 31;(b) G. Becker, H. Schmidt, G. Uhl, W. Uhl, *Inorg. Synth.* **1990**, *27*, 243.
[11] A. Mack, E. Pierron, T. Allspach, U. Bergstrasser, M. Regitz, *Synthesis* **1998**, 1305.
[12] F. G. N. Cloke, J. C. Green, J. R. Hanks, J. F. Nixon, J. L. Suter, *J. Chem. Soc., Dalton Trans.* **2000**, 3534.

[13] E. Urnezius, W. W. Brennessel, C. J. Cramer, J. E. Ellis, P. V. Schleyer, *Science* **2002**, *295*, 832.

[14] P. L. Arnold, F. G. N. Cloke, P. B. Hitchcock, J. F. Nixon, *J. Am. Chem. Soc.* **1996**, *118*, 7630.

[15] A. Elvers, F. W. Heinemann, B. Wrackmeyer and U. Zenneck, *Chem. Eur. J.*, **1999**, *5*, 3143.

[16] J. F. Bai, A. V. Virovets and M. Scheer, *Science*, **2003**, *300*, 781.

[17] P. Binger, T. Wettling, R. Schneider, F. Zurmuhlen, U. Bergstrasser, J. Hoffmann, G. Maas, M. Regitz, *Angew. Chem. Int. Ed.* **1991**, *30*, 207.

[18] G. Becker, W. Becker, R. Knebl, H. Schmidt, U. Weeber, M. Westerhausen, *Nova Acta Leopold.* **1985**, *59*, 55.

[19] C. Müller, D. Vogt, *Dalton Trans.* **2007**, 5505.

[20] M. D. Francis, C. Holtel, C. Jones, R. P. Rose, *Organometallics* **2005**, *24*, 4216.

[21] S. B. Clendenning, P. B. Hitchcock, J. F. Nixon, *Chem. Commun.* **1999**, 1377.

[22] (a) F. Preuss, E. Fuchslocher, E. Leber, W. Towae, *Z. Naturforsch. B* **1989**, *44*, 271; (b) F. Preuss, W. Towae, *Z. Naturforsch. B* **1981**, *36*, 1130.

[23] F. Tabellion, A. Nachbauer, S. Leininger, C. Peters, F. Preuss, M. Regitz, *Angew. Chem. Int. Ed.* **1998**, *37*, 1233.

[24] C. W. Tate, P. B. Hitchcock, G. A. Lawless, Z. Benkoö, L. Nyulászi, J. F. Nixon, C. R. Chim. 2010, 13, 1063–1072; N S. Townsend, M. Green, and C. A. Russell Organometallics, 2012, DOI: 10.1021/om201265n

12
P-chiral Ligands

Jérôme Bayardon and Sylvain Jugé

Institut de Chimie Moléculaire, Université de Bourgogne, Dijon, France

12.1 Introduction

The asymmetric reactions and catalysis are now commonly used for the enantioselective synthesis of chiral substances [1], including at the industrial stage [2]. In this field, mono- or diphosphines are by far the most widely used chiral ligands in asymmetric catalysis [3]. This class of ligand is widely used because various structural changes are relatively easy to make by electrophilic or nucleophilic reactions of P(III)- or P(IV)-precursors, on the phosphorus atom or in the α or β position [3f,4,5]. A further interest of chiral phosphorus ligands is due to the fact that the chirality can be located on the carbon skeleton, the phosphorus atom itself, or both, thereby increasing the number of stereoisomers which can be used for the development of a catalytic asymmetric process [6–8]. Noteworthy, as for all these applications the formation of a product in a given configuration is required, it is recommended to have at our disposal methodologies able to produce separately both enantiomers of a chiral ligand, in order to satisfy a given stereoselectivity or property. Anyway, the enantiomeric pure organophosphorus ligands can be obtained by resolution of a racemic mixture [9], by hemisynthesis from the chiral pool [10], by stereoselective synthesis [4], or still by asymmetric catalysis [11].

Since the pioneering work of William Knowles (Nobel Prize in Chemistry, 2001) [12], interest in organophosphorus ligands bearing chirality on the phosphorus atom has greatly expanded due to the development of new stereoselective synthetic methods, particularly those using phosphine borane chemistry [13].

A representative selection of highly enantioselective asymmetric catalysis by transition metal complexes and P-chiral phosphines is summarized in Table 12.1.

In coordination chemistry, P-chiral diphosphinomethanes such as **9** have been recently used for the preparation of the first chiral palladium clusters **10** (Scheme 12.1) [22]. The study of electrochemical properties and the EPR analysis of these clusters have demonstrated the formation of the first radical $[Pd_3(dppm^*)_3(CO)]^{+\bullet}$ in

Phosphorus(III) Ligands in Homogeneous Catalysis: Design and Synthesis, First Edition.
Edited by Paul C. J. Kamer and Piet W. N. M. van Leeuwen.

Table 12.1 *P-chiral ligands in asymmetric catalysis.*

Ligands	Asymmetric catalysis	References
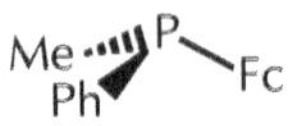 **1**	Reductive coupling (Ni)	[14]
 **2**	Hydrogenation (Rh) Addition 1–4 (Cu) Hydrosilylation (Rh)	[15]
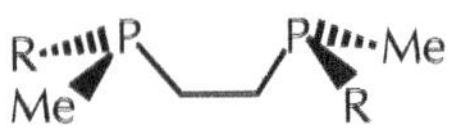 **3**	Hydrogenation (Rh) Allylation (Pd)	[16]
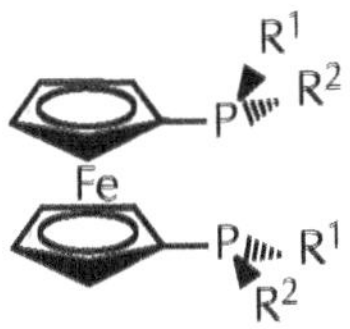 **4**	Hydrogenation (Rh) Allylation (Pd)	[17]
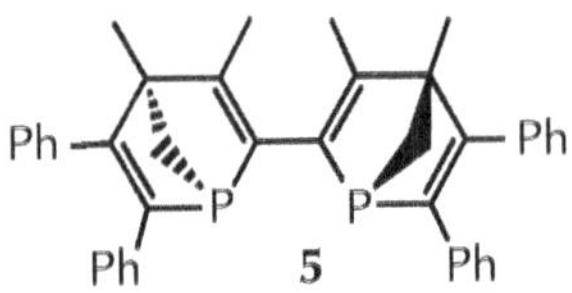 **5** BIPNOR	Hydrogenation (Rh) Heck (Pd) Isomerisation (Rh)	[18]
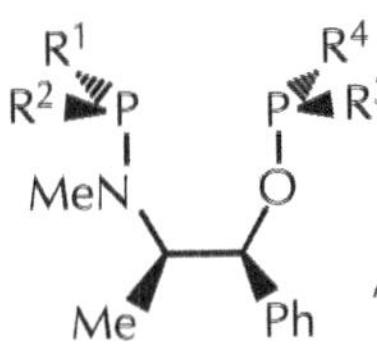 **6** AMPP*	Hydrogenation (Rh) Allylation (Pd) Hydroformylation (Rh)	[8,19]
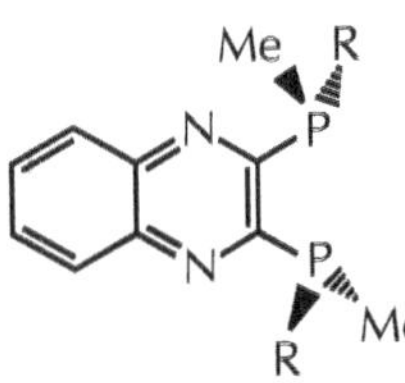 **7** QuinoxP*	Hydrogenation (Ru) Addition 1–4 (Rh) Allylation (Pd)	[20]
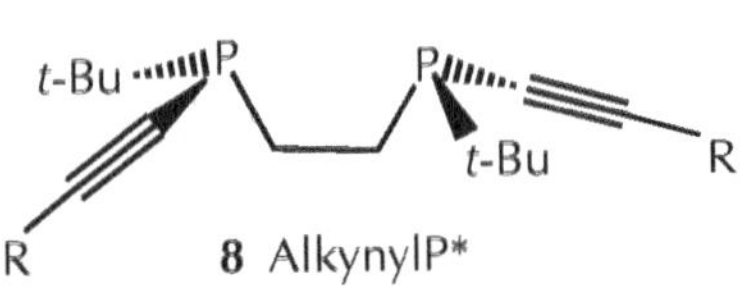 **8** AlkynylP*	Addition 1–4 (Rh) Allylation (Pd)	[21]

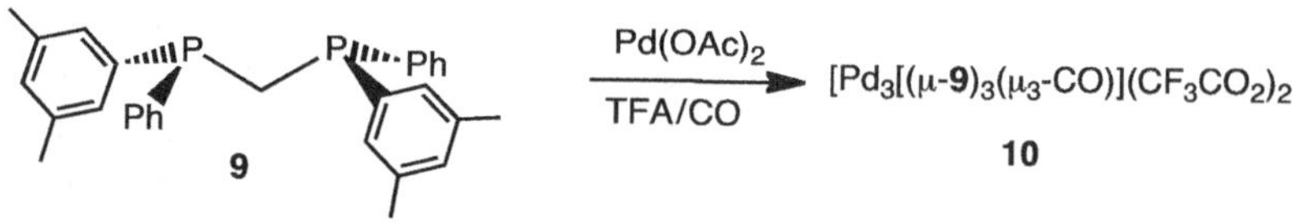

9

10

Scheme 12.1

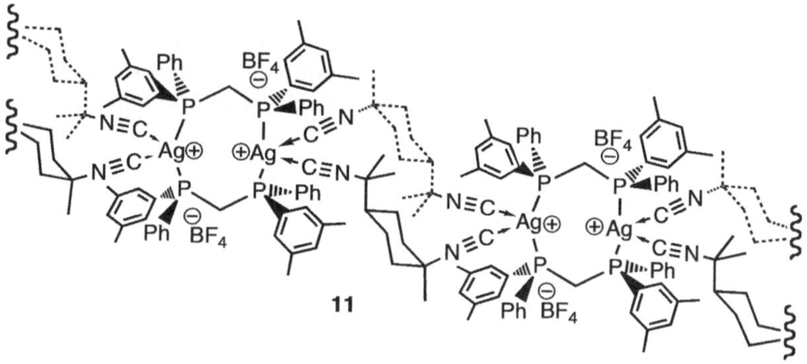

11

Scheme 12.2

a highly structured chiral environment, driven by a molecular orbital of C_3 symmetry [23]. Such paramagnetic chiral transition metal complexes currently attract much interest because their applications are relevant to models of metalloenzyme, molecular magnets, and radical activators, and also as components for nanoelectronics.

In the field of materials, P-chiral phosphines were recently used for the preparation of a new class of chiral coordination polymers, using metals like platinum [24], copper, or silver [25] as the macrocomplex **11** (Scheme 12.2).

The interest of P-chiral ligands has greatly expanded in recent years, due to the development of new, efficient, stereoselective synthetic methods, using mainly menthol, borneol, prolinol, ephedrine, sparteine, or α-arylethylamine as chiral auxiliaries. In addition, the use of borane as P(III)-protecting group has also caused a significant breakthrough in asymmetric synthesis of P-chiral ligands because their complex derivatives are easily purified and allow numerous clean reactions. Finally, to date the designing of P-chiral organophosphorus compounds has undergone much progress, allowing researchers to prepare various kind of ligands, functionalized or not, in a given configuration. The best methodologies for designing P-chiral ligands are reported in this chapter, according to the chiral auxiliary used.

12.2 Designing P-chiral ligands using alcohols as chiral auxiliaries

The stereoselective synthesis of P-chiral compounds can be achieved using natural chiral alcohols such as (−)-menthol **12**, (−)-cholesterol [26], (−)-borneol **13**, glucofuranoses [27], cinchonine, or cinchonidine [28]. However, the more efficient methods were achieved using (−)-menthol **12** and (−)-borneol **13**, which are summarized here (Scheme 12.3). It should be noted that the (+)-menthol and (+)-borneol are also commercially available, although they are much more expensive than their (−)-enantiomers.

The (−)-menthol methodology described by Mislow *et al.* is based on the diastereoselective synthesis of the menthyl phosphinates **16**, which are used for the synthesis of phosphine oxides **17**, and then P-chiral phosphines **18** (Scheme 12.4) [29].

This methodology gives rise to the phosphinates (*S*)-**16**, but the access to the minor epimer (*R*)-**16** with a good diastereomeric purity remains complicated, and it is certainly preferable to use (+)-menthol to afford the enantiomeric ligands. The reaction of phosphinate **16** with Grignard reagents led to the tertiary phosphine oxide **17**, with complete inversion of configuration at the phosphorus atom, then to the phosphine **18**, by deoxygenation with trichlorosilane as reductive reagent (Scheme 12.4) [30]. The efficiency of this methodology was demonstrated by the preparation and use of numerous menthylphosphinate **16** bearing various substituents, as reported in Table 12.2.

Knowles has used this method for the synthesis of DIPAMP **20** starting from (*S*)-methylphenylphosphinate **16a** (Scheme 12.5) [12]. Thus, the reaction of **16a** with *o*-anisylmagnesium bromide afforded the corresponding phosphine oxide (*S*)-**19** (PAMPO), which was then coupled by deprotonation, then oxidation in the presence of Cu(II). After final reduction of the diphosphine dioxide with trichlorosilane, the (*R*,*R*)-DIPAMP

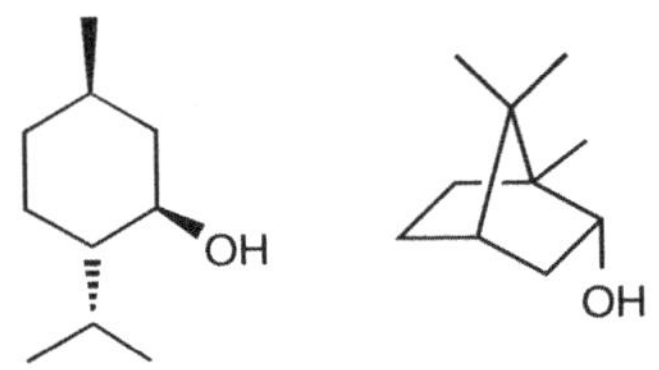

(–)-Menthol **12** (–)-Borneol **13**

Scheme 12.3

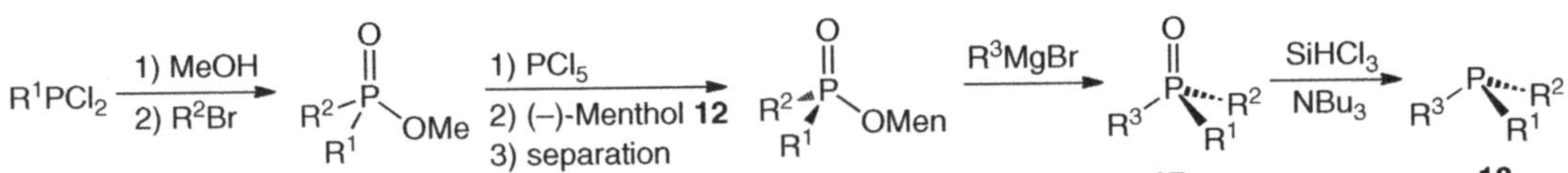

14 (±)-**15** (*S*)-**16** **17** **18**

Scheme 12.4

Table 12.2 *List of substituents for the menthylphosphinates **16**.*

16	R^1	R^2	References
a	Ph	Me	[29a]
b	2-Np	Ph	[29a]
c	Me	*c*-Hex	[29a]
d	Ph	*n*-Pr	[29a]
e	Ph	*i*-Pr	[29c]
f	Me	Et	[29b]
g	Me	4-*t*-BuPh	[31]
h	Ph	CH_2Ph	[32]
i	*o*-An	Ph	[33]
j	Me	*p*-An	[34]
k	Me	*o*-H_2NPh	[34]
l	Me	*p*-Me_2NPh	[34]
m	Ph	Et	[35]

20 was obtained with an overall yield of 7% over eight steps (Scheme 12.5a) [12]. An efficient synthesis of the diphosphine oxide **22** was also described by Imamoto, starting from the same (*S*)-menthylphosphinate **16a** (Scheme 12.5b) [36]. This method is based on the oxidative coupling into diphosphinate **21**, followed by a reductive P–O bond cleavage of the menthyloxy groups with lithium 4,4′-di-*tert*-butyldiphenylide (LiDBB). Finally, the resulting diphosphide intermediate led stereoselectively to the P-chiral diphosphine oxide **22**, upon quenching with methyl iodide (Scheme 12.5b).

An alternative methodology has been performed using the borneol **13** as asymmetric inductor [37]. Thus, Schmidt *et al.* have described the preparation of the bornyl methylphenyl phosphinate **24** in a 80:20 ratio, owing to a kinetic resolution starting from phosphoryl chloride **23** (Scheme 12.6). After separation of the major diastereomer by chromatography, the bornyl phosphinate **24** was used to afford the (*R*)-PAMPO **19** in 83% yield by reaction with the *o*-anisyl Grignard reagent. Contrary to the Mislow method, (*R*)-PAMPO **19** was reduced to free phosphine, then complexed by BH_3. DMS to perform quantitatively the phosphine borane complex **25** (PAMP.BH_3). The methylphosphine borane **25** is coupled to the corresponding diphosphine diborane, which is decomplexed by Me_2NH to afford the (*R,R*)-DIPAMP **20** in 86% yield (Scheme 12.6) [37].

On the other hand, the use of menthyl H-phosphinates **26** has been independently described by Han and Buono for the synthesis of secondary phosphine oxides **27** (SPO) [38]. SPOs are useful as P-chiral ligands [39] or building blocks for the synthesis of phosphine oxide derivatives **28** (Scheme 12.7) [40]. The addition of two equivalents of an organolithium reagent to the starting phosphinate **26** affords stereospecifically a phosphinite

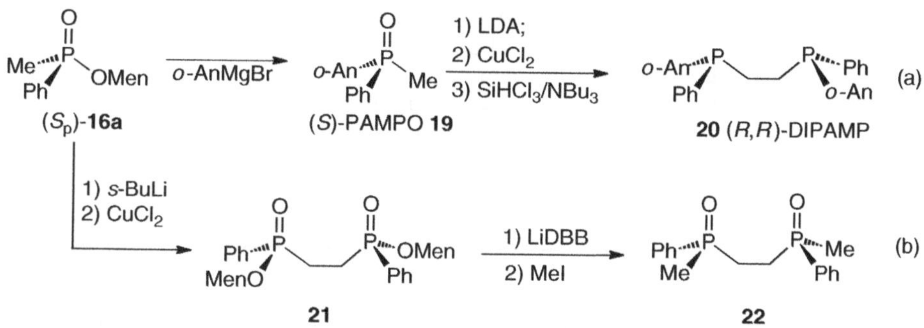

Scheme 12.5

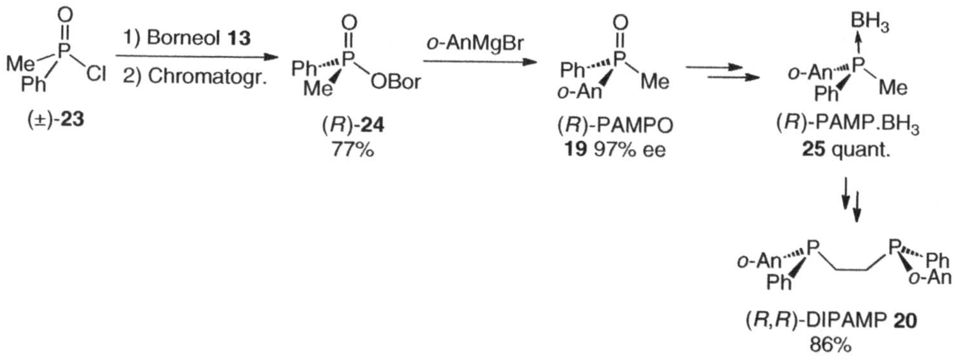

Scheme 12.6

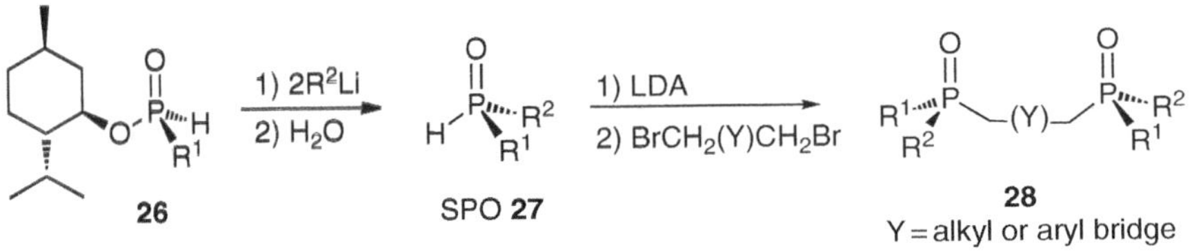

Scheme 12.7

Table 12.3 *SPO **27** synthesized from menthyl H-phosphinates **26**.*

27	R^1	R^2	ee (%)
a	Ph	Me	97
b	Ph	*n*-Bu	99
c	Ph	*i*-Pr	97
d	Ph	*t*-Bu	99
e	Ph	$TMSCH_2$	96
f	Ph	All	97
g	Ph	Bn	97
h	Mes	Ph	94
i	Bn	*n*-Bu	92
j	Bn	*t*-Bu	97
k	Ph	*o*-Tol	98
l	Ph	2,4-$(MeO)_2$Ph	92
m	Ph	1-Np	99
n	Ph	2-Np	92
o	Ph	3-thiophenyl	99
p	Ph	2-Pyr	92

salt intermediate, giving rise to the corresponding enantiopure SPO **27**, after hydrolysis. The efficiency of this methodology was demonstrated by the stereoselective synthesis of various SPOs **27**, as reported in Table 12.3.

This methodology has been applied for one-pot synthesis of enantio-enriched phosphinous acid boranes **29**, after direct trapping of the intermediate with BH_3. SMe_2 followed by protonation (Scheme 12.8) [38b]. Phosphinous acid boranes **29** can be further used to prepare either one or the other enantiomer of the secondary phosphine **32**, by stereoselective reduction or substitution of the phosphinite borane derivatives **30** or **31**, respectively (Scheme 12.8) [41]. Noteworthy, the secondary phosphine boranes **32** which also could be prepared using the sparteine method (*vide infra*), is a useful building block for the stereoselective synthesis of P-chiral ligands.

As an alternative to the Mislow route, Imamoto *et al.* have extended the synthesis of (*S*,*S*)-DIPAMP **20** starting from enantiopure menthylphosphinite boranes **34**, obtained by separation of a diastereomeric mixture (Scheme 12.9a) [33,42]. Following this work, Imamoto proposed the first method of enantioselective synthesis of P-chiral phosphides **36** obtained by stereoselective P–O bond cleavage of the menthyl phosphinite borane **34** with lithium naphthalene [42b–e]. The resulting phosphide borane **36** can be either protonated or alkylated sequentially, giving rise to the secondary or tertiary phosphine boranes **37** and **38**, respectively (Scheme 12.9b).

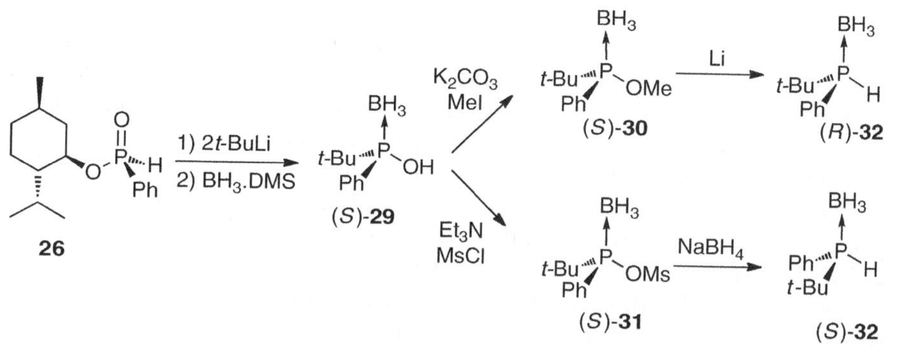

Scheme 12.8

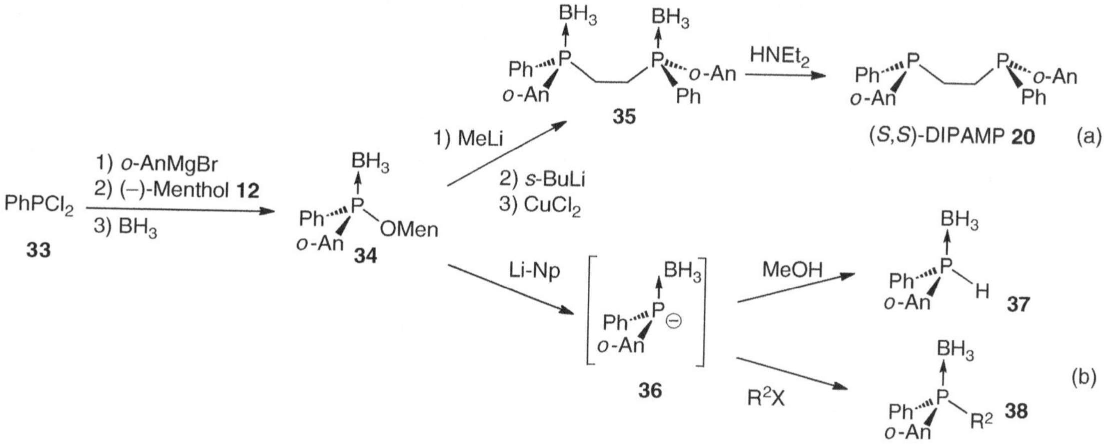

Scheme 12.9

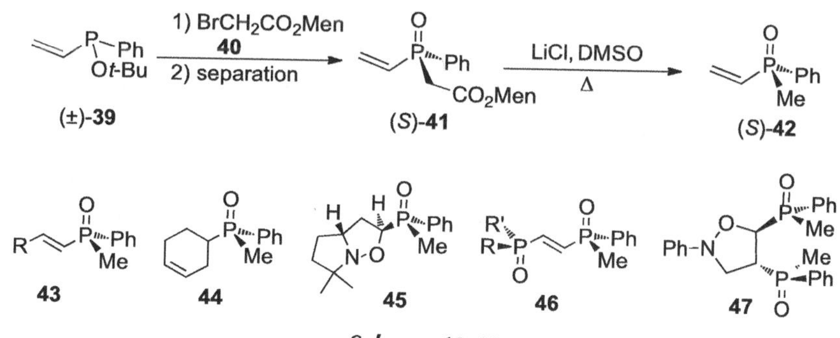

Scheme 12.10

In a different approach, efficient asymmetric synthesis of methylvinylphosphine oxide **42** was reported by Pietrusiewicz, using menthyl bromoacetate **40** as chiral reagent (Scheme 12.10) [43]. The vinyl phosphine oxide **42** is a useful precursor for the synthesis of various kinds of P-chiral phosphorus compounds such as **43–47**, by Michael reactions, Diels–Alder reactions, 1,3-dipolar addition, and/or Heck or metathesis couplings (Scheme 12.10) [44].

The principle of this synthesis is based on the Michaelis–Arbuzov rearrangement of the racemic *t*-butyl phosphinite **39** upon reaction with menthyl bromoacetate **40** affording the corresponding phosphine oxide derivative **41**. Fortuitously, one diastereomer spontaneously crystallized, providing the enantiomerically pure (*S*)-**42** after decarbo-alkoxylation using Krapcho conditions (Scheme 12.10). This procedure is well suited for a large-scale preparation of methylphenylvinylphosphine oxide **42** [43].

A similar procedure using menthyl chloroacetate **49** was described for the synthesis of P-chiral mono- and diphosphines **51** and **52** bearing bulky substituents, such as *t*-Bu, Mes, and 8-MeO-α-Np [42a,45]. In this case, the menthyl phosphine oxide derivatives **50a–h** were prepared by reaction of the secondary phosphine oxide with menthyl chloroacetate **49** (Scheme 12.11). After fractional crystallization, the diastereomerically pure **50a–h** were obtained and decarboxylated in basic conditions to afford the corresponding methyl phosphine oxide **51a–h**, which were obtained in moderate overall yields. Methyl phosphine oxides **51** could be used for the preparation of bulky P-chiral 1,2-diphosphinoethane **52** by bridge formation with a classical C–C bond coupling (Scheme 12.11).

Finally, the use of (–)-menthol **12** has been recently reported by Gilheany *et al.* for an easy deracemization of tertiary P-chiral methylphenylphosphine **53** (Scheme 12.12) [46]. The process is based on the stereoselective reaction of the racemic phosphine **53** with a mixture of (–)-menthol **12** and hexachloroacetone, leading to the corresponding phosphine oxide **51** with good to excellent enantiomeric excess (ee) (Table 12.4). The stereoselectivity of the reaction was explained by the dynamic kinetic interconversion of the alkyloxyphosphorane intermediates **54** leading to the phosphine oxide **51** by elimination of menthyl chloride (Scheme 12.12).

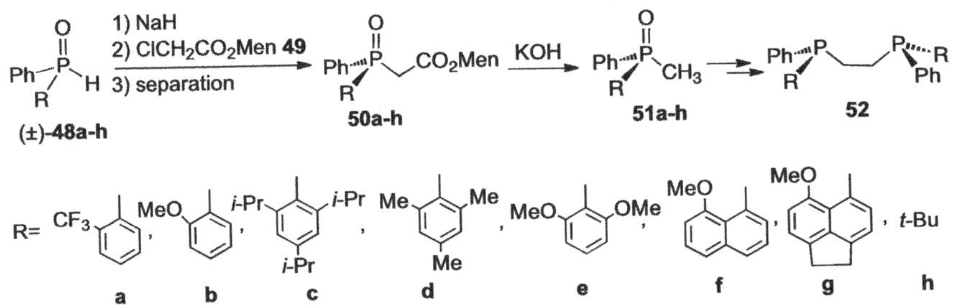

Scheme 12.11

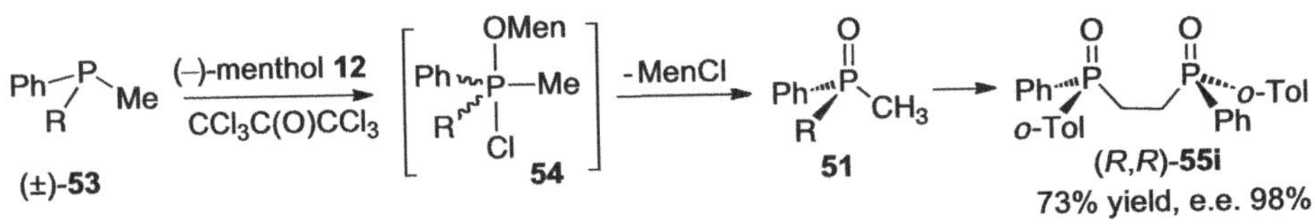

Scheme 12.12

*Table 12.4 Deracemization of the phosphines **53** into their oxide derivatives **51**.*

51	R	Yield	(Absolute configuration) ee (%)
a	*o*-An	90	(*R*)-50
h	*t*-Bu	—	(*S*)-48
i	*o*-Tol	91	(*R*)-80
j	2-*i*-PrPh	—	(*R*)-80
k	*o*-ClPh	—	71

The efficiency of this methodology was demonstrated by the diastereoselective preparation of the 1,2-bis(phenyl-*o*-tolylphosphino)ethane dioxide **55i** from **51i**, in 73% yield and with 98% ee, after recrystallization (Scheme 12.12).

12.3 Designing P-chiral ligands using amino alcohols as chiral auxiliaries

From the 1980s, an efficient strategy has emerged based on the stereoselective ring cleavage of a diheterophosphacycloalkane by electrophilic or nucleophilic attack, leading to actual strategy using 1,3,2-oxazaphospholidine **57**. The principle is based on two key steps (Scheme 12.13):

- Diastereoselective preparation of 1,3,2-oxazaphospholidine derivative **57** from β-amino alcohols **56**
- Regio- and stereoselective ring opening leading to the aminophosphine derivative **58**

Finally, the ring-opening product **58** is transformed into various *P*-chiral compounds such as phosphines **60** by subsequent reactions involving first electrophilic and then nucleophilic reagents (Scheme 12.13).

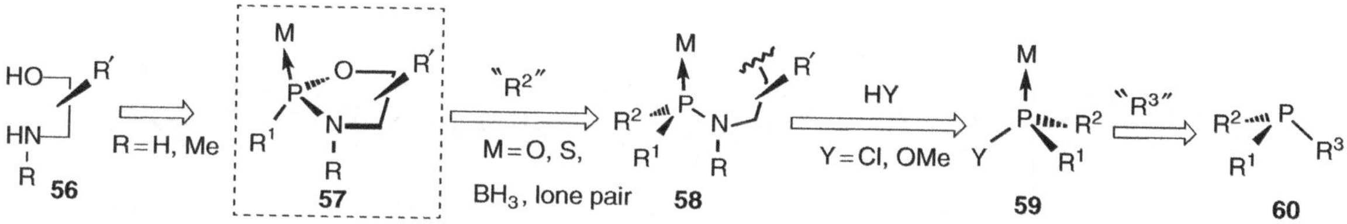

Scheme 12.13

To date, this strategy has been performed with 1,3,2-oxazaphospholidine **57** prepared from numerous natural or easily obtainable β–amino alcohols, such as ephedrine **56a**, Ψ-ephedrine **56b** [47], prolinol **56c**, 2-azabicyclo[3.3.0]octan-3-yl-methanol **56d** [48], glycidol derivatives **56e** [48], and or aminoindan-2-ol **56f** [49] (Scheme 12.14; Table 12.5).

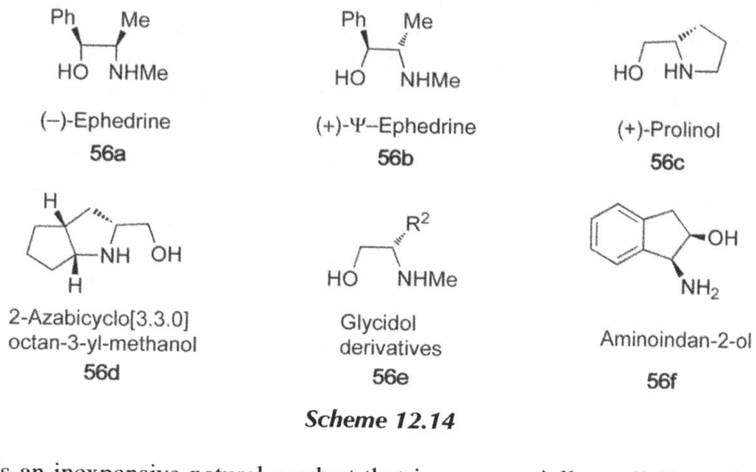

Scheme 12.14

Ephedrine **56a** is an inexpensive natural product that is commercially available at the same price under both enantiomeric forms. The other β–amino alcohols **56b–f** were either commercially available or prepared by reduction of the corresponding amino acids or from industrial waste materials. According to this strategy, the main oxazaphospholidines and their derivatives used for the asymmetric synthesis of P-chiral compounds are reported in Table 12.5.

Table 12.5 *Main 1,3,2-oxazaphospholidine **57** derived from ephedrine **56a** or prolinol **56c**.*

β-aminoalcohols	1,3,2-Oxazaphospholidines		M	References
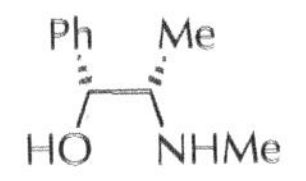 (–)-Ephedrine **56a**	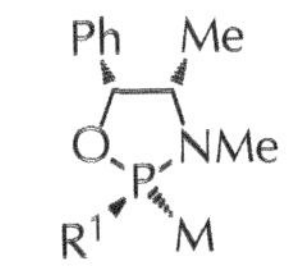 R^1 = alkyl, aryl, Ad, Fc	**57a**	Lone pair	[50,51]
		57b	O	[52,53]
		57c	S	[54]
		57d	BH_3	[52b,52c,53,55, 56,57]
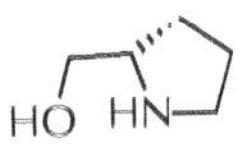 (+)-Prolinol **56c**	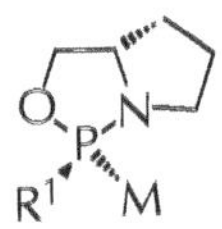 R^1 = Me, aryl	**57e**	Lone pair	[48,51,58]
		57f	O	[59]
		57g	S	[59]
		57h	BH_3	[48b,60]

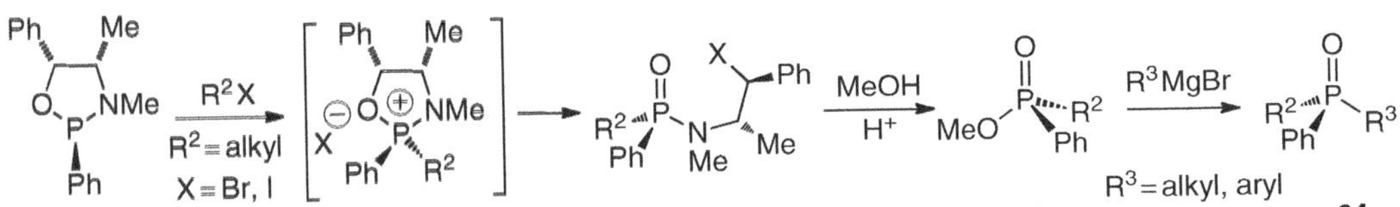

57a **61** **62** **63** **64**

Scheme 12.15

12.3.1 Synthesis starting from tricoordinated 1,3,2-oxazaphospholidines

The use of P(III)-oxazaphospholidine **57a** for the synthesis of *P*-chiral phosphine oxides, according to the defined strategy, was first reported by Jugé *et al.* [50]. In this approach, the ring-opening step was induced by the electrophilic reaction of an alkyl halide, allowing the P–O bond cleavage via a Michaelis–Arbuzov (MA) type rearrangement (Scheme 12.15). The MA rearrangement presents a great advantage to form a phosphorus–carbon bond under neutral conditions. The reaction of the diastereomerically pure P(III)-oxazaphospholidines **57a** (R^1 = Ph) with an alkyl halide R^2X afforded the phosphinamide derivatives **62** (Scheme 12.15). The MA rearrangement proceeds in two successive steps: firstly, a quaternization of the heterocycle by the alkyl halide leading to phosphonium salt **61**; and, secondly, a rearrangement affording the ring-opening product by C(5)–O bond cleavage. In this way, the reaction with methyl iodide led to product **62** (R^2 = Me) in 85% yield and 85% diastereomeric excess (de) (Scheme 12.15). However, the stereoselectivity strongly depends on the alkyl halide used and the experimental conditions. The monitoring of the reaction with methyl iodide or benzyl bromide by ^1H, ^{13}C, and ^{31}P NMR clearly showed the epimerization of the phosphonium intermediate **61**, limiting unfortunately this synthetic approach [50b].

The acidic methanolysis of diastereomerically pure phosphinamide **62** (R^2 = Me), which is obtained after recrystallization, afforded the corresponding methyl phosphinate **63** in 85% yield and with 96% ee (Scheme 12.15). Finally, the reaction of **63** with Grignard reagents led stereospecifically to the corresponding phosphine oxides **64**.

Another example of MA rearrangement as a key step was also reported, using the P(III)-oxazaphospholidine **57e** for the synthesis of the hybrid ligand **66** (Scheme 12.16) [58]. The reaction of **57e** with benzyl bromide gave rise to the ring-opening product **65**. Then subsequent substitution by lithium diphenylphosphide afforded the resulting P-chiral phosphine oxide–phosphine ligand **66** (Scheme 12.16).

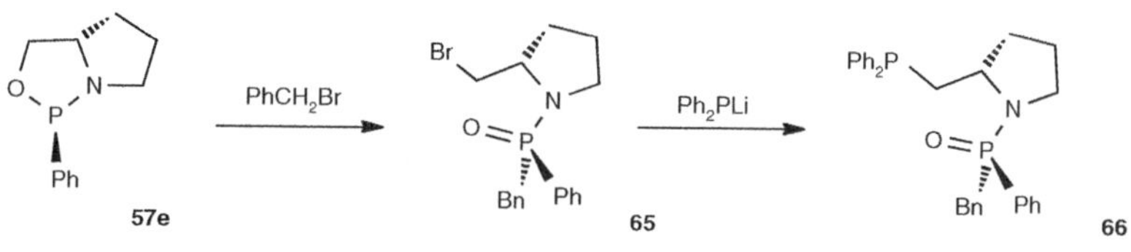

Scheme 12.16

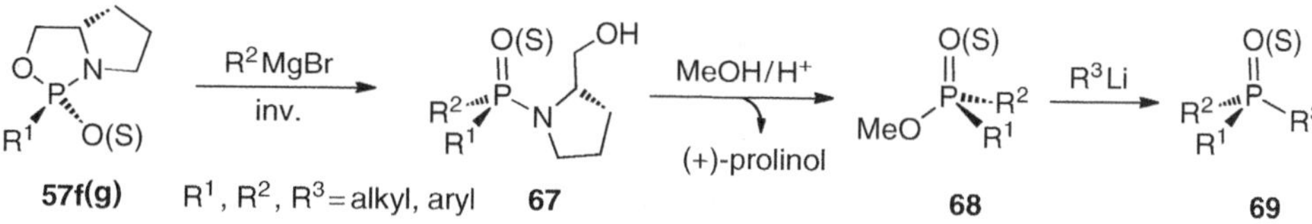

Scheme 12.17

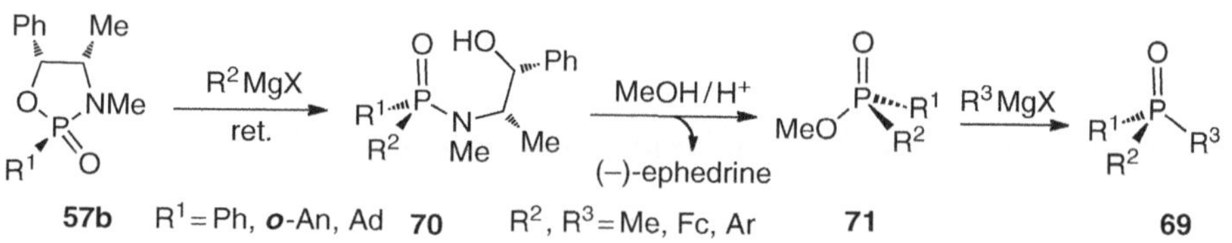

Scheme 12.18

12.3.2 Synthesis starting from tetracoordinated 1,3,2-oxazaphospholidines

Starting from oxazaphospholidine 2-oxo (or 2-thio), another strategy based on the ring opening induced by reaction with an organometallic reagent was possible (Scheme 12.17).

In pioneering works, Inch *et al.* have shown that the reaction of phenyllithium with 2-sulfide-oxazaphospholidine **57c** (R^1=Me, M=S) led to the ring-opening products by P–O bond cleavage, and major retention of configuration at the phosphorus atom [54]. Following this strategy, a convenient route for the preparation of P-chiral methyl (thio)phosphinates **68**, useful for the synthesis of chiral phosphine derivatives **69**, was reported starting from the oxo- (or sulfide) derivatives **57f** and **57g** (Scheme 12.17) [61]. In this case, the ring-opening step was induced by reaction of the Grignard reagent with the oxazaphospholidine derivatives **57f** (or **57g**) leading to (thio)phosphinamides **67**, by P–O bond cleavage and with inversion of configuration at the P-center (Scheme 12.17). It should be noted that the stereochemistry of the ring opening depends on the mono- or bicyclic oxazaphospholidines **57** used [54,61]. Acidic methanolysis of the (thio)phosphinamide **67** afforded the corresponding P-chiral (thio)phosphinates **68** with inversion of the configuration at the P-center, in moderate to good yields and with ee ranging between 73% and 99% (Scheme 12.17). This last step allowed the recovery of (+)-prolinol which can be reused.

This method was also performed for the synthesis of P-chiral phosphine oxides **69** starting from 2-oxo-1,3,2-oxazaphospholidines **57b** [52a–c]. Thus, the reaction of oxazaphospholidine **57b** with a Grignard reagent, afforded the ring-opened product **70** by P-O bond cleavage and with major retention of configuration at the P-center (Scheme 12.18). The acid methanolysis of the crystallized phosphinamide **70** into phosphinate **71**, followed by the reaction with a second Grignard reagent led to phosphine oxides **69** in overall yields up to 73%. When *o*-anisyl or 2-naphthyl are employed as R^1 substituent of the starting oxazaphospholidines **57b**, this method allowed the preparation of the corresponding P-chiral phosphine oxides **69** with excellent ee (>94%) (Scheme 12.18). However, the ring-opening step was not fully stereoselective when R^1 is a bulky

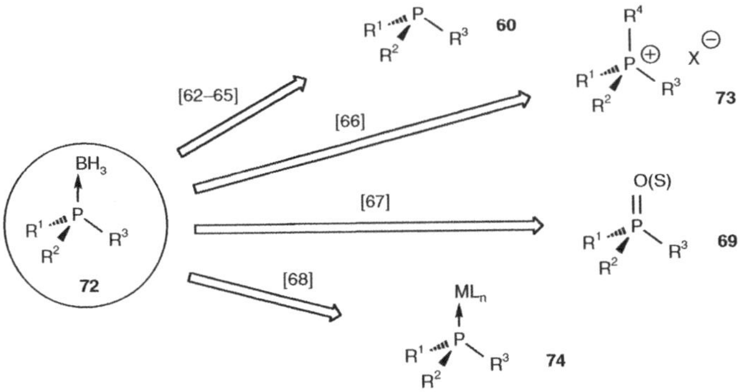

Scheme 12.19

substituent such as adamantyl, and in this case an epimeric mixture of phosphinamide **70** was obtained [52c]. Once again, the chiral auxiliary (*i.e.*,(−)-ephedrine) is recovered from the methanolysis step, allowing its reuse.

12.3.3 Synthesis starting from 1,3,2-oxazaphospholidine borane complexes

12.3.3.1 Interest of the borane–phosphorus complex chemistry

The use of borane as protecting group of the phosphorus atom enabled significant progress in organophosphorus chemistry and for their asymmetric synthesis, as was mentioned in this chapter [13]. The borane complexes **72** are stable, odorless, and often crystalline, and give clean reactions on the P-center or in the α- or β-position of one of the phosphorus substituents [13]. On the other hand, the borane decomplexation into P(III)-chiral compounds **60** is stereospecifically achieved using an amine [62], a strong acid [63], an alcohol (ROH) [64], or an olefin [65], and with complete retention of configuration on the phosphorus center (Scheme 12.19). In addition, the borane complexes **72** could be directly used in organic synthesis or in coordination chemistry for the stereoselective preparation of P-chiral organophosphorus compounds, which cannot be easily prepared by alternative approaches or obtained in their trivalent form (Scheme 12.19). Using this methodology, P-chiral quaternary phosphonium salts **73** [66], (thio)phosphorylated derivatives **69** [67], and transition metal complexes **74** [68] become accessible (Scheme 12.19).

12.3.3.2 Ephedrine method

The synthesis of P-chiral compounds using ephedrine as asymmetric inductor is based on two key steps: diasteroselective preparation of the oxazaphospholidine borane complex **57d**, and regio- and stereoselectivering opening by reaction with an organolithium reagent, to afford the aminophosphine borane **76** (Scheme 12.20) [56]. Subsequent stereoselective reactions of the aminophosphine borane **76** by methanolysis (or reaction with HCl), followed by organometallic R^3Li and decomplexation of the borane, led to free phosphine **60** (Scheme 12.20).

To date, the ephedrine method is one of the most efficient for the synthesis of P-chiral ligands because the starting complex **57d** bears only one of the three substituents of the final P(III)-product, and offers many possibilities for the design of P-chiral ligands [56].

The oxazaphospholidine borane **57da** ($R^1 = Ph$) is prepared as a crystallized pure diasteromer on a multigram scale, by reaction of ephedrine with the bis(dimethylamino)phosphine **75**, followed by complexation with BH_3.DMS (Scheme 12.20) (a typical procedure is given in this chapter). The synthesis of oxazaphospholidine borane complexes **57d** bearing substituents R^1 such as alkyl, aryl, ferrocenyl, and so on could also be achieved

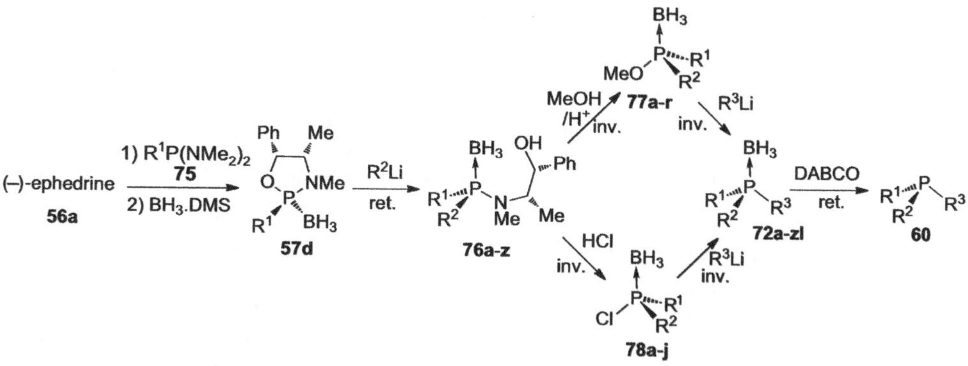

Scheme 12.20

by reaction of the ephedrine with PCl_3, then with an organometallic reagent and finally complexation with borane [52c,52d,55j]. However, in this case, it must be noted that lower diastereoselectivity and chemical yields were obtained than with the former method using diaminophosphine **75**. The starting complex **57da–de** then reacts with alkyl or aryl lithium reagents, to afford regio- and stereoselectively the corresponding ring-opening products **76a–y** by P–O bond cleavage of the heterocyclic ring (Scheme 12.20) (a typical procedure is given in this chapter). Various R^2 substituents such as alkyl, cycloalkyl, aryl, or ferrocenyl could be introduced on the P-center of the aminophosphine boranes, which are usually obtained in good or excellent yields as a single detectable diastereomer (d.r. > 98:2) (Table 12.6). The stereochemistry of the ring-opening step was established by X-ray analysis of several aminophosphine boranes **76**, and this step proceeds with retention of configuration at the P-center [48b,55h,55i]. However, the stereoselectivity is lower for the reaction of the borane complex **57da** with allyl- or ferrocenyllithium, since the aminophosphine boranes **76b**, **76s** were obtained with diastereomeric ratios of 80:20 and 91:9, respectively (Table 12.6, entries 2 and 19). In addition, mesityl oxazaphospholidine borane **57dd** (R^1 = Mes) did not react with MeLi, but reacted with PhLi giving the corresponding aminophosphine borane **76y** in 9% yield (entry 25). Furthermore, no reaction of **57de** occurred with bulky organolithium reagents bearing substituents such as mesityl, 2,4,5-trimethoxyphenyl, or 9-anthracenyl (entry 26).

12.3.3.3 Methyl phosphinite boranes as P-chiral electrophilic building blocks

The acidic methanolysis of the aminophosphine boranes **76** affords the phosphinite borane derivatives **77** and the ephedrine salt by P–N bond cleavage (Scheme 12.20) (a typical procedure is given in this chapter) [56]. This reaction, which proceeds with inversion of configuration at the P-center, is quite general and provides the methyl phosphinite boranes **77** in high yields and with excellent enantiomeric purities (Table 12.7). In the case of bulky R^2 substituents such as *t*-butyl, the methanolysis did not proceed and this lack of reactivity was explained by the steric hindrance around the phosphorus atom.

The reaction of the methyl phosphinite boranes **77** with an organolithium reagent affords the corresponding tertiary phosphine boranes **72** in good yields and with high enantiomeric excesses (>90% ee) (Scheme 12.20). This last reaction proceeds also with inversion of the configuration at the P-center, and occurs with various alkyl or aryllithium reagents, except with too hindered ones as the mesityl group (Table 12.7). The phosphine boranes **72** were easily obtained enantiomerically pure by recrystallization, and were then decomplexed with retention of the configuration into the free phosphines **60** using DABCO [62b] or another borane-removing method [63–65] (Scheme 12.20). The efficiency of this method allows preparing tertiary phosphines **60** on a large scale, in overall yields reaching 50%.

Table 12.6 *Synthesis of the aminophosphine boranes* **76** *starting from 1,3,2-oxazaphospholidineborane complexes* **57d**.

Entry	Starting complex **57d**		Aminophosphine borane **76**				References
		R^1		R^2	Yield (%)	Diastereomeric ratio	
1			**a**	Me	95	97:3	[7,14,55,62c]
2			**b**	All	77	8:2[a]	[69]
3			**c**	*n*-Bu	80	97:3	[7,14,55]
4			**d**	*t*-Bu	82	>98:2	[47,55i]
5			**e**	Neophyl	72	>98:2	[14]
6			**f**	*c*-Hex	88	>98:2	[14,55i]
7			**g**	Vinyl	96	>98:2	[70]
8			**h**	*o*-An	95	97:3	[14,17,47,55,62]
9			**i**	*p*-An	80	>98:2	[14]
10	**a**	Ph	**j**	3,5-diMe-4-PrOPh	57	>98:2	[8c]
11			**k**	*o*-Tol	92	>98:2	[14,55i]
12			**l**	*m*-Xyl	60	>98:2	[8c]
13			**m**	Calix[4]arenyl	84	>98:2	[8c]
14			**n**	*o*-[MeO(CH$_2$)$_2$]Ph	57	>98:2	[71]
15			**o**	*o*-biPh	85	>98:2	[14,17,55i]
16			**p**	1-Np	87	>98:2	[7,17,55i,76]
17			**q**	2-Np	89	>98:2	[17,55i]
18			**r**	9-Phenanthryl	86	>98:2	[17]
19			**s**	Fc	84	91:9	[17d]
20			**t**	*o*-HOPh	92	>98:2	[78e]
21			**u**	1-HO-2-Np	92	>98:2	[78e]
22			**v**	*i*-Pr	80	>98:2	[55j]
23	**b**	*p*-An	**w**	1-Np	81	95:5	[57b]
24	**c**	4-CF$_3$Ph	**x**	1-Np	72	95:5	[57b]
25	**d**	Mes	**y**	Ph	9	>98:2	[72]
26	**e**	Ad	**z**	Ph	—	3:1	[52c]

[a] Starting from racemic complex **57da**.

The ephedrine method and the methyl phosphinite borane intermediates have been widely used for the preparation of highly efficient *P*-chiral diphosphine ligands such as **79–82**, for the asymmetric catalyzed reactions by transition metal complexes (Scheme 12.21) [75,76,78].

12.3.3.4 *Chlorophosphine boranes as P-chiral electrophilic building blocks*

Despite the numerous tertiary phosphines **60** synthesized from methyl phosphinite boranes **77**, the lack of reactivity with soft nucleophiles such as cyclopentadienyl anion or phenolates remains a limitation (Scheme 12.20). For this reason, the chlorophosphine boranes **78** have recently emerged as better electrophilic *P*-chiral building blocks than the methyl phosphinite boranes **77**.

The configurational stability of tricoordinate *P*-chiral chlorophosphines is low, and these compounds slowly racemize at room temperature [79]. This racemization was explained by traces of HCl involving reversible protonation of the phosphorus atom, with a concerted back-side attack of a chloride anion to the

Table 12.7 *Preparation of methyl phosphinite boranes* ***77*** *and phosphine boranes* ***72***.

Methyl phosphinite borane **77**					Phosphine borane **72**				
	R^1	R^2	Yield (%)	References		R^3	Yield (%)	ee (%)	References[a]
a	Ph	Me	95	[14,55, 62c,72]	**a**	*n*-Bu	91	—	[55]
					b	*t*-Bu	66	92	[72]
					c	2-Pic	77	—	[62c]
					d	2,6-Lut	70	—	[62c]
					e	*o*-An	50	95	[55]
					f	2-Np	90	85	[55]
					g	Fc	58	83	[14]
b	Ph	*n*-Bu	62	[14]	**h**	Fc	90	77	[14]
c	Ph	Neophyl	65	[14]	**i**	Fc	79	98	[14]
d	Ph	*c*-Hex	29	[14]	**j**	Fc	66	97	[14]
e	Ph	*o*-An	95	[14,17, 47,55, 52c,62c]	**k**	Me	95	95	[55]
					l	*t*-Bu	92	98	[52c,77c]
					m	Ad	37	—	[52c]
					n	2-Pic	70	—	[62c]
					o	2,6-Lut	55	—	[62c]
					p	3,4-diMeOPh	80	—	[47]
					q	Fc	92	98	[52c,14]
f	Ph	*p*-An	77	[14]	**r**	Fc	85	95	[14]
g	Ph	*o*-Tol	90	[14,73]	**s**	Me	95	95	[73]
					t	Fc	70	83	[14]
h	Ph	*o*-[MeO(CH$_2$)$_2$]Ph	64	[71]	**u**	2-Np	28		[71]
i	Ph	*o*-biPh	77	[14,17, 74,75]	**v**	Me	—	—	[74]
					w	*i*-Pr	—	—	[74]
					x	Fc	69	98	[14,75]
j	Ph	1-Np	84	[17,74, 75,76]	**y**	Me	96	93	[74,76]
					z	*i*-Pr	61	99	[74]
					f	Fc	> 66	—	[75]
k	Ph	2-Np	94	[55c,17]	**za**	Me	90	95	[77a]
					zb	*o*-An	—	95	[77c]
l	Ph	9-Phenanthryl	64	[17,77]	**zc**	Me	44	99	[74]
m	Ph	*o*-HOPh	95	[78e]	**zd**	Me	94	—	[78e]
n	Ph	1-HO-2-Np	96	[78e]	**ze**	Me	92	—	[78e]
o	Ph	Fc	31	[17]	—	—	—	—	—
p	*p*-An	1-Np	84	[57b]	—	—	—	—	—
q	4-CF$_3$Ph	1-Np	75	[57b]	—	—	—	—	—
r	Mes	Ph	55	[72]	—	—	—	—	—

[a]The synthesis was performed starting from the (+)-ephedrine in the following references [14,17,55i,77a,78e].

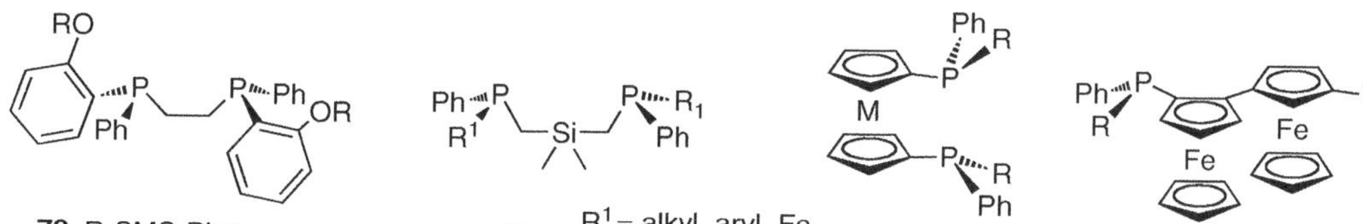

79 R-SMS-Phos
R = H, *c*-Hex, *i*-Pr, *t*-Bu...

80 R^1 = alkyl, aryl, Fc
M = Fe, Ru

81

82

Scheme 12.21

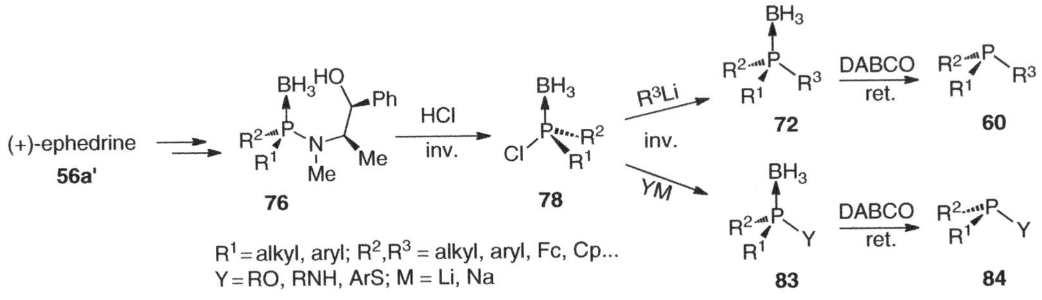

Scheme 12.22

Table 12.8 *Synthesis of P-chiral organophosphorus borane complexes **72** and **83** from chlorophosphine boranes **78**.*

Entry	Chlorophosphine borane **78**			Reagents	Organophosphorus borane complexes			
		R^1	R^2	R^3Li or YM		Yields (%)	ee (%)	References[a]
1	**a**	Ph	Me	*o*-AnLi	(*S*)-**72e**	80	90	[55i]
2	**a**			CpNa	**83a**	83	85	[55g]
3	**a**			2-BrPhONa	**83b**	71	72	[65a]
4	**a**			1-Br-2-NpONa	**83c**	66	91	[64a]
5	**a**			PhSLi	**83d**	87	60	[55i]
6	**a**			*o*-Ph(OLi)$_2$	**83e**	86	—	[55i]
7	**a**			2-BrPhNHLi	**83f**	53	99	[77e]
8	**b**	Ph	*o*-An	MeLi	(*R*)-**72e**	90	98	[55i]
9	**b**			1-Np	**72zf**	75	97	[77c]
10	**b**			2-Np	**72f**	69	98	[77c]
11	**b**			*o*-biPh	**72zg**	81	77	[77c]
12	**b**			*o*-Tol	(*S*)-**72zh**	53	—	[77c]
13	**b**			FcLi	(*R*)-**72q**	71	95	[55j]
14	**b**			*m*-Xyl	(*S*)-**72zi**	79	99	[55j]
15	**b**			4-MePhONa	**83g**	85	97	[55i]
16	**b**			2-BrPhONa	**83h**	70	80	[64a]
17	**b**			1-Br-2-NpONa	**83i**	88	99	[64a]
18	**b**			*m*-Ph(OLi)$_2$	**83j**	58	96	[77d]
19	**b**			1,1′-biPh(OLi)$_2$	**83k**	70	96	[77c]
20	**b**			2-BrPhNHLi	**83l**	98	78	[77e]
21	**b**			$(CH_2NH)_2Na_2$	**83m**	89	96	[55i]
22	**c**	Ph	*o*-Tol	MeLi	**72s**	61	99	[55i]
23				*o*-AnLi	(*R*)-**72zh**	92	99	[55i]
24	**d**	Ph	*o*-biPh	MeLi	**72v**	41	99	[55i]
25	**d**			*c*-Hex	**72zj**	62	97	[77c]
26	**e**	Ph	1-Np	MeLi	**72y**	50	0	[55i]
27	**f**	Ph	2-Np	MeLi	**72za**	46	81	[55i]
28	**g**	Ph	Fc	*o*-AnLi	(*S*)-**72q**	79	98	[55j]
29	**h**	Ph	*m*-Xyl	*o*-AnLi	(*R*)-**72zi**	82	99	[55j]
30	**i**	Ph	*c*-Hex	MeLi	**72zk**	41	80	[55i]
31	**j**	Ph	*i*-Pr	MeLi	(*R*)-**72zl**	68	89	[55j]
32	**j**	*i*-Pr	Ph	MeLi	(*S*)-**72zl**	86	96	[55j]

[a] Most of these syntheses were performed starting from the (+)-ephedrine **56a′**.

P–Cl bond, leading to the reversible formation of an achiral dichloropentacoordinated intermediate [80]. Complexed to the borane moiety, the corresponding chlorophosphine boranes **78** have a configurational stability and give similar reactivity [55i].

Chlorophosphine boranes **78** are easily obtained via the ephedrine methodology, performing acidolysis of the aminophosphine boranes with dry HCl in toluene solution (Scheme 12.22) (a typical procedure is given in this chapter). Therefore, the P–N bond cleavage is efficiently achieved with inversion of configuration at the P-center. The acidolysis step depends on the steric hindrance of the substituents on the phosphorus atom, the excess of HCl, the concentration of reagents, and the reaction time [55i]. The enantiomeric purity of the chlorophosphine boranes **78** was determined by chiral HPLC on the corresponding phosphine boranes **72** resulting from their reaction with an organolithium reagent (Scheme 12.22; Table 12.8).

The chlorophosphine boranes **78** are efficient electrophilic building blocks for the synthesis of new classes of P-chiral compounds, which bear substituents such as cyclopentadienyl, aryloxy, amino, or sulphide and are very difficult to synthesize in another way (Scheme 12.22) [55i]. Thus, various carbon, oxygen, nitrogen, or sulfur nucleophiles react with the freshly prepared chlorophosphine boranes **78** to afford the corresponding products **83** with inversion of configuration at the P-center (a typical procedure is given in this chapter). The results are reported in Table 12.8.

12.3.3.5 Designing P-chiral aminophosphine phosphinites (AMPP)*

Using the ephedrine methodology, the preparation of a library of P-chiral AMPP* **87** ligands could be prepared easily by changing the substituents on either the aminophosphine or the phosphinite moiety (Scheme 12.23) [7,8]. These P-chiral ligands have widely demonstrated their efficiency in hydroformylation, allylation, and hydrogenation catalyzed by transition metal complexes, affording asymmetric induction in the range of 75–100% ee [7,8,81].

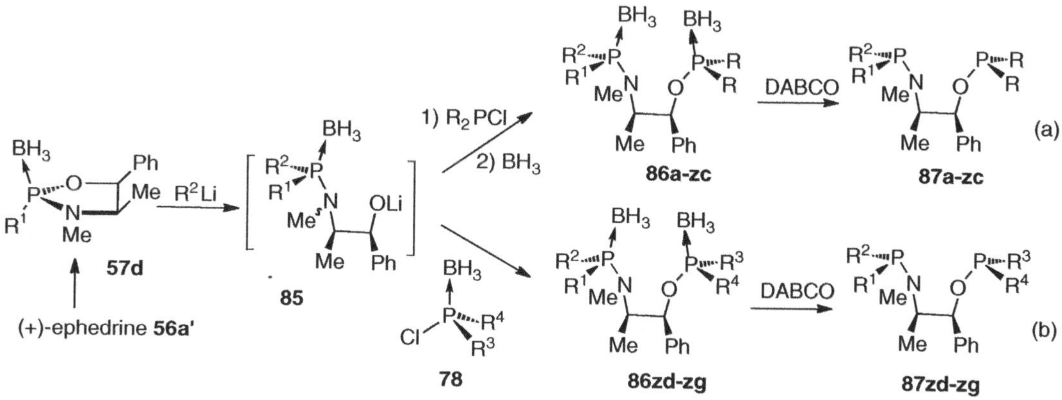

Scheme 12.23

Thus, by reaction with an organolithium reagent, the starting complex **57da** (R^1=Ph) leads to a ring-opened product **85** which is then trapped with a chlorophosphine and subsequently borane (Scheme 12.23a), or the P-chiral chlorophosphine borane **78** (Scheme 12.23b), to afford the corresponding AMPP* as diborane complex **86**. The aminophosphine–phosphinite diboranes **86** are isolated by chromatography as stable complexes (Table 12.9). The borane of these complexes is easily removed by reaction with DABCO to give high yields (>90%) of the corresponding free AMPP* ligands **87**, which are isolated by filtration over a short column of alumina (a typical procedure is given in this chapter). The structure of the aminophosphine–phosphinite diboranes **86** derived from the ephedrine are summarized in Table 12.9.

Table 12.9 *Structure of the aminophosphine-phosphinite diboranes **86** derived from ephedrine.*

		Aminophosphine-phosphinite borane complexes **86** ($R^1 = Ph$)				
Entry		R^2	R (or R^3)	R (or R^4)	Yield[a] (%)	References
1	**a**	Ph	Ph	Ph	63	[7a,8a,b]
2	**b**	Me	Ph	Ph	79	[7a,8a,b]
3	**c**	Me	*p*-Tol	*p*-Tol	82	[7a]
4	**d**	Me	3,5-(CF_3)Ph	3,5-(CF_3)Ph	96	[7a]
5	**e**	*n*-Bu	Ph	Ph	78	[7a]
6	**f**	"	3,5-(CF_3)Ph	3,5-(CF_3)Ph	96	[7a]
7	**g**	*o*-An	Ph	Ph	62	[8a,b]
8	**h**	*o*-An	OPh	OPh	30	[8b]
9	**i**	*o*-An	*c*-Hex	*c*-Hex	67	[8b]
10	**j**	1-Np	Ph	Ph	62	[7a,8a,b]
11	**k**	2-Np	Ph	Ph	78	[8a,b]
12	**l**	*t*-Bu	Ph	Ph	64	[8a,b]
13	**m**	*o*-biPh	Ph	Ph	40	[8b]
14	**n**	*o*-biPh	3,5-$(Me)_2$Ph	3,5-$(Me)_2$Ph	42	[8d]
15	**o**	*o*-biPh	1-Np	1-Np	64	[8d]
16	**p**	Fc	Ph	Ph	18[b]	[8b]
17	**q**	*o*-MEMPh	Ph	Ph	66	[8b]
18	**r**	*c*-Hex	Ph	Ph	38	[8b]
19	**s**	2-Pyr	Ph	Ph	54	[8d]
20	**t**	*o*-O*i*Pr-Ph	Ph	Ph	90	[8d]
21	**u**	*o*-OBn-Ph	Ph	Ph	87	[8d]
22	**v**	3,4-$(Me)_2$Ph	Ph	Ph	25[b]	[8d]
23	**w**	3,5-$(Me)_2$Ph	Ph	Ph	34	[8c]
24	**x**	2,4-$(t\text{-}Bu)_2$-3-OMe-Ph	Ph	Ph	83	[8d]
25	**y**	Calix[4]arenyl	Ph	Ph	62	[8c]
26	**z**	3,5-$(Me)_2$-4-OPr-Ph	Ph	Ph	43	[8c]
27	**za**	*o*-biNp	Ph	Ph	41	[8d]
28	**zb**	2′,6′-(OMe)-biPh	Ph	Ph	41	[8d]
29	**zc**	3,5-$(t\text{-}Bu)_2$Ph	Ph	Ph	71	[8d]
30	**zd**	Ph	*o*-An	Ph	55	[8a,b]
31	**ze**	"	Ph	*o*-An	55	[8a,b]
32	**zf**	*o*-An	Ph	*o*-An	65	[8a,b]
33	**zg**	"	*o*-An	Ph	40	[8a,b]

[a] Isolated yield. [b] Not optimized.

12.3.3.6 *P-chiral o-hydroxyaryl phosphines*

Recently, chiral organophosphorus derivatives bearing an *o*-hydroxyaryl substituent have received a lot of interest because these compounds could be used as either ligands [82] or organocatalysts in various asymmetric reactions [83,84]. The stereoselective synthesis of mono- and di-*o*-hydroxyarylphosphines, such as **79** (H-SMS-Phos), has been recently described by M. Stephan *et al.* using the ephedrine methodology and an *o*-lithiated phenolate (or naphtoate) as organometallic reagent (Table 12.6, entries 20 and 21) [78e].

On the other hand, the *o*-hydroxyarylphosphine **89** could also be obtained from the *o*-bromoarylphosphinite boranes **83b,c,h,i**, previously prepared by reaction of the chlorophosphine boranes **78** with the

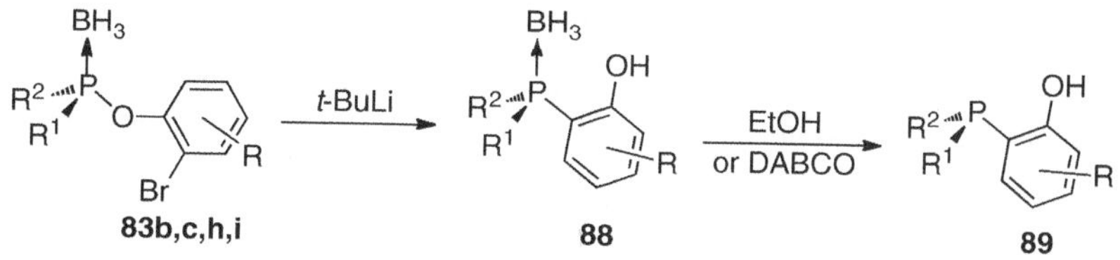

Scheme 12.24

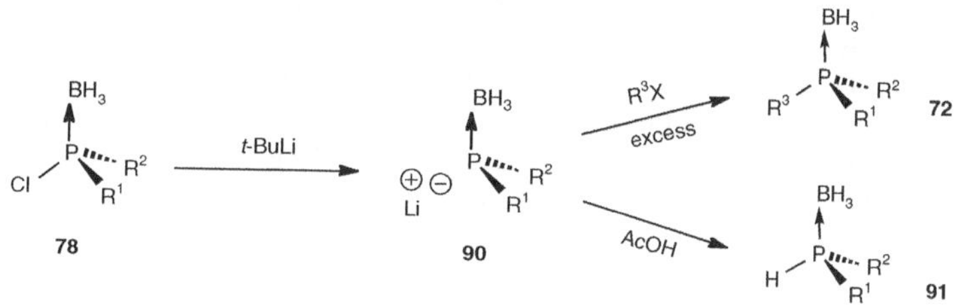

Scheme 12.25

corresponding phenolates (Table 12.8, entries 3, 4, 16, and 17; Scheme 12.24). Thus, the reaction of *o*-bromoarylphosphinite boranes **83** with *t*-BuLi, affords the anion in *ortho*-position by halogen–metal exchange, which leads to the *o*-hydroxyarylphosphine boranes **88** in good to excellent yields via a Fries-like rearrangement (Scheme 12.24) [64a]. This reaction proceeds stereospecifically with retention of configuration at the P-center, and provides the P-chiral *o*-hydroxyarylphosphine **89** after decomplexation of the borane (Scheme 12.24).

12.3.3.7 P-chiral secondary phosphine boranes

An unexpected metal–halide exchange of the chlorophosphine boranes **78** with *t*-BuLi was recently described, affording the corresponding P-chiral phosphide boranes **90** [55j]. The trapping at low temperature with an excess of alkyl halide affords the corresponding phosphine boranes **72** in good to excellent yields and with ee up to 99% (Scheme 12.25). The stereochemistry of the reaction proceeds with retention of configuration during the metal–halide exchange and the following alkylation step. When the phosphide boranes **90** were protonated by acetic acid, the corresponding P-chiral secondary phosphine boranes **91** were obtained in good yields and with ee up to 97% (Scheme 12.25). The results are reported in Table 12.10.

The deprotonation of the P-chiral secondary phosphine borane **91**, with one equivalent of base at low temperature, leads stereospecifically to the corresponding P-chiral phosphide borane **90**, which can be used for the synthesis of phosphine boranes **72** by reaction with an alkyl halide R^3X (Scheme 12.25) [55j,85]. The phosphine boranes **72** were easily obtained enantiomerically pure by recrystallization and then decomplexed with retention of the configuration to give the free phosphines **60** as shown in Scheme 12.22. It should be noted that under the conditions of borane decomplexation, the free secondary phosphines are obtained as racemates.

12.3.3.8 P-chiral 1,2-diphosphinobenzenes

Recently, a new strategy for the stereoselective synthesis of P-chiral 1,2-diphosphinobenzene **96** was proposed, starting from the secondary phosphine boranes **91** and the 1,2-dibromobenzene **92** (Scheme 12.26) [86]. The key step of this synthesis is based on the reaction of the secondary phosphine boranes **91** with

Table 12.10 Preparation of secondary and tertiary P-chiral phosphine boranes **72** and **91** from chlorophosphineboranes **78**.

Entry	Chlorophosphineboranes **78**[a]			Reagents	Phosphine boranes[a]		
		R[1]	R[2]			Yields (%)	ee(%)
1	**b**	Ph	*o*-An	MeI	(*S*)-**72e**	75	92
2	**b**	Ph	*o*-An	BnBr	**72zm**	75	95
3	**b**	Ph	*o*-An	AllBr	**72zn**	88	85
4	**b**	Ph	*o*-An	AcOH	**91a**	98	97
5	**c**	Ph	*o*-Tol	MeI	**72s**	95	99
6	**c**	Ph	*o*-Tol	AcOH	**91b**	83	94
7	**d**	Ph	2-biPh	MeI	**72v**	55	93
8	**g**	Ph	Fc	MeI	**72g**	73	90
9	**g**	Ph	Fc	BnBr	**72zo**	59	87
10	**g**	Ph	Fc	AcOH	**91c**	60	90
11	**i**	Ph	*c*-Hex	MeI	**72zk**	70	93
12	**i**	Ph	*c*-Hex	AcOH	**91d**	91	95
13	**j**	Ph	*i*-Pr	MeI	(*S*)-**72zl**	68	89
14	**j**	*i*-Pr	Ph	MeI	(*R*)-**72zl**	71	96
15	**j**	*i*-Pr	*i*-Pr	AcOH	**91e**	41	95

[a]All compounds were prepared starting from the (+)-ephedrine except entry 14, prepared from (–)-ephedrine [55j].

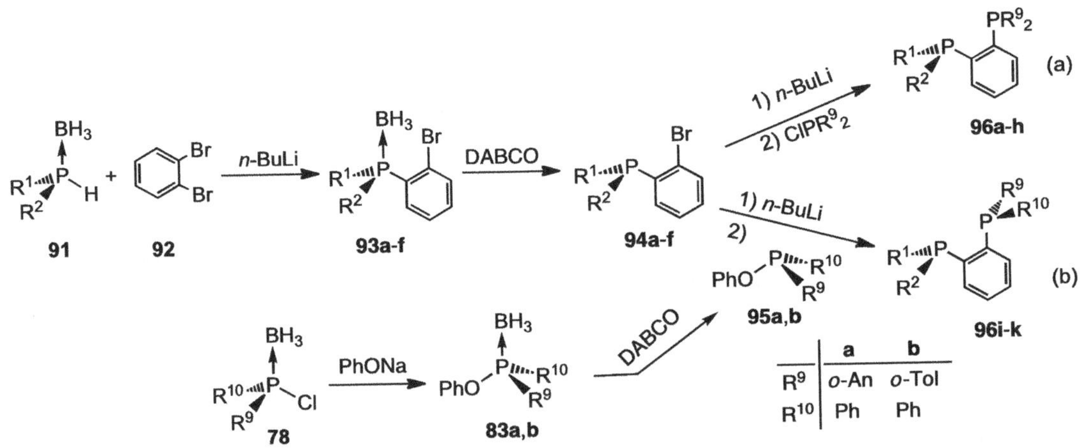

Scheme 12.26

1,2-dibromobenzene **92**, which affords the corresponding *o*-bromophenylphosphine boranes **93a–f** in good to excellent yields (Scheme 12.26; Table 12.11). The reaction proceeds by deprotonation of **91** and addition of the corresponding phosphide borane to the benzyne generated *in situ* by a metal–halide or elimination of 1,2-dibromobenzene **92**.

So, the free *o*-bromophenylphosphines **94a–c**, obtained by decomplexation of **93a–c** with DABCO, react after metal–halogen exchange with a chlorophosphine to yield stereospecifically the corresponding P-chiral 1,2-diphosphinobenzenes **96a–h** in 37–70% isolated yields (Scheme 12.26a; Table 12.11, entries 1–8). When the free *o*-bromophenylphosphines **94d–f** react after metal–halogen exchange with the P-chiral phenylphosphinites **95a,b**, the diphosphinobenzenes **96i–k** are obtained in 52–56% yields in high enantiomeric purities

Table 12.11 Preparation of o-bromophenylphosphines **93** and 1,2-diphosphinobenzenes **96**.

Entry			o-bromophenylphosphine boranes **93**				1,2-diphosphinobenzènes **96**			
			R^1	R^2	Yield (%)[b]		R^9	R^{10}	Yield (%)[b]	ee (%)[a]
1	**a**	(*S*)-	Ph	*o*-An	53	**a**	Ph	—	70	99
2	**a**	(*S*)-	Ph	*o*-An	53	**b**	*c*-Hex	—	47	99
3	**a**	(*S*)-	Ph	*o*-An	53	**c**	*i*-Pr	—	43	99
4	**a**	(*S*)-	Ph	*o*-An	53	**d**	*o*-Tol	—	37	99
5	**a**	(*S*)-	Ph	*o*-An	53	**e**	*p*-Tol	—	52	99
6	**a**	(*S*)-	Ph	*o*-An	53	**f**	*p*-CF_3Ph	—	58	99
7	**b**	(*S*)-	Ph	Fc	50	**g**	Ph	—	56	99
8	**c**	(*R*)-	Ph	*i*-Pr	48	**h**	Ph	—	54	98
9	**d**	—	Ph	Ph	75	**i**	*o*-Tol	Ph	54	99
10	**e**	(*S*)-	Ph	*o*-An	53	**j**	*o*-An	Ph	52	99
11	**f**	(*R*)-	Ph	*i*-Pr	58	**k**	*o*-An	Ph	56	99

[a] Determined by HPLC on chiral column. [b] Isolated yields.

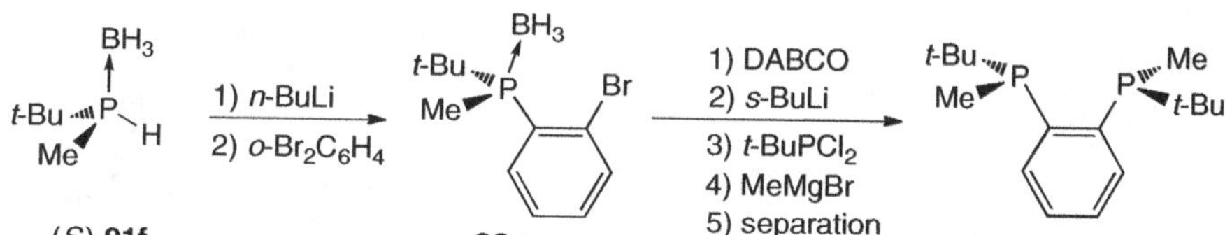

Scheme 12.27

(ee >99%) (Scheme 12.26b, entries 9–11). The P-chiral arylphosphinites **95a,b** were previously prepared by reaction of sodium phenolate with the corresponding chlorophosphine boranes **78**, followed by decomplexation with DABCO (Scheme 12.26). Noteworthy is that this method allows one to prepare the P-chiral 1,2-diphosphinobenzene **96** with the chirality located on either one or both P-centers (Scheme 12.26) [86].

This method was recently applied by Imamoto *et al.* for the stereoselective synthesis of the P-chiral 1,2-diphosphinobenzene (*R,R*)-BenzP* **96l** starting from the secondary phosphine borane (*S*)-**91f** (Scheme 12.27) [87]. Thus, the reaction of the *t*-butylmethylphosphine borane **91f** with the *n*-butyllithium reagent and the 1,2-dibromobenzene **92** affords stereospecifically the corresponding *o*-bromophosphine borane **93g** in 65% yield (Scheme 12.27). After decomplexation of the borane and subsequent reactions with *s*-BuLi, *t*-butyldichlorophosphine, and then methylmagnesium bromide, the (*R,R*)-BenzP* **96l** was readily separated as a crystalline solid from the *meso*-diphosphine epimer (Scheme 12.27) [87].

12.3.3.9 Strategies for the enantiodivergent synthesis of P-chiral ligands

At last, the synthesis of both enantiomers of P-chiral ligands is easily achieved following the ephedrine methodology, using either the (+)- or (–)- commercially available enantiomer of **56a** as chiral auxiliary. More interestingly, both enantiomers of a P-chiral phosphine ligand **60** could also be obtained starting from the same complex **57d**, prepared as example from (–)-ephedrine **56a**, by changing the order of addition of the organolithium reagents (*i.e.*, R^2Li then R^3Li *vs* R^3Li then R^2Li; Scheme 12.28) [55e].

On the other hand, the unexpected metal–halide exchange reaction between *t*-butyllithium and chlorophosphine boranes **78** afforded the corresponding P-chirogenic phosphide boranes **90** with retention of

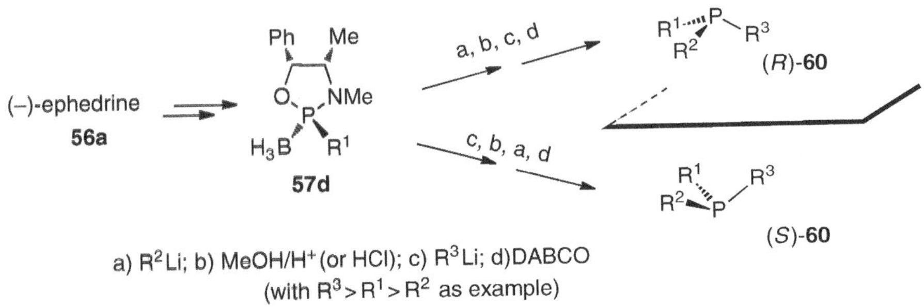

Scheme 12.28

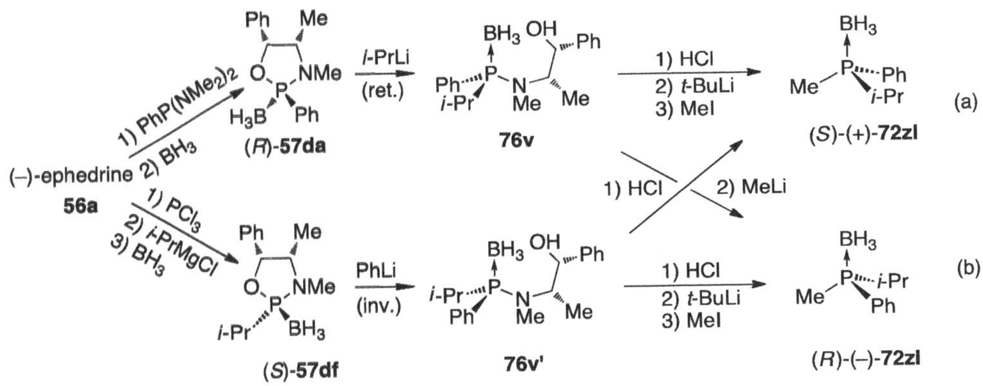

Scheme 12.29

the configuration at the phosphorus atom (Scheme 12.25). Subsequent addition of alkyl halide in excess gave the corresponding phosphine boranes **72** in good to excellent yields and with ee up to 99% (Scheme 12.25; Table 12.10). This synthetic pathway was successfully applied for the enantioselective preparation of the phosphine borane (*S*)-**72zl** via the aminophosphine **76v**, by reaction of the phosphide borane intermediate with methyl iodide (Scheme 12.29a; Table 12.10, entry 13)[55j]. Noteworthy, the phosphine borane (*S*)-**72zl** obtained by this pathway is the enantiomer of the one prepared by acidolysis with HCl of the aminophosphine borane **76v**, followed by reaction with methyl lithium (Scheme 12.29; Table 12.8, entry 31).

The synthesis of both enantiomers of the phosphine borane **72zl** was also achieved starting from the phenyl-*i*-propylaminophosphine borane **76v'**, readily obtained from the 2-*i*-propyloxazaphospholidine borane **57df**. This new starting complex (*S*)-**57df** was prepared in a diastereomerically pure form by subsequent reactions of PCl_3 with (–)-ephedrine **56a**, followed by *i*-propylmagnesium chloride and finally with BH_3·DMS (Scheme 12.29) [55j]. The reaction of the oxazaphospholidine borane **57df** with phenyllithium afforded the phenyl-*i*-propylaminophosphine borane **76v'** by P–O bond cleavage of the heterocycle (Scheme 12.29b). The X-ray structure of compound **76v'**, which has been established, proves the stereochemistry of the ring-opening process with inversion of configuration on the phosphorus center (Scheme 12.29b) [55j]. After acidolysis with HCl of the epimer **76v'** followed by metal–halide exchange with *t*-BuLi of the chlorophosphine borane intermediate, and finally trapping with MeI, the (*R*)-methylphenyl-*i*-propylphosphine borane **72zl** was obtained in 71% yield and with 96% ee (Scheme 12.29b; Table 12.10, entry 14). Finally, the (*S*)-methylphenyl-*i*-propylphosphine borane **72zl** was also obtained from the *i*-propylaminophosphine borane **72v'** by acidolysis with HCl and subsequent reaction with MeLi (Scheme 12.29; Table 12.8, entry 32).

12.4 Designing of P-chiral ligands using amines as chiral auxiliaries

12.4.1 Sparteine as chiral auxiliary

In 1995, Evans described the use of (–)-sparteine **97** as chiral auxiliary for the stereoselective deprotonation of the prochiral dimethylphosphine sulfide or borane **98** (Scheme 12.30) [88]. The P-chiral phosphine borane α-carbanion **99** reacted with benzophenone to afford the corresponding β-hydroxy derivatives **100** in 80–98% yield with ee up to 88%. In addition, the oxidative coupling of the α-carbanion in the presence of Cu(II) salt, or its grafting with a dichlorosilane, led stereospecifically to the diphosphine diboranes **101** and **102**, respectively (Scheme 12.30).

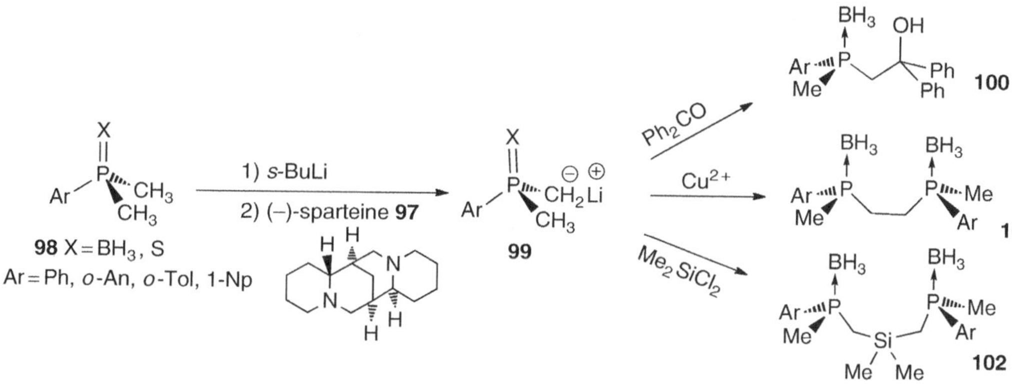

Scheme 12.30

This procedure has been largely used and developed by Imamoto to prepare efficient P-chiral ligands, such as MiniPHOS **2**, BisP* **3**, and QuinoxP***7**, using dimethylphosphine boranes **103** bearing bulky R^1 substituents (Scheme 12.31; Table 12.12) [16]. Noteworthy, some of these ligands are commercially available and give very high stereoselectivities in numerous asymmetric catalyzed reactions by transition metal complexes (see Table 12.1). Thus, efficient P-chiral 1,2-di(methylphosphino)ethane ligands **3** (BisP*) for asymmetric catalysis by transition metal complexes were prepared by oxidative coupling of the corresponding enantiomerically pure P-chiral α-carbanion **104** (Scheme 12.31a) [16]. In addition, the synthesis of dissymmetrical P-chiral 1,2-di(methylphosphino)ethane ligands **108** was also achieved from the carbanion **104** by subsequent reactions via the β-mesylate phosphine borane **106**, which is substituted by the P-chiral phosphide **107** and then decomplexed (Scheme 12.31b) [5a,89]. On the other hand, the P-chiral carbanion **104** can also react with molecular oxygen to afford the α-hydroxymethylphosphine borane **109**, useful to prepare the secondary phosphine boranes **91f,g** in high yields and with ee up to 99%, by an oxidative decarboxylation process (Scheme 12.31c) [90]. On the other hand, the secondary phosphine boranes **91h–k** have been prepared from the β-hydroxyphosphine borane **100** by a reduction of the pivalate ester with the Np–Li reagent [91]. Finally, the reaction of the secondary phosphine boranes **91f,g** with 2,3-dichloroquinoxaline afforded the P-chiral diphosphine ligands QuinoxP***7** (Scheme 12.31c) [20,92]. At last, the stereoselective synthesis of P-chiral diphosphinomethane **2** (MiniPHOS) was achieved by reaction of the carbanion **104** with dichlorophosphine, followed by addition of methyl magnesium bromide and borane, separation of the meso isomer, and finally decomplexation (Scheme 12.31d) [15].

However, as only one enantiomer of the sparteine **97** is available, this methodology allows the synthesis of a defined P-chiral compound in only one configuration. For example, only the BisP* (*S,S*)-**3a** (R^1 = *t*-Bu), can be obtained starting from the α-carbanion **104** prepared using the sparteine **97** (Scheme 12.31a). However, in order to obtain the (*R,R*)-**3a**, Imamoto developed a modified strategy which is shown in Scheme 12.32 [93].

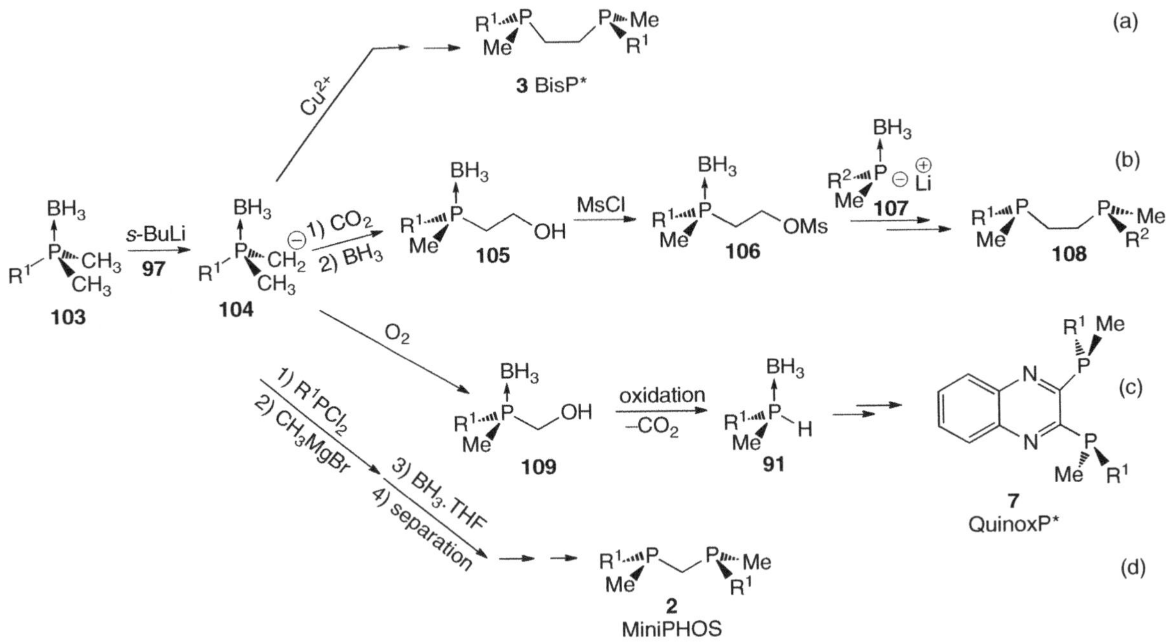

Scheme 12.31

Table 12.12 Substituents described for the synthesis of methylphosphine borane derivatives using sparteine **97** *as an asymmetric inductor.*

	R[1]	R[2]	References
2,3,7,91f	*t*-Bu	—	[15,16,20,92]
3	Et_3C	—	[16]
2,3,7,91g	Ad	—	[16,20,92]
2,91h	Ph	—	[15,91]
2	*i*-Pr	—	[15]
91i	*o*-Tol	—	[91]
91j	2-*i*-PrPh	—	[91]
91k	*o*-An	—	[91]
3	1-Me-*c*-Hex	—	[16]
3	*c*-Pe	—	[16]
2,3	*c*-Hex	—	[15,16]
108	Ad	*t*-Bu	[5a,89]
108	Ad	*c*-Pe	[5a,89]
108	Ad	*c*-Hex	[5a,89]
108	Ad	*i*-Pr	[5a,89]
108	Ad	Me	[5a,89]
108	Ad	Ph	[5a,89]
108	*t*-Bu	Ad	[5a,89]
108	*t*-Bu	*c*-Hex	[5a,89]
108	*t*-Bu	*i*-Pr	[5a,89]
108	*c*-Hex	*c*-Pe	[5a,89]

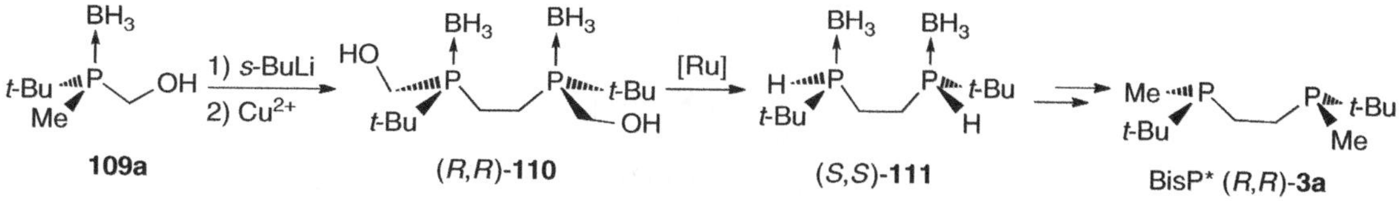

Scheme 12.32

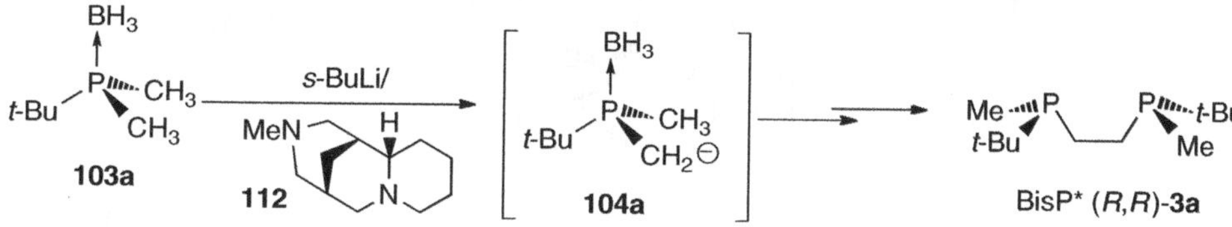

Scheme 12.33

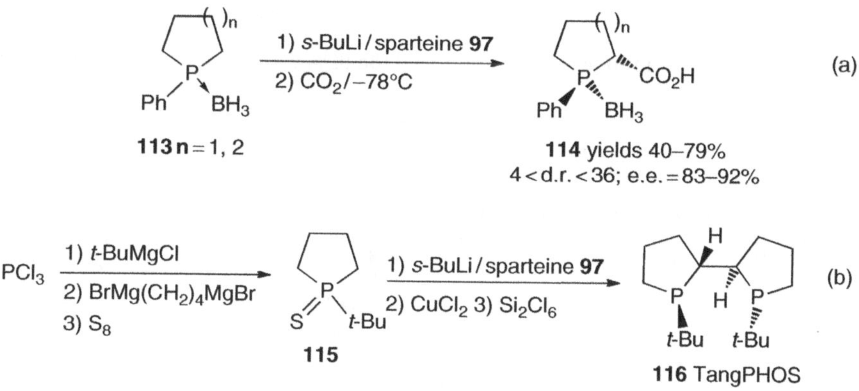

Scheme 12.34

The enantiomerically enriched α-hydroxyphosphine borane **109a** was subjected to double deprotonation, followed by copper-promoted oxidative coupling, to afford the diphosphine (*R,R*)-**110**. The ruthenium-catalyzed oxidative one-carbon degradation of **110** afforded the bis-secondary phosphine (*S,S*)-**111**, which reacts after deprotonation with methyl iodide to lead finally to the desired BisP* enantiomer (*R,R*)-**3a** (Scheme 12.32).

Alternative strategies were also developed in order to improve the synthesis of both enantiomers of ethylene-bridged P-chiral diphosphine ligands using secondary phosphine–boranes or prochiral dimethylphosphine sulfide [94]. Thus, by changing the reaction conditions (kinetic or thermodynamic control), the (-)-sparteine methodology can provide access to the opposite P-chiral enantiomer [95]. Recently, the synthesis of the opposite enantiomer of a P-chiral diphosphine such as (*R,R*)-**3a**, starting from dimethyl phosphine borane **103a**, was shown using the (+)-sparteine surrogate **112** (Scheme 12.33) [96,97].

Sparteine **97** was also used for the synthesis of P-chiral phospholane derivatives **114** and **116** by stereoselective deprotonation of the prochiral heterocycles **113** or **115**, respectively (Scheme 12.34) [98,99]. Thus, the reaction of the phosphacycloalkane borane complexes **113** with *s*-BuLi afforded the chiral carbanions in α-position of the P-center, which are trapped with CO_2 to yield the carboxylic acid derivatives **114** with a

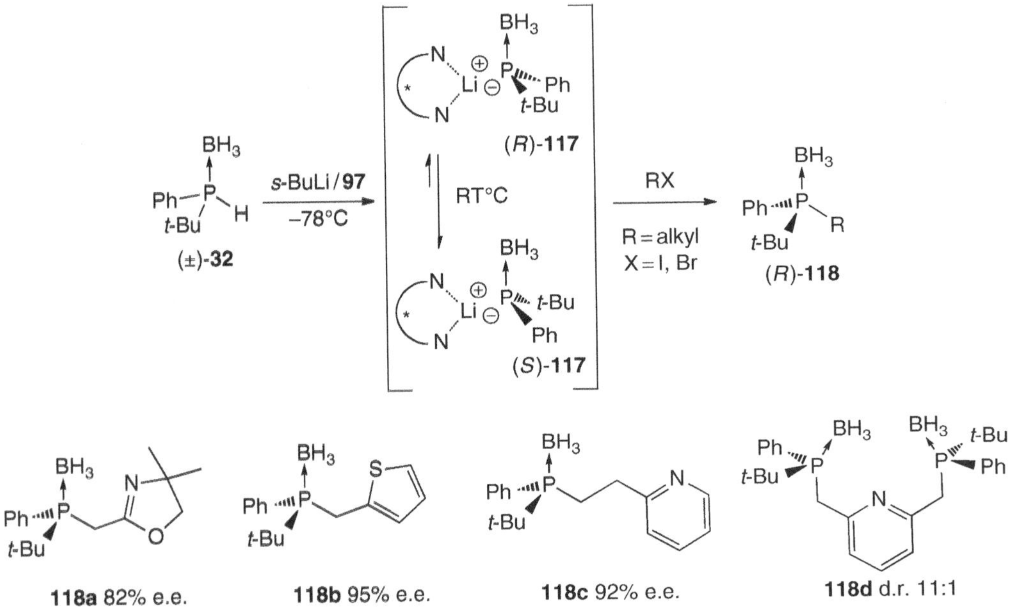

Scheme 12.35

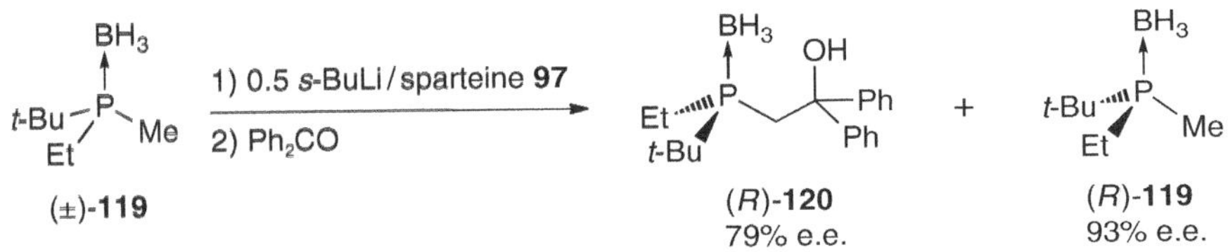

Scheme 12.36

good diastereomeric ratio (*trans/cis* >4) and ee up to 92% (Scheme 12.34a) [98]. In the case of the anion derived from the phospholane sulfide **115**, prepared in three steps from PCl_3, the oxidative coupling with a cupric salt, followed by the desulfuration using hexachlorodisilane, led to the bisphospholane ligand TangPHOS **116** (Scheme 12.34b) [99]. It must be noted that the TangPHOS ligand **116** affords very high asymmetric induction in the rhodium-catalyzed hydrogenation of various α- and β-arylenamide substrates [99].

In 1998, Livinghouse reported the use of (−)-sparteine **97** for the synthesis of P-chiral phosphines **118**, based on dynamic thermodynamic resolution of the secondary *t*-butylphenylphosphine borane (±)-**32** (Scheme 12.35) [85]. After deprotonation, the phosphide borane complex **117** was epimerized in the presence of (−)-sparteine at 25°C, and the major diasteromer (*S*)-**117** was then trapped at −78°C with an alkyl halide to afford the corresponding phosphine boranes **118** with good to excellent stereoselectivities (Scheme 12.35). Thus, this method was used to prepare various P-chiral phosphine borane complexes bearing chelating heterocyclic groups or pincer ligands, such as **118a–d**, with enantioselectivities up to 92% ee and a diastereomeric ratio superior to 11:1.

More recently, the kinetic resolution of P-chiral methylphosphine borane **119** using *s*-BuLi/(-)-sparteine **97** as chiral base was investigated by O'Brien (Scheme 12.36) [100]. The reaction of the racemic **119** with 0.5 equivalent of base and (−)-sparteine **97** afforded the enantiomerically enriched α-carbanion by deprotonation of the methyl substituent, which is then trapped by benzophenone to give the corresponding β-hydroxy phosphine borane **120** with 79% ee. The remaining enantiomer **119** was then recovered with 93% ee (Scheme 12.36).

12.4.2 α-Arylethylamines as chiral auxiliaries

α-Phenylethylamine **121** is a cheap chiral amine, available in both enantiomeric forms, which is currently used as chiral auxiliary in numerous asymmetric reactions. Several stereoselective syntheses of P-chiral organophosphorus compounds have been described using the α-phenylethylamine **121** as chiral auxiliary (Scheme 12.37) [101,102].

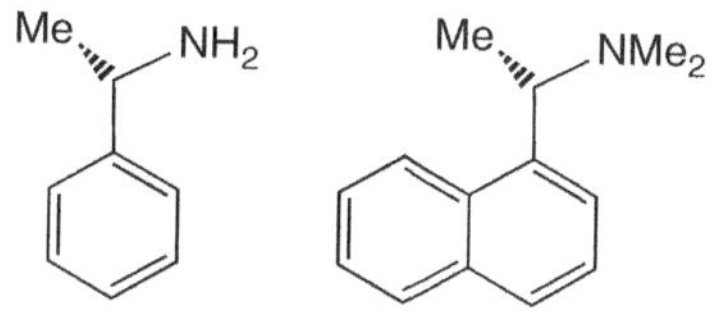

Scheme 12.37

The stereoselective synthesis of P-chiral organophosphorus compounds using the palladacycle **123** derived from N, N-dimethyl-α-naphtylethylamine **122** was widely investigated by Leung *et al.* (Scheme 12.38) [103,104]. The Pd-complex **123** reacted with the 1-phenyl-3,4-dimethylphosphole **124** (DMPP) to afford the corresponding monomeric complex **125a**, which led to the adduct **127** by Diels–Alder reaction with the ethylvinyl ketone **126** (Scheme 12.38).

The reaction occurred with major *endo*-cycloaddition, giving a mixture of diastereomers **127** which can be separated to afford the optically pure ketophosphine (R_p)- or (S_p)-**128**, after decomplexation with 1,2-bis(diphenylphosphino)ethane (dppe; Scheme 12.38a). When the Pd-complex **125a** reacted with $AgClO_4$, prior to the cycloaddition with ethylvinyl ketone **126**, the reaction was slower and the *exo*-cycloaddition took place affording the diastereomerically pure complex (R_p)-**127**, and then the enantiomerically pure ketophosphine (S_p)-**128** was obtained after treatment with dppe (Scheme 12.38b). Accordingly, the slower *exo*-cycloaddition process showed a better stereoselectivity than the faster *endo* one.

This method was further developed for the synthesis of unusual bicyclic P-chiral mono- and diphosphines, such as **131**, **133**, or **135** (Scheme 12.39) [104]. Thus, when two equivalents of the racemic vinylphosphines **129** were used as dienophilic reagent with the cationic palladacycle **125b**, the *exo*-adduct **130** was diastereoselectively obtained in 80% yield, together with the unreacted (*R*)-methylphenylvinyl phosphine **129** (87% yield; 100% optical purity) (Scheme 12.39). Subsequent reactions of the palladacycle **130** afforded stereospecifically the P-chiral diphosphine **131**. In the case of the α, β-unsaturated keto ester **132** or the 2-methylene-3-quinuclidinone **134** as dieneophilic reagents, the cycloaddition with the palladacycle **125b**, followed by decomplexation, afforded the enantiomerically pure heterofunctionalized P-chiral phosphines **133** or **134**, respectively (Scheme 12.39). The latter case of Diels–Alder reaction exhibited remarkable stereochemical and electronic directing effects, due to the nitrogen atom which is finally located in the *exo*-position (Scheme 12.39).

12.5 Conclusion

The best method for the design of P-chiral ligands is based on the use of chiral auxiliaries such as menthol, borneol ephedrine, prolinol, sparteine, or α-arylethylamine. Noteworthy, the use of borane as P(III)-protecting group has afforded significant breakthroughs for the asymmetric synthesis of P-chiral

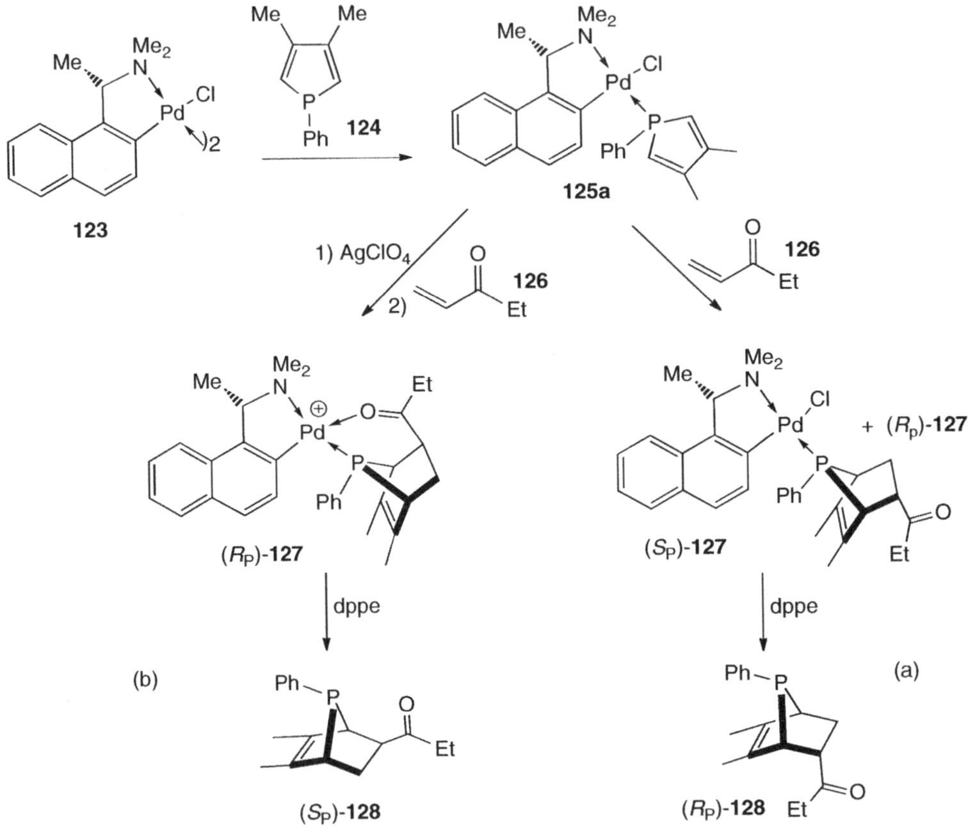

Scheme 12.38

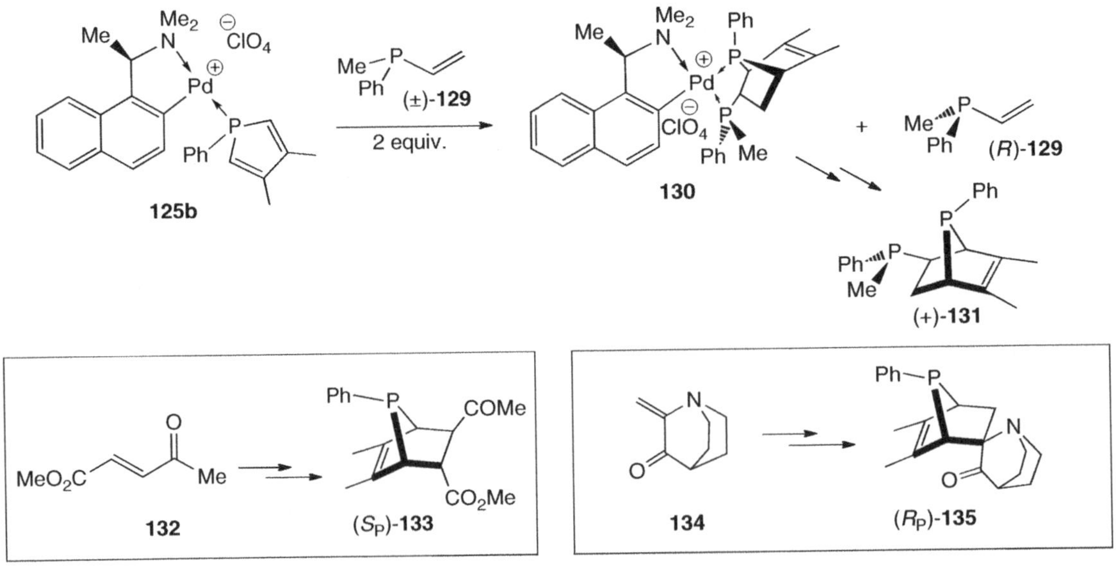

Scheme 12.39

organophosphorus compounds, because their borane complexes are stable and easy to handle and purify, whereas the free P(III)-ligands are quantitatively recovered by decomplexation. Among the best chiral auxiliaries, indeed ephedrine and sparteine (especially for one enantiomeric series) could be successfully employed for the preparation of numerous classes of P-chiral phosphorus compounds, functionalized or not, by clean reactions involving P*–C, P*–H, P*–Cl, P*–O, P*–N bond formation.

12.6 Experimental procedures

Typical procedure for the preparation of oxazaphospholidine borane complex ***57da***: A 1 L three-necked, round-bottomed flask was equipped with a mechanical stirrer, a nitrogen inlet, and a short-path distillation head fitted with a dropping condenser. The flask was charged with 500 mL of toluene, 0.10 mol of (-)-ephedrine, and 0.10 mol of freshly distilled bis(dimethylamino)phenyl phosphine. The solution was stirred at 105°C for 5 h, under a gentle flow of nitrogen to remove the dimethylamine formed. The formation of the oxazaphospholidine was monitored by ^{31}P NMR (δ=+142ppm). The flask was then cooled to room temperature, and BH_3.DMS (10 ml/10 M or 50 ml/2 M in toluene) added under stirring. After 4 h, the solvent was removed under vacuum, and the crude complex crystallized upon cooling. Recrystallization from i-PrOH gave 20 g of oxazaphospholidine borane complex **57da** in two crops. After concentration of the filtrate, the residue was purified on a short column of silica gel, eluting with toluene. After recrystallization, 22 g of the complex **57da** were obtained in addition (77% yield).

Typical procedure for the preparation of aminophosphine boranes ***76***: In a 100 mL three-necked flask, equipped with a magnetic stirrer and an argon inlet, 15 mmol of the oxazaphospholidine borane complex **57da** were dissolved in 15 mL of anhydrous tetrahydrofuran (THF). The mixture was cooled at –78°C, and 30 mmol of the organolithium reagent previously prepared were slowly added. The resulting mixture was stirred and warmed to 0°C until the starting material had completely reacted. The reaction was monitored by thin-layer chromatography (TLC) over silica (toluene as eluent), and was finally hydrolyzed at 0 °C with 5 mL of water. The THF was removed under reduced pressure, and the aqueous layer was extracted several times with dichloromethane. The combined organic phases were dried over magnesium sulfate, and the solvent was removed. The residue was purified on a short column of silica gel, using a mixture of toluene–AcOEt 95:5 as eluent, to afford the aminophosphine boranes **76** in excellent yields of up to 95%. The aminophosphine boranes **76** can be recrystallized using a mixture of hexane–isopropanol 7:3.

Typical procedure for the preparation of methyl phosphinite boranes ***77***: A 25 mL round-bottomed flask equipped with a magnetic stirrer and an argon inlet was charged with methanol (10 mL), 1 mmol of aminophosphine borane **76**, and 1 mmol of concentrated sulfuric acid (98%). The mixture was stirred at room temperature for 1–3 days, depending on the aminophosphine borane **76**, and then filtered under reduced pressure on a Büchner funnel through silica gel. The solvent was then removed and the crude residue purified by column chromatography with a mixture of toluene–ethyl acetate as eluent to afford the corresponding methyl phosphinite boranes **77** with yield varying from 31% to 95% (Table 12.7).

Typical procedure for the preparation of chlorophosphine boranes ***78***: In a 50 mL two-necked flask, equipped with a magnetic stirrer, an argon inlet, and a rubber septum, was introduced 2 mmol of aminophosphine borane **76** (in toluene solution or not). A solution of HCl in toluene (0.38 M) (2–15 equiv., depending on the aminophosphine borane **76**) was next added under stirring at room temperature. After stirring at room temperature (from 1 to 48 h; reaction monitored by TLC), the precipitate of ephedrine hydrochloride was filtered off with a Millipore 4 μm filter, and the excess of HCl was removed by several vacuum–argon cycles. The toluene solution of chlorophosphine borane **78** was then used for synthetic applications without further purification.

Typical procedure for the preparation of tertiary phosphine boranes ***72*** *and free P-chiral phosphines* ***60***: In a 50 mL two-necked flask, equipped with a magnetic stirrer, an argon inlet, and a rubber septum was introduced a solution of chlorophosphine borane **78** in toluene. To this solution cooled at –78°C was added the organolithium reagent (2.5 equiv.). The reaction mixture was allowed to warm to room temperature during 1 h, then hydrolyzed with water. The organic phase was removed, the aqueous layer was extracted with dichloromethane, and the combined organic phases were dried over $MgSO_4$. After evaporation of the solvent, the residue was purified by column chromatography with a mixture of toluene–ethyl acetate as eluent to afford the corresponding tertiary phosphine boranes with good yields (41–92%; Table 12.8). Their recrystallization in a mixture of hexane–i-PrOH or hexane–dichloromethane gives the enantiomerically pure phosphine boranes **72**.

*Decomplexation into free phosphines **60***: A 50 mL round-bottomed flask equipped with a magnetic stirrer and an argon inlet was charged with 0.5 mmol of tertiary phosphine borane **72** and 1.5 mmol of DABCO. 3 mL of dry toluene was added, and the resulting solution was stirred under argon at 50°C overnight. The solvent was removed under vacuum, and the crude product was rapidly filtered on a neutral alumina column using degassed eluent to afford the corresponding P(III) tertiary phosphine **60**.

*Typical procedure for the preparation of AMMP*diborane complexes **86** and free AMMP* **87***: A 100 mL round-bottomed flask equipped with a magnetic stirrer and an argon inlet was charged with 3.5 mmol of oxazaphospholidine borane **57da** (R^1 = Ph) and 6 mL of THF. 7 mmol of a previously prepared organolithium reagent was added slowly at –78 °C under stirring. The temperature reaction was slowly warmed to 0°C, and the stirring maintained until oxazaphospholidine complex **57da** had disappeared completely. 7 mmol of chlorophosphine (or chlorophosphine borane **78**) were added and the mixture was stirred 2 h and warmed to room temperature. 35 mmol of borane dimethylsulfide (BH_3.DMS) were added and the mixture was stirred overnight. BH_3.DMS and THF were evaporated under reduced pressure, and the residue was hydrolyzed at room temperature and then extracted with CH_2Cl_2. The organic extracts were dried over anhydrous $MgSO_4$, and the solvent was removed. The residue was purified by chromatography on silica gel using a mixture of toluene–petroleum ether as eluent affording the corresponding aminophosphine–phosphinite diboranes **87** with moderate to good yield (18–90%, Table 12.9). These AMPP*diborane complexes **86** can also be recrystallized in a mixture of CH_2Cl_2–hexane.

Decomplexation into free AMPP **87***: A 50 mL round-bottomed flask equipped with a magnetic stirrer and an argon inlet was charged with 0.5 mmol of aminophosphine phosphinite diborane **88** and 3 mmol of DABCO. 3 mL of dry toluene was added, and the resulting solution was stirred under argon at 50 °C overnight. The solvent was removed under vacuum, and the crude product was rapidly filtered on a neutral alumina column using degassed eluent to afford the corresponding aminophosphine phosphinite AMPP* **87**.

*Typical procedure for the preparation of secondary phosphine boranes **91***: In a 50 mL two-necked flask, equipped with a magnetic stirrer, an argon inlet, and a rubber septum, was introduced 6 mmol of a solution of chlorophosphine borane **78** in toluene. To this solution cooled at –78°C was added dropwise 12 mmol of t-BuLi (1.6 M in pentane). After stirring for 5 min at –78 °C, 6.5 mL of dry THF was added dropwise and the stirring was maintained for 3 min. At –78 °C, 3.1 mL of acetic acid was added, followed by 20 mL of water. The cooling bath was removed, and the solution was stirred to room temperature for 20 min. The organic phase was removed, the aqueous layer was extracted with dichloromethane, and the combined organic phases were dried over $MgSO_4$. After evaporation of the solvent, the residue was purified by column chromatography with a mixture of petroleum ether-ethyl acetate as eluent to afford the corresponding secondary phosphine boranes **91** in good yields and good enantiomeric purities (see Table 12.10).

*Typical procedures for the preparation of o-bromophenylphosphine borane **93** and 1,2-diphosphinobenzene **96***

*Preparation of o-bromophenylphosphine borane **93***: A 50 mL round-bottomed flask equipped with a magnetic stirrer and an argon inlet was charged with 0.83 mmol of secondary phosphine borane **91** and dry THF (2 mL). To this solution was added dropwise under argon at –78°C, 0.83 mmol of n-BuLi (1.6 M in hexane). The resulting solution was stirred at this temperature for 1 h, and 1.16 mmol of

1,2-dibromobenzene **92** was then added dropwise followed by 0.17 mmol of n-BuLi (1.6 M in hexane). After 1 h at –78 °C, the reaction mixture was quenched with water (10 mL) and extracted with methylene chloride (3 × 10 mL). The organic phases were dried over $MgSO_4$ and filtered, and the solvent evaporated, giving a residue which was purified by column chromatography on silica gel and/or by recrystallization to afford the corresponding *o*-bromophenylphosphine borane **93** in good yield (see Table 12.11).

*Decomplexation of the o-bromophenylphosphine borane **93** into **94***: A 50 mL round-bottomed flask equipped with a magnetic stirrer and an argon inlet was charged with 0.5 mmol of o-bromophenylphosphine borane ***93*** and 3 mmol of DABCO. 3 mL of dry toluene was added, and the resulting solution was stirred under argon at room temperature overnight. The solvent was removed under vacuum, and the crude product was purified by column chromatography on silica gel using degassed eluent to afford the corresponding *o*-bromophenylphosphine **94**.

*Preparation of 1,2-diphosphinobenzene **96***: A 50 mL round-bottomed flask equipped with a magnetic stirrer and an argon inlet was charged with 0.54 mmol of o-bromophenylphosphine **94** and dry THF (2 mL). 0.59 mmol of n-BuLi (1.6 M in hexane) was added dropwise at –78 °C under argon, and the resulting solution was stirred at this temperature for 1 h. At this time, 0.65 mmol of chlorophosphine (or P-chiral phenylphosphinite **95**) in solution in 1 mL of THF was added slowly at –78 °C. After stirring at room temperature overnight, the solution was quenched with water and extracted with dichloromethane. The organic phases were dried over $MgSO_4$, and the solvent evaporated to give a residue which was purified by chromatographic column and/or by recrystallization to afford the corresponding 1,2-diphosphinobenzene **96**.

References

[1] (a) I. Ojima, *Catalytic Asymmetric Synthesis*, 2nd ed., Wiley-VCH, **2000**; and (b) E. N. Jacobsen, A. Pfaltz, H. Yamamoto (Eds.), *Comprehensive Asymmetric Catalysis*, Springer, **2004**.

[2] (a) H.-U. Blaser, F. Spindler, M. Studler, *Appl. Cat. A: Gen.* **2001**, *221*, 119; (b) H.-U. Blaser, *Adv. Synth. Catal.* **2002**, *344*, 17; (b) H.-U. Blaser, *Chem. Commun.* **2003**, 293; (c) H.-U. Blaser, C. Malan, B. Pugin, F. Spindler, H. Steiner, M. Studer, *Adv. Synth. Catal.***2003**, *345*, 103;(d) H.-U. Blaser, E. Schmidt, *Asymmetric Catalysis on Industrial Scale*, Wiley-VCH, **2003**; (e) J. M. Hawkins, J. N. Watson, *Angew. Chem. Int. Ed.* **2004**, *43*, 3224; (f) H.-U. Blaser, B. Pugin, F. Spindler, M. Thommen, *Acc. Chem. Res.* **2007**, *40*, 1240; (g) N. B. Johnson, I. C. Lennon, P. H. Moran, J. A. Ramsden, *Acc. Chem. Res.* **2007**, *40*, 1291; and (h) L. A. Saudan, *Acc. Chem. Res.* **2007**, *40*, 1309.

[3] (a) W. Tang, X. Zang, *Chem. Rev.* **2003**, *103*, 3029; (b) H. Shimidzu, I. Nagasaki, T. Saito, *Tetrahedron* **2005**, *61*, 5405; (c) R. G. Arrayas, J. Adrio, J. C. Carretero, *Angew. Chem. Int. Ed.* **2006**, *45*, 7674; (d) C. Jäkel, R. Paciello, *Chem. Rev.* **2006**, *106*, 2912; (e) W. Zhang, Y. Chi, X. Zhang, *Acc. Chem. Res.* **2007**, *40*, 1278; (f) A. Börner (Ed.), *Phosphorus Ligands in Asymmetric Catalysis*, Vol. 3, Wiley-VCH, **2008**; and (g) G. C. Hargaden, P. J. Guiry, *Chem. Rev.* **2009**, *109*, 2505.

[4] (a) H. B. Kagan, M. Sasaki, in F. R. Hartley (Ed.), *The Chemistry of Organophosphorus Compounds*, John Wiley& Sons, **1990**, Vol. 1; (b) M. K. Pietrusiewicz, M. Zablocka, *Chem. Rev.* **1994**, *94*, 1375; (c) O. I. Kolodiazhnyi, *Tetrahedron: Asymmetry* **1998**, *9*, 1279; (d) M. Berthod, G. Mignani, G. Woodward, M. Lemaire, *Chem. Rev.* **2005**, *105*, 1801; and (e) A. Grabulosa, J. Granell, G. Muller, *Coord. Chem. Rev.* **2007**, *251*, 25.

[5] (a) A. Ohashi, S.-I. Kikuchi, M. Yasutake, T. Imamoto, *Eur. J. Org. Chem.* **2002**, 2535; and (b) K. V. L. Crépy, T. Imamoto, *Top. Curr. Chem.* **2003**, *229*, 1.

[6] (a) U. Nagel, B. Rieger, *Organometallics* **1989**, *8*, 1534; and (b) U. Nagel, C. Roller, *Z. Naturforsch. B: Chem. Sci.* **1998**, *53*, 267.

[7] (a) R. Ewalds, E. B. Eggeling, A. C. Hewat, P. C. J. Kamer, P. W. N. M. Van Leeuwen, D. Vogt, *Chem. Eur. J.* **2000**, *6*, 1496; and (b) R. Den Heeten, B. H. G. Swennenhuis, P. W. N. M. van Leeuwen, J. G. de Vries, P. C. J. Kamer, *Angew. Chem. Int. Ed.* **2008**, *47*, 1.

[8] (a) D. Moulin, C. Darcel, S. Jugé, *Tetrahedron: Asymmetry* **1999**, *10*, 4729; (b) C. Darcel, D. Moulin, J. C. Henry, M. Lagrelette, P. Richard, P. D. Harvey, S. Jugé, *Eur. J. Org. Chem.* **2007**, 2078; (c) N. Khiri, E. Bertrand,

M.-J. Ondel-Eymin, Y. Rousselin, J. Bayardon, P. D. Harvey, S. Jugé, *Organometallics* **2010**, *29*, 3622; and (d) J. Bayardon, C. Darcel, *et al.* to be published.

[9] For representative examples of chiral ligands prepared by resolution of racemic, see (a) H. Takaya, K. Mashima, K. Koyano, M. Yagi, H. Kumobayashi, T. Taketomi, S. Akutagawa, R. Noyori, *J. Org. Chem.* **1986**, *51*, 629; (b) R. Schmidt, J. Foricher, M. Cereghetti, P. Schönholzer, *Helv. Chim. Acta* **1974**, *74*, 370; and (c) S. Duprat de Paule, S. Jeulin, V. Ratovelomanana-Vidal, J. P. Genêt, N. Champion, G. Deschaux, P. Dellis, *Org. Process. Res. Dev.* **2003**, *7*, 399.

[10] For representative examples of ligands derived from chiral pool, see (a) H. B. Kagan, T.-P. Dang, *J. Am. Chem. Soc.* **1972**, *94*, 6429; (b) G. Helmchem, A. Pfaltz, *Acc. Chem. Res.* **2000**, *33*, 336; and (c) E. Belanger, M.-F. Pouliot, J.-F. Paquin, *Org. Lett.* **2009**, *11*, 2201.

[11] For an example of a chiral ligand synthesized by asymmetric synthesis, see M. J. Burk, J. E. Feaster, W. A. Nugent, R. L. Harlow, *J. Am. Chem. Soc.* **1993**, *115*, 10125.

[12] B. D. Vineyard, W. S. Knowles, M. J. Sabacky, G. L. Bachma, D. J. Weinkauff, *J. Am. Chem. Soc.* **1977**, *99*, 5946.

[13] (a) M. Ohff, J. Holz, M. Quirmbach, A. Börner, *Synthesis* **1998**, 1391; (b) J. M. Brunel, B. Faure, M. Maffei, *Coord. Chem. Rev.* **1998**, *178–180*, 665; and (c) B. Carboni, L. Monnier, *Tetrahedron* **1999**, *55*, 1197.

[14] E. A. Colby, T. F. Jamison, *J. Org. Chem.* **2003**, *68*, 156.

[15] Y. Yamanoi, T. Imamoto, *J. Org. Chem.* **1999**, *64*, 2988.

[16] (a) T. Imamoto, J. Watanabe, Y. Wada, H. Masuda, H. Yawada, H. Tsuruta, S. Matsukawa, K. Yamaguchi, *J. Am. Chem. Soc.* **1998**, *120*, 1635; (b) T. Imamoto, *Pure Appl. Chem.* **2001**, *73*, 373; and (c) N. Oohara, K. Katagiri, T. Imamoto, *Tetrahedron: Asymmetry* **2003**, *14*, 2171.

[17] (a) U. Nettekoven, M. Widhalm, P. C. J. Kamer, P. W. N. M. Van Leeuwen, *Tetrahedron: Asymmetry* **1997**, *8*, 3185; (b) U. Nettekoven, P. C. J. Kamer, P. W. N. M. Van Leeuwen, M. Wildhalm, A. Spek, M. Lutz, *J. Org. Chem.* **1999**, *64*, 3996; (c) F. Maienza, M. Wörle, P. Steffanut, A. Mezzetti, F. Spindler, *Organometallics* **1999**, *18*, 1041; and (d) U. Nettekoven, M. Widhalm, H. Kalchhauser, P. C. J. Kamer, P. W. N. M. Van Leeuwen, M. Lutz, A. L. Spek, *J. Org. Chem.* **2001**, *64*, 759.

[18] (a) F. Mathey, F. Mercier, F. Robin, L. Ricard, *J. Organomet. Chem.* **1998**, *557*, 117; (b) T. Faitg, J. Soulie, J.-Y. Lallemand, F. Mercier, F. Mathey, *Tetrahedron* **1999**, *56*, 101; and (c) M. Siutkowki, F. Mercier, L. Ricard, F. Mathey, *Organometallics* **2006**, *18*, 2585.

[19] R. Ewalds, E. B. Eggeling, A. C. Hewat, P. C. J. Kamer, P. W. N. M. Van Leeuwen, D. Vogt, *Chem. Eur. J.* **2000**, *6*, 1496.

[20] (a) T. Imamoto, K. Sugita, K. Yoshida, *J. Am. Chem. Soc.* **2005**, *127*, 11934; and (b) T. Imamoto, M. Nishimura, A. Koide, K. Yoshida, *J. Org. Chem.* **2007**, *72*, 7413.

[21] T. Imamoto, Y. Saitoh, A. Koide, T. Ogura, *Angew. Chem. Int. Ed.* **2007**, *46*, 8636.

[22] C. Salomon, D. Fortin, C. Darcel, S. Jugé, P. D. Harvey, *J. Clust. Sci.* **2009**, *20*(2), 267.

[23] C. Salomon, S. Dal Molin, D. Fortin, Y. Mugnier, R. T. Boeré, S. Jugé, P. D. Harvey, *J. Chem. Soc., Dalton Trans.* **2010**, *39*, 10068.

[24] Y. Ouchi, Y. Morisaki, T. Ogoshi, Y. Chujo, *Chem. Asian J.* **2007**, *2*, 397.

[25] C. Salomon, D. Fortin, N. Khiri, S. Jugé, P. D. Harvey, *Eur. J. Inorg. Chem.* **2011**, 2597.

[26] (a) A. Nudelman, D. J. Cram, *J. Am. Chem. Soc.* **1968**, *90*, 3869; and (b) A. Nudelman, D. J. Cram, *J. Org. Chem.* **1971**, *36*, 335.

[27] (a) O. I. Kolodiazhnyi, E. V. Grishkun, *Tetrahedron: Asymmetry* **1996**, *7*, 967; (b) A. Benabra, A. Alcudia, N. Khiar, I. Fernandez, F. Alcudia, *Tetrahedron: Asymmetry* **1996**, *7*, 3353; (c) M. Oliana, F. King, P. N. Horton, M. B. Hursthouse, K. K. Hii, *J. Org. Chem.* **2006**, *71*, 2472.

[28] (a) W. Chodkiewicz, D. Jore, W. Wodzki, *Tetrahedron Lett.* **1979**, *20*, 1069; (b) W. Chodkiewicz, D. Jore, A. Pierrat, W. Wodzki, *J. Organomet. Chem.* **1979**, *174*, C21; and (c) W. Chodkiewicz, *J. Organomet. Chem.* **1984**, *273*, C55.

[29] (a) O. Korpium, R. A. Lewis, J. Chickos, K. Mislow, *J. Am. Chem. Soc.* **1968**, *90*, 4842; (b) R. A. Lewis, K. Mislow, *J. Am. Chem. Soc.* **1969**, *91*, 7009; and (c) W. B. Farnham, R. A. Lewis, R. K. Murray Jr., K. Mislow, *J. Am. Chem. Soc.* **1970**, *92*, 5809.

[30] For a review on the stereoselective synthesis and reduction of P-chiral phosphine oxides, see M. K. Pietrusiewicz, M. Zablocka, *Chem. Rev.* **1994**, *94*, 1375.

[31] E. P. Kyba, *J. Am. Chem. Soc.* **1976**, *98*, 4805.

[32] T. L. Emmick, R. L. Letsinger, *J. Am. Chem. Soc.* **1968**, *90*, 3459.
[33] T. Oshiki, T. Imamoto, *J. Am. Chem. Soc.* **1992**, *114*, 3975.
[34] L. Horner, B. Schlotthauer, *Phosphorus, Sulfur* **1978**, *4*, 155.
[35] M. Mikolajczyk, J. Omelanczuk, W. Perlikowska, *Tetrahedron* **1979**, *35*, 1531
[36] Y. Koide, A. Sakamoto, T. Imamoto, *Tetrahedron Lett.* **1991**, *32*, 3375.
[37] U. Schmidt, B. Riedl, H. Griesser, C. Fitz, *Synthesis* **1991**, *8*, 655.
[38] (a) A. Leyris, J. Bigeault, D. Nuel, L. Giordano, G. Buono, *Tetrahedron Lett.* **2007**, *48*, 5247; (b) D. Moraleda, D. Gatineau, D. Martin, L. Giordano, G. Buono, *Chem. Commun.* **2008**, 3031; (c) L.-B. Han, Q. Xu, International Patent WO2008032531; and (d) Q. Xu, C.-Q. Zhao, L.-B. Han, *J. Am. Chem. Soc.* **2008**, *130*, 12648.
[39] (a) X.-B. Jiang, A. J. Minnaard, B. Hessen, B. L. Feringa, A. L. Duchateau, J. G. O. Andrien, J. A. F. Boogers, J. G. de Vries, *Org. Lett.* **2003**, *5*, 1503; (b) N. V. Dubrovina, A. Börner, *Angew. Chem. Int. Ed.* **2004**, *43*, 5883; (c) X.-B. Jiang, M. van der Berg, A. J. Minnard, B. L. Feringa, J. G. de Vries, *Tetrahedron: Asymmetry* **2004**, *15*, 2223; and (d) H. Landert, F. Spindler, A. Wyss, H.-U. Blaser, B. Pugin, Y. Ribourduoille, B. Gschwend, B. Ramalingam, A. Pfaltz, *Angew. Chem. Int. Ed.* **2010**, *49*, 6873.
[40] (a) J. Zhang, Y. Xu, G. Huang, H. Guo, *Tetrahedron Lett.* **1988**, *29*, 1955; (b) Y. Xu, H. Wei, J. Zhang, H. Guo, *Tetrahedron Lett.* **1989**, *30*, 1955; (c) R. K. Haynes, T.-L. Au-Yeung, W.-K. Chan, W.-L. Lam, Z.-Y Li, L.-L. Yueng, A. S. C. Chan, P. Li, M. Koen, C. R. Mitchell, S. C. Vonwiller, *Eur. J. Org. Chem.* **2000**, 3205; (d) T. Hirai, L.-B. Han, *Org. Lett.* **2007**, *9*, 53; and (e) F. Alonso, I. P. Beletskaya, M. Yus, *Chem. Rev.* **2004**, *104*, 3079.
[41] M. Stankevic, M. K. Pietrusiewicz, *J. Org. Chem.* **2007**, *72*, 816.
[42] (a) T. Imamoto, T. Oshiki, T. Onozawa, T. Kusumoto, K. Sato, *J. Am. Chem. Soc.* **1990**, *112*, 5244; (b) T. Oshiki, T. Hikosaka, T. Imamoto, *Tetrahedron Lett.* **1991**, *32*, 3371; (c) T. Imamoto, T. Yoshizawa, K. Hirose, Y. Wada, H. Masuda, K. Yamaguchi, H. Seki, *Heteroatom Chem.* **1995**, *6*, 99; (d) Y. Wada, T. Imamoto, H. Tsuruta, K. Yamaguchi, I. D. Gridnev, *Adv. Synth. Cat.* **2004**, *346*, 777; and (e) Y. Takahashi, Y. Yamamoto, K. Katagiri, H. Danjo, K. Yamaguchi, T. Imamoto, *J. Org. Chem.* **2005**, *70*, 9009.
[43] (a) R. Bodalski, E. Rutkowska-Olma, M. K. Pietrusiewicz, *Tetrahedron* **1980**, *36*, 2353; (b) M. K. Pietrusiewicz, M. Zablocka, J. Monkiewicz, *J. Org. Chem.* **1984**, *49*, 1522; (c) M. K. Pietrusiewicz, M. Zablocka, *Tetrahedron Lett.* **1988**, *29*, 1991; and (d) Z. Pakulski, O. M. Demchuk, J. Frelek, R. Luboradzki, M. K. Pietrusiewicz, *Eur. J. Org. Chem.* **2004**, 3913.
[44] (a) M. K. Pietrusiewicz, M. Zablocka, *Tetrahedron Lett.* **1988**, *29*, 937; (b) M. K. Pietrusiewicz, M. Zablocka, *Tetrahedron Lett.* **1988**, *29*, 1987; (c) M. K. Pietrusiewicz, W. Wisniewski, M. Zablocka, *Tetrahedron* **1989**, *45*, 337; (d) A. Brandi, S. Cicchi, A. Goti, M. K. Pietrusiewicz, M. Zablocka, W. Wisniewski, *J. Org. Chem.* **1991**, *56*, 4383; (e) M. K. Pietrusiewicz, M. Kuznikowski, M. Koprowski, *Tetrahedron: Asymmetry* **1993**, *4*, 2143; (f) O. M. Demchuk, M. K. Pietrusiewicz, A. Michrowska, K. Grela, *Org. Lett.* **2003**, *5*, 3217; and (g) N. Vinokurov, M. K. Pietrusiewicz, S. Frynas, M. Wiebcke, H. Butenschön, *Chem. Commun.* **2008**, 5408.
[45] T. Imamoto, K. Sato, C. R. Johnson, *Tetrahedron Lett.* **1985**, *26*, 783.
[46] E. Bergin, C. T. O'Connor, S. B. Robinson, E. M. McGarrigle, C. P. O'Mahony, D. G. Gilheany, *J. Am. Chem. Soc.* **2007**, *129*, 9566.
[47] A. J. Rippert, A. Linden, H.-J. Hansen, *Helv. Chim. Acta* **2000**, *83*, 311.
[48] (a) P. Cros, G. Buono, G. Peiffer, D. Denis, A. Mortreux, F. Petit, *New J. Chem.* **1987**, *11*, 573; and (b) V. Peper, K. Stingl, H. Thümler, W. Saak, D. Haase, S. Pohl, S. Jugé, J. Martens, *Liebigs Annalen* **1995**, 2123.
[49] T. Léon, A. Riera, X. Verdaguer, *J. Am. Chem. Soc.* **2011**, *133*, 5740.
[50] (a) S. Jugé, Procédé de préparation de phosphinamides, application et nouveaux produits, French Patent and Extensions 2518100, **1981**; and (b) S. Jugé, M. Wakselman, M. Stephan, J. P. Genêt, *Tetrahedron Lett.* **1990**, *31*, 4443.
[51] W. J. Richter, *Chem. Ber.* **1984**, *117*, 2328; and (b) S. Jugé, J. P. Genêt, *Tetrahedron Lett.* **1989**, *30*, 2783.
[52] (a) J. M. Brown, J. V. Carey, M. J. H. Russell, *Tetrahedron* **1990**, *46*, 4877; (b) J. V. Carey, M. D. Barker, J. M. Brown, M. J. H. Russell, *J. Chem. Soc., Perkin Trans.* **1993**, 831; (c) J. M. Brown, J. C. P. Laing, *J. Organomet. Chem.* **1997**, *529*, 435; and (d) D. Vinci, N. Mateus, X. Wu, F. Hancock, A. Steiner, J. Xiao, *Org. Lett.* **2006**, *8*, 215.
[53] S. K. Sheehan, M. Jiang, L. McKinstry, T. Livinghouse, D. Garton, *Tetrahedron* **1994**, *50*, 6155.
[54] C. R. Hall, T. D. Inch, I. W. Lawston, *Tetrahedron Lett.* **1979**, *29*, 2729.
[55] (a) S. Jugé, J. P. Genêt, Procédé de preparation de derives de l'acide phosphineux et produit intermédiaire utilisé à cet effet, French Patent 2603286, **1986**; (b) S. Achi, J. P. Genêt, S. Jugé, P. Potin, *Phosphorus, Sulfur, and Silicon*

1987, *30*, 688; (c) S. Jugé, J. P. Genêt, Composés organiques du phosphore complexé par du bore, French Patent and Extensions 2649109, **1989**; (d) S. Jugé, M. Stephan, S. Achi, J. P. Genêt, *Phosphorus, Sulfur and Silicon* **1990**, *49–50*, 267; (e) S. Jugé, M. Stephan, J. A. Laffitte, J. P. Genêt, *Tetrahedron Lett.* **1990**, *31*, 6357; (f) S. Jugé, M. Stephan, J. P. Genêt, S. Halut-Desportes, S. Jeannin, *Acta Crystallographica* **1990**, *C46*, 1869; (g) S. Jugé, M. Stéphan, R. Merdès, J. P. Genêt, S. Halut-Desportes, *J. Chem. Soc., Chem. Commun.* **1993**, 531; (h) E. B. Kaloun, R. Merdès, J. P. Genêt, J. Uziel, S. Jugé, *J. Organomet. Chem.* **1997**, *529*, 455; (i) C. Bauduin, D. Moulin, E. B. Kaloun, C. Darcel, S. Jugé, *J. Org. Chem.* **2003**, *68*, 4293; and (j) F. Chaux, S. Frynas, H. Laureano, C. Salomon, G. Morata, M.-L. Auclair, M. Stephan, R. Merdès, P. Richard, M.-J. Ondel, J. C. Henry, J. Bayardon, C. Darcel, S. Jugé, *Compte-Rendus Chimie* **2010**, *13*, 1213.

[56] For reviews on the ephedrine method, see (a) C. Darcel, J. Uziel, S. Jugé, in *Phosphorus Ligands in Asymmetric Catalysis Synthesis and Applications*, A. Börner (Ed.), Wiley-VCH, **2008**; and (b) S. Jugé, *Phosphorus, Sulfur and Silicon and the Related Compounds* **2008**, *183* (2–3), 233.

[57] (a) J. M. Johansson, N. Kann, K. Larsson, *Acta Crystallographica* **2004**, *E60*, 287; (b) U. Nettekoven, P. C. J. Kamer, M. Widhalm, P. W. N. M. Van Leeuven, *Organometallics* **2000**, *19*, 4596; and (c) F. Maienza, F. Spindler, M. Thommen, B. Pugin, C. Malan, A. Mezzetti, *J. Org. Chem.* **2002**, *67*, 5239.

[58] B. Faure, A. Archavlis, G. Buono, *J. Chem. Soc., Chem. Comm.* **1989**, 805.

[59] T. Koizumi, R. Yanada, H. Takagi, H. Hirai, E. Yoshii, *Tetrahedron Lett.* **1981**, *22*, 477.

[60] J. M. Brunel, O. Chiodi, B. Faure, F. Fotiadu, G. Buono, *J. Organomet. Chem.* **1997**, *529*, 285.

[61] T. Koizumi, R. Yanada, H. Takagi, H. Hirai, E. Yoshii, *Tetrahedron Lett.* **1981**, *22*, 571.

[62] (a) T. Imamoto, T. Kusumoto, N. Suzuki, K. Sato, *J. Am. Chem. Soc.* **1985**, *107*, 5301; (b) H. Brisset, Y. Gourdel, P. Pellon, M. Le Corre, *Tetrahedron Lett.* **1993**, *34*, 4523; and (c) H. Yang, N. Lugan, R. Mathieu, *Organometallics* **1997**, *16*, 2089.

[63] (a) G. Knühl, P. Sennhenn, G. Helmchen, *J. Chem. Soc., Chem. Commun.* **1995**, 1845; and (b) L. McKinstry, J. J. Overberg, C. Soubra-Ghaoui, D. S. Walsh, K. A. Robins, T. T. Toto, J. L. Toto, *J. Org. Chem.* **2000**, *65*, 2261.

[64] (a) D. Moulin, S. Bago, C. Bauduin, C. Darcel, S. Jugé, *Tetrahedron: Asymmetry* **2000**, *11*, 3939; (b) M. Schröder, K. Nozaki, T. Hiyama, *Bull. Chem. Soc. Jpn.* **2004**, *77*, 1931; and (c) M. van Overschelde, E. Vervecken, S. G. Mohdha, S. Cogen, E. van der Eycken, J. van der Eycken, *Tetrahedron* **2009**, *65*, 6410.

[65] A. Pelter, R. Rosser, S. Mills, *J. Chem. Soc., Chem. Commun.* **1981**, 1014.

[66] J. Uziel, N. Riegel, B. Aka, P. Figuière, S. Jugé, *Tetrahedron Lett.* **1997**, *38*, 3405.

[67] (a) T. Imamoto, K. Hirose, H. Amano, H. Seki, *Main Group Chemistry* **1996**, *1*, 331; and (b) J. Uziel, C. Darcel, D. Moulin, C. Bauduin, S. Jugé, *Tetrahedron: Asymmetry* **2001**, *12*, 1441.

[68] (a) N. Brodie, S. Jugé, *Inorg. Chem.* **1998**, *37*, 2438; and (b) C. Darcel, E. B. Kaloun, R. Merdès, D. Moulin, N. Riegel, S. Thorimbert, J. P. Genêt, S. Jugé, *J. Organomet. Chem.* **2001**, *624*, 333.

[69] M. Schuman, M. Trevitt, A. Redd, V. Gouverneur, *Angew. Chem. Int. Ed.* **2000**, *39*, 2491.

[70] J. A. Ramsden, J. M. Brown, M. B. Hursthouse, A. I. Karalulov, *Tetrahedron: Asymmetry* **1994**, *5*, 2033.

[71] M. B. Tollefson, J. J. Li, P. Beak, *J. Am. Chem. Soc.* **1996**, *118*, 9052.

[72] (a) R. M. Stoop, C. Bauer, P. Setz, M. Wörle, T. Y. H. Wong, A. Mezzetti, *Organometallics* **1999**, *18*, 5691; (b) F. Maienza, F. Spindler, M. Thommen, B. Pugin, C. Malan, A. Mezzetti, *J. Org. Chem.* **2002**, *67*, 5239; and (c) F. Maienza, F. Santoro, F. Spindler, C. Malan, A. Mezzetti, *Tetrahedron: Asymmetry*, **2002**, *13*, 1817.

[73] H. Lam, P. N. Horton, M. B. Hursthouse, D. J. Aldous, K. K. Hii, *Tetrahedron Lett.* **2005**, *6*, 145.

[74] A. Grabulosa, G. Muller, J. I. Ordinas, A. Mezzetti, M. A. Maestro, M. Font-Bardia, X. Solans, *Organometallics* **2005**, *24*, 4961.

[75] U. Nettekoven, M. Widhalm, P. C. J. Kamer, P. W. N. M. van Leeuwen, K. Mereiter, M. Lutz, A. L. Spek, *Organometallics* **2000**, *19*, 2299.

[76] R. M. Stoop, A. Mezzetti, F. Spindler, *Organometallics* **1998**, *17*, 668.

[77] (a) M. Stephan, PhD thesis, Université P. & M. Curie, Paris, **1991**; (b) R. Merdès, PhD thesis, Université P. & M. Curie, Paris, **1993**; (c) C. Bauduin, PhD thesis, Université Paris Sud, Orsay, **2001**; (d) M.-L. Auclair, PhD thesis, Université de Bourgogne, Dijon, **2004**; and (e) G. Morata, PhD thesis, Université de Bourgogne, Dijon, **2005.**

[78] (a) M. Stephan, S. Sterk, B. Modec, B. Mohar, *J. Org. Chem.* **2007**, *72*, 8010; (b) B. Zupancic, B. Mohar, M. Stephan, *Adv. Synth. Catal.* **2008**, *350*, 2024; (c) B. Zupancic, M. Mohar, M. Stephan, *Org. Lett.* **2010**, *12*,

1296; (d) B. Zupancic, M. Mohar, M. Stephan, *Org. Lett.* **2010**, *12*, 3022; and (e) M. Stephan, B. Modec, B. Mohar, *Tetrahedron Lett.* **2011**, *52*, 1086.

[79] (a) L. Horner, M. Jordan, *Phosphorus, Sulfur* **1980**, *8*, 235; and (b) J. Omelanczuk, *J. Chem. Soc., Chem. Commun.* **1992**, 1718.

[80] S. Humbel, C. Bertrand, C. Darcel, C. Bauduin, S. Jugé, *Inorg. Chem.* **2003**, *42*, 420.

[81] J. I. Jarbo, A. Lledos, D. Vogt, C. Bo, *Chem. Eur. J.* **2006**, *12*, 1457.

[82] (a) M. Quirmbach, A. Kless, J. Holtz, V. Tararov, A. Börner, *Tetrahedron: Asymmetry* **1999**, *10*, 1803; and (b) J. M. Brunel, O. Legrand, G. Buono, *Tetrahedron: Asymmetry* **1999**, *10*, 1979.

[83] (a) J. M. Brunel, T. Contantieux, O. Legrand, G. Buono, *Tetrahedron Lett.* **1998**, *39*, 2961; (b) O. Legrand, J. M. Brunel, G. Buono, *Tetrahedron Lett.* **1998**, *39*, 9419; (c) G. Buono, O. Chiodi, M. Wills, *Synlett* **1999**, *4*, 377; and (d) O. Legrand, J. M. Brunel, G. Buono, *Tetrahedron Lett.* **2000**, *41*, 2105.

[84] (a) M. Shi, L.-H. Chen, C.-Q . Li, *J. Am. Chem. Soc.* **2005**, *127*, 3790; and (b) Y.-Q. Jiang, Y.-L. Shi, M. Shi, *J. Am. Chem. Soc.* **2008**, *130*, 7202.

[85] (a) B. Wolfe, T. Livinghouse, *J. Am. Chem. Soc.* **1998**, *120*, 5116; and (b) H. Heath, B. Wolfe, T. Livinghouse, S. K. Bae, *Synthesis* **2001**, 2341.

[86] S. Jugé, J. Bayardon, H. Lauréano, J. C. Henry, F. Colobert, F. Leroux, E. Rémond, *P-chirogenic organophosphorus compounds*, International Patent pending, **2011**.

[87] K. Tamura, M. Sugiya, K. Yoshida, A. Yanagisawa, T. Imamoto, *Org. Lett.* **2010**, *12*, 4400.

[88] A. R. Muci, K. R. Campos, D. A. Evans, *J. Am. Chem. Soc.* **1995**, *117*, 9075.

[89] (a) A. Ohashi, T. Imamoto, *Tetrahedron Lett.* **2001**, *42*, 1099; and (b) A. Ohashi, T. Imamoto, *Org. Lett.* **2001**, *3*, 373.

[90] K. Nagata, S. Matsukawa, T. Imamoto, *J. Org. Chem.* **2000**, *65*, 4185.

[91] B. Wolfe, T. Livinghouse, *J. Org. Chem.* **2001**, *66*, 1514.

[92] T. Imamoto, A. Kumada, K. Yoshida, *Chem. Lett.* **2007**, *36*, 500.

[93] K. V. L. Crepy, T. Imamoto, *Tetrahedron Lett.* **2002**, *43*, 7735.

[94] (a) T. Imamoto, K. Tamura, T. Ogura, Y. Ikematsu, D. Mayama, M. Sugiya, *Tetrahedron: Asymmetry* **2010**, *21*, 1522; (b) T. Imamoto, *J. Synth. Org. Chem. Jpn.* **2007**, *66*, 18; and (c) J. J. Gammon, P. O'Brien, B. Kelly, *Org. Lett.* **2009**, *11*, 5022.

[95] J. J. Gammon, V. H. Gessner, G. R. Barker, J. Granander, A. C. Whitwood, C. Strohmann, P. O'Brien, B. Kelly, *J. Am. Chem. Soc.* **2010**, *132*, 13922.

[96] (a) M. J. Dearden, M. J. McGrath, P. O'Brien, *J. Org. Chem.* **2004**, *69*, 5789; (b) C. Genet, S. J. Canipa, P. O'Brien, S. Taylor, *J. Am. Chem. Soc.* **2006**, *128*, 9336; and (c) J. Granander, F. Secci, S. J. Canipa, P. O'Brien, B. Kelly, *J. Org. Chem.* **2011**, *76*, 4794.

[97] M. J. Johansson, L. Schwartz, M. Amedjkouh, N. Kann, *Tetrahedron: Asymmetry* **2004**, *15*, 3531.

[98] S. Kobayashi, N. Shiraishi, W. Lam, K. Manabe, *Tetrahedron Lett.* **2001**, *42*, 7303.

[99] W. Tang, X. Zhang, *Angew. Chem. Int. Ed.* **2002**, *41*, 1612.

[100] J. Granader, F. Secci, P. O'Brien, B. Kelly, *Tetrahedron: Asymmetry* **2009**, *20*, 2432.

[101] (a) M. Mikolajczyk, J. Drabowicz, J. Omelanczuk, E. Fluck, *J. Chem. Soc., Chem. Commun.* **1975**, 382; (b) A. J. Macpherson, D. J. H. Smith, *J. Chem. Res.* **1984**, 32; and (c) B. S. Purnanand Batra, G. Lal, *Tetrahedron Lett.* **1994**, *35*, 4641.

[102] (a) O. I. Kolodiazhnyi, E. V. Gryshkun, N. V. Andrushko, M. Freytag, P. G. Jones, R. Schmutzler, *Tetrahedron: Asymmetry* **2003**, *12*, 181; and (b) E. V. Gryshkun, N. V. Andrushko, O. I. Kolodiazhnyi, *Phosphorus, Sulfur and Silicon* **2004**, *179*, 1027.

[103] P.-H. Leung, S.-K. Loh, J. J. Vittal, A. J. P. White, D. J. Williams, *J. Chem. Soc., Chem. Commun.* **1997**, 1987.

[104] (a) P.-H. Leung, A. Liu, K. F. Mok, *Tetrahedron: Asymmetry* **1999**, *10*, 1309; (b) P.-H. Leung, H. Lang, J. P. White, D. J. Williams, *Tetrahedron: Asymmetry* **1998**, *9*, 2961; and (c) Y. Song, J. J. Vittal, S.-H. Chan, P.-H. Leung, *Organometallics* **1999**, *18*, 650; (d) P.-H. Leung, *Acc. Chem. Res.* **2004**, *37*, 169; and (e) M. Yuan, S. A. Pullarkat, Y. Li, Z.-Y. Lee, P.-H. Leung, *Organometallics* **2010**, *29*, 3582.

13

Phosphatrioxa-adamantane Ligands

Paul G. Pringle[1] and Martin B. Smith[2]

[1]*School of Chemistry University of Bristol, Bristol, United Kingdom*
[2]*Department of Chemistry, Loughborough University, Loughborough, United Kingdom*

13.1 Introduction

A database search in January 2011 of the phospha-adamantane cage substructure shown in Figure 13.1 (denoted CgP here) revealed 75 articles, and even though the first report of the parent secondary phosphine CgPH was in 1961, 80% of the articles have appeared since 2001. Remarkably, over half of these articles are patents that are mainly concerned with applications of CgP ligands in a variety of homogeneous catalysis processes including carbonylation, hydroformylation, hydrocyanation, Buchwald–Hartwig amination and Suzuki coupling. The interest in CgP-containing ligands has led to the commercial availability of CgPH, CgPPh and even a CgP ligand immobilised on a silica support.

As will be evident from this chapter, the synthesis of these apparently exotic cage ligands is normally straightforward, high yielding and from readily available reagents. The CgP ligands are generally air stable, thermally robust and resistant to aqueous acids and bases. Infrared (IR) and nuclear magnetic resonance (NMR) spectroscopic measurements, along with X-ray crystal structure determinations of many metal

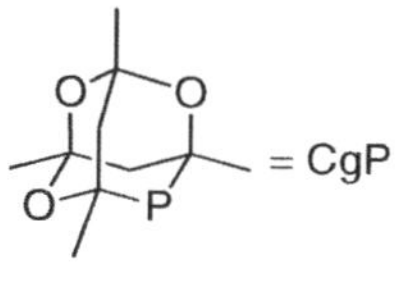

Figure 13.1

Phosphorus(III) Ligands in Homogeneous Catalysis: Design and Synthesis, First Edition.
Edited by Paul C. J. Kamer and Piet W. N. M. van Leeuwen.

complexes of CgP-containing ligands, have revealed that the CgP substituent can be categorised as very bulky (comparable with a P^tBu_2 group) and electron poor (comparable with a $P(OR)_2$ group). The adamantane cage imposes a very rigid structure on the ligand and compresses the C–P–C angle close to 90°. This unusual combination of steric bulk, rigidity, small C–P–C angle and phosphonite-like electronic properties not only makes these CgPR ligands fascinating to investigate, but also may be the source of their striking track record for producing highly effective catalysts for several processes.

Buckler and Epstein discovered the acid-catalysed condensation reaction of PH_3 with acetylacetone to give CgPH (Eq. 13.1) and suggested a mechanism for the cage formation [1,2].

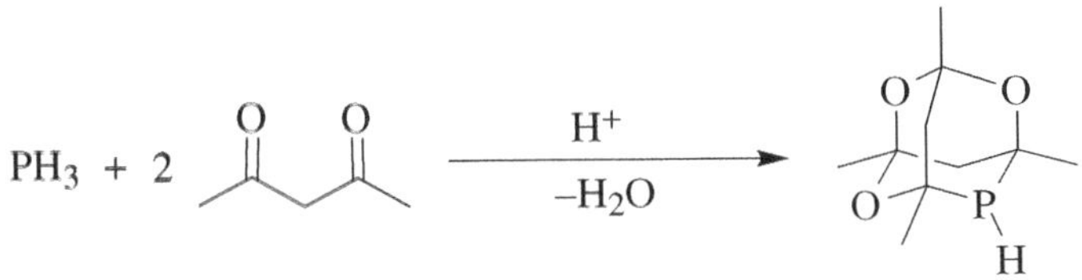

(Eq. 13.1)

In addition, it was shown [1] that primary phosphines RPH_2 (R=Ph, n-C_8H_{17}, i-C_4H_9) react with acetylacetone to give the corresponding CgPR and, in the case of $PhPH_2$, the di- and tri-phospha-adamantanes shown in Eq. 13.2.

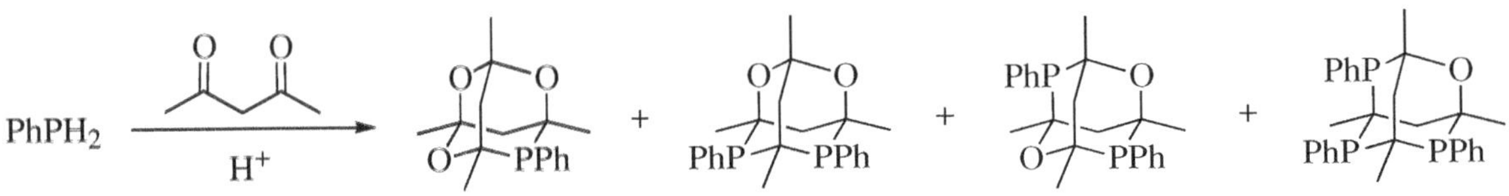

(Eq. 13.2)

The organophosphorus chemistry of CgPH was developed as summarised in Scheme 13.1 [1–3] and the phosphonium salts were of interest as flame retardants [4].

The first report of catalysis with a CgP ligand was made in 1968 when CgPH was claimed to support the Co-catalysed hydroformylation of propene and hexene [5]. It took 30 years for the second claim of catalysis by CgP-containing ligands to be made; in 1998, the bidentate $CgPCH_2CH_2CH_2PCg$ ligand was reported to give efficient Pd catalysts for methoxycarbonylation of alkenes and Rh catalysts for hydroformylation [6]. This report drew attention to CgP ligands and led to interest in synthesis of monodentate and bidentate CgP ligands described in this chapter.

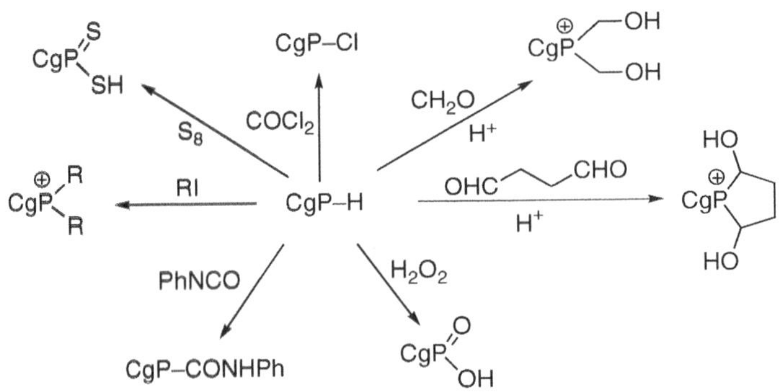

Scheme 13.1

13.2 Synthesis of phosphatrioxa-adamantanes

There are several versatile routes that have been used to synthesise CgP-containing ligands. These are itemised below, and detailed procedures for some of these routes are given in Section 13.4.

1. *Buckler–Epstein reaction (Scheme 13.2).* The acid-catalysed conversion of PH_3 or RPH_2 to a CgPR where R = hydrogen, alkyl or an aryl substituent has been applied to the synthesis of monophosphines [1] and diphosphines [7]. The good yields and high purity of the products obtained using this method belie the complexity of the cage construction that is taking place. The key to its success is probably the precipitation of the cage product from the aqueous medium in which the reaction is carried out. A limited number of other 1,3-diketones (e.g. $CF_3COCH_2COCF_3$ [8] and $CH_3CH_2COCH_2COCH_2CH_3$ [9]) can be used in place of acetylacetone in this reaction to give the corresponding phosphatrioxa-adamantanes, but to date these have attracted little attention as ligands for catalysis.
2. *Nucleophilic source of CgP (Scheme 13.3).* The air-stable compound $CgPH(BH_3)$ is readily deprotonated to give the powerful nucleophile $CgP(BH_3)Li$ which has been shown to react with a range of halocarbons to give monophosphines [10] and diphosphines [11].
3. *Electrophilic source of CgP (Scheme 13.4).* The halophosphines CgPCl and CgPBr readily react with organolithium or Grignard reagents to give tertiary phosphines [10] and with alkoxides or aryloxides to give phosphinites [12].
4. *Pd-catalysed arylations (Scheme 13.5).* The reaction of bromoarenes with CgPH catalysed by palladium complexes (e.g. $[Pd(PPh_3)_4]$) gives good yields of CgPAr ligands [13]. The reaction is tolerant of a variety of functionalised bromoarenes including bromopyridine.
5. *Radical P–H additions (Scheme 13.6).* The radical addition of CgPH to alkenes is a route that has been used to make CgP-containing alkyl phosphines which incorporate a functionality [14].
6. *Mannich-type condensations (Scheme 13.7).* The variant of the Mannich reaction shown in this chapter has been used to make CgP-containing aminophosphines and has been extended to make aminodiphosphines [15].

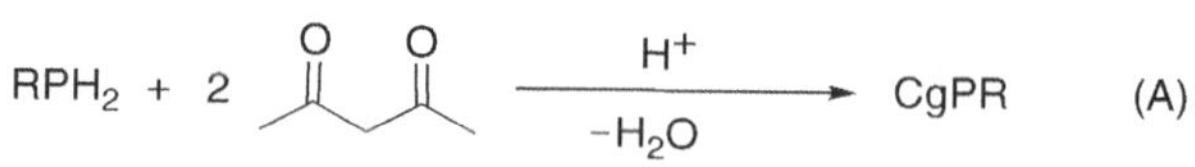

Scheme 13.2

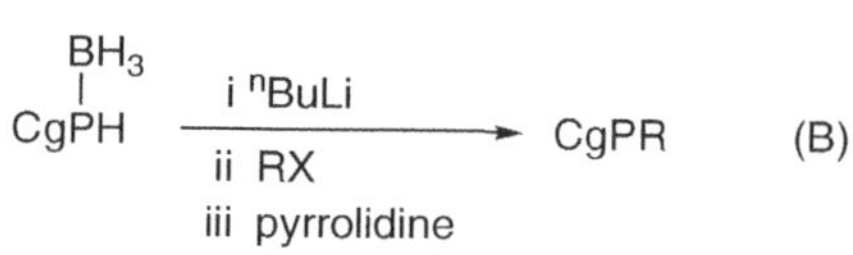

Scheme 13.3

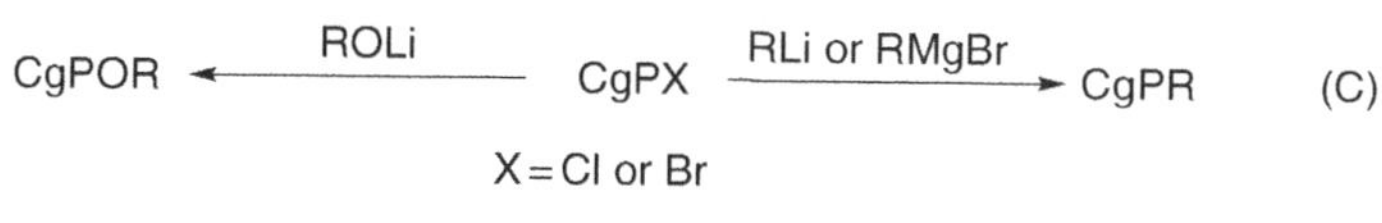

Scheme 13.4

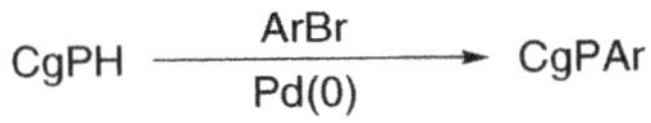

(D)

Scheme 13.5

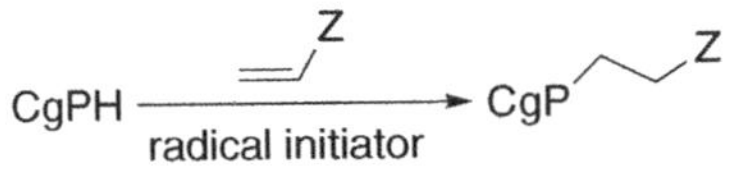

(E)

Scheme 13.6

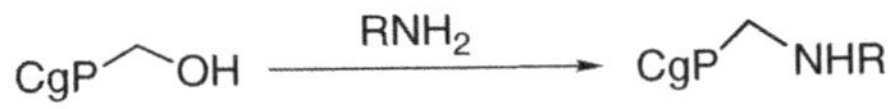

(F)

Scheme 13.7

Table 13.1 *Synthesis of ligands of the type CgPR.*

R	Route	Ref.
H; C_6H_5; $(CH_2)_7CH_3$; $CH_2CH(CH_3)_2$	A	[1]
c-C_6H_{11}	A	[16]
$CH(CH_3)_2$	A	[10]
$CH_2C_6H_5$; $CH(CH_3)_2$; $(CH_2)_{19}CH_3$	B	[10]
CH_3; $CH(CH_3)_2$	C	[10]
OCH_3, OCH_2CF_3; OC_6H_5; OC_6H_4Me-2; $OC_6H_2Me_3$-2,4,6; $OC_6H_3{}^tBu_2$-2,4	C	[12]
2-BuOC_6H_4; 3-MeOC_6H_4; 4-MeOC_6H_4; 2-$Me_2NC_6H_4$; 3-$Me_2NC_6H_4$; 4-$Me_2NC_6H_4$; 4-MeCOC_6H_4; 2,4-$(MeO)_2C_6H_3$; 1-naphthyl	D	[13]
2-BrC_6H_4	D	[17]
C_6H_5; 2-MeC_6H_4	D	[18]
$(CH_2)_{13}CH_3$	E	[18]
$CH_2CH_2PO(OEt)_2$	E	[19]
$CH_2CH_2Si(OMe)_3$	E	[20]
C_6H_5; 4-MeC_6H_4	F	[15]

The CgP-containing monodentate and bidentate ligands that have been reported to date are collected in Tables 13.1 and 13.2 with the route adopted for their synthesis. The CgPR ligands have C_1-symmetry and thus exist as enantiomers (labelled α and β in Figure 13.2). In general, the ligands described here are derived from racemic CgP precursors, and therefore diphosphines containing two CgP units will be mixtures of two diastereoisomers, *rac* and *meso*. The presence of these isomers is easily detected by the two signals in the ^{31}P NMR spectra which can be separated by several ppm; in the case of $CgPCH_2CH_2CH_2PCg$, the *rac* and *meso* isomers were readily separated by crystallisation (see Section 13.4).

Table 13.2 *Synthesis of CgP-containing diphos ligands.*

	Route	Ref.
$CgPCH_2CH_2PCg$; $CgPCH_2CH_2CH_2PCg$	A	[7]
1,2-$(CgPCH_2)_2C_6H_4$	B	[11,21]
1,2-$(CgPCH_2)_2C_5H_3Fe(C_5H_5)$	B	[22,23]
1,2-$(CgPOCH_2)_2C_6H_4$; 2,2′-$(CgPOC_6H_4)_2$	C	[12]
1-CgP-2-${}^tBu_2PCH_2C_6H_4$	D	[24]
$CgPCH_2CH_2PPh_2$	E	[14]
$CgPCH_2NPhCH_2PPh_2$; $CgPCH_2N(4\text{-}MeC_6H_4)CH_2PPh_2$	F	[15]

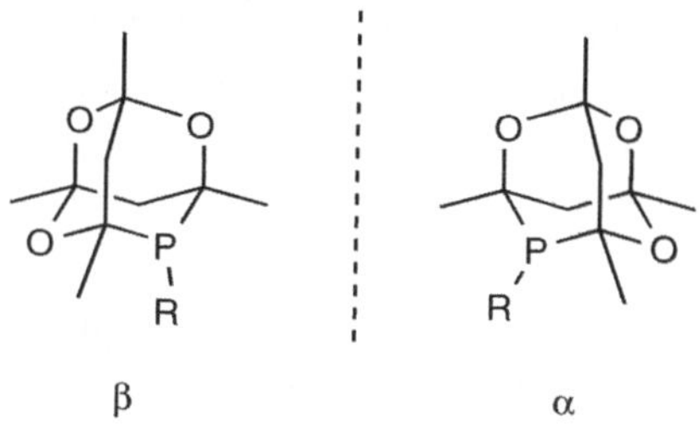

Figure 13.2

13.3 Catalysis supported by phosphatrioxa-adamantane ligands

Monodentate and bidentate ligands containing a ${}^tBu_2P$ group have made an enormous impact on homogeneous catalysis [25]. From the crystal structures of metal complexes of CgP ligands, it is apparent that the CgP and ${}^tBu_2P$ groups have similar steric bulk [7]. It is this similarity that may explain why it has been found that ligands based on CgPR and ${}^tBu_2PR$ both give efficient catalysts for some reactions, for example alkoxycarbonylation (Section 3.1) and C–C and C–N coupling reactions (Section 3.3). The dissimilarity between CgP and ${}^tBu_2P$ groups is in their σ-donor and π-acceptor properties. A CgP group has been shown to have electronic properties similar to those of a $(RO)_2P$ group [9,12], and it is this feature that may lie behind the success of CgPR ligands in the catalysis of reactions such as hydroformylation and hydrocyanation. Stability to air and moisture is a general property of CgP ligands and is a considerable advantage that they have over their ${}^tBu_2PR$ or $(RO)_2PR$ analogues.

The range of reactions that complexes of CgP-containing ligands have been shown to catalyse augurs well for further applications in the future.

13.3.1 Alkoxycarbonylation

The methoxycarbonylation of ethene to give methyl propionate (Eq. 13.3) is a recently commercialised Pd-catalysed process in which the bulky xylenyl diphosphine 1,2-$({}^tBu_2PCH_2)_2C_6H_4$ is the ancillary ligand that is used [26]. Two other ligands that give highly active and selective catalysts are the propane derivative ${}^tBu_2PCH_2CH_2CH_2P{}^tBu_2$ [27] and the ferrocene derivative 1,2-$({}^tBu_2PCH_2)_2C_5H_3Fe(C_5H_5)$ [28]. The ligands $\mathbf{L_1}$ [6,29], $\mathbf{L_2}$ [24] and $\mathbf{L_3}$ [30] have all been reported to give Pd catalysts that have efficiencies comparable with the ${}^tBu_2P$ analogues for ethene methoxycarbonylation (Eq. 13.3). Since the CgP and ${}^tBu_2P$ substituents are similar sterically but not electronically, it appears that it is ligand steric bulk that is more important in determining the catalyst performance for this important reaction.

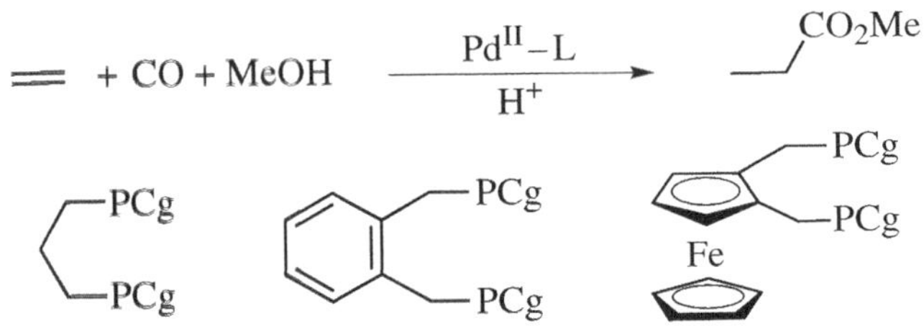

$\mathbf{L_1}$ $\mathbf{L_2}$ $\mathbf{L_3}$

(Eq. 13.3)

The *meso* isomer of $\mathbf{L_1}$ gives a significantly more active catalyst than the *rac* isomer [6,29], and in the presence of polyamines, the Pd catalyst based on $\mathbf{L_1}$ is stabilised [31].

With internal alkenes, tandem isomerisation–methoxycarbonylation is efficiently catalysed by Pd–$\mathbf{L_1}$ complexes to give linear products selectively, for example a mixture of internal tetradecenes gave the linear ester (Eq. 13.4) and methyl-3-pentenoate gave dimethyl adipate (Eq. 13.5) [6,29].

$$CH_3(CH_2)_nCH=CH(CH_2)_{10-n}CH_3 + CO + MeOH \xrightarrow[H^+]{Pd^{II}-\mathbf{L_1}} CH_3(CH_2)_{12}CO_2Me$$

(Eq. 13.4)

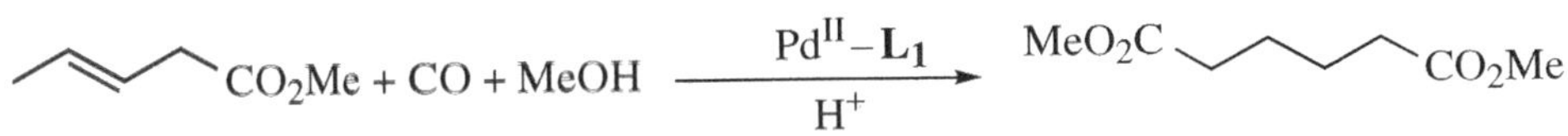

(Eq. 13.5)

When the Pd-catalysed carbonylation of ethene was carried out in the presence of water instead of methanol, propionic acid was obtained in 100% selectivity (Eq. 13.6). The rates of CO gas uptake using $\mathbf{L_1}$ (2500 mol/molPd h^{-1}) and $\mathbf{L_2}$ (10 000 mol/molPd h^{-1}) were superior to the 800 mol/molPd h^{-1} obtained with 1,2-(^{t}Bu$_2$PCH$_2$)$_2$C$_6$H$_4$ [11].

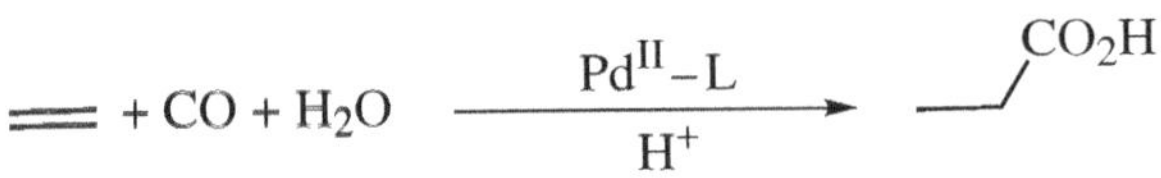

(Eq. 13.6)

In another variant, when the carbonylation was carried out in the presence of propionic acid instead of methanol, propanoic anhydride was obtained in 100% selectivity (Eq. 13.7). In this case, the rates of CO gas uptake were greater using *meso*-$\mathbf{L_1}$ (2500 mol/molPd h^{-1}) than for *rac*-$\mathbf{L_1}$ (800 mol/mol h^{-1}) and the rates with $\mathbf{L_2}$ (1 000 mol/molPd h^{-1}) were greater than with the analogue 1,2-(^{t}Bu$_2$PCH$_2$)$_2$C$_6$H$_4$ (120 mol/molPd h^{-1}) [32].

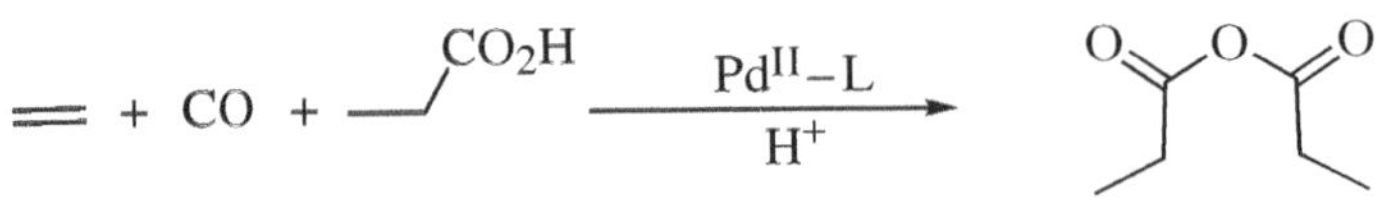

(Eq. 13.7)

The carbonylation of butadiene to pentenoic acid (Eq. 13.8) is catalysed by Pd complexes of **L**$_2$, **L**$_4$ and **L**$_5$ but at inferior rates to those obtained with 1,2-(^{t}Bu$_2$PCH$_2$)$_2$C$_6$H$_4$ [24,33].

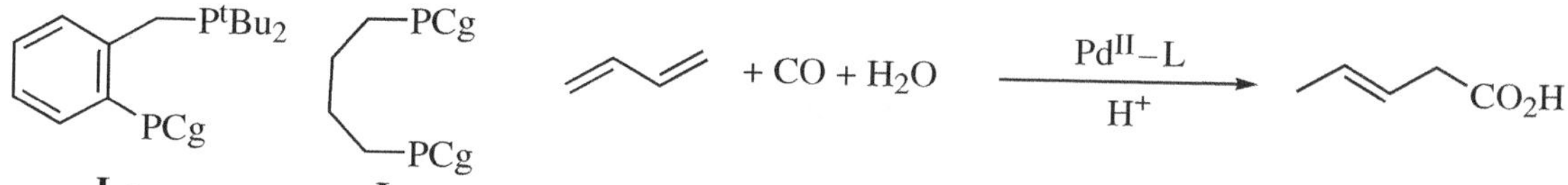

L$_4$ **L**$_5$

(Eq. 13.8)

Alkene carbonylation catalysis is dominated by Pd, but it was reported [29] that Pt complexes of **L**$_1$ also catalysed the methoxycarbonylation of ethylene to methyl propionate (Eq. 13.3) in 100% selectivity, albeit at a low rate (500 mol/molPt h^{-1}) compared with the Pd analogue (8000 mol/molPd h^{-1}). Moreover, Pt complexes of **L**$_1$ and **L**$_6$ catalyse the carbonylation of acetylene to give n-propyl acrylate (Eq. 13.9) in up to 90% selectivity (with succinate and malonate diesters as the by-products) and rates of 2000 mol/molPt h^{-1} [34]. Propyne and phenyl acetylene were similarly methoxycarbonylated to give mixtures of regioisomeric unsaturated esters [34].

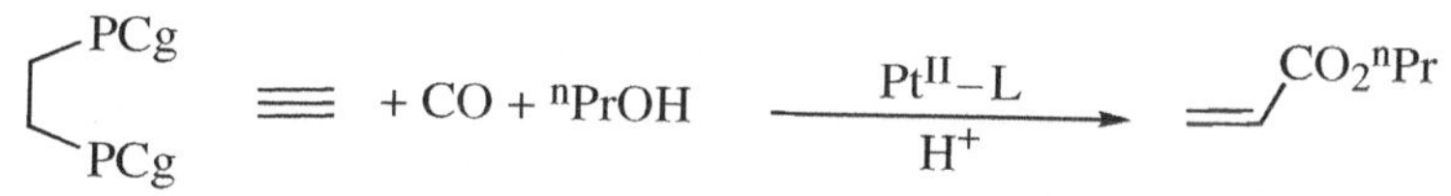

(Eq. 13.9)

L$_6$

13.3.2 Hydroformylation and hydrocyanation

Some of the bidentates that were used in the carbonylations described in this chapter have been shown to give complexes of Rh that are hydroformylation catalysts. For example, **L**$_1$ gave a Rh catalyst for the hydroformylation of 1-octene whose chemoselectivity and regioselectivity depended strongly on the ratio of Rh:**L**$_1$. Under the same conditions, a 1:1 ratio of Rh:**L**$_1$ gave 12% conversion to nonanals in a 1:1 n:iso ratio (the rest of the product was internal octenes), while 95% conversion to nonanals in a 97:1 n:iso ratio was obtained when a 1:10 ratio of Rh:**L**$_1$ was employed [35]. This indicates that dissociation of the bulky **L**$_1$ readily occurs in these systems.

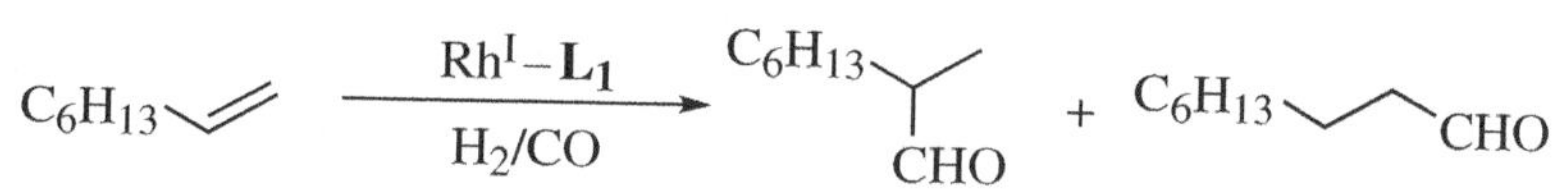

(Eq. 13.10)

Rhodium complexes of **L**$_2$ catalyse the hydroformylation of methyl 3-pentenoate to a mixture of isomeric aldehydes, predominantly (90%) the isomers shown in Eq. 13.11 [11].

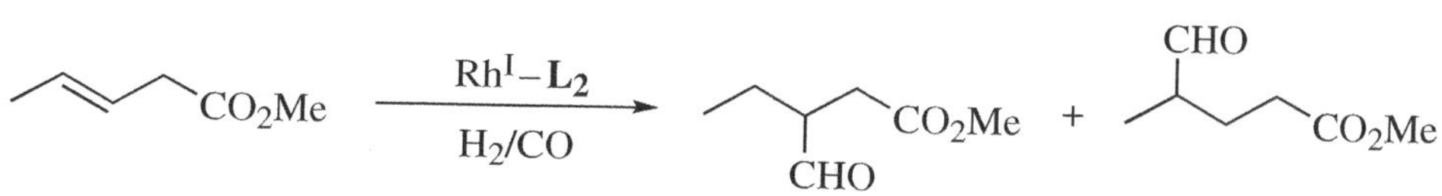

(Eq. 13.11)

Monodentate CgPR has been extensively researched for Rh-catalysed hydroformylation. Under mild conditions (60 ppm Rh, 10 bar syngas, 100°C, 4 h) using Rh:L of 1:50, the phosphines CgPCy and CgP(n-C$_8$H$_{17}$) gave quantitative conversion of octene to nonanals in a 1:1 n:iso ratio (Eq. 13.10) [36]. Even the secondary

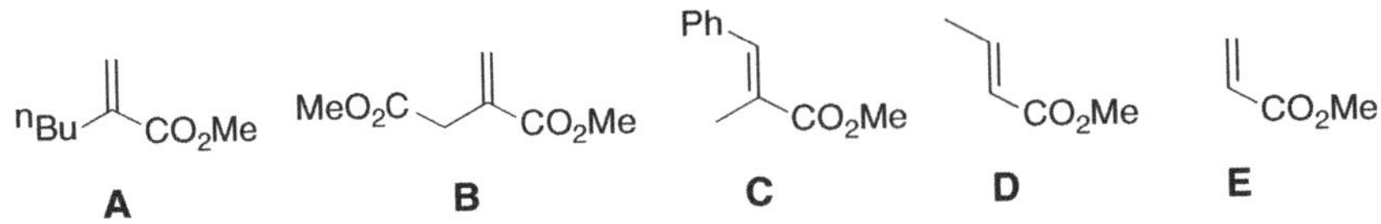

Scheme 13.8

phosphine CgPH gave a catalyst for octene hydroformylation under the same conditions, albeit of lower activity (29% conversion after 4 h).

The hydroformylation of formaldehyde to give glycol aldehyde (Eq. 13.12) is of interest as the product is readily hydrogenated to give glycol. The monodentates CgP(n-C_8H_{17}), CgP(n-$C_{20}H_{41}$) and CgP($CH_2CH_2CONMe_2$) all gave excellent catalysts with conversions and chemoselectivities of up to 90% [37].

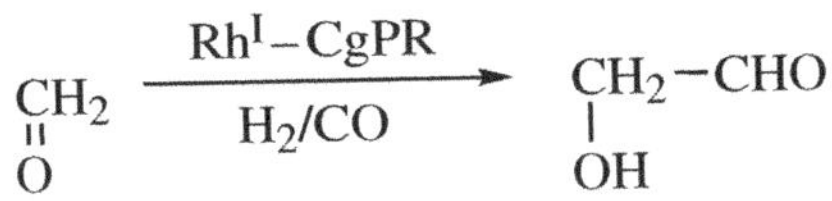

(Eq. 13.12)

The ligand CgPPh (**L**$_7$) has been extensively used in Rh-catalysed hydroformylation with the main focus on the hydroformylation of 1,1-disubstituted alkenes to give quaternary aldehydes in unprecedentedly high selectivity using comparatively mild conditions for the catalysis (50 bar, 50°C). Thus the hydroformylation of methyl atropate (Eq. 13.13) gives the quaternary aldehyde and linear aldehyde in greater than 100:1 ratio [9,38].

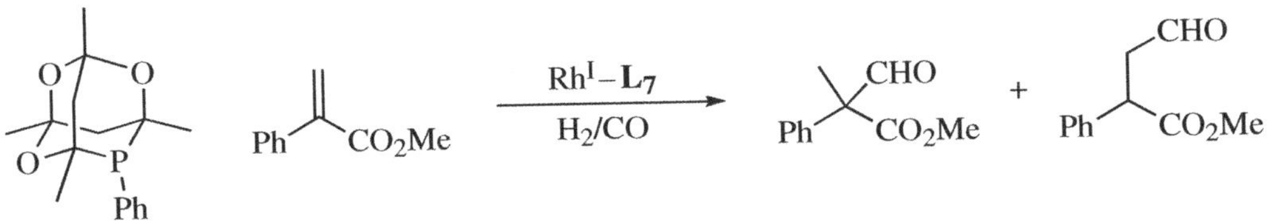

L$_7$

(Eq. 13.13)

The reaction shown in Eq. 13.13 has been extended to other 1,1-disubstituted, 1,2-disubstituted and trisubstituted alkenes (such as in Scheme 13.8A–D), and in each case **L**$_7$ produces a catalyst that gives excellent chemoselectivity and regioselectivity for aldehyde α to the carboxymethyl group [38]. Methyl acrylate (**E**) is also efficiently and regioselectively hydroformylated even in the absence of solvent [39].

The monophosphinites CgPOR (R = alkyl or aryl) and the diphosphinites CgPOZOPCg (Z = 1,2-$(CH_2)_2C_6H_4$ or 2,2-biphenyl) have been reported to give Rh catalysts for the hydroformylation of hexene [40]. More importantly, the same phosphinite ligands have been shown to be effective for nickel(0)-catalysed hydrocyanation of 3-pentenenitrile (Eq. 13.14), with **L**$_8$ giving the most efficient catalyst [12].

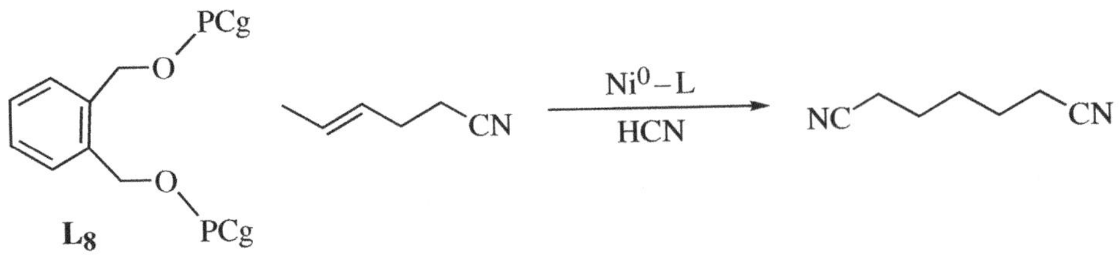

L$_8$

(Eq. 13.14)

13.3.3 Pd-catalysed coupling reactions

Palladium-catalysed C–C, C–N and C–O couplings are ubiquitous in synthetic organic chemistry, and bulky, electron-rich phosphines such as P^tBu_3 have been widely used [25]. In recent years, CgPPh (**L**$_7$) has been shown to give Pd catalysts for some C–C and C–N couplings that are more efficient than P^tBu_3 analogues. It seems once again that it is ligand sterics rather than electronics that determine the catalyst performance. Ligand **L**$_7$ has the advantages over P^tBu_3 of being air stable and easily recovered from the reaction mixture.

The complex generated from $[Pd_2(dba)_3]$ and 2 equiv. of **L**$_7$ has been shown to catalyse a range of Suzuki–Miyaura reactions (Eq. 13.15) in excellent yields including with difficult chloroarene substrates [18]. The crystal structure of $[Pd(\mathbf{L}_7)_2(dba)]$ has been determined, and in solution NMR evidence shows that this complex dissociates dba to give highly unsaturated species [41], similar to the analogous $[Pd(P^tBu_2Ph)_2]$ [42]. Suzuki couplings have also been carried out in the solid phase [43], catalysed by Pd(0)–**L**$_7$.

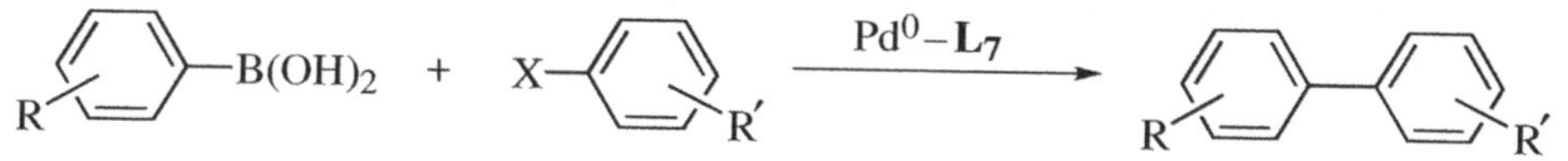

(Eq. 13.15)

A series of CgPAr ligands were screened, and the conditions optimised for the Suzuki coupling are shown in Eq. 13.16 where β elimination to give 1-dodecene is the side reaction to be avoided. Ligand **L**$_9$ was found to be the best, giving 96% selectivity for dodecylbenzene [13]. The same ligand **L**$_9$ was also shown [13] to be effective in a variety of couplings of alkyl halides or tosylates with 9-BBN derivatives (Eq. 13.17). A key step in the synthesis of the anti-hypertension drug Nebivol is an intramolecular version of the coupling shown in Eq. 13.17 catalysed by Pd(0)–**L**$_7$ [44].

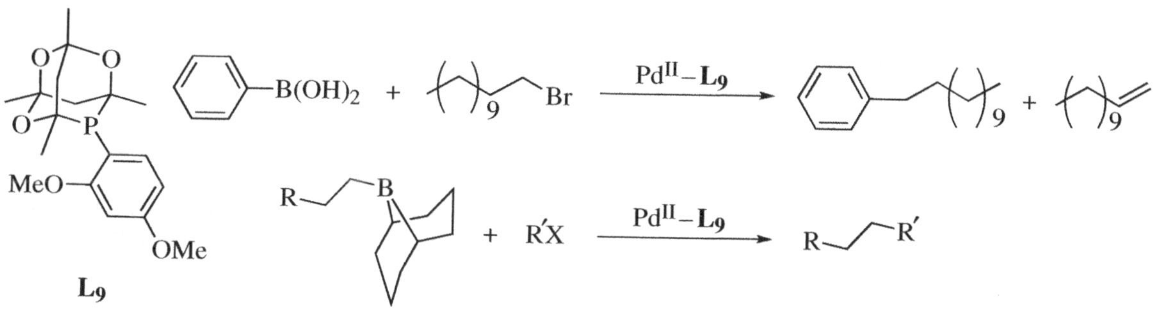

(Eq. 13.16)
(Eq. 13.17)

The phosphatrioxa-adamantane unit has been covalently attached to silica via the reaction shown in Eq. 13.18. Immobilised ligand **L**$_{10}$ combined with Pd(II) or Pd(0) compounds is a very efficient and recyclable catalyst for a variety of Suzuki couplings [20].

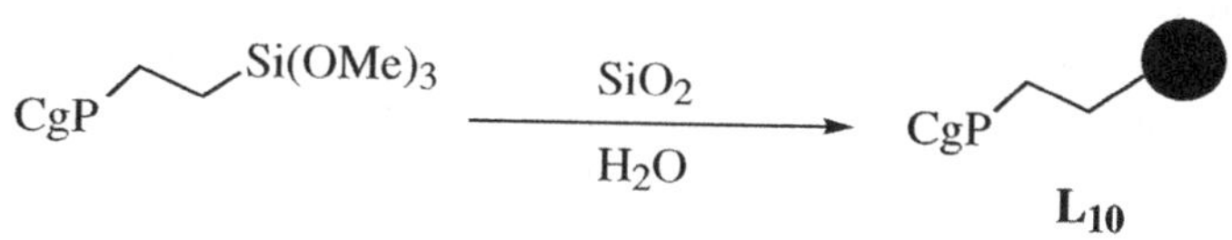

(Eq. 13.18)

Sonogashira coupling (Eq. 13.19) with a range of iodoarenes and α-arylation of ketones (Eq. 13.20) are catalysed by [Pd($\mathbf{L}_7$)$_2$(dba)] catalysts [41].

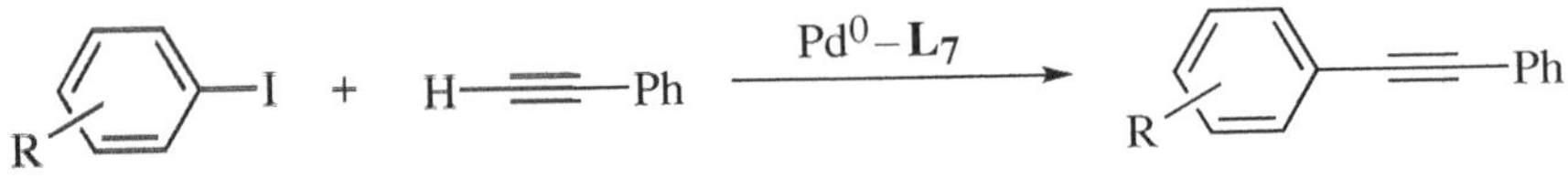

(Eq. 13.19)

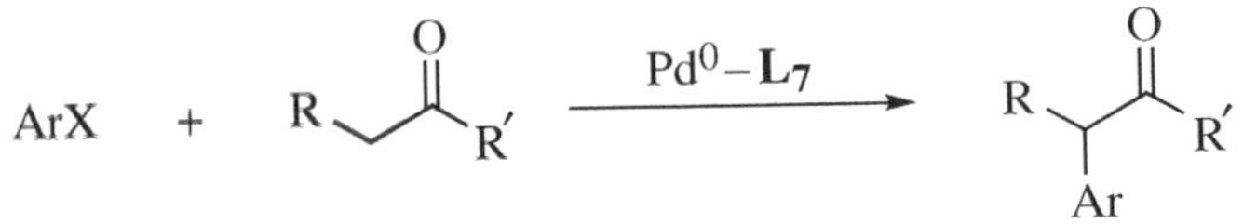

(Eq. 13.20)

The Buchwald–Hartwig aminations shown in Eq. 13.21 were shown [45] to be efficiently catalysed by [$Pd_2(dba)_3$] / $\mathbf{L}_7$ (ratio 1:2); the primary or secondary amines could be alkyl or aryl, and the arene electrophile could be ArI, ArBr or ArCl. This reaction has attracted industrial interest for the manufacture of diarylamines and triarylamines [46].

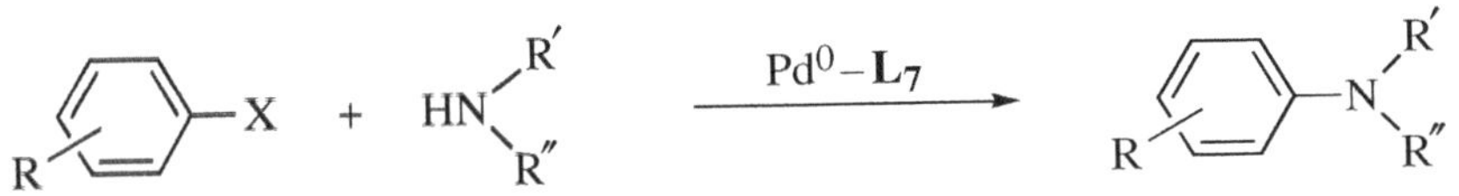

(Eq. 13.21)

The carbonylative cross-coupling of aryl halides with water, alcohols or amines to produce carboxylic acids, esters and amides (Eq. 13.22) is each catalysed by $Pd(OAc)_2$ / $\mathbf{L}_7$ (ratio 1:1) [47]. Conversion of an ArI to an $ArCO_2Me$ was used as a key step in a synthesis of the antibiotic Platensimysin. An intramolecular variant of this reaction has been used to make seven-membered cyclic amides [48].

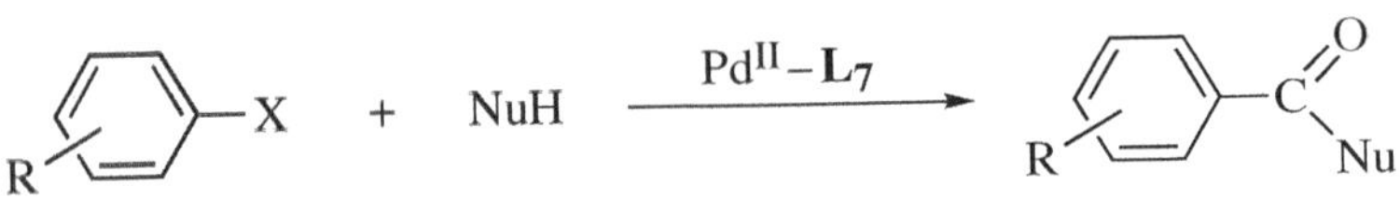

(Eq. 13.22)

The cross-coupling of a distannane with a haloarene (Eq. 13.23) is catalysed by $Pd(OAc)_2$ / $\mathbf{L}_9$ (ratio 1:2), and a variant of this reaction has been used as a key step in the synthesis of some antitumour radiopharmaceuticals [49].

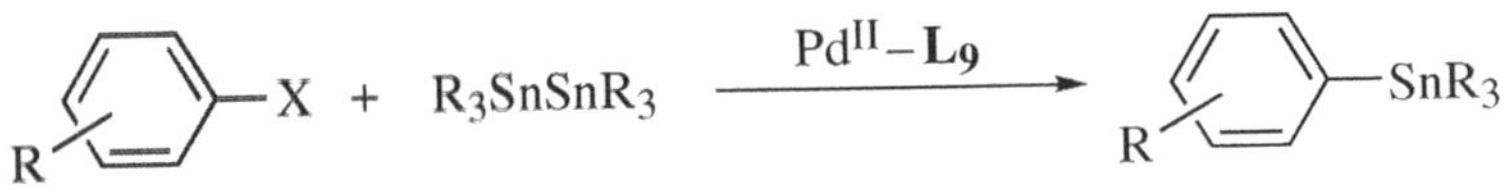

(Eq. 13.23)

13.3.4 Asymmetric hydrogenation

A diastereomeric mixture of $CgPCH_2CH_2CH_2PCg$ ($\mathbf{L}_1$, the structure of which has been evocatively likened to headphones [50]) was used to make optically active octahedral complexes [RuHCl($\mathbf{L}_1$)(diamine)] where the diamine = (*1R,2R*)-1,2-diaminocyclohexane) or (*1R,2R*)-1,2-diphenylethylenediamine. These complexes were very active for the hydrogenation of acetophenone (Eq. 13.24), but the optical yield was negligible (< 10% ee).

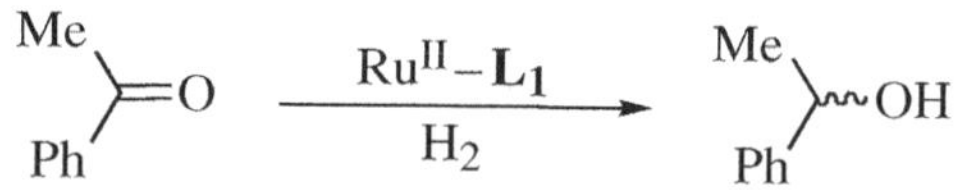

(Eq. 13.24)

The secondary phosphine CgPH is the simplest CgP-containing ligand that could be used in catalysis. The enantiomers of CgPH have been resolved by a classical procedure involving fractional crystallisation of diastereomeric quinine salts of the acid CgP(O)OH [19]. The complexes $[Rh(\beta\text{-}CgPH)_2(cod)]BF_4$ are enantioselective catalysts for the asymmetric hydrogenation of MAA, MAC and DMI (Eq. 13.25). The high optical yields that were observed (88–95% ee) were surprising in view of the apparently 'weak' chirality of the CgPH ligand and the anticipated C_2 symmetry of the complex [51]. The explanation may lie in the observed restricted Rh–P rotation leading to highly asymmetric (C_1 symmetry) rotamers which might dominate the catalysis.

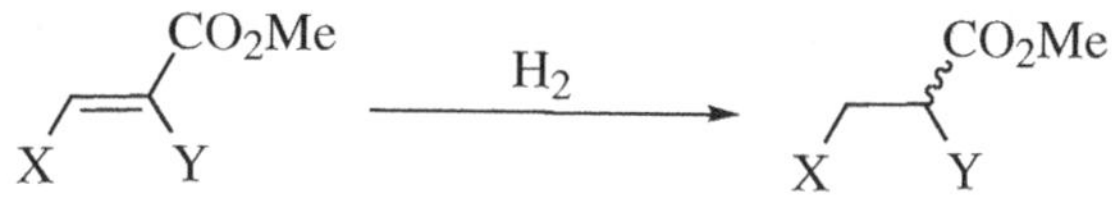

(Eq. 13.25)

MAA X = H Y = NHAc
MAC X = Ph Y = NHAc
DMI X = H Y = CH_2CO_2Me

13.4 Experimental procedures for phosphatrioxa-adamantanes ligands

The synthesis of each of the compounds CgPH, $CgPH(BH_3)$ and CgPBr is described in this section since these are the starting materials for many other CgP-containing ligands. The most widely used CgP ligands in catalysis are $CgPCH_2CH_2CH_2PCg$ (**L_1**) and CgPPh (**L_7**), and their syntheses are also described here. Note that CgPH, CgPPh and a silica-supported phospatrioxa-adamantane ligand are commercially available [52]. All reactions below are carried out in dried solvents under a nitrogen atmosphere.

13.4.1 Preparation of CgPH

A three-necked, 1 L flask was equipped with a large magnetic stirrer bar, a thermometer, a gas inlet fitted to a mineral oil bubbler and then a gas manifold to allow for N_2–PH_3 admission (fine gas bubbles were obtained using a sintered glass bulb attached to the bottom of a standard glass inlet), a gas outlet fitted to a mineral oil bubbler followed by a bleach trap and then another mineral oil bubbler. The flask was charged with acetylacetone (150 cm^3, 1.46 mol) and aqueous HCl (300 cm^3, 5 M) and then N_2 was passed through the solution for 18 h to ensure the elimination of oxygen. The reaction vessel was cooled in an ice–water bath to 0°C, and then gaseous PH_3 was then passed through the solution for 7 h at such a rate that it was absorbed by the reaction (monitored by observation of the oil bubblers). *Caution: PH_3 gas is highly toxic and an explosion hazard. It should be handled with extreme care.* Nitrogen was again bubbled through the mixture for 16 h to ensure the elimination of any PH_3 before workup. A very sensitive test for the presence of PH_3 is to use filter paper which has been dipped in a solution of $AgNO_3$ – it turns black in the presence of minute quantities of PH_3. The white crystalline solid was then filtered off in air, washed with water (3 × 200 cm^3) and dried under reduced pressure to give CgPH (78.5 g, 50% yield). ^{31}P NMR ($CDCl_3$): δ_P −49.5.

13.4.2 Preparation of CgPH(BH_3)

A solution of BH_3.THF (37.3 cm^3, 1 M in THF, 37.3 mmol) was added slowly over 30 min to a stirred, ice-cold solution of CgPH (8.00 g, 37.3 mmol) in THF (20 cm^3). The reaction mixture was then allowed to warm to room temperature, and the solvent removed under reduced pressure to give a white powder. The solid was dissolved in hot THF (6 cm^3), and the resulting solution allowed to cool slowly to room temperature and then put in the freezer (−10°C) to give the crystalline white solid CgPH(BH_3) (8.22 g, 96%). ^{31}P NMR ($CDCl_3$): δ_P 1.5 (br. q, ^1J(PB) 39 Hz).

13.4.3 Preparation of CgPBr

A solution of Br_2 (1.83 mmol, 8.40 cm^{-3}, 0.218 M in CH_2Cl_2) was added dropwise over 10 min to a cooled (0°C) solution of CgPH (0.368 g, 1.70 mmol) in CH_2Cl_2 (10 cm^3). After 1 h the solvent was removed to give a yellow-orange solid. Toluene (10 cm^3) was added to the solid and the suspension filtered. Removal of the toluene then yielded the white powder CgPBr (0.39 g, 78%). ^{31}P NMR ($CDCl_3$): δ_P 54.9. CgPBr can also be made *in situ* and used immediately [12].

13.4.4 Preparation of CgP$CH_2CH_2CH_2$PCg (L_1)

To a solution of 2,4-pentanedione (11.4 cm^3, 111 mmol) in N_2-saturated aqueous HCl (40 cm^3, 5 M) 1,3-diphosphinopropane [53] (2.0 cm^3, 18.5 mmol) was added dropwise over 5 min. After 1 h a white solid began to form and the mixture was stirred for 3 days. The white solid product was filtered off and washed with water (2×20 cm^3) and cold MeOH (2×10 cm^3) and then dried under reduced pressure to give a 1:1 mixture of *rac*- and *meso*-$\mathbf{L_1}$ (5.24 g, 60%). ^{31}P NMR ($CDCl_3$): δ_P −29.4 and −30.2. Recrystallisation of the mixture from boiling MeOH (4 cm^3 per 0.1 g of $\mathbf{L_1}$) gave a solid that was 95% *rac*-$\mathbf{L_1}$ in 80% recovery. From the filtrates, samples rich (80%) in *meso*-$\mathbf{L_1}$ were recovered.

13.4.5 Preparation of CgPPh (L_7)

To a solution of 2,4-pentanedione (2.60 cm^3, 25.1 mmol) in aqueous HCl (10 cm^3, 12 M), phenylphosphine (1.0 cm^3, 9.1 mmol) was added dropwise over 5 min. After 4 days, water (10 cm^3) was added to the solution and a white precipitate formed. The white solid was filtered off and washed with water (2×10 cm^3) and cold MeOH (2×5 cm^3) and then dried under reduced pressure to give $\mathbf{L_7}$ (2.31 g, 86%). ^{31}P NMR ($CDCl_3$): δ_P −23.6.

References

[1] M. Epstein, S. A. Buckler, *J. Am. Chem. Soc.* **1961**, *83*, 3279.
[2] M. Epstein, S. A. Buckler, US Patent US 3026327, **1962**.
[3] (a) S. A. Buckler, M. Epstein, *J. Org. Chem.* **1962**, *27*, 1090; and (b) S. A. Buckler, M. Epstein, French Patent Fr. 1348669, **1963**.
[4] H. C. Gillham, A. E. Sherr, US Patent US 3309425 **1967**.
[5] J. F. Deffner, E. R. Tucci, J. V. Ward, H. I. Thayer, H. J. Elder, French Patent Fr. 1534510, **1968**.
[6] J. Suykerbuyk, E. Drent, P. G. Pringle, World Patent WO9842717, **1998**.
[7] V. Gee, A. G. Orpen, H. Phetmung, P. G. Pringle, R. I. Pugh, *Chem. Commun.* **1999**, 901.
[8] G. Bekiaris, E. Lork, W. Offerman, G. V. Röschenthaler, *Chem. Ber.* **1997**, *130*, 1547.
[9] R. A. Baber, M. L. Clarke, K. M. Heslop, A. C. Marr, A. G. Orpen, P. G. Pringle, A. Ward, D. E. Zambrano-Williams, *Dalton Trans.* **2005**, 1079.

[10] J. H. Downing, J. Floure, K. Heslop, M. F. Haddow, J. Hopewell, M. Lusi, H. Phetmung, A. G. Orpen, P. G. Pringle, R. I. Pugh, D. E. Zambrano-Williams, *Organometallics* **2008**, *27*, 3216.
[11] E. Drent, R. H. van der Made, P. G. Pringle, World Patent WO03070370, **2003**.
[12] I. S. Mikhel, M. Garland, J. Hopewell, S. Mastroianni, C. L. McMullin, A. G. Orpen, P. G. Pringle, *Organometallics* **2011**, *30*, 974.
[13] T. Brenstrum, D. A. Gerristma, G. M. Adjabeng, C. S. Frampton, J. Britten, A. J. Robertson, J. McNulty, A. Capretta, *J. Org. Chem.* **2004**, *69*, 7635.
[14] T. J. Cunningham, M. R. J. Elsegood, P. F. Kelly, M. B. Smith, P. M. Staniland, *Eur. J. Inorg. Chem.* **2008**, 2326.
[15] G. M. Brown, M. R. J. Elsegood, A. J. Lake, N. M. Sanchez-Ballester, M. B. Smith, T. S. Varley, K. Blann, *Eur. J. Inorg. Chem.* **2007**, 1405.
[16] C.-A. Carraz, D. W. Stephan, *Organometallics* **2000**, *19*, 3791.
[17] T. J. Cunningham, M. R. J. Elsegood, P. F. Kelly, M. B. Smith, P. M. Staniland, *Dalton Trans.* **2010**, *39*, 5216.
[18] G. Adjabeng, T. Brenstrum, J. Wilson, C. Frampton, A. Robertson, J. Hillhouse, J. McNulty, A. Capretta, *Org. Lett.* **2003**, *5*, 953.
[19] J. Hopewell, P. Jankowski, C. L. McMullin, A. G. Orpen, P. G. Pringle, *Chem. Commun.* **2010**, *46*, 100.
[20] M. Guinó, A. C. Sullivan, J. R. H. Wilson, *J. Mol. Catal. A: Chem.* **2008**, *293*, 25.
[21] G. R. Eastham, M. Waugh, P. G. Pringle, T. Fanjul Solares, *World Patent*, **2010**, WO2010001174.
[22] G. R. Eastham, *World Patent*, **2005**, WO2005118519.
[23] I. R. Butler, P. K. Baker, G. R. Eastham, K. M. Fortune, P. N. Horton, M. B. Hursthouse, *Inorg. Chem. Comm.*, **2004**, *7*, 1049.
[24] E. Drent, R. Ernst, W. W. Jager, *World Patent*, **2004**, WO2004103948.
[25] (a) P. W. N. M. van Leeuwen in *Homogeneous Catalysis: Understanding the Art*, Springer-Verlag, **2004**; and (b) J. F. Hartwig in *Organotransition Metal Chemistry: From Bonding to Catalysis*, University Science Books, **2009**.
[26] W. Clegg, G. R. Eastham, M. R. J. Elsegood, R. P. Tooze, X. L. Wang, K. Whiston, *Chem. Commun.* **1999**, 1877.
[27] E. Drent, E. Kragtwijk, Eur. Patent EP0495548, **1992**.
[28] G. R. Eastham, World Patent WO2004024322, **2004**.
[29] R. I. Pugh, E. Drent, P. G. Pringle, *Chem. Commun.* **2001**, 1476.
[30] (a) G. Eastham, World Patent WO2004024322, **2004**; (b) G. Eastham, World Patent WO2005079981, **2005**; and (c) G. Eastham, World Patent WO2005118519, **2005**.
[31] (a) M. Slany, M. Schäfer, M. Röper, German Patent DE10228293, **2001**; and (b) M. Schäfer, M. Slany, E. Zeller, M. Röper, M. Schultz, G. Grimm, World Patent WO0226382, **2002**.
[32] E. Drent, R. H. van der Made, R. I. Pugh, World Patent WO03070679, **2003**.
[33] X. Sava, M. Slany, M. Röper, *World Patent*, **2003**, WO03031457.
[34] E. Drent, P. G. Pringle, R. I. Pugh, World Patent WO0128972, **2001**.
[35] W. Ahlers, M. Slany, World Patent WO0185662, **2001**.
[36] W. Ahlers, World Patent WO0185661, **2001**.
[37] K. Q. Almeida-Leñero, E. Drent, R. van Ginkel, R. I. Pugh, World Patent WO2005058788, **2005**.
[38] M. L. Clarke, G. R. Roff, *Chem. Eur. J.* **2006**, *12*, 7978.
[39] M. L. Clarke, G. J. Roff, *Green Chem.* **2007**, *9*, 792.
[40] S. Mastroianni, I. Mikhel, P. Pringle, World Patent WO2010102962, **2010**.
[41] G. Adjabeng, T. Brenstrum, C. S. Frampton, A. J. Robertson, J. Hillhouse, J. McNulty, A. Capretta, *J. Org. Chem.* **2004**, *69*, 5082.
[42] M. Matsumoto, H. Yoshioka, K. Nakatsu, T. Yoshida, S. Otsuka, *J. Am. Chem. Soc.* **1974**, *96*, 3322.
[43] S. A. Ohnmacht, T. Brenstrum, K. H. Bleicher, J. McNulty, A. Capretta, *Tetrahedron Lett.* **2004**, *45*, 5661.
[44] R. Volpicelli, P. Maragni, L. Cotarca, J. Foletto, F. Massaccesi, World Patent WO2008064827, **2008**.
[45] D. Gerristma, T. Brenstrum, J. McNulty, A. Capretta, *Tetrahedron Lett.* **2004**, *45*, 8319.
[46] (a) T. P. Bender, J. A. Coggan, US Patent US2008125607, **2008**; (b) T. P. Bender, J. A. Coggan, US Patent US7345203, **2008**; (c) T. P. Bender, E. L. Moore, J. A. Coggan, F. P. Lee, D. Borbely, Eur. Patent EP1949957, **2008**; and (d) J. A. Coggan, US Patent US20090131720, **2009**.
[47] J. McNulty, J. J. Nair, A. Capretta, *Tetrahedron Lett.* **2009**, *50*, 4087.

[48] Q. Yang, H. Cao, A. Robertson, H. Alper, *J. Org. Chem.* **2010**, *75*, 6297.
[49] J. W. McIntee, C. Sundararajan, A. C. Donovan, M. S. Kovacs, A. Capretta, J. F. Valliant, *J. Org. Chem.* **2008**, *73*, 8236.
[50] A. Hadzovic, A. J. Lough, R. H. Morris, P. G. Pringle, D. E. Zambrano-Williams, *Inorg. Chim. Acta* **2006**, *359*, 2864.
[51] P. Jankowski, C. L. McMullin, I. D. Gridnev, A. G. Orpen, P. G. Pringle, *Tetrahedron: Asym.* **2010**, *21*, 1206.
[52] CgPH is available from Cytec Specialty Chemicals, Woodland Park, NJ, US. CgPPh is available from Sigma-Aldrich, St. Louis, MO, US. Silica-supported CgPR ligands are available from PhosphonicsS Ltd., Abingdon, UK.
[53] L Maier, *Helv. Chim. Acta*, **1966**, *49*, 842.

14

Calixarene-based Phosphorus Ligands

Angelica Marson[1], Piet W. N. M. van Leeuwen[2], and Paul C. J. Kamer[3]

[1]*Clariant International Ltd., Muttenz, Switzerland*
[2]*Institute of Chemical Research of Catalonia (ICIQ), Tarragona, Spain*
[3]*EaStCHEM, School of Chemistry, University of St Andrews, St Andrews, United Kingdom*

14.1 Introduction

The history of calixarenes started in 1872, when Adolph von Baeyer reported for the first time the reactions between phenols and aldehydes, but at that time many of the products could not yet be identified [1]. It took almost 70 years before Zinke and Ziegler assigned cyclic tetrameric structures to these products [2], and more than a century (1978) before Gutsche introduced the name calixarenes [3]. The use of this word was suggested because of their resemblance in shape to a type of Greek vase called a 'calix crater', thus describing the so-called *cone* conformation that these macrocycles often adopt. The calixarene syntheses reported by Gutsche in the 1970s showed that careful control of reaction conditions can provide good yields of pure compounds of various ring sizes, leading to the many applications realised at the beginning of the twenty-first century. Today, the easy one-step preparation of the cyclic tetramer (calix[4]arene), cyclic hexamer (calix[6]arene) and cyclic octamer (calix[8]arene) from *p-tert*-butyl-phenol and formaldehyde [4] makes these materials extremely attractive for a variety of studies including, *inter alia*, complexation of cations, anions and metals, ion transport phenomena, enzyme model building, organic purification, demulsifiers, sensor construction and, last but not least, catalysis [5].

Calixarenes are particularly attractive as ligands for metal ions because of their chemical architecture. Several important features of calixarene scaffolds can be exploited to design and develop new ligand systems for individual applications: (1) the calixarene framework can be regarded as a platform with two rims that can be independently modified with different functional groups; (2) the cavity size for intercalation of the metal ion can be modified by changing the number of arene units within the calixarene ring (i.e. the value of n in the calix[n]arene); (3) the interior of the cavity represents a space able to entrap or confine reactive fragments bound to the

Phosphorus(III) Ligands in Homogeneous Catalysis: Design and Synthesis, First Edition.
Edited by Paul C. J. Kamer and Piet W. N. M. van Leeuwen.

transition metal; (4) the calixarene framework has sufficient flexibility to adapt its conformation to a preferred cavity size and shape; (5) both the number and the geometrical arrangement of the ligating groups can be varied to accommodate the particular coordination properties of the metal ion; (6) separate chemical modifications to each rim can be used to incorporate ligating groups on one rim and groups that confer desired solubility properties on the other; and (7) the selective incorporation of different functional groups on each rim can be used to synthesise calixarenes that can selectively complex dissimilar metals on these rims.

The non-centrosymmetric structure of calixarenes and the ease with which enantiomeric congeners can be synthesised make calixarenes attractive as synthons for a wide range of ligand applications.

The general class of cyclophanes consists of any circular arrangement of arene rings linked through bridges of an arbitrary number of atoms. In particular, $[1_n]$*meta*cyclophanes constitute one of the subclasses in which all arene rings are linked via the *meta* positions by methylene groups (Figure 14.1). The trivial name calixarenes denotes both the characteristic shape of these macrocycles and the presence of arene rings. The number of arene rings is given in brackets between calix and arene (e.g. calix[4]arene) (Figure 14.1). Although most of the calixarenes are synthesised from *para*-substituted phenols and, thus, are characterised by the presence of hydroxyl groups, this is not essential for the classification. If the calixarene framework is functionalised, substituents are normally denoted as a prefix before calix.

Modifications of the calixarene scaffold can be carried out mainly at two regions, namely, the 'lower rim', which contains the phenolic hydroxyl groups (formation of ethers and esters) and the 'upper rim', which is defined by the *para*-positions (electrophilic substitution and *ipso*-substitution). The word 'rim' denotes the two regions in the vase-like three-dimensional structure (Figure 14.2) and can be easily visualised when all the OH groups are pointing downwards (i.e. *cone* conformation), but in other cases this term can be

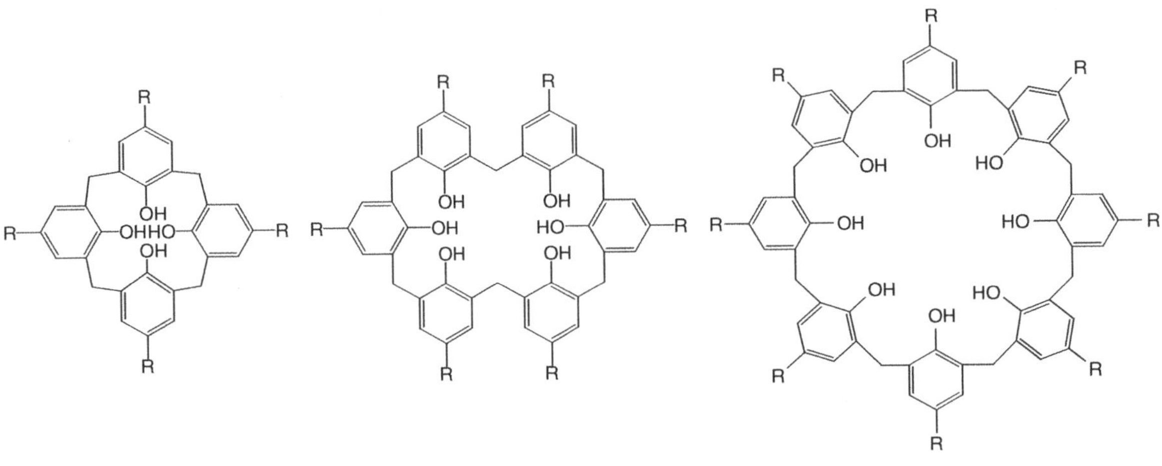

Figure 14.1 *General structure of calix[4]-, calix[6]- and calix[8]arene.*

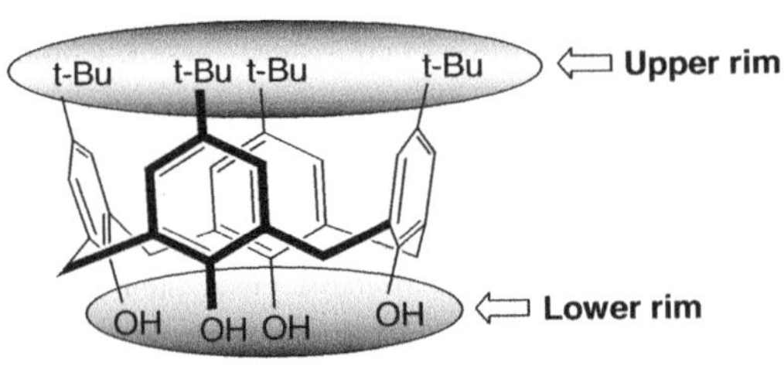

Figure 14.2 *Three-dimensional representation of calix[4]arene (cone conformation).*

misleading. Additionally, reactions may be carried out at the methylene bridges, they may comprise the aromatic system of the phenolic units as a whole (oxidation and hydrogenation) or they may lead to the replacement of the OH-function by other groups.

The most simple fashion in which a calixarene can bind to a metal centre is via one or more phenolic oxygen atoms. In most cases, however, the coordination chemistry of calixarenes results from the presence of functional groups arranged around the calixarene periphery (i.e. attached either to the phenolic oxygen atoms or to the carbon atoms at the *para* positions). Many publications have been devoted to the functionalisation of the phenolic oxygen atoms of calixarenes frameworks, which can be achieved either by direct acylation or by performing a Williamson type of alkylation, allowing straightforward introduction of different groups that may be functionalised further. If calixarenes functionalised at specific oxygen atoms are needed, both the degree of alkylation and the site(s) that will be modified need to be controlled. This can be achieved by using carefully selected reaction conditions, notably by controlling the stoichiometry and by employing an optimal base–solvent couple.

14.2 Conformational properties

General conformational notation is adopted for calixarenes on the basis of the orientation of the aryl units, which can be *up* (*u*) or *down* (*d*) relative to the horizontal plane of the calixarene skeleton, defined by the bridging methylene groups ($ArCH_2Ar$). The *up* (*u*) notation refers to the aryl units with the *para*-substituent pointing upwards (upper rim) and the phenolic oxygen atom pointing downwards (lower rim), whereas *down* (*d*) refers to the exact opposite orientation. Thus, the **u**,u,u,u (all *up* with **u** representing the aryl ring taken as reference) arrangement for calix[4]arenes indicates the *cone* conformation generally adopted by these cavities (*vide infra*). Furthermore, due to the presence of different substituents on the upper and/or lower rim of the cavity, the aryl groups are often not perpendicularly oriented with respect to the calixarene reference plane. When the orientation of the aryl unit approaches that of the reference plane, the group can be designated as *out* (*o*) (the *para*-substituent points outwards the cavity) or *in* (*i*) (the *para* substituent points inwards the cavity). This can be clearly visualised in terms of the dihedral angle Φ defined by the plane of the aryl fragment and by the reference plane of the calixarene skeleton (Figure 14.3), which quantifies the inclination of aromatic moieties that can be remarkably 'open' in distorted calixarene frameworks.

Thus, the *up* and *down* notation can be expanded to *up* and *out* (*uo*), *down* and *out* (*do*), *up* and *in* (*ui*) and *down* and *in* (*di*).

All calixarenes containing free intra-annular OH groups are conformationally mobile in solution at room temperature. Not surprisingly, the degree of mobility varies from one ring system to another.

It was recognised by Cornforth that calix[4]arenes can exist in four discrete conformations, according to the relative orientation of the four phenoxy groups [6]. Later these four conformations have been named by Gutsche as *cone*, *partial cone*, *1,2 alternate* and *1,3 alternate* (Figure 14.4).

Interconversion between different conformations can take place by rotation of the aryl groups around the axis that passes through the *meta* carbon atoms bonded to the bridging methylene units. When the

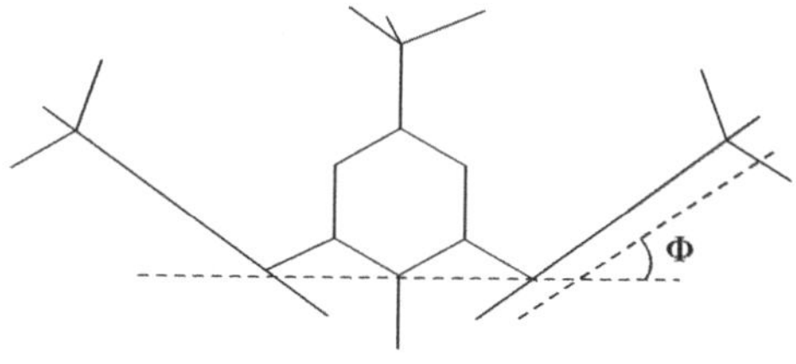

Figure 14.3 Inclination of the aromatic units in calixarene frameworks (dihedral angle *Φ*).

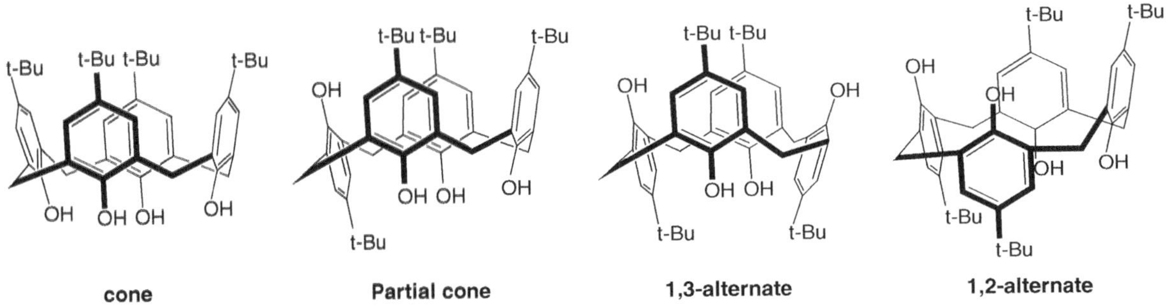

Figure 14.4 The four conformations of calix[4]arenes.

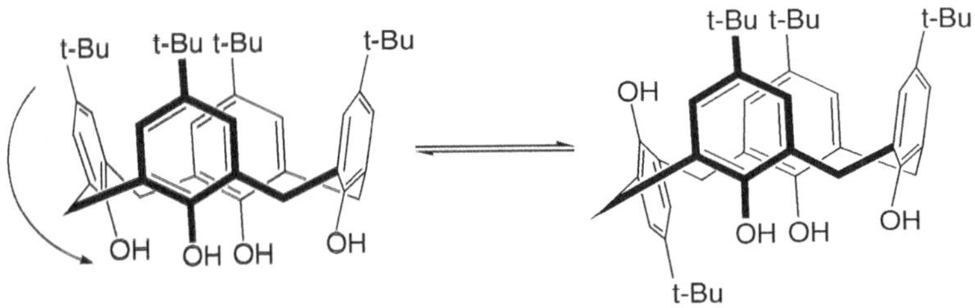

Figure 14.5 The 'hydroxyl through the annulus' rotation in calix[4]arenes.

para-positions of the aryl units bear bulky substituents (like in the *p-tert*-butyl calix[4]arenes), the interconversion exclusively involves the passage of the hydroxyl groups through the calixarene ring and, hence, the process is called the 'hydroxyl through the annulus' rotation (Figure 14.5).

Stable conformations of calix[4]arenes can be obtained by a proper blocking of the 'through the annulus' rotation via the introduction of pendant alkyl–aryl groups onto the phenolic positions (*cone* or *partial cone* with exceptional 1,3 alternate conformations) [7]. When the upper rim is functionalised (*para*-substitution) the energy barrier to inversion varies only slightly, while, due to the complexating abilities of the calixarene cavity, larger differences can be observed upon changing the solvent. This is a result of the disruption of the intramolecular hydrogen bonding at the lower rim of the cavity, which is the primary force responsible for the preference of the calixarene to assume a cone conformation [8].

Any of the four basic conformations depicted in Figure 14.4 represents a special three-dimensional arrangement with different complexation behaviour and, hence, with different potential applications as a molecular scaffold and/or useful building blocks in supramolecular chemistry.

Calix[6]arenes, the cyclic hexamers, can exist in eight *up-down* conformations but more conformations are possible when other orientations of the aryl rings are considered. Due to the larger ring size, calix[6]arenes have higher conformational flexibility than calix[4]arenes in nonpolar solvents. Calix[6]arenes in the **u**,u,u,u,u,u (**u** represents the aryl unit chosen as a reference) arrangement [9] may adopt two extreme conformations, which have been referred to as 'pinched' and 'winged' cones (Figure 14.6) [10]. They differ in the position of the methylene carbon atoms relative to the cavity formed by the aromatic rings. In the winged conformation, all six bridging methylene groups point outwards from the macrocycle annulus, while in the pinched conformation two opposite methylene carbons point inwards. In both conformations, the six hydroxyl groups form two clusters of three OH groups each in which circular flip-flop hydrogen bonding might be possible.

The inversion barrier for calix[6]arenes is lower than for calix[4]arenes, but the effects of nonpolar solvents show the same order as that in the cyclic tetramers. Also for lower rim substituted calix[6]arenes, the barrier

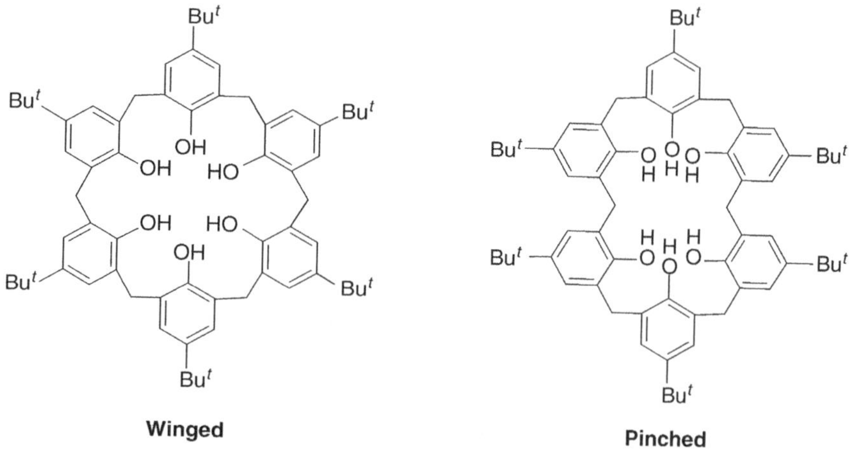

Figure 14.6 The winged and pinched conformations for calix[6]arenes.

to conformational inversion depends mainly on the nature of the substituent at the upper rim. This is clear when assuming that the inversion involves the passage of the oxygen end rather than the *para* substituent end of the aryl moieties through the calixarene ring.

Other ring sizes, like in calix[5]arenes and calix[8]arenes [11], show similar or increased fluxional behaviour respectively.

14.3 Calixarene-based phosphorus ligands

Since the first use, more than two decades ago, of calixarene-derived ligands for the complexation of transition metal ions, a considerable number of fundamental studies have been undertaken demonstrating the suitability of these ligands for the construction of a large variety of mononuclear and polynuclear metal complexes [12]. It has been shown that the presence of transition metal centres bound to calixarene matrices may seriously perturb the shape and structural properties of such macrocycles. Nevertheless, many aspects of the coordination chemistry of calixarenes still remain to be explored. Due to their suitability to form inclusion complexes with organic substrates, the potential of molecules combining catalytic centres with a calixarene cavity can be exploited for novel shape-selective catalysts and asymmetric catalysis.

A large variety of phosphorus-containing calixarenes, mainly derived from calix[4]arene frameworks, have been synthesised and the potential of these compounds as complexating agents for transition metals as well as their application in homogeneous catalysis have received increasing attention [12,13]. Calixarenes bearing trivalent phosphorus centres can be divided into two classes: those in which the phosphorus atom is connected via a phenolic oxygen atom at the lower rim of a calixarene cavity, and those in which the phosphorus atom is attached to the upper rim of the calixarene matrix, most often at the 4-position by replacement of the *tert*-butyl group with a reactive linker or with phosphorus directly.

14.3.1 Phosphines and phosphinites

Since the first report of transition metal complexes of phosphorus-modified calixarenes by Floriani *et al.* [14], the synthesis and design of calixarene-based phosphorus donor ligands have received fast-growing attention. The most straightforward modification of calixarene is via substitution of one or more of the

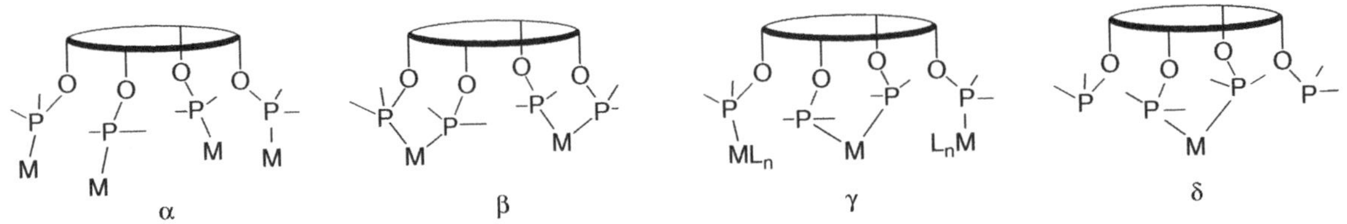

Figure 14.7 The four coordination modes of tetra-functionalised cone-calix[4]arenes.

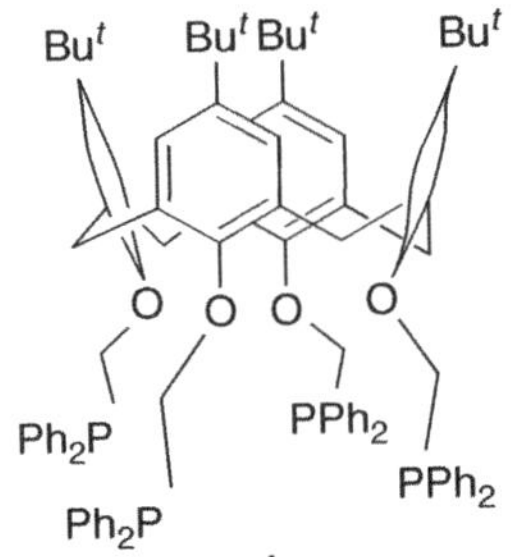

1

Figure 14.8 Complex **1**.

hydroxy functionalities at the lower rim. Attachment of a trivalent phosphorus atom to a phenolic oxygen of calixarenes may readily be achieved by treatment of the calixarene with a base, followed by reaction with halogenated phosphorus compounds. The number of phosphino groups that may be introduced mainly depends on the basicity of the deprotonating agent. Instead of using the chlorophosphine–base route, it is sometimes advantageous to perform the phosphination with a dialkylaminophosphine. In this case, the amino group plays the role of both the deprotonating agent and the leaving group.

Full functionalization of calix[4]arenes provides tetradentate ligands which can support several different coordination modes. Several tetra-functionalised *cone*-shaped calix[4]arenes have been reported in literature, and their coordinative properties investigated by different research groups. Four coordination modes were found for this type of ligands: (**α**) each phosphorus ligand is monocoordinated to a single metal centre [15]; (**β**) the calixarene behaves as a double chelator, with each chelation involving two proximal phosphorus atoms [16]; (**γ**) the calixarene acts simultaneously as a chelating ligand via two distal phosphorus atoms and as a bridging ligand via the other two P atoms [17]; and (**δ**) the calixarene behaves as a *trans* chelator and the other two distal phosphorus atoms remain uncoordinated (Figure 14.7) [18].

An example of a lower-rim functionalised calix[4]arene with four pendant CH_2PPh_2 groups (**1**) was reported by Matt and co-workers (Figure 14.8). Reaction of compound **1** with cyclopalladated dimer [Pd(*o*-$C_6H_4CH_2NMe_2$)Cl]$_2$ afforded a tetranuclear complex in which every phosphine group was monocoordinated to one palladium centre [19]. Similar calix[4]arene-based diphosphine and diphosphinite ligands, tetrasubstituted with diphenylphosphinoxyethoxy or diphenylphosphoxyethoxy moieties, respectively, have also been reported and their coordinating abilities with Pt(II), Pd(II) and Rh(I) investigated [20]. The tetraphosphine formed a dinuclear complex with both Pt(II) and Pd(II). The former was obtained exclusively in the *cis* configuration, and the latter as a mixture of *cis* and *trans* geometries. Reaction of the phosphinite with [$MCl_2(PhCN)_2$] (M = Pt or Pd) in a 1:2 ratio produced the corresponding dinuclear complex where pairs of adjacent phosphorus atoms chelate to the metal centre in a *cis* geometry. Reaction with [Rh(nbd)Cl]$_2$ afforded di- or tetranuclear complexes, depending on the amount of rhodium precursor used.

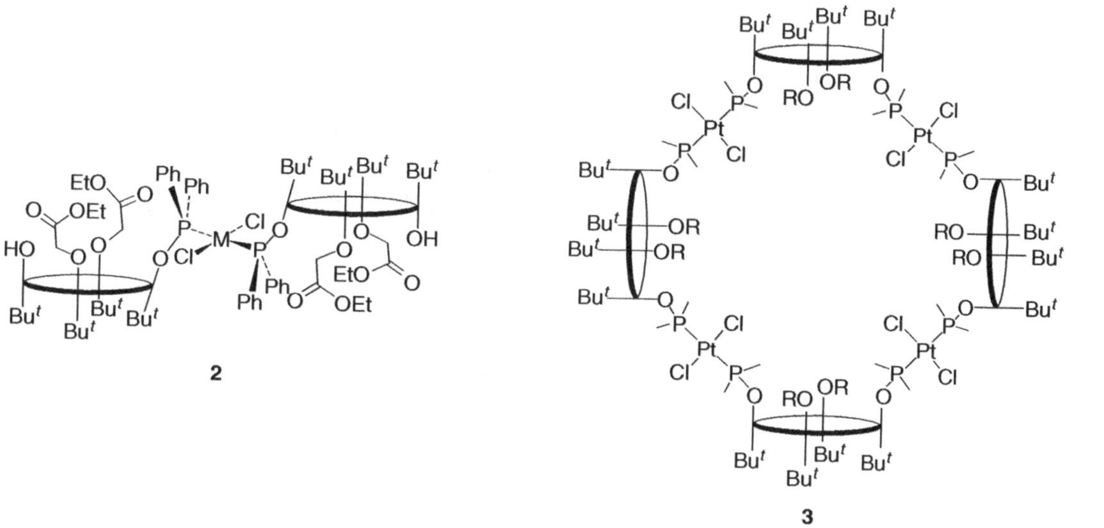

Figure 14.9 Complexes **2** and **3**.

Calixarene ligands containing less than four phosphino groups have also been prepared which require selective substitution of one or more hydroxy groups of calixarenes. Several mono- and diphosphino calix[4]arenes in a *cone* conformation have been synthesised and their complexation abilities studied in the presence of transition metal centres [21]. It was determined that the nature of the substituent plays a critical role in determining the final structure of the complexes. Also the preparation of terdentate ligands in the form of a tri(phosphinito)-calixarene complex has been reported [15]. Calix[4]arene-derived mono- and diphosphinite were obtained from their 1,3-alternating diester precursors [22]. Use of strong bases such as *n*-butyllithium gave facile formation of the diphosphinite ligands, whereas employing a weaker base as triethylamine provided the monodentate ligand [21].

Monophosphinito–diester calix[4]arene was found to form *trans* $[MCl_2L_2]$ (M = Pd(II), Pt(II) and L = ligand (**2**) complexes exclusively, while the 1,3-diphosphinito–diester calix[4]arene was found to act as bridging ligand, affording tetrametallic species with a *trans* stereochemistry of the chlorides (**3**) (Figure 14.9). Unsymmetrical substituted calixarenes yield dimers, while symmetrical ones form tetramers.

The steric effect in the coordination of metal ions has been studied by dealkylating the upper rim of calixarenes substituted with phosphine and ester groups at the phenolic sites [23]. Results were unclear as to whether the absence of *tert*-butyl groups obviated the formation of tetrameric species, though the *trans* stereochemistry was maintained. Interestingly, the steric bulk of the *trans*-phosphinite ligands changes the dimensions of the calixarene ring by pinching two of the rings together. This change in diameter of the calixarene could result in unique complexating or catalytic properties.

Analogous diphenylphosphino derivatives were prepared in a similar three-step procedure [24]. Chelating behaviour of 1,3-di(phosphinomethyl)calix[4]arenes towards group 9 metals was found in complexes **4** (M = Rh, X = BF_4; M = Ir, X = SbF_6) and **5** (Figure 14.10) [24]. Reaction of $[RhCl(nbd)]_2$ with $AgBF_4$ followed by the addition of the calix[4]arene yields the monomeric complex **5**, in which the Rh(I) atom is chelated by two phosphorus atoms in a *cis* geometry. Formation of a *cis* complex was also observed in the presence of $[PtCl_2(cod)]$. The metal centre in these complexes was found to be located above the hemispherical cavity formed by the four functional groups to minimise steric interactions between the phosphine and the amide groups.

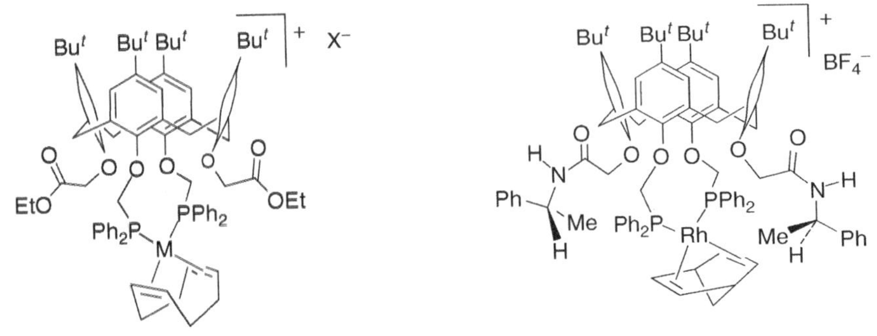

4

5

Figure 14.10 Complexes **4** and **5**.

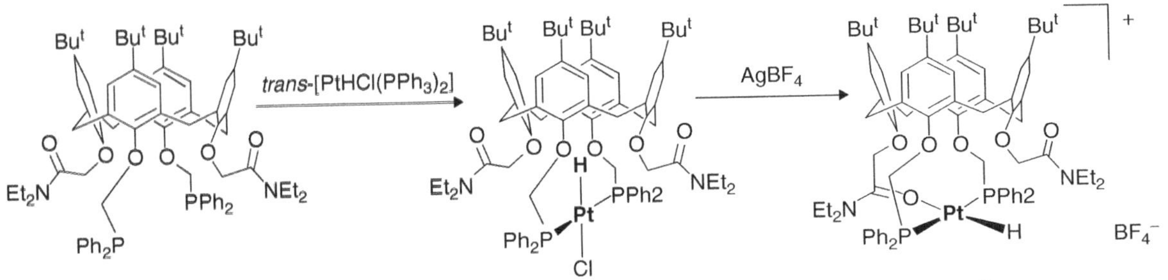

6

7

8

Scheme 14.1 Formation of calix[4]arene-based platinum hydride complexes.

The first calixarene complex engulfing a Pt–H bond (**7**) was synthesised by reaction of *trans*-[PtH(Cl)$(PPh_3)_2$] with calix[4]arene-based diphosphine ligand **6** (Scheme 14.1) [25]. The *trans* chelating behaviour of the ligand was inferred from nuclear magnetic resonance (NMR) spectroscopy and X-ray diffraction study. This investigation also established that, at least in the solid state, the hydrido ligand of **7** is trapped within the cavity defined by the four calixarene substituents, with the Pt–H bond pointing towards the centre of the cavity. Directional change of the Pt–H vector was forced by reacting **7** with $AgBF_4$.

This resulted in the formation of the cationic complex (**8**), in which the calixarene behaves as a hemispherical P,O,P tridentate. Molecular models clearly showed that the Pt–H bond lies approximately parallel to the calixarene reference plane and points towards the exterior of the cone.

Further studies by Wieser *et al.* led to the synthesis of *trans*-platinum hydride complexes of bis(diphenylphosphinomethoxy)-substituted calix[4]arenes with various functionalities at the other two phenolic oxygens [26]. This resulted in the modification of the chemical properties and the environment of the encapsulated hydride.

Large triphosphinocalix[6]arene ligands (**9**) were prepared by Tsuji *et al.* [27] from their alternating trimethoxy-protected precursors [28]. Complexation of this ligand with Rh(I) or Ir(I) provided nanocapsules with a ligand-to-metal ratio of 2:3, which were capable of encapsulating solvent molecules such as dichloromethane. The calix[6]arene moiety of both the free ligand and the metal complexes adopted the pinched-cone conformation. The terdentate ligand **9** proved to be extremely efficient for the stabilization of gold nanoparticles by preventing sintering to larger aggregates, both in solution and on support [29]. The steric bulk of the ligand influenced the accessible surface of the metal nanoparticle resulting in a relatively large

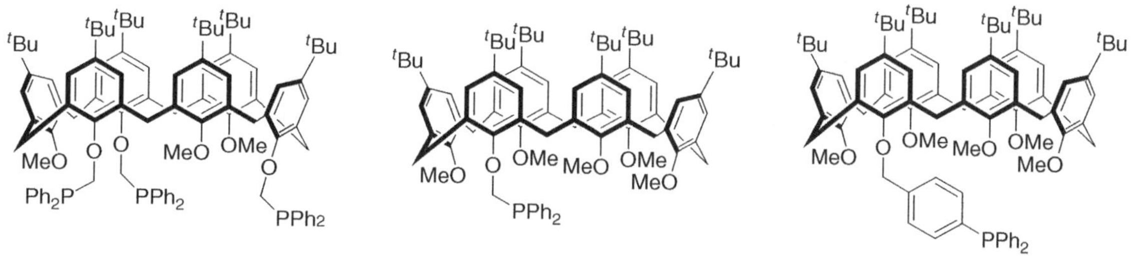

9 10 11

Figure 14.11 Complexes **10** and **11**.

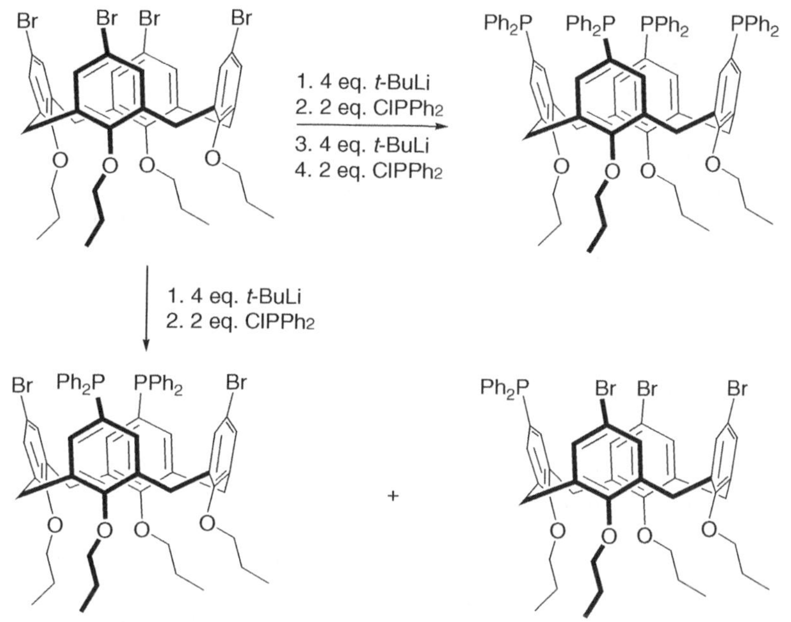

Scheme 14.2 Introduction of phosphines at upper rim according to Matt et al.

exposed surface area of the gold nanoparticle. In combination with the increased stability and uniform particle size distribution, this might lead to useful catalytic applications.

Tsuji *et al.* prepared monophosphanylcalix[6]arene ligands **10** and **11** (Figure 14.11) [30] starting from the analogous pentamethoxyprotected precursors [31]. X-ray crystal structure analysis revealed that **10** adopted a 1,2,3 alternate structure in the solid state, but NMR experiments showed that both **10** and **11** were in the cone conformation in solution. Both systems were active in Rh-catalysed hydroformylation of alkenes, but no special influence of the calixarene cavity on catalyst performance was observed.

An alternative approach to the direct functionalization of the lower-rim hydroxyl functions is substitution at the upper-rim *para*-position. This can easily be accomplished by $AlCl_3$-catalysed transalkylation of *p–tert–*butylcalix[4]arene [32] followed by bromination [33]. A series of calix[4]arenes carrying diphenylphosphino groups at the upper rim was prepared by Matt *et al.* [34]. They introduced single or multiple phosphine donors starting from 5,11,17,23-tetrabromo-25,26,27,28-tetrapropoxycalix[4]arene via lithiation followed by reaction with chlorodiphenylphosphine (Scheme 14.2). Reaction of the monophosphine-modified

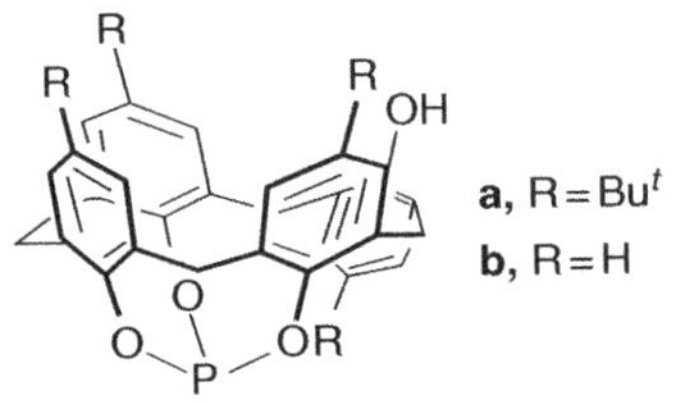

12

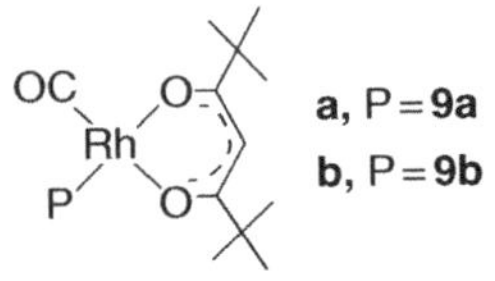

13

Figure 14.12 Complexes **12** and **13**.

calixarene with $[RuCl_2(p\text{-cymene})]_2$ afforded the monophosphine ruthenium(II)dichloride complex with the *p*-cymene ligand encapsulated inside the calixarene basket. A related bis-phosphine ligand based on a calix[4]arene scaffold resulted in the formation a trans-coordinated dinuclear complex with $PdCl_2$(cod) under dilute conditions. X-ray crystal structure analysis revealed that the two phosphinocalix[4]arene molecules were linked by two $PdCl_2$ fragments [35].

Tsuji *et al.* introduced a methylene linker between the calixarene cavity and the phosphine donors [36] starting from tetrachloromethylcalix[4]arene [37]. The resulting tetradentate ligand formed a complex with four $RuCl_2$(p-cymene) moieties in a 1,3 alternating mode. Related bidentate phosphine and phosphite ligands were prepared by Matt *et al.* by formylation at the two remaining *para*-positions of 5,17-bis(tert-butyl)-25,26,27,28-tetrakis(2-ethoxyethoxy)calix[4]arene with $SnCl_4$–Cl_2CHOCH_3. Subsequent reduction with $LiAlH_4$ gave the benzylic alcohol, which could be easily converted to the benzylic chloride with thionyl chloride. These products conveniently provided bidentate phosphite and phosphine ligands, respectively [38].

14.3.2 Phosphites and phosphonites

One of the earliest monophosphites (**12a**) was synthesised by Khasnis *et al.* in 1994, by using *p-tert*-butyl-calix[4]arene as starting material [39]. At that moment, compound **12a** was assumed to exist in a *cone* conformation, but in 2000 Cobley *et al.* proved by NMR spectroscopy and X-ray analysis that the calixarene-based monophosphite adopts a *partial cone* conformation [40].

In the same paper, *p-tert*-butyl-calix[4]arene-based monophosphite **9a** and the corresponding debutylated derivative **12b** were studied for their ability to complex rhodium precursors. Addition of $[Rh(acac)(CO)_2]$ to a CH_2Cl_2 solution containing the appropriate ligand afforded complexes **13**, which were tested as catalysts in the hydroformylation of 1-hexene (Figure 14.12). Although high aldehyde selectivity was obtained, the substrate conversion was low and the regioselectivity poor. Similar results were obtained with both ligands **12a** and **12b**, in keeping with the characteristic high steric bulk of the calix[4]arene scaffold, which, as proposed, might promote the formation of unsaturated rhodium species in solution.

Calix[4]arenes, which contain a μ_3-*O,O,O* bridging phosphorus atom at the lower rim of the cavity and an auxiliary function on the fourth phenolic oxygen, can exist in six different conformations (Figure 14.13), and the following notation can be used to differentiate among them: (*a*) u-u-o-**u**, (*b*) u-u-o-**d**, (*c*) u-u-u-**u**, (*d*) u-u-u-**d**, (*e*) u-o-u-**u** and (*f*) u-o-u-**d**.

Conformations *e* and *f* are relatively unfavourable and *a*–*d* are of comparable energy. Before 2000 only conformations *b* and *e* were known, as calix[4]arene monophosphates and calix[4]arene monoarsenites, with an *exo* orientation for P=O moiety and an *endo* orientation for the arsenic lone pair [39,41].

Simultaneous with the unsubstituted calix[4]-based monophosphites studied by Cobley *et al.*, eight different monophosphites (**11a–h**) built on monofunctionalised calix[4]arene cavities were reported by Parlevliet *et al.* (Scheme 14.3) [42]. Two possible synthetic routes were explored and, depending on the order in which the phosphorylation step and the *O*-alkylation (acylation) step were followed, different conformations could be obtained. When the phosphorylation reaction is carried out on a monofunctionalised calix[4]-matrix, the

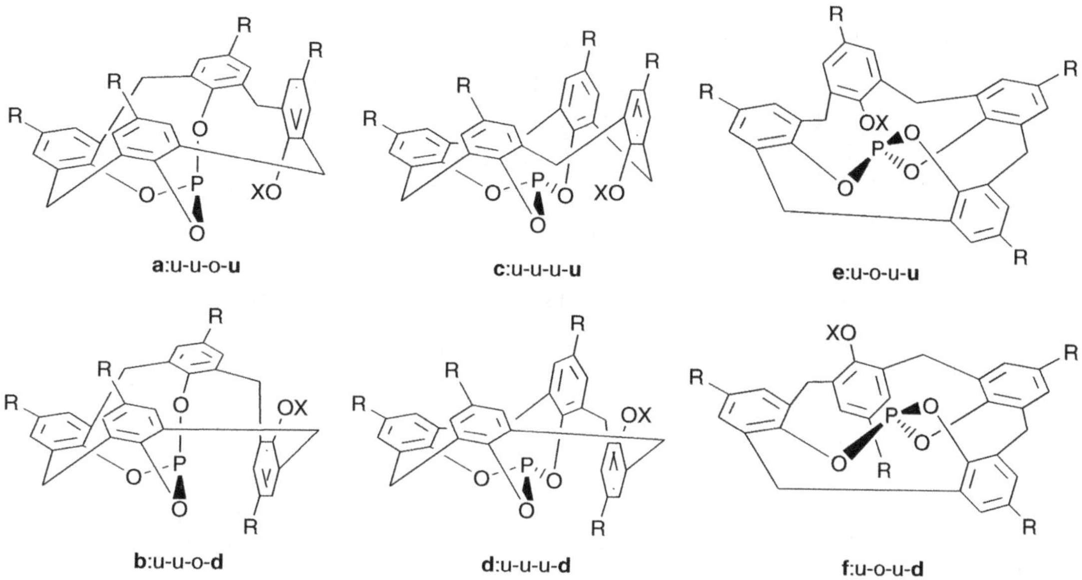

Figure 14.13 Possible conformations of monoalkylated calix[4]-based monophosphites. OX containing aryl printed in bold.

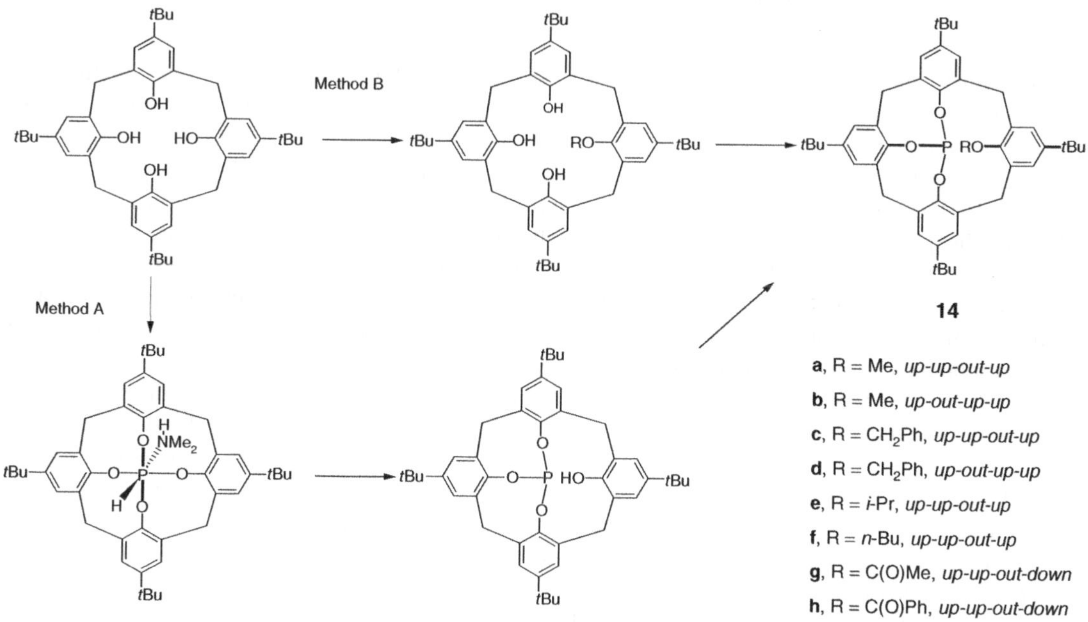

Scheme 14.3 Synthetic routes to monoalkylated calix[4]-based monophosphites (**14**).

resulting phosphite is characterised by an overall *cone* conformation of the cavity. On the contrary, when the two steps are reversed, the phosphite product shows a *partial cone* conformation. In the first case, due to the two possible orientations of the phosphorus lone pair (*exo* and *endo*, respectively), compounds with structure *a* or *e* can form, whereas in the last synthetic sequence only conformation *b* was observed.

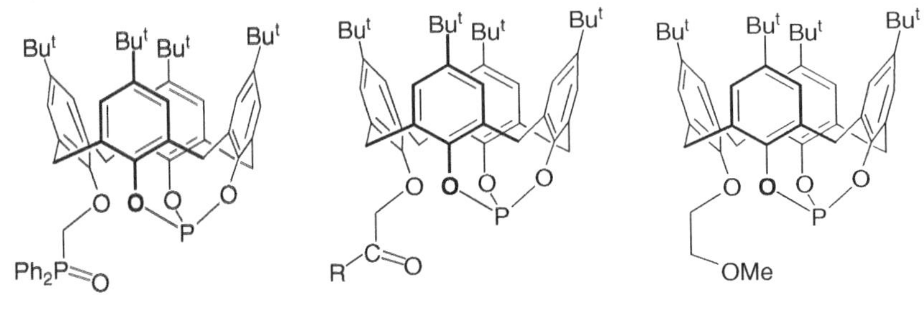

Figure 14.14 Complexes **15**, **16** and **17**.

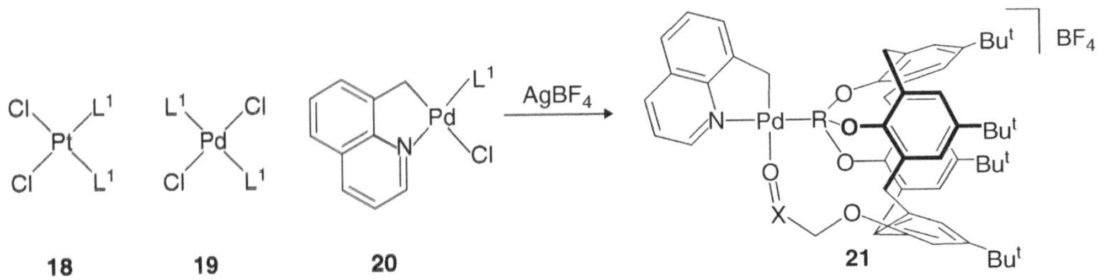

Figure 14.15 Complexes **18**, **19**, **20** and **21**.

When the fourth phenolic oxygen of a calix[4]arene-based μ_3-bridging phosphite is functionalised with a second donor atom-containing moiety, the resulting bidentate structure may show chelating behaviour towards transition metal centres.

Steyer *et al.* reported heterofunctionalised calix[4]arene monophosphites (**15–17**; Figure 14.14) as P,O-ligands, in which the auxiliary function is introduced at the lower rim of the cavity prior to P-functionalisation [43]. All the phosphites adopt a *cone* conformation, but partial flattening of the calixarene core may be accomplished as a consequence of the strain imposed by the very short μ_3-*O,O,O* capping unit.

In general, ligands **15** and **16** were found to coordinate to transition metal centres in a monodentate fashion, leaving the secondary function as a noncoordinated pendant arm. Only in one case, chelating behaviour was observed with the formation of 12-membered palladacycles.

Reaction of **15** with [PtCl$_2$(cod)] afforded the *cis* complex **18** in high yield. X-ray diffraction studies confirmed the *cis* stereochemistry of this complex. and ^{1}H NMR data indicated that the two calix[4]arene units are no longer C_s-symmetric. This observation reflects the important steric crowding of the two calixarene-fragments, which prevents the *cis*-arranged phosphites to freely rotate around the M–P bonds. Because of the highly distorted structure with respect to calix[4]arenes containing uncapped phenolic units, both *cone*-calixarenes adopt an *up-up-out-up* conformation. Reaction of **15** with [PdCl$_2$(cod)] yielded complex *trans*-**19** quantitatively. Due to the large separation between the two calixarene fragments, in this complex there is no barrier to rotation around the M–P bond.

When reacted with [Pd(8-mq)Cl]$_2$ (8-mqH=8-methylquinoline), ligand **15** showed again monocoordination. In order to force bonding of the auxiliary groups, complex **20** was treated with silver tetrafluoroborate, affording the cationic species **21** (Figure 14.15) in which the phosphite–phosphine oxide behaves as a chelating ligand.

Stronger chelators were reported by Marson *et al.*, who reported the synthesis of the first hybrid P,N-ligand **24** based on a calix[4]arene platform (Scheme 14.4) [44]. The cone conformation of the calixarene cavity

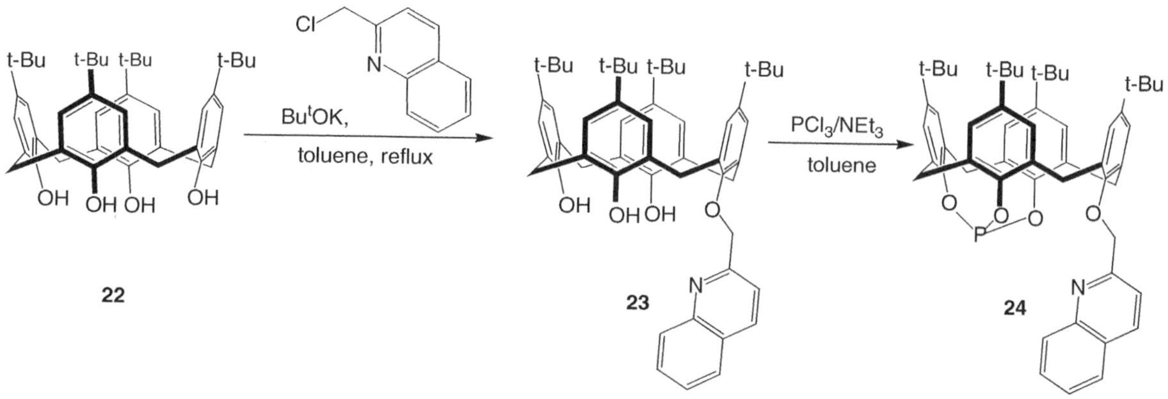

Scheme 14.4 Synthetic route to calix[4],quinoline-based P,N-ligand (**24**).

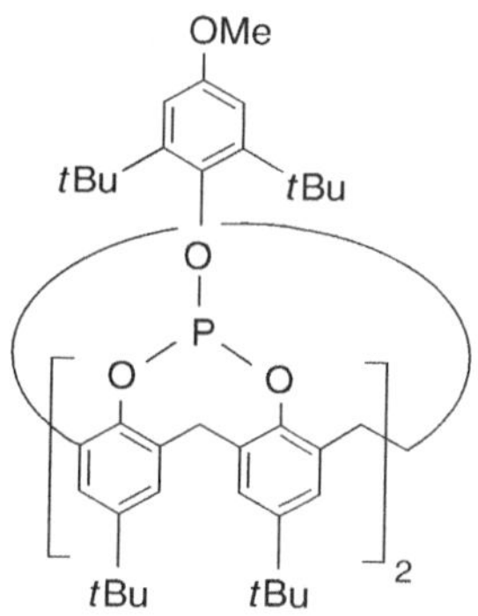

Figure 14.16 Complex **25**.

rendered the bidentate P,N-ligand suitable as a potential chelating system. In the presence of [PdCl(CH_3)(cod)] and $[Rh(CO)_2Cl]_2$ metal precursors, it readily formed 12-membered chelate P,N-rings, while with [$PtCl_2$(cod)] and with [Rh(acac)$(CO)_2$] showed monodentate coordination exclusively via the phosphorus atom, affording a 2:1 and a 1:1 complex, respectively.

In 1999, Paciello *et al.* synthesised the first chelating diphosphite ligand using a calix[4]arene cavity as a backbone and two P(OAr) fragments (Ar=2,6-di-*tert*-butyl-4-methoxyphenyl) bridging two proximal oxygen atoms of the lower rim (**25**; Figure 14.16) [45]. The resulting molecular pincer has a channel type of structure, in which the binding sites of the two phosphorus atoms are further blocked by the sterically demanding aryloxy substituents. The ligand afforded a highly active and selective rhodium hydroformylation catalyst towards linear aldehydes.

Other calix[4]-based diphosphite and diphosphonite ligands have been reported by Kunze *et al.* and subjected to detailed complex formation analysis [46]. In this type of structures, two $P(OAr)_2$ units are linked to two distal oxygen atoms of the calix[4]arene cavity while the other two oxygens remain unsubstituted or bear methyl groups (**26–28**; Figure 14.17).

In contrast to diphosphite **26**, which is characterised by only one singlet in the ^{31}P NMR spectrum, four major signals were observed for diphosphite **27**. Due to the complexity of the ^{1}H and ^{13}C NMR spectra, a detailed explanation of single diastereomers and conformers could not be given.

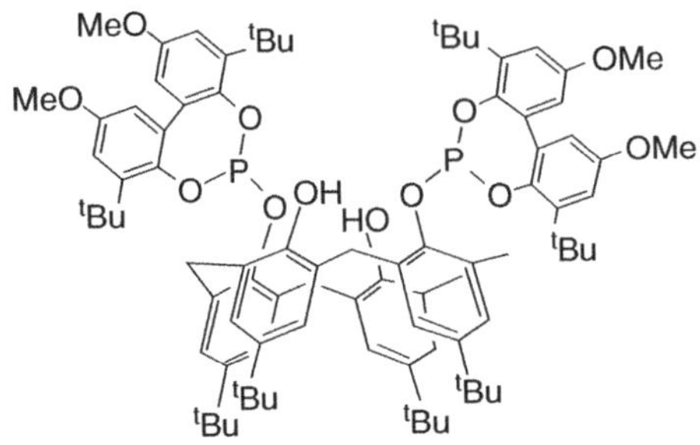

26

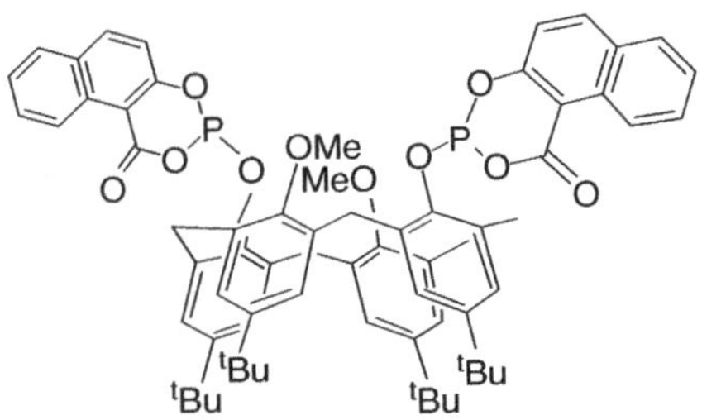

27

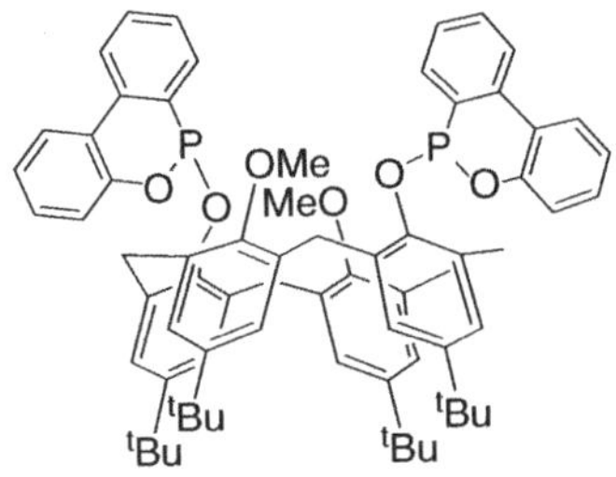

28

Figure 14.17 *Complexes **26**, **27** and **28**.*

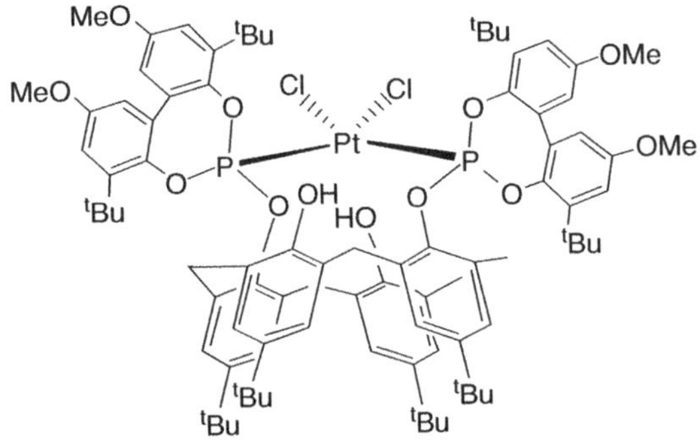

29

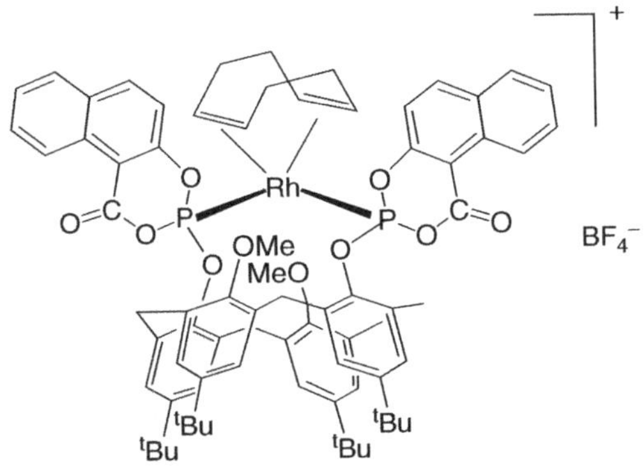

30

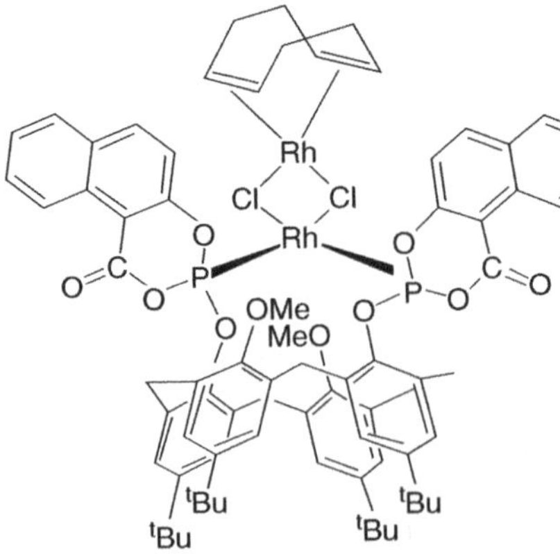

31

Figure 14.18 *Complexes **29**, **30** and **31**.*

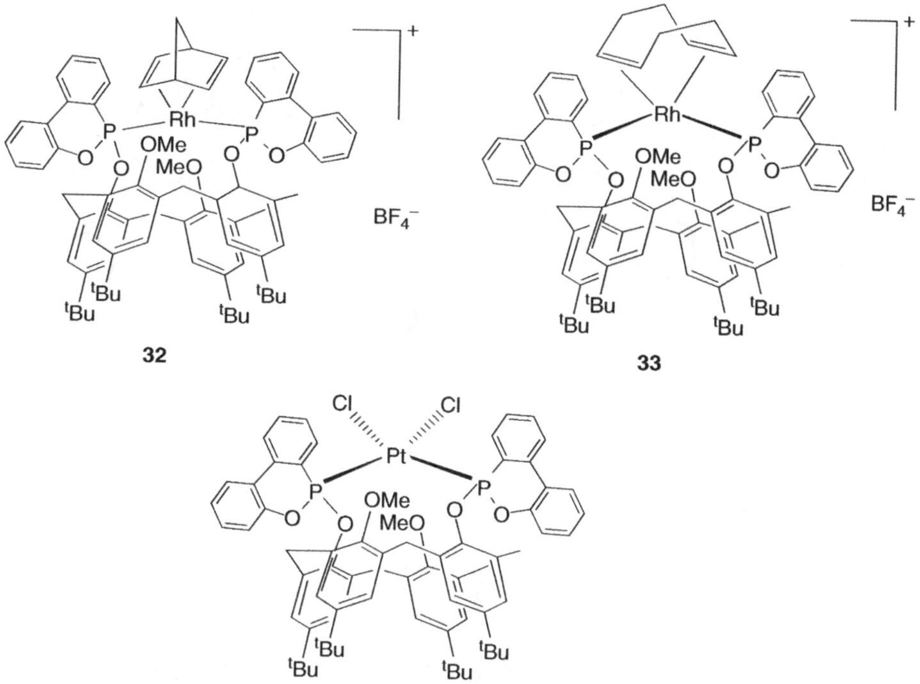

Figure 14.19 Complexes **32**, **33** and **34**.

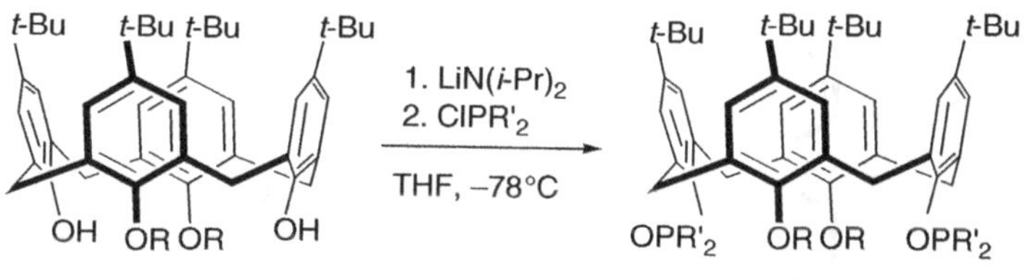

Scheme 14.5 Synthesis of 1,3 alternating calix[4]arene-based diphos(phin)ites

The coordinative properties of diphosphite ligands **26** and **27** towards platinum and rhodium precursors were studied. NMR spectroscopy proved the *cis*-arrangement of the diphosphite ligands around the metal centres and the exclusive *cone* conformation adopted by the calix[4]-cavity (complexes **29–31**; Figure 14.18). A notable additional feature was the complete loss of conformational flexibility upon complexation, which was proved by the ^{1}H, ^{13}C and ^{31}P NMR data.

Regarding diphosphonite **28**, two signals of about the same intensity were found in the ^{31}P NMR spectrum. Due to the nonstereoselective manner of the synthesis, two pairs of enantiomers were formed: (S,R)/(R,S), which are identical and constitute the *meso* form, and (S,S)/(R,R). Reaction with [Rh(nbd)(acac)], [Rh(cod)(acac)] and [$PtCl_2$(cod)] resulted in complexes **32**, **33** and **34**, respectively (Figure 14.19). Diastereomeric complexes were formed, and ^{31}P NMR data showed again *cis* coordination of the ligand around the metal centres.

A series of 1,3-diphosphites **35** and phosphinites **36** was prepared by Matt *et al.* from 1,3-protected *p–tert–*butylcalix[4]arenes (Scheme 14.5) [47]. The ligands were found to form a conical structure which is probably

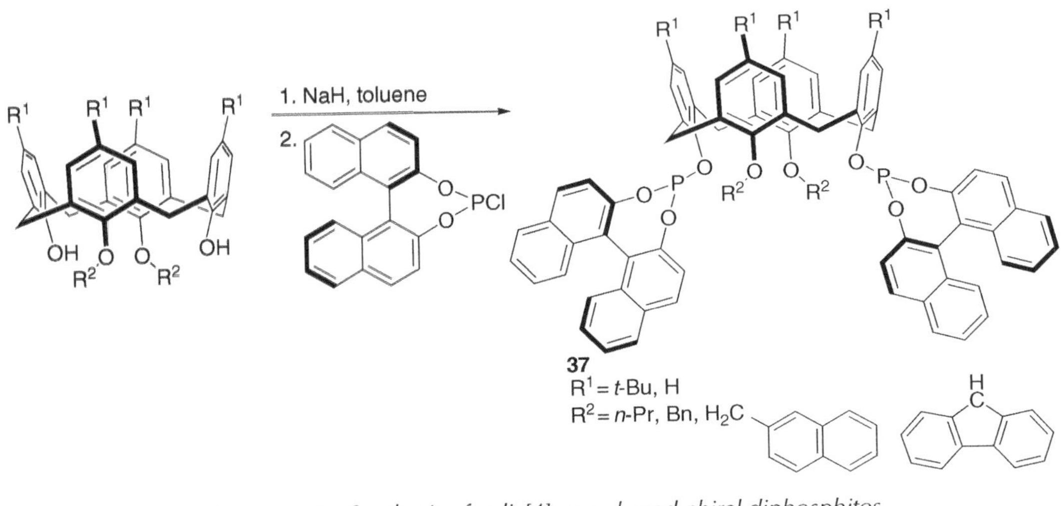

Scheme 14.6 *Synthesis of calix[4]arene-based chiral diphosphites.*

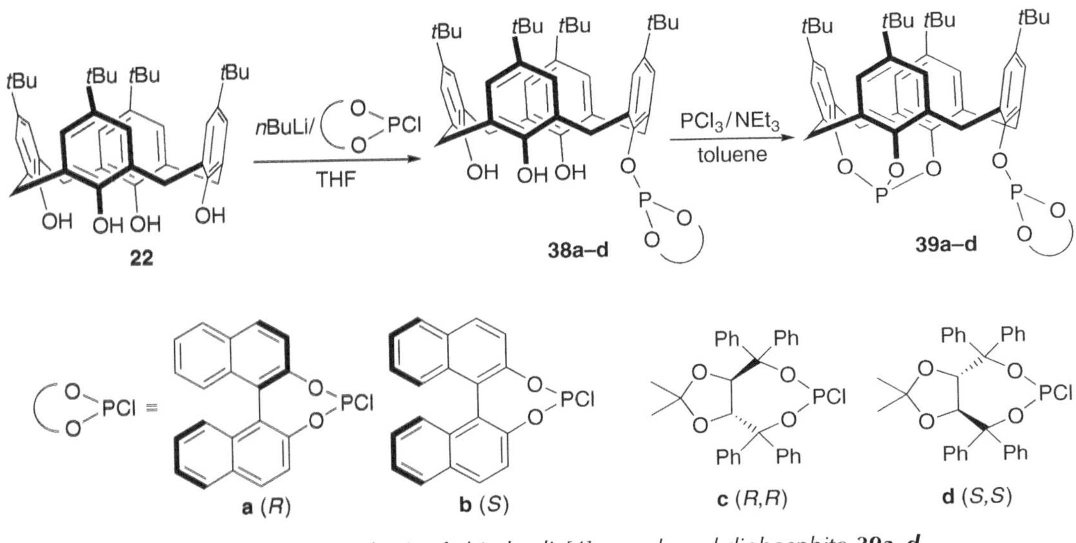

Scheme 14.7 *Synthesis of chiral calix[4]arene-based diphosphite* ***39a–d****.*

causing the formation of chelating complexes with Rh(I) and Pd(II) despite the large distance between the phosphorus donors which are separated by 12 bonds.

The good chelating properties and well-defined conical structures made these ligand systems excellent candidates for the development of chiral analogues. Matt *et al.* prepared a series of chiral calix[4]arene-based ligands by introduction of bisnaphthol-based phosphites at 1,3 alternating free hydroxy groups (Scheme 14.6) [48–50]. The synthesis was straightforward by deprotonation of the two remaining free hydroxyl groups followed by addition of the phosphorochloridite. The well-structured cone formation of these ligands provides a suitable scaffold for the development of new bulky ligands leading to shape-selective catalysts.

An alternative approach was followed by Marson *et al.*, who prepared enantiopure calix[4],BINOL- and calix[4],TADDOL-containing diphosphites through a two-step synthesis from *p–tert–*butylcalix[4]arene

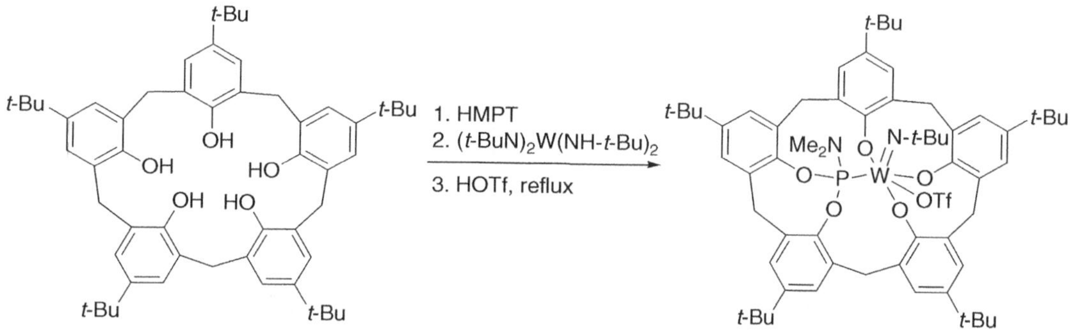

Scheme 14.8 *Phosphoramidite ligand derived from* p–tert–*butylcalix[5]arene exhibiting tetradentate behaviour.*

(Scheme 14.7) [51]. The sequence in which the two steps were followed was crucial in order to obtain diphosphites in the appropriate cone conformation as was previously observed by Matt *et al.* [35] and Parlevliet *et al.* [36]. The first step was the introduction of the chiral phosphite moiety by treatment of *p–tert–*butylcalix[4]arene with one equivalent of *n*-BuLi in THF at room temperature, followed by reaction with one equivalent of chiral phosphorochloridite. The selectivity of this step proved to be dependent on the steric bulk of the phosphorochloridite; the more sterically hindered fragments (i.e. TADDOL-PCl (**c**,**d**)) afford mainly monofunctionalised products, while with the bare BINOL derivatives (**a**,**b**) the two fractions are obtained in a 1:1 ratio. Fortunately, purification of the intermediates **38** was straightforward. Subsequent treatment with an excess of PCl_3–NEt_3 gave the pure chiral diphosphite products (**39a–d**).

Lattman *et al.* exploited the hydroxy groups of calix[5]arene both for direct coordination to a metal and to accommodate a phosphoramidite ligand [52]. The phosphoramidite ligand was conveniently prepared by reaction between hexamethylphosphorus triamide (HMPT) with *p–tert–*butylcalix[5]arene. After complexation of tungsten to the lower-rim oxygens, the phosphorus donor is forced into close proximity of the metal by the cone structure of the calix[5]arene backbone. After ligand exchange the calix[5]arene behaves as a tetradentate ligand with three hard oxide donor atoms and one soft phosphorus donor (Scheme 14.8).

The *p–tert–*butylcalix[5]arene moiety provides a good balance between flexibility and rigidity of the cavity, which was exploited by Lattman *et al.* to prepare a series of structurally well-defined mono- and bidentate phosphorus donor ligands (Scheme 14.9) [53]. Bidentate ligands **40a** and **40b** were easily prepared by addition of two equivalents of HMPT or dichlorophenylphosphine, respectively. Interestingly, monodentate ligand **41** could be made to react further to form phosphites in both an intra- and an intermolecular fashion. The monomeric phosphite **24** can be employed as a monodentate ligand in a similar way as **41**, whereas the dimeric ligand **43** forms a rigid tubular structure which is capable of trapping metals deep inside the cavity.

Aleksiuk *et al.* synthesised the first example of phosphorus-functionalised calix[6]arene containing two μ_3-bridging phosphorus atoms at the lower rim of the cavity (Figure 14.20) [54]. A *syn-* and an *anti-*conformation were proposed for the resulting diphosphate **44**. Similar isomeric structures have been also postulated by Parlevliet *et al.* for the calix[6]-based diphosphite **45** reported in 1996 [55]. Besides the conformation of the backbone (*syn* or *anti* depending on the mutual orientation of the phosphite units with respect to the main plane of the molecule), the phosphorus lone pair can adopt an *exo* or an *endo* orientation with respect to the calix[6]arene cavity. X-ray analysis showed that the P=O moieties in the calix[6]arene diphosphate and the phosphorus lone pairs in the analogous diphosphite were, in both cases, exclusively *exo* orientated. Chelating behaviour was observed for the *syn* conformer of the calix[6]-based diphosphite. *Cis*-complexes were formed with palladium and platinum, and the stereochemistry confirmed by X-ray structure analysis.

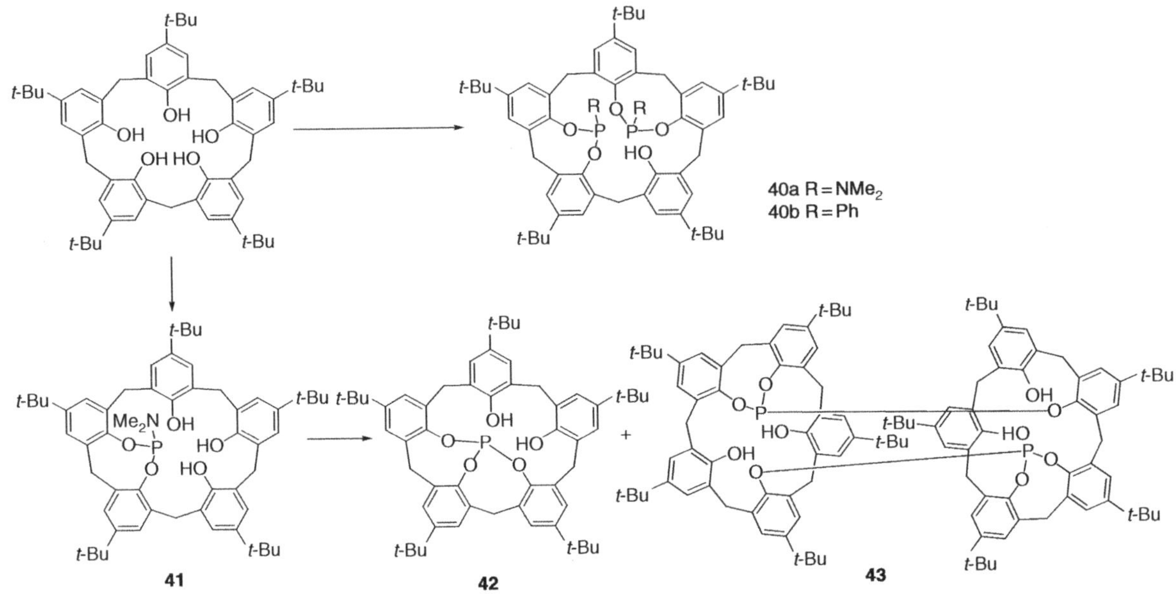

Scheme 14.9 *Synthesis of calix[5]arene-based mono- and bidentate ligands.*

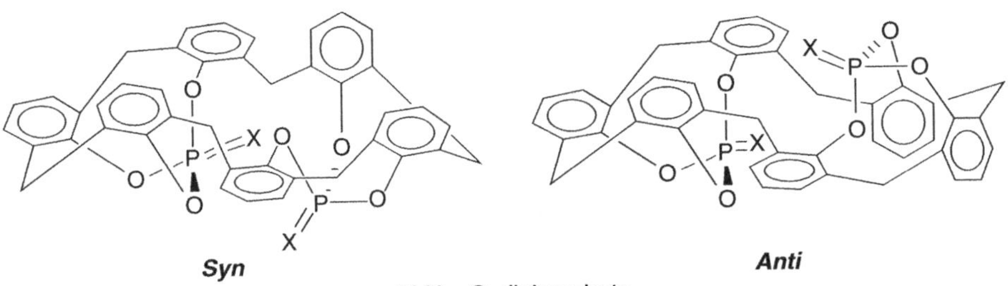

44 X = O, diphosphate
45 X = lone pair, diphosphite

Figure 14.20 Syn *and* anti *conformations for phosphorus-modified calix[6]arenes.*

14.4 Applications in homogeneous catalysis

Already in the mid-1990s, Pt(II) and Rh(I) complexes of phosphine and the phosphinite prepared by Matt *et al.* were employed in the hydroformylation of styrene [19]. The activity and the chemoselectivity of the phosphine-based catalysts were much higher than those of the phosphinite-based catalysts. Both the rhodium–phosphine and the rhodium–phosphinite were lower in activity than conventional catalysts. Hydroformylation of styrene, performed with the chiral rhodium complex **5** as catalyst, produces a branched-to-linear aldehyde ratio of 95:5, which is not an unusual value for a Rh–phosphine catalyst [24]. The rather low reaction rate was explained in terms of the partial encapsulation of the metal centre, hence preventing close approach of the substrate. Under the explored reaction conditions, no chiral induction was observed.

Parlevliet *et al.* investigated the effect of the different calix[4]arene phosphite conformation on the catalyst performances of the corresponding metal complexes in the rhodium-catalysed hydroformylation of 1-octene [42]. The three classes of conformers (*a*, *b* and *e*; Figure 14.8) showed totally different rates in catalysis, although all were fast. Ligands with an *e* conformation formed the slowest catalyst due to the environment of the phosphorus lone pair that for these compounds is sterically congested. As reported, this may hamper coordination of the ligand to the metal centre and, as a result, catalyst formation is probably slow. Moreover, steric hindrance in the complex would impede substrate coordination, slowing down the reaction. If the phosphorus lone pair suffers the least from steric crowding, catalyst formation is fast and the catalyst is easily accessible for the alkene substrate. The introduction of hemilabile groups at the fourth arene afforded lower activities than those obtained with the monophosphites reported by Parlevliet *et al.*, which was attributed to the possible formation of P,O-chelate complexes during catalysis [43].

The first example of great increase of selectivity in a catalytic reaction induced by the calixarene structure was reported by Paciello *et al.* [45]. In the rhodium-catalysed hydroformylation of 1-octene, their chelating diphosphites led to nonanal with surprisingly high regioselectivities but also to octane and considerable amounts of isomerised octene. Furthermore, activity and regioselectivity of the reaction could be tuned in a controlled fashion by variation of the substituents on the aryl group of the phosphite moieties.

An early application of calixarenes in asymmetric catalysis was reported by Matt *et al.*, who functionalized *p-tert*-butylcalix[4]arenes at the lower rim with phosphinite ligands in an alternating way [56]. The resulting ligands assumed a cone conformation and the pendant chiral groups induced moderate to good enantioselectivity in rhodium-catalysed asymmetric hydrogenation of dimethyl itaconate and palladium-catalysed asymmetric allylic alkylation of 1,3-diphenylprop-2-enyl acetate with dimethyl malonate. In subsequent studies, they introduced chirality in the calix[4]arene-based phosphite ligands using binol as chiral entity. These ligands **37** induced high regioselectivity for the linear product in the palladium-catalysed allylic alkylation of 3-phenylallyl acetate with dimethyl malonate and in the rhodium-catalysed hydroformylation of 1-octene [48]. These ligands also induced high linearities in the hydroformylation of challenging substrates such as styrene and 2-octene [49]. The linearities could be further improved by the addition of surfactants, and remarkably the best surfactant was [tetra(*p*-sulfonato)]-(tetra-*n*-butoxy)-calix[4]arene [50]. The high linearities render these chiral ligands not very suitable for asymmetric catalysis, but moderate enantioselectivities up to 61% were obtained in hydroformylation of norbornene [49].

The first example of chiral calix[4]arene-modified ligands that induce high enantioselectivity in metal-catalysed asymmetric reactions was reported by Marson *et al.* [50]. Chiral calixarene-based diphosphite ligands **39a–d** induced high enantiomeric excesses (up to 94%) in the rhodium-catalysed asymmetric hydrogenation of prochiral olefins with good activities. Even higher enantioselectivities were obtained by Jugé *et al.*, who reported a P-chirogenic aminophosphinephosphinite (AMP*P) ligand supported on the upper rim of a calix[4]arene moiety [57].

Achiral calix[4]arenes functionalized with phosphines at the upper rim have been extensively studied by Matt *et al.* The upper-rim monophosphinated calixarenes showed remarkable chelating behaviour in their complexes with $[RuCl_2(p\text{-cymene})]_2$ coordinating with phosphorus to the ruthenium metal, whereas the arenes of the calix cavity interacted with the p-cymene [58]. The calix[4]arene-based monophosphines provided very fast palladium catalysts for the Suzuki–Miyaura cross-coupling reaction [59]. The catalysts based on the calixarenyl–phosphines showed comparable activity as the state-of-the-art Buchwald type of ligands [60]. This excellent catalyst performance was explained by supramolecular stabilization of a monoligand $[Pd^0L(ArBr)]$ species of the Pd-bound arene inside the calixarene cavity via π–π interactions similar to the ruthenium-p-cymene complexes.

The related upper-rim functionalized diphosphines showed remarkable high activity in the Kumada–Corriu cross-couplings, which was attributed to the large natural bite induced by the calixarene structure [61]. The ability to support enlarged bite angles of the calixarene-based diphosphines was also thought to induce high activity and selectivity in the nickel-catalysed dimerization of propene [62]. Remarkably, Matt *et al.* reported that similar calix[4]arene-based monophosphines showed excellent performance in the nickel-catalysed Kumada–Tamao–Corriu cross-coupling, which was attributed to the formation of a monoligand Ni(0) intermediate [63].

Whereas most catalytic applications of calixarenes are based on calix[4]arenes, *syn*-calix[6]arene diphosphite was successfully applied in the palladium-catalysed CO–ethylene copolymerisation reaction [64]. The positive results achieved were attributed to the unique combination of the following features: (1) a chelate coordination with effective C_{2v} symmetry, (2) a bite angle being slightly larger than 90° and (3) a rigid backbone, chemically stable in the presence of a coordinative unsaturated cationic metal centre.

14.5 Experimental procedures

*General procedure for preparation of calix[4]arene-based monophosphites (**38a–d**)*: *p-tert*-butyl-calix[4]arene **22** (1.54 mmol) was suspended in 100 ml of dry THF, and a 2.5 M solution of n-BuLi/hexane (0.62 ml, 1.54 mmol) was added dropwise at room temperature (RT). The mixture was stirred for about 2 h and then cooled down to –78°C. A solution of the appropriate chlorophosphite ((R)-2,2′-dihydroxy-1,1′-binaphthyl (**a**), (S)-2,2′-dihydroxy-1,1′-binaphthyl (**b**), (R,R)-2,2-dimethyl-α,α,α′,α′-tetraphenyldioxolane-4,5-dimethanol (**c**) and (S,S)-2,2-dimethyl-α,α,α′,α′-tetraphenyldioxolane-4,5-dimethanol (**d**)) (1.54 mmol) in 15 ml of dry THF was added dropwise, and the mixture was then allowed to reach RT. After stirring at RT for 20 h, the THF was evaporated to dryness and dry dichloromethane was added to the residue. The salts were removed by filtration through a layer of SiO_2 and washed with dichloromethane. The filtrate and washing solutions were then evaporated under vacuum, giving calix[4]arene-based monophosphites **2a–d** as white residues that were used in the next step without further purification. Yields: 0.54÷0.77 mmol, 35÷50% (on the basis of the ^{1}H NMR spectra).

*General procedure for preparation of chiral calix[4]arene-based diphosphites (**39a–d**)*: The crude isolated from the first step was azeotropically dried twice with dry toluene and then dissolved in 50 ml of dry toluene. Distilled NEt_3 (18.65 mmol) was added and the solution was cooled to –78°C. 4.58 mmol of PCl_3 in toluene (15 ml) were then added dropwise, and the mixture was allowed to come to RT whilst stirring under argon atmosphere. After 20 h at 80°C, the salts formed were removed by filtration through a layer of SiO_2 and washed with toluene. The filtrate and washing solutions were then evaporated under vacuum. The residue was subjected to chromatographic separation on the Chromatotron® affording diphosphites **3a–d** in pure form as white powders.

*Synthesis of calix[6]arene diphosphite (**45**)*: A suspension of 4.86 g (5 mmol) of calix[6]arene [65] was dissolved in 100 ml of anhydrous THF/DMF (4/1, v/v) and added dropwise to a suspension of 0.90 g (30 mmol) NaH (80%) in 50 ml of anhydrous THF under inert atmosphere. After stirring for 1.5 hours at 40–45°C, the red suspension was added dropwise simultaneously and at the same rate with a solution of 0.89 ml (10 mmol) of PCl_3 in 100 ml of anhydrous THF, to 50 ml of refluxing THF. After refluxing for 2 h, the reaction mixture was allowed to cool to RT, and most of the solvent was evaporated. The resulting suspension was taken up in 100 ml of CH_2Cl_2 and washed quickly with 100 ml of water. The water layer was washed four times with 100 ml of CH_2Cl_2, the combined CH_2Cl_2 layers were dried over $MgSO_4$ and filtered, and the solvent was evaporated. The resulting solid was purified by column chromatography over silica gel with CH_2Cl_2 as eluent followed by recrystallization from hexane, yielding 1–2 g (20–40%) pure *syn*-**1** as a white powder. The *anti*-**45 s** could not be obtained in pure form, and therefore assignment of the structure was only done on the basis of NMR studies in the presence of *syn*-**45**.

5,17-Bis(diphenylphosphino)-25,26,27,28-tetrapropoxycalix[4]arene [34]: A 1.72 M solution of *t*-BuLi-pentane (2.8 cm^3, 4.8 mmol) was added to a cold (–78°C) solution of 5,11,17,23-tetrabromo-25,26,27,28-tetrapropoxycalix[4]arene [66] (2.169 g, 2.38 mmol) in THF (50 ml). After 1 h, Ph_2PCl (0.8 ml, 1.053 g, 4.76 mmol) was added dropwise and the solution was stirred for 10 h at –78°C. The lithiation step was then repeated using the same amounts of *t*-BuLi. The reaction was quenched with 3 ml of MeOH. The solution was allowed to warm up to RT was evaporated to dryness, and the residue was dissolved in CH_2Cl_2. Addition of EtOH afforded a white precipitate. Yield: 1.875 g, 82%.

^{1}H NMR ($CDCl_3$): δ 7.37–7.33 (20 H, PPh_2), 7.06 (d, 4 H, m-H of OArP, $^3J_{PH}$=8 Hz), 6.29 and 6.11 (AB, 6 H, m and p-H of OAr, ^3J=8 Hz), 4.41 and 3.06 (AB, 8 H, $ArCH_2Ar$, $^2J_{AB}$=13.0 Hz), 4.03 (pseudo t, 4 H, OCH2, $^3J\approx$8 Hz), 3.63 (t, 4 H, OCH_2, ^3J=8 Hz), 2.00 (m, 4 H, OCH_2CH_2), 1.81 (m, 4 H, OCH_2CH_2), 1.06 (t, 6 H, CH_3, ^3J=7.5 Hz), 0.91 (t, 6 H, CH_3, ^3J=7.5 Hz). ^{13}C{^{1}H} NMR ($CDCl_3$): δ 159.80 and 155.10 (2 s, arom. Cq–O), 137.25–122.02 (arom. C's), 77.00 and 76.60 (2 s, OCH_2), 30.95 (s, $ArCH_2Ar$), 23.51 and 23.15 (2 s, CH_2CH_3), 10.82 and 9.93 (2 s, CH_3). ^{31}P{^{1}H} NMR ($CDCl_3$): δ .6.4 (s, PPh_2). Found: C, 80.18; H, 6.74. Calc. for $C_{64}H_{66}O_4P_2$ (Mr=961.18) C, 79.98; H, 6.92%.

References

[1] (a) A. V. Bayer, *Ber. Dtsch. Chem. Ges.* **1872**, *5*, 25; (b) A. V. Bayer, *Ber. Dtsch. Chem. Ges.* **1872**, *5*, 280; and (c) A. V. Bayer, *Ber. Dtsch. Chem. Ges.* **1872**, *5*, 1094.

[2] A. Zinke and E. Ziegler, *Wiener Chem. Ztg.* **1944**, *47*, 151.

[3] C. D. Gutsche, R. Muthukrishman, *J. Org. Chem.* **1978**, *43*, 4905.

[4] C. D. Gutsche, M. Iqbal, *Org. Synth.* **1990**, *68*, 234.

[5] Z. Asfari, V. Böhmer, J. Harrowfield, J. Vicens, *Calixarenes*, Kluwer Academic, **2001**, and references he]rein.

[6] E. D. Cornforth, K. T. Morgan, R. J. W. Rees, *Tetrahedron* **1973**, *29*, 1659.

[7] C. D. Gutsche, L. J. Bauer, *Tetrahedon Lett.* **1981**, *22*, 4763.

[8] C. D. Gutsche, L. J. Bauer, *J. Am. Chem. Soc.* **1985**, *107*, 6052.

[9] The all-*u* conformation is also referred to as the *cone* conformation.

[10] (a) M. A. Molins, P. M. Nieto, C. Sanchez, P. Prados, J. de Mendoza, M. Pons, *J. Org. Chem.* **1992**, *57*, 6924; (b) R. G. Janssen, J. P. M. van Duynhoven, W. Verboom, G. J. van Hummel, S. Harkema, D. N. Reinhoudt, *J. Am. Chem. Soc.* **1996**, *118*, 3666; and (c) W. P. van Hoorn, F. C. J. M. van Veggel, D. N. Reinhoudt, *J. Org. Chem.* **1996**, *61*, 7180.

[11] C. D. Gutsche, R. Muthuhrishnan, *J. Org. Chem.* **1978**, *43*, 4905.

[12] For a review about calixarene ligands in transition metal chemistry, see C. Wieser, C. B. Dieleman, D. Matt, *Coord. Chem. Rev.* **1997**, *165*, 93, and references herein.

[13] For a review on the coordination chemistry of phosphorylated calixarenes, see S. D. Alexandratos, S. Natesan, *Ind. Eng. Chem. Res.* **2000**, *39*, 3998.

[14] C. Floriani, D. Jacoby, A. Chiesi-Villa, C. Guastini, *Angew. Chem. Int. Ed. Engl.* **1989**, *28*, 1376.

[15] W. Xu, R. J. Puddephatt, L. Manojlovic-Muir, K. W. Muir, C. S. Frampton, *J. Incl. Phenom.* **1994**, *19*, 277.

[16] D. Jacoby, C. Floriani, A. Chiesi-Villa, C. Rizzoli, *J. Chem. Soc., Dalton Trans.* **1993**, 813.

[17] C. Floriani, D. Jacoby, A. Chiesi-Villa, C. Guastini, *Angew. Chem., Int. Ed. Engl.* **1989**, *28*, 1376.

[18] I. Neda, H. J. Plinta, R. Sonnenburg, A. Fischer, P. G. Jones, R. Schmultzler, *Chem. Ber.* **1995**, *128*, 267.

[19] C. Dieleman, C. Loeber, D. Matt, A. De Cian, J. Fischer, *J. Chem. Soc., Dalton Trans.* **1995**, 3097.

[20] Z. Csok, G. Szalontai, G. Czira, L. Kollar, *J. Organomet. Chem.* **1998**, *570*, 23.

[21] C. Loeber, D. Matt, P. Briard, D. Grandjean, J. *Chem. Soc.*, Dalton Trans. **1996**, 513.

[22] (a) P. J. Dijkstra, J. A. J. Brunink, K.-E. Bugge, D. N. Reinhoudt, S. Harkema, R. Ungaro, F. Ugozzoli, E. Ghidini, *J. Am. Chem. Soc.* **1989**, *111*, 7567; (b) E. Ghidini, F. Ugozzoli, R. Ungaro, S. Harkema, A. A. El-Fad1, D. N. Reinhoudt, *J. Am. Chem. Soc.* **1990**, *112*, 6979; (c) K. Iwamoto, K. Araki, S. Shinkai, *J. Org. Chem.* **1991**, *56*, 4955; and (d) F. Arnaud-Neu, E. M. Collins, M. Deasy, G. Ferguson, S. J. Harris, B. Kaitner, A. J. Lough, M. A. McKervey, E. Marques, B. L. Ruhl, M.-J. Schwing-Weill, E. M. Seward, *J. Am. Chem. Soc.* **1989**, *111*, 8681.

[23] P. Faidherbe, C. Wieser, D. Matt, A. Harriman, A. De Cian, J. Fischer, *Eur. J. Inorg. Chem.* **1998**, 451.

[24] (a) C. Loeber, D. Matt, A. De Cian, J. Fischer, J. Organomet *Chem.* **1994**, *475*, 297; (b) C. Loeber, C. Wieser, D. Matt, A. De Cian, J. Fischer, L. Toupet, *Bull. Soc. Chim. Fr.* **1995**, *132*, 166; and (c) J. Gragno, C. Loeber, D. Matt, P. D. Harvey, *Inorg. Chim. Acta* **1996**, *242*, 137.

[25] C. Wieser, D. Matt, L. Toupet, H. Bourgeois, J. P. Kintzinger, *J. Chem. Soc., Dalton Trans.* **1996**, 4041.

[26] C. Wieser, D. Matt, J. Fischer, A. Harriman, *J. Chem. Soc., Dalton Trans.* **1997**, 2391.

[27] Y. Obora, Y. K. Liu, L. H. Jiang, K. Takenaka, M. Tokunaga, Y. Tsuji, *Organometallics* **2005**, *24*, 4.
[28] R. G. Janssen, V. Verboom, D. N. Reinhoudt, A. Casnati, M. Freriks, A. Pochini, F. Ugozzoli, R. Ungaro, P. M. Nieto, M. Carramolino, F. Cuevas, P. Prados, J. de Mendoza, *Synthesis* **1993**, 380.
[29] J.-M. Ha, A. Solovyov, A. Katz, *Langmuir* **2009**, *18*, 10548.
[30] Y. Obora, Y. K. Liu, S. Kubouchi, M. Tokunaga, Y. Tsuji, *Eur. J. Inorg. Chem.* **2006**, 222.
[31] J. de Mendoza, M. Carramolino, F. Cuevas, P. M. Nieto, P. Prados, D. N. Reinhoudt, W. Verboom, R. Ungaro, A. Casnati, *Synthesis* **1994**, 47.
[32] (a) C. D. Gutsche, J. A. Levine, *J. Am. Chem. Soc.* **1982**, *104*, 2652; and (b) C. D. Gutsche, L.-G. Lin, *Tetrahedron* **1986**, 1633.
[33] C. D. Gutsche, P. F. Pagoria, *J. Org. Chem.* **1985**, *50*, 5795.
[34] M. Lejeune, C. Jeunesse, D. Matt, N. Kyritsakas, R. Welter, J.-P. Kintzinger, *J. Chem. Soc., Dalton Trans.* **2002**, 1642.
[35] K. Takenaka, Y. Obora, L. H. Jiang, Y. Tsuji, *Organometallics* **2002**, *21*, 1158.
[36] K. Takenaka, Y. Obora, Y. Tsuji, *Inorg. Chim. Acta* **2004**, *357*, 3895.
[37] M. Almi, A. Arduini, A. Casnati, A. Pochini, R. Ungaro, *Tetrahedron* **1989**, *45*, 2177.
[38] I. A. Bagatin, D. Matt, H. Thonnessen, P. G. Jones, *Inorg. Chem.* **1999**, *38*, 1585.
[39] D. V. Khasnis, J. M. Burton, J. D. McNeil, C. J. Santini, H. Zhang, M. Lattman, *Inorg. Chem.* **1994**, *33*, 2657.
[40] C. J. Cobley, D. D. Ellis, A. G. Orpen, P. G. Pringle, *J. Chem. Soc., Dalton Trans.* **2000**, 1101.
[41] (a) S. Shang, D. Khasnis, H. Zhang, A. C. Small, M. Fan, M. Lattman, *Inorg. Chem.* **1995**, *34*, 3610; and (b) D. V. Khasnis, J. M. Burton, J. D. McNeill, H. Zhang, M. Lattman, *Phosphorus, Sulfur Silicon Relat. Elem.* **1993**, *75*, 253.
[42] F. J. Parlevliet, C. Kiener, J. Fraanje, M. Lutz, A. L. Spek, P. C. J. Kamer, P. W. N. M van Leeuwen, *J. Chem. Soc., Dalton Trans.* **2000**, 1113.
[43] S. Steyer, C. Jeunesse, D. Matt, R. Welter, M. Wesolek, *J. Chem. Soc., Dalton Trans.* **2002**, 4264.
[44] A. Marson, J. E. Ernsting, M. Lutz, A. L. Spek, P. W. N. M. van Leeuwen, P. C. J. Kamer, *Dalton Trans.* **2009**, 621.
[45] R. Paciello, L. Siggel, M. Röper, *Angew. Chem. Int. Ed.* **1999**, *38*, 1920.
[46] C. Kunze, D. Selent, I. Neda, M. Freytag, P. G. Jones, R. Schmutzer, W. Baumann, A. Börner, *Z. Anorg. Allg. Chem.* **2002**, *628*, 779.
[47] S. Steyer, C. Jeunesse, J. Harrowfield, D. Matt, *Dalton Trans.* **2005**, 1301.
[48] D. Sémeril, C. Jeunesse, D. Matt, L. Toupet, *Angew. Chem. Int. Ed.* **2006**, *45*, 5810.
[49] D. Sémeril, D. Matt, L. Toupet, *Chem. Eur. J.* **2008**, *14*, 7144.
[50] L. Monnereau, D. Sémeril, D. Matt, L. Toupet, *Adv. Synth. Catal.* **2009**, *351*, 1629.
[51] A. Marson, Z. Freixa, P. C. J. Kamer, P. W. N. M. van Leeuwen, *Eur. J. Inorg. Chem.* **2007**, 4587.
[52] M. Fan, H. Zhang, M. Lattman, *Chem. Commun.* **1998**, 99.
[53] P. Sood, M. Koutha, M. Fan, Y. Klichko, H. Zhang, M. Lattman, *Inorg. Chem.* **2004**, *43*, 2975.
[54] F. Grynszpan, O. Aleksiuk, S. E. Biali, *J. Chem. Soc., Chem. Commun.*, **1993**, 13.
[55] F. J. Parlevliet, A. Olivier, W. G. J. de Lange, P. C. J. Kamer , A. L. Spek, H. Kooijman, P. W .N .M. van Leeuwen, *J. Chem. Soc., Chem. Commun.* **1996**, 583.
[56] C. Dieleman, S. Steyer, C. Jeunesse, D. Matt, *J. Chem. Soc., Dalton Trans.*, **2001**, 2508.
[57] N. Khiri, E. Bertrand, M.-J. Ondel-Eymin, Y. Rousselin, J. Bayardon, P. D. Harvey, S. Jugé, *Organometallics* **2010**, *29*, 3622.
[58] S. Sameni, M. Lejeune, C. Jeunesse, D. Matt, R. Welter, *Dalton Trans.*, **2009**, 7912.
[59] L. Monnereau, D. Sémeril, D. Matt, L. Toupet, *Chem. Eur. J.* **2010**, *16*, 9237.
[60] J. P. Wolfe, R. A. Singer, B. H. Yang, S. L. Buchwald, *J. Am. Chem. Soc.* **1999**, *121*, 9550.
[61] D. Sémeril, M. Lejeune, C. Jeunesse, D. Matt, *J. Mol. Catal. A* **2005**, *239*, 257.
[62] M. Lejeune, D. Sémeril, C. Jeunesse, D. Matt, P. Lutz, L. Toupet, *Adv. Synth. Catal.* **2006**, *348*, 881.
[63] L. Monnereau, D. Sémeril, D. Matt, *Chem. Commun.* **2011**, *47*, 6626.
[64] F. J. Parlevliet, M. A. Zuideveld, C. Kiener, H. Kooijman, A. L. Spek, P. C. J. Kamer, P. W. N. M. van Leeuwen, *Organometallics* **1999**, *18*, 3394.
[65] C. D. Gutsche, B. Dhawan, M. Leonis, D. Stewart, *Org. Syn.* **1990**, *68*, 238.
[66] M. Larsen, M. Jørgensen, *J. Org. Chem.* **1996**, *61*, 6651.

15

Supramolecular Bidentate Phosphorus Ligands

Jarl Ivar van der Vlugt and Joost N. H. Reek

Supramolecular and Homogeneous Catalysis, van 't Hoff Institute for Molecular Sciences, University of Amsterdam, Amsterdam, The Netherlands

15.1 Introduction: general design principles

Ligand design and development are the most important tools to obtain effective transition metal catalysts. The properties of a homogeneous catalyst can be tailored to very specific conversions by proper design of the ligand. In past decades, this has resulted in the development of numerous ligands that provide good catalysts for various reactions. To assist ligand design on rational bases, interesting general concepts have been developed. Important parameters that describe ligand properties include the cone angle and the bite angle.Tolman introduced the cone angle (Θ) as a means of describing the steric properties of monodentate ligands. The electronic properties of phosphine ligands were also first quantified by Tolman, who introduced the χ parameter [1]. This χ value was determined from the CO frequency of $[Ni(L)(CO)_3]$ complexes for a large set of ligands to quantify the electronic donor properties of the ligands. Much later, the importance of the geometry of bidentate ligands was recognized and the "natural bite angle" was introduced as a way of describing the properties of bidentate chelating ligands [2]. The natural bite angle β is defined as the angle at which two donor atoms of a chelating bidentate ligand "bite" into a transition metal; this can be calculated by molecular modeling of the backbone.A strongly related parameter is the flexibility range, which is defined as the accessible range of bite angles within less than 12.6 kJ/mol excess strain energy from the calculated bite angle. Although these parameters are very helpful and computational techniques are becoming increasingly important, one cannot easily predict the activity and selectivity of a transition metal catalyst. This is especially true for chiral catalysts developed for asymmetric conversions, and therefore catalyst development (for asymmetric reactions) relies to a great extent on trial and error and

Phosphorus(III) Ligands in Homogeneous Catalysis: Design and Synthesis, First Edition.
Edited by Paul C. J. Kamer and Piet W. N. M. van Leeuwen.

sophisticated guesses. As a consequence, combinatorial approaches and high-throughput experimentation have become increasingly important. Several elegant screening techniques for combinatorial catalysis have been developed [3], but there is only limited technology available to prepare large catalyst libraries with sufficient diversity. Although some interesting modular bidentate ligands have been reported [4], the library size generally remains relatively small. An interesting breakthrough is the revival of the use of monodentate ligands, simultaneously reported by Reetz, Feringa and de Vries, and Pringle [5], and more recently the successful use of mixtures of monodentate ligands [6]. These approaches lead much easier to large libraries of ligands because these monodentates are generally easier to prepare. The mixture approach does lead to large catalyst libraries, but because the catalysis is performed by a mixture of catalysts, hits can be found only if the catalysis is dominated by the heterocomplex. In addition, the complexes based on monodentate ligands generally have a less well-defined "pocket" for the substrate, as more orientations are possible for the ligands to coordinate to the metal.

Recently a new class of ligands has been introduced that potentially combines the advantages of bidentate and monodentate ligands. They have been introduced as supramolecular bidentate ligands (SUPRAphos) or self-assembled ligands, and the key point is that they form via self-assembly of two functionalized monodentate ligands. By simply mixing monodentate ligands functionalized with proper complementary binding motifs, well-defined metal complexes form in situ. We believe that this new generation of bidentate ligands, based on supramolecular building blocks, is a breakthrough in the field of catalysis. It combines the easy synthesis of the building blocks with the control over coordination space, and in addition, large libraries of catalysts can be generated from a relatively small number of building blocks.

There are two principle strategies that can be followed (Scheme 15.1):

1. A template can be used that contains binding sites for the selective assembly of two (different) ligands.
2. The ligands can be functionalized with complementary binding motifs.

An important difference between these two strategies is the number of components involved. Approach A leads to a three-component assembly, and as a consequence, the system is more complicated, but it does easily produce large ligand libraries ($10\times10\times10=1000$ members based on 30 compounds). Approach B is simpler, but more building blocks are required to arrive at large ligand libraries ($30\times30=900$ members based on 60 compounds).

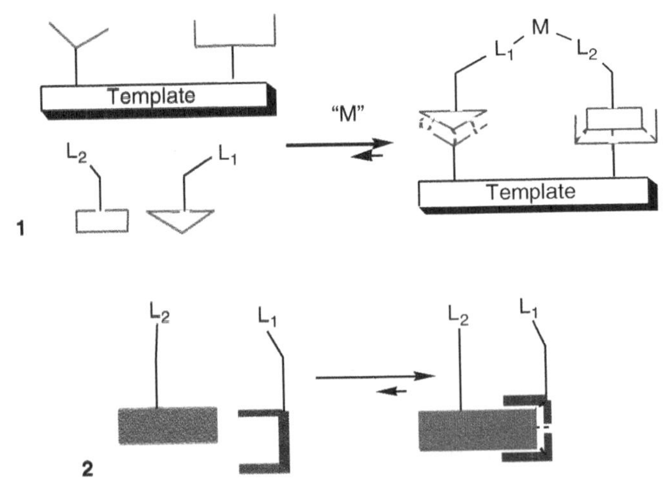

Scheme 15.1 *Schematic representation of two methods to form chelating bidentate ligands by assembly: (1) via a template, and (2) via complementary binding motifs directly attached to the ligand.*

This chapter will discuss recent developments in the field of homogeneous catalysis with supramolecular bidentate phosphorus ligands, including their design, synthesis, assembly, and performance in selected applications.

We divide three types of supramolecular ligand concepts to assemble a bidentate ligand from monodentate components: (1) assembly based on H-bonding interactions, (2) assembly based on ionic interactions, and (3) assembly based on weak coordinative forces. For the latter type, various types of building motifs can be chosen, each offering distinct possibilities. All methods have been established now as viable methodologies for homogeneous catalysis, even under forcing reaction conditions, thus enabling a new, versatile type of catalyst design to be implemented for practical applications.

15.2 Construction of bidentate phosphorus ligands via self-assembly

15.2.1 H bonding

Despite being very weak interactions (typically in the order of 3 kcal/mole), H bonds provide interesting motifs as they are highly directional, scalable (with binding strength increasing with the number of bonds), and commonly generated from a relatively large number of functional groups, allowing various synthetic approaches, including tautomeric and mesomeric forms, as well as self-complementarity versus cross-complementarity principles. Therefore, this type of interaction is very well suited for the fabrication of supramolecular bidentate ligand systems from its basic monodentate components. Functional group tolerance may be an issue during complexation, although this is generally not so significant for late transition metals, and also hydrogen bonds are solvent dependent which may limit the application of such systems. Interestingly, we will see these limitations are not severe as one intuitively expects, as hydrogen bond motifs are also preorganized via the metal and thus easily compete with solvents.

By making proper use of self-association of hydroxypyridine compounds functionalized with diphenylphosphino-groups, the group of Breit was able to induce selective chelation using two monodentate phosphorus ligand building blocks [7]. Molecule **1** exists in two tautomeric forms, which can self-assemble by forming two hydrogen bonds (Scheme 15.2). As a consequence, the two phosphine moieties are also aligned

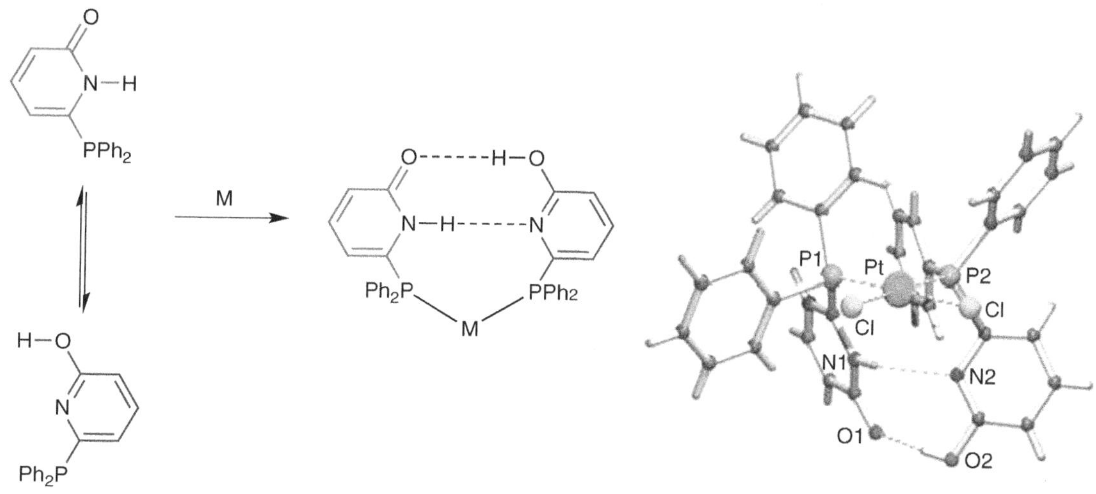

Scheme 15.2 *Left: Self-complementary assembly of a bidentate ligand through H-bonding interactions between two tautomers of a monodentate phosphine. Right: Molecular structure of a $PtCl_2$ complex. Reprinted from I. Usui, et al., 2008, with permission of the American Chemical Society.*

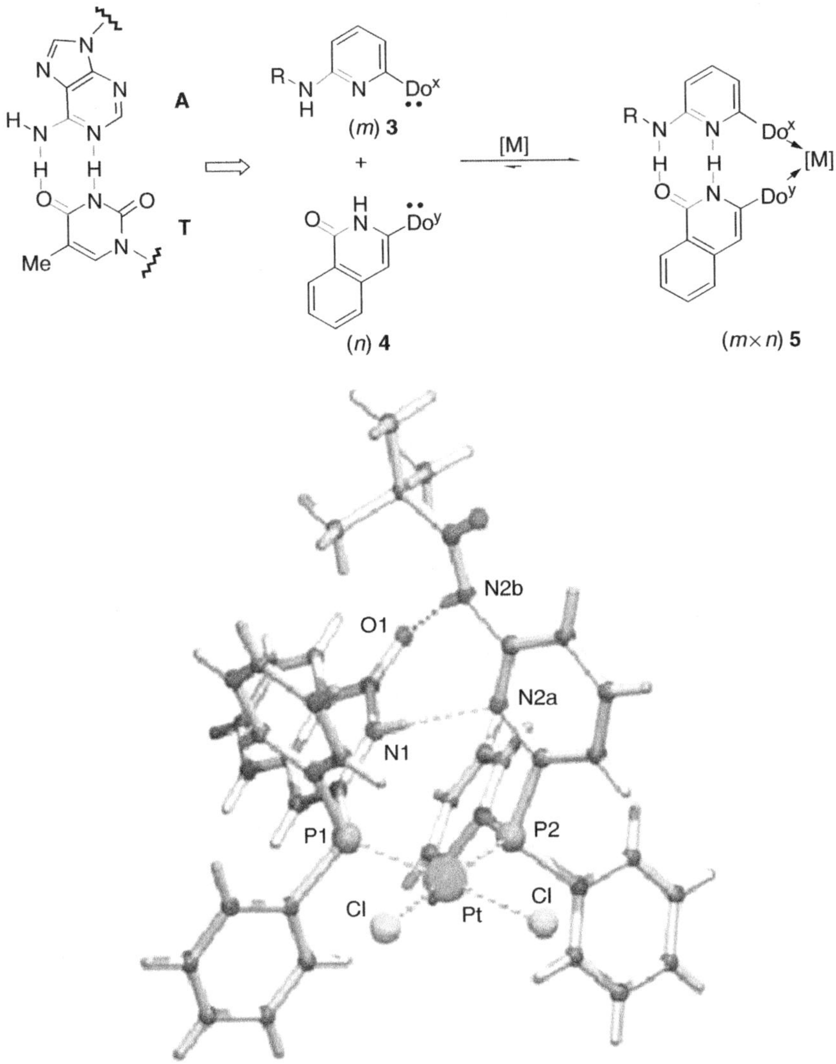

Scheme 15.3 *Top: Formation of a supramolecular bidentate diphosphine ligand through complementary self-assembly of two monophosphine fragments. Bottom: Molecular structure of $PtCl_2$ complex. Reprinted from L. Pignataro, et al., 2010, with permission of John Wiley & Sons.*

in such a way that bidentate coordination to a metal ion is possible, with both phosphines in a mutual *cis* fashion. This has been demonstrated by X-ray analysis for Pt^{II} complexes.

These systems proved to give high activities and remarkable selectivities in several types of reactions, including the Rh-catalyzed hydroformylation of alkenes [8]. Regioselectivities reminiscent of traditional bidentate ligand systems such as Xantphos were achieved with these self-assembled ligands, while the complexes formed in situ by just adding the monodentate building blocks. It was demonstrated that the selectivity for the linear aldehyde was retained up to 100°C, indicating that the bidentate character of the supramolecular assembly remains intact at elevated temperatures. In addition, a wide range of solvents

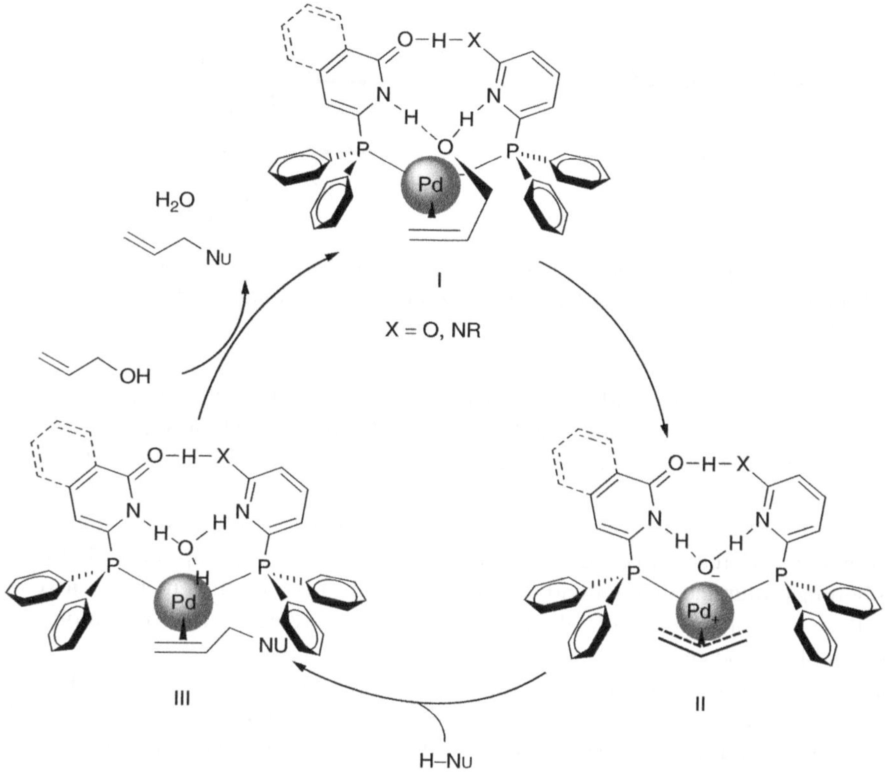

Scheme 15.4 *Proposed catalytic cycle for Pd-catalyzed amination of allyl alcohol via a ligand-assisted bifunctional mechanism. Reprinted from B. Breit, et al., 2003, with permission of the American Chemical Society.*

has been explored, and even in protic solvents the performance of the catalyst indicated that the bidentate character of the ligand was retained. The transformation to practical organic synthesis was demonstrated by performing the reaction at room temperature, at ambient pressure converting a wide range of substrates.

A recent spectroscopic and computational study on the 6-diphenylphosphanylpyridin-2(*1*H)-one system supported the hypothesis that H bonding is a deciding and directing factor during coordination of these formally monodentate ligands and that the H bonds remain intact under catalytic conditions [9].

In follow-up work, Breit and coworkers also developed very elegant and highly efficient supramolecular bidentate ligand classes inspired by the adenosine–tyrosine base pairing present in DNA nucleotides (Scheme 15.3) [10,11]. In this manner, the natural self-assembly mechanisms have been translated to synthetic supramolecular systems. From a small library of building blocks, a series of supramolecular bidentate ligands was evaluated in hydroformylation reactions, and it was demonstrated that activity and selectivity could be easily optimized to levels beyond those displayed by traditional bidentate ligands. Remarkably, Rh-catalyzed hydroformylation was shown to be possible even in MeOH, which is typically believed to be a detrimental solvent for H-bonding interactions due to its competing nature, leading to breakdown of the supramolecular structure [12]. Similar high reactivities and selectivities, thanks to bidentate coordination of two monodentate ligands, were observed in the Ni-catalyzed hydrocyanation of alkenes [13]. The designed H-bonding interactions, besides enabling chelate coordination of two monodentate ligand fragments, also proved very efficient to orient and pre-activate functionalized substrates such as allylic alcohols, which opened up new reactivity for the allylation of *N* heterocycles using Pd complexes (Scheme 15.4) [14].

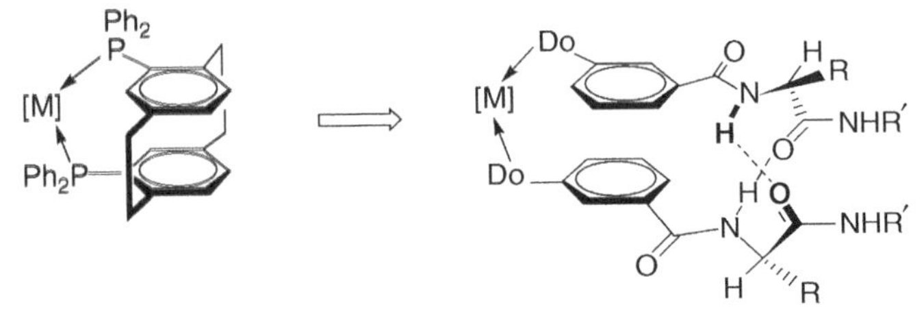

Figure 15.1 *Translation of classic, covalent Phanephos structure, derived from a paracyclophane backbone, to a self-assembled chiral scaffold SupraPhanePhos. Reprinted from A. C. Laungani, et al., 2008, with permission of the Royal Society of Chemistry.*

Libraries of chiral aminopyridine and isoquinoline self-complementary phosphonite ligands were shown to give selective catalysts for the asymmetric hydrogenation of C=C double bonds in acrylate, cinnamate, and itaconate substrates [15]. The exclusive formation of heterobidentate metal complexes allows for novel catalyst-screening approaches. Breit reported a deconvolution strategy: mixtures of catalysts derived from a library of self-assembled bidentate ligands were screened simultaneously in a small number of subgroups, with nine combinations per reactor. By selecting the best performing subgroup for further deconvolution in new smaller subgroups (four and then individual), the best catalyst is identified in an iterative fashion. Using this technique, they were able to identify three highly active and selective hydrogenation catalysts from a library of 120 possible ligand combinations in just 17 experiments [16].

These ligand systems were also applied in the Ru-catalyzed hydration of nitriles [17], and they were proposed to involve the intimate interplay of both the self-assembled diphosphine ligand as well as substrate preorientation and activation using the H-bonding motifs present in the ligand scaffolds. This is one of the examples in which the functional group of a ligand plays a dual role: self-assembly to form the bidentate and activation of the substrate.

Selective heterocombinations of peptidyl-functionalized phosphines and phosphites could be used to mimic the well-known rigid Phanephos structure – derived from the rigid cyclophane backbone – using H bonding as directional tool. This scaffold, dubbed SupraPhanePhos (Figure 15.1), was successfully applied in the Rh-catalyzed asymmetric hydrogenation of benchmark substrates with moderate to good enantioselectivities [18]. Note that the cyclophane structure itself exhibits planar chirality, while the chirality of this supramolecular analogue is primarily induced by chiral auxiliary groups of the peptidyl phosphite fragment, and planar chirality is inferred mainly from the helical arrangement of the peptide side chains, as evidenced by X-ray crystallographic analysis (Figure 15.2).

Our own group has developed several approaches based on hydrogen bonding for supramolecular bidentate ligands. We have developed monodentate phosphorus ligands with appended urea moieties that are capable of forming self-complementary H bond interactions [19]. These systems allow the formation of chelating diphosphine structures in a supramolecular fashion, as we have demonstrated for both Pd and Rh complexes. The rhodium complexes were applied in asymmetric hydrogenation of alkenes. Initial results show that the chirality in the bridge of UREAPhos has an important effect on the outcome of the reaction, which is interesting as it provides a tool for generating a library with large diversity. In the subsequent work in this area, it was demonstrated that these ligands could be prepared in an automated fashion (using a ChemSpeed robot), facilitating the rapid preparation of these ligands. In combination with the supramolecular assembly approach, this easy preparation of building blocks provides a powerful strategy for combinatorial catalysis.

The coordination chemistry to various rhodium precursors was also studied in detail. Interestingly, upon reaction of two equivalents of a simple ligand building block (i.e., $Ph_2PCH_2CH_2NHC(O)NHPh$ (**L1**)) with the precursor $Rh(acac)(CO)_2$, an ABX pattern was observed in the ^{31}P NMR spectrum, indicating the presence of two inequivalent phosphorus atoms coordinated to the Rh center. This was corroborated by an X-ray

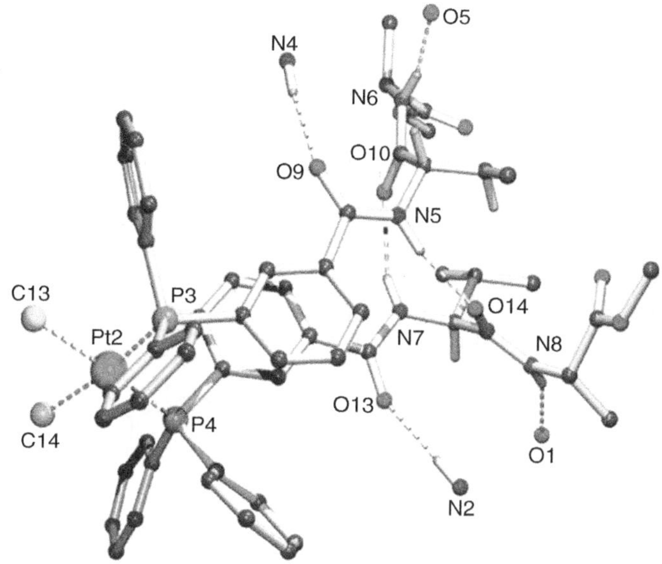

Figure 15.2 *Molecular structure of $PtCl_2$-complex featuring a supramolecular diphosphine comprised of two peptide-based monophosphines. Reprinted from A. C. Laungani, et al., 2008, with permission of the Royal Society of Chemistry.*

crystallographic study, showing that the N atom of one urea group was deprotonated by the acac fragment and coordinates to the Rh, whereas the second urea group is hydrogen bonded to this ligand (Scheme 15.5 and Figure 15.3). However, upon addition of two equivalents of this same ligand to the cationic rhodium precursor $[Rh(nbd)_2]BF_4$ (nbd=2,5-norbornadiene), only one doublet could be discerned by ^{31}P NMR spectroscopy ($^1J_{RhP}$=152 Hz), indicating the formation of a bis-ligated complex with both equivalent P atoms in a *cis* geometry. The urea protons in this complex are involved in hydrogen bonding, according to the chemical shifts observed in the corresponding ^1H NMR spectrum. Use of an equimolar ratio of ligand to cationic Rh resulted in selective P–O coordination of the ureaphosphine ligand, whereas with neutral Rh precursors the ligand becomes deprotonated to induce P–N coordination. Hence, this new kind of ligand structure allows for various different types of coordination, which significantly increases the potential applications in catalysis. Furthermore, it has been shown that this is a very powerful tool to construct ligand libraries because of the modular nature of the assembling components. We subsequently demonstrated that these various coordination modes translate nicely in different catalytic behavior and showed the ureaphosphine to act as an adaptive ligand scaffold [20].

The group of Breit used a related strategy of phosphine side group functionalization to induce H bonding by preparing peptide-appended phosphines that could aggregate and form self-assembled bidentate ligands [21]. Both self-complementary homo- and cross-complementary heteroligated systems were obtained under the proper conditions and using Rh as metal template. These systems have been evaluated in the asymmetric hydroformylation of styrene, showing moderate activity. Furthermore, a combination of weak and strong H-bonding interactions for the metal-induced construction of a self-assembled β-turn mimic was examined (Figure 15.4) [22].

In a search for simple and accessible building blocks, we presented LEUPhos as a new ligand building block which was synthesized by a simple condensation reaction between enantiopure (*S*)-2,2′-dihydroxy-1,1′-binaphthyl ((*S*)-binol) and PCl_3, followed by the addition of L-leucine methyl ester [23]. We subsequently showed that mixing this functionalized phosphoramidite ligand – bearing a leucine amino acid as the N terminus – with an urea-appended phosphine, and in the presence of $[Rh(cod)_2]BF_4$ as the Rh precursor, leads to the exclusive formation of the corresponding heterocomplex according to ^{31}P NMR spectroscopy,

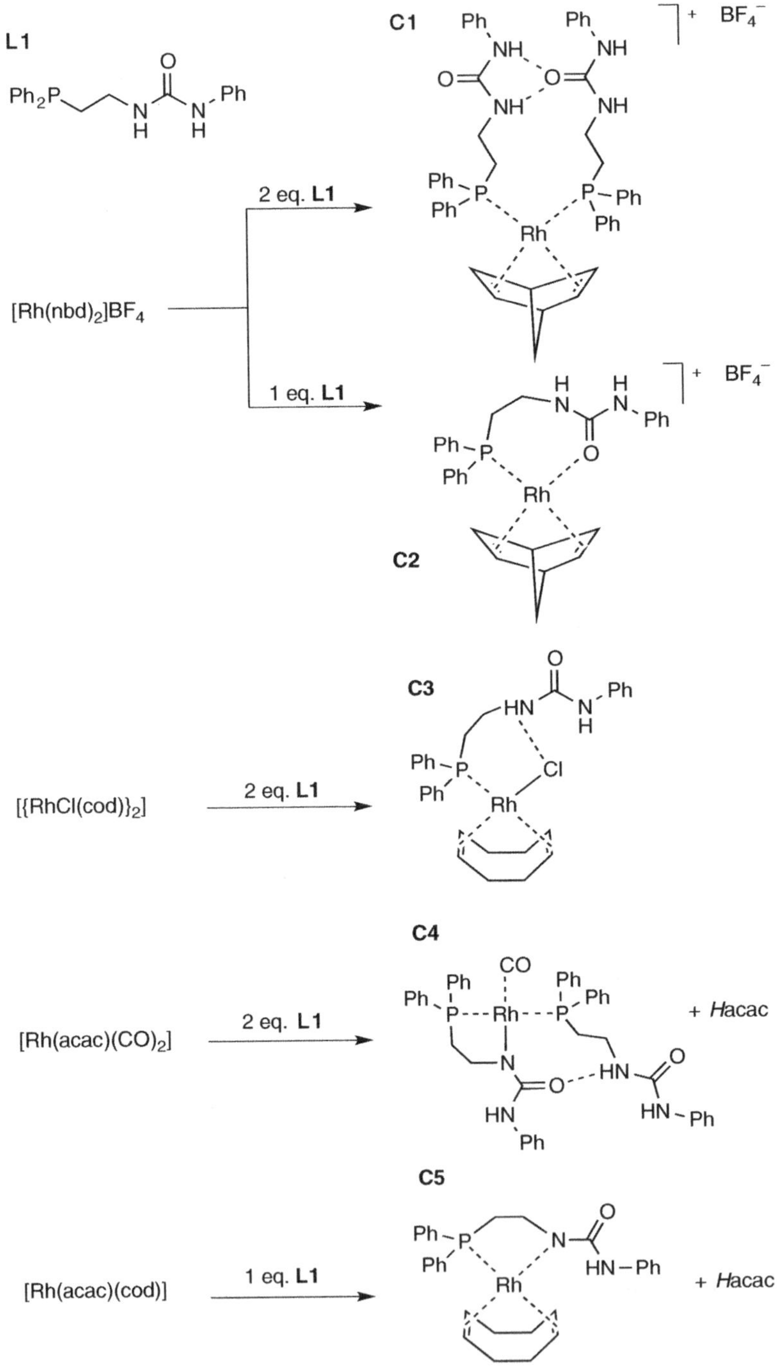

Scheme 15.5 *Diverse coordination behaviour of Ureaphosphine **L1** with rhodium.*

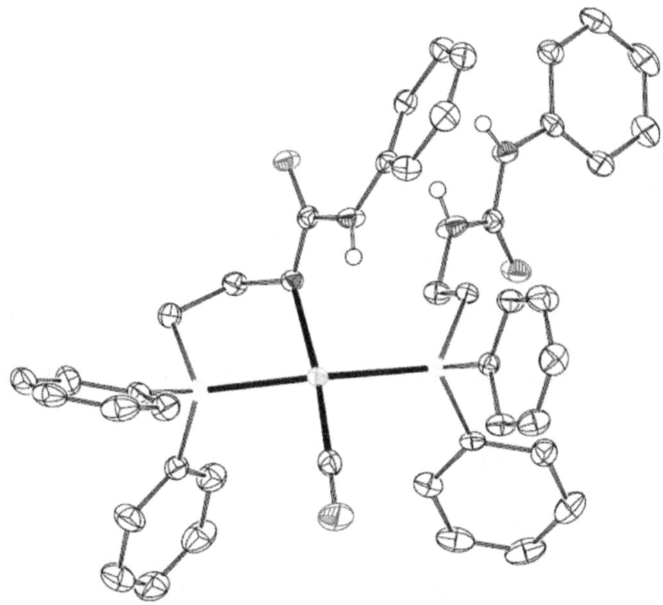

Figure 15.3 Molecular structure of Rh(**L1**)(CO).

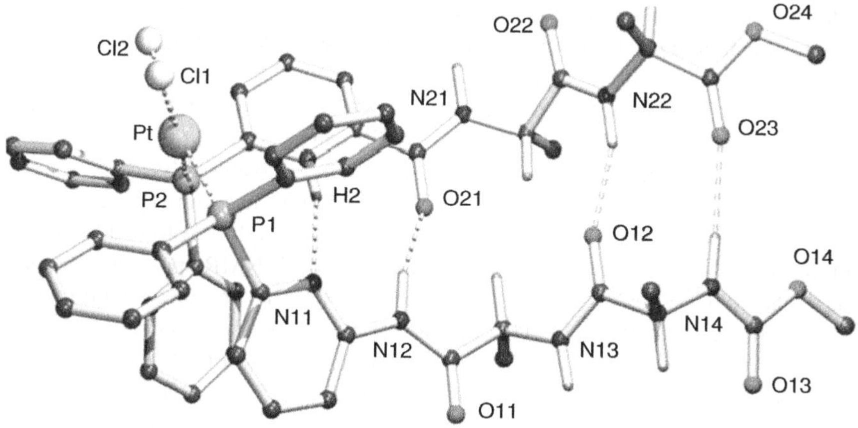

Figure 15.4 Overview of ligand library based on chiral peptides. Reprinted from A. C. Laungani, et al., 2008, with permission of John Wiley & Sons Ltd.

due to the favorable interactions between the H bond donor (carboxylic acid) and the H bond acceptor (urea C=O) (Scheme 15.6). Control experiments with ligands that are not able to form a hydrogen bond between the two building blocks showed formation of homoligated analogues alongside the heterocomplex. The supramolecular heterocomplex was characterized by nuclear magnetic resonance (NMR) and infrared (IR) spectroscopy, and its structure was calculated using density functional theory (DFT) calculations. The computed structures show that the single hydrogen bond interaction is more favorable (by 7.5 kcal mol^{-1}) than the double hydrogen bond interaction between the urea–NH group and the ester carbonyl group. IR spectroscopic measurements confirmed the formation of the hydrogen bond between the P–NH unit and

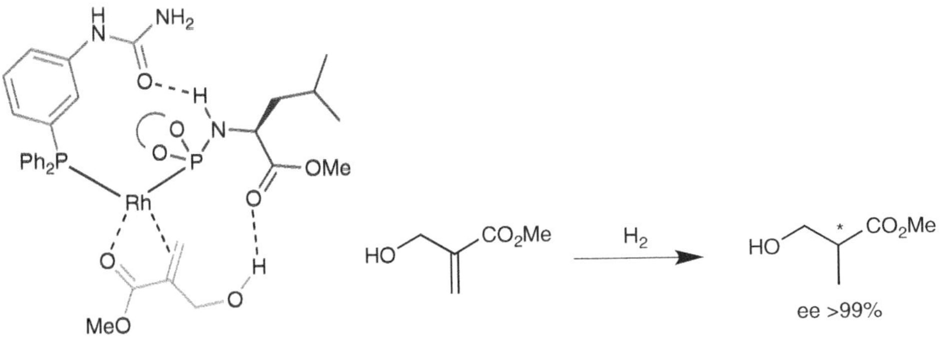

Scheme 15.6 *Supramolecular diphosphorus assembly involving LeuPhos, which also plays a role in substrate activation.*

the urea carbonyl group. Similar experiments with reference compounds such as Monophos, methylated derivatives of LEUPhos, or unfunctionalized phosphines confirm the specific need for H bond interaction to induce heteroligation. This result is interesting as it provides a simple tool to make relatively pure heterocomplexes without using an excess of either ligand. In addition, these ligands will have some bidentate character as the interaction between the two building blocks reduces the conformational freedom in these systems.

Upon introduction of a functionalized alkene substrate (i.e., phenyl acetamido cinnamate or the Roche ester) [24], it could be shown that the H-bonding pattern changes to allow for selective substrate preorganization and coordination. The asymmetric hydrogenation of these substrates was very efficiently catalyzed using this supramolecular ligand assembly, resulting in enantiomeric excess (ee) of up to 99%. Competition experiments with substrates incapable of H bonding showed that these weak bonding interactions are essential to induce high activity and selectivity to the catalyst system.

In search for new binding motifs, we prepared sulfonamide-based monophosphorus ligands (both phosphines or phosphoramidites), coined METAMORPhos, which were prepared by a simple condensation reaction between appropriate sulfonamides and chlorophosphines or phosphochloridites. These ligands were initially designed as novel ligands with a hydrogen bond motif (donor–acceptor type) close to the phosphorus ligand [25]. Upon closer inspection of the NMR signature of the diphenylphosphine derivative **X**, we realized that two different tautomers exist in $CDCl_3$, as is evident from the ^{31}P NMR spectrum, that slowly exchange on the NMR time scale (Scheme 15.7). The two peaks δ 33.98 and 6.67 ppm, for tautomeric forms **Xa** and **Xb**, respectively, show distinctly different P–H couplings of 7 and 490 Hz, respectively, indicative of the P^{III} and HP^{V} states. No coalescence was observed in the temperature window of 323 to 223 K; however, the ratio between **Xa** and **Xb** changed dramatically from ~4.4 at 323 K to ~0.85 at 223 K.

Upon addition of two equivalents of METAMORPhos **X** to $Rh(acac)(CO)_2$, rather than the formation of a typical *cis*-diphosphorus complex, an unusual AB pattern was observed in the ^{31}P NMR spectrum, indicative of the presence of two different phosphorus ligands. Furthermore, the large P–P coupling constant ^2J of 335 Hz is a strong indication for a mutual *trans* orientation. Combined with the observation that one equivalent of Hacac is produced and the presence of only one CO band in the IR spectrum, we postulated that a POP supramolecular pincer ligated Rh complex is formed. One of the METAMORPhos ligands is deprotonated (by $acac^-$) *and* has mesomerized to create a P–N=S–O^- fragment X^-, with the phosphorus and the oxido group coordinating to Rh. The second, neutral METAMORPhos ligand binds only via the phosphorus atom, while the –NH group acts as hydrogen bond donor to the oxido fragment to create the pseudo-pincer arrangement. This hypothesis was supported by solution NMR and IR spectroscopy and the structure was calculated with DFT. We have also prepared chiral ligand **Y**, which showed only one signal in the ^{31}P NMR spectrum, consistent with its existence in only one tautomeric form, due to the lower basicity of the phosphorus atom.

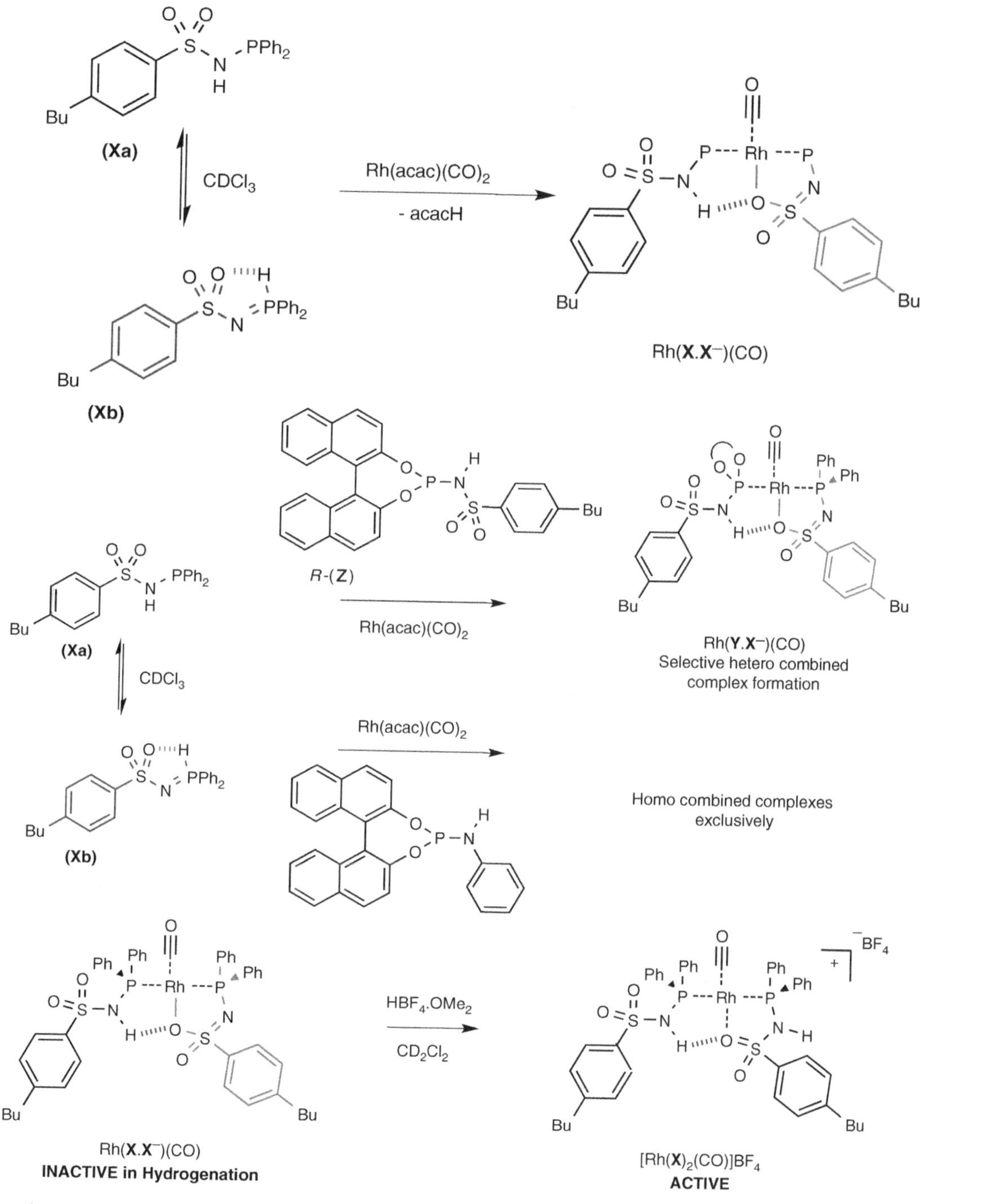

Scheme 15.7 *Coordination chemistry and adaptive character of METAMORPhos ligand* ***X*** *and formation of hetero- and homocomplexes.*

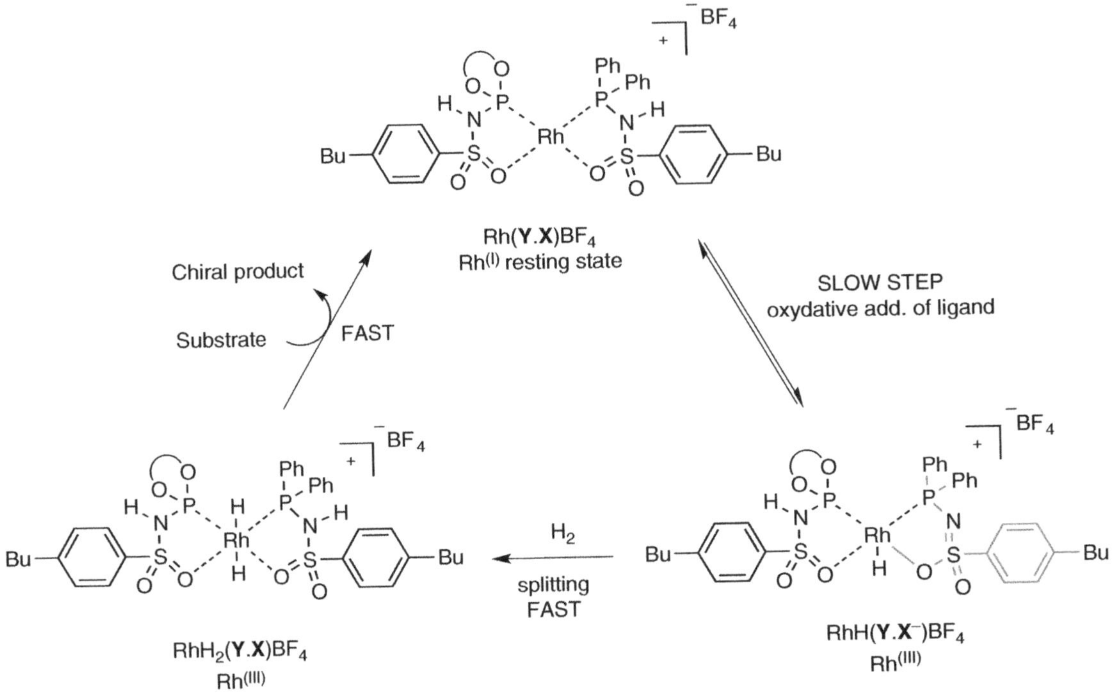

Scheme 15.8 *Proposed catalytic cycle for asymmetric hydrogenation via a ligand-assisted H_2 activation step.*

Upon mixing these two types of METAMORPhos ligands **X** and **Y**, we succeeded in selectively obtaining the heterocomplex, owing to the two distinct features necessary for formation of these species (i.e., a ligand–hydrogen bond acceptor in the anionic form and a hydrogen bond donor in the neutral state). As the organic rest group on the sulfonamide group is identical in both cases, the relative NH acidity of the aminophosphine versus the phosphoramidite is key to this remarkable synthetic selectivity.

We were interested in determining whether the typical METAMORPhos coordination would also result in new catalytic properties for its rhodium complexes. As expected, the neutral Rh^{I} complexes **A** and **B** were found to be completely inactive for the Rh-catalyzed asymmetric hydrogenation of methyl 2-acetamido acrylate (MAA). Interestingly, these neutral complexes could be transformed into their cationic analogs by simple protonation of the ligand with $HBF_4{\cdot}OMe_2$, which leads to active complexes for the hydrogenation reaction (Scheme 15.7). Alternatively, the catalytically relevant cationic species could be generated by coordination of the respective ligand set to $[Rh(nbd)_2]BF_4$.

Applying these novel systems in the Rh-catalyzed hydrogenation of functionalized alkenes, we established that the different catalysts all showed different behavior. Strikingly, the most active heteroligated rhodium complex, based on ligands **X** and **Y**, showed very unusual kinetic behavior, with zero order in both substrate and H_2; this is unprecedented to the best of our knowledge. Based on the cooperative coordination behavior of these sulfonamide–phosphorus ligands and the observed kinetics described in this chapter, we have proposed a new mechanism for the asymmetric hydrogenation that is specific for this particular METAMORPhos catalyst system (Scheme 15.8). As the reaction rate only depends on the catalyst concentration, the rate-determining step in this mechanism must be intramolecular, involving only the Rh complex, which suggests that "oxidative addition" (proton transfer) of the neutral, most acidic METAMORPhos ligand **X** onto the Rh center is the essential step. This reactive cationic Rh^{III} hydride species rapidly splits H_2 in a heterolytic fashion, with ligand **X** acting as base, to yield the corresponding cationic Rh^{III} dihydride species. Subsequent decoordination of the oxygen atoms of the ligands creates the vacant sites required for alkene substrate coordination, while the cationic

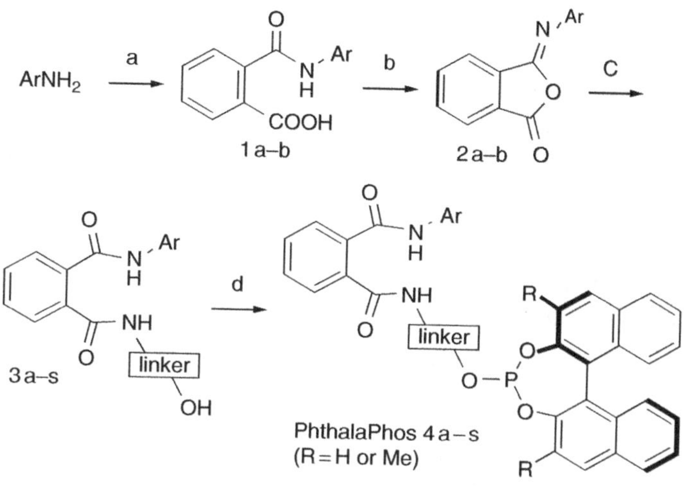

Scheme 15.9 Synthesis of PhthalaPhos. Reprinted from B. Breit, et al., 2005, with permission of John Wiley & Sons Ltd.

Rh^{I} resting state is regenerated after hydride migration and reductive elimination of the product. Heterolytic splitting of molecular hydrogen could also take place after substrate coordination and the first migration reaction, but because these steps all occur after the rate-determining step we cannot distinguish between these different sequences on the basis of our kinetic data. This intramolecular oxidative addition of METAMORPhos to the rhodium center, combined with ligand adaptivity to activate H_2, is unique and could become important for other types of reactions as well. An alternative explanation of the unusual kinetic behavior is that the reductive elimination is the rate-determining step of the reaction, which would be very unique for this reaction.

We have since showed that METAMORPhos ligands can also act as monoanionic bridging ligands, displaying other kinds of remarkable coordination behavior, but as this is not supramolecular in nature, we have chosen not to discuss this in detail in this chapter [26].

Related to these H-bonding approaches, the group of Gennari recently published a family of PhthalaPhos ligands that are also amenable to H bonding due to the presence of phthalic acid side groups on phosphite frameworks (Scheme 15.9) [27]. Supramolecular self-assembly was observed during Rh coordination, and this also resulted in the highly efficient asymmetric hydrogenation of acetamide and acrylate substrates. Similar attempts using combinations of acid- and base-appended phosphites did not show the same level of selective heterocomplex formation [28]. The synthesis of these ligands follows a straightforward route starting from the desired aniline, which is converted to the hydroxyl-functionalized phthalamide precursor in three steps prior to introduction of the chiral phosphite unit.

Similar as proposed by our group, the two functionalized phosphite ligands were suggested to self-assemble thanks to the presence of H-bond donor and acceptor moieties in the phthalamide groups (Figure 15.5). Good results in the asymmetric hydrogenation of benchmark substrates were obtained, with high ee. The role of hydrogen bonding was assessed by NMR and IR spectroscopic investigations of a representative substrate complex, and the results corroborated by DFT calculations. It was postulated that self-assembly of these monodentate ligands led to a reduced degree of conformational freedom for the substrate, which might explain the high stereoselectivities observed. This stimulated additional catalytic screening for the hydrogenation of acetamides and acrylates.

The group of Ding has also reported on supramolecular bidentate phosphorus ligands using H bond motifs between the NH groups of two phosphoramidite ligands to create a supramolecular bidentate ligand that proved very effective, active, and selective in the asymmetric hydrogenation of β-aryl itaconic acid

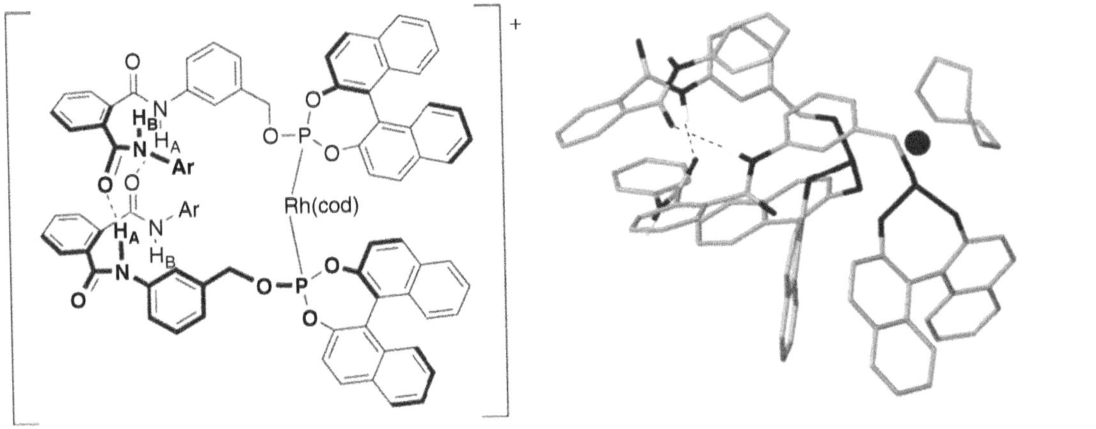

Figure 15.5 Coordination of Rh-complex of PhthalaPhos and the molecular structure of this species. Reprinted from B. Breit, et al., 2005, with permission of John Wiley & Sons Ltd.

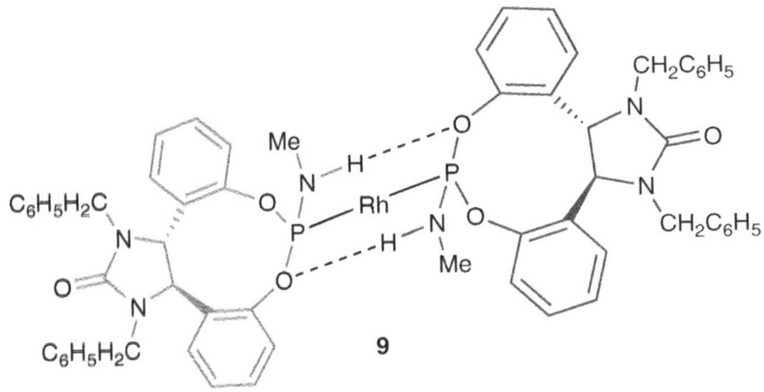

9

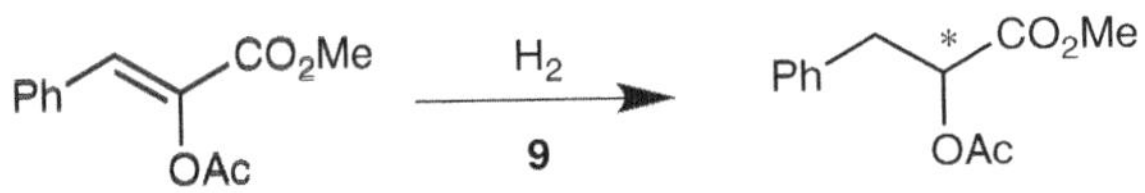

Scheme 15.10 Supramolecular interactions between two phosphoramidite ligands in a Rh complex.

derivatives (Scheme 15.10) [29]. The analogous unfunctionalized monodentate derivatives, lacking any H-bonding motif, proved inactive.

Fan and coworkers synthesized a phosphite ligand functionalized with a crown ether, which is capable of forming a pseudo-rotaxane assembly by means of ionic- and hydrogen-bonding interactions with the ammonium group of a modified phosphine ligand [30]. This assembly proved to be active and selective in the asymmetric hydrogenation reaction of α-dehydroamino acid esters.

15.2.2 Metal template assembly

Dative coordination bonds between a σ-donor and a metal-based acceptor system, although stronger than typical H bond interactions, can still be classified as weak when compared with traditional organometallic, ionic, or covalent interactions that typify the toolbox of synthetic chemists. Porphyrins are currently used as

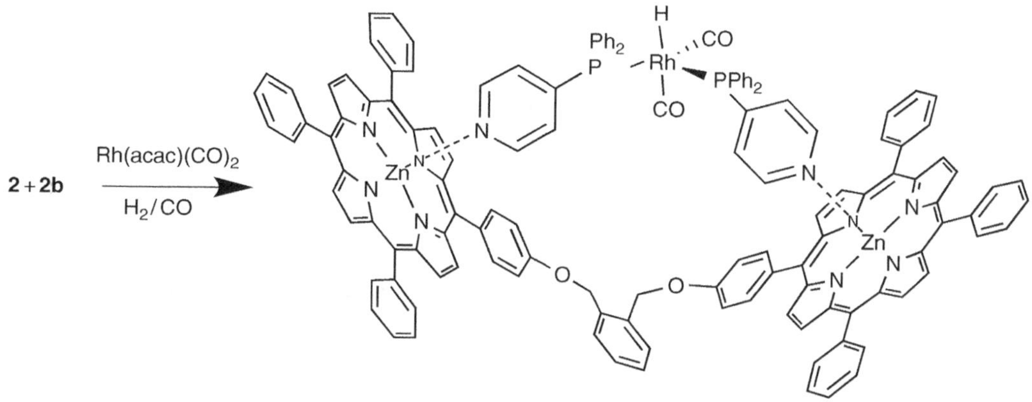

Scheme 15.11 *Metal-templated formation of a diphosphine ligand.*

multipurpose components in many different types of supramolecular assemblies, and their photochemical and catalytic properties generally play a crucial role. In particular, the pyridine–M^{II}–porphyrin (M = Zn, Ru, Co) motif has been extensively exploited. The association of nitrogen ligands with porphyrins has been studied in great detail, and currently pyridines and cyclic amines such as 1,4-diazabicyclo[2.2.2]octane (dabco) are most widely used as components of functional supramolecular assemblies, with association constants between 10^3 and 10^5. Using these coordination interactions, it is also feasible to generate bidentate ligands by mixing amine- or pyridine-based ligands with porphyrin-functionalized ligand building blocks or bisporphyrin templates, which we started to explore when we entered this field.

The use of template ligands – that is, molecules that have the ability to form moderately strong coordination complexes to, for example, zinc(II)–porphyrins via specific nitrogen donor atoms whilst also bearing soft-donor atoms (e.g., phosphines) that do not interact with the axial site at Zn but that can coordinate to other transition metal fragments – has allowed for new and modular strategies to form supramolecular bidentate ligands as well as encapsulated transition metal catalysts. Herein, we will focus on the former aspect; the interested reader is referred to recent reports for the design and application of these sterically encumbered monophosphorus ligand systems and the corresponding implications for catalysis [31].

Recently, a new class of bidentate ligands, resulting from the self-assembly of two monodentate ligands, was introduced. Specific zinc(II) porphyrin–pyridine interactions were used as templating scaffold and as assembly motif to generate bidentate phosphorus-based ligands. The first report dealt with the use of a bisporphyrin, featuring a flexible bis(benzyloxy)ether linkage, acting as scaffold for the supramolecular anchoring of two pyridylphosphine fragments (Scheme 15.11) [32]. The highly directional Zn–N bond, and the V-shaped bisporphyrin platform, lead to the possibility of mutual *cis*-coordination of both phosphines to a Rh center, effectively forming a supramolecular bidentate ligand. These supramolecular ligands were evaluated in the Rh-catalyzed hydroformylation of terminal alkene and styrene.The system was shown to be active and it was demonstrated that it displayed bidentate behavior, although selectivities and activities were only moderate. Later, the same concept was shown to work for Sn-porphyrins and carboxylic acid–functionalized phosphorus ligands [33].

Related approaches have also been developed to generate bidentate ligands from monodentate scaffolds using metal templation. For instance, tris(zinc(II) porphyrin)phosphite ligands are accessible by reaction of PCl_3 with three equivalents of the desired Zn(II) porphyrin, for example 5-(3-hydroxyphenyl)-10,15,20-tris(phenyl)-zinc(II) porphyrin. The resulting monomeric phosphite ligand thus contains three macrocyclic Zn-units that can function as anchoring points for the formation of multicomponent assemblies [34].

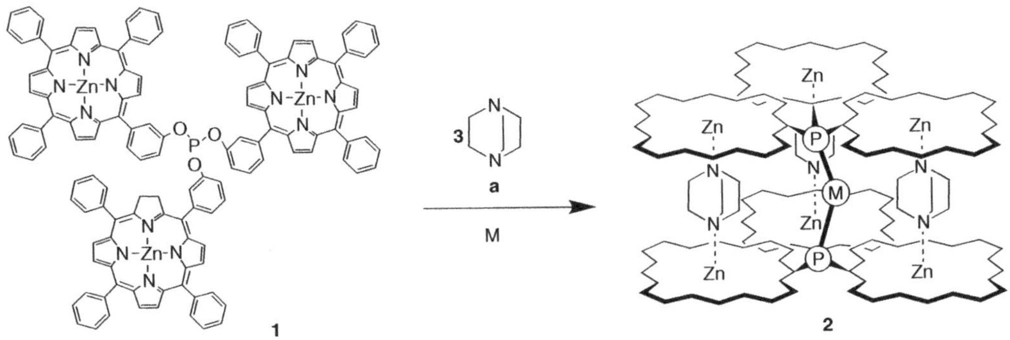

Scheme 15.12 Formation of a supramolecular diphosphite ligand through multiple Zn–amine interactions.

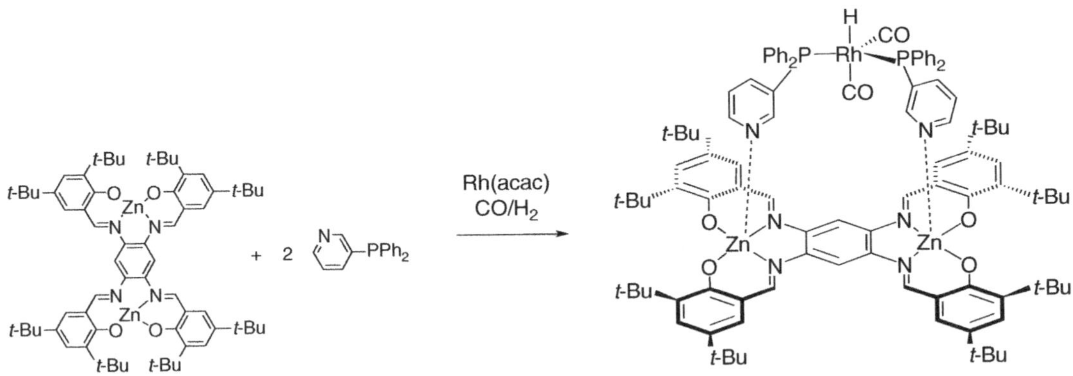

Scheme 15.13 Salphen building blocks used as a template for the formation of a supramolecular diphosphine ligand.

To restrict fluxional flexibility of the system and to tune the bite angle of the resulting ligand system, we employed rigid ditopic struts to link the supramolecular units. For example, 1,4-diazabicyclo[2.2.2]octane (dabco) can be used as templating ligand, since it has been shown previously to form strong, well-defined 1:2 complexes with zinc(II) porphyrins (Scheme 15.12) [35].

The bidentate ligand system held together by three bridging ditopically coordinated template molecules showed high regioselectivity and very high activity, typical for phosphite-based ligands – in the rhodium-catalyzed hydroformylation of 1-octene. Interestingly, three dabco templates were required to create selective catalysts, probably to enforce rigidity in the supramolecular ligand.

Using the same type of templating interaction, but employing pyridyl-functionalized monotopic phosphorus units and bis-salphen frameworks as an anchoring platform, we demonstrated the construction of other kinds of supramolecular bidentate phosphorus ligands (Scheme 15.13). Judicious choice of the templating and ligating scaffolds allows for tuning of the ligand "bite angle" [36]. Interestingly, mixing bulky phosphoramidite building blocks and phosphine building blocks resulted in selective heterocomplex formation, only in the presence of the bis–zinc(II) salphen template. In absence of such template, only homocomplexes were formed when Rh(acac)$(CO)_2$ was used as a metal precursor. The templated heteroligands induced relatively high ee in the asymmetric hydroformylation of styrene, although the activities were rather low [37].

A class of supramolecular ligands, termed Supraphos,was developed, and the supramolecular formation is based on the same zinc(II)–nitrogen interaction. These ligands consist of two components – a

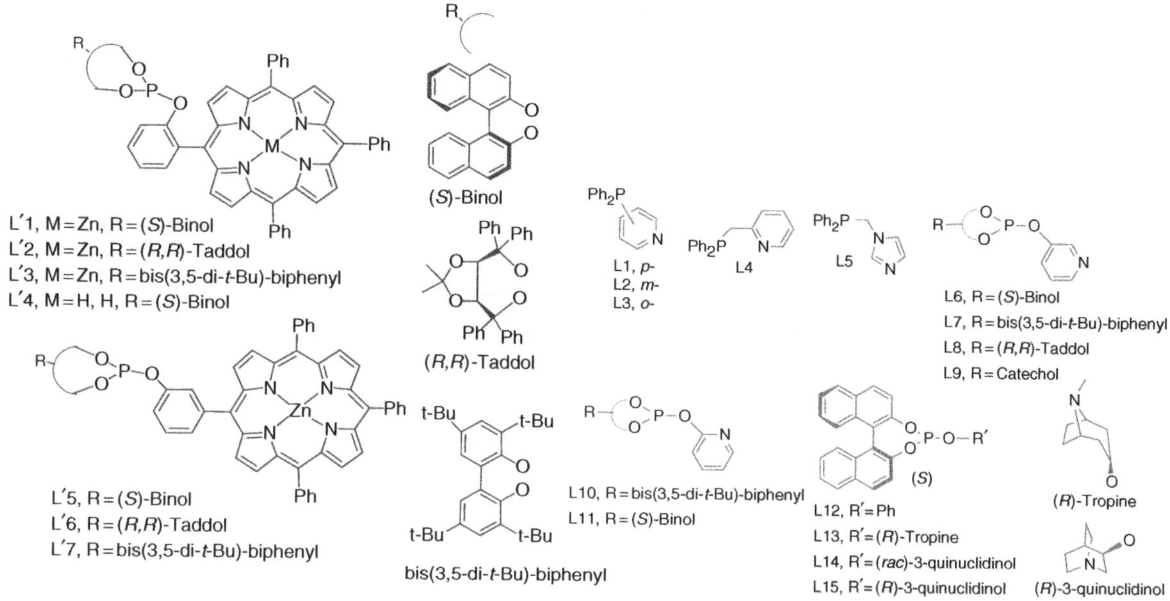

Figure 15.6 *Diol and phosphine building blocks for a combinatorial library of supramolecular bidentate phosphorus ligands.*

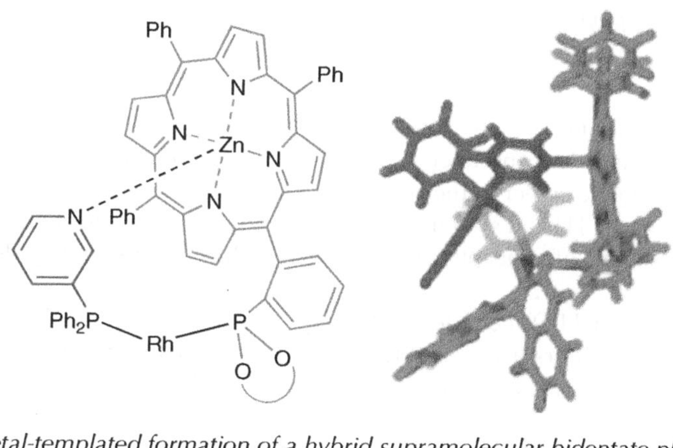

Figure 15.7 *Metal-templated formation of a hybrid supramolecular bidentate phosphorus ligand.*

pyridylphosphine and a porphyrin-appended phosphite or phosphoramidite – that form bidentate ligands via selective and reversible coordination of the pyridyl nitrogen donor atom to the zinc atom of the porphyrin (Figures 15.6 and 15.7). Importantly, the phosphine does not coordinate to the zinc, providing a strict orthogonal interaction between the supramolecular template and the phosphorus unit. These SUPRAPhos ligands have been used as chelating ligands for palladium and rhodium [38]. We demonstrated that based on 14 building blocks, a library of 48 (8×6) bidentate ligands could be formed by just mixing the different building blocks. The thus-formed catalyst gave different outcomes in the asymmetric allylic alkylation reaction, providing first proof of principle of the strategy. This stimulated us to increase the library of building blocks to create 400 supramolecular bidentate ligands with around 40 building blocks [39]. A part of this

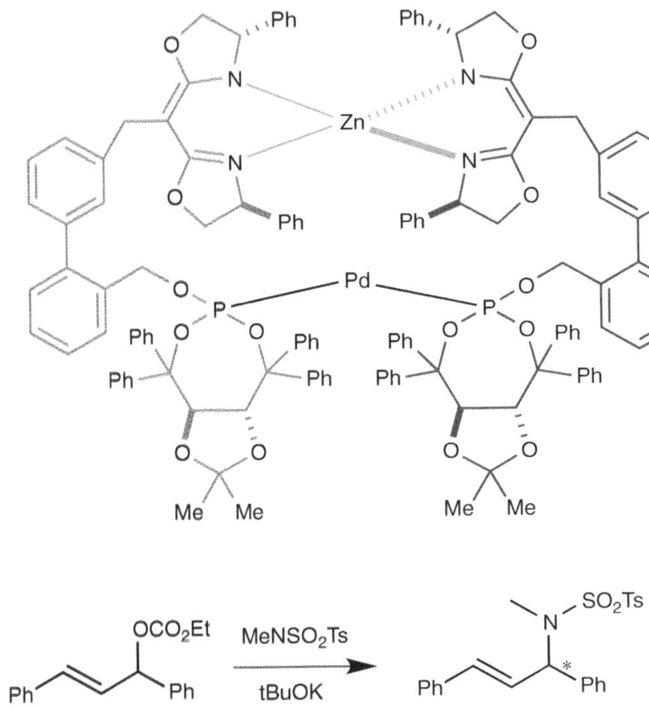

Scheme 15.14 Second-sphere supramolecular interactions to generate a bidentate phosphorus ligand.

library was screened for the asymmetric hydrogenation of a challenging cyclic enamide substrate, and one of the ligand members produced the product in 100% conversion and with the highest ee ever reported (94%), clearly demonstrating the power of the concept [40]. Later we also identified palladium catalysts for the dynamic kinetic resolution [41] and systems for the regioselective hydroformylation of styrene [42], but further exploration of these type of concepts was continued with cheaper building blocks that could lead to commercial applications [43].

Another early player in the field of metal-templated supramolecular bidentate ligation has been the group of Takacs. The approach taken differs from the others described here in that bifunctional ligands were utilized that consisted of a bidentate donor site for a templating, catalytically inactive metal (e.g., Zn^{II}) and a monodentate phosphorus group (most often a phosphite), connected via a spacer that can be varied in a modular fashion (Scheme 15.14). These monodentate ligand systems can be turned into chelating diphosphorus ligands, capable of coordinating to catalytically relevant metal centers, by selective formation of a heteroleptic assembly around a Zn(II) center of the secondary groups. The modular approach allowed them to prepare large combinations of different bisoxazoline-functionalized ligands. Interestingly, the donor atoms did not change in this approach, but were only fixed in space differently by changing the spacer and the backbone. In the first instance, this strategy was used in the palladium-catalyzed allylic amination, involving a library of 50 different catalysts, which led to the discovery of very selective catalysts [44]. Next, it was demonstrated that the same ligand library provided selective catalysts for the rhodium-catalyzed asymmetric hydrogenation [45] as well as for the rhodium-catalyzed asymmetric hydroboration [46].

The group of Gudat contributed to the field of supramolecular bidentate ligands by designing catechol-functionalized phosphine ligands [47]. The presence of both soft (P) and hard (O^- when deprotonated) donor

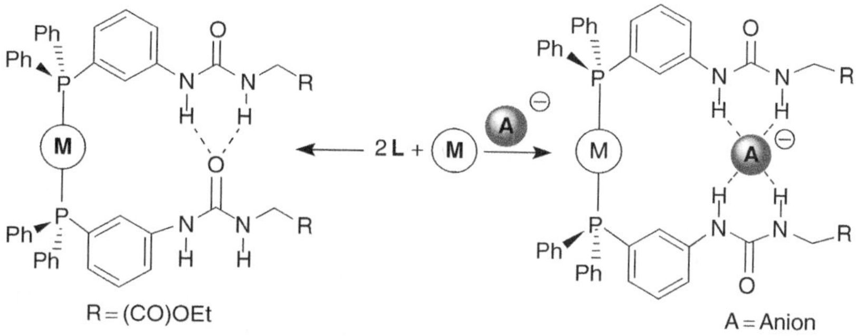

Scheme 15.15 Ion templation as a means to generate bidentate phosphorus ligands.

groups allows the targeted assembly of bidentate phosphorus ligands. Using main-group elements such as Ga or Zn as templating units, coordinating to the two catecholate moieties, creates a pocket.

15.2.3 Ion templation

We [48] and the group of Love [49] have independently introduced novel phosphine ligands containing appended urea groups. These self-complementary hydrogen bond motifs enable the formation of supramolecular bidentate ligands in a similar fashion as previously described for both H-bonding and metal templation concepts. Urea-based phosphine ligand (m-$Ph_2PC_6H_4NH(CO)NHCH_2(CO)OEt$) was synthesized by converting m-fluoroaniline into m-(diphenylphosphino)aniline with $KPPh_2$. Subsequent reaction with ethyl isocyanatoacetate resulted in the urea-appended phosphine–ligand. Two equivalents of this ligand *P* were added to CODPdMeCl to generate the metal complex $PdMeCl(P)_2$, for which a singlet was observed at 32.7 ppm in the $^{31}P\{^1H\}$ NMR spectrum, indicating that the phosphine ligands are coordinated in a mutual *trans* configuration to Pd. In addition, a $^3J_{P\text{-}H}$ value of 6.0 Hz was observed for the Pd–CH_3 resonance. Infrared spectroscopy showed that the urea functional groups are involved in intramolecular hydrogen bonding even at relatively low concentrations (<14 mM), likely due to preorganization of the urea groups by coordination to the palladium metal center. Computational studies (DFT and B3LYP) confirmed that the intramolecular hydrogen-bonded complex is energetically more favorable than the corresponding complex lacking H bonds by 9 kcal/mol. As urea groups are frequently employed to construct receptors for anion recognition, we envisaged that our new functionalized ligand might also bind anions (Scheme 15.15; Figure 15.8). The addition of various equivalents of tetra-n-butylammonium chloride (nBu_4NCl) led to binding, as indicated by monitoring the NH resonances of the urea fragments using ^1H NMR spectroscopy as well as following the changes in the same fragments by IR spectroscopy.With Δν(N–H) values of 113 and 70, respectively, the hydrogen bond between the ureas and the chloride anion of the corresponding bisphosphine–palladium complex is clearly stronger than the self-association of the urea moieties in ligand *P*, likely due to the anionic nature of the chloride. In addition to the increase in strength, the number of hydrogen bonds involved is larger (4 vs. 2) in the anion-templated system, enabling the chloride anion to compete with the intramolecular hydrogen-bonded complex (Scheme 15.16).

The combined data clearly suggest that chloride binds in the bis-urea pocket at the expense of the hydrogen bonds between the urea functionalities,thereby generating a Pd complex with an anion-templated bidentate chelating ligand. Interestingly, the ligands in the resulting complex are also in a *trans* configuration at the metal center, as evident by the singlet that is present in the $^{31}P\{^1H\}$ NMR spectrum. This is the first example of a *trans*-coordinating, anion-templated, hydrogen-bonded phosphine ligand system. Upon addition of a

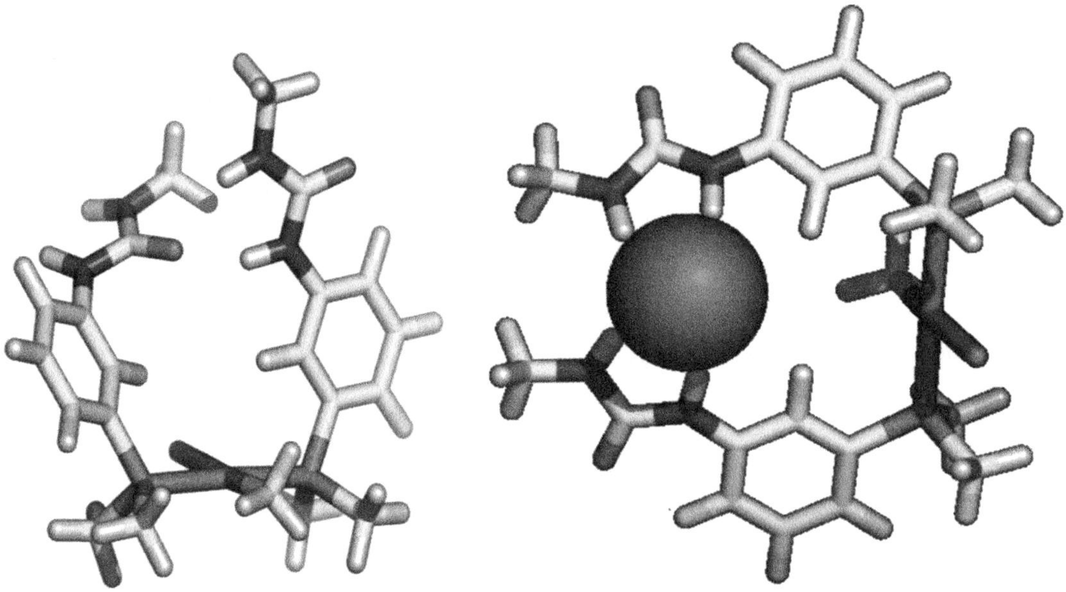

Figure 15.8 Molecular modeling of species depicted in Scheme 15.15.

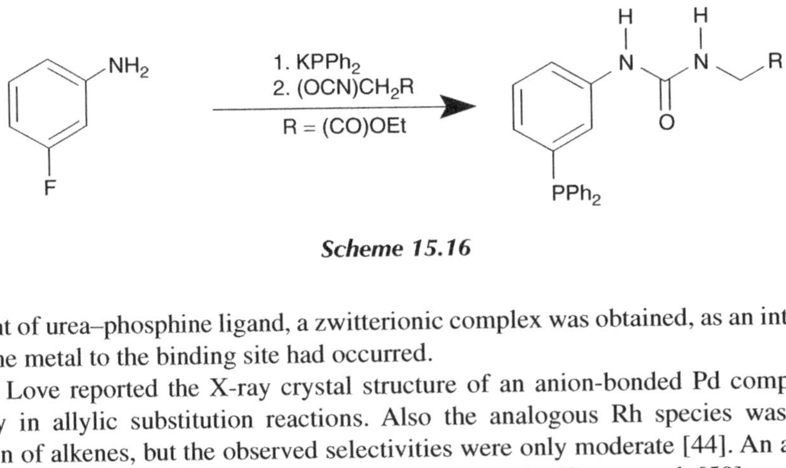

Scheme 15.16

further equivalent of urea–phosphine ligand, a zwitterionic complex was obtained, as an intramolecular chloride shift from the metal to the binding site had occurred.

The group of Love reported the X-ray crystal structure of an anion-bonded Pd complex as well as its catalytic activity in allylic substitution reactions. Also the analogous Rh species was evaluated in the hydroformylation of alkenes, but the observed selectivities were only moderate [44]. An attempt to develop asymmetric induction by using chiral anions did not lead to any significant result [50].

15.3 Conclusions

Supramolecular bidentate phosphorus ligands have recently been introduced as a new class of ligands that form by the self-assembly of ligand building blocks through specific interactions. It has been demonstrated that hydrogen bonds, ionic interactions, and metal–ligand interactions can all facilitate the assembly process. Multicomponent catalyst assemblies are easily accessible and modular by nature, and a limited number of building blocks quickly makes a multitude of combinations possible. It is these properties that make the approach of multicomponent catalyst assemblies ideally suited for combinatorial strategies, and as such the application thereof on a larger scale is just a matter of time. Although still early in its development, already a number of reactions have been covered and several supramolecular catalysts have emerged with properties that surpass those of traditional homogeneous catalysts, showing the strength and potential of such multicomponent catalyst assemblies. Interestingly, the number of supramolecular bidentate ligands grows exponentially with the number of available building blocks, which clearly shows the power of the supramolecular approach; therefore this approach is generally associated with combinatorial routes to rapid catalyst discovery [51]. The dynamic character inherent to the class of supramolecular ligands also brings about new opportunities, including adaptive and responsive catalyst systems, and supramolecular interactions between ligands and substrates may also lead to new reactivity and/or selectivity. These endeavors will be the subject of research in many labs in the next decade.

15.4 Experimental procedures

15.4.1 General remarks

All experiments were performed under a dry, inert atmosphere of nitrogen using standard Schlenk techniques, unless stated otherwise. Solvents were dried and distilled under nitrogen prior to use. Toluene was distilled from sodium, diethylether, and tetrahydrofuran (THF) from sodium–benzophenone and hexanes from sodium–benzophenon–triethyleneglycol dimethyl ether ("triglyme") Dichloromethane, ethylacetate, and acetonitrile were distilled from CaH_2. Dimethylformamide was distilled from $CaSO_4$. NMR spectra were recorded on a Varian Mercury 300 or a Varian Inova 500 spectrometer. $CDCl_3$ was used as a solvent; shifts are given relative to TMS (^1H, $^{13}C\{^1H\}$) and 85% H_3PO_4 ($^{31}P\{^1H\}$). Optical rotations were measured on a Perkin Elmer 241 Polarimeter using a sodium lamp (289 nm), unless stated otherwise. Elemental analyses were performed at Kolbe Mikroanalytisches laboratorium in Mülheim (Germany). High-resolution mass spectra were recorded at the Department of Mass Spectrometry at the University of Amsterdam using FAB^+ ionization on a JEOL JMS SX/SX102A four-sector mass spectrometer with 3-nitrobenzyl alcohol as the matrix.

15.4.2 Synthesis of UREAPhos (Scheme 15.17)

General procedure for preparation of urea-alcohols: A solution of 5 mmol of isocyanate (1 equiv.) 10 ml in DCM was added dropwise to a vigorously stirred solution of 5 mmol amino alcohol (1 equiv.) in 20 ml DCM kept at 0°C. The reaction mixture was allowed to warm up to room temperature and was left stirring for 1 h. The formed precipitate was filtered off and washed twice with hexanes. Remaining solvents were evaporated *in vacuo* to obtain the product almost quantitatively.

General procedure for the preparation of urea-amines: A solution of 5 mmol of isocyanate (1 equiv.) 10 ml in toluene/hexanes (1/1) was added dropwise to a vigorously stirred solution of 10 mmol diamine (2 equiv.) in 20 ml toluene/hexanes (1/1) kept at 0°C. The reaction mixture was allowed to warm up to

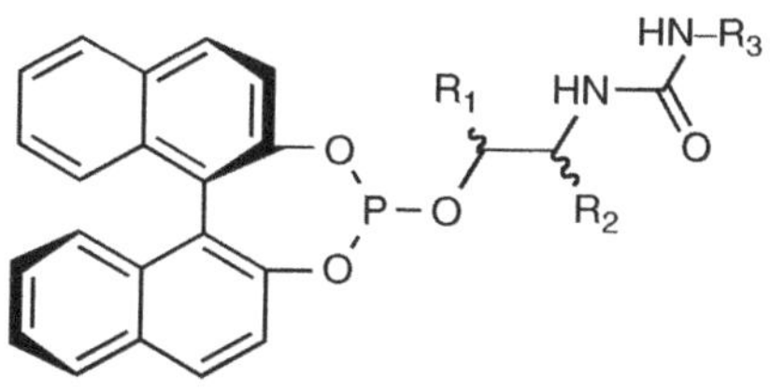

Scheme 15.17

room temperature and was left stirring for 1 h. The formed precipitate was filtered off and washed twice with hexanes. Remaining solvents were evaporated *in vacuo* to obtain the product. The unreacted diamine was recovered from the filtrate.

General procedure for the preparation of urea-phosphoramidite and -phosphite ligands: A solution of 1 mmol phosphorochloridite (1 equiv.) dissolved in 10 ml THF was added dropwise to a vigorously stirred solution of 1 mmol of urea-alcohol or urea-amine (1 equiv., dried via co-evaporation with toluene (3×)) and an excess of triethylamine in 20 ml THF kept at 0°C. The turbid reaction mixture was subsequently warmed up to room temperature and left stirring for 12 h. After filtration over a small layer of neutral alumina oxide and evaporation of solvents, the product was obtained almost quantitatively. If necessary, the product was purified by column chromatography (eluent: DCM/EtOAc 4/1). After evaporation of solvents *in vacuo*, the last traces of EtOAc were removed by co-evaporation with toluene.

General procedure for the preparation of ureaphosphines: A solution of 5 mmol of isocyanate (1 equiv.) in DCM was added dropwise to a vigorously stirred solution of 5 mmol aminophosphine (1 equiv.) in 20 ml DCM kept at 0°C. The reaction mixture was allowed to warm up to room temperature and was left stirring for 2 h. After evaporation of solvents *in vacuo*, the product was obtained almost quantitatively.

Multigram synthesis ligand UreaPhos: In a 3 L flask, 430 g Amberlyst A-21 and 0.505 mol of urea-alcohol **U4** (97.86 g, 1 equiv.) were dried by co-evaporation with toluene (3×). 500 ml THF was added as a solvent. The reaction mixture was cooled in an ice bath, and a solution of phosphorochloridite of (*R*)-2,2′-bisnaphthol (0.505 mol, 1 equiv.) in toluene was added slowly. The reaction mixture was stirred overnight at 45°C. The reaction mixture was filtered, and the Amberlyst A-21 washed with THF. Solvents were evaporated *in vacuo* to give the crude product. The product was purified by recrystallization from DCM/PE. White powder. Yield 33% (0.16 mol, 83 g).

15.4.3 Synthesis of METAMORPhos

4-Butylbenzene-1-sulfonamide-diphenylphosphine (1): METAMORPhos (Scheme 15.18): Commercially available 4-butylbenzene-1-sulfonamide (9.376 mmol, 1 equiv.) was placed in a Schlenk flask under argon. 10 mL of toluene was dropped on it, leading to a blurred suspension which was stirred at room temperature for 5 min. The solvent was then evaporated. The process was repeated twice.The compound was then dissolved in THF (20 mL) and Et_3N (25 mmol), leading to a clear colorless solution. Distilled chlorodiphenylphosphine (9.376 mmol, 1.0 equiv.) was added dropwise under strong magnetic stirring at room temperature. The immediate formation of salt was observed as a white suspension. The suspension was left to stir overnight at room temperature. The suspension was then filtered under argon atmosphere, and the resulting clear solution was evaporated to a white solid. The product was redissolved in 10 mL of Et_2O, leading to a clear solution. 10 mL of pentane was added dropwise to the solution, causing a white solid to

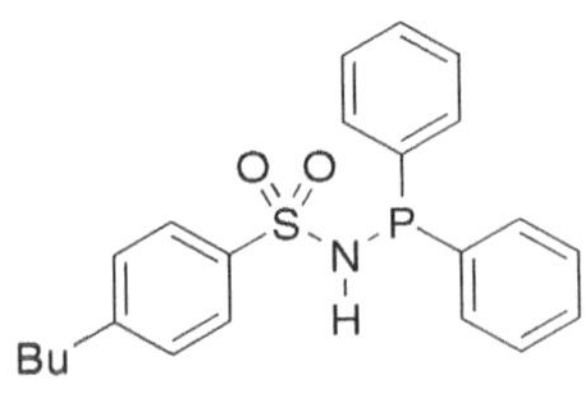

$C_{22}H_{24}NO_2PS$
Molecular weight: 397.47

Scheme 15.18

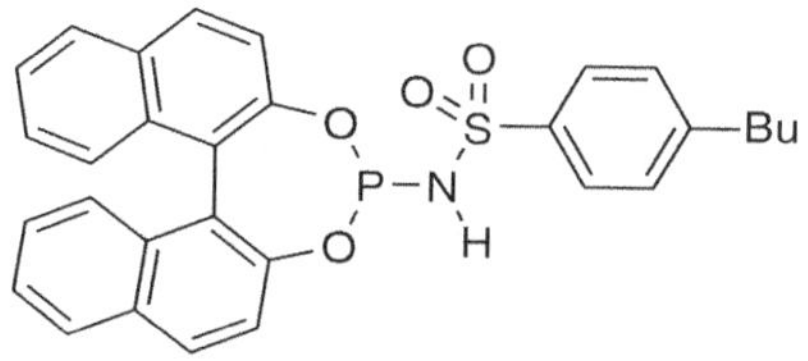

$C_{30}H_{26}NO_4PS$
Molecular weight: 527.57

Scheme 15.19

crash out. The organic layer was syringed out, and the white solid obtained was washed twice with 10 mL pentane. The product was filtered over a short silica gel column with CH_2Cl_2 as eluant.The fractions containing the product were united and evaporated, yielding a white solid: ligand 1, with a quantitative yield.

4-Butylbenzene-1-sulfonamide-R-(+)-1,1′-Bi-2-naphthol-phosphoramidite (2) (Scheme 15.19): Commercially available *R*-(+)-1,1′-Bi-2-naphtol (9.647 mmol, 1.03 equiv.) was placed in a Schlenk flask under argon. 10 mL of toluene was added to it, leading to a suspension which was stirred for 10 min at room temperature. The solvent was then evaporated, and the process repeated twice. 36 mL of toluene was finally added to the white solid along with PCl_3 (12 mL, distilled), leading to a suspension. The mixture was gently heated up to 80°C, yielding a clear colorless solution which was left to stir overnight. The solution was then cooled down at room temperature and evaporated, leading to a white sticky foam.The product was redissolved in toluene (20 mL) and evaporated again. The sticky foam obtained was then dissolved in 10 mL THF, leading to a clear colorless solution.

In a second Schlenk flask was placed commercially available 4-butylbenzene-1-sulfonamide (9.376 mmol, 1 equiv.) under argon atmosphere. The amide was then submitted to 10 mL of toluene, leading to a suspension stirred at room temperature for 10 min. The solvent was evaporated, and the process was repeated twice. The amide was then dissolved in THF (25 mL) and Et_3N (57 mmol), leading to a clear colorless solution. The solution of *R*-(+)-1,1′-Bi-2-naphtol-PCl previously prepared was added dropwise to the amide solution under strong magnetic stirring at room temperature. The suspension obtained was left to stir overnight at room temperature. The suspension was then filtered under argon atmosphere, and the clear colorless solution collected was evaporated, leading to a white foam. The product was redissolved in CH_2Cl_2 (10 mL), and re-evaporated to a white foam. 10 mL of toluene was then added, leading to a thick solution. Within a few seconds, however, a white solid broke out under the form of what appeared to be small white crystals. An extra 50 mL of toluene was added, leading to a suspension which was filtered under argon atmosphere. The white product obtained was recovered from the filter by redissolving in CH_2Cl_2, and the clear solution obtained was evaporated. Ligand **2** was obtained with a quantitative yield as a thin white powder. $^{31}P\{^1H\}$ NMR (202.3 MHz, $CDCl_3$, 298 K): δδ: 170.142 (broad s). Ligand **2** does not show the slow tautomerization process observed in the case of the diphenylphophine analogue : Ligand **1**.

^1H NMR (500 MHz, $CDCl_3$, 223 K): δ 10.35 (broad s, 1 H, NH), 7.97 (d, 1 H, J=8.5 Hz, Ar. H), 7.92 (d., 7.5 Hz, 2 H, Ar. H), 7.83 (d, J=8 Hz, 1 H, Ar. H), 7.65 (d, J=8.5 Hz, 1 H, Ar. H), 7.49 (d., J=9 Hz, 1 H, Ar. H), 7.46–7.23 (Ar), 6.515 (d, J=8.5 Hz, 1 H, Ar. H), 2.69 (m., 2 H, (C_3H_7)–CH_2–Ar), 2.62 (broad m, 3 H, splitted CH_3–CH_2–N of remaining Et_3N (1 equiv.)), 2.00 (broad m, 3 H, split CH_3–CH_2–N of remaining Et_3N (1 equiv.)), 1.60 (m, 2 H, (C_2H_5)–CH_2–CH_2–Ar), 1.33 (m, 2 H, CH_3–CH_2–(C_2H_4)–Ar), 0.915 (t, J=7.5 Hz, 3 H, CH_3–(C_3H_6)–Ar), 0.62 (broad m, 9 H, CH_3–CH_2–N of remaining Et_3N (1equiv.)). ^{13}C NMR (125.7 MHz, $CDCl_3$, 223 K): δ 149.6 (C_{quat}), 145.4 (C_{quat}), 144.5 (C_{quat}), 132.8 (d, J_{C-P}=3 Hz, C_{quat}), 131.3 (C_{quat}), 130.4 (C_{quat}), 130.1 (CH), 129.0 (CH), 128.6 (CH), 128.3 (CH), 128.3 (CH), 128.0 (CH),

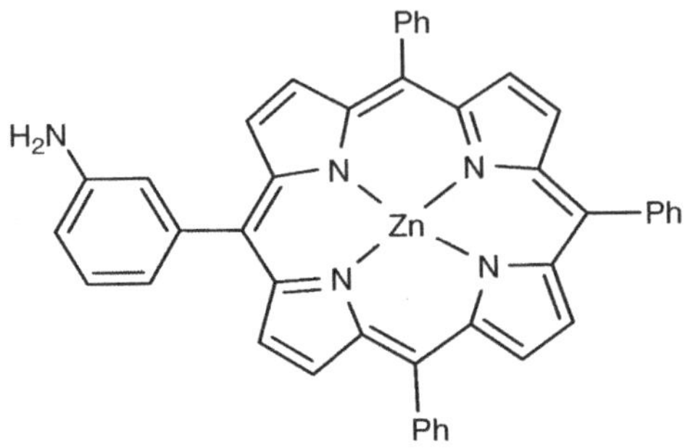

Scheme 15.20

126.8 (CH), 126.4 (CH), 126.3 (CH), 126.0 (CH), 126.0 (CH), 126.0 (CH), 125.8 (CH), 124.6 (CH), 124.3 (CH), 124.2 (CH), 123.8 (d. $J_{C\text{-}P}$=2Hz, C_{quat}), 122.3 (CH), 44.5 (s, CH_3–CH_2–N of remaining Et_3N (1equiv.)), 35.5 (s., (C_3H_7)–CH_2–Ar), 33.4 (s, (C_2H_5)–CH_2–CH_2–Ar), 22.2 (s, CH_3–CH_2–(C_2H_4)–Ar), 13.9 (s, CH_3–(C_3H_6)–Ar), 8.0 (s, CH_3–CH_2–N of remaining Et_3N (1equiv.)).

4-Pyridylphosphine was prepared following a literature procedure [52].

15.4.4 Synthesis of supraphos

Synthesis of 5-(3-aminophenyl)-10,15-20-tris(phenyl)-zinc(II) porphyrin (Scheme 15.20): A solution of *meso*-(3-aminophenyl)triphenylporphyrin (387 mg, 0.61 mmol) and $Zn(OAc)_2 \cdot 2H_2O$ (337 mg, 1.54 mmol) in a CH_2Cl_2–MeOH mixture was stirred for 16 h at room temperature, after which part of the solvent was evaporated. During evaporation of the CH_2Cl_2, methanol was added gradually in order to precipitate the product. After leaving the product to precipitate overnight, it was filtrated and washed with methanol. The product was used without further purification. Yield: 326 mg (77%).

^{1}H NMR (300 MHz, $CDCl_3$ + 2 drops MeCN-d3): δ 8.78–8.70 (m, 4 H), 8.52 (d, 1 H, J=4.20 Hz), 8.23–8.15 (m, 4 H), 8.09–8.02 (m, 3 H), 7.94 (d, 1 H, J=7.50 Hz), 7.65–7.44 (m, 13 H), 7.14 (d, 1 H, J=7.20 Hz)

Synthesis of 5-(2-aminophenyl)-10,15-20-tris(phenyl)-zinc(II) porphyrin: A solution of *meso*-(2-aminophenyl)triphenylporphyrin (388 mg, 0.62 mmol) and $Zn(OAc)_2 \cdot 2H_2O$ (780 mg, 3.55 mmol) in a CH_2Cl_2–MeOH solvent mixture (2:1, 40 mL), was stirred for 16 h at room temperature, after which part of the solvent was evaporated. During evaporation of the CH_2Cl_2, methanol was added gradually in order to precipitate the product. After precipitation overnight, the product was isolated by filtration and washed with methanol. The product was used without further purification. Yield: 278 mg (65%). ^{1}H NMR (300 MHz, $CDCl_3$): δ 9.03 (d, 2 H, J=4.50 Hz), 8.90 (d, 2 H, J=5.10 Hz), 8.57 (d, 2 H, J=4.20 Hz), 8.49 (br. d, 1 H, J=6.00 Hz), 8.33 (br. d, 1 H, J=6.90 Hz), 8.24 (br. s, 2 H), 7.97 (d, 2 H, J=7.20 Hz), 7.87–7.62 (m, 12 H), 7.37–7.34 (m, 1 H), 6.89–6.87 (m, 2 H), 4.04 (br. s, 2 H) ppm. ^{13}C NMR (300 MHz, $CDCl_3$): δ 150.6, 150.4, 149.9, 148.7, 143.6, 143.3, 140.7, 134.9, 134.7, 134.6, 134.4, 132.7, 132.2, 130.4, 129.9, 127.8, 127.7 127.5, 126.9, 126.8, 126.7, 126.6, 121.5, 121.2, 121.0, 119.7, 116.6, 112.9 ppm. Anal.calculated for $C_{44}H_{29}N_5Zn$ + 2.5 eq. H_2O: C 71.59, H 4.64, N 9.49; found: C 71.93, H 4.39, N 8.36.

Synthesis of (S)-(1,1′-binaphthyl-2,2′-diyl)(5-(phenyl-3-yl)-10,15,20-tris(phenyl) zinc(II) porphyrin) phosphoramidite (Scheme 15.21): 5-(3-aminophenyl)-10,15,20-tris(phenyl)-zinc(II) porphyrin (0.47 mmol, 326 mg) was azeotropically dried with toluene (3×2 ml). After dissolving in 30 mL THF, it was cooled to –50°C, after which dipea (4.70 mmol, 0.82 mL) was added. Freshly prepared (S)-2,2′-binaphtol phosphorus chloridite (0.47 mmol, 165 mg), dissolved in 10 mL THF, was added dropwise. The mixture was stirred

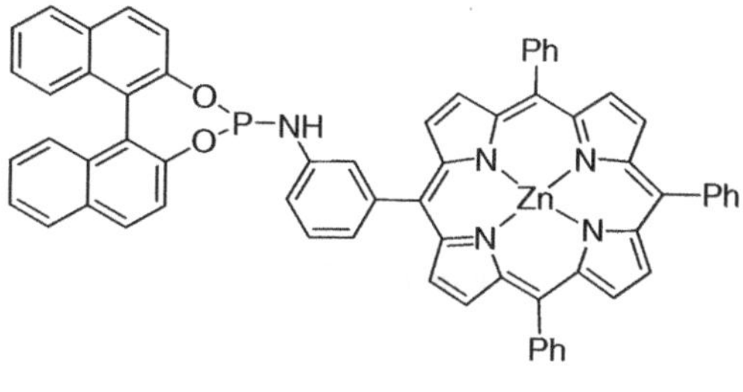

Scheme 15.21

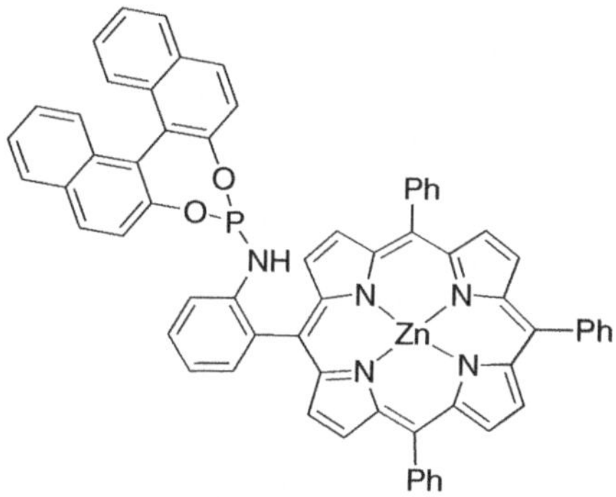

Scheme 15.22

overnight, allowing it to warm to room temperature. The solvent was evaporated, and the product dissolved in 5 mL CH_2Cl_2 and filtrated over a short column of basic alumina (4 cm), using dichloromethane as eluent. Evaporation of the solvent yielded the ligand as a purple solid. Yield: 213 mg (45%).

^{1}H NMR (300 MHz, $CDCl_3$): δ 9.07–9.05 (m, 2 H), 9.01–8.96 (m, 6 H), 8.28–8.21 (m, 6 H), 7.95 (d, 1 H, J=5.40 Hz), 7.91 (d, 1 H, J=8.00 Hz), 7.87 (d, 1 H, J=7.50 Hz), 7.85–7.83 (m, 2 H), 7.79–7.74 (m, 7 H), 7.67 (d, 1 H, J=8.00 Hz), 7.61 (t, 1 H, J=8.00 Hz), 7.54 (d, 1 H, J=7.50 Hz), 7.45–7.40 (m, 4 H), 7.35 (d, 1 H, J=9.00 Hz), 7.31 (d, 1 H, J=8.50 Hz), 7.27–7.24 (m, 2 H), 7.13 (t, 1 H, 7.50 Hz), 7.07 (t, 1 H, J=7.50 Hz), 5.44 (s, 1 H) ppm. ^{31}P NMR (300 MHz, $CDCl_3$): δ 147.8 ppm ^{13}C NMR (300 MHz, $CDCl_3$): δ 150.5, 150.4, 150.3, 149.0, 147.3, 144.4, 143.2, 139.7 (d, J=16.2 Hz), 134.7 (d, J=24.9 Hz), 133.8, 132.9 (J=16.2 Hz), 132.2, 131.8, 131.3, 130.7 (d, J=30.0 Hz), 130.1 (d, J=28.3 Hz), 129.0, 128.7, 128.5, 128.3, 127.8, 127.6, 127.1, 126.9, 126.6, 125.3, 125.1, 124.9, 124.4, 123.9, 122.5, 121.8, 121.7, 121.4, 121.3, 120.7, 117.3 ppm. MS (FAB) for $C_{64}H_{40}N_5O_2PZn$ calculated: 1007.2368; observed: 1007.2396.

Synthesis of (S)-(1,1′-binaphthyl-2,2′-diyl)(5-(phenyl-2-yl)-10,15,20-tris(phenyl) zinc(II) porphyrin) phosphoramidite (Scheme 15.22): 5-(2-aminophenyl)-10,15-20-tris(phenyl)-zinc(II) porphyrin (0.40 mmol, 278 mg) was azeotropically dried with toluene (3 × 2 ml). After dissolving in 30 mL THF, it was cooled to –50°C, after which dipea (4.01 mmol, 0.70 mL) was added. Freshly prepared (S)-2,2′-binaphtol phosphorus chloridite (0.40 mmol, 141 mg), dissolved in 10 mL THF, was added dropwise. The mixture was stirred overnight, allowing it to warm to room temperature. The solvent was evaporated, and the product dissolved in 5 mL CH_2Cl_2 and filtrated over a very short column of basic alumina (5 cm), using dichloromethane as eluent. Evaporation of the solvent yielded the ligand as a purple solid. Yield: 169 mg (42%).

^{1}H NMR (300 MHz, $CDCl_3$): δ 8.97–8.83 (m, 8 H), 8.25–8.02 (m, 6 H), 7.88–7.66 (m, 14 H), 7.41–7.36 (m, 1 H), 7.19 (d, 1 H, J=6.90 Hz), 7.14 (d, 1 H, J=8.70 Hz), 6.97 (d, 1 H, J=6.60 Hz), 6.87 (d, 1 H, J=8.40 Hz),

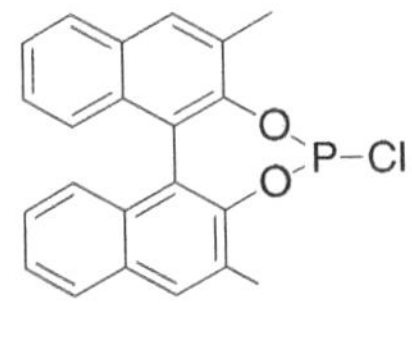

Scheme 15.23

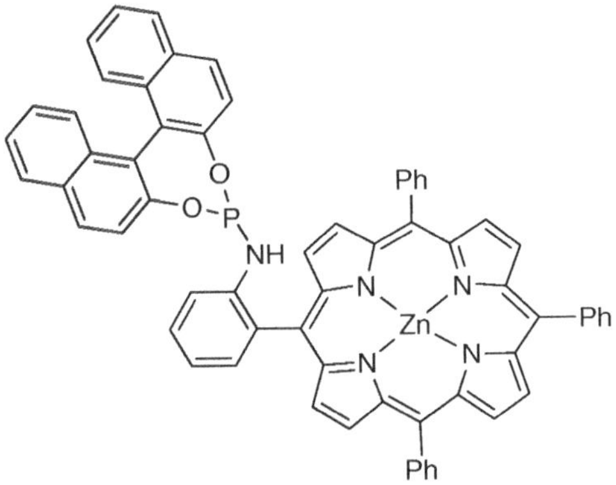

Scheme 15.24

6.58–6.53 (m, 3 H), 6.07 (d, 1 H, J=8.10 Hz), 5.97 (d, 1 H, J=8.70 Hz), 4.79 (s, 1 H) ppm. ^{31}P NMR (300 MHz, $CDCl_3$): δ 146.4 ppm ^{13}C NMR (300 MHz, $CDCl_3$): δ 150.54, 150.46, 150.3, 150.11, 150.06, 148.3, 146.3, 143.2, 143.0, 142.6 (d, $J_{(C,P)}$=17.1 Hz), 135.2, 134.7, 132.9, 132.8, 132.5, 132.2, 131.7, 131.5, 131.4, 130.2, 129.8, 129.6, 128.4, 128.3, 127.8, 127.7, 127.6, 127.3, 126.8, 126.0 (d, $J_{(C,P)}$=20.8 Hz), 125.2 (d, $J_{(C,P)}$=26.9), 124.0, 122.6, 121.7, 121.4, 121.2 (d, $J_{(C,P)}$=6.12 Hz), 120.4, 116.6 (d, $J_{(C,P)}$=22.0 Hz), 114.7 ppm.

Synthesis of (R)-(+)-3,3′-dimethyl-2,2′-bisnaphtol phosphorochloridite (Scheme 15.23): (R)-(+)-3,3′-dimethylbinol (3.42 g, 10.9 mmol) was azeotropically dried with toluene (3×3 mL), dissolved in 30 mL CH_2Cl_2, and added dropwise to a solution of PCl_3 (9.48 mL, 109 mmol) and Et_3N (3.02 mL, 21.7 mmol) in 80 mL CH_2Cl_2 at –60°C. When the addition of (R)-3,3′-dimethylbinol was completed, the cooling bath was removed and the mixture was stirred for 3 h at room temperature. The reaction mixture was concentrated *in vacuo* and redissolved in 30 mL toluene, after which 60 mL hexanes were added. The precipitated salts were removed via filtration under nitrogen. After evaporation of the solvent, the product was obtained as a white-yellowish foam in quantitative yield. ^{1}H NMR (300 MHz, $CDCl_3$): δ 7.91 (br. s. 2 H), 7.89 (br. s., 2 H), 7.51–7.43 (m, 2 H), 7.35–7.30 (m, 2 H), 7.27–7.22 (m, 2 H), 2.66 (s, 6 H). ^{31}P NMR (300 MHz, $CDCl_3$): d=175.3 ppm.

Synthesis of (R)-(3,3′-dimethyl-1,1′-binaphthyl-2,2′-diyl)(5-(phenyl-2-yl)-10,15,20-tris(phenyl) zinc(II) porphyrin) phosphite (Scheme 15.24): To a mixture of 5-(2-hydroxyphenyl)-10,15-20-tris(phenyl)-zinc(II) porphyrin (319 mg, 0.46 mmol), azeotropically dried with toluene (3×2 mL), and dipea (0.80 mL, 4.60 mmol) in 20 mL CH_2Cl_2 was added dropwise a solution of (R)-3,3′-dimethyl-2,2′-bisnaphtol phosphorus chloridite in 3 mL CH_2Cl_2 at –78°C. After addition of the PCl compound, the mixture was stirred 15 min at –78°C and 3 h at room temperature. The solvent was evaporated, and the crude product dissolved in a minimum amount of CH_2Cl_2 and purified over a short column (±3 cm) of basic alumina. After evaporation of the solvent, the product was obtained as a purple solid. Yield: 224 mg (47%). ^{1}H NMR (300 MHz,

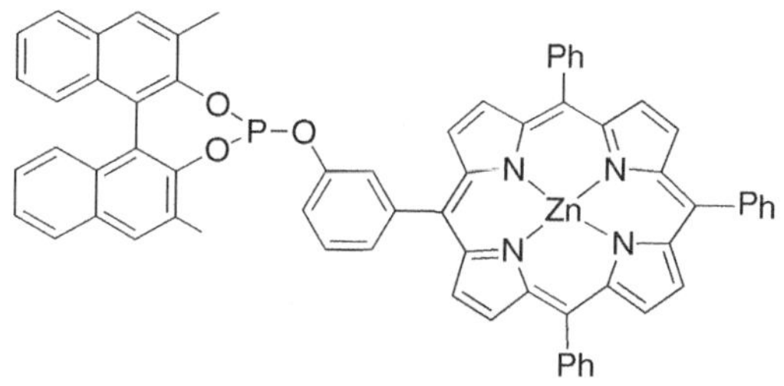

Scheme 15.25

$CDCl_3$): δ 9.00–8.93 (m, 4 H), 8.89 (d, 1 H, J = 9.00 Hz), 8.87 (d, 1 H, J = 9.00 Hz), 8.75 (d, 1 H, J = 4.80 Hz), 8.67 (d, 1 H, J = 4.50), 8.35–8.34 (br. m, 2 H), 8.25–8.17 (m, 3 H), 8.04 (d, 1 H, J = 7.20 Hz), 7.86–7.50 (m, 14 H), 7.23–7.11 (m, 2 H), 6.67 (t, 1 H, 7.65 Hz), 6.45 (d, 1 H, 8.40 Hz), 5.93–5.83 (m, 3 H), 5.76 (br. s., 2 H), 2.18 (s, 6 H). ^{31}P NMR (121.3 MHz, $CDCl_3$): δ 143.7. ^{13}C NMR (125.7 MHz, $CDCl_3$): δ 151.5 (d, J = 3.52 Hz), 150.7, 150.33, 150.30 (d, J = 2.39 Hz), 150.25 (d, J = 2.39 Hz), 150.0 (d, J = 8.04 Hz), 146.8 (d, J = 5.78 Hz), 145.8 (d, J = 3.52 Hz), 143.3 (d, J = 6.40 Hz), 143.1, 136.3, 135.7, 134.8, 134.7, 134.6, 132.4, 132.1, 132.0, 131.7, 131.4, 131.2, 129.9, 129.54, 129.49, 129.3, 129.2, 128.9, 128.5, 127.7 (d, J = 2.89 Hz), 127.5, 127.4, 126.8, 126.7 (d, J = 4.02 Hz), 126.6, 125.5, 125.1 (d, J = 9.30 Hz), 124.9 (d, J = 6.28 Hz), 123.9 (d, J = 5.15 Hz), 123.4, 123.1 (d, J = 5.15 Hz), 121.5, 121.4 (d, J = 2.89 Hz), 121.1, 121.0, 119.8 (d, J = 10.4 Hz). MS (FAB) for $C_{66}H_{43}N_4O_3PZn$ calculated: 1036.2521; observed: 1036.2526. Anal.calculated for $C_{66}H_{43}N_4O_3PZn$ + 1 equiv. H_2O: C75.18, H4.30, N5.31; found: C74.77, H4.65, N5.39.

Synthesis of (R)-(3,3′-dimethyl-1,1′-binaphthyl-2,2′-diyl)(5-(phenyl-3-yl)-10,15,20-tris(phenyl) zinc(II) porphyrin) phosphite (Scheme 15.25): To a mixture of 5-(3-hydroxyphenyl)-10,15-20-tris(phenyl)-zinc(II) porphyrin (319 mg, 0.46 mmol), azeotropically dried with toluene (3 × 2 mL), and dipea (0.80 mL, 4.60 mmol) in 20 mL CH_2Cl_2 was added dropwise a solution of (R)-3,3′-dimethyl-2,2′-bisnaphtol phosphorus chloridite in 3 mL CH_2Cl_2 at –78°C. After addition of the PCl-compound, the mixture was stirred 15 min at –78°C and 3 h at room temperature. The solvent was evaporated, the crude product dissolved in a minimum amount of CH_2Cl_2 and purified over a short column (±3 cm) of basic alumina. After evaporation of the solvent, the product was obtained as a purple solid. Yield: 229 mg (48%).^{1}H NMR (300 MHz, $CDCl_3$): δ 9.00–8.93 (m, 7 H), 8.28–8.08 (m, 5 H), 8.08–8.03 (m, 2 H), 7.82–7.68 (m, 13 H), 7.64–7.61 (m, 2 H), 7.40–7.35 (m, 2 H), 7.30–7.25 (m, 2 H), 7.22–7.14 (m, 3 H), 7.08–7.03 (m, 1 H), 2.65 (d. $J_{(H,P)}$ = 11.1 Hz). ^{31}P NMR (121.3 MHz, $CDCl_3$): δ 142.5. ^{13}C NMR (75.5 MHz, $CDCl_3$): δ 150.56 (s), 150.49 (s), 150.16 (s), 147.86 (s), 147.79 (s), 146.48 (s), 146.45 (s), 145.00 (s), 143.09 (s), 134.72 (s), 132.43 (s), 132.33 (s), 132.28 (s), 132.01 (s), 131.92 (s), 131.76 (s), 131.59 (s), 131.37 (s), 130.84 (s), 130.29 (s), 130.0 (s), 127.99 (s), 127.77 (s), 127.62 (s), 127.16 (d, $J_{(C,P)}$ = 3.70 Hz), 126.81 (s), 126.37 (d, $J_{(C,P)}$ = 8.61 Hz), 125.62 (s), 125.44 (s), 125.37 (s), 125.05 (s), 124.94 (s), 124.86 (s), 123.03 (s), 121.60 (s), 121.48 (s), 119.84 (s), 119.22 (d, $J_{(C,P)}$ = 8.53 Hz). MS (FAB) for $C_{66}H_{43}N_4O_3PZn$ calculated 1036.2521 obsd. 1036.2526. anal. calculated for $C_{66}H_{43}N_4O_3PZn$: C76.48, H4.18, N5.41; found: C76.55, H4.10, N5.40.

Synthesis of 3-[bis-(2,6-dimethyl-phenyl)-phosphanyl]-pyridine (Scheme 15.26): Bis(2,6-dimethylphenyl) chlorophosphine was prepared according to literature procedure [31]. *n*-Buthyllithium (5.28 mL of a 2.5 M solution in hexanes, 13.2 mmol) and TMEDA (1.99 mL, 1.53 g, 13.2 mmol) were stirred for 15 min at room temperature. After cooling to –70°C, 50 mL of a THF-Et_2O-pentane solution (1/1/1) was added. The mixture was cooled to –120°C and 3-bromopyridine (1.60 mL, 2.61 g, 16.5 mmol), dissolved in 10 mL of a 1/1/1 mixture of THF-Et_2O-pentane, was added slowly. The mixture was stirred for 10 minutes at –120°C, after which bis(2,6-dimethylphenyl)chlorophosphine, dissolved in THF, was added dropwise. The resulting reaction mixture was stirred overnight and allowed to warm to room temperature. The orange solution

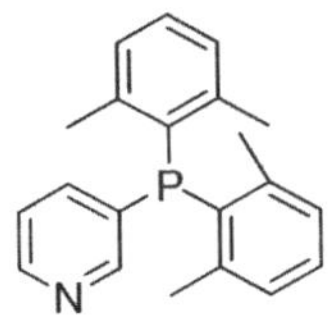

Scheme 15.26

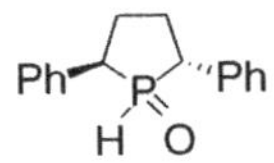

Scheme 15.27

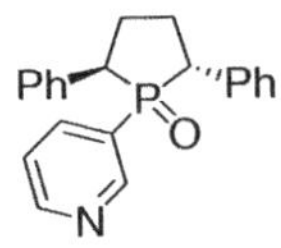

Scheme 15.28

was washed once with brine, dried over $MgSO_4$, and filtrated (all under nitrogen atmosphere). After adding some silica, the mixture was concentrated *in vacuo* (and simultaneously impregnated on silica). The product was purified by column chromatography using PE 40–65 with 10% Et_3N as an eluent. The product was further purified by means of chromatographic separation with Chromatotron®, using CH_2Cl_2 as an eluent. When the impurity was removed, 1% MeOH was added to the eluent. (R_f on silica with 1% MeOH/CH_2Cl_2: 0.37). Yield: 1.2 g (23%). ^{1}H NMR (300 MHz, $CDCl_3$): δ 8.60 (br. s, 1 H), 8.50 (dd, 1 H, J = 4.65 Hz, J = 1.65 Hz), 7.59 (br. t, 1 H, J = 6.30 Hz), 7.20–7.14 (m, 3 H), 7.03–6.99 (m, 4 H), 2.13 (s, 12 H). ^{31}P NMR (121.5 MHz, $CDCl_3$): δ −26.4 ppm. ^{13}C NMR (75.5 MHz, $CDCl_3$): δ 154.7 (d, $J_{(C,P)}$ = 28.1 Hz), 149.1 Hz (s), 142.9 (d, $J_{(C,P)}$ = 15.9 Hz), 141.0 (d, $J_{(C,P)}$ = 18.3 Hz), 133.9 (s, $J_{(C,P)}$ = 17.0 Hz), 132.6 (s, $J_{(C,P)}$ = 19.6 Hz), 129.6 (s), 129.3 (d, 34.1 Hz), 123.4 (d, $J_{(C,P)}$ = 3.70 Hz). Anal. calculated for $C_{21}H_{22}NP$: C78.97, H6.94, N4.39; found: C78.74, H6.87, N4.46.

Synthesis of (R,R)-2,5-diphenyl-phospholane 1-oxide (Scheme 15.27): 1-Oxo-2,5-diphenyl-1λ^5-phospholan-1-ol was prepared according to literature procedure [25]. (*R,R*)-1-Oxo-2,5-diphenyl-1λ^5-phospholan-1-ol (1.22 g, 4.51 mmol) was three times azeotropically dried with toluene and suspended in THF. After cooling to 0°C, oxalylchloride (1.55 mL, 2.29 g, 18.1 mmol) was added. After 5 min stirring at 0°C, the cooling bath was removed and the mixture stirred for 2 h at room temperature. The solvent and excess oxalyl chloride were removed by evaporation, and the product was dissolved in Et_2O. After 5 minsonification in the ultrasound bath, $LiAlH_4$ (0.17 g, 4.51 mmol) was added in portions at 0°C. After stirring overnight, all the phosphinicacid is reduced to the hydride, as judged from ^{31}P NMR (singlet −19 ppm). The workup was performed under air, thereby oxidizing the product. 1.0 M aqueousHCl solution was added, followed by separation of the layers. The aqueous layer was once extracted with EtOAc. Afterward, the combined organic layers were twice washed with saturated aqueous K_2CO_3 solution and once with saturated NaCl solution. The organic layer was dried over $MgSO_4$, filtrated, and concentrated *in vacuo*. The white solid was crystallized from EtOAc and immediately used for the next step. Yield: 1.06 g (92%).

Synthesis of 3-(1-oxo-2,5-diphenyl-phospholan-1-yl)-pyridine (Scheme 15.28): $Pd(OAc)_2$ (43.6 mg, 0.19 mmol) and 1,3-Bis(diphenylphosphino)propane (dppp) (88.2 mg, 0.21 mmol) were put in a dry, degassed Schlenk flask and degassed by three vacuum–nitrogen cycles. The catalyst precursor was dissolved in 10 mL DMF and stirred for 15 min at room temperature. 3-iodopyridine (478 mg, 2.33 mmol)

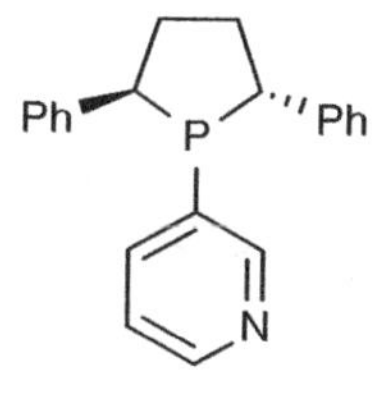

Scheme 15.29

was added, and the mixture was stirred for an additional 15 min. A solution of (R,R)-2,5-diphenylphospholane 1-oxide (498 mg, 1.94 mmol), three times azeotropically dried with toluene, and dipea in 5 mL DMF was added to the catalyst solution. The resulting dark red solution was stirred for 16 h at 105°C. After cooling to room temperature, the DMF was evaporated. The crude product was redissolved in dichloromethane, and washed twice with 5% $NaHCO_3$ (aqueous) and once with brine. The organic layer was dried over $MgSO_4$, filtrated, and evaporated. After a very fast purification over a small layer of silica, to remove main impurities, the product was purified by means of chromatographic separation with Chromatotron®, usingEtOAC/PE 3/1 as an eluent.

Analysis data (S,S): ^{1}H NMR (300 MHz, $CDCl_3$): δ 8.63 (br. s, 1 H), 8.54 (br. s, 1 H), 7.82–7.76 (m, 1 H), 7.31–7.25 (m, 5 H), 7.12–7.02 (m, 6 H), 4.01–3.88 (m, 1 H), 3.57–3.74 (m, 1 H), 2.80–2.52 (m, 3 H), 2.33–2.23 (m, 1 H). ^{31}P NMR (121.5 MHz, $CDCl_3$): δ 52.9 ppm. ^{13}C NMR (75.5 MHz, $CDCl_3$): δ 151.7 (s), 150.6 (s), 150.5 (s), 140.4 (br. s), 135.5 (s), 135.2 (s), 129.0 (br. s), 127.6 (s), 127.3 (d, $J_{(C,P)}$ = 4.76 Hz), 127.1 (s), 123.6 (d, $J_{(C,P)}$ = 7.32 Hz), 51.1 (d, $J_{(C,P)}$ = 62.5 Hz), 47.3 (d, $J_{(C,P)}$ = 62.3 Hz), 31.9 (d, $J_{(C,P)}$ = 7.40 Hz), 27.7 (d, $J_{(C,P)}$ = 8.53 Hz). anal. calculated for $C_{21}H_{20}NOP$: C 75.66, H 6.05, N 4.20, found: C 75.75, H 6.08, N 4.04.

Analysis data (R,R): ^{1}H NMR (300 MHz, $CDCl_3$): δ 8.61 (d, J = 3.30 Hz), 8.51 (dd, 1 H, J = 4.65 Hz, J = 2.25 Hz), 7.80–7.73 (m, 1 H), 7.31–7.17 (m, 5 H), 7.12–7.02 (m, 6 H), 3.96–3.85 (m, 1 H), 3.53–3.46 (m, 1 H), 2.71–2.52 (m, 3 H), 2.32–2.23 (m, 1 H). ^{31}P NMR (121.5 MHz, $CDCl_3$): δ 52.6 ppm. ^{13}C NMR (75.5 MHz, $CDCl_3$): δ 152.4 (s), 151.3 (s), 151.2 (s), 139.8 (s), 135.4 (d, $J_{(C,P)}$ = 3.77 Hz), 135.1 (d, $J_{(C,P)}$ = 6.03 Hz), 129.1 (d, $J_{(C,P)}$ = 5.28 Hz), 129.0 (d, $J_{(C,P)}$ = 1.51 Hz, 128.9 (br.s), 127.6 (d, $J_{(C,P)}$ = 2.26 Hz), 127.3 (d, $J_{(C,P)}$ = 4.52 Hz), 127.1 (d, $J_{(C,P)}$ = 3.02 Hz), 123.3 (d, $J_{(C,P)}$ = 7.32 Hz), 51.1 (d, $J_{(C,P)}$ = 62.3 Hz), 47.3 (d, $J_{(C,P)}$ = 62.3 Hz), 31.9 (d, $J_{C,P)}$ = 7.32 Hz), 27.7 (d, $J_{(C,P)}$ = 9.74 Hz).

Synthesis of R,R-3-(2,5-diphenyl-phospholan-1-yl)-pyridine (Scheme 15.29): *(R,R)*-3-(1-oxo-2,5-diphenyl-phospholan-1-yl)-pyridine (300 mg, 0.90 mmol) was azeotropically dried with toluene (3 × 2 mL). $PhSiH_3$ (± 4 mL) was added, and the mixture was heated to reflux and stirred under reflux for 16 h. After evaporation, the product was dissolved in CH_2Cl_2 and purified using Chromatotron using CH_2Cl_2 asa eluent. Yield: 234 mg (84%).

^{1}H NMR (300 MHz, $CDCl_3$): δ 8.40 (d, 2 H, J = 4.80 Hz), 7.43–7.21 (m, 6 H), 7.12–6.91 (m, 6 H), 4.11–4.03 (m, 1 H), 3.97–3.85 (m, 1 H), 2.82–2.75 (m, 1 H), 2.47–2.38 (m, 1 H), 2.36–2.25 (m, 1 H), 2.17–2.05 (m, 1 H) ppm. ^{31}P NMR (121.5 MHz, $CDCl_3$): δ 15.9 ppm. ^{13}C NMR (75.5 MHz, $CDCl_3$ δ 153.4 (d, J = 17.1 Hz), 149.6, 144.1 (d, J = 18.3 Hz), 141.1 (d, J = 19.6 Hz), 138.1, 132.3 (d, J = 34.2 Hz), 129.0, 128.4, 128.1 (d, J = 9.74 Hz), 127.8 (d, J = 3.70 Hz), 126.5 (d, J = 2.42 Hz), 126.2 (d, J = 2.42 Hz), 122.9 (d, J = 6.12 Hz), 48.6 (d, J = 17.1 Hz), 46.8 (d, J = 15.9 Hz), 37.1, 32.9 (d, J = 3.62 Hz). Anal. calculated for $C_{21}H_{20}NP$ + 33% oxide: C 78.16, H 6.25, N 4.34; found: C 78.37, H 6.28, N 3.95.

Tert-butylphenylphosphine borane was prepared according to literature procedure [32].

Tert-butylmethylphenylphosphine borane was prepared according to literature procedure [23].

Synthesis of (R)-tert-butylphenyl(carboxylic acid)phosphine borane (Scheme 15.30): Tert-butylmethylphenylphosphine borane (1.00 g, 5.15 mmol) was dried azeotropically (3 × 4 ml) and dissolved in 40 mL Et_2O. *tert*-buthyllithium (4.47 ml of a 1.5 M solution in pentane, 6.70 mmol) was added at –78°C.

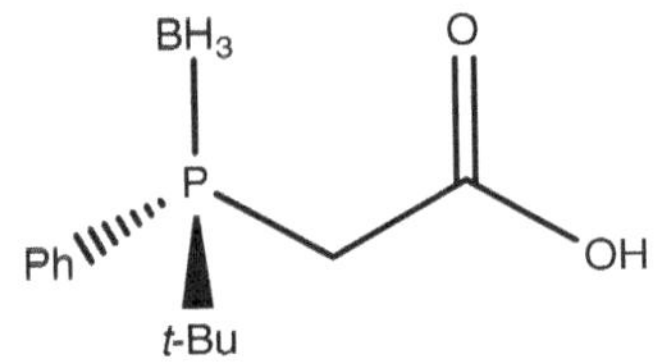

Scheme 15.30

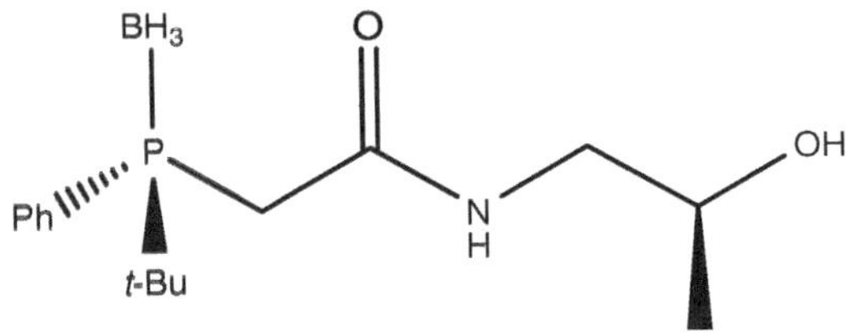

Scheme 15.31

After 1 h stirring at this temperature, CO_2 gas was bubbled through the solution. For that purpose, a round-bottomed flask was filled with CO_2 (s) and connected to a column with sicapent© in order to dry the released gas. The gas was brought to the flask via a polyflow® tube. The CO_2 was bubbled through for 1 h at –78°C, and for 1 h at room temperature (40 mL Et_2O was added to keep the solvent level constant). CO_2 bubbling was stopped, and the mixture was stirred at room temperature for 16 h. Then the mixture was poured out in ice water with a few ml of 37% aqueous HCl. The layers were separated, and the aqueous layer was twice extracted with chloroform. 100 mL 0.5 M Na_2CO_3 (aqueous) was added to the combined organic layers, resulting in deprotonation of the product. As a consequence, the deprotonated acid moves to the aqueous layer. The layers are separated, after which 100 mL 4 M HCl (aqueous) and 200 mL $CHCl_3$ were added to the aqueous layer. The protonated product immediately moves to the organic layer, which was collected. The aqueous layer was three times extracted with $CHCl_3$. The combined organic layers were dried over Na_2SO_4, filtrated, and concentrated *in vacuo*. The product was crystallized from toluene, yielding 1.0 g (87%) of the acid as a white solid. $[a]_D$=+ 50.1 (c=0.974 in $CHCl_3$).^{1}H NMR (300 MHz, $CDCl_3$): δ 10.33 (br. s, 1 H), 7.78–7.71 (m, 2 H), 7.53–7.41 (m, 3 H), 3.23 (t, 1 H, $J_{(P,H)}$=13.2 Hz), 3.00 (dd, 1 H, $J_{(P,H)}$=13.2 Hz, $J_{(H,H)}$=9.3 Hz), 1.12 (d, 9 H, $J_{(P,H)}$=14.4 Hz). 0.8–0.2 (br. s, 3 H). ^{31}P NMR (121.5 MHz, $CDCl_3$): δ 33.4 (d, $J_{P,B}$=61.2 Hz).^{13}C NMR (75.5 MHz, $CDCl_3$): δ 173.3 (s), 133.6 (d, $J_{(C,P)}$=8.53 Hz), 131.9 (d, $J_{(C,P)}$=2.49 Hz), 128.6 (d, $J_{(C,P)}$=9.82 Hz), 125.1 (d, $J_{(C,P)}$=48.8 Hz), 30.3 (d, $J_{(C,P)}$=31.8 Hz), 27.6 (d, $J_{(C,P)}$=25.7 Hz), 25.5 (d, $J_{(C,P)}$=2.42 Hz). Anal. calculated for $C_{12}H_{20}BO_2P$: C60.54, H8.47; found: C 60.42, H 8.42.

Synthesis of 2-[(R)-boronatophenyl(tert-butyl)phosphanyl)-N-((S)-2-hydroxy-propyl)-acetamide (Scheme 15.31): *tert*-Butylphenyl(carboxylic acid)phosphine borane (250 mg, 1.11 mmol) was azeotropically dried (2×, toluene), dissolved in 15 mL THF, and cooled to 0°C. 1-Hydroxybenzotriazole (HOBt) (301 mg, 2.23 mmol) and 1-(3-dimethylaminopropyl)-3-ethylcarbodiimide hydrochloride (EDCI) (256 mg, 1.34 mmol) were added, and the resulting mixture was stirred for 30 min at room temperature. After cooling to 0°C, (S)-(+)-1-amino-2-propanol (100 mg, 1.34 mmol) was added, after which the cooling was removed and the mixture stirred overnight at room temperature. 10 ml 1 M HCl (aqueous) was added to the reaction mixture, followed by 50 mL $CHCl_3$. The layers were separated, and the aqueous layer was twice extracted with chloroform. The combined organic layers were washed once with saturated $NaHCO_3$ (aqueous) and once with brine. After drying on Na_2SO_4, the solution was filtrated and concentrated *in vacuo*. Column chromatography (silica, eluent: pure EtOAc, R_f=0.32) yielded 219 mg (67%) of the product as a white solid. $[a]_D$= +70.4 (c=0.100 in $CHCl_3$);

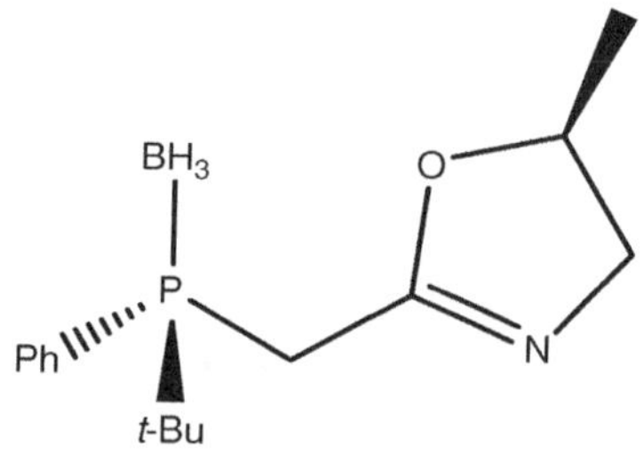

Scheme 15.32

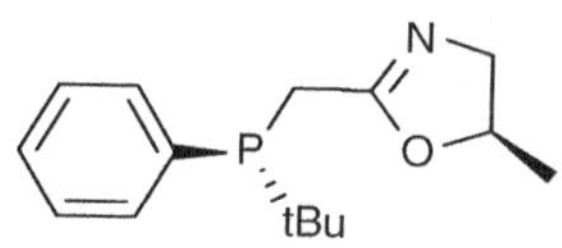

Scheme 15.33

^{1}H NMR (300 MHz, $CDCl_3$): δ 7.77–7.71 (m, 2 H), 7.53–7.41 (m, 3 H), 6.74 (br.s, 1 H), 3.65–3.59 (m, 1 H), 3.37–3.29 (m, 1 H), 3.15 (t, 1 H, J = 14.3 Hz), 3.02–2.97 (m, 1 H), 2.88–2.79 (m, 1 H), 2.65 (br. s, 1 H), 1.10 (d, 9 H, $J_{(P,H)}$ = 14.7 Hz), 0.95 (d, 3 H, J = 6.30 Hz), 0.90–0.20 (br, 3 H, BH_3). ^{31}P NMR (121.5 MHz, $CDCl_3$): δ 30.2 (br. d, $J_{(P,B)}$ = 66.3 Hz). ^{13}C APT NMR (75.5 MHz, $CDCl_3$): δ 166.8 (s), 133.7 (d, J = 8.61 Hz), 132.0 (s), 128.6 (d, J = 9.74 Hz), 124.7 (d, J = 51.3 Hz), 66.7 (s), 47.6 (s), 30.1 (d, J = 30.6), 28.9 (d, 25.7 Hz), 25.3 (s), 20.5 (s). Anal. calculated for $C_{15}H_{27}BNO_2P$: C61.04, H 9.22, N 4.75; found: C 61.18, H 8.78, N 4.71.

Synthesis of 2-[(R)-boronatophenyl(tert-butyl)phosphino]methyl-(3 S)-3-methyl-2-oxazoline (Scheme 15.32): 2-[(R)-boronatophenyl(tert-butyl)phosphanyl)-N-((S)-2-hydroxy-propyl)-acetamide (0.82 g, 2.77 mmol) was azeotropically dried with toluene (3 × 3 mL) and dissolved in 50 mL CH_2Cl_2. The reaction mixture was cooled to 0°C, and Et_3N (1.9 mL, 13.6 mmol, 4.9 equiv.) and mesityl chloride (1.1 mL, 13.8 mmol, 5 equiv.) were added. After 30 min stirring, the ice bath was removed and the reaction mixture was stirred for 13 h at room temperature. The reaction was quenched by the addition of saturated $NaHCO_3$ (aqueous) and diluted with $CHCl_3$, and the layers were separated. The aqueous layer was twice extracted with $CHCl_3$. The combined organic layers were washed with brine and dried over Na_2SO_4. After filtration, the solvent was evaporated under reduced pressure, after which the product was purified by column chromatography over basic alumina. Eluent: petroleum ether:ethylacetate 1:1, upgrade to pure ethylacetate. R_f = 0.34; yield = 664 mg (86%).

$[a]_D$ = + 35.1 (c = 0.480 in $CHCl_3$); ^{1}H NMR (300 MHz, $CDCl_3$): δ 7.80 (t, 2H, J = 8.4 Hz), 7.54–7.42 (m, 3H), 4.63–4.55 (m, 1H), 3.89–3.80 (m, 1H), 3.27–3.18 (m, 2H), 3.05–2.95 (m, 1H), 1.14 (d, 9H, J = 13.8 Hz), 1.02 (d, 3H, 6.60 Hz), 0.82–0.20 (br., 3H). ^{31}P NMR (121.5 MHz, $CDCl_3$): δ 33.5 ppm (br. d, J = 60.0 Hz). ^{13}C NMR (75.5 MHz, $CDCl_3$): δ 161.8, 133.9 (d, J = 8.67 Hz), 131.7 (d, J = 2.39 Hz), 128.4 (d, J = 9.30 Hz), 125.7 (d, J = 48.4 Hz), 61.4, 30.2 (d, J = 30.7 Hz), 25.6, 21.3 (d, J = 29.4 Hz), 21.0. Anal. calculated for $C_{15}H_{25}BNOP$: C 65.00, H 9.09, N 5.05; found: C 64.77, H 8.98, N 5.02.

Synthesis of 2-[(R)-phenyl(tert-butyl)phosphino]methyl-(3S)-3-methyl-2-oxazoline (Scheme 15.33): 2-[(R)-Boronatophenyl(tert-butyl)phosphino]methyl-(3S)-3-methyl-2-oxazoline (155.2 mg, 0.59 mmol) was azeotropically dried with toluene (3 × 2 mL) and dissolved in 10 mL CH_2Cl_2. The solution was cooled to –20°C, after which tetrafluoroboric acid–dimethyl ether complex (0.61 mL, 5.90 mmol) was added. The mixture was stirred for 16 h at room temperature. Degassed saturated $NaHCO_3$ aqueous solution was added, and the mixture was stirred for an additional hour. The layers were separated, and the water layer was three times extracted with Et_2O. The combined organic layers were dried on Na_2SO_4, filtrated, and concentrated

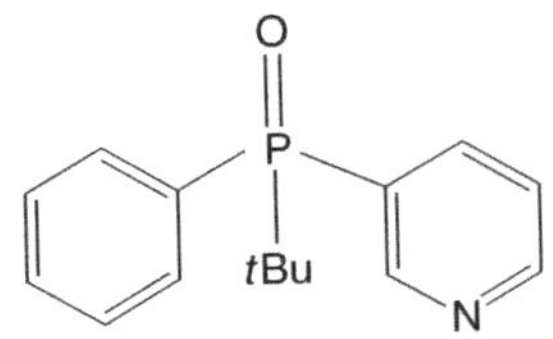

Scheme 15.34

in vacuo. The crude product was purified over a small column of basic alumina (ca. 5 cm), using Et_2O as an eluent. ^{31}P NMR (121.5 MHz, $CDCl_3$): δ 47.5 ppm.

(Rac)-3-(tert-butyl-phenyl-phosphinoyl)-pyridine (Scheme 15.34): In a dry, degassed Schlenk, $Pd(OAc)_2$ (135 mg, 0.60 mmol) and dppb (282 mg, 0.66 mmol) were dissolved in 15 ml freshly distilled DMF. After 5 min stirring, 3-iodopyridine (1.48 g, 7.22 mmol) was added, after which the mixture was stirred for 10 min at room temperature. *tert*-Butyl-phenyl-phosphine (1.0 g, 6.02 mmol) and dipea (4.2 ml, 3.11 g, 24.1 mmol), dissolved in 5 ml DMF, were added, after which the mixture was stirred at 105°C for 16 h. The solution was allowed to cool down to room temperature, and the DMF was evaporated. The resulting brown-red oil was dissolved in chloroform, and washed twice with 5% aqueous $NaHCO_3$ and once with brine (under air; oxidation of phosphine). The organic layer was dried on $MgSO_4$, filtrated, and concentrated *in vacuo*. The crude product (oxide) was purified by means of chromatographic separation with Chromatotron®, equipped with a silica plate, EtOAc–CH_2Cl_2 1:1 with 1% Et_3N as a eluent. Yield: 0.97 g (62%) of an off-white solid.[1] H NMR (300 MHz, $CDCl_3$): δ 9.12 (m, 1H), 8.73 (m, 1H), 8.31 (m, 1H), 7.92 (m, 2H), 7.49 (m, 4H), 1.25 (d, $J_{(H,P)}$ = 15.3 Hz, 9H) ppm. ^{31}P NMR (300 MHz, $CDCl_3$): δ 38.2 ppm. ^{13}C NMR (125.7 MHz, $CDCl_3$): δ 152.2 (d, $J_{(C,P)}$ = 1.76 Hz), 152.1 (d, $J_{(C,P)}$ = 9.80 Hz), 140.8 (d, $J_{(C,P)}$ = 5.78 Hz), 132.2 (d, $J_{(C,P)}$ = 8.67 Hz), 132.2 (d, $J_{(C,P)}$ = 2.89 Hz), 130.4 (d, $J_{(C,P)}$ = 92.4 Hz), 128.8 (d, $J_{(C,P)}$ = 10.9 Hz), 128.3 (d, $J_{(C,P)}$ = 85.5 Hz), 123.8 (d, $J_{(C,P)}$ = 7.54 Hz), 34.3 (d, $J_{(C,P)}$ = 71.6 Hz), 25.2 ppm. Anal. calculated for $C_{15}H_{18}NOP$: C 69.48, H 7.00, N 5.40; found: C 69.58, H 8.44, N 5.20.

The enantiomers were separated at DSM, using preparative HPLC equipped with an AD column (l = 25 cm, eluent *n*-heptane/*tert*-butanol 72.5/27.5, 1 mL/min). t_1 = 15.5 min (enantiomeric purity: 98.5±0.5%); t_2 = 20.3 min (enantiomeric purity: 97.6%).

Elemental analysis after separation of enantiomers:

Enantiomer 1: Anal. calculated for $C_{15}H_{18}NOP$: C 69.48, H 7.00, N 5.40; found: C 69.33, H 6.93, N 5.32.
Enantiomer 2: Anal. calculated for $C_{15}H_{18}NOP$: C 69.48, H 7.00, N 5.40; found: C 69.34, H 6.96, N 5.30.

*3-(tert-butyl-phenyl-phosphanyl)-pyridine, both enantiomers **B5** and **B6** (Scheme 15.35)*: The phosphine–oxide (120 mg, 0.46 mmol) was dissolved in 5 mL $PhSiH_3$, heated until reflux and stirred for 40 h at reflux conditions. Then the solvent was removed *in vacuo* and the product was purified over a short silica column (±5 cm), using EtOAc–PE 1/1 mixture as a solvent.

Analysis data:

Enantiomer 1: ^1H NMR (300 MHz, $CDCl_3$): δ 8.77 (br. s., 1 H), 8.56 (d, 1 H, J = 5.10 Hz), 7.83–7.81 (m, 1H), 7.58–7.53 (m, 2H), 7.38–7.35 (m, 3 H), 7.29–7.25 (m, 1 H), 1.19 (d, 9 H, J = 12.9 Hz) ppm. ^{31}P NMR (300 MHz, $CDCl_3$): δ 12.8 ppm. Anal. calculated for $C_{15}H_{18}NO_2P + H_2O$: C 64.97, H 7.27, N 5.05, found: C65.18, H 6.75, N 4.82.

Enantiomer 2: ^1H NMR (300 MHz, $CDCl_3$): δ 8.78 (br. s., 1 H), 8.57 (d, 1H, J = 4.80 Hz), 7.87–7.82 (m, 1 H), 7.62–7.52 (m, 2 H), 7.44–7.26 (m, 4 H), 1.20 (d, 9 H, J = 12.9 Hz) ppm. ^{31}P NMR (300 MHz, $CDCl_3$):

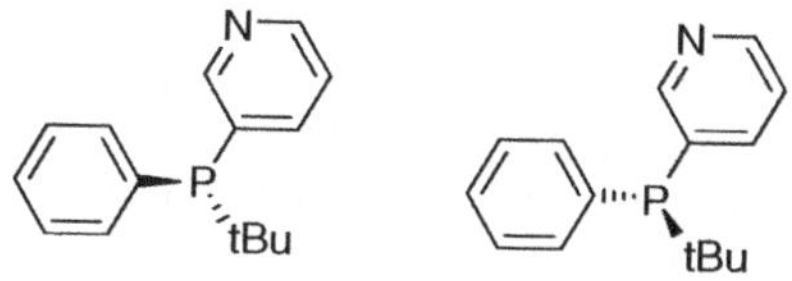

B5 B6

Scheme 15.35

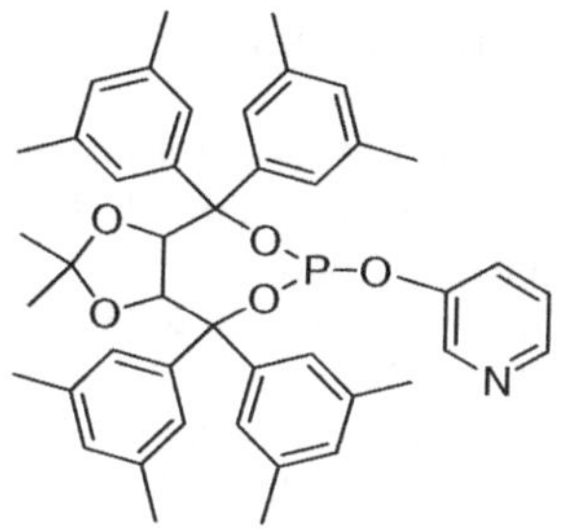

Scheme 15.36

δ 12.8 ppm. ^{13}C NMR (125.7 MHz, CDCl3): δ 154.1, 153.8 (d, J=9.81 Hz), 149.8, 148.8, 142.5, 136.2, 134.6 (d, J=37.9 Hz), 129.5 (d, J=29.4 Hz), 128.4, 123.5, 31.1, 28.8 (d, J=14.7 Hz) ppm.

Synthesis of (S,S)-3-[4,4,8,8-tetrakis-(3,5-dimethyl-phenyl)-2,2-dimethyl-tetrahydro-[1,3]dioxolo[4,5-e][1,3,2]dioxaphosphepin-6-yloxy]-pyridine (Scheme 15.36): 3-hydroxypyridine (0.58 g, 6.13 mmol), azeotropically dried with toluene (3×5 ml), and triethylamine (0.94 ml, 6.74 mmol) were dissolved in dichloromethane (20 ml), and the solution was cooled to –50°C. Freshly prepared (1*S*,7*S*)-4-chloro-9,9-dimethyl-2,2,6,6-tetra(3,5-dimethylphenyl)-3,5,8,10-tetraoxa-4-phosphabicyclo-[5.3.0]decane (3.94 g, 6.13 mmol), dissolved in dichloromethane (20 ml) was added dropwise. The cooling bath was removed, and the solution was allowed to warm to room temperature; stirring was continued for 2 h. The solvent was evaporated, after which the product was extracted with a mixture of 20 ml toluene and 60 ml hexanes. After filtration the solvent was removed *in vacuo*, giving **D7** (3.31 g, 4.72 mmol, 77%) as white foam. ^{1}H NMR ($CDCl_3$, 300 MHz): δ 8.26 (dd, 1H, J=4.65 Hz, 1.50 Hz), 7.90 (br. d, 1H, J=2.70 Hz), 7.28–7.26 (m, 1H), 7.21–7.15 (m, 5H), 7.10–7.02 (m, 5H), 6.97–6.91 (m, 3H), 6.84 (br. d, 2H, J=9.60 Hz), 5.45 (d, 1H, J=7.80 Hz), 5.06 (dd, J=8.25 Hz, 0.60 Hz), 2.32 (d, 15H, J=11.4 Hz), 2.24 (s, 9H), 0.85 (s, 3H), 0.73 (s, 3H); ^{31}P NMR (121.5 MHz, $CDCl_3$): δ 126.45; ^{13}C NMR (75.5 MHz, $CDCl_3$): δ 149.9, 149.2, 149.1, 145.9, 144.3, 142.5, 142.4, 141.1, 141.0, 137.6, 137.4, 136.7, 136.6, 129.7, 129.3, 129.2, 129.1, 128.5, 127.1, 127.0, 126.9, 126.7, 125.6, 125.2, 125.1, 123.5, 113.2, 86.9, 86.8, 85.9, 85.8, 82.6, 82.5, 80.1, 27.0, 26.7, 21.9, 21.8; anal. calculated for $C_{44}H_{48}NO_5P$: C 75.30, H 6.89, N 2.00; found: C 75.18, H 6.77, N 1.96.

References

[1] C. A. Tolman, *Chem. Rev.* **1977**, *77*, 313.

[2] (a) P. W. N. M. van Leeuwen, P. C. J. Kamer, J. N. H. Reek, P. Dierkes, *Chem. Rev.* **2000**, *100*, 2741; (b) P. C. J. Kamer, P. W. N. M. van Leeuwen, J. N. H. Reek, *Acc. Chem. Res.* **2001**, *34*, 895; and (c) C. P. Casey, G. T. Whiteker, *Isr. J. Chem.* **1990**, *30*, 299.

[3] J. Meeuwissen, M. Kuil, A. M. van der Burg, A. J. Sandee, J. N. H. Reek, *Chem. Eur. J.* **2010**, *16*, 10272, and references therein.

[4] (a) J. I. van der Vlugt, J. M. Bonet, A. M. Mills, A. L. Spek, D. Vogt, *Tetrahedron Lett.* **2003**, *44*, 4389; (b) J. I. van der Vlugt, M. M. P. Grutters, A. M. Mills, H. Kooijman, A. L. Spek, D. Vogt, *Eur. J. Inorg. Chem.* **2003**, 4361; (c) J. I. van der Vlugt, J. Ackerstaff, T. W. Dijkstra, A. M. Mills, H. Kooijman, A. L. Spek, A. Meetsma, H. C. L. Abbenhuis, D. Vogt, *Adv. Synth. Catal.* **2004**, *346*, 399; (d) J. I. van der Vlugt, R. Sablong, A. M. Mills, A. L. Spek, D. Vogt, *Organometallics* **2004** *23*, 3177; (e) E. J. Zijp, J. I. van der Vlugt, D. M. Tooke, A. L. Spek, D. Vogt, *Dalton Trans.* **2005**, 512; (f) T. P. Clark, C. R. Landis, S. L. Freed, J. Klosin, K. A. Abboud, *J. Am. Chem. Soc.* **2005**, *127*, 5040; (g) J. Wassenaar, J. N. H. Reek, *Dalton Trans.* **2007**, 3750; and (h) S. H. Chikkali, R. Bellini, G. Berthon-Gelloz, J. I. van der Vlugt, B. de Bruin, J. N. H. Reek, *Chem. Commun.* **2010**, *46*, 1244.
[5] (a) M. T. Reetz, G. Mehler, *Angew. Chem. Int. Ed.* **2000**, *39*, 3889; (b) M. van den Berg, A. J. Minnaard, E. P. Schudde, J. van Esch, A. H. M. de Vries, J. G. de Vries, B. L. Feringa, *J. Am. Chem. Soc.* **2000**, *122*, 11539; and (c) C. Claver, E. Fernandez, A. Gillon, K. Heslop, D. J. Hyett, A. Martorell, A. G. Orpen, P. G. Pringle, *Chem. Commun.* **2000**, 961.
[6] M. T. Reetz, T. Sell, A. Meiswinkel, G. Mehler, *Angew. Chem. Int. Ed.* **2003**, *42*, 790. See also D. Peña, A. J. Minnaard, J. A. F. Boogers, A. H. M. de Vries, J. G. de Vries, B. L. Feringa, *Org. Biomol. Chem.* **2003**. *1*, 1087.
[7] B. Breit, W. Seiche, *J. Am. Chem. Soc.* **2003**, *125*, 6608.
[8] W. Seiche, A. Schuschkowski, B. Breit, *Adv. Synth. Catal.* **2005**, *347*, 1488.
[9] U. Gellrich, J. Huang, W. Seiche, M. Keller, M. Meuwly, B. Breit, *J. Am. Chem. Soc.* **2011**, *133*, 964.
[10] B. Breit, W. Seiche, *Angew. Chem. Int. Ed.* **2005**, *44*, 1640.
[11] M.-N. Birkholz, N. V. Dubrovina, H. Jiao, D. Michalik, J. Holz, R. Paciello, B. Breit, A. Börner, *Chem. Eur. J.* **2007**, *13*, 5896.
[12] C. Waloch, J. Wieland, M. Keller, B. Breit, *Angew. Chem. Int. Ed.* **2007**, *46*, 3037.
[13] M. de Greef, B. Breit, *Angew. Chem. Int. Ed.* **2009**, *48*, 551.
[14] I. Usui, S. Schmidt, M. Keller, B. Breit, *Org. Lett.* **2008**, *10*, 1207.
[15] M. Weis, C. Waloch, W. Seiche, B. Breit, *J. Am. Chem. Soc.* **2006**, *128*, 4188.
[16] J. Wieland, B. Breit, *Nature Chem.* **2010**, *2*, 832.
[17] (a) F. Chevallier, B. Breit, *Angew. Chem. Int. Ed.* **2006**, *45*, 1599; and (b) T. Šmejkal, B. Breit, *Organometallics* **2007**, *26*, 2461.
[18] A. C. Laungani, B. Breit, *Chem. Commun.* **2008**, 844.
[19] A. J. Sandee, A. M. van der Burg, J. N. H. Reek, *Chem. Commun.* **2007**, 864.
[20] J. Meeuwissen, R. Detz, A. J. Sandee, B. de Bruin, M. A. Siegler, A. L. Spek, J. N. H. Reek, *Eur. J. Inorg. Chem.* **2010**, 2992.
[21] A. C. Laungani, J. M. Slattery, I. Krossing, B. Breit, *Chem. Eur. J.* **2008**, *14*, 4488.
[22] A. C. Laungani, M. Keller, J. M. Slattery, I. Krossing, B. Breit, *Chem. Eur. J.* **2009**, *15*, 10405.
[23] (a) P.-A. R. Breuil, F. W. Patureau, J. N. H. Reek, *Angew. Chem. Int. Ed.* **2009**, *47*, 2162; and (b) P.-A. R. Breuil, J. N. H. Reek, *Eur. J. Org. Chem.* **2009**, 6225.
[24] J. Wassenaar, M. Kuil, J. N. H. Reek, *Adv. Synth. Catal.* **2008**, *350*, 1610.
[25] F. W. Patureau, M. Kuil, A. J. Sandee, J. N. H. Reek, *Angew. Chem. Int. Ed.* **2008**, *47*, 3180.
[26] F. W. Patureau, S. de Boer, M. Kuil, J. Meeuwissen, P.-A. R. Breuil, M. A. Siegler, A. L. Spek, A. J. Sandee, B. de Bruin, J. N. H. Reek, *J. Am. Chem. Soc.* **2009**, *131*, 6683.
[27] L. Pignataro, S. Carboni, M. Civera, R. Colombo, U. Piarulli, C. Gennari, *Angew. Chem. Int. Ed.* **2010**, *49*, 6633.
[28] L. Pignataro, B. Lynikaite, J. Cvengroš, M. Marchini, U. Piarulli, C. Gennari, *Eur. J. Org. Chem.* **2009**, 2539.
[29] Y. Liu, C. A. Sandoval, Y. Yamaguchi, X. Zhang, Z. Wang, K. Kato, K. Ding, *J. Am. Chem. Soc.* **2006**, *128*, 14212.
[30] Y. Li, Y. Feng, Y.-M. He, F. Chen, J. Pan, Q.-H.Fan, *Tetrahedron Lett.* **2008**, *49*, 2878.
[31] (a) V. F. Slagt, J. N. H. Reek, P. C. J. Kamer, P. W. N. M. van Leeuwen, *Angew. Chem. Int. Ed.* **2001**, *40*, 4271; (b) A. W. Kleij, M. Kuil, D. M. Tooke, A. L. Spek, J. N. H. Reek, *Inorg. Chem.* **2005**, *44*, 7696; (c) M. Kuil, T. Soltner, P. W. N. M. van Leeuwen, J. N. H. Reek, *J. Am. Chem. Soc.* **2006**, *128*, 11344; (d) A. W. Kleij, J. N. H. Reek, *Chem. Eur. J.* **2006**, *12*, 4218; (e) V. F. Slagt, P. Kaiser, A. Berkessel, M. Kuil, A. M. Kluwer, P. W. N. M. van Leeuwen, J. N. H. Reek, *Eur. J. Inorg. Chem.* **2007**, 4653; (f) A. M. Kluwer, R. Kapre, F. Hartl, M. Lutz, A. L. Spek, A. M. Brouwer, P. W. N. M. van Leeuwen, J. N. H. Reek, *Proc. Natl. Acad. Sci. USA* **2009**, *106*, 10460; and (g) R. Bellini, S. H. Chikkali, G. Berthon-Gelloz, J. N. H. Reek, *Angew. Chem. Int. Ed.* **2011**, *50*, 7342. See also J. Flapper, J. N. H. Reek, *Angew. Chem. Int. Ed.* **2007**, *46*, 8590.

[32] V. F. Slagt, P. W. N. M. van Leeuwen, J. N. H. Reek, *Chem. Commun.* **2003**, 2474.
[33] V. F. Slagt, P. W. N. M. van Leeuwen, *Dalton Trans.* **2007**, 2302.
[34] V. F. Slagt, P. W. N. M. van Leeuwen, *Angew. Chem. Int. Ed.* **2003**, *42*, 5619.
[35] C. C. Mak, N. Bampos, J. K. M. Sanders, *Angew.Chem. Int. Ed.* **1998**, *37*, 3020.
[36] M. Kuil, P. E. Goudriaan, A. W. Kleij, D. M. Tooke, A. L. Spek, P. W. N. M. van Leeuwen, J. N. H. Reek, *Dalton Trans.* **2007**, 2311.
[37] M. Kuil, P. E. Goudriaan, P. W. N. M. van Leeuwen, J. N. H. Reek, *Chem. Commun.* **2006**, 4679.
[38] V. F. Slagt, M. Röder, P. C. J. Kamer, P. W. N. M. van Leeuwen, J. N. H. Reek, *J. Am. Chem. Soc.* **2004**, *125*, 4056.
[39] P. E. Goudriaan, X.-B. Jiang, M. Kuil, R. Lemmens, P. W. N. M. van Leeuwen, J. N. H. Reek, *Eur. J. Org. Chem.* **2008**, 6079.
[40] X.-B. Jiang, L. Lefort, P. E. Goudriaan, A. H. M. de Vries, P. W. N. M. van Leeuwen, J. G. de Vries, J. N. H. Reek, *Angew. Chem. Int. Ed.* **2006**, *45*, 1223.
[41] X.-B. Jiang, P. W. N. M. van Leeuwen, J. N. H. Reek, *Chem. Commun.* **2007**, 2287.
[42] P. E. Goudriaan, M. Kuil, X.-B. Jiang, P. W. N. M. van Leeuwen, J. N. H. Reek, *Dalton. Trans.* **2009**, 1801.
[43] J. N. H. Reek, M. Röder, P. E. Goudriaan, P. C. J. Kamer, P. W. N. M. van Leeuwen, V. F. Slagt, *J. Organomet. Chem.* **2005**, *690*, 4505.
[44] J. M. Takacs, D. S. Reddy, S. A. Moteki, D. Wu, H. Palencia, *J. Am. Chem. Soc.* **2004**, *126*, 4494.
[45] J. M. Takacs, K. Chaiseeda, S. A. Moteki, D. S. Reddy, D. Wu, K. Chandra, *Pure Appl. Chem.* **2006**, *78*, 501.
[46] S. A. Moteki, J. M. Takacs, *Angew.Chem. Int. Ed.* **2008**, *47*, 894.
[47] (a) S. H. Chikkali, D. Gudat, F. Lissner, M. Niemeyer, T. Schleid, M. Nieger, *Chem. Eur. J.* **2008**, *14*, 482; (b) S. Chikkali, D. Gudat, *Eur. J. Inorg. Chem.* **2006**, 3005; (c) S. H. Chikkali, D. Gudat, F. Lissner, M. Nieger, T. Schleid, *Dalton Trans.* **2007**, 3906; and (d) S. Chikkali, D. Gudat, M. Niemeyer, *Chem. Commun.* **2007**, 981.
[48] L. K. Knight, Z. Freixa, P. W. N. M. van Leeuwen, J. N. H. Reek, *Organometallics* **2006**, *25*, 954.
[49] P. A. Duckmanton, A. J. Blake, J. B. Love, *Inorg.Chem.* **2005**, *44*, 7708.
[50] For another recent example of anion-steering catalysis, see P. Dydio, W. I. Dzik, B. de Bruin, J. N. H. Reek, *Angew. Chem. Int. Ed.* **2011**, *50*, 396.
[51] (a) C. Gennari, U. Piarulli, *Chem. Rev.* **2003**, *103*, 3071; (b) J. G. de Vries, A. H. M. de Vries, *Eur. J. Org. Chem.* **2003**, 799; (c) T. Satyanarayana, H. B. Kagan, *Adv. Synth. Catal.* **2005**, *347*, 737; (d) J. G. de Vries, L. Lefort, *Chem. Eur. J.* **2006**, *12*, 4722; and (e) C. Jäkel, R. Paciello, *Chem. Rev.* **2006**, *106*, 2912.
[52] A. M. Kluwer, I. Ahmad, J. N. H. Reek, *Tetrahedron Lett.* **2007**, *48*, 2999.

16

Solid-phase Synthesis of Ligands

Michiel C. Samuels[1], Bert H. G. Swennenhuis[2], and Paul C. J. Kamer[1]

[1] *EaStCHEM, School of Chemistry, University of St Andrews, St Andrews, United Kingdom*
[2] *Department of Chemistry, Texas A&M University at Qatar, Doha, Qatar*

16.1 Introduction

Homogeneous catalysis plays an increasingly important role in chemical synthesis, both in the production of fine chemicals and in the bulk industry [1]. Notwithstanding its huge success, there is still a pressing need for novel highly active, selective, stable, and reusable catalysts. By far the most important and widespread ligands for (asymmetric) transition-metal catalysis are (chiral) phosphorus-based ligands [1]. The electronic and steric properties of these ligands often have a pronounced influence on the activity, stability, and selectivity of a catalyst. Consequently, ligand design has evolved as a powerful tool in the development of superior catalysts. Despite considerable progress in the last few decades, it is often not possible to rationally design ligands, and thus the development of new catalysts often relies on "trial and error." This in turn necessitates the synthesis and screening of large families of ligands [2], but due to the reactivity of phosphorus compounds (e.g., the moisture sensitivity of phosphites), this can be a laborious and time-consuming process.

One way to speed up this process can be the application of solid-phase synthesis (SPS). Originally developed for the synthesis of peptides [3], SPS is based on the covalent attachment of a reactive functionality onto a polymeric support, which can also facilitate a combinatorial approach. The formed immobilized functionality can then serve as the starting point for a sequence of reactions, leading to the desired supported compound. The primary advantage of the solid-phase approach is the ease of purification, as the support can be separated from the reaction mixture by simple filtration or decantation. Consequently, the use of large excesses of reagents, which can be practical as well as beneficial for the yield, does not complicate the purification of the ligands. Furthermore, the reagents can be neutralized *after* the filtration step, which can be convenient if reactive reagents like organometallic compounds are used, but can be essential if supported compounds with reactivities similar to those of the reagents are formed. Moreover, as a result

Phosphorus(III) Ligands in Homogeneous Catalysis: Design and Synthesis, First Edition.
Edited by Paul C. J. Kamer and Piet W. N. M. van Leeuwen.

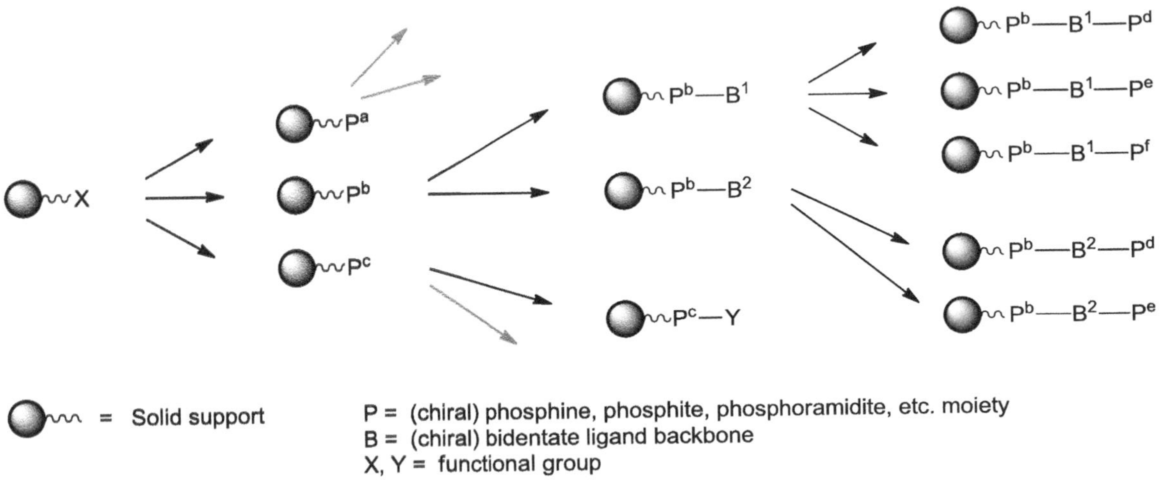

Scheme 16.1 *Application of solid-phase synthesis for the combinatorial synthesis of ligands.*

of the immobilization on the support, intermolecular reactions can be reduced and immobilization can stabilize reactive intermediates and thus reduce the formation of side products.

Due to these features, SPS facilitates the use of combinatorial approaches for the synthesis of large libraries of compounds in a fast and efficient manner. When applied to the synthesis of ligands, these combinatorial techniques can facilitate the formation of ligand families and thus SPS can be a powerful tool in the development of novel catalysts. Scheme 16.1 depicts the general approach of the solid-phase synthesis of phosphorus-based ligands. Reaction of a functional group on the support with a suitable phosphorus reagent allows the loading of this reagent on the support. Subsequently, this intermediate can be converted to the desired supported ligand in a sequence of reactions. Similar to the synthesis of nonsupported ligands, structural diversity can be accomplished by using various reagents in one or more phases of the synthesis, allowing the modular synthesis of mono- and bidentate ligands with various electronic, steric, and stereo-chemical properties.

Supported ligands have proven their efficiency in several transition metal-catalyzed reactions [4]. One of the main disadvantages of homogeneous catalysis is the difficult recovery of the catalysts, which prohibits recycling of the valuable metals and ligands and can lead to contamination of the product with toxic transition metals. Consequently, it is no surprise that the development of techniques to separate the catalyst from the reaction mixture has received considerable attention [5]. The use of supported ligands allows the formation of supported catalysts, which can be easily separated from the reaction mixture by a simple filtration procedure and can subsequently be reused. Good mixing of the catalyst and the reactants can be maintained by the application of supports with excellent swelling properties (*vide infra*). The approach thus combines the advantages of homogeneous catalysis such as good control over catalyst properties via ligand design and with the easy catalyst separation and recycling of heterogeneous catalysis.

Despite the advantages of SPS, examples of the combinatorial synthesis of families of phosphorus ligands by multistep solid-phase reactions are still relatively scarce, which can be attributed mainly to the lack of efficient synthetic methodologies [6]. Consequently, the majority of the reviews on polymer-supported ligands discuss the application of this class of ligands in catalysis [2g,4c,4f,4i–k,5b], while the synthetic approaches have received less attention. Below, we will first outline some practical issues associated with SPS, followed by an overview of the various synthetic methodologies developed to generate libraries of supported ligands, without claiming to discuss all polymer-bound ligands that have appeared in literature.

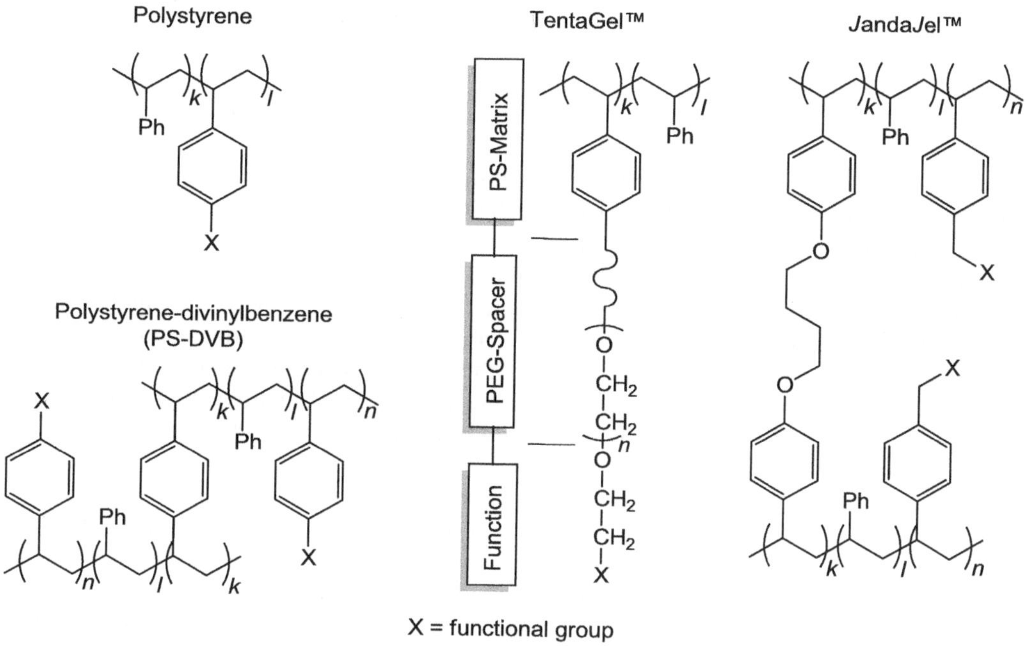

Figure 16.1 *General structures of several polystyrene-based supports.*

Since most of the standard purification techniques such as crystallization or column chromatography cannot be used, the formation of supported side products should be suppressed, necessitating all synthetic steps to be optimized to proceed quantitatively. In general, conditions used for the synthesis of nonsupported ligands cannot be simply applied to the synthesis of the supported analogues. Frequently the type of solvent and the applied reaction temperatures need to be adjusted to get good conversion, which often can be attributed to the specific swelling properties of the support (*vide infra*). Furthermore, although organometallic chemistry can often be successfully applied to SPS [7], the high reactivity of some reagents may prohibit their use in combination with a support, necessitating the development of alternative routes.

Solid supports (often referred to as resins) have been initially developed for and are still extensively used in peptide synthesis, but SPS has increasingly been applied to other fields of chemistry, and consequently a large variety of supports is (commercial) available [4j]. Linear non-cross-linked polystyrenes (PS; Figure 16.1) are soluble in most commonly used organic solvents, but precipitate in protic solvents like water and methanol and are therefore classified as *soluble* supports. If styrene is co-polymerized with divinylbenzene (DVB), cross-linked polymers are formed (PS-DVB; Figure 16.1), which are insoluble in all solvents and therefore are referred to as *insoluble* supports. For these resins, good swelling in the reaction medium is of crucial importance, as this impacts the accessibility of the functional groups and consequently the reaction rates and yields. For PS-DVB the swelling is generally very good in THF, toluene, dioxane, and CH_2Cl_2, while the resin contracts in solvents like diethyl ether, methanol, and water. This difference in swelling is being utilized in the purification of the support, by alternating washing cycles with solvents in which the support swells and in which it contracts. One of the most commonly used supports is Merrifield resin, a copolymer of styrene, chloromethylstyrene, and DVB, with a chloromethyl reactivity ($X=CH_2Cl$; Figure 16.1) [3].

The type and percentage of cross-linking and the type and loading of the functional group determine the swelling properties of the support as well [4j]. For instance, JandaJel™ [4d,4e,8] is a support based on a flexible chemically robust polytetrahydrofuran cross-linker, resulting in a polymer with excellent swelling properties in a broad range of solvents. The presence of polyethyleneglycol (PEG) linkers on the polystyrene backbone, as for instance in the TentaGel™ resins (Figure 16.1) [9], also enhances the swelling properties and makes this resin almost solvent independent. These PEG-based linkers can, however, react with strong bases and organometallic reagents, resulting in "PEG leakage" [8]. Moreover, due to the higher molecular weight of these types of support, the loading of functional groups (expressed as mmol per gram resin) can drop considerably in comparison to PS-DVB.

In addition to commercial available supports, straightforward (co-)polymerization processes allow the creation of "custom-made" supports with relative ease. Jana *et al.* [10], for instance, co-polymerized styrene with a dihydroxy-functionalized styrene in order to synthesize polymer-bound bidentate phosphites, for which the dihydroxy moiety serves as the backbone of the ligand. Bergbreiter *et al.* [11] developed soluble polymeric supports of which the solubility was temperature dependent (i.e., they were highly soluble at low temperatures but insoluble at room temperature). This enabled tuning of the hydrogenation activity of a supported rhodium catalyst by altering the temperature of the reaction mixture. Takaishi *et al.* [12] tuned the properties of polymeric bis-diphenylphosphine ligands by application of hydroxyethyl methacrylate in the co-polymerization process to create an insoluble resin which, unlike typical PS-based supports, has good swelling properties in alcohol and other polar solvents.

An important requirement for the application of a specific polymer as a support is the stability of both the ligand and the intermediates toward intermolecular side reactions. Several factors influence the occurrence of these: the nature of the ligand, the applied reaction conditions such as solvent and temperature, and the loading and flexibility of the support. Insoluble supports with a high degree of cross-linking may supply more site isolation than high-loading soluble or flexible supports, and consequently the latter may be unsuitable for some types of ligands and/or reactions. For instance, aryl exchange in high-loading polymer-bound triphenol phosphites can give nonsupported phosphites and polymers in which the phosphite moiety acts as a cross-linker. This not only leads to ligand degradation but also can yield polymers with reduced swelling properties and can result in metal leaching during catalysis. Replacement of the phenols with a chelating bis-phenoxy group can prevent this reaction [13].

The identification of the supported compounds is often difficult, as standard analysis techniques like nuclear magnetic resonance (NMR) and mass spectrometry are commonly unsuitable. However, supports with excellent swelling properties may allow characterization by ^{13}C- and ^{1}H-NMR spectrometry. Gel phase ^{31}P-NMR spectroscopy, on the other hand, can be used in most cases to monitor the reaction progress, to identify the phosphorus species present on the support, and to determine the purity of the product [14]. These measurements can be performed in common NMR tubes, as long as the resin is well swollen and is brought into the focus of the magnetic field within the spectrometer, making solvents on which the supports float (such as dichloromethane for PS-DVB) unsuitable. Generally the resonances of immobilized compounds have broader line widths than those of nonsupported analogues, but the broadening of the signals will depend strongly on the swelling of the used support. Elemental analysis can be used to determine the loading of phosphorus on the support and thus to determine the yield of a transformation.

16.2 Insoluble supports in ligand synthesis

The majority of immobilized ligands are synthesized by a single reaction between a support bearing a suitable reactive group and a phosphorus reagent (Scheme 16.2) [4g,6b,6c,6h,15]. The commercially available polystyrene-bound triphenylphosphines (PS-PPh$_2$) are among the most commonly applied supported ligands and can be formed via several routes. In 1971, Camps *et al.* [16] reported the first preparation by co-polymerization of styrene with *p*-styryldiphenylphosphine (3:1) and 2% divinylbenzene as the cross-linking agent, but more common is the synthesis by the reaction of bromopolystyrene with lithium diphenylphosphide (path **I**) [17] or by lithiation of bromopolystyrene followed by addition of chlorodiphenylphosphine (path **II**)

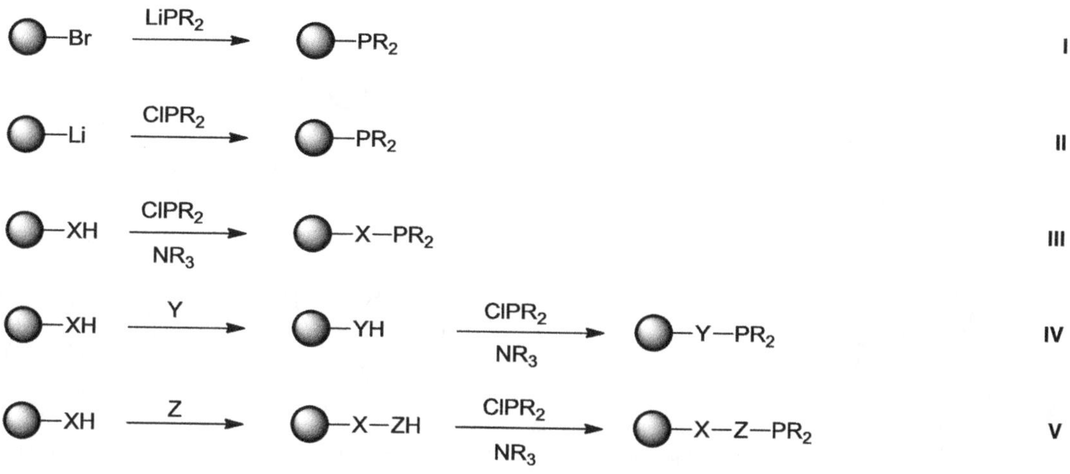

Scheme 16.2 *Commonly applied routes toward supported phosphorus ligands.*

[18]. The high reactivity of the used organometallic reagents can, however, lead to side reactions. For instance, it was observed that not all bromo groups were cleanly converted to the phosphine moiety, as some were simply reduced [16], while unwanted attack of alkyllithium on the C=C bonds of cross-linked polystyrene can lead to contamination and reduced swelling properties [4f]. Structural diversity can be achieved in routes **I** and **II** by the application of structurally different phosphorus reagents. For instance, Le Drain and coworkers [15c,19] synthesized series of eight supported diarylphosphines and six supported *tert*-butylarylphosphines by the straightforward reaction of various lithium phosphides ($LiPAr_2$ and LiP(*t*-Bu)Ar, respectively; Ar=Ph, Tolyl, Naphthyl, etc.) with Merrifield resin in THF at room temperature. Although 99% of the chlorine atoms were consumed in the reaction, only ~80% of the theoretical maximum of phosphino groups was introduced on the resin. The reaction of lithium *tert*-butylphenylphosphide with *p*-bromopolystyrene proved considerably less efficient, as 98% of the bromine groups were consumed, but only 25% were converted to a phosphino moiety [15c].

Immobilized phosphoramidites and phosphites are generally prepared by reaction of supports containing, respectively, amino- or hydroxy-functionalities with chlorophosphorus compounds in the presence of a tertiary amine as a base (path **III**). Similar to the supported phosphines, structural diverse ligands can be prepared by application of different phosphorus reagents.

An alternative approach to create families of ligands is by modifying the steric and electronic properties of the functional group on the support (path **IV** and **V**, Scheme 16.2). The group of Waldmann [6c], for instance, created a library of 78 phosphoramidites by applying a modular approach based on varying three functionalities (Figure 16.2, **A**, **B**, and **C**). The diversity of the ligands was mainly based on the use of 28 different R^1 groups (block **B**), which were systematically varied and included aliphatic and aromatic acyl groups, sulfonamides, urea and thiourea groups, and phosphorus substituents. Moreover, by changes in the amino functionality on the support (block **A**, X=C, N, O, S) and by the use of four BINOL derivatives (block **C**, R^2=H, Me, Ph, Br), the library of ligands was extended even more. Portnoy and coworkers [6i] synthesized a variety of amino-functionalized supports to create a library of 40 borane-protected α-aminophosphines (Figure 16.2, top right). Variations were made by connecting different linkers at the *ortho*-, *meta*-, and *para*-positions and 12 different electron-rich, electron-poor, and sterically crowed aromatic amines. Additionally, with the use of

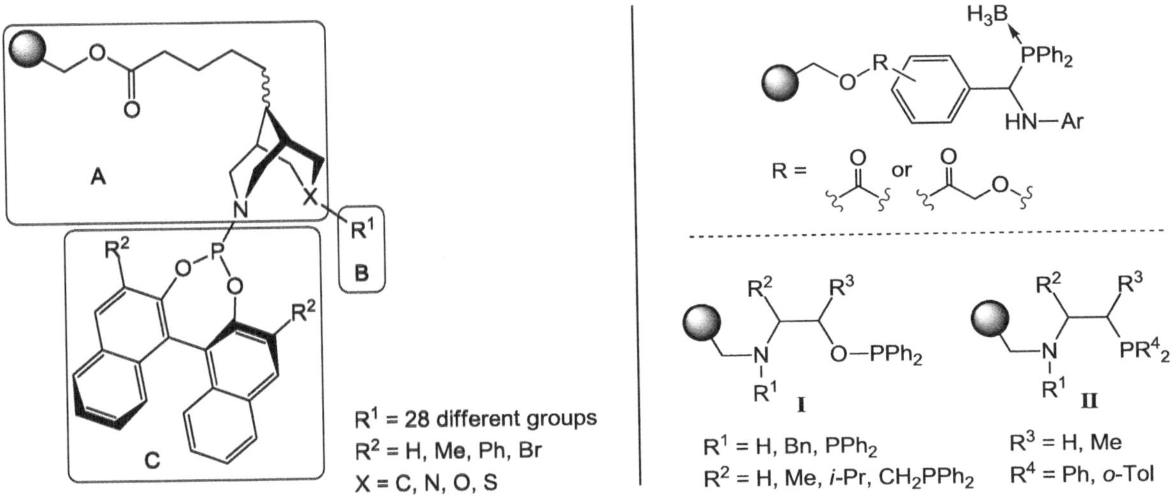

Figure 16.2 *General structure of ligands synthesized by Waldmann (left) and Portnoy (right).*

various aminoalcohol linkers, the formation of a family of phosphinite (bidentate) ligands was achieved (Figure 16.2, bottom right, **I**). Moreover, chlorodehydroxylation of the same aminoalcohol linkers allowed the synthesis of a series of amino–phosphine ligands (Figure 16.2, bottom right, **II**) [6d].

Similarly, the groups of Gilbertson [6a,20] and of Meldal [6e,21] created libraries of immobilized ligands by varying the peptide chain that links the phosphorus with the support. Examples of these types of ligands will be discussed in more detail in Chapter 19.

To prepare a library of 15 immobilized (chiral) monodentate ligands with various electronic and steric properties, Kamer and coworkers [6h,22] varied both the functional group on the (commercial) available polystyrene supports, as well as the chlorophosphorus reagent (Figure 16.3). The main advantage of SPS is the ease of purification, which allows the use of excesses of reagents and/or reagents containing impurities. Kamer *et al.* [6h] utilized this feature in the synthesis of the supported phosphites and phosphoramidites (Figure 16.3), as the use of an excess of partially hydrolyzed (~30%) chlorophosphites did not compromise the workup or the purity of the immobilized ligand. The synthesis of chlorophosphites from phenols and PCl_3 frequently leads to mixtures containing additional species, such as dichlorophosphites. The authors showed that the presence of a phosphite does not hinder the reaction between a hydroxy-functionalized polystyrene and a chlorophosphite and that, by pushing the equilibrium toward the formation of a mixture of chlorophosphite and phosphite, the formation of dichlorophosphite, and thus of resin-bound by-products, could be prevented effectively. The importance of the purity of the support was also discussed. Reaction of the *N*-methylaminomethylated and aminomethylated polystyrene supports with chlorophosphites yielded not only the expected supported phosphoramidites but also a second species. ^{31}P-NMR spectroscopy suggested that alcohol functionalities were also present on the resins, leading to supported phosphites. Since these were covalently bound to the polymer, they could not be separated from the targeted supported phosphoramidites.

In the routes in Scheme 16.2, the structural diversity is created on the phosphorus *prior* to the coupling with the support or on the support *prior* to the coupling with the phosphorous reagent. Examples in which structural diverse ligands are created from a single, polymer-supported, phosphorus-containing precursor (i.e., the "bottom-up" approach) are considerably rarer. This is surprising, as by synthesizing the ligand on the resin from the start of the reaction sequence, the advantages of SPS can be fully exploited during all steps of the sequence. Polystyrene-bound chlorophosphorus reagents allow maximum structural diversity, as they can be reacted with a large variety of reagents. The group of Kamer [6h] showed that these can be easily

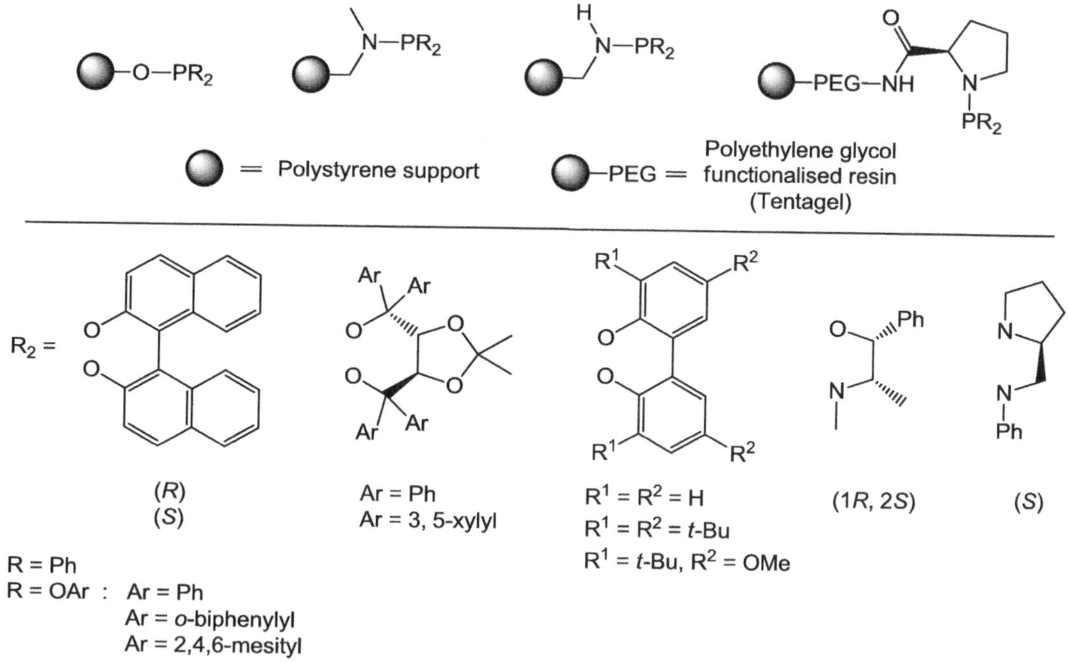

Figure 16.3 *Supported monodentate ligands synthesized by Kamer* et al.

accessed using relatively simple commercially available reagents and supports and that they are excellent starting materials for the synthesis of supported ligands (Figure 17.3). The synthesis of a supported dichlorophosphite and a chlorophosphoramidite and the conversion of these intermediates to, respectively, a phosphite and a phosphoramidite are described at the end of the chapter.

The synthesis of supported chiral phosphorus ligands received considerable attention, but the chirality in the majority of reported ligands originates from a BINOL moiety. Kamer and coworkers [6g] reported a rare example of the SPS of a family of *P*-stereogenic ligands. Adjusting a method developed by Jugé and coworkers [23] to polystyrene support enabled the parallel synthesis of supported *P*-stereogenic bidentate ligands. By reacting *para*-lithiopolystyrene **1b** [18] with oxazaphospholidine boranes **2a–f**, six different *P*-stereogenic immobilized aminophosphine boranes (**3a–f**) were synthesized (Scheme 16.3). Analysis by gel phase ^{31}P- NMR showed that these chiral intermediates were formed in excellent yields and purities and with high diastereoselectivities (diastereomeric ratios >96:4). More precisely, the reaction of the oxazaphospholidine boranes (**2a–f**) with the support proceeds with stereoselectivities which are comparable to those of the solution phase reactions. This last fact is particularly significant, as supported chiral ligands can only be applied in the development of efficient enantioselective catalysts, if the optical purity of the formed ligands is not compromised by the application of SPS. In fact, the optical purity should ideally be superior to that obtained with solution phase routes, as further enrichment using, for example, crystallization is not possible with SPS. The hydroxy moiety in aminophosphines **3a–f** allowed the straightforward synthesis of 10 bidentate aminophosphine–phosphinite (**6**) and aminophosphine–phosphite (**7**) ligands, by reaction with the appropriate chlorophosphorus reagents, followed by removal of the borane with $HNEt_2$. All ligands were obtained with a (stereo-)chemical purity of more than 92%. Synthesis of immobilized bidentate aminophosphine–phosphite **7b** (R^1=*o*-anisyl) via aminophosphine borane **3b** as a representative example of the approach can be found at the end of this chapter.

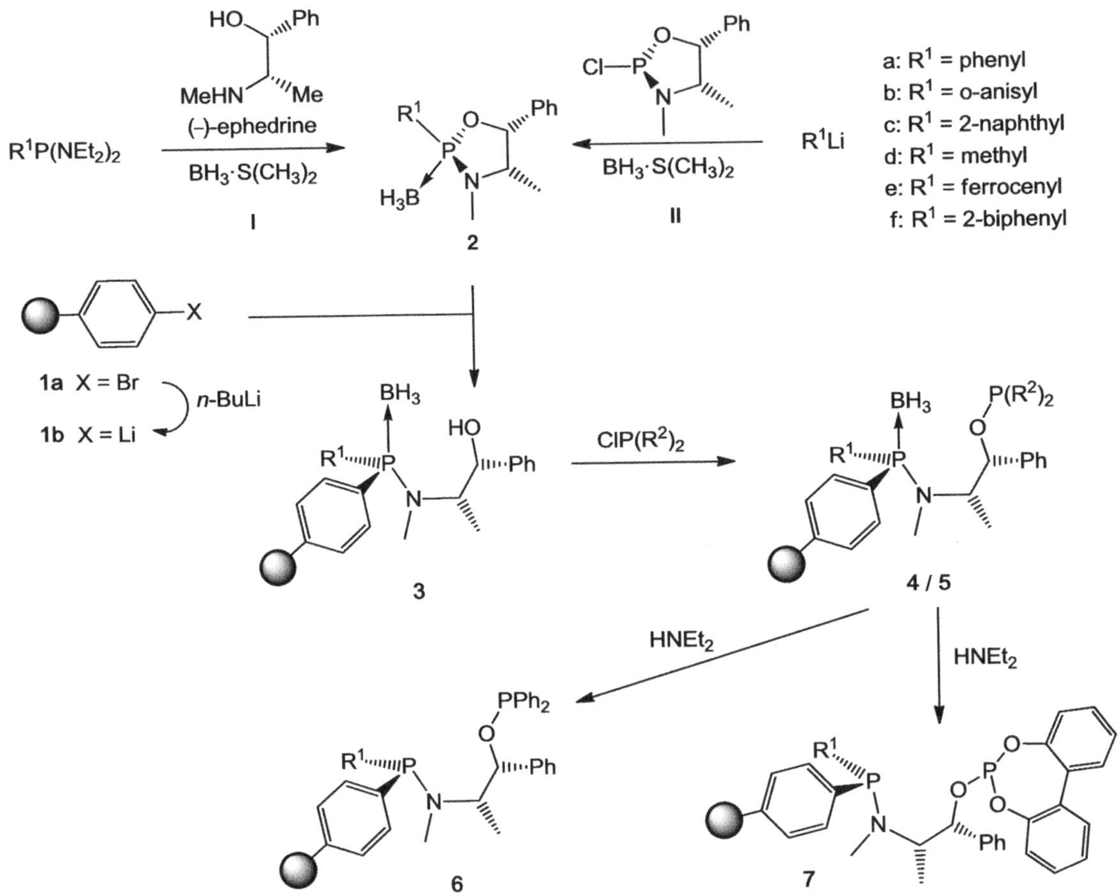

Scheme 16.3 Immobilized aminophosphine–phosphinite and aminophosphine–phosphite ligands.

16.3 Soluble polymeric supports

Soluble polymers have received considerable attention as supports for immobilized ligands and catalysts [24], and a number of reviews on this subject have appeared in recent years [4e,4i,25]. They are attractive alternatives to insoluble polymers, as they do not possess some of the limitations associated with the insolubility of the support (*vide supra*). A key difference is that soluble supports allow the synthesis to be carried out under homogeneous conditions, while the phase separation and the associated purification of the immobilized ligand can occur after the reaction. Consequently, the reaction conditions used for the synthesis of nonsupported ligands can often be applied to the synthesis of soluble polymer-bound analogues, as long as the support is inert to the applied reagents. However, other restrictions of synthesis on solid support still apply; the reactions have to be complete because polymer-bound side products cannot be removed. The solubility of the polymers is highly dependent on the nature of the polymer and the type of solvent, and this property can be used to separate and purify the soluble ligand. Several separation techniques have been developed, but solid–liquid and liquid–liquid techniques are the most commonly applied. An accessible approach is the utilization of polyethyleneglycol as support, as it is soluble in water and in common solvents such as toluene, DMF, and acetonitrile, but it precipitates in diethyl ether and hexane and this allows the

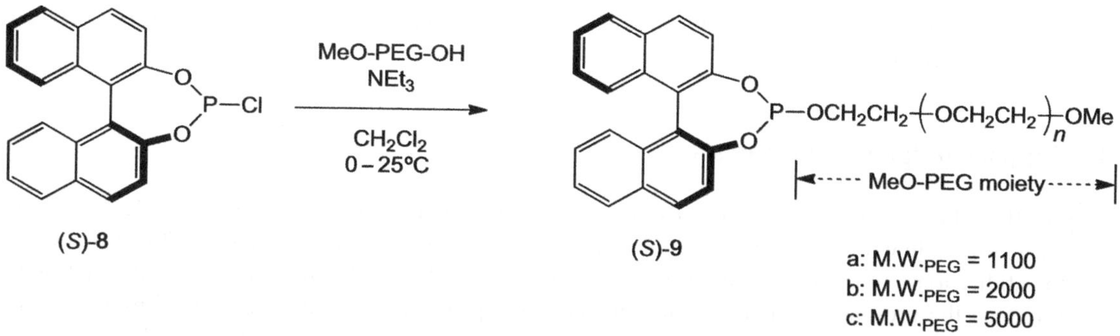

Scheme 16.4 Synthesis of chiral MeO–PEG–monophosphites.

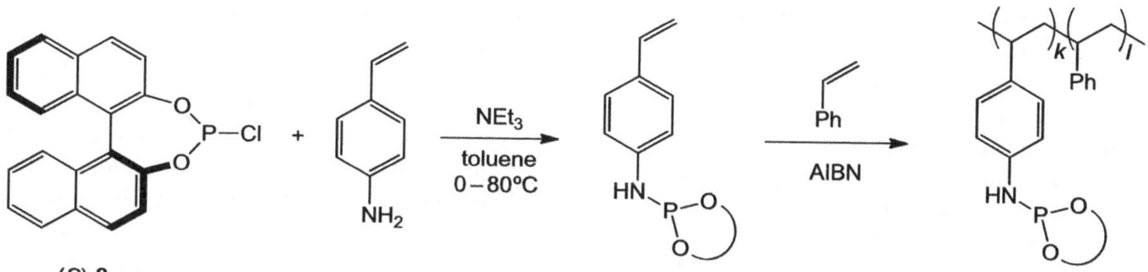

Scheme 16.5 Supported phosphoramidite by co-polymerization.

simple separation of PEG from the reaction mixture by the addition of suitable solvent to precipitate the polymer, followed by filtration [26]. A variety of soluble polymers with a wide range of properties have been developed, and several are commercially available. The reported phosphorus ligands are mainly supported di- or triphenylphosphines and BINOL-derived phosphites which are generally prepared by a simple single-step reaction between a chlorophosphorus reagent and the polymer [4e,4i,25d]. For instance, the group of Zheng [25e] reported the synthesis of MeOPEG–monophosphites (**9**) by treatment of the BINOL-based chlorophosphite **8** with an equimolar amount of commercially available PEG–monomethyl ether (MeOPEG-OH) in CH_2Cl_2 at 0 – 25°C, in the presence of triethylamine (Scheme 16.4). The soluble ligand was separated from the formed ammonium salt (NEt_3HCl) by filtration. Subsequent addition of diethyl ether to the filtrate resulted in precipitation of the polymer. Whittall *et al.* [15a] synthesized several related phosphites by reacting the commercial available PS-PEG$_{600}$-OH, TentaGel™-OH, MeO-PEG$_{2000}$-OH, MeO-PEG$_{5000}$-OH, and HO-PEG$_{4600}$-OH supports with chlorophosphite **8**. The use of trioctylamine as the base turned out to be essential for the purification of the ligands, as trioctylamine hydrochloride is soluble in diethyl ether and therefore can be separated from the ligands by selective precipitation of the latter.

Doherty *et al.* [27] investigated an alternative approach to synthesize soluble polymer-bound ligands (Scheme 16.5). Instead of treatment of a polymer containing a suitable functionality with a phosphorus reagent, the supported ligand (**12**) was formed by co-polymerization of a phosphoramidite-functionalized monomer (**11**) with styrene. The solubility properties of the polymer allowed a simple purification procedure (precipitation and washing), yielding the ligand in high purity. Note that the properties of the resulting polymer can be tuned by varying the type and the relative amounts of the monomers used in the co-polymerization, while cross-linking agents can be applied to create related insoluble supports [16]. A related soluble polymer-bound ligand was formed by co-polymerizing a quinoline-based phosphoramidite with styrene, and both polymers displayed different activities and enantioselectivities in rhodium-catalyzed asymmetric hydrogenations.

16.4 Supported ligands in catalysis

The application of polymer-supported ligands in catalysis has been extensively reviewed [2g,4c,4f,4i–k,5b], and therefore only a few representative examples will be discussed here. The key feature of supported catalysts is the ability to recover (by a simple filtration procedure) and reuse them after a catalytic reaction [4]. This can reduce the presence of metal contaminants in the product and allows the recycling of valuable metals and ligands. Polymer-bound ligands are particularly attractive when the excellent swelling properties of the support allow good mixing of the catalyst and the reactants. The fact that a PEG-based *soluble* polymer-supported chiral monodentate phosphite developed by Whittall *et al.* [15a] yields one of the most active rhodium catalysts for the asymmetric hydrogenation of dimethyl itaconate demonstrates this. It is, however, more commonly encountered that supported catalysts are less active and selective than the nonsupported analogues. The polystyrene-bound BINOL-based phosphite synthesized by Swennenhuis *et al.* [22] (Figure 16.3), for instance, induced 11% enantiomeric excess (ee) in the rhodium-catalyzed hydrogenation of dimethyl itaconate, which is in sharp contrast to the 97% ee observed for the nonsupported analogue. Despite this commonly encountered disadvantage, polymer-bound ligands can speed up the development of novel catalysts, as SPS allows the rapid synthesis of ligand libraries [6c,6d,6g,6j,19b,28]. Gilbertson and coworkers [6j] synthesized a family of 96 supported peptide-based phosphine ligands and tested these in a palladium-catalyzed asymmetric allylic alkylation reaction. While 77 members of the library gave ≥60% ee, for only one ligand 80% ee was observed, showing that in order to find highly efficient catalysts, it is often essential to screen large libraries of ligands, which can be efficiently created by SPS. In addition, screening libraries of supported ligands can provide essential information about the relationships between ligand structure and catalyst performance. Applying the family of 10 supported chiral bidentate ligands developed by den Heeten *et al.* [6g] (Scheme 16.3) in the rhodium-catalyzed hydrogenation of methyl α-acetamidocinnamate revealed that the nature of the R^1 and R^2 moieties have a decisive influence on the enantioselectivity and the absolute chirality of the product. Application of the biphenol-based aminophosphine–phosphite ligands leads to predominantly the *S*-enantiomer, while the aminophosphine–phosphines lead to the hydrogenation product with the *R* absolute configuration.

The application of SPS in catalysis research is not restricted to finding efficient supported catalysts. Waldmann and coworkers [6c] showed that the screening of *supported* ligands can be an excellent tool to identify the most efficient *nonsupported* ligand. For this purpose, a library of 78 polymer-bound phosphoramidites was synthesized by varying the electronic and steric properties of amino-functionalized resins. In the enantioselective copper-catalyzed conjugate addition to enones, the observed ee values ranged from 3% to 67%. A selection of the most efficient supported ligands was synthesized as their soluble nonsupported analogues. Application of these homogeneous systems in catalysis revealed that the polymer-bound phosphoramidites mirrored the performance of the analogue ligands in solution. Note that in numerous studies, however, no straightforward correlation between the performances of the supported and nonsupported analogues was found, which can most likely be attributed to the specific microenvironment created by the support. For instance, Pittman *et al.* [29] observed higher linear-to-branched selectivities in the hydroformylation of 1-pentene, catalyzed by DVB-cross-linked polymer-attached carbonylhydrotris(triphenylphosphine)rhodium $[(PS\text{-}PPh_3)_3RhH(CO)]$ compared with the nonsupported analogue, which was attributed to the high ligand catalyst concentration within the swollen resin.

Related to this is the fact that the type of solvent can have a decisive influence on the catalytic performance of polymer-supported catalysts, which often can be attributed to the swelling of the support in the applied reaction medium and thus on the microenvironment of the catalyst [30]. An often encountered problem is that a solvent system, in which the swelling properties of the support are optimal, may not be an ideal solvent for the catalyst. To overcome this, a support should be selected that suits the preferred solvent system of the catalyst and this is particularly important if the goal is to find efficient nonsupported catalysts through screening libraries of supported ligands. Gilbertson and Yamada studied the performance of palladium allylic allylation catalysts immobilized on various polymeric supports in several commonly used solvents [30a]. The support–solvent combinations displayed a variety of activities and enantioselectivities, and for the support–solvent combinations providing good swelling properties, the observed ee was comparable to that of the nonsupported catalyst in that specific solvent. For the combinations where there is an incompatibility between the solvent and the support, no clear correlation between the performances of the supported and nonsupported analogues could be found.

16.5 Solid-phase synthesis of nonsupported ligands

As discussed in this chapter, in several catalytic systems the nonsupported ligands proved to be superior over their supported analogues, as they yielded more active and selective catalysts. Cleaving the ligands from the support after synthesis is an attractive approach, as it allows utilizing the advantages of solid-phase-based synthesis, while yielding nonsupported ligands. Surprisingly, this methodology has received limited attention, as the majority of supported ligands have been synthesized with the aim to create supported catalysts. The general strategy is based on the use of a functionality (C, Scheme 16.6) that allows loading of phosphorus-based reagents on the solid phase (**I**), while it also permits cleavage of the product from the support at a desired stage of the synthetic sequence (**II**). Since analysis of solid-supported compounds is not always straightforward, this methodology can assist in the identification of products ($\mathbf{II}_a$), as well as intermediates ($\mathbf{II}_b$) that are formed during the reactions, as the cleaved compounds can be analyzed using standard techniques. A wide variety of linkers, the structural motif which temporarily joins the polymeric support and the ligand, have been developed, but the suitability of these in a specific synthetic route is highly dependent on the type of reagents and the reaction conditions applied [31]. The linker must allow attachment and cleavage of the phosphorus species under conditions mild enough not to affect the ligand, while it should tolerate the chemistry involved in the synthesis of the supported ligand. To the best of our knowledge, only amine-functionalized resins have so far been used in the solid-phase synthesis of nonsupported phosphorus-based ligands. The methodologies exploit the acid lability of P–N bonds, allowing the controlled release of the intermediates or ligands under relatively mild conditions.

Kamer *et al.* [6h] applied phosphoramidites linked to the solid support via an amido functionality as synthetic intermediates for phosphites (Scheme 16.7). Acid-catalyzed substitution of the P–N bond of the supported phosphoramidites (**15**) with alcohol yielded monophosphites (**16**) liberated from the resin (**13**). For this purpose, five supported phosphoramidites were formed *in situ*, by reacting amino-functionalized resins (**13**) with different chlorophosphites (**14**). A subsequent reaction with various alcohols in the presence of an acid activator allowed the parallel synthesis of a family of 20 (chiral) phosphites (**16**) in high overall yields

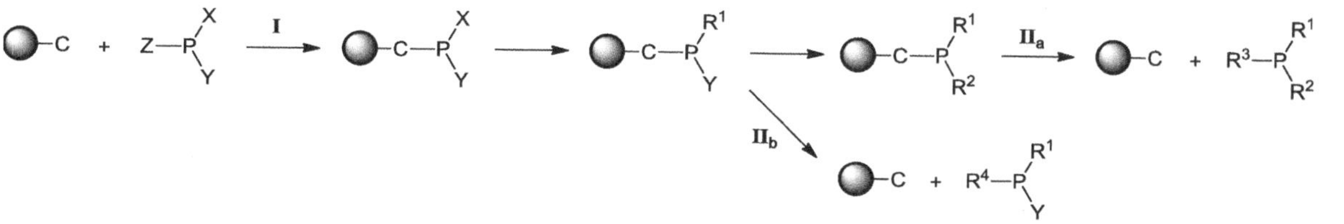

Scheme 16.6 Solid-phase synthesis of nonsupported ligands.

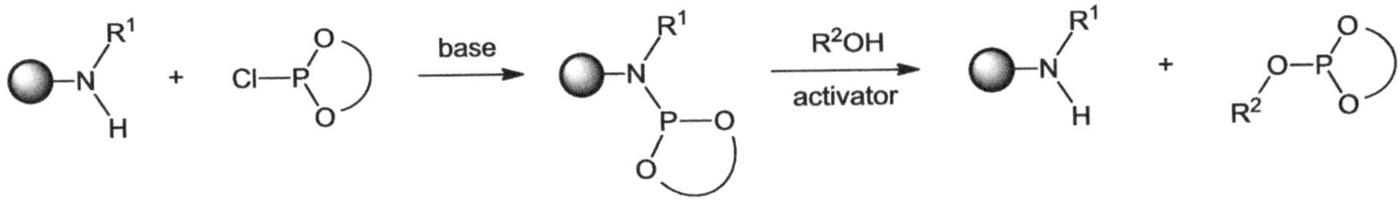
13 14 15 13 16

Scheme 16.7 *Solid-phase synthesis of phosphites.*

(82–91%) and purities (>97%). The cleavage of the P–N bond regenerates the amine-functionalized resin (**13**), and the authors showed that this can be reused with only a slight drop in yield (from 88% to 85%) between successive runs. The synthesis of 6-methoxydibenzo[d,f][1,3,2]dioxaphosphepine via this approach is described at the end of this chapter.

The leaving group capacity of the amino-functionalized resin and the nature of the alcohol and substituents on the phosphorus all influence the efficiency of the cleavage reaction. For instance, bis-aryl phosphites required a good leaving group like pyrrolidine as linker, which has a higher basicity ($pK_a = 11.3$) compared to a *N*-methyl-benzylamine moiety ($pK_a = 9.7$), whereas both linkers could be used for the synthesis of aliphatic TADDOL-based phosphites. Equally important is the acidity of the activator. In the absence of an acid activator, no reaction was observed. With 1*H*-tetrazole, *N*-methyl-imidazolinium trifluoroacetate, or pyridinium tetrafluoroborate, the conversion to product was slow or incomplete and the required prolonged reaction times promoted side product formation via oxidation and hydrolysis. *N*- methyl-anilinium trifluoroacetate (TAMA) was found to be the most suitable activator, although one equivalent was needed to allow a fast product formation, which necessitates the purification of the phosphite by flash chromatography after cleaving. In the conventional synthetic route, phosphites are synthesized by reaction of a chlorophosphite with an alcohol in the presence of a base, followed by chromatography to purify the ligand. By introducing supported phosphoramidites as intermediates, which are formed from the same chlorophosphites, an advantage over the traditional route is achieved, namely, the higher stability of the supported phosphoramidites in comparison to the chlorophosphites. A second advantage is that the supported phosphoramidites can be easily purified, and therefore this route does allow the usage of partially hydrolysed chlorophosphites without complicating the purification of the phosphite.

In the previous approach, two of the three substituents on the phosphorus atom in the nonsupported ligand originate from the type of chlorophosphite reagent that is applied in the synthesis of the supported phosphoramidite. The structural diversity of ligands that can be synthesized from one type of supported intermediate is, therefore, somewhat limited. Li *et al.* [6b] developed an efficient solid-phase route that allowed the formation of a large variety of nonsupported bidentate ligands, with an immobilized bis-chlorophosphine as the key intermediate (Scheme 16.8). Reaction of secondary amine-functionalized polystyrene support **17**, formed by treating Merrifield resin with an excess of tert-butylamine, with the commercially available 2-bis(dichlorophosphanyl)ethane (**18**) in the presence of triethylamine yielded supported bis-chlorophosphine **19**. Only one phosphorus atom was linked to the resin, and the authors attributed the high selectivity to the steric impediments of the *tert*-butyl group and the use of an excess of **18**. Intermediate **19** allows great structural diversity, as due to the reactive P–Cl bonds, three of the substituents on the two phosphorus atoms can be introduced by SPS. Polymer-bound **19** can be converted cleanly under mild conditions to intermediates **20** and **21**, by reaction with organomagnesium and organolithium reagents or metal alkoxides, respectively. The inertness of the P–N bond to these reagents ensures that the ligands remain bonded to the resin.

Cleavage of the bidentate ligands from the resin can be achieved with a variety of reagents, which further expands the structural diversity of ligands that can be synthesized. Reaction of **20** with alcohols or thioalcohols leads to phosphine-phosphinites (**22**) and phosphine–thiophosphines (**23**), respectively, while reaction with PCl_3 leads to chlorophosphines (**24**). Similarly, nonsupported phosphinites–chlorophosphinites (**25**) can be formed by

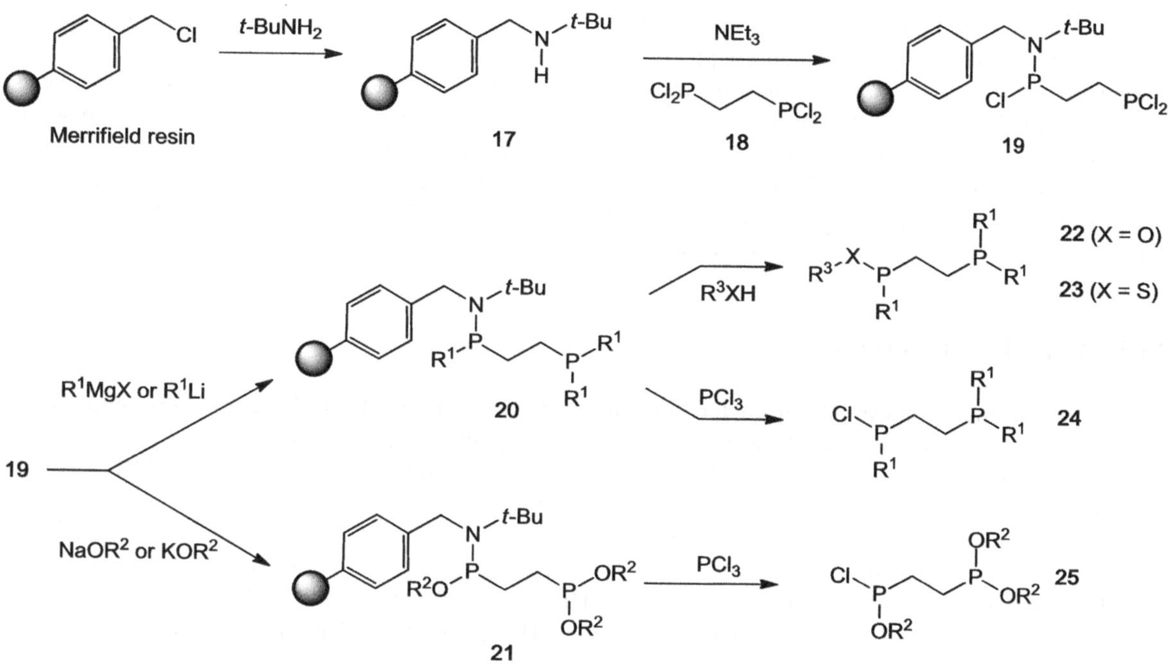

Scheme 16.8 *Solid-phase synthesis of C1 symmetric bidentate ligands.*

reaction of **21** with PCl_3. The method allowed the synthesis of a family of 15 different bidentate ligands in good yields (24–75%) and with a high degree of purity (≥95%) from a single supported intermediate (**19**). Moreover, it allows the synthesis of C1 symmetric ligands, which are generally difficult to prepare by standard solution chemistry. The utility of the process is evidenced by the synthesis of several ligands on a multigram scale.

16.6 Conclusions and outlook

Polymeric-supported ligands have proven their efficiency in several transition metal-catalyzed reactions, as they allow the formation of supported catalysts and thereby facilitate the recovery and reusability of the catalysts. Good mixing of the catalyst and the reactants is maintained by the excellent swelling properties of the polymeric support. These properties also allow polymer-bound ligands to be purified by simple filtration and washing procedures, which is attractive from a synthetic point of view, as phosphorus intermediates are in general quite reactive, resulting in good conversions, but separation and purification are often the problems in homogeneous systems. Moreover, it allows the use of excesses or even impure reagents. Consequently, application of solid-phase synthesis can facilitate the combinatorial synthesis of libraries of ligands in an efficient manner and thus can be a powerful tool in catalysis research. The approach can be utilized to create supported ligands, but also allows the synthesis of nonsupported ligands by cleaving them from the support in the final step of the synthetic sequence. Interestingly, the utilization of polymer-supported ligands in homogeneous catalysis has received considerably more attention than the development of efficient synthetic routes to form the ligands. Although a variety of *soluble* and *insoluble* polymeric supports have been applied in the synthesis of phosphorus-based ligands, the majority of these were synthesized by a single reaction between a support bearing a suitable reactive group and a phosphorus reagent. Structural diversity thus originates from the type of phosphorus-based reagent and/or from the support and not from the application of solid-phase synthetic techniques. Examples in which ligands are created by multistep reactions from a single solid-bound phosphorus-containing precursor are particularly rare. This can be ascribed to the lack of efficient synthetic protocols, which in turn can be ascribed to the fact that conditions used for the synthesis of nonsupported ligands are not always suitable for the synthesis of supported analogues. Multistep solid-phase routes can not only greatly extend the range of supported ligands which can be formed from a single supported intermediate but also allow the exploitation of the advantages of solid-phase synthesis during all steps of the synthetic sequence. Consequently, the development of efficient and flexible solid-phase protocols is a prerequisite to the successful utilization of solid supports in ligand synthesis and catalysis research.

16.7 Experimental procedures

Procedure for the synthesis of supported phosphite 28 via dichlorophosphite 27[6h] (Scheme 16.9):
Polymer-bound 4-hydroxyphenol **26** (0.12 mmol, 130 mg, polystyrene 1% cross-linked with divinylbenzene, 50–100 mesh) was washed with THF (3 × 2 mL) and subsequently suspended in 3 mL of THF. *i*-Pr_2NEt (0.76 mmol, 0.20 mL, 6.3 eq.) and PCl_3 (2.3 mmol, 0.2 mL, 19 eq.) were added, and argon was bubbled through the mixture several times for 5 min periods. After 3 h, the suspension was filtered and resin **27** was washed three times with 3 mL of THF. $\delta\,^{31}P\{^1H\}$ (THF-C_6D_6 8:1): 180 ppm. Subsequently, supported dichlorophosphite **27** was suspended in 3 mL of THF followed by the addition of *i*-Pr_2NEt (0.76 mmol, 0.2 mL, 6.3 eq.) and 2,4,6-trimethyl-phenol (0.31 mmol, 41 mg, 2.6 eq.), and argon was bubbled through the mixture several times for 5 min periods. After 3 h, the suspension was filtered. Washing with subsequently THF, CH_2Cl_2, and Et_2O (2 × 2 mL each) and drying with a gentle flow of argon, followed by drying under vacuum, yielded resin **28**. $\delta\,^{31}P\{^1H\}$ (THF–C_6D_6 8:1): 141 ppm. Elemental analysis: P 1.90%.

Procedure for the synthesis of supported phosphoramidite 30 [6h] (Scheme 16.10):
Polymer-bound 4-hydroxyphenol **26** (polystyrene, 50–100 mesh) 1% cross-linked with divinylbenzene (0.18 mmol, 200 mg) was washed with THF (2 × 2 mL) and subsequently suspended in 3 mL of THF. NEt_3

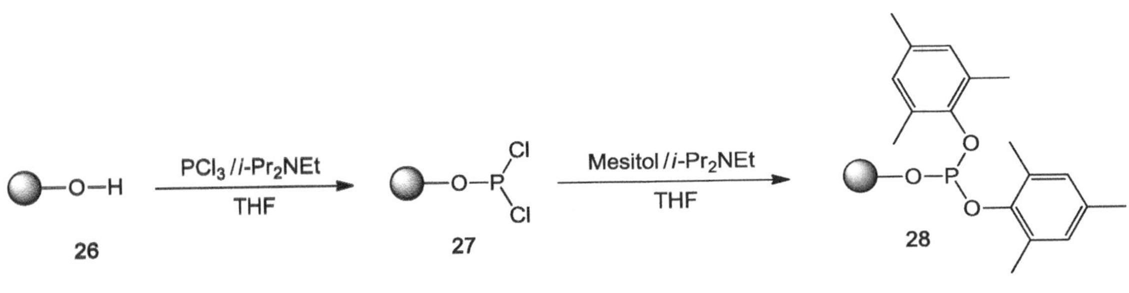

Scheme 16.9 Synthesis of supported phosphite 28 via dichlorophosphite 27 [6h].

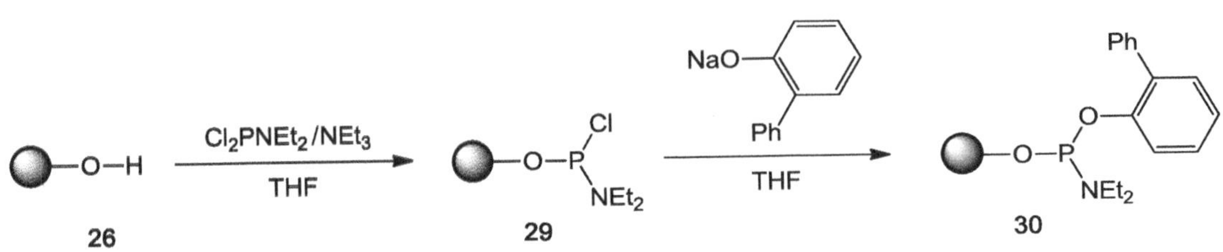

Scheme 16.10 Synthesis of supported phosphoramidite 30 [6h].

(1.4 mmol, 0.20 mL, 7.8 eq.) and Cl_2PNEt_2 (1.4 mmol, 0.2 mL, 7.7 eq.) were added, and argon was bubbled through the mixture several times for 5 min periods. After 3 h, the suspension was filtered and resin **29** was washed with THF (3×3 mL) and Et_2O (3 mL). $\delta^{31}P\{^1H\}$ (THF): 168 ppm. Subsequently, supported chlorophosphoramidite **29** was suspended in 3 mL of THF followed by the addition of 2-phenylphenol sodium salt (0.54 mmol, 3 eq.) in 4 mL of THF, which was prepared in a separate flask by allowing 22 mg of sodium hydride (60% dispersion in mineral oil, 0.54 mmol) to react with 2-phenylphenol (0.54 mmol, 92 mg) in THF for 30 min. Argon was bubbled through the reaction mixture several times for 5 min periods, and after 3 h the suspension was filtered. Washing with subsequently THF (4×5 mL), CH_2Cl_2, Et_2O, CH_2Cl_2, and Et_2O (2×2 mL each), drying with a gentle flow of argon, followed by drying under vacuum, yielded resin **30**. $\delta^{31}P\{^1H\}$ (THF, unlocked): 143 ppm. Elemental analysis: P 1.55, N 0.73%.

Procedure for the synthesis of supported aminophosphine borane 3b [6g] (Scheme 16.11): 4-Bromopolystyrene (**1a**, 0.5 g, 1.9 mmol/g) was swollen in toluene (20 mL), and *n*-BuLi (2.5 M in hexanes, 2.4 eq.) was added [18]. The reaction mixture was gently stirred at 60°C for 3 h. After cooling to room temperature, the liquid phase was removed. The lithiated resin (**1b**) was washed twice with toluene (20 mL). 30 mL of toluene was added, followed by a solution of (2*R*,4*S*,5*R*)-3,4-dimethyl-2-(2-methoxyphenyl)-5-phenyl-[1,3,2]-oxazaphospholidine 2-borane **2b** [32] (449.1 mg, 1.43 mmol, 1.5 eq.) in toluene (5 mL), and the reaction mixture was gently stirred for 16 h at room temperature. The liquid phase was removed, and the resin was washed with 5 mL portions of subsequently THF, THF–water (1:1), THF, CH_2Cl_2, Et_2O, CH_2Cl_2, and Et_2O. Finally, white resin **3b** was dried using a gentle flow of argon and under reduced pressure. $\delta^{31}P\{^1H\}$ (THF, D_2O inner tube) = 70.0 (br). Elemental analysis: P 1.23%.

Procedure for the synthesis of supported aminophosphine–phosphite 7b [6g] (Scheme 16.12): Resin-bound aminophosphine borane **3b** (50 mg) was washed with THF (5 mL) and allowed to swell in THF (5 mL) for 1 h. Subsequently, *N*-methyl morpholine (0.5 mL) and 2,2′-biphenol phosphorochloridite (1.3 eq.) were added. The reaction mixture was gently stirred for 16 h at room temperature. The liquid

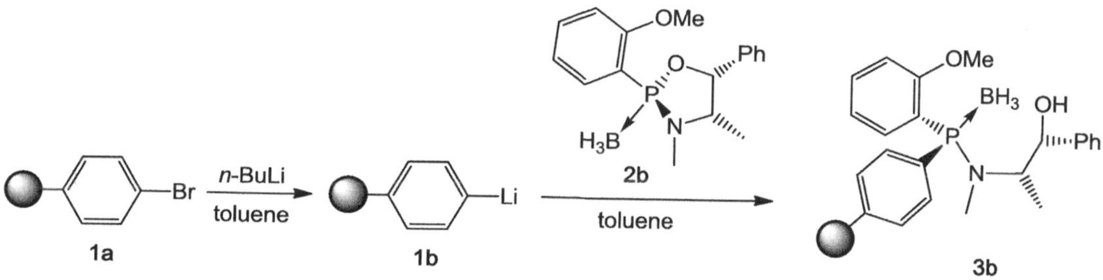

Scheme 16.11 Synthesis of supported aminophosphine borane 3b [6g].

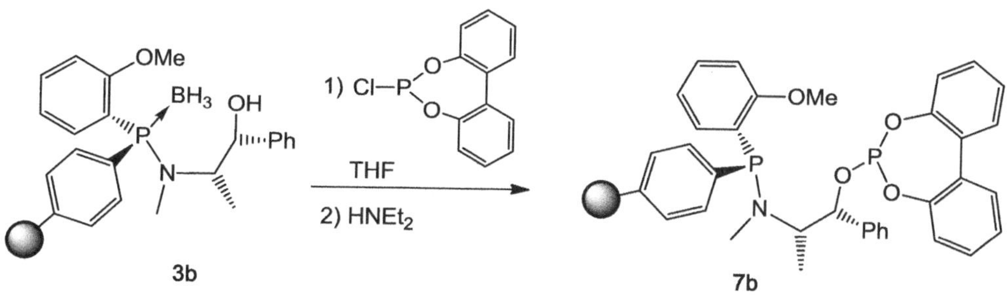

Scheme 16.12 Synthesis of supported aminophosphine–phosphite 7b [6g].

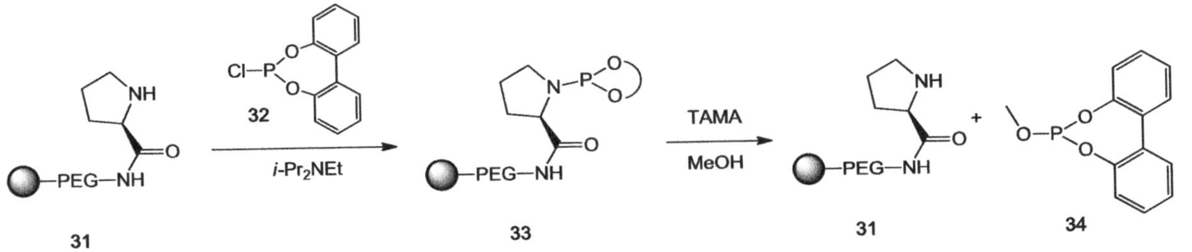

Scheme 16.13 Synthesis of 6-methoxydibenzo[d,f][1,3,2]dioxaphosphepine 34 via SPS.

phase was removed, and resin **5b** was washed with subsequently 5 ml THF, CH_2Cl_2, Et_2O, CH_2Cl_2, and Et_2O. In order to remove the borane, the resin was suspended in Et_2NH (2 mL) and heated to 50°C. After 16 h, the solution phase was removed and resin **7b** was washed with 1 mL portions of subsequently CH_2Cl_2, Et_2O, CH_2Cl_2, and Et_2O, and dried using a gentle flow of argon and under reduced pressure. $\delta^{31}P\{^1H\}$ (THF, D_2O inner tube) = 149.2 (br), 59.5 (br).

Procedure for the synthesis of 6-methoxydibenzo[d,f][1,3,2]dioxaphosphepine 34 via SPS (Scheme 16.13): 500 mg of resin **31**, derived from TentaGel™-S-NH_2 functionalized with a proline moiety 6 h (0.127 mmol), was washed with 5 mL of CH_2Cl_2 and allowed to swell in 10 mL of CH_2Cl_2 for 1 h. Subsequently 6-chlorodi benzo[d,f][1,3,2]dioxaphosphepine (**32**, 253 mg, 1.01 mmol, 8 equiv.) and *i*-Pr_2NEt (0.40 mL, 1.14 mmol, 9 equiv.) were added. After bubbling argon through the suspension for 1 h, the resin was filtered and washed with THF, CH_2Cl_2, and Et_2O (2×4 mL each). $\delta^{31}P\{^1H\}$ (C_6D_6):151 ppm. Resin **33** was suspended in 10 mL of THF, and subsequently methanol (8 µl, 0.19 mmol, 1.5 equiv.) and TAMA (45 mg, 0.13 mmol, 1 equiv.) were added. Argon was bubbled through the mixture for 30 min, which was subsequently filtered; the resin was washed with THF (3×5 mL), and the combined organic extracts were concentrated *in vacuo* to afford the crude product. Purification by flash chromatography (with CH_2Cl_2–hexanes (1/5, v/v)) furnished title compound **34** in 28.1 mg yield (0.114 mmol, 90%) as a colorless oil. δ^1H (300 MHz, $CDCl_3$) (ppm): 7.29–6.95 (m, 8 H, C*H*); 3.73(d, 3 H, CH_3, $^3J_{P\text{-}H}$ = 7.2 Hz). $\delta^{31}P\{^1H\}$ (300 MHz, $CDCl_3$): 139.2 ppm.

References

[1] P. W. N. M. van Leeuwen, *Homogeneous Catalysis: Understanding the Art, Kluwer Academic*, **2004**.

[2] (a) M. T. Reetz, *Angew. Chem., Int. Ed.* **2001**, *40*, 284; (b) J. G. de Vries, A. H. M. de Vries, *Eur. J. Org. Chem.* **2003**, 799; (c) C. Jäkel, R. Paciello, *Chem. Rev.* **2006**, *106*, 2912; (d) J. P. Stambuli, S. R. Stauffer, K. H. Shaughnessy, J. F. Hartwig, *J. Am. Chem. Soc.* **2001**, *123*, 2677; (e) O. Lavastre, F. Bonnette, L. Gallard, *Curr. Opin. Chem. Biol.* **2004**, *8*, 311; (f) L. Lefort, J. A. F. Boogers, A. H. M. de Vries, J. G. de Vries, *Org. Lett.* **2004**, *6*, 1733; and (g) C. Gennari, U. Piarulli, *Chem. Rev.* **2003**, *103*, 3071.

[3] R. B. Merrifield, *J. Am. Chem. Soc.* **1963**, *85*, 2149.

[4] (a) D. C. Sherrington, A. P. Kybett, Editors, *Supported Catalysts and Their Applications*, Royal Society of Chemistry, Cambridge, **2001**; (b) B. M. L. Dioos, I. F. J. Vankelecom, P. A. Jacobs, *Adv. Synth. Catal.* **2006**, *348*, 1413; (c) M. Guino, K. K. Hii, *Chem. Soc. Rev.* **2007**, *36*, 608; (d) B. Clapham, T. S. Reger, K. D. Janda, *Tetrahedron* **2001**, *57*, 4637; (e) T. J. Dickerson, N. N. Reed, K. D. Janda, *Chem. Rev.* **2002**, *102*, 3325; (f) N. E. Leadbeater, *Curr. Med. Chem.* **2002**, *9*, 2147; (g) N. E. Leadbeater, M. Marco, *Chem. Rev.* **2002**, *102*, 3217; (h) C. A. McNamara, M. J. Dixon, M. Bradley, *Chem. Rev.* **2002**, *102*, 3275; (i) D. E. Bergbreiter, J. Tian, C. Hongfa, *Chem. Rev.* **2009**, *109*, 530; (j) J. Lu, P. H. Toy, *Chem. Rev.* **2009**, *109*, 815; and (k) N. Kann, *Molecules* **2010**, *15*, 6306.

[5] (a) R. T. Baker, W. Tumas, *Science* **1999**, *284*, 1477; (b) *Chem. Rev.* **2002**, *102*, no. 10; and (c) D. J. Cole-Hamilton, *Science* **2003**, *299*, 1702.

[6] (a) S. R. Gilbertson, S. E. Collibee, A. Agarkov, *J. Am. Chem. Soc.* **2000**, *122*, 6522; (b) G. Y. Li, P. J. Fagan, P. L. Watson, *Angew. Chem., Int. Ed.* **2001**, *40*, 1106; (c) O. Huttenloch, E. Laxman, H. Waldmann, *Chem. Eur. J.* **2002**, *8*, 4767; (d) A. Mansour, M. Portnoy, *Tetrahedron Lett.* **2003**, *44*, 2195; (e) C. A. Christensen, M. Meldal, *Chem. Eur. J.* **2005**, *11*, 4121; (f) Y. Kobayashi, D. Tanaka, H. Danjo, Y. Uozumi, *Adv. Synth. Catal.* **2006**, *348*, 1561; (g) R. den Heeten, B. H. G. Swennenhuis, P. W. N. M. van Leeuwen, J. G. de Vries, P. C. J. Kamer, *Angew. Chem., Int. Ed.* **2008**, *47*, 6602; (h) B. H. G. Swennenhuis, R. Chen, P. W. N. M. van Leeuwen, J. G. de Vries, P. C. J. Kamer, *Org. Lett.* **2008**, *10*, 989; (i) B. Bar-Nir Ben-Aroya, M. Portnoy, *Tetrahedron* **2002**, *58*, 5147; (j) A. Agarkov, S. J. Greenfield, T. Ohishi, S. E. Collibee, S. R. Gilbertson, *J. Org. Chem.* **2004**, *69*, 8077.

[7] (a) H. Graden, N. Kann, *Curr. Org. Chem.* **2005**, *9*, 733; and (b) G. Reginato, M. Taddei, *Il Farmaco* **2002**, *57*, 373.

[8] (a) P. H. Toy, K. D. Janda, *Tetrahedron Lett.* **1999**, *40*, 6329; and (b) P. H. Toy, T. S. Reger, P. Garibay, J. C. Garno, J. A. Malikayil, G.-Y. Liu, K. D. Janda, *J. Comb. Chem.* **2001**, *3*, 117.

[9] R. Quarrell, T. D. W. Claridge, G. W. Weaver, G. Lowe, *Mol. Diversity* **1996**, *1*, 223.

[10] R. Jana, J. A. Tunge, *Org. Lett.* **2009**, *11*, 971.

[11] D. E. Bergbreiter, L. Zhang, V. M. Mariagnanam, *J. Am. Chem. Soc.* **1993**, *115*, 9295.

[12] N. Takaishi, H. Imai, C. A. Bertelo, J. K. Stille, *J. Am. Chem. Soc.* **1976**, *98*, 5400.

[13] (a) T. Jongsma, M. Fossen, G. Challa, P. W. N. M. van Leeuwen, *J. Mol. Catal.* **1993**, *83*, 17; and (b) T. Jongsma, H. van Aert, M. Fossen, G. Challa, P. W. N. M. van Leeuwen, *J. Mol. Catal.* **1993**, *83*, 37.

[14] (a) F. Bardella, R. Eritja, E. Pedroso, E. Giralt, *Bioorg. Med. Chem. Lett.* **1993**, *3*, 2793; and (b) D. E. Bergbreiter, Y.-C. Yang, *J. Org. Chem.* **2010**, *75*, 873.

[15] (a) W. Chen, S. M. Roberts, J. Whittall, *Tetrahedron Lett.* **2006**, *47*, 4263; (b) A. Mandoli, M. Calamante, B. L. Feringa, P. Salvadori, *Tetrahedron: Asymmetry* **2003**, *14*, 3647; and (c) S. Schweizer, J.-M. Becht, C. Le Drian, *Tetrahedron* **2010**, *66*, 765.

[16] C. R. Harrison, P. Hodge, B. J. Hunt, E. Khoshdel, G. Richardson, *J. Org. Chem.* **1983**, *48*, 3721.

[17] H. M. Relles, R. W. Schluenz, *J. Am. Chem. Soc.* **1974**, *96*, 6469.

[18] M. J. Farrall, J. M. J. Frechet, *J. Org. Chem.* **1976**, *41*, 3877.

[19] (a) S. Schweizer, J.-M. Becht, C. Le Drian, *Org. Lett.* **2007**, *9*, 3777; and (b) S. Schweizer, J.-M. Becht, C. Le Drian, *Adv. Synth. Catal.* **2007**, *349*, 1150.

[20] (a) S. R. Gilbertson, X. Wang, *Tetrahedron* **1999**, *55*, 11609; and (b) A. Agarkov, S. R. Gilbertson, *Tetrahedron Lett.* **2005**, *46*, 181.

[21] (a) J. M. Benito, C. A. Christensen, M. Meldal, *Org. Lett.* **2005**, *7*, 581; and (b) C. A. Christensen, M. Meldal, *J. Comb. Chem.* **2007**, *9*, 79.

[22] B. H. G. Swennenhuis, R. F. Chen, P. W. N. M. van Leeuwen, J. G. de Vries, P. C. J. Kamer, *Eur. J. Org. Chem.* **2009**, 5796.

[23] (a) D. Moulin, C. Darcel, S. Juge, *Tetrahedron: Asymmetry* **1999**, *10*, 4729; and (b) C. Darcel, D. Moulin, J.-C. Henry, M. Lagrelette, P. Richard, P. D. Harvey, S. Juge, *Eur. J. Org. Chem.* **2007**, 2078.

[24] E. Bayer, V. Schurig, *Chemtech* **1976**, *6*, 212.

[25] (a) P. Wentworth Jr, A. M. Vandersteen, K. D. Janda, *Chem. Commun.* **1997**, 759; (b) F. Sieber, P. Wentworth Jr., J. D. Toker, A. D. Wentworth, W. A. Metz, N. N. Reed, K. D. Janda, *J. Org. Chem.* **1999**, *64*, 5188; (c) P. H. Toy, K. D. Janda, *Acc. Chem. Res.* **2000**, *33*, 546; (d) D. E. Bergbreiter, *Chem. Rev.* **2002**, *102*, 3345; and (e) X. P. Hu, J. D. Huang, Q. H. Zeng, Z. Zheng, *Chem. Commun.* **2006**, 293.

[26] D. E. Bergbreiter, S. D. Sung, *Adv. Synth. Catal.* **2006**, *348*, 1352.

[27] S. Doherty, E. G. Robins, I. Pal, C. R. Newman, C. Hardacre, D. Rooney, D. A. Mooney, *Tetrahedron: Asymmetry* **2003**, *14*, 1517.

[28] O. Lavastre, J. P. Morken, *New J. Chem.* **2002**, *26*, 745.

[29] C. U. Pittman, R. M. Hanes, *J. Am. Chem. Soc.* **1976**, *98*, 5402.

[30] (a) S. R. Gilbertson, S. Yamada, *Tetrahedron Lett.* **2004**, *45*, 3917; (b) H. Nakano, K. Takahashi, Y. Suzuki, R. Fujita, *Tetrahedron: Asymmetry* **2005**, *16*, 609; and (c) A. Leyva, H. García, A. Corma, *Tetrahedron* **2007**, *63*, 7097.

[31] (a) B. J. Backes, J. A. Ellman, *Curr. Opin. Chem. Biol.* **1997**, *1*, 86; (b) I. W. James, *Tetrahedron* **1999**, *55*, 4855; and (c) A. C. Comely, S. E. Gibson, *Angew. Chem., Int. Ed.* **2001**, *40*, 1012.

[32] C. H. Schwalbe, G. Chopra, S. Freeman, J. M. Brown, J. V. Carey, *J. Chem. Soc.-Perkin Trans. 2* **1991**, 2081.

17

Biological Approaches

René den Heeten[1], Paul C. J. Kamer[2], and Wouter Laan[2]

[1]*Huntsman Polyurethanes, Rotterdam, The Netherlands*
[2]*EaStCHEM, School of Chemistry, University of St Andrews, St Andrews, United Kingdom*

17.1 Introduction

The concepts of molecular recognition underlying biocatalysis and biomolecular assembly have stimulated the development of a variety of intriguing approaches for the construction of bio-inspired ligand systems for transition metal-catalysed reactions. Besides the use of biomacromolecules as a source of chirality in asymmetric reactions, these endeavours open the potential for second-sphere ligand–substrate interactions, which may lead to enzyme-like control over regio-, chemo-, enantio- and substrate selectivities in homogeneous catalysis. Moreover, the advances in parallel solid-phase synthesis and molecular biology make it increasingly facile to generate and screen large biomolecule-based ligand libraries. This chapter will discuss the various strategies developed for the synthesis of phosphine-containing peptides, DNA and proteins.

17.2 Peptide-based phosphine ligands

17.2.1 Solid-phase synthesis using phosphine-containing amino acids

Because methods for the parallel solid-phase synthesis of peptides are well established, peptide synthesis using phosphine-containing amino acids is an attractive approach for the combinatorial synthesis of functionalized chiral ligands. To be compatible with standard peptide-coupling conditions, the amino acid building blocks are typically synthesized with the Fmoc or Boc-protecting group on the amine. In phosphine amino acids, the phosphorous is generally protected as the sulfide to prevent oxidation during the synthesis of the amino acids as well as during the ensuing steps involved in the solid-phase synthesis of phosphine-containing peptides.

Phosphorus(III) Ligands in Homogeneous Catalysis: Design and Synthesis, First Edition.
Edited by Paul C. J. Kamer and Piet W. N. M. van Leeuwen.

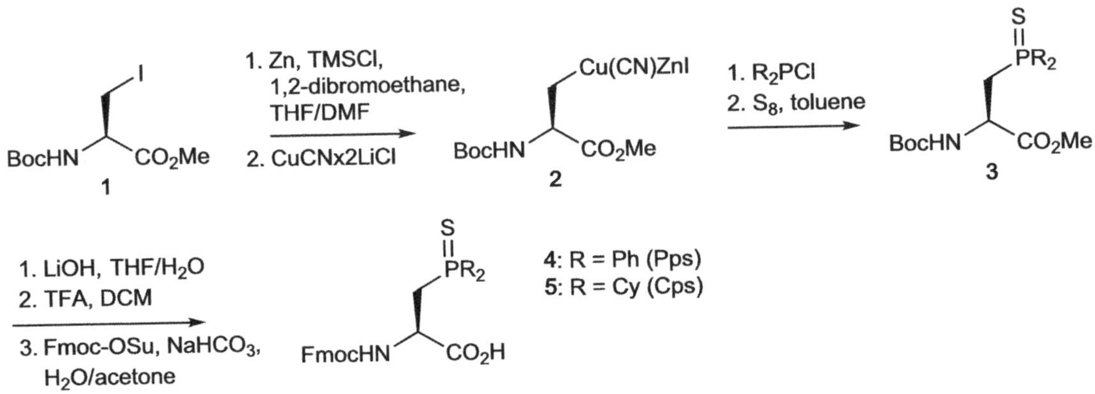

Scheme 17.1 *Synthesis of phosphine-containing alanine derivatives.*

17.2.1.1 *Synthesis of phosphine-containing amino acids*

In 1994, Gilbertson and co-workers pioneered the synthesis of phosphine-containing amino acids by synthesizing a number of phosphino-alanine derivatives, which are readily incorporated into peptides using standard peptide-coupling chemistry. The original approach made use of Evans chiral oxazolidone chemistry. Although this method provided a number of different phosphine-containing amino acids in good yields, the procedure was lengthy and entailed several chromatographic steps, limiting the wider applicability of the approach. Therefore a more expedient alternative route was developed. The commercially available 3-iodo-alanine methylester **1** (available in both enantiomeric forms) was converted to a Knochel cuprate [1], which was then reacted with a variety of aromatic and alkyl phosphine chlorides. Protective group manipulation afforded several phosphino amino acids, suitable for peptide-synthesis, in good yields (Scheme 17.1) [2].

Although application of this chemistry to the synthesis of phosphino-alanines functionalized at the β-carbon provided the desired products, the yields were rather low. For these types of amino acids, the original Evans oxazoline chemistry proved the most effective. Reaction conditions were found which provided access to both diastereoisomers, which could be separated by column chromatography [3].

In 1998, an alternative synthetic route towards the Boc-protected phenylphosphinoserine (Pps), starting from serine, was reported by Burgess and co-workers [4]. Although amenable to large-scale synthesis, the overall yield of this route is considerably lower than that of Gilbertson's procedure.

None of the routes outlined above are readily extended to phosphine-containing amino acids with linkages different from methylene. To address this issue, Lammertsma and co-workers introduced a procedure that involves the enantioselective alkylation of a chiral nickel Schiff base template **7** (Belekon's template), which allowed for the synthesis of compounds **12** and **13** with varying alkyl chains between the phosphine group and the amino acid part (Scheme 17.2) [5].

Gilbertson also developed a method for the conversion of the aromatic amino acids tyrosine and hydroxyphenylglycine into phosphines (Scheme 17.3). Conversion of the OH group of the protected amino acids **14** and **15** to their triflates was achieved by the reaction with *N*-phenyltrifluoromethanesulfonamide and diisopropyl-ethylamine. Palladium-catalysed coupling of the triflates with diphenylphosphine gave the phosphine amino acids **16** and **17** (protected as the sulfide) in good yields. The nature of the amine-protecting group was found to be critical: whereas the initially used acetyl group led to racemisation of the α-carbon of the amino acid, changing to the Boc-group prevented this. This reaction is also suitable for the direct conversion of tyrosine residues in larger peptides to phosphine ligands: running the reaction on a tyrosine-containing pentapeptide yielded the phosphine-containing peptide in 78% yield [6]. This illustrates the possibility to use this chemistry to convert the phenols of naturally occurring tyrosine-containing proteins to phosphines.

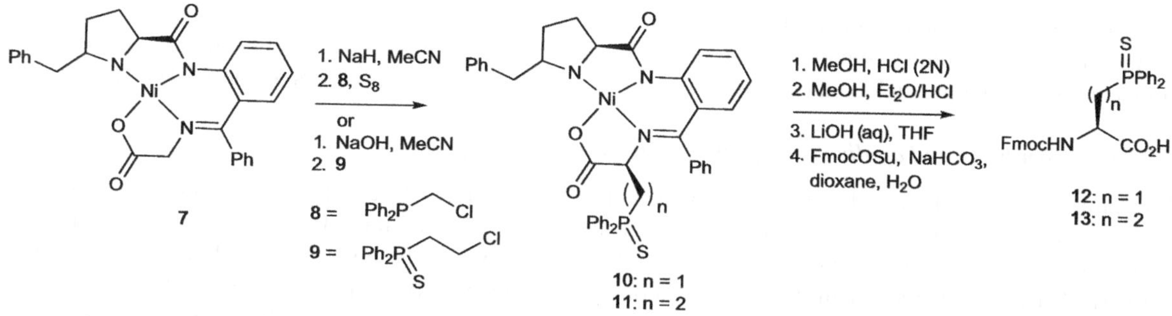

Scheme 17.2 *Synthesis of phosphino amino acids using Belekon's template.*

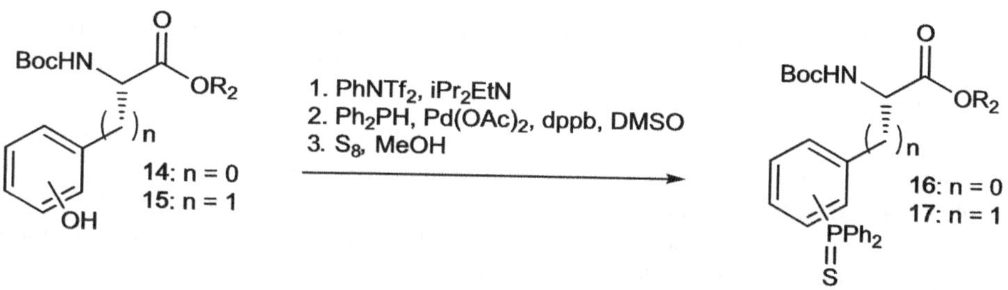

Scheme 17.3 *Conversion of aromatic amino acids into phosphines as developed by Gilbertson.*

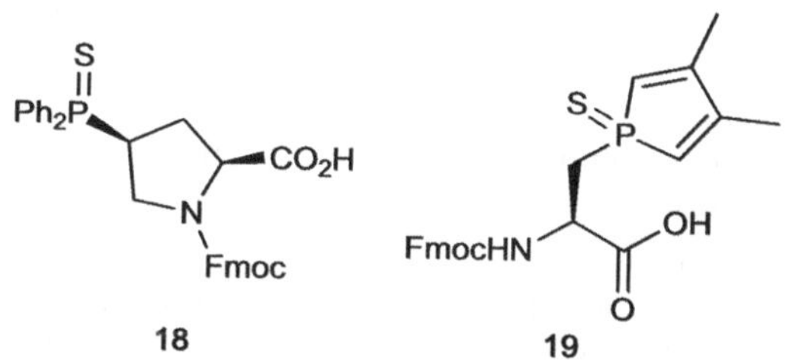

Scheme 17.4 *Phosphine-containing proline (**18**) and phosphole-containing amino acid (**19**).*

Kraatz and Pletsch synthesized similar (4-diphenylphosphanyl) derivatives of D- and L-phenylalanine using a palladium-catalysed cross-coupling reaction of diphenylphosphine and the corresponding iodo-aromatic amino acid precursor. In this case, the use of the acetyl group for protection of the amine did not lead to the racemisation of the α-carbon during synthesis [7].

Stelzer and co-workers reported the synthesis of *ortho-* and *para-*phosphinophenyl derivatives of glycine and alanine by nucleophilic phosphination of 2- and 4-fluorophenylglycine and -alanine with Ph(R)PK (R=Me, Ph), although no evidence of enantiopurity of the products was provided [8].

Gilbertson also reported the synthesis of a phosphine derivative of proline. Starting from *trans*-hydroxyproline, phosphine-modified proline **18** (Scheme 17.4) was obtained using a laborious synthetic route in which a major obstacle was the protection of the acid function from attack by the highly nucleophilic phosphide anion. In search for a shorter synthetic route, oxazoline turned out to be an excellent group for the protection of the acid moiety during the nucleophilic addition [9].

Le Floch and co-workers reported the synthesis of a readily available amino acid building block for the preparation of phosphole-containing peptides. Nucleophilic substitution of a phospholide anion onto protected 3-iodoalanine yielded an amino acid with an appended phosphole (**19**, Scheme 17.4) [10].

17.2.1.2 Synthesis and application of phosphine-containing peptides

Phosphine amino acids have been employed to synthesize a variety of peptide-based diphosphine ligands using the standard Fmoc solid-phase peptide synthesis strategy. To facilitate the isolation and screening of ligands in catalysis, it is desirable to form the metal complexes while the ligands are still bound to the solid support. Whereas reduction with the heterogeneous Raney nickel is typically used for the deprotection of ligands in solution, immobilized ligands require a homogeneous reagent system. Unprotected phosphines bound to the solid support can be accessed via the methylation of the phosphine-sulfides with methyl triflate, followed by transfer of the sulphur to the phosphorous of hexamethylphosphorous triamide (HMPT) [11].

Phosphine-containing helix and β-turn structures Gilbertson choose the helical and β-turn secondary structure elements as structurally stable peptide scaffolds, which facilitates the orientation of the introduced phosphorous ligands in the desired chelating coordination mode.

By predominantly using alanine and placing α-aminoisobutyric acid (Aib) in a number of locations throughout the sequence, a peptide can be enforced into a helical secondary structure. To obtain chelating helical diphosphine-peptides, the phosphine amino acids were positioned in an *i*, *i+4* relationship, which places the ligands on the same side of the peptide. The secondary structure of a helical dodecapeptide-based diphosphine-ligand was confirmed by nuclear magnetic resonance (NMR) and X-ray analysis [12].

Using these design principles, libraries of over 100 helical diphoshine-peptides containing the phosphine–amino acids Pps and Cps (cyclohexylphosphinoserine) were synthesized on the spatially addressable Mimotopes/Chiron Multipin™ Multiple Peptide Synthesis system, which allows the synthesis of discrete isolated compounds in a 96-pin format (Scheme 17.5). The solid-phase anchored ligands were screened in the rhodium-catalysed hydrogenation of methyl 2-acetamidoacrylate. The obtained maximum enantioselectivity was low (38%), and the selectivities varied widely depending on the solvent used, as well as on whether the ligands were polymer bound or in solution. These issues were attributed to peptide chain interactions. Nevertheless, these studies demonstrated that the method is a viable approach to the combinatorial synthesis and testing of new chiral phosphine ligands [13].

More recently, Gilbertson *et al.* reported the synthesis of phosphine libraries based on a β-turn motif. Placing the phosphines in an *i*, *i+3* relationship flanking a proline and D-amino acid affords peptides which adopt a β-turn secondary structure, which brings the ligands close enough to coordinate metals in a chelating fashion. These ligands exhibited good to high selectivities in the palladium-catalysed allylic alkylation of

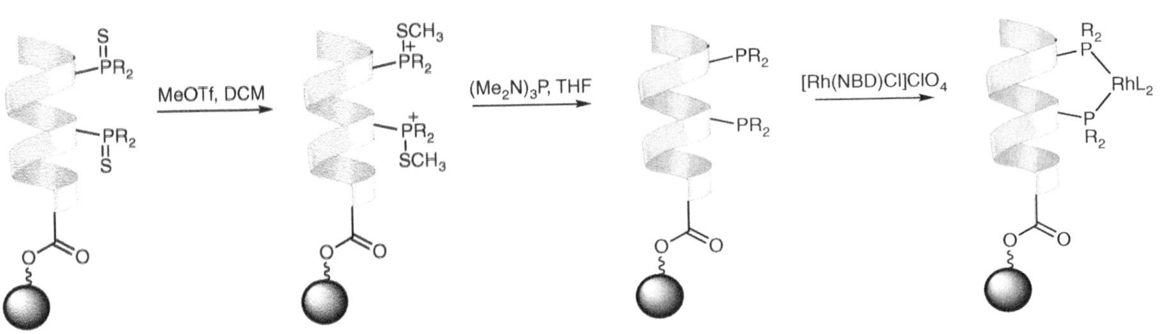

Scheme 17.5 Synthesis of polymer-bound rhodium complexes of helical diphosphine-peptides.

cyclic allyl acetates with dimethylmalonate, providing the product in up to 95% ee, and a good correlation between the immobilized and solution phase ligands was observed [2,14].

Phosphine-containing cyclic decapeptides Lammertsma *et al.* reported bisphosphine analogues of Gramicidin S, a naturally occurring cyclic decapeptide containing a rigid structure enforced by two β-turns [5]. These peptides were synthesized on solid support by the replacement of the two ornithines in the natural sequence with phosphine-containing amino acids. X-ray crystal structures of two of these cyclic peptides containing sulphur-protected phosphines show that the phosphine amino acids are in a favourable conformation for diphosphine complex formation. No application in catalysis has been reported to date.

17.2.2 Functionalisation of peptides with phosphines

An alternative strategy for the synthesis of phosphine-containing peptides is to synthesize the peptide first, followed by its functionalization with a phosphine moiety. As the introduction of the phosphine is typically the final synthetic step, this strategy may avoid a phosphine protection and deprotection sequence, and generally facilitates the synthesis and handling of the ligands. A number of different approaches for the derivatization of peptides have been reported.

17.2.2.1 Phosphinomethylation of amines

When peptides containing primary and/or secondary amino groups are synthesized, phosphine moieties can be introduced by phosphinomethylation of the amines using hydroxymethylphosphines. Hydroxymethylphosphines can be easily obtained *in situ*, by heating a neat mixture of paraformaldehyde and a secondary phosphine [15].

Meldal and co-workers used this strategy for the solid-phase synthesis of peptide-based diphosphine ligands. Resin-bound peptides containing primary or secondary amines (incorporated into the backbone of the peptide scaffold by reductive alkylation) were reacted with diphenylphosphinomethanol. The resin-bound ligands were applied in the palladium-catalysed asymmetric allylic substitution of 1,3-diphenylallyl acetate and dimethylmalonate obtaining moderate enantioselectivities (up to 21%) [16]. In a more recent study, the methodology was employed for the solid-phase synthesis of peptide-based P,S-ligand systems derived from cysteine that gave the product of the same Pd-catalysed reaction with enantioselectivities up to 60% [17]. Although the enantioselectivities obtained to date are only moderate, these approaches can in principle be applied to the synthesis and screening of libraries of ligands.

Aiming to address the issue of metal leaching in catalysis, Arya *et al.* reported the synthesis of phosphine-containing peptide-based dendrimers. Following the solid-phase synthesis of peptide arms containing free amines, reaction of the amines with diphenylphosphinomethanol afforded the diphosphine-modified dendrimers. Their rhodium complexes were applied in the hydroformylation of a variety of olefins, resulting in very good conversions and high branched-to-linear ratios. Moreover, the catalysts still showed very good activity after five cycles [18].

17.2.2.2 Phosphine modification of peptides via imine or amide formation

Hoyveda disclosed a modular strategy for the synthesis of dipeptide-based ligands. Following the synthesis of structurally diverse dipeptides using standard peptide-coupling methods, phosphine derivatives **23** are easily prepared via Schiff base formation between 2-(diphenylphosphino)benzaldehyde and the free terminal amino group of the peptides [19]. Ligand diversity was achieved by the choice of the two amino acid fragments as well as the protecting group R. The analogous amide derivatives **24** are likewise readily obtained by employing 2-(diphenylphosphino)benzoic acid (Scheme 17.6) [20]. Whereas in initial studies the ligands were synthesized in solution, later ligands were also prepared on solid support and subsequently cleaved before use. These ligands were applied in the copper-catalysed conjugate addition of dialkyl zinc reagents to various electrophiles, in general providing high enantioselectivities. Thus far, the application of the ligands attached to the solid support has not been reported.

Similar *ortho*-peptidyl phosphine ligands were developed by Breit and Laungani via DCC–NEt_3 mediated coupling of 2-(diphenylphosphino)benzoic acid to mono-, di-, and tripeptides, which afforded enantioselectivities above 97% ee in the conjugate addition of dialkylzinc reagents to cyclic enones [21]. Breit and co-workers exploited the propensity of peptides to associate through noncovalent interactions for the supramolecular assembly of bidentate ligands (Scheme 17.7). Various ligands were obtained by coupling of 3-(diphenylphosphino)benzoic acid to dipeptides using DCC–DMAP, whereas a phosphite-modified peptide **26** was obtained via reaction of binaphthol–PCl with a hydroxybenzoyl peptide. These peptides allowed for the rhodium-templated formation of bidentate ligands via the hydrogen-bonding mediated self-assembly of the peptides into homodimeric helical structures, of which some showed excellent enantioselectivities in hydrogenation of methyl *N*-acetyl-dehydroamino acids (up to 99% ee) [22].

The approach was extended to heterodimeric complexes by utilizing complementary N-linked peptidyl phosphines, which were synthesized via the stepwise modification of 6-diphenylphospanyl-2-aminopyridine with two amino acids. The rhodium-templated assembly of N- and C-linked peptide ligands afforded β-sheet structures **27**, which gave higher selectivities (up to 38% ee) in the hydroformylation of styrene than helical homodimeric combinations [23].

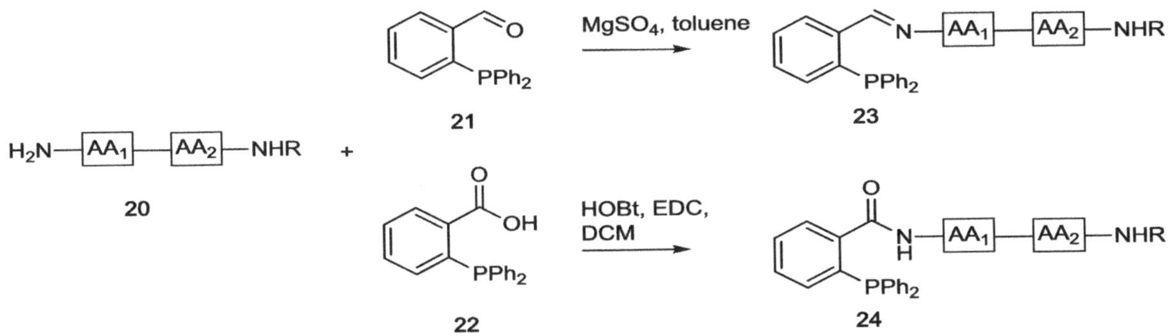

Scheme 17.6 Modular synthesis of phosphine-peptides via imine or amide bond formation.

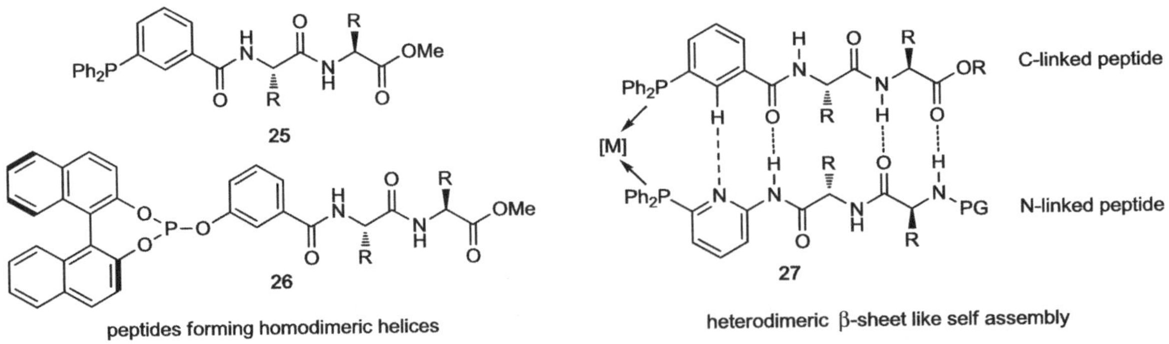

Scheme 17.7 Self-assembling ligand-modified peptides.

Wills *et al.* reported the functionalisation of the free amino groups of supported polyleucine, polylysine and leucine–lysine copolymers with *p*-diphenylphosphino benzoic acid using EDC–HOBt.[24] The thus-obtained ligands were tested in palladium-catalysed allylic substitution of diphenylallyl acetate, but no ee was observed. It was assumed that the phosphine-functionalized side chains in the polymers were too remote from the stereogenic centres to be capable of transferring chiral information from the support to the products.

In 2001, Landis and co-workers reported the facile synthesis of mono- and bis-3,4-diazaphospholanes. The asymmetry of the phospholane ring in combination with the diversity of amino acid fragments provided ligand libraries successfully applied in palladium-catalysed asymmetric allylic alkylation reactions [25] and rhodium-catalysed enantioselective hydroformylation reactions [26]. Furthermore, Landis employed the advantages of solid-phase synthesis in the formation of a library of resin-bound peptide-modified diazaphospholanes. Enantioselectivities up to 92% were reached in the Pd-catalysed allylic alkylation of 1,3-dimethylallyl acetate with dimethyl malonate [27].

17.3 Oligonucleotide-based phosphine ligands

Besides providing the biological blueprint for natural enzymes, DNA is also an attractive chiral biopolymer for the development of functionalized phosphine ligands. In addition to being a source of chirality, it has become evident in recent years that oligonucleotide sequences can function as highly selective binding sites for small organic molecules [28]. Methods for the solid-phase synthesis of oligonucleotides are well established [29]. Due to incompatibility between DNA synthesis conditions and ligand synthesis, transition metal-binding ligands are generally introduced after DNA synthesis and purification to an incorporated unnatural nucleotide.

17.3.1 Covalent anchoring of phosphines to DNA

Jäschke *et al.* employed the elegant 'convertible nucleoside approach' to prepare internally amino-modified oligonucleotides. Oligonucleotides (19- and 23-mers) containing a triazole-modified nucleotide were treated with various diamines, resulting in the site-selective introduction of linkers of varying length. A carboxylic acid containing diphosphines and a carboxylic acid containing phosphinooxazoline were activated by N-hydroxysuccinimide (NHS) in the presence of EDC, and the NHS esters were then allowed to react with the amine-containing oligonucleotides via amide bond formation. The coupling products **28** (Scheme 17.8) were purified by reversed-phase HPLC. No catalysis with these systems has been reported to date [30].

A similar approach was used by Kamer *et al.* to prepare phosphine-modified mono- and oligonucleotides [31]. Sonogashira coupling between the commercially available 5-iodo-2′-deoxyuridine (IdU) and

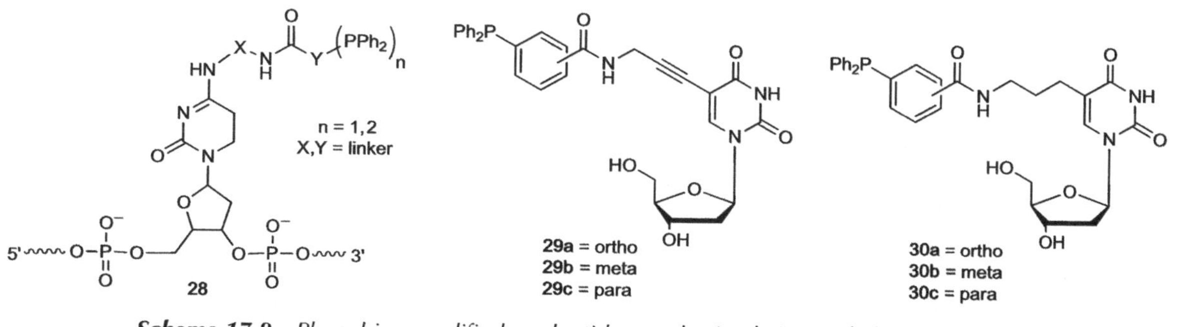

Scheme 17.8 *Phosphine-modified nucleotides synthesized via amide bond formation.*

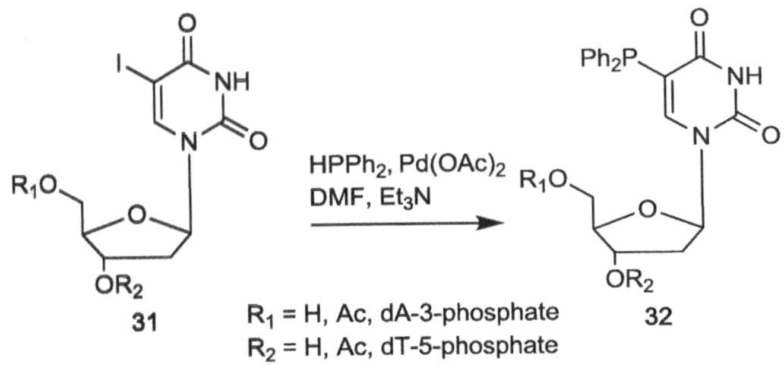

Scheme 17.9 *Phosphine-modified nucleotides synthesized via Pd catalyzed coupling.*

N-propynyltrifluoroacetamide, followed by deprotection of the amine with aqueous ammonia, provided an amino-modified deoxyuridine. Reduction of the propynyl linker to a single bond by hydrogenation over Pd–C (prior to deprotection) afforded a more flexible and reactive amino group on the deoxyuridine. Following Jäschke's approach, carboxylic acid–containing (di)phosphines were converted to NHS esters in the presence of EDC and reacted with the functionalized bases, affording seven different phosphine-modified deoxyuridines, with yields being significantly higher when the hydrogenated amino-modified deoxyuridine was employed. Application of the same approach to oligonucleotides (15–16 mers) containing the hydrogenated amino-modified deoxyuridine afforded several phosphine-modified oligonucleotides, which were purified by ethanol precipitation. A preliminary catalytic study was performed with ligands **29** and **30** (Scheme 17.8). Allylic alkylation of 1,3-diphenylallyl acetate with dimethyl malonate and allylic amination with benzylamine as nucleophile resulted in 100% conversion using ligands **29b**, **29c**, **30b** and **30c**, while ligands **29a** and **30a** gave less than 10% conversion. Unfortunately, no enantioselectivity was obtained.

Others have also introduced phosphines into oligonucleotides using amide bond formation, but only using terminally amine-functionalized DNA and not for catalytic purposes [32].

Kamer and co-workers also reported an alternative approach to phosphine-modified nucleotides [33]. The Pd-catalysed coupling of diphenylphosphine to IdU led to the functionalized nucleotide 5-diphenylphosphino-2′-deoxyuridine (dppdU) in high yield (Scheme 17.9). Functionalization of supported IdU-containing trinucleotides was also achieved, although higher palladium loadings were required. The nucleotides were used as ligands for the palladium-catalysed asymmetric allylic amination. Good enantiomeric excess was obtained (up to 82% for the S enantiomer) with the mononucleotide ligand dppdU. The functionalized trinucleotides also showed good conversion, albeit at the expense of the enantioselectivity.

17.4 Phosphine-based artificial metalloenzymes

The catalytic proficiency of enzymes is dependent on highly organized protein substrate interactions. The introduction of a transition metal catalyst in a protein scaffold can afford a hybrid system in which the protein environment is expected to dictate the selectivity, thus affording the potential to achieve enzyme-like control over the second coordination sphere of the catalyst. The performance of the hybrids can be optimized by synergistic tuning of both the structure of the protein and the organometallic catalyst, a strategy for which the term 'chemogenetic optimization' has been coined [34]. Structural variants of a protein scaffold can nowadays easily be generated using advanced mutagenesis strategies, making it possible to evolve systems by the combinatorial screening of large numbers of mutant proteins. For the anchoring of phosphine-ligands to proteins, two complementary strategies have been pursued: supramolecular assembly and site-selective covalent modification.

17.4.1 Supramolecular anchoring of phosphines to proteins

17.4.1.1 Avidin–biotin

The most extensively developed approach for the synthesis of phosphine-containing proteins as artificial metalloenzymes is based on the avidin–biotin system. Avidin is a remarkably robust tetrameric protein, which binds the vitamin biotin or its chemically modified derivatives with such high affinity that association is considered to be irreversible ($K_a \sim 10^{15}\,M^{-1}$). Biotin contains a valeric acid side chain, which provides a convenient handle for the synthesis of biotin conjugates. These features make the biotin–avidin couple a very attractive system for the supramolecular anchoring of transition metal catalysts to a protein host.

In 1978, Wilson and Whitesides [35] exploited this to embed a phosphine-based rhodium catalyst in a protein environment. A biotinylated phosphine ligand (**Biot-1**) was synthesized by amide bond formation between biotin *N*-hydroxysuccinimide (Biotin-NHS) and bis(2-(diphenylphosphino)ethyl)amine (Scheme 17.10), and subsequently used to create a biotinylated rhodium–diphosphine complex. Simple mixing of this catalyst with a solution of avidin afforded a supramolecular assembled artificial metalloenzyme, which was applied in the asymmetric hydrogenation of *N*-acetamidoacrylic acid, providing the product *N*-acetylalanine in 41% ee (S) with quantitative conversion.

In 1999, Chan and co-workers reported the use of the same synthetic strategy to obtain enantiopure biotinylated Pyrphos–rhodium complexes. These complexes were used in conjunction with avidin for the hydrogenation of itaconic acid. Depending on the configuration of the ligand and the reaction conditions,

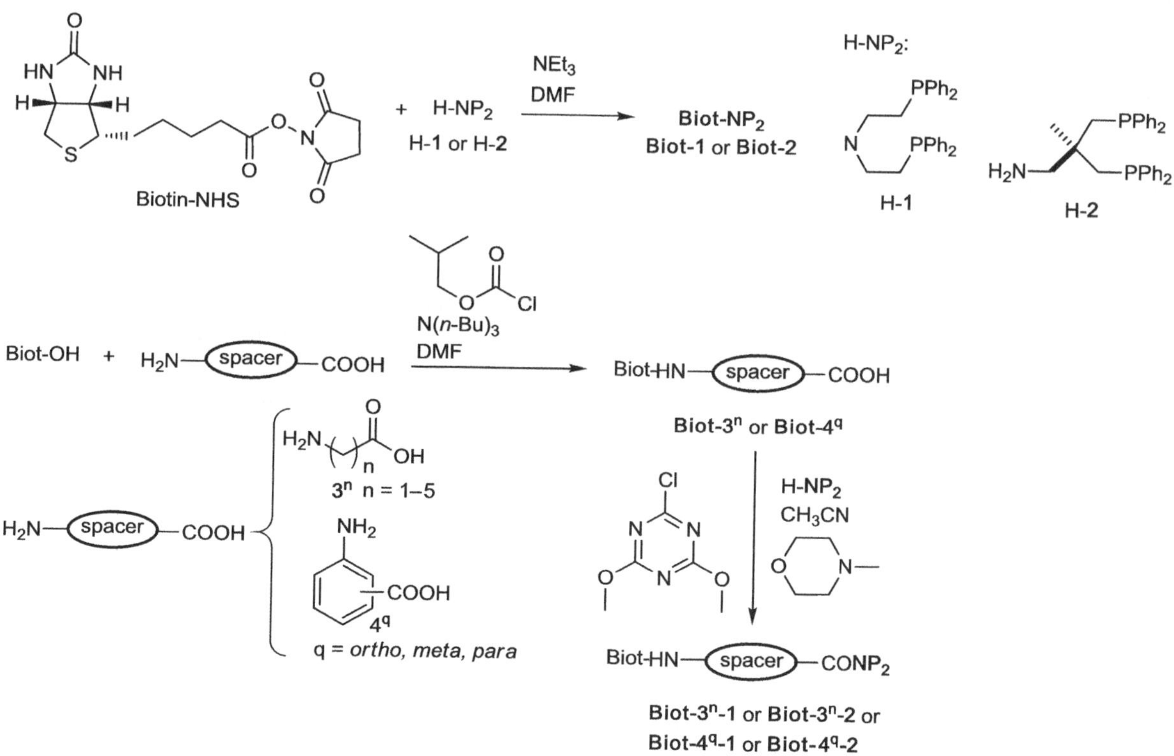

Scheme 17.10 Synthesis of biotinylated diphosphine ligands for anchoring to (strept)avidin.

the absolute configuration of the hydrogenation product ranged from 48% (*R*) to 26% (*S*), and the results indicated a positive effect of the presence of the protein on the enantioselectivity [36].

Inspired by these reports, Ward and co-workers have substantially expanded the avidin–biotin platform for artificial metalloenzyme development [37]. Firstly, the protein streptavidin was included as host protein. Streptavidin possesses a similar affinity for biotin ($K_a \sim 10^{14}\,M^{-1}$) as avidin, but streptavidin contains a deeper binding pocket. Secondly, to allow the chemical optimization of the performance of the artificial metalloenzymes, a range of biotinylated phosphine ligands were developed via amide bond coupling reactions (Scheme 17.10) [38].

Biotinylated amidodiphosphine ligands devoid of a spacer, **Biot-1** and **Biot-2**, were obtained by coupling of amino-diphosphines to biotin–NHS (Scheme 17.10). Ligands containing achiral amino acid spacers between the biotin and the phosphines were synthesized by coupling of biotin to unprotected linear α,ω-amino acids, $\mathbf{3}^n$ (n=1–5), as well as arylamino amino acids, $\mathbf{4}^q$ (q=*ortho*, *meta*, *para*), using isobutylchloroformate as the activating agent. This was followed by the coupling of the resulting biotinylated amino acid to aminodiphoshine ligands, which was best achieved using 2-chloro-4,6-dimethoxy-1,3,5-triazine (CDMT) as the activating agent in anhydrous acetonitrile.

Ligands containing enantiopure chiral amino acid spacers were obtained by coupling both enantiomers of the Boc-protected amino acids phenylalanine and proline to aminodiphoshine H-**1**, followed by deprotection and coupling to biotin [39].

The screening of various combinations of protein mutants and ligands yielded substrate-specific and either (*R*) or (*S*) selective rhodium-catalysts which hydrogenated α-acetamidoacrylic and α-acetamidocinnamic acid with enantioselectivities up to 96%. In some cases the artificial metalloenzymes displayed a higher activity in the reduction of acetamidoacrylic acid than the corresponding hydrogenation by the rhodium–ligand complex alone [40]. Second coordination sphere effects were assumed to be responsible for this acceleration. Introduction of enantiopure amino acid spacers between the biotin anchor and the diphosphine moiety considerably improved the activity and stability of the artificial metalloenzymes in rhodium-catalysed hydrogenation reactions. In addition, these so-called second-generation artificial hydrogenases displayed improved organic solvent tolerance [39]. Besides the use of artificial metalloenzymes for enantioselective hydrogenation, Ward and co-workers demonstrated in 2008 that the biotin–avidin technology could be extended to palladium-catalysed asymmetric allylic alkylation. With a combination of chemical and genetic optimization, the product of the substitution reaction of 1,3-diphenylpropenyl acetate and dimethyl malonate was obtained with enantiomeric excesses ranging from 90% (*R*) to 82% (*S*) [41].

Reetz *et al.* subjected the system composed of streptavidin and the $Rh(COD)BF_4$ complex of the ligand introduced by Whitesides (**Biot-1**) to directed evolution, a Darwinistic approach fundamentally different from Ward's chemogenetic engineering [42]. Starting from 23% ee for the (*R*) product of the hydrogenation of methyl-α-acetamidoacrylate, three rounds of screening afforded catalysts displaying 65% ee (*R*) or 7% ee (*S*). This demonstrated that it is possible to apply this methodology to tailor the enantioselectivity of an artificial metalloenzyme, although it was noted that it is a very labour-intensive procedure [42].

17.4.1.2 Antibodies

An alternative method for the anchoring of a transition metal catalyst into a protein based on molecular recognition was developed by Harada and co-workers, who elicited antibodies against an achiral diphosphine–rhodium complex (Scheme 17.11) as the hapten [43]. Following activation of the carboxylic acid group of the complex with *N,N'*-carbonyldiimidazole (CDI), the complex was coupled to the carrier proteins keyhole limpet hemocyanin (KLH) and bovine serum albumin (BSA). These protein–hapten conjugates were then employed to elicit antibodies, providing four monoclonal antibodies for the complex. One antibody–hapten complex provided 98% ee in the hydrogenation of α-acetamidoacrylic acid, albeit with poor conversion.

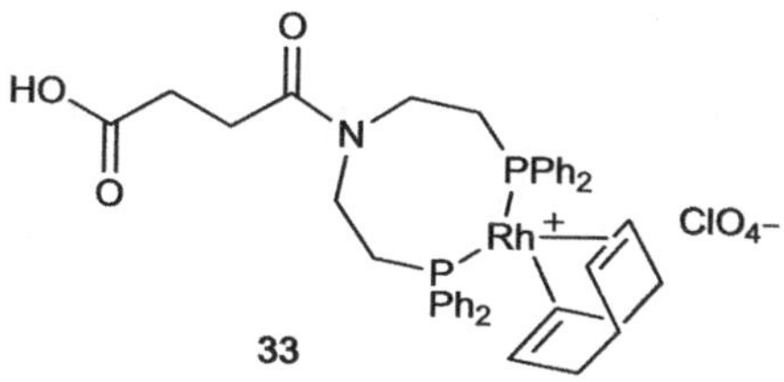

33

Scheme 17.11 Rhodium–diphosphine complex used as hapten to generate antibodies.

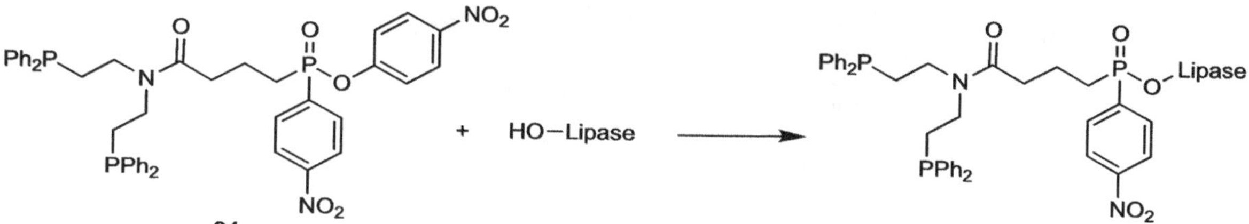

34

35

Scheme 17.12 Covalent anchoring of a diphosphine ligand to the active-site serine of lipase via a phosphonate inhibitor.

Notably, no conversion was obtained for the larger substrates α-acetamidocinnamic acid and α-acetamido-4-methylpentenoic acid, suggesting that the antibody enforces substrate-specific access to the catalytic centre.

17.4.2 Covalent anchoring of phosphines

The pronounced nucleophilicity of certain amino acid side chains (typically serine, lysine or cysteine) can be exploited for the site-selective covalent modification of proteins using suitable electrohilic moieties. The serine residue in the active site of lipases can be selectively modified with compounds containing a *para*-nitrophosphonate group, a known class of covalent inhibitors for these enzymes. Reetz *et al.* exploited this to couple the diphosphine ligand **34** to a lipase (Scheme 17.12). However, the obtained hybrids were unstable due to basic hydrolysis of the phosphonate moiety, and the catalytic activity of the native lipase was restored within 24 h [44].

The unique reactivity of its nucleophilic thiol side chain and its low natural abundance make cysteine a very attractive target for site-selective covalent anchoring of compounds to proteins. De Vries and co-workers modified papain at cysteine 25 with a monodentate phosphite ligand using phenacyl bromide as the anchor. To overcome problems with cofactor solubility in the aqueous medium required for the protein, the ligand was functionalized with two triethyleneglycol tails. After the introduction of the ligand, the modified protein was treated with an excess of $[Rh(COD)_2]BF_4$ and purified by gel filtration. The artificial metalloenzyme was tested in the hydrogenation of methyl α-acetamido acrylate and turned out to be active but not enantioselective. It was concluded that papain allows the catalytic centre too much conformational freedom to ensure a structurally defined metal complex and that other, more robust proteins may be more suitable as scaffolds [45].

The nucleophilicity of phosphines (which are more nucleophilic than phosphites) precludes the use of the most common electrophiles employed for cysteine selective bioconjugation, requiring the development of alternative coupling strategies. For example, the widely employed maleimide moiety reacts with phosphines to yield phosphorous ylides [46]. Laan *et al.* reported the use of imidazolides as alternative electrophiles for

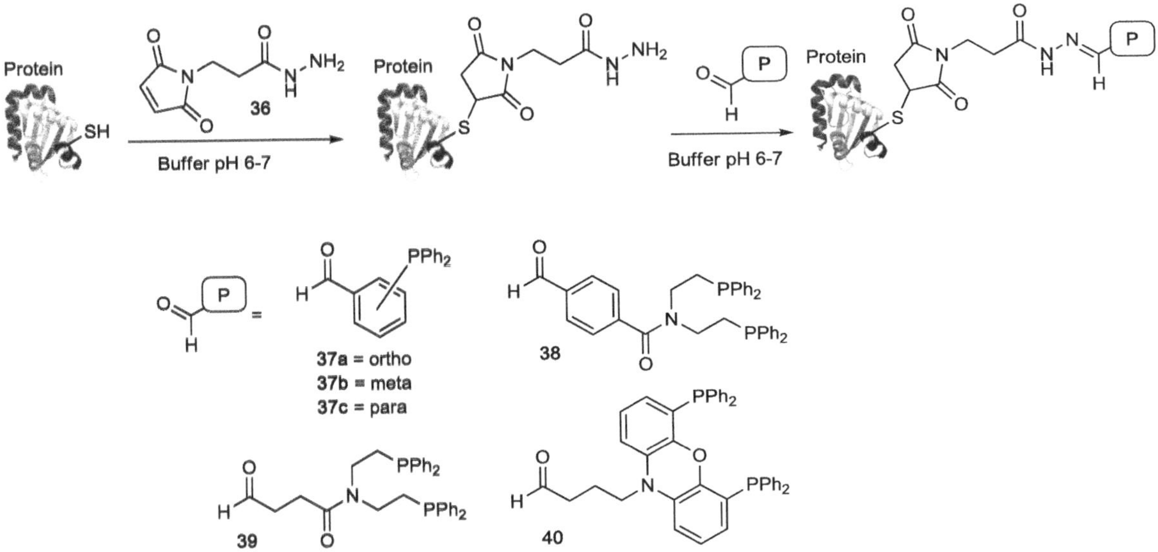

Scheme 17.13 Cysteine-selective covalent anchoring of phosphines via a hydrazone formation.

the covalent anchoring of several phosphines and phosphine–transition metal complexes to the unique cysteine of photoactive yellow protein (PYP) [47]. Unfortunately, the imidazolide methodology lacked the desired chemoselectivity when applied to other protein scaffolds, resulting in mixed cysteine and lysine acylation [48].

A generally applicable methodology for the conjugation of phosphines to single cysteine-containing proteins was developed by Deuss *et al.*, combining two well-known and highly selective bioconjugation techniques: Michael addition of cysteine thiol to maleimide and hydrazone formation between a hydrazide and aldehyde (Scheme 17.13) [48]. Using the commercially available cross-linker N-(β-maleimidopropionic acid) hydrazide (BMPH, **36**), a hydrazide moiety is conjugated to the cysteine. Hydrazines react selectively with carbonyls like ketones and aldehydes to form stable hydrazone bonds. This reaction is widely employed for bioconjugation (e.g. to detect carbonyls in proteins) and for modification of carbonyl-containing proteins with reporter groups. Modification of the hydrazine-modified proteins with a variety of phosphino-aldehydes (**37–40**) afforded proteins containing a free phosphorous ligand bound via a hydrazone linkage. The versatility of this approach was demonstrated by the coupling of a variety of phosphine ligands to three structurally different proteins. Moreover, incubation of one of the phosphine-modified proteins with $[Rh(acac)(CO)_2]$ resulted in the formation of a metalloprotein. This development holds great promise for the use of a wide range of protein structures as hosts for the preparation of structurally diverse phosphine-bearing artificial metalloenzymes.

17.5 Conclusions and outlook

This chapter has outlined the various strategies for the synthesis of phoshine-modified biomolecules pursued to date. The diversity of systems and applications reported to date underlines the potential of these approaches. It is evident that the application of advanced biomolecular techniques such as site-directed mutagenesis and directed evolution in combination with traditional synthetic tools is a very powerful approach for optimising the performance of these systems. Continuous progress in the computational design and modelling techniques

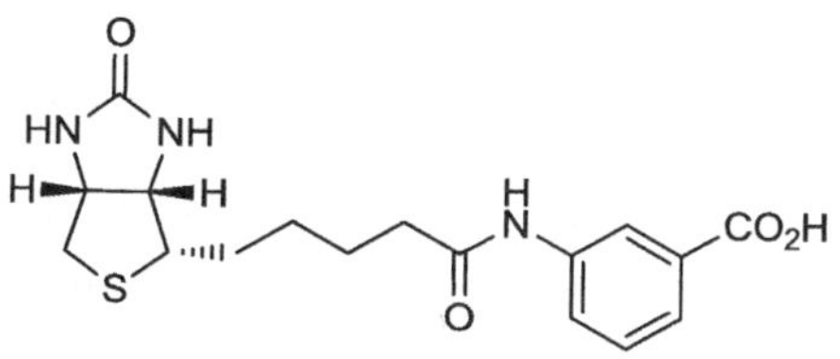

Biot–4m

Scheme 17.14 Biot–4meta.

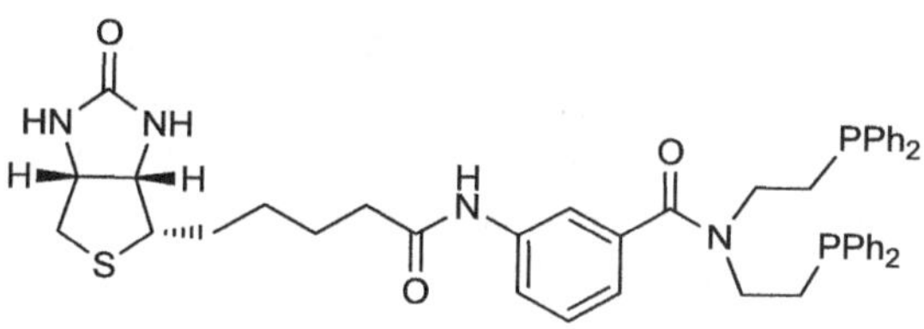

Biot–4m–1

Scheme 17.15 Biot–4m–1.

for both DNA and polypeptide structures will lead to further insight and possibilities of these systems. With these powerful tools available and considering the increasing research efforts in the field we anticipate that further advances in the development of these hybrid catalyst systems will lead to effective solutions for demanding transition metal–catalysed chemical reactions.

17.6 Representative synthetic procedures

17.6.1 Artificial hydrogenases based on the biotin–streptavidin technology (adapted from [38a])

Representative procedure for the synthesis of biotinylated amino acid precursors: Biot–4meta (Scheme 17.14): Isobutyl chloroformate (0.32 mL, 2.47 mmol) was added to a solution of biotin (500 mg, 2.05 mmol) in DMF (40 mL) containing tri-*N*-butylamine (0.64 mL, 2.69 mmol). After 10 min at room temperature (RT), the mixture was slowly added at 5°C to a suspension of *m*-aminobenzoicacid (4.1 mmol) in DMF (40 mL). After stirring at 5°C for 2 h, the solvent was distilled under vacuum and the crude precipitate was dissolved in warm aqueous ethanol (1:1, 36 mL total). The mixture was acidified with 2.0 N HCl to pH 2 and kept at 0°C for 12 h. The resulting precipitate was filtered, washed with water and dried under vacuum. Yield: 70%.

^{1}H NMR (400 MHz, [D$_6$] DMSO, 20°C): δ = 1.30–1.70 (m, 6 H; CH_2), 2.34 (m, 2 H; CH_2), 2.61 (d, J = 12.4 Hz, 1 H; CH_2–S), 2.83 (dd, J = 12.4, 5.0 Hz, 1 H; CH_2–S), 3.16 (m, 1 H; CH–S), 4.16 (m, 1 H; CH–N), 4.32 (m, 1 H; CH–N), 6.38 (br, 1 H; NH), 6.46 (br, 1 H; NH), 7.42 (m, 1 H; ArH), 7.61 (m, 1 H; ArH), 7.84 (m, 1 H; ArH), 8.25 (m, 1 H; ArH), 10.15 (br, 1 H; NH), 12.80 (br, 1 H; CO_2H); ^{13}C NMR (100.6 MHz, [D$_6$] DMSO, 20°C): δ = 25.9 (CH_2), 29.0 (CH_2), 29.1 (CH_2), 37.1 (CH_2), 40.7 (CH_2–S), 56.3 (CH–S), 60.1 (CH–N), 61.9 (CH–N), 120.6 (Ar–H), 124.0 (Ar–H), 124.6 (Ar–H), 129.8 (Ar–H), 132.1 (Ar–C), 140.4 (Ar–C), 163.6 (N–CO–N), 168.1 (CO–N), 172.3 (CO_2H); ESI–MS for $C_{17}H_{21}N_3O_4S$: 362.2 [M–H]$^-$.

Synthesis of Biot–4m–1 (Scheme 17.15): The biotinylated amino acid **Biot–4m** (0.7 mmol), the diphosphino-amine (H–**1**·HCl) (0.7 mmol) and CDMT (0.135 g, 0.77 mmol) were mixed in degassed acetonitrile (20 mL), and *N*-methylmorpholine (0.27 mL, 2.5 mmol) was added. The slurry was stirred overnight under reflux. After cooling to RT, water (50 mL) was added. This solvent mixture initially dissolved the suspension but quickly generated a thick slurry as the product precipitated. After 3 h stirring, the solvent was decanted and the oil was dried under vacuum to afford a white foam which was purified by column chromatography on silica gel 60 F_{254} using chloroform:ethanol (7:1) as eluent. Yield: 29%.

^{1}H NMR (500 MHz, $CDCl_3$, 20°C): δ = 1.20–1.72 (m, 6 H; CH_2), 2.10 (m, 2 H; CH_2), 2.24 (m, 2 H; CH_2), 2.35 (m, 2 H; CH_2), 2.53 (d, J = 12.6 Hz, 1 H; CH_2–S), 2.67 (dd, J = 12.6, 5.0 Hz, 1 H; CH_2–S), 2.92 (m, 1 H; CH–S), 3.20 (m, 2 H; CH_2), 3.46 (m, 2 H; CH_2), 4.07 (m, 1 H; CH–N), 4.27 (m, 1 H; CH–N), 6.70–7.70 (m, 24 H; ArH); ^{13}C NMR (125.7 MHz, $CDCl_3$, 20°C): δ = 24.5 (CH_2), 25.4 (d, J = 12.5 Hz; CH_2), 26.4 (d, J = 15.0 Hz; CH_2), 27.1 (CH_2), 27.2 (CH_2), 35.6 (CH_2), 39.4 (CH_2–S), 41.9 (d, J = 22.5 Hz; CH_2), 45.9 (d, J = 23.7 Hz; CH_2), 54.6 (CH–S), 59.4 (CH–N), 60.8 (CH–N), 116.0–131.7 (Ar–H), 135.6–138.4 (Ar–C), 163.4 (N–CO–N), 170.6 (CO–NH), 171.3 (CO–N); ^{31}P NMR (161.9 MHz, $CDCl_3$, 20°C): δ = –20.4, –20.8; ESI–MS for $C_{45}H_{48}N_4O_3P_2S$: 803.3 (15) $[M+H+O]^+$, 787.3 (100) $[M+H]^+$.

Hydrogenation protocol: Considering the very small amounts of catalyst involved, we found that the reproducibility of the results critically depends on the care taken in the preparation of the catalytic runs. All organic and aqueous solutions were degassed by flushing nitrogen through the solutions for *at least* 3 h. All operations were performed in a glove box. Acetate, pH 4.0, 0.1 M final concentration.

The metal source $[Rh(COD)_2]BF_4$ (4 mg, 10 μmol) was dissolved in DMSO, and 1.56 μmol of this solution was added to **Biot–4^{meta}–1** (2.02 μmol, 1.3 equivalents with respect to the metal source). The solution was stirred at RT for 15 min. This solution can be stored and reused for a few days without any noticeable loss of activity and of selectivity.

A glass tube (ca. 3 mL) was placed in an autoclave and charged with an acetamidoacrylic acid solution in water (260 μL of a 23.85 mM solution, 6.20 μmol). The acetate buffer pH 4.0 solution (100 μL of a 1 M solution) was added and the volume adjusted with water to 900 μL. The protein solution in water was then added (100 μL of a 0.207 mM solution, 0.0207 μmol of the tetramer). The catalyst solution in DMSO (100 μL, 0.062 μmol) was added last. The resulting mixture was vortexed.

The glass autoclave containing up to 21 samples was closed, purged under vacuum and charged with H_2. After stirring for 15 h on an orbital shaker at the desired temperature, the reaction was quenched by adjusting the pH to 2 with 2.0 N aqueous HCl solution. The aqueous solution was continuously extracted with ethyl acetate for 2 h, and the organic phase was evaporated to a minimum volume (0.5 mL). The hydrogenation product (acetamidoalanine) was converted *in situ* to its methyl ester using trimethysulfonium hydroxyde before GC analysis on Heptakis-2,3,6-perethyl-β-cyclodextrin, 22 m×0.32 mm column: He carrier gas: 3.3 mL/min, inlet injector 280°C, oven 97°C, FID detector 230°C. Retention time: (*S*)-*N*-acetylacetamidoalanine: 4.6 min; and (*R*)-*N*-acetylacetamidoalanine: 5 min.

17.6.2 Site-selective covalent modification of proteins with phosphines via hydrazone linkage (adapted from [48])

General remarks: 3-Maleimidopropionic acid hydrazide hydrochloride **36** is obtained from Shanghai SpeedChemical Co., Ltd, and (*o*-diphenylphosphino)benzaldehyde **37a** from Aldrich. Spin-filter concentrators with a molecular weight cut-off of 10 kDa (Amicon® Millipore Ultra®, Ultracel 10 K, Millipore®, Carrigtwohill, Co. Cork, Ireland) are used to concentrate and/or wash protein solutions. All aqueous solutions were degassed by flushing nitrogen through the solutions for *at least* 3 h.

Conjugation of maleimidopropionic acid hydrazide 36 to SCP-2L V83C: A 35 μM solution of protein in buffer (20 mM MES 50 mM NaCl pH 6 for SCP-2L V83C) was incubated with 400 μM 3-maleimidopropionic acid hydrazide hydrochloric acid 36 added from a 50 mM stock solution in the corresponding buffer. This mixture is stirred at RT for 16 h, after which it is transferred to a centrifugal concentrator. Concentration of the protein to 250 µl followed by dilution to 15 ml with buffer (20 mM MES 50 mM NaCl pH 6) was repeated three times to remove excess hydrazide. The conjugation product was analysed by LCMS(ES$^+$). A sample of the washed conjugation product to which a known amount of unmodified protein was added was also subjected to LCMS(ES$^+$) analysis. This analysis showed that the unmodified protein shows more efficient ionization, thus confirming that the observed signal ratios do reflect more than 95% conversion.

Coupling of phosphine aldehyde 37a to hydrazide-modified SCP-2L V83C: Phosphine aldehyde 37a (3 eq.) was added from a 50 mM stock solution in degassed DMF to a solution of 36-modified protein in buffer (20 mM MES 50 mM NaCl pH 6 for SCP-2L V83C) under inert atmosphere. The reaction was mixed for 16 h at RT, after which the excess phosphine was removed by centrifugation and the supernatant washed three times with the appropriate buffer in a centrifugal concentrator. Again, analysis of relative ionization efficiencies of unmodified and modified protein confirmed the high conversion observed by LCMS(ES$^+$) analysis. Often peaks corresponding to the modification of the protein with an oxidized phosphine are observed in the LCMS(ES$^+$) spectra. As the LCMS(ES$^+$) analysis is performed without exclusion of oxygen, this is likely the result of oxidation during the analysis procedure. This is supported by the fact that no significant amount of oxidized product is observed in ^{31}P-NMR experiments. For ^{31}P-NMR experiments the modified protein solution was concentrated to 2 mM after which 400 µl of the protein solution was mixed with 100 µl D_2O in an NMR tube under argon. The ^{31}P-spectrum was recorded on a Bruker Avance 500 spectrometer over 16 h.

17.7 Acknowledgments

We thank Professor Thomas R. Ward for providing us with a representative procedure for the synthesis of artificial hydrogenases based on the biotin–streptavidin technology.

References

[1] P. Knochel, R. D. Singer, *Chem. Rev.* **1993**, *93*, 2117.
[2] S. J. Greenfield, A. Agarkov, S. R. Gilbertson, *Org. Lett.* **2003**, *5*, 3069.
[3] A. Agarkov, S. Greenfield, D. J. Xie, R. Pawlick, G. Starkey, S. R. Gilbertson, *Biopolymers* **2006**, *84*, 48.
[4] A. M. Porte, W. A. van der Donk, K. Burgess, *J. Org. Chem.* **1998**, *63*, 5262.
[5] S. Burck, S. G. A. van Assema, B. Lastdrager, J. C. Slootweg, A. W. Ehlers, J. M. Otero, B. Dacunha-Marinho, A. L. Llamas-Saiz, M. Overhand, M. J. van Raaij, K. Lammertsma, *Chem. Eur. J.* **2009**, *15*, 8134.
[6] S. R. Gilbertson, G. W. Starkey, *J. Org. Chem.* **1996**, *61*, 2922.
[7] H. B. Kraatz, A. Pletsch, *Tetrahedron: Asymmetry* **2000**, *11*, 1617.
[8] M. Tepper, O. Stelzer, T. Hausler, W. S. Sheldrick, *Tetrahedron Lett.* **1997**, *38*, 2257.
[9] S. R. Gilbertson, R. V. Pawlick, *Angew. Chem. Int. Ed.* **1996**, *35*, 902.
[10] S. van Zutphen, V. J. Margarit, G. Mora, P. Le Floch, *Tetrahedron Lett.* **2007**, *48*, 2857.
[11] S. R. Gilbertson, X. F. Wang, G. S. Hoge, C. A. Klug, J. Schaefer, *Organometallics* **1996**, *15*, 4678.
[12] S. R. Gilbertson, G. H. Chen, J. Kao, A. Beatty, C. F. Campana, *J. Org. Chem.* **1997**, *62*, 5557.
[13] S. R. Gilbertson, X. F. Wang, *Tetrahedron* **1999**, *55*, 11609.
[14] (a) A. Agarkov, S. J. Greenfield, T. Ohishi, S. E. Collibee, S. R. Gilbertson, *J. Org. Chem.* **2004**, *69*, 8077; and (b) S. R. Gilbertson, S. E. Collibee, A. Agarkov, *J. Am. Chem. Soc.* **2000**, *122*, 6522.
[15] J. B. H. Hellmann, H. Birkner, O. Schumacher, *Liebigs Ann. Chem.* **1962**, *659*, 49.
[16] C. A. Christensen, M. Meldal, *Chem. Eur. J.* **2005**, *11*, 4121.
[17] C. A. Christensen, M. Meldal, *J. Comb. Chem.* **2007**, *9*, 79.
[18] P. Arya, G. Panda, N. V. Rao, H. Alper, S. C. Bourque, L. E. Manzer, *J. Am. Chem. Soc.* **2001**, *123*, 2889.

[19] (a) S. J. Degrado, H. Mizutani, A. H. Hoveyda, *J. Am. Chem. Soc.* **2002**, *124*, 13362; and (b) S. J. Degrado, H. Mizutani, A. H. Hoveyda, *J. Am. Chem. Soc.* **2001**, *123*, 755.
[20] A. W. Hird, A. H. Hoveyda, *Angew. Chem. Int. Ed.* **2003**, *42*, 1276.
[21] B. Breit, A. C. Laungani, *Tetrahedron: Asymmetry* **2003**, *14*, 3823.
[22] A. C. Laungani, B. Breit, *Chem. Commun.* **2008**, 844.
[23] A. C. Laungani, J. M. Slattery, I. Krossing, B. Breit, *Chem. Eur. J.* **2008**, *14*, 4488.
[24] C. W. Edwards, M. R. Shipton, M. Wills, *Tetrahedron Lett.* **2000**, *41*, 8615.
[25] T. P. Clark, C. R. Landis, *J. Am. Chem. Soc.* **2003**, *125*, 11792.
[26] T. P. Clark, C. R. Landis, S. L. Freed, J. Klosin, K. A. Abboud, *J. Am. Chem. Soc.* **2005**, *127*, 5040.
[27] C. R. Landis, T. P. Clark, *Proc. Natl. Acad. Sci. USA* **2004**, *101*, 5428.
[28] (a) A. D. Ellington, J. W. Szostak, *Nature* **1990**, *346*, 818; (b) C. Tuerk, L. Gold, *Science* **1990**, *249*, 505; and (c) R. Stoltenburg, C. Reinemann, B. Strehlitz, *Biomol. Eng* **2007**, *24*, 381.
[29] C. B. Reese, *Org. Biomol. Chem.* **2005**, *3*, 3851.
[30] M. Caprioara, R. Fiammengo, M. Engeser, A. Jaschke, *Chem. Eur. J.* **2007**, *13*, 2089.
[31] M. Nuzzolo, A. Grabulosa, A. M. Z. Slawin, N. J. Meeuwenoord, G. A. van der Marel, P. C. J. Kamer, *Eur. J. Org. Chem.* **2010**, 3229.
[32] (a) R. M. Franzini, E. T. Kool, *ChemBioChem* **2008**, *9*, 2981; (b) K. Furukawa, H. Abe, J. Wang, M. Uda, H. Koshino, S. Tsuneda, Y. Ito, *Org. Biomol. Chem.* **2009**, *7*, 671; and (c) K. Sakurai, T. M. Snyder, D. R. Liu, *J. Am. Chem. Soc.* **2005**, *127*, 1660.
[33] L. Ropartz, N. J. Meeuwenoord, G. A. van der Marel, P. W. N. M. van Leeuwen, A. M. Z. Slawin, P. C. J. Kamer, *Chem. Commun.* **2007**, 1556.
[34] D. F. Qi, C. M. Tann, D. Haring, M. D. Distefano, *Chem. Rev.* **2001**, *101*, 3081.
[35] M. E. Wilson, G. M. Whitesides, *J. Am. Chem. Soc.* **1978**, *100*, 306.
[36] C. C. Lin, C. W. Lin, A. S. C. Chan, *Tetrahedron: Asymmetry* **1999**, *10*, 1887.
[37] T. R. Ward, *Acc. Chem. Res.* **2011**, *44*, 47.
[38] (a) M. Skander, N. Humbert, J. Collot, J. Gradinaru, G. Klein, A. Loosli, J. Sauser, A. Zocchi, F. Gilardoni, T. R. Ward, *J. Am. Chem. Soc.* **2004**, *126*, 14411; and (b) J. Collot, J. Gradinaru, N. Humbert, M. Skander, A. Zocchi, T. R. Ward, *J. Am. Chem. Soc.* **2003**, *125*, 9030.
[39] M. Skander, C. Malan, A. Ivanova, T. R. Ward, *Chem. Commun.* **2005**, 4815.
[40] U. E. Rusbandi, C. Lo, M. Skander, A. Ivanova, M. Creus, N. Humbert, T. R. Ward, *Adv. Synth. Catal.* **2007**, *349*, 1923.
[41] J. Pierron, C. Malan, M. Creus, J. Gradinaru, I. Hafner, A. Ivanova, A. Sardo, T. R. Ward, *Angew. Chem. Int. Ed.* **2008**, *47*, 701.
[42] M. T. Reetz, J. J. P. Peyralans, A. Maichele, Y. Fu, M. Maywald, *Chem. Commun.* **2006**, 4318.
[43] H. Yamaguchi, T. Hirano, H. Kiminami, D. Taura, A. Harada, *Org. Biomol. Chem.* **2006**, *4*, 3571.
[44] M. T. Reetz, M. Rentzsch, A. Pletsch, M. Maywald, *Chimia* **2002**, *56*, 721.
[45] L. Panella, J. Broos, J. F. Jin, M. W. Fraaije, D. B. Janssen, M. Jeronimus-Stratingh, B. L. Feringa, A. J. Minnaard, J. G. de Vries, *Chem. Commun.* **2005**, 5656.
[46] E. Hedaya, S. Theodoropulos, *Tetrahedron* **1968**, *24*, 2241.
[47] (a) R. den Heeten, B. K. Munoz, G. Popa, W. Laan, P. C. J. Kamer, *Dalton Transactions* **2010**, *39*, 8477; and (b) W. Laan, B. K. Munoz, R. den Heeten, P. C. J. Kamer, *ChemBioChem* **2010**, *11*, 1236.
[48] P. J. Deuss, G. Popa, C. H. Botting, W. Laan, P. C. J. Kamer, *Angew. Chem. Int. Ed.* **2010**, *49*, 5315.

18

The Design of Ligand Systems for Immobilisation in Novel Reaction Media

Paul B. Webb[1] **and David J. Cole Hamilton**[2]

[1] *Sasol Technology UK Ltd, St Andrews, United Kingdom*
[2] *EaStCHEM, School of Chemistry, University of St Andrews, St Andrews, United Kingdom*

18.1 Introduction

There is an ever-increasing demand to minimise the environmental impact of the chemicals industry and this can only be achieved if we move increasingly towards commercial operations that exhibit 100% atom economy with high activity and high product selectivity. The ideal catalytic reaction would also be operated in a continuous flow process, where the catalyst remains under its optimum operating conditions at all times, there would be no volatile organic solvents, no waste and no by-products. It is unlikely that such demanding atom economy and selectivity criteria can be met by heterogeneously catalysed reactions and as such the future for a greener chemicals industry must lie with homogeneous catalysis. Homogeneous catalysts are typically more active than their heterogeneous counterparts, a consequence of all reaction sites being equally accessible. Through careful control of ligand structure it is also possible to fine-tune the chemoselectivity, regioselectivity and/or enantioselectivity of a reaction. Yet despite these great advantages the commercialisation of otherwise very attractive homogeneously catalysed reactions has been hindered because of one major technical challenge; the separation of catalyst from product can be difficult to achieve. This problem arises because the most commonly employed method of separation, distillation, typically requires elevated temperatures. Most homogeneous catalysts are thermally labile, often decomposing below the temperatures required for product distillation. Other conventional separation strategies, such as chromatography or extraction, can also lead to catalyst loss because the conditions of separation are very different from those of catalysis itself where the catalyst can be stabilised by the presence of reactant gases for instance. The catalyst separation problem must therefore be addressed if the transition to more environmentally benign processes is ever to be realised.

Phosphorus(III) Ligands in Homogeneous Catalysis: Design and Synthesis, First Edition.
Edited by Paul C. J. Kamer and Piet W. N. M. van Leeuwen.

The issue of catalyst recycle and the ability to operate as a continuous process remain two of the biggest challenges facing the chemical industry and have led to extensive research efforts into methods that combine the ease of catalyst recovery associated with heterogeneous systems with the more desirable activity and selectivity obtained with homogeneous catalysts. The separation of catalyst from product can be effected by a number of distinct approaches that can be broadly classed as (1) heterogenisation, the anchoring of catalyst to a support, and (2) biphasic catalysis, the immobilisation of catalyst in a solvent that is immiscible with the reaction product under certain conditions [1]. Within each category there are several variants. For example, the support for heterogenisation can be an oxidic material commonly used in heterogeneous catalysis or take the form of a soluble entity, such as a polymer [2] or dendritic species [3], where separation is effected by ultrafiltration. There are also a number of solvent systems that can be utilised in the biphasic approach to catalyst separation, the most recognised being aqueous biphasic [4], fluorous biphasic [5], ionic liquids [6], supercritical carbon dioxide [7] and combinations thereof. The search for alternative reaction media is addressed not only from a catalyst separation perspective but also as a means to improve the environmental impact of chemical processes [8]. According to our earlier definition, the ideal catalytic reaction would not rely on volatile organic compounds (VOCs) as solvents. Ideally, we would like to avoid using solvents altogether but when a solvent is required water has many positive attributes; it is nontoxic, nonvolatile and environmentally benign to say the least. In the context of green chemistry, supercritical carbon dioxide [9] and ionic liquids [6a,10] also present interesting reaction media. These so-called neoteric, green solvents not only provide potentially cleaner and safer alternatives but also possess physical properties that enable access to operations not possible with common VOCs and this will be described in later sections of this chapter.

Research into the catalyst separation problem has grown and diversified to such an extent that today it represents a scientific discipline in its own right and is the subject of many excellent books and reviews [1–10]. In this chapter we aim to describe strategies that have been developed to enable catalyst separation focusing specifically on novel reaction media; heterogenisation, separation by size exclusion filtration, thermoregulated systems and so on will not be described in detail here. Catalysts designed for immobilisation or solubilisation in novel reaction media are rarely developed from first principles. The generally adopted approach is to modify existing ligands that have been designed for use in conventional solvents to display specific activity and selectivity behaviours for the reaction of interest. There are a number of strategies that can be employed to enable immobilisation depending on the nature of the reaction medium. Ideally, in all cases such modifications should not alter the catalytic properties exhibited by the parent ligand, and catalysts derived therefrom, because this would negate the use of a specific ligand framework in the first place. An important consideration in ligand design is therefore to ensure that functionalisation with moieties that impart desired solubility characteristics remain sufficiently isolated from the phosphorus centre so that properties such as steric crowding, bite angle and electronics are not adversely affected. This is of particular importance in the construction of fluorous soluble catalysts, although in some cases beneficial effects of the highly electron-withdrawing nature of fluorine are observed.

If we consider the number of P(III) ligands that have been developed for use in conventional solvents, and the broad range of modifications that can be applied to effect immobilisation in several different reaction media, the number of possible permutations and combinations is vast. It would therefore be impossible to give full justification to all P(III) ligands that have been developed for applications in alternative solvent systems. It is far more informative, and of practical use, to focus on general strategies for immobilisation with worked examples. Thus, synthetic methodologies will be described for different compound classes and associated synthetic challenges highlighted where necessary using examples chosen purely on the basis of literature prevalence and ease of preparation. Synthetic strategies will therefore be demonstrated using the ubiquitous PPh_3 and chiral BINAP scaffolds. Since its discovery in 1980, the number of patents and publications dedicated to the synthesis and application of BINAP is impressive. BINAP is one of the most widely employed ligands in asymmetric catalysis, it is currently used in several industrial processes and is also commercially available. It is not surprising, therefore, that separation and recycling strategies are often explored with this highly successful ligand [11], hence its inclusion as an example throughout this chapter.

18.2 Aqueous biphasic catalysis

Two-phase catalysis provides an elegant solution to the catalyst separation problem and was first applied industrially in the Shell higher olefin process (SHOP) [12]. In the SHOP process, the polymerisation of ethylene occurs over a polar 1,4-butanediol phase, containing an organonickel catalyst, to yield apolar α-olefin products that separate as a second liquid phase and are easily removed. The basic requirement of any biphasic system is the existence of a miscibility gap between catalyst-containing reaction solution and reaction products under certain conditions. It is not surprising, therefore, that water represents an ideal solvent for biphasic catalysis owing to its general immiscibility with organic compounds. In aqueous, homogeneous, two-phase catalysis the active catalyst is dissolved in water so that the reactants and reaction products, which are ideally organic and relatively nonpolar, can be separated off after the reaction is complete. Commercialisation of the aqueous biphasic Ruhrchemie–Rhône-Poulenc process (RCHRP) in 1984 signified a major breakthrough in catalyst separation and recycling technologies [13]. The most important aspect of the RCHRP process, which converts propylene to *n*-butyraldehyde, is the tailoring of Rh-based organometallic complexes as catalysts. Aryl- and alkyl-substituted tertiary phosphines are among the most commonly employed ligands in homogeneous catalysis, but they are hydrophobic, as in general are the products of catalysis. The RCHRP catalyst is rendered water-soluble through modification of the P(III) ligand (i.e. tri(*m*-sulfonyl)triphenylphosphine TPPTS (**6**) is employed as a hydrophilic analogue of PPh_3). Numerous water-soluble ligands have been extensively tested for this process but only TPPTS has proven economically viable [14]. Nonetheless, fuelled largely by the industrial success of the RCHRP process, extensive research into aqueous biphasic catalysis has led to an array of alternative means for introducing hydrophilicity in P(III) ligands. In principle, any hydrothermally stable, polar moiety can be utilised to impart water solubility. The functionalisation of P(III) ligands with hydrophilic phosphonium [15], phosphonate [16], ammonium [17], carboxylate [18], carbohydrate [19] and polyether [20] groups has been demonstrated. The number of examples within each of these classes is copious; virtually every imaginable ligand framework has been modified in some way with hydrophilic groups to generate water-soluble catalysts. Figure 18.1 does not give

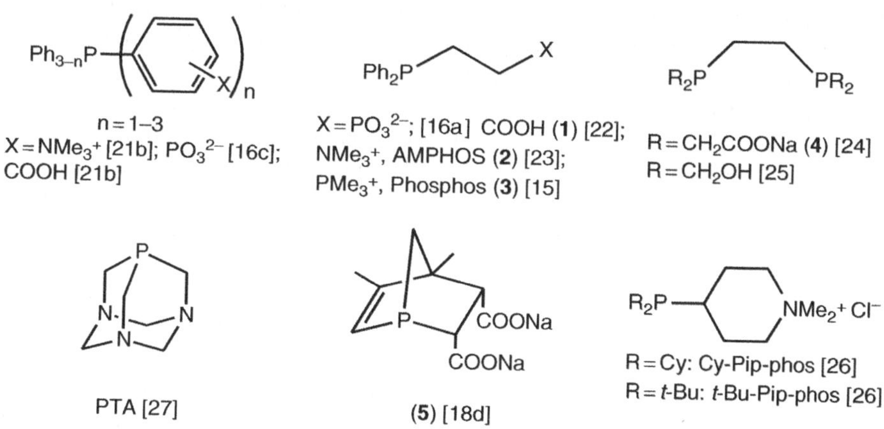

Figure 18.1 *Examples of hydrophilic phosphines developed for chemistry in water. References are given to literature describing synthesis of the associated ligand. For sulphonated phosphines see Figure 18.2.*

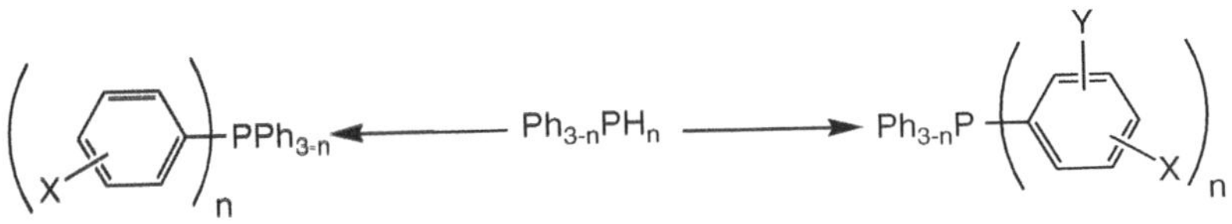

X, Y = Me, OH, NH_2, COOH, COOMe, SO_3Na

Scheme 18.1

justice to the extent of research in this field and serves merely to illustrate the number of approaches to enhancing hydrophilicity and the diversity of ligand structures that have been modified for aqueous phase catalysis. Several strategies have been developed to construct ligands depicted in Figure 18.1 and here we highlight only the routes that demonstrate broad application. Of particular note is the Pd-catalysed P–C coupling reaction of substituted aryl halides with primary or secondary phosphines developed by Stelzer [17a,17b,21]. In principle, the aryl halide can contain any polar substituent, which will impart the desired hydrophilicity, in *ortho-*, *meta-* or *para-* positions with respect to the halogen of the aromatic ring system.

The coupling procedure also works for multiply substituted aryl halides containing either electron-withdrawing or electron-donating substituents. It becomes immediately apparent that a comprehensive library of water-soluble phosphines can be constructed easily using this simple and generic methodology that is also typically high yielding. The versatility of this approach is depicted in Scheme 18.1, which demonstrates the number of hydrophilic phosphines that can be prepared.

Phosphines with carboxylic groups were some of the earliest ligands developed for use in water. The first example, described by Mann *et al.* (**1**) [22], dates back to 1952 and was prepared by cyanoethylation of diphenylphosphine followed by nitrile saponification. Ligands of structure type (**1**) can also be obtained by oxidation of borane-protected hydroxyalkylphosphines with chromic anhydride in sulphuric acid [28] or *via* standard preparative methods involving alkylation of alkali–metal organophosphides [29]. The phosphine analogue of ethylene diaminetetraacetic acid (**4**) [24] is perhaps the most important example of phosphino carboxylic acid ligands owing to its application in catalysis [18a]. Also of note is the phosphanorbornene (**5**) synthesised by Mercier *et al.* that exhibits high water solubility (>300 g dm^{-3}) [18d].

Quaternisation of aminoalkyl or aminoaryl substituents offers another convenient method for introducing hydrophilicity. The most significant example from this family of water-soluble ligands is AMPHOS (**2**). This prototypical cationic phosphine is prepared simply by the quaternisation of $Ph_2P(CH_2)_2NMe_2$ [23,30]. Protection of the more reactive phosphorus centre by borane, oxidation or complexation is a necessary first step with subsequent reduction or decomplexation yielding the desired ligand. This preparative route can be used to generate several AMPHOS analogues and is easily extended to other ligand frameworks [17d,17e,31]. For example, Tóth *et al.* have applied this methodology to the preparation of hydrophilic, chiral ligands based on the DIOP and CHIRAPHOS scaffolds [17d,17e]. Phosphonium analogues of (**2**), known as Phosphos (**3**), are accessible through quaternisation of terminal PMe_3 with ω-chloroalkyl-diphenylphosphine or alkylation of $LiPPh_2$ with ω-haloalkylphosphonium salts $[X–(CH_2)_n–PMe_3]^+X^-$ [15]. The halides are then converted to more stable nitrates by ion exchange or salt metathesis.

Despite the diverse range of possibilities for transforming ligands into water-soluble entities, and the large volume of work dedicated to the field, sulfonation undoubtedly remains the most recognised immobilisation strategy in aqueous biphasic catalysis. However, appending the sulfonate moiety to P(III) ligands is not without its complications. Direct sulfonation with fuming sulphuric acid, originally described by Chatt in 1957 [32] and much later by Kuntz in 1975 for the preparation of TPPTS (**6**) [33], is the most commonly employed method for introducing sulfonate functionality and has been applied to various ligand frameworks (see Figure 18.2). Irrespective of its frequent application, this comparatively simple synthetic procedure poses

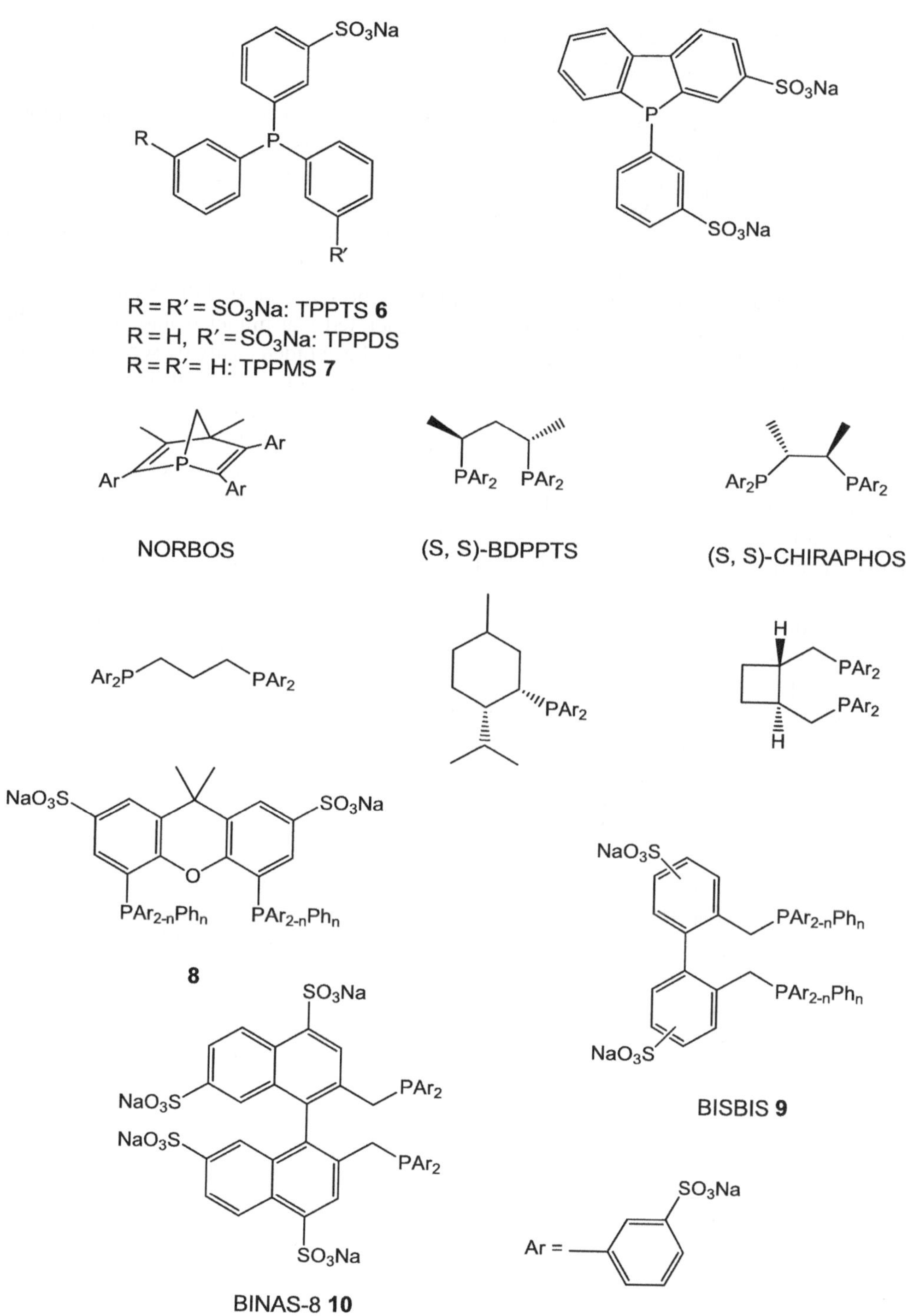

Figure 18.2 Examples of water-soluble phosphines obtained by direct sulfonation with oleum.

several challenges. Firstly, the formation of phosphine oxides is a serious disadvantage but can be suppressed by working in a superacidic reaction medium comprising boric acid and anhydrous sulphuric acid [34]. Minimising oxidation can also be achieved through the use of sodium oxalate or basic aluminium oxide [35], the careful control of pH [36] and operation at low temperatures especially during work-up. At low pH the phosphine is protected against oxidation through formation of phosphonium species. Controlling the pH between 2 and 4 retains phosphorus in this protonated form, while sulfonic acid is converted during neutralisation giving rise to $HP(m\text{-}C_6H_4SO_3Na)_3^+$ and restricting oxidation.

Product isolation from the sulfonation medium presents the main difficulty associated with the direct sulfonation route. Several work-up procedures, including fractional crystallisation [14], extraction re-extraction [37] and chromatography [38], have been applied in the synthesis of TPPTS alone but in all cases it is the considerable amounts of waste sodium sulphate generated that hinder isolation of the phosphine product. The production of large volumes of inorganic salts can be avoided easily by separation of the phosphine from the reaction solution as its free acid prior to neutralisation. Several approaches have been adopted with early work focusing on the separation of phosphine *via* extraction from the hydrolysed reaction mixture with a hydrophobic amine, typically triisooctylamine [34,37,39]. This method is described in our first synthesis example for TPPTS in Section 18.6, although we have found the extraction process to be difficult to reproduce on several occasions. The origin of this irreproducibility may lie in the free SO_3 content of the post-sulfonation mixture. During studies of a range of sulfonation–separation methods, Tic concluded that the efficacy of extraction is directly related to the free SO_3 content of the reaction solution; extraction is more efficient at low SO_3–phosphine ratios [40]. Recrystallisation of the free acid of the sulfonated phosphine directly from the sulfonation mixture has been demonstrated by Mul *et al.* as an effective work-up of sulfonated bidentate ligands [41]. Subsequent neutralisation of the solid precipitate obtained yields the desired phosphine and avoids neutralisation of the acidic reaction solution in its entirety. Sulfonated Xantphos (**8**), for example, was obtained by dilution of the sulfonation medium with water at −10°C, which resulted in immediate precipitation of the phosphine as a sticky white solid. Filtration and washing yielded the acidic product as a fine white powder, which is easily converted to the sodium salt by dilution in an aqueous NaOH solution, containing stoichiometric amounts of sodium, and subsequent freeze drying to remove co-produced water. Mul also notes the importance of removing all traces of sulphuric acid so that neutralisation of the sulphonic acid groups can be stoichiometrically controlled and the production of $NaHSO_4$ and Na_2SO_4 avoided as a consequence. On a final note, the choice of solvent at the washing stage is particularly important because nonpolar solvents can lead to gel formation. This highly efficient means of phosphine isolation has also been applied to the synthesis of ligands (**9** and **10**) shown in Figure 18.2.

Finally, direct sulfonation also suffers from poor regioselectivity. In the preparation of TPPTS, for instance, mono- and disubstituted species are always co-produced. Careful selection of reaction temperature can improve matters [42] but on the whole regioselective control is not offered by direct sulfonation procedures. An alternative approach is to introduce the sulfonate moiety during construction of the ligand framework. For example, the cross-coupling of haloarylsulfonic acids and primary or secondary phosphines according to the method developed by Stelzer (Scheme 18.1) provides easy access to sulfonated species [17a,17b,21]. Nevertheless, this route does require the handling of air-sensitive primary and secondary phosphines, and as such there will be little value in applying the P–C coupling approach to the synthesis of TPPMS (**7**), for example, which can be prepared on a large scale by direct sulfonation. More importantly, the synthesis of analogous PAr_3 systems according to the P–C coupling route will require the use of highly toxic and flammable PH_3 gas as the phosphorus precursor. This is not easily handled and is perhaps best avoided if at all possible. The P–C coupling strategy works well for the triarylphosphine framework but is not easily adaptable to the preparation of more sophisticated bidentate ligand systems. In this regard the preparation of sulfonated bidentate ligands according to the work-up method described by Mul *et al.* appears to be the most attractive approach [41]. Hydrophilic ligands and their application in aqueous phase transition metal-catalysed reactions have been reviewed recently by Shaughnessy [43]. The synthesis of the trisodium salt of 3,3′,3″-phosphinetriylbenzenesulfonic acid (TPPTS) (**6**) is chosen as our first synthesis example owing to its industrial importance and prevalence in the open literature. The procedure given in Section 18.6 is described by Herrmann *et al.* in *Inorganic Synthesis* [39]. The preparation of sulfonated Xantphos and BINAP is also described in Section 18.6 according to the synthetic procedures of van Leeuwen [44] and Davis [45], respectively.

18.3 Fluorous biphasic catalysis

The concept of fluorous biphasic catalysis was first reported in the seminal work of Horváth and Rábai [46], although earlier studies by Vogt [47] based on the same concept are often overlooked. Both groups independently recognised the potential of utilising fluorocarbons in a new form of biphasic catalysis based on the temperature-dependent phase behaviour of fluorocarbon–hydrocarbon mixtures. Fluorous biphasic catalysis operates on the basic principle that the catalyst partitions preferentially towards a fluorous phase and that a mutual miscibility of all phases exists under reaction conditions. Upon cooling, a two-phase system is rapidly established enabling facile separation of product from the catalyst-containing reaction solution. The operation of fluorous biphasic catalysis is similar in nature to the industrially applied aqueous biphasic approach described in Section 18.2. An important distinction, however, is that fluorous biphasic systems are truly homogeneous under reaction conditions and as such mass transport limitations are rarely encountered. Horváth and Rábai demonstrated the concept using the hydroformylation of 1-decene in a fluorous biphasic system comprising perfluoro(methylcyclohexane) and toluene [46]. The Rh-based catalyst was rendered preferentially soluble in the fluorous phase through a fluorinated phosphine $P[CH_2CH_2(CF_2)_5CF_3]$ (**11**). Following this pioneering work, research into fluorous biphasic catalysis expanded significantly and has now been applied to a wide range of reactions including hydrogenation [48], hydroboration [49] and hydroformylation [50]. By virtue of the ability to obtain a single phase under reaction conditions with a broad range of substrates, and a tolerance towards hydrolytically sensitive materials, the fluorous biphasic approach offers greater flexibility over analogous aqueous biphasic systems, although the solvents and ligands employed are undoubtedly more expensive. The field has been extensively reviewed [51] and is the subject of several books and book chapters [5a,52]. The intention here is to provide details concerning general strategies for the transformation of P(III) ligands into fluorous-phase soluble species.

The most important consideration for catalyst design is simple: the P(III) ligand must contain fluorine to impart preferential solubility of the corresponding complex in the fluorous phase, but exactly how much fluorine is necessary? This important question is a matter of debate with early claims that a fluorine content of more than 60 wt% is required [53]. More accurately, efficient immobilisation occurs when the organic–metal core is completely encapsulated by fluorine atoms and accordingly the specific wt% of fluorine incorporation required will be dictated by ligand structure. This is exemplified by the work of Hope and co-workers who demonstrated that ligands displaying subtle structural differences with identical composition (in terms of wt% F) can exhibit markedly different solubility behaviour [54]. It is also worth noting that preferential solubility of a perfluorinated ligand does not always confer greater solubility of the corresponding complex [54].

A second important design consideration is the potentially negative effect of highly electron-withdrawing F atoms on catalyst behaviour. Ideally, we would like to confer fluorous phase solubility on a catalyst without affecting the desirable activity–selectivity behaviours for which it was chosen in the first place. This has been achieved, with varying degrees of success, through incorporation of spacer groups that serve to shield phosphorus from the electron-withdrawing effects of fluorine. The need for such spacer groups was recognised from the inception of the fluorous biphasic concept with early perfluoroalkyl ligands containing $(CH_2)_n$ units as electronic insulators [46,50a,55]. Theoretical calculations suggest that two methylene groups are sufficient to isolate P from the effects of F but in practice this is not always the case [51c]. Aryl-based fluorous ligands are also often developed with additional spacer groups between the aryl ring and perfluorinated ponytail owing to the incomplete shielding offered by the ring itself. Methylene-based spacers are most commonly employed, although alkoxy, $O(CH_2)_n$ [56] and silyl substituents [48e,57] are also known to exhibit desired insulating properties. For example, ligands of the generic structure $P(4\text{-}C_6H_4SiMe_{3-n}(CH_2CH_2C_6F_{13})_n)_3$ are shown to be electronically similar to PPh_3 [57a,57b].

The number of strategies that can be employed to achieve immobilisation in a fluorous solvent is somewhat limited because the basic necessity that the catalyst must contain fluorine can only be achieved through either ligand design or the use of ionic catalyst systems containing heavily fluorinated anions such as tetrakis{3,5-bis(perfluorohexyl)phenyl}borate [48f,58]. The diversity in reported ligands developed for fluorous biphasic catalysis arises from variations in the extent of fluorination, the length of fluorous 'ponytails', the nature of spacer units and the ligand framework employed rather than the means of immobilisation itself. Figure 18.3 illustrates the diverse range of perfluorinated ligands that have been studied according to several basic classes with reference specifically to literature describing synthetic procedures. Thus, perfluorinated ligands are broadly described here as (1) perfluoroalkyl, (2) perfluoroaryl and (3) perfluoroaryl systems comprising electronic insulator groups between the fluorinated ponytail and the aryl ring. There are obviously variants within each generic structure. For instance, unsymmetrical ligands, where at least one ponytail of the phosphine is structurally different from the others, and 'split ponytail' systems based on $(CH_2)_n$ [59] spacer units are not shown. However, the classification given is intended to aid description of general synthetic strategies and to provide an historical perspective for the development of fluorous P(III) ligands.

The fluorous biphasic concept was first demonstrated with an alkyl phosphine (**11**) that was prepared by the reaction of 1H, 1H, 2H-perfluoro-1-octene with PH_3 [46]. This radical induced reaction of olefin with phosphine formed the basis of early synthetic procedures [50a,55]. Unsymmetrically substituted trialkyl phosphines can also be generated through this route [55b]. In an additional first step, a perfluoroalkyl iodide is combined with $LiPH_2$ yielding a primary phosphine from which the unsymmetrical ligand is subsequently prepared. Alternative approaches to perfluoroalkyl phosphines, that do not require the use of hazardous PH_3, include the coupling of perfluoroalkyl iodides with PCl_3 *via* the corresponding Rieke [67] or Grignard reagents [60,61]. It is worth noting that the organo-zinc route is not necessarily extendable to systems containing longer perfluoroalkyl substituents [51g]. The aforementioned synthetic procedures can easily be applied to the preparation of mixed aryl–alkyl phosphines of the generalised form $PPh_{3-n}(C_2H_4C_6F_{13})_n$ using the appropriate P(III) reagents. For instance, (**12**) can be prepared from the reaction of Ph_2PCl with $C_6F_{13}C_2H_4MgI$. More recent advances have seen the development of a modular synthetic protocol that enables the construction of ligands of the general form $P[(CH_2)_AR_{FX}][(CH_2)_BR_{FY}][(CH_2)_CR_{FZ}]$ where A, B, C=3 and 4; and X, Y, Z=4, 6 and 8 [68]. This synthetic approach is based on the alkylation of 2-cyanoethyl substituted phosphines with $R_{FX}(CH_2)_AI$ to yield a phosphonium iodide. Subsequent removal of a 2-cyanoethyl group by sodium methoxide gives 3-methoxypropionitrile and the corresponding phosphine. Through consecutive alkylation–dealkylation steps, this synthetic methodology allows independent control of the number of methylene spacers and the length of fluorous chain comprising each ponytail. Thus, the construction of a broad range of perfluoroalkyl ligands can be readily achieved.

The main difficulty associated with the preparation of trialkyl phosphines is arguably their air sensitivity. Research in the field therefore moved predominantly towards the use of more stable triaryl-based systems. The *para*-substituted perfluoroalkyl triphenylphosphine (**13**) was first prepared in 1997 using a copper-mediated cross-coupling between a perfluoroalkyl iodide and an aryl halide [60]. Subsequent low-temperature lithiation of the perfluoroalkylated aryl bromide followed by reaction with phosphorus chloride reagents

Type I			References
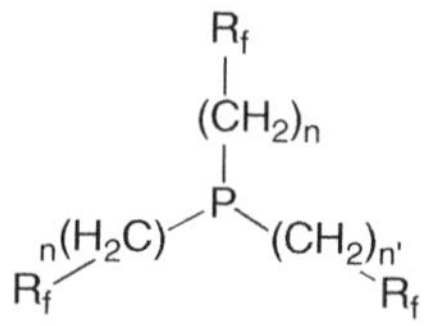	n = n′ = 2		
	$R_f = C_6F_{13}$	**(11)**	[46,55a,60]
	C_8F_{17}		[55a]
	$C_{10}F_{21}$		[55a]
	n = n′ = 3–5		
	$R_f = C_8F_{17}$		[55b]
$Ph_{3-n}P(C_2H_4R_f)_n$	n = 1, 2		
	$R_f = C_6F_{13}$	**(12)** (n = 1)	[60,61]
Type II			
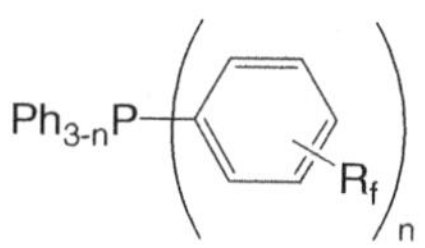	n = 1–3		
	$R_f = p\text{-}C_6F_{13}$	**(13)** (n = 3)	[60,62]
	$p\text{-}C_8F_{17}$		[62,63]
	$m\text{-}C_6F_{13}$		[51f,g]
	$m\text{-}C_8F_{17}$		[63]
Type III			
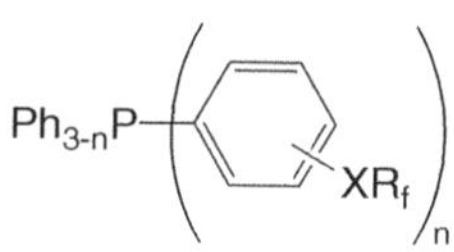	$X = C_2H_4$		
	$R_f = p\text{-}C_6F_{13}$	**(14)**	[64]
	$p\text{-}C_8F_{17}$	**(15)**	[63,65]
	$X = C_3H_6$		
	$R_f = p\text{-}C_6F_{13}$		[66]
	$m\text{-}C_6F_{13}$		[64b]
	$p\text{-}C_8F_{17}$		[66]
	$p\text{-}C_{10}F_{21}$		[66]
	$X = OCH_2, O(CH_2)_3$	**(16)**	
	$R_f = p\text{-}C_7F_{15}$		[56b]
	$p\text{-}C_8F_{17}$		[56b]
	$X = SiMe_{3-m}(CH_2CH_2)_m$		
	n = 3, m = 1–3		
	$R_f = C_6F_{13}$		[57a,57c]
	C_8F_{17}		[57a,57c]
	n = 1–3, m = 1–3		
	$R_f = C_4F_9$		[57c]
	C_6F_{13}		[57c]
	C_8F_{17}		[57c]
	$C_{10}F_{21}$		[57c]

Figure 18.3 *Basic fluorinated P(III) ligands classed according to several distinct structural motifs.*

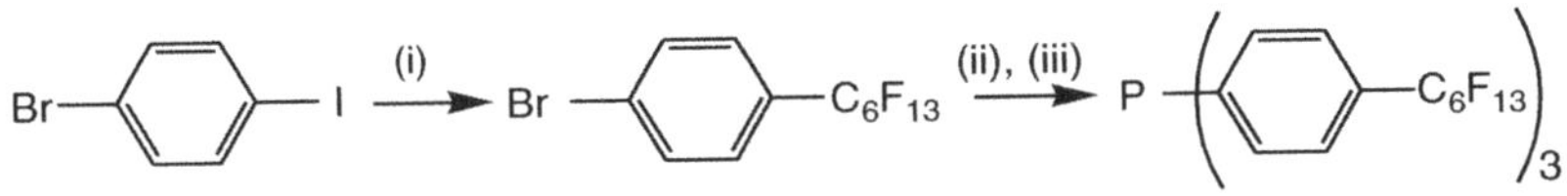

Scheme 18.2

Note: (i) $C_6F_{13}I$, Cu, 2,2′-bipyridine, DMSO, 70°C; (ii) *n*-BuLi, ether, –78°C; and (iii) PCl_3, ether, –78°C.

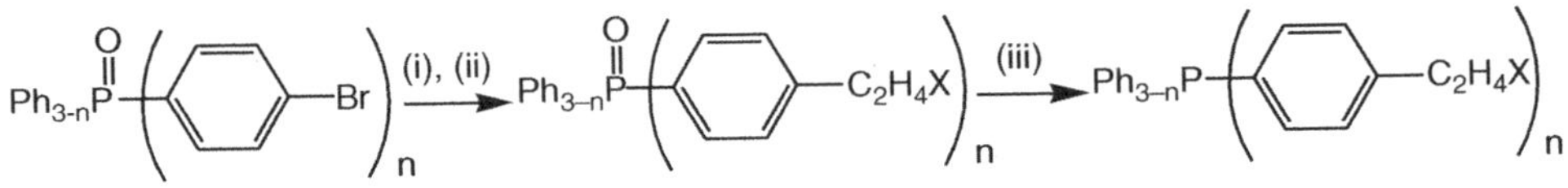

Scheme 18.3

Note: (i) 1.3 eq. alkene, 1.3 eq. NaOAc, 0.1% palladacycle, in DMF, 125°C; (ii) H_2 (10 bar), 10% Pd–C, in EtOAc, 25°C; and (iii) 5 eq. $HSiCl_3$, 5–10 eq. NEt_3, in toluene, 120°C.

according to Scheme 18.2 provides an easily accessible route to a broad range of fluorinated ligands. This synthetic methodology can also be used to generate *ortho*- and *meta*-substituted ligands and fluorinated analogues of chelating bidentates such as dppe [69] and BINAP [48g,70]. Ligands with mixed perfluoroalkyl substituents are also accessible *via* the perfluoroalkyl-derivatised chlorophosphine intermediate $ClP(C_6H_4Rf)_2$. In all cases the main by-product of the coupling reaction is the bis-substituted arene, which is easily separated by distillation [60]. Yields are typically high with the exception of coupling reactions involving 2-C_6H_4BrI, which suffers suppressed conversion as a consequence of steric interactions. Further evidence of the effects of sterically demanding *ortho*-substitution is provided by the reaction of PCl_3 with 2-$LiC_6H_4C_6F_{13}$, which does not lead to completion but rather yields $PCl(2\text{-}C_6H_4C_6F_{13})_2$ [51g].

Spectroscopic investigations of metal complexes formed from perfluorinated ligands reveal that the aryl unit is a better electronic insulator than C_2H_4 but still does not completely isolate phosphorus from the effects of fluorine [71]. Type II ligands are therefore typically less basic than PPh_3 and do not represent direct replacements. If the chemical properties of a specifically chosen ligand framework are to be retained, the electronic effects of perfluorinated ponytails must be isolated from the P centre, as previously discussed in this chapter. The incorporation of $(CH_2)_n$ groups between aryl ring and perfluorinated ponytail is a commonly employed means of increasing insulation effects. The first example of perfluorinated aryl phosphines containing C_2H_4 spacer groups can be attributed to Leitner and co-workers who at the time were developing CO_2-phillic catalysts [64b,72]. Using PPh_3 and dppe frameworks $C_2H_4C_6F_{13}$ **(14)** and $C_2H_4C_8F_{17}$ **(15)** groups were introduced in *meta*- and *para*-positions by the coupling of the mono Grignard of aryl dibromides with commercially available perfluoroalkyl iodides. Lithiation, followed by reaction with chlorophosphine reagents, yielded a range of new phosphines which were successfully employed for catalysis in supercritical CO_2. A generic and economical route to ligands of type III structure, based on the Pd-catalysed Heck olefination of haloarylphosphine oxides, was introduced by Chen and co-workers [64a]. This simple three-step synthetic route (Scheme 18.3) involves (1) C–C coupling of haloarylphosphine oxide with an olefinic substrate, (2) hydrogenation of the oxide-substituted olefin and (3) reduction to the desired P(III) product. The olefinic substrate can contain virtually any functionality and accordingly the route outlined in Scheme 18.3 can also be applied to the preparation of hydrophilic ligand systems.

An alternative synthetic strategy described by Gladysz employs the Wittig reaction for introduction of a $(CH_2)_n$ unit between the aryl ring and perfluorocarbon ponytail [73]. In a first step fluorous phosphonium salts are prepared and reacted with bromobenzaldehyde to yield the olefinic Wittig product. Subsequent hydrogenation, performed under mild conditions to prevent overreduction, gives fluorous bromoarenes $BrC_6H_4(CH_2)_3Rf$ in high yields. Lithiation followed by reaction with PCl_3 generates a mixture of the target phosphine and its oxide. Full conversion to the oxide by H_2O_2 is then carried out to simplify subsequent purification (i.e. the mixture is finally reduced in its entirety with $Cl_3SiH–Et_3N$). Although providing an additional methylene unit, this synthetic strategy is clearly more involved than the Heck-based route described by Chen [64a].

Perfluorinated aryl ligands containing alkoxy spacer units (**16**) have also been developed with the aim of enhancing the electronic isolation of P and are accessible *via* commercially available tris(*p*-methoxyphenyl) phosphine [56b]. The multistep reaction sequence involves (1) conversion to the phosphine oxide and demethylation with BBr_3 to afford tris(*p*-hydroxyphenyl)phosphine oxide, (2) reaction of the phosphine oxide with $Rf(CH_2)_xOTf$ and (3) a final reduction step using $HSiCl_3$ to give the desired product. As our final example, the silyl methylene grouping, $SiMe_{(3-x)}(CH_2CH_2)_x$, has also been exploited as a spacer unit. The silyl linkage not only provides effective insulation but also enables the number of fluorinated ponytails to be increased dramatically. Using this so-called 'split ponytail' approach up to 9 C_2H_4Rf groups can be appended to the PPh_3 framework following a relatively straightforward procedure [57a]. This strategy has been extended to the preparation of fluorous derivatives of dppe containing up to 12 tails per molecule [48e].

The copper-catalysed cross-coupling reaction illustrated in Scheme 18.2 can also be used to generate perfluoralkyl-substituted phenols from which a series of phosphinite, phosphonite and phosphite ligands can be prepared. Similarly, ligands of the generalised structure $PPh_{3-n}(OCH_2CH_2C_6F_{13})_n$ are easily formed *via* direct reaction of commercially available perfluoroalcohols with phosphorus chlorides in the presence of a base [60]. The additional oxygen atom facilitates the preparation of *ortho*-substituted ligands through alleviation of the steric constraints described earlier. Nevertheless, fluorous trialkyl and triaryl phosphites have received comparatively little attention.

Bidentate phosphines belonging to each of the preceding categories have also been reported but are much less plentiful than fluorous monodentate ligands [48e,60,64b,74]. It turns out that many perfluorinated bidentates actually exhibit poor solubility in fluorous solvents, perhaps because it is difficult to meet the requirement of complete encapsulation by fluorine for larger ligand frameworks. Perfluorinated analogues of BINAP, for example, are insoluble in perfluorinated cycloalkane solvents such as perfluoro-1,3-dimethylcyclohexane [70]. However, perfluorinated BINAP is appreciably soluble in dense CO_2, and for this reason more detailed synthesis discussion is included in the section relating to catalysis in supercritical fluids [75].

18.4 Ionic liquids as reaction media

Reducing the environmental impact of chemical processes, through reductions in energy consumption and fugitive emissions, remains a primary objective in the chemical industry. The replacement of environmentally harmful, organic solvents provides an effective means to achieve some of these goals. Ionic liquids (ILs), organic salts that are liquid at ambient conditions, offer great potential as environmentally benign alternatives to VOCs, primarily because of their immeasurably low vapour pressures, a property that is not exhibited by classical solvents. Importantly, ionic liquids possess many of the attributes required of a reaction medium not least in their ability to dissolve an extended range of solutes. In fact, the accolade of the most diverse reaction medium is perhaps attributable to the ionic liquid family. The origins of this great diversity lies not only in the sheer number of ILs that possess all of the desired properties of a reaction solvent but also in the ability to tailor their physical and chemical properties to meet the specific demands of a reaction system. Through careful choice of ion pairs it is possible to tune not only solvent characteristics but also physical properties including density, viscosity, polarity, acidity, heat capacity and so on. It is this great versatility that has led to the designation of ILs as 'designer solvents'. Relationships between the structure of an IL and its properties are a key aspect in the development of IL-based processes but are beyond the scope of this chapter. Suffice to say that an appreciation of IL properties, and how they are governed, is necessary if these solvents are to be exploited to their full potential in catalysis. The reader is referred to more specialised literature on the subject [6a].

The recognition of ionic liquids as neoteric green solvents has prompted a fast-growing area of research into their application in chemistry with efforts focused predominantly on their use as alternative reaction media. In terms of commercial viability, the use of ILs in the development of catalyst separation technologies is of particular importance. When the substrate shows significant solubility in the IL and the product forms a second immiscible phase, separation is simply achieved by decanting. Distillation also provides a viable means of separation, owing to the immeasurably low vapour pressure of the solvent itself, but is limited to solutes of high volatility and/or catalyst systems of high thermal stability. Products can also be isolated from the IL by solvent extraction, an approach that is particularly attractive when water can be used as the extractant in conjunction with hydrophobic ionic liquids. However, distillation and aqueous extraction cannot be applied to all solutes, in particular lipophillic and high-boiling organics. Volatile organic solvents that are immiscible with the IL solvent can of course be used for liquid–liquid extraction, but this somewhat negates the use of an ionic liquid in the first place. Biphasic systems based on ILs as a stationary reaction phase with CO_2 as a transport vector help to fill this gap and this will be described further in Section 18.5.

Liquid–liquid biphasic catalysis relies on the solubility of substrates and limited solubility of products in the reaction medium. The advantage of being able to control the solvent characteristics of an IL becomes immediately apparent; it broadens the potential scope of biphasic approaches to the catalyst separation problem. Effective immobilisation of the transition metal catalyst in the IL solvent is an obvious prerequisite for biphasic operations. Most ILs provide the ideal reaction medium for catalysis because they are weakly coordinating and readily dissolve transition metal complexes. However, the ability to dissolve the catalyst is only part of the solution to the separation problem; it is equally important to ensure that the catalyst is completely insoluble in the product phase or extracting solvent so that it is not lost at the separation stage. In 1995 Chauvin and Olivier-Bourbigou reported on the Rh-catalysed hydroformylation of 1-pentene in an IL solvent [76]. Significant leaching of the catalyst into the product was observed when PPh_3 was employed as a ligand. Leaching could be suppressed by switching to TPPTS (**6**) or TPPMS (**7**), but this was reported to result in appreciable decomposition as evidenced by comparatively low reaction rates. Given that poor selectivity was also observed, it is likely that the catalyst species derived from the sulphonated phosphines were actually poorly soluble in the IL or perhaps not formed at all. Similar observations were made by Sellin *et al.* who also described immobilisation of a Rh hydroformylation catalyst in an IL using TPPMS as the ligand [77]. The rate of reaction was lower than expected and this was attributed to a low solubility of the Rh complex. Simple replacement of the Na counterion of TPPMS by the imidazolium cation comprising the IL solvent dramatically improved complex solubility and re-established expected reaction rates. As with any biphasic system the effectiveness of product separation and catalyst recycling is dictated by catalyst partitioning (i.e. the catalyst must be sufficiently soluble in the reaction medium for catalysis to occur and insoluble in the product phase or extracting solvent to prevent leaching). The nature of ligand modification necessary to impart preferential solubility in an IL is, of course, dependent on the nature of the IL solvent itself, which in turn is dictated largely by the solubility demands of reagents. Here, the great diversity in IL structure poses a problem for it gives rise to great flexibility but, unfortunately, makes discussions relating to generic synthetic methodologies of P(III) ligands designed for IL solvents more difficult. Nonetheless, there are still commonly employed strategies for the immobilisation of P(III) ligands, and catalysts derived therefrom, in IL solvents (see Figure 18.4).

In some cases it is sufficient to employ ligands modified with hydrophilic groups to achieve the required catalyst solubility combined with acceptable catalyst retention. A classic example is provided by (**19**) that was originally developed to impart water solubility on PPh_3-based catalysts and is prepared easily according to the method outlined in Scheme 18.1 [78]. The Pd-catalysed coupling of *m*-iodophenylguanidine with $PhPH_2$ first yields the iodide salt [17b] from which (**19**) is prepared *via* anion exchange with NH_4PF_6 in

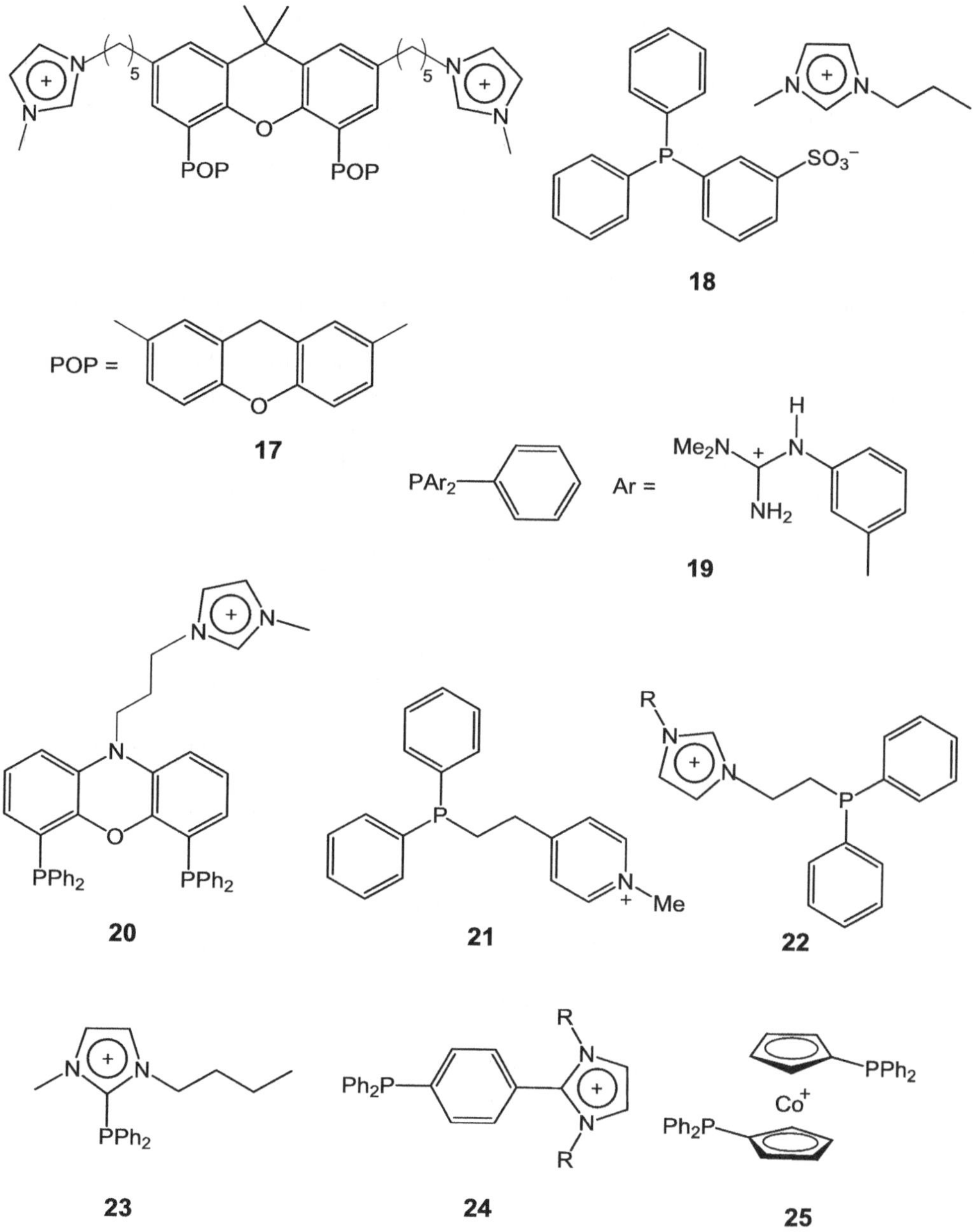

Figure 18.4 *Ionophilic ligands developed for immobilisation in ionic liquids.*

aqueous solution. The guanadinium moiety provides exceptional water solubility and is also effective for enhancing dissolution in IL solvents [79]. This strategy has been extended to the IL immobilisation of ligands of the Xantphos family known for their superior regioselectivity in the hydroformylation reaction [79]. Wasserscheid and Stelzer prepared the guanadinium-modified Xantphos for the hydroformylation of 1-octene in hexafluorophosphate ionic liquids. The expected high regioselectivity was preserved with good catalyst retention. Dupont and co-workers have also described Rh-catalysed hydroformylation of olefins in ILs using

Xantphos-based ligands [80]. In this example, the sulfonated analogue of Xantphos (**8**) (Figure 18.2) yields far superior catalyst retention when compared to Xantphos itself, which displays high preferential solubility in the organic phase. The sulfonated analogue of BINAP (**10**) has also been exploited as an immobilising ligand by the group of Yuan who studied asymmetric hydroformylation and hydrogenation reactions [81]. Again, the use of a ligand originally designed to impart solubility in water proved successful for catalyst recycling in ILs; successive reaction and extraction cycles did not lead to loss of activity, enantioselectivity or regioselectivity. Similarly, a phosphonic acid derivative of BINAP has been employed in the Ru-catalysed asymmetric hydrogenation of β-keto esters in ILs with excellent catalyst retention [82].

The effectiveness of water-soluble ligands in IL biphasic catalysis is largely dependent on the nature of the solvent itself; the more lipophilic its character, the lower the expected solubility of highly hydrophilic species. Of course, there are many factors that will contribute to solute–solvent interactions, including the structure and size of the ligand framework and the number and type of appended hydrophilic groups, not to mention the modifying effects of dissolved substrates and products. Nonetheless, the starting point for ligand design must be the nature of the solvent itself because this in turn is dictated by the solubility demands of the reaction system. For instance, the dissolution of appreciable concentrations of high molecular weight and low polarity substrates in an IL requires tuning of solvent characteristics; in essence, the ionic solvent must become less ionic. The commonly employed IL [BMIM]PF_6 is a relatively poor solvent for the hydroformylation of 1-octene in biphasic systems comprising the IL and $scCO_2$ [83]. Simply increasing the alkyl chain length of the imidazolium cation from C_4 to C_8 has a profound effect on catalysis because as the solubility of alkene in the IL is enhanced it partitions more favourably towards the phase where the catalyst resides [83]. Hydrophilic ligands will not necessarily provide sufficient catalyst solubility in ILs designed to solvate low-polarity substrates such as long-chain alkenes. However, exchange of the cation of any water-soluble phosphine with that comprising the IL solvent provides a simple and generic strategy for catalyst immobilisation. Importantly, this approach enables entry into IL-based processes starting from well-understood, water-soluble ligands that have been developed over many years. In addition to the aforementioned example, this highly effective method of immobilisation has been adopted in several subsequent studies but despite its simplicity is still under-explored in our opinion [83,84].

Enhancing catalyst solubility can also be achieved by attaching cations that comprise the IL solvent directly onto the ligand framework, as illustrated by examples (**17**, **20–24**) in Figure 18.4. Phosphines with 2-imidazolium moieties (**23**) can be derived directly from commercially available ionic liquids such as [BMIM]PF_6 according to Scheme 18.4 [85]. The imidazolium salt is first rigorously dried to allow lithiation at low temperature. Reaction of the intermediate formed with chlorodiphenylphosphine gives the target 2-imidazolium phosphine in high yield. Ligands of this structure do not provide direct ionophilic substitutes for PPh_3, although this can be addressed by insertion of a phenylene spacer between phosphorus and the heterocyclic substituent (**24**). However, the preparation of (**24**) is considerably more involved, proceeding through a route that requires isolation of the neutral phosphine. The neutral analogue of (**24**) is formed in a two-step process involving a Kusugi–Stille-type coupling followed by nucleophilic phosphination. Firstly, 2-(4-fluorophenyl)-1-methylimidazole is generated by the Pd-catalysed coupling of 4-fluoroiodobenzene and 2-tributylstannyl-1-methylimidazole. Nucleophilic phosphination with potassium diphenylphosphide then gives the neutral phosphine. Conversion to the imdazolium salt is achieved in a final step involving *N*-quaternisation with dimethylsulphate or trimethyloxonium tetrafluoroborate, for example. At this stage, care must be taken to avoid the formation of phosphonium species.

Imidazolium-containing ligands of type (**22**) structure are also accessible *via* *N*-quaternisation of the neutral, parent phosphine [86]. In a first step, primary or secondary phosphines are added to the exocyclic C=C double bond of *N*-vinylimidazole in the presence of a strong base (Scheme 18.5). An intermediate tertiary phosphine bearing the 2-(1-imidazolyl)ethyl substituent is formed in high yield. Subsequent *N*-quaternisation first requires protection of phosphorus against the alkylating agent and can be achieved through sulfurisation.

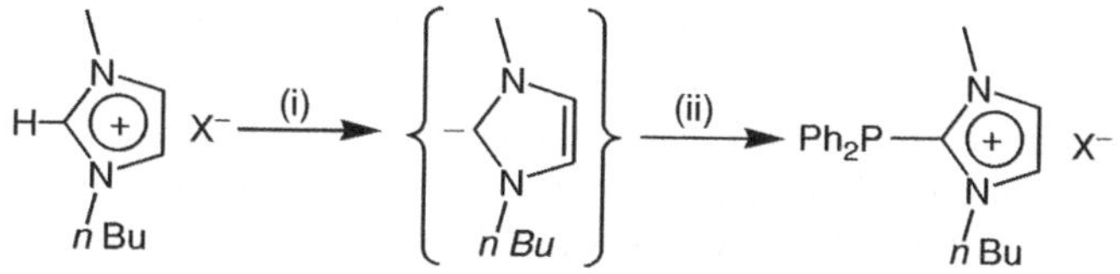

Scheme 18.4

Note: (i) *n*-BuLi, CH_2Cl_2, –70°C; (ii) Ph_2PCl, CH_2Cl_2, –70°C.

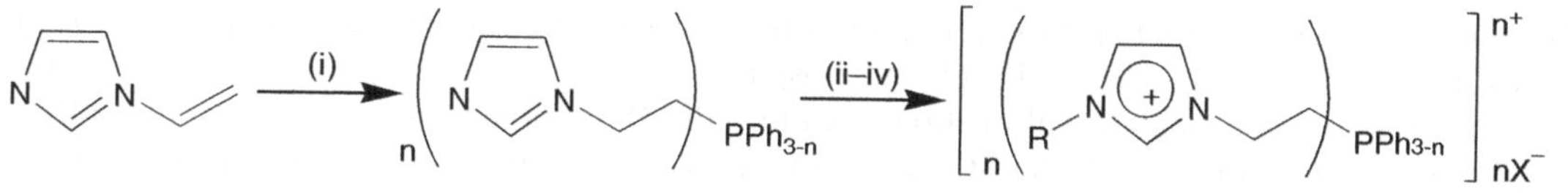

Scheme 18.5

Note: (i) $Ph_{3-n}PH_n$, *t*BuOK; (ii) S_8; (iii) for example, MeI or $Et_3O^+PF_6^-$; (iv) Raney Ni.

N-quaternisation of the corresponding sulphide, with MeI or $Et_3O^+PF_6^-$ for example, followed by deprotection with Raney nickel yields the desired imidazolium phosphine. In principle, this synthetic methodology can be applied to any PH-functional phosphine enabling the preparation of a broad range of P(III) ligands, containing peripheral 1-imidazolium groups, from commercially available 1-vinylimidazole. This method can also be adapted for the generation of pyridinium analogues (**21**) [87]. More recently, Consorti, Dupont and co-workers described an alternative route to ligands of this generic structure that does not require phosphorus protection and deprotection steps [88]. The trick is to avoid *N*-quaternisation in the presence of phosphorus through allyl or vinyl imidazolium salt precursors. The addition of secondary phosphines to the allyl or vinyl imidazolium salt, in the presence of an appropriate radical initiator, yields an imidazolium-tagged ligand. Thus, the radical chain addition method enables one-step generation of ionophilic phosphines in what is typically a high-yielding reaction. The success of this synthetic approach appears to be largely dictated by choice of radical initiator.

Appending imidazolium or pyridinum tags onto P(III) ligands is perhaps the most popular approach for effecting immobilisation in ILs and has been applied to a broad range of ligand structures including ferrocenyl diphosphine [89], triphos [90] and xanthene (**17**, **20**) [91] frameworks. In all cases effective catalyst retention is observed but the preparative routes to these ligands are not always straightforward. Incorporating the IL solvent's anion, rather than its cation, into catalyst structure provides a synthetically less challenging approach to immobilisation. Ligands composed of the cobaltcenium backbone (**25**) are an important example of this particular class of ionophilic ligands because they represent the first tailor-made phosphines for immobilisation in ILs. Cobaltcenium species can be prepared by oxidation of 1,1′-bis(diphenylphosphino) cobaltocene with C_2Cl_6 followed by anion exchange with NH_4PF_6, for example, in acetone [92]. The efficacy of (**25**) in biphasic catalysis was studied using the hydroformylation of 1-octene. When compared to PPh_3, TPPTS and several neutral bidentate ligands, a clear increase in the *n*:*i* product ratio was observed. More importantly, the recovered reaction solution of (**25**) could be re-used without loss of activity or selectivity.

To summarise, several approaches can be adopted to immobilise catalysts in IL solvents but in all cases the basic principle of 'like dissolves like' is met by incorporating charge separation into the catalyst structure. In some cases, hydrophilic ligands developed for aqueous biphasic catalysis can be utilised without further modification. When such ligands are insufficiently soluble, exchanging a fragment of the ionic moiety with that comprising the IL solvent provides a straightforward solution. Imidazolium or pyridinium fragments can also be incorporated directly into the ligand backbone but the associated synthetic procedures can be more challenging. On a final note, it is important to recognise that ionic liquid purity is an unavoidable consideration in catalytic applications. Small amounts of impurities can have a profound effect on the properties of an IL and in turn its usefulness as a reaction medium [93]. Non-volatility obviously precludes purification of ILs by distillation in all but a few exceptional cases [94] and accordingly the final product is often isolated following a series of extractions. Purification is particularly important when IL preparation involves halide intermediates. It is imperative that all traces of halides are removed because of the generally accepted negative effect on catalysis. Details concerning the preparation of ionic liquids can be found in several excellent books and reviews [6a,95].

In keeping with our theme of demonstrating synthetic methodologies using BINAP and PPh_3 frameworks, the preparation of 4,4′-phosphorylated 2,2′-bis(diphenylphosphanyl)-1,1′-binaphthyl and 1-propyl-3-methylimidazolium diphenyl(3-sulfonatophenyl)phosphine (**18**) is given in Section 18.6 as examples. Of the known sulfonated phosphines, monosulfonated triphenylphosphine (**7**) is by far the simplest to prepare owing to its relatively facile separation from the sulfonation mixture. TPPMS is perhaps unique in that it has sufficiently low solubility in water, compared to Na_2SO_4, to enable recrystallisation directly from its post-neutralisation reaction solution. This overcomes the necessity for complex work-up procedures often encountered in the preparation of sulfonated phosphines. The procedure described by Joo is easy to follow and reproducible on a relatively large scale [96]. The important aspects of this synthetic route are (1) to ensure that the solution temperature is kept low during both oleum addition and eventual neutralisation with NaOH and (2) to exclude oxygen particularly during the final heating stage. This does not necessarily require Schlenk line manipulations; it is sufficient to bubble constantly a stream of inert gas through the reaction solution as it is heated in an Erlenmeyer flask. Despite the simplicity of its preparation, TPPMS has seen relatively little application in aqueous biphasic catalysis compared to its more heavily sulfonated analogues. However, TPPMS is extremely useful for the preparation of IL-soluble catalysts, hence its inclusion in Section 18.6 as a synthesis example.

18.5 Supercritical fluids as solvents in single and multiphasic reaction systems

Supercritical fluids (SCFs) are a curious hybrid of liquid and gas, fluids with an identity crisis that display properties characteristic of both states of matter. Pure compounds or mixtures become supercritical when heated and pressurised beyond their critical parameters. Above its critical temperature a substance cannot be liquefied by compression alone and as a consequence density, and thermodynamically related properties, can be varied continuously from gas-like to liquid-like values without the emergence of a phase boundary. A fluid in its supercritical state can therefore have appreciable density, sufficient enough to exhibit liquid-like solvent properties, whilst maintaining gas-like mass transport characteristics. The existence of the critical point and the characteristic signatures of a critical point phase transition have been known since the end of the nineteenth century. Only in the last two decades or so, however, have SCFs attracted interest as environmentally acceptable replacements for processes that currently rely on the use of organic solvents. The ability to dissolve species in a compressed fluid is the most important property in SCF technology. It is the emergence of much increased solvating capacity near the critical point that has led to the vast research into SCFs as alternative solvents in a wide variety of applications. Importantly, the range of intermediate densities, that are inaccessible at subcritical conditions, gives rise to tuneable solvent strength and provides a degree of control that cannot be achieved through the use of normal liquid solvents.

Carbon dioxide is the most studied SCF owing to its non-toxic, non-flammable nature and readily accessible critical parameters ($T_c = 31.1\,°C$, $p_c = 73.8\,bar$, $d_c = 0.437\,g\,cm^{-3}$). The wide scope of applications involving $scCO_2$ has arisen not only because of its environmentally benign nature but also as a consequence of advantageous physical properties that are not exhibited by subcritical solvents. From a catalysis perspective, these advantages include a complete miscibility with permanent gases such as hydrogen, high rates of molecular diffusion and heat transfer, the aforementioned tuneable solvent characteristics and the potential to operate in a single phase comprising multiple components. It is this unique combination of gas-like and liquid-like behaviour that provides access to new methods of catalyst immobilisation. For instance, tuneable solvent strength can be exploited to separate product and catalyst in a process referred to as 'catalysis and extraction using supercritical solutions' (CESS) [64c,75]. Alternatively, supercritical CO_2 can also be employed as a transport vector for substrates and products in continuous flow processes where the catalyst is heterogenised on a solid support [97] or in a second liquid phase and remains insoluble in CO_2 at all times. There are subtle variants of these two generic scCO_2-based separation strategies but all have in common the basic criterion that the catalyst must be either soluble or insoluble in CO_2. The ability to generate soluble and insoluble catalysts is therefore the most important consideration in catalyst design. Solubility in a SCF is determined by both chemical (molecular interaction) and physical (vapour pressure) contributions resulting in complex solubility behaviour that manifests itself in retrograde condensation and crossover effects. It is not always appreciated that the conditions of criticality for a mixture can differ significantly from those of the pure fluid. A fluid mixture can lower its free energy by splitting into two phases of different composition and will therefore tend to phase-separate before its compressibility becomes infinite. The nature of solute–solvent interactions and phase behaviour in the highly compressible, supercritical regime governs the design of SCF processes but is a complex topic beyond the scope of this text. For the purpose of this chapter, it is sufficient to consider only the basic principles that govern solubility in scCO_2 and to appreciate that solubility can be enhanced by the incorporation of CO_2-philic groups.

As we might expect, supercritical fluids are not 'super' solvents when compared to denser subcritical liquids but how does their solvent strength compare? Supercritical carbon dioxide is often said to display solvent characteristics similar to *n*-hexane but this is a common misconception based entirely on similarities between their dielectric constants. Vapour pressure is perhaps the most important indicator of solubility in a dense gas; the more volatile a substance the more soluble it is likely to be (in the limit a gas is totally miscible with the fluid) [98]. Compounds with convenient critical temperatures, up to 100°C for example, are usually nonpolar and, as a consequence, the chemical solvation of polar materials in these fluids is often poor. It is hardly surprising, therefore, that compounds with high lattice energies, such as metals and their salts, are insoluble in scCO_2. This is an important observation when considering scCO_2 as a transport vector for biphasic operations and this will be described further in this chapter. In general, the solubilisation of a metal ion requires that the metal charge be shielded from CO_2 so that what the fluid 'sees' is ideally a CO_2-philic shell. So-called 'CO_2-philic' species have been developed for the efficient transport of insoluble or poorly soluble materials into CO_2, assisting the development of a host of CO_2-based processes. The most well-known and effective CO_2-philes are fluorocarbons. The high affinity of CO_2 towards fluorinated species has been known for some time. For example, there has been extensive research into the use of perfluorinated compounds as potential blood substitutes [99]. A unique property of these compounds is their ability to dissolve abnormally large quantities of gas, in particular CO_2. As the fluorine content of these compounds is increased, the solubility of CO_2 in them also increases. The high solubility of CO_2 in fluorinated compounds is also observed in polymeric materials. The solubility of CO_2 in Teflon, for example, supersedes that of many other polymers despite its higher degree of crystallinity [100], observations that suggest a strong chemical interaction between CO_2 and CF bonds. If CO_2 shows high solubility in fluorinated compounds, as a consequence of strong chemical interaction, then we would expect the reverse to also be true, that is, fluorinated compounds will exhibit high solubility in CO_2. This is exactly what is observed, and the use of fluorinated compounds in preference to their nonfluorinated analogues has been exploited in a diverse range of scCO_2 applications from dispersion polymerisation [101] to metal extraction [102].

The effects of fluorination have also been employed extensively in the development of CO_2-soluble homogeneous catalysts. Burk and Tumas used the fluorinated anion BArf, $[B\{3,5\text{-}(CF_3)_2C_6H_3\}_4]^-$, to increase the CO_2 solubility of Rh catalysts for asymmetric hydrogenation reactions [103]. This simple and effective approach has been adopted by others but is of course restricted to ionic complexes [48f,64c,104]. Other early developments in the field focused on catalysts derived from simple fluorinated analogues of PPh_3, in particular tris[3,5-bis(trifluoromethyl)phenyl]phosphine (**26**). Tumas and co-workers demonstrated that Pd catalysts derived from (**26**) gave higher conversions in Heck and Stille reactions performed in CO_2, when compared to PPh_3-based catalysts, and attributed the higher activity to enhanced catalyst solubility [105]. Substitution of PPh_3 for commercially available (**26**) and structurally similar ligands is now a commonly adopted approach for achieving catalyst solubility in CO_2 [106]. However, PPh_3 does not always yield catalysts with the required activity and selectivity properties. This can be addressed through a more generic approach to solubilisation involving the incorporation of perfluoroalkyl chains into ligand structure. Leitner and co-workers first demonstrated this approach with the preparation of perfluoroalkylated ligands based on PPh_3 (**14**) and dppe (**27**) frameworks [64b,74]. Whereas Rh(hfpd)(dppe) showed no measurable solubility in CO_2, even at high fluid densities, substitution at the aryl *para*-position with perfluoro alkyl chains significantly increased complex solubility. Following the pioneering work of Leitner, the use of perfluoroalkyl ligands to impart solubility in CO_2 has been widely adopted [61,107]. It is therefore perhaps no surprise to learn that the first fluorous, chiral ligand (**29**) was actually developed for use in $scCO_2$ [108]. Leitner and Franciò used the chiral phosphine–phosphite ligand (**29**), an analogue of (R,S)-BINAPHOS, in the rhodium-catalysed asymmetric hydroformylation of vinyl arenes in both liquid and $scCO_2$. Higher regio- and enantioselectivity were observed compared to reactions performed in benzene with catalysts derived from (R,S)-BINAPHOS. From a series of control experiments, the origin of these beneficial effects could be traced to the presence of fluorous ponytails rather than the use of CO_2 as a reaction medium [108]. In principle, any P(III) ligand developed for fluorous biphasic catalysis can be utilised to impart solubility in $scCO_2$. In fact, the scope of perfluorinated ligands that can be successfully employed in $scCO_2$ is arguably broader than applications in fluorous solvents because conditions relating to the extent of F incorporation are somewhat relaxed. This is demonstrated by ligand (**28**) that is found to be insoluble in fluorous solvents but is sufficiently soluble in CO_2 to give active catalysis.

The high cost of fluorinated compounds and their potent global-warming potential are major concerns and may be determining factors for process commercialisation. So what are the alternatives? From earlier arguments relating to volatility as an indicator of solubility, complexes derived from alkyl phosphines, such as PMe_3 and PEt_3, are expected to be more soluble in CO_2 and can indeed be used without further modification [109]. However, this approach to achieving solubility allows the solvent system to dictate chemistry and is not always a viable option when alkylphosphines do not deliver the desired catalytic properties. Unfortunately, aryl-based systems are notoriously difficult to dissolve in CO_2 and require modification to achieve appreciable solubility [110]. Nevertheless, nonfluorinated substituents known to enhance the solubility of aryl phosphines are comparatively rare. One approach is to switch to tri(2-furyl)phosphine (**30**) [111], which can be used as a CO_2-soluble substitute in some cases, but this by no means offers a systematic and generic approach to the solubility problem. Other effective CO_2-philes that have been used to impart catalyst solubility include peracetylated sugars (**31**) [111c,112] and siloxane groups (**32**) [113]. Siloxanes have long been known to exhibit favourable solubility characteristics in CO_2 [114] and were exploited by Saffarzadeh-Matin *et al.* to modify the aryl-based phosphine (**32**) depicted in Figure 18.5 [113]. The siloxane-tagged ligand is generated from the brominated ester of a poly(dimethylsiloxane) monocarbinol [115]. Lithiation of the bromide at low temperature followed by reaction with chlorophosphines gives (**32**) in good yield. The origins of the known CO_2-philic character of peracetylated sugars can be traced back to the development of hydrophilic amphiphiles as stabilisers of dispersion polymerisation reactions in CO_2 [116]. Later studies demonstrated that peracetoxy-alkyl chains impart CO_2 solubility to otherwise insoluble amides [117]. This beneficial effect of peracetylated sugars was subsequently adopted in the preparation of ligand (**31**), which proved to be more soluble in CO_2 than PPh_3 itself [111c]. Structurally related peracetylated beta-cyclodextrins have also been employed to impart solubility on P(III) ligands through a complexation process (i.e. the cyclodextrin is not covalently attached to the ligand but is instead present as a solvent modifier) [112]. Finally, trimethylsilyl-substituted PPh_3 has been shown to enhance complex solubility in CO_2 but the effect is insignificantly small [118].

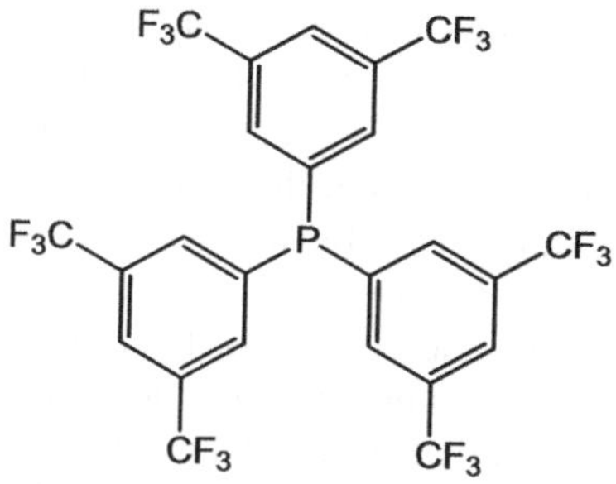

26

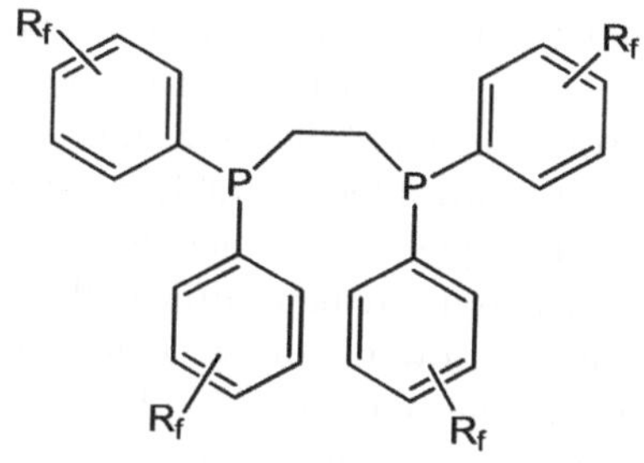

27

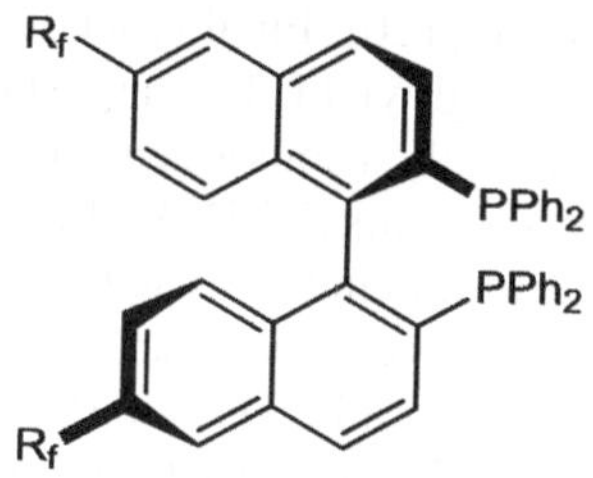

28

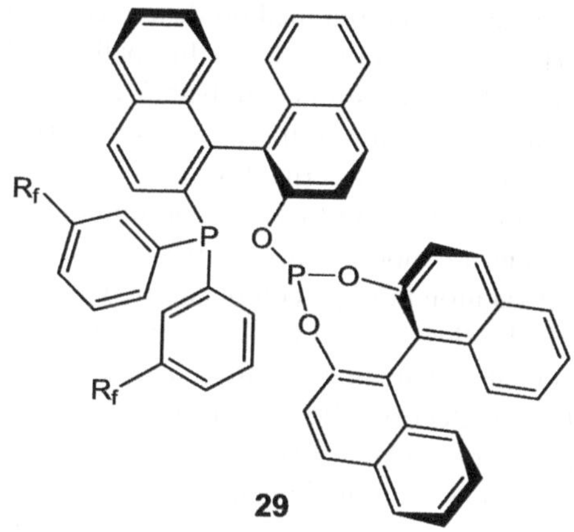

29

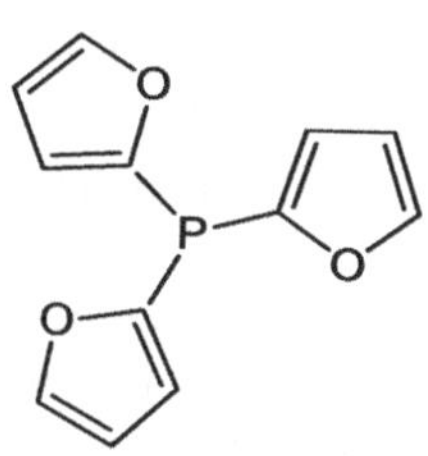

30

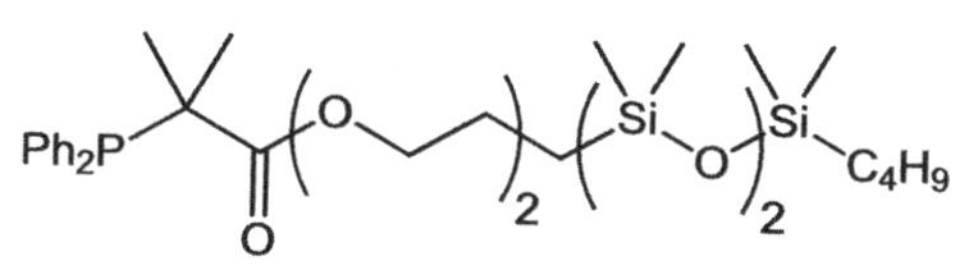

32

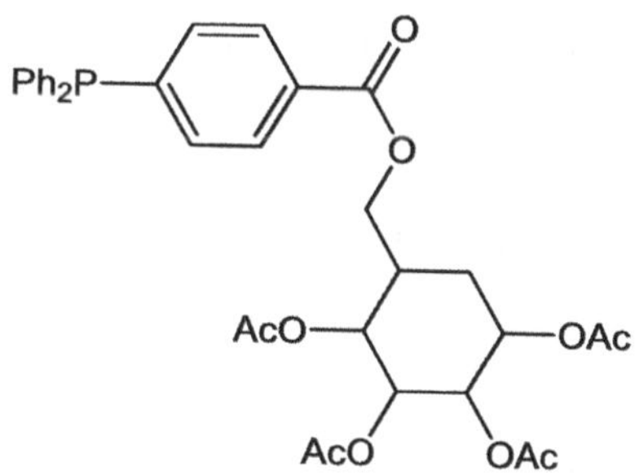

31

Figure 18.5 *Examples of ligands employed for catalysis in* $scCO_2$*.* R_f *represents a perfluoroalkyl chain.*

It is apparent that strategies for enhancing solubility in CO_2 through catalyst design are somewhat limited and appending perfluoroalkyl chains to ligand frameworks is perhaps the only generic and consistently proven approach. Synthetic methodologies for the preparation of perfluorinated ligands were described in Section 18.4. In Section 18.6 we also give an example of a fluorous chiral ligand (**28**) that was actually developed for use in CO_2. This example is chosen because it also allows us to highlight important aspects regarding the functionalisation of BINAP. The preparation of perfluorinated BINAP is synthetically more challenging than perfluorination of the PPh_3 framework also described in this chapter. It is imperative that any modification does not affect the desirable electronic and steric properties of the ligand. Modification of BINAP, for immobilisation purposes, is therefore often achieved through the binaphthyl framework so that the additional functional groups remain sufficiently isolated from the P centre. Furthermore, such modifications are typically directed towards the 4, 5 and 6 positions because these are least likely to alter ligand properties. The synthetic route employed for the incorporation of fluorous ponytails on the binaphthyl backbone is dependent on the targeted site of modification. Attachment of a fluoroalkyl chain to the 4,4′ positions of BINAP has been demonstrated using the copper-mediated coupling of a fluoroalkyl iodide and 4,4′-dibromo BINAP oxide (BINAPO) [119]. The preceeding bromination of BINAP oxide is highly regioselective and achieved easily according to a method described by Köckritz and Kant [120]. Final reduction of the oxide with $PhSiH_3/HSiCl_3$ yields the 4,4′-perfluoroalkyl-BINAP product in high yield. The largest class of modified BINAP ligands are substituted in the 5,5′ positions. This is undoubtedly a consequence of the possibility for electrophilic substitution at the 5 position of BINAPO. 5,5′ Perfluoroalkyl BINAP has been prepared using an identical strategy described for the 4,4′ substituted analogue although the route to the 5,5′-dibromo intermediate differs [119]. In this case, bromination is performed in the presence of an iron catalyst [121]. The 4,4′ positions are more reactive in basic conditions but the presence of a Lewis acid ($FeBr_3$) is believed to cause deactivatation through complexation with the phosphine oxide. The selectivity of bromination then shifts towards the 'less deactivated' 5,5′ positions. The 5,5′-dibromo-BINAPO species thus obtained provides an excellent starting material for a large class of 5,5′-substituted BINAP ligands. The 6,6′ positions are not accessible *via* electrophilic substitution but are nonetheless intensely studied owing to the reactivity of the BINOL precursor. Unlike BINAP, the 6,6′ positions of BINOL are more reactive so that electrophilic substitutions such as bromination and Friedel–Crafts acylation can still be effected. In a final step, the aryl phosphine is introduced *via* phosphination of the modified BINOL. This more involved route is used as our example for the preparation of perfluorinated BINAP. More in-depth experimental details with full analyses at each stage are available in reference [70]. As a final point, we note that CESS does not always give effective separation of product and catalyst because it relies on highly selective precipitation from CO_2 or highly selective extraction with CO_2. When both catalyst and product are CO_2 soluble, this can be difficult to achieve and a more appropriate strategy might be to use catalysts which are insoluble in $scCO_2$ but are still active [122].

18.5.1 Biphasic systems based on CO_2

Biphasic catalysis is arguably the most successful approach to the homogeneous catalyst separation problem. Nevertheless, liquid–liquid biphasic systems still suffer from the disadvantage that separation first requires some of the catalyst solution to be removed from the reactor. Besides the obvious negative impact of not having the entire catalyst inventory involved in active catalysis at any one time, the conditions of separation are often very different from those of the reaction so that decomposition may occur. Our early definition of the ideal reaction system specified continuous operation with separation under conditions identical to those of the reaction itself. In this scenario all of the catalyst remains in its active state at all times and decomposition associated with normal modes of separation is circumvented. The ability to operate in a continuous flow mode also has a considerable effect on reactor volumes required for a given product throughput, offering further advantages in terms of safety and reduced capital costs. There is one very good reason why this ideal scenario is difficult to achieve for it relies on fulfilment of a somewhat paradoxical requirement; the substrates and catalyst must reside in the same phase for reaction to occur, yet simultaneously the product and catalyst must reside in different phases for separation purposes. This is impossible to achieve in liquid–liquid biphasic catalysis, hence the need for some of the catalyst solution to be removed before separation can be effected. However, the situation is very different if one of the phases is a gas that cannot dissolve the reaction medium or its immobilised catalyst. Ionic liquids and $scCO_2$ provide the ideal biphasic couple to achieve these goals. Brennecke and Beckman made the important observation that CO_2 is highly soluble in ionic liquids (up to 0.6 mole fraction), whilst many ionic liquids remain insoluble in CO_2 [123]. Immiscibility gaps are key to the success of biphasic catalysis and later work by the group of Brennecke demonstrated that two phase immiscibility regions exist for a number of ionic liquids [124] even under conditions where the phase envelope of organic liquid–CO_2 systems normally begins to close [125]. For biphasic operations it is equally important that this desirable phase behaviour does not become disrupted by additional components representing substrates and products. The ability to extract a wide range of organic molecules from ionic liquids using $scCO_2$, without affecting the phase behaviour necessary for two-phase catalysis, was also demonstrated by Brennecke [126]. Importantly, the separation of involatile solutes, with boiling points too high to allow separation by distillation alone, was also shown. These ground-breaking studies led to extensive research into $scCO_2$ and ILs as new solvent systems for biphasic catalysis. The combination of $scCO_2$ and IL is the perfect biphasic environment because it provides the potential to fulfil all of the criteria that defines the ideal catalytic reaction described in the introduction to this chapter. The separation of product from catalyst and reaction solvent can be carried out under a single set of conditions, with all of the aforementioned advantages, simply because one of the phases is a dense gas in which the reaction medium is insoluble. A further advantage arises when the supercritical mixture is expanded downstream because products are obtained free of any reaction solvent, which dissipates as a low-pressure gas.

The success of multiphase catalysis is ultimately dictated by the efficiency of catalyst retention. In the design of $scCO_2$-IL biphasic systems the catalyst must be rendered insoluble in the mobile $scCO_2$ phase and sufficiently soluble in the stationary IL phase. According to our earlier arguments, the easiest way to achieve these targets is to develop a catalyst whose vapour pressure is immeasurably low under the conditions of reaction. Thus, by virtue of its ionic character any ligand designed for effective immobilisation in an IL should exhibit an inherently low solubility in CO_2. This is not to say that all ionic species are insoluble in SCFs because we know that ionic catalysts comprising highly fluorinated anions of low charge density (BArF) are CO_2 soluble. Nevertheless, all of the ligands described in Section 18.4 should in principle provide a good starting point for catalyst design. Complexes derived from (**18**), for example, show remarkable retention in $scCO_2$-IL biphasic systems developed for the hydroformylation of long-chain alkenes [83]. Over periods of up to 72 h of operation the product recovered down-stream was found to contain ~0.012 ppm Rh. Even when we consider that the detected Rh also included some physical entrainment that occurred during start-up, this level of leaching is equivalent to 1 g Rh in every 40 tonnes of product under industrially relevant turnovers. This highly successful approach to catalyst separation has been extensively explored and is described in several good reviews [127]. In recent developments the SCF-IL concept has been adapted by supporting a thin film of the IL within pores in silica. The catalysts, obtained as free-flowing powders, can then be used in conventional heterogeneous flow reactors and use very much less IL [128].

On a final note, the condensed phase does not have to be an IL to achieve the required phase behaviour for continuous flow homogeneous catalysis. Biphasic systems based on CO_2 with a second phase comprising water [129], fluorous solvents [130], poly(ethyleneglycol) [131] or the reagents and products [84a] themselves have all been demonstrated. In each case, the same basic principles apply (i.e. the catalyst should partition favourably towards the stationary phase and a permanent separation between mobile and stationary phases must exist). When water is employed as the reaction solvent there are additional factors to consider. For instance, the application of this particular biphasic couple is restricted to products and catalysts that can tolerate low pH owing to the acidity of water in the presence of compressed CO_2. With acidity issues taken into account the remaining factors that govern catalyst design come as no great surprise; the catalyst must be insoluble in CO_2 but sufficiently soluble in water to achieve active catalysis. Thus, any hydrophilic ligand described in Section 18.2 will provide an ideal candidate for the water–$scCO_2$ biphasic mode of operation. Multiphasic catalysis involving $scCO_2$ is described in detail in several excellent books and reviews [1a,132].

18.6 Experimental section

18.6.1 Trisodium salt of 3,3′,3″-phosphinetriylbenzenesulfonic acid (TPPTS) (6) (Reproduced with permission from [39] © John Wiley & Sons, Ltd)

$$P(C_6H_5)_3 + SO_3 \xrightarrow{H_2SO_4} P(C_6H_4SO_3H)_3$$

$$P(C_6H_4SO_3H)_3 + 3N(C_8H_{17})_3 \longrightarrow [HN(C_8H_{17})_3]_3[P(C_6H_4SO_3)_3]$$

$$[HN(C_8H_{17})_3]_3[P(C_6H_4SO_3)_3] + 3NaOH \longrightarrow P(C_6H_4SO_3Na)_3 + 3N(C_8H_{17})_3 + 3H_2O$$

Method A: Oleum (96 g, 30 wt% SO_3) is placed in a 250 mL flask equipped with a stirrer, thermometer and dropping funnel and cooled in an external ice bath to an internal temperature of 15°C. Over a period of 1 h triphenylphosphine (10.5 g, 40 mmol) and oleum (32 g, 30 wt% SO_3) are added with stirring. The reaction temperature is kept between 15°C and 20°C. After the addition of oleum and triphenylphosphine has been completed, the mixture is stirred for 24 h at 20°C. Subsequently, the content of the flask is added to a 1 L flask containing water (300 g) having a temperature of about 10°C. During the addition the internal temperature is kept between 20°C and 40°C by intensive external cooling. The homogeneous sulfonation mixture is placed in a 1 L flask and stirred with a mixture of triisooctyl amine (47.7 mL, 110 mmol) and toluene (180 mL). After the addition has been completed, the reaction mixture is stirred for 30 min and left to separate for a further 30 min.[a] The lower phase is separated and discarded. Subsequently, to the reaction product aqueous sodium hydroxide solution (5%) is added while stirring. At a pH of 5.50, the addition of sodium hydroxide is interrupted. The water phase is separated and discarded. The addition of base is then continued until the mixture reaches a pH of 6.0–6.3. The aqueous solution which contains mainly TPPTS, minor amounts of TPPDS and their oxides is separated. By concentration of the aqueous solution until the beginning of crystallisation, followed by filtration, washing with methanol and drying, the sodium salt of 3,3′,3″-phosphinetriylbenzenesulfonic acid can be obtained as a white solid with a purity of at least 95%. For highest purities, the solid is dissolved in water and transferred to a Sephadex G-15 column for chromatography.

Method B: In a 100 mL, two-necked, round-bottomed flask equipped with a magnetic stirrer and dropping funnel, boric acid (14.1 g, 228 mmol) is dissolved in concentrated H_2SO_4 (96 wt%) in a vacuum (10^{-2} Torr) while vigorously stirring. Triphenyphosphine (4.0 g, 15.25 mmol) is added and after complete dissolution oleum (67 mL, 65 wt% SO_3) is slowly dropped over a period of 30 min into the solution. The temperature of the reaction mixture should not exceed 20°C. The mixture is stirred for 2 h at room temperature (RT) and afterwards hydrolysed by means of a degassed water–ice mixture (240 g). Triisooctylamine (54 g, 152 mmol), dissolved in toluene (153 mL), is added to the hydrolysis mixture. After vigorous stirring for 15 min, the previously colourless organic phase is slightly yellow and the aqueous phase is colourless.[a] In a separation funnel the lower water phase is removed and the organic phase is washed with water (3 × 50 mL). The water phases are collected and discarded. While stirring, a solution of sodium hydroxide in water (7.5 molar) is added slowly to the mixture until a pH of 7.0, as measured by a standard glass electrode, is reached. The aqueous phase is separated and worked up as described in procedure A. The resulting solid contains 10 wt% TPPDS and TPPMS without oxides. For higher purities, the solid is dissolved in water and transferred to a column for chromatography on Sephadex G-15. Anal. calculated for $C_{18}H_{18}O_{12}S_3PNa_3$, 622.47: C, 34.73; H, 2.91; O, 30.84; S, 15.45; Na, 11.08; P, 4.98. Found: C, 34.84; H, 2.80; O, 30.62; S, 15.73; Na, 11.24; P, 4.75. Properties: TPPTS is a white solid which is slightly air sensitive. The ^{31}P NMR (161.8 MH_z, D_2O) exhibits a singlet at $\delta = -5.1$ ppm. In-depth NMR analysis is given by Bartik *et al.* [133].

Note

[a] The efficacy of phosphine extraction with triisooctylamine may well be dictated by the free SO_3 content of the reaction solution following the sulfonation procedure as previously discussed. Oleum solutions readily lose SO_3 on standing. Older batches of oleum can therefore contain lower than expected SO_3 levels and this may also cause a degree of irreproducibility in direct sulfonation procedures. Removing SO_3 *in vacuo* following the sulfonation reaction may drastically improve extraction efficiency according to the results presented by Tic [40].

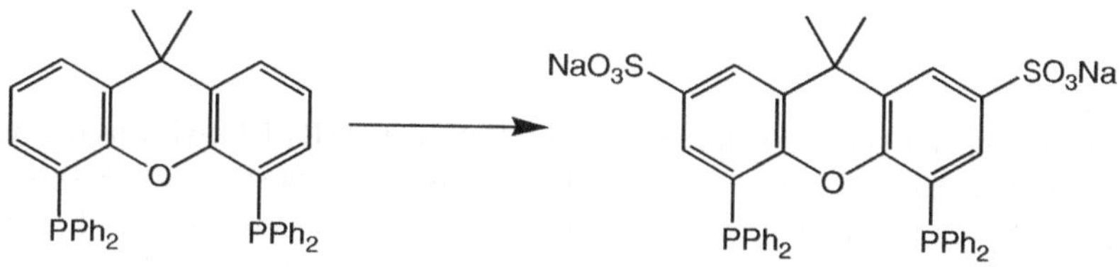

Scheme 18.6

18.6.2 2,7-bis(SO_3Na)-Xantphos (8) (Scheme 18.6) (Reproduced with permission from [44] © Elsevier)

In a typical experiment, Xantphos (1 g, 1.73 mmol) was added slowly to oleum (2.9 mL, 25% SO_3) in 4 h at 5°C. After the addition was complete, the resulting brownish solution was warmed to RT. After 16 h, 15 mL degassed ice water was added slowly to the solution which resulted in a white suspension. Another 20 mL of water was added, and the resulting yellowish solution was poured into a vigorously stirred solution of tri-isooctylamine (54 mmol, 2.3 mL) in 10 mL of toluene. The toluene layer was washed twice with water, and a solution of (NaOH 6.25 M) was slowly added until pH 12 was reached. The water layer was decanted and neutralized with a 3 M H_2SO_4 solution. Evaporation gave a white solid which was poured into refluxing MeOH. After 30 min, the white suspension was carefully filtered and the resulting clear solution was evaporated to dryness resulting in a pale yellow solid. Further purification was accomplished by refluxing the solid in EtOH which after decantation and evaporation gave a white solid.[a] Yield: 1.03 g (1.31 mmol; 76%). Anal. calculated for $C_{39}H_{30}Na_2O_7P_2S_2$: C, 58.3; H, 4.0; S, 8.2%. Found: C, 57.8; H, 4.2; S, 8.2. Properties: ^{1}H NMR (CD_3OD; ppm): δ 8.00 (s, 2 H, ArH), 7.28–7.18 (m, 22 H, ArH), 1.69 (s, 6 H, $C(CH_3)_2$). ^{13}C{H} NMR (CD_3OD; ppm): δ 154.6 (t, *J*. = 9.3 Hz), 142.1, 138.0 (t, *J*. = 6.0 Hz), 135.2 (t, *J*. = 10.1 Hz), 131.1, 129.9, 129.6, 128.3, 128.0, 126.0, 36.2 (*C*$(CH_3)_3$), 32.1 (CH_3). ^{31}P{^{1}H} NMR (CD_3OD; ppm): δ −14.7. Exact mass (FAB): *m*/*z* 783.7165 [M + H]. Calculated for $C_{39}H_{30}Na_2O_7P_2S_2$: 783.7168. M.p. >300°C (decomposition).

Note

[a] Subsequently reported preparations employed the recrystallisation procedure of Mul *et al.* described in this chapter [41].

18.6.3 Sulfonated BINAP (10) (Scheme 18.7) (Reproduced with permission from [45] © Royal Society of Chemistry)

The sulfonation of (R)-BINAP was accomplished by dissolving (R)-BINAP (1 g) in concentrated sulfuric acid (3.5 mL) with subsequent dropwise addition over 2–3 h of fuming sulfuric acid (15 ml, 40 wt% SO_3) and stirring for 3 days at 10°C under argon. The reaction was quenched by pouring the solution into 100 mL of ice-cooled water and the product neutralized by dropwise addition of aqueous sodium hydroxide (50 wt% NaOH) until a pH of 7 was reached. The volume of the solution was reduced to 30 mL *in vacuo* and methanol

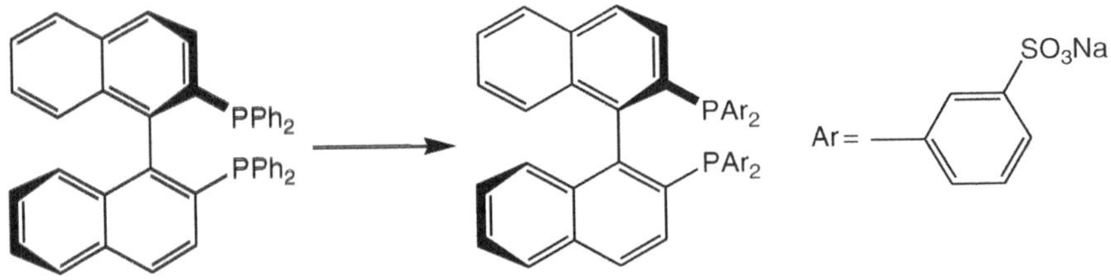

Scheme 18.7

(100 mL) added to precipitate any sodium sulphate present. Sulfonated (R)-BINAP was recovered as a solid by vacuum drying the filtered methanolic solution. Anal. calculated for BINAP(SO_3–Na)$_4$: S/P:Na/P:C/Na=2:2:11; found, S/P:Na/P:C/Na=2.34:2.02:11.31. Properties: This result is in agreement with the ^{31}P NMR spectrum that shows a single major resonance at δ ca. −11.1 with a second small peak around δ −12.7 (relative intensities are 86:14). The ^{33}S NMR spectrum of the oxidized form of this sample (in D_2O) indicates that no sodium sulphate is present, to at least the detection limits of NMR (150 000 scans employed) and that no further information on the sample composition is possible because the sulphate groups on the BINAP give a broad peak around δ −13 (relative to ammonium sulphate). NMR data are consistent with the premise that only the four phenyl rings are sulfonated in the major species obtained in 85% abundance. The remaining species is likely to be penta- or hexa-sulfonated BINAP with one or two extra sulfonated group(s) on the naphthyl ring(s).

18.6.4 Synthesis of Tris(1H,1H,2H,2H-perfluorooctyl)phosphine (11) (Reproduced with permission from [60] © Royal Society of Chemistry)

A solution of $CF_3(CF_2)_5C_2H_4I$ (23.7 g, 0.05 mmol) in diethyl ether (70 cm^3) was added dropwise to magnesium turnings (1.44 g, 0.06 mol) suspended in diethyl ether (15 cm^3) with two drops of 1,2-dibromoethane. The mixture was refluxed for 2 h and the solution decanted into a second flask under nitrogen. Phosphorus trichloride (2.2 g, 0.016 mol) in diethyl ether (75 cm^3) was added dropwise to the solution over 2 h, and the mixture was hydrolysed with degassed aqueous ammonium chloride. The organic layer was separated and dried over $MgSO_4$ and the solvent removed *in vacuo*. The resulting brown solid was dissolved in the minimum amount of diethyl ether and passed quickly through an alumina column eluting with light petroleum (bp 40–60°C), the solvent was removed *in vacuo* and the white solid was heated (60°C) under dynamic vacuum (0.01 mmHg) for 2 h to remove all traces of fluorinated starting material to afford the product as a colourless solid–liquid, mp 24–26°C (8.6 g, 50%). Anal. calculated for $C_{24}H_{12}F_{39}P$: C, 26.9; H, 1.1; P, 2.9. Found: C, 26.7; H, 1.1; P, 2.5. Properties: ^{1}H NMR (250 MHz, $CDCl_3$) δ 2.3 (6H, br t, $^3J_{HH}$ 10, PCH_2), 2.1 (6H, um, CH_2CF_2); ^{19}F NMR (235 MHz, $CDCl_3$) δ −81.0 (3F, t, $^3J_{FF}$ 14, CF_3), −114.3 (2F, um, $C^{\alpha}F_2$), −121.7 (2F, um, $C^{\beta}F_2$), −123.0 (2F, um, $C^{\delta}F_2$), −123.7 (2F, um, $C^{\varepsilon}F_2$), −126.1 (2F, um, $C^{\gamma}F_2$); *m/z* (EI) 1072 (M^+, 86%), 1053 (67), 953 (4), 803 (4), 739 (21) and 656 (46).

Note

In NMR assignments, um refers to unresolved multiplet.

18.6.5 Synthesis of Tris (4-tridecafluorohexylphenyl)phosphine (13) (Reproduced with permission from [134] © Wiley-VCH Verlag GmbH & Co. KGaA)

Step A. 4-(Tridecafluorohexyl)bromobenzene: A 500 mL, three-necked, round-bottomed flask equipped with a magnetic stirring bar, pressure-equalising dropping funnel, thermometer and condenser was purged for

10 min with dinitrogen. The flask was charged with 4-bromoiodobenzene (25.0 g, 88.34 mmol), copper bronze (12.36 g, 193.13 mmol),[a] 2,2′-bipyridine (0.98 g, 6.36 mmol), DMSO (200 mL)[c] and fluorobenzene (150 mL). The pressure-equalising dropping funnel was charged with perfluoro-*n*-hexyl iodide (39.42 g, 88.39 mmol) and fluorobenzene (50 mL). The flask was purged with dinitrogen for a further 20 min whilst the suspension was stirred vigorously. The mixture was then heated to exactly 70°C and the perfluoro-*n*-hexyl iodide added dropwise over 7 h, with the temperature of the mixture being maintained at exactly 70°C throughout. The mixture was then stirred for a further 120 h at 70°C. After cooling the reaction mixture to room temperature, it was added to water (250 mL) and diethyl ether (500 mL) in a 1 L conical flask. The solids that precipitate were removed by filtration through a Buchner funnel and washed with two 50 mL portions of diethyl ether. The organic layer was separated in a 1 L separating funnel and washed five times with 250 mL water, dried over magnesium sulphate, filtered and concentrated under reduced pressure using a rotary evaporator to give a brown oil. This oil was heated in a Kugelröhr oven at 65°C at 0.01 mmHg to remove residual 4-bromoiodobenzene and, subsequently, the temperature of the oven was increased to 100°C to distill the 4-(tridecafluoro)bromobenzene and 1,4 bis(tridecafluoro)benzene as a clear oil away from the residual copper complexes which remained in the base bulb as a dark brown tar. The clear oil was then distilled in a microfine distillation apparatus to give 28.56 g (68%) of 4-(tridecafluoro)bromobenzene as a colourless oil (bp 89–92°C at 10 mbar). NMR analysis showed that this sample is >98% pure and contained less than 2% of 1,4-bis(tridecafluoro)benzene. Material of this purity is acceptable for use in step B. Anal. calculated for $C_{12}H_4BrF_{13}$: C, 30.3; H, 0.8; F, 52.0. Found: C, 30.2; H, 0.8; F, 51.7. Properties: ^{1}H NMR (250 MHz, $CDCl_3$) δ 7.90 (2 H, d, $^3J_{HH}$ = 8.5 Hz, ArH), 7.70 (2 H, d, $^3J_{HH}$ = 8.5 Hz, ArH); ^{19}F NMR (235 MHz, $CDCl_3$) δ −81.80 (3 F, t, $^4J_{FF}$ = 9.3 Hz, CF), −111.72 (2 F, t, $^4J_{FF}$ = 14.6 Hz, α-CF_2), −122.10 (2 F, m, CF_2), −122.63 (2 F, m, CF_2), −123.63 (2 F, m, CF_2), −126.90 (2 F, m, CF_2); *m/z* (EI) 474/6 (M^+).

Step B. Tris(4-tridecafluorohexylphenyl)phosphine: A 500 mL, three-necked round-bottomed flask equipped with a magnetic stirring bar, pressure-equalising dropping funnel, low-temperature thermometer and Rotaflo stopcock adaptor was attached to a Schlenk line. After flame drying under high vacuum, the flask was cooled and filled with dinitrogen. The flask was charged with 4-(tridecafluorohexyl)bromobenzene (24.0 g, 50.53 mmol) and degassed again under high vacuum for 20 min. The apparatus was again filled with nitrogen before adding dry diethyl ether (300 mL) to the main flask. Dry diethyl ether (30 mL) and a 1.6 M hexane solution of *n*-butyllithium (31.5 mL, 50.40 mmol) were both added to the dropping funnel and the flask was cooled to −78°C using a dry ice–acetone bath. The *n*-butyllithium solution was added dropwise over 60 min with the internal temperature never allowed to warm above −75°C. The mixture was then stirred for 5 h at −78°C. The dropping funnel was rinsed with a 10 mL portion of dry diethyl ether and then charged with dry diethyl ether (30 mL) and phosphorus trichloride (2.29 g, 16.72 mmol). The phosphorus trichloride solution was added dropwise over 60 min with the internal temperature never allowed to warm above −75°C. The solution was then allowed to warm to RT and stirred overnight. A 10% ammonium hydroxide solution (100 mL) was then added[b] and the mixture was stirred for 20 min. The organic solution was then transferred to a flame-dried Schlenk flask under dinitrogen containing magnesium sulphate *via* a cannula. After stirring for 20 min, the solution was transferred to a flame-dried Schlenk flask under dinitrogen and the solvent removed *in vacuo* to give a yellow oil. This oil was distilled in a Kugelröhr oven to give a clear oil (bp 210°C at 0.01 mmHg). This was transferred to a glove box where it slowly solidified to give a white solid (10.98 g, 56%). Anal. calculated for $C_{36}H_{12}F_{39}P$: C, 35.5; H, 1.0; P, 2.5. Found: C, 35.8; H, 0.9; P, 2.5. Properties: NMR analysis showed that this sample is >98% pure, mp 63–64°C. ^{1}H NMR (250 MHz, $CDCl_3$) δ 7.53 (6 H, d, $^3J_{HH}$ = 8.0 Hz, 3–C_6H_4), 7.35 (6 H, vt, $^3J_{HH}$ = $^3J_{HP}$ = 8.0 Hz, 2-C_6H_4); ^{19}F NMR (235 MHz, $CDCl_3$) δ −81.32 (9 F, t, $^4J_{FF}$ = 9.6 Hz, CF_3), −111.48 (6 F, t, $^4J_{FF}$ = 14.2 Hz, α-CF2), −121.90 (6 F, m, CF_2), −122.24 (6 F, m, CF_2), -123.24 (6 F, m, CF_2), −126.59 (6 F, m, CF_2); ^{31}P{^{1}H} NMR (101 Hz, $CDCl_3$) δ −6.3; *m/z* (FAB) 1216 (M^+).

Notes

[a] It is essential that copper bronze is used for this stage, not copper powder.
[b] The ammonium hydroxide solution was degassed prior to addition by stirring under vacuum for a few seconds and then backfilling with dinitrogen. This procedure was repeated six times. Alternatively, the aqueous ammonium hydroxide solution is degassed by bubbling dinitrogen through the solution for 20 min.
[c] The copper coupling reaction is sensitive to the quantity of DMSO present; if it is too low, the reaction will not proceed [135].

18.6.6 (Meta-sulfonatophenyl)diphenylphosphine, sodium salt (monosulfonated triphenylphosphine, TPPMS) (7) (Reproduced with permission from [96] © John Wiley & Sons, Ltd)

$$P(C_6H_5)_3 + SO_3 \xrightarrow{H_2SO_4} P(C_6H_5)_2(C_6H_4\text{-}m\text{-}SO_3H)$$

$$P(C_6H_5)_2(C_6H_4\text{-}m\text{-}SO_3H) + NaOH + H_2O \longrightarrow P(C_6H_5)_2(C_6H_4\text{-}m\text{-}SO_3Na.2H_2O)$$

In a 100 mL Erlenmeyer flask chilled in ice water, finely ground PPh_3 (20 g, 76.3 mmol) was dissolved in fuming sulfuric acid (50 mL, 20%, contains 230 mmol free SO_3). The phosphine was added in small portions, and the flask covered with a watchglass and gently shaken to dissolve most of the PPh_3 before the next addition. Complete dissolution of PPh_3 requires approximately 30 min. The solution was then left to warm to RT and placed on top of a boiling water (steam) bath where it was kept for 75 min with occasional swirling. After cooling to RT, the solution was carefully poured onto crushed ice (400 g) in a 1 L beaker. A milky emulsion was obtained with some gummy material which stuck to the wall of the beaker and to the glass stirring rod.[a] This emulsion was further cooled in ice water and carefully neutralized by slow addition of aqueous NaOH solution (~140 mL, 50%) until it reached pH 3–4, yielding a white precipitate. The resulting solution was allowed to stand at room temperature for 30 min, after which the precipitate was filtered with gentle suction on a Buchner funnel and, finally, as much of the remaining moisture was pressed out with a suitable flat glass stopper. This raw product contained TPPMS, nonsulfonated starting material and some sodium sulphate. The crude product was dissolved in hot water (800 mL), placed in an Erlenmeyer flask on top of a boiling water bath and flushed with a stream of argon. During this time the milky emulsion slowly cleared, and in 1–3 h a clear, or only very slightly cloudy, solution was obtained while the nonsulfonated PPh_3 collected on the bottom of the flask.[b] This solution was left to cool until the PPh_3 melt solidified and was then quickly filtered through a Celite (Hyflo Super-cell) pad. Typically, 4 g and 15 mmol of crystalline PPh_3 is recovered, which can be used in subsequent preparations. The filtrate was allowed to cool under argon and then placed in a refrigerator overnight. The product crystallized as small shiny white flakes and was dried over P_4O_{10}. Yield: 12.5 g, 51% (based on reacted PPh_3). This material contained 2–5% PPh_3, was virtually free (HPLC) from phosphine oxide (OTPPMS) and multiply sulfonated by-products and was suitable for subsequent preparation of water-soluble metal complexes *via* ligand exchange of PPh_3-containing starting materials. Analytically pure TPPMS was obtained by recrystallization of 10 g of this compound from 100 mL water under argon. Yield: 5.7 g, 57% (overall yield 29%). Anal. calculated for $C_{18}H_{18}O_5PSNa$: C, 53.99; H, 4.53: Found: C, 54.33; H, 4.60. Properties: The solubility of the sodium salt of (meta-sulfonatopheny1)diphenylphosphine, TPPMS in water is approximately 12 g/L at room temperature. It dissolves slightly in cold ethanol, but is soluble at elevated temperatures. It is virtually insoluble in acetone and aliphatic, aromatic or chlorinated hydrocarbons, but is soluble at RT in tetrahydrofuran (THF). The compound crystallizes with two waters of crystallization; however, the anhydrous form can also be obtained. Dry TPPMS is stable to air but is oxidized rapidly when wet, especially in basic aqueous solutions. It is a highly surfactant compound and forms aggregates and micelles in neutral aqueous solutions.

Notes

[a] The checker observed a clear solution.

[b] In two separate preparations, the checkers did not observe separation of PPh_3. On the many occasions we have prepared TPPMS according to this method, we also never observed separation of PPh_3 or the appearance of an emulsion following hydrolysis of the reaction mixture.

18.6.7 1-Propyl-3-methylimidazolium diphenyl(3-sulfonatophenyl)-phosphine ([PrMIM][TPPMS]) (18) (Reproduced with permission from [83] © American Chemical Society)

$$[PPh_2P(C_6H_4SO_3)]Na + [PrMIM]Cl \longrightarrow [PPh_2P(C_6H_4SO_3)][PrMIM] + NaCl$$

1-Propyl-3-methylimidazolium chloride [83] (14.31 g, 89.1 mmol) was added to a rapidly stirred solution of sodium diphenyl-(3-sulfonatophenyl)phosphine dihydrate (25.74 g, 64.3 mmol) in THF, resulting in immediate formation of a fine white precipitate. The solution was stirred at room temperature for a further 24 h and filtered through diatomaceous earth, and the solvent removed *in vacuo*. The resulting residue was taken up into dichloromethane and again filtered, and the solvent removed *in vacuo*. The crude product, which contains an excess of the imidazolium chloride, was left at –10°C to afford (**18**) as colourless rhomboidal crystals (15.28 g, 32.8 mmol, 50.9%) mpt. 88–90°C. Anal. calculated for $C_{25}H_{27}O_3N_2PS$: C 64.36 H 5.83 N 6.00. Found: C 64.42 H 5.64 N 6.14. Properties: IR (CsI): $\tilde{\nu}$=3156 (w), 3097 (m), 3047 (m), 2969 (w), 2883 (w), 1580 (m), 1479 (m), 1461 (w), 1434 (m), 1394 (w), 1309 (w), 1202 (s, br), 1139 (m), 1091 (m), 1070 (w), 1031 (s), 995 (w), 890 (w), 796 (m), 784 (m), 749 (s), 698 (s), 673 (m), 617 (s), 595 (w), 555 (w) cm^{-1}. ^1H NMR (300 MHz, CD_2Cl_2, 298 K): δ=0.99 (3 H, t, CH_3), 1.92 (2 H, sext, CH_2CH_3), 3.97 (3 H, s, NCH_3), 4.20 (2 H, t, NCH_2), 7.43 (14 H, m, $Ph_2PC_6H_4SO_3$), 7.92, 7.96 (2 H, 2×d, NC(H)C(H)N), 9.71 (1 H, s, NC(H)N) ppm. ^{13}C NMR (75.4 MHz, CD_2Cl_2, 298 K): δ=148.1 (d, CSO_3, J_{pc}=6.9 Hz), 138.2 (s, NCN), 138.0 (d, *ipso*-C_6H_5, J_{pc}=12.7 Hz), 137.5 (d, *ipso*-$C_6H_4SO_3$, J_{pc}=11.5 Hz), 134.6 (d, *o*-$C_6H_4SO_3$, J_{pc}=14.9 Hz), 134.2 (d, *o*-C_6H_5, J_{pc}=19.6 Hz), 131.3 (d, *o*-$CCSO_3$, J_{pc}=25.3 Hz), 129.3 (s, *p*-C_6H_5), 129.0 (d, *m*-C_6H_5, J_{pc}=6.9 Hz), 128.8 (d, *m*, $C_6H_4SO_3$, J_{pc}=5.8 Hz), 126.8 (s, *p*-$C_6H_4SO_3$), 123.8, 122.4 (2×s, NCCN), 51.6 (s, NCH_2), 36.5 (s, NCH_3), 23.9 (s, CH_2), 10.9 (s, CH_3) ppm. ^{31}P NMR (121.4 MHz, CD_2Cl_2, 298 K) δ=–3.90.

18.6.8 4,4′-Phosphorylated 2,2′-Bis(diphenylphosphanyl)-1,1′-binaphthyl (Reproduced with permission from [120] © Wiley-VCH Verlag GmbH & Co. KGaA)

Step A: 4,4′-dibromo-2,2′-bis(diphenylphosphoryl)-1,1′-binaphthyl: (*R*)-BINAP oxide (15.06 g, 23 mmol) was dissolved in CH_2Cl_2 (340 mL). While the solution was stirred, Br_2 (11.03 g, 69 mmol) and pyridine (1.82 g, 23 mmol) were added. The stirring was continued at room temperature for 20 h. Thereafter the organic phase was extracted with 1 M aqueous sodium hydrogen sulfite (320 mL), brine and saturated sodium hydrogen carbonate solution. Drying over Na_2SO_4 was followed by filtration. After evaporation of the solvent, the crude oily product mainly consisted of the 4-bromosubstituted product. Yield: 16.83 g (consisting of 62 mol% of 4-bromo, 18 mol% of 4,4′-dibromo and 20 mol-% of (*R*)-BINAP oxide, determined by 31P NMR spectroscopy). To obtain the bis-brominated product, the bromination procedure that we have described was repeated twice and the crude products of the preceding brominations were used as starting materials. The dichloromethane solution was reduced in volume, and the crude oily product was dried *in vacuo* and used for the following phosphorylation. Yield: 18.43 g (consisting of 86 mol-% of the 4,4′-dibromo and 14 mol-% of 4-bromo products determined by ^{31}P NMR spectroscopy).

Step B: (R)-(1)-4,4′-Bis(diethoxyphosphoryl)-2,2′-bis(diphenylphosphoryl)-1,1′-binaphthyl (5): Crude A (2.00 g), diethyl phosphite (1.22 g, 10 mmol), tetrakis(triphenylphosphine)–palladium (0.58 g, 0.5 mmol) and triethylamine (1.11 g, 11 mmol) were dissolved in absolute toluene (5 mL) and stirred for 40 h at 90°C.

After cooling, the solution was diluted with CH_2Cl_2 (80 mL), extracted three times with H_2O (30 mL) and dried with Na_2SO_4. The solvent was evaporated and the crude product chromatographed on a silica gel column (eluent: ethyl acetate–hexane–ethanol, 7:3:1). After chromatography, the product was isolated as a glassy mass. Yield: 1.03 g (53%).

Step C: (R)-(1)-4,4′-Bis(diethoxyphosphoryl)-2,29-bis(diphenylphosphanyl)-1,1′-binaphthyl (7): Compound B (0.93 g, 1 mmol) and phenylsilane (6 mL) were refluxed for 65 h. The mixture was cooled down and kept in the refrigerator overnight. Colourless crystals precipitated. These were filtered off, washed with *n*-hexane, and dried *in vacuo*. Yield: 0.50 g (56%).

Step D: (R)-(1)-2,2′-Bis(diphenylphosphanyl)-4,4′-bisphosphono-1,1′-binaphthyl: Compound C (2.00 g, 2.23 mmol) was stirred with bromotrimethylsilane (5 mL) at room temperature for 24 h. Volatile components were evaporated *in vacuo* and the residue was dissolved in a mixture of THF (5 mL) and water (2 mL), stirred for 1 h, and concentrated *in vacuo*. Yield: 1.74 g (quantitative). Properties: ^{1}H NMR (CD_3OD): δ=6.64 (d, J = 8.5, 2 H), 6.80 (dt, J = 8.1, 1.3, 2 H), 7.34 (dt, J = 8.1, 1.3, 2 H), 8.15 (d, J = 16.2, 2 H), 8.62 (d, J = 8.5, 2 H, $CH_{naphthyl}$), 6.96–6.99 (m, 8 H), 7.02–7.08 (m, 10 H), 7.11 (t, J = 7.2, 7.3, 2 H, CH_{phenyl}). ^{13}C NMR (CD_3OD): δ= 129.1 (d, J = 163.5), 129.6 (d, J = 4.6), 130.1, 130.3, 130.6 (m), 131.2, 135.1 (m), 135.8 (m), 136.5 (m), 138.9 (m), 139.6 (m), 151.5 (m). ^{31}P NMR: δ=14.4 (phosphonic acid), -15.7 (phosphine). (*m/z*)=782 [M$^+$]. $[\alpha]^D_{20}$=92.4 (DMSO, c=1.0).

18.6.9 Synthesis of (R)-6,6′-bis(perfluorohexyl)-2,2′ bis (diphenylphosphino)-1,1′-binaphthyl ((R)-Rf-BINAP) (28) (Scheme 18.8) (Reproduced with permission from [70] © Wiley-VCH Verlag GmbH & Co. KGaA)

Step A. (R)-6,6′-dibromo-2,2′-diacetoxy-1,1′-binaphthyl: A 250 mL, round-bottomed flask equipped with a magnetic stirrer bar was charged with (*R*)-6,6′-dibromo-1,1′-bi-2-naphthol (5.0 g, 11.2 mmol), triethylamine (9.4 mL, 66.6 mmol), (4-*N,N*dimethylamino) pyridine (DMAP; 100 mg, 10 mmol), acetic anhydride (2.1 mL, 22.4 mmol) and dichloromethane (100 mL). A reflux condenser was fitted to the flask, and the mixture was heated to reflux for 1 h with vigorous stirring. Upon cooling, 1 M hydrochloric acid (100 mL)

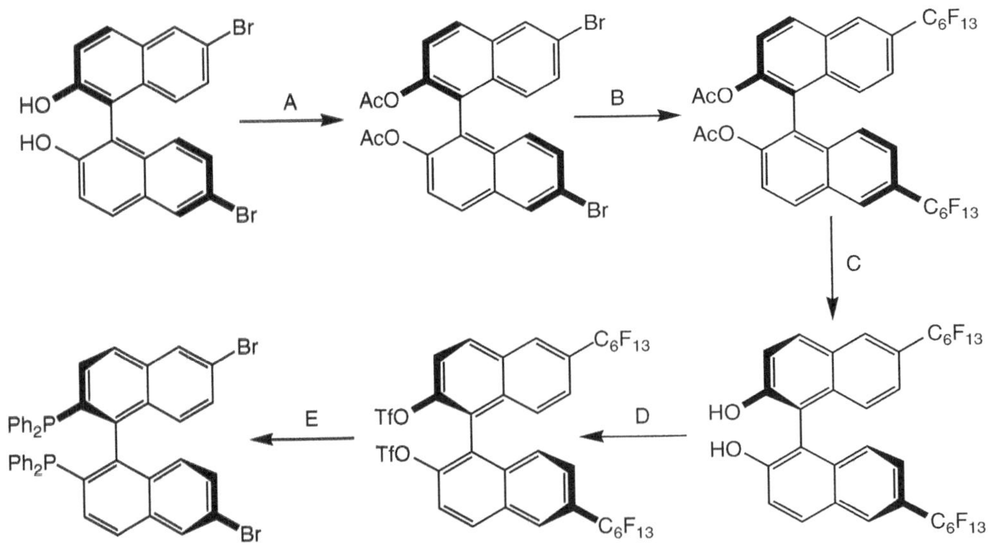

Scheme 18.8

was slowly added and the mixture transferred to a separating funnel. The organic layer was separated and washed with saturated sodium carbonate solution (100 mL), brine (100 mL) and water (100 mL) before being dried over anhydrous magnesium sulphate, filtered and concentrated under reduced pressure using a rotary evaporator to yield the crude product as a yellow semi-solid. Recrystallisation from ethanol yields 8.4 g (77%) of (*R*)-6,6′-dibromo-2,2′-diacetoxy-1,1′-binaphthyl as white needles.

Step B. (R)-6,6′-bis(tridecafluorohexyl)-2,2′-diacetoxy-1,1′-binaphthyl: A 250 mL, round-bottomed flask was equipped with a magnetic stirrer bar, thermometer and condenser and purged for 10 min with dinitrogen. The flask was charged with (*R*)-6,6′ dibromo-2,2′-diacetoxy-1,1′-binaphthyl (3.5 g, 6.6 mmol), perfluoro-*n*-hexyl iodide (8.9 g, 20 mmol), copper powder (2.6 g, 39.7 mmol), 2,2′-bipyridine (0.2 g, 1.5 mmol), fluorobenzene (50 mL) and DMSO (100 mL). The flask was purged with dinitrogen for a further 15 min before heating the resulting mixture at 80°C for 72 h with constant stirring. The mixture was then transferred to a 1 L conical flask, and diethyl ether (300 mL) and water (200 mL) were added. The suspension was filtered using a Buchner funnel and the solids were washed with diethyl ether (2 × 25 mL). The filtrate was transferred to a 1 L separating funnel and the organic layer was separated. This was washed with 1 M hydrochloric acid (100 mL) and five portions of water (100 mL), dried over anhydrous magnesium sulphate, filtered and concentrated under reduced pressure using a rotary evaporator to give 4.4 g (66%) of (*R*)-6,6′-bis(tridecafluorohexyl)-2,2′-diacetoxy-1,1′-binaphthyl as a yellow oil. The sample was shown to be >95% pure by ^{1}H NMR. Material of this purity was acceptable for use in the third step. The yellow oil can be solidified by addition of hexane (30 mL) followed by removal of the solvent under reduced pressure using a rotary evaporator three times, followed by prolonged drying under high vacuum.

Step C. (R)-6,6′-bis(tridecafluorohexyl)-1,1′-bi-2-naphthol: A 100 mL, round-bottomed flask, equipped with a magnetic stirrer bar, was charged with (*R*)-6,6′-bis(tridecafluorohexyl)-2,2′-diacetoxy-1,1′-binaphthyl (2.1 g, 2 mmol) and ethanol (65 mL). The resulting suspension was stirred rapidly, and sodium ethoxide (0.34 g, 5 mmol) was added. The resulting brown solution was stirred for 1 h at room temperature. The magnetic stirrer bar was removed from the mixture and the ethanol was then removed under reduced pressure using a rotary evaporator. Dichloromethane (50 mL) was added to the brown oil and the liquid was transferred to a 250 mL separating funnel. The organic solution was washed with hydrochloric acid (100 mL, 1 *Mol.* dm^{-3}) and water (100 mL), dried over anhydrous magnesium sulphate, filtered and concentrated under reduced pressure using a rotary evaporator to yield a yellow-brown oil. Methanol (30 mL) was added to the oil and the solution was filtered. The solvent was removed again using a rotary evaporator to yield a yellow oil. This oil was triturated with hexane and recrystallised from hexane to give 1.6 g (90%) of (*R*)-6,6′-bis(tridecafluorohexyl)-1,1′-bi-2-naphthol as a yellow solid.

Step D. (R)-6,6′-bis(tridecafluorohexyl)-2,2′-di-trifluoromethane-sulfonyloxy-1,1′-binaphthyl: A 100 mL, three-necked, round-bottomed flask equipped with a magnetic stirrer was flame-dried under high vacuum. Once cool, the flask was filled with dinitrogen. (*R*)-6,6′-bis(tridecafluorohexyl)-1,1′-bi-2-naphthol (2.0 g, 2.2 mmol) was added to the flask against a positive flow of dinitrogen and the removed glass stopper was replaced with a rubber septum. Dry and degassed dichloromethane (50 mL) was added to the flask through the septum *via* a cannula and dry and degassed pyridine (0.15 mL, 3.2 mmol) was added *via* syringe. The mixture was cooled in an ice bath to 0°C and trifluoromethanesulfonic anhydride (0.44 mL, 2.6 mmol) was added dropwise *via* syringe through the septum. The resulting orange solution containing a white precipitate was allowed to warm to room temperature over 4 h under nitrogen. Hydrochloric acid (50 mL, 1 *Mol.* dm^{-3}) was added and the mixture transferred to a separating funnel. The organic layer was separated, washed with saturated sodium carbonate solution (50 mL) followed by water (50 mL), dried over anhydrous magnesium sulphate, filtered and concentrated under reduced pressure using a rotary evaporator. The crude brown product was recrystallised from hexane to yield 1.2 g (95%) of (*R*)-6,6′-bis(tridecafluorohexyl)-2,2′-di-trifluoromethanesulfonyloxy-1,1′-binaphthyl as a white crystalline solid.

Step E. (R)-6,6′-bis(tridecafluorohexyl)-2,2′-bis(diphenylphosphino)-1,1′-binaphthyl: A 100 mL, three-necked, round-bottomed flask equipped with a magnetic stirrer bar and reflux condenser was flame-dried under high vacuum. Once cool, the flask was filled with dinitrogen. Freshly prepared [$NiCl_2$(dppe)] (44 mg, 8.3×10^{-5} mol) was added to the flask against a positive flow of dinitrogen and the removed glass stopper was replaced with a rubber septum. A 100 mL Schlenk flask was equipped with a magnetic stirrer bar and flame-dried under high vacuum. Upon cooling, this flask was also filled with dinitrogen. (*R*)-6,6′-bis(tridecafluorohexyl)-2,2′-di-trifluoromethane-sulfonyloxy-1,1′-binaphthyl (0.84 mmol) was added to the Schlenk flask against a positive flow of dinitrogen and a rubber septum was placed in the mouth of the flask. Dry and degassed DMF (50 mL) was then added to each flask *via* a cannula. To the round-bottomed flask, diphenylphosphine (120 mL, 1.1 mmol) was added through the septum *via* syringe and the mixture of phosphine and nickel stirred at 100°C for 30 min. The solution of (*R*)-6,6′-bis(tridecafluorohexyl)-2,2′-di-trifluoromethanesulfonyloxy-1,1′-binaphthyl in the Schlenk flask was then added to the phosphine *via* a cannula. The septum was removed and 1,4-diazabicyclo[2,2,2]octane (DABCO, 0.4 g, 3.3 mmol) was added against a positive flow of dinitrogen and the septum was replaced. The resulting green solution was stirred at 110°C for 2 h. An additional portion of diphenylphosphine (120 mL, 1.1 mmol) was then added *via* syringe, the septum was replaced with a stopper and the solution stirred at 110°C for a further 72 h. Upon cooling, the DMF was removed under high dynamic vacuum at 70°C by trap-to-trap distillation using a round-bottomed flask trap cooled in liquid nitrogen. The resulting dark brown solid was stirred in methanol (50 mL) for 30 min and filtered to yield an off-white crude product. This was recrystallised from dichloromethane/methanol to yield 0.4 g (41%) of (*R*)-6,6′-bis(tridecafluorohexyl)-2,2′-bis(diphenyl-phosphino)-1,1′-binaphthyl as a white powder, mp 241–244°C. Anal. calculated for $C_{56}H_{30}F_{26}P_2$: C, 53.4; H, 2.4. Found: C, 53.3; H, 2.3. Properties: ^{1}H NMR (250 MHz, $CDCl_3$) $\delta = 6.69$ (2 H, d, $^3J_{HH} = 9.1$ Hz, ArH), 6.85 (2 H, d, $^3J_{HH} = 8.9$ Hz, ArH), 7.02 (8 H, m, ArH), 7.25 (12 H, m, ArH), 7.59 (2 H, d, $^3J_{HH} = 8.5$ Hz, ArH), 8.00 (2 H, d, $^3J_{HH} = 8.5$ Hz, ArH), 8.05 (2 H, s, ArH); ^{19}F NMR (235 MHz, $CDCl_3$) δ −81.18 (6 F, t, $^4J_{FF} = 10.2$ Hz, CF_3), −110.68 (4 F, m, α-CF_2), −121.77 (4 F, m, CF_2), −121.77 (4 F, m, CF_2), −123.15 (4 F, m, CF_2), −126.49 (4 F, m, CF_2); ^{31}P{1H} NMR (101 MHz, $CDCl_3$) δ −13.56 (s); *m/z* (ES^+) 1259 $(MH)^+$ (5%); $[\alpha]_D$ 104.2 (C_6H_6, *c* 0.1).

References

[1] (a) B. Cornils, W. A. Herrmann, I. T. Horvath, W. Leitner, S. Mecking, H. Olivier-Bourbigou, D. Vogt, *Multiphase Homogeneous Catalysis*, Wiley-VCH, Weinheim, **2005**; (b) D. J. Cole-Hamilton, *Science* **2003**, *299*, 1702–1706.

[2] (a) D. E. Bergbreiter, J. H. Tian, C. Hongfa, *Chem Rev* **2009**, *109*, 530–582; (b) D. E. Bergbreiter, S. D. Sung, *Adv Synth Catal* **2006**, *348*, 1352–1366.

[3] E. de Jesus, J. C. Flores, *Ind Eng Chem Res* **2008**, *47*, 7968–7981.

[4] (a) B. Cornils, W. A. Herrmann, *Aqueous-Phase Organometallic Catalysis: Concepts and Applications*, 2nd ed., Wiley-VCH, Weinheim, **2004**; (b) F. Joo, *Accounts Chem Res* **2002**, *35*, 738–745.

[5] (a) J. A. Gladysz, D. P. Curran, I. T. Horvath, *Handbook of Fluorous Chemistry*, Wiley-VCH, Weinheim, **2004**; (b) J. Rabai, Z. Szlavik, I. T. Horvath, in *Handbook of Green Chemistry and Technology*, Blackwell Science Ltd., **2002**, pp. 502–523.

[6] (a) P. Wasserscheid, T. Welton, *Ionic Liquids in Synthesis*, 2nd ed., Wiley-VCH, Weinheim, **2007**; (b) H. Olivier-Bourbigou, L. Magna, D. Morvan, *Appl Catal a-Gen* **2010**, *373*, 1–56; (c) J. Dupont, R. F. de Souza, P. A. Z. Suarez, *Chem Rev* **2002**, *102*, 3667–3691; (d) F. van Rantwijk, R. A. Sheldon, *Chem Rev* **2007**, *107*, 2757–2785; (e) X. Han, D. W. Armstrong, *Accounts Chem Res* **2007**, *40*, 1079–1086.

[7] (a) P. G. Jessop, W. Leitner, *Chemical Synthesis Using Supercritical Fluids*, Wiley-VCH, Weinheim, **1999**; (b) G. Brunner, *Supercritical Fluids as Solvents and Reaction Media*, Elsevier, **2004**; (c) P. G. Jessop, T. Ikariya, R. Noyori, *Chem Rev* **1999**, *99*, 475–493.

[8] (a) D. J. Adams, P. J. Dyson, S. J. Tavener, *Chemistry in Alternative Reaction Media*, John Wiley & Sons, **2003**; (b) R. Sheldon, *Chem Commun* **2001**, 2399–2407; (c) R. A. Sheldon, *Green Chem* **2005**, *7*, 267–278; (d) R. A. Sheldon, *Nato Sci Ser Ii Math* **2008**, *246*, 1–28; (e) R. A. Sheldon, *Chem Commun* **2008**, 3352–3365; (f) H. R. Hobbs, N. R. Thomas, *Chem Rev* **2007**, *107*, 2786–2820; (g) P. T. Anastas, M. M. Kirchhoff, *Accounts Chem Res* **2002**, *35*, 686–694.

[9] W. Leitner, *Accounts Chem Res* **2002**, *35*, 746–756.

[10] (a) T. Welton, *Coordin Chem Rev* **2004**, *248*, 2459–2477; (b) M. J. Earle, K. R. Seddon, *Pure Appl Chem* **2000**, *72*, 1391–1398.

[11] M. Berthod, G. Mignani, G. Woodward, M. Lemaire, *Chem Rev* **2005**, *105*, 1801–1836.

[12] W. Keim, T. M. Shryne, R. S. Bauer, H. Chung, P. W. Glockner, H. van Zwet, SHELL International Research, DE-P 2054009, **1969**.

[13] (a) E. Wiebus, B. Cornils, *Chem-Ing-Tech* **1994**, *66*, 916–923; (b) B. Cornils, *Org Process Res Dev* **1998**, *2*, 121–127; (c) C. W. Kohlpaintner, R. W. Fischer, B. Cornils, *Appl Catal A-Gen* **2001**, *221*, 219–225.

[14] B. Cornils, E. G. Kuntz, *J Organomet Chem* **1995**, *502*, 177–186.

[15] E. Renaud, R. B. Russell, S. Fortier, S. J. Brown, M. C. Baird, *J Organomet Chem* **1991**, *419*, 403–415.

[16] (a) S. Ganguly, J. T. Mague, D. M. Roundhill, *Inorg Chem* **1992**, *31*, 3500–3501; (b) A. M. Herring, B. D. Steffey, A. Miedaner, S. A. Wander, D. L. Dubois, *Inorg Chem* **1995**, *34*, 1100–1109; (c) T. L. Schull, J. C. Fettinger, D. A. Knight, *J Chem Soc Chem Comm* **1995**, 1487–1488; (d) S. Lelievre, F. Mercier, F. Mathey, *J Org Chem* **1996**, *61*, 3531–3533; (e) T. L. Schull, J. C. Fettinger, D. A. Knight, *Inorg Chem* **1996**, *35*, 6717–6723.

[17] (a) O. Herd, A. Hessler, M. Hingst, P. Machnitzki, M. Tepper, O. Stelzer, *Catal Today* **1998**, *42*, 413–420; (b) P. Machnitzki, M. Tepper, K. Wenz, O. Stelzer, E. Herdtweck, *J Organomet Chem* **2000**, *602*, 158–169; (c) A. Hessler, S. Kucken, O. Stelzer, J. Blotevogelbaltronat, W. S. Sheldrick, *J Organomet Chem* **1995**, *501*, 293–302; (d) I. Toth, B. E. Hanson, *Organometallics* **1993**, *12*, 1506–1513; (e) I. Toth, B. E. Hanson, *Tetrahedron-Asymmetry* **1990**, *1*, 895–912.

[18] (a) A. Jegorov, J. Podlaha, *Catal Lett* **1991**, *9*, 9–14; (b) H. Schumann, V. Ravindar, L. Meltser, W. Baidossi, Y. Sasson, J. Blum, *J Mol Catal a-Chem* **1997**, *118*, 55–61; (c) V. Ravindar, H. Schumann, H. Hemling, J. Blum, *Inorg Chim Acta* **1995**, *240*, 145–152; (d) F. Mercier, F. Mathey, *J Organomet Chem* **1993**, *462*, 103–106.

[19] (a) M. T. Reetz, J. Rudolph, *Tetrahedron-Asymmetry* **1993**, *4*, 2405–2406;(b) M. T. Reetz, S. R. Waldvogel, *Angew Chem Int Edit* **1997**, *36*, 865–867; (c) M. Beller, J. G. E. Krauter, A. Zapf, S. Bogdanovic, *Catal Today* **1999**, *48*, 279–290; (d) M. Beller, J. G. E. Krauter, A. Zapf, *Angew Chem Int Edit* **1997**, *36*, 772–774.

[20] (a) T. Malmstrom, C. Andersson, *J Mol Catal a-Chem* **1999**, *139*, 259–270; (b) E. Valls, J. Suades, R. Mathieu, N. Lugan, *Eur J Inorg Chem* **2003**, 3047–3053; (c) E. Valls, A. Solsona, J. Suades, R. Mathieu, F. Comelles, C. Lopez-Iglesias, *Organometallics* **2002**, *21*, 2473–2480; (d) E. Valls, J. Suades, R. Mathieu, J. F. Piniella, A. Alvarez-Larena, *J Organomet Chem* **2001**, *626*, 139–144; (e) E. Valls, J. Suades, R. Mathieu, *Organometallics* **1999**, *18*, 5475–5483; (f) E. Valls, J. Suades, B. Donadieu, R. Mathieu, *Chem Commun* **1996**, 771–772; (g) J. Powell, M. R. Gregg, A. Kuksis, C. J. May, S. J. Smith, *Organometallics* **1989**, *8*, 2918–2932; (h) J. Powell, A. Kuksis, C. J. May, P. E. Meindl, S. J. Smith, *Organometallics* **1989**, *8*, 2933–2941; (i) J. Powell, M. Gregg, A. Kuksis, P. Meindl, *J Am Chem Soc* **1983**, *105*, 1064–1065; (j) A. Varshney, M. L. Webster, G. M. Gray, *Inorg Chem* **1992**, *31*, 2580–2587; (k) A. Varshney, G. M. Gray, *Inorg Chem* **1991**, *30*, 1748–1754.

[21] (a) O. Herd, K. P. Langhans, O. Stelzer, N. Weferling, W. S. Sheldrick, *Angew Chem Int Edit* **1993**, *32*, 1058–1059; (b) O. Herd, A. Hessler, M. Hingst, M. Tepper, O. Stelzer, *J Organomet Chem* **1996**, *522*, 69–76; (c) D. J. Brauer, M. Hingst, K. W. Kottsieper, C. Liek, T. Nickel, M. Tepper, O. Stelzer, W. S. Sheldrick, *J Organomet Chem* **2002**, *645*, 14–26.

[22] F. G. Mann, I. T. Millar, *Journal of the Chemical Society* **1952**, 4453–4457.

[23] R. T. Smith, R. K. Ungar, L. J. Sanderson, M. C. Baird, *Organometallics* **1983**, *2*, 1138.

[24] J. Podlahova, J. Podlaha, *Collection of Czechoslovak Chemical Communications* **1980**, *45*, 2049–2053.

[25] V. S. Reddy, D. E. Berning, K. V. Katti, C. L. Barnes, W. A. Volkert, A. R. Ketring, *Inorg Chem* **1996**, *35*, 1753–1757.

[26] (a) D. M. Lynn, B. Mohr, R. H. Grubbs, L. M. Henling, M. W. Day, *J Am Chem Soc* **2000**, *122*, 6601–6609; (b) R. B. DeVasher, L. R. Moore, K. H. Shaughnessy, *J Org Chem* **2004**, *69*, 7919–7927.

[27] D. J. Daigle, Pepperma .Ab, S. L. Vail, *J Heterocyclic Chem* **1974**, *11*, 407–408.

[28] P. Pellon, *Tetrahedron Lett* **1992**, *33*, 4451.

[29] W. Keim, R. P. Schultz, *J Mol Catal* **1994**, *92*, 21–33.

[30] R. T. Smith, M. C. Baird, *Transition Metal Chemistry* **1981**, *6*, 197.

[31] (a) G. Peiffer, S. Chhan, A. Bendayan, B. Waegell, J. P. Zahra, *Journal of Molecular Catalysis* **1990**, *59*, 1–9; (b) U. Nagel, E. Kinzel, *Chemische Berichte* **1986**, *119*, 1731–1733.

[32] S. Ahrland, J. Chatt, N. R. Davies, A. A. Williams, *Nature Chemistry* **1957**, *179*, 1187–1188.

[33] E. Kuntz, Rhone-Poulenc Recherche, FR 2.314.910, **1975**.

[34] W. A. Herrmann, G. P. Albanese, R. B. Manetsberger, P. Lappe, H. Bahrmann, *Angew Chem Int Edit* **1995**, *34*, 811–813.

[35] H. Jansch, S. Kannenberg, G. Boche, *Eur J Org Chem* **2001**, 2923–2926.

[36] S. Hida, P. J. Roman, A. A. Bowden, J. D. Atwood, *J Coord Chem* **1998**, *43*, 345–348.

[37] W. A. Herrmann, C. W. Kohlpaintner, H. Bahrmann, W. Konkol, *J Mol Catal* **1992**, *73*, 191–201.

[38] (a) W. A. Herrmann, J. A. Kulpe, W. Konkol, H. Bahrmann, *J Organomet Chem* **1990**, *389*, 85–101; (b) L. Lecomte, J. Triolet, D. Sinou, J. Bakos, B. Heil, *J Chromatogr* **1987**, *408*, 416–419.

[39] W. A. Herrmann, C. W. Kohlpaintner, B. E. Hanson, X. Kang, *Inorg Synth* **1998**, *32*, 8–25.

[40] W. J. Tic, *Sep Sci Technol* **2007**, *42*, 125–135.

[41] W. P. Mul, K. Ramkisoensing, P. C. J. Kamer, J. N. H. Reek, A. J. van der Linden, A. Marson, P. W. N. M. van Leeuwen, *Adv Synth Catal* **2002**, *344*, 293–298.

[42] B. M. Bhanage, S. S. Divekar, R. M. Deshpande, R. V. Chaudhari, *Org Process Res Dev* **2000**, *4*, 342–345.

[43] K. H. Shaughnessy, *Chem Rev* **2009**, *109*, 643–710.

[44] M. S. Goedheijt, P. C. J. Kamer, P. W. N. M. van Leeuwen, *J Mol Catal a-Chem* **1998**, *134*, 243–249.

[45] K. T. Wan, M. E. Davis, *J Chem Soc Chem Comm* **1993**, 1262–1264.

[46] I. T. Horvath, J. Rabai, *Science* **1994**, *266*, 72–75.

[47] M. Vogt, Technische Hochschule Aachen **1991**.

[48] (a) D. Rutherford, J. J. J. Juliette, C. Rocaboy, I. T. Horvath, J. A. Gladysz, *Catal Today* **1998**, *42*, 381–388; (b) E. G. Hope, R. D. W. Kemmitt, D. R. Paige, A. M. Stuart, *J Fluorine Chem* **1999**, *99*, 197–200; (c) B. Richter, B. J. Deelman, G. van Koten, *J Mol Catal a-Chem* **1999**, *145*, 317–321; (d) B. Richter, A. L. Spek, G. van Koten, B. J. Deelman, *J Am Chem Soc* **2000**, *122*, 3945–3951; (e) E. de Wolf, A. L. Spek, B. W. M. Kuipers, A. P. Philipse, J. D. Meeldijk, P. H. H. Bomans, P. M. Frederik, B. J. Deelman, G. van Koten, *Tetrahedron* **2002**, *58*, 3911–3922; (f) J. van den Broeke, E. de Wolf, B. J. Deelman, G. van Koten, *Adv Synth Catal* **2003**, *345*, 625–634; (g) D. J. Birdsall, E. G. Hope, A. M. Stuart, W. P. Chen, Y. L. Hu, J. L. Xiao, *Tetrahedron Lett* **2001**, *42*, 8551–8553.

[49] (a) J. J. J. Juliette, I. T. Horvath, J. A. Gladysz, *Angew Chem Int Edit* **1997**, *36*, 1610–1612; (b) J. J. J. Juliette, D. Rutherford, I. T. Horvath, J. A. Gladysz, *J Am Chem Soc* **1999**, *121*, 2696–2704.

[50] (a) I. T. Horvath, G. Kiss, R. A. Cook, J. E. Bond, P. A. Stevens, J. Rabai, E. J. Mozeleski, *J Am Chem Soc* **1998**, *120*, 3133–3143; (b) A. Aghmiz, C. Claver, A. M. Masdeu-Bulto, D. Maillard, D. Sinou, *J Mol Catal a-Chem* **2004**, *208*, 97–101; (c) Y. L. Huang, E. Perperi, G. Manos, D. J. Cole-Hamilton, *J Mol Catal a-Chem* **2004**, *210*, 17–21; (d) E. Perperi, Y. L. Huang, P. Angeli, G. Manos, C. R. Mathison, D. J. Cole-Hamilton, D. A. Adams, E. G. Hope, *Chem Eng Sci* **2004**, *59*, 4983–4989; (e) D. F. Foster, D. Gudmunsen, D. J. Adams, A. M. Stuart, E. G. Hope, D. J. Cole-Hamilton, G. P. Schwarz, P. Pogorzelec, *Tetrahedron* **2002**, *58*, 3901–3910; (f) D. J. Adams, D. J. Cole-Hamilton, E. G. Hope, P. J. Pogorzelec, A. M. Stuart, *J Organomet Chem* **2004**, *689*, 1413–1417; (g) D. Bonafoux, Z. H. Hua, B. H. Wang, I. Ojima, *J Fluorine Chem* **2001**, *112*, 101–108; (h) T. Mathivet, E. Monflier, Y. Castanet, A. Mortreux, J.-L. Couturier, *Tetrahedron* **2002**, *58*, 3877–3888; (i) T. Mathivet, E. Monflier, Y. Castanet, A. Mortreux, J. L. Couturier, *Cr Chim* **2002**, *5*, 417–424.

[51] (a) G. Pozzi, I. Shepperson, *Coordin Chem Rev* **2003**, *242*, 115–124; (b) B. Cornils, *Angew Chem Int Edit* **1997**, *36*, 2057–2059; (c) I. T. Horvath, *Accounts Chem Res* **1998**, *31*, 641–650; (d) M. Cavazzini, F. Montanari, G. Pozzi, S. Quici, *J Fluorine Chem* **1999**, *94*, 183–193; (e) R. H. Fish, *Chemistry - A European Journal* **1999**, *5*, 1677–1680; (f) E. G. Hope, A. M. Stuart, *J Fluorine Chem* **1999**, *100*, 75–83; (g) P. Bhattacharyya, B. Croxtall, J. Fawcett, D. Gudmunsen, E. G. Hope, R. D. W. Kemmitt, D. R. Paige, D. R. Russell, A. M. Stuart, D. R. W. Wood, *J Fluorine Chem* **2000**, *101*, 247–255; (h) C. C. Tzschucke, C. Markert, W. Bannwarth, S. Roller, A. Hebel, R. Haag, *Angew Chem Int Edit* **2002**, *41*, 3964–4000; (i) A. P. Dobbs, M. R. Kimberley, *J Fluorine Chem* **2002**,

118, 3–17; (j) E. de Wolf, G. van Koten, B. J. Deelman, *Chem Soc Rev* **1999**, *28*, 37–41; (k) J. A. Gladysz, *Science* **1994**, *266*, 55–56.
[52] I. T. Horvath, in *Multiphase Homogeneous Catalysis, Vol. 1* (Eds.: B. Cornils, W. A. Herrmann, I. T. Horvath, W. Leitner, S. Mecking, H. Olivier-Bourbigou, D. Vogt), Wiley VCH, **2005**, pp. 339–404.
[53] M. Cavazzini, A. Manfredi, F. Montanari, S. Quici, G. Pozzi, *Chem Commun* **2000**, 2171–2172.
[54] D. J. Adams, D. J. Cole-Hamilton, E. G. Hope, P. J. Pogorzelec, A. M. Stuart, *J Organomet Chem* **2004**, *689*, 1413–1417.
[55] (a) L. J. Alvey, D. Rutherford, J. J. J. Juliette, J. A. Gladysz, *J Org Chem* **1998**, *63*, 6302–6308; (b) L. J. Alvey, R. Meier, T. Soos, P. Bernatis, J. A. Gladysz, *Eur J Inorg Chem* **2000**, 1975–1983.
[56] (a) D. Sinou, D. Maillard, A. Aghmiz, A. M. M. i-Bulto, *Adv Synth Catal* **2003**, *345*, 603–611; (b) D. Sinou, G. Pozzi, E. G. Hope, A. M. Stuart, *Tetrahedron Lett* **1999**, *40*, 849–852; (c) C. M. Haar, J. Huang, S. P. Nolan, J. L. Petersen, *Organometallics* **1998**, *17*, 5018–5024.
[57] (a) B. Richter, E. de Wolf, G. van Koten, B. J. Deelman, *J Org Chem* **2000**, *65*, 3885–3893; (b) E. de Wolf, A. J. M. Mens, O. L. J. Gijzeman, J. H. van Lenthe, L. W. Jenneskens, B. J. Deelman, G. van Koten, *Inorg Chem* **2003**, *42*, 2115–2124; (c) E. de Wolf, E. Riccomagno, J. J. M. de Pater, B. J. Deelman, G. van Koten, *J Comb Chem* **2004**, *6*, 363–374; (d) J. J. M. de Pater, C. E. P. Maljaars, E. de Wolf, M. Lutz, A. L. Spek, B. J. Deelman, C. J. Elsevier, G. van Koten, *Organometallics* **2005**, *24*, 5299–5310.
[58] J. van den Broeke, B. J. Deelman, G. van Koten, *Tetrahedron Lett* **2001**, *42*, 8085–8087.
[59] M. Wende, F. Seidel, J. A. Gladysz, *J Fluorine Chem* **2003**, *124*, 45–54.
[60] P. Bhattacharyya, D. Gudmunsen, E. G. Hope, R. D. W. Kemmitt, D. R. Paige, A. M. Stuart, *J Chem Soc-Perkin Trans. 1* **1997**, 3609–3612.
[61] M. A. Carroll, A. B. Holmes, *Chem Commun* **1998**, 1395–1396.
[62] W. P. Chen, J. L. Xiao, *Tetrahedron Lett* **2000**, *41*, 3697–3700.
[63] S. Schneider, W. Bannwarth, *Angew Chem Int Edit* **2000**, *39*, 4142–4145.
[64] (a) W. P. Chen, L. J. Xu, J. L. Xiao, *Org Lett* **2000**, *2*, 2675–2677; (b) S. Kainz, D. Koch, W. Baumann, W. Leitner, *Angew Chem Int Edit* **1997**, *36*, 1628–1630; (c) S. Kainz, A. Brinkmann, W. Leitner, A. Pfaltz, *J Am Chem Soc* **1999**, *121*, 6421–6429; (d) Q. S. Zhang, Z. Y. Luo, D. P. Curran, *J Org Chem* **2000**, *65*, 8866–8873.
[65] S. Darses, M. Pucheault, J. P. Genet, *Eur J Org Chem* **2001**, 1121–1128.
[66] T. Soos, B. L. Bennett, D. Rutherford, L. P. Barthel-Rosa, J. A. Gladysz, *Organometallics* **2001**, *20*, 3079–3086.
[67] (a) F. Langer, K. Puntener, R. Sturmer, P. Knochel, *Tetrahedron-Asymmetry* **1997**, *8*, 715–738; (b) S. Beneficemalouet, H. Blancou, A. Commeyras, *J Fluorine Chem* **1985**, *30*, 171–187.
[68] G. Vlad, F. Richter, I. T. Horvath, *Org Lett* **2004**, *6*, 4559–4561.
[69] J. Fawcett, S. Friedrichs, J. H. Holloway, E. G. Hope, V. McKee, M. Nieuwenhuyzen, D. R. Russell, G. C. Saunders, *J Chem Soc Dalton* **1998**, 1477–1484.
[70] D. J. Adams, E. G. Hope, A. M. Stuart, A. J. West, in *Handbook of Fluorous Chemistry*, Wiley-VCH Verlag GmbH & Co, **2004**, pp. 431–435.
[71] (a) J. Fawcett, E. G. Hope, R. D. W. Kemmitt, D. R. Paige, D. R. Russell, A. M. Stuart, D. J. ColeHamilton, M. J. Payne, *Chem Commun* **1997**, 1127–1128; (b) M. A. Guillevic, A. M. Arif, I. T. Horvath, J. A. Gladysz, *Angew Chem Int Edit* **1997**, *36*, 1612–1615; (c) M. A. Guillevic, C. Rocaboy, A. M. Arif, I. T. Horvath, J. A. Gladysz, *Organometallics* **1998**, *17*, 707–717.
[72] D. Koch, W. Leitner, *J Am Chem Soc* **1998**, *120*, 13398–13404.
[73] (a) C. Rocaboy, D. Rutherford, B. L. Bennett, J. A. Gladysz, *J Phys Org Chem* **2000**, *13*, 596–603; bT. Soos, B. L. Bennett, D. Rutherford, L. P. Barthel-Rosa, J. A. Gladysz, *Organometallics* **2001**, *20*, 3079–3086.
[74] (a) E. de Wolf, B. Richter, B. J. Deelman, G. van Koten, *J Org Chem* **2000**, *65*, 5424–5427; (b) E. G. Hope, R. D. W. Kemmitt, A. M. Stuart, *J Chem Soc Dalton* **1998**, 3765–3770; (c) B. L. Bennett, J. M. Hoerter, J. F. Houlis, D. M. Roddick, *Organometallics* **2000**, *19*, 615–621.
[75] (a) Y. L. Hu, D. J. Birdsall, A. M. Stuart, E. G. Hope, J. L. Xiao, *J Mol Catal a-Chem* **2004**, *219*, 57–60; (b) D. J. Adams, W. P. Chen, E. G. Hope, S. Lange, A. M. Stuart, A. West, J. L. Xiao, *Green Chem* **2003**, *5*, 118–122.
[76] Y. Chauvin, L. Mussmann, H. Olivier, *Angewandte Chemie-International Edition* **1995**, *34*, 2698–2700.
[77] M. F. Sellin, P. B. Webb, D. J. Cole-Hamilton, *Chem Commun* **2001**, 781–782.

[78] A. Hessler, O. Stelzer, H. Dibowski, K. Worm, F. P. Schmidtchen, *J Org Chem* **1997**, *62*, 2362–2369.

[79] P. Wasserscheid, H. Waffenschmidt, P. Machnitzki, K. W. Kottsieper, O. Stelzer, *Chem Commun* **2001**, 451–452.

[80] J. Dupont, S. M. Silva, R. F. de Souza, *Catal Lett* **2001**, *77*, 131–133.

[81] (a) C. X. Deng, G. N. Ou, J. R. She, Y. Z. Yuan, *J Mol Catal a-Chem* **2007**, *270*, 76–82; (b) J. R. She, L. M. Ye, J. Zhu, Y. Z. Yuan, *Catal Lett* **2007**, *116*, 70–75.

[82] H. L. Ngo, A. G. Hu, W. B. Lin, *Chem Commun* **2003**, 1912–1913.

[83] P. B. Webb, M. F. Sellin, T. E. Kunene, S. Williamson, A. M. Z. Slawin, D. J. Cole-Hamilton, *J Am Chem Soc* **2003**, *125*, 15577–15588.

[84] (a) P. B. Webb, D. J. Cole-Hamilton, *Chem Commun* **2004**, 612–613; (b) P. B. Webb, T. E. Kunene, D. J. Cole-Hamilton, *Green Chem* **2005**, *7*, 373–379; (c) A. C. Frisch, P. B. Webb, G. Zhao, M. J. Muldoon, P. J. Pogorzelec, D. J. Cole-Hamilton, *Dalton T* **2007**, 5531–5538; (d) C. P. Mehnert, R. A. Cook, N. C. Dispenziere, E. J. Mozeleski, *Polyhedron* **2004**, *23*, 2679–2688.

[85] D. J. Brauer, K. W. Kottsieper, C. Liek, O. Stelzer, H. Waffenschmidt, P. Wasserscheid, *J Organomet Chem* **2001**, *630*, 177–184.

[86] K. W. Kottsieper, O. Stelzer, P. Wasserscheid, *J Mol Catal a-Chem* **2001**, *175*, 285–288.

[87] F. Favre, H. Olivier-Bourbigou, D. Commereuc, L. Saussine, *Chem Commun* **2001**, 1360–1361.

[88] C. S. Consorti, G. L. P. Aydos, G. Ebeling, J. Dupont, *Org Lett* **2008**, *10*, 237–240.

[89] (a) R. Sebesta, M. Meciarova, V. Polackova, E. Veverkova, I. Kmentova, E. Gajdosikova, J. Cvengros, R. Buffa, V. Gajda, *Collection of Czechoslovak Chemical Communications* **2007**, *72*, 1057–1068; (b) R. Sebesta, F. Bilcik, *Tetrahedron-Asymmetry* **2009**, *20*, 1892–1896.

[90] (a) J. Andrieu, M. Azouri, *Inorg Chim Acta* **2007**, *360*, 131–135; (b) J. Andrieu, L. Harmand, M. Picquet, *Polyhedron* **2010**, *29*, 601–605.

[91] (a) R. P. J. Bronger, S. M. Silva, P. C. J. Kamer, P. W. N. M. van Leeuwen, *Chem Commun* **2002**, 3044–3045; (b) R. P. J. Bronger, S. M. Silva, P. C. J. Kamer, P. W. N. M. van Leeuwen, *Dalton T* **2004**, 1590–1596.

[92] C. C. Brasse, U. Englert, A. Salzer, H. Waffenschmidt, P. Wasserscheid, *Organometallics* **2000**, *19*, 3818–3823.

[93] K. R. Seddon, A. Stark, M.-J. Torres, *Pure Appl Chem* **2000**, *72*, 2275–2287.

[94] M. J. Earle, J. M. S. S. Esperanca, M. A. Gilea, J. N. C. Lopes, L. P. N. Rebelo, J. W. Magee, K. R. Seddon, J. A. Widegren, *Nature* **2006**, *439*, 831–834.

[95] (a) T. Welton, *Chem Rev* **1999**, *99*, 2071–2083; (b) P. Wasserscheid, W. Keim, *Angew Chem Int Edit* **2000**, *39*, 3772–3789; (c) B. Clare, A. Sirwardana, D. R. MacFarlane, *Top Curr Chem* **2009**, *290*, 1–40.

[96] F. Joo, J. Kovacs, A. Katho, A. C. Benyei, T. Decuir, D. J. Darensbourg, *Inorg Synth* **1998**, *32*, 1–8.

[97] N. J. Meehan, A. J. Sandee, J. N. H. Reek, P. C. J. Kamer, P. W. N. M. van Leeuwen, M. Poliakoff, *Chem Commun* **2000**, 1497–1498.

[98] J. A. Darr, M. Poliakoff, *Chem Rev* **1999**, *99*, 495–541.

[99] (a) J. G. Riess, M. Leblanc, *Pure Appl Chem* **1982**, *54*, 2383–2406; (b) C. Arlen, M. Leblanc, J. G. Riess, *J Fluorine Chem* **1985**, *29*, 246.

[100] J. H. Aubert, *J Supercrit Fl* **1998**, *11*, 163–172.

[101] A. I. Cooper, *J Mater Chem* **2000**, *10*, 207–234.

[102] (a) C. M. Wai, S. Wang, *J Chromatogr* **1997**, *785*, 369–383; (b) C. M. Wai, S. F. Wang, J. J. Yu, *Anal Chem* **1996**, *68*, 3516–3519.

[103] M. J. Burk, S. G. Feng, M. F. Gross, W. Tumas, *J Am Chem Soc* **1995**, *117*, 8277–8278.

[104] (a) B. Guzel, M. A. Omary, J. P. Fackler, A. Akgerman, *Inorg Chim Acta* **2001**, *325*, 45–50; (b) K. Burgemeister, G. Francio, H. Hugl, W. Leitner, *Chem Commun* **2005**, 6026–6028.

[105] D. K. Morita, D. R. Pesiri, S. A. David, W. H. Glaze, W. Tumas, *Chem Commun* **1998**, 1397–1398.

[106] (a) D. R. Palo, C. Erkey, *Ind Eng Chem Res* **1998**, *37*, 4203–4206; (b) D. R. Palo, C. Erkey, *Ind Eng Chem Res* **1999**, *38*, 3786–3792; (c) D. R. Palo, C. Erkey, *Ind Eng Chem Res* **1999**, *38*, 2163–2165; (d) D. R. Palo, C. Erkey, *Organometallics* **2000**, *19*, 81–86; (e) T. Davis, C. Erkey, *Ind Eng Chem Res* **2000**, *39*, 3671–3678; (f) S. Fujita, S. Fujisawa, B. M. Bhanage, M. Arai, *Tetrahedron Lett* **2004**, *45*, 1307–1310; (g) S. Fujita, S. Fujisawa, B. M. Bhanage, Y. Ikushima, M. Arai, *Eur J Org Chem* **2004**, 2881–2887; (h) A. C. J. Koeken, M. C. A. van Vliet, L. J. P. van den Broeke, B. J. Deelman, J. T. F. Keurentjes, *Adv Synth Catal* **2006**, *348*, 1553–1559;

(i) A. C. J. Koeken, M. C. A. van Vliet, L. J. P. van den Broeke, B. J. Deelman, J. T. E. Keurentjes, *Adv Synth Catal* **2008**, *350*, 179–188; (j) C. T. Estorach, A. M. Masdeu-Bulto, *Catal Lett* **2008**, *122*, 76–79; (k) C. T. Estorach, A. Orejon, A. M. Masdeu-Bulto, *Green Chem* **2008**, *10*, 545–552; (l) Y. Shimoyama, M. Sonoda, K. Miyazaki, H. Higashi, Y. Iwai, Y. Arai, *J Supercrit Fl* **2008**, *44*, 266–272; (m) S. Fujita, K. Yuzawa, B. M. Bhanage, Y. Ikushima, M. Arai, *J Mol Catal a-Chem* **2002**, *180*, 35–42.

[107] (a) A. M. B. Osuna, W. P. Chen, E. G. Hope, R. D. W. Kemmitt, D. R. Paige, A. M. Stuart, J. L. Xiao, L. J. Xu, *J Chem Soc Dalton* **2000**, 4052–4055; (b) S. Lange, A. Brinkmann, P. Trautner, K. Woelk, J. Bargon, W. Leitner, *Chirality* **2000**, *12*, 450–457; (c) H. Hori, C. Six, W. Leitner, *Macromolecules* **1999**, *32*, 3178–3182; (d) C. A. G. Carter, R. T. Baker, S. P. Nolan, W. Tumas, *Chem Commun* **2000**, 347–348.

[108] (a) G. Francio, W. Leitner, *Chem Commun* **1999**, 1663–1664; (b) G. Francio, K. Wittmann, W. Leitner, *J Organomet Chem* **2001**, *621*, 130–142.

[109] (a) P. G. Jessop, T. Ikariya, R. Noyori, *Nature* **1994**, *368*, 231–233; (b) P. G. Jessop, Y. Hsiao, T. Ikariya, R. Noyori, *J Am Chem Soc* **1996**, *118*, 344–355; (c) I. Bach, D. J. Cole-Hamilton, *Chem Commun* **1998**, 1463–1464; (d) T. R. Early, R. S. Gordon, M. A. Carroll, A. B. Holmes, R. E. Shute, I. F. McConvey, *Chem Commun* **2001**, 1966–1967.

[110] K. D. Wagner, N. Dahmen, E. Dinjus, *J Chem Eng Data* **2000**, *45*, 672–677.

[111] (a) N. Shezad, R. S. Oakes, A. A. Clifford, C. M. Rayner, *Tetrahedron Lett* **1999**, *40*, 2221–2224; bN. Shezad, A. A. Clifford, C. T. Rayner, *Green Chem* **2002**, *4*, 64–67; (c) C. D. Ablan, D. Sheppard, E. J. Beckman, M. M. Olmstead, P. G. Jessop, *Green Chem* **2005**, *7*, 590–594.

[112] (a) G. Filardo, M. Di Blasi, A. Galia, A. Ponchel, H. Bricout, A. D. Sayede, E. Monflier, *J Supercrit Fl* **2006**, *36*, 173–181; (b) C. T. Estorach, M. Gimenez-Pedros, A. M. Masdeu-Bulto, A. D. Sayede, E. Monflier, *Eur J Inorg Chem* **2008**, 2659–2663; (c) A. Galia, E. C. Navarre, O. Scialdone, M. Ferreira, G. Filardo, S. Tilloy, E. Monflier, *J Phys Chem B* **2007**, *111*, 2573–2578; (d) A. Galia, E. C. Navarre, O. Scialdone, G. Filardo, E. Monflier, *J Supercrit Fl* **2009**, *49*, 154–160.

[113] S. Saffarzadeh-Matin, C. J. Chuck, F. M. Kerton, C. M. Rayner, *Organometallics* **2004**, *23*, 5176–5181.

[114] M. A. McHugh, V. J. Krukonis, *Supercritical Fluid Extraction*, 2nd ed., Butterworth-Heinemann, **1994**.

[115] K. Huan, L. Bes, D. M. Haddleton, E. Khoshdel, *J Polym Sci Pol Chem* **2001**, *39*, 1833–1842.

[116] W. J. Ye, J. M. DeSimone, *Ind Eng Chem Res* **2000**, *39*, 4564–4566.

[117] V. K. Potluri, J. H. Xu, R. Enick, E. Beckman, A. D. Hamilton, *Org Lett* **2002**, *4*, 2333–2335.

[118] F. Montilla, A. Galindo, V. Rosa, T. Aviles, *Dalton T* **2004**, 2588–2592.

[119] M. Berthod, G. Mignani, M. Lemaire, *Tetrahedron-Asymmetry* **2004**, *15*, 1121–1126.

[120] M. Kant, S. Bischoff, R. Siefken, E. Grundemann, A. Kockritz, *Eur J Org Chem* **2001**, 477–481.

[121] M. Berthod, C. Saluzzo, G. Mignani, M. Lemaire, *Tetrahedron-Asymmetry* **2004**, *15*, 639–645.

[122] M. F. Sellin, D. J. Cole-Hamilton, *J Chem Soc Dalton* **2000**, *11*, 1681–1683.

[123] L. A. Blanchard, D. Hancu, E. J. Beckman, J. F. Brennecke, *Nature* **1999**, *399*, 28–29.

[124] L. A. Blanchard, Z. Y. Gu, J. F. Brennecke, *J Phys Chem B* **2001**, *105*, 2437–2444.

[125] R. Dohrn, G. Brunner, *Fluid Phase Equilibria* **1995**, *106*, 213–282.

[126] L. A. Blanchard, J. F. Brennecke, *Ind Eng Chem Res* **2001**, *40*, 287–292.

[127] (a) D. J. Cole-Hamilton, T. E. Kunene, P. B. Webb, in *Multiphase Homogeneous Catalysis, Vol. 2* (Eds.: B. Cornils, W. A. Herrmann, I. T. Horvath, W. Leitner, S. Mecking, H. Olivier-Bourbigou, D. Vogt), Wiley-VCH, **2005**, pp. 688–713; (b) S. V. Dzyuba, R. A. Bartsch, *Angew Chem Int Edit* **2003**, *42*, 148–150.

[128] (a) A. Riisager, K. M. Eriksen, P. Wasserscheid, R. Fehrmann, *Catal Lett* **2003**, *90*, 149–153; (b) A. Riisager, R. Fehrmann, S. Flicker, R. van Hal, M. Haumann, P. Wasserscheid, *Angew Chem Int Edit* **2005**, *44*, 815–819; (c) A. Riisager, R. Fehrmann, M. Haumann, P. Wasserscheid, *Top Catal* **2006**, *40*, 91–102; (d) A. Riisager, R. Fehrmann, M. Haumann, P. Wasserscheid, *Eur J Inorg Chem* **2006**, 695–706; (e) U. Hintermair, Z. X. Gong, A. Serbanovic, M. J. Muldoon, C. C. Santini, D. J. Cole-Hamilton, *Dalton T* **2010**, *39*, 8501–8510; (f) U. Hintermair, T. Hofener, T. Pullmann, G. Francio, W. Leitner, *Chemcatchem* **2010**, *2*, 150–154; (g) U. Hintermair, G. Y. Zhao, C. C. Santini, M. J. Muldoon, D. J. Cole-Hamilton, *Chem Commun* **2007**, 1462–1464.

[129] (a) R. J. Bonilla, B. R. James, P. G. Jessop, *Chem Commun* **2000**, 941–942; (b) B. M. Bhanage, Y. Ikushima, M. Shirai, M. Arai, *Chem Commun* **1999**, 1277–1278; (c) B. M. Bhanage, Y. Ikushima, M. Shirai, M. Arai,

Tetrahedron Lett **1999**, *40*, 6427–6430; (d) G. B. Jacobson, J. Brady, J. Watkin, W. Tumas, *Abstr Pap Am Chem S* **1999**, *217*, U786–U786.

[130] K. N. West, J. P. Hallett, R. S. Jones, D. Bush, C. L. Liotta, C. A. Eckert, *Ind Eng Chem Res* **2004**, *43*, 4827–4832.

[131] D. J. Heldebrant, P. G. Jessop, *J Am Chem Soc* **2003**, *125*, 5600–5601.

[132] (a) M. Araia, S. Fujitaa, M. Shirai, *J Supercrit Fl* **2009**, *47*, 351–356; (b) C. M. Gordon, W. Leitner, in *Catalyst Separation, Recovery and Recycling* (Eds.: D. J. Cole-Hamilton, R. P. Tooze), Springer-Verlag, Berlin Heidelberg, **2006**, pp. 215–234.

[133] T. Bartik, B. Bartik, B. E. Hanson, T. Glass, W. Bebout, *Inorg Chem* **1992**, *31*, 2667–2670.

[134] E. G. Hope, A. M. Stuart, in *Handbook of Fluorous Chemistry*, Wiley-VCH Verlag GmbH & Co, **2004**, pp. 247–256.

[135] Personal Communication with Dr Alison M. Stuart, University of Leicester.

Index

Phosphorus(III) Ligands in Homogeneous Catalysis: Design and Synthesis, First Edition.
Edited by Paul C. J. Kamer and Piet W. N. M. van Leeuwen.

Printed and bound by CPI Group (UK) Ltd, Croydon, CR0 4YY
16/09/2026
14963390-0008